# TABLE OF INTEGRALS

**b.** $\displaystyle \int \frac{dx}{x\sqrt{ax+b}} = \frac{1}{\sqrt{b}} \ln\left|\frac{\sqrt{ax+b}-\sqrt{b}}{\sqrt{ax+b}+\sqrt{b}}\right| + C \qquad [b>0]$

**9.** $\displaystyle \int \frac{\sqrt{ax+b}}{x}\, dx = 2\sqrt{ax+b} + b\int \frac{dx}{x\sqrt{ax+b}}$

**10.** $\displaystyle \int \frac{\sqrt{ax+b}}{x^2}\, dx = -\frac{\sqrt{ax+b}}{x} + \frac{a}{2}\int \frac{dx}{x\sqrt{ax+b}} + C$

**11.** $\displaystyle \int \frac{dx}{x^2\sqrt{ax+b}} = -\frac{\sqrt{ax+b}}{bx} - \frac{a}{2b}\int \frac{dx}{x\sqrt{ax+b}} + C$

**Forms with $\sqrt{a^2+u^2}$:**

**12.** $\displaystyle \int \frac{du}{\sqrt{a^2+u^2}} = \ln\left|u+\sqrt{a^2+u^2}\right| + C$

**13.** $\displaystyle \int \sqrt{a^2+u^2}\, du = \frac{u}{2}\sqrt{a^2+u^2} + \frac{a^2}{2}\ln(u+\sqrt{a^2+u^2}) + C$

**14.** $\displaystyle \int u^2\sqrt{a^2+u^2} = \frac{u(a^2+2u^2)\sqrt{a^2+u^2}}{8} - \frac{a^4}{8}\ln(u+\sqrt{a^2+u^2}) + C$

**15.** $\displaystyle \int \frac{\sqrt{a^2+u^2}}{u}\, du = \sqrt{a^2+u^2} - a\ln\left|\frac{a+\sqrt{a^2+u^2}}{u}\right| + C$

**16.** $\displaystyle \int \frac{\sqrt{a^2+u^2}}{u^2}\, du = \ln(u+\sqrt{a^2+u^2}) - \frac{\sqrt{a^2+u^2}}{u} + C$

**17.** $\displaystyle \int \frac{u^2}{\sqrt{a^2+u^2}}\, du = -\frac{a^2}{2}\ln(u+\sqrt{a^2+u^2}) + \frac{u\sqrt{a^2+u^2}}{2} + C$

**18.** $\displaystyle \int \frac{du}{u\sqrt{a^2+u^2}} = -\frac{1}{a}\ln\left|\frac{a+\sqrt{a^2+u^2}}{u}\right| + C$

**19.** $\displaystyle \int \frac{du}{u^2\sqrt{a^2+u^2}} = -\frac{\sqrt{a^2+u^2}}{a^2 u} + C$

**Forms with $\sqrt{a^2-u^2}$:**

**20.** $\displaystyle \int \frac{du}{\sqrt{a^2-u^2}} = \sin^{-1}\left(\frac{u}{a}\right) + C$

*Continued on overleaf*

# TABLE OF INTEGRALS

**21.** $\displaystyle \int \sqrt{a^2 - u^2}\, du = \frac{u}{2}\sqrt{a^2 - u^2} + \frac{a^2}{2}\sin^{-1}\left(\frac{u}{a}\right) + C$

**22.** $\displaystyle \int u^2 \sqrt{a^2 - u^2}\, du = \frac{a^4}{8}\sin^{-1}\left(\frac{u}{a}\right) - \frac{1}{8}u\sqrt{a^2 - u^2}(a^2 - 2u^2) + C$

**23.** $\displaystyle \int \frac{\sqrt{a^2 - u^2}}{u}\, du = \sqrt{a^2 - u^2} - a\ln\left|\frac{a + \sqrt{a^2 - u^2}}{u}\right| + C$

**24.** $\displaystyle \int \frac{\sqrt{a^2 - u^2}}{u^2}\, du = -\sin^{-1}\left(\frac{u}{a}\right) - \frac{\sqrt{a^2 - u^2}}{u} + C$

**25.** $\displaystyle \int \frac{u^2}{\sqrt{a^2 - u^2}}\, du = \frac{a^2}{2}\sin^{-1}\left(\frac{u}{a}\right) - \frac{1}{2}u\sqrt{a^2 - u^2} + C$

**26.** $\displaystyle \int \frac{du}{u\sqrt{a^2 - u^2}} = -\frac{1}{a}\ln\left|\frac{a + \sqrt{a^2 - u^2}}{u}\right| + C$

**27.** $\displaystyle \int \frac{du}{u^2\sqrt{a^2 - u^2}} = -\frac{\sqrt{a^2 - u^2}}{a^2 u} + C$

**Forms with $\sqrt{u^2 - a^2}$:**

**28.** $\displaystyle \int \frac{du}{\sqrt{u^2 - a^2}} = \ln\left|u + \sqrt{u^2 - a^2}\right| + C$

**29.** $\displaystyle \int \frac{du}{(\sqrt{u^2 - a^2})^n} = \frac{u(\sqrt{u^2 - a^2})^{2-n}}{(2-n)a^2} - \frac{n-3}{(n-2)a^2}\int \frac{du}{(\sqrt{u^2 - a^2})^{n-2}} \qquad n \neq 2$

**30.** $\displaystyle \int \sqrt{u^2 - a^2}\, du = \frac{u}{2}\sqrt{u^2 - a^2} - \frac{a^2}{2}\ln\left|u + \sqrt{u^2 - a^2}\right| + C$

**31.** $\displaystyle \int (\sqrt{u^2 - a^2})^n\, du = \frac{u(\sqrt{u^2 - a^2})^n}{n+1} - \frac{na^2}{n+1}\int (\sqrt{u^2 - a^2})^{n-2}\, du, \qquad n \neq -1$

**32.** $\displaystyle \int u^2 \sqrt{u^2 - a^2}\, du = \frac{u}{8}(2u^2 - a^2)\sqrt{u^2 - a^2} - \frac{a^4}{8}\ln\left|u + \sqrt{u^2 - a^2}\right| + C$

**33.** $\displaystyle \int \frac{\sqrt{u^2 - a^2}}{u}\, du = \sqrt{u^2 - a^2} - a\sec^{-1}\left|\frac{u}{a}\right| + C$

**34.** $\displaystyle \int \frac{\sqrt{u^2 - a^2}}{u^2}\, du = \ln\left|u + \sqrt{u^2 - a^2}\right| - \frac{\sqrt{u^2 - a^2}}{u} + C$

**35.** $\displaystyle \int \frac{u^2}{\sqrt{u^2 - a^2}}\, du = \frac{a^2}{2}\ln\left|u + \sqrt{u^2 - a^2}\right| + \frac{u}{2}\sqrt{u^2 - a^2} + C$

*This table is continued on the back endpapers.*

# CALCULUS

# CALCULUS

Robert Seeley

University of Massachusetts at Boston

HARCOURT BRACE JOVANOVICH, PUBLISHERS

and its subsidiary, Academic Press

San Diego   New York   Chicago   Austin   Washington, D.C.
London   Sydney   Tokyo   Toronto

ISBN: 0-15-505681-6

Library of Congress Catalog Card Number: 89-84683

Printed in the United States of America

# PREFACE

Mathematics, and particularly calculus, provides an essential language and intellectual framework for science. With the explosive growth of science and technology comes a growing need for mastery of calculus, at least as a language, and better yet, as a way of thinking.

The wide variation in backgrounds and talents of calculus students presents a real challenge both to the instructor and to the textbook used. This book responds by exposing the main ideas of calculus as quickly as possible, before elaborating all the details. For example, a chapter on the applications of calculus to problems of graphing, optimization, and motion precedes any real discussion of limits. The idea is to exploit the students' geometric intuition, which is generally much better developed than their facility in calculation.

In the course of class testing over the years, I have found only good consequences of this approach. Students easily grasp the basic ideas of the applications of calculus and, with the understanding of this background, develop the maturity and motivation necessary for confronting the technical questions of limits. The same philosophy determines the treatment of infinite series and other topics. The textbook presents infinite series as a means of calculation *before* developing the abstract convergence tests.

Such minor reordering of traditional topics does not imply a lax treatment of the subject. Theoretical questions are treated with the prevailing standards of rigor; they are just treated somewhat later than usual. The difficult parts of the theory—such as the proofs of the Intermediate Value Theorem and Extreme Value Theorem and the justification of term-by-term operations on infinite series—are postponed to an appendix. The $\varepsilon - \delta$ definition of function limits is also in the appendix because very few students are ready for it early in the course. But the $\varepsilon - N$ definition of sequence limits is in the main body of the text in Chapter 11; by that point the definition seems natural.

The transcendental functions are presented in the order that is currently standard. The trigonometric functions are discussed early and provide a variety of examples for limits and derivatives. The exponential and logarithmic functions follow integrals, but could actually be covered at any time after Chapter 4. (Limits involving these functions are treated by l'Hopital's Rule in Chapter 4.)

Differential equations, discussed as slope fields, are introduced as soon as possible, beginning with antiderivatives in Chapter 2. They provide significant examples and problems throughout the text. The final chapter presents complex numbers and their use in solving constant coefficient equations, particularly the damped oscillator equation.

The textbook uses several pedagogical features to reinforce topic presentations:

- *An extensive art program* illustrates concepts and problems with two- or full-color figures.

- *Examples*, essential to illustrate and clarify general principles, are used extensively. They are clearly set off from the text discussions and are worked in great detail. Students should be encouraged to do their own work in similar detail to clarify thinking and avoid careless errors.

- Section *summaries* precede the problem sets for that section so that students can easily review as they work the problems.

- The *problems* are grouped in "A," "B," and "C" sets in order of increasing maturity. "A" problems are straightforward applications of the text material, "B" problems require more initiative or originality to work, and "C" problems are for students who show greater mathematical sophistication.

- *Review problems* end each chapter.

- *Answers* to problems marked with an asterisk (∗) are given at the end of the book. (Answers to all the problems are given in the Instructor's Manual.)

## ACKNOWLEDGMENTS

A book such as this is not created single-handedly. I owe a great debt to my students for their patience in using class notes and for their suggestions. My family, too, has contributed a good share of patience. My son Karl did much of the word processing, while Joe provided answers to many problems. I am grateful to my colleagues, particularly to Matt Gaffney for a multitude of suggestions and conversations about the book, to Steve Parrott for reading all of one version, and to Paul Salomaa for providing many answers; but most of all I am grateful to Dennis Wortman, who undertook the major task of writing the Instructor's Manual.

The book would not have been published without the support of several people. John Parker of Harcourt Brace Jovanovich kept an interest as the book developed. It has had essential support from a sequence of editors. Shelley Langman, Wesley Lawton, Ted Buchholz, and Richard Wallis presided over its development and provided support for the class testing, and Mike Johnson brought it into the world. The conscientious and cooperative book team included Robert Watrous, manuscript editor; Karen Denhams, production editor; Martha Gilman, designer; Vicki Kelly, art editor; Lesley Lenox, production manager; and Pamela Whiting, associate editor.

Comments from reviewers have, of course, been essential. They all have my grateful thanks: Charles C. Alexander, University of Mississippi; Glen D. Anderson, Michigan State University; Thomas F. Banchoff, Brown University; Douglas B. Crawford, College of San Mateo; Daniel S. Drucker, Wayne State University; Roger W. Hansell, University of Connecticut; James E. Hodge, Angelo State University; Kendell Hyde, Weber State College; Eleanor L. Kendrick, San Jose City College; Eleanor Killam, University of Massachusetts,

Amherst; Matthew Liu, University of Wisconsin, Stevens Point; John Montgomery, University of Rhode Island; David Price, Tarrant County Junior College; Robert G. Russell, West Valley College; Donald Schmidt, University of Northern Colorado; and Keith L. Wilson, Oklahoma City Community College.

Robert Seeley

## SUPPLEMENTS

### For Students

*Student Solutions Manual* by Dennis Wortman contains solutions for those problems that have answers in the text. (These are marked in the text with an asterisk.) The solutions are worked out with unusual care and completeness, and will be particularly instructive to those who want this extra help.

### For Instructors

*Instructor's Manual with Solutions* by Dennis Wortman contains the worked-out solutions for *all* the problems in the text.

*Computerized Testbank* (Micro-Pac Genie) by Microsystems Software, Ltd., is the most complete test-generating and author system with graphics on the market. This system is available for the IBM PC, XT or compatible system, and the Macintosh.

*Testbank* is also available in a printed version.

## COMPUTER SUPPLEMENTS FOR INSTRUCTORS AND STUDENTS

*CALCULUS* 3.0 from True Basic is intended for users who wish to expand their basic understanding. Easy-to-use features include a simple, menu-driven interface and context-sensitive pop-up HELP and GLOSSARY screens. A dialog box is used to enter numbers and functions and to execute problems; a graph box displays plotted functions.

*Calculus* encourages free exploration of topics including the concept of limits, tangents, l'Hopital's Rule, as well as exploring and visualizing parametric equations and differential equations. It is available for Apple Macintosh, IBM PC, and IBM PS/2.

*INTERACTIVE CALCULUS* from Math Lab allows students to experience complex mathematical ideas in a graphic and intuitive setting. Concepts such as limits, differentiation, and integration are numerically and graphically performed in a single keystroke. It is available for Apple II, IBM PC, and IBM PS/2.

*CALCAIDE*, a computer software program by Elizabeth Chang, Hood College, is a microcomputer disk for IBM PC and IBM PS/2 systems with interactive color graphing and numerical computation programs for the mainstream calculus course.

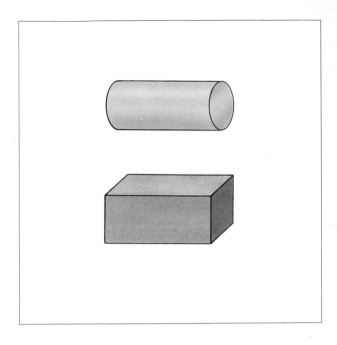

# CONTENTS

## CHAPTER 1
## BACKGROUND    1

1.1   Coordinates    2
1.2   Straight Lines and Slope    9
1.3   Functions    18
1.4   Optimization    25
      Review Problems    28

## CHAPTER 2
## THE DERIVATIVE OF A POLYNOMIAL    31

2.1   The Derivative and the Tangent Line    32
2.2   Quadratic Functions    40
2.3   The Derivative of a Polynomial    45
      Appendix: The Derivative of $x^n$. Factoring Polynomials    51
2.4   Graphing with the Derivative    53
2.5   The Second Derivative. Concavity    61
2.6   Newton's Method (optional)    66
2.7   Velocity and Acceleration    70
2.8   Antiderivatives    77
      Review Problems    86

## CHAPTER 3
## LIMITS, CONTINUITY, AND DERIVATIVES    89

3.1   Limits    90
3.2   Some Trigonometric Limits. The Trapping Theorem    99
3.3   Limits as $x \to +\infty$ or $x \to -\infty$    111
3.4   Continuity. The Intermediate Value Theorem    120
3.5   Derivatives. The Product and Quotient Rules    131
3.6   Derivatives of the Trigonometric Functions    142
3.7   Composite Functions. The Chain Rule    148
3.8   Implicit Differentiation. The Derivative of $x^{m/n}$    157
      Review Problems    174

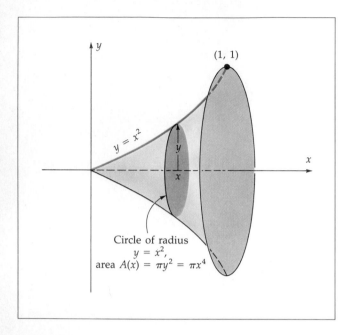

Circle of radius
$y = x^2$,
area $A(x) = \pi y^2 = \pi x^4$

CHAPTER 4
MORE APPLICATIONS OF
   DERIVATIVES   177

4.1  Rate of Change. The Linear
       Approximation   178
4.2  Related Rates   186
4.3  Parametric Curves and
       l'Hopital's Rule   194
4.4  Fermat's Principle and the
       Law of Refraction
       (optional)   202
4.5  The Lorenz Curve
       (optional)   208
       Review Problems   210

CHAPTER 5
THE MEAN VALUE
   THEOREM   213

5.1  The Extreme Value
       Theorem and the Critical
       Point Theorem   214
5.2  The Mean Value Theorem
       220
5.3  Applications to Graphing
       and Optimization   226
       Review Problems   232

CHAPTER 6
INTEGRALS   235

6.1  The Integral as Signed Area
       236
6.2  Summation Notation. Limits
       of Riemann Sums   245
6.3  The Integral of a
       Derivative. The First
       Fundamental Theorem of
       Calculus   252
6.4  The Integral of a Rate of
       Change. Trapezoid Sums
       258
6.5  Properties of Integrals.
       Mean Value   265
6.6  The Second Fundamental
       Theorem of Calculus
       275.
6.7  Indefinite Integrals   280
6.8  Differentials and
       Substitution   285
       Review Problems   292

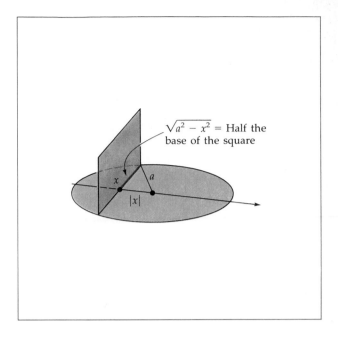

$\sqrt{a^2 - x^2}$ = Half the base of the square

CHAPTER 7
SOME APPLICATIONS OF
   INTEGRALS   295

7.1   Areas by Integration   296
7.2   Volumes by Cross Sections
       301
7.3   Volumes by Cylindrical
       Shells. Flow in Pipes
       312
7.4   Energy and Work   318
7.5   Improper Integrals   325
       Review Problems   332

CHAPTER 8
EXPONENTIALS,
   LOGARITHMS, AND
   INVERSE FUNCTIONS
   335

8.1   The Natural Exponential
       Function   336
8.2   Inverse Functions   346
8.3   Natural Logarithms   352
8.4   Other Bases. Logarithmic
       Differentiation.
       Indeterminate Forms 361
8.5   Exponential and
       Logarithmic Integrals
       364
8.6   Proportional Growth and
       Decay   366
8.7   Separable Differential
       Equations   371
8.8   Inverse Trigonometric
       Functions   381
8.9   Oscillations (optional)
       390
8.10  The Hyperbolic Functions
       400
       Review Problems   404

CHAPTER 9
TECHNIQUES OF
   INTEGRATION   407

9.1   Integration by Parts   408
       Appendix: The Differential
       Equation $\dfrac{dy}{dt} + by = f(t)$
       (optional)   415
9.2   Powers and Products of
       Sines and Cosines   417
9.3   Powers of Tangent and
       Secant   424
9.4   Trigonometric Substitution
       429
9.5   Some Algebraic Methods
       436
9.6   Integrating Rational
       Functions by Partial
       Fractions   442
9.7   Rationalizing Substitutions
       452
       Review Problems   454

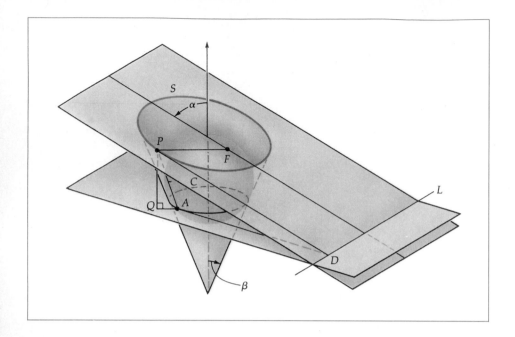

# Contents

CHAPTER 10
MORE APPLICATIONS OF
    INTEGRALS   457

10.1   Arc Length   458
10.2   Area of a Surface of
       Revolution   463
10.3   Center of Mass   467
10.4   Probability Density
       Functions   472
10.5   Numerical Integration
       480
       Appendix: Proving the
       Remainder Estimates
       490
       Review Problems   494

CHAPTER 11
INFINITE SEQUENCES AND
    SERIES   495

11.1   Taylor Polynomials
       496
       Appendix: The
       Remainder for Taylor
       Polynomials   506
11.2   Limit of a Sequence
       Defined   508
11.3   More on Sequences   519
11.4   Infinite Series   528
11.5   Taylor Series   536
11.6   Operations on Power
       Series   542
11.7   Estimating Remainders
       (optional)   549
11.8   Convergence Tests. The
       Integral Test   553
11.9   The Comparison Test
       562
11.10  Alternating Series   567
11.11  Absolute and Conditional
       Convergence   570
11.12  The Ratio Test. The Radius
       of Convergence   574
       Review Problems   581

CHAPTER 12
CONIC SECTIONS. POLAR
    COORDINATES   583

12.1   Parabolas. Completing the
       Square   584
12.2   Ellipses   591
12.3   Hyperbolas   598
12.4   General Quadratics.
       Rotation of Axes   606
12.5   Polar Coordinates   614
12.6   Conic Sections in Polar
       Coordinates.
       Eccentricity   620
12.7   Area in Polar Coordinates
       628
       Review Problems   634

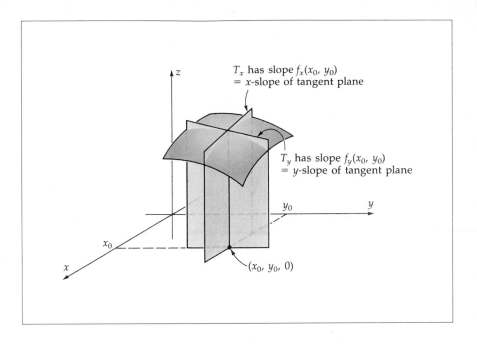

$T_x$ has slope $f_x(x_0, y_0)$
= $x$-slope of tangent plane

$T_y$ has slope $f_y(x_0, y_0)$
= $y$-slope of tangent plane

$(x_0, y_0, 0)$

CHAPTER 13
PLANE VECTORS   635

13.1   The Vector Space $R^2$
          636
13.2   Parametric and Vector
          Equations   645
13.3   Limits and Derivatives.
          Velocity and
          Acceleration   651
13.4   Arc Length   659
13.5   Planetary Motion   665
          Review Problems   671

CHAPTER 14
SPACE COORDINATES
     AND VECTORS   675

14.1   Space Coordinates   676
14.2   The Vector Space $R^3$. The
          Dot Product   680
14.3   The Cross Product.
          Determinants   689
14.4   Lines and Planes   698
14.5   Vector Functions and
          Space Curves   706
14.6   Quadric Surfaces.
          Cylinders   716
          Review Problems   724

CHAPTER 15
FUNCTIONS OF TWO
     VARIABLES   727

15.1   Graphs and Level Curves
          728
15.2   Continuity and Limits
          736
15.3   Partial Derivatives. The
          Tangent Plane and the
          Gradient   741
15.4   The Linear Approximation.
          Tangency   748
15.5   The Derivative along a
          Curve. The Directional
          Derivative   757
15.6   Higher Derivatives. Mixed
          Partials. $C^k$ Functions
          768
15.7   Critical Points. The
          Second Derivative Test
          772
15.8   The Least Squares Line
          (optional)   782
          Review Problems   785

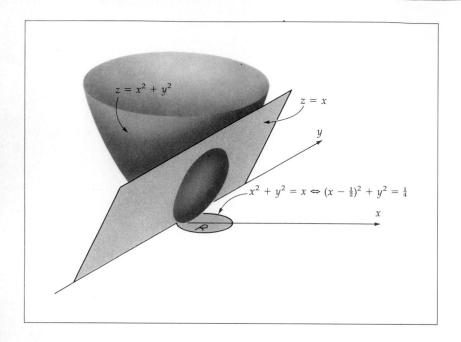

CHAPTER 16
FUNCTIONS OF SEVERAL
VARIABLES. FURTHER
TOPICS   787

16.1   Functions of Three
Variables   788
16.2   The General Chain Rule
795
16.3   Implicit Functions   801
16.4   Exact Differential
Equations (optional)
807
16.5   Lagrange Multipliers
(optional)   813
Review Problems   824

CHAPTER 17
MULTIPLE INTEGRALS
825

17.1   Double Integrals   826
17.2   More General Regions
835
17.3   Mass. Center of Mass.
Moment of Inertia
844
17.4   Double Integrals in Polar
Coordinates   850
17.5   Bivariate Probability
Densities (optional)
856
17.6   Surface Area   860
17.7   Triple Integrals   865
17.8   Cylindrical and Spherical
Coordinates   872
17.9   General Coordinate
Systems. The Jacobian
(optional)   879
Review Problems   886

CHAPTER 18
LINE AND SURFACE
INTEGRALS   887

18.1   Work and Line Integrals
888
18.2   Gradient Fields. Curl   898
18.3   Green's Theorem   908
18.4   Surface Integrals. Gauss'
Theorem   916
18.5   Stokes' Theorem   926
Review Problems   932

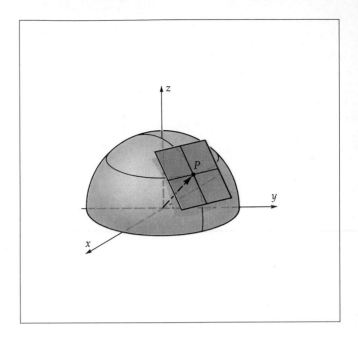

CHAPTER 19
COMPLEX NUMBERS AND
DIFFERENTIAL EQUATIONS
935

19.1 Complex Numbers 936
19.2 Multiplication of Complex
Numbers 940
19.3 Complex Functions of a
Real Variable 946
19.4 Damped Oscillations
951
19.5 Forced Oscillations 958

APPENDIX A
PROOFS OF THE HARD
THEOREMS 963

A.1 The Definition of Limit
964
A.2 Some Limit Theorems
971
A.3 Continuity 975
A.4 The Intermediate Value
Theorem 978
A.5 The Extreme Value
Theorem 980
A.6 Uniform Convergence.
Term-by-term
Operations on Series
983

APPENDIX B
REVIEW OF EXPONENTS
AND LOGARITHMS 991

APPENDIX C
REVIEW OF
TRIGONOMETRY 999

SELECTED ANSWERS 1009

INDEX 1045

# 1

# BACKGROUND

# 1.1

## COORDINATES

We begin by reviewing basic ideas and notation concerning coordinate systems. (Some of you know all of this, and all of you know some of it; that's the nature of a review.) Then we use coordinates to investigate two "optimization" problems, typical of those to be solved by calculus later, in Chapters 2–4.

### The Number Line

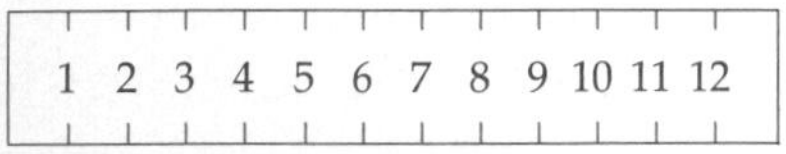

**FIGURE 1**
Humble ruler.

The humble ruler (Fig. 1) suggests the basic premise of analytic geometry: *The points of a straight line can be identified with numbers.* This appealing idea gave rise to some deep questions: What is a line? What is a number? Fortunately the questions have been answered[1], and we accept this premise as the basis for uniting algebra and geometry.

To identify numbers with points on a horizontal line, choose a unit of distance and an origin. Then label each point according to its distance from the origin, with negative numbers to the left and positive ones to the right (Fig. 2). You now have a **coordinate axis**, or a **number line**. The axis is the line itself; the number associated with a point on the line is a coordinate.

**FIGURE 2**
Number line.

On the coordinate axis, the distance from point $x$ to the origin is the **absolute value of $x$**, denoted $|x|$ (Fig. 3):

$$|-2| = 2, \qquad |2| = 2, \qquad |-3| = 3, \qquad |0| = 0.$$

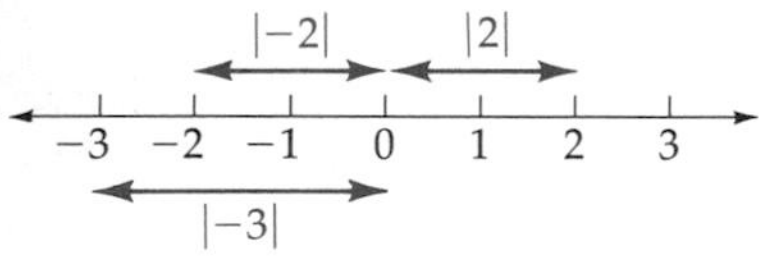

The absolute value $|x|$ gives the "size" of $x$ but not its sign; that is, both 2 and $-2$ have the same absolute value 2. In algebraic terms, the absolute value is defined by

$$|x| = \begin{cases} x & \text{if } x \geq 0 \\ -x & \text{if } x < 0. \end{cases}$$

**FIGURE 3**
$|x| = $ distance from $x$ to 0.

That $|x| = x$ when $x \geq 0$ is clear; but to understand the definition when $x < 0$, try for example $x = -2$:

$$|-2| = 2 = -(-2)$$

so indeed $|x| = -x$ when $x$ is the negative number $-2$.

With a little thought and experimentation, you can understand two properties of the absolute value:

$$|xy| = |x| \cdot |y|$$

and

$$|x + y| \leq |x| + |y|.$$

---

[1] One of the best answers was given by Richard Dedekind in 1872; *see* James R. Newman *The World of Mathematics*, Vol. 1, pp 525–36.

**FIGURE 4**
$a < b$ means $a$ is to the left of $b$.

Numbers are *ordered* according to their positions on the line, as shown in Figure 4:

$$a < b \text{ means } a \text{ is to the left of } b.$$

So $-100 < 2$; in words, $-100$ is less than 2. To compare size without regard to sign, use the absolute value; the inequality

$$|2| < |-100|$$

says that 2 is "smaller" than $-100$.

The *difference* of two numbers $x_1$ and $x_2$ is denoted $\Delta x$:

$$\Delta x = x_2 - x_1.$$

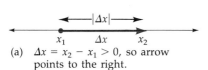

(a) $\Delta x = x_2 - x_1 > 0$, so arrow points to the right.

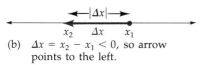

(b) $\Delta x = x_2 - x_1 < 0$, so arrow points to the left.

**FIGURE 5**
Arrows for $\Delta x$ point *from $x_1$ to $x_2$*.

(The symbol $\Delta$ is the Greek capital letter delta, which corresponds to our letter $D$, for difference. Thus $\Delta x$ is the *difference* of two values of $x$, not the product of a number $\Delta$ times a number $x$.) To form the difference, you need to know which number is first and which is second, that is, you need an *ordered* pair: $x_1$ first and $x_2$ second. Then $\Delta x$ is *positive* if $x_2 > x_1$, and *negative* if $x_2 < x_1$. As in Figure 5 we indicate $\Delta x$ by an arrow from $x_1$ to $x_2$; it points to the right if $\Delta x > 0$, and to the left if $\Delta x < 0$. The length of the arrow is the absolute value $|\Delta x|$, which gives the *distance* between $x_1$ and $x_2$:

$$\text{distance from } x_1 \text{ to } x_2 = |x_2 - x_1| = |\Delta x|.$$

Segments on the line are called **intervals** of numbers. It can be crucial to know whether the endpoints are included or not. Square brackets indicate an included endpoint, parentheses an excluded one (Fig. 6). For example,

$$[a, b) = \text{all numbers } x \text{ between } a \text{ and } b, \text{ including } a \text{ and excluding } b.$$

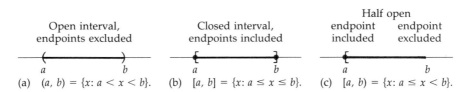

(a) $(a, b) = \{x: a < x < b\}$.   (b) $[a, b] = \{x: a \le x \le b\}$.   (c) $[a, b) = \{x: a \le x < b\}$.

**FIGURE 6**
Intervals: Open, closed, half open.

If both endpoints are included, the interval is called **closed**; if both are excluded, it is **open**.

To describe intervals, and sets in general, we use the concise "set-builder notation." With this, the interval $[a, b)$ is described as

$$\{x: a \le x < b\}$$

Typical member of set. | Condition determining whether $x$ is in the set.

Inside the braces, { }, use any convenient symbol as the typical member of the set. After labeling this typical member, put a colon (:) and state the conditions for membership. For example,

$$(1, 2] = \{t: 1 < t \le 2\}$$
$$= \text{the set of all numbers } t \text{ such that } 1 < t \text{ and } t \le 2.$$

Likewise, $[-1, 3] = \{z: -1 \le z \le 3\}$.

The "interval" $(1, 0) = \{x: 1 < x < 0\}$ is empty, since there are *no* numbers, $x$, which are both greater than 1 and less than 0.

**Coordinates in the plane** require two axes, one horizontal and one vertical (Fig. 7). Their point of intersection is called the **origin**. A number scale is chosen on each axis, with 0 at the origin; when convenient, the scale is the same on each axis. On the horizontal axis, numbers increase from left to right; and on the vertical axis, they increase from bottom to top.

Given these scales, an ordered pair of numbers $(x, y)$ is plotted as in Figure 7. Imagine the vertical line through point $x$ on the horizontal axis, and the horizontal line through point $y$ on the vertical axis; plot $(x, y)$ at the intersection of these two lines.

Given two points $P_1 = (x_1, y_1)$ and $P_2 = (x_2, y_2)$, we have the differences

$$\Delta x = x_2 - x_1 \quad \text{and} \quad \Delta y = y_2 - y_1.$$

Draw them as arrows parallel to the axes, giving a route from $P_1$ to $P_2$ as in Figure 8. The $\Delta x$ arrow points to the right if $\Delta x > 0$, and the $\Delta y$ arrow points up if $\Delta y > 0$. Figure 8 shows various possibilities.

The ratio $\dfrac{\Delta y}{\Delta x}$ is called the **slope** of the segment between $P_1$ and $P_2$.

We will say more about this important ratio later on.

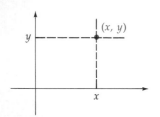

**FIGURE 7**
Coordinates in the plane.

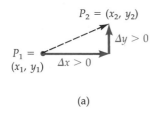

(a)

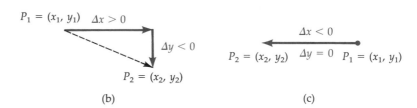

(b)                    (c)

**FIGURE 8**
$\Delta x$ and $\Delta y$.

**The distance** between $P_1$ and $P_2$ is denoted $|P_1 P_2|$. It can be computed from the Pythagorean Theorem (Fig. 9):

$$|P_1 P_2|^2 = |\Delta x|^2 + |\Delta y|^2 = (\Delta x)^2 + (\Delta y)^2$$

so

$$|P_1 P_2| = \sqrt{(\Delta x)^2 + (\Delta y)^2}.$$

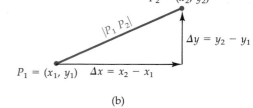

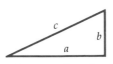

Pythagoras: $c^2 = a^2 + b^2$

(a)                    T                    (b)

**FIGURE 9**
Distance formula $|P_1 P_2| = \sqrt{(\Delta x)^2 + (\Delta y)^2}$ (Slope from $P_1$ to $P_2$ is $\Delta y / \Delta x$)

The symbol $\sqrt{\ }$ stands for the *nonnegative* square root:

$$\sqrt{9} = 3, \text{ not } \pm 3.$$

To give the two square roots, use the $\pm$ sign. For example,

$$x^2 = 2 \quad \text{means} \quad x = \pm\sqrt{2}.$$

## Circles

The circle of radius $r$ with center $C = (a, b)$ consists of all points $P = (x, y)$ at distance $r$ from the center $C$, as in Figure 10. Apply the Pythagorean Theorem to the right triangle in Figure 10, to find

$$(x - a)^2 + (y - b)^2 = r^2. \tag{1}$$

This is the standard equation of the circle: *Every point $(x, y)$ on the circle satisfies equation (1), and conversely every $(x, y)$ satisfying (1) lies on the circle.* The circle is called the **graph** of the equation; it is the set

$$\{(x, y): (x - a)^2 + (y - b)^2 = r^2\}.$$

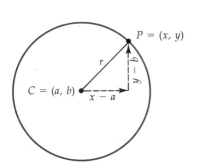

**FIGURE 10**
Circle $(x - a)^2 + (y - b)^2 = r^2$.

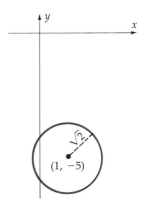

**FIGURE 11**
$(x - 1)^2 + (y + 5)^2 = 2$.

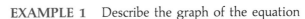

**EXAMPLE 1**   Describe the graph of the equation

$$(x - 1)^2 + (y + 5)^2 = 2. \tag{2}$$

*SOLUTION*   What choices of $a$, $b$, and $r$ in the general circle equation (1) will make it equivalent to the particular example (2)? Clearly $a = 1$, and almost as clearly $b = -5$, and $r = \sqrt{2}$; using these choices in the general equation (1), you get

$$(x - 1)^2 + (y - (-5))^2 = (\sqrt{2})^2,$$

which is equivalent to (2). So the graph is the circle with center $(1, -5)$ and radius $\sqrt{2}$ drawn in Figure 11.

Many practical problems require the distance formula; the next two examples will be studied more completely in Chapter 3.

Alphton

Betaville

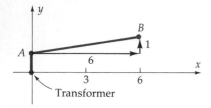

High voltage transmission line

**FIGURE 12**
Two towns near a high voltage
power transmission line.

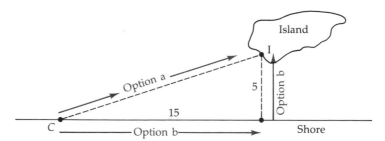

**FIGURE 13**
One possible location for the
transformer in Example 2.

**EXAMPLE 2**    Two towns, Alphton and Betaville, are near a long-distance high voltage line, positioned as in Figure 12. Electricity is to be provided to both towns from a single transformer station, $T$, somewhere along the line. Where should you place $T$ so as to minimize the total length of power line from $T$ to towns $A$ and $B$?

**SOLUTION**    First of all, set up a coordinate system as in Figure 13. The problem is, where to put $T$. One reasonable place is at the origin, as shown. Then you need a line to $A$, and another from $A$ to $B$. The first line is 1 mile long, and the line from $A$ to $B$ is

$$|AB| = \sqrt{6^2 + 1^2} = \sqrt{37} \text{ miles long;}$$

the total is $1 + \sqrt{37}$, about 7.08 miles.

Another reasonable place for $T$ would be at the three-mile mark along the transmission line, with a separate line to each town from there; you can compute the total length in that case and compare it to the 7.08 already found. It turns out, however, that neither of these "trial and error" solutions is the *optimal* one, which can be found by a geometric analysis suggested in the problems.

**EXAMPLE 3**    An island harbor is 5 miles offshore, and 15 miles down the coast from Sea City (Fig. 14). Cars drive to a terminal, where they take a ferry to the island harbor $I$. Assuming that cars travel 50 mph, and the ferry travels 16 mph, how long does it takes to go from the city to the island?

Island

Option a

5

Option b

15

C

Option b

Shore

**FIGURE 14**
Routes from Sea City (at $C$) to island harbor (at $I$). *Option a*) Terminal in the city; entire trip by ferry. *Option b*) Terminal opposite island; 15 miles by car and 5 by ferry.

**SOLUTION**    The answer depends on where the ferry terminal is.

***Option a***    Suppose the terminal is in the city. Then the whole trip is by ferry, at 16 mph. The distance traversed is

$$D = \sqrt{15^2 + 5^2} = \sqrt{250} = 5\sqrt{10} \text{ miles.}$$

At 16 mph, the trip will take

$$\frac{5\sqrt{10} \text{ miles}}{16 \text{ mi/hr}} = 0.988 \text{ hours.}$$

*Option b*   Suppose that the terminal is opposite the island. Then there are 15 miles by car, and 5 miles by ferry. The car part of the trip takes

$$\frac{15 \text{ miles}}{50 \text{ mi/hr}} = 0.3 \text{ hours,}$$

and the ferry part takes

$$\frac{5 \text{ miles}}{16 \text{ mi/hr}} = 0.3125 \text{ hours,}$$

so the total trip takes 0.6125 hours. The route is longer than with option a, 20 miles vs. $5\sqrt{10} \approx 16$ miles; but it is quicker, because the car is much faster than the ferry.

*Further Options*   There are slightly quicker routes; see the problems. The quickest of all is found in Chapter 3 using calculus.

## SUMMARY

*Absolute Value (see Fig. 3):*

$$|x| = \text{distance from } x \text{ to } 0 = \begin{cases} x & \text{if } x \geq 0 \\ -x & \text{if } x < 0. \end{cases}$$

$$|xy| = |x| \cdot |y|$$
$$|x + y| \leq |x| + |y|$$

*Intervals:*   See Figure 6.

*Distance Formula (see Fig. 9):*

$$|P_1 P_2| = \sqrt{(x_2 - x_1)^2 + (y_2 - y_1)^2} = \sqrt{(\Delta x)^2 + (\Delta y)^2}.$$

## PROBLEMS

### A

1. Given $x$, find $-x$ and $|x|$. Show that $|-x| = |x|$, $(-x)^2 = x^2$, and $|x|^2 = x^2$. Note that $-x$ is not necessarily negative!
   a) $x = 2$
   b) $x = -3$
   c) $x = -\sqrt{3}$
   d) $x = \sqrt{2}$

2. For the given pair of numbers, verify that $|xy| = |x| \cdot |y|$ and $|x + y| \leq |x| + |y|$.
   a) $x = 2, y = -3$
   b) $x = -5, y = -7$
   c) $x = -5, y = 7$

3. Plot $x_1$, $x_2$, and indicate $\Delta x$ by an appropriate arrow, as in Figure 5.
   *a) $x_1 = 1, x_2 = 3$
   b) $x_1 = 6, x_2 = 4$
   *c) $x_1 = -1, \Delta x = 2$ (find $x_2$)
   d) $x_2 = 0, \Delta x = -1$ (find $x_1$)

4. Describe the following sets in interval notation.
   *a) $\{x: 1 < x \le 3\}$   b) $\{x: -1 \le x < 1\}$
   *c) $\{t: 0 \le t \le 5\}$   d) $\{x: |x| < 2\}$
   *e) $\{x: |x| \le 1\}$

5. Describe the following in set notation $\{x: \text{condition on } x\}$.
   *a) $[-1, 2)$   b) The interval $(0, 5)$
   *c) All solutions $x$ of the equation
      $$x^3 + 3x + 1 = 0$$
   d) All numbers $x$ outside the interval $[-1, 2)$

6. List all members of the set $\{x: x^2 = 4\}$.

*7. Plot $P_1$ and $P_2$, and indicate $\Delta x$ and $\Delta y$ by appropriate arrows, as in Figure 8. Compute the slope and the distance between $P_1$ and $P_2$.
   a) $(1, 3)$ and $(2, 5)$
   b) $(3, 1)$ and $(5, 2)$
   c) $(1, -3)$ and $(2, -5)$
   d) $(-1, 3)$ and $(-2, 5)$
   e) $(-1, -3)$ and $(-2, -5)$
   f) $(2, 5)$ and $(1, 3)$

8. From the examples in problem 7, explain what happens to the slope when:
   i) For each point, $x$ and $y$, coordinates are exchanged (compare 7a and 7b).
   ii) Each $x$ coordinate is reversed in sign (compare 7a and 7d).
   iii) Each $y$ coordinate is reversed in sign (compare 7a and 7c).
   iv) All coordinates are reversed in sign.
   v) The roles of $P_1$ and $P_2$ are exchanged (compare 7a and 7f).

9. Find the center and radius of the graph of the given equation. Sketch the circle, and give the coordinates of five distinct points on it. (Check that these points satisfy the equation of the circle.)
   *a) $(x + 2)^2 + (y - 3)^2 = 4$
   b) $x^2 + y^2 = 1$
   c) $(x - 1)^2 + (y + 2)^2 = 3$

*10. In Example 2 and Figure 13, suppose that the transformer station is at the point $T = (3, 0)$, with one power line directly from $T$ to $A$ and another from $T$ to $B$. Is the total length of the line more, or less, than the length for the arrangement in Example 2?

*11. In Example 3 and Figure 14, suppose that the ferry terminal is 12 miles down the coast from $C$. Compute the time required to traverse this route, and compare it to the two routes in Example 3.

12. A power cable is to be run from a power transformer, $T$, to a house, $H$, on the opposite side of the street, 80 feet down the street from $T$ as in Figure 15; the street is 30 feet wide. Suppose that it costs \$20 a foot to lay cable alongside the street, and \$50 a foot to lay it under the street.

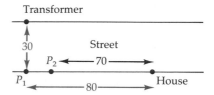

**FIGURE 15**
A route for cable from $T$ to $H$.

Compute the total cost to lay it:
   *a) Directly from $T$ to $H$.
   *b) Directly across the street to $P_1$, then along the street to $H$.
   c) Diagonally from $T$ to point $P_2$, 70 feet from $H$, then along the street to $H$.
   (In Chapter 3 you can find the cheapest of all possible routes.)

13. Draw a sketch illustrating the distance formula $|P_1P_2| = \sqrt{(\Delta x)^2 + (\Delta y)^2}$, and explain why the formula is true.

**B**

14. In Example 2, the optimal placement of $T$ for shortest total transmission line can be determined geometrically. Imagine a town $B'$ as in Figure 16, directly across the transmission line from $B$.
   a) What is the optimal placement of $T$ to serve towns $A$ and $B'$? Explain.
   b) What is the optimal placement of $T$ to serve $A$ and $B$? Explain.

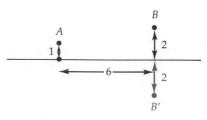

**FIGURE 16**
Imagine a town $B'$ opposite $B$.

# 1.2

## STRAIGHT LINES AND SLOPE

We review the equations of straight lines and the important idea of *slope*, which is central in calculus.

A *vertical line* in the $xy$-plane is parallel to the $y$-axis as shown in Figure 1. It intersects the $x$-axis at some point, say, point $(a, 0)$; the number $a$ is called the **$x$-intercept** of the line. As you move up and down the line, $y$ changes but $x$ remains the same, always $x = a$. Thus the vertical line with $x$-intercept $a$ is the set

$$\{(x, y): x = a\}.$$

For *nonvertical lines* the fundamental property is this: *The slope between any pair of points on line L is the same as the slope between any other pair on L.* That is, in Figure 2

$$\frac{\Delta y}{\Delta x} = \frac{\overline{\Delta y}}{\overline{\Delta x}}$$

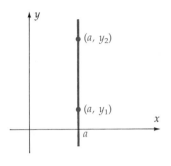

**FIGURE 1**
Vertical line $x = a$.

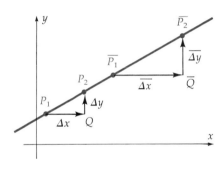

**FIGURE 2**
On a straight line $\dfrac{\Delta y}{\Delta x} = \dfrac{\overline{\Delta y}}{\overline{\Delta x}}$.

because triangles $P_1QP_2$ and $\overline{P_1QP_2}$ are similar. The common value of all these ratios is called the **slope of the line**.

The equation of a nonvertical line $L$ can be derived from this "equal slope" principle, as shown in Figure 3. The line cuts the $y$-axis at some point $P_0$ whose $y$ coordinate we label $b$. This is called the **$y$-intercept** of the line. Let $P = (x, y)$ be any other point on the line, and let $m$ be the slope of the line. Then

$$m = \frac{\Delta y}{\Delta x} = \frac{y - b}{x - 0} = \frac{y - b}{x}.$$

Solve for $y$ to get

$$y = mx + b. \tag{1}$$

*This equation is satisfied by every point $P = (x, y)$ on the line.* We just proved this for points $P$ different from $P_0$, and it holds for $P_0 = (0, b)$ as well, since

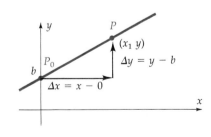

**FIGURE 3**
Line of slope $m = \dfrac{\Delta y}{\Delta x}$, $y$ intercept $b$.

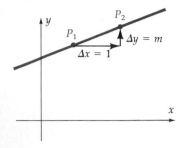

**FIGURE 4**
$m$ = increase of $y$ per unit increase of $x$.

(1) is true with $x = 0$ and $y = b$. Conversely, one can show that *every point satisfying (1) lies on the line.* So (1) is the **equation of the line**, and the line in turn is the **graph of this equation**. This is called the **slope-intercept** equation of the line; the coefficient $m$ in (1) is the slope of the line, and $b$ is the $y$-intercept.

The slope of the line gives the *rate of increase of $y$ per unit increase in $x$.* To see this, take two points $P_1$ and $P_2$ on $L$ with $\Delta x = x_2 - x_1 = 1$ (Fig. 4). Then

$$m = \frac{\Delta y}{\Delta x} = \frac{\Delta y}{1} = \Delta y$$

as in the figure. Thus when $x$ increases by 1 ($\Delta x = 1$) then $y$ changes by $m$ ($\Delta y = m$).

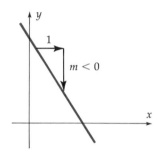

**FIGURE 5**
Line with negative slope.

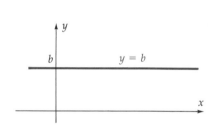

**FIGURE 6**
Slope 0, $y = b$.

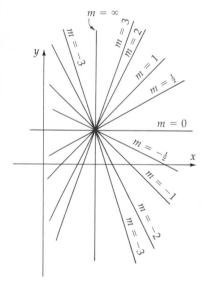

**FIGURE 7**
Lines with various slopes.

Geometrically, the slope $m$ describes how the line is tipped, or inclined. If $m$ is positive, then the line is tipped up from left to right (Fig. 4); in other words, left to right is uphill. If $m < 0$, then left to right is downhill (Fig. 5). If $m = 0$, the line is horizontal (Fig. 6) and its equation $y = 0 \cdot x + b$ reduces to $y = b$. If $m$ is large (positive or negative), the line is steeply inclined (Fig. 7).

For the steepest of all possible lines, a vertical line $L$ as in Figure 1, the slope is undefined. Every point $P$ on $L$ has the same $x$ coordinate, $x = a$, so in the ratio $\Delta y / \Delta x$ the denominator $\Delta x$ is zero,

$$\Delta x = x_2 - x_1 = a - a = 0.$$

Since lines which are *nearly* vertical can have arbitrarily large slope, it is natural to say that a *vertical* line has infinite slope, $m = \infty$.

Usually, the equation of a line with specified properties can be obtained from the equal slopes principle.

---

**EXAMPLE 1** Sketch the line $L$ through point $(1, 2)$ with slope $-1/2$, and write its equation.

**SOLUTION (Fig. 8)** Plot the given point $(1, 2)$, then move one unit to the right and $1/2$ unit down (since slope $m = -1/2$). This gives a second point $(2, 1\frac{1}{2})$ on the line. Sketch the line $L$ through these two points.

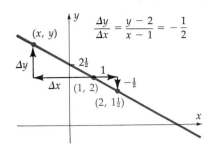

**FIGURE 8**
Line of slope $-\frac{1}{2}$ through (1, 2).

To find the equation of $L$, let $(x, y)$ be any point on $L$ (Fig. 8). The slope between $(x, y)$ and the given point (1, 2) is $-1/2$:

$$\frac{\Delta y}{\Delta x} = \frac{y - 2}{x - 1} = -\frac{1}{2}.$$

Solve for $y$:

$$y - 2 = -\tfrac{1}{2}(x - 1) = -\tfrac{1}{2}x + \tfrac{1}{2}$$

$$y = -\tfrac{1}{2}x + 2\tfrac{1}{2}.$$

Compare this to the standard form $y = mx + b$; the slope is $m = -\frac{1}{2}$ as required, and the $y$-intercept is $2\frac{1}{2}$ (Fig. 8).

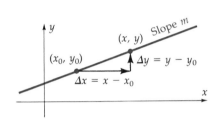

**FIGURE 9**
$$m = \frac{\Delta y}{\Delta x} = \frac{y - y_0}{x - x_0}.$$

The method in Example 1 can be applied in general, to obtain an equation for the line through a given point $P_0 = (x_0, y_0)$ with given slope $m$. A point $P = (x, y)$ is on this line if and only if the slope between $P$ and $P_0$ equals $m$ (Fig. 9):

$$\frac{\Delta y}{\Delta x} = \frac{y - y_0}{x - x_0} = m$$

so

$$y - y_0 = m(x - x_0).$$

This is the **point-slope equation** for the line.

**EXAMPLE 2**   Sketch the line $L$ through (1, 2) and (2, 1), and write its equation.

*SOLUTION (Fig. 10)*   Plot the two points and sketch the line through them. To find the equation, use the "equal slopes" principle. First compute the slope between the two given points:

$$m = \frac{\Delta y}{\Delta x} = \frac{1 - 2}{2 - 1} = -1.$$

Then let $(x, y)$ be any point on $L$; the slope between $(x, y)$ and either of the given points, say (1, 2), equals $-1$:

$$\frac{\Delta y}{\Delta x} = \frac{y - 2}{x - 1} = -1.$$

Solve for $y$:

$$y - 2 = (-1)(x - 1) = -x + 1,$$

$$y = -x + 3.$$

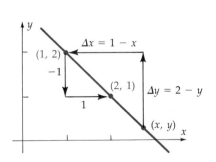

**FIGURE 10**
Line through (1, 2) and (2, 1).

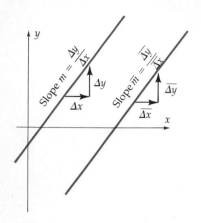

**FIGURE 11**
Parallel lines, $m = \bar{m}$.

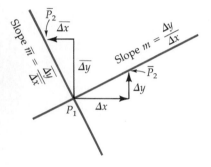

**FIGURE 12**
Perpendicular lines.

## Parallel and Perpendicular Lines

Since slope measures how a line is tipped, you would expect that *parallel lines have the same slope*. This is true. In Figure 11 the two triangles are similar, since corresponding sides are parallel; hence the ratio $\Delta y / \Delta x$ in one triangle equals the corresponding ratio in the other:

$$\frac{\Delta y}{\Delta x} = \frac{\overline{\Delta y}}{\overline{\Delta x}}$$

that is, $m = \bar{m}$. Conversely, *two lines with equal slope are parallel*. For if the slopes are equal, then the triangles in Figure 11 can be shown to be similar, so the lines are indeed parallel.

The slopes of two perpendicular lines have a more complicated relation, which can be deduced from Figure 12. The segment $P_1 P_2$ is rotated $90°$ to produce $P_1 \overline{P_2}$. Then $\Delta x$ rotates into $\overline{\Delta y}$, so these segments have the same length; and both point in the positive direction, so

$$\overline{\Delta y} = \Delta x.$$

Further, $\overline{\Delta x}$ and $\Delta y$ have the same length, but $\Delta y$ is positive (it points *up*) and $\overline{\Delta x}$ is negative (it points to the *left*) so

$$\overline{\Delta x} = -\Delta y.$$

Hence

$$\bar{m} = \frac{\overline{\Delta y}}{\overline{\Delta x}} = \frac{\Delta x}{-\Delta y} = -\frac{\Delta x}{\Delta y} = -\frac{1}{\Delta y / \Delta x} = -\frac{1}{m}.$$

This gives the relation between the slopes of nonvertical perpendicular lines:

$$\bar{m} = -1/m.$$

Vertical lines, of course, are perpendicular to horizontal ones.

---

### EXAMPLE 3
**a.** Show that the graph of $2x + 3y = 1$ is a line, and find its slope.
**b.** Find an equation of the line through the origin and perpendicular to the graph of $2x + 3y = 1$.

### SOLUTION
**a.** To show that the graph of $2x + 3y = 1$ is a line, rewrite the given equation in $y = mx + b$ form, by solving for $y$:

$$y = -\tfrac{2}{3}x + \tfrac{1}{3}.$$

The graph of this is a line of slope $m = -2/3$.
**b.** The required perpendicular line has slope

$$\bar{m} = \frac{-1}{m} = \frac{-1}{-2/3} = \frac{1}{2/3} = \frac{3}{2}.$$

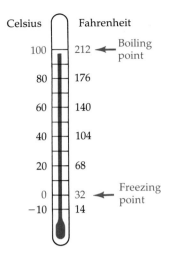

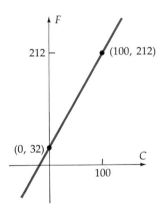

**FIGURE 13**
Celsius and Fahrenheit temperature scales.

**FIGURE 14**
Fahrenheit vs. Celsius.

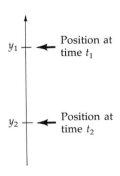

**FIGURE 15**
Positions of an object moving on a vertical line.

It is required to pass through the origin, so the $y$-intercept is 0, and the equation of the line is

$$y = \tfrac{3}{2} x + 0 = \tfrac{3}{2} x.$$

## Linear Relations

The equation

$$y = mx + b$$

gives a relation between $x$ and $y$. Since the graph of the equation is a straight line, the relation is called **linear**. There are many concrete examples of linear relations, and in each case the slope has a specific meaning as a **rate of change**.

**EXAMPLE 4: Temperature Scales**    The Celsius and Fahrenheit temperature scales illustrated in Figure 13 are linearly related. Let C stand for degrees Celsius, and F for degrees Fahrenheit. The freezing point of water is $0°$ in Celsius and $32°$ in Fahrenheit; the boiling point is $100°$ in Celsius, $212°$ in Fahrenheit. So the points $(0, 32)$ and $(100, 212)$ are on the graph of the relation between C and F (Fig. 14). The slope of the graph is

$$m = \frac{\Delta F}{\Delta C} = \frac{212 - 32}{100 - 0} = 1.8.$$

This gives the rate of change of Fahrenheit with respect to Celsius—each increase of $1°$ Celsius corresponds to an increase of $1.8°$ Fahrenheit. So the relation between Fahrenheit and Celsius is that of a line with slope 1.8; the F-intercept is 32 (Fig. 14), so the equation of the line is

$$F = 32 + (1.8)C = 32 + \tfrac{9}{5}C.$$

**EXAMPLE 5: Motion Along a Line**    Suppose an object moves along the $y$-axis. Consider any time interval, from $t_1$ to $t_2$. At time $t_1$ the object is at some position $y_1$, and by time $t_2$ it has moved to another position $y_2$ (Fig. 15). The **average velocity** during this time interval is the ratio

$$\frac{\text{change of position}}{\text{change of time}} = \frac{\Delta y}{\Delta t}.$$

If this ratio has the *same value* $v$ for all time intervals, then the object is moving with constant velocity $v$; this constant $v$ is the rate of change of position with respect to time.

Now think of the graph of position $y$ (on the vertical axis) versus time (on the horizontal axis, Figure 16). At time $t = 0$, the object is at some *initial position*, $y_0$. If the motion has constant velocity $v$ then

$$\frac{\Delta y}{\Delta t} = v$$

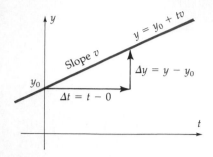

**FIGURE 16**
$$\frac{\Delta y}{\Delta t} = \frac{y - y_0}{t - 0} = v.$$

for every time interval $[t_1, t_2]$. Take the time interval $[0, t]$ shown in Figure 16; then

$$\frac{\Delta y}{\Delta t} = \frac{y - y_0}{t} = v.$$

Solve for $y$:

$$y = y_0 + vt.$$

So the velocity $v$ is the slope of the graph of position $y$ against time $t$; and the initial position $y_0$ is the $y$-intercept (Fig. 16).

---

*NOTE*   The formula $y = y_0 + vt$ is valid *only when the velocity $v$ is constant!* Motion with variable velocity is discussed in Chapter 2.

---

**EXAMPLE 6: Fixed Costs and Marginal Costs**   In producing $x$ items of a product there are usually certain *fixed costs*, $F$ (such as the cost of tools), that are the same no matter how many items are produced; and in addition there are costs for material and so on, which vary with the quantity produced. For a very simple production process, it is reasonable to suppose that these variable costs, $V$, are proportional to $x$, the number of items produced; that is, there is a *constant of proportionality $M$* such that

$$V = Mx.$$

Then the total cost of producing $x$ items is the sum of the fixed and the variable costs,

$$C = F + Mx.$$

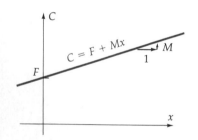

**FIGURE 17**
$C$ = cost of producing $M$ items;
$F$ = fixed cost, $M$ = marginal cost.

The graph of this is a straight line (Fig. 17); the $C$-intercept $F$ gives the fixed costs, and the slope $M$ is the cost of producing each additional unit. This slope is the so-called "marginal cost" of production, the rate of change of the total cost, $C$, with respect to the number of items produced, $x$.

A similar formula applies to car rental prices. There is a fixed charge, call it $F$ dollars per day, and a mileage charge of $M$ dollars per mile. The cost of renting the car to drive $x$ miles in one day is the sum of the fixed charge and the mileage charge,

$$C = F + Mx.$$

---

## Linear Interpolation

Most graphs considered in calculus are curved; but if you look at just a small section, the graph is nearly straight (Fig. 18) and can be approximated by a straight line. This is the basis of linear interpolation.

---

**EXAMPLE 7**   Figure 18 is a graph of the "cumulative normal probability distribution" commonly used in probability and statistics. The usual table for this distribution gives probability values $y$ to four decimal places, for each

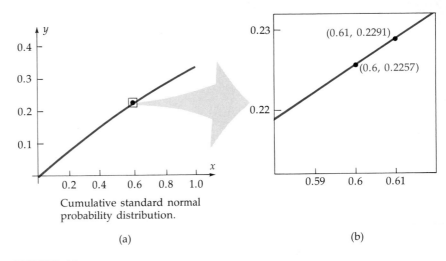

**FIGURE 18**
A small piece of the graph, magnified, is practically a straight line.

$x$ to two decimals; for example

| $x$ | $y$ |
|------|--------|
| 0.60 | 0.2257 |
| 0.61 | 0.2291 |

Suppose that you are making a very refined calculation requiring probability $y$ when $x = 0.603$. The table above gives two points on the graph:

$$(x_1, y_1) = (0.6, 0.2257) \quad \text{and} \quad (x_2, y_2) = (0.61, 0.2291).$$

You can approximate the graph between these two points by the straight line passing through them. The equation of that line is

$$y = 0.34(x - 0.6) + 0.2257.$$

In particular, when $x = 0.603$, the corresponding probability is approximately

$$y = 0.34(0.003) + 0.2257 = 0.22672.$$

Since the table entries have been rounded to four places, round the result to four places; and since it is an approximation, not an exact value, use the "approximately equals" symbol $\approx$:

$$\text{when} \quad x = 0.603 \quad \text{then} \quad y \approx 0.2267.$$

## SUMMARY

If $P_1 = (x_1, y_1)$ and $P_2 = (x_2, y_2)$ are on a line of slope $m$, then

$$\frac{\Delta y}{\Delta x} = \frac{y_2 - y_1}{x_2 - x_1} = m.$$

The slope is the rate of change of $y$ with respect to $x$.

The equation of a line of slope $m$ and $y$-intercept $b$ (see Fig. 2) is

$$y = mx + b.$$

The equation of the line through $(x_0, y_0)$ with slope $m$ (see Fig. 9) is

$$y - y_0 = m(x - x_0).$$

Parallel lines are lines of equal slope.
Perpendicular lines have slopes related by

$$m_2 = -1/m_1.$$

## PROBLEMS

### A

1. Let $L$ be a line of slope $m$, and suppose that the scales on the horizontal and vertical axes are equal.
   a) What is the angle between $L$ and the $x$-axis if $m = 1$? If $m = -1$?
   *b) If the angle between $L$ and the $x$-axis is $60°$, what is $m$?

2. Plot two points which lie on the given line, and sketch the graph. Show all $x$- and $y$-intercepts, and determine the slope.
   a) $y = x$
   b) $y = -x$
   c) $y = 2x - 1$
   d) $y = -\frac{1}{2}x + 3$
   e) $y = 3$
   f) $x = 3$
   *g) $x + y = 1$
   h) $2x + y = 1$

3. Write the equation of the line through the given point $P_0$ with given slope $m$. Sketch the line.
   *a) $(1, 1)$, $m = 1$
   *b) $(1, 1)$, $m = -2$
   *c) $(0, 0)$, $m = -1$
   d) $(2, 4)$, $m = -1/2$
   e) $(5, 4)$, $m = 0$

4. Plot the given points, sketch the line through them, and find its slope and equation.
   *a) $(0, 0)$ and $(1, 2)$
   *b) $(2, 3)$ and $(3, 2)$
   *c) $(2, 1)$ and $(0, 0)$
   d) $(1, -1)$ and $(1.1, 1.01)$
   e) $(5, 4)$ and $(-4, 5)$
   f) $(5, 4)$ and $(-5, -4)$

5. Write the equation of the line through $(-1, 2)$ and parallel to the given line.
   *a) $y = \frac{1}{2}x + 2$
   *b) $2x + 3y = 2$
   *c) $y = 5$
   d) $x = 3$

6. Write the equation of the line through $(1, 5)$ and perpendicular to the given line.
   *a) $y = 2x - \frac{1}{2}$
   *b) $3x + 2y = 1$
   c) $y = 100$
   d) $x = -\pi$

7. The three points $A = (0, 0)$, $B = (1, 2)$, and $C = (2, 0)$ are the vertices of a triangle.
   a) Plot the points and sketch the triangle.
   *b) Find the equation of the line through $B$, parallel to the line through $A$ and $C$.

8. The position of a particle moving on the axis is given by $x = -1 + 2t$.
   *a) Where is it at time 0?
   *b) When is it at position 0?
   c) What is its velocity?
   d) Sketch the graph of $x$ versus $t$.

### B

9. In the triangle $ABC$, where $A = (0, 0)$, $B = (1, 2)$, $C = (2, 0)$, find:
   *a) The equation of the line through $A$, perpendicular to the line through $B$ and $C$.
   b) The intersection of this perpendicular with the line through $B$ and $C$.
   c) The point where the perpendicular through $A$ intersects the analogous perpendicular through $B$.

10. For the triangle in problem 9, show that the perpendiculars through $A$, $B$, and $C$ all meet in a common point.

*11. a) While visiting Rome, you find that your temperature is $40°C$. Are you ill?
   b) Is $°F$ ever equal to $°C$? Ever less?

12. In the Kelvin temperature scale, the freezing point of water is $273°K$, and the boiling point is $373°K$. This scale is linearly related to Celsius and to Fahrenheit. Let K denote the temperature in the Kelvin scale.
    a) Express K in terms of C (degrees Celsius)
    b) Express F (degrees Fahrenheit) in terms of K

13. Gnome Copy Service charges 4 cents a page for photocopying; for photo-offset the charge is $2.50 for a photo plate of each page, and 2 cents for each copy made from the plate.
    a) Write the formula for the cost of photocopying $n$ copies of a page. [Call this cost $X(n)$.]
    b) Write the formula for the cost of $n$ copies by the photo-offset process. [Call this cost $P(n)$.]
    c) For what range of $n$ is photocopying cheaper, and for what range is photo-offset cheaper?

*14. In 1980, a minicar could be rented for $10/day, plus $0.10/mile, plus $0.05/mile for gas.
    a) What was the total cost of driving $x$ miles in one day?
    b) What was the *cost per mile* of driving $x$ miles in one day?

*15. A certain new piece of equipment costs $5,000 and is considered to be worn out and useless after ten years of operation. According to the "straight-line" depreciation method, the graph of value $V$ versus time $t$ is a straight line, with $V = 5,000$ when $t = 0$ and $V = 0$ when $t = 10$. Write the relation between $V$ and $t$, for $0 \le t \le 10$. What is the slope of the graph? Why is it negative?

*16. A normal probability table gives

| $x$ | $y$ |
|------|--------|
| 0.13 | 0.0517 |
| 0.14 | 0.0557 |

By linear interpolation, approximate $y$ when $x = 0.136$.

*17. A table of natural logarithms gives

$$\ln(7.00) = 1.94501, \qquad \ln(7.10) = 1.96009.$$

Approximate $\ln(7.07)$ by linear interpolation.

18. Suppose that it takes a certain plane 1,445 gallons of fuel to fly a 400-mile trip, and 1,077 gallons to fly a 260-mile trip. Assume that the distance flown, $D$, is linearly related to the gallons of fuel consumed, $G$.
    a) Write the relation expressing $G$ in terms of $D$.
    b) Interpret the slope and the $G$-intercept of the graph. (Note: It takes a certain amount of fuel to get airborne and a certain amount to land.)

19. A certain state income tax rate is 5% of all income over $1,000; the first $1,000 is tax-free, but the excess over $1,000 is taxed at 5%.
    a) Write a formula for the amount of tax, $T$, on an income of $I$ dollars, $I \ge 1,000$.
    b) Sketch the graph of the tax, $T$, versus income $I$, for $0 \le I \le 2,000$.

20. Suppose the tax in problem 19 is raised for incomes over $20,000; the state still taxes 5% of each dollar between 1,000 and 20,000, but 10% of each dollar over 20,000. Write a formula for the total tax, $T$, on an income of $I$ dollars, for each range:
    a) $0 \le I \le 1,000$.
    b) $1,000 \le I \le 20,000$.
    c) $20,000 \le I$.
    Sketch the graph of the total tax $T$ versus income $I$.

21. Let $P_1 = (1, 2)$ and $P_2 = (-1, -1)$. A third point $P$ is sought such that the triangle $P_1 P P_2$ is isosceles, with $P_1 P_2$ as its base; that is, $|PP_1| = |PP_2|$. Find the equation that must be satisfied by the coordinates $(x, y)$ of $P$. By algebraic reduction, show that this is the equation of a straight line, $L$, and that $L$ is perpendicular to $P_1 P_2$.

**C**

22. Suppose that segment $P_1 P_2$ is perpendicular to $P_2 P_3$. Use the distance formula to prove the Pythagorean theorem:

$$|P_1 P_3|^2 = |P_1 P_2|^2 + |P_2 P_3|^2.$$

(The distance formula was based on the Pythagorean theorem for triangles with legs parallel to the coordinate axes. Taking that formula as a definition of distance, this problem proves the Pythagorean theorem for right triangles with legs in any position.)

## 1.3

### FUNCTIONS

We need one more preliminary: the concept of **function**.

A function is like an abstract machine (Fig. 1) that accepts certain inputs; to each input $x$, it assigns a definite output $f(x)$ [read "$f$ of $x$"]. Think of "$f(\ )$" as the machine; insert $x$ in the slot formed by the parentheses, and you get $f(x)$.

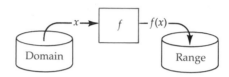

**FIGURE 1**
Function machine.

A hand calculator contains several functions, called "sin", "cos", "ln", and so on. You enter a number $x$ (the input), press the appropriate function key $f$, and read the output $f(x)$. For example, enter "3.2", press "ln", and read the output

$$\ln(3.2) = 1.16315081.$$

(Later chapters will show how this function is defined and computed.)

Many functions arise in geometry. For instance, given the edge-length $x$ of a square, its area is determined: $A = x^2$. The area is called a **function of the edge-length**; the function is defined by the equation

$$f(x) = x^2.$$

The input is the edge-length $x$, and the output is the area $A = f(x) = x^2$.

There are countless physical examples: The position of an object is a function of time; so are its velocity, temperature, radioactivity, and so on. In biology, the size of a given population is a function of time. In economics, for a given commodity, the quantity that can be sold is a function of the price $P$.

The **domain** of a function $f$ consists of all the possible inputs (Fig. 1). The output $f(x)$ is called the **value** of $f$ at $x$. The set of all these outputs is called the **range of values** or, simply, the **range** of $f$.

The domain may be given explicitly (Example 1), or else may be determined implicitly by the formula for $f$ (Examples 2–4), or by the context in which $f$ arises (as in the examples in Sec. 1.4). In most cases, it is easy and worthwhile to determine the domain precisely. The range, on the other hand, may be difficult to determine.

---

**EXAMPLE 1**  A function is defined by

$$f(x) = \sqrt{x}, \qquad x \geq 0. \tag{1}$$

The "$x \geq 0$" says that the domain of $f$ consists of all numbers $x \geq 0$. The formula "$f(x) = \sqrt{x}$" gives the output $f(x)$ for each input $x$ in that domain. To

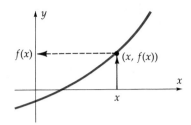

**FIGURE 2**
Graph of $f$ is $\{(x, y): y = f(x)\}$.

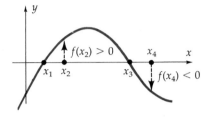

**FIGURE 3**
Graph of $f$. $x_1$ and $x_3$ are zeroes of $f$.

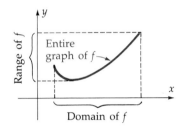

**FIGURE 4**
Domain and range.

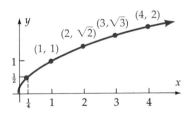

**FIGURE 5**
Graph of $f(x) = \sqrt{x}$.

find the output for any particular input, say $x = 4$, replace $x$ by 4 on both sides of the defining formula (1):

$$f(4) = \sqrt{4} = 2.$$

Similarly, $f(3) = \sqrt{3} = 1.732 \cdots$. Table 1 lists a few sample outputs.

**Table 1**

| Input $x$ | 0 | 1/4 | 1 | 2 | 9/4 | 3 | 4 |
|---|---|---|---|---|---|---|---|
| Output $\sqrt{x}$ | 0 | 1/2 | 1 | $\sqrt{2}$ | 3/2 | $\sqrt{3}$ | 2 |

The range of this function is all numbers $y \geq 0$; each such $y$ is the output corresponding to the input $y^2$:

$$y = \sqrt{y^2} = f(y^2).$$

**The Graph** of a function consists of all ordered pairs $(x, f(x))$, formed using all possible inputs $x$ in the domain of $f$ (Fig. 2). It is the graph of the equation $y = f(x)$, and can be described as the set

$$\{(x, y): y = f(x)\}.$$

The function value $f(x)$ gives the vertical distance from point $x$ on the horizontal axis up (or down) to the graph (Figs. 2 and 3). When $f(x) > 0$ the graph is $f(x)$ units *above* the axis; when $f(x) < 0$ the graph is $|f(x)|$ units *below* the axis. Points where the graph crosses the horizontal axis are called **zeroes of $f$**, or **$x$-intercepts**, or **horizontal intercepts**. They are the solutions, $x$, of the equation $f(x) = 0$.

The *domain* of $f$ can be visualized from the graph (Fig. 4). It consists of all points on the $x$-axis which lie directly below or above some point of the graph. These points form what is called the **projection** of the graph on the $x$-axis. Similarly, the *range* is the projection of the graph on the $y$-axis.

**EXAMPLE 1 (continued)**   To get the graph of $f(x) = \sqrt{x}$, plot the entries in Table 1 taking the input as first coordinate and the output as the second (Fig. 5). These points suggest a curve, which we draw as reasonably and smoothly as possible. This is a good representation of that part of the graph which fits in the sketch. The complete graph continues infinitely far to the right; we indicate this by an arrow at the right end of the part sketched.

**EXAMPLE 2**   $f(x) = x(10 - x)$. No domain is specified; but the formula $x(10 - x)$ gives a real number for every real $x$, so take the domain of $f$ to be all real numbers.

The zeroes of $f$ are the solutions of $f(x) = 0$:

$$x(10 - x) = 0, \quad \text{so } x = 0 \text{ or } x = 10. \tag{2}$$

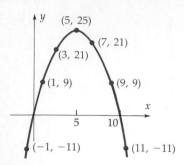

**FIGURE 6**
$f(x) = x(10 - x)$.

Table 2 includes these $x$ values, and selected other inputs $x$ with the corresponding outputs $f(x)$.

**Table 2**

| Input $x$ | $-1$ | 0 | 1 | 3 | 5 | 7 | 9 | 10 | 11 |
|---|---|---|---|---|---|---|---|---|---|
| Output $x(10 - x)$ | $-11$ | 0 | 9 | 21 | 25 | 21 | 9 | 0 | $-11$ |

Plot the resulting points as in Figure 6, and draw a smooth curve through them; this is a sketch of the graph.

---

*NOTE 1*  In Figure 6 it appears that the maximum value of $f(x) = x(10 - x)$ is $f(5) = 25$. This can be proved by completing the square (see Section 12.1):

$$x(10 - x) = -[x^2 - 10x] = -[x^2 - 10x + 5^2 - 5^2]$$
$$= -(x - 5)^2 + 5^2 = 25 - (x - 5)^2.$$

If $x \neq 5$, then $(x - 5)^2 > 0$, hence $f(x) = 25 - (x - 5)^2 < 25$, which proves that the maximum value of $f$ is indeed $25 = f(5)$.

The next chapter gives an easier way to find the maximum, using calculus.

*NOTE 2*  The line marked (2) used the "product principle": *A product equals zero if and only if one or another of the factors equals zero.* Thus

$$x(10 - x) = 0 \quad \text{if and only if} \quad x = 0 \text{ or } 10 - x = 0$$

which gives the solutions $x = 0$ and $x = 10$.

*NOTE 3*  The phrase "if and only if" has a convenient abbreviation as a double arrow, $\Leftrightarrow$. Thus the product principle can be stated as

$$ab = 0 \Leftrightarrow a = 0 \text{ or } b = 0.$$

We use this to indicate concisely the relation between two equivalent statements or equations.

A *single* arrow $\Rightarrow$ means "implies"; for example

$$x > 2 \Rightarrow x > 1$$
$$x = 0 \Rightarrow x(10 - x) = 0.$$

*NOTE 4*  If the domain of $f$ is not stated explicitly, we take it to be all real numbers which make sense in the formula for $f$. The two main operations which do not make sense are:

(i)  Division by zero.

(ii)  Taking the square root of a negative number, or any even root of a negative number.

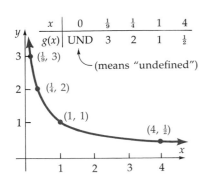

**FIGURE 7**

Graph of $g(x) = \dfrac{1}{\sqrt{x}}$.

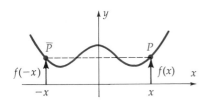

**FIGURE 8**
Even function: $f(-x) = f(x)$. $P$ and $\overline{P}$ are symmetric with respect to the $y$ axis.

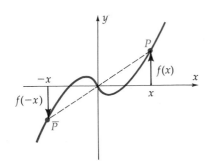

**FIGURE 9**
Odd function: $f(-x) = -f(x)$. $P$ and $\overline{P}$ are symmetric with respect to the origin.

**EXAMPLE 3**  $g(x) = \dfrac{1}{\sqrt{x}}$. Here the domain consists of all real numbers $x$ such that $\dfrac{1}{\sqrt{x}}$ makes sense as a real number. $\sqrt{x}$ is not defined when $x < 0$, and $\dfrac{1}{\sqrt{x}}$ is not defined when $\sqrt{x} = 0$ (because division by zero is not a legitimate algebraic operation). This rules out all $x \leq 0$, leaving only $x > 0$; thus the domain of $g$ consists of all numbers $x > 0$.

The function $g$ has no zeroes, since

$$\frac{1}{\sqrt{x}} = 0 \quad \text{implies} \quad 1 = 0 \cdot \sqrt{x} = 0,$$

which is impossible. Figure 7 shows a table and a sketch of the graph. Apparently the range of $g$ is all numbers $y > 0$.

**Odd and even functions** have certain symmetries which show up in their graphs. A function $f$ is called **even** if

$$f(-x) = f(x) \tag{3}$$

for every $x$ in the domain of $f$. The graph of an even function is symmetric about the vertical axis, as illustrated in Figure 8. The even powers have this symmetry; for example

$$(-x)^4 = (x)^4,$$

so $f(x) = x^4$ is an even function, and its graph is symmetric about the vertical axis.

The *odd* functions are those such that

$$f(-x) = -f(x) \tag{4}$$

for every $x$ in the domain of $f$. Naturally, the odd powers have this property:

$$(-x)^3 = -(x)^3.$$

The graph of an odd function is symmetric with respect to the origin (Fig. 9).

When you see an equation like (3) or (4), keep in mind that the function notation $f(x)$ does *not* mean that a number $f$ is multiplying a number $x$! So familiar rules for multiplication, such as

$$a(-b) = -a(b)$$
$$a(b + c) = a(b) + a(c)$$

do *not* generally apply to functions! Usually $f(-x)$ is different from $-f(x)$; only for *odd* functions are they equal, for all $x$. And $f(x + y)$ is rarely equal to $f(x) + f(y)$; for instance, $\sqrt{x + y}$ is not equal to $\sqrt{x} + \sqrt{y}$, except when $x = 0$ or $y = 0$ (see problem 17).

Most functions are neither even nor odd. For instance, with $f(x) = \sqrt{x}$ as in Example 1, the domain contains no points with $x < 0$, and the graph has no apparent symmetry (Fig. 5).

---

**EXAMPLE 4**    Is $f(x) = x(10 - x)$ even, odd, or neither?

*SOLUTION*    Compute a sample pair of values:

$$f(1) = 9, \qquad f(-1) = -11.$$

Since $f(-1) \neq f(1)$, equation (3) fails when $x = 1$, so $f$ is not even; likewise, $f(-1) \neq -f(1)$, so equation (4) fails, and $f$ is not odd either. The graph (Fig. 6) is symmetric with respect to neither the $y$-axis, nor the origin.

---

**EXAMPLE 5**    Is $f(x) = \dfrac{x}{1 + x^2}$ even, odd, or neither?

*SOLUTION*    Compute a sample pair of values:

$$f(1) = \tfrac{1}{2}, \qquad f(-1) = -\tfrac{1}{2}.$$

So $f(-1) = -f(1)$ as in condition (4) for an odd function; maybe this function is odd. To be sure, determine whether $f(-x) = -f(x)$ for *every* $x$ in the domain. To compute $f(-x)$, replace $x$ by $(-x)$ in the formula for $f(x)$:

$$f(-x) = \frac{(-x)}{1 + (-x)^2} = \frac{-x}{1 + x^2} = -\frac{x}{1 + x^2}$$
$$= -f(x).$$

We have proved this equality for *every* $x$ in the domain of $f$ and can therefore conclude that the function $f$ is odd. The graph in Figure 10 is indeed symmetric about the origin.

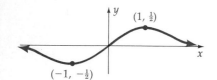

**FIGURE 10**

$f(x) = \dfrac{x}{1 + x^2}$ is odd.

---

Example 5 illustrates how to use function notation; given an expression for $f(x)$, you obtain $f(\text{anything})$ by replacing "$x$" by "(anything)" in the defining expression. For example, if $f(x) = x^2 + 2$ then

$$f(\text{anything}) = (\text{anything})^2 + 2;$$
$$f(a) = a^2 + 2; \qquad f(mx + b) = (mx + b)^2 + 2.$$

If $g(x) = \sqrt{x}$ then

$$g(a + 1) = \sqrt{a + 1}, \quad \text{and} \quad g(f(x)) = \sqrt{f(x)} = \sqrt{x^2 + 2},$$

with $f(x) = x^2 + 2$ as above.

Don't just listen to the sound of the words; look carefully at the symbols, so you can distinguish $\sqrt{a + 1}$ from $\sqrt{a} + 1$.

## SUMMARY

A function $f$ assigns a definite output $f(x)$ to each input $x$ in its domain. The *domain* is the set of all possible inputs, and the set of all resulting outputs $f(x)$ is the *range*. The number $f(x)$ is the *value* of $f$ at $x$.

The *graph* of $f$ (see Fig. 2) is the set $\{(x, y): y = f(x)\}$.

The *zeroes* of $f$ are the solutions of $f(x) = 0$, where the graph crosses the x-axis (see Fig. 3). These are also called the *x-intercepts* of the graph. The *y-intercept* is the value $f(0)$, where the graph crosses the y-axis.

The function $f$ is *even* if $f(-x) = f(x)$, and *odd* if $f(-x) = -f(x)$, for every $x$ in the domain of $f$ (see Figs. 8 and 9).

## PROBLEMS

### A

1. Let $f(x) = \dfrac{1}{\sqrt{4 - x^2}}$. Compute the following; if any is undefined, answer "UND".
   *a) $f(0)$     b) $f(2)$     *c) $f(-3)$
   *d) $f(-x)$    e) $f(a)$     f) $f(h + 1)$
   g) $f(a + h)$  h) $f(2x)$

2. Determine the domain and zeroes of the following functions. Graph each one, showing the zeroes and plotting points for the suggested values of $x$. From your graph, try to determine the range.
   a) $f(x) = 2x + 1$
   *b) $f(x) = \sqrt{1 + x}$;     $x = -1, -3/4, 0, 3$.
   *c) $g(x) = \dfrac{1}{\sqrt{1 + x}}$;     $x = -8/9, 0, 3$
   *d) $h(x) = \dfrac{1}{x - 1}$;     $x = 0, 1, 9/10, 11/10, 2$
   *e) $j(x) = x(x + 1)$;     $x = -2, -1/2, 1$
   f) $k(x) = x\sqrt{x + 1}$;     $x = -1/2, 3$
   g) $l(x) = x^2(1 - x^2)$;   $x = \pm 1/2, \pm 2$

*3. Sketch the graph of $f(x) = |x|$, for $-2 \le x \le 2$. Show any zeroes and any symmetry. (Plot points for $x = 0, \pm\frac{1}{2}, \pm 1, \pm 1\frac{1}{2}, \pm 2$.)

*4. For each of the graphs in Figure 11, indicate the *domain* and the *range* of the function. (In Figure 11c each small open circle indicates a point not on the graph. In Figure 11d the arrows indicate that the graph continues infinitely far; the dashed lines indicate that it comes up closer and closer to the lines $y = \pi$ and $y = -\pi$.)

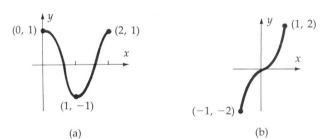

(a)                    (b)

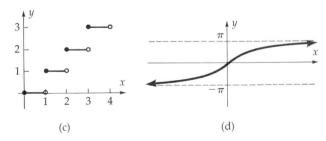

(c)                    (d)

**FIGURE 11**

5. Classify the following functions as even, odd, or neither.
   *a) $x$                  b) $\sqrt{x}$
   *c) $x\sqrt{x}$            d) $\dfrac{x^2}{1 + x^2}$
   *e) $\dfrac{x^3}{1 + x^2}$     f) $\dfrac{x^2}{1 + x}$
   *g) $\dfrac{\sqrt{x}}{1 + x^2}$   h) $2x^2 + 3x - 5$

*i)  $2x^2 - 5$        j)  $2x^3 + 3x^2 + 3x - 5$
k)  $2x^3 + 3x$       l)  $f(x) = 1$, for all $x$

6.  From the graphs in Figure 12, decide which of the functions shown are even, which are odd, and which are neither.

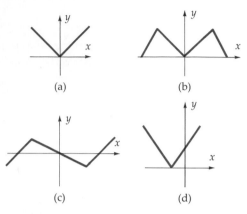

(a)                              (b)

(c)                              (d)

**FIGURE 12**
Even, odd, or neither?

7.  Let $f(x) = 1 + x$ and $g(x) = x^3$. Compute
*a)  $f(g(x))$.          *b)  $g(f(x))$.
c)  $g(x + 1)$.         *d)  $f(x - 1)$.
*e)  $8g\left(\dfrac{x}{2}\right)$.       *f)  $g(\text{ice})$.

8.  Let $f(x) = x^2 + 2x$ and $g(x) = x^2$, for all $x$. Show that:
a)  $f(x) - g(x) = 2x$.
b)  $f(x) = g(x + 1) - 1$.

9.  For the function $f$ in Figure 13, find:
a)  $f(2)$ and $f(3)$.
b)  The slope of the line through $P = (2, f(2))$ and $\hat{P} = (3, f(3))$.
c)  The equation of the line through $P$ and $\hat{P}$.

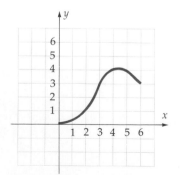

**FIGURE 13**
Graph of $f$ for problem 10.

10.  Let $f$ be the function in Figure 13, and $g(x) = f(x^2 + 1)$.
*a)  Compute $g(2)$.
b)  Compute $g(1)$ and $g(-1)$.
c)  Is $g$ even, odd, or neither?

**B**

11.  Express the distance from the origin to the point $(x, x^2)$ on the graph of $y = x^2$ as a function of $x$ (call the distance $D(x)$.)

12.  Consider all right triangles having one leg of length 2 cm. Denote by $x$ the length of the other leg. Express the length of the hypotenuse as a function of $x$. What is the domain of this function?

*13.  A man 7 feet tall is walking on a level street $x$ feet from the base of an 18-foot high streetlight. Express the length, $s$, of his shadow as a function of $x$.

*14.  Two functions are *equal* if they have *precisely* the same graph. Let

$$f(x) = \frac{x^2 - 9}{x - 3}$$

$$g(x) = x + 3 \qquad (x \neq 3)$$

$$h(x) = x + 3 \qquad (\text{for all } x).$$

Does $f = g$? $f = h$? $g = h$?

15.  Give an example of a function $f$ with domain $[0, 1]$ and with its graph symmetric in the $x$-axis. How many such functions are there?

16.  Are there any functions with domain $(-\infty, +\infty)$ which are both even and odd? How many?

17.  Prove that $\sqrt{x + y} < \sqrt{x} + \sqrt{y}$, unless $x = 0$ or $y = 0$. (Hint: Square $\sqrt{x} + \sqrt{y}$.)

**C**

18.  Suppose that $f$ is any function with domain $(-\infty, +\infty)$.
a)  Does the function $g$ defined by $g(x) = f(x) + f(-x)$ have any special symmetry?
b)  Does the function $h$ defined by $h(x) = f(x) - f(-x)$ have any special symmetry?
c)  Can every $f$ with domain $(-\infty, +\infty)$ be written as the sum of an even function and an odd function?

# 1.4

## OPTIMIZATION

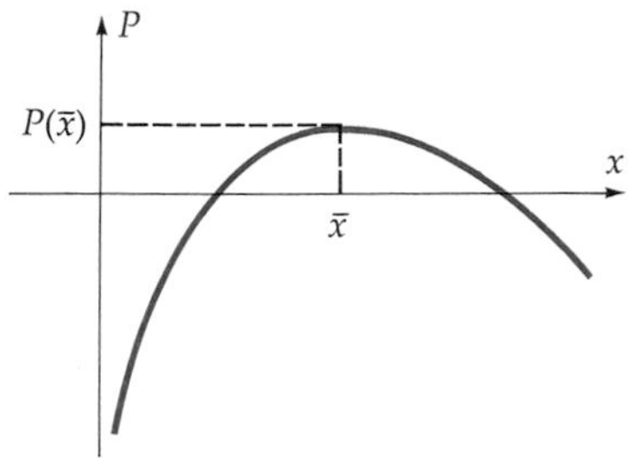

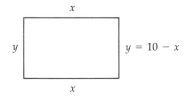

**FIGURE 1**
$P$ = profit from selling widgets priced at $x$ per widget.

Everyone wants "optimal results"; functions can help to achieve them. Consider, for example, the hypothetical "profit curve" in Figure 1, giving the net profit $P(x)$ to be made by selling as many widgets as possible at price $x$. If the price $x = 0$, there is a huge loss, representing the unrecovered cost of producing many widgets. The situation improves as the price rises; but when it gets too high, then few widgets are sold and the profit declines. Somewhere between these extremes is an optimal price $\bar{x}$, giving maximum profit $P(\bar{x})$. (This is optimal from the manufacturer's point of view; a consumer group might take a different view of "optimal price.")

This section gives examples in which a specific function can be set up to describe a given situation; then a nearly optimal result can be obtained by careful graphing. Chapters 2 and 3 will show how calculus perfects this technique by finding the optimal solution exactly, and in many cases more easily.

---

**EXAMPLE 1**   Given 20 meters of fence, design a rectangular enclosure containing the maximum possible area.

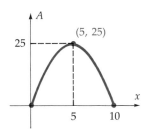

**FIGURE 2**
Fence with perimeter 20:
$2x + 2y = 2x + 2(10 - x) = 20$.

**SOLUTION**   You might say "Make it square," but could you explain why? Analyze the situation carefully. Figure 2 shows the rectangle, with sides labeled $x$ and $y$. Since the enclosure uses 20 meters of fence, the perimeter is

$$2x + 2y = 20.$$

Thus if the base $x$ is chosen, the other dimension $y$ is determined:

$$2y = 20 - 2x, \quad \text{or} \quad y = 10 - x.$$

You can now express the area enclosed, $A = xy$, as a function of $x$ alone:

$$A = x(10 - x).$$

The idea now is to *graph A versus x*, and find the maximum value of $A$ (Fig. 3). Since $A = x(10 - x)$, we graph the function

$$f(x) = x(10 - x), \qquad 0 \le x \le 10.$$

The restriction $0 \le x$ is needed since, in this context, $x$ is a length and cannot be negative. Likewise $10 - x$ is a length, so $10 - x \ge 0$ and $x \le 10$.

This function, $f$, has the same formula as the one in Example 2, Section 1.3—only the domain is different. As noted in that example, the maximum value of $A = f(x)$ is 25, and it occurs with $x = 5$. So the rectangle of maximum area has base $b = 5$ and height $10 - x = 10 - 5 = 5$; it is a square, as expected.

**FIGURE 3**
$A = x(10 - x), 0 < x < 10.$

---

**EXAMPLE 2**   Design a bin with square base, rectangular sides, and open top, containing 8 m³ and with minimum surface area.

**SOLUTION**   You might say "Make it a $2 \times 2 \times 2$ cube," but can you be sure? Again, analyze the situation.

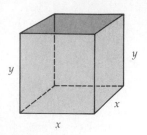

**FIGURE 4**
Topless bin.

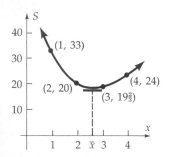

**FIGURE 5**

Surface $S = x^2 + \dfrac{32}{x}$.

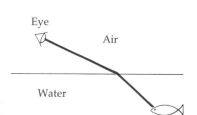

**FIGURE 6**
Refracted light ray.

Figure 4 shows the bin, with square base ($x$ by $x$) and rectangular sides ($x$ by $y$). The volume is given as 8, so we write

$$V = (\text{area of base})(\text{height}) = x^2 y = 8$$

and conclude that $y = 8/x^2$. Thus, once $x$ is chosen, then the requirement that $V = 8$ determines $y = 8/x^2$; and this in turn determines the surface area:

$$
\begin{aligned}
S &= x^2 + 4xy \quad (\text{area of base, } x \text{ by } x, \text{ plus four sides, each } x \text{ by } y)\\
&= x^2 + 4x(8/x^2) \quad (\text{since } y = 8/x^2)\\
&= x^2 + 32/x.
\end{aligned}
$$

For minimum surface area, we need the input $x$ that *minimizes* $S = x^2 + 32/x$; that is, we want the minimum value of the function

$$f(x) = x^2 + 32/x, \qquad x > 0.$$

Plotting points for $x = 1, 2, 3, 4$ gives the sketch in Figure 5. It appears that the minimum of $S$ occurs at some point $\bar{x}$ between $x = 2$ and $x = 3$. In any case, the minimum is *not* at $x = 2$, so the "optimal" bin is *not* a $2 \times 2 \times 2$ cube!

You could get the minimum more accurately by trying values between $x = 2$ and $x = 3$; after a lot of work, you would conclude that $x = 2.52$ is about the best choice.

In the mid-seventeenth century, Pierre de Fermat solved optimization problems by a "method of tangents" that he had devised. How does it work? Look back at the "optimal" $\bar{x}$ in Figure 5. Notice that at the corresponding point on the graph, the tangent line is horizontal and has zero slope. Fermat's method gives an equation whose solution is precisely the point $\bar{x}$ where the slope of the tangent equals 0; solving this equation gives the desired minimum value.

The most significant application made by Fermat was his explanation, in 1662, of the *law of refraction* describing the behavior of light rays passing from one medium to another as illustrated in Figure 6. (See Sec. 4.4 for details and a derivation of the law.)

Fermat's method of computing tangents was developed by Isaac Newton and Gottfried Leibniz into the general theory that we now call *differential calculus*, the subject of our next several chapters.

## PROBLEMS

### A

1. With 20 meters of fence, a rectangular pen is to be made along the side of a long barn; three sides require fence, and the fourth does not (Fig. 7).
   *a) Express $y$ in terms of $x$, given that 20 meters are to be used.
   *b) Express the area in the pen as a function of $x$.

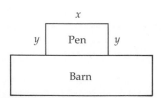

**FIGURE 7**

c) Graph the function in part b, showing intercepts.

*d) What dimensions $x$ and $y$ appear to yield maximum area? Is the optimal pen a square?

2. Repeat the previous problem, but with two *adjacent* pens made from 20 meters of fence (Fig. 8).

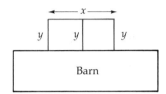

**FIGURE 8**

*3. A box with square bottom of edge $x$, and height $h$, is closed at the top. It is to contain exactly 8 cubic meters.

a) Express the required surface area $S$ as a function of $x$ alone. What is the appropriate domain of this function?

b) Sketch the graph of $S$ versus $x$ by plotting points for several values of $x$, and guess optimal values for $x$ and $h$.

4. The bottom of the bin in Example 2 costs $0.10/m^2$, and the sides $0.05/m^2$. Express the cost, $C$, as a function of $x$. Plot a few points, sketch the graph, and guess an approximate optimal value for $x$.

*5. a) The volume of the can in Figure 9 is $\pi r^2 h$, and its surface area is

$$S = 2\pi r^2 + 2\pi rh.$$

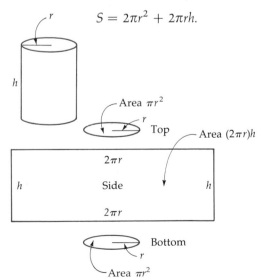

**FIGURE 9**
Surface area of a can.

It is to contain $100$ cm$^3$. Use this to express $S$ as a function of $r$ alone (eliminate $h$). What is the domain of the function, that is, for what values of $r$ is the function relevant?

b) To design a can containing $100$ cm$^3$, and requiring the least material to construct, you would look for the minimum of $S$ as a function of $f$. Plot points for $r = 1, 2, 3, 4, 5$, and guess an approximate optimal value of $r$.

6. The "demand function" for a certain product is

$$A(x) = 800,000(1 - x).$$

That is, when the price is $x$ dollars/unit, then $A(x)$ units can be sold. The *gross revenue* is then $R(x) = \left[ x \dfrac{\text{dollars}}{\text{unit}} \right] \cdot [A(x) \text{ units}] = x\,A(x)$ dollars. Sketch the graph of $R$ vs. $x$, and try to determine what price yields the greatest gross revenue.

*7. An island harbor $I$ is 5 miles offshore and 15 miles down the coast from city $C$ as shown in Figure 10. A ferry terminal is to be located $x$ miles down the coast from point $O$ on the shore opposite $I$. Assume that cars average 50 mph on land, and the high speed ferry travels 20 mph.

a) Express the time it takes to go from $C$ to $I$ as a function $f(x)$. What values of $x$ are relevant to the problem?

b) To optimize, would you look for a maximum value of $f$, or a minimum value? Sketch the graph of $f$, plotting points for $x = 0, 2, 4, 10, 15$; and estimate an optimal value for $x$.

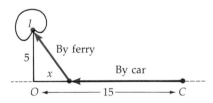

**FIGURE 10**

8. In Figure 11, telephone cable is to be laid from point $A$ on one side of the road to point $B$ on the

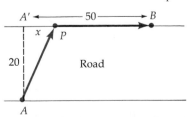

**FIGURE 11**

other side. The portion under the road, from $A$ to $P$, costs \$20/meter to lay, while the portion along the road, from $P$ to $B$, costs only \$10/meter.

a) Express the total cost $C$ of laying the cable as a function of $x$, the distance from $A'$ (opposite $A$) to $P$.

b) Sketch the graph of $C$ versus $x$, computing $C$ for $x = 2, 4, 6, 8, 10, 15$; and estimate an optimal choice of $x$.

## B

9. Along a highway are four cities (Fig. 12): $A$ (pop. 50,000), $B$ (pop. 10,000), $C$ (pop. 40,000) and $D$ (pop. 20,000). An oil depot somewhere along the highway is to be supplied by a pipeline. Assuming that each thousand citizens require 0.2 megagallons of oil a year:

a) Express the amount of oil transport required, in miles times megagallons per year, as a function of the distance $x$ from $A$ to the depot. (Hint: If $0 \le x \le 20$, then the depot is between $A$ and $B$, and the amount of transport required is

| miles | | megagallons | |
|-------|-------|-------|-------|
| $(x)$ | times | $(50)(.2)$ | $= 10x$ |

to supply city $A$

| | | | |
|-------|-------|-------|-------|
| $(20 - x)$ | times | $(10)(.2)$ | $= 2(20 - x)$ |

to supply city $B$

| | | | |
|-------|-------|-------|-------|
| $(60 - x)$ | times | $(40)(.2)$ | $= 8(60 - x)$ |

to supply city $C$

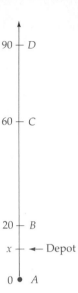

**FIGURE 12**
Depot along the road.

| | | | |
|-------|-------|-------|-------|
| $(90 - x)$ | times | $(20)(.2)$ | $= 4(90 - x)$ |

to supply city D,

a grand total of:

$$880 - 4x \quad \text{mile—megagallons per year.}$$

Make the corresponding calculations for $20 \le x \le 60$ and for $60 \le x \le 90$.)

b) Sketch the graph.

c) Find a position for the depot that minimizes the transport required.

---

# REVIEW PROBLEMS    CHAPTER 1

*1. Graph.
a) $y = x + 1$.    b) $y = |x|$.
c) $y = |x + 1|$.

*2. Write the following sets in interval notation.
a) $\{x: 0 < x \le 2\}$    b) $\{x: |x| \le 1\}$
c) $\{t: |t + 1| < 2\}$

*3. Write the following in set notation $\{x: \text{condition on } x\}$.
a) The interval $[-1, 1)$
b) The set of all solutions of $x^3 + 3x + 1 = 0$

*4. Write a formula for the distance between $(1, -1)$ and $(3, x)$.

*5. Write the equation of the line with the given properties and sketch it
a) with $y$-intercept $-1$ and slope 2.
b) with slope $-1$, passing through $(2, 5)$.
c) passing through $(1, 3)$ and $(3, 1)$.

*6. Write the equation of the line passing through the origin and
a) parallel to the line $x + 2y = 1$.
b) perpendicular to the line $x + 2y = 1$.

**\*7.** Let $A = (1, -1)$ and $B = (2, 1)$. Using the distance formula, write out the equation satisfied by all points $P = (x, y)$ such that $|PA| = |PB|$.

**\*8.** Determine the domain of

a) $\dfrac{x^2 - 1}{(x - 1)(x + 2)}$.

b) $\sqrt{4 - x^2}$.

c) $\dfrac{x - 1}{\sqrt{x^2 - 4}}$.

**\*9.** Determine whether $f$ is even, odd, or neither.

a) $f(x) = x + 1/x$

b) $f(x) = x^2 + 1/x^2$

c) $f(x) = \dfrac{x^2}{x + 1}$

d) $f(x) = \sqrt{1 + x^2}$

e) $f(x) = \sqrt{1 + x}$

f) $f(x) = |2x|$

g) $f(x) = |x + 2|$

**\*10.** Let $f(x) = x^3 + 2x$ and $g(x) = \sqrt{1 + x}$. Compute

a) $f(x + 1)$.

b) $f(x) + 1$.

c) $f(g(x))$.

d) $g(f(x))$.

e) $f(x + h) - f(x)$.

**\*11.** A lifeguard runs at 9 m/sec along the $x$-axis from the origin to the point $(x, 0)$, then swims at 4 m/sec to a drowning swimmer at $(5, 2)$. How long does it take?

**\*12.** A topless box is to have a square bottom costing $1/m^2$ and sides costing $2/m^2$. The volume is to be 4 $m^3$. Express the cost of material as a function of the height of the box; call the height $h$.

# 2

# THE DERIVATIVE
# OF A POLYNOMIAL

The central concept in calculus is the *derivative* of a function. We define the derivative in an intuitive geometric way, show how to compute the derivative of a polynomial, and then sample its various uses and interpretations.

## 2.1

## THE DERIVATIVE AND THE TANGENT LINE

Suppose that $f$ is a function with a nice smooth graph, such as a polynomial. Pick a point $P = (x, f(x))$ on that graph, and imagine a line drawn through $P$, *tangent* to the graph at $P$ as in Figure 1. The slope of this tangent line is called the **derivative of $f$ at $x$**, denoted $f'(x)$:

$f'(x) =$ the slope of the line tangent to the graph of $f$
at the point $(x, f(x))$.

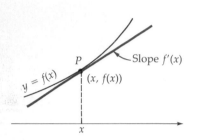

**FIGURE 1**
$f'(x) =$ Slope of tangent line at $(x, f(x))$.

The function sketched in Figure 2 appears to have a tangent line at every point; at the point $(0, f(0))$, for example, there is a tangent with slope roughly equal to $1/2$, so at that point the derivative of $f$ is about $1/2$:

$$f'(0) \approx 1/2.$$

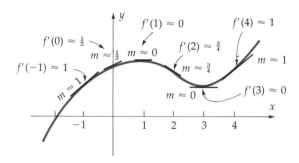

**FIGURE 2**
Graphical estimation of $f'(x)$.

By sketching more tangents and estimating their slopes we find, approximately, the values shown in Figure 2.

This graphical method gives a rough approximation of the derivative, but we need a precise way to compute the slope of the tangent line. The key idea is due originally to Pierre de Fermat. Suppose we are computing the slope of the tangent at a point $P = (x, f(x))$, (Fig. 3). Take a second point $\hat{P} = (\hat{x}, f(\hat{x}))$ on the graph. The straight line cutting the graph at $P$ and $\hat{P}$ is called a **secant**.[1] Now, move $\hat{P}$ gradually over to $P$ along the graph, like a bead sliding on a wire. As $\hat{P}$ approaches $P$, the secant line approximates the tangent line more and more closely. If we follow the *slope* of the secant line very carefully, it will lead us to the slope of the tangent line.

---

[1] "Secant" is Latin for "cutting," and "tangent" means "touching."

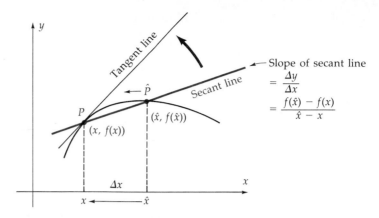

**FIGURE 3**
Secant line approximating tangent line, as $\hat{x} \to x$.

**EXAMPLE 1**   For $f(x) = x^2$, compute $f'(1)$.

**SOLUTION**   $f'(1)$ is the slope of the line tangent to the graph at point $P = (1, 1)$. Compute the slope of the secant line through $P$ and a nearby point $\hat{P} = (\hat{x}, f(\hat{x}))$ on the graph in Figure 4; this is

$$\frac{\Delta y}{\Delta x} = \frac{f(\hat{x}) - f(1)}{\hat{x} - 1} = \frac{\hat{x}^2 - 1^2}{\hat{x} - 1}.$$

Factor the numerator,

$$\frac{\Delta y}{\Delta x} = \frac{(\hat{x} - 1)(\hat{x} + 1)}{\hat{x} - 1},$$

and cancel the factors $\hat{x} - 1$:

$$\frac{\Delta y}{\Delta x} = \hat{x} + 1. \tag{1}$$

Now move $\hat{P} = (\hat{x}, f(\hat{x}))$ over to $P = (1, 1)$; as you do so, $\hat{x}$ moves over to 1. Figure 5 shows in detail a few steps along the way. In Figure 5a, $\hat{x} = 1.5$,

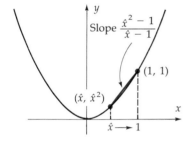

**FIGURE 4**
$f(x) = x^2$; computing $f'(1)$.

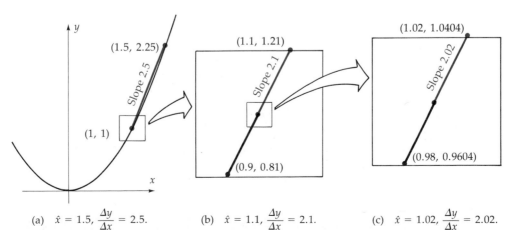

(a)   $\hat{x} = 1.5$, $\dfrac{\Delta y}{\Delta x} = 2.5$.        (b)   $\hat{x} = 1.1$, $\dfrac{\Delta y}{\Delta x} = 2.1$.        (c)   $\hat{x} = 1.02$, $\dfrac{\Delta y}{\Delta x} = 2.02$.

**FIGURE 5**

and formula (1) for the slope of the secant line gives

$$\frac{\Delta y}{\Delta x} = \hat{x} + 1 = 1.5 + 1 = 2.5.$$

In Figure 5b, $\hat{x} = 1.1$ is much closer to 1, and the slope of the corresponding secant line is

$$\frac{\Delta y}{\Delta x} = \hat{x} + 1 = 1.1 + 1 = 2.1.$$

In Figure 5c, $\hat{x} = 1.02$ is much closer yet, and

$$\frac{\Delta y}{\Delta x} = \hat{x} + 1 = 2.02.$$

To get the slope of the tangent line itself, move $\hat{P}$ all the way over to $P$, and set $\hat{x} = 1$ in the formula $\dfrac{\Delta y}{\Delta x} = \hat{x} + 1$, obtaining $1 + 1 = 2$. So the slope of the tangent line at $(1, 1)$ is exactly equal to 2. This is the *derivative* $f'(1)$:

$$f'(1) = 2.$$

---

The inherent difficulty in computing the slope of the tangent line at $P = (x, f(x))$ is this: To compute the slope of a line, we generally use two points; but on the tangent line we are given only one point, $P$. However, the tangent depends not only on the point $P$ itself, but also on the nearby points, which determine how the graph is "tipped" at $P$. Fermat's idea is to exploit those nearby points $\hat{P} = (\hat{x}, f(\hat{x}))$. The slope of the line through $P$ and $\hat{P}$ is

$$\frac{\Delta y}{\Delta x} = \frac{f(\hat{x}) - f(x)}{\hat{x} - x}. \tag{2}$$

As $\hat{P}$ approaches $P$ along the curve then $\hat{x}$ approaches $x$ (see Fig. 3); and the quotient in formula (1), the slope of the secant, approaches $f'(x)$, the slope of the tangent at $P$. We indicate this briefly by writing

$$\frac{f(\hat{x}) - f(x)}{\hat{x} - x} \to f'(x) \qquad \text{as } \hat{x} \to x.$$

The arrow ($\to$) is read as "approaches."

The quotient $\dfrac{f(\hat{x}) - f(x)}{\hat{x} - x}$ is called a **difference quotient** for $f$, and the derivative $f'(x)$ is called its "limit value" as $\hat{x} \to x$. For many functions this limit value can be calculated with elementary algebra.[2]

---

[2] The concept of "limit values" or "limits" is crucial for a thorough study of derivatives. We take this up in Chapter 3 and again, at a deeper level, in the Appendix. For the moment, however, we skate quickly over this thin ice, to see what lies beyond.

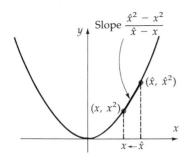

**FIGURE 6**
Computing $f'(x)$ for $f(x) = x^2$.

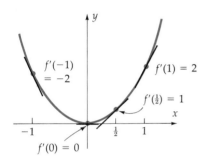

**FIGURE 7**
Sample values of $f'(x)$, for $f(x) = x^2$.

**EXAMPLE 2**   For $f(x) = x^2$, compute the derivative $f'(x)$.

*SOLUTION*   The difference quotient is (Fig. 6)

$$\frac{\Delta y}{\Delta x} = \frac{f(\hat{x}) - f(x)}{\hat{x} - x} = \frac{\hat{x}^2 - x^2}{\hat{x} - x}. \tag{3}$$

Factor and simplify:

$$\frac{\Delta y}{\Delta x} = \frac{(\hat{x} - x)(\hat{x} + x)}{\hat{x} - x}$$

$$= \hat{x} + x. \tag{4}$$

As $\hat{x} \to x$ then $\hat{x} + x \to x + x = 2x$, so $f'(x) = 2x$. The whole calculation can be written briefly as

$$\frac{\Delta y}{\Delta x} = \frac{f(\hat{x}) - f(x)}{\hat{x} - x} = \frac{\hat{x}^2 - x^2}{\hat{x} - x}$$

$$= \hat{x} + x$$

$$\to 2x \qquad [\text{as } \hat{x} \to x]$$

$$= f'(x).$$

Figure 7 illustrates sample values of the derivative $f'(x) = 2x$:

$$f'(-1) = 2(-1) = -2, \qquad f'(0) = 2 \cdot 0 = 0, \text{ etc.}$$

When we sketch lines with these slopes at the appropriate points on the graph, they really do appear to be tangent to the graph.

The secret to calculating $f'(x)$ in these examples is the algebraic simplification in going from (3) to (4). You have to let $\hat{x} \to x$; if you do this in (3), you get

$$\frac{\hat{x}^2 - x^2}{\hat{x} - x} \to \frac{x^2 - x^2}{x - x} = \frac{0}{0},$$

which is an undefined algebraic expression. But once the difference quotient (3) is simplified to (4), the denominator $\hat{x} - x$ has been cancelled against a like factor $\hat{x} - x$ in the numerator, and there is no longer any difficulty:

$$\hat{x} + x \to 2x \qquad \text{as } \hat{x} \to x.$$

This cancellation is always possible when $f$ is a polynomial, so in that case Fermat's method produces a simple formula for the derivative.

The necessary algebraic steps are simplified by a change in notation. Since $\Delta x = \hat{x} - x$ (Fig. 3), then $\hat{x} = x + \Delta x$, and the difference quotient is

$$\frac{\Delta y}{\Delta x} = \frac{f(\hat{x}) - f(x)}{\hat{x} - x} = \frac{f(x + \Delta x) - f(x)}{\Delta x}.$$

As $\hat{P}$ moves over to $P$, $\hat{x} \to x$, so $\Delta x = \hat{x} - x \to 0$. Thus the derivative is calculated as follows:

$$\frac{\Delta y}{\Delta x} = \frac{f(x + \Delta x) - f(x)}{\Delta x} \to f'(x) \qquad \text{as } \Delta x \to 0. \qquad (5)$$

**EXAMPLE 3** Compute the derivative of $f(x) = 3x^2 - x$, using the notation in (5).

*SOLUTION* The difference quotient is

$$\frac{\Delta y}{\Delta x} = \frac{f(x + \Delta x) - f(x)}{\Delta x} = \frac{[3(x + \Delta x)^2 - (x + \Delta x)] - [3x^2 - x]}{\Delta x}.$$

You can't set $\Delta x = 0$, for then the denominator would be 0. In order to see what happens as $\Delta x \to 0$, *rewrite* the difference quotient so that a factor $\Delta x$ appears in the numerator, to cancel the $\Delta x$ in the denominator. Multiply out:

$$\frac{\Delta y}{\Delta x} = \frac{3x^2 + 6x(\Delta x) + 3(\Delta x)^2 - x - (\Delta x) - 3x^2 + x}{\Delta x}.$$

All terms without a factor $\Delta x$ cancel, and

$$\frac{\Delta y}{\Delta x} = \frac{6x(\Delta x) + 3(\Delta x)^2 - \Delta x}{\Delta x} = \frac{\Delta x(6x + 3\Delta x - 1)}{\Delta x} = 6x + 3\Delta x - 1.$$

Now you can see what happens as $\Delta x \to 0$:

$$\frac{\Delta y}{\Delta x} = 6x + 3\Delta x - 1$$

$$\to 6x - 1 \qquad \text{as } \Delta x \to 0.$$

So $f'(x) = 6x - 1$.

### Equation of the Tangent Line

The derivative makes it easy to write the equation of the line tangent to the graph of $f$ at any given point $Q$ on the graph. Suppose that $Q = (a, f(a))$. The tangent line passes through $(a, f(a))$, and *its slope is precisely* $f'(a)$, the derivative $f'(x)$ *evaluated at the point* $x = a$; so the equation of the tangent line is (Fig. 8)

$$\frac{y - f(a)}{x - a} = f'(a)$$

or

$$y = f(a) + f'(a)(x - a).$$

Tangent line has slope $f'(a)$

$(x, y)$

$\Delta y = y - f(a)$

$(a, f(a))$ $\Delta x$

$\dfrac{y - f(a)}{x - a} = f'(a)$

$a$ $x$

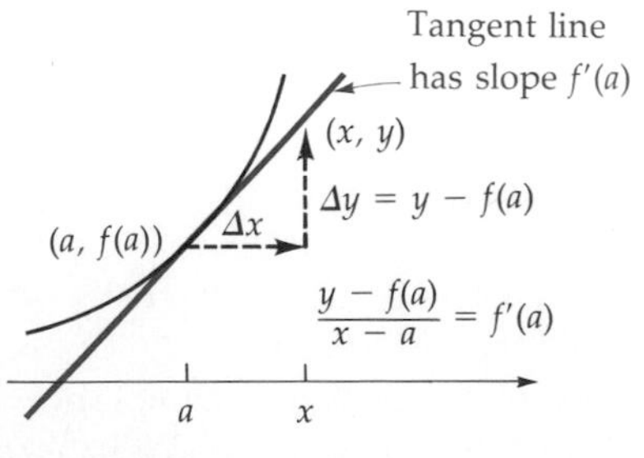

**FIGURE 8**
Equation of line tangent to graph of $f$ at $(a, f(a))$.

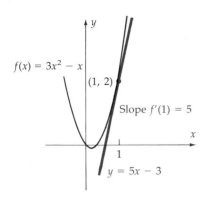

$f(x) = 3x^2 - x$

(1, 2)

Slope $f'(1) = 5$

1

$y = 5x - 3$

**FIGURE 9**
Tangent line at (1, 2).

**EXAMPLE 4**   Write the equation of the line tangent to the graph of $f(x) = 3x^2 - x$ at the point (1, 2).

**SOLUTION**   From Example 3, the derivative of $f$ is $f'(x) = 6x - 1$, so the slope of the tangent line at the point (1, 2) is

$$f'(1) = 6 \cdot 1 - 1 = 5.$$

The tangent line is therefore the line through (1, 2) with slope 5 (Fig. 9). Its equation is

$$\frac{y - 2}{x - 1} = 5$$

or

$$y = 2 + 5(x - 1) = 5x - 3.$$

**WARNING**   A common error is to use $f'(x)$ instead of $f'(a)$ in the equation of the tangent line, writing (in Example 4)

E R R O R $\Rightarrow$   $\dfrac{y - 2}{x - 1} = f'(x) = 6x - 1,$   or   $y = 2 + (6x - 1)(x - 1).$   $\Leftarrow$ E R R O R

This is not correct; you need the slope of the tangent line *at the point* $(1, f(1))$, which is $f'(1)$, not $f'(x)$. Notice that the equation obtained in this erroneous way is not the equation of a line at all; it cannot be written in the form $y = mx + b$ with *constants m and b.*

## Numerical Approximation of the Derivative

If function $f$ is not a polynomial, it may not be practical to cancel the factor $\Delta x = \hat{x} - x$ in the difference quotient by any algebraic trick. We will eventually see how to take derivatives of more general functions; but in any case, you can *approximate* the derivative $f'(x)$ by difference quotients:

$$f'(x) \approx \frac{f(x + \Delta x) - f(x)}{\Delta x}   \qquad \text{when } \Delta x \text{ is small.}$$

**EXAMPLE 5**   Use a calculator to approximate the derivative of the square root function $g(x) = \sqrt{x}$ at the point $x = 1$.

**SOLUTION**   When $\Delta x$ is small,

$$g'(1) \approx \frac{g(1 + \Delta x) - g(1)}{\Delta x} = \frac{\sqrt{1 + \Delta x} - 1}{\Delta x}.$$

With an eight-place calculator, it would be reasonable to take $\Delta x = 0.0000001$:

$$g'(1) \approx \frac{\sqrt{1.0000001} - 1}{0.0000001} = \frac{0.00000005}{0.0000001} = 0.5.$$

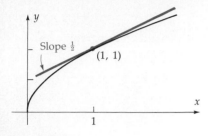

**FIGURE 10**
$f(x) = \sqrt{x}; f'(1) = 1/2.$

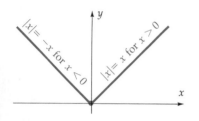

**FIGURE 11**
$f(x) = |x|$ has no tangent line at (0, 0).

Equally reasonable is $\Delta x = -0.0000001$:

$$g'(1) \approx \frac{\sqrt{0.9999999} - 1}{-0.0000001} = 0.5.$$

These calculations suggest that $g'(1) = 1/2$; this is correct (see problem 8c, or wait for Sec. 3.5). Figure 10 shows the graph of $g$ and the tangent at (1, 1) with slope 1/2.

---

Until now we have considered only functions with smooth graphs. For some functions, the graph is not smooth at every point, and accordingly $f'(x)$ does not exist for every $x$. A simple example is the absolute value function in Figure 11

$$f(x) = |x|.$$

The graph is not smooth at $x = 0$, where it has a corner. Suppose that we try to compute $f'(0)$. From the figure, the difference quotient for $f'(0)$ is

$$\frac{f(\hat{x}) - f(0)}{\hat{x} - 0} = \begin{cases} +1 & \text{if } \hat{x} > 0 \\ -1 & \text{if } \hat{x} < 0. \end{cases}$$

As $\hat{x} \to 0$, these quotients do not approach any *one* value (they approach two different values, 1 and $-1$). In such a case, we say that $f'(0)$ *does not exist, and the graph does not have a tangent line at* (0, 0).

This may contradict your feeling about tangents—in a way, the $x$-axis might be called a tangent to the graph at (0, 0), since it touches the graph there without crossing it—so we have to explain what we mean by "tangent line."

Consider a point $P$ on a smooth curve, such as a circle, or the graph of $y = x^2$ (Fig. 5). If you magnify a small piece of the curve near $P$ as in Figures 5b and 5c, it becomes straighter, just as the round earth looks flat when you view a relatively small part of it. As you magnify the curve more and more, a straight line emerges; *this is precisely the tangent line!*

The idea of successive magnifications is consistent with Fermat's method of difference quotients. With each magnification, the point $\hat{P} = (\hat{x}, f(\hat{x}))$ in the

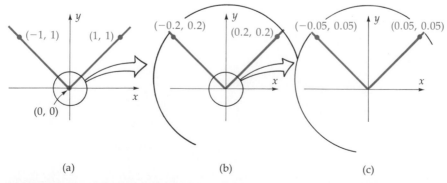

(a)                    (b)                    (c)

**FIGURE 12**
Magnifying graph of $f(x) = |x|$ near (0, 0).

figure corresponds to a value of $\hat{x}$ closer to $x$. As you continue magnifying, the numerical difference $\Delta x = \hat{x} - x$ becomes vanishingly small, and the difference quotient $\Delta y / \Delta x$ approaches the slope of the emerging straight line, the tangent line.

If you apply the same process to the graph of the absolute value $f(x) = |x|$ (Fig. 12), at every stage there is a right angle in the graph; no straight line emerges, so there is no tangent line.

## SUMMARY

*Derivative of f (see Fig. 3):*

$$\frac{\Delta y}{\Delta x} = \frac{f(\hat{x}) - f(x)}{\hat{x} - x} \to f'(x) \qquad \text{as } \hat{x} \to x.$$

$$\frac{\Delta y}{\Delta x} = \frac{f(x + \Delta x) - f(x)}{\Delta x} \to f'(x) \qquad \text{as } \Delta x \to 0.$$

*Equation of Line Tangent to Graph of f at (a, f(a))   (see Fig. 7):*

$$y = f(a) + f'(a)(x - a)$$

## PROBLEMS

(NOTE: The answer to problem 4b is used in the next section.)

### A

1.  For the given function:
    i)   Write the difference quotient

    $$\frac{\Delta y}{\Delta x} = \frac{f(x + \Delta x) - f(x)}{\Delta x},$$

    and illustrate it with a sketch like that in Figure 6, appropriately labelled;
    ii)   Simplify the difference quotient and cancel $\Delta x$;
    iii)   Compute $f'(x)$, by letting $\Delta x \to 0$;
    iv)   Compute $f'(2)$ and $f'(-1)$.
    *a)   $f(x) = x^2 + 2x$
    *b)   $f(x) = \frac{1}{2}x^2 + 1$
    c)   $f(x) = 2x^2 + 3x + 1$
    *d)   $f(x) = 3x + 1$
    e)   $f(x) = 3$ (a constant function)
    *f)   $f(x) = 2x - x^2$

*2.  For each function in problem 1, write the equation of the tangent line at $(2, f(2))$.

3.  Compute $g'(x)$. Write the equation of the tangent line at $(3, g(3))$.
    a)   $g(x) = 2x + 1$
    *b)   $g(x) = x - x^2$

*4.  Compute $f'(x)$; $m$, $b$, $a$, $c$ are constants.
    a)   $f(x) = mx + b$
    b)   $f(x) = ax^2 + bx + c$

5.  Using a calculator or tables, estimate the derivative of the given function at the given point (see Example 5).
    *a)   $f(x) = 1/x$; $f'(1)$
    *b)   $f(x) = \sqrt{x}$; $f'(4)$
    *c)   $f(x) = \sin x$; $f'(0)$ ($x$ in radians; on some calculators, round-off errors create inaccuracies in computing $\sin(\Delta x)$ for $\Delta x < 0.00001$, say.)
    d)   $f(x) = \cos x$; $f'(0)$ ($x$ in radians)
    e)   $f(x) = \log_{10} x$; $f'(1)$
    f)   $f(x) = \ln x$; $f'(1)$

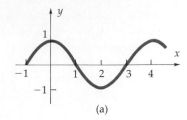

(a)

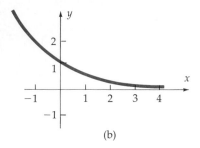

(b)

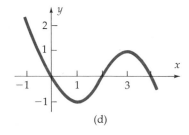

(c)

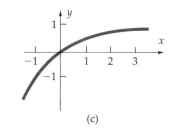

(d)

**FIGURE 13**
Graphs for problem 6.

6. Figure 13 shows graphs of four functions $f$. For each, estimate $f'(x)$ for $x = -1, 0, 1, 2, 3$. (The $x$- and $y$-axes have equal scales.)

**B**

7. For each function in problem 6, sketch the graph of $f'(x)$ for $-1 \le x \le 3$.

8. Find $f'(x)$ exactly. Write the equation of the tangent line at $(1, 1)$.
   a) $f(x) = 1/x$     b) $f(x) = 1/x^2$
   c) $f(x) = \sqrt{x}$
      [Hint: $\hat{x} - x = (\sqrt{\hat{x}} - \sqrt{x})(\sqrt{\hat{x}} + \sqrt{x})$]

9. a) Draw a sketch illustrating the "two-sided" difference quotient

$$\frac{\Delta y}{\Delta x} = \frac{f(x + \Delta x) - f(x - \Delta x)}{2\Delta x}.$$

   Why is the denominator $2\Delta x$?

   b) Show that if $f(x)$ is a quadratic $ax^2 + bx + c$, then the two-sided difference quotient gives $f'(x)$ exactly, even when $\Delta x \ne 0$. (The two-sided difference quotient is generally a more accurate approximation than the usual one; but when $\Delta x \to 0$, both give the same result for $f'(x)$.)

**C**

10. Find the equation of the line tangent to the graph of $y = x^{1/3}$ at the point $(0, 0)$.

11. What is the relation between the graph for $g(x) = \sqrt{x}$ in Figure 10 and the graph for $f(x) = x^2$ in Figure 4? What is the relation between the derivatives $g'(1) = 1/2$ and $f'(1) = 2$? Explain the relation.

## 2.2

# QUADRATIC FUNCTIONS

We'll use the derivative to graph quadratics, draw their tangents, and analyze parabolic mirrors.

A function $f$ is called **quadratic** if it has the form

$$f(x) = ax^2 + bx + c,$$

where $a$, $b$, and $c$ are constants with $a \ne 0$. The derivative is given by

$$f'(x) = 2ax + b \tag{1}$$

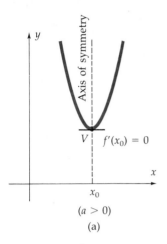

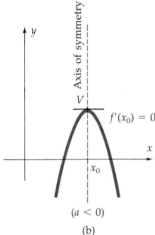

**FIGURE 1**
$f(x) = ax^2 + bx + c.$

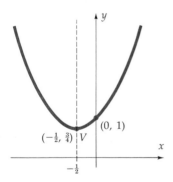

**FIGURE 2**
$f(x) = x^2 + x + 1.$

(problem 4b, Sec. 2.1). For example, the derivative of $f(x) = 3x^2 + 5x + 1$ is

$$f'(x) = 2 \cdot 3x + 5 = 6x + 5.$$

## Graphing

The graph of a quadratic function is a parabola with a vertical axis of symmetry (Figure 1—for a proof, see problems 19 and 20, or Section 12.1). The parabola opens up if $a > 0$ and down if $a < 0$. The point where the graph crosses the axis of symmetry is called the **vertex**. You can see that at the vertex, the tangent line is horizontal and the slope is zero. So you graph the quadratic like this: Compute $f'(x)$ and find the number $x_0$ where $f'(x_0) = 0$. Then $(x_0, f(x_0))$ is the vertex of the graph, and the vertical line $x = x_0$ is the axis of symmetry! Plot the vertex and one or two other points, and the graph is clear.

---

**EXAMPLE 1** Sketch the graph of $f(x) = x^2 + x + 1$.

*SOLUTION* The derivative is

$$f'(x) = 2x + 1.$$

The slope is zero when $2x + 1 = 0$, that is, $x = -1/2$. When $x = -1/2$, then $y = f(-1/2) = (-1/2)^2 - 1/2 + 1 = 3/4$; so the vertex is

$$V = (-1/2, 3/4).$$

Another point on the graph is the $y$-intercept, where the graph crosses the $y$-axis, at

$$f(0) = 0^2 + 0 + 1 = 1.$$

Plot $V$ and the intercept, sketch a parabola symmetric in the vertical line through $V$, and you have the graph in Figure 2.

---

Any $x$-intercepts of $f(x) = ax^2 + bx + c$ can be found either by factoring or by using the quadratic formula (proved in Section 12.1)

$$ax^2 + bx + c = 0 \Leftrightarrow x = \frac{-b \pm \sqrt{b^2 - 4ac}}{2a}.$$

An example of factoring is:

$$x^2 - x - 2 = 0 \Leftrightarrow (x - 2)(x + 1) = 0$$
$$\Leftrightarrow x = 2 \text{ or } x = -1.$$

An example requiring the quadratic formula is:

$$2x^2 - x - 2 = 0 \Leftrightarrow x = \frac{1 \pm \sqrt{1 + 16}}{2 \cdot 2} = \frac{1 \pm \sqrt{17}}{4}$$

as shown in Figure 3.

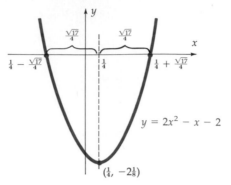

**FIGURE 3**

$$2x^2 - x - 2 \Leftrightarrow x = \frac{1 + \sqrt{17}}{4}.$$

## Parabolic Mirrors

Take the graph of $y = ax^2$, revolve it about the $y$-axis, and silver the inside of the resulting bowl-shaped figure. You now have a parabolic mirror with an interesting and useful property (Fig. 4): *All light rays entering the parabola parallel to its axis of symmetry are reflected to the same point F on the y-axis.* That point is called the **focus** of the parabola.

The reflecting property also works in reverse—any ray leaving the focus is reflected along a line parallel to the axis of symmetry. Microwave transmission exploits this (Fig. 5): microwaves are transmitted from the focus $T$ of one parabolic reflector, which converts them into a parallel beam aimed at a second reflector, which then focuses the beam on a receiver $R$. Further applications are in sonar, reflecting telescopes, satellite transmission, headlights, searchlights, and so on.

The proof of the reflecting property is based on Figure 6 and uses our newfound ability to determine the tangent line. A light ray (the "incident ray") entering the parabola parallel to the $y$-axis is reflected at point $P = (x_0, ax_0^2)$. We seek the point $F$ where the reflected ray crosses the $y$-axis.

By the law of reflection, the angle $\alpha$ between the incident ray and the tangent line $T$ equals the angle $\beta$ between $T$ and the reflected ray:

$$\alpha = \beta.$$

Moreover, since the incident ray is parallel to the $y$-axis, angles $\alpha$ and $\gamma$ are equal. Hence $\beta = \gamma$, so triangle $PFI$ is isosceles, and it follows that

$$|PF| = |FI|. \tag{2}$$

Our aim is to discover the coordinates of the point $F$ where the reflected ray crosses the $y$-axis. To do that, we find the coordinates of point $I$, then use equation (2).

*To find I,* write the equation of the tangent line at $P$. The slope of this line is

$$f'(x_0) = 2ax_0.$$

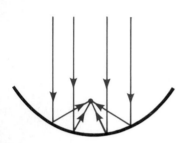

**FIGURE 4**
Cross section of parabolic mirror.

**FIGURE 5**
Microware transmission scheme.

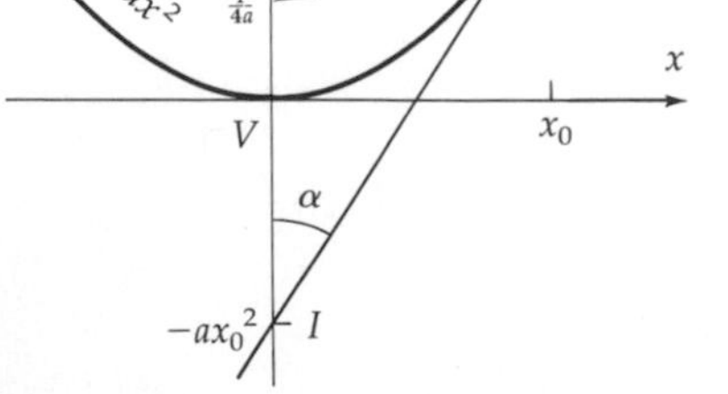

**FIGURE 6**
Reflection in a parabola.

The equation of the tangent at $P$ is thus

$$\frac{y - ax_0^2}{x - x_0} = 2ax_0,$$

which gives

$$y = 2ax_0(x - x_0) + ax_0^2$$

or

$$y = 2ax_0x - ax_0^2.$$

The $y$-intercept in this equation is $-ax_0^2$, so the desired point $I$ is $(0, -ax_0^2)$, as shown in Figure 6.

    *Now, to find F:* Since $F$ lies on the $y$-axis, its coordinates can be labeled $(0, p)$ for some number $p$; we will determine $p$. Equation (2) gives

$$|PF|^2 = |FI|^2$$

which by the distance formula is

$$(x_0 - 0)^2 + (ax_0^2 - p)^2 = (p - (-ax_0^2))^2 + 0.$$

Algebra gives

$$x_0^2 + (ax_0^2)^2 - 2ax_0^2 p + p^2 = p^2 + 2pax_0^2 + (ax_0^2)^2.$$

Leaving $x_0^2$ on the left, transpose and combine all other terms:

$$x_0^2 = 4pax_0^2.$$

Cancel $x_0^2$ and solve for $p$:

$$p = 1/4a.$$

Hence, *any incident ray parallel to the y-axis, reflected at any point P whatsoever, crosses the y-axis at the point $F = (0, 1/4a)$. This point $F$ is the same for all rays; it is independent of the point of reflection $P = (x_0, ax_0^2)$.* This is the reflecting property of the parabola that we set out to prove.

## SUMMARY

The graph of $f(x) = ax^2 + bx + c$ is a parabola if $a \neq 0$. The vertex $V$ is where $f'(x) = 0$; the axis of symmetry is the vertical line through $V$; and any $x$-intercepts can be found by factoring, or by using the quadratic formula:

$$ax^2 + bx + c = 0 \Leftrightarrow x = \frac{-b \pm \sqrt{b^2 - 4ac}}{2a}.$$

## PROBLEMS

NOTE: Problems 7–9 are useful preparation for the next section.

### A

1. Graph the given parabola by finding its vertex. Show the axis of symmetry and at least three points on the graph. Show any intercepts with the coordinate axes.
   *a) $y = x^2 - x + 1$
   *b) $y = x^2 - 3x - 1$
   c) $y = 3x^2 - 2x + 1$
   *d) $y = x(10 - x)$

2. Find the equation of the line tangent to the given graph at the given point. Sketch the graph and the tangent line.
   *a) $y = 5x^2 - x$, $P = (2, 18)$
   b) $y = x(3 - x)$, $Q = (1, 2)$

*3. With 120 meters of fence, a rectangular pen is to be made along a straight river, using the river as one unfenced side. What dimensions yield maximum area?

4. Repeat the previous problem with $L$ meters of fence. Show that the optimal pen is half a square.

5. A rectangular building is to be constructed on a triangular lot, as in Figure 7.

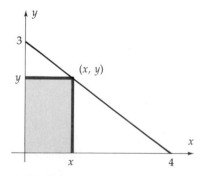

**FIGURE 7**

a) Determine the equation of the line containing the hypotenuse of the lot.
*b) Determine dimensions $x$ and $y$ to maximize the area on the ground floor.

### B

6. A rectangular building is to be constructed on a triangular lot as in Figure 8. Determine dimensions $l$ and $w$ to maximize the area on the ground floor. (Express $l$ and $w$ in terms of the $x$ in the figure.) Compare this result to that in problem 5.

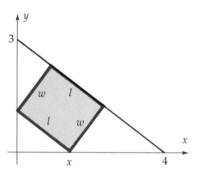

**FIGURE 8**

7. a) Check that $b^3 - a^3 = (b - a)(b^2 + ba + a^2)$.
   *b) For $f(x) = x^3$, compute $f'(x)$, using $\dfrac{\Delta y}{\Delta x} = \dfrac{f(\hat{x}) - f(x)}{\hat{x} - x}$.

8. a) Complete the factorization $b^4 - a^4 = (b^2 - a^2)(b^2 + a^2) = ?$
   b) For $f(x) = x^4$, compute $f'(x)$, using $\dfrac{\Delta y}{\Delta x}$ as in problem 7b.

9. a) Show that $b^4 - a^4 = (b - a)(b^3 + b^2a + ba^2 + a^3)$.
   b) Prove an analogous formula for $b^5 - a^5$.
   c) For $f(x) = x^5$, compute $f'(x)$.

10. A ray of light is *outside* a parabola, but headed for the focus. What is the direction of the reflected ray? Why?

11. a) Show that the line $y - 1 = m(x - 1)$ intersects the graph of $y = x^2$ at point $P = (1, 1)$.
    b) Find *all* intersections of the line and graph in part a. Show that there are two distinct intersections unless $m = 2$. (Hence the tangent line to the graph of $y = x^2$ is the only

nonvertical line which intersects the graph at just one point.)

12. Repeat the previous problem for an arbitrary point $P = (x_0, x_0^2)$ on the graph of $y = x^2$, using the line $y - x_0^2 = m(x - x_0)$; show that the line intersects the graph in two points unless $m = 2x_0$. (This suggests a method to compute derivatives without using Fermat's idea; but it does not work well for functions more complicated than quadratics.)

13. The parabola $y = x^2$ divides the plane into three parts $R_0$, $R_1$, $R_2$, as follows: if $P$ is in $R_0$, then *no* line through $P$ is tangent to the parabola; through any point $P$ in $R_1$ passes exactly one line tangent to the parabola; and through any point $P$ in $R_2$ pass two such lines. Describe $R_0$, $R_1$, $R_2$.

14. Determine a quadratic function whose graph passes through the three given points, or explain why no such function exists. (Give a geometric explanation, or an algebraic one, or preferably both.)
    *a)  $(-1, 0), (0, 0), (1, 0)$
    *b)  $(0, 3), (1, 2), (2, 4)$
    c)  $(1, 0), (3, 1), (7, 3)$
    d)  $(1, 0), (1, 1), (2, 1)$

15. At the end of this section, in deriving the coordinate $p = 1/4a$ of the focus (page 43), we divided by $x_0$, thus tacitly assuming that $x_0 \neq 0$. Complete the derivation by describing the path of the reflected ray when $x_0 = 0$.

16. A large parabolic reflector is 4 meters in diameter and 1.5 meters deep at the center. In an appropriate coordinate system, find the equation $y = ax^2$ of its cross section, and find the focus.

17. Let $x_1$ and $x_2$ be zeroes of $ax^2 + bx + c$.
    a)  Show that $x_1 + x_2 = -b/a$.
    b)  Express the product $x_1 x_2$ in terms of $a$, $b$, and $c$.
    c)  Show that $ax^2 + bx + c = a(x - x_1)(x - x_2)$.

18. In Figure 6, write the equation of the perpendicular bisector of the segment $IP$, and find its intersection with the $y$-axis.

19. a)  Find the vertex $(x_0, y_0)$ of the graph of $y = ax^2 + bx + c$. ($x_0$ and $y_0$ will depend on $a$, $b$, and $c$.)
    b)  Show that $y = a(x - x_0)^2 + y_0$.
    c)  Suppose that $a > 0$. Explain directly from part b why $(x_0, y_0)$ is the lowest point on the graph. (Use the fact that $(x - x_0)^2 > 0$ unless $x = x_0$.)
    d)  Suppose that $a < 0$, and explain why $(x_0, y_0)$ is the highest point on the graph.
    e)  Deduce from part b that the graph is symmetric with respect to the vertical line $x = x_0$.

20. In Figure 6, draw the parallelogram having $PF$ and $FI$ as two of its sides and $PI$ as diagonal. Denote by $D$ the vertex of this parallelogram opposite $F$. Show that:
    a)  $D$ lies on the line $y = -1/4a$. (This line is called the *directrix* of the graph.)
    b)  $|PF| = |PD|$. (This equation gives a geometric characterization of the graph and shows that it is a *parabola*—see Sec. 12.1.)

## 2.3

## THE DERIVATIVE OF A POLYNOMIAL

Our next goal is to simplify the calculation of derivatives. We'll introduce some standard notations and then develop a few easy formulas that give the derivative of any polynomial at a glance.

**Operator notation** for the derivative is a capital "$D$", meaning "the derivative of":

$$Df(x) = f'(x)$$

For example, $Dx^2 = 2x$, $D(3x^2 + x) = 6x + 1$.

**Leibniz notation** for the derivative of $f$ is

$$\frac{df}{dx} = f'(x)$$

or, if $y = f(x)$,

$$\frac{dy}{dx} = f'(x).$$

The expression $dy/dx$ is not really a fraction, since neither $dy$ nor $dx$ has been defined; but it does recall the difference quotient $\Delta y/\Delta x$ that we use in defining the derivative. With Leibniz notation

$$\frac{dx^2}{dx} = 2x, \qquad \frac{d(3x^2 + x)}{dx} = 6x + 1.$$

Using a different letter for the variable $x$ calls for a corresponding change in the Leibniz notation:

$$\frac{dt^2}{dt} = 2t, \qquad \frac{d(8z^2 - 3z + 1)}{dz} = 16z - 3.$$

Why all these notations? Each has its advantages; we use whichever is most convenient in the given context.

Now we'll prove the derivative formulas. We begin with the "power function" $f(x) = x^n$, then show how to handle monomials $cx^n$ and finally polynomials $c_n x^n + c_{n-1} x^{n-1} + \cdots + c_1 x + c_0$.

**The Power Rule** gives the derivative of any power $x^n$. You already know the derivative when $n = 2$:

$$\frac{dx^2}{dx} = 2x. \tag{1}$$

As a further hint to the general power rule, we compute $\dfrac{dx^4}{dx}$. If $f(x) = x^4$ then

$$\frac{f(\hat{x}) - f(x)}{\hat{x} - x} = \frac{\hat{x}^4 - x^4}{\hat{x} - x}$$

$$= \frac{[\hat{x}^2 - x^2] \cdot [\hat{x}^2 + x^2]}{\hat{x} - x}$$

$$= \frac{(\hat{x} - x)(\hat{x} + x)[\hat{x}^2 + x^2]}{\hat{x} - x}$$

$$= (\hat{x} + x)(\hat{x}^2 + x^2)$$

$$\to (2x)(2x^2) = 4x^3 \qquad \text{as } \hat{x} \to x.$$

Thus

$$\frac{dx^4}{dx} = 4x^3.$$

This suggests the general formula

$$\frac{dx^n}{dx} = nx^{n-1}.$$

(2)

This is the **Power Rule**, valid for every *constant* power $n$; it is proved for positive integers in the Appendix to this section and for more general powers later on. In the special case $n = 2$, the power rule gives

$$\frac{dx^2}{dx} = 2x^{2-1} = 2x^1 = 2x$$

in accord with (1). And when $n = 1$, it gives

$$\frac{dx^1}{dx} = 1 \cdot x^{1-1} = 1 \cdot x^0 = 1,$$

which is correct; the graph of $y = x^1 = x$ is a straight line of slope 1.

## The Derivative of a Polynomial

Every polynomial is built up from powers $x^n$ by two operations—multiplying by constants and adding functions. (For example, $2x^2 + x$ is obtained by multiplying the power function $x^2$ by 2, then adding the function $x$.) We will prove two general formulas relating derivatives to these operations.

Suppose that $f$ is a function with derivative $f'$, that is,

$$\frac{f(\hat{x}) - f(x)}{\hat{x} - x} \to f'(x) \qquad \text{as } \hat{x} \to x.$$

(3)

Let $c$ be any constant. Then the function $cf(x)$ has a derivative, computed as follows:

$$\frac{cf(\hat{x}) - cf(x)}{\hat{x} - x} = \frac{c[f(\hat{x}) - f(x)]}{\hat{x} - x}$$

$$= c\frac{f(\hat{x}) - f(x)}{\hat{x} - x}$$

$$\to cf'(x) \qquad \text{as } \hat{x} \to x$$

because of (3). So *the derivative of cf is c times the derivative of f*:

$$(cf)' = c \cdot f'$$

or

$$D(cf) = cDf$$

or

$$\frac{dcf(x)}{dx} = c\frac{df(x)}{dx}.$$

For example,

$$D(5x^3) = 5Dx^3 = 5 \cdot 3x^2$$

$$\frac{d(-x^5)}{dx} = \frac{d(-1 \cdot x^5)}{dx} = -1\frac{dx^5}{dx} = -5x^4.$$

Next, consider the sum of two functions $f$ and $g$. Suppose that both have derivatives; thus in addition to (3),

$$\frac{g(\hat{x}) - g(x)}{\hat{x} - x} \to g'(x) \qquad \text{as } \hat{x} \to x. \tag{4}$$

Then the sum $f + g$ has a derivative, computed as follows:

$$\frac{[f(\hat{x}) + g(\hat{x})] - [f(x) + g(x)]}{\hat{x} - x} = \frac{f(\hat{x}) + g(\hat{x}) - f(x) - g(x)}{\hat{x} - x}$$

$$= \frac{[f(\hat{x}) - f(x)] + [g(\hat{x}) - g(x)]}{\hat{x} - x}$$

$$= \frac{f(\hat{x}) - f(x)}{\hat{x} - x} + \frac{g(\hat{x}) - g(x)}{\hat{x} - x}$$

$$\to f'(x) + g'(x) \qquad \text{as } \hat{x} \to x$$

because of (3) and (4). This shows that *the derivative of a sum is the sum of the derivatives*:

$$(f + g)' = f' + g' \tag{5}$$

or

$$D(f + g) = Df + Dg$$

or

$$\frac{d[f(x) + g(x)]}{dx} = \frac{df(x)}{dx} + \frac{dg(x)}{dx}.$$

For example,

$$D(x^2 + x^3) = 2x^1 + 3x^2 = 2x + 3x^2$$

$$\frac{d(t^5 + 1)}{dt} = 5t^4 + 0 = 5t^4.$$

For the *difference* of two functions, the derivative is

$$(f - g)' = f' - g'.$$

This can be proved just like formula (5) for sums.

For a sum with three terms, you have

$$(f + g + h)' = (f + (g + h))'$$
$$= f' + (g + h)' \quad \text{[by (5), with } g \text{ replaced by } g + h]$$
$$= f' + g' + h' \quad \text{[by (5) again]}.$$

Clearly, *the derivative of any finite sum of functions is the sum of the derivatives of the functions.*

Combining all these rules, you can handle any polynomial.

---

## EXAMPLE 1

$$D(2x^5 - 3x^2 + 2) = D(2x^5) + D(-3x^2) + D2$$
$$\qquad\qquad\qquad\qquad\qquad \text{[since } D(f + g + h) = Df + Dg + Dh]$$
$$= 2Dx^5 - 3Dx^2 + 0 \quad \text{[since } D(cf) = c \cdot Df]$$
$$= 2 \cdot 5x^4 - 3 \cdot 2x \quad \text{[since } Dx^n = nx^{n-1}]$$
$$= 10x^4 - 6x.$$

---

In Leibniz or operator notation, the value of a derivative at a particular point $x_0$ is denoted like this:

$$\frac{df(x)}{dx}\bigg|_{x_0} \quad \text{or} \quad \frac{dy}{dx}\bigg|_{x_0} \quad \text{or} \quad \frac{dy}{dx}\bigg|_{x = x_0} \quad \text{or} \quad Dy\bigg|_{x = x_0}.$$

---

## EXAMPLE 2

$$\frac{d(x^2 + 1)}{dx}\bigg|_3 = (2x + 0)\bigg|_3 = 2 \cdot 3 = 6.$$

---

## EXAMPLE 3

$$D(x^2(x^2 - 1))\bigg|_{x = 2} = D(x^4 - x^2)\bigg|_{x = 2} = 4x^3 - 2x\bigg|_{x = 2} = 28.$$

---

*WARNING*   To do Example 3, multiply out the product $x^2(x^2 - 1)$ and apply the rule for sums; you cannot simply take the derivative of each factor. Thus $D(x^2(x^2 - 1))$ does *not* equal $(2x)(2x - 0) = 4x^2$; indeed, the derivative of $x^2(x^2 - 1)$ is really $4x^3 - 2x$. The trouble is that the derivative of a product is *not* the product of the derivatives; in general,

$$(fg)' \neq f'g'.$$

The correct way to handle products is given in Section 3.3.

REMARK   The formulas in this section were explained informally. A more thorough discussion of the principles involved is given in Chapter 3 and a still more thorough one in Appendix A at the end of the book.

## SUMMARY

$$f'(x) = Df(x) = \frac{df(x)}{dx}$$

$$Dx^n = nx^{n-1}$$

$$D(cf) = cDf$$

$$D(f + g) = Df + Dg$$

$$D(f - g) = Df - Dg$$

## PROBLEMS

### A

Compute the following derivatives (problems 1–12):

*1.  $D(x^3 + x^2 - x - 1)$.

2.  $D(2x^5 - 5x + 6)$.

*3.  $\left.\dfrac{d(3x^{100} - 2x^{50} + 11)}{dx}\right|_{x=0}$.

*4.  $\left.\dfrac{d(100x^3 - 50x^2 + 2)}{dx}\right|_{x=1}$.

*5.  $\left.\dfrac{d(-16t^2 + 5t + 1)}{dt}\right|_{t=0}$.

*6.  $\left.\dfrac{d(-4.9t^2 + 3.1t + 2.6)}{dt}\right|_{t=1}$.

*7.  $f'(-3)$ if $f(x) = x^3(x^2 + 2)$.

8.  $g'(1)$ if $g(x) = x(1 + x^3)$.

*9.  $\dfrac{d(2x - 1)^2}{dx}$.

10.  $\dfrac{d(t^2 - 1)^2}{dt}$.

11.  $D(3(-x + 2)^2)$.

12.  $D(3^2(x^2 - 1))$.

*13.  Compute the equation of the line tangent to the graph of $f(x) = x^3$ at each of the given points.

Plot each point carefully; sketch the graph of $f$ and the graph of each tangent line. Note that one tangent line *crosses* the graph at the point of tangency, and each of the others intersects the graph in two points.

a)  $(0, 0)$    b)  $(1/2, 1/8)$
c)  $(-1, -1)$

### B

14.  Show that $Dx^n = nx^{n-1}$ is true in the following cases:
a)  $n = -1$.
b)  $n = -2$.
c)  $n = 1/2$ [$a - b = (\sqrt{a} - \sqrt{b})(\sqrt{a} + \sqrt{b})$].

15.  A polynomial is *even* if it involves only even powers of $x$, and *odd* if it involves only odd powers.
a)  What can be said about the derivative of an even polynomial?
b)  Sketch any graph that is symmetric about the $y$-axis, and try to see geometrically the corresponding symmetry in the graph of the derivative.
c)  What can be said about the derivative of an odd function? Illustrate your answer with a graph and verify it for polynomials.

## 2.3

## APPENDIX:
## THE DERIVATIVE OF $x^n$.
## FACTORING POLYNOMIALS

We compute the derivative of $x^n$ from the difference quotient

$$\frac{\Delta y}{\Delta x} = \frac{\hat{x}^n - x^n}{\hat{x} - x}$$

by factoring the numerator, which has the form $b^n - a^n$. The factorization $b^2 - a^2 = (b - a)(b + a)$ is familiar. You can check that

$$b^3 - a^3 = (b - a) \cdot \underbrace{(b^2 + ab + a^2)}_{3 \text{ terms}}$$

and

$$b^4 - a^4 = (b - a) \cdot \underbrace{(b^3 + ab^2 + a^2b + a^3)}_{4 \text{ terms}}.$$

We will show that for any positive integer $n$,

$$b^n - a^n = (b - a) \cdot \underbrace{(b^{n-1} + ab^{n-2} + \cdots + a^{n-2}b + a^{n-1})}_{n \text{ terms}}. \tag{1}$$

The dots on the right-hand side indicate terms not written out, which can be guessed from those that *are* written. The powers on $b$ descend, while those on $a$ ascend; thus the term after $ab^{n-2}$ is $a^2b^{n-3}$, and the one before $a^{n-2}b$ is $a^{n-3}b^2$.

Formula (1) can be proved by multiplying out the right-hand side, aligning like terms, and noting the cancellations:

$$b(b^{n-1} + ab^{n-2} + \cdots\cdots + a^{n-2}b + a^{n-1})$$
$$-a(b^{n-1} + ab^{n-2} + \cdots\cdots + a^{n-2}b + a^{n-1})$$
$$= b^n + ab^{n-1} + \cdots\cdots + a^{n-2}b^2 + a^{n-1}b$$
$$\quad - ab^{n-1} - a^2b^{n-2} - \cdots\cdots - a^{n-1}b - a^n$$
$$= b^n - a^n.$$

Now we can compute the derivative of $x^n$. Formula (1) gives

$$\frac{\hat{x}^n - x^n}{\hat{x} - x} = \frac{(\hat{x} - x)(\hat{x}^{n-1} + x\hat{x}^{n-2} + \cdots + x^{n-2}\hat{x} + x^{n-1})}{\hat{x} - x}$$
$$= (\hat{x}^{n-1} + x\hat{x}^{n-2} + \cdots + x^{n-2}\hat{x} + x^{n-1})$$
$$\rightarrow (x^{n-1} + x^{n-1} + \cdots + x^{n-1} + x^{n-1}) \qquad \text{as } \hat{x} \rightarrow x$$
$$= nx^{n-1}$$

since there are precisely $n$ terms in the parentheses. This proves the Power Rule

$$\frac{dx^n}{dx} = nx^{n-1}.$$

*REMARK* Formula (1) has many applications, including three basic facts about polynomials:

---

**THE REMAINDER THEOREM**

If $P$ is a polynomial of degree $n$, and $a$ is a number, then there is a polynomial $Q_a$ of degree $n - 1$ such that

$$P(x) - P(a) = (x - a)Q_a(x).$$

---

**EXAMPLE 1** For $P(x) = x^2 + 2x + 2$,

$$P(x) - P(a) = x^2 + 2x + 2 - [a^2 + 2a + 2] = (x - a)[(x + a) + 2].$$

Here $Q_a(x) = (x + a) + 2$.

---

**THE FACTOR THEOREM**

If $P$ is a polynomial of degree $n$, then $P(a) = 0$ if and only if $(x - a)$ is a factor of $P(x)$. Precisely, $P(a) = 0$ if and only if

$$P(x) = (x - a)Q_a(x)$$

for a polynomial $Q_a$ of degree $n - 1$.

---

**EXAMPLE 2** For $P(x) = x^3 + x - 2$, $P(1) = 0$, so $P(x) = (x - 1)Q_1(x)$; in fact

$$P(x) = (x - 1)(x^2 + x + 2).$$

---

**COROLLARY**

A polynomial of degree $n$ has at most $n$ distinct zeroes.

---

The problems below outline the proofs.

## PROBLEMS

**B**

**1.** Show that the derivative of $x^{-n}$ is $-nx^{-n-1}$, for $n = 1, 2, 3, \ldots$.

**2.** Show that

**a)** $1 + x + \cdots + x^{n-1} = \dfrac{1 - x^n}{1 - x}$ if $x \neq 1$.

**b)** $1 + x + \cdots + x^n = \dfrac{1 - x^{n+1}}{1 - x}$ if $x \neq 1$.

**c)** $1 + \dfrac{1}{2} + \cdots + \left(\dfrac{1}{2}\right)^n = 2 - \left(\dfrac{1}{2}\right)^n$.

**d)** $1 + \dfrac{1}{3} + \cdots + \left(\dfrac{1}{3}\right)^n = \dfrac{3}{2} - \dfrac{1}{2}\left(\dfrac{1}{3}\right)^n$.

**e)** $1 + (.1) + \cdots + (.1)^n = \dfrac{10}{9} - \dfrac{1}{9}(0.1)^n$.

**3.** Suppose that $P(x) = c_2 x^2 + c_1 x + c_0$ is any quadratic polynomial; the symbols $c_2$, $c_1$, and $c_0$ are the coefficients. Prove the Remainder Theorem:

$$P(x) - P(a) = (x - a)Q_a(x),$$

where in this case $Q_a(x) = c_2(x + a) + c_1$.

**C**

**4.** Prove the Remainder Theorem for an arbitrary polynomial of degree $n$,

$$P(x) = c_n x^n + c_{n-1} x^{n-1} + \cdots + c_1 x + c_0.$$

(Factor $x^n - a^n$, $x^{n-1} - a^{n-1}$, and so on.)

**5.** Suppose that $P(x) - P(a) = (x - a)Q_a(x)$. Relate $Q_a(a)$ to the derivative $P'(a)$.

**6.** Prove the Factor Theorem. (Use the Remainder Theorem.)

**7.** The quadratic formula proves that every quadratic polynomial has at most two distinct zeroes. Prove that a cubic polynomial has at most three distinct zeroes. (Use the Factor Theorem.)

**8.** Prove that a polynomial of degree $n$ has at most $n$ distinct zeroes.

## 2.4

# GRAPHING WITH THE DERIVATIVE

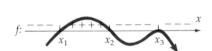

**FIGURE 1**
Sign of $f$.

Given a function $f$, how can you tell what its graph looks like? Two crucial bits of information are the *sign of $f$* and the *sign of the derivative $f'$*.

Figure 1 illustrates an efficient way to determine where a polynomial $f$ is positive and where it is negative. Where $f(x) > 0$, the graph is above the $x$-axis, and where $f(x) < 0$, it is below. Since the graph of a polynomial is a smooth, unbroken curve, it cannot cross from one side to the other without intersecting the axis; these intersections occur at the zeroes of $f$. Thus, to determine the intervals where $f > 0$ and those where $f < 0$, the first step is to *determine all the zeroes of $f$*. In the interval between two successive zeroes the graph does not cross the axis, so $f$ cannot change sign; thus the value of $f$ at just one point in each interval determines its sign throughout that interval. The second step is therefore to *evaluate $f$ at any convenient point in each interval between successive zeroes*.

In practice it may be difficult or impossible to determine precisely *all* the zeroes; in that case there is unfortunately no way to determine exactly where $f$ changes sign.

---

**EXAMPLE 1** Determine the sign of $f(x) = x^3 - 2x^2 + x$ for each real $x$.

**SOLUTION** Find all the zeroes, in this case by factoring:

$$f(x) = x^3 - 2x^2 + x = x(x^2 - 2x + 1) = x(x - 1)^2.$$

So $f(x) = 0$ when $x = 0$ or $x = 1$. This determines three intervals on the axis, $(-\infty, 0)$, $(0, 1)$, and $(1, +\infty)$:

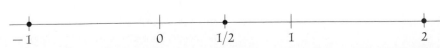

Choose a point in each interval, and evaluate $f$ there:

$$f(-1) = -1(-1 - 1)^2 = -4 < 0$$

$$f(1/2) = \tfrac{1}{2}(\tfrac{1}{2} - 1)^2 = \tfrac{1}{8} > 0$$

$$f(2) = 2(2 - 1)^2 = 2 > 0.$$

These values determine the sign of $f$ throughout each interval:

$f(-1) < 0$ $f(\tfrac{1}{2}) > 0$ $f(2) > 0$

$- - - - - - - - - -$ $+ + + + + + +$ $+ + + + + + + +$

$$-1 \qquad 0 \qquad 1/2 \qquad 1 \qquad 2$$

Thus the graph of $f$ is below the axis for $x < 0$, above it for $0 < x < 1$, and again above the axis for $x > 1$, as indicated in Figure 2. Notice that at one of the zeroes, $f$ does not actually change sign.

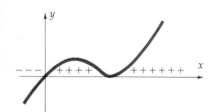

**FIGURE 2**
Sign of $x^3 - 2x^2 + x$.

The sign of the *derivative $f'$* is often more important than the sign of $f$ itself. Geometrically, $f'(x)$ is the slope of the tangent to the graph at the point $(x, f(x))$; we say more briefly that it is the slope of the graph at $(x, f(x))$. In Figure 3 you can see:

*On an interval where $f' > 0$, the graph slopes up to the right, and the function values increase as you read the graph from left to right. Where $f' < 0$ the function values decrease, reading from left to right.*

What you see in Figure 3 is valid in general; a proof is given in Chapter 5.

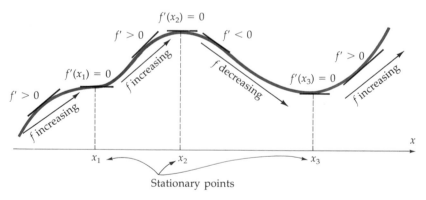

**FIGURE 3**
$f$ increasing where $f' > 0$; $f$ decreasing where $f' < 0$.

If $f$ is a polynomial then so is $f'$, and the sign of $f'$ can change only at the zeroes of $f'$. These zeroes are called **stationary points** for $f$; at these points the graph is neither rising nor falling. In graphing, a crucial step is to *determine all the stationary points and the corresponding function values*; then *determine the sign of $f'$* in the intervals determined by those points. Plot the points on the graph for each stationary point, along with a few others, and draw a graph consistent with the sign of $f'$. Reading the graph from left to right, you see: Where the sign of $f'$ changes from $+$ to $-$ there is a "peak," as at $x_2$ in Figure 3; where it changes from $-$ to $+$, there is a "pit," as at $x_3$ in Figure 3.

Notice that $f'$ does not *necessarily* change sign at a stationary point—see point $x_1$ in Figure 3.

---

**EXAMPLE 2**   Graph $f(x) = -x^3 + 6x + 1$.

**SOLUTION**   The zeroes of $f$ are hard to determine, so we go on and compute $f'$:

$$f'(x) = -3x^2 + 6.$$

Find *all* zeroes of $f'$:

$$-3x^2 + 6 = 0 \Leftrightarrow x^2 = 2 \Leftrightarrow x = \pm\sqrt{2}.$$

Mark these on a line, and determine the sign of $f'$ in the resulting intervals; the values

$$f'(-2) = -3(-2)^2 + 6 = -6 < 0$$

$$f'(0) = 6 > 0, \qquad f'(2) = -6 < 0$$

give

Sign of $f'$:
$$- - - - - \mid + + + + + + + \mid - - - - - $$
$$-2 \qquad -\sqrt{2} \qquad 0 \qquad \sqrt{2} \qquad 2$$

Trend of $f$:
$$\searrow \searrow \searrow \searrow \mid \nearrow \nearrow \nearrow \nearrow \nearrow \nearrow \mid \searrow \searrow \searrow \searrow$$
$$\text{"pit"} \qquad\qquad \text{"peak"}$$

Make a table for $f$ including the stationary points (where $f' = 0$) and a few others; plot the points and draw the graph (Figure 4). For each stationary point, draw a small horizontal segment to indicate that the slope of the graph

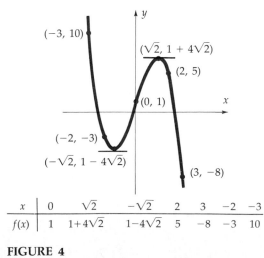

| $x$ | 0 | $\sqrt{2}$ | $-\sqrt{2}$ | 2 | 3 | $-2$ | $-3$ |
|-----|---|-----------|------------|---|---|------|------|
| $f(x)$ | 1 | $1+4\sqrt{2}$ | $1-4\sqrt{2}$ | 5 | $-8$ | $-3$ | 10 |

**FIGURE 4**
$f(x) = -x^3 + 6x + 1.$

is zero at that point. To the left of $x = -\sqrt{2}$, the graph slopes down to the right, since $f' < 0$ there; between $x = -\sqrt{2}$ and $x = \sqrt{2}$, it slopes up to the right; and for $x > \sqrt{2}$, it slopes down to the right again.

At the beginning, it was hard to find the zeroes of $f$. The graph in Figure 4 clarifies the picture. There is one zero just to the left of $x = -2$, another just to the left of $x = 0$, and a third between 2 and 3. Plotting the stationary points of a cubic will always reveal how many zeroes there are and their approximate locations.

**Optimization problems**    such as those in Section 1.4 can be reduced to finding the maximum or minimum point on some graph. These points can be found if the sign of $f'$ is known everywhere.

**EXAMPLE 3**    A cylindrical can is to be made, using 24 in$^2$ of material. What dimensions give maximum volume? What are the proportions (height/diameter) of this optimal can? How much of the 24 in$^2$ is used on the top and bottom, and how much on the sides?

*SOLUTION*    You have to consider all possible cans with the given surface area 24 in$^2$ and select the one of maximum volume. Let the radius of a typical can be $r$ and the height be $h$ (Fig. 5). The top and bottom are discs of radius $r$; the area of each is $\pi r^2$. The side is made of a rectangle $2\pi r$ long (the perimeter of the can) and $h$ high (Fig. 6); its area is $2\pi rh$. The total area of the material in this typical can (two discs and a rectangle) is therefore

$$S = 2(\pi r^2) + (2\pi r)h.$$

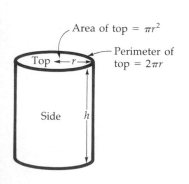

Area of top $= \pi r^2$

Perimeter of top $= 2\pi r$

Top $\leftarrow r \rightarrow$

Side    $h$

**FIGURE 5**

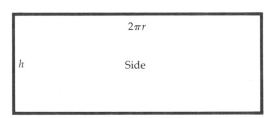

$2\pi r$

$h$          Side

**FIGURE 6**

This is to equal 24 in$^2$:

$$24 = 2\pi r^2 + 2\pi rh, \quad \text{or} \quad 12 = \pi r^2 + \pi rh. \tag{1}$$

So you cannot choose $r$ and $h$ independently; if you choose a value for $r$, then $h$ is determined by solving (1):

$$h = \frac{12 - \pi r^2}{\pi r}. \tag{2}$$

To maximize the volume $V$, you need the formula

$$V = (\text{area of base}) \times (\text{height}) = \pi r^2 h.$$

And $h$ is determined by $r$, via (2), so

$$V = \pi r^2 \frac{(12 - \pi r^2)}{\pi r} = r(12 - \pi r^2)$$

$$= 12r - \pi r^3. \tag{3}$$

This expresses the volume $V$ resulting from any possible choice of radius $r$. To maximize $V$, you want the highest point on the graph of $V$ versus $r$, for $r \geq 0$ ($r < 0$ is ruled out, since the radius $r$ cannot be negative). At the high point, $dV/dr = 0$, so compute from (3)

$$\frac{dV}{dr} = 12 - 3\pi r^2.$$

This equals zero if

$$12 = 3\pi r^2 \Leftrightarrow r^2 = 4/\pi \Leftrightarrow r = \pm 2/\sqrt{\pi}.$$

Determine the sign of $dV/dr$:

$$\frac{dV}{dr}:$$

$$- - - - \mid + + + + + + + + + + \mid - - - - -$$
$$\frac{-2}{\sqrt{\pi}} \qquad\qquad \frac{2}{\sqrt{\pi}}$$

$$V: \quad \searrow\searrow\searrow\searrow \mid \nearrow\nearrow\nearrow\nearrow\nearrow\nearrow\nearrow\nearrow \mid \searrow\searrow\searrow$$

The graph rises for $-\dfrac{2}{\sqrt{\pi}} < r < \dfrac{2}{\sqrt{\pi}}$ and then descends, so the high point

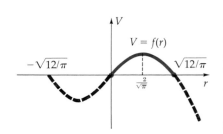

**FIGURE 7**
Volume as a function of radius.

for $r > 0$ is with $r = 2/\sqrt{\pi}$, as shown in Figure 7. The part for $r < 0$ is dashed, since it has no significance in this problem. To make the figure reasonably accurate, we show the zeroes of $V$, at $r = 0$ and $r = \pm\sqrt{12/\pi}$. The part for $r > \sqrt{12/\pi}$ is also irrelevant; those values of $r$ make $h < 0$, as you can see from equation (2).

Now you know that $V$ is maximum when the radius $r = 2/\sqrt{\pi}$. Determine the corresponding height $h$ from (2):

$$h = \frac{12 - \pi r^2}{\pi r} = \frac{12 - \pi \cdot 4/\pi}{\pi(2/\sqrt{\pi})} = \frac{8}{2\sqrt{\pi}} = \frac{4}{\sqrt{\pi}}.$$

The solution has two interesting features. First, the *proportions* of the can (height/diameter) are

$$\frac{h}{2r} = \frac{4/\sqrt{\pi}}{2 \cdot 2/\sqrt{\pi}} = 1.$$

For this optimal can, height = diameter. Second, *exactly one-third* of the 24 in$^2$ of material is used to form the top and bottom:

$$2\pi r^2 = 2\pi(2/\sqrt{\pi})^2 = 8 \text{ in}^2.$$

The other two-thirds forms the sides.

## SUMMARY

Where $f' > 0$, $f$ is *increasing* from left to right.

Where $f' < 0$, $f$ is *decreasing* from left to right.

A *zero* of $f'$ is called a *stationary point* of $f$.

## PROBLEMS

### A

1. Sketch a graph of $f$ consistent with the given data.

   a)

   b)

   | $x$ | $-3$ | $-2$ | $-1$ | $0$ | $1$ | $2$ | $3$ |
   |------|------|------|------|-----|-----|-----|-----|
   | $f(x)$ | $0$ | $-1$ | $0$ | $1$ | $0$ | $-1$ | $0$ |

   $f'$:

2. Graph the following functions. Find all stationary points and determine the sign of $f'$ everywhere. Find the "peaks" and "pits." Find the intercepts algebraically if possible; if not, approximate them by reading your graph.

   *a)  $x^3 - x$          *b)  $x^3 - x^2$
   c)  $x^3 + x^2$          *d)  $x^3 + 3x^2 - 2$
   e)  $x^3 - 3x^2 + 1$     *f)  $x(x^2 - 1)$
   g)  $x^4 - 8x^2 + 16$    h)  $t^3 - t^4$
   i)  $x^4 + x$

3. The graphs in Figure 8i–iv give the *derivatives* of the functions in a–d, but in the wrong order. Which derivative matches which function? Why?

*4. A box with square bottom of edge $x$ and height $h$ is open at the top. The surface area is to be 300 cm². What dimensions $x$ and $h$ give maximum volume? What proportion of the total surface is on the bottom?

*5. The material for the bottom of a topless box costs \$0.02/cm², and for the sides, \$0.01/cm². If the total cost is to be \$6, what dimensions $x$ and $h$ give maximum volume? What proportion of the \$6 is spent on the material for the bottom?

6. The material for the top and bottom of a cylindrical can costs \$0.005/in², and for the sides,

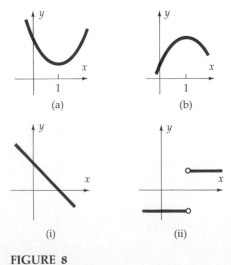

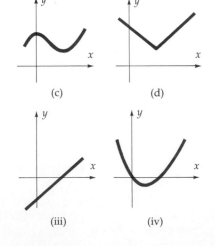

**FIGURE 8**

$0.002/\text{in}^2$. If the can is to cost $0.03, what dimensions yield the maximum volume? What fraction of the total cost is spent on top and bottom together?

*7. A cylindrical cup is to have 15 cm$^2$ of surface area. What dimensions give maximum volume? What fraction of the 15 cm$^2$ is used on the sides, and what fraction on the bottom?

*8. By postal regulations, a parcel post package may have a length plus girth of no more than 84 in. (In Figure 9, the length is the longest dimension, labeled $x$, and the girth is the perimeter perpendicular to the length, $2y + 2z$.) What dimensions meeting these requirements give maximum volume for a rectangular box? (Assume that the cross-section is square, so $y = z$ in Figure 9.)

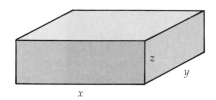

**FIGURE 9**

9. (The following "box problem" is modified from one suggested by Kay Dundas of Hutchinson Community College.) Figures 10 and 11 show two different ways to cut and fold a rectangular piece of cardboard to make a box; the seams will be taped. In each case, assume that the box is made from a $12 \times 12$ square.

a) In Figure 10, find $x$ to give maximum volume.

b) In Figure 11, find $x$ to give maximum volume.

c) Which method yields a box with more volume?

**B**

10. There is a theory of the cough that gives the velocity of air in the throat, $V$, as a function of the radius of the throat, $r$. The relation is

$$V = ar^2(r_0 - r),$$

where $a$ is a constant, and $r_0$ is the radius of the throat in its relaxed state (also a constant).

a) Graph $V$ as a function of $r$, showing intercepts and stationary points.

b) The idea of coughing is to clear the throat by forcing air through it at high velocity. What throat radius $r$ produces maximum velocity, and what is the maximum velocity? (Apparently, during a cough the throat commonly contracts about 33%. How does this compare to the "optimal" result that you computed? Source: UMAP Module 211 "The Human Cough" by Philip Tuchinsky.)

11. Suppose that a polynomial $f$ has two distinct zeroes. Do you feel that it must necessarily have at least one stationary point? If so, why? What if $f$ has three distinct zeroes?

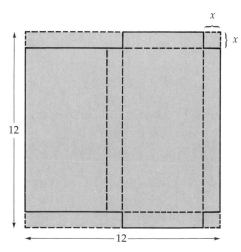
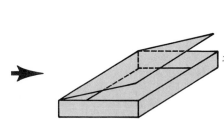

**FIGURE 10**
Make a box with lid.

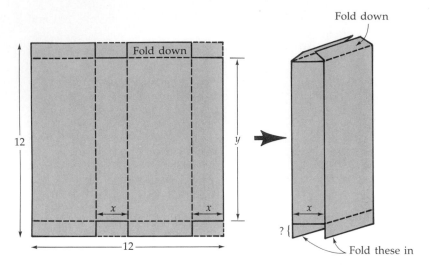

**FIGURE 11**
Another box.

12. How many real-valued solutions are there to the following equations? Approximate each solution to the nearest integer.
  *a) $x^4 + x^3 - 5x^2 - 6 = 0$
  b) $x^4 + 10x^2 + 22 = 0$
  c) $x^{11} + x^3 + x + 4 = 0$

13. In economic theory, the *profit* realized in producing and selling quantity $q$ is

$$P(q) = R(q) - C(q),$$

where $R(q)$ is the revenue (receipts from sales), and $C(q)$ the cost of production.
  a) Suppose that at a certain level of production $\bar{q}$, it is found that $P'(\bar{q}) > 0$. In order to increase profits, should production be raised or lowered?
  b) Suppose that at some other level $q^*$, $R'(q^*) < C'(q^*)$. Should production be raised or lowered?
  c) Suppose that, at level $\tilde{q}$, profit is maximized. What is $P'(\tilde{q})$? What is the relation between $R'(\tilde{q})$ and $C'(\tilde{q})$? (In economics, the derivatives $C'$, $R'$, and $P'$ are called the *marginal cost*, *marginal revenue*, and *marginal profit*, denoted MC, MR, and MP.)

14. Bytes of information are stored on a memory disk, each byte occupying a point on the disk as shown in Figure 12. For easy retrieval, they are located on circular arcs and lined up on radial lines. The minimum distance between arcs is $a$, and the minimum distance between bytes on a given arc is $b$; since the bytes are aligned, those

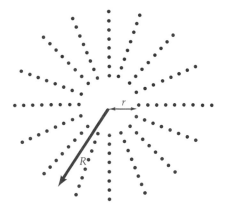

**FIGURE 12**
Memory sites on a disk.

on the inner circle are spaced $b$ units apart, and those on the outer circles are farther apart.
  a) Let $r$ be the radius of the inner circle and $R$ the radius of the outer circle. Suppose that $2\pi r$ is a multiple of $b$ and $R - r$ is a multiple of $a$. Show that the number of bytes is

$$N = \frac{2\pi r}{b}\left(\frac{R - r}{a} + 1\right).$$

  b) When $a$ and $b$ are small compared to $r$ and $R - r$, the formula in part a is a good approximation even if $2\pi r$ is *not* a multiple of $b$ and $R - r$ is *not* a multiple of $a$. Assuming that formula, determine, for a given outer radius $R$, the inner radius $r$ that maximizes the number $N$.

## 2.5

## THE SECOND DERIVATIVE. CONCAVITY

The derivative of a function $f$ is a new function $f'$. If you take the derivative of this new function, you get what is called the **second derivative** of $f$, denoted $f''$.

---

EXAMPLE 1

$$f(x) = x^3 + 3x^2 + 4x$$

$$f'(x) = 3x^2 + 6x + 4$$

$$f''(x) = \text{derivative of } f' = 6x + 6.$$

---

With operator notation, the first derivative is $Df$, and the second derivative is $DDf = D^2f$. With Leibniz notation the symbol $\dfrac{d}{dx}$ replaces $D$, and $D^2$ is replaced by $\dfrac{d^2}{dx^2}$, an abbreviation for $\left(\dfrac{d}{dx}\right)^2$. Thus the second derivative of $f(x)$ is

$$\frac{d^2}{dx^2} f(x) \quad \text{or} \quad \frac{d^2 f(x)}{dx^2}.$$

This is read as "$d$ square $f$ $d$ $x$ square."

---

EXAMPLE 2

$$y = 2x^4, \qquad \frac{dy}{dx} = 8x^3, \qquad \frac{d^2y}{dx^2} = 24x^2$$

$$D^2(x^3 + 3x^2 + 4x) = D[Dx^3 + D(3x^2) + D(4x)]$$
$$= D[3x^2 + 6x + 4] = 6x + 6.$$

---

Notice the difference between $\dfrac{d^2y}{dx^2}$ and $\left(\dfrac{dy}{dx}\right)^2$. With $y = 2x^4$, Example 2 gives

$$\frac{d^2y}{dx^2} = 24x^2$$

but

$$\left(\frac{dy}{dx}\right)^2 = (8x^3)^2 = 64x^6.$$

In the last case, it is the function $dy/dx$ that is squared; in the first case the *operation* $d/dx$ is "squared," by performing it twice.

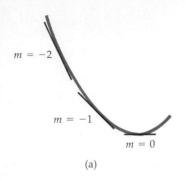

$m = -2$

$m = -1$

$m = 0$

(a)

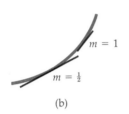

$m = 1$

$m = \frac{1}{2}$

(b)

**FIGURE 1**

$f'' > 0$; slope function *increases* from left to right; graph lies *above* its tangent, concave *up*.

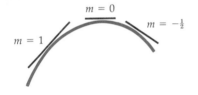

$m = 0$

$m = 1$

$m = -\frac{1}{2}$

**FIGURE 2**

$f'' < 0$; slope function *decreases* from left to right; graph lies *below* its tangent, concave *down*.

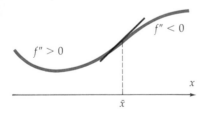

$f'' < 0$

$f'' > 0$

$x$

$\bar{x}$

**FIGURE 3**

$\bar{x}$ is an inflection point.

The **third derivative** of $y = f(x)$ is denoted $f'''(x)$, or $D^3 f(x)$, or $\dfrac{d^3 y}{dx^3}$.

The $n^{\text{th}}$ derivative is

$$f^{(n)}(x) \quad \text{or} \quad D^n f(x) \quad \text{or} \quad \frac{d^n y}{dx^n}.$$

For example, if $f(x) = x^4$ then

$$f'(x) = Dx^4 = \frac{dx^4}{dx} = 4x^3$$

$$f''(x) = D^2 x^4 = \frac{d^2 x^4}{dx^2} = 4 \cdot 3x^2$$

$$f^{(3)}(x) = D^3 x^4 = \frac{d^3 x^4}{dx^3} = 4 \cdot 3 \cdot 2x.$$

## Concavity

Recall the basic principle of graphing: Where the derivative $f'$ is positive, the function values $f(x)$ are increasing as you read the graph from left to right, that is,

$$f' > 0 \Rightarrow f \text{ is increasing.}$$

Now, $f''$ is the derivative of $f'$, so the same relation holds between these two: Where $f''$ is positive, the values of $f'(x)$ are increasing, that is

$$f'' > 0 \Rightarrow f' \text{ is increasing.}$$

Figure 1 shows graphs where $f'' > 0$; reading each graph from left to right, the algebraic value of the slope increases. As a result, the graph lies *above* each of its tangent lines; the graph is called **concave up**.

Where $f''$ is negative, there $f'$ is decreasing as in Figure 2, and the graph lies *below* each tangent line; the graph is called **concave down**.

At a point where $f''$ changes sign (Fig. 3) the graph *crosses* its tangent line; this is called an **inflection point**. If $f$ is a polynomial, inflection points can occur only at zeroes of $f''$. (But note that $f''$ does not *necessarily* change sign at a zero; and if $f''$ does not change sign, there is no inflection point.) The theoretical justification of this is given in Chapter 5.

---

**EXAMPLE 3**  Graph $f(x) = x^3 - 3x$.

**SOLUTION**  $f'(x) = 3x^2 - 3 = 0$ when $x = \pm 1$. The signs of $f'$ are found to be

$f'$:  $\underline{\qquad + + + + \quad | \quad - - - - - - \quad | \quad + + + + +\qquad}$ .
$\qquad\qquad\qquad\qquad\quad -1 \qquad\qquad\qquad 1$

The second derivative and its signs are

$f''(x) = 6x$:  $\underline{\qquad - - - - - - \quad | \quad + + + + + + +\qquad}$ .
$\qquad\qquad\qquad\qquad\qquad\qquad 0$

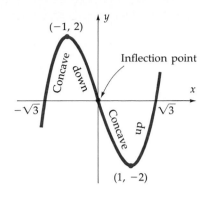

**FIGURE 4**
$y = x^3 - 3x$.

The graph has stationary points at $x = \pm 1$. For $x < 0$, $f'' < 0$, and the graph is concave down; for $x > 0$ it is concave up; and $x = 0$ gives an inflection point $(0, 0)$ (Fig. 4).

**EXAMPLE 4** Graph $f(x) = \dfrac{x^3}{12} + \dfrac{x^2}{4} + \dfrac{x}{3}$, using the first and second derivative.

**SOLUTION** Find the zeroes of $f$, $f'$, and $f''$, and check the sign of each in between these points. The zeroes of $f$ are the solutions of $f(x) = 0$:

$$0 = \frac{x^3}{12} + \frac{x^2}{4} + \frac{x}{3} = \frac{x}{12}(x^2 + 3x + 4).$$

The quadratic $x^2 + 3x + 4$ has no zeroes (check this with the quadratic formula), so $f(x) = 0$ only for $x = 0$. Checking signs on each side of $x = 0$ shows

$$f(-1) = -\tfrac{1}{12} + \tfrac{1}{4} - \tfrac{1}{3} = -\tfrac{1}{6} < 0,$$

while $f(1) = \tfrac{2}{3} > 0$, so the sign of $f$ is

$$f: \quad \frac{\overset{\text{intercept}}{\underset{0}{- - - - - \mid + + + + +}}}{}$$

Turning to $f'(x) = \dfrac{x^2}{4} + \dfrac{x}{2} + \dfrac{1}{3}$, this polynomial has no zeroes (check that) and $f'(0) = \dfrac{1}{3} > 0$, so $f'(x) > 0$ everywhere:

$$f': \quad \overline{+ + + + + + + + + + + +}.$$

The entire graph slopes up to the right.

Finally, $f''(x) = \tfrac{1}{2}x + \tfrac{1}{2} = \tfrac{1}{2}(x + 1)$ is zero only for $x = -1$, and has signs

$$f'': \quad \frac{- - - - - \mid + + + + + + + +}{\underset{-1}{}}.$$

Since $f''$ actually changes sign at $-1$, this is an inflection point.

Now sketch the graph, in two steps:

**Step 1** Plot the various points calculated, and the points where $f = 0$, $f' = 0$, or $f'' = 0$. If these are few, plot some others as well. In this example we use:

inflection ⟶          ⟵ intercept

| $x$    | $-2$          | $-1$          | $0$ | $1$           |
|--------|---------------|---------------|-----|---------------|
| $f(x)$ | $-\tfrac{1}{3}$ | $-\tfrac{1}{6}$ | $0$ | $\tfrac{2}{3}$ |

**Step 2** Between and beyond these points, fill in the graph, *consistent with the known signs of f, f', and f''*. The result is Figure 5.

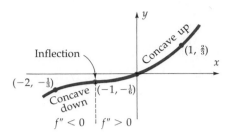

**FIGURE 5**

Inflection point for $f(x) = \dfrac{x^3}{12} + \dfrac{x^2}{4} + \dfrac{x}{3}$.

## SUMMARY

Where $f'' > 0$ the graph is *concave up*; it lies above its tangent line.

Where $f'' < 0$ it is *concave down*; it lies below its tangent line.

Where $f''$ changes sign, the graph has an *inflection point*; it crosses its tangent line.

## PROBLEMS

### A

1. Take the second derivative.
   *a) $f(x) = 2x + 1$
   b) $f(x) = 3x^2 - 2x + 2$
   *c) $g(t) = -t^3 + 2t$
   d) $h(z) = 4z^4 - 5z^3 + 3$
   *e) $f(x) = x^5 - 8x^2 + 2$
   f) $y = 11x^5 + 2x^2 - 1$
   *g) $s = t^2$
   h) $z = y^2 - 5y + 4$

2. For the given graph, locate *stationary points*, *inflection points*, the intervals where $f'' > 0$, and the intervals where $f'' < 0$.
   *a) The graph in Figure 6
   *b) The graph in Figure 7
   c) The graph in Figure 8

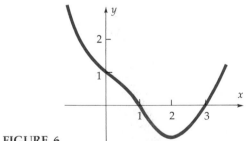

**FIGURE 6**

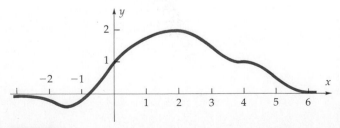

**FIGURE 7**

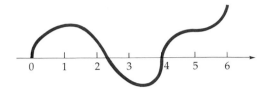

**FIGURE 8**

**3.** Find the stationary point and determine whether it is a "peak," a "pit," or an inflection point. Draw a general conclusion about the monomials $x^n$.
a) $x^2$   b) $x^3$   c) $x^4$   d) $x^5$
e) $x^n$

**4.** Sketch the graphs, using first and second derivatives. Where practical, find the intercepts. Plot at least four points on each curve.
*a) $y = -x^3 + 6x + 1$
*b) $y = (x + 1)^2(x - 2)$
*c) $y = x^3 - x^4$
d) $y = \frac{1}{5}x^5 - \frac{2}{3}x^3 + x$
e) $y = x^4 - 32x$
f) $y = \dfrac{x^5}{5} - 4x^2 + 4$

**5.** Compute the second derivative of the general quadratic function $f(x) = ax^2 + bx + c$, and relate the constant $a$ to the concavity of the graph.

**6.** The graphs in Figure 9i–iii give the second derivative $f''$ for the functions $f$ graphed in a–c, but in the wrong order. Which $f''$ corresponds to which $f$? Why?

**7.** An ant walking along the graph of $f$ from left to right finds itself turning to the right. Is $f'' > 0$, or $f'' < 0$? Draw a sketch explaining your answer.

**B**

**8.** Suppose that $f(x) = ax^3 + bx^2 + cx + d$ is a cubic with three real zeroes $x_1$, $x_2$, and $x_3$; then $f(x)$ can be factored as $a(x - x_1)(x - x_2)(x - x_3)$. Show that $f$ has an inflection point $(\bar{x}, f(\bar{x}))$ with $\bar{x}$ the average of the three zeroes.

Problems 9–11 investigate the symmetry of the graph of the general cubic function.

**9.** Show that the graph of $x^3 - 3x$ is symmetric about its point of inflection.

**10.** a) For the cubic $x^3 + 3x^2 + 2x + 2$, find the point of inflection.
b) Let $(h, k)$ be the point of inflection found in part a. Show that $f(h + z) - k = k - f(h - z)$.
c) Describe the symmetry of the graph (see Fig. 10).

**11.** Let $(h, k)$ be the point of inflection of $f(x) = ax^3 + bx^2 + cx + d$.
a) Express $(h, k)$ in terms of $a$, $b$, $c$, and $d$.
b) Show that $f(h + z) - k = k - f(h - z)$.
c) Describe the symmetry of the graph (see Fig. 10).

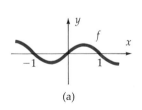

(a)

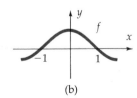

(b)

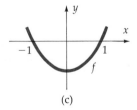

(c)

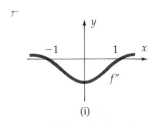

(i)

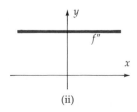

(ii)

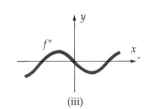

(iii)

**FIGURE 9**

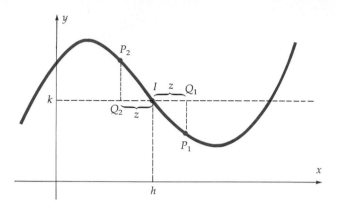

**FIGURE 10**
Graph of a cubic; $I = (h, k) =$ Inflection point.

# 2.6

## NEWTON'S METHOD (optional)

Solving equations is a basic part of mathematics. For some types there is a simple solution; for example, to solve

$$ax^2 + bx + c = 0$$

we have the quadratic formula. But this is a very special case—generally there is no specific formula to solve an equation of the form $f(x) = 0$. Newton proposed a method that *approximates* the solutions, often with great efficiency.

Suppose that you are trying to solve the equation $f(x) = 0$, looking for the intercept $\bar{x}$ in Figure 1. Take a point $x_0$ somewhere near $\bar{x}$. Draw the tangent line at $(x_0, f(x_0))$ and follow it to the x-axis. Newton's first approximation to $\bar{x}$ is the point $x_1$ where that tangent crosses the x-axis. You can then repeat the process; draw the tangent at $(x_1, f(x_1))$ and find the second approximation $x_2$ in Figure 2. Repeat if necessary, until the desired accuracy is achieved.

The formula for the first approximation $x_1$ is derived from Figure 1. The slope of the tangent line at $(x_0, f(x_0))$ is $f'(x_0)$. Hence from the figure

$$f'(x_0) = \frac{\Delta y}{\Delta x} = \frac{f(x_0) - 0}{x_0 - x_1}.$$

Solve for $x_1$:

$$x_1 = x_0 - \frac{f(x_0)}{f'(x_0)}. \tag{1}$$

Repeating the process with $x_1$ in place of $x_0$ gives a second approximation

$$x_2 = x_1 - \frac{f(x_1)}{f'(x_1)}.$$

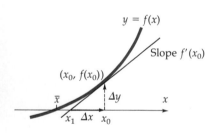

**FIGURE 1**
$x_1$ is Newton's approximation to $\bar{x}$, a solution of $f(\bar{x}) = 0$.

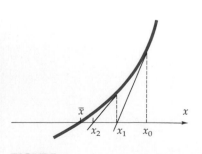

**FIGURE 2**
Two successive approximations by Newton's method.

As in Figure 2, $x_2$ is usually a better approximation than $x_1$. You can go on to compute further approximations $x_3$, $x_4$, and so on with the general formula

$$x_{n+1} = x_n - \frac{f(x_n)}{f'(x_n)}.$$

Thus each new approximation is given by

$$x_{\text{new}} = x_{\text{old}} - \frac{f(x_{\text{old}})}{f'(x_{\text{old}})}.$$

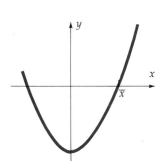

**FIGURE 3**
$f(x) = x^2 - 2$, $\bar{x} = \sqrt{2}$.

**EXAMPLE 1**  Solve $x^2 - 2 = 0$.

**SOLUTION**  Of course, $x = \sqrt{2}$ is a solution; this is the $\bar{x}$ in Figure 3. In applying Newton's method to the equation $x^2 - 2 = 0$, we construct approximations to the square root of 2.

Take as an initial approximation $x_0 = 1.4 = 7/5$, since $(1.4)^2 = 1.96$ is close to 2. Then

$$f(x_0) = x_0^2 - 2 = \tfrac{49}{25} - 2 = -\tfrac{1}{50}$$

and

$$f'(x_0) = 2x_0 = \tfrac{14}{5},$$

so Newton's formula (1) gives

$$x_1 = x_0 - \frac{f(x_0)}{f'(x_0)} = \frac{7}{5} - \frac{-1/50}{14/5} = \frac{99}{70} = 1.41428 \cdots.$$

Repeat the process with $x_1 = 99/70$ (using an eight-digit calculator):

$$f(x_1) = (x_1)^2 - 2 = .000204$$

$$f'(x_1) = 2x_1 = \tfrac{99}{35}$$

$$x_2 = x_1 - \frac{f(x_1)}{f'(x_1)} = \frac{99}{70} - \frac{.000204}{99/35} = 1.4142136 \cdots,$$

which is good to six decimal places. The next approximation $x_3$ would be much more accurate; you need a twelve-digit calculator to do it justice.

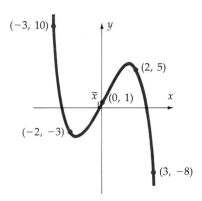

**FIGURE 4**
$f(x) = -x^3 + 6x + 1$. (Vertical scale compressed for convenience.)

**EXAMPLE 2**  The cubic $f(x) = -x^3 + 6x + 1$ has three zeroes (Figure 4; this was graphed in Example 2, Section 2.4). Approximate the middle zero (Fig. 5).

**SOLUTION**  A convenient initial approximation is $x_0 = 0$. You get

$$f(x_0) = 1, \qquad f'(x_0) = 6$$

hence

$$x_1 = x_0 - \frac{f(x_0)}{f'(x_0)} = 0 - \frac{1}{6} = -.1666 \cdots.$$

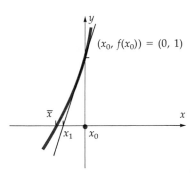

**FIGURE 5**
First approximation to $\bar{x}$. (Vertical scale compressed.)

Repeating:

$$f(x_1) = \tfrac{1}{216}$$

$$f'(x_1) = -3x_1^2 + 6 = \tfrac{71}{12}$$

$$x_2 = x_1 - \frac{f(x_1)}{f'(x_1)} = -\frac{1}{6} - \frac{1/216}{71/12} = -\frac{107}{639} = -.1674491\cdots.$$

You can't check the accuracy of $x_2$ by comparing it to some "official" decimal expansion of $\bar{x}$—perhaps nobody has computed this particular $\bar{x}$ before! This $x_2$ *appears* to be good to three or more decimal places, since it agrees with $x_1$ up to three places. To determine the accuracy rigorously, we try to bracket $\bar{x}$ between two points where $f$ has opposite sign. First compute

$$f(x_2) = f(-.1674491) = 0.0000005\cdots > 0.$$

Since $f'(x) = -3x^2 + 6 > 0$ in the region where we are working, the function $f$ is increasing there. Since $f(x_2) > 0$ and $f$ is increasing, we conclude that

$$\bar{x} < x_2 = -.1674491.$$

Thus to bracket $\bar{x}$, we need a nearby point $\bar{x}_2 < x_2$ where $f(\bar{x}_2) < 0$. We try $\bar{x}_2 = -.1674492$ and find that

$$f(\bar{x}_2) = -0.00000005\cdots < 0.$$

Thus

$$-.1674492 < \bar{x} < -.1674491$$

and we have determined $\bar{x}$ to almost seven decimals.

---

*REMARK* Before using Newton's method, it is a good idea to have a reasonable sketch of the graph near the desired zero $\bar{x}$. Otherwise, you might unwittingly approximate the wrong zero of $f$ (Fig. 6a), or oscillate back and forth (Fig. 6b), or find larger and larger "approximations" to a nonexistent zero (Fig. 6c).

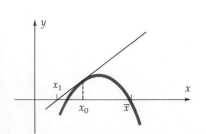

(a)  Approximating wrong zero.

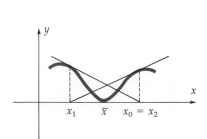

(b)  Oscillating approximations.

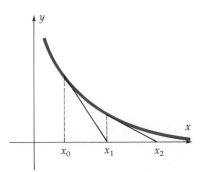

(c)  Successive "approximations" grow infinitely large.

**FIGURE 6**
Possible difficulties with Newton's method.

## SUMMARY

Newton's method approximates solutions $\bar{x}$ of $f(\bar{x}) = 0$, by the formula

$$x_{\text{new}} = x_{\text{old}} - \frac{f(x_{\text{old}})}{f'(x_{\text{old}})}.$$

The success of the method requires a reasonable starting value $x_0$.

## PROBLEMS

### A

1. From an appropriate sketch (such as Figure 2), deduce the formula for Newton's second approximation, $x_2$, in terms of $x_1$.

2. Find Newton's approximations $x_1$ and $x_2$ to a solution of the given equation in parts a–d. Start with the suggested value of $x_0$. Sketch the graph in each case, showing $(x_0, f(x_0))$, the tangent line, and $(x_1, f(x_1))$. If you have a calculator, find $x_3$ for at least one of the parts a–d.
   *a) $-x^3 + 6x + 1 = 0$, $x_0 = -2$ (This approximates a root different from the one in Example 2.)
   b) $x^2 - 10 = 0$, $x_0 = 3$ (You are approximating $\sqrt{10} = 3.1622777 \cdots$. How accurate is $x_1$? $x_2$?)
   *c) $x^3 - 2 = 0$, $x_0 = 1$
   d) $x^5 - 2 = 0$; choose $x_0$ yourself

### B

3. a) Graph $f(x) = x^3 - 3x + 1$, and approximate the smallest positive and largest negative zero, to the nearest 0.1. [If $f(x_0 - 0.1)$ and $f(x_0 + 0.1)$ have opposite signs, then $x_0$ approximates a zero within 0.1.]
   *b) Approximate the largest positive zero, to the nearest 0.01.
   c) Approximate the largest negative zero, to the nearest 0.0001.

4. Suppose that the zero $\bar{x}$ and the successive approximations $x_0$, $x_1$, $x_2$, all lie in an interval I where $f' > 0$ and $f'' > 0$. Are $x_1$ and $x_2$ greater, or less, than $\bar{x}$? Is $x_2$ greater, or less, than $x_1$? (Consider the two possibilities $x_0 < \bar{x}$, and $x_0 > \bar{x}$.)

5. [Adapted from W. C. Rheinholdt, UMAP unit #264.] The van der Waals equation of state for an "imperfect" gas is

$$\left(p + \frac{a}{v^2}\right)(v - b) = RT, \qquad (*)$$

where $p$ is the pressure; $a$ and $b$ are constants; different for different gases; $v$ is the volume per mole; $R = 0.082054$ is a constant independent of the gas, and $T$ is the temperature in degrees Kelvin. For carbon dioxide $a = 3.592$, $b = 0.04267$. Assume that the gas is carbon dioxide and that $p = 1$ (normal atmospheric pressure) and $T = 300°K$ (27°C).
   a) Rewrite equation (*) as a cubic equation for $v$.
   b) Show that there is just one solution, between $v = 24$ and $v = 25$.
   c) Approximate the solution of (*) to three significant figures. (Start with $x_0 = 24.5$, and apply Newton's method once. You are finding the volume per mole of carbon dioxide at 1 atmosphere pressure, at 27°C.)

6. What happens when you apply Newton's method to the following equations? (Determine your answer graphically.)
   a) $x^2 + 1 = 0$, with $x_0 = 1$
   b) $\dfrac{1}{x} = 0$, with $x_0 = 1$
   c) $x^2 - 1 = 0$, with $x_0 = 1$
   d) $x^{1/3} = 0$, with $x_0 \neq 0$

7. Show that Newton's method for $x^2 - a = 0$ gives

$$x_{\text{new}} = \frac{(x_{\text{old}})^2 + a}{2x_{\text{old}}} = \frac{1}{2}\left[x_{\text{old}} + \frac{a}{x_{\text{old}}}\right]$$

as approximations to $\sqrt{a}$. Use this so-called "divide and average" formula to approximate $\sqrt{3}$.

8. Derive a formula for approximations to $\sqrt[3]{a}$ analogous to the one in Problem 7 for $\sqrt{a}$. Use it to approximate $\sqrt[3]{10}$.

## 2.7

## VELOCITY AND ACCELERATION

The first application of calculus to physical problems was Isaac Newton's study of motion. Newton strode onto a stage set by Galileo, showed that velocity is really a derivative, and laid the foundations of a theory relating the fall of an apple to the paths of the moon and planets, a theory which today predicts the motion of satellites and spacecraft.

The longer an apple (or anything else) falls, the farther it goes. Galileo was first to tell how far the apple falls in a given time. We will show how Galileo's law could be discovered, then go on to define the essential concepts in the study of motion: the *velocity* and *acceleration*.

Galileo's law is based on experiments something like the one illustrated in Figure 1a. (According to some rumors, Galileo used a convenient leaning tower in Pisa.) A heavy stone is dropped from height 64 feet, and its height is recorded every half second. Let $f(t)$ stand for the height after it has dropped for $t$ seconds as in Figure 1b. From Figure 1a, $f(0) = 64$, $f(.5) = 60$, $f(1) = 48$, $f(1.5) = 28$, $f(2) = 0$.

The stone apparently gains speed as it falls, since the gaps between successive positions grow. The key to analyzing the motion is the **average speed**. The distance fallen in the first $t$ seconds is $64 - f(t)$ (Fig. 1b). The average speed in the first $t$ seconds is

$$\frac{\text{distance}}{\text{time}} = \frac{64 - f(t) \text{ feet}}{t \text{ seconds}} = \frac{64 - f(t)}{t}\frac{\text{ft}}{\text{sec}}.$$

The following table shows these averages:

64 $t = 0$
60 $t = 0.5$
48 $t = 1$
28 $t = 1.5$
0 $t = 2$

(a)

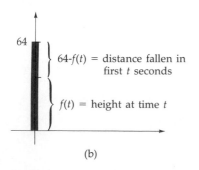

64

$64 - f(t) =$ distance fallen in first $t$ seconds

$f(t) =$ height at time $t$

(b)

**FIGURE 1**

| $t$ | .5 | 1 | 1.5 | 2 |
|---|---|---|---|---|
| $\dfrac{64 - f(t)}{t}$ | $\dfrac{64 - 60}{.5}$ $= 8$ | $\dfrac{64 - 48}{1}$ $= 16$ | $\dfrac{24 - 28}{1.5}$ $= 24$ | $\dfrac{64}{2}$ $= 32$ |

The average speeds 8, 16, 24, 32, show a clear pattern: Each entry is 16 times the corresponding time $t$ in the first row. Thus for times $t = .5$, 1, 1.5, and 2,

$$\frac{64 - f(t)}{t} = 16t. \qquad (1)$$

It is an easy guess, supported by further experiment, that this same relationship (1) holds throughout the fall, for $0 \le t \le 2$. Taking this for granted, solve equation (1) for $f(t)$, and you get

$$f(t) = 64 - 16t^2, \qquad 0 \le t \le 2.$$

This gives the height $f(t)$ of a stone $t$ seconds after it is dropped from height 64 feet. If it were dropped at time 0 from height $h$ feet, the formula would be

$$f(t) = h - 16t^2, \qquad 0 \le t \le t_0, \tag{2}$$

where $t_0$ is the time that it lands. This is Galileo's formula for the height of any dense object (like a stone) $t$ seconds after it is dropped from height $h$. It applies near the surface of the earth, but not out in space, nor near the moon, nor in a satellite.

## Velocity

To study motion further, we need the concept of **average velocity**, which differs from average speed in a subtle but important way. The average speed is

$$\frac{\text{distance traversed}}{\text{time elapsed}};$$

it tells how fast an object is moving. The average velocity is

$$\frac{\text{change in position}}{\text{change in time}};$$

this tells how fast the object is moving *and in which direction*, as we now show.

For an object moving up or down, the position is usually measured along a vertical $y$-axis, directed upward. Suppose the position at time $t$ is denoted $y = f(t)$. Then the average velocity from time $t_1$ to a later time $t_2$ is

$$\frac{\text{change in position}}{\text{change in time}} = \frac{f(t_2) - f(t_1)}{t_2 - t_1} = \frac{\Delta y}{\Delta t}.$$

If the object has moved up during that time as in Figure 2, then $f(t_2) > f(t_1)$, so $\Delta y > 0$ and the average velocity is positive. If the object has moved down as in Figure 3, then $\Delta y < 0$ so the average velocity is negative. Thus, with the $y$-axis pointing up, *positive* velocity corresponds to *upward* motion, and *negative* velocity to *downward* motion.

Now we come to the central concept in the study of motion, the **instantaneous velocity** at a particular time $t$, denoted $v(t)$. This can be approximated by the average velocity over a short time interval, that is, by

$$\frac{\Delta y}{\Delta t} = \frac{f(t_2) - f(t)}{t_2 - t}.$$

Make the time interval shorter and shorter, so that $t_2 \to t$. Then the average velocity $\Delta y/\Delta t$ should approach the instantaneous velocity $v(t)$, that is,

$$\frac{\Delta y}{\Delta t} = \frac{f(t_2) - f(t)}{t_2 - t} \to v(t) \qquad \text{as } t_2 \to t. \tag{3}$$

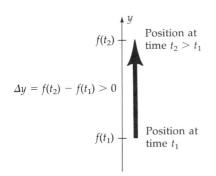

**FIGURE 2**
Object has moved up, $\Delta y > 0$.

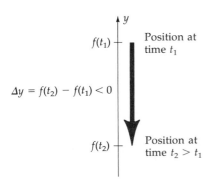

**FIGURE 3**
Object has moved down, $\Delta y < 0$.

Here we recognize the difference quotient and the derivative:

$$\frac{f(t_2) - f(t)}{t_2 - t} \rightarrow f'(t) \qquad \text{as } t_2 \rightarrow t. \tag{4}$$

Comparing (3) and (4) shows: *The velocity is the derivative of the position function $y = f(t)$,* that is,

$$v(t) = f'(t) = \frac{dy}{dt}.$$

For example, with Galileo's law $y = h - 16t^2$, the velocity is

$$v = \frac{dy}{dt} = -32t.$$

At $t = 0$, when the object is dropped, the velocity equals zero. With each passing second, the velocity changes by $-32$ ft/sec. The heavy stone in Figure 1a landed after two seconds of fall, so its velocity at the time of impact was

$$v(2) = -64 \text{ ft/sec.}$$

The minus sign tells the *direction* at impact, that is, downward; and 64 is the *speed* of impact.

Thus, the velocity tells *how fast* the position is changing, and *in which direction*. It is called the **rate of change of position**.

---

**EXAMPLE 1** A baseball is hit straight up. At time $t$, its height is $y = 4 + 96t - 16t^2$. When is it going up? Down? How high does it go? When does it land ($y = 0$)? How fast is it going at time $t = 0$?

**SOLUTION** The velocity is

$$v = \frac{dy}{dt} = 96 - 32t.$$

Determine the sign of $v$, by finding the time, $t$, when $v = 0$:

$$96 - 32t = 0, \qquad t = \tfrac{96}{32} = 3.$$

You can easily check the following signs:

$$v: \quad \underline{\begin{array}{c} + + + + + + \ \bigg| \ - - - - - - - - - - - \\ 3 \end{array}}.$$

Hence the ball goes up for $0 \le t \le 3$ and comes down after that as shown in Figure 4. It therefore achieves its maximum height, $y_{max}$, at time $t = 3$:

$$y_{max} = 4 + 96 \cdot 3 - 16 \cdot 3^2 = 148 \text{ ft.}$$

It lands when $y = 0$, so solve that equation for t:

$$0 = y = 4 + 96t - 16t^2 = 4(1 + 24t - 4t^2),$$

which is equivalent to

$$0 = 4t^2 - 24t - 1$$

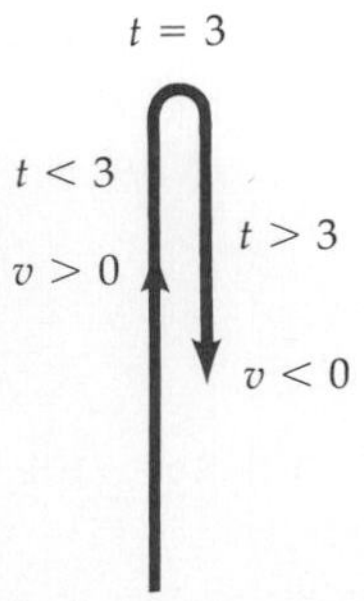

**FIGURE 4**
$v > 0$ for $t < 3$; $v < 0$ for $t > 3$.

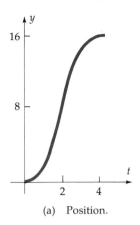

(a) Position.

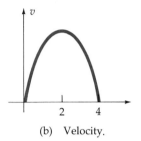

(b) Velocity.

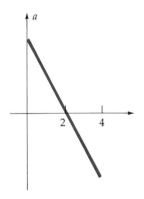

(c) Acceleration.

**FIGURE 5**
Position, velocity, and acceleration for elevator.

or

$$t = \frac{24}{8} \pm \frac{\sqrt{(24)^2 + 4 \cdot 4}}{8} = 3 \pm \frac{\sqrt{37}}{2}.$$

Which $t$ value do you want? The ball reached maximum height at $t = 3$, so it lands *after* $t = 3$, at $t = 3 + \sqrt{37}/2$, not $3 - \sqrt{37}/2$. The velocity at that time is

$$v\left(3 + \frac{\sqrt{37}}{2}\right) = 96 - 32\left(3 + \frac{\sqrt{37}}{2}\right) = -16\sqrt{37} = -97.3 \cdots.$$

At time 0 the velocity is

$$v(0) = 96 - 32 \cdot 0 = 96,$$

positive, since the ball is going up at $t = 0$.

## Acceleration

The rate of change of the velocity with respect to time is called the **acceleration**, denoted $a$. In other words, the acceleration is the *derivative of the velocity*,

$$a = \frac{dv}{dt}.$$

Since $v = \frac{dy}{dt}$, then $a = \frac{d}{dt}\left(\frac{dy}{dt}\right) = \frac{d^2y}{dt^2}$; the acceleration is the *second derivative* of the position function.

In Galileo's law, $y = h - 16t^2$, you find $v = \frac{dy}{dt} = -32t$, and $a = \frac{dv}{dt} = -32$. The acceleration is constant; this is called **uniformly accelerated motion**. When the acceleration is due only to gravity, the motion is described as "free fall."

**EXAMPLE 2** Figure 5a shows the position function for an elevator rising 16 ft. in 4 seconds; it is at ground level when $t = 0$, then rises more and more rapidly until $t = 2$, then slows down to a stop at $y = 16$ when $t = 4$. The velocity $v$ in Figure 5b is always $\geq 0$, since the elevator is rising; but $v = 0$ at the start ($t = 0$) and stop ($t = 4$). The acceleration $a$ in Figure 5c is the derivative of the velocity; thus $a > 0$ when $v$ is increasing; $a$ decreases to 0 when the velocity reaches its maximum (at $t = 2$); and $a$ becomes negative when the velocity decreases ($2 < t < 4$).

## Remarks about Units

Position $y$ has units of length. Hence the *average velocity* $\frac{\Delta y}{\Delta t}$ has units of $\frac{\text{length}}{\text{time}}$, and so does the *instantaneous velocity* $v$. This is reflected in the

Leibniz notation

$$v = \frac{dy}{dt} = \frac{d(\text{length})}{d(\text{time})}.$$

The *acceleration*

$$a = \frac{dv}{dt} = \frac{d(\text{length/time})}{d(\text{time})}$$

has units of (length/time)/time, or length/time$^2$.

In English units, velocity is ft/sec, or mi/hr, or mi/min, or other units, depending on circumstances. Hence acceleration is (ft/sec)/sec = ft/sec$^2$; or mi/hr$^2$, and so on.

In metric units, velocity is cm/sec, or km/hr, and so on depending on circumstances; then acceleration is cm/sec$^2$, or km/hr$^2$.

## Acceleration and Speed

In "free fall," the constant negative acceleration $a = -32$ means that the velocity $v$ is decreasing at 32 ft/sec/sec. But the *speed* $|v|$ may be increasing! In Figure 6, when $t < 0$, the object rises; $v = -32t > 0$, and the speed $|v| = v = -32t$ is decreasing along with the velocity. But when $t > 0$, then the object is falling; $v = -32t < 0$, and the speed $|v| = 32t$ *increases* at 32 ft/sec/sec. Generally, when $v$ and $a$ have the *same* sign, then the acceleration is in the same direction as the motion, thereby speeding it up; but when $v$ and $a$ have *opposite* sign, the acceleration is opposite to the motion, and slows it down.

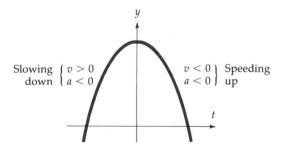

**FIGURE 6**
$y = h - 16t^2$.

## SUMMARY

*Average Velocity* over an interval is

$$\frac{\text{change in position}}{\text{change in time}}.$$

*Velocity* is the derivative of position.

*Acceleration* is the derivative of velocity.
If $y = f(t)$ gives position, then

$$v_{av} = \frac{\Delta y}{\Delta t} = \frac{f(t_2) - f(t_1)}{t_2 - t_1}$$

$$v = \frac{dy}{dt} = f'(t)$$

$$a = \frac{dv}{dt} = \frac{d^2y}{dt^2} = f''(t).$$

## PROBLEMS

### A

1. Let $y = 64 - 16t^2$. Compute $v(2)$, and compare it to the average velocity over the time intervals from $t_1$ to $t_2$.
   **a)** $t_1 = 1, t_2 = 2$   **b)** $t_1 = 1.9, t_2 = 2$
   ***c)** $t_1 = 1.99, t_2 = 2$

2. Let $y = 10 - 4.9t^2$. Compute $v(1)$ and compare it to the average velocity over the time interval from $t_1$ to $t_2$, where:
   **a)** $t_1 = .9, t_2 = 1$.
   ***b)** $t_1 = .9, t_2 = 1.1$.
   **c)** $t_1 = 1, t_2 = 1.1$.
   **d)** $t_1 = .99, t_2 = 1.01$.

3. Compute the velocity and acceleration for the following motion. When is the object moving up, and when is it moving down? What is its maximum height $y$? What is the velocity at time $t = 0$? When does it land and with what velocity?
   ***a)** $y = 6 + 16t - 16t^2$
   ***b)** $y = t(1 - t^2), t \geq 0$
   **c)** $y = t(1 - t)^2, t \geq 0$

4. An object moves on the $x$-axis with the given position function. At what times is it stopped? When does it move to the right? When to the left? How far to the left does it get, for $t \geq 0$?
   ***a)** $x = (2t - 6)^2$
   ***b)** $x = t^4 - 18t^2 + 25$
   **c)** $x = (t + 1)^2(t + 2)$

5. In problem 3a, sketch the graph of $y$ versus $t$, for $0 \leq t \leq t_0$ where $t_0$ is the "landing time." Sketch the graph of $v$ versus $t$ and the graph of $a$ versus $t$.

***6.** In metric units, Galileo's law is $y = h - 490t^2$ cm, for the height of an object $t$ seconds after it is dropped from height $h$ cm. Find the velocity in cm/sec. Find the acceleration in cm/sec$^2$.

7. An elevator has height $y = 3t^5 - 15t^4 + 20t^3$ ft at time $t$, $0 \leq t \leq 2$ seconds. Compute the velocity and acceleration for $0 \leq t \leq 2$. Graph $y$, $v$, and $a$, showing the maximum and minimum of each for $0 \leq t \leq 2$.

***8.** Which graph of $y$ in Figure 7 a–d corresponds to which graph of $v$ in Figure 8 i–iv? Explain your choices.

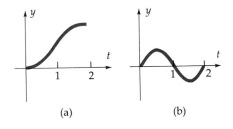

(a)                    (b)

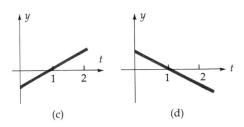

(c)                    (d)

**FIGURE 7**
Position functions.

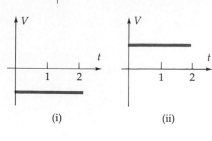

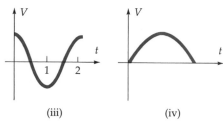

**FIGURE 8**
Velocity functions.

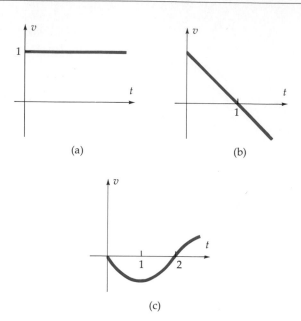

**(a)**

**(b)**

**(c)**

**FIGURE 10**
Velocity functions.

9. For each graph of $y$ versus $t$ in Figure 9, sketch a reasonable version of the graph of $v$ versus $t$.

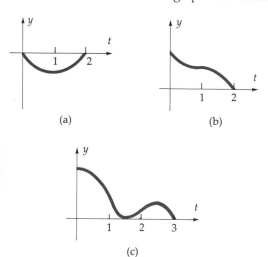

(a)

(b)

(c)

**FIGURE 9**
Position functions.

10. For each graph of $v$ versus $t$ in Figure 10, sketch a reasonable version of the graph of $y$ versus $t$. Make $y = 0$ when $t = 0$.

11. For the given motion, determine when the object is speeding up, and when it is slowing down.
    a) $y = 64 + 10t - 16t^2$     b) $x = 3t - t^3$

**B**

12. A piece of ice slides down a $30°$ inclined slide as shown in Figure 11. It has slid down $x$ feet

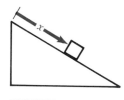

**FIGURE 11**
Inclined slide.

at time $t$, as given in this table:

| $t$ | 0 | .5 | 1 | 1.5 |
|---|---|---|---|---|
| $x$ | 0 | 2 | 8 | 18 |

a) Compute the average velocity $\dfrac{x}{t}$, for $t = 0.5$, $t = 1$, and $t = 1.5$.

b) The averages in part a should suggest that $x/t$ is a linear function of $t$; find this linear function, and find $x$ as a function of $t$.

c) Find the instantaneous velocity $\dfrac{dx}{dt}$ at $t = 1$; at $t = 1.5$. (Galileo performed experiments like this; there were no really accurate devices for timing short intervals, so he used inclined ramps to "dilute" the action of gravity and slow down the motion.)

**FIGURE 12**
Dropping a stone in a deep Mayan well.

13. Two explorers in the Yucatan discover a large deep well as in Figure 12. To see how far down the surface of the water is, they drop stones, timing the fall at 1.75 seconds. The stones are dropped from their hands, 3 feet above the surface of the earth. How deep is the well?

14. The Rovers leave the ruins of Palenque and head south into the jungle of Chiapas. They find a high waterfall pouring over a cliff and landing in a pool. By timing globs of water and occasional branches carried in the stream, they find that it takes 2.5 seconds of free fall from the top of the cliff to the pool. How high is the waterfall? What is the velocity of the falling water as it hits the pool?

15. During a "slam dunk," Dr. J. (Julius Erving, the first master of the slam dunk) raised himself about 3 feet off the floor. How long did it take him to get back to the floor from the highest point of the jump? With what velocity did he land? (This is the negative of his take-off velocity.)

# 2.8
## ANTIDERIVATIVES

What we have done so far is to take derivatives: Given a function, find the slope of its graph; given a position function, find the velocity. Now we look at the reverse problem: Given the slope for each $x$, find the function; given the velocity at every time, find the position.

The given information is a function, call it $f$. We are looking for another function, $F$, such that

$$F' = f.$$

The desired $F$ is called an **antiderivative** of the given $f$.

Consider first the special case that the prescribed derivative is 0 at every point:

$$F'(x) = 0 \qquad \text{for every } x,$$

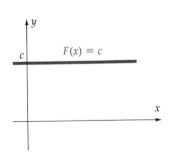

**FIGURE 1**
Constant function has zero slope at every point.

so the graph of $F$ has zero slope at every point. An obvious solution for $F$ is a *constant* function (Fig. 1)

$$F(x) \equiv C,$$

and it is difficult to imagine any other type of solution. In fact, in Chapter 5 it is proved that *only* the constant functions have slope identically equal to zero:

---

**THEOREM 1**

Any function whose derivative equals zero throughout an interval $I$ is constant on $I$:

$$F'(x) = 0 \text{ throughout } I \Rightarrow F(x) = C \text{ on } I,$$

for some constant $C$.

---

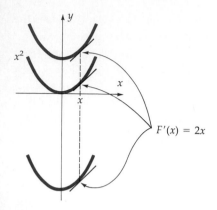

**FIGURE 2**
On each curve $F(x) = x^2 + C$, the slope is the same: $F'(x) = 2x$.

Now suppose that the prescribed derivative is a nonzero function, say $f(x) = 2x$. In this particular case, $x^2$ is an antiderivative, since $Dx^2 = 2x$. Another antiderivative is $x^2 + 3$, since

$$D(x^2 + 3) = 2x + 0 = 0.$$

In fact, for any constant $C$, the function $x^2 + C$ is an antiderivative of $2x$:

$$D(x^2 + C) = 2x.$$

Figure 2 illustrates the situation. The function $F(x) = x^2$ satisfies $F'(x) = 2x$, that is, the slope at $x$ equals $2x$. If you add a constant $C$, you simply shift the entire graph up or down; this does not affect the slope at $x$.

Does the formula $F(x) = x^2 + C$ give *all* the antiderivatives of the function $f(x) = 2x$? Yes it does, as the following argument shows.

Suppose that $F(x)$ is *any* antiderivative of $2x$:

$$DF(x) = 2x. \tag{1}$$

Consider the difference between $F(x)$ and the known antiderivative $x^2$. The derivative of this difference is

$$\begin{aligned} D[F(x) - x^2] &= DF(x) - Dx^2 \\ &= 2x - 2x \quad \text{[by (1)]} \\ &= 0. \end{aligned}$$

So the derivative of $F(x) - x^2$ is zero, for every $x$. It follows from Theorem 1 that the function $F(x) - x^2$ equals a constant $C$,

$$F(x) - x^2 = C,$$

and this implies that

$$F(x) = x^2 + C.$$

Therefore *every antiderivative of $f(x) = 2x$ has the form $F(x) = x^2 + C$*, for some constant $C$.

The same argument applies to any function $f$. Suppose that $F_0$ is one antiderivative of $f$ on some interval $I$; then every antiderivative of $f$ on the same interval $I$ has the form

$$F(x) = F_0(x) + C$$

for some constant $C$. Geometrically, this means that if the slope $F'(x)$ is prescribed throughout an interval, then the graph is determined, except for shifts up or down (Fig. 3).

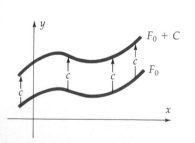

**FIGURE 3**
$F_0' = f$, and $(F_0 + C)' = f$.

## Notation for Antiderivatives

To indicate that

$$D(x^2 + C) = 2x$$

we write

$$x^2 + C = D^{-1}(2x).$$

The symbol $D^{-1}$ stands for the reverse of the derivative operator $D$, and $D^{-1}f(x)$ stands for the antiderivatives of $f(x)$.

**Formulas for Antiderivatives** follow from the corresponding formulas for derivatives. For example

$$D^{-1}x^2 = \tfrac{1}{3}x^3 + C \qquad [\text{since } D(\tfrac{1}{3}x^3) = \tfrac{1}{3} \cdot 3x^2 = x^2]$$
$$D^{-1}x^5 = \tfrac{1}{6}x^6 + C \qquad [\text{since } D(\tfrac{1}{6}x^6) = x^5]$$
$$D^{-1}t^3 = \tfrac{1}{4}t^4 + C \qquad [\text{since } D(\tfrac{1}{4}t^4) = t^3]$$

and generally for $n = 0,\ 1,\ 2,\ 3,\ \ldots,$

$$D^{-1}x^n = \frac{1}{n+1}x^{n+1} + C. \tag{2}$$

This is just the power rule in reverse.

The antiderivative of a sum $f + g$ is the sum of the antiderivatives. For if $DF = f$ and $DG = g$, then

$$D(F + G) = DF + DG = f + g.$$

We indicate this by the formula

$$D^{-1}(f + g) = D^{-1}f + D^{-1}g. \tag{3}$$

Similarly, for a constant multiple, $cf$,

$$D^{-1}(cf) = cD^{-1}f. \tag{4}$$

---

**EXAMPLE 1**

$$D^{-1}(2x + 1) = D^{-1}(2x) + D^{-1}(1)$$
$$= x^2 + x + C.$$

We don't add a constant $C$ to each term of the sum—it is sufficient to add just one at the end, for a simple reason: The function $F_0(x) = x^2 + x$ is one antiderivative of $2x + 1$, so every other antiderivative has the form

$$F_0(x) + C = x^2 + x + C$$

for some constant $C$.

---

In the study of motion, the velocity $v$ is the derivative of the position $y$; so $y$ is an antiderivative of $v$,

$$y = D^{-1}v.$$

And the acceleration is the derivative of the velocity, so $v$ is an antiderivative of $a$:

$$v = D^{-1}a.$$

---

**EXAMPLE 2**    Given:

$$\text{the position } y = 40 \text{ at } \textit{one particular time } t = 0; \qquad (5)$$

$$\text{the velocity } v(t) = -32t \text{ for } \textit{all } t \geq 0; \qquad (6)$$

find $y$ for all times $t \geq 0$.

**SOLUTION**    In general, $y = D^{-1}v$. In the present case, $v = -32t$ so

$$\begin{aligned}
y &= D^{-1}(-32t) \\
&= (-32)D^{-1}t && \text{[by (4)]} \\
&= (-32)\left[\frac{1}{2}t^2 + C\right] && \text{[by (2)]} \\
&= -16t^2 - 32C.
\end{aligned}$$

This has the form "$-16t^2 + $ constant"; we can denote $-32C$ by $\bar{C}$ and write

$$y = -16t^2 + \bar{C}. \qquad (7)$$

So condition (6) determines $y$, except for the constant $\bar{C}$. To determine $\bar{C}$, use the data from (5) [$y = 40$ when $t = 0$] in each side of (7):

$$40 = -16 \cdot 0^2 + \bar{C} = \bar{C}.$$

So $\bar{C} = 40$, and you can rewrite (7) as

$$y = -16t^2 + 40.$$

The solution is valid only for $t \geq 0$, since $v$ is given only for $t \geq 0$.

---

Example 2 solves a simple "navigation" problem: An object moving on a line has a known position at time 0 and known velocity for all times $t \geq 0$. You then deduce the exact position at all times $t \geq 0$.

If instead of the velocity, the *acceleration* is given, then you have to take antiderivatives twice—once to find the velocity, and again to find the position.

---

**EXAMPLE 3**    The acceleration of a dense object in free fall near the surface of the earth (that is, acted on by gravity alone) is a constant, $-9.8$ meters/sec$^2$. Find the position of the object at time $t$.

**SOLUTION**    It is given that $Dv = a = -9.8$, so

$$\begin{aligned}
v(t) = D^{-1}a = D^{-1}(-9.8) \\
= -9.8t + C.
\end{aligned}$$

The constant $C$ is the velocity at time 0:

$$v(0) = (-9.8)0 + C = C.$$

The velocity at time 0 is generally denoted $v_0$, so $C = v_0$, and

$$v = -9.8t + v_0, \qquad \text{where } v_0 = v(0).$$

Knowing $v$, you can find its antiderivative $y$:

$$
\begin{aligned}
y = D^{-1}v &= D^{-1}(-9.8t + v_0) \\
&= (-9.8)D^{-1}t + D^{-1}v_0 \\
&= (-9.8)\tfrac{1}{2}t^2 + v_0 t + \bar{C} \\
&= -4.9t^2 + v_0 t + \bar{C},
\end{aligned}
\tag{8}
$$

where $\bar{C}$ is a new constant. When $t = 0$ then $y = \bar{C}$ from (8). That is, $\bar{C}$ is the position at time 0, generally denoted $y_0$. Thus

$$y = -4.9t^2 + v_0 t + y_0. \tag{9}$$

So, Galileo's free fall formula follows from the fact that the acceleration is constant!

## Slope Fields

The process of reconstructing $F$ from its derivative $F'$ can be understood geometrically by sketching a "slope field" for $F$, as in this final example.

**EXAMPLE 4**   Find a curve $y = F(x)$ such that

$$\frac{dy}{dx} = F'(x) = \frac{1}{x+2}, \tag{10}$$

$$y = 0 \qquad \text{when } x = -1. \tag{11}$$

**SOLUTION**   You are not yet prepared to find a specific formula for the antiderivative of (10); nevertheless, you can make a reasonable sketch of its graph by sketching the given information (10) and (11), as in Figure 4. When

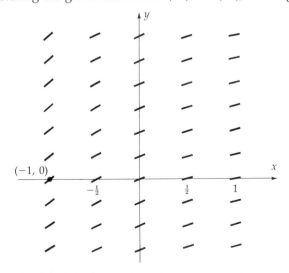

**FIGURE 4**

Slope field for $F'(x) = \dfrac{1}{x+2}$.

$x = 0$, the slope prescribed by (10) is

$$F'(0) = \frac{1}{0 + 2} = \frac{1}{2}.$$

Indicate this by sketching short segments of slope 1/2 all up and down the $y$-axis. You don't yet know where the desired curve crosses the $y$-axis; but wherever that may be, the curve will be parallel to the segments sketched there. Similarly, at $x = 1$ the prescribed slope is

$$F'(1) = \frac{1}{1 + 2} = \frac{1}{3};$$

so draw short segments of slope 1/3 all up and down the line $x = 1$. Follow the same procedure for the lines $x = \pm 1/2$ and $x = -1$, getting the "slope field" for $F$, shown in Figure 4. This pictures only a small part of the information in equation (10); to show it completely would require short segments all along *every* vertical line from $x = -1$ to $x = 1$! But even this small part gives a good idea of the possible curves satisfying (10).

The second condition, (11), is much easier to understand:

$$y = 0 \qquad \text{when } x = -1$$

says simply that the graph passes through the point $(-1, 0)$. Add this point in Figure 4, and you have a geometric view of the given information.

Now to sketch the desired solution, start at the point $(-1, 0)$ [to satisfy (11)] and sketch a smooth curve that crosses each vertical line at the angle indicated by the slope field [to satisfy (10)]. The result is Figure 5.

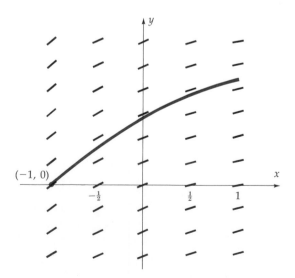

**FIGURE 5**

Solution curve for $F'(x) = \dfrac{1}{x + 2}$, $-1 \leq x \leq 1$, $F(-1) = 0$.

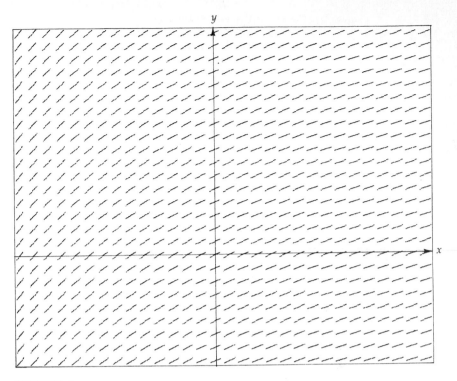

**FIGURE 6**

Slope field for $F'(x) = \dfrac{1}{x+2}$.

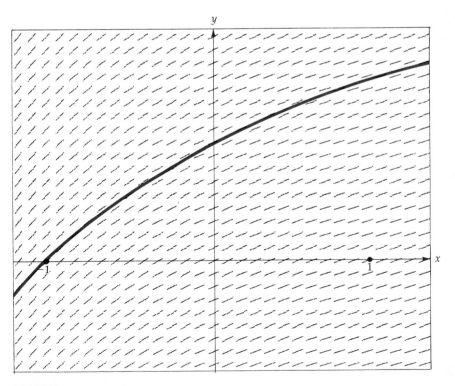

**FIGURE 7**

Solution curve $y = F(x)$ with $F'(x) = \dfrac{1}{x+2}$ and $F(-1) = 0$.

There is software for personal computers that will draw the slope field for a given derivative $F'(x)$, and the solution curve $y = F(x)$ passing through a given point. Figure 6 shows the slope field for $F'(x) = 1/(x + 2)$ drawn by the program "Diffs" (Bridge Software, P.O. Box 118, New Town Branch, Boston, MA 02258); the axes were added by hand. Figure 7 shows the solution curve with $F(-1) = 0$. Figure 8 shows slopes for $F'(x) = 2x$, and several solution curves $y = x^2 + C$.

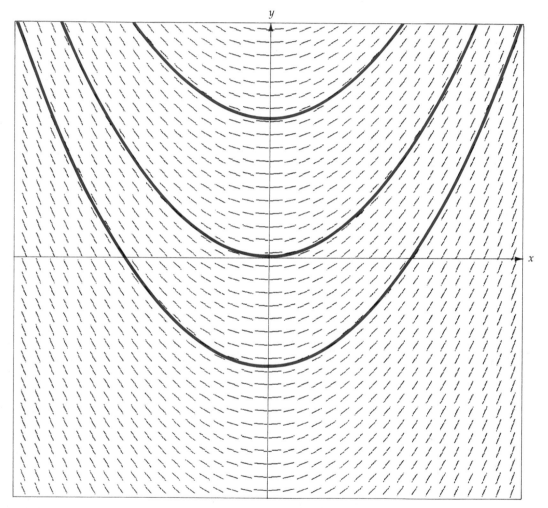

**FIGURE 8**
Slope field for $F'(x) = 2x$ with several solution curves $y = x^2 + c$.

## SUMMARY

$D^{-1}f$ denotes the *antiderivatives* of $f$, the functions $F$ such that $F' = f$.

If $F_0$ is one particular antiderivative on an interval $I$, then on that interval

$$D^{-1}f = F_0 + C.$$

General formulas:

$$D^{-1}x^n = \frac{x^{n+1}}{n+1} + C, \qquad n = 0, 1, 2, \dots$$

$$D^{-1}(cf) = cD^{-1}f$$

$$D^{-1}(f + g) = D^{-1}f + D^{-1}g.$$

Position is an antiderivative of velocity.

Velocity is an antiderivative of acceleration.

## PROBLEMS

### A

1. Find all antiderivatives for the given function on the entire real line.
   *a) $x^2 + 1$
   b) $2x^3 - x$
   c) $(x + 1)(x - 1)$
   d) $x(x - 1)^2$
   *e) $-32t + 10$
   f) $8t^3 - 3t$
   *g) $u^{100}$ (Find a function of $u$ whose derivative is $u^{100}$.)

2. Find the function $F$, given that
   *a) $F'(x) = x^2$, $F(0) = 0$.
   *b) $F'(x) = x(x + 1)$, $F(1) = 2$.
   c) $F'(x) = x(1 - x)(2 - x)$, $F(0) = 2$.

*3. A cat, seeing its image in a mirror, starts creeping toward it and accelerates toward the mirror with acceleration $a(t) = t^2$ ft/sec$^2$. The mirror is at position $x = 3$ ft, and at time $t = 0$ the cat is motionless at position $x_0 = 0$ ft. Give the cat's velocity and position at time $t$. When does the cat reach the mirror and with what velocity? (Since the cat moves horizontally, we use $x$ for the position and $dx/dt$ for the velocity.)

4. a) The acceleration due to gravity near the surface of the moon is about 1.6 meters/sec$^2$ (in absolute value). Derive the formula for free fall on the surface of the moon, analogous to formula (9) for the earth.
   b) Do the same for Mars, where the acceleration is about 3.6 meters/sec$^2$.

*5. A basketball player jumps up at time $t = 0$, with initial height $y_0 = 1$ meter (to her center of gravity) and velocity $v_0 = 4$ m/sec. What is her height at time $t$, until she comes back to the floor? What is her maximum height? How long is she in the air? (She lands again when $y = 1$, since $y$ is height to the center of gravity, not the soles of the feet.)

*6. A basketball player on the moon jumps up with the same initial velocity $v_0 = 4$ m/sec and initial height $y_0 = 1$ as in problem 5. How high does he go? How long is he in the air? (See problem 4 for the acceleration on the moon.)

7. A space-landing module cuts its power at time $t = 0$, when it is 3 meters above the surface of Mars and is falling at 1 meter/sec. Find its height at time $t$. When does it land? With what velocity? (See problem 4b for the acceleration on Mars.)

8. For the given derivative $F'$, sketch carefully the slope field for $F$, showing slopes along the suggested vertical lines. Sketch a graph of $F$ passing through the given point $P_0$.
   *a) $F'(x) = \frac{1}{x}$; $x = \frac{1}{4}, \frac{1}{2}, 1, \frac{3}{2}, 2$; $P_0 = (1, 0)$
   b) $F'(x) = \frac{1}{x}$; $x = -\frac{1}{4}, -\frac{1}{2}, -1, -\frac{3}{2}, -2$; $P_0 = (-1, 1)$
   *c) $\frac{dy}{dx} = \frac{1}{1 + x^2}$; $x = 0, \pm\frac{1}{2}, \pm 1, \pm\frac{3}{2}$; $P_0 = (0, -1)$
   d) $\frac{dy}{dx} = 1 - \frac{1}{x^2}$; $x = \frac{1}{4}, \frac{1}{2}, 1, \frac{3}{2}, 2$; $P_0 = (1, -1)$

9. a) Sketch carefully the slope field for $F$, given the derivative $F'(x) = x^2 - 1$; show slopes along the vertical lines $x = 0, \pm 1/2, \pm 1, \pm 3/2, \pm 2$. Sketch the graph of $F$, given that $F(1) = 2$.

b) *Calculate* the antiderivative for part a and sketch it in the usual way by plotting stationary and inflection points. Compare the result to the curve in part a.

10. A particle moves on the $x$-axis with velocity as follows:

| $t$ | 0 | 0.1 | 0.2 | 0.3 | 0.4 | 0.5 | 0.6 | 0.7 |
|---|---|---|---|---|---|---|---|---|
| $v$ | $-1$ | $-.7$ | $-.3$ | 0 | 0.3 | 0.5 | 0.7 | 1 |

a) Sketch a slope field for position $x$ as a function of $t$.
b) Given $x = 2$ when $t = 0$, sketch a graph of $x$ as a function of $t$.

11. An elevator is designed so that the acceleration is $a(t) = 15t(t - 1)(t - 2)$ m/sec$^2$ for $0 \leq t \leq 2$. It starts at $t = 0$ with $v_0 = 0$ and $y_0 = 0$.
*a) Find $v(t)$ and $y(t)$ for $0 \leq t \leq 2$.
b) Graph $a$, $v$, and $y$ by plotting points for $t = 0, 1, 2$; show the appropriate signs for $a$, the stationary points for $v$, and the concavity for $y$. Show that $v(t) > 0$ for $0 < t < 2$, and $v(0) = v(2) = 0$. Why is this good? Note also that $a(0) = a(2) = 0$; why is that good?

c) How high does the elevator go?
*d) Find the maximum acceleration and minimum acceleration.

**B**

12. Multiply the position $y(t)$ in problem 11 by an appropriate constant, so that the elevator rises exactly 3 meters in 2 seconds. How is acceleration $a$ affected?

13. Determine a motion $y(t)$, similar to the one in problem 11, but taking 4 seconds to rise 3 meters. Thus $a(0) = a(4) = 0$, $a(t) > 0$ for $0 < t < 2$, $a(t) < 0$ for $2 < t < 4$; $v(0) = v(4) = 0$; $y(0) = 0$, $y(4) = 3$. Determine the maximum velocity and acceleration for this motion.

14. Suppose the acceleration in problem 11 is replaced by $a(t) = t^2(t - 1)(t - 2)^2$ for $0 \leq t \leq 2$, with $a(t) = 0$ for $t < 0$ and $t > 2$. Show that $a'(0) = 0$ and $a'(2) = 0$. Graph $a(t)$ for $-1 < t < 3$ and compare it to acceleration $a$ in problem 11. Which motion is smoother?

## REVIEW PROBLEMS    CHAPTER 2

*1. Compute $f'(x)$ using difference quotients. Make a sketch to illustrate the quotient you use.
a) $f(x) = 2x^2 - x$    b) $f(x) = 1/(2x + 1)$
*2. Write the equation of the line tangent to the graph of $f$ at the given point.
a) $f(x) = 2x^3 - x^2$ at $(-1, -3)$
b) $f(x) = 1/(2x + 1)$ at $(0, 1)$
*3. Graph, showing stationary points, inflection points, and concavity.
a) $f(x) = x^2 - x - 1$
b) $f(x) = x^3 + x^2 + 1$
c) $f(x) = x^4 - 2x^2 + 1$
*4. A cylindrical can is made of sides costing 2 cents/in$^2$ and a top and bottom costing 5 cents/in$^2$. For 30 cents in material costs, what dimensions yield a can with maximal volume?
*5. How many solutions are there for the following?
a) $5x^3 - 5x + 1 = 0$
b) $x^3 + x + 1 = 0$

c) $x^3 - 3x + 2 = 0$
d) $x^4 - 3x^2 + 2x = 0$
*6. (From optional section on Newton's method)
a) Approximate all solutions of $x^3 + x + 1 = 0$ by Newton's method, with error $< 0.0001$.
b) Demonstrate that the error is $< 0.0001$.
*7. The position at time $t$ of an object moving on the $x$-axis is given by $x = 2t^3 - 3t^2 + 1$. Find velocity $v$ and acceleration $a$. When is the object moving to the right? When to the left? When is it momentarily stationary? What is the minimum velocity for $0 \leq t \leq 2$? When is it speeding up?
*8. Given $F'(x) = x^4 - x$ and $F(1) = 3$, determine $F(x)$ for all $x$.
*9. Find position $y$, given that velocity $v = -3t + 2$, and $y = 1$ when $t = 0$.
*10. Find position $y$, given that acceleration $a = 2t$, velocity $v(0) = 2$, and $y = 0$ when $t = 1$.

11. Sketch the slope field and the solution curve $y = F(x)$ for $-2 \leq x \leq 2$, given that $F'(x) = (1+x^2)^{1/2}$, $F(0) = 0$.

12. A particle moves on the $y$-axis with velocity $v = dy/dt$ as follows:

| $t$ | 0 | .2 | .4 | .6 | .8 | 1.0 | 1.2 | 1.4 | 1.6 | 1.8 | 2.0 |
|---|---|---|---|---|---|---|---|---|---|---|---|
| $v(t)$ | 1 | 1.3 | 1.4 | 1.3 | 1 | .5 | 0 | $-.3$ | 0 | 0 | 0 |

Sketch the slope field for $y$. Sketch $y$, given $y = 1$ when $t = 0$.

## SUPPLEMENTARY PROBLEMS    CHAPTER 2

1. Find the point(s) on the parabola $y = x^2$ closest to (0, 1). [Hint: Minimize the *square* of the distance from $(x, x^2)$ to (0, 1).]

2. A beam of length $L$ has one end clamped in position, while the other end rests on a wall. The beam sags from the straight line between its ends by an amount

$$z = k(2x^4 - 5Lx^3 + 3L^2x^2), \qquad 0 \leq x \leq L,$$

where $k$ is a constant depending on the nature of the beam. At what point does the maximum sag occur? Judging from your answer, is the clamped end at $x = 0$, or at $x = L$? Determine which end is clamped in another way, by computing $dz/dx$ at $x = 0$ and $x = L$.

3. A baseball is hit at 3 feet above ground level, and rises straight up to a height of 150 feet, then falls back to earth. What is its initial velocity?

4. A car traveling 60 mph is stopped, with constant deceleration, in 200 feet.
   a) What was the rate of deceleration?
   b) How fast was it going 100 feet before the stopping point?

5. The following data give the position of a falling body at time $t$:

| $t$ (secs) | 0 | 0.1 | 0.2 | 0.3 | 0.4 | 0.5 | 0.6 |
|---|---|---|---|---|---|---|---|
| $y$ (meters) | 2 | 1.95 | 1.80 | 1.56 | 1.22 | 0.78 | 0.24 |

   a) Use difference quotients to estimate velocity $v$ at times $t = 0.25$ and 0.35.
   b) Estimate acceleration $a$ at time $t = 0.3$.

6. Find a cubic $f(x) = ax^3 + bx^2 + cx + d$ with $f(0) = 1$, $f'(0) = 1$, $f''(0) = 1$, and $f^{(3)}(0) = 1$.

7. Find a cubic with $f(0) = 1$, $f'(0) = 1$, $f''(1) = 2$, and $f''(-1) = 3$.

8. Approximate the stationary points of $y = x^4 + x^2 + x$ by Newton's method, with error $<0.01$. Demonstrate that the error is $<0.01$.

9. Suppose that $\bar{x}$ is a zero of $f$, and $\bar{x}$ together with the successive Newton approximations $x_0$, $x_1$, $x_2$ all lie in interval $I$ where $f' < 0$ and $f'' > 0$. Are $x_1$ and $x_2$ greater or less than $\bar{x}$? Is $x_1$ greater or less than $x_2$?

10. a) Find the minimum value of $f(x) = Ax^2 + 2Bx + C$, given $A > 0$.
    b) Suppose that $Ax^2 + 2Bx + C \geq 0$ for all $x$. Show that $B^2 \leq AC$.
    c) Suppose that $a_1, b_1, a_2, b_2, a_3, b_3, \ldots, a_n, b_n$ are any real numbers, and define $f(x) = (a_1x + b_1)^2 + \cdots + (a_nx + b_n)^2$. Explain why $f(x) \geq 0$ for all $x$.
    d) Combining parts c and b, deduce the *Schwarz inequality*

$$(a_1^2 + a_2^2 + \cdots + a_n^2)(b_1^2 + b_2^2 + \cdots + b_n^2)$$
$$\geq (a_1b_1 + a_2b_2 + \cdots + a_nb_n)^2.$$

    e) Under what conditions on the numbers $a_1, \ldots, a_n, b_1, \ldots, b_n$ does the function in part b have a minimum value 0? Under what conditions does the left-hand side of the Schwarz inequality *equal* the right-hand side?

# 3

# LIMITS,
# CONTINUITY,
# AND
# DERIVATIVES

## LIMITS

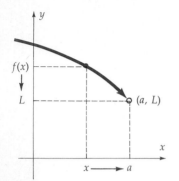

**FIGURE 1**
$$\lim_{x \to a^-} f(x) = L.$$

Underlying derivatives is the concept of *limit*. Like most things in calculus, limits have a geometric aspect (describing certain features of graphs) and an algebraic aspect, providing the rules for computing limits and derivatives. We begin with the geometric aspect.

In Figure 1, pick a point on the graph of $f$ just to the left of the line $x = a$. Then follow the graph to the right as it approaches that line; it heads for a particular point $(a, L)$, like a purposeful bee headed for a flower at $(a, L)$. This means that as $x$ approaches $a$ from the left, the function values $f(x)$ approach the number $L$. We write

$$\lim_{x \to a^-} f(x) = L$$

and say "the limit of $f(x)$ as $x$ approaches $a$ from the left equals $L$." Notice that in this case the "target point" $(a, L)$ is not on the graph. The limit as $x \to a^-$ is completely determined by the function values $f(x)$ for $x$ *less* than $a$; you ignore the value $f(a)$, which may not be defined, or may have any value whatsoever, without affecting the limit as $x$ *approaches a*.

In Figure 2, as $x$ approaches 0 from the left, the graph heads for the point $(0, -1)$; thus

$$\lim_{x \to 0^-} \frac{x}{|x|} = -1.$$

In the same figure, as $x$ approaches 0 from the *right* the graph heads for the point $(0, 1)$. We say then that "the limit of $x/|x|$ as $x$ approaches 0 from the right equals 1," and write

$$\lim_{x \to 0^+} \frac{x}{|x|} = 1.$$

The square root function, $f(x) = \sqrt{x}$, gives another example of "limit from the right." As you see from the graph in Figure 3

$$\lim_{x \to 0^+} \sqrt{x} = 0.$$

A table of values shows the same thing—as $x$ approaches 0 from the right then $\sqrt{x}$ approaches 0:

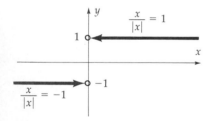

**FIGURE 2**
$$\lim_{x \to 0^-} \frac{x}{|x|} = -1, \quad \lim_{x \to 0^+} \frac{x}{|x|} = 1.$$

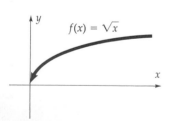

**FIGURE 3**
$$\lim_{x \to 0^+} \sqrt{x} = 0.$$

| $x$ | 1 | 0.1 | 0.01 | 0.001 | 0.0001 | $10^{-6}$ | $10^{-2n}$ |
|---|---|---|---|---|---|---|---|
| $\sqrt{x}$ | 1 | $0.316 \cdots$ | 0.1 | $0.0316 \cdots$ | 0.01 | $10^{-3}$ | $10^{-n}$ |

It may happen that as $x \to a$ from the right then $f(x)$ does not approach any single number; for example in Figure 4, as $x \to 0^+$, then $1/x$ grows larger

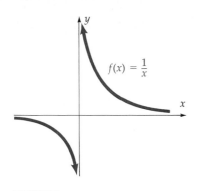

**FIGURE 4**

$\lim\limits_{x \to 0^+} \dfrac{1}{x}$ does not exist.

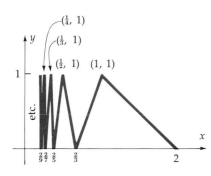

**FIGURE 5**
The graph of $f$ rises to the line $y = 1$ at $x = 1, \frac{1}{2}, \frac{1}{3}, \frac{1}{4}, \frac{1}{5}, \ldots$, and returns to $y = 0$ at $x = \frac{2}{3}, \frac{2}{5}, \frac{2}{7}, \frac{2}{9}, \ldots$.
$\lim\limits_{x \to 0^+} f(x)$ does not exist. (The part of the graph for $0 < x < \frac{1}{5}$ cannot be shown to this scale, but it continues to oscillate in the same fashion.)

and larger without bound. In such a case we say that "the limit does not exist":

$$\lim_{x \to 0^+} \frac{1}{x} \quad \text{does not exist.}$$

Figure 5 shows another case where the limit does not exist—the graph oscillates back and forth, like a frantic bee that knows not where to land. Since the graph does not head for any one particular point $(0, L)$ as $x \to 0^+$, no limit $L$ exists. (Examples like this are not common but, as you see, they are a possibility to be reckoned with.)

In Figure 6 the graph heads for the point $(a, L)$ from *both* sides; both $\lim\limits_{x \to a^-} f(x) = L$ and $\lim\limits_{x \to a^+} f(x) = L$. In that case we write

$$\lim_{x \to a} f(x) = L$$

and say "the limit of $f(x)$ as $x$ approaches $a$ equals $L$." A very simple example is the general linear function (Fig. 7)

$$f(x) = mx + b.$$

As $x \to a$ from either side, the graph heads for the point $(a, ma + b)$, so

$$\lim_{x \to a} (mx + b) = ma + b.$$

A more instructive example is

$$f(x) = \frac{x^2 - 1}{x - 1}, \qquad x \neq 1. \tag{1}$$

The point $x = 1$ is excluded from the domain of $f$ because of the denominator $x - 1$. But when $x \neq 1$, then

$$\frac{x^2 - 1}{x - 1} = \frac{(x - 1)(x + 1)}{x - 1} = x + 1, \qquad (x \neq 1).$$

So the graph of the function $f$ defined in (1) is just the straight line $y = x + 1$

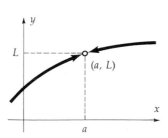

**FIGURE 6**
$\lim\limits_{x \to a} f(x) = L.$

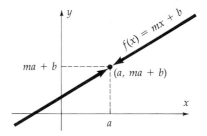

**FIGURE 7**
$\lim\limits_{x \to a} (mx + b) = ma + b.$

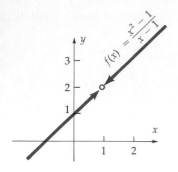

**FIGURE 8**

$$\lim_{x \to 1} \frac{x^2 - 1}{x - 1} = 2.$$

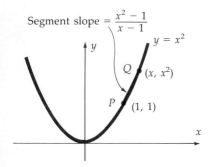

**FIGURE 9**

Slope of graph at $P$ is $\lim_{x \to 1} \dfrac{x^2 - 1}{x - 1}$.

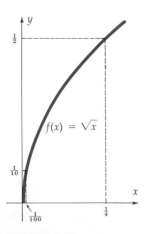

**FIGURE 10**

$$\lim_{x \to 0^+} \sqrt{x} = 0.$$

*with the point* $(1, 2)$ *deleted,* as shown in Figure 8. The graph has no point with $x = 1$; but as $x \to 1$ from either side, the graph heads for the point $(1, 2)$, and the function values $f(x)$ approach the limit 2. That is,

$$\lim_{x \to 1} f(x) = 2.$$

One might wonder, "Why write the quotient $\dfrac{x^2 - 1}{x - 1}$ instead of the simpler expression $x + 1$?" We encounter this quotient quite naturally in computing the slope of the graph of $y = x^2$ at the point $P = (1, 1)$ in Figure 9; the slope of the segment between $P$ and any other point $Q = (x, x^2)$ on the graph is

$$\frac{\Delta y}{\Delta x} = \frac{x^2 - 1}{x - 1}, \qquad x \ne 1,$$

which is precisely the function $f(x)$ in (1). As $x \to 1$, then $Q \to P$ and $\dfrac{\Delta y}{\Delta x}$ approaches the slope of the graph of $y = x^2$ at $(1, 1)$; so the slope of the graph at that point is

$$\lim_{x \to 1} \frac{x^2 - 1}{x - 1} = 2.$$

This slope is precisely the derivative of the function $g(x) = x^2$ at $x = 1$. Thus *the derivative is a certain limit;* that is why we are studying limits.

Turning now to the algebraic aspect of limits, we need to formulate the concept more precisely. Keeping in mind the previous examples, we make the following attempt at a definition:[1]

---

**LIMIT FROM THE RIGHT**

$\lim\limits_{x \to a^+} f(x) = L$ if you can keep $f(x)$ arbitrarily close to $L$, by requiring that $x$ be sufficiently close to $a$, but greater than $a$.

---

According to this, it is true that $\lim\limits_{x \to 0^+} \sqrt{x} = 0$; you *can* keep $\sqrt{x}$ arbitrarily close to 0 by requiring that $x$ be sufficiently close to 0, but greater than 0. For instance (Fig. 10) you can keep $\sqrt{x} < 1/2$ by requiring $0 < x < 1/4$; keep $\sqrt{x} < 1/10$ by requiring $0 < x < 1/100$; keep $\sqrt{x} < 10^{-6}$ by requiring $x < 10^{-12}$; and so on.

For the situation back in Figure 5, the attempted definition shows that the limit fails to exist; there is *no* number $L$ such that the function values remain arbitrarily close to $L$ when $x$ is required to be sufficiently close to 0. For no matter how closely you require $x$ to remain near 0 (but greater than 0), the function values $f(x)$ assume all values between 0 and 1, but do not remain close to *any one* number $L$.

---

[1] The formal, rigorous version of this definition is stated and discussed in Appendix A.1.

For the left-hand limit, $\lim_{x \to a^-} f(x) = L$, the definition above is changed; the requirement that $x$ be "greater than $a$" is replaced by "less than $a$." And for the two-sided limit, $\lim_{x \to a} f(x) = L$, we require the same limit from both sides:

$$\lim_{x \to a} f(x) = L \quad \text{means} \quad \lim_{x \to a^+} f(x) = L \quad \text{and} \quad \lim_{x \to a^-} f(x) = L.$$

Remember that in computing the limit as $x$ approaches $a$, we leave $a$ itself completely out of consideration; the function value $f(a)$ (which may or may not be defined) has no effect on the *limit* as $x$ *approaches $a$*.

Next, we consider some essential algebraic relations for limits:

---

**THEOREM 1**

Let lim stand for any one of $\lim_{x \to a^+}$, $\lim_{x \to a^-}$, or $\lim_{x \to a}$. Then:

$$\lim(f + g) = (\lim f) + (\lim g)$$

$$\lim(fg) = (\lim f)(\lim g)$$

$$\lim(cf) = c \lim f \qquad \text{(for any constant $c$)}$$

$$\lim(f/g) = (\lim f)/(\lim g) \qquad \text{(if $\lim g \neq 0$)}$$

provided, in each case, that $\lim f$ and $\lim g$ exist.

---

Each of these relations has a commonsense explanation. Consider for example the limit of a product, and interpret "lim" as "$\lim_{x \to a}$." It is assumed that $\lim_{x \to a} f(x)$ has a certain value $L$ and $\lim_{x \to a} g(x)$ has a certain value $M$. That is, as $x \to a$, $f(x)$ approaches $L$ and $g(x)$ approaches $M$. It then stands to reason that the product $f(x)g(x)$ approaches $L \cdot M$. In terms of the attempted definition of limit, you can keep $f(x)g(x)$ arbitrarily close to $L \cdot M$ by keeping $f(x)$ arbitrarily close to $L$ and $g(x)$ arbitrarily close to $M$.

The rules for sums, quotients, and constant multiples can be similarly explained. (These explanations are not *proofs*, however—proofs require the formal definition of limit given in the Appendix.)

The rule for sums is stated for just two terms, but it then follows for three; $f + g + h$ can be considered the sum of two terms, $f$ and $(g + h)$, so

$$\begin{aligned}
\lim(f + g + h) &= \lim(f + (g + h)) \\
&= \lim f + \lim(g + h) \qquad \text{(Theorem 1)} \\
&= \lim f + [\lim g + \lim h] \qquad \text{(Theorem 1 again)} \\
&= \lim f + \lim g + \lim h.
\end{aligned}$$

In fact, for any finite sum

$$\lim[f_1 + f_2 + \cdots + f_n] = \lim f_1 + \lim f_2 + \cdots + \lim f_n.$$

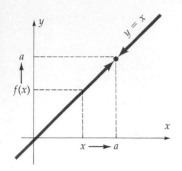

**FIGURE 11**
$\lim\limits_{x \to a} x = a.$

Similarly, for products with many factors

$$\lim[f_1 \cdot f_2 \cdot \,\cdots\, \cdot f_n] = (\lim f_1)(\lim f_2) \cdots (\lim f_n).$$

In all these formulas it is assumed that the limits on the right-hand side exist. To apply them we must start with known limits and build up. One known limit is shown in Figure 11,

$$\lim_{x \to a} x = a.$$

From this it follows that

$$\lim_{x \to a} x^n = a^n; \qquad (2)$$

for

$$\lim_{x \to a} x^n = \lim_{x \to a} (x \cdot x \cdot \,\cdots\, \cdot x)$$

$$= \left[\lim_{x \to a} x\right]\left[\lim_{x \to a} x\right] \cdots \left[\lim_{x \to a} x\right]$$

$$= a \cdot a \cdot \,\cdots\, \cdot a$$

$$= a^n$$

as claimed.

Next, consider any polynomial

$$P(x) = c_n x^n + c_{n-1} x^{n-1} + \cdots + c_1 x + c_0.$$

---

**THEOREM 2**

For any polynomial $P$, $\lim\limits_{x \to a} P(x) = P(a).$

---

**PROOF**

$$\lim_{x \to a} P(x) = \lim_{x \to a} (c_n x^n + c_{n-1} x^{n-1} + \cdots + c_1 x + c_0)$$

$$= \lim_{x \to a} c_n x^n + \lim_{x \to a} c_{n-1} x^{n-1} + \cdots + \lim_{x \to a} c_1 x + \lim_{x \to a} c_0$$

(limit of a sum)

$$= c_n \lim_{x \to a} x^n + c_{n-1} \lim_{x \to a} x^{n-1} + \cdots + c_1 \lim_{x \to a} x + c_0$$

(constant multiples)

$$= c_n a^n + c_{n-1} a^{n-1} + \cdots + c_1 a + c_0 \qquad \text{[using (2)]}$$
$$= P(a).$$

In words: To evaluate the limit of a polynomial $P(x)$ as $x \to a$, simply replace $x$ by $a$ in the formula for $P$. For example,

$$\lim_{x \to 2} (x^3 + 5x - 4) = 2^3 + 5 \cdot 2 - 4 = 14.$$

The same is true of any rational function $P/Q$ (where $P$ and $Q$ are polynomials), at points $a$ where $Q(a) \neq 0$. For by Theorem 2, $\lim\limits_{x \to a} P(x) = P(a)$

and $\lim\limits_{x \to a} Q(x) = Q(a) \neq 0$; hence

$$\lim_{x \to a} \frac{P(x)}{Q(x)} = \frac{\lim\limits_{x \to a} P(x)}{\lim\limits_{x \to a} Q(x)} \qquad \text{(limit of a quotient)}$$

$$= \frac{P(a)}{Q(a)}.$$

For example,

$$\lim_{x \to 2} \frac{x^2 - 1}{x - 1} = \frac{\lim\limits_{x \to 2} (x^2 - 1)}{\lim\limits_{x \to 2} (x - 1)}$$

$$= \frac{2^2 - 1}{2 - 1} = \frac{3}{1} = 3$$

as Figure 8 confirms.

When the denominator $Q(a)$ *does* equal zero, then the formula for the limit of a quotient does not apply, and further analysis is needed, as in the limit that we computed in Figure 8,

$$\lim_{x \to 1} \frac{x^2 - 1}{x - 1} = 2.$$

---

**EXAMPLE 1**   A function $f$ is defined by

$$f(x) = x + 1 \qquad \text{if } x < 1$$

$$= \frac{x^2 - 2x}{x + 1} \qquad \text{if } x \geq 1.$$

Evaluate $\lim\limits_{x \to 1^-} f(x)$, $\lim\limits_{x \to 1^+} f(x)$, and $\lim\limits_{x \to 1} f(x)$.

**SOLUTION**   The limit as $x$ approaches 1 from the left uses only function values $f(x)$ with $x < 1$. For such $x$, the definition of $f$ is

$$f(x) = x + 1, \qquad x < 1.$$

Hence

$$\lim_{x \to 1^-} f(x) = \lim_{x \to 1^-} (x + 1) = 1 + 1 = 2.$$

Similarly, for the limit as $x \to 1$ from the right, use the definition of $f$ for $x > 1$:

$$\lim_{x \to 1^+} f(x) = \lim_{x \to 1^+} \frac{x^2 - 2x}{x + 1} = \frac{1 - 2}{2} = -1/2.$$

Since the left-hand limit differs from the right-hand limit, the two-sided limit $\lim\limits_{x \to a} f(x)$ does not exist.

## SUMMARY

$$\lim_{x \to a} f(x) = L \Leftrightarrow \lim_{x \to a^-} f(x) = L \quad \text{and} \quad \lim_{x \to a^+} f(x) = L.$$

### Algebra of Limits:

$$\lim(f + g) = (\lim f) + (\lim g)$$

$$\lim(fg) = (\lim f) \cdot (\lim g)$$

$$\lim(cf) = c \lim f$$

$$\lim(f/g) = (\lim f)/(\lim g) \qquad \text{if } \lim g \neq 0$$

provided in each case that $\lim f$ and $\lim g$ exist.
For a polynomial $P$,

$$\lim_{x \to a} P(x) = P(a).$$

For a rational function $P/Q$

$$\lim_{x \to a} \frac{P(x)}{Q(x)} = \frac{P(a)}{Q(a)} \qquad \text{if } Q(a) \neq 0.$$

## PROBLEMS

### A

In problems 1–9, evaluate the required limit, or explain why it does not exist.

1. For $f(x)$ in Figure 12, determine

   *a) $\displaystyle\lim_{x \to 2^-} f(x)$.    b) $\displaystyle\lim_{x \to 2^+} f(x)$.

   *c) $\displaystyle\lim_{x \to 2} f(x)$.    d) $\displaystyle\lim_{x \to 3} f(x)$.

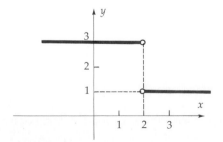

**FIGURE 12**

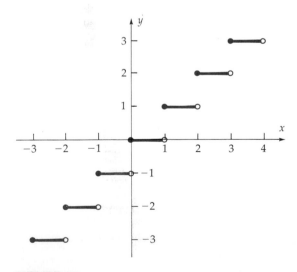

**FIGURE 13**
Step function $s(x)$.

2. For the "step function" $s(x)$ in Figure 13, determine

   *a) $\displaystyle\lim_{x \to 0^+} s(x)$.    b) $\displaystyle\lim_{x \to 0^-} s(x)$.

   *c) $\displaystyle\lim_{x \to 0} s(x)$.    d) $\displaystyle\lim_{x \to 3^-} s(x)$.

   e) $\displaystyle\lim_{x \to 3^+} s(x)$.

3. For the function graphed in Figure 14, determine

*a) $\lim\limits_{x \to -1^-} f(x)$.  b) $\lim\limits_{x \to -1^+} f(x)$.

*c) $\lim\limits_{x \to -1} f(x)$.  d) $\lim\limits_{x \to 1^+} f(x)$.

*e) $\lim\limits_{x \to 1^-} f(x)$.  f) $\lim\limits_{x \to 1} f(x)$.

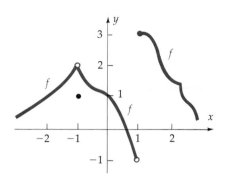

**FIGURE 14**

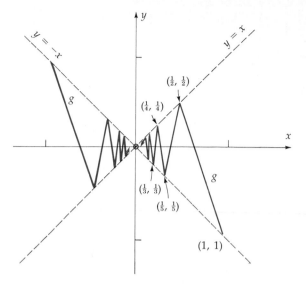

**FIGURE 16**
The graph of $g$ oscillates between $y = x$ and $y = -x$, as indicated, and $g(x)$ is defined for all $x \neq 0$. (The part of the graph for $-\frac{1}{8} < x < \frac{1}{8}$ cannot be sketched at this scale.)

4. For the function $f$ in Figure 15, determine

*a) $\lim\limits_{x \to 0^-} f(x)$.  *b) $\lim\limits_{x \to 0^+} f(x)$.

c) $\lim\limits_{x \to 0} f(x)$.

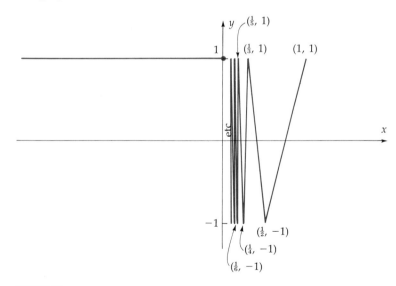

**FIGURE 15**
$f(x) = 1$, $x < 0$. At $x = \frac{1}{3}, \frac{1}{5}, \frac{1}{7}, \ldots, f(x) = 1$; at $x = \frac{1}{2}, \frac{1}{4}, \frac{1}{6}, \frac{1}{8}, \ldots, f(x) = -1$.

5. For the function $g$ graphed in Figure 16, determine

a) $\lim\limits_{x \to 0^-} g(x)$.  b) $\lim\limits_{x \to 0^+} g(x)$.

*c) $\lim\limits_{x \to 0} g(x)$.

6. In 1987, the postage $P(x)$ on a letter weighing $x$ ounces was 22 cents for $0 < x \leq 1$, and 39 cents for $1 < x \leq 2$. Determine

a) $\lim\limits_{x \to 0^+} P(x)$.  b) $\lim\limits_{x \to 1^-} P(x)$.

c) $\lim\limits_{x \to 1^+} P(x)$.    d) $\lim\limits_{x \to 1} P(x)$.

e) $\lim\limits_{x \to 2^-} P(x)$.

*7.  Let $f(x) = x^2 + 1$    if $x \le 2$
            $= x - 1$    if $x > 2$.
Find the limit, or explain why it does not exist.

a)  $\lim\limits_{x \to 2^-} f(x)$    b)  $\lim\limits_{x \to 2^+} f(x)$

c)  $\lim\limits_{x \to 2} f(x)$    d)  $\lim\limits_{x \to -2} f(x)$

8.  Let $f(x) = x^2 + x$,    if $x \le 0$
            $= x^2 - x$,    if $0 < x < 1$
            $= 2$,    if $x \ge 1$.

Find the limit, or explain why it does not exist.

a)  $\lim\limits_{x \to 0^-} f(x)$    b)  $\lim\limits_{x \to 0^+} f(x)$

c)  $\lim\limits_{x \to 1^-} f(x)$    d)  $\lim\limits_{x \to 1^+} f(x)$

e)  $\lim\limits_{x \to 0} f(x)$    f)  $\lim\limits_{x \to 1} f(x)$

g)  $\lim\limits_{x \to -2} f(x)$

9.  For $f(x) = 1/x$,    $x > 0$
            $= 2x$,    $x < 0$
determine

*a)  $\lim\limits_{x \to 0^+} f(x)$.    *b)  $\lim\limits_{x \to 0^-} f(x)$.

c)  $\lim\limits_{x \to 0} f(x)$.    d)  $\lim\limits_{x \to 1^-} f(x)$.

10.  Use a calculator to guess, by experimenting, the following limits. (Some do not exist.)

*a)  $\lim\limits_{x \to 1} \dfrac{x^3 - 1}{x - 1}$    *b)  $\lim\limits_{x \to 1} \dfrac{x^3 + 1}{x - 1}$

*c)  $\lim\limits_{x \to 1} \dfrac{\sqrt{x} - 1}{x - 1}$    d)  $\lim\limits_{x \to 2} \dfrac{x^2 - 4}{x^3 - 8}$

11.  Sketch the graph of $\dfrac{|x + 1|}{x + 1}$, plotting several points for $x > -1$ and several for $x < -1$. Then evaluate the limits (if they exist).

a)  $\lim\limits_{x \to -1^-} \dfrac{|x + 1|}{x + 1}$    b)  $\lim\limits_{x \to -1^+} \dfrac{|x + 1|}{x + 1}$

*c)  $\lim\limits_{x \to -1} \dfrac{|x + 1|}{x + 1}$    *d)  $\lim\limits_{x \to 1} \dfrac{|x + 1|}{x + 1}$

12.  Evaluate the limits by sketching the graphs.

a)  $\lim\limits_{x \to 1} |x - 1|$    b)  $\lim\limits_{x \to 1} \dfrac{|x - 1|}{x - 1}$

c)  $\lim\limits_{x \to 2} \dfrac{x^2 - 4}{x - 2}$ (The graph is a straight line with one point missing.)

d)  $\lim\limits_{x \to 4} \dfrac{x^2 - 4}{x - 2}$

13.  Evaluate the limits, making algebraic simplifications where needed.

*a)  $\lim\limits_{x \to 1} (x^2 + 2x + 3)$

b)  $\lim\limits_{x \to 2} \dfrac{x + 1}{x - 1}$

*c)  $\lim\limits_{x \to 1} \dfrac{x - 1}{x^2 - 1}$

d)  $\lim\limits_{x \to 2} \dfrac{x}{|x|}$

*e)  $\lim\limits_{x \to 0} \dfrac{3/x^2}{2 + (1/x^2)}$

f)  $\lim\limits_{x \to -1} (x^3 + 1)$

g)  $\lim\limits_{x \to 1} \left( \dfrac{1}{x - 1} - \dfrac{2}{x^2 - 1} \right)$

h)  $\lim\limits_{x \to 0^+} \dfrac{1}{1 + 1/x}$

i)  $\lim\limits_{x \to 0} \dfrac{x^3 - 2x}{x}$

14.  Use a calculator with a "ln" key to guess the following limits by "experimenting."

*a)  $\lim\limits_{x \to 1} \dfrac{\ln x}{x - 1}$    *b)  $\lim\limits_{x \to 0^+} x \ln x$

c)  $\lim\limits_{x \to 2} \dfrac{\ln x - \ln 2}{x - 2}$    d)  $\lim\limits_{x \to 0^+} \sqrt{x} \ln x$

15.  Use a calculator with an $x^y$ key to guess the following limits by "experimenting."

*a)  $\lim\limits_{x \to 0^+} x^x$    b)  $\lim\limits_{x \to 0^+} x^{1/x}$

c)  $\lim\limits_{x \to 0^+} x^{\sqrt{x}}$

**B**

16.  Does there exist a function $f$ with $f(x) > 0$ for all $x \ne 0$, yet with $\lim\limits_{x \to 0} f(x) = 0$? If "no," say why not; if "yes," give an example of such a function.

17. Rewrite the given expression in such a way that the limit becomes easy to evaluate. Evaluate it.

   a) $\lim\limits_{x \to 1} \dfrac{x^3 - 1}{x - 1}$

   b) $\lim\limits_{x \to 4} \dfrac{\sqrt{x} - 2}{x - 4}$

   $[\sqrt{a} - \sqrt{b} = (a - b)/(\sqrt{a} + \sqrt{b})]$

   c) $\lim\limits_{x \to 1} \dfrac{x - 1}{x^2 - 1}$   d) $\lim\limits_{x \to 1} \dfrac{x^3 - 1}{x^2 - 1}$

18. Suppose that $\lim\limits_{x \to 0^+} f(x) = L$. What can be said about $\lim\limits_{x \to 0^-} f(x)$ if:

   a) $f$ is an even function?
   b) $f$ is an odd function?

19. Suppose that $-1 < f(x) < 1$ for all $x$, and that $\lim\limits_{x \to 0} f(x)$ exists. What can be said about $\lim\limits_{x \to 0} f(x)$?

---

## 3.2

# SOME TRIGONOMETRIC LIMITS. THE TRAPPING THEOREM

The trigonometric functions are even more important than their geometric origins suggest. They arise naturally in periodic phenomena (such as oscillations and crystal structure) and are pressed into service in almost every branch of mathematics.

Our immediate aim is to review the definitions of $\sin \theta$ and $\cos \theta$, and their graphs; to derive a limit relation needed in computing their derivatives; and to illustrate some interesting limit phenomena.

The sine and cosine are defined in terms of arc length on the unit circle, the graph of $x^2 + y^2 = 1$ (Fig. 1). Given a real number $\theta$, start at the point $P_0 = (1, 0)$ and go $\theta$ units of arc length *counterclockwise* around the unit circle, arriving at a point $P_\theta = (x, y)$. Then by definition

$$x = \cos \theta \quad \text{and} \quad y = \sin \theta.$$

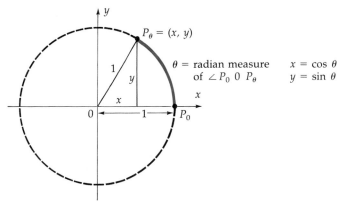

**FIGURE 1**
Sin $\theta$ and cos $\theta$ defined using unit circle $x^2 + y^2 = 1$.

The number $\theta$ is the radian measure of the angle $P_0 O P_\theta$, so these are trigonometric functions in *radian* measure.

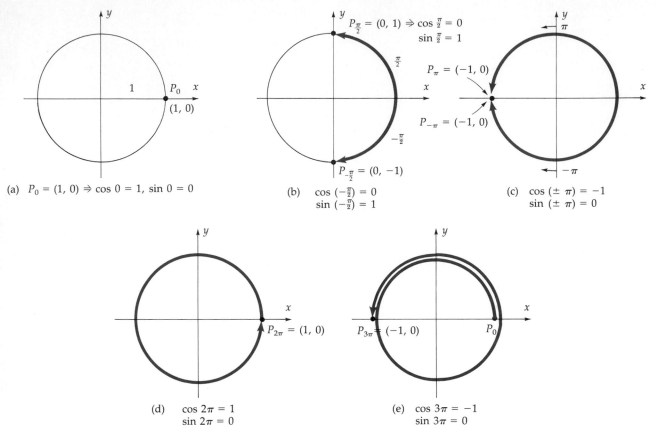

(a) $P_0 = (1, 0) \Rightarrow \cos 0 = 1, \sin 0 = 0$

(b) $\cos\left(-\frac{\pi}{2}\right) = 0$
$\sin\left(-\frac{\pi}{2}\right) = 1$

(c) $\cos(\pm\,\pi) = -1$
$\sin(\pm\,\pi) = 0$

(d) $\cos 2\pi = 1$
$\sin 2\pi = 0$

(e) $\cos 3\pi = -1$
$\sin 3\pi = 0$

**FIGURE 2**
Computing $\cos\theta$ and $\sin\theta$ from the definition.

Figure 2 illustrates some easily computed values of $\sin\theta$ and $\cos\theta$. When $\theta = 0$, you are still at $P_0 = (1, 0)$ (Fig. 2a) so

$$\cos 0 = 1, \qquad \sin 0 = 0.$$

When $\theta = \pi/2$, you have gone around a quarter circle (Fig. 2b); for, the full unit circle (radius $r = 1$) has length $2\pi r = 2\pi$, and $\pi/2$ is one fourth of this. So $P_{\pi/2} = (0, 1)$ and

$$\cos\frac{\pi}{2} = 0, \qquad \sin\frac{\pi}{2} = 1.$$

When $\theta = -\pi/2$, you go $-\pi/2$ units counterclockwise, that is, $\pi/2$ units in the opposite direction (Fig. 2b again) and arrive at $P_{-\pi/2} = (0, -1)$; so

$$\cos\left(-\frac{\pi}{2}\right) = 0, \qquad \sin\left(-\frac{\pi}{2}\right) = -1.$$

Figures 2c and 2d are self-explanatory. In Figure 2e where $\theta = 3\pi$, you go once around the circle (arc length $2\pi$) and continue $\pi$ units more; thus $P_{3\pi} = P_\pi =$

$(-1, 0)$, and $\cos(3\pi) = \cos \pi = -1$, $\sin(3\pi) = \sin \pi = 0$. More generally,

$$\cos(\theta + 2\pi) = \cos \theta, \qquad \sin(\theta + 2\pi) = \sin \theta.$$

The functions are called **periodic**, with period $2\pi$; hence their use to study periodic phenomena.

The graphs in Figure 3 use function values such as those just computed. The maximum of $\cos \theta$ is 1, and the minimum $-1$; for $\cos \theta$ is the $x$ coordinate of a point on the unit circle, and there $-1 \le x \le 1$. Similarly, $-1 \le \sin \theta \le 1$. The graphs show the periodicity; each has a certain basic shape in the interval $[0, 2\pi]$ which is repeated ad infinitum in both directions.

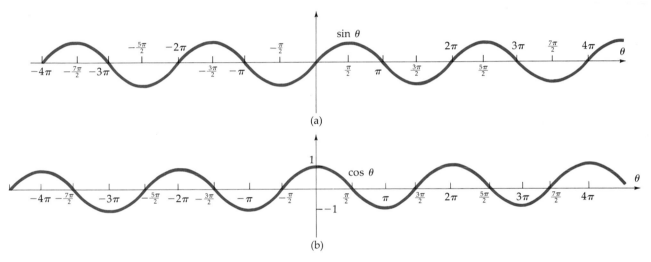

(a)

(b)

**FIGURE 3**
Graphs of $\sin \theta$ and $\cos \theta$.

It is important to keep these graphs in mind, particularly the exact location of the maxima, minima, and zeroes for both $\cos \theta$ and $\sin \theta$. Notice also that $\cos \theta$ is an *even* function, while $\sin \theta$ is *odd*.

## THE LIMIT $\lim\limits_{\theta \to 0} \dfrac{\sin \theta}{\theta}$

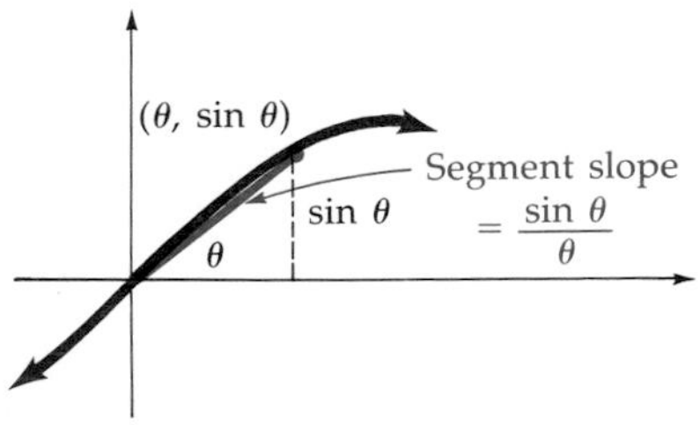

Now we focus on the graph of $\sin \theta$ near the origin and compute the slope of the graph at that point (Fig. 4). The slope of the segment from the origin to the point $(\theta, \sin \theta)$ on the graph is

$$\frac{\Delta y}{\Delta \theta} = \frac{\sin \theta - 0}{\theta - 0} = \frac{\sin \theta}{\theta}.$$

The slope of the graph at the origin is the limit of this quotient as $\theta \to 0$,

$$\lim_{\theta \to 0} \frac{\sin \theta}{\theta}. \tag{1}$$

**FIGURE 4**

Slope at origin $= \lim\limits_{\theta \to 0} \dfrac{\sin \theta}{\theta}$.

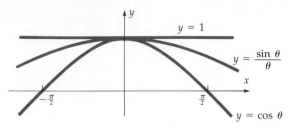

**FIGURE 5**

$$\cos \theta \le \frac{\sin \theta}{\theta} \le 1.$$

As an aid in computing the limit, we sketch the graph of the function $\frac{\sin \theta}{\theta}$ by plotting a number of points as in Figure 5; for comparison, we include the graphs of $y = \cos \theta$ and $y = 1$. It appears that

$$\cos \theta < \frac{\sin \theta}{\theta} < 1, \quad \text{for} \quad 0 < |\theta| < \frac{\pi}{2}, \tag{2}$$

and we will show that this is correct. Moreover, as in the figure,

$$\lim_{\theta \to 0} \cos \theta = \cos 0 = 1,$$

and the function $\frac{\sin \theta}{\theta}$ is "trapped" between $\cos \theta$ and 1. We will conclude (as Figure 5 suggests) that

$$\lim_{\theta \to 0} \frac{\sin \theta}{\theta} = 1.$$

How can we be sure that $\sin \theta$ is trapped? The second inequality in (2),

$$\frac{\sin \theta}{\theta} < 1 \tag{3}$$

is explained in Fig. 6, for the case where $0 < \theta < \pi/2$. The arc from $P_0$ to $P$ has length $\theta$; so by definition, $\sin \theta$ is the second coordinate of point $P$, as shown in the figure. Then the cord $PP' = 2 \sin \theta$, while the arc $PP_0P' = 2\theta$. Since the chord in a circle is less than the arc that it cuts off, we conclude that

$$2 \sin \theta < 2\theta, \qquad 0 < \theta < \frac{\pi}{2}.$$

Divide this by $2\theta$ to obtain (3), for $0 < \theta < \frac{\pi}{2}$.

Again in Figure 6, segment $PQ$ is drawn tangent to the circle, hence perpendicular to $OP$; so $OPQ$ is a right triangle, and $|PQ| = \tan \theta$, as labeled.

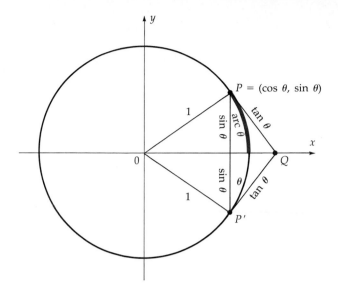

**FIGURE 6**
$2 \sin \theta < 2\theta < 2 \tan \theta$.

The circular arc $PP'$ is shorter than the broken-line path $PQP'$, so

$$2\theta < 2 \tan \theta.$$

Multiply both sides by $\dfrac{\cos \theta}{2\theta}$, use $\tan \theta = \dfrac{\sin \theta}{\cos \theta}$, and you find

$$\cos \theta < \frac{\sin \theta}{\theta}, \qquad 0 < \theta < \frac{\pi}{2}. \tag{4}$$

Together with (3), this shows that

$$\cos \theta < \frac{\sin \theta}{\theta} < 1 \tag{5}$$

for $0 < \theta < \dfrac{\pi}{2}$. To prove the same inequalities for $\theta < 0$, recall that $\sin \theta$ is an odd function, and hence $\dfrac{\sin \theta}{\theta}$ is even:

$$\frac{\sin(-\theta)}{-\theta} = \frac{-\sin \theta}{-\theta} = \frac{\sin \theta}{\theta}.$$

Thus the graph of $\dfrac{\sin \theta}{\theta}$ is symmetric in the vertical axis, as shown in Figure 5. Since (5) is true for $0 < \theta < \dfrac{\pi}{2}$, it must therefore be true for $-\dfrac{\pi}{2} < \theta < 0$ as well.

Summing up: $\dfrac{\sin \theta}{\theta}$ is trapped between $\cos \theta$ and 1, as in (2) (see Figure 5) and we conclude that

$$\lim_{\theta \to 0} \frac{\sin \theta}{\theta} = 1. \qquad \text{(6)}$$

This is the basis of all other limits and derivatives for the trigonometric functions.

Figures 7 and 8 provide a way to remember this basic limit. In Figure 7 you see that

$$\frac{\sin \theta}{\theta} = \frac{2 \sin \theta}{2\theta} = \frac{\text{chord } PP'}{\text{arc } PP'},$$

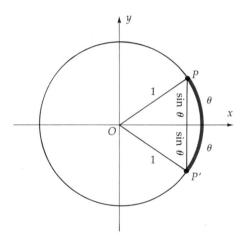

**FIGURE 7**
$$\frac{\sin \theta}{\theta} = \frac{2 \sin \theta}{2\theta} = \frac{\text{chord } PP'}{\text{arc } PP'}.$$

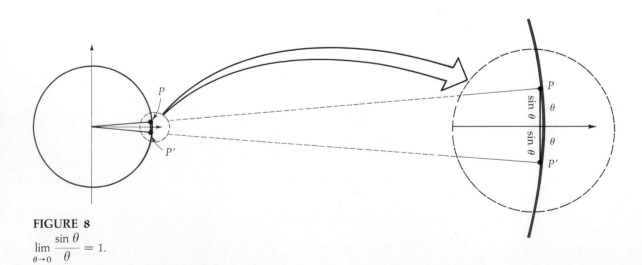

**FIGURE 8**
$$\lim_{\theta \to 0} \frac{\sin \theta}{\theta} = 1.$$

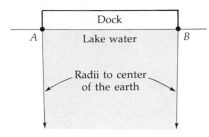

**FIGURE 9**
Chord $AB$ indistinguishable from arc $AB$.

where chord $PP'$ is the straight line distance from $P$ to $P'$, and arc $PP'$ is the length of the circular arc from $P$ to $P'$. As the angle $\theta$ becomes very small (Fig. 8) the chord $2 \sin \theta$ is very much like the arc $2\theta$, and it seems that *as the angle $\theta$ tends to 0, the ratio (chord length)/(arc length) tends to* 1. This is precisely the limit in (6).

To visualize an angle much smaller even than the one in Figure 8, imagine a dock in a calm lake (Fig. 9). The waterline along the dock is really an arc in a huge circle of radius about 4,000 miles, subtending at the center of the earth an extremely small angle $\theta$. The length of the arc from $A$ to $B$ is almost exactly the same as the length of the straight line segment (that is, the chord) from $A$ to $B$; hence the ratio (chord length)/(arc length) is almost exactly equal to 1.

Many limits can be reduced to (6) by a simple change of variable.

**EXAMPLE 1** Evaluate $\displaystyle\lim_{x \to 0} \frac{\sin(2x)}{x}$.

**SOLUTION** Let $2x = \theta$. Then as $x \to 0$, $\theta \to 0$ as well. You find that

$$\lim_{x \to 0} \frac{\sin(2x)}{x} = \lim_{\theta \to 0} \frac{\sin \theta}{\theta/2} \qquad (2x = \theta \Rightarrow x = \theta/2)$$

$$= \lim_{\theta \to 0} 2\,\frac{\sin \theta}{\theta} = 2 \lim_{\theta \to 0} \frac{\sin \theta}{\theta}$$

$$= 2.$$

Still other limits are evaluated by recognizing $\dfrac{\sin \theta}{\theta}$ as a factor.

**EXAMPLE 2**

$$\lim_{\theta \to 0} \frac{\sin^2 \theta}{\theta} = \lim_{\theta \to 0} \frac{\sin \theta}{\theta} \cdot \sin \theta = \lim_{\theta \to 0} \frac{\sin \theta}{\theta} \cdot \lim_{\theta \to 0} \sin \theta$$

$$= 1 \cdot 0 = 0.$$

$$\lim_{x \to 0} \frac{\tan \pi x}{x} = \lim_{\theta \to 0} \frac{\tan \theta}{\theta/\pi} \qquad (\theta = \pi x)$$

$$= \pi \cdot \lim_{\theta \to 0} \frac{\sin \theta}{\theta} \cdot \frac{1}{\cos \theta} \qquad \left(\text{since } \tan \theta = \frac{\sin \theta}{\cos \theta}\right)$$

$$= \pi \cdot 1 \cdot 1 = \pi.$$

## The Trapping Theorem

In our calculation of $\displaystyle\lim_{\theta \to 0} \left(\frac{\sin \theta}{\theta}\right)$, an essential idea was to trap the function $\dfrac{\sin \theta}{\theta}$ between two functions with known limits. We will need this method

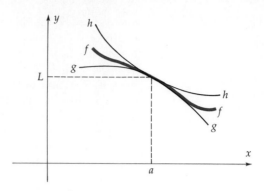

**FIGURE 10**
Trapping Theorem:
$\lim g = L = \lim h \Rightarrow \lim f = L.$

again from time to time; its use is justified by the following theorem illustrated in Figure 10 and proved in Appendix A:

**THE TRAPPING THEOREM**

Suppose that

$$g \leq f \leq h$$

and that

$$\lim_{x \to a} g(x) = \lim_{x \to a} h(x) = L.$$

Then also

$$\lim_{x \to a} f(x) = L.$$

The same holds if $\lim_{x \to a}$ is replaced by any of the other forms of limit.

We conclude with two examples of functions that make infinitely many oscillations as they approach the origin; one has a limit as $x \to 0$, the other does not.

**EXAMPLE 3**    Sketch the graph of $f(x) = \cos \dfrac{1}{x}$, and determine its behavior as $x \to 0$.

**SOLUTION**    You can plot the maximum and minimum values of $f$ without taking derivatives, as follows. The function $\cos \theta$ graphed in Figure 3 has

a maximum value of 1 when $\theta = 0, \pm 2\pi, \pm 4\pi, \ldots,$
a minimum value of $-1$ when $\theta = \pm \pi, \pm 3\pi, \pm 5\pi, \ldots.$

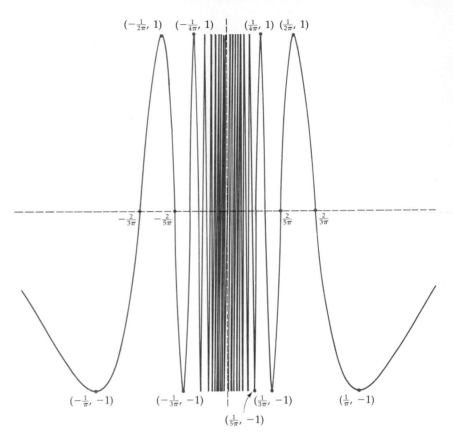

**FIGURE 11**
Graph of $f(x) = \cos(1/x)$, $-\frac{1}{2} \le x < 0$ and $0 < x \le \frac{1}{2}$. (After about five peaks, the oscillations are too close together to be accurately graphed.)

Hence $\cos \dfrac{1}{x}$ has

a maximum value of 1 when $\dfrac{1}{x} = \pm 2\pi, \pm 4\pi, \ldots$

$$\left(\text{thus when } x = \frac{\pm 1}{2\pi}, \frac{\pm 1}{4\pi}, \ldots\right) \text{ and}$$

a minimum value of $-1$ when $\dfrac{1}{x} = \pm \pi, \pm 3\pi, \ldots$

$$\left(\text{thus when } x = \frac{\pm 1}{\pi}, \frac{\pm 1}{3\pi}, \frac{\pm 1}{5\pi}, \ldots\right).$$

Figure 11 shows the corresponding points on the graph. As $x \to 0$, the function oscillates more and more rapidly between the extremes $+1$ and $-1$ and does not approach any one point on the $y$-axis. So $\displaystyle\lim_{x \to 0} \cos \dfrac{1}{x}$ does not exist.

**EXAMPLE 4** Sketch the graph of $g(x) = x \cos \dfrac{1}{x}$, and determine

$\displaystyle\lim_{x \to 0} x \cos \dfrac{1}{x}$, if it exists.

*SOLUTION* Since $\cos \dfrac{1}{x}$ varies between $+1$ and $-1$, $x \cos \dfrac{1}{x}$ varies

between $+x$ and $-x$, as in Figure 12. Precisely, $x \cos \dfrac{1}{x} = x$ when $\cos \dfrac{1}{x} = 1$,

thus when $x = \dfrac{\pm 1}{2\pi}, \dfrac{\pm 1}{4\pi}, \ldots$ as in the previous example; for these values of $x$,

the graph of $y = x \cos \dfrac{1}{x}$ touches the line $y = x$. Likewise $x \cos \dfrac{1}{x} = -x$

when $\cos \dfrac{1}{x} = -1$, or $x = \dfrac{\pm 1}{\pi}, \dfrac{\pm 1}{3\pi}, \dfrac{\pm 1}{5\pi}, \ldots$, and for these values of $x$,

the graph touches the line $y = -x$. In particular, for $x > 0$,

$$-x \le x \cos \dfrac{1}{x} \le x.$$

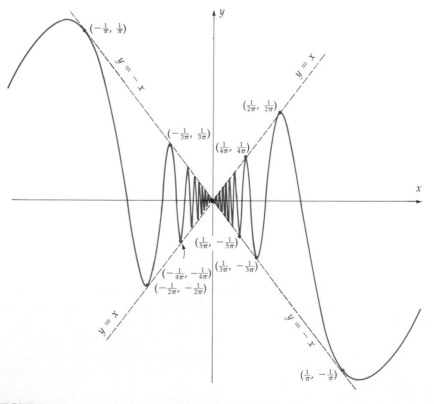

**FIGURE 12**
Graph of $g(x) = x \cos \frac{1}{x}$, trapped between $y = x$ and $y = -x$. (As $x \to 0$, it
is impossible to show all the oscillations.) (Bridge Software "Curves")

Since $\lim\limits_{x \to 0^+} x = 0$ and $\lim\limits_{x \to 0^+} x = 0$, it follows from the Trapping Theorem that

$$\lim_{x \to 0^+} x \cos \frac{1}{x} = 0.$$

Similarly, $\lim\limits_{x \to 0^-} x \cos \dfrac{1}{x} = 0$.

## Some Graphing Hints

The function $\cos \dfrac{1}{x}$ in Example 4 has the form $\cos(g(x))$, with $g(x) = 1/x$. This type of function can generally be graphed by the method in that example. Since $\cos \theta = 1$ when $\theta = 0, \pm 2\pi, \pm 4\pi, \ldots$, then $\cos(g(x)) = 1$ when

$$g(x) = 0, \pm 2\pi, \pm 4\pi, \ldots;$$

by solving these equations, you find the points where $\cos(g(x))$ reaches its maximum value of 1. Likewise, the minimum value is achieved where

$$g(x) = \pm \pi, \pm 3\pi, \pm 5\pi, \ldots$$

and the zeroes occur where

$$g(x) = \pm \frac{\pi}{2}, \pm \frac{3\pi}{2}, \ldots.$$

A similar method applies to functions of the form $\sin(g(x))$.

In Example 5 we have the more complicated form $h(x)\cos(g(x))$. Since $-1 \le \cos(g(x)) \le 1$, the graph of $h(x)\cos(g(x))$ lies between the graphs of $-h(x)$ and $h(x)$; so you graph these two functions and find the points where the graph of $h(x)\cos(g(x))$ touches them by solving the equations $\cos(g(x)) = \pm 1$.

## SUMMARY

$$\lim_{\theta \to 0} \frac{\sin \theta}{\theta} = 1$$

*The Trapping Theorem (see Fig. 10):*

If $\ g \le f \le h$, and $\lim g = \lim h = L$, then $\lim f = L$.

*Graphs of sin θ and cos θ (see Fig. 3):*   Note the maxima, minima, zeroes, period, and symmetries.

## PROBLEMS

### A

*1. a) Graph sin $2x$, $-\pi \le x \le \pi$, showing maxima, minima, and zeroes. (sin $2x = 0$ when $2x = 0, \pm\pi, \pm 2\pi, \ldots$, thus when $x = 0, \pm\pi/2, \pm\pi, \ldots$; use a similar method to solve sin $2x = 1$ and sin $2x = -1$.)

b) Compute $\lim\limits_{x \to 0} \dfrac{\sin 2x}{x}$. What does this tell about the graph in part a?

c) Graph $\dfrac{\sin 2x}{x}$, $0 < |x| \le \pi$, showing all zeroes, all points where sin $2x = \pm 1$, and the limit as $x \to 0$.

2. a) Graph $\sin\left(\dfrac{x}{3}\right)$, $-3\pi \le x \le 3\pi$, showing maxima, minima, and zeroes.

b) Compute $\lim\limits_{x \to 0} \dfrac{\sin \dfrac{x}{3}}{x}$. What does this tell about the graph in part a?

c) Graph $\dfrac{\sin \dfrac{x}{3}}{x}$, $0 < |x| \le 3\pi$, showing all zeroes, points where $\sin \dfrac{x}{3} = \pm 1$, and the limit as $x \to 0$.

3. Evaluate.

*a) $\lim\limits_{\theta \to 0} \dfrac{\sin \theta}{2\theta}$

*b) $\lim\limits_{\theta \to 0} \dfrac{\theta}{\sin \theta}$

c) $\lim\limits_{\theta \to 0} \dfrac{\sin^2 \theta}{\theta}$

d) $\lim\limits_{\theta \to 0} \dfrac{\tan \theta}{\theta}$ $\left(\tan \theta = \dfrac{\sin \theta}{\cos \theta}\right)$

e) $\lim\limits_{\theta \to 0} \dfrac{\theta^2}{\sin \theta}$

4. Evaluate.

*a) $\lim\limits_{t \to 0} \dfrac{\sin 5t}{t}$

*b) $\lim\limits_{t \to 0} \dfrac{t}{\sin 3t}$

c) $\lim\limits_{t \to 0} \dfrac{\sin 5t}{7t}$

d) $\lim\limits_{t \to 0} \dfrac{\sin^2 13t}{t}$

5. Evaluate.

a) $\lim\limits_{x \to 0} \dfrac{\sin(x^2)}{x^2}$

*b) $\lim\limits_{x \to 0} \dfrac{\sin(x^2)}{x}$

c) $\lim\limits_{x \to 0} \dfrac{x^2}{\sin x}$

*6. a) Graph $\sin\left(\dfrac{1}{x}\right)$, showing several maximum and minimum points.

b) Is $\sin \dfrac{1}{x}$ odd, even, or neither?

c) Does $\lim\limits_{x \to 0} \sin \dfrac{1}{x}$ exist?

7. a) Graph $x \sin \dfrac{1}{x}$, showing several points where $\sin \dfrac{1}{x} = \pm 1$.

b) Is $x \sin \dfrac{1}{x}$ odd, even, or neither?

c) Does $\lim\limits_{x \to 0} x \sin \dfrac{1}{x}$ exist?

8. Evaluate.

*a) $\lim\limits_{x \to 0} x^2 \sin\left(\dfrac{1}{x}\right)$

b) $\lim\limits_{x \to 0} x \cos \dfrac{1}{x^2}$

c) $\lim\limits_{x \to 0} (\sin x)\left(\cos \dfrac{1}{x}\right)$

9. Evaluate.

*a) $\lim\limits_{x \to 0} \dfrac{1}{x} \sin x$

b) $\lim\limits_{x \to 0} \sqrt{x} \cos \dfrac{1}{x}$

*c) $\lim\limits_{x \to 0} x\left[1 + \cos \dfrac{2}{x}\right]$

d) $\lim\limits_{x \to 0} \left[1 + \sin \dfrac{1}{x^2}\right]$

10. With a calculator, compute
*a) $\sin(1°)$ (the sine of one degree).
*b) $\sin(1)$ (the sine of one radian).

c) $\dfrac{\sin 1}{1}$.

d) $\dfrac{\sin(0.1)}{0.1}$.

e) $\dfrac{\sin(0.01)}{0.01}$.

**11.** The radius of the earth is about $6.38 \times 10^6$ meters. Compute the angle in radians subtended at the center of the earth by a horizontal dock 10 meters long. (You can assume that the edge of the dock is the arc of a very large circle. What fraction of the full circle of $2\pi$ radians is subtended by the dock?)

**B**

**\*12.** Evaluate.

**a)** $\displaystyle \lim_{\theta \to 0} \frac{1 - \cos \theta}{\theta}$

(Hint: $(1 - \cos \theta)(1 + \cos \theta) = ?$)

**b)** $\displaystyle \lim_{\theta \to 0} \frac{1 - \cos \theta}{\theta^2}$

**13.** Suppose that $1 - |x| \le f(x) \le 1 + |x|$. Evaluate $\displaystyle \lim_{x \to 0} f(x)$.

**14.** Our proof that $\displaystyle \lim_{\theta \to 0} \frac{\sin \theta}{\theta} = 1$ assumed that $\displaystyle \lim_{\theta \to 0} \cos \theta = 1$. Verify this. (Hint: One way is to show from Figure 6 that $0 \le 1 - \cos \theta \le |\theta|$ if $|\theta| \le \pi/2$.)

---

# 3.3

## LIMITS AS $x \to +\infty$ OR $x \to -\infty$

Figure 1 shows the graph of $f(x) = 1/x$; you can see that as $x$ grows large and positive, the function values $1/x$ grow small. The following calculations show that as $x$ grows extremely large, far beyond the confines of the sketch, the function values become vanishingly small:

| $x$ | 1 | 10 | 100 | 1000 | 10,000 ... |
|-----|---|----|-----|------|------------|
| $1/x$ | 1 | 1/10 | 1/100 | 1/1000 | 1/10,000 ... |

In such a case we write

$$\lim_{x \to +\infty} 1/x = 0. \tag{1}$$

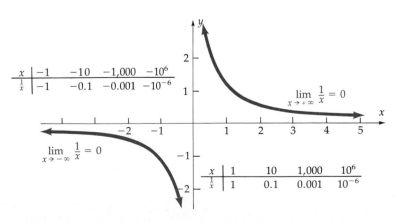

FIGURE 1
$y = 0$ is an asymptote of the graph of $f(x) = 1/x$.

Similarly, considering large negative values of $x$, you see that as $x \to -\infty$ then $1/x \to 0$:

$$\lim_{x \to -\infty} 1/x = 0. \tag{2}$$

In general, we write

$$\lim_{x \to +\infty} f(x) = L$$

if the function values $f(x)$ approach $L$ as $x$ grows infinitely large and positive (Fig. 2); and

$$\lim_{x \to -\infty} f(x) = L$$

if $f(x)$ approaches $L$ as $x$ grows infinitely large and negative. In either case, the horizontal line $y = L$ is called an **asymptote** of the graph.

The limit theorems in Section 3.1 and the Trapping Theorem (Sec. 3.2) apply to these types, $\lim\limits_{x \to +\infty}$ and $\lim\limits_{x \to -\infty}$. Using those theorems, many examples can be deduced from the limits (1) and (2).

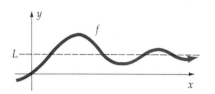

**FIGURE 2**
$\lim\limits_{x \to +\infty} f(x) = L.$

---

### EXAMPLE 1

$$\lim_{x \to +\infty} 1/x^2 = \lim_{x \to +\infty} \left( \frac{1}{x} \cdot \frac{1}{x} \right) = \left( \lim_{x \to +\infty} \frac{1}{x} \right)\left( \lim_{x \to +\infty} \frac{1}{x} \right)$$
$$= 0 \cdot 0 = 0.$$

Similarly, $\lim\limits_{x \to +\infty} \dfrac{1}{x^3} = 0$, and generally $\lim\limits_{x \to +\infty} \dfrac{1}{x^n} = 0$ for $n = 1, 2, 3, 4,$ and so on.

---

### EXAMPLE 2

$$\lim_{x \to +\infty} \frac{1}{1 - 1/x^2} = \frac{\lim(1)}{\lim(1 - 1/x^2)}$$
$$= \frac{1}{\lim(1) - \lim(1/x^2)}$$
$$= \frac{1}{1 - 0} = 1,$$

where "lim" stands for "$\lim\limits_{x \to +\infty}$" throughout.

---

*WARNING*    *The symbol $\infty$ is used in a descriptive sense. We write* $\lim\limits_{x \to +\infty}$ and speak of the "limit as $x$ approaches infinity," as if there were some object labeled $+\infty$ attached to the far right of the number line. This is harmless, as long as we don't begin to think of $+\infty$ as an algebraic part of the number system.

Similarly, it is natural to write

$$\lim_{x \to +\infty} x^2 = +\infty. \tag{3}$$

This "equation" has a descriptive meaning: As $x$ grows large and positive, then also $x^2$ grows large and positive, without bound. But an equation such as (3) does *not* mean that $\lim_{x \to +\infty} x^2$ exists in the sense required by the theorem about the limit of a sum, product, or quotient; an equation such as $\lim(fg) = (\lim f)(\lim g)$ is guaranteed to apply only if $\lim f$ and $\lim g$ exist as *finite numbers*, not as "discriptive symbols" such as $+\infty$ or $-\infty$. So to respect the conditions in the limit theorem, we say that $\lim_{x \to +\infty} x^2$ *does not exist*; equation (3) is merely descriptive, and does not imply that $\lim_{x \to +\infty} x^2$ actually exists.

---

**EXAMPLE 3**  Determine $\lim_{x \to +\infty} \dfrac{2x}{x + 2}$, if it exists.

*SOLUTIONS*  In this case $\lim_{x \to +\infty} 2x$ does not exist, so you cannot apply the limit theorem for $\lim(f/g)$ with $f(x) = 2x$ and $g(x) = x + 2$. But even though the limit cannot be evaluated in that particular way, there may be other ways.

To gain some insight into this example, start with an "experimental" approach—compute function values $f(x)$ for large positive values of $x$:

| $x$ | 0 | 100 | 1000 | 10000 |
|---|---|---|---|---|
| $f(x) = \dfrac{2x}{x + 2}$ | $1\frac{2}{3}$ | 1.96 | 1.996 | 1.9996 |

It seems that

$$\lim_{x \to +\infty} \frac{2x}{x + 2} = 2.$$

You can verify this algebraically. Divide numerator and denominator by $x$:

$$\frac{2x}{x + 2} = \frac{2x/x}{(x + 2)/x}$$

$$= \frac{2}{1 + (2/x)}.$$

As $x \to \infty$ then the term $(2/x)$ becomes practically equal to 0, and the quotient

$$\frac{2}{1 + (2/x)}$$

becomes practically equal to

$$\frac{2}{1 + 0} = 2.$$

You can reach the same conclusion in a more formal way; rewriting the fraction as we did makes it possible to apply the theorem on the algebra of limits:

$$\lim_{x \to +\infty} \frac{2x}{x + 2} = \lim_{x \to +\infty} \frac{2}{1 + (2/x)} \qquad \text{(algebra)}$$

$$= \frac{\lim 2}{\lim(1 + (2/x))} \qquad \text{(limit of a quotient)}$$

$$= \frac{2}{1 + 2 \lim(1/x)} \qquad \text{(limit of a sum)}$$

$$= \frac{2}{1 + 0} \qquad \left(\lim_{x \to +\infty} \frac{1}{x} = 0\right)$$

$$= 2.$$

Likewise, as $x$ grows infinitely large and negative,

$$\lim_{x \to -\infty} \frac{2x}{x + 2} = \lim_{x \to -\infty} \frac{2}{1 + (2/x)} = 2.$$

This is confirmed by the graph in Figure 3; the line $y = 2$ is a horizontal asymptote, approached both as $x \to +\infty$ and as $x \to -\infty$.

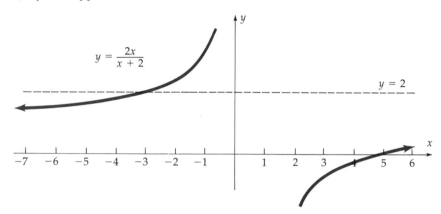

**FIGURE 3**

$$\lim_{x \to +\infty} \frac{2x}{x + 2} = 2.$$

**REMARK**   The function $\dfrac{2x}{x + 2}$ in this example describes the *electrical resistance* of the circuit in Figure 4. Electricity flowing from $A$ to $B$ encounters a fixed resistance of 2 ohms in the upper branch and a variable resistance of $x$ ohms in the lower branch. According to basic circuit theory, the resistance $R$ of the entire circuit is given by

$$\frac{1}{R} = \frac{1}{2} + \frac{1}{x} = \frac{x + 2}{2x}.$$

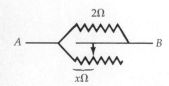

**FIGURE 4**
Circuit with a variable resistance of $x$ ohms.

Hence $R = \dfrac{2x}{x + 2}$, $x \geq 0$, and the relation between $x$ and $R$ is given by the part of the graph in Figure 3 for $x \geq 0$. When $x = 0$, there is no resistance, which is reasonable; for when $x = 0$, the lower branch of the circuit offers no resistance, and all the electrical current flows there. As $x$ increases, the resistance of the circuit as a whole grows (the graph shows this). When $x$ is huge, most of the current flows in the upper branch, where it meets a resistance of 2 ohms, so the resistance of the entire circuit is nearly 2 ohms; that is, $\lim_{x \to +\infty} \dfrac{2x}{x + 2} = 2$, as shown in Figure 3.

Example 3 illustrates a standard procedure in evaluating the limit of a rational function as $x \to \pm\infty$, namely: *Divide numerator and denominator by the highest power of $x$ that appears in the denominator*; then apply the limit theorems.

**EXAMPLE 4**  Evaluate $\lim_{x \to +\infty} \dfrac{x^2 - 4}{x^3 - 8}$, if it exists.

*SOLUTION*  Divide numerator and denominator by $x^3$, the highest power of $x$ that appears in the denominator:

$$\lim_{x \to +\infty} \frac{x^2 - 4}{x^3 - 8} = \lim_{x \to +\infty} \frac{(x^2 - 4)/x^3}{(x^3 - 8)/x^3}$$

$$= \lim_{x \to +\infty} \frac{(1/x) - (4/x^3)}{1 - (8/x^3)}$$

$$= \frac{\lim(1/x) - \lim(4/x^3)}{1 - \lim 8/x^3}$$

$$= \frac{0 - 0}{1 - 0} = \frac{0}{1}$$

$$= 0.$$

The limit exists and equals 0. Similarly, $\lim_{x \to -\infty} \dfrac{x^2 - 4}{x^3 - 8} = 0$; the $x$-axis is a horizontal asymptote, approached both as $x \to +\infty$ and as $x \to -\infty$.

For rational functions, the limit as $x \to \pm\infty$ can be found by a simple shortcut: In the numerator, drop all terms but the one with the highest power; do the same in the denominator; combine the remaining terms, and the limit as $x \to \pm\infty$ is clear. Thus in Example 3,

$$\lim_{x \to \pm\infty} \frac{2x}{x + 2} = \lim_{x \to \pm\infty} \frac{2x}{x} \quad \text{(drop lower powers)}$$

$$= \lim_{x \to \pm\infty} 2 = 2.$$

In Example 4,

$$\lim_{x \to \pm\infty} \frac{x^2 - 4}{x^3 - 8} = \lim_{x \to \pm\infty} \frac{x^2}{x^3} \qquad \text{(drop lower powers)}$$

$$= \lim_{x \to \pm\infty} \frac{1}{x} = 0.$$

As $x \to \pm\infty$, the highest power in numerator and denominator becomes relatively more significant than the lower powers; and in taking the *limit* as $x \to \pm\infty$, the lower powers can be ignored. That is what is verified by the formal method in Examples 3 and 4.

---

**EXAMPLE 5**   Evaluate $\lim\limits_{x \to +\infty} \dfrac{2x^3 - 3x + 1}{x^2 + 2}$.

*Shortcut Solution:*

$$\lim_{x \to +\infty} \frac{2x^3 - 3x + 1}{x^2 + 2} = \lim_{x \to +\infty} \frac{2x^3}{x^2} \qquad \text{(drop lower powers)}$$

$$= \lim_{x \to +\infty} 2x = +\infty.$$

*Analysis of the Solution:*

$$\lim_{x \to +\infty} \frac{2x^3 - 3x + 1}{x^2 + 2} = \lim_{x \to +\infty} \frac{x^3(2 - 3/x^2 + 1/x^3)}{x^2(1 + 2/x^2)}$$

$$= \lim_{x \to +\infty} x \cdot \frac{(2 - 3/x^2 + 1/x^3)}{(1 + 2/x^2)}.$$

As $x \to +\infty$, the fraction $\dfrac{2 - 3/x^2 + 1/x^3}{1 + 2/x^2}$ has the limit 2; and the other factor $x$ tends to $+\infty$; so the product tends to $+\infty$.

---

## Dealing with the "Limits" $+\infty$ and $-\infty$

Even though $+\infty$ and $-\infty$ have only a symbolic meaning, they can to some extent be manipulated algebraically. Table 1 lists several important com-

**Table 1**

| | Given | | Deduced | | | |
|---|---|---|---|---|---|---|
| | $\lim f$ | $\lim g$ | $\lim(f + g)$ | $\lim(fg)$ | $\lim(f/g)$ | $\lim(g/f)$ |
| 1. | 0 | $+\infty$ | $+\infty$ | indeterminate | 0 | indeterminate |
| 2. | 0 | $-\infty$ | $-\infty$ | indeterminate | 0 | indeterminate |
| 3. | finite $> 0$ | $+\infty$ | $+\infty$ | $+\infty$ | 0 | $+\infty$ |
| 4. | finite $> 0$ | $-\infty$ | $-\infty$ | $-\infty$ | 0 | $-\infty$ |
| 5. | $+\infty$ | $+\infty$ | — | — | — | — |
| 6. | $+\infty$ | $-\infty$ | — | — | — | — |
| 7. | $-\infty$ | $-\infty$ | — | — | — | — |

binations and some of the results. Not all the rows are filled in—the blanks are left as problems for you to think out.

This complicated table is not to be memorized! Its purpose is to list major possibilities and to develop a reliable intuitive way to analyze them.

Here are the reasons for the entries in row 1: *Suppose that* $\lim\limits_{x \to +\infty} f(x) = 0$ *and* $\lim\limits_{x \to +\infty} g(x) = +\infty$. Then when $x$ is very large, $f(x)$ is near 0 while $g(x)$ is very large and positive, so $f(x) + g(x)$ is very large and positive; thus $\lim(f + g) = +\infty$.

On the other hand, the limit of the product $fg$ is not determined by $\lim f = 0$ and $\lim g = +\infty$. For example, each of the following limits has the form $\lim(fg)$ with $\lim f = 0$ and $\lim g = +\infty$:

$$\lim_{x \to +\infty} \left[ \frac{1}{x^2} \cdot x \right] = \lim_{x \to +\infty} \frac{1}{x} = 0,$$

$$\lim_{x \to +\infty} \left[ \frac{1}{x} \cdot x \right] = \lim_{x \to +\infty} [1] = 1$$

$$\lim_{x \to +\infty} \left[ \frac{1}{x} \cdot x^2 \right] = \lim_{x \to +\infty} [x] = +\infty.$$

Thus if $\lim f = 0$ and $\lim g = +\infty$, then $\lim(fg)$ can be determined only by a further study of the particular product $fg$ at hand. This is the meaning of the entry "indeterminate" in the table.

As for the quotient $f/g$, if $\lim f = 0$ and $\lim g = +\infty$, then $\lim(f/g) = 0$; for dividing a very small number $f(x)$ by a very large number $g(x)$ yields a *very* small quotient $f(x)/g(x)$. On the other hand, the limit $\lim(g/f)$ is indeterminate; for if $g(x)$ is very large positive while $f(x)$ is near 0, then $g(x)/f(x)$ might be very large positive [if $f(x) > 0$] or very large negative [if $f(x) < 0$] or even undefined [if $f(x) = 0$]. This explains all the entries in row 1; the rest are left as problems.

## Oscillations as $x \to \pm\infty$

As $x \to +\infty$, the function $\sin\theta$ (Fig. 5) oscillates forever back and forth between $+1$ and $-1$, and does not approach a unique limiting value; so $\lim\limits_{\theta \to +\infty} \sin\theta$ does not exist. On the other hand, $\lim\limits_{\theta \to +\infty} \dfrac{\sin\theta}{\theta}$ does exist (Fig. 6); for when $\theta$ is very large then $\dfrac{1}{\theta}$ is nearly 0, while $|\sin\theta| \leq 1$, so the

**FIGURE 5**
$\lim\limits_{\theta \to +\infty} \sin\theta$ does not exist.

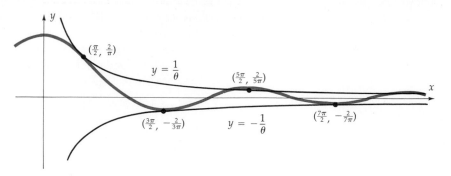

**FIGURE 6**

$$-\frac{1}{\theta} \le \frac{\sin \theta}{\theta} \le \frac{1}{\theta} \Rightarrow \lim_{\theta \to +\infty} \frac{\sin \theta}{\theta} = 0.$$

product $\frac{1}{\theta} \sin \theta$ is nearly 0. This conclusion is rigorously justified by the Trapping Theorem; $-1 \le \sin \theta \le 1$, and for $\theta > 0$ you can multiply each term by $1/\theta$, finding

$$-\frac{1}{\theta} \le \frac{\sin \theta}{\theta} \le \frac{1}{\theta}.$$

So the graph of $\frac{\sin \theta}{\theta}$ is trapped between the graphs of $-\frac{1}{\theta}$ and $\frac{1}{\theta}$ as in Figure 6. Since $\lim\limits_{\theta \to +\infty} -\frac{1}{\theta} = 0$ and $\lim\limits_{\theta \to +\infty} \frac{1}{\theta} = 0$, it follows from the Trapping Theorem that $\lim\limits_{\theta \to +\infty} \frac{\sin \theta}{\theta} = 0$. This means that the line $\theta = 0$ is an asymptote of the graph; in this case, the graph actually crosses the asymptote infinitely often.

## PROBLEMS

### A

1. Determine $\lim\limits_{x \to +\infty}$ and $\lim\limits_{x \to -\infty}$ for the function in the given figure, or explain why they do not exist.

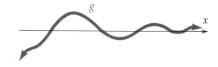

b) **FIGURE 8**

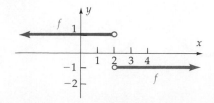

*a)  **FIGURE 7**

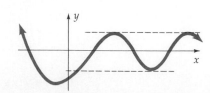

*c)  **FIGURE 9**

In problems 2–23, compute the limit as $x \to +\infty$ and $x \to -\infty$, or explain why it does not exist. Check that your answer is reasonable by computing function values for some large values of $x$.

**2.** $\dfrac{1 + 2/x}{2 - 1/x}$

**\*3.** $\dfrac{x + 2}{2x - 1}$

**4.** $\dfrac{1}{x + x^3}$

**\*5.** $\dfrac{1}{x} - \dfrac{1}{x^2}$

**6.** $\dfrac{x + 2}{2x^2 - 1}$

**\*7.** $\dfrac{x^2 + 2}{2x - 1}$

**8.** $\dfrac{x^4 + 1}{x^4 - 1}$

**\*9.** $\dfrac{2x^2 + 3x + 1}{3x^2 + 2x + 2}$

**10.** $\dfrac{x^3 + 5x - 1}{2 - 6x + 3x^3}$

**\*11.** $\dfrac{1 + x^4}{x^2(1 + x^2)}$

**\*12.** $\dfrac{x^2}{x^3 + 1}$

**13.** $\dfrac{(x + 1)^2 - x^2}{x + 1}$

**\*14.** $\dfrac{x}{|x|}$  (Think of the graph.)

**15.** $\dfrac{x - 1}{|x - 1|}$  ($|x - 1| = x - 1$ for $x > 1$; what is it for $x < 1$?)

**\*16.** $\dfrac{x}{|x + 1|}$

**17.** $\lim\limits_{\theta \to -\infty} \dfrac{\cos \theta}{\theta}$

**\*18.** $\lim\limits_{\theta \to +\infty} 100 \dfrac{\sin(\theta^2)}{\theta}$

**19.** $\lim\limits_{x \to +\infty} \dfrac{10 + \sin 2x}{x^2}$

**\*20.** $\lim\limits_{x \to +\infty} x \sin \dfrac{1}{x}$

**21.** $\lim\limits_{x \to +\infty} \cos \dfrac{1}{x}$

**22.** $\lim\limits_{x \to -\infty} x \cos \dfrac{1}{x}$

**23.** $\lim\limits_{x \to -\infty} x \sin\left(\dfrac{1}{x^2}\right)$

**24.** By "experimenting" with a calculator, guess the following limits.

**\*a)** $\lim\limits_{x \to +\infty} \dfrac{1}{\sqrt{x}}$

**\*b)** $\lim\limits_{x \to -\infty} \dfrac{1}{\sqrt[3]{x}}$

**\*c)** $\lim\limits_{x \to +\infty} \dfrac{\sqrt{x + 1}}{\sqrt{x + 2}}$

**\*d)** $\lim\limits_{x \to +\infty} \dfrac{\sqrt{x + 1}}{x + 2}$

**e)** $\lim\limits_{x \to +\infty} (\sqrt{x + 1} - \sqrt{x})$

**f)** $\lim\limits_{x \to +\infty} (\sqrt{x^2 + 1} - x)$

**25.** A certain rental car costs \$30/day and \$.30/mile, including gas. Compute:
   a)  The cost of driving it $x$ miles in one day.

**b)**  The cost per mile of driving it $x$ miles in one day.

**c)**  The limit as $x \to +\infty$ of the cost per mile of driving it $x$ miles in one day.

**26.** In Table 1, explain the entries in
   **a)**  row 2.      **b)**  row 3.      **c)**  row 4.

**B**

**27.** In Table 1, determine the correct entries, and explain them, for
   **a)**  row 5.      **b)**  row 6.      **c)**  row 7.

**28.** (This problem shows why $\dfrac{+\infty}{+\infty}$ is called an "indeterminate form," and $+\infty$ cannot be treated as an ordinary number.)

**a)**  Given any number $m > 0$, show that there are functions $f$ and $g$ such that

$$\lim_{x \to +\infty} f(x) = +\infty, \ \lim_{x \to +\infty} g(x) = +\infty \quad (3)$$

and

$$\lim_{x \to +\infty} \dfrac{f(x)}{g(x)} = m.$$

Thus, in a sense, $\dfrac{+\infty}{+\infty} = m$, for any $m \geq 0$!

(Hint: Try linear functions $f$ and $g$.)

**b)**  Show that there are functions $f$ and $g$ satisfying equation (3), and

$$\lim_{x \to +\infty} \dfrac{f(x)}{g(x)} = 0, \qquad \lim_{x \to +\infty} \dfrac{g(x)}{f(x)} = +\infty.$$

Thus, in a sense, $\dfrac{+\infty}{+\infty}$ could also be 0 or $+\infty$!

**29.** Rewrite the following expressions in such a way that their limits become easy to evaluate. [Note: $\sqrt{a} - \sqrt{b} = (a - b)/(\sqrt{a} + \sqrt{b})$].

**\*a)** $\lim\limits_{x \to +\infty} (\sqrt{x + 1} - \sqrt{x})$

**\*b)** $\lim\limits_{x \to +\infty} (\sqrt{1 + x^2} - x)$

**c)** $\lim\limits_{x \to +\infty} (\sqrt{1 + x + x^2} - x)$

**30.** If $P(x) = c_n x^n + \cdots + c_0$ is any polynomial, evaluate $\lim\limits_{x \to \pm\infty} P(1/x)$.

# 3.4

## CONTINUITY. THE INTERMEDIATE VALUE THEOREM

The simplest graphing method is to plot a few points and draw a continuous, unbroken curve through them. Obviously, this is legitimate only where the graph is in fact continuous—it is essential to know any discontinuities that may occur.

In this section we define continuity in terms of limits, illustrate typical discontinuities, and then discuss the Intermediate Value Theorem for continuous functions. This theorem justifies the general "sign change principle": A function $f$ can change sign only at a zero or at a discontinuity.

**Continuity** is illustrated in Figure 1. For $f$ to be continuous at a given point $a$ on the $x$-axis, first of all, $f(a)$ must be defined. Second, as $x$ approaches $a$ from either side, the function values $f(x)$ must approach $f(a)$; that is,

$$\lim_{x \to a} f(x) = f(a).$$

Think of $\lim_{x \to a} f(x)$ as the function value you would expect at $x = a$; and $f(a)$ is the value you actually get when $x = a$. Then continuity means that "what you expect = what you get." In other words, $f(a)$ is the natural continuation of the values of $f(x)$ just to the right and left of $x = a$.

If $f(a)$ is the natural continuation *considering only values $f(x)$ with $x$ to the right* of $a$, then $f$ is called *continuous from the right* (Fig. 2). Continuity from the left is similarly defined as in Figure 3. Summing up:

---

**DEFINITION**

A function $f$ is *continuous at* $a$ (Fig. 1) if $f(a)$ is defined, and $\lim_{x \to a} f(x)$ exists, and

$$\lim_{x \to a} f(x) = f(a).$$

$f$ is *continuous from the right* at $a$ (Fig. 2) if

$$\lim_{x \to a^+} f(x) = f(a)$$

and *continuous from the left* at $a$ (Fig. 3) if

$$\lim_{x \to a^-} f(x) = f(a).$$

---

Any polynomial $P$ is continuous at every point $a$, because (from Sec. 3.1)

$$\lim_{x \to a} P(x) = P(a).$$

This means that the graph of a polynomial consists of a single unbroken curve.

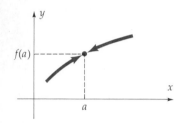

**FIGURE 1**
$f$ is continuous at $a$, $\lim_{x \to a} f(x) = f(a)$.

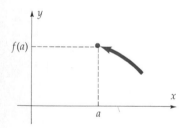

**FIGURE 2**
$f$ is continuous from the right at $a$, $\lim_{x \to a^+} f(x) = f(a)$. $f(x)$ may or may not be defined for $x < a$.

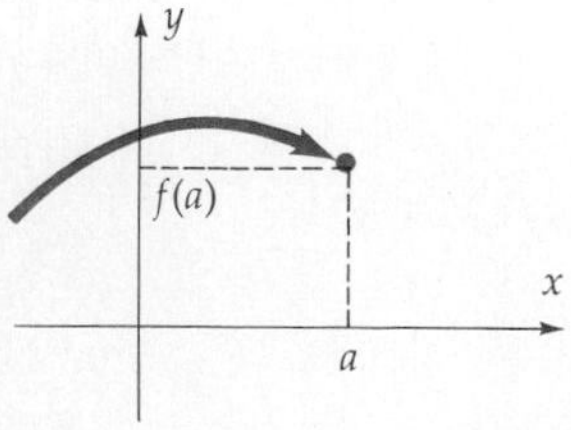

**FIGURE 3**
$f$ is continuous from the left at $a$, $\lim_{x \to a^-} f(x) = f(a)$.

Any rational function $P/Q$ is continuous at every point $a$ where it is defined, since (from Sec. 3.1)

$$\lim_{x \to a} \frac{P(x)}{Q(x)} = \frac{P(a)}{Q(a)}$$

wherever the denominator $Q(a) \neq 0$. So the graph of a rational function is a curve broken only at the points where $Q(a) = 0$.

Our first example displays a typical discontinuity.

**EXAMPLE 1**   $f(x) = \dfrac{1}{x}$ is a rational function; the denominator is 0 only at $x = 0$, so this is its only discontinuity. The graph in Figure 4 shows that as $x \to 0$ from the right, the function values $1/x$ grow large and positive without bound, since $1/x$ is very large when $x$ is very small. Describe this by writing

$$\lim_{x \to 0^+} \frac{1}{x} = +\infty.$$

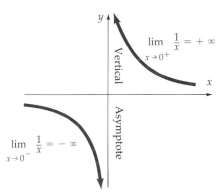

**FIGURE 4**

$f(x) = \dfrac{1}{x}$ is discontinuous at $x = 0$.

As $x \to 0$ from the left, then $1/x$ grows large and negative, so write

$$\lim_{x \to 0^-} \frac{1}{x} = -\infty.$$

REMEMBER: The symbols $+\infty$ and $-\infty$ cannot be treated like ordinary numbers, and the rules for the limit of a sum, product, or quotient do not apply when any of the "limits" is $+\infty$ or $-\infty$.

In Figure 4, the $y$-axis is called a **vertical asymptote** of the graph. In general, the line $x = a$ is a vertical asymptote if the function values $f(x)$ grow infinitely large, either positive or negative, as $x \to a$ from one side or the other.

Another typical discontinuity is a *jump* as in Figure 5; $f$ has a jump discontinuity at $a$ if the left-hand limit $\lim_{x \to a^-} f(x)$ and the right-hand limit $\lim_{x \to a^+} f(x)$

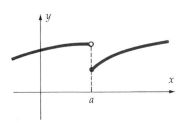

**FIGURE 5**

Jump discontinuity,

$\lim_{x \to a^-} f(x) \neq \lim_{x \to a^+} f(x).$

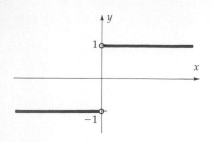

**FIGURE 6**
$f(x) = \dfrac{x}{|x|}$ has a jump discontinuity
at $x = 0$.

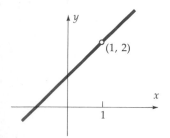

**FIGURE 7**
$f(x) = \dfrac{x^2 - 1}{x - 1}$, $x \neq 1$ has a "hole"
at $x = 1$.

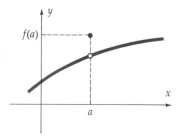

**FIGURE 8**
"Blip" at $x = a$: $f(a) \neq \lim\limits_{x \to a} f(x)$.

exist, but are not equal. A simple example is shown in Figure 6:

$$f(x) = \frac{x}{|x|} \qquad x \neq 0.$$

As in Example 1, the discontinuity arises at a zero of the denominator. But in this case it is *not* a vertical asymptote; the zero in the denominator is "cancelled" by a zero in the numerator at the same point. We call this a "$\frac{0}{0}$ case." Such a quotient $f(x)/g(x)$ with $f(a) = g(a) = 0$ can produce not only jumps, but most other varieties of discontinuity as well, as the remaining examples and problems will show.

Another common sort of discontinuity is a "hole," a point $a$ where $\lim\limits_{x \to a} f(x)$ exists, but $f(a)$ is not defined. The function in Figure 7

$$f(x) = \frac{x^2 - 1}{x - 1}, \qquad x \neq 1$$

has a "hole" in its graph at $x = 1$, as we showed in Section 3.1. This is another "$\frac{0}{0}$ case"; both numerator and denominator equal 0 at $x = 1$.

Figure 8 shows a "blip" discontinuity, a point $a$ where $\lim\limits_{x \to a} f(x)$ exists,

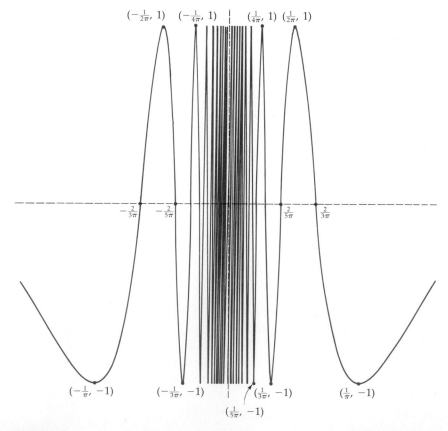

**FIGURE 9**
$f(x) = \cos(1/x)$ has an oscillatory discontinuity at $x = 0$.

but this limit is different from $f(a)$. This type is rare and does not arise naturally in elementary calculus.

Finally, there are *oscillatory* discontinuities as in Figure 9 where $\lim\limits_{x \to a} f(x)$ does not exist because the graph oscillates back and forth, and the function values $f(x)$ do not approach one specific finite limit $L$.

*NOTE* The terms *jump* and *oscillatory* are standard; "hole" and "blip" are not.

## Combinations of Continuous Functions

The theorems on limits show that sums, products, and quotients of continuous functions are themselves continuous, except at a zero denominator. Precisely:

> **THEOREM**
>
> If $f$ and $g$ are continuous at $a$, then so are the sum $f + g$, product $fg$, and also the quotient $f/g$ if $g(a) \neq 0$.

**PROOF** Since $f$ and $g$ are assumed continuous at $a$, then

$$\lim_{x \to a} f(x) = f(a) \quad \text{and} \quad \lim_{x \to a} g(x) = g(a).$$

Hence

$$\lim_{x \to a} [f(x) + g(x)] = \lim_{x \to a} f(x) + \lim_{x \to a} g(x) \qquad \text{(limit of a sum)}$$
$$= f(a) + g(a)$$

and it follows that $f + g$ is continuous at $a$. The proofs for the product $fg$ and quotient $f/g$ are virtually the same.

**The sign change principle,** an essential tool in graphing, is justified by a fundamental fact about continuous functions:

> **THE INTERMEDIATE VALUE THEOREM**
>
> If $f$ is continuous at every point of the interval $[a, b]$, and if $\bar{y}$ is any given number between $f(a)$ and $f(b)$, then there is a number $\bar{x}$ in $(a, b)$ such that
>
> $$f(\bar{x}) = \bar{y}.$$

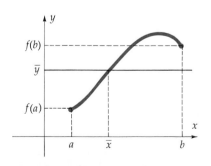

**FIGURE 10**
Intermediate Value Theorem.

Figure 10 illustrates the theorem. Since $\bar{y}$ is between $f(a)$ and $f(b)$, and the graph is continuous, it must intersect the horizontal line $y = \bar{y}$ at some point $P = (\bar{x}, \bar{y})$; then $f(\bar{x}) = \bar{y}$. The proof is interesting, but requires subtle properties of the real numbers; we leave it to Appendix A.

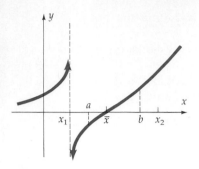

**FIGURE 11**
Sign changes only at an intercept or a discontinuity.

From this theorem follows a useful corollary:

---

**COROLLARY**

If $f$ has no discontinuities and no zeroes in a certain interval $I = (x_1, x_2)$, then $f$ must be either positive throughout $I$, or negative throughout $I$.

---

For if, on the contrary, $f$ were negative at some point $a$ in $I$ and positive at another such point $b$ as in Figure 11, then by the Intermediate Value Theorem there would be a point $\bar{x}$ between $a$ and $b$ where $f(\bar{x}) = 0$; and this would contradict the assumption that $f$ has no zeros in the interval $I$.

The corollary can be rephrased as a *sign change principle*:

*A function can change sign only at a zero or a discontinuity.*

---

**EXAMPLE 2**   Determine the sign of $\dfrac{2x}{x + 2}$, for every real $x$.

**SOLUTION**   $f$ can change sign only at a zero or discontinuity. To find the discontinuities: $f$ is the ratio of two polynomials, so it is continuous everywhere except where the denominator equals 0, that is, at $x = -2$. As for the zeroes,

$$\frac{2x}{x + 2} = 0 \Leftrightarrow 2x = 0 \quad \text{and} \quad x \neq -2$$

$$\Leftrightarrow x = 0.$$

So there are two possible sign changes:

$$\text{at } x = -2 \qquad \text{(a discontinuity)},$$

$$\text{at } x = 0 \qquad \text{(a zero of } f).$$

Mark these points on a line

$$f: \quad \underline{\qquad\qquad\overset{|}{-2}\qquad\qquad\qquad\overset{|}{0}\qquad\qquad}$$

and check the sign of $f$ at one point in each of the resulting intervals, say at $-3, -1$, and $1$:

$$f(-3) = \frac{-6}{-3 + 2} = 6 > 0$$

and similarly

$$f(-1) = -2 < 0, \qquad f(1) = \tfrac{2}{3} > 0.$$

Thus the sign of $f$ is

$$f: \quad \underline{\overset{+\ +\ +\ +\ +}{\qquad}\Big|\overset{-\ -\ -\ -\ -\ -\ -\ -}{\qquad}\Big|\overset{+\ +\ +\ +\ +\ +}{\qquad}}$$
$$\qquad\qquad -2 \qquad\qquad\qquad 0$$

which means that

$$\frac{2x}{x+2} > 0 \qquad \text{for } x < -2 \text{ or } x > 0,$$

$$\frac{2x}{x+2} < 0 \qquad \text{for } -2 < x < 0.$$

**EXAMPLE 3**   Examine the nature of the discontinuities of $f(x) = \dfrac{2x}{x+2}$.

*SOLUTION*   The only discontinuity is at $x = -2$, where the denominator is 0. To see what it is like you can "experiment" by computing function values close to $-2$. Approach $-2$ from the left:

| $x$ | $-2.1$ | $-2.01$ | $-2.001$ |
|---|---|---|---|
| $\dfrac{2x}{x+2}$ | 42 | 402 | 4002 |

The function values grow larger and larger, suggesting that the limit from the left is $+\infty$:

$$\lim_{x \to -2^-} \frac{2x}{x+2} = +\infty.$$

Approach $-2$ from the right:

| $x$ | $-1.9$ | $-1.99$ | $-1.999$ |
|---|---|---|---|
| $\dfrac{2x}{x+2}$ | $-38$ | $-398$ | $-3998$ |

Apparently $\displaystyle\lim_{x \to -2^+} \frac{2x}{x+2} = -\infty$.

An alternate approach is to think about the fraction $\dfrac{2x}{x+2}$. When $x$ is near $-2$, then the numerator is nearly $-4$, and the denominator is very small; so the function values are "near $+\infty$ or $-\infty$" when $x$ is near $-2$. You can decide between $+\infty$ and $-\infty$ by consulting the signs in Example 2. When $-2 < x < 0$, then $f(x)$ is negative; so approaching $-2$ from the right $f(x)$ must be "near $-\infty$":

$$\lim_{x \to -2^+} \frac{2x}{x+2} = -\infty.$$

When $x \to -2$ from the left, then $\dfrac{2x}{x+2} > 0$, so $\displaystyle\lim_{x \to -2^-} \frac{2x}{x+2} = +\infty$.

Figure 12 shows a graph of $\dfrac{2x}{x+2}$. The limits as $x \to \pm\infty$ were worked out in Section 3.3.

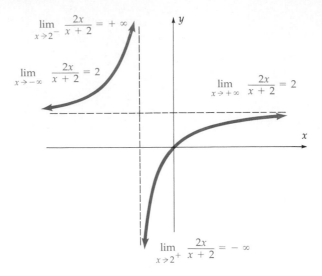

**FIGURE 12**

$$f(x) = \frac{2x}{x + 2}, \ x \neq 2.$$

*REMARK* Figure 12 shows a function that is *increasing* for $x < -2$, and for $x > -2$ (with a "huge decrease" right at $x = -2$). In Section 3.5 we will be able to verify this by taking the derivative, and show that $f'(x) > 0$ except at $x = -2$.

## Discovering and Analyzing Vertical Asymptotes

A vertical asymptote will occur for any quotient $g/h$ at a point $a$ where $g$ and $h$ are continuous with

$$g(a) \neq 0 \quad \text{but} \quad h(a) = 0.$$

We call this a $\dfrac{``\neq 0"}{0}$ case. It gives rise to a vertical asymptote because, when $x$ is very near $a$, then the numerator $g(x)$ is very near the nonzero number $g(a)$, while the denominator $h(x)$ is very near 0. Thus, dividing $g(x)$ by the very small $h(x)$ produces a large quotient $g(x)/h(x)$.

The precise nature of a vertical asymptote (whether the graph goes up to $+\infty$ or down to $-\infty$) can be determined from the *signs* of $g/h$.

**EXAMPLE 4** Examine the discontinuities of $f(x) = \dfrac{1}{1 - (1/x)}$.

*SOLUTION* The function $f(x)$ is not defined for two values for $x$:

$x = 0,$ since then $1/x$ is undefined, and

$x = 1,$ since then the denominator $1 - 1/x = 0$.

At $x = 1$ there is a vertical asymptote, since the denominator in $f(x)$ is 0 at that point, while the numerator is not; we have only to decide whether the limits in question are $+\infty$ or $-\infty$. Checking signs shows

$$\text{sign of } f: \quad \underset{0}{\underbrace{+\;+\;\;\bigg|}} \;\; \underset{}{\underbrace{-\;-\;-\;-\;-\;-\;-\;-\;-}} \; \underset{1}{\bigg|\;+\;+\;+\;+}$$

So as $x \to 1$ from the left, $f(x)$ remains negative and we find

$$\lim_{x \to 1^-} f(x) = -\infty.$$

For like reasons,

$$\lim_{x \to 1^+} f(x) = +\infty.$$

At the other discontinuity $x = 0$, experiments with function values suggest that this is *not* a vertical asymptote; $f(x)$ is not large even when $x$ is very small. So try taking the limit, after rewriting the formula for $f$:

$$\lim_{x \to 0} \frac{1}{1 - (1/x)} = \lim_{x \to 0} \frac{x}{x - 1} \quad \text{(multiply numerator and denominator by } x\text{)}$$

$$= \frac{\lim x}{\lim(x - 1)} = \frac{0}{-1} = 0.$$

Since $\lim\limits_{x \to 0} f(x)$ exists, but $f(0)$ is not defined, the discontinuity at $x = 0$ is a "hole."

To complete the picture, check for horizontal asymptotes:

$$\lim_{x \to \pm \infty} \frac{1}{1 - 1/x} = \frac{\lim 1}{\lim(1 - 1/x)} = 1.$$

The graph in Figure 13 shows the horizontal asymptote $y = 1$, and the two discontinuities.

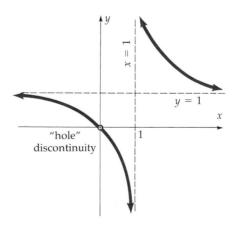

**FIGURE 13**

$$f(x) = \frac{1}{1 - (1/x)}, \; x \neq 0, \; x \neq 1.$$

### The Trigonometric Functions

$$\tan \theta = \frac{\sin \theta}{\cos \theta}, \qquad \cot \theta = \frac{\cos \theta}{\sin \theta}, \qquad \sec \theta = \frac{1}{\cos \theta}, \qquad \csc \theta = \frac{1}{\sin \theta}$$

have vertical asymptotes at each zero of their denominators. We can determine the main features of the graph of $\tan \theta$ from the signs of $\sin \theta$ and $\cos \theta$, as follows:

$\sin \theta$:

$$- \ 0 + + + \ 0 - - - - \ 0 + + + \ 0 - - - - - \ 0 + + + \ 0$$

$-2\pi \qquad -\pi \qquad 0 \qquad \pi \qquad 2\pi \qquad 3\pi$

$\cos \theta$:

$$+ + + \ 0 - - - - \ 0 + + + \ 0 - - - \ 0 + + + \ 0 - - -$$

$-\dfrac{3\pi}{2} \qquad -\dfrac{\pi}{2} \qquad \dfrac{\pi}{2} \qquad \dfrac{3\pi}{2} \qquad \dfrac{5\pi}{2}$

$\tan \theta$:

$$- \ 0 + \infty - \ 0 + \infty - \ 0 + \infty - \ 0 + \infty - \ 0 + \infty - - \ 0 -$$

$-2\pi \ -\dfrac{3\pi}{2} \ -\pi \ -\dfrac{\pi}{2} \ 0 \ \dfrac{\pi}{2} \ \pi \ \dfrac{3\pi}{2} \ 2\pi \ \dfrac{5\pi}{2} \ 3\pi$

Where $\sin \theta$ and $\cos \theta$ have the *same sign*, $\tan \theta$ is positive, and otherwise negative. Where the numerator $\sin \theta$ is 0, so is $\tan \theta$ (since the denominator $\cos \theta$ is *not* 0 at these points); where the denominator $\cos \theta$ is 0, $\tan \theta$ has a vertical asymptote, indicated by "$\infty$" on its sign chart. Figure 14 shows the graph.

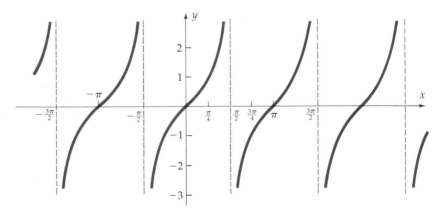

**FIGURE 14**
$y = \tan \theta$.

## SUMMARY

A function $f$ is

> *continuous at a* (see Fig. 1) *if* $\lim\limits_{x \to a} f(x) = f(a)$,
>
> *continuous from the right* (see Fig. 2) *if* $\lim\limits_{x \to a^+} f(x) = f(a)$,
>
> *continuous from the left* (see Fig. 3) *if* $\lim\limits_{x \to a^-} f(x) = f(a)$.

Every polynomial is continuous everywhere.

Every rational function is continuous wherever defined.

**Intermediate Value Theorem (Fig. 10):** If $f$ is continuous for all $x$ in $[a, b]$, and $\bar{y}$ lies between $f(a)$ and $f(b)$, then there is a number $\bar{x}$ in $(a, b)$ where $f(\bar{x}) = \bar{y}$.

**Sign Change Principle:** A function $f$ can change sign only at a zero or a discontinuity.

**Theorem:** If $f$ and $g$ are continuous at $a$, then so are $f + g$, $fg$, and also $f/g$ if $g(a) \neq 0$.

## PROBLEMS

### A

1. By inspecting Figure 15, determine the following limits; use $+\infty$ and $-\infty$ where appropriate.

   *a) $\displaystyle\lim_{x \to -\infty} f(x)$    *b) $\displaystyle\lim_{x \to -3^-} f(x)$

   *c) $\displaystyle\lim_{x \to -3^+} f(x)$    *d) $\displaystyle\lim_{x \to -1^-} f(x)$

   e) $\displaystyle\lim_{x \to -1} f(x)$    f) $\displaystyle\lim_{x \to 2} f(x)$

   g) $\displaystyle\lim_{x \to 3} f(x)$    h) $\displaystyle\lim_{x \to 4} f(x)$

   i) $\displaystyle\lim_{x \to 4^-} f(x)$    j) $\displaystyle\lim_{x \to +\infty} f(x)$

3. For each given function, identify all zeroes and points of discontinuity.

   *a) $\dfrac{1}{x + 1}$    *b) $\dfrac{1}{x^2}$

   *c) $\dfrac{x}{x + 2}$    *d) $\dfrac{x}{x^2 + 1}$

   e) $\dfrac{x}{x^2 - 1}$    f) $\dfrac{x}{x^2 - 4}$

   *g) $\dfrac{x - 1}{1 - 2/x}$    *h) $\dfrac{1}{x} - \dfrac{1}{x + 1}$

   *i) $\dfrac{1}{x} + \dfrac{1}{x + 1}$    j) $\dfrac{1}{x} - \dfrac{1}{x^2} + \dfrac{1}{x + 1}$

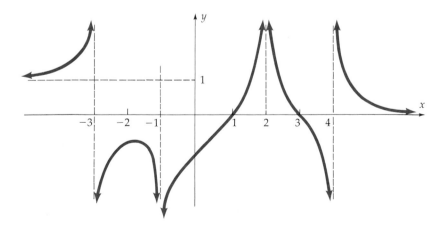

**FIGURE 15**
Graph of $f$ for problem 1.

   k) $\dfrac{|x|}{x - 1}$    l) $\dfrac{1 - 1/x}{1 + 1/x}$

*2. For each graph in Figure 16, find all discontinuities, and identify each as a vertical asymptote, a jump, a "hole," a "blip," or oscillatory.

*4. In problem 3, for each part a–l determine the sign of $f$ everywhere. (Answers for a, b, c, d, g, h, i.)

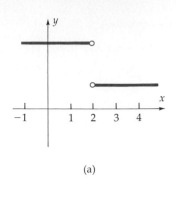

(a)

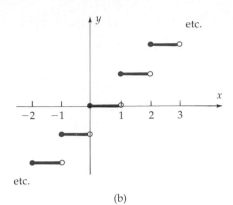

etc.

(b)

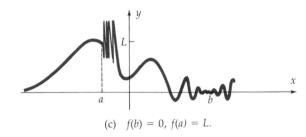

(c) $f(b) = 0$, $f(a) = L$.

**FIGURE 16**

*5. In problem 3, for each part a–l determine $\lim_{x\to a^-} f(x)$ and $\lim_{x\to a^+} f(x)$, at each discontinuity

a. (Answers for a, b, c, d, g, h, i.)

*6. In problem 3, for each part a–l determine $\lim_{x\to-\infty} f(x)$ and $\lim_{x\to+\infty} f(x)$. (Answers for a, b, c, d, g, h, i.)

7. In problem 3, for each part a–l sketch a graph consistent with all the information in problems 3–6.

8. Determine the zeroes, discontinuities, and signs, and sketch an appropriate graph.

a) $\cot\theta \quad \left[ = \dfrac{\cos\theta}{\sin\theta} \right]$

b) $\sec\theta \quad \left[ = \dfrac{1}{\cos\theta} \right]$

c) $\csc\theta \quad \left[ = \dfrac{1}{\sin\theta} \right]$

d) $\tan 2\theta$

e) $\sec(\pi x)$

9. Determine all zeroes and classify all discontinuities of $\dfrac{\sin\theta}{\theta}$.

10. Determine the sign of $\dfrac{\theta}{\sin\theta}$ for $-3\pi < \theta < 3\pi$. Classify all discontinuities and sketch a graph consistent with this information.

11. Describe all discontinuities of the function

$$\begin{aligned} f(x) &= 1 - x^2, & x \le 0 \\ &= 1 & 0 < x < 1 \\ &= (x-1)^2 & 1 \le x < 2 \\ &= 2 & x = 2 \\ &= 1 & x > 2. \end{aligned}$$

*12. The acceleration of a rocket at time $t$ is

$$\begin{aligned} a(t) &= 0, & t \le 0 & \quad\text{(at rest)} \\ &= 10 + t, & 0 < t \le 2 & \quad\text{(motor firing)} \\ &= 16 - t^2, & 2 < t \le 3 & \\ && & \text{(gradual fuel depletion)} \\ &= -9.8, & 3 < t & \quad\text{(free fall).} \end{aligned}$$

Describe all discontinuities in $a(t)$.

13. A switch is "off" until time $t = 0$, when it is turned on and instantaneously a current of 3 amps begins to flow. Describe the discontinuity in the graph of current versus time.

14. The force of gravity between point masses $m_1$ and $m_2$ is $F = \gamma m_1 m_2 / r^2$, where $\gamma$ is a constant

and $r$ is the distance between the masses. Describe any asymptotes of the graph of $F$ versus $r$.

15. In 1987, the Post Office charged $0.22 for the first ounce and $0.17 for each succeeding ounce (or part thereof), up to a total of 12 ounces. (For more than 12 ounces, the postage depended on the destination of the letter.) Sketch the graph of postage $p$ versus weight of letter $x$ for $0 < x \le 12$. Describe its discontinuities. Is it *continuous from the right* for every $x > 0$? Continuous from the left? Why, or why not?

**B**

16. A function $f$ is defined by

$$f(x) = \begin{cases} x^2, & x \le 1 \\ c, & x > 1 \end{cases}.$$

Determine the constant $c$ so that $f$ is continuous everywhere.

17. A function is defined by

$$f(x) = x^2, \qquad x \le -1$$
$$= mx + b, \qquad -1 < x < 1$$
$$= 2x^2, \qquad x \ge 1.$$

Determine $m$ and $b$ so that $f$ is continuous everywhere.

18. All the following functions give examples of the "$\frac{0}{0}$ case" at $x = 0$. Determine the nature of each discontinuity and the limits as $x \to 0^-$ and $x \to 0^+$.

a) $\dfrac{x}{x}$    b) $\dfrac{5x}{x}$    c) $\dfrac{x}{|x|}$    d) $\dfrac{x}{x^2}$

e) $\dfrac{x}{x^3}$    f) $\dfrac{x^2}{x}$    g) $\dfrac{x^3}{x}$

## 3.5

## DERIVATIVES. THE PRODUCT AND QUOTIENT RULES

The rest of this chapter gives general techniques of differentiation. In this section we prove two important formulas, one for the derivative of a product, $fg$, and one for the derivative of a quotient, $f/g$. We begin by relating derivatives to limits and continuity.

The derivative was introduced in Chapter 2 as the "limit value" of the difference quotients, as in Figure 1:

$$\frac{f(\hat{x}) - f(x)}{\hat{x} - x} \to f'(x) \qquad \text{as } \hat{x} \to x.$$

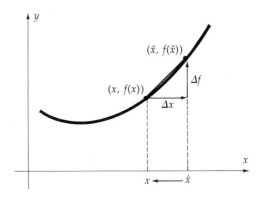

**FIGURE 1**

$$f'(x) = \lim_{\Delta x \to 0} \frac{\Delta f}{\Delta x} = \lim_{\hat{x} \to x} \frac{f(\hat{x}) - f(x)}{\hat{x} - x}.$$

Stated formally:

---

### DEFINITION

The derivative of a function $f$ is defined by

$$f'(x) = \lim_{\hat{x} \to x} \frac{f(\hat{x}) - f(x)}{\hat{x} - x} \tag{1}$$

for every number $x$ such that the limit exists.

---

We often abbreviate $\hat{x} - x$ as $\Delta x$ and $f(\hat{x}) - f(x)$ as $\Delta f$. (Fig. 1). As $\hat{x} \to x$, then $\Delta x = \hat{x} - x \to 0$ and vice versa, so (1) can be written

$$f'(x) = \lim_{\Delta x \to 0} \frac{\Delta f}{\Delta x}. \tag{2}$$

In Figure 1, $\Delta x = \hat{x} - x$ is the change from $x$ to $\hat{x}$, and $\Delta f = f(\hat{x}) - f(x)$ is the resulting change in $f$.

If the limit in (1) or (2) exists, then the function $f$ is called **differentiable at $x$**—the term refers to the difference quotient in (1) and (2). Computing derivatives is called **differentiating**.

### Differentiability Related to Continuity

In Figure 2, $f$ has the "smoothest" graph, and $h$ has the "roughest." These differences can be expressed in terms of continuity and derivatives; in Figure 2a, $f$ has a derivative at $a$; in Figure 2b, $g$ has no derivative at $a$ (see Sec. 2.2) but is continuous there; and in Figure 2c, $h$ is not even continuous at $a$.

This suggests a hierarchy of function behavior: Discontinuity is bad, continuity is better, and having a derivative is better yet. Is it true, then, that every differentiable function is automatically continuous? Yes:

---

### THEOREM 1

If $f'(a)$ exists, then $f$ is continuous at $a$; that is

$$\lim_{x \to a} f(x) = f(a).$$

---

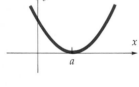

(a)  $f(x) = (x - a)^2$

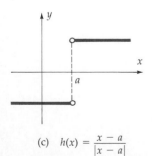

(b)  $g(x) = |x - a|$

(c)  $h(x) = \dfrac{x - a}{|x - a|}$

**FIGURE 2**

**PROOF**  We express $f(x)$ in terms of a difference quotient:

$$f(x) = f(a) + [f(x) - f(a)]$$
$$= f(a) + \frac{f(x) - f(a)}{x - a} (x - a).$$

Hence

$$\lim_{x \to a} f(x) = \lim_{x \to a} f(a) + \left[ \lim_{x \to a} \frac{f(x) - f(a)}{x - a} \right] \cdot \left[ \lim_{x \to a} (x - a) \right]$$
$$= f(a) + f'(a) \cdot 0$$
$$= f(a)$$

and this proves that $f$ is continuous at $a$.

Now we develop the derivative formulas:

### The Power Rule

$$Dx^n = nx^{n-1}$$

was shown in Section 2.3 for $n = 0, 1, 2, \ldots$. At the end of this section, it will be proved for $n = -1, -2, \ldots$; ultimately we will see that the formula is true for *every* real number $n$.

**The derivative of a sum** is the sum of the derivatives:

$$(f + g)' = f' + g'.$$

This, too, was shown in Section 2.3.

**The derivative of a product** is *not* the product of the derivatives! For example,

$$D(x^2 \cdot x^3) \neq (Dx^2) \cdot (Dx^3) = 2x \cdot 3x^2 = 6x^3$$

because in fact

$$D(x^2 \cdot x^3) = Dx^5 = 5x^4.$$

To get the correct formula for the derivative of a product $fg$, go back to the limit (2) defining the derivative:

$$(fg)' = \lim_{\Delta x \to 0} \frac{\Delta(fg)}{\Delta x},$$

where $\Delta(fg) = (fg)(\hat{x}) - (fg)(x) = f(\hat{x})g(\hat{x}) - f(x)g(x)$. Figure 3 shows the relation of $\Delta(fg)$ to $\Delta f$ and $\Delta g$. Rectangles I, II, and III together have area $f(\hat{x})g(\hat{x})$, while rectangle I alone has area $f(x)g(x)$. So $\Delta(fg)$, the difference of these, is the sum of areas II and III; reading the dimensions of each rectangle from Figure 3, you find

$$\Delta(fg) = (\Delta f)g(x) + (\Delta g)f(\hat{x}). \tag{3}$$

Now, assuming that $f'(x)$ and $g'(x)$ exist, we find the derivative of $fg$:

$$
\begin{aligned}
(fg)'(x) &= \lim_{\Delta x \to 0} \frac{\Delta(fg)}{\Delta x} \\
&= \lim_{\Delta x \to 0} \frac{(\Delta f)g(x) + (\Delta g)f(\hat{x})}{\Delta x} && \text{[by (3)]} \\
&= \lim_{\Delta x \to 0} \left[ \frac{\Delta f}{\Delta x} g(x) + \frac{\Delta g}{\Delta x} f(\hat{x}) \right] && \text{(algebra)} \\
&= \left( \lim_{\Delta x \to 0} \frac{\Delta f}{\Delta x} \right) g(x) + \lim_{\Delta x \to 0} \frac{\Delta g}{\Delta x} \lim_{\Delta x \to 0} f(\hat{x}) && \text{(Limit Theorem).}
\end{aligned}
$$

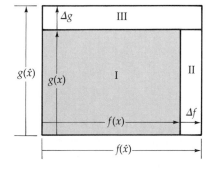

**FIGURE 3**
$\Delta(fg) = \text{II} + \text{III} = (\Delta f)g(x) + (\Delta g)f(\hat{x}).$

At this point we use the definition of the derivative

$$\lim_{\Delta x \to 0} \frac{\Delta f}{\Delta x} = f'(x) \quad \text{and} \quad \lim_{\Delta x \to 0} \frac{\Delta g}{\Delta x} = g'(x)$$

and Theorem 1: If $f'(x)$ exists, then $\lim_{\hat{x} \to x} f(\hat{x}) = f(x)$. Here $\hat{x} \to x$ is the same as $\Delta x \to 0$, so

$$\lim_{\Delta x \to 0} f(\hat{x}) = f(x).$$

With these three limits we complete the above calculation of $(fg)'(x)$, finding

$$(fg)'(x) = f'(x)g(x) + g'(x)f(x).$$

This is the *Product Rule*. In operator notation

$$D(fg) = (Df)g + fDg$$

and in Leibniz notation,

$$\frac{d(fg)}{dx} = \frac{df}{dx}g + f\frac{dg}{dx}.$$

In words: The derivative of a product $fg$ is the sum of two terms: in one term, differentiate $f$ and leave $g$ alone; in the other term, leave $f$ alone and differentiate $g$. Figure 3 suggests why there are two terms—they correspond to the two rectangles II and III that form the difference $\Delta(fg)$.

---

**EXAMPLE 1**

$$\frac{d(x^2 \cdot x^3)}{dx} = \frac{dx^2}{dx} \cdot x^3 + x^2 \cdot \frac{dx^3}{dx}$$

$$= 2x \cdot x^3 + x^2 \cdot 3x^2 = 2x^4 + 3x^4$$

$$= 5x^4,$$

which agrees with $\dfrac{dx^2 \cdot x^3}{dx} = \dfrac{dx^5}{dx} = 5x^4$.

---

## Multiplication by a Constant

If $c$ is a constant, then $\dfrac{dc}{dx} = 0$, so by the product rule

$$\frac{dcf}{dx} = \frac{dc}{dx} \cdot f + c\frac{df}{dx}$$

$$= 0 \cdot f + c\frac{df}{dx}.$$

Thus when one of the factors is constant, the derivative of the product has just one term:

$$\frac{dcf}{dx} = c\,\frac{df}{dx}, \quad \text{or} \quad (cf)' = c \cdot f'.$$

**EXAMPLE 2**

$$\frac{d5x^2}{dx} = 5\,\frac{dx^2}{dx} = 5 \cdot 2x = 10x.$$

### Products of Three or More Factors

The derivative of a product $fgh$ can be found by two applications of the Product Rule. First, apply it to the product of $(fg)$ by $h$,

$$(fgh)' = ((fg)(h))' = (fg)'h + (fg)h',$$

and then to $fg$,

$$(fgh)' = (f'g + fg')h + (fg)h'$$
$$= f'gh + fg'h + fgh'.$$

The end result is the sum of three terms; in each term, one of the three factors is differentiated, and the other two are left alone. For example,

$$\frac{d(x \cdot x \cdot x)}{dx} = \frac{dx}{dx} \cdot x \cdot x + x \cdot \frac{dx}{dx} \cdot x + x \cdot x \cdot \frac{dx}{dx}$$
$$= 1 \cdot x^2 + x \cdot 1 \cdot x + x \cdot x \cdot 1$$
$$= 3x^2.$$

(You can probably imagine the result of differentiating a product with four or more factors.)

**The derivative of a reciprocal $1/g$** is a straightforward computation:

$$\left(\frac{1}{g}\right)'(x) = \lim_{\Delta x \to 0} \frac{\Delta(1/g)}{\Delta x} = \lim_{\Delta x \to 0} \frac{\dfrac{1}{g(\hat{x})} - \dfrac{1}{g(x)}}{\Delta x}$$

$$= \lim_{\Delta x \to 0} \frac{1}{\Delta x}\left[\frac{1}{g(\hat{x})} - \frac{1}{g(x)}\right] = \lim_{\Delta x \to 0} \frac{1}{\Delta x}\left[\frac{g(x) - g(\hat{x})}{g(\hat{x})g(x)}\right]$$

$$= \lim_{\Delta x \to 0} \frac{-1}{g(\hat{x})g(x)}\,\frac{g(\hat{x}) - g(x)}{\Delta x}. \tag{4}$$

If $g'(x)$ exists, we thus find the derivative of the reciprocal:

$$\left(\frac{1}{g}\right)'(x) = \frac{-1}{g(x)^2}\,g'(x) \tag{5}$$

or

$$\frac{d(1/g)}{dx} = \frac{-1}{g^2}\,\frac{dg}{dx}.$$

**EXAMPLE 3**

$$\frac{d}{dx}\frac{1}{x^2+1} = \frac{-1}{(x^2+1)^2}\frac{d(x^2+1)}{dx} = \frac{-1}{(x^2+1)^2}2x = \frac{-2x}{(x^2+1)^2}.$$

**The derivative of a quotient** $f/g$ can now be derived by writing $f/g$ as a product:

$$\frac{f}{g} = f \cdot \frac{1}{g}.$$

The derivative of this product is the sum of two terms,

$$\left[f \cdot \frac{1}{g}\right]' = (f')\left(\frac{1}{g}\right) + (f)\left(\frac{1}{g}\right)' \qquad \text{[now use (5)]}$$

$$= (f')\left(\frac{1}{g}\right) + (f)\left[\frac{-1}{g^2}g'\right] = \frac{f'}{g} - \frac{fg'}{g^2}$$

$$= \frac{f'g}{g^2} - \frac{fg'}{g^2} = \frac{f'g - fg'}{g^2}.$$

This proves the Quotient Rule:

$$\left(\frac{f}{g}\right)' = \frac{f'g - fg'}{g^2}.$$

In operator notation,

$$D\left(\frac{f}{g}\right) = \frac{gDf - fDg}{g^2}. \tag{6}$$

It is crucial to remember which term gets the minus sign—the term containing the derivative $Dg$ of the denominator. In reading formula (6)

$$g \text{ dee } f - f \text{ dee } g$$
$$\text{all over } g^2$$

you begin and end with the denominator, $g$.

**EXAMPLE 4**

$$D\left[\frac{2x}{x+2}\right] = \frac{(x+2)D(2x) - 2xD(x+2)}{(x+2)^2}$$

$$= \frac{(x+2)\cdot 2 - 2x\cdot 1}{(x+2)^2}$$

$$= \frac{4}{(x+2)^2}.$$

The derivative is positive everywhere (except at $x = -2$) so the function is increasing for $x < -2$, and increasing again for $x > -2$. This confirms the graph sketched previously in Section 3.4, Figure 12.

---

**The derivative of a quotient $f/c$** with *constant* denominator $c$ does not require the Quotient Rule:

$$\left(\frac{f}{c}\right)' = \left(\frac{1}{c}f\right)' = \frac{1}{c}f' = \frac{f'}{c}.$$

For example,

$$\frac{d(x^2/3)}{dx} = \frac{1}{3}\frac{dx^2}{dx} = \frac{1}{3}2x = \frac{2x}{3}.$$

**The derivative of $x^n$ for $n = -1, -2, \ldots$**

The derivative of $x^{-2}$ can be deduced from the formula for the derivative of a reciprocal:

$$\frac{dx^{-2}}{dx} = \frac{d\left[\dfrac{1}{x^2}\right]}{dx} = \frac{-1}{(x^2)^2}\frac{dx^2}{dx}$$

$$= \frac{-2x}{x^4} = -2x^{-3}.$$

This follows the pattern of the Power Rule

$$\frac{dx^n}{dx} = nx^{n-1}$$

with $n = -2$. Similarly, the Power Rule holds for every negative integer $n$. To prove it, write $n = -m$, where $m$ is a *positive* integer. Then

$$\frac{dx^n}{dx} = \frac{dx^{-m}}{dx} = \frac{d(1/x^m)}{dx}$$

$$= \frac{-1}{(x^m)^2}mx^{m-1} = \frac{-mx^{m-1}}{x^{2m}}$$

$$= -mx^{-m-1} = nx^{n-1}$$

since $n = -m$.

The easiest way to differentiate the reciprocal of a power such as $1/x^5$ is to rewrite it as a negative power:

$$\frac{d(1/x^5)}{dx} = \frac{dx^{-5}}{dx} = -5x^{-6} = \frac{-5}{x^6}$$

---

**EXAMPLE 5**   Graph $y = 2x^2 + \dfrac{1}{2x}$

**SOLUTION** It simplifies the differentiation to write $\dfrac{1}{2x} = \dfrac{1}{2} \cdot \dfrac{1}{x} = \dfrac{1}{2} \cdot x^{-1}$.

Thus

$$\frac{dy}{dx} = \frac{d2x^2}{dx} + \frac{d\,\dfrac{1}{2x}}{dx} \qquad \text{(derivative of a sum)}$$

$$= 4x + \frac{d[\frac{1}{2}x^{-1}]}{dx} = 4x + \frac{1}{2}\frac{dx^{-1}}{dx}$$

$$= 4x + \frac{1}{2}(-1)x^{-2} = 4x - \frac{1}{2x^2}.$$

The second derivative is

$$\frac{d^2y}{dx^2} = \frac{d4x}{dx} - \frac{1}{2}\frac{dx^{-2}}{dx} = 4 - \frac{1}{2}(-2)x^{-3} = 4 + x^{-3}.$$

Now determine the signs of $y$, $dy/dx$, and $d^2y/dx^2$. $y$ is discontinuous at $x = 0$ and has a vertical asymptote there. For zeroes,

$$y = 0 \Leftrightarrow 2x^2 + \frac{1}{2x} = 0 \Leftrightarrow 4x^3 = -1 \Leftrightarrow x = \frac{-1}{\sqrt[3]{4}}.$$

Find the signs:

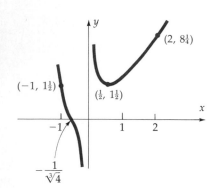

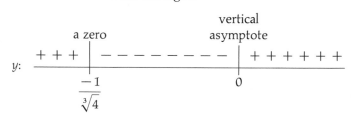

A similar check (left as an exercise) gives

**FIGURE 4**

$y = 2x^2 + \dfrac{1}{2x}$.

Make a table of values, including the intercept, stationary point, and inflection point, and sketch a graph consistent with all the above data as in Figure 4.

| $x$ | $-1$ | $\dfrac{-1}{\sqrt[3]{4}}$ | $0$ | $\dfrac{1}{2}$ | $2$ |
|---|---|---|---|---|---|
| $y$ | $1\frac{1}{2}$ | $0$ | UND | $1\frac{1}{2}$ | $8\frac{1}{4}$ |

## SUMMARY

*Definition of the Derivative:*

$$f'(x) = \lim_{\hat{x} \to x} \frac{f(\hat{x}) - f(x)}{\hat{x} - x} = \lim_{\Delta x \to 0} \frac{\Delta f}{\Delta x}.$$

*Theorem:* If $f'(a)$ exists, then $f$ is continuous at $a$:

$$\lim_{x \to a} f(x) = f(a).$$

*Power Rule:* $\quad Dx^n = nx^{n-1} \quad$ for every integer $n$.

*Product Rule:* $\quad (fg)' = f'g + fg'$

$$(cf)' = cf' \quad \text{if } c \text{ is constant.}$$

*Reciprocals:* $\quad \left(\dfrac{1}{g}\right)' = \dfrac{-1}{g^2} g'.$

*Quotient Rule:* $\quad \left(\dfrac{f}{g}\right)' = \dfrac{gf' - fg'}{g^2}$

$$\left(\frac{f}{c}\right)' = \left(\frac{1}{c}f\right)' = \frac{1}{c}f' = \frac{f'}{c} \quad \text{if } c \text{ is constant.}$$

## PROBLEMS

### A

1. Given that $f(a) = 1$, $f'(a) = -1$, $g(a) = 2$, $g'(a) = 3$, compute
   *a) $(3f)'(a)$.
   b) $(f + g)'(a)$.
   *c) $(fg)'(a)$.
   d) $(f/g)'(a)$.
   *e) $(g/f)'(a)$.
   *f) $(f^2)'(a)$.
   g) $(f^2g)'(a)$.

2. Differentiate the following (i) by the Product Rule and (ii) by multiplying out and then differentiating. Show that the two results are equivalent.
   a) $x^5 \cdot x^7$
   b) $x^m \cdot x^n$
   c) $(x^2 + 1)(x - x^3)$
   d) $(3x^2 + 5x + 4)(x^{100} + x^2)$

3. Explain why $D\left(\dfrac{f}{c}\right) = \dfrac{Df}{c}$, if $c$ is a constant. Use this simplified rule in the following cases:

   *a) $D\left(\dfrac{x^2 + 3x + 1}{5}\right)$.

   b) $\dfrac{d}{dt} \dfrac{5t^2 + 6t + 1}{2 \cdot 10^8}$.

4. Compute the derivative.

   *a) $\dfrac{x}{x + 1}$      *b) $\dfrac{x}{3}$

   *c) $\dfrac{3}{x}$      d) $\dfrac{5}{x + 1}$

   *e) $\dfrac{7}{1 - x}$      f) $\dfrac{1}{(1 + x)^2}$

   g) $\dfrac{x + 1}{x - 1}$      *h) $x^2 + \dfrac{1}{x^2}$

*i)  $\dfrac{1}{x}(x+1)^2$    *j)  $-3x^{-100}$

k)  $3x^2 - (5/x)$    *l)  $\dfrac{1}{x^2+1}$

**\*5.** For the functions in problem 4, compute the second derivative. (Answer to part l only.)

**6.** Compute the derivative.

*a)  $\dfrac{2x-3}{3x-2}$    *b)  $\dfrac{x^3+3x+2}{x^2-1}$

c)  $\dfrac{3-x^2}{1+x^2}$    *d)  $x^3\left(1-\dfrac{3}{x+1}\right)$

e)  $\dfrac{t+2}{t^2+t+1}$    f)  $\dfrac{x+1}{x+2}(2x+3)$

**7.** Rewrite the following functions so as to differentiate them *without* the Quotient Rule; then differentiate.

*a)  $\dfrac{5}{x^2}$    b)  $\dfrac{x^2+1}{x}\;[=x+1/x]$

*c)  $\dfrac{x+1}{x^2}$    d)  $\dfrac{1+1/x}{x^3}$

**8.** Find the equation of the line tangent to the graph of the given function at the given point.

*a)  $\dfrac{x}{1+x}$  at $(1,\frac{1}{2})$

b)  $y = \dfrac{x}{1-x}$  at $(\frac{1}{2},1)$

**9.** Graph the following, showing regions of increase and decrease, convexity and concavity, inflection points, and vertical asymptotes.

*a)  $x-\dfrac{1}{x}$    b)  $x^2-\dfrac{1}{x}$    *c)  $x+\dfrac{1}{x^2}$

d)  $x^2-\dfrac{1}{x^2}$    *e)  $\dfrac{1}{x^2+1}$    f)  $\dfrac{x}{x^2+1}$

**10.** The current, $I$, from a battery is given by $I = \dfrac{E}{r+R}$, where $E$ is the battery voltage, $r$ the internal resistance of the battery, and $R$ the resistance of the electrical circuit connected to the battery. $E$ and $r$ are constants, determined by the nature of the battery; $R$ varies, depending on the circuit to which the battery is connected. The energy used by the battery is $W = I^2 R$. For what value of $R$ is $W$ a maximum? What is $\lim\limits_{R\to+\infty} W$?

The following problems 11–16 are slightly different from the optimization problems in Chapter 2. There we asked "What rectangle with given perimeter has maximum area?" Now we ask "What rectangle with given area has minimum perimeter?" Phrased this way, the problem leads to a rational function, not a polynomial.

**11.** A rectangle is to have area 9 m². What is the minimum possible perimeter?

**\*12.** A rectangular area is to be fenced, enclosing 1,000 m². One side, facing the road, requires fencing at \$5/m, and the other three sides require fencing at \$1/m.
  a) What dimensions require the least cost of fencing?
  b) What fraction of the cost is spent on the two sides parallel to the road?

**13.** A rectangular area is to be fenced, enclosing $A$ m². The fencing for one side costs \$$a$ per meter, and for the other three sides costs \$$b$ per meter.
  a) Find the dimensions that require minimum cost of fence.
  *b) What fraction of the cost is spent on the two sides perpendicular to the side that costs \$$b$ per meter?

**\*14.** A flat-roofed shed is to be constructed, with walls 10 feet high and a rectangular base. The roof costs \$5/ft², the floor \$3/ft², one side costs \$4/ft², and the other three sides cost \$1/ft². The floor area is to be 400 ft². What dimensions give minimum cost?

**15.** A rectangular basin is to be constructed, with a square base and open top, containing 1000 m³. The bottom costs \$5/m² and the sides cost \$2/m². What dimensions give minimum cost? What fraction of the cost is spent on the bottom?

**\*16.** A cylindrical can is to contain a certain given volume $V$. What dimensions require the minimum surface area? Find the "proportions" of this optimal can, that is, the ratio diameter/height.

**17.** The van der Waals gas equation is

$$\left(P+\dfrac{n^2 a}{V^2}\right)(V-bn) = nRT.$$

This is a refinement of the "ideal gas law" $PV = nRT$. $P$ is the pressure, $V$ the volume, and $T$ the temperature of a certain quantity of gas in equilibrium; the constants $a$, $b$, $n$, and $R$ describe the number and nature of the molecules of gas in that quantity. For this problem, suppose that $T$ is held constant, but $P$ and $V$ may vary.

a) Express $P$ as a function of $V$, and compute the derivative $dP/dV$.

b) Suppose that $a = b = n = R = T = 1$. Sketch the graph of $P$ versus $V$, for $V > 0$, showing all vertical and horizontal asymptotes. Show that $P < 0$ for $0 < V < 1$, and $P > 0$ for $V > 1$.

c) Still with $a = b = n = R = T = 1$, show that $dP/dV = 0$ at just one point, in $0 < V < 1$. Is this consistent with your graph in part b?

d) Drop the assumption in parts b and c, that all the constants equal 1, and express the asymptotes in terms of these constants. Show that $P < 0$ for $0 < V < bn$. (The van der Waals equation does not apply in this interval.)

**18.** a) Describe the derivative of the product of four factors $f_1f_2f_3f_4$.

b) Prove your answer to part a.

**B**

**19.** Prove that a cubic $y = k(x - a)(x - b)(x - c)$ with three zeroes $a$, $b$, $c$ has a point of inflection at the average of $a$, $b$, and $c$.

**20.** a) Prove that $(fg)'' = f''g + 2f'g' + fg''$.

b) Prove an analogous formula for $(fg)'''$.

**21.** a) Prove that $\dfrac{df^2}{dx} = 2f\dfrac{df}{dx}$. (Write $f^2 = f \cdot f$ and use the Product Rule.)

b) Using part a, prove that $\dfrac{df^3}{dx} = 3f^2\dfrac{df}{dx}$.

c) Prove that $\dfrac{df^4}{dx} = 4f^3\dfrac{df}{dx}$.

d) Guess a formula for $df^n/dx$.

e) Compute $\dfrac{d(2x + 1)^3}{dx}$ by the formula in part b. Recompute it by first multiplying out $(2x + 1)^3$.

f) Compute $\dfrac{d(2x + 1)^{10}}{dx}$.

**22.** a) Assume that $\dfrac{df^n}{dx} = nf^{n-1}\dfrac{df}{dx}$ for a certain

number $n$. Prove that
$$\frac{df^{n+1}}{dx} = (n + 1)f^n\frac{df}{dx}.$$

b) Explain why the formula in part a is actually true for every $n = 1, 2, 3, 4, \ldots$.

c) Compute $\dfrac{d(2x + 1)^{100}}{dx}$. [Imagine multiplying out $(2x + 1)^{100}$ in order to take the derivative *without* the formula in part a!]

**C**

**23.** Suppose that $f$ and $g$ have derivatives throughout an interval $I$, and $g(x) \neq 0$ for every $x$ in $I$. Prove that $f$ is a constant multiple of $g$ if and only if $fg' - f'g = 0$ throughout $I$. (The expression $fg' - f'g$ is called the *Wronskian* of $f$ and $g$.)

**24.** a) Prove that $\lim\limits_{x \to a} \dfrac{f(x) - f(a)}{g(x) - g(a)} = \dfrac{f'(a)}{g'(a)}$ if $f'(a)$ exists and $g'(a) \neq 0$.

b) Prove that $\lim\limits_{x \to a} \dfrac{f(x)}{g(x)} = \dfrac{f'(a)}{g'(a)}$ if the right-hand side is defined *and* $f(a) = g(a) = 0$. (This is one form of a method for evaluating limits known as "l'Hôpital's Rule".)

**25.** Evaluate the following, using the formula in problem 24b.

a) $\lim\limits_{x \to 2} \dfrac{x^2 - 4}{x^3 - 8}$

b) $\lim\limits_{x \to 1} \dfrac{x - 1}{x^2 - 1}$

c) $\lim\limits_{x \to -1} \dfrac{x^3 + x^2 + x + 1}{x^2 - 1}$

**26.** a) Suppose that $P$ is a polynomial divisible by $(x - a)^2$, that is $P(x) = (x - a)^2Q(x)$ for some polynomial $Q$. Prove that $P'(x)$ is divisible by $x - a$.

b) Prove that if $P$ is divisible by $(x - a)^3$ then $P'$ is divisible by $(x - a)^2$ and $P''$ is divisible by $x - a$.

**27.** a) If $f'(a)$ exists then the difference quotient $\dfrac{f(x) - f(a)}{x - a}$ has a "hole" discontinuity at $x = a$; for the quotient is not defined when $x = a$, but $\lim\limits_{x \to a} \dfrac{f(x) - f(a)}{x - a} = f'(a)$ exists.

Show that there is a unique function $Q_a(x)$ such that

$$\left. \begin{aligned} f(x) &= f(a) + (x - a)Q_a(x) \quad \text{for all } x \\ \lim_{x \to a} Q_a(x) &= Q_a(a). \end{aligned} \right\} \quad (*)$$

b) Let $f$ be any function. Suppose that there is a function $Q_a(x)$ satisfying the two conditions (*). Prove that $f'(a)$ exists and equals $Q_a(a)$.

28. A function $f$ is defined by

$$\begin{aligned} f(x) &= x^2, && x \le 1 \\ &= mx + b, && x > 1. \end{aligned}$$

Determine constants $m$ and $b$ so that $f$ and $f'$ are continuous everywhere. Be careful to show that the limit defining $f'(1)$ exists.

## 3.6

# DERIVATIVES OF THE TRIGONOMETRIC FUNCTIONS

To compute the derivatives of $\sin \theta$ and $\cos \theta$, we need two limits. One is

$$\lim_{h \to 0} \frac{\sin h}{h} = 1, \tag{1}$$

which gives the slope of the graph of $\sin \theta$ at the origin (Figure 1); this is proved in Section 3.2. The other limit is

$$\lim_{h \to 0} \frac{\cos h - 1}{h} = 0, \tag{2}$$

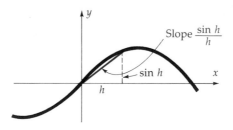

**FIGURE 1**

Slope of graph of $\sin \theta$ at origin is $\lim\limits_{h \to 0} \dfrac{\sin h}{h} = 1$.

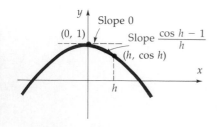

**FIGURE 2**

$\lim\limits_{h \to 0} \dfrac{\cos h - 1}{h} = $ [slope of graph of $\cos \theta$ at $\theta = 0$] = 0.

giving the slope of the graph of $\cos \theta$ at $\theta = 0$ (Fig. 2); it is clear from the graph that this slope is 0. (A more formal proof of (2) is suggested in Section 3.2, problem 12.)

We also need the *addition formulas* for the sine and cosine (reviewed in Appendix C):

$$\sin(\theta + \varphi) = \sin \theta \cos \varphi + \cos \theta \sin \varphi \tag{3}$$

$$\cos(\theta + \varphi) = \cos \theta \cos \varphi - \sin \theta \sin \varphi. \tag{4}$$

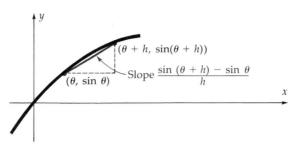

**FIGURE 3**

$$\sin'(\theta) = \lim_{h \to 0} \frac{\sin(\theta + h) - \sin \theta}{h}.$$

With all this, we compute the derivative $\sin'(\theta)$ as the limit of the difference quotient in Figure 3:

$$\sin'(\theta) = \lim_{h \to 0} \frac{\sin(\theta + h) - \sin \theta}{h}.$$

The addition formula (3) gives

$$\sin'(\theta) = \lim_{h \to 0} \frac{\sin \theta \cdot \cos h + \cos \theta \cdot \sin h - \sin \theta}{h}.$$

Separate the $\sin \theta$ terms from the $\cos \theta$ terms:

$$\sin'(\theta) = \lim_{h \to 0} \left[ \frac{\sin \theta \cdot \cos h - \sin \theta}{h} + \frac{\cos \theta \cdot \sin h}{h} \right]$$

$$= \lim_{h \to 0} \sin \theta \frac{\cos h - 1}{h} + \lim_{h \to 0} \cos \theta \frac{\sin h}{h}$$

$$= 0 + \cos \theta$$

by (2) and (1). Thus the derivative of $\sin \theta$ is

$$\sin'(\theta) = \cos \theta.$$

The same method gives the derivative of the cosine, which turns out to be

$$\cos'(\theta) = -\sin \theta.$$

You could almost guess the formula $\sin'(\theta) = \cos \theta$ from the graphs in Figure 4: Where the graph of $\sin \theta$ is rising (for example, with $0 < \theta < \pi/2$), $\cos \theta$ is positive; where the graph of $\sin \theta$ is horizontal (as at $\theta = \pi/2$), $\cos \theta = 0$.

The *second* derivative of $\sin \theta$ is

$$\frac{d^2 \sin \theta}{d\theta^2} = \frac{d \cos \theta}{d\theta} = -\sin \theta.$$

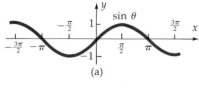

(a)

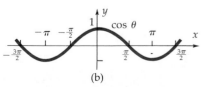

(b)

**FIGURE 4**

Thus where $\sin\theta$ is positive, the second derivative $-\sin\theta$ is negative; where the graph of $\sin\theta$ is above the axis, it is concave down (Fig. 4). The cosine has the same property.

**EXAMPLE 1**

$$\frac{d\, x \cdot \sin x}{dx} = \frac{dx}{dx} \cdot \sin x + x \cdot \frac{d\sin x}{dx} \qquad \text{(Product Rule)}$$

$$= 1 \cdot \sin x + x \cdot \cos x.$$

$$\frac{d^2(x \cdot \sin x)}{dx^2} = \frac{d}{dx}(\sin x + x \cdot \sin x)$$

$$= \cos x + \left[\frac{dx}{dx} \cdot \cos x + x \cdot \frac{d\cos x}{dx}\right]$$

$$= \cos x + \cos x + x(-\sin x)$$

$$= 2\cos x - x \cdot \sin x.$$

**The other trigonometric functions** are defined in terms of $\sin\theta$ and $\cos\theta$:

$$\tan\theta = \frac{\sin\theta}{\cos\theta}, \qquad \cot\theta = \frac{\cos\theta}{\sin\theta},$$

$$\sec\theta = \frac{1}{\cos\theta}, \qquad \csc\theta = \frac{1}{\sin\theta}. \tag{5}$$

Their derivatives are found by using the Quotient Rule. For the tangent:

$$\frac{d\tan\theta}{d\theta} = \frac{d\,\dfrac{\sin\theta}{\cos\theta}}{d\theta} = \frac{\cos\theta\,\sin'(\theta) - \sin(\theta)\,\cos'(\theta)}{\cos^2\theta}$$

$$= \frac{\cos^2\theta - (\sin\theta)(-\sin\theta)}{\cos^2\theta}$$

$$= \frac{\cos^2\theta + \sin^2\theta}{\cos^2\theta}.$$

We simplify this with the Pythagorean identity (Fig. 5)

$$\sin^2\theta + \cos^2\theta = 1.$$

Thus

$$\frac{d\tan\theta}{d\theta} = \frac{1}{\cos^2\theta} = \left(\frac{1}{\cos\theta}\right)^2$$

$$= \sec^2\theta.$$

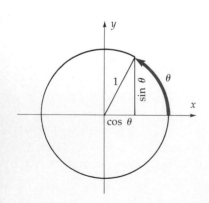

**FIGURE 5**
$\cos^2\theta + \sin^2\theta = 1.$

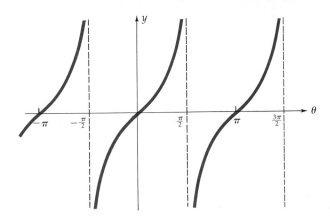

**FIGURE 6**
$y = \tan \theta$.

This shows that $\dfrac{d \tan \theta}{d\theta} > 0$, so throughout each interval where it is defined, the graph is rising from left to right (Fig. 6).

You can work out the formulas for the remaining derivatives yourself, and complete the following list:

$$\tan'(\theta) = \sec^2\theta \qquad\qquad \cot'(\theta) = -\csc^2\theta$$
$$\sec'(\theta) = \sec\theta\tan\theta \qquad \csc'(\theta) = -\csc\theta\cot\theta$$

The formulas for $\tan'(\theta)$ and $\sec'(\theta)$ should be memorized; the remaining two are the same, except with a minus sign, and with each function replaced by its "cofunction."

---

**EXAMPLE 2**

$$\frac{d}{dx}\frac{\sec x}{x} = \frac{x[\sec \cdot \tan x] - [\sec x] \cdot 1}{x^2}$$

$$= \sec x \cdot \frac{x \cdot \tan x - 1}{x^2}$$

---

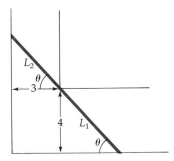

**FIGURE 7**
$L(\theta) = L_1 + L_2$.

**EXAMPLE 3**  A door 9 feet long is to be carried on edge around a corner from a corridor 4 feet wide into a corridor 3 feet wide. Will it fit?

*SOLUTION*  This can be answered by determining the *minimum* length $L(\theta) = L_1 + L_2$ in Figure 7. From the figure

$$\frac{4}{L_1} = \sin\theta, \qquad \frac{3}{L_2} = \cos\theta$$

so

$$L_1 = \frac{4}{\sin \theta} = 4 \csc \theta, \qquad L_2 = \frac{3}{\cos \theta} = 3 \sec \theta$$

and

$$L(\theta) = L_1 + L_2 = 4 \csc \theta + 3 \sec \theta, \qquad 0 < \theta < \pi/2.$$

The first and second derivatives are

$$L'(\theta) = -4 \csc \theta \cot \theta + 3 \sec \theta \tan \theta$$

$$L''(\theta) = 4(\csc \theta \cot^2\theta + \csc^3\theta) + 3(\sec \theta \tan^2\theta + \sec^3\theta).$$

In the relevant range $0 < \theta < \pi/2$, the second derivative is positive, so the graph is concave up. Hence (Fig. 8) any zero of $L'(\theta)$ will give the desired minimum of $L$. Solve $L'(\theta) = 0$:

$$3 \sec \theta \tan \theta = 4 \csc \theta \cot \theta$$

$$\frac{4}{3} = \frac{\sin^3\theta}{\cos^3\theta} = \tan^3\theta$$

$$\tan \theta = (4/3)^{1/3}.$$

A calculator gives $\sec \theta \approx 1.4871$, $\csc \theta \approx 1.3511$, so the minimum length is

$$L(\theta) = 4 \csc \theta + 3 \sec \theta \approx 4(1.3511) + 3(1.4871) = 9.8656.$$

The door will fit, with less than a foot to spare.

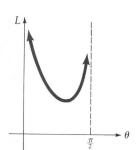

**FIGURE 8**
Graph of $L$ is concave up.

## SUMMARY

*Derivatives:*   $\dfrac{d \sin \theta}{d\theta} = \cos \theta$ $\qquad\qquad$ $\dfrac{d \cos \theta}{d\theta} = -\sin \theta$

$\qquad\qquad\quad$ $\dfrac{d \tan \theta}{d\theta} = \sec^2\theta$ $\qquad\qquad$ $\dfrac{d \cot \theta}{d\theta} = -\csc^2\theta$

$\qquad\qquad\quad$ $\dfrac{d \sec \theta}{d\theta} = \sec \theta \cdot \tan \theta$ $\qquad\quad$ $\dfrac{d \csc \theta}{d\theta} = -\csc \theta \cdot \cot \theta$

*Addition Formulas:*   $\sin(\theta + \varphi) = \sin \theta \cdot \cos \varphi + \cos \theta \cdot \sin \varphi$

$\qquad\qquad\qquad\quad$ $\cos(\theta + \varphi) = \cos \theta \cdot \cos \varphi - \sin \theta \cdot \sin \varphi$

*Relations:*

$$\tan \theta = \frac{\sin \theta}{\cos \theta}, \qquad \cot \theta = \frac{\cos \theta}{\sin \theta}, \qquad \sec \theta = \frac{1}{\cos \theta}, \qquad \csc \theta = \frac{1}{\sin \theta}$$

## PROBLEMS

### A

**1.** Prove the formula for the derivative of cos θ.

**2.** Take the derivative.

  **\*a)** $\sin \theta + \cos \theta$    **b)** $3 \tan \theta$

  **\*c)** $\dfrac{\cot \theta}{5}$    **d)** $3 \csc \theta + 5 \sec \theta$

  **\*e)** $\dfrac{10 \csc \theta}{\theta}$    **f)** $\dfrac{\theta^2 + 2}{\tan \theta}$

  **g)** $\theta(\tan \theta + \sec \theta)$

  **\*h)** $\dfrac{\csc \theta + \cot \theta}{\theta^2}$

**3.** Take first and second derivatives.

  **\*a)** $x \cdot \cos x$    **b)** $x^2 \cdot \sin x$

  **\*c)** $\cos x \cdot \sin x$    **d)** $\dfrac{\sin \theta}{\theta}$

  **\*e)** $\dfrac{5\theta}{\sin \theta}$    **f)** $50 \csc \theta$

  **g)** $3 \sec \theta$

**4.** For the given function $f(t)$, find $f''$ and $f^{(4)}$. Find a simple relation between $f$ and $f''$, and an even simpler one between $f$ and $f^{(4)}$.

  **\*a)** $2 \cos(t)$

  **b)** $3 \sin(t)$

  **c)** $2 \cos(t) + 3 \sin(t)$

**5.** The two hallways in Example 3 have been re-measured; it is found that in fact their widths are only 2.9 feet and 3.8 feet. What is the longest door that can be carried around the corner?

**\*6.** A trough with a trapezoidal cross section is to be made with three boards, each 1 foot wide (Fig. 9). The carrying capacity is proportional to the area of the cross section.

  **a)** Express the cross-sectional area $A$ as a function of $\theta$.

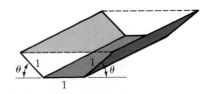

**FIGURE 9**

**b)** Determine $\theta$ so as to maximize $A$. (Hint: Express $dA/d\theta$ as a function of cos $\theta$ alone. If necessary, review the values of cos $\theta$ for special values of $\theta$, Appendix C.)

**7.** A hiker at $H$ in Figure 10 is headed for cabin $C$, 6 miles away. Both $H$ and $C$ are one mile from a straight road. The hiker can go 2 mi/hr in the woods and 4 mi/hr on the road.

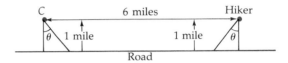

**FIGURE 10**
Hiker's dilemma.

  **a)** Determine the sine of the angle $\theta$ in the figure that gives the quickest route from $H$ to the road and then to $C$. Is this route quicker than the straight line route from $H$ to $C$?

  **b)** Several years later, the underbrush has grown up so much that the hiker goes only 1 mi/hr in the woods. Redo part a for this situation.

**8.** A box of width $w$ is to be carried around a right angle corner between two halls of width $h$.

  **a)** In Figure 11, express $L$ as a function of $\theta$.

  **b)** Show that the maximum possible length of such a box is computed from Figure 11 with $\theta = 45°$. [Hint: Rewrite the equation $dL/d\theta = 0$ as (polynomial in $\sin \theta$ and $\cos \theta) = 0$; $\sin \theta - \cos \theta$ factors out of this polynomial.]

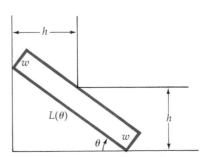

**FIGURE 11**

**B**

9. Use the definition of derivative to compute the derivative.
   *a)  $\sin 2\theta$      b)  $\cos 5\theta$
   c)  $\sin(a\theta)$ ($a$ is constant)      d)  $\cos(a\theta)$

10. Use the previous problem to compute the derivative.
   *a)  $\tan(a\theta)$    b)  $\cot(a\theta)$
   c)  $\sec(a\theta)$    d)  $\csc(a\theta)$

11. You might wonder, how can $\sin x$ and $\cos x$ be computed? This problem and the following two show how they can be accurately approximated. Prove the following for $x > 0$.
   a)  $1 - \cos x \geq 0$
   b)  $x - \sin x > 0$ (Hint: $x - \sin x = 0$ when $x = 0$; show that $x - \sin x$ is an increasing function for $x \geq 0$.)
   c)  $\dfrac{x^2}{2} - 1 + \cos x > 0$
   d)  $\dfrac{x^3}{3!} - x + \sin x > 0$   $(3! = 3 \cdot 2 \cdot 1 = 6)$
   e)  $\dfrac{x^4}{4!} - \dfrac{x^2}{2} + 1 - \cos x > 0$

   $$(4! = 4 \cdot 3 \cdot 2 \cdot 1 = 24)$$

   f)  $\dfrac{x^5}{5!} - \dfrac{x^3}{3!} + x - \sin x > 0$

   $$(5! = 5 \cdot 4 \cdot 3 \cdot 2 \cdot 1 = 120)$$

12. From problem 11, $x - \dfrac{x^3}{3!} < \sin x < x - \dfrac{x^3}{3!} + \dfrac{x^5}{5!}$ for $x > 0$. Approximate:
   a)  $\sin 1$.    *b)  $\sin(\frac{1}{2})$.    c)  $\sin(0.1)$.
   [You are computing the sine of $x$ radians, not $x$ degrees: $\pi$ radians $= 180°$, 1 radian $= \left(\dfrac{180}{\pi}\right)° \approx (57.3)°$, $\frac{1}{2}$ radian $\approx (28.65)°$.]

13. From problem 10, $1 - \dfrac{x^2}{2} < \cos x < 1 - \dfrac{x^2}{2} + \dfrac{x^4}{4!}$, for $x > 0$.
   a)  Is this true for $x < 0$?
   b)  Sketch graphs of $1 - \dfrac{x^2}{2}$ and $1 - \dfrac{x^2}{2} + \dfrac{x^4}{4!}$, for $|x| \leq 3$.
   c)  Approximate $\cos(\frac{1}{2})$.

## 3.7

## COMPOSITE FUNCTIONS. THE CHAIN RULE

The last (but not least!) of the general rules for derivatives is the *Chain Rule*, giving the derivative of a so-called "composite function."

### Composition

Two functions $f$ and $g$ can be combined by using the output of $g$ as the input for $f$:

$$x \xrightarrow{g} g(x) \xrightarrow{f} f(g(x)). \tag{1}$$

The result is called the **composition** of $f$ and $g$, denoted $f \circ g$:

$$f \circ g(x) = f(g(x)).$$

The composite $f \circ g$ is computed in two steps—*first $g$, then $f$.*

---

**EXAMPLE 1**   $y = \sin(2x + 1)$ is a composite function, computed in two steps:

$$x \xrightarrow{g} 2x + 1 \xrightarrow{f} \sin(2x + 1) = y.$$

We labeled the function at the first step "$g$" and at the second step "$f$," to match the scheme in (1). Given $x$, you first compute

$$g(x) = 2x + 1,$$

then take the sine of this output; so the second function $f$ is the sine. We commonly use the letter $u$ for the output of the first function; in this example, $u = 2x + 1$. This output is also the input of the second function $f$; in this example the second function is the sine, so

$$f(u) = \sin(u).$$

To check these identifications, form the composition

$$f(g(x)) = f(2x + 1) \qquad \text{(since } g(x) = 2x + 1\text{)}$$
$$= \sin(2x + 1) \qquad \text{(since } f(u) = \sin u\text{)},$$

which is indeed the function originally given.

---

**EXAMPLE 2**   $y = (x^2 + 1)^5$ is a composite:

$$x \underset{g}{\to} x^2 + 1 \underset{f}{\to} (x^2 + 1)^5.$$

The first step is $g(x) = x^2 + 1$; the second step is raising $x^2 + 1$ to the fifth power, so $f(u) = u^5$. As a check we find with these choices of $g$ and $f$ that

$$f(g(x)) = f(x^2 + 1) = (x^2 + 1)^5.$$

---

**EXAMPLE 3**   The price of beef, $B$, is a function of the price of corn, $C$:

$$B = f(C).$$

The price of corn is itself a function of the price of fertilizer, $F$:

$$C = g(F).$$

And the price of fertilizer is a function of the price of oil, $O$:

$$F = h(O).$$

Through this chain of relations, the price of beef is a function of the price of oil:

$$B = f(C) = f(g(F)) = f(g(h(O))).$$

Any increase in the price of oil will increase, in turn, the prices of fertilizer, corn, and beef.

---

**EXAMPLE 4**   An object moves along the graph of a function $f$, so its $y$-coordinate is a function of its $x$-coordinate:

$$y = f(x).$$

Because the object is moving, the $x$-coordinate is a function of time $t$,

$$x = g(t).$$

The $y$-coordinate is also a function of $t$, and

$$y = f(x) = f(g(t)).$$

## Continuity

If the two functions $f$ and $g$ in a composition are continuous, then so is $f \circ g$; for as $x \to a$, then $g(x) \to g(a)$, so $f(g(x)) \to f(g(a))$. (The rigorous form of this argument is in Appendix A.) For instance, the function $\sin(2x + 1)$ in Example 1 is continuous, since the linear function $g(x) = 2x + 1$ is continuous, and so is the sine.

In Section 3.3 we saw that sums, products, and quotients of continuous functions are continuous wherever defined. Since the same is true of compositions, it follows that *any* function of $x$, given by a single formula involving the operations of addition, multiplication, division, the trigonometric functions, roots, and compositions of these, is continuous wherever defined. For such functions, discontinuities can arise only where division by zero is required.

**Limits** can be taken "inside" a continuous function, in the sense that

$$\lim_{x \to a} f(g(x)) = f(\lim_{x \to a} g(x))$$

if $f$ is continuous at the point $\lim_{x \to a} g(x)$. The argument for this is the same as for the continuity of the composite function $f \circ g$. For example,

$$\lim_{x \to a} \cos(g(x)) = \cos(\lim_{x \to a} g(x)).$$

The same holds for limits as $x \to \infty$ and for one-sided limits. For example, since the square root is continuous,

$$\lim_{x \to 1^-} \sqrt{1 - x^2} = \sqrt{\lim_{x \to 1^-} (1 - x^2)} = \sqrt{0} = 0.$$

Now we come to our main topic.

**The Chain Rule** expresses the derivative of a composite function $f(g(x))$ in terms of the derivatives of $f$ and $g$. The rule can be discovered as follows. By definition, the derivative of the composite is the limit

$$\frac{d}{dx} f(g(x)) = \lim_{\hat{x} \to x} \frac{f(g(\hat{x})) - f(g(x))}{\hat{x} - x}.$$

Assuming that $g(\hat{x}) \neq g(x)$, rewrite this as

$$\frac{d}{dx} f(g(x)) = \lim_{\hat{x} \to x} \frac{f(g(\hat{x})) - f(g(x))}{g(\hat{x}) - g(x)} \cdot \frac{g(\hat{x}) - g(x)}{\hat{x} - x}. \tag{2}$$

By definition

$$\lim_{\hat{x} \to x} \frac{g(\hat{x}) - g(x)}{\hat{x} - x} = g'(x)$$

and (as we explain more fully below)

$$\lim_{\hat{x} \to x} \frac{f(g(\hat{x})) - f(g(x))}{g(\hat{x}) - g(x)} = f'(g(x)). \tag{3}$$

Use these two limits in (2), and you find that *the derivative of the composite* $f(g(x))$ *is the product of the derivatives at each step:*

$$\frac{d}{dx} f(g(x)) = f'(g(x)) \cdot g'(x). \tag{4}$$

Now we go back to explain the limit (3). We introduce a variable $u$ with

$$u = g(x), \qquad \hat{u} = g(\hat{x}).$$

We assume that $g'(x)$ exists; then $g$ is continuous at $x$, so $g(\hat{x}) \to g(x)$ as $\hat{x} \to x$; that is

$$\hat{u} \to u \qquad \text{as } \hat{x} \to x.$$

We assume also that $f'(u)$ exists; then

$$\lim_{\hat{x} \to x} \frac{f(g(\hat{x})) - f(g(x))}{g(\hat{x}) - g(x)} = \lim_{\hat{u} \to u} \frac{f(\hat{u}) - f(u)}{\hat{u} - u} = f'(u)$$

$$= f'(g(x))$$

as claimed in (3).

Before writing (3), we assumed that $g(\hat{x}) - g(x) \neq 0$. However, the Chain Rule can be proved without this awkward assumption (see problem 26).

---

**EXAMPLE 5**   Compute $\dfrac{d \sin(2x + 1)}{dx}$.

*SOLUTION*   Apply (4) with

$$g(x) = 2x + 1, \qquad g'(x) = 2 \tag{5}$$

and

$$f(u) = \sin u, \qquad f'(u) = \cos u. \tag{6}$$

The result is

$$\frac{d\,\sin(2x\,+\,1)}{dx} = f'(g(x)) \cdot g'(x)$$

$$= \cos(g(x)) \cdot g'(x) \qquad \text{[by (6)]}$$

$$= \cos(2x\,+\,1) \cdot 2 \qquad \text{[by (5)]}.$$

## Peeling the Onion

Think of the composite $f(g(x))$ as an onion, with $f(\cdot\,\cdot\,\cdot)$ the outer layer and $g(x)$ the inner. To take the derivative, differentiate the outer layer and multiply by the derivative of the inner layer; this gives $f'(g(x)) \cdot g'(x)$ as in (4).

When the outer layer is a power, $f(u) = u^n$, then $f'(u) = nu^{n-1}$, so $f'(g(x)) = ng(x)^{n-1}$, and

$$\frac{d}{dx}\,g(x)^n = n \cdot g(x)^{n-1} \cdot g'(x)$$

$$= n \cdot g(x)^{n-1}\,\frac{dg}{dx}.$$

## EXAMPLE 6

$$\frac{d(x^2\,+\,1)^{-2}}{dx} = (-2)(x^2\,+\,1)^{-2-1} \cdot \frac{d(x^2\,+\,1)}{dx} = -2(x^2\,+\,1)^{-3}(2x)$$

$$= \frac{-4x}{(x^2\,+\,1)^3},$$

following the pattern

$$\frac{d(g(x))^n}{dx} = n \cdot g(x)^{n-1} \cdot \frac{dg(x)}{dx}.$$

You can check this derivative by applying the quotient rule to $\dfrac{d}{dx}\dfrac{1}{(x^2\,+\,1)^2}$; try it. Which method is easier?

**Leibniz notation** for the Chain Rule is very suggestive. Let $u = g(x)$ and $y = f(u) = f(g(x))$. Then

$$\frac{dy}{dx} = \frac{df(g(x))}{dx}, \qquad \frac{dy}{du} = f'(u) = f'(g(x)), \quad \text{and} \quad \frac{du}{dx} = g'(x).$$

With this notation, the Chain Rule (4) is

$$\frac{dy}{dx} = \frac{dy}{du} \cdot \frac{du}{dx}. \tag{7}$$

If the symbols $\frac{dy}{dx}$, $\frac{dy}{du}$, and $\frac{du}{dx}$ were really fractions, this equation would be true by elementary algebra—just cancel $du$ on the right-hand side! In fact, $\frac{dy}{du}$ and $\frac{du}{dx}$ are not really fractions, so (7) cannot be justified this way; but the Chain Rule shows that the symbols behave as if they were fractions, and this is a useful way to remember the rule.

It is a good idea to remember the basic derivative formulas with the Chain Rule "built in":

$$\frac{du^n}{dx} = n \cdot u^{n-1} \cdot \frac{du}{dx},$$

$$\frac{d \sin u}{dx} = \cos u \cdot \frac{du}{dx}, \qquad \frac{d \cos u}{dx} = -\sin u \cdot \frac{du}{dx},$$

and so on. (A complete list of these basic formulas to date is in the Summary.)

Some "onions" have several layers, each requiring the Chain Rule.

---

**EXAMPLE 7**

$$\frac{d \cos[(x^2 + 1)^3]}{dx} = -\sin[(x^2 + 1)^3] \frac{d(x^2 + 1)^3}{dx} \qquad \left[ \frac{d \cos u}{dx} = -\sin u \frac{du}{dx} \right]$$

$$= -\sin[(x^2 + 1)^3] \cdot 3(x^2 + 1)^2 \cdot \frac{d(x^2 + 1)}{dx}$$

$$\left[ \frac{du^n}{dx} = n \cdot u^{n-1} \cdot \frac{du}{dx} \right]$$

$$= -\sin[(x^2 + 1)^3] \cdot 3(x^2 + 1)^2 \cdot 2x$$

$$= -6x(x^2 + 1)^2 \sin[(x^2 + 1)^3].$$

---

**EXAMPLE 8**

$$\frac{d \cos^5 2t}{dt} = ?$$

***SOLUTION***  Recall that $\cos^5 \theta$ means $(\cos \theta)^5$. Compute

$$\frac{d[\cos 2t]^5}{dt} = 5(\cos 2t)^4 \cdot \frac{d \cos 2t}{dt}$$

$$= 5(\cos 2t)^4 \left( -\sin 2t \frac{d2t}{dt} \right)$$

$$= 5(\cos 2t)^4 (-\sin 2t)(2)$$

$$= -10(\sin 2t)(\cos 2t)^4.$$

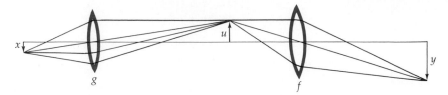

**FIGURE 1**
$u = m_g x,\; y = m_f u = m_f m_g x.$

The intuitive idea behind the Chain Rule is illustrated by a compound lens system as shown in Figure 1. The first lens $g$ has a magnification factor which we denote by $m_g$; a source of length $x$ yields an image of length

$$u = m_g \cdot x. \tag{8}$$

The second lens has a magnification factor $m_f$; a source of length $u$ yields an image of length

$$y = m_f \cdot u. \tag{9}$$

For the two lenses together, the source of length $x$ yields an image of length $y$, and (8) and (9) give

$$y = m_f \cdot u = m_f \cdot m_g \cdot x; \tag{10}$$

so the magnification factor for the compound lens is the product of the factors for each component.

Phrased in terms of derivatives, $\dfrac{du}{dx} = m_g$ (since the factor $m_g$ in (8) is a constant) while from (9) $\dfrac{dy}{du} = m_f$, and from (10)

$$\frac{dy}{dx} = m_f \cdot m_g = \frac{dy}{du} \cdot \frac{du}{dx}$$

precisely as in the Chain Rule.

# SUMMARY

*The Composition* of two functions $f$ and $g$ is denoted $f \circ g$ and defined by

$$(f \circ g)(x) = f(g(x)).$$

If $g$ is continuous at $a$, and $f$ is continuous at $g(a)$, then $f \circ g$ is continuous at $a$.

If $f$ is continuous at $\lim\limits_{x \to a} g(x)$, then

$$\lim_{x \to a} f(g(x)) = f\left( \lim_{x \to a} g(x) \right).$$

**The Chain Rule:** $\dfrac{df(g(x))}{dx} = f'(g(x))g'(x),$

or

$$\frac{dy}{dx} = \frac{dy}{du}\frac{du}{dx}.$$

**Special Cases:** $\dfrac{du^n}{dx} = n \cdot u^{n-1} \cdot \dfrac{du}{dx}$

$$\frac{d \sin u}{dx} = \cos u \cdot \frac{du}{dx} \qquad \frac{d \cos u}{dx} = -\sin u \frac{du}{dx}$$

$$\frac{d \tan u}{dx} = \sec^2 u \cdot \frac{du}{dx} \qquad \frac{d \cot u}{dx} = -\csc^2 u \cdot \frac{du}{dx}$$

$$\frac{d \sec u}{dx} = \sec u \cdot \tan u \cdot \frac{du}{dx} \qquad \frac{d \csc u}{dx} = -\csc u \cdot \cot u \cdot \frac{du}{dx}.$$

## PROBLEMS

**A**

1. Let $f(u) = 2u + 3$ and $g(x) = 5x - 2$. Compute $f \circ g(x)$. Show that it is linear with slope 10. Relate this to the slopes of $f$ and $g$, and compare the result with the Chain Rule.

2. Let $f(x) = 1/x$; $g(x) = x^2 + 1$; $h(x) = \sqrt{x}$, $x \geq 0$. Represent the following as compositions, using these three functions.

   *a) $\dfrac{1}{x^2 + 1}$    *b) $\left(\dfrac{1}{x}\right)^2 + 1$

   *c) $\sqrt{\dfrac{1}{x}}$    *d) $\dfrac{1}{\sqrt{x}}$

   e) $\sqrt{x^2 + 1}$    f) $x + 1, x \geq 0$

3. Represent the following as compositions, $f \circ g$, of appropriate functions.

   *a) $\dfrac{1}{2x + 1}$       b) $2\sqrt{x^2 - 1}$

   *c) $(x^2 + 2x + 2)^{1/4}$    d) $\dfrac{1}{1 + \sqrt{x}}$

   *e) $\sqrt{\sin \theta}$    f) $\sec^2 \theta + 1$

4. Find $\dfrac{dy}{dx}$ by the Chain Rule, and express your answer in terms of $x$ alone. Then go back, express $y$ directly as a function of $x$, take the derivative, and compare it to the Chain Rule answer.

   *a) $y = u^2, \quad u = 2x + 1$

   b) $y = 2u + 1, \quad u = x^2$

   c) $y = \dfrac{1}{u}, \quad u = x^3 + 1$

5. Find $dy/dx$; express your answer in terms of $x$ alone.

   *a) $y = \cos u, \quad u = 1 + x^2$

   b) $y = 1 + u^2, \quad u = \tan x$

   *c) $y = \dfrac{1}{1 + v}, \quad v = \dfrac{1}{1 + x}$

   d) $y = \sec u, \quad u = v^3 + 1, \quad v = \csc(x - 1)$

6. Compute the derivative.
   *a)  $(2x + 1)^3$
   *b)  $(1 - x)^2$
   *c)  $(x^2 - 1)^5$
   d)  $\tan(2x^2 - 1)$
   *e)  $(x^3 - x + 1)^4$
   f)  $\sec(r^2 - x^2)$   ($r$ is constant.)
   g)  $x \sin(x^2 + 1)$   (Product Rule!)

7. Take first and second derivatives.
   *a)  $y = x \sin x$
   b)  $y = x \cos (2x)$
   *c)  $x = \cos^2 t$
   d)  $x = \cos^2 3t$
   *e)  $y = \cos(\sin x)$
   f)  $y = \cos(2x^2 + 1)$
   *g)  $y = \tan(3x)$
   h)  $x = 5 \sec 2\theta$
   i)  $y = \tan^2(x^2 + 1)$
   j)  $z = \csc 3\theta + 2 \cot 3\theta$

   *k)  $y = (\sin x)/x$
   l)  $\dfrac{x}{\sin x}$

   m)  $y = x^2 \cos 2x$

8. In a quotient with a constant numerator you can use the $du^n/dx$ formula instead of the Quotient Rule: $\dfrac{d(c/u^m)}{dx} = \dfrac{dcu^{-m}}{dx}$. Take the following derivatives both ways.

   *a)  $\dfrac{2}{x^2 + 1}$
   *b)  $\dfrac{5}{(x^2 + 1)^3}$
   c)  $\dfrac{1}{\sin^2\theta}$

9. Let $f$ and $g$ be linear; $g(x) = mx + b$, $f(u) = nu + c$. Show that $f\circ g(x)$ is linear, and determine its slope. Relate this to the Chain Rule.

10. Suppose that $f$ is an amplifier which, taking in a signal of strength $u$, puts out a signal of strength $au$; the constant $a$ is the *amplification factor*. Suppose that $g$ is another amplifier, with amplification factor $b$: taking in a signal of strength $x$, $g$ puts out a signal of strength $bx$. Schematically,

$$x \to \boxed{g} \to bx \qquad u \to \boxed{f} \to au.$$

Now hook the two amplifiers together, feeding the output of $g$ into $f$:

$$x \to \boxed{g} \to \boxed{f} \to ?$$

What is the amplification factor for this combination?

11. The position of an oscillating particle is given by $x = A \sin(\omega t + \phi)$, where $A$, $\omega$, and $\phi$ are constants.
   a)  Compute the velocity and acceleration of the particle.
   b)  If $A = 2$ cm, $\phi = \pi/2$, $\omega = \pi/3$ rads/sec, find the velocity of the particle when $t = 4$.

12. Take the first and second derivatives. In each case, find a relation between $y$ and $d^2y/dt^2$.
   *a)  $y = \sin(3t + 4)$
   b)  $y = 4 \cos(t - \pi)$
   c)  $y = -\sin(\pi - 3t)$
   d)  $y = 2 \cos 5t - 3 \sin 5t$
   e)  $y = A \cos(\omega t + \phi)$ ($A$, $\omega$, $\phi$ are constants)
   f)  $y = A \cos \omega t + B \sin \omega t$ ($\omega$, $A$, $B$ are constants)

13. Graph $y = (x^2 - 1)^4$, using first and second derivatives. [$(x^2 - 1)^2$ is a factor of $d^2y/dx^2$.]

14. Graph $y = x + \sin x$, using first and second derivatives. Show where the graph crosses the line $y = x$.

**B**

*15. Let $f(t) = 8 \sin(3t + 1)$. Find $f^{(4)}(t)$; $f^{(6)}(t)$; $f^{(48)}(t)$; $f^{(81)}(t)$; $f^{(402)}(t)$; $f^{(4023)}(t)$.

16. Let $s(t) = \sin(t$ degrees$)$ and $c(t) = \cos(t$ degrees$)$; for example, $s(30) = 1/2$ and $c(90) = 0$. Show that $s'(t) = (\pi/180)c(t)$ and $c'(t) = -(\pi/180)s(t)$. (The factor $\pi/180$ here explains why radian measure, not degrees, is appropriate for the calculus of trigonometric functions.)

17. Differentiate the addition formula $\sin(\theta + \varphi) = \sin \theta \cos \varphi + \cos \theta \sin \varphi$ with respect to $\theta$, keeping $\varphi$ constant. What results? What if you differentiate again?

18. Find $f'(2)$, given that $g(1) = 2$, $g'(1) = 4$, and $(f\circ g)'(1) = 6$.

19. An object moves along the curve $y = f(x)$. Its $x$-coordinate is a known function of $t$. Express $\dfrac{dy}{dt}$ in terms of $\dfrac{dy}{dx}$ and $\dfrac{dx}{dt}$.

20. For $f(x) = \dfrac{\sin x}{x}$ compute (if it exists):
   a)  $\lim\limits_{x \to +\infty} f(x)$
   b)  $\lim\limits_{x \to +\infty} f'(x)$

21. Let $f(x) = x \cdot \sin \dfrac{1}{x}$. Compute (if it exists):
   a)  $\lim\limits_{x \to 0} f(x)$
   b)  $\lim\limits_{x \to 0} f'(x)$

22. Prove that $\dfrac{df(2x)}{dx} = 2f'(2x)$
   a)  By the Chain Rule.
   b)  By direct use of the definition of derivative as a limit.

**23.** Prove that $\dfrac{df(x+1)}{dx} = f'(x+1)$ two ways, as in the previous problem. What is the relation between the graphs of $y = f(x)$ and $y = f(x+1)$?

**C**

**24.** A function $f$ is related to its inverse $g$ by $f(g(x)) = x$. Deduce that $g'(x) = 1/f'(g(x))$, if $g'(x)$ and $f'(g(x))$ exist.

**25.** Suppose that $f$ is a differentiable function such that $f(ax) = f(a) + f(x)$ for every $a > 0$ and $x > 0$. Prove that

a) $f'(x) = af'(ax)$.

b) $f'(x) = f'(1)/x$, for all $x > 0$.

**26.** (Proof of the Chain Rule) One of the problems in Section 3.4 gave this: If $f'(a)$ exists then

$$f(x) = f(a) + (x - a)Q_a(x) \qquad \text{(i)}$$

where

$$\lim_{x \to a} Q_a(x) = Q_a(a) \qquad \text{(ii)}$$

and

$$Q_a(a) = f'(a). \qquad \text{(iii)}$$

Conversely, if there is a function $Q_a$ with properties (i) and (ii), then $f'(a)$ exists and equals $Q_a(a)$. Use this to prove the Chain Rule. [Hint: In (i) replace $x$ by $g(x)$ and $a$ by $g(a)$.]

**27.** a) Let $u$ be a differentiable function of $x$. Write the Chain Rule for $\dfrac{d^2 u^n}{dx^2}$.

b) Write the Chain Rule for $(f \circ g)''$, that is, $\dfrac{d^2 f(g(x))}{dx^2}$.

## 3.8

# IMPLICIT DIFFERENTIATION. THE DERIVATIVE OF $x^{m/n}$

Implicit differentiation is an ingenious way to compute derivatives that might be hard or impossible to obtain directly.

As a simple example, we differentiate the square root function

$$f(x) = \sqrt{x}, \qquad x \geq 0.$$

Square both sides:

$$f(x)^2 = x.$$

Now *differentiate each side with respect to $x$.* Since $\dfrac{df(x)^2}{dx} = 2f(x)\dfrac{df}{dx}$, you get

$$2f(x)\frac{df}{dx} = \frac{dx}{dx} = 1.$$

Hence

$$\frac{df}{dx} = \frac{1}{2f(x)}.$$

Since $f(x) = \sqrt{x}$, you find

$$\frac{d\sqrt{x}}{dx} = \frac{1}{2\sqrt{x}}. \qquad \text{(1)}$$

This type of calculation is usually done with a briefer notation, replacing $f(x)$ by $y$. The steps are then

$$y = \sqrt{x}$$

$$y^2 = x \qquad \text{(square)}$$

$$\frac{dy^2}{dx} = \frac{dx}{dx} \qquad \text{(derivative of each side with respect to } x)$$

$$2y \frac{dy}{dx} = 1 \qquad \text{(use the Chain Rule)}$$

$$\frac{dy}{dx} = \frac{1}{2y} \qquad \text{(solve for } dy/dx)$$

leading again to equation (1), since $y = \sqrt{x}$. The main point is to differentiate each side of $y^2 = x$ *with respect to* $x$; the necessary derivative of $y^2$ is not $\dfrac{dy^2}{dy} = 2y$, but rather

$$\frac{dy^2}{dx} = 2y \frac{dy}{dx}.$$

The same method gives the derivative of $y = x^r$ for any rational power $r = m/n$, with $m$ and $n$ both integers. In the equation

$$y = x^{m/n} \tag{2}$$

raise each side to the power $n$:

$$y^n = (x^{m/n})^n = x^{(m/n) \cdot n} = x^m.$$

Now differentiate each side *with respect to* $x$:

$$ny^{n-1} \frac{dy}{dx} = mx^{m-1}.$$

Solve for the derivative

$$\frac{dy}{dx} = \frac{m}{n} \frac{x^{m-1}}{y^{n-1}} = \frac{m}{n} x^{m-1} y^{1-n}$$

and use equation (2) to substitute $x^{m/n}$ for $y$:

$$\frac{dx^{m/n}}{dx} = \frac{m}{n} \cdot x^{m-1} \cdot (x^{m/n})^{1-n} = \frac{m}{n} \cdot x^{m-1} \cdot x^{(m/n)(1-n)}$$

$$= \frac{m}{n} \cdot x^{m-1} \cdot x^{(m/n)-m} = \frac{m}{n} x^{(m/n)-1}.$$

This verifies the Power Rule for rational powers.

**EXAMPLE 1**   Sketch the graph of $y = x^{3/4}$.

**SOLUTION**

**Domain**   The power $x^{3/4}$ means $(x^{1/4})^3$; and $x^{1/4}$, the fourth root of $x$, is defined only for $x \geq 0$. So the domain is $x \geq 0$; in the graph, you should check the limiting values of $y$ and $\dfrac{dy}{dx}$ as $x \to 0^+$.

**Derivatives**

$$y' = \frac{3}{4}x^{(3/4)-1} = \frac{3}{4}x^{-1/4} = \frac{3}{4x^{1/4}} > 0 \qquad \text{(graph rising)},$$

$$y'' = \frac{3}{4}\left(-\frac{1}{4}\right)x^{-5/4} = \frac{-3}{16x^{5/4}} < 0 \qquad \text{(concave down)}.$$

**Limits**

$$\lim_{x \to 0^+} x^{3/4} = \left[\lim_{x \to 0^+} x^3\right]^{1/4} = 0^{1/4} = 0,$$

$$\lim_{x \to 0^+} y' = \lim_{x \to 0^+} \frac{3}{4x^{1/4}} = +\infty$$

since the denominator $4x^{1/4} \to 0$ as $x \to 0$. Since $y' \to \infty$ as $x \to 0^+$, the graph is infinitely steep at the origin, and is tangent to the $y$-axis, as in Figure 1.

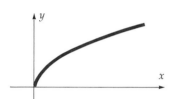

**FIGURE 1**
$y = x^{3/4}$. Graph is tangent to $y$-axis at $(0, 0)$.

## Implicit Differentiation

The rational power function

$$y = x^{m/n} \tag{3}$$

is equivalent to the equation

$$y^n = x^m. \tag{4}$$

In (3), $y$ is expressed *explicitly* as a function $f(x) = x^{m/n}$. Equation (4) implies the same relation between $x$ and $y$, and thus defines $y$ *implicitly* as a function of $x$. We found $dy/dx$ by differentiating the relation (4). This is called **implicit differentiation**; in principle, it can be applied to any equation defining $y$ implicitly as a function of $x$.

A simple example is the equation of the unit circle in Figure 2:

$$x^2 + y^2 = 1. \tag{5}$$

Think of this as defining $y$ as a function of $x$. Since the left-hand side is identically equal to the right-hand side, the derivatives must be the same:

$$\frac{d}{dx}(x^2 + y^2) = \frac{d}{dx}(1) = 0,$$

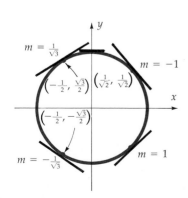

**FIGURE 2**
Tangents to $x^2 + y^2 = 1$.

which gives by the Chain Rule

$$2x + 2y\frac{dy}{dx} = 0.$$

Solve for $dy/dx$:

$$dy/dx = -x/y. \tag{6}$$

This formula contains both $x$ and $y$; if you replace them by the coordinates of a point on the circle (5), then you get the slope of the graph at that point as in Figure 2:

| point $(x, y)$ | $(0, 1)$ | $\left(\dfrac{1}{\sqrt{2}}, \dfrac{1}{\sqrt{2}}\right)$ | $\left(\dfrac{-1}{2}, \dfrac{\sqrt{3}}{2}\right)$ | $\left(\dfrac{-1}{2}, \dfrac{-\sqrt{3}}{2}\right)$ | $\left(\dfrac{1}{\sqrt{2}}, \dfrac{-1}{\sqrt{2}}\right)$ |
|---|---|---|---|---|---|
| slope $\dfrac{dy}{dx} = -\dfrac{x}{y}$ | $0$ | $-1$ | $1/\sqrt{3}$ | $-1/\sqrt{3}$ | $1$ |

Implicit differentiation requires only the familiar rules of differentiation, carefully applied. Thus if $y$ is a function of $x$, then

$$\frac{d}{dx}(x^3y^5) = 3x^2y^5 + x^3 \cdot \frac{dy^5}{dx} = 3x^2y^5 + x^3 \cdot 5y^4 \cdot \frac{dy}{dx},$$

$$\frac{d}{dx}\sin(xy) = \cos(xy)\frac{d(xy)}{dx} = \cos(xy)\left[1 \cdot y + x\frac{dy}{dx}\right]$$

$$\frac{d}{dx}\left(\frac{x}{y}\right) = \frac{1 \cdot y - x \cdot dy/dx}{y^2}.$$

The next example was apparently devised by Descartes as a challenge to Fermat, to see whether he could construct tangents to such a curve. (He could.)

---

**EXAMPLE 2**  The graph of

$$x^3 + y^3 = 3xy \tag{7}$$

is called a *folium* (leaf) *of Descartes*. The point $(\frac{3}{2}, \frac{3}{2})$ lies on this curve, since (7) is satisfied when $x = 3/2$, $y = 3/2$. Find the slope of the curve at that point.

*SOLUTION*  In equation (7), think of $y$ as a function of $x$. Then the left-hand side of (7) is a function of $x$, and its derivative is

$$3x^2 + 3y^2\frac{dy}{dx}.$$

The right-hand side is also a function of $x$, and its derivative is (by the Product Rule)

$$\frac{d3xy}{dx} = \left(\frac{d3x}{dx}\right)y + 3x\frac{dy}{dx} = 3y + 3x\frac{dy}{dx}.$$

Equation (7) says that the functions $x^3 + y^3$ and $3xy$ are equal; so their derivatives are equal,

$$3x^2 + 3y^2 \frac{dy}{dx} = 3y + 3x \frac{dy}{dx}.$$

Solve for $dy/dx$—put all $dy/dx$ terms on the left side and the remaining terms on the right:

$$3y^2 \frac{dy}{dx} - 3x \frac{dy}{dx} = 3y - 3x^2$$

$$(y^2 - x)\frac{dy}{dx} = y - x^2 \tag{8}$$

$$\frac{dy}{dx} = \frac{y - x^2}{y^2 - x}. \tag{9}$$

This gives the slope at any point $(x, y)$ that is on the curve (provided, of course, that the denominator in (9) is not zero). At $(\frac{3}{2}, \frac{3}{2})$, it gives

$$\frac{dy}{dx} = \frac{(3/2) - (9/4)}{(9/4) - (3/2)} = -1.$$

We indicate this in Figure 3 by drawing a short segment of slope $-1$ through the point $(\frac{3}{2}, \frac{3}{2})$.

Formula (9) for the slope can be used to find the points where the curve (7) is horizontal. Let $P = (x, y)$ be such a point. Then $P$ is on the curve, so $(x, y)$ satisfies (7); and $dy/dx = 0$ at $P$, so by (9), $y = x^2$. Thus $(x, y)$ satisfies the simultaneous equations

$$x^3 + y^3 = 3xy, \quad \text{and} \quad y = x^2.$$

To solve these, use the second equation $y = x^2$ in the first:

$$x^3 + (x^2)^3 = 3x(x^2)$$

$$x^3 + x^6 - 3x^3 = 0$$

$$x^3(x^3 - 2) = 0.$$

There are two solutions for $x$:

$$x_1 = 0, \qquad x_2 = 2^{1/3}.$$

To get the corresponding $y_1$ and $y_2$, use $y = x^2$ once more:

$$y_1 = x_1^2 = 0, \qquad y_2 = (x_2)^2 = 2^{2/3}.$$

Finally, check the results. Both $(0, 0)$ and $(2^{1/3}, 2^{2/3})$ lie on the given curve $x^3 + y^3 = 3xy$; the second point gives 0 slope, from (9):

$$\frac{dy}{dx} = \frac{2^{2/3} - (2^{1/3})^2}{2^{4/3} - 2^{1/3}} = 0.$$

But the point $(0, 0)$ gives $dy/dx = 0/0$ in (9), and this does not determine the slope. *Implicit differentiation fails at any point that gives a zero denominator in the expression for $dy/dx$.*

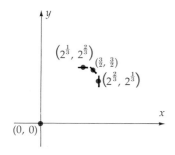

**FIGURE 3**
Four points on the folium $x^3 + y^3 = 3xy$.

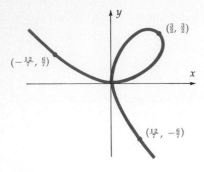

**FIGURE 4**
Folium of Descartes, $x^3 + y^3 = 3xy$.

Vertical tangents are found in a similar way: solve $dx/dy = 0$. This gives the point $(2^{2/3}, 2^{1/3})$.

Figure 3 shows the four points found and the slope at three of them; this gives a hint as to the nature of the complete curve shown in Figure 4. From that figure, you can see why there was some confusion about the value of $dy/dx$ at the origin; the curve crosses itself, and there is no well-defined slope.

**Higher implicit derivatives** are found as in the next example.

**EXAMPLE 3**   Find $\dfrac{d^2y}{dx^2}$ for the folium of Descartes in Example 2.

*SOLUTION*   It would be natural to differentiate the derivative $dy/dx$ in formula (9), but it is easier to differentiate formula (8), using the Product Rule and Chain Rule as necessary:

$$\frac{d}{dx}\left[(y^2 - x)\frac{dy}{dx}\right] = \frac{d}{dx}(y - x^2),$$

so by the Product Rule

$$\frac{d(y^2 - x)}{dx} \cdot \frac{dy}{dx} + (y^2 - x)\frac{d}{dx}\left(\frac{dy}{dx}\right) = \frac{dy}{dx} - 2x,$$

$$\left(2y\frac{dy}{dx} - 1\right)\cdot\frac{dy}{dx} + (y^2 - x)\frac{d^2y}{dx^2} = \frac{dy}{dx} - 2x.$$

Solve for $\dfrac{d^2y}{dx^2}$:

$$\frac{d^2y}{dx^2} = 2(y^2 - x)^{-1}\left[-y\left(\frac{dy}{dx}\right)^2 + \frac{dy}{dx} - x\right].$$

Thus at the point $(\tfrac{3}{2}, \tfrac{3}{2})$, where $\dfrac{dy}{dx} = -1$,

$$\frac{d^2y}{dx^2} = 2\left(\frac{9}{4} - \frac{3}{2}\right)^{-1}\left[-\frac{3}{2} - 1 - \frac{3}{2}\right] = -\frac{32}{3}.$$

This verifies that at $(\tfrac{3}{2}, \tfrac{3}{2})$ the curve is concave down, as shown in Figure 4.

Implicit differentiation raises an interesting question: If we cannot solve the given equation to express $y$ explicitly as a function of $x$, is it legitimate to treat $y$ *as if it were* a function of $x$? The answer: It *is* legitimate, except at points where the resulting formula for $dy/dx$ calls for division by zero. The circle $x^2 + y^2 = 1$ illustrates the situation (see Fig. 5). This equation *can* be solved for $y$:

$$x^2 + y^2 = 1 \Leftrightarrow y^2 = 1 - x^2 \Leftrightarrow y = \pm\sqrt{1 - x^2}.$$

On the upper semicircle in Figure 5, $y = \sqrt{1 - x^2}$, and on the lower semicircle, $y = -\sqrt{1 - x^2}$. Take any point $(x_0, y_0)$ on the upper semicircle, with

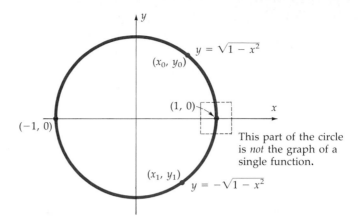

**FIGURE 5**

$y_0 > 0$; then all points on the circle near $(x_0, y_0)$ satisfy the equation $y = \sqrt{1 - x^2}$, giving $y$ as a differentiable function of $x$. Likewise, all the points on the circle near any given point $(x_1, y_1)$ with $y_1 < 0$ lie on the graph of $y = -\sqrt{1 - x^2}$. Near such points, it *is* legitimate to treat $y$ as a function of $x$. Not so, however, near the point $(1, 0)$ on the $x$-axis. The part of the circle near this point is not the graph of *any* function, differentiable or not; for any vertical line just to the left of $(1, 0)$ intersects the circle in two points, which would be impossible on the graph of a function. Similarly, near $(-1, 0)$ the circle is not the graph of a single function.

Now, recall the formula (6) obtained by implicit differentiation in this case:

$$\frac{dy}{dx} = -\frac{x}{y}.$$

This gives a value for $dy/dx$ at each point of the circle *except where $y = 0$*, since then $x/y$ is not defined. And the points on the circle where $y = 0$ are $(1, 0)$ and $(-1, 0)$, precisely those near which the circle does not coincide with the graph of a function.

This is the situation in general: Near any point on the given curve where the formula for $dy/dx$ does not call for division by zero, the curve is in fact given by a differentiable function $y = f(x)$, so the method of implicit differentiation is justified. For example, on the folium of Descartes in Figure 6 we found

$$\frac{dy}{dx} = \frac{y - x^2}{y^2 - x}. \tag{10}$$

Take any point $P_0$ on the curve where the denominator $y^2 - x \neq 0$ (Fig. 6). If we look only in some small circle about $P_0$, that part of the folium is the graph of a single function $y = f(x)$, and the slope is given by solving (10). Other parts of the curve may be given by different functions, but there too, (10) gives the slope, as long as the factor $y^2 - x \neq 0$.

The same Figure 6 shows another point $P_1 = (2^{2/3}, 2^{1/3})$ where the situation is different. At $P_1$, the tangent line is vertical, and near $P_1$ the curve lies to the left of this tangent. No matter how small a circle you draw around

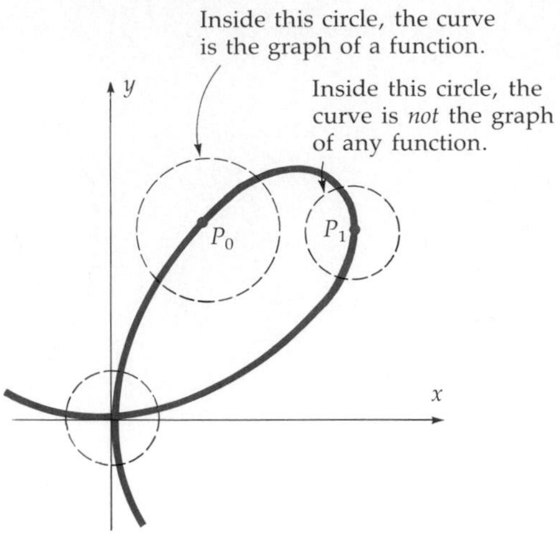

Inside this circle, the curve
is the graph of a function.

Inside this circle, the
curve is *not* the graph
of any function.

**FIGURE 6**

$P_1$, the part of the folium inside that circle is not the graph of any function; for each $x < 2^{2/3}$, there are two distinct values of $y$.

At the origin, the situation is even worse. Draw any circle about the origin, however small, and look only inside that circle; for any $x$ just to the right of 0 the folium gives *three* distinct values of $y$.

At points $P_1$ and $O$ there is *no* guarantee that the folium can be represented as the graph of a single function $y = f(x)$; for at those points, the factor $(y^2 - x)$ in the denominator of (10) is zero.

Our final example gives some hints for dealing with derivatives and limits where roots are involved.

---

**EXAMPLE 4** Sketch the graph of $y = x/\sqrt{x^2 + 1}$, showing stationary points, concavity, and asymptotes.

**SOLUTION** To apply the Power Rule, rewrite the function as $y = x(x^2 + 1)^{-1/2}$. Then

$$y' = (x^2 + 1)^{-1/2} + x\left(-\frac{1}{2}\right)(x^2 + 1)^{-3/2}(2x)$$
$$= (x^2 + 1)^{-1/2} - x^2(x^2 + 1)^{-3/2}.$$

To simplify this, clear the negative exponents; multiply both sides by $(x^2 + 1)^{3/2}$:

$$(x^2 + 1)^{3/2}y' = (x^2 + 1)^{3/2 - 1/2} - x^2(x^2 + 1)^{-3/2 + 3/2}$$
$$= (x^2 + 1) - x^2 = 1.$$

So

$$y' = (x^2 + 1)^{-3/2} > 0 \qquad \text{(graph always rising)}$$

and

$$y'' = -\tfrac{3}{2}(x^2 + 1)^{-5/2}(2x) = -3x(x^2 + 1)^{-5/2}.$$

For $x > 0$, $y'' < 0$ and the graph is concave down; for $x < 0$ it is concave up.

The function $x/\sqrt{x^2 + 1}$ is defined by a single formula valid for all $x$ (since $x^2 + 1 > 0$), so it is continuous everywhere, and its graph has no vertical asymptotes. As $x \to +\infty$,

$$\lim_{x \to +\infty} \frac{x}{\sqrt{x^2 + 1}} = \lim \frac{(1/x)x}{(1/x)\sqrt{x^2 + 1}}$$

$$= \lim \frac{1}{\sqrt{1 + 1/x^2}} = \frac{1}{\sqrt{\lim(1 + 1/x^2)}} = 1.$$

The crucial step here is

$$\frac{1}{x}\sqrt{x^2 + 1} = \frac{\sqrt{x^2 + 1}}{x} = \frac{\sqrt{x^2 + 1}}{\sqrt{x^2}}$$

$$= \sqrt{\frac{x^2 + 1}{x^2}} = \sqrt{1 + 1/x^2}.$$

This is valid when $x > 0$, for then $x = \sqrt{x^2}$. But when $x \to -\infty$ then $x < 0$ and $x = -\sqrt{x^2}$; in this case, computing as above show that $\displaystyle \lim_{x \to -\infty} \frac{x}{\sqrt{x^2 + 1}} = -1$. Figure 7 gives the graph.

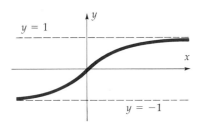

**FIGURE 7**

$y = \dfrac{x}{\sqrt{x^2 + 1}}.$

## PROBLEMS

### A

1.  Which are odd? Even? Neither?
    *a)  $x^{1/3}$    *b)  $x^{2/3}$
    *c)  $x^{3/2}$     d)  $x^{5/3}$

2.  Under what conditions on $m$ and $n$ is $x^{m/n}$ odd? Even? Neither?

3.  Graph, showing domain, intervals of increase and decrease, concavity, and limits of $y$ and $dy/dx$ as $x \to 0^+$, $x \to +\infty$, and (where applicable) $x \to 0^-$, $x \to -\infty$.
    a)  $x^{1/3}$     b)  $x^{2/3}$     c)  $x^{4/3}$
    d)  $x^{3/2}$     e)  $x^{3/5}$     f)  $x^{-1/5}$
    g)  $x^{-5/4}$

4.  Compute and simplify the first and second derivatives (as in Example 4).
    *a)  $x\sqrt{x^2 + 1}$    *b)  $\sqrt[3]{x^2 - 1}$
    c)  $\dfrac{x}{(x^2 + 1)^{1/4}}$

5.  Sketch the graph. Use $y'$ and $y''$, and show domain and asymptotes.

    *a)  $y = (x + 3)(x^2 + 1)^{-1/2}$
    b)  $y = x(x^2 - 1)^{-1/2}$
    *c)  $y = x(1 - x^2)^{-1/2}$

6.  For the given equation (i) differentiate each side with respect to $x$, and (ii) find $\dfrac{dy}{dx}$.

    *a)  $x^2 + 3xy^4 + y = 11$
    b)  $x^2 + x^2y + y^4 = 9$
    c)  $(x + y)(y - 2x) + 9 = 0$
    *d)  $(x^2 + 2y^2)^{13} = y$
    e)  $x^{2/3}y + y^{2/3}x = x^3$
    *f)  $y = \sin(xy)$
    g)  $xy = \tan(x^2 + y^2)$

*7.  Find the equation of the tangent line at the point $(2, 1)$ on each of the curves in problem 6 parts a–c. (Answer provided for part a.)

8.  In the folium of Descartes $x^3 + y^3 = 3xy$ (Fig. 4):
    *a)  Compute the slope at $(\frac{-12}{7}, \frac{6}{7})$. Write the equation of the tangent line there.

**b)** Compute the slope at $(\frac{6}{7}, \frac{-12}{7})$. Why is this slope the reciprocal of the slope in part a?

9. Find $dy/dx$.

   **a)** $x^3 + y^3 + 3xy = 0$ (Another folium of Descartes)

   **\*b)** $\dfrac{1}{x^2} - \dfrac{1}{y^2} = 1$ ["Bullet nose" of Schoute (1885)]

   **c)** $x^{2/3} + y^{2/3} = 1$ [Astroid of Roemer (1694)]

10. For the curve in the corresponding part of problem 9, write the equation of the tangent line at the given point.

   **a)** $\left(-\dfrac{3}{2}, -\dfrac{3}{2}\right)$    **\*b)** $\left(\dfrac{1}{\sqrt{3}}, \dfrac{1}{\sqrt{2}}\right)$

   **c)** $(1, 0)$

11. For the curve in the corresponding part of problem 9, compute $y''$ at the given point.

   **a)** $\left(-\dfrac{3}{2}, -\dfrac{3}{2}\right)$    **\*b)** $\left(\dfrac{1}{\sqrt{3}}, \dfrac{1}{\sqrt{2}}\right)$

   **c)** $(1, 0)$

12. In problem 9b solve for $y$ explicitly as a function of $x$, then differentiate. Evaluate $dy/dx$ for $x = 1/\sqrt{3}$ and compare it with the slope found in problem 9b.

**\*13.** On the folium of problem 9a, find all points where $dy/dx = 0$ and all points where there is a vertical tangent.

14. The graph of $(x^2 + y^2 - 2y)^2 = x^2 + y^2$ is a *limaçon of Pascal* (1650) as shown in Figure 8. Find the equation of the tangent line at $(\sqrt{3}, 1)$. Find all points where there is a horizontal tangent.

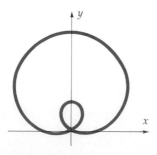

**FIGURE 8**
A limaçon of Pascal.

15. The graph of $(x^2 + y^2)^2 = x^2 - y^2$ is a *lemniscate of Bernoulli* (1664) (Fig. 9).

   **a)** Write the equation of the tangent line at $(\sqrt{3}/5, \sqrt{2}/5)$.

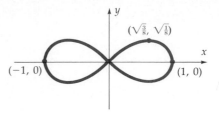

**FIGURE 9**
A lemniscate of Bernoulli.

   **b)** Find the points where the tangent line is horizontal and where it is vertical.

   **c)** There are three exceptional points $P$, such that in no circle around $P$ can the curve be represented as the graph of a function $y = f(x)$. Which points are they? What does your formula for $dy/dx$ produce at these points?

   **d)** Near which points is the curve not the graph of a function $x = g(y)$? Find a formula for $dx/dy$, and see what it produces at these points.

16. Why is it useless to find the slope of $x^4 + y^4 = -1$ by implicit differentiation?

17. The graph of $x^2 + xy + y^2 = 4$ is the ellipse shown in Figure 10. Find the slope of the ellipse at each point where it crosses the $y$-axis. Note that the tangents at these points are parallel.

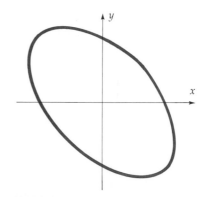

**FIGURE 10**
Ellipse $x^2 + xy + y^2 = 4$.

18. Find $dy/dx$ on the curve $x^2 + 2xy + y^2 = 1$. From your answer, what kind of curve is it?

**B**

19. Show that the line tangent to $\dfrac{x^2}{a^2} + \dfrac{y^2}{b^2} = 1$ at

the point $(x_0, y_0)$ is

$$\frac{xx_0}{a^2} + \frac{yy_0}{b^2} = 1.$$

20. Show that the line tangent to $\dfrac{x^2}{a^2} - \dfrac{y^2}{b^2} = 1$ at $(x_0, y_0)$ is

$$\frac{xx_0}{a^2} - \frac{yy_0}{b^2} = 1.$$

21. For the astroid $x^{2/3} + y^{2/3} = 1$, show that $y'' > 0$ in quadrants I and II and that $y'' < 0$ in quadrants III and IV.

22. The curve $x^2 + xy + y^2 = 3$ has two points on the line $x = 1$. Find the slope of the curve at each of these points.

23. a) Find where the curves $xy = 2$ and $x^2 - y^2 = 3$ intersect and show that they are perpendicular at those points.
    b) Consider any two curves $xy = c$ and $x^2 - y^2 = k$, where $c$ and $k$ are constants. Show

that at any point where these curves intersect, they are perpendicular.

24. The folium $x^3 + y^3 = 3xy$ intersects the line $y = 2x$ at the origin and at a second point $P$. Find $P$ and the slope of the folium at $P$.

25. For the "bullet nose" $x^{-2} - y^{-2} = 1$, show that $dy/dx > 0$ in the first and third quadrants, and $dy/dx < 0$ in the second and fourth quadrants.

C

26. To find points on the folium $x^3 + y^3 = 3xy$ of Example 2, look for the intersections of this curve with the line $y = mx$ for varying slopes $m$.

    a) Show that these intersections are $(0, 0)$ and
    $$\left( \frac{3m}{1 + m^3}, \frac{3m^2}{1 + m^3} \right).$$
    b) Plot the points for $m = 0, 1/2, 1, 2, 3, -2,$ and $-\frac{1}{2}$.
    c) The curve crosses the line $y = -x$ at $x = 0$. Does it cross this line at any other point?

## SUMMARY CHAPTER 3

*Limit Concepts:*

$$\lim_{x \to a^+} f(x); \qquad \lim_{x \to a^-} f(x); \qquad \lim_{x \to a} f(x) \qquad \text{(Sec. 3.1)}$$

$$\lim_{x \to +\infty} f(x) \qquad \lim_{x \to -\infty} f(x) \qquad \text{(Sec. 3.3)}$$

*Limit Theorems (Sec. 3.1):*  If $\lim f$ and $\lim g$ exist then

$$\lim(f + g) = \lim f + \lim g$$

$$\lim(cf) = c \cdot \lim f$$

$$\lim(fg) = (\lim f) \cdot (\lim g)$$

$$\lim\left(\frac{f}{g}\right) = \frac{\lim f}{\lim g} \qquad \text{if } \lim g \neq 0.$$

*Basic Trigonometric Limit (Sec. 3.2):*

$$\lim_{\theta \to 0} \frac{\sin \theta}{\theta} = 1.$$

*Trapping Theorem (Sec. 3.2):*  If $f \leq g \leq h$ and $\lim\limits_{x \to a^+} f = \lim\limits_{x \to a^+} h$ then also

$\lim\limits_{x \to a^+} g = \lim\limits_{x \to a^+} f$. The same holds for the other four types of limits.

*Continuity (Sec. 3.5):*  $f$ is

continuous at $a$ if $f(a) = \lim_{x \to a} f(x)$;

continuous from the right at $a$ if $f(a) = \lim_{x \to a^+} f(x)$;

continuous from the left at $a$ if $f(a) = \lim_{x \to a^-} f(a)$.

Every function defined by a single formula using only powers, roots, trigonometric functions, sums, products, quotients, or composition, is continuous wherever defined. (If the domain is a closed interval $[a, b]$, "continuous at $a$" means "continuous from the right", and "continuous at $b$" means "continuous from the left".)

*Intermediate Value Theorem (Sec. 3.5):*  If $f$ is continuous at every $x$, $a \leq x \leq b$, and if $\bar{y}$ lies between $f(a)$ and $f(b)$, then there is a point $\bar{x}$ in $[a, b]$ where $f(\bar{x}) = \bar{y}$.

*The Derivative (Sec. 3.6):*

$$f'(x) = \lim_{\Delta x \to 0} \frac{f(x + \Delta x) - f(x)}{\Delta x}.$$

*Theorem:*  If $f'(a)$ exists, then $f$ is continuous at $a$: $f(a) = \lim_{x \to a} f(x)$.

*Derivative Formulas:*

$$(f + g)' = f' + g' \qquad (cf)' = cf'$$

$$(fg)' = f'g + fg' \qquad \text{(Product Rule)}$$

$$\left(\frac{f}{g}\right)' = \frac{f'g - fg'}{g^2} \qquad \text{(Quotient Rule)}$$

$$\frac{df(g(x))}{dx} = f'(g(x)) \cdot g'(x), \quad \text{or} \quad \frac{dy}{dx} = \frac{dy}{du} \cdot \frac{du}{dx} \qquad \text{(Chain Rule)}$$

$$\frac{du^r}{dx} = r \cdot u^{r-1} \frac{du}{dx} \qquad \text{(Power Rule)}$$

$$\frac{d \sin(u)}{dx} = \cos(u) \frac{du}{dx} \qquad\qquad \frac{d \cos(u)}{dx} = -\sin(u) \frac{du}{dx}$$

$$\frac{d \tan(u)}{dx} = \sec^2(u) \frac{du}{dx} \qquad\qquad \frac{d \cot(u)}{dx} = -\csc^2(u) \frac{du}{dx}$$

$$\frac{d \sec(u)}{dx} = \sec(u) \tan(u) \frac{du}{dx} \qquad \frac{d \csc(u)}{dx} = -\csc(u) \cot(u) \frac{du}{dx}.$$

# SUMMARY OF GRAPHING TECHNIQUES

We here combine the graphing techniques from Chapters 1–3 and make suggestions for recording the information visually as it is derived. In any given case, it may be impractical to carry out all these steps; use discretion! But the starred items are particularly important.

*Step 1. Domain and Possible Discontinuities.* In the $xy$ plane, lightly cross out vertical strips through points on the $x$-axis which are not in the domain. Draw a dashed vertical line through isolated "holes" in the domain and through any points where the defining formula for $f$ changes.

*Step 2. Symmetry.* Odd? Even? Periodic?

Step 3. Zeroes and Sign of f. Plot zeroes; cross out "forbidden areas" above or below axis (see the examples below).

*Step 4. For First Derivative f':*
    **a)** Find zeroes and discontinuities.
    **b)** Find $f(c)$ at all points $c$ in the domain of $f$ with $f'(c) = 0$, or $f'(c)$ undefined.
    **c)** Determine sign of $f'$. On an auxiliary $x$-axis, mark intervals of increase and decrease of $f$ with appropriate arrows.

Step 5. For Second Derivative f'':
    **a)** Find zeroes and discontinuities.
    **b)** Find $f(c)$ for all points $c$ in the domain of $f$ with $f''(c) = 0$ or $f''(c)$ undefined.
    **c)** Determine sign of $f''$. On the auxiliary $x$-axis, mark intervals where $f$ is concave up or down.

*Step 6.* At each discontinuity $a$ and at endpoints of intervals forming the domain of $f$, determine $\lim\limits_{x \to a^+} f$ and $\lim\limits_{x \to a^-} f$ as appropriate. (This determines the nature of any vertical asymptotes, jumps, "holes," "blips," and so on.)

*Step 7.* $\lim\limits_{x \to \pm \infty} f(x)$ for horizontal asymptotes.

Step 8. $\lim\limits_{x \to a^\pm} f'(x)$ at points $a$ in domain where $f'$ not defined. Plot additional points as desired.

---

**EXAMPLE 1**    Graph $f(x) = \dfrac{x}{\sqrt{x^2 - 1}}$.

*SOLUTION*

*Step 1. Domain and Possible Discontinuities.* $f$ is defined only when $x^2 - 1 > 0$, that is, $x^2 > 1$, or $|x| > 1$. This rules out the shaded region in Figure 1.

*Step 2. Symmetry.* $f(-x) = \dfrac{-x}{\sqrt{(-x)^2 - 1}} = -\dfrac{x}{\sqrt{x^2 - 1}} = -f(x)$, so $f$ is *odd.*

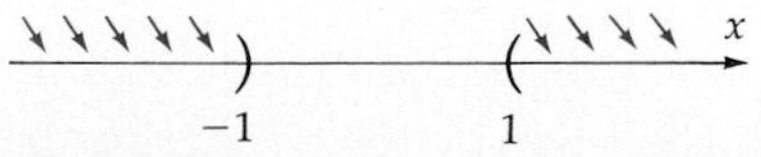

*Step 3. Zeroes, and Sign of f.* $\dfrac{x}{\sqrt{x^2 - 1}} = 0 \Leftrightarrow x = 0$; but 0 is not in the domain of $f$, so $f$ has no zeroes. The defining formula for $f$ shows that $f(x) < 0$ for $x < 1$, and $f(x) > 0$ for $x > 1$; this rules out further regions as in Figure 2.

*Step 4.*
$$f'(x) = (x^2 - 1)^{-1/2} + x(-\tfrac{1}{2})(x^2 - 1)^{-3/2}(2x)$$
$$= (x^2 - 1)^{-3/2}[(x^2 - 1) - x^2] = -(x^2 - 1)^{-3/2}.$$

No graph here;
$f(x)$ not defined

**FIGURE 1**

**FIGURE 2**

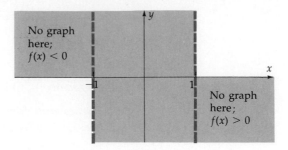

**FIGURE 3**

So $f'(x) < 0$ at all points of the domain of $f$, giving Figure 3. (Note that $f'$ is even, while $f$ is odd.)

*Step 5.* $f''(x) = \frac{3}{2}(x^2 - 1)^{-5/2}(2x) = 3x(x^2 - 1)^{-5/2}$ has no zeroes in the domain of $f$; $f''(x) < 0$ for $x < 1$ and $f''(x) > 0$ for $x > 1$. Add to Figure 3 appropriate marks indicating the concavity; the result is Figure 4.

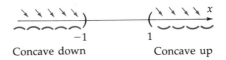

Concave down          Concave up

**FIGURE 4**

*Step 6.* As $x \to 1^+$, $f(x) = \dfrac{x}{\sqrt{x^2 - 1}} \dfrac{\text{"}\to 1\text{"}}{0}$, so $f(x)$ becomes infinite and there is a vertical asymptote at $x = 1$. The pattern of signs in Figure 2 shows that $\lim_{x \to 1^+} f(x) = +\infty$; since $f$ is odd, it follows from this that $\lim_{x \to -1^-} f(x) = -\infty$.

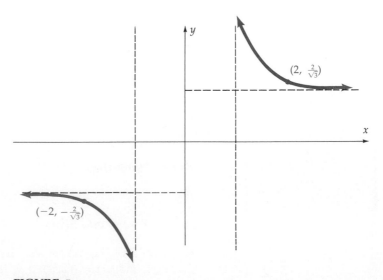

**FIGURE 5**

$$f(x) = \frac{x}{\sqrt{x^2 - 1}}.$$

*Step 7.*

$$\lim_{x \to +\infty} \frac{x}{\sqrt{x^2 - 1}} = \lim_{x \to +\infty} \frac{1}{\frac{1}{x}\sqrt{x^2 - 1}} = \lim_{x \to +\infty} \frac{1}{\sqrt{1/x^2}\sqrt{x^2 - 1}}$$

$$= \lim_{x \to +\infty} \frac{1}{\sqrt{1 - 1/x^2}} = 1.$$

In this calculation we could write $1/x = \sqrt{1/x^2}$ because $x > 0$ as $x \to +\infty$. As $x \to -\infty$ then $x < 0$, so $1/x = -\sqrt{1/x^2}$, and a calculation as above shows that $\lim_{x \to -\infty} f(x) = -1$. (This could also be deduced from the symmetry of the graph.)

*Step 8.* $f'(a)$ is defined at all points of the domain of $f$.

*Step 9.* $f(\pm 2) = \dfrac{\pm 2}{\sqrt{3}}$.

The graph in Figure 5 displays all these features.

---

**EXAMPLE 2** Graph $f(x) = (x^2 - 1)^{1/3}$.

*SOLUTION*

*Step 1.* **The Domain of $f$** is all $x$ (every real number has a cube root), and $f$ is continuous everywhere.

*Step 2.* $f(-x) = [(-x)^2 - 1]^{1/3} = [x^2 + 1]^{1/3} = f(x)$, so $f$ is *even*.

*Step 3.* $(x^2 - 1)^{1/3} = 0 \Leftrightarrow x^2 - 1 = 0 \Leftrightarrow x = \pm 1$. Check signs in each of the three intervals $(-\infty, -1), (-1, 1), (1, +\infty)$ to find the pattern of excluded regions indicated in Figure 7.

*Step 4.* $f'(x) = \dfrac{2x}{3}(x^2 - 1)^{-2/3} = \dfrac{2x}{3(x^2 - 1)^{2/3}}$. (Notice that $f'$ is odd, while $f$ is even.)

    a) $f'(0) = 0$; $f'(x)$ is discontinuous at $x = \pm 1$.
    b) $f(0) = -1$, $f(\pm 1) = 0$.
    c) $(x^2 - 1)^{2/3} = [(x^2 - 1)^{1/3}]^2 \ge 0$, so the signs are

$$f'(x) = \frac{2x}{3(x^2 - 1)^{2/3}}: \qquad \begin{array}{ccccc} - - - - - - & | & - - - - - & | & + + + + + & | & + + + + + + \\ & -1 & & 0 & & 1 & \end{array}$$

*Step 5.*

$$f''(x) = \frac{2}{3}(x^2 - 1)^{-2/3} + \frac{2x}{3}\left(-\frac{2}{3}\right)(x^2 - 1)^{-5/3}(2x)$$

$$= -\frac{2}{9}(x^2 - 1)^{-5/3}(x^2 + 3), \qquad \text{an even function, like } f.$$

    a) $f''$ never 0; $f''$ discontinuous at $x = \pm 1$.
    b) $f(\pm 1) = 0$.

c)  $f''$ can change sign only at $x = \pm 1$; check the signs:

$f''$:  $\underline{\hspace{4cm}}$

$- - - - - - - - - - \mid + + + + + \mid + + + + + \mid - - - - - -$

$\qquad\qquad\qquad -1 \qquad\qquad 0 \qquad\qquad 1$

Figure 6 displays the information in steps 4 and 5 together.

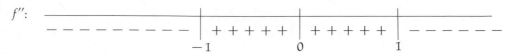

**FIGURE 6**

**Step 6.**  No discontinuities.

**Step 7.**  $\displaystyle\lim_{x \to \pm\infty} (x^2 - 1)^{1/3} = +\infty.$

**Step 8.**  $f'(x)$ is undefined at $x = \pm 1$. There

$$\lim_{x \to 1} f'(x) = \lim_{x \to 1} \frac{2x}{3(x^2 - 1)^{2/3}} = +\infty, \quad \lim_{x \to -1} \frac{2x}{3(x^2 - 1)^{2/3}} = -\infty.$$

Hence the graph is tangent to the vertical lines $x = 1$ and $x = -1$.

**Step 9.**  Additional points on the graph: $f(\pm 3) = 2$.

Figure 7 shows the graph.

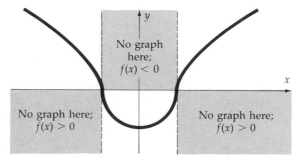

**FIGURE 7**
$f(x) = (x^2 - 1)^{1/3}$.

**EXAMPLE 3**  Graph $f(x) = \dfrac{1}{1 - \tan x}$.

**SOLUTION**

**Step 1. Domain**  $\tan x = \dfrac{\sin x}{\cos x}$ is undefined when $\cos x = 0$, thus when

$x = \pm\dfrac{\pi}{2}, \pm\dfrac{3\pi}{2}, \pm\dfrac{5\pi}{2}, \ldots$ . Further, $f(x)$ is undefined when $\tan x = 1$, or

$x = \dfrac{\pi}{4}, \dfrac{\pi}{4} \pm \pi, \dfrac{\pi}{4} \pm 2\pi, \ldots$ . All these are points of discontinuity for $f$.

*Step 2.* $\tan(x + \pi) = \tan x$ (see Appendix C) so $f$ has period $\pi$. Graph it for $0 \le x \le \pi$ and extend the graph by periodicity.

*Step 3.* $f(x) = \dfrac{1}{1 - \tan x}$ has no zeroes, so $f$ can change sign only at the discontinuities. For $0 \le x \le \pi$, Figure 8 shows the pattern of "forbidden areas" (and also the concavity found in step 5).

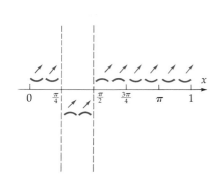

**FIGURE 8**

*Step 4.* $f'(x) = \dfrac{\sec^2 x}{(1 - \tan x)^2} > 0$ whenever defined. $f'$ is discontinuous at the same points as $f$.

*Step 5.*

a) $f''(x) = \dfrac{2 \sec^2 x}{(1 - \tan x)^3} (1 + \tan x) = 0$ when $\tan x = -1$, $x = \dfrac{3\pi}{4} \pm \pi$, $\dfrac{3\pi}{4} \pm 2\pi$.

b) $f(3\pi/4) = 1/2$.

c) Figure 8 shows the concavity.

*Step 6.*

$$\lim_{x \to \pi/2} \frac{1}{1 - \tan x} = \lim_{x \to \pi/2} \frac{\cos x}{\cos x - \sin x} = \frac{0}{0 - 1} = 0.$$

As $x \to \pi/4$ then $1 - \tan x \to 0$, so $\dfrac{1}{1 - \tan x} \to \pm\infty$. Since $f' > 0$ everywhere, $\lim\limits_{x \to \pi/4^-} f(x) = +\infty$ and $\lim\limits_{x \to \pi/4^+} f(x) = -\infty$ as shown in Figure 9.

*Step 7.* $\lim\limits_{x \to \pm\infty} f(x)$ does not exist, since $f$ is periodic and not constant.

*Step 8.* $f'(a)$ is defined at all points of the domain.

*Step 9.* $f(0) = f(\pi) = 1$.

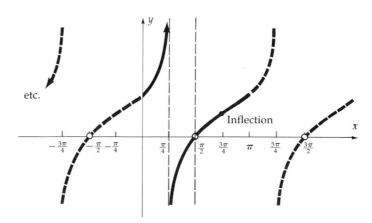

**FIGURE 9**

$$f(x) = \frac{1}{1 - \tan x}.$$

Figure 9 shows the graph. The part for $0 \leq x \leq \pi$ is determined by the preceding data; the rest of the graph is obtained by repeating that part with period $\pi$.

# REVIEW PROBLEMS    CHAPTER 3

## A

1.  Describe all discontinuities of the given function as vertical asymptotes, "holes," "jumps," or "oscillatory." At each discontinuity, evaluate the one- and two-sided limits, if they exist. Determine any limits as $x \to +\infty$ or $x \to -\infty$.

*a)  $f(x) = \dfrac{1 - 2/x}{x - 2}$

*b)  $f(x) = \dfrac{|1 - 2/x|}{x - 2}$

*c)  $g(x) = \dfrac{x - 1}{x^2 - 1}$

d)  $g(x) = \dfrac{x^3 - 1}{x^3 + 1}$

e)  $g(x) = \dfrac{x^2 + 4}{x - 2} + 1/x$

f)  $f(x) = \dfrac{\sin x}{x}$

g)  $g(x) = \tan 2x \left[ = \dfrac{\sin 2x}{\cos 2x} \right]$

h)  $h(\theta) = \sec 2\theta$

i)  $f(x) = \dfrac{\sqrt{x^4 + 1}}{x^3}$

j)  $g(x) = \dfrac{(x^2 + 1)^{1/2}}{(x^3 + 1)^{1/3}}$

k)  $s(x) = \sin(\pi/x)$

l)  $\varphi(x) = x \sin(1/x)$

2.  Define the derivative $f'(x)$. Illustrate the definition with an appropriately labeled figure.

3.  Compute the derivative of the given function.

*a)  $\varphi(t) = \dfrac{t}{\sqrt{t + 1}}$

*b)  $r(x) = \sin(x + \sqrt{2x + 1})$

*c)  $s(z) = \sqrt{z + \sqrt{z}}$

*d)  $f(\theta) = \cos^2(5\theta + 1)$

*e)  $\alpha(t) = \tan(3t) + \cot(3t)$

f)  $\beta(t) = t^{-1}(\sec 2t + \csc 2t)$

*g)  $\dfrac{\sin 2\theta}{\theta}$

4.  Compute $dy/dx$.

*a)  $y = \dfrac{x}{1 + x^2}$

*b)  $y = \sqrt{\dfrac{x}{1 + x^2}}$

c)  $y = (x + 1)^{1/3}(x^2 + 1)^{1/4}$

*d)  $x^5 + y^5 + x^2y^2 = 0$

e)  $x \cos(xy) = y \sin(xy)$

*5.  Write the equation of the line tangent to the given curve at point $(1, 3)$.

a)  $y = \dfrac{x^2 + 2}{2 - x^2}$    b)  $x^3 + xy + y^3 = 31$

6.  *a)  Find $g'(1)$, given that $g(1) = 2$, $f'(2) = 3$, and $(f \circ g)'(1) = 4$.

b)  Find $g'(3)$, given that $g(3) = 2$, $f(3) = 5$, $f'(3) = 4$, and $(fg)'(3) = -7$.

*7.  In Figure 1, four "$x$ by $x$" corners are to be cut

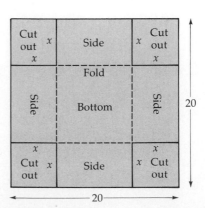

**FIGURE 1**

out and the sides folded up to create an open box. What value of $x$ gives maximum volume?

*8. A cylindrical pot with aluminum sides and copper bottom is to hold 1000 cm$^3$. If copper costs four times as much as aluminum, what dimensions require the least cost of material?

9. Sketch the graph of the given function.

*a) $\dfrac{3x^2 - 1}{x^3}$  *b) $\dfrac{x^3}{3x^2 - 1}$

c) $\dfrac{4x}{\sqrt{x^2 + 15}}$  d) $\dfrac{x^2 + 1}{x^2 - 2}$

e) $x + \dfrac{32}{x^2}$  *f) $2x - \tan x$

g) $\theta + \sin \theta$  h) $t + \sin 2t$

10. Sketch the graph of a function with the given properties, or explain why no such function exists.

*a) $f(x) > 0$ for all $x \neq 1$, and $\lim\limits_{x \to 1} f(x) = 0$.

*b) $f(x) > 0$ for all $x$, and $\lim\limits_{x \to 1} f(x) = -1$.

*c) $f$ is increasing and continuous for all $x$, with $\lim\limits_{x \to +\infty} f(x) = -1$ and $\lim\limits_{x \to -\infty} f(x) = +1$.

*d) $f'(x) > 0$ everywhere except at $x = 0$, and $\lim\limits_{x \to +\infty} f(x) = -1$, $\lim\limits_{x \to -\infty} f(x) = 1$.

e) $f'(x) > 0$ for all $x > 0$, and $\lim\limits_{x \to 0^+} f(x) = -\infty$.

f) $f'(x) > 0$ for all $x > 0$, and $\lim\limits_{x \to +\infty} f(x) = -\infty$.

g) $f$ is continuous everywhere, with no zeroes, and $\lim\limits_{x \to -\infty} f(x) = -1$, $\lim\limits_{x \to +\infty} f(x) = 1$.

h) $f$ is defined everywhere, with no zeroes, and $\lim\limits_{x \to -\infty} f(x) = 1$, $\lim\limits_{x \to +\infty} f(x) = -1$.

11. Prove the formula for
a) $(fg)'$.  b) $(1/g)'$.
c) $(f/g)'$.  d) $(f \circ g)'$.

12. Which, if any, of the following is an antiderivative of $\cos 2x$?
a) $\sin 2x$  b) $-\sin 2x$
c) $\frac{1}{2} \sin 2x$  d) $-2 \sin 2x$
e) $-\cos(x^2)$

13. Which, if any, of the following is an antiderivative of $(x^2 + 1)^{1/2}$?
a) $\frac{2}{3}(x^2 + 1)^{3/2}$  b) $(\frac{1}{3}x^3 + x)^{1/2}$
c) $x(x^2 + 1)^{-1/2}$

14. Which, if any, of the following is an antiderivative of $\sin(x^2)$?

a) $2x \cos(x^2)$  b) $\sin \dfrac{x^3}{3}$

c) $\cos \dfrac{x^3}{3}$  d) $-\cos \dfrac{x^3}{3}$

e) $-\cos(x^2)$  f) $\dfrac{-1}{2x} \cos(x^2)$

**B**

15. a) Suppose that $f'(0)$ exists. Evaluate $\lim\limits_{x \to 0^+} \dfrac{f(x) - f(-x)}{2x}$. Make a sketch illustrating the limit.

b) If the limit in part a exists, does $f'(0)$ necessarily exist? If yes, explain why; if no, illustrate by an example.

16. The Remainder Theorem for polynomials says that for any polynomial $P$, and any number $r$, there is a polynomial $Q$ such that

$$P(x) = P(r) + (x - r)Q(x).$$

Express $Q(r)$ in terms of the derivative $P'$.

# 4

# MORE APPLICATIONS OF DERIVATIVES

# 4.1

# RATE OF CHANGE. THE LINEAR APPROXIMATION

In applications of calculus, the derivative represents a *rate of change*. If $y$ is a function of $x$, any change $\Delta x$ in $x$ gives rise to a corresponding change $\Delta y$ in $y$ (Fig. 1). The ratio $\Delta y / \Delta x$ is called the **average rate of change** of $y$ with respect to $x$. The limit of these ratios, the derivative

$$\frac{dy}{dx} = \lim_{\Delta x \to 0} \frac{\Delta y}{\Delta x} \tag{1}$$

is called simply the **rate of change** of $y$ with respect to $x$. For example, if $y$ is position at time $t$, then $\Delta y / \Delta t$ is the average rate of change of position with respect to time, and $dy/dt$, the velocity, is the (instantaneous) rate of change.

The limit relation (1) means that when $\Delta x$ is near 0, then $\Delta y / \Delta x$ is a good approximation to $dy/dx$, and vice versa. This is indicated by

$$\frac{\Delta y}{\Delta x} \approx \frac{dy}{dx}.$$

Hence, multiplying by $\Delta x$,

$$\Delta y \approx \frac{dy}{dx} \cdot \Delta x. \tag{2}$$

Geometrically, $\Delta y$ is the actual change in $y$, and $\dfrac{dy}{dx} \cdot \Delta x$ is the change along the tangent line (Fig. 2); when $\Delta x$ is small, these two changes are nearly the same. We call $\dfrac{dy}{dx} \cdot \Delta x$ the **linear approximation of** $\Delta y$.

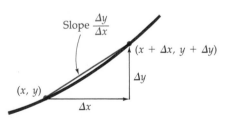

**FIGURE 1**

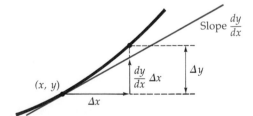

**FIGURE 2**
$\dfrac{dy}{dx} \Delta x \approx \Delta y.$

**EXAMPLE 1**    Let position $y$ be a function of time $t$, so $dy/dt$ is the rate of change of position with respect to time (the velocity). Over a small time interval $\Delta t$, the change in position $\Delta y$ is approximately

$$\Delta y \approx \frac{dy}{dt} \cdot \Delta t.$$

If you are traveling 55 mph at twelve noon, then your change in position during the first second after noon will be very nearly

$$\Delta y \approx \frac{dy}{dt} \cdot \Delta t$$

$$= 55 \cdot \frac{1}{3600} \approx .015 \text{ miles} \qquad \left( \Delta t = 1 \text{ second} = \frac{1}{3600} \text{ hours} \right)$$

$$= 80\tfrac{2}{3} \text{ feet.}$$

If the speed is constant, this is the *exact* change in position; if the speed varies a little in that one second, this is the *approximate* change.

---

**EXAMPLE 2**   (Population growth) If $P$ is the size of a certain population and $t$ is time, then $dP/dt$ is the rate of change of the population with respect to time. If the population is, say, the mass of bacteria in a certain culture, and right now $dP/dt = 3$ milligrams per day, then the change in population during the next hour is nearly

$$\Delta P \approx \frac{dP}{dt} \cdot \Delta t$$

$$= 3 \cdot \frac{1}{24} = \frac{1}{8} \text{ milligrams,}$$

since the change in time is $\Delta t = 1$ hour $= \frac{1}{24}$ day. (You must express $\Delta t$ in days, since $dP/dt$ is given in mg/day.)

---

*REMARK*   Leibniz notation is often the first choice in applications, because it helps interpret the meaning of the derivative for the particular problem at hand. For example, if $y$ denotes position along a line, the symbol $dy/dt$ stands for the derivative of $y$ with respect to time $t$; writing this symbol implies that $y$ is a function of $t$. Moreover, the symbol gives the proper units—if $y$ is in meters and $t$ in seconds then the derivative has units of meters/sec, just as the "fraction" $dy/dt$ suggests. Analogously, if $P$ stands for numbers of people and $t$ is time in years, then $dP/dt$ has units of "people per year."

---

**EXAMPLE 3**   The area of a square of side $x$ is

$$A = x^2.$$

If the square is enlarged by a small amount $\Delta x$, then the change in area is

$$\Delta A \approx \frac{dA}{dx} \cdot \Delta x$$

$$= 2x \cdot \Delta x.$$

This approximation includes the two thin rectangles in Figure 3 and neglects the tiny square of area $(\Delta x)^2$. When $\Delta x$ is small, the square is negligible compared to the rectangles.

In this example, the derivative $dA/dx$ is the rate of change of area with respect to edge length.

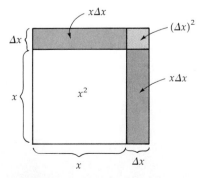

**FIGURE 3**
$\Delta A = 2x\,\Delta x + (\Delta x)^2 \approx 2x\,\Delta x.$

## Differential Notation

The linear approximation is sometimes expressed with the notation in Figure 4. The small change in $x$ is denoted by $dx$, and the linear approximation by $dy$:

$$dy = \frac{dy}{dx} \cdot dx. \tag{3}$$

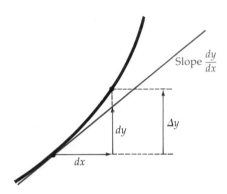

**FIGURE 4**

Differential approximation: $\Delta y \approx dy = \dfrac{dy}{dx}\, dx.$

The symbol $dy$ is called the **differential** of $y$; it represents the change along the tangent line corresponding to a change $dx$ in $x$. In principle, formula (3) defines a differential approximation $dy$ for any given $dx$; but the approximation is useful only when $dx$ is small.

**The linear approximation of function values** is illustrated in Figure 5. Suppose that a given function $f$ and its derivative $f'$ are easy to compute at some particular point $x$ but hard to compute at some nearby point $x + \Delta x$. Then you can *approximate* $f(x + \Delta x)$ as in the figure:

$$\Delta y = f(x + \Delta x) - f(x) \approx f'(x)\, \Delta x$$

so

$$f(x + \Delta x) \approx f(x) + f'(x)\, \Delta x. \tag{4}$$

Think of $x$ as a "base point" where $f(x)$ and $f'(x)$ are known; then (4) approximates $f$ at the nearby point $x + \Delta x$.

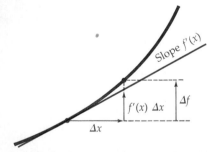

**FIGURE 5**
$\Delta f = f(x + \Delta x) - f(x) \approx f'(x)\, \Delta x.$

**EXAMPLE 4**   Approximate $\sqrt{10}$.

*SOLUTION*   The function involved is

$$f(x) = \sqrt{x}.$$

You can easily compute $f(9) = \sqrt{9} = 3$, so take as base point

$$x = 9.$$

Then (4) approximates the desired value $\sqrt{10} = f(10)$ if you set

$$10 = x + \Delta x = 9 + \Delta x, \qquad \text{so } \Delta x = 1.$$

Having chosen $f$, $x$, and $\Delta x$ in (4), go ahead and work out the right-hand side: $f'(x) = \dfrac{1}{2\sqrt{x}}$, and (4) gives

$$\sqrt{9 + 1} \approx \sqrt{9} + \frac{1}{2\sqrt{9}} \cdot 1$$

$$= 3 + \frac{1}{6} = 3.166 \ldots$$

This is slightly larger than the value given by a calculator, $\sqrt{10} = 3.162 \ldots$. If you sketch the graph of $y = \sqrt{x}$ and its tangent at $(9, 3)$, you can see why this linear approximation is larger than the actual value.

---

Of course, with calculators so cheap, you are unlikely to need numerical approximations like this, unless stranded on a desert island (and then, who would care about $\sqrt{10}$?). The linear approximation is nevertheless significant; a good illustration shows up in Einstein's theory of "special relativity."

In that theory, the mass $m$ of an object varies with its velocity $v$. If the object at rest has mass $m_0$, then at velocity $v$ it has mass

$$m = \frac{m_0}{\sqrt{1 - \dfrac{v^2}{c^2}}} = m_0\left(1 - \frac{v^2}{c^2}\right)^{-1/2}$$

where $c$ is the velocity of light in vacuum. Moreover, by another famous Einsteinian equation, the *energy* of a moving object is

$$E = mc^2$$

$$= m_0\left(1 - \frac{v^2}{c^2}\right)^{-1/2} c^2. \tag{5}$$

When the velocity $v$ is small compared to $c$, Einstein's theory should be closely related to Newton's. In fact it is; at low velocities, the two theories attribute nearly the same amount of energy to the effects of motion. We show this by approximating $(1 - v^2/c^2)^{-1/2}$, using the linear approximation of the function

$$f(x) = x^{-1/2}$$

with $x = 1$ and $\Delta x = -v^2/c^2$. Since $v$ is small compared to $c$, then $\Delta x$ will be very small, and the linear approximation should be good. The derivative of $f$ is

$$f'(x) = -\tfrac{1}{2}x^{-3/2}$$

and the linear approximation (4) gives

$$(x + \Delta x)^{-1/2} \approx x^{-1/2} - \tfrac{1}{2}x^{-3/2}\, \Delta x.$$

Set $x = 1$ and $\Delta x = -v^2/c^2$:

$$(1 - v^2/c^2)^{-1/2} \approx 1^{-1/2} - \frac{1}{2}\, 1^{-3/2}(-v^2/c^2)$$

$$= 1 + \frac{1}{2}\frac{v^2}{c^2}.$$

Use this in equation (5) and find that the energy is

$$E \approx m_0 \left( 1 + \frac{1}{2}\frac{v^2}{c^2} \right) c^2$$

$$= m_0 c^2 + \frac{1}{2}\, m_0 v^2.$$

The first term $m_0 c^2$ is the energy of the object at rest, since $m_0$ is the mass at rest; so the second term $\tfrac{1}{2}m_0 v^2$ is the energy due to the velocity $v$. *This is precisely the formula which in Newtonian theory gives the kinetic energy*, the energy due to motion! So at low velocities, Einstein's energy formula is nearly the same as Newton's.

### Relative Error

In approximations, the meaning of "small" depends on circumstances. An error of 1 gram in measuring the mass of the earth (about $5.98 \times 10^{27}$ gm) would be awesomely small; but an error of 1 gm in measuring the mass of an electron (about $9.11 \times 10^{-28}$ gm) would be ridiculously large.

Thus the accuracy of a measurement is best described by the **relative error**, the error divided by the true value, or (since the true value may not be exactly known) the error divided by the measured value:

$$\text{Relative error in measurement} = \frac{\text{error}}{\text{measured value}}.$$

If we denote by $\Delta x$ the error in finding measurement $x$, then

$$\text{relative error} = \frac{\Delta x}{x}.$$

---

**EXAMPLE 5** Suppose that the radius $r$ of a ball bearing is measured as $(5 \pm .01)$ mm. This means that the measured value is $r = 5$, and the error $\Delta r$ is between $-0.1$ and $0.1$:

$$|\Delta r| \le 0.01.$$

Then the relative error is, in absolute value,

$$\left|\frac{\Delta r}{r}\right| \le \frac{.01}{5} = 0.002.$$

**EXAMPLE 6**   The volume of the ball bearing in Example 5 is to be determined from the measurement of its radius as $(5 \pm .01)$ mm. Estimate the relative error in thus determining the volume.

*SOLUTION*   The volume of a sphere of radius $r$ is $V = \frac{4}{3}\pi r^3$, so

$$\Delta V \approx \tfrac{4}{3}\pi(3r^2)\,\Delta r = 4\pi r^2\,\Delta r,$$

and the relative error is

$$\frac{\Delta V}{V} = \frac{4\pi r^2\,\Delta r}{(4/3)\pi r^3} = 3\,\frac{\Delta r}{r}.$$

From Example 5, $\dfrac{|\Delta r|}{r} \le .002$, so

$$\frac{|\Delta V|}{V} = 3\,\frac{|\Delta r|}{r} < 3(.002) = .006.$$

So for small errors, the relative error in the volume is about three times the relative error in the radius. (What do you suppose is the relation between relative error in surface area and relative error in radius?)

## SUMMARY

*Rate of Change:*

$$\frac{\Delta y}{\Delta x} = \text{average rate of change of } x \text{ with respect to } y.$$

$$\frac{dy}{dx} = \lim_{\Delta x \to 0} \frac{\Delta y}{\Delta x} = \text{rate of change of } y \text{ with respect to } x.$$

*Linear Approximation:*

$$\Delta y \approx \frac{dy}{dx}\,\Delta x \qquad\qquad \text{(Fig. 2)}$$

$$dy = \frac{dy}{dx}\,dx \qquad\qquad \text{(Fig. 4)}$$

$$f(x + \Delta x) \approx f(x) + f'(x)\,\Delta x. \qquad \text{(Fig. 5)}$$

*Relative Error:*   If $\Delta x$ represents the error in a measurement $x$, then the relative error is $\dfrac{\Delta x}{x}$.

# PROBLEMS

## A

Recall: A sphere of radius $r$ has volume $V = \frac{4}{3}\pi r^3$ and surface area $S = 4\pi r^2$.

1. Approximate
   *a) $\sqrt{15}$ using $\sqrt{16} = 4$.
   *b) $\sqrt{50}$.
   c) $\sqrt{55}$ using $\sqrt{49} = 7$.
   Compare your approximations to the values given by a calculator or a table. Why is your answer to part c less accurate than for part b?

2. Certain dice are designed with an edge of 1 cm. The dice actually constructed have an edge of 0.99 cm.
   *a) Estimate the difference in volume between the actual dice and the designed ones.
   *b) Compute the difference exactly.
   c) Compare the relative error in the volume to the relative error in the edge.

3. A cubical crystal with edge 0.5 cm grows, so that the new edge is 0.505 cm.
   a) Estimate the increase in volume, using derivatives.
   b) Compute the exact increase in volume.
   c) Compare the relative increase in volume to the relative increase in the edge.

4. A cube has volume $V = x^3$. Suppose that $x$ is increased by a small increment $\Delta x$.
   a) Show that $\Delta V \approx 3x^2\, \Delta x$.
   b) Show that $3x^2\, \Delta x$ is the volume of the three slabs on the faces of the cube in Figure 6.
   c) Show that $\Delta V = 3x^2\, \Delta x + 3x(\Delta x)^2 + (\Delta x)^3$ exactly.

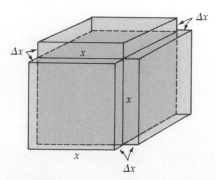

**FIGURE 6**

d) Try to visualize $3x(\Delta x)^2$ and $(\Delta x)^3$ as volumes which, added to the original cube and the three slabs in Figure 6, would fill out the cube of edge $x + \Delta x$. (When $\Delta x$ is small, these additional volumes are negligible compared to the three slabs.)

5. A cubical box has edge of length $x$. Its outside edge is painted with a layer of paint of thickness $h$, forming a new slightly larger cube (ignore any rounding at the edges). The box has six faces, each of area $x^2$, so the volume of paint required is about $6x^2h$. Identify this as the linear approximation with an appropriate $\Delta x$.

*6. Circular disks are designed with a radius of 8 cm and constructed with an actual radius of 7.9 cm.
   a) Estimate the difference in area between the constructed and the designed disks.
   b) Compare the relative error in area to the relative error in radius.

7. A path 1 meter wide is laid around a large circular garden bed of diameter 60 meters. The area covered by the path is approximately

$$(\text{circumference}) \times (\text{width}) = (60\pi)(1).$$

Show that this is precisely the linear approximation to the increment

$$\Delta A = (\text{Area of path and garden}) - (\text{Area of garden}).$$

Is the approximation less than or greater than the actual increment?

*8. A certain stately pleasure dome is a hemisphere of diameter 20 meters. It is to be covered with gold leaf 0.3 mm thick. Estimate the volume of gold required, in two ways:
   a) Use the surface area of the dome.
   b) Use a linear approximation of $\Delta V$ as the radius is increased by 0.3 mm.

9. The radius of a certain hemispherical dome is measured as $(10.3 \pm .05)$ meters. Estimate or compute:
   a) The resulting error in determining the surface area.
   b) The relative error in measuring the radius.

c) The relative error in determining the surface area. Compare this to the relative error in determining the radius.

*10. The height of a cliff is determined by measuring its angle of elevation, viewed from a distance of 20 meters (Fig. 7). The angle is measured as $30°$. Estimate:

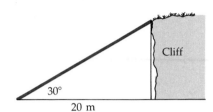

**FIGURE 7**

a) The height $y$.
b) The error in computing the height due to an error of $2°$ in measuring the angle of elevation. (Recall that the trigonometric derivative formulas are for radian measure.)

11. The force between a proton and an electron at distance $r$ meters is about $(2.3) \times 10^{-29} r^{-2}$ newtons. (A force of one newton is about the weight of 100 grams at the surface of the earth.) In a hydrogen atom, a proton and an electron are separated by about $5.3 \times 10^{-11}$ meters.
a) Estimate the force binding the electron to the proton.
b) Suppose there is a relative error of .05 in measuring the distance $r$; what is the relative error in measuring the force?

12. A section of stove pipe has length 1 m, inner diameter 20 cm, and walls $\frac{1}{2}$ mm thick. Compute the volume of the walls:
a) Precisely, using $V = \pi r^2 h$ for the volume of a cylinder.
b) Approximately, thinking of $V$ as a function of $r$, with $h = 1$.
c) Relate the approximation in b) to the surface area of the inside of the pipe.

13. Given (i) $y$ as a function of $x$, and (ii) $x$ and $dx$, compute the differential approximation $dy = \frac{dy}{dx} dx$.

*a) $y = \frac{1}{x}$, $x = 1$, $dx = 0.1$

*b) $y = x^n$, $x = 1$, $dx = 0.1$

c) $y = \frac{1}{x}$, $x = 10$, $dx = 1$

*14. The New York *Times* reported (7/17/82) that the rate of decline of the Polish economy was slowing. If $P(t)$ represents the amount produced at time $t$, what were they saying about $P'(t)$ and $P''(t)$?

15. Let $C(t)$ be the average cost of typical goods at time $t$. If the rate of inflation is slowing, what is being said about $C'$ and $C''$?

**B**

16. Show that if $y = \sqrt{x}$, then the exact increment $\Delta y$ due to an increment $\Delta x$ is

$$\Delta y = \frac{1}{\sqrt{x + \Delta x} + \sqrt{x}} \Delta x.$$

Compare this to $\frac{dy}{dx} \cdot \Delta x$. Why are the two nearly the same when $\Delta x$ is small?

17. The radius $r$ of a sphere is changed by a small quantity $\Delta r$. Compare the relative change in $r$ to the resulting relative change in the
*a) Volume. b) Surface area.
c) Diameter.

18. A cube of side $x$ is changed by a small amount $\Delta x$. Compare the percent change in $x$ to the percent change in
a) The volume. *b) The surface area.

19. The edge, $x$, of a square is changed by a small amount $\Delta x$. Compare the percent change in $x$ to the percent change in
*a) The perimeter. b) The area.

20. The radius of a circle is changed by a small amount $\Delta r$. Compare the relative change in $r$ to the relative change in
a) The circumference. b) The area.

21. The rate of flow through a pipe of radius $R$ is $F = k/R^4$; $k$ is a constant, depending on the viscosity of the fluid and the pressure gradient in the pipe. The radius is changed by a small amount $\Delta R$. What is the relation between the relative change in $R$ and the resulting relative change in $F$?

## 4.2

### RELATED RATES

Imagine two quantities changing with time. Any relation between these quantities implies a relation between their rates of change; given one rate, the other can often be determined.

---

**EXAMPLE 1**   A spherical hailstone is accumulating ice; the radius is growing at the rate of 2 mm/sec. How fast is the volume increasing?

*SOLUTION*   The hailstone is a sphere, so its volume is

$$V = \tfrac{4}{3}\pi r^3, \tag{1}$$

where $r$ is the radius. The radius is growing at the rate of 2 mm/sec; that is, the rate of change of $r$ with respect to time is 2,

$$\frac{dr}{dt} = 2.$$

The example asks how fast the volume is increasing; what is the rate of change of volume with respect to time,

$$\frac{dV}{dt} = ?$$

To answer, differentiate the relation (1) between $V$ and $r$, treating both $V$ and $r$ as functions of $t$:

$$\frac{dV}{dt} = \frac{dV}{dr}\frac{dr}{dt} = \left(\frac{4}{3}\pi \cdot 3r^2\right)\left(\frac{dr}{dt}\right)$$

$$= (4\pi r^2) \cdot 2 \qquad \left(\frac{dr}{dt} = 2 \text{ is given}\right)$$

$$= 8\pi r^2.$$

This is the required rate of increase of volume.

Think of the calculation like this: A small increase $\Delta t$ in time causes an increase in $r$, $\Delta r \approx \dfrac{dr}{dt}\Delta t = 2\,\Delta t$. This increase in $r$ causes an increase in $V$,

$$\Delta V = \underbrace{(4\pi r^2)}_{dV/dr}\,\Delta r \approx 4\pi r^2 \cdot \underbrace{2\,\Delta t}_{\Delta r \,\approx\, 2\,\Delta t} = 8\pi r^2\,\Delta t.$$

So the change in $V$ is nearly $8\pi r^2$ times the change in $t$; that is, the rate of change of $V$ with respect to $t$ is $8\pi r^2$.

Notice that the factor $dV/dr = 4\pi r^2$ is precisely the surface area of the hailstone; increasing the thickness by $\Delta r = 2\,\Delta t$ adds a layer whose volume is nearly $4\pi r^2(2\,\Delta t)$.

The problem in Example 1 is a typical "related rates" problem. The rate of change of the volume $V$ is related to the rate of change of the radius $r$, because $V$ itself is related to $r$. The two main steps in such a problem are:

**(1)**   Write the relation between the quantities involved.

**(2)**   Take the derivative of this relation (usually with respect to time $t$) using any relevant derivative formulas, particularly the chain rule.

Then:

**(3)**   Substitute any given numerical quantities into your formula.

**(4)**   Solve for the desired rate.

In the next example, there is an equally important preliminary "zeroth" step:

**(0)**   Draw an appropriate sketch and label the essential parts.

---

**EXAMPLE 2**   A train goes whistling down a straight track at 72 km/hr, or 20 m/sec. An observer is 10 meters from the track. Find the rate of change of distance between observer and whistle:
   **a.**   When the whistle is at the point $P$ directly opposite the observer.
   **b.**   When the whistle is 100 meters before point $P$.
   **c.**   When the whistle is 100 meters beyond point $P$.

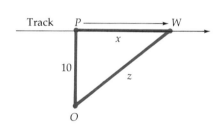

Track

**FIGURE 1**

**SOLUTION**   Draw a sketch and label the parts (Fig. 1). Think of the track as a coordinate line with origin at $P$. Draw the observer O 10 meters away from $P$. (If the observer were moving, the distance $|OP|$ would be labeled with a letter, not a number.) Let $x$ be the position of the whistle $W$, and label the distance $|OW|$ by some convenient letter, say $z$.

Now write the relation between $x$ and $z$. From the right triangle in Figure 2,

$$z^2 = 10^2 + x^2. \tag{2}$$

**FIGURE 2**

The given information is the velocity of the train, which is $dx/dt$. The desired information is the rate of change of the distance $|OW| = z$; this rate is $dz/dt$. You get the relation between the two rates by taking the derivative of each side of (2) with respect to time, treating $x$ and $z$ as functions of $t$. By the chain rule

$$\frac{dz^2}{dt} = \frac{dz^2}{dz}\frac{dz}{dt} = 2z\frac{dz}{dt}.$$

So differentiating (2) gives

$$2z\frac{dz}{dt} = 0 + 2x\frac{dx}{dt}.$$

This is the relation between the two rates:

$$\frac{dz}{dt} = \frac{x}{z}\frac{dx}{dt}. \tag{3}$$

Now you can answer the three questions:

**a.** When $W$ is at $P$, then $x = 0$, so the rate of change of distance $z = |OW|$ is

$$\frac{dz}{dt} = \frac{0}{z}\frac{dx}{dt} = 0,$$

no matter what the velocity $dx/dt$ is.

**b.** When $W$ is 100 meters before $P$, then $x = -100$ (Fig. 2) and

$$z = \sqrt{(10)^2 + (100)^2} = \sqrt{10100}.$$

So from (3)

$$\frac{dz}{dt} = \frac{-100}{\sqrt{10100}}\frac{dx}{dt}$$

$$= \frac{-10}{\sqrt{101}} \cdot 20 \text{ m/sec}$$

$$= -19.90 \text{ m/sec}$$

since the train's velocity $dx/dt$ is given as 20 m/sec. The derivative $dz/dt$ is negative, since the train is approaching and distance $z$ is decreasing.

**c.** When $W$ is 100 meters beyond $P$ then $x = 100$, $z = \sqrt{10100}$, and

$$\frac{dz}{dt} = \frac{100}{\sqrt{10100}}\frac{dx}{dt}$$

$$= \frac{10}{\sqrt{101}} \cdot 20 \text{ m/sec}$$

$$= 19.90 \text{ m/sec}.$$

*REMARK*  These rates are related to the *Doppler effect* on the pitch of sound emitted by a moving object. The pitch heard by an observer is raised when the emitter is approaching the observer, and lowered when it is receding. The train is first approaching (so the pitch is raised) then receding (so the pitch is lowered). See problems 30–32 for more details.

---

*WARNING*  When given specific numerical values for a changing variable, *do not use these values too soon.* In Example 2, if you started out with Figure 2, you would get

$$z^2 = 10^2 + 100^2 \tag{4}$$

and it would be impossible to say anything about $dz/dt$. This relation (4) is valid at *only one time t*, so it makes no sense to take its derivative with respect to $t$. In each problem, the figure will include certain parts that change in time, and others that don't.

*All parts that change should be labeled with a letter, not a number.*

---

**EXAMPLE 3**  A chemical solution is seeping through a conical filter; the angle between the side of the cone and its axis is $45°$. At a certain time $t$, the depth of fluid in the filter is 15 mm, and it is dropping at the rate of 3 mm/sec. At that instant, how fast is fluid passing through the filter?

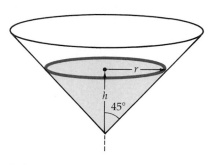

**FIGURE 3**
Depth of fluid labeled with a *letter h*, since it varies as the level drops.

*SOLUTION*   The quantity of fluid is measured by its volume $V$, say in mm$^3$. The desired rate at which it passes through the filter is then $dV/dt$. The given information is the depth of fluid, and the rate of change of the depth. The relevant figure is a cone (Fig. 3). The depth is labeled by a variable $h$, representing the depth at any time (not by the constant 15, giving the depth at just one instant). Label the radius $r$ and use the formula for the volume of a cone

$$V = \tfrac{1}{3}\pi r^2 h.$$

In the given cone with a $45°$ angle, $r = h$ (at all times!) so

$$V = \tfrac{1}{3}\pi h^2 h = \tfrac{1}{3}\pi h^3.$$

This is the relation between volume $V$ and depth $h$. Differentiate both sides *with respect to t*:

$$\frac{dV}{dt} = \frac{1}{3}\pi 3h^2 \frac{dh}{dt} = \pi h^2 \frac{dh}{dt}. \tag{5}$$

Now that you have the relation between the rates, you can use the given data $h = 15$, $dh/dt = -3$ (why minus?):

$$\frac{dV}{dt} = \pi(15)^2(-3) = -675\pi \text{ mm}^3/\text{sec}.$$

**EXAMPLE 4**   A ball rolls down a chute shaped like the graph of $y = x^3$ (Fig. 4). At a certain time $t_0$, the ball is at point $(1, 1)$, and its $y$ coordinate is decreasing at 2 m/sec. How fast is its $x$ coordinate decreasing?

*SOLUTION*   Since the ball rolls down the chute, its coordinates $(x, y)$ at every time $t$ satisfy the equation

$$y = x^3. \tag{6}$$

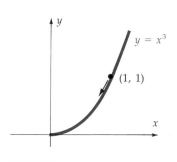

**FIGURE 4**

Both $x$ and $y$ vary with time, and the relation between their derivatives is found by differentiating (6) with respect to $t$. Think of $x$ and $y$ as functions of $t$, and use the power rule on $x^3$:

$$\frac{dy}{dt} = 3x^2 \frac{dx}{dt}. \tag{7}$$

At time $t_0$, $x = 1$, $y = 1$, and $dy/dt = -2$ ($y$ is *decreasing* at the rate of 2 m/sec). Substitute this in (7) and get

$$-2 = 3 \cdot 1^2 \frac{dx}{dt}$$

$$\frac{dx}{dt} = -2/3.$$

So $x$ is decreasing at 2/3 m/sec. The rate of decrease in $x$ is less than the rate of decrease in $y$, because at the point $(1, 1)$ the chute points more down than to the left.

**EXAMPLE 5**    A crankshaft turning at 1000 rpm (revolutions per minute) is connected to a piston (Fig. 5). How fast is the piston moving when $\theta = \pi/2$? When $\theta = 0$?

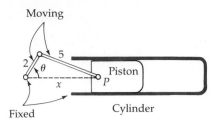

**FIGURE 5**
Piston attached to crankshaft.

*SOLUTION*    Each time the crankshaft revolves, the angle $\theta$ in Fig. 5 increases by $2\pi$ radians. Since it revolves 1000 times per minute, $\theta$ increases by $2000\pi$ radians per minute:

$$\frac{d\theta}{dt} = 2000\pi \ \frac{\text{rad}}{\text{min}}.$$

The position of the piston pin $P$ is given by the distance $x$ from $P$ to the center of the crankshaft. The question "How fast is the piston moving?" asks for $\frac{dx}{dt}$.

The relation between $x$ and $\theta$ is given by the law of cosines (Fig. 6):

$$5^2 = 2^2 + x^2 - 2 \cdot 2x \cos \theta, \quad \text{or} \quad 25 = 4 + x^2 - 4x \cos \theta.$$

Differentiate this equation with respect to $t$:

$$0 = 0 + \frac{dx^2}{dt} - 4 \frac{d(x \cos \theta)}{dt}$$

$$= 2x \frac{dx}{dt} - 4 \frac{dx}{dt} \cos \theta - 4x \left( -\sin \theta \frac{d\theta}{dt} \right).$$

So

$$0 = 2x \sin \theta \frac{d\theta}{dt} + (x - 2 \cos \theta) \frac{dx}{dt} \tag{8}$$

and

$$\frac{dx}{dt} = \frac{2x \sin \theta \dfrac{d\theta}{dt}}{2 \cos \theta - x} = \frac{4000\pi x \sin \theta}{2 \cos \theta - x} \tag{9}$$

since $\frac{d\theta}{dt} = 2000\pi$. When $\theta = \pi/2$, then $\sin \theta = 1$, $\cos \theta = 0$, and by the Pythagorean Theorem $x = \sqrt{25 - 4} = \sqrt{21}$, so

$$\theta = \frac{\pi}{2} \Rightarrow \frac{dx}{dt} = \frac{4000\pi \sqrt{21}}{-\sqrt{21}} = -4000\pi \ \text{cm/min}$$

$$\approx -20.94 \ \text{m/sec},$$

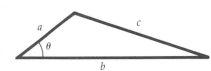

**FIGURE 6**
Law of cosines (Appendix C):
$c^2 = a^2 + b^2 - 2ab \cos \theta.$

about twice as fast as a good sprinter. When $\theta = 0$ then $\sin \theta = 0$, so (9) gives $d\theta/dt = 0$.

**EXAMPLE 6**   Compute the acceleration of the piston in Example 5, when $\theta = 0$.

**SOLUTION**   To compute $d^2x/dt^2$, it would be natural to differentiate formula (9) for $dx/dt$; but since that involves a quotient, it is easier (and more interesting) to differentiate (8) with respect to $t$, and solve for $d^2x/dt^2$. As usual, the product and chain rules are needed:

$$0 = 2 \frac{dx}{dt} \sin \theta \frac{d\theta}{dt} + 2x \left( \cos \theta \frac{d\theta}{dt} \right) \frac{d\theta}{dt} + 2x \sin \theta \frac{d^2\theta}{dt^2}$$

$$+ \left( \frac{dx}{dt} + 2 \sin \theta \frac{d\theta}{dt} \right) \frac{dx}{dt} + (x - 2 \cos \theta) \frac{d^2x}{dt^2}. \tag{10}$$

As in Example 5, the crankshaft goes around at a constant rate, so $\dfrac{d\theta}{dt} = 2000\pi$

is constant, and thus $\dfrac{d^2\theta}{dt^2} = 0$. When $\theta = 0$ then $x = 7$ (imagine $\theta = 0$ in

Figure 5) and $\dfrac{dx}{dt} = 0$ (see Example 5) so (10) gives

$$0 = 2 \cdot 7(2000\pi)^2 + (7 - 2) \frac{d^2x}{dt^2}.$$

Hence,

$$\frac{d^2x}{dt^2} = -\frac{14(2000\pi)^2}{5} \text{ cm/min}^2 \approx -307 \text{ m/(sec)}^2,$$

about 30 times the acceleration due to gravity.

# PROBLEMS

In various of these problems, you need the following formulas (proved in Chapters 6 and 9):

*Volumes*

Cone:   $V = \frac{1}{3}\pi r^2 h$

Sphere:   $V = \frac{4}{3}\pi r^3$

Pyramid:   $V = \frac{1}{3}(\text{area of base}) \times (\text{height})$

*Surface areas*

Cone:   $S = 2\pi r \sqrt{r^2 + h^2}$

Sphere:   $S = 4\pi r^2$

**A**

1.   A balloon is being blown up at the constant rate of 13 cm$^3$/sec.
   *a)   How fast does the radius $r$ increase, when $r = 3$?
   b)   Does the radius increase faster when $r$ is small or when $r$ is large? Why?

*2.   For the balloon in problem 1, how fast is the surface area increasing when $r = 3$? Does it increase faster when $r$ is small or when $r$ is large?

3. A rocket is going straight up. At a certain time its height is 100 meters, and it is rising at the rate of 15 m/sec. A camera on the ground, 50 meters from the launch pad, is constantly aimed at the rocket.

   *a) How fast is the distance between camera and rocket changing at the given time?

   b) How fast is the angle of elevation $\theta$ changing (Fig. 7)?

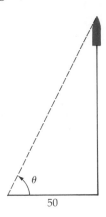

**FIGURE 7**
Angle of elevation $\theta$.

4. A man $M$ is standing 200 meters from the nearest point $P$ on a straight train track, where a train is going 100 km/hr. The pitch of the train whistle $W$, as heard by the man, depends on the rate of change of the distance between $M$ and $W$ (with respect to time). What is this rate of change, when the train is

   *a) 200 meters down the track from $P$, and approaching?

   b) At $P$?

   c) 300 meters from $P$ and going away?

5. Two parallel train tracks are a quarter of a mile apart. Passenger $A$ is going 40 miles per hour on one track, and passenger $B$ is going 60 miles per hour, in the same direction, on the other track. What is the rate of change of the distance between them when

   a) They are directly opposite each other?

   *b) $A$ is an eighth of a mile in front of $B$?

   c) $B$ is an eighth of a mile in front of $A$?

*6. Two cars are on roads that intersect at right angles, each car headed for the intersection. At what rate is the straight-line distance between them decreasing if car $A$ is one mile from the intersection and going 60 miles per hour, while

car $B$ is one-half mile from the intersection and going 80 miles per hour?

7. Two cars are on roads intersecting at right angles. Car $A$ is two miles north of the intersection, approaching it at 40 mph; car $B$ is one mile east of the intersection, driving away from it at 60 mph. Are the cars getting closer, or farther apart? How rapidly? (Consider the straight-line distance between them, not the distance along the roads.)

8. Ships Ajax and Belle leave Halifax at noon. Ajax sails south at 15 knots and Belle sails east at 10 knots. At what rate does the straight line distance between them increase at 12:30? At 3:00? (A knot is one nautical mile per hour. Give your answer in knots.)

*9. A stone is moving along a parabolic arc, the graph of $y = 9 - x^2$. From observing its shadow, it is known that the $x$ coordinate is increasing at 3 m/sec. How fast is it rising when $x = -2$? $x = 0$? $x = 1$? $x = 2$?

10. A boat is moving toward a dock at a constant speed 2 m/sec. A line from the boat to the dock is held taut; the dock end is 1 m above the boat end. How fast is the line reeled in when the boat is

    a) 10 m from the dock?

    b) 2 m from the dock?

*11. A tractor lifts a load of hay by pulling a rope over a pulley, as in Figure 8. Given the following data, find the rate at which the load of hay is rising:

   Length of rope, tractor hitch to pulley: 15 m.
   Height of pulley above tractor hitch: 9 m.
   Height of hay above tractor hitch: 3 m.
   Speed of tractor: 2 m/sec.

**FIGURE 8**
Tractor raising hay.

12. A man 6 feet tall walks away from a high wall, at 5 ft/sec, toward a light at ground level, 20 feet from the wall. How fast does his shadow grow when he is
    *a) 15 feet away from the light?
    b) 10 feet away?
    c) 5 feet away?

13. A blown-out undersea oil well is spewing oil at 25 m³ per second, forming a circular slick about 1 cm thick on the surface of the calm sea. Compute the rate of increase of the area of the slick, and of its radius $r$, when
    *a) $r = 10$ m.  b) $r = 1000$ m.

14. A container is shaped like the cone in Figure 9. Water flows in at the constant rate of 2 cm³/min. How fast is the water level rising when the container is filled to height $h$ cm? (Your answer, of course, will depend on $h$, and on the angle $\alpha$ between the axis and the side of the cone.)

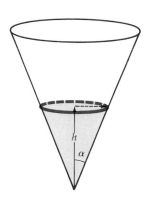

FIGURE 9

15. Sawdust is dumped on a conical pile at the rate of 5 m³/hr. The "angle of repose" of the sawdust is 45°; that is, the sides of the cone always make a 45° angle with the ground. How fast does the top of the pile rise when it is
    *a) 2 m high?  b) 10 m high?

16. A trough 2 m long has a triangular cross section, 0.3 m high and 0.3 m across at the top. Water evaporates at such a rate that the level drops 1 mm/hr. At what rate does the trough lose water (in m³/hr) when
    *a) The trough is full?
    b) The water is 0.1 m deep?
    (1 mm is 0.001 m.)

17. Imagine the ancient Egyptians building a solid pyramid with a square base 500 m on each side;

it will be 300 m high when it is finished. They first build the square base, and gradually raise the level; after each day, they have a truncated pyramid whose volume is 1000 m³ greater than the previous day's. How fast is the level of the structure rising, in m/day, when it is
    *a) 10 m high?  b) 200 m high?

18. The mast of a certain ship is a cylinder 10 m long. Ice is forming on the mast, so that the thickness grows 1 cm/hr. At what rate is the volume of ice growing when the icy mast is 12 cm in diameter? (Neglect ice collecting on the top of the mast.)

*19. In a freezing rain, ice is forming on a telephone line, and the thickness of ice increases at the rate of $\frac{1}{8}$ in/hr. Consider a 1-foot long section of wire on which ice has formed to produce a cylinder of radius $r_0$ inches. At what rate is the volume of ice increasing on this 1-foot section of wire?

20. In Example 6, compute the acceleration $d^2x/dt^2$ when
    a) $\theta = \pi/2$.  b) $\theta = \pi$.

*21. On a sunny June 21 in Quito, Ecuador, a building 80 m high casts a shadow 100 m long.
    a) How fast is the shadow lengthening? (At how many degrees per hour does the sun set at the equator?)
    b) What is the acceleration of the tip of the shadow?

22. A beacon light 200 m offshore rotates twice per minute, forming a spot of light that moves along a straight wall along the shore. How fast is the spot of light moving when it is
    a) At point O, directly opposite the beacon?
    b) 200 m from point O?

23. In the previous problem, compute the acceleration of the spot of light at the given positions.

24. From the shore, a man observes a rotating beacon. The light rotates once per minute, and the spot of light crosses a wall directly behind the observer at a rate of 10 m per second. The wall is perpendicular to the line from the observer to the beacon. How far is the beacon from the wall?

25. (Calculator) An airplane is flying east at 10,000 meters, aided by a tail wind of unknown speed. Its airspeed is 225 m/sec (about 500 mph). It sights a beacon directly in its line of flight and determines that the angle of elevation is 0.48 rad and increasing at 0.0055 rad/sec. Determine, in

m/sec and in mph:
a) The speed of the plane with respect to the ground.
b) The speed of the tail wind.

**26.** For an "ideal gas" at constant temperature, pressure $P$ and volume $V$ satisfy $PV = k$ (a constant).
a) Suppose that at some time $t_0$, $P = 1$ atm, $V = 1$ liter, and $P$ increases at 0.1 atm/min. How fast is $V$ changing? Is it increasing or decreasing?
b) Repeat the calculation if $P = 2$ and $V = 1$ at time $t_0$.
c) Repeat it if $P = 1$ and $V = 2$ at time $t_0$.

**27.** When a gas is adiabatically compressed (that is, there is no change in heat) then pressure and volume satisfy $PV^{1.4} = k$ (a constant). At time $t_0$, $P = 3$, $V = 30$ cm³, and $V$ is decreasing at a rate of 2 cm³/min. How fast is $P$ changing? Increasing or decreasing?

**B**

**28.** An observer at O sees two headlights $H_1$ and $H_2$ approaching. Familiar with cars, he can make a reasonable assumption about the spacing $s$ between the headlights. His eyes give him the angle $\theta$ formed by $H_1$ and $H_2$ at O, and also $d\theta/dt$. With this information, is it mathematically possible to determine
a) The distance $r$ between himself and the headlights? If yes, give a formula for $r$.
b) The rate $v$ at which the headlights are approaching him? If yes, give a formula for $v$.

**29.** In the previous problem, do *not* assume that $s$ is known, even approximately.

a) Determine the *relative rate* at which the headlights are approaching, that is, $\dfrac{1}{r} \cdot \dfrac{dr}{dt}$.
b) If $\theta = 0.1$ (rad) and $\dfrac{d\theta}{dt} = 0.2$ rad/sec, about how soon will the headlights be 5% closer?

**30.** The pitch of a sound is determined by its frequency. In the Doppler effect, the frequency $f_r$ received by the observer is related to the frequency $f_e$ emitted by the transmitter. If $S$ is the speed of sound and $V$ is the rate at which the emitter is going away from the receiver, the relation is

$$f_r = \frac{S}{S + V} f_e.$$

In normal atmospheric conditions, $S$ is about 330 m/sec.
a) What velocity $V$ would double the frequency (that is, raise the pitch one octave)?
b) What velocity would cut the frequency in half (lower it one octave)?

**31.** In Example 2, compute $\displaystyle\lim_{x \to +\infty} \frac{dz}{dt}$ and $\displaystyle\lim_{x \to -\infty} \frac{dz}{dt}$. Relate these limits to the velocity of the train.

**32.** An observer 10 meters from a track hears a train whistle approaching in the distance. As it passes and disappears in the opposite direction, the frequency she hears is reduced to 3/4 of its initial value (that is, the pitch changes by the interval known as a "fourth"). Assuming the speed of sound is 330 m/sec, find the speed of the train. (See problems 30 and 31.)

## 4.3

# PARAMETRIC CURVES AND L'HÔPITAL'S RULE[1]

Imagine a small particle moving in the plane. At any time $t$, it has a certain position $(x, y)$; since the particle moves, its coordinates $x$ and $y$ are functions of time $t$.

---

[1] This section can be postponed. L'Hôpital's Rule is not used again until Chapter 8.

**EXAMPLE 1**  A stone is thrown, and its coordinates at time $t$ seconds are

$$x = 10t, \qquad y = 2 + 10t - 4.8t^2. \tag{1}$$

Figure 1 shows the position at times $t = 0, \frac{1}{2}, 1, \frac{3}{2}, 2$; the path appears to be a parabola. To find the exact relation between $x$ and $y$, solve $x = 10t$ for $t$, and substitute the result in the given equation for $y$; you get $t = x/10$, so

$$y = 2 + 10(x/10) - 4.8(x/10)^2 = 2 + x - (0.048)x^2. \tag{2}$$

So the path really is a parabola.

The equations (1) are called **parametric equations** of the curve (2), with $t$ as parameter.

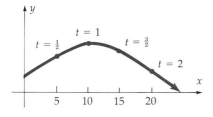

**FIGURE 1**
$x = 10t; \ y = 2 + 10t - 4.8t^2.$

Whenever $x$ and $y$ are functions of $t$, the point $(x, y)$ traces a curve in the plane. The *slope* of the curve, $\dfrac{dy}{dx}$, can be computed from the derivatives $\dfrac{dx}{dt}$ and $\dfrac{dy}{dt}$; the chain rule gives $\dfrac{dy}{dt} = \dfrac{dy}{dx} \cdot \dfrac{dx}{dt}$, so

$$\frac{dy}{dx} = \frac{dy/dt}{dx/dt}. \tag{3}$$

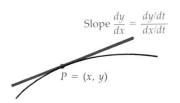

**FIGURE 2**
Slope of a parametric curve.

Thus the slope of the tangent line in Figure 2 is $\dfrac{dy/dt}{dx/dt}$.

**EXAMPLE 1 (continued)**  The parametric equations (1) give

$$\frac{dx}{dt} = 10, \qquad \frac{dy}{dt} = 10 - 9.8t.$$

Hence,

$$\frac{dy}{dx} = \frac{dy/dt}{dx/dt} = \frac{10 - 9.8t}{10} = 1 - (0.98)t.$$

On the other hand, differentiating (2) directly gives

$$\frac{dy}{dx} = 1 - (0.098)x.$$

These two expressions for $dy/dx$ agree, since $x = 10t$.

Parametric equations will be studied further in Chapter 14. But now we use them to study a simple method of evaluating limits called l'Hôpital's Rule.

## L'Hôpital's Rule

Many limit problems involve quotients $N/D$, where the limits of the numerator $N$ and denominator $D$ are known. The easiest case is where the denominator has a nonzero limit; then

$$\lim \frac{N}{D} = \frac{\lim N}{\lim D}$$

by the simple theorem on the limit of a quotient (see Sec. 3.1).

Next is the "$\frac{\neq 0}{0}$ case," where the denominator has a zero limit but the numerator does not. This produces a vertical asymptote which can be analyzed as in Section 3.3.

There remains the "$\frac{0}{0}$ case," where both numerator $N$ and denominator $D$ have limit zero. This case is called **indeterminate**, since the limit of the fraction $N/D$ is not determined by the limits of $N$ and $D$. The following are all examples of the $0/0$ case, and each has a different limit:

$$\lim_{t \to 0} \frac{t^2}{t} = 0; \qquad \lim_{t \to 0} \frac{5t}{t} = 5; \qquad \lim_{t \to 0^+} \frac{t}{t^2} = \lim_{t \to 0^+} \frac{1}{t} = +\infty.$$

Most indeterminate cases cannot be handled by such obvious algebraic cancellations. Fortunately there is another method, known as *l'Hôpital's Rule*. Figure 3 shows how it works in the case of a one-sided limit, $\lim_{t \to a^+}$. Suppose that numerator $N$ and denominator $D$ are functions of $t$ and that we are in the $0/0$ case, where

$$\lim_{t \to a^+} N(t) = 0, \qquad \lim_{t \to a^+} D(t) = 0. \tag{4}$$

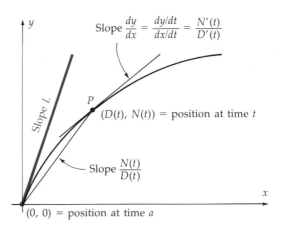

**FIGURE 3**

The "$\frac{0}{0}$ case": $L = \lim \dfrac{N}{D} = \lim \dfrac{N'}{D'}$.

Imagine that the functions $D$ and $N$ give the coordinates of a point moving in the plane, so that at time $t$ its position is

$$x = D(t), \qquad y = N(t).$$

Condition (4) says that as $t \to a^+$, the point $P = (D(t), N(t))$ moves to the origin. The quotient $N/D$ is precisely the slope of the secant line between the origin and $P$ (Fig. 3). As $t \to a^+$, then $P$ approaches the origin. Recalling how the slope of the secant line approaches the slope of the tangent line, we expect that

$$\lim_{t \to a^+} \frac{N(t)}{D(t)} = \text{slope of the curve at the origin.} \qquad (5)$$

Now, there is another way to compute the slope of the curve at the origin. On the curve, $x = D(t)$ and $y = N(t)$, so the slope of the curve at $P$ is

$$\frac{dy}{dx} = \frac{dy/dt}{dx/dt} = \frac{N'(t)}{D'(t)}.$$

The limit of this slope as $t \to a^+$ should also give the slope at the origin:

$$\lim_{t \to a^+} \frac{N'(t)}{D'(t)} = \text{slope of curve at the origin.} \qquad (6)$$

You would naturally conclude that the limit in (6) equals the limit in (5). The careful statement of this principle is:

---

**L'HÔPITAL'S RULE (0/0 case)**

Suppose that $\lim N = 0$ and $\lim D = 0$. Then

$$\lim \frac{N}{D} = \lim \frac{N'}{D'}$$

provided that $\lim \dfrac{N'}{D'}$ exists. Moreover, if $\lim \dfrac{N'}{D'} = +\infty$ or $-\infty$, then the same is true of $\lim \dfrac{N}{D}$.

---

Here lim can stand for $\lim_{t \to a^+}$, $\lim_{t \to a^-}$, $\lim_{t \to a}$, $\lim_{t \to +\infty}$ or $\lim_{t \to -\infty}$. The proof of the rule is given in Chapter 5.

---

**EXAMPLE 2**  $\displaystyle \lim_{t \to 1} \frac{4t^2 - t^4 - 3}{t^2 - 1}$ is the 0/0 case, since

$$\lim_{t \to 1} (4t^2 - t^4 - 3) = 0 \quad \text{and} \quad \lim_{t \to 1} (t^2 - 1) = 0.$$

So l'Hôpital's Rule applies:

$$\lim_{t \to 1} \frac{4t^2 - t^4 - 3}{t^2 - 1} = \lim_{t \to 1} \frac{8t - 4t^3}{2t} = \frac{4}{2} = 2.$$

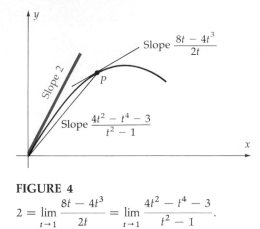

**FIGURE 4**

$$2 = \lim_{t \to 1} \frac{8t - 4t^3}{2t} = \lim_{t \to 1} \frac{4t^2 - t^4 - 3}{t^2 - 1}.$$

Figure 4 shows the curve traced by the point $P = (N(t), D(t)) = (t^2 - 1, 4t^2 - t^4 - 3)$ and the slope of the curve at the origin.

---

**EXAMPLE 3**   $\lim\limits_{t \to 2} \dfrac{t^3 - 1}{t^2 - 1}$ is an "easy case," since the limit of the denominator is $\lim\limits_{t \to 2} t^2 - 1 = 3 \neq 0$. So

$$\lim_{t \to 2} \frac{t^3 - 1}{t^2 - 1} = \frac{8 - 1}{4 - 1} = \frac{7}{3}. \tag{7}$$

l'Hôpital's Rule does not apply here, since it is *not* the 0/0 case. If you had used the rule in this example, you would have found

$$\begin{matrix} E \\ R \\ R \\ O \\ R \end{matrix} \quad \Rightarrow\Rightarrow\Rightarrow \quad \lim_{t \to 2} \frac{t^3 - 1}{t^2 - 1} = \lim_{t \to 2} \frac{3t^2}{2t^2} = \frac{12}{8} = \frac{3}{2}, \quad \Leftarrow\Leftarrow\Leftarrow \quad \begin{matrix} E \\ R \\ R \\ O \\ R \end{matrix}$$

which is not the correct result given in (7).

---

Figure 5 shows why l'Hôpital's Rule applies only in the indeterminate case. If $\lim N$ and $\lim D$ are *not* both 0, then the moving point approaches some position $P_0$ different from the origin. In this case, the limit of $N/D$ does not generally give the slope of the curve at $P_0$, so it does *not* equal the limit of $N'/D'$.

**Infinite Limits**

Suppose that in our fraction $N/D$, either $N$ or $D$ grows large without bound, that is $\lim N = \pm\infty$ or $\lim D = \pm\infty$. Again, consider three cases:

i.  $\dfrac{\text{finite}}{\infty}$, that is $\lim N$ is finite but $\lim D$ is $+\infty$ or $-\infty$. Then in the

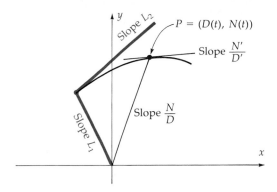

**FIGURE 5**

When $\lim_{t\to a^+} N(t) \neq 0$ or $\lim_{t\to a^+} D(t) \neq 0$, then $\lim_{t\to a^+} \dfrac{N(t)}{D(t)} = L_1$ can differ from $\lim_{t\to a^+} \dfrac{N'(t)}{D'(t)}$.

fraction $N/D$, the numerator remains bounded while the denominator grows infinitely large, so

$$\lim \frac{N}{D} = 0.$$

ii.   $\dfrac{\infty}{\text{finite}}$, that is, $\lim N = +\infty$ or $-\infty$ while $\lim D$ is finite (including zero). In this case, clearly, $\lim N/D$ is $+\infty$ or $-\infty$, depending on the sign of $N$ and of $D$.

iii.   $\dfrac{\infty}{\infty}$, the indeterminate case. Here we have:

---

**L'HÔPITAL'S RULE ($\frac{\infty}{\infty}$ case)**

If $\lim N = +\infty$ or $-\infty$ and $\lim D = +\infty$ or $-\infty$, then

$$\lim \frac{N}{D} = \lim \frac{N'}{D'}$$

provided that $\lim \dfrac{N'}{D'}$ exists, or is itself $+\infty$.

---

Figure 6 illustrates this version of the rule. A point moves in the plane as before. Since $N$ and $D$ become infinite, the point moves "off to infinity" in some way. The ratio $N/D$ is the slope of the line from the origin to the moving point $P$, and the ratio $N'/D'$ is the slope of the line tangent to the path at $P$. Suppose, for instance, that this slope has limit 1. Then as the point moves off to infinity, its own "compass heading" comes closer and closer to northeast. In that case, says l'Hôpital's rule, the line from the origin to the moving point will also aim closer and closer to the northeast as the point moves off to infinity.

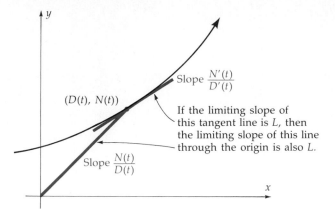

**FIGURE 6**

$$\lim \frac{N}{D} = \lim \frac{N'}{D'} \text{ in the } {}^{''}\frac{\infty}{\infty} \text{ case }{}^{''}.$$

---

**EXAMPLE 4** $\displaystyle\lim_{t \to +\infty} \frac{2t + 1}{t - 1}$ is an $\dfrac{\infty}{\infty}$ case, so

$$\lim_{t \to +\infty} \frac{2t + 1}{t - 1} = \lim_{t \to +\infty} \frac{2}{1} = 2.$$

---

*REMARK 1* The variable in these limits does not have to be $t$—we used that just so we could think of $N(t)/D(t)$ in terms of a moving point. L'Hôpital's Rule could say just as well:

$$\text{If } \lim f(x) = 0 \quad \text{and} \quad \lim g(x) = 0 \quad \text{then} \quad \lim \frac{f(x)}{g(x)} = \lim \frac{f'(x)}{g'(x)}$$

provided the latter limit exists.

*REMARK 2* If necessary, you can reapply the rule, *as long as you still have an indeterminate case.*

---

**EXAMPLE 5**

$$\lim_{x \to -\infty} \frac{x^2 + 2x + 1}{5x^2 - 3x + 2} \qquad \text{(an } \tfrac{\infty}{\infty} \text{ case)}$$

$$= \lim_{x \to -\infty} \frac{2x + 2}{10x - 3} \qquad \text{(again an } \tfrac{\infty}{\infty} \text{ case)}$$

$$= \lim_{x \to -\infty} \frac{2}{10} = \frac{1}{5}.$$

## SUMMARY

**L'Hôpital's Rule:** If $\lim N/D$ is either of the indeterminate cases $\frac{0}{0}$ or $\frac{\infty}{\infty}$, then

$$\lim \frac{N}{D} = \lim \frac{N'}{D'}$$

provided that $\lim \dfrac{N'}{D'}$ exists.

## PROBLEMS

### A

**1.** Determine which "case" the limit is, and evaluate, using l'Hôpital's Rule where appropriate.

**a)** $\displaystyle\lim_{t \to 0} \frac{\sin(t^2)}{3t}$

**b)** $\displaystyle\lim_{\theta \to 0} \frac{\tan 5\theta}{\sin 2\theta}$

**\*c)** $\displaystyle\lim_{x \to 1} \frac{x^2 - 1}{x - 1}$

**\*d)** $\displaystyle\lim_{x \to 0} \frac{x^2 - 1}{x - 1}$

**\*e)** $\displaystyle\lim_{x \to 2} \frac{x^3 - 8}{x^2 - 4}$

**\*f)** $\displaystyle\lim_{x \to +\infty} \frac{x^2 + 1}{x + 1}$

**\*g)** $\displaystyle\lim_{x \to -\infty} \frac{x + 1}{2x + 1}$

**\*h)** $\displaystyle\lim_{x \to +\infty} \frac{x + 1}{x^2 + 1}$

**\*i)** $\displaystyle\lim_{x \to +\infty} \frac{x^2 + 1}{1 + x + 3x^2}$

**\*j)** $\displaystyle\lim_{x \to -2} \frac{x^4 - 16}{x^3 + 8}$

**\*k)** $\displaystyle\lim_{x \to 2} \frac{x^4 - 16}{x^3 + 8}$

**\*l)** $\displaystyle\lim_{x \to -2} \frac{x^3 + 2x^2 + 4x + 8}{x^3 + 8}$

**m)** $\displaystyle\lim_{\theta \to 0} \frac{\cos 2\theta - 1}{\theta^2}$

**n)** $\displaystyle\lim_{\theta \to \pi/2} \frac{\cos 2\theta}{\theta - \pi/2}$

**o)** $\displaystyle\lim_{\theta \to \pi/2} \frac{\sin 2\theta}{\cos \theta}$

**\*2.** A ball flies through the air; its position at time $t$ is

$$x = 10t, \qquad y = 2 + 20t - 4.9t^2,$$

where $y =$ height above the ground. Find

**a)** The slope of the ball's path when $t = 0$.

**b)** The tangent of the angle at which the ball strikes the ground.

**\*3.** The position of a point $(x, y)$ at time $t$ is

$$x = t + \frac{1}{t}, \qquad y = t - \frac{1}{t}.$$

**a)** Determine the position of the point when $t = 2$, and the slope of the path of motion at that position.

**b)** Determine the time when the point is at $(2\frac{1}{2}, -1\frac{1}{2})$, and the slope of the curve at that point.

**c)** Show that the point moves on the curve $x^2 - y^2 = 4$ (a hyperbola [see Sec. 12.3]).

**4.** The position of a point $(x, y)$ at time $t$ is

$$x = 2 \cos \pi t, \qquad y = 3 \sin \pi t.$$

**a)** Where is it at $t = 1/4$? What is the slope of the path of motion there?

**b)** When is it at $(0, 3)$? What is the slope of the path there?

**c)** Show that the point moves on the curve $\dfrac{x^2}{4} + \dfrac{y^2}{9} = 1$ (an ellipse [see Sec. 12.2]).

**5. a)** Can $\displaystyle\lim_{x \to +\infty} \frac{\sin x}{x}$ be evaluated by l'Hôpital's Rule?

**b)** Evaluate $\displaystyle\lim_{x \to +\infty} \frac{\sin x}{x}$.

**B**

6.  Let $C$ be any constant. Find functions $N$ and $D$ such that $\lim \dfrac{N}{D}$ has the indeterminate form $\frac{0}{0}$, and $\lim\limits_{x \to 0} N/D = C$.

7.  Find functions $N$ and $D$ such that $\lim\limits_{x \to 0^+} N/D$ has the indeterminate form $0/0$, and $\lim\limits_{x \to 0^+} N/D = +\infty$.

8.  A point $P$ moves in the plane, with position at time $t$
    $$x = \frac{3t}{1 + t^3}, \qquad y = \frac{3t^2}{1 + t^3}.$$

    a)  Show that the point moves on the "folium" $x^3 + y^3 = 3xy$ of Example 2, Section 3.8 (Fig. 7).

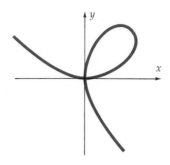

**FIGURE 7**
Folium of Descartes $x^3 + y^3 = 3xy$.

    b)  Plot the points for $t = 0$, $1/2$, $1$, $2$, $3$, $-2$, $-\frac{1}{2}$.
    c)  What is the limiting position of $P$ as $t \to +\infty$? As $t \to -\infty$?
    d)  What slope is determined by the parametric equations at $t = 0$? What is the limiting slope as $t \to +\infty$? As $t \to -\infty$?

## 4.4

## FERMAT'S PRINCIPLE AND THE LAW OF REFRACTION (optional)

One of the first applications of the ideas of calculus was Pierre de Fermat's deduction of the law of refraction. In fact, he did it before calculus had officially been invented! In this section we give some background to the problem, then the solution, and finally speculate on its role in the birth of calculus.

### Background

From infancy we have to explain to ourselves the phenomena of light. Why can't we see one object hidden behind another? What makes shadows? What happens in a mirror? How do prisms and lenses work? We all come to terms with these questions in our own way. For a scientist, the hope is to answer them with a single principle, the simpler the better.

The first two phenomena—hiding and shadows—are easily explained if you assume that light consists of infinitesimal particles (photons) that travel along straight lines, called "light rays." The hidden object can't be seen (Fig. 1) because the photons emanating from it are intercepted by the screen. An object casts a shadow (Fig. 2) because photons from the light source are intercepted by the object.

In reflection (Fig. 3), the photons seem to bounce off the mirror; the eye responds as if the photons had come to it along a straight line, and reports an object at $A'$ instead of at $A$. Experiments show that the angles $\alpha$ and $\beta$ in Figure 3 are equal, that is, "the angle of incidence equals the angle of

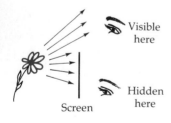

Visible here

Hidden here

Screen

**FIGURE 1**

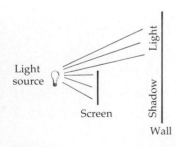

Light source

Screen

Light

Shadow

Wall

**FIGURE 2**

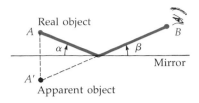

**FIGURE 3**

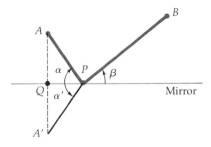

**FIGURE 4**
General route from $A$ to mirror to $B$.

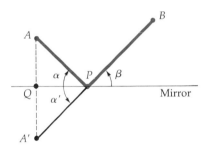

**FIGURE 5**
Shortest route from $A$ to mirror to $B$.

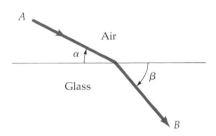

**FIGURE 6**

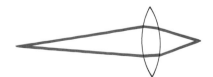

**FIGURE 7**
Bending of light by a lens.

reflection." This can be explained in various ways. One of the simplest was given (in about 75 A.D.) by Heron of Alexandria. Heron hypothesized that in going from $A$ to the mirror to $B$, the photon would take the *shortest* of all such possible routes—"nature does nothing in vain." How does this explain the equal angles? In Figure 4, each possible route $APB$ has the same length as the corresponding route $A'PB$, where $A'$ is the point opposite to $A$. (By "opposite" we mean that segments $AQ$ and $A'Q$ have equal length, and both are perpendicular to the mirror. Thus in Figure 4, $\alpha = \alpha'$, since triangles $APQ$ and $A'PQ$ are congruent.) Now, the shortest route $A'PB$ is the straight line (Fig. 5) and on this route $\beta = \alpha'$. Hence, $\beta = \alpha$ for the shortest route $APB$.

Heron's "shortest route" principle also explains the straight line motion of light—the shortest route between two points is indeed a straight line.

**Refraction** is more mysterious than reflection, and it violates the "shortest route" principle. Light rays passing from one medium to another are broken ("refracted") at the boundary between the two media (Figs. 6 and 7). This phenomenon can be seen at work in lennses, and in the "bent straw" effect (Fig. 8). Prisms and rainbows display their colors because each color is refracted through a different angle.

Ptolemy of Alexandria (about 150 A.D.) described the phenomenon in a qualitative way: The ray in the denser medium is closer to perpendicular, that is, $\beta > \alpha$ in Figure 6. But it was not until 1620 that Willibrord Snell (and independently René Descartes) discovered the law of refraction known as **Snell's law**: As the angle $\alpha$ for the incident ray varies, the ratio of the cosines

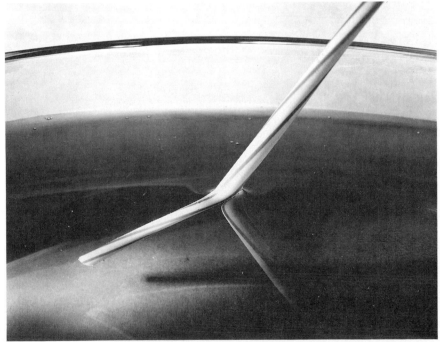

**FIGURE 8**
"Bent straw" effect.

of $\alpha$ and $\beta$ remains a constant, which is generally denoted by $n$:

$$\frac{\cos \alpha}{\cos \beta} = n. \tag{1}$$

The "index of refraction" $n$ is independent of $\alpha$ and $\beta$, but does depend on the two media involved—for example, the "$n$" for air and glass is different from the "$n$" for air and water.

Experiments supported Snell's law, but the explanation given by its discoverers created more controversy than conviction. Finally, in 1662, Pierre de Fermat gave a good explanation. First of all, he assumed that light travels with a finite velocity, and slower in a dense medium (like glass) than in a thin medium (like air). Second, he hypothesized that light takes the *quickest* route; this is known as **Fermat's least time principle**.

In a single homogeneous medium, the quickest route is simply the shortest one. But in Figure 6, the bent route from $A$ to $B$ might be quicker than the straight line; the light rays spend a little more time in the faster medium (air) but less in the slower medium (glass). The problem is to discover the optimal route, a problem which light rays solve in their own mysterious way. We will solve it in the usual calculus way.

### Snell's Law Deduced from Fermat's Principle

Figure 9 illustrates the situation. Light travels from point $A$ (in air) to point $B$ (in glass); the boundary between air and glass is flat. The speed of light in air is $s_1$ and in glass it is $s_2 < s_1$. By Fermat's principle, light takes the quickest route from $A$ to $B$, following a straight line to the air-glass boundary, then another straight line to $B$.

To find the quickest route, we minimize the time spent. Let $T_1$ be the time spent in air, and $T_2$ the time in glass; we must minimize the sum

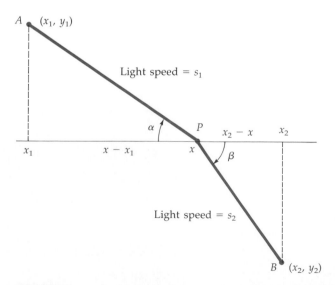

**FIGURE 9**
Deriving Snell's law.

$T = T_1 + T_2$. In each medium the speed is constant, so

$$\text{distance} = (\text{speed})(\text{time}).$$

Hence, $|AP| = s_1 T_1$ and $|PB| = s_2 T_2$, and

$$T = T_1 + T_2 = \frac{1}{s_1}|AP| + \frac{1}{s_2}|PB|.$$

Now we can suppose that the air-glass boundary is the $x$-axis, and denote by $(x, 0)$ the point $P$ where the ray crosses the axis (Fig. 9). Then time $T$ is a function of $x$, and we seek the minimum by solving $dT/dx = 0$. The speeds $s_1$ and $s_2$ are constant, so

$$\frac{dT}{dx} = \frac{1}{s_1}\frac{d|AP|}{dx} + \frac{1}{s_2}\frac{d|PB|}{dx}.$$

We therefore seek the point where

$$\frac{1}{s_1}\frac{d|AP|}{dx} + \frac{1}{s_2}\frac{d|PB|}{dx} = 0. \tag{2}$$

To find $d|AP|/dx$, let $A = (x_1, y_1)$ (Fig. 9). Then

$$|AP| = \sqrt{(x - x_1)^2 + y_1^2}$$
$$\frac{d|AP|}{dx} = \frac{2(x - x_1)}{2\sqrt{(x - x_1)^2 + y_1^2}}$$
$$= \frac{x - x_1}{|AP|}$$
$$= \cos\alpha \text{ in Figure 9.}$$

Similarly, $|PB| = \sqrt{(x - x_2)^2 + y_2^2}$, and you find

$$\frac{d|PB|}{dx} = \frac{x - x_2}{|PB|} = -\frac{x_2 - x}{|PB|} = -\cos\beta.$$

Put these in equation (2):

$$\frac{1}{s_1}\cos\alpha + \frac{1}{s_2}(-\cos\beta) = 0$$

which can be rewritten as

$$\frac{\cos\alpha}{\cos\beta} = \frac{s_1}{s_2}. \tag{3}$$

This is the law of refraction (1) given above. The constant $n$ in (1) is the *ratio of speeds $s_1/s_2$.*

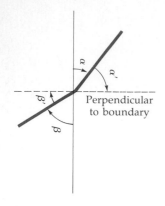

**FIGURE 10**
$\cos \alpha = \sin \alpha'$
$\cos \beta = \sin \beta'$

*REMARK 1* The law of refraction is usually stated for the angles $\alpha'$ and $\beta'$ in Figure 10, between the light rays and the line perpendicular to the boundary. These are the complements of $\alpha$ and $\beta$. Recall from trigonometry that

$$\cos \alpha = \sin \alpha', \qquad \cos \beta = \sin \beta'.$$

Put these in (3) to get Snell's law in its usual form:

$$\frac{\sin \alpha'}{\sin \beta'} = \frac{s_1}{s_2}.$$

*REMARK 2* Calculus is supposed to have been invented by Newton in 1666; so how did Fermat solve his least time problem in 1662? He says: "For this purpose I had to appeal to my method of maxima and minima which expedites the solution of this sort of problem with much success." This method, and his closely related tangent construction, are described in "The Historical Development of Calculus" by C. H. Edwards, Jr. Fermat worked with the formulas for two nearby points on the curve, performed algebraic simplifications, then set the difference between the two points equal to zero. Here lay the germ of our concept of the derivative; Newton said "I had the hint of this method [of derivatives] from Fermat's way of drawing tangents, and by applying it to abstract equations, directly and invertedly, I made it general." Newton seems to say that Fermat applied his method to particular examples (the law of refraction was surely the most interesting), but Newton saw it as a method that could be applied to functions in general, with streamlined rules of calculation—essentially, the rules developed in Chapter 3. Once these rules were given, he could then apply them "invertedly," that is, take antiderivatives. He saw, too, that velocity is a derivative, so these ideas could be used to study motion. Newton deserves all the credit he is given, but Fermat's contribution was significant.

*REMARK 3* The success of the least time principle may have led to a much grander thought: that the whole universe can be explained as the solution of a vast optimization problem. In fact, Leibniz proposed that God had considered all "possible" worlds, and selected the "best" one. This is not the place to argue the merits of that idea—Voltaire did it well in his story of Candide.

## PROBLEMS

### A

1. Prove the law of reflection, using calculus. In Figure 5, let $A = (a, b)$, $B = (c, d)$, $P = (x, 0)$, and show that $\cos \beta = \cos \alpha$ precisely when $|AP| + |PB|$ is minimum.

*2. Cars travel from city $C$, to ferry $T$, to island $I$ (Fig. 11). Exactly where should the ferry terminal

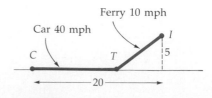

**FIGURE 11**

be placed to get the quickest route? (Imitate the proof of Snell's law.)

3. A power cable is to be laid from $A$ to $B$ on opposite banks of a river 500 meters wide (Fig. 12). Along the bank it costs \$20/meter to lay the cable, and under the river \$40/meter.
   *a) Describe a route of minimum cost.
   b) Describe *all* routes of minimum cost.
   (In this problem, cost plays the role that time plays in the refraction problem.)

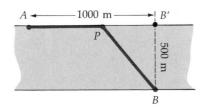

**FIGURE 12**

4. Redo problem 3 if point $B$ is only 500 meters downstream from $A$.

*5. A pumping station $P$ at $(x, 0)$ on the $x$-axis is to serve a small city $A$ and a large city $B$ (Fig. 13). It costs \$100/meter for the pipe from $P$ to $A$, and \$200/meter for the pipe from $P$ to $B$. What is the relation between $\alpha$ and $\beta$ if the total cost is to be minimum?

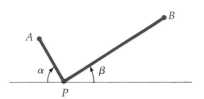

**FIGURE 13**

6. a) Take $|AP|$ as in Figure 9. Prove that $d^2|AP|/dx^2 = y_1^2/|AP|^3$.
   b) Prove: $\dfrac{d^2T}{dx^2} = \dfrac{y_1^2}{s_1|AP|^3} + \dfrac{y_2^2}{s_2|PB|^3}$. (Hence the graph of $T$ versus $x$ is concave up and has a unique minimum, at the point characterized by Snell's law.)

7. In Figure 14, describe the path of the ray of light after it leaves the glass. What is the relation between its path below the glass and the path above the glass?

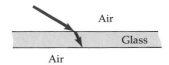

**FIGURE 14**

8. The eye contains a lens, with a watery fluid in place of glass. Why is it that you can't focus properly under water without goggles? Why do the goggles help? (See the previous problem.)

**B**

9. Let $n_{ag}$ be the index of refraction in equation (1) for light passing from air to glass, and $n_{aw}$ the index for light passing from air to water. Show that $\dfrac{n_{ag}}{n_{aw}} = n_{wg}$, where $n_{wg}$ is the index from water to glass. (This formula can be checked by simple experiments.)

10. In Figure 10, where Snell's law is satisfied, it is possible for $\alpha'$ to be $\pi/2$. The corresponding angle $\beta'$ is called the *critical angle*. Express the sine of the critical angle in terms of the speed of light above and below the boundary.

11. Relate the critical angle described in problem 10 to the solution of problem 2 and of problem 3.

**C**

12. Snell's law is a beautiful geometrical result, but it doesn't give the actual path of the ray directly. Let $A = (-1, 1)$, $B = (1, -1)$, and let the index of refraction be 1.5 (the value for air and ordinary glass). In Figure 15, find the point $x$ giving the route prescribed by Snell's law, to two decimals. (You can use Newton's method to solve a fourth degree polynomial equation.)

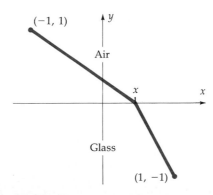

**FIGURE 15**

## 4.5

# THE LORENZ CURVE (optional)

In economics and political science, the distribution of wealth in a given population is described by a graph called the **Lorenz curve.**[2] Imagine a large population ranked according to wealth. For every number $p$ between 0 and 1, the Lorenz function $w = f(p)$ gives the fraction of the total wealth owned by the poorest $100p$ percent of the population. If, say, the poorest 90 percent has only 40 percent of the wealth, then $f(.90) = .40$.[3] Combining this with similar data for every other percentage gives the entire curve in Figure 1.

The derivative of the Lorenz function

$$\frac{dw}{dp} = f'(p)$$

has an interesting interpretation which can be discovered by taking difference quotients. Consider the segment of the population ranging between fractions $p_1$ and $p_2$ (Fig. 1). The entire population up to level $p_2$ enjoys a fraction $w_2$ of the wealth, and the part up to level $p_1$ enjoys a fraction $w_1$ of it, so the segment between $p_1$ and $p_2$ enjoys the fraction

$$\Delta w = w_2 - w_1$$

of the wealth. These people represent a fraction

$$\Delta p = p_2 - p_1$$

of the whole population. So

$$\frac{\Delta w}{\Delta p} = \frac{\text{fraction of wealth}}{\text{fraction of population}}.$$

If this ratio equals 1, the segment has just a fair share of the wealth; if it is greater than 1, this segment is "advantaged," and if less than 1, "disadvantaged."

If you take a very small segment, then $\Delta w/\Delta p$ is practically the same as the derivative $dw/dp$. So the derivative $dw/dp$ is a measure of "advantage"; population segments where $dw/dp < 1$ are disadvantaged, and those where $dw/dp > 1$ are advantaged (Fig. 2).

Since the population is ranked by wealth, the measure of advantage $dw/dp$ must be an increasing function of $p$; hence every Lorenz curve is *concave up.*

In the usual case $dw/dp$ is less than 1 at $p = 0$, and greater than 1 at $p = 1$. Somewhere in between you would expect that $dw/dp = 1$. The smallest value of $p$ where this occurs is called the "equal share coefficient"

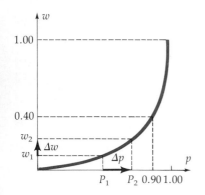

**FIGURE 1**
A Lorenz curve. The population segment $\Delta p$ shares fraction $\Delta w$ of the wealth.

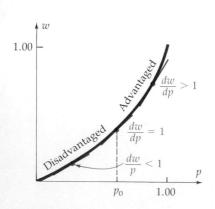

**FIGURE 2**
$p_0 = $ ESC (Equal Share Coefficient).

---

[2] P. A. Samuelson, *Economics*, 2nd ed. (New York: McGraw-Hill, 1951), pp. 70–71.

[3] The fraction 0.90 is equivalent to 90 percent; in general, the fraction $p$ of the population is equivalent to $100p$ percent of the population.

(ESC) (Fig. 2). If the ESC = $p_0$, then $dw/dp < 1$ everywhere to the left of $p_0$, so $p_0$ percent of the population is disadvantaged.

We mentioned that a Lorenz curve must be concave up; it must also have $w = 0$ when $p = 0$, and $w = 1$ when $p = 1$ (why?). Any curve with these three properties is a potential Lorenz curve.

**EXAMPLE 1**    Show that the function

$$w = p^5, \qquad 0 \le p \le 1$$

is a possible Lorenz function. Find the ESC.

*SOLUTION*    When $p = 0$, then $w = 0$; and when $p = 1$, then $w = 1$, as required. Moreover, the derivative

$$\frac{dw}{dp} = 5p^4$$

is an increasing function of $p$, so the graph is indeed concave up. The derivative equals 1 when $p^4 = 1/5$, or

$$p = \left[\frac{1}{5}\right]^{1/4} \approx 0.669.$$

The ESC is 0.669; about two thirds of the population is disadvantaged, a possible source of political unrest.

For more detail on this topic, see UMAP units 60 and 61, from COMAP, 60 Lowell St., Arlington, MA 02148. The data in Table 1 are from this source.

**Table 1    Distribution of Farmland**

| Fraction of farmers | Fraction of land (1964) | | |
|---|---|---|---|
| | Bolivia | Denmark | United States |
| 0 | 0 | 0 | 0 |
| .10 | 0 | .06 | .025 |
| .20 | 0 | .12 | .05 |
| .30 | 0 | .18 | .075 |
| .40 | 0 | .24 | .10 |
| .50 | .10 | .30 | .13 |
| .60 | .16 | .36 | .18 |
| .70 | .22 | .45 | .22 |
| .80 | .30 | .54 | .28 |
| .90 | .40 | .70 | .42 |
| 1.00 | 1.00 | 1.00 | 1.00 |

Source: Reference to UMAP Module #60, published by COMAP, Inc., 60 Lowell St., Arlington, MA 02174.

## PROBLEMS

### A

1. Show that $w = p^4$ defines a possible Lorenz function. Find the ESC.

2. Table 1 gives data for three Lorenz curves for the distribution of farmland among farmers in Bolivia, Denmark, and the United States. Sketch the three curves and on each, find (approximately) the point where $dw/dp = 1$.

*3. Figure 3 shows two Lorenz curves.
   a) Which has the higher ESC?
   b) Could a case be made that one has just as equal a distribution as the other?

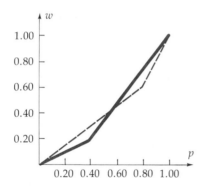

**FIGURE 3**
Two Lorenz curves.

4. Can a Lorenz curve have the shape in Figure 4? Why, or why not?

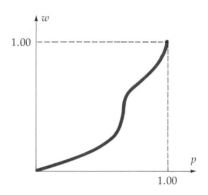

**FIGURE 4**
Can this be a Lorenz curve?

### B

5. Why must a Lorenz curve have $w = 0$ when $p = 0$? Why must $w = 1$ when $p = 1$?

6. If everyone had an equal share, there would be complete equality of wealth. What would be the Lorenz function in that case? What would it be in the other extreme, complete inequality?

7. a) For what real values of $r$ is the function $f_r(p) = p^r$ a Lorenz function?
   b) Find the ESC for $f_r(p) = p^r$, for the appropriate values of $r$ (as determined in part a).
   c) Graph $f_r(p)$ for $r = 1$, $\frac{3}{2}$, 2, and 3; show the ESC on each graph. Which of these is most equitable, and which least?

## REVIEW PROBLEMS    CHAPTER 4

*1. Approximate $(30)^{1/5}$, using $(32)^{1/5} = 2$ and an appropriate linear approximation.

2. Let $S$ denote the size of a certain glacier at time $t$. What are the signs of $dS/dt$ and $d^2S/dt^2$ if the glacier is
   *a) Shrinking more and more rapidly?
   b) Shrinking more and more gradually?
   c) Growing faster and faster?
   d) Growing more and more slowly?

*3. The period $T$ of small oscillations of a pendulum of length $L$ is

$$T = 2\pi\sqrt{L/g},$$

where $g$ is the acceleration due to gravity. Suppose that $g$ is to be determined by measuring the period $T$, with a possible (small) error $\Delta T$.
   a) What is the corresponding error $\Delta g$ in determining $g$?

**b)** Is $|\Delta g|$ greater with a long, or with a short pendulum?

**\*4.** A point $P$ moves on the circle $x^2 + y^2 = 25$.

When $P = (-4, 3)$ then $\dfrac{dx}{dt} = 4$.

**a)** Is $P$ moving clockwise, or counterclockwise?

**b)** What is $\dfrac{dy}{dt}$ at that time?

**5.** A point $P$ moves on the circle $x^2 + y^2 = 16$, with $\dfrac{dy}{dt} = x$.

**a)** Is it moving clockwise, or counterclockwise?
**\*b)** What is $dx/dt$?

**\*6.** A painter on a 10 ft ladder sees the top resting on a vertical wall, 8 ft from the ground, and *slipping down* at $\frac{1}{2}$ ft/sec! How rapidly is the bottom of the ladder being pulled away from the wall by a sinister prankster?

**\*7.** A car going 20 m/sec on a straight and level highway passes directly beneath a balloon moving vertically. Two seconds later, the distance from car to balloon is 50 m, and increasing at 6 m/sec. How high is the balloon? Is it rising or descending? How rapidly?

**\*8.** A conical cup with diameter 10 cm and height 10 cm is being filled at 7 cm$^3$/sec. How rapidly is the fluid level rising just as the cup becomes full?

**\*9.** A helicopter flying at 50 m/sec 300 m above a highway sights an approaching car. The angle of elevation $\theta$ between car and helicopter is 45°, and increasing at 10°/sec.
**a)** How fast is the car going?
**b)** Assuming that car and helicopter move at constant velocity, find $d^2\theta/dt^2$.

**\*10.** A pipeline is to be laid from an oil well $w$ miles off a straight shore to a refinery $r$ miles down the coast. The cost underwater is \$$U$/mile, and along the shore is \$$S$/mile, where $S < U$. The cheapest pipe route goes $x$ miles along the shore, then at an angle from the shore to the well. Determine $x$ to minimize the cost of laying pipe.

**\*11.** Evaluate.

**a)** $\displaystyle \lim_{x \to 0} \frac{2 \sin 5x}{\sin 3x}$

**b)** $\displaystyle \lim_{x \to 0} \tan 5x \cdot \cot 3x$

**c)** $\displaystyle \lim_{x \to 0} \frac{\sin^2(\sqrt{x})}{x}$

**d)** $\displaystyle \lim_{x \to +\infty} \frac{x}{x + 3\sqrt{x}}$

**e)** $\displaystyle \lim_{x \to 0} \frac{\sec 2x - 1}{x^2}$

**f)** $\displaystyle \lim_{x \to +\infty} \frac{\sin 3x}{1 + x^2}$

# 5

# THE MEAN VALUE
# THEOREM

Much of our intuition about calculus depends on thinking of the set of real numbers as a line, like an infinite piece of string. But how, exactly, are the real numbers to be defined? What properties can actually be proved? Our definition of the trigonometric functions presents similar difficulties, for it assumes that every real number $\theta$ corresponds to a definite point on the unit circle, $\theta$ units of arc length from the point (1, 0). How is the arc length to be defined? Once defined, can the correspondence be proved? These questions were answered in the nineteenth century (though not easily), and it is now possible to develop calculus on widely accepted logical foundations. Much of this structure is subtle and takes time to master; a happy exception is the Mean Value Theorem, the main subject of this brief chapter. Assuming earlier parts of the theory, it can be proved without appeal to pictures or intuition; yet it can be easily understood and interpreted at an intuitive level; and it gives simple proofs of other results that are less accessible to intuition.

This chapter presents first those "earlier parts" necessary for the proof, then the Mean Value Theorem itself, and finally its consequences for graphing and for antiderivatives.

## 5.1

### THE EXTREME VALUE THEOREM AND THE CRITICAL POINT THEOREM

In an optimization problem you look for maximum or minimum values of an appropriate function $f$ on an interval $I$. In principle, these points can be found by carefully graphing $f$, but this is often laborious. The two theorems in this section, when they apply, provide a quicker way to optimize.

To begin, we state the exact meaning of "maximum."

---

**DEFINITION 1**

A number $M$ is called the **maximum** of $f$ on an interval $I$ if

(i)   $M = f(\bar{x})$ for *some* point $\bar{x}$ in $I$, and
(ii)  $M \geq f(x)$ for *all* points $x$ in $I$.

In this case we write

$$M = \max_{I} f.$$

The definition of **minimum** is similar. A number which is either a maximum or a minimum is called an **extreme value**.

---

For example in Figure 1, where $f(x) = 3 - x^2$ on the interval $I = [-1, 1]$, you see that there is a maximum value

$$3 = f(0)$$

and a minimum value

$$2 = f(-1) = f(1).$$

Both 3 and 2 are *extreme values* of $f$ on $[-1, 1]$.

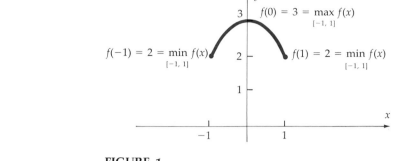

**FIGURE 1**
Maximum and minimum of $f(x) = 3 - x^2$ on $[-1, 1]$.

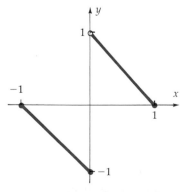

**FIGURE 2**
$f(x) = x$ has neither maximum nor minimum on infinite interval $(-\infty, \infty)$.

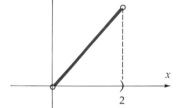

**FIGURE 3**
$f(x) = x$ has neither maximum nor minimum on the *open* interval $(0, 2)$.

**FIGURE 4**
$f$ has no maximum on $[-1, 1]$.

It can easily happen that there *are* no extreme values. For instance in Figure 2, where the function grows infinitely large positive and negative, there is no maximum and no minimum.

For the function $f(x) = x$ on the interval $I = (0, 2)$ (Fig. 3), speaking loosely, you might say that the maximum of $f$ on $I$ is 2. But $M = 2$ does not satisfy the definition; condition (i) fails, since there is *no* point $\bar{x}$ in the interval $(0, 2)$ where $f(\bar{x}) = 2$. In this case, *f has no maximum on I.* Figure 4 shows another case where $f$ has no maximum on $I$.

These figures illustrate three conditions under which there might be no maximum value: The function $f$ may be discontinuous somewhere in the interval $I$ (Fig. 4); the interval $I$ may not include its endpoints (Fig. 3); or the interval may be infinite (Fig. 2). *These are the only conditions under which a function may fail to have a maximum or minimum on I.*

---

### THEOREM 1   THE EXTREME VALUE THEOREM

Suppose that $I = [a, b]$ is a closed finite interval, and $f$ is continuous at every point of $I$. Then $f$ has a maximum and a minimum value on $I$.

---

In this theorem, continuity at the endpoints of $I$ is defined by means of the appropriate one-sided limits:

$$\lim_{x \to a^+} f(x) = f(a) \quad \text{and} \quad \lim_{x \to b^-} f(x) = f(b).$$

For example, $\sqrt{x}$ is continuous on $[0, 1]$.

The theorem depends on deep properties of the set of real numbers, and its proof is left to Appendix A; but the example in Figure 4 illustrates the essential role of continuity.

Suppose now that you are seeking the extreme points of a continuous function $f$ on a closed finite interval. By the Extreme Value Theorem, you know that they exist. Figures 5 and 6 show where they may occur. In Figure 5, the minimum and maximum occur at points $c$ where $f'(c) = 0$. In Figure 6, the minimum is at a "kink" in the graph, a point $c$ where $f'(c)$ does not exist; and the maximum occurs at an endpoint. We will prove that these are the only possibilities for extreme points.

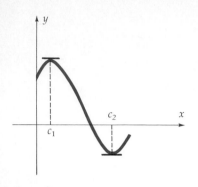

**FIGURE 5**
$f'(c) = 0$ at the maximum and minimum points.

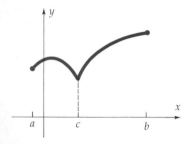

**FIGURE 6**
$f'(c)$ does not exist at the minimum; maximum occurs at endpoint $b$.

---

**THEOREM 2**

*If $f(\overline{x})$ is a maximum or a minimum value of $f$ on $I$, then either*

$$\overline{x} \text{ is an endpoint of } I,$$

or

$$f'(\overline{x}) = 0,$$

or

$$f'(\overline{x}) \text{ does not exist.}$$

---

The points $c$ where $f'(c) = 0$ are called **stationary points** (see Sec. 2.4). These, together with those where $f'(c)$ does not exist, are called **critical points**; along with the endpoints, they are the places to look when seeking extreme values.

We call Theorem 2 the *Critical Point Theorem*. Combined with the Extreme Value Theorem, it gives our "quicker way to optimize."

---

**EXAMPLE 1**    Find the extreme values of $f(x) = x^2 - 2x$ on $-1 \le x \le 2$.

**SOLUTION**    $f$ is continuous on the finite interval $-1 \le x \le 2$, including its endpoints; so the extreme values exist, and you may use the Critical Point Theorem to find them. In this case $f'(x)$ exists at every point, so the critical points are the solutions of

$$0 = f'(x) = 2x - 2.$$

Thus, $x = 1$ is the only critical point. The extreme values are therefore among the points

$$x = -1 \text{ (endpoint)}, \qquad x = 1 \text{ (critical point)}, \qquad x = 2 \text{ (endpoint)}.$$

Compute $f$ at these points:

$$f(-1) = 3 \qquad f(1) = -1 \qquad f(2) = 0.$$

Since the maximum of $f$ on $[-1, 2]$ is one of these three function values, it must be the largest one:

$$\max_{[-1,2]} f(x) = f(-1) = 3.$$

And the smallest must be the minimum:

$$\min_{[-1,2]} f(x) = f(1) = -1.$$

---

**EXAMPLE 2**    Find the extreme values of $f(x) = x + \dfrac{1}{x}$ on $(0, 4]$.

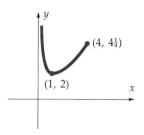

**FIGURE 7**

$f(x) = x + \dfrac{1}{x}, \ 0 < x \le 4.$

*SOLUTION*

$$f'(x) = 1 - \frac{1}{x^2} = 0 \qquad \text{when } x = \pm 1.$$

By the Critical Point Theorem, the candidates for extreme values of $f$ on the interval $(0, 4]$ are

$$x = 1 \ \text{(critical point)} \quad \text{and} \quad x = 4 \ \text{(endpoint)}.$$

(The endpoint $x = 0$ is not a candidate, because it is not in the interval $(0, 4]$.) You find that $f(1) = 2$ and $f(4) = 4\frac{1}{4}$, and might be tempted to conclude that $x = 1$ is a minimum point with $\min_{(0,4]} f = f(1) = 2$; while $x = 4$ is a maximum point with $\max_{(0,4]} f = 4\frac{1}{4}$. However, the second conclusion is false—the graph (Fig. 7) shows that *there is no maximum value!* The trouble is that the interval $(0, 4]$ is not closed, so the Extreme Value Theorem does not apply. Similar problems could occur if the interval were not finite, or the function $f$ were not continuous.

---

Example 2 illustrates a limitation of the Critical Point Theorem—it provides a guaranteed maximum or minimum only for continuous functions on closed finite intervals. Theorems covering other cases will be proved as corollaries of the Mean Value Theorem.

### Proof of the Critical Point Theorem (Theorem 2)

Suppose that $f(\bar{x})$ is the maximum of $f$ on $I$. If $\bar{x}$ is an endpoint of $I$, or if $f'(\bar{x})$ does not exist, then the conclusion of the theorem is true. Consider the one remaining case: $\bar{x}$ is *not* an endpoint, and $f'(\bar{x})$ *does* exist. By the definition of the derivative,

$$f'(\bar{x}) = \lim_{x \to \bar{x}} \frac{f(x) - f(\bar{x})}{x - \bar{x}}.$$

Since the limit exists, it equals the limit from the left and the limit from the right:

$$\lim_{x \to \bar{x}} \frac{f(x) - f(\bar{x})}{x - \bar{x}} = \lim_{x \to \bar{x}^-} \frac{f(x) - f(\bar{x})}{x - \bar{x}} = \lim_{x \to \bar{x}^+} \frac{f(x) - f(\bar{x})}{x - \bar{x}}.$$

Since $\bar{x}$ is not an endpoint, the interval $I$ contains all the points $x$ in some small interval to the left of $\bar{x}$. At such points $x - \bar{x} < 0$, and $f(x) - f(\bar{x}) \le 0$ since $f(\bar{x})$ is the maximum of $f$ on $I$. Thus (Fig. 8)

$$\frac{f(x) - f(\bar{x})}{x - \bar{x}} \ge 0.$$

Hence,

$$f'(\bar{x}) = \lim_{x \to \bar{x}^-} \frac{f(x) - f(\bar{x})}{x - \bar{x}} \ge 0. \tag{1}$$

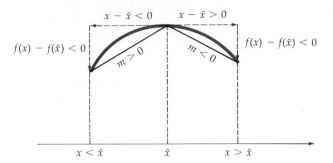

**FIGURE 8**

$$\frac{f(x) - f(\overline{x})}{x - \overline{x}} \geq 0 \text{ if } x < \overline{x}$$

$$\frac{f(x) - f(\overline{x})}{x - \overline{x}} \leq 0 \text{ if } x > \overline{x}.$$

Similarly, for points $x$ just to the right of $\overline{x}$, $\dfrac{f(x) - f(\overline{x})}{x - \overline{x}} \leq 0$ (Fig. 8), hence

$$f'(\overline{x}) = \lim_{x \to \overline{x}^+} \frac{f(x) - f(\overline{x})}{x - \overline{x}} \leq 0. \tag{2}$$

The inequalities (1) and (2) leave only one possibility: $f'(\overline{x}) = 0$. A similar argument applies if $f(\overline{x})$ is a minimum, and this completes the proof of the Critical Point Theorem.

## SUMMARY

**Definition 1:**   A *maximum value* of a function $f$ on an interval $I$ is a function value $f(\overline{x})$, where $\overline{x}$ is in $I$, and $f(\overline{x}) \geq f(x)$ for every $x$ in $I$.

   A *minimum value* on $I$ is a value $f(\underline{x})$, where $\underline{x}$ is in $I$ and $f(\underline{x}) \leq f(x)$ for every $x$ in $I$.

   An *extreme value* is either a maximum or a minimum value.

**Definition 2:**   A *critical point* of $f$ is a point $c$ in the domain of $f$ where either $f'(c) = 0$, or $f'(c)$ does not exist.

**The Critical Point Theorem:**   If $f(\overline{x})$ is an extreme value for $f$ on an interval $I$, then $\overline{x}$ is either an endpoint of $I$, or $\overline{x}$ is a *critical point*.

**The Extreme Value Theorem:**   If $f$ is continuous on a closed finite interval $[a, b]$, then $f$ has a *maximum value* and a *minimum value* on $[a, b]$. That is, there are numbers $\overline{x}$ and $\underline{x}$ in $[a, b]$ such that

$$f(\overline{x}) \geq f(x) \geq f(\underline{x}) \qquad \text{for all } x \text{ in } I.$$

## PROBLEMS

### A

1. For $f$ defined on the given interval, find all critical points and the maximum and minimum.
   *a)   $x^2 - x$ on $[0, 1]$
   b)   $2x - x^3$ on $[-1, 1]$
   *c)   $|x - 1|$ on $[0, 2]$
   d)   $x^{4/3}$ on $[-1, 1]$
   *e)   $x^{1/3}$ on $[-8, 8]$

2. Determine whether the Extreme Value Theorem applies. In any case, try to find the maximum and minimum of the given function on the given interval, if they exist. Give reasons justifying your answer.
   *a)   $f(x) = x^2$ on $[0, 2]$
   *b)   $f(x) = x^2$ on $(0, 2]$
   *c)   $f(x) = \dfrac{1}{1 + x^2}$ on $[0, 1]$
   *d)   $f(x) = \dfrac{1}{1 + x^2}$ on $[0, \infty)$
   e)   $f(x) = 2\sqrt{x(1 - x)}$ on $[0, 1]$
   f)   $f(x) = \begin{cases} x, & 0 \leq x < 1 \\ x - 1, & 1 \leq x \leq 2 \end{cases}$
   g)   $f(x) = |x|$ on $[-1, 1]$

   Solve the following optimization problems by finding the maximum or minimum of an appropriate function. Which extreme values can be found using the Extreme Value Theorem and the Critical Point Theorem?

3. a)   A rectangle is to contain 1 m$^2$. What dimensions minimize the perimeter?
   b)   A rectangle is to have perimeter 4 m. What dimensions maximize the area?

4. A rectangle with base on the $x$-axis has its upper corners on the graph of $y = 4 - x^2$.
   a)   Find the maximum area of such a rectangle.
   b)   Find the minimum area of such a rectangle. (You may treat a line segment as an "extreme case" of a rectangle.)

5. A wire of length $L$ is cut into two pieces, one of which is bent into a circle, the other into a square.
   a)   Find the minimum possible sum of the areas of the two figures.
   b)   Is there a maximum possible sum, assuming that you really cut it into two pieces?
   c)   Is there a maximum possible sum, if it is allowed to use the entire wire either for the circle or for the square?

### B

*6. In each of the following cases, sketch an example of a continuous function $f$ on the interval $I = (-\infty, +\infty)$ with the given properties, or explain why no such function exists. (In the case of a sketch, no explicit formula for $f$ is required.)
   a)   $f$ has both maximum and minimum on $I$.
   b)   $f$ has a maximum on $I$, but no minimum.
   c)   $-1 \leq f(x) \leq 1$ for every $x$, but $f$ has no maximum or minimum on $I$.

*7. In each of the following cases, sketch an example of a function $f$ defined on the given interval $I$ with the given properties, or explain why no such function exists.
   a)   $I = [0, 1]$; the function values are unbounded, that is, there is no number $L$, which is $\geq |f(x)|$ for every $x$.
   b)   $I = [0, 1]$; $f$ is continuous and the function values are unbounded.
   c)   $I = (0, 1)$; $f$ is continuous and the function values are unbounded.
   d)   $I = [0, +\infty)$; $f$ is continuous, $-1 \leq f(x) \leq 1$, but the function has neither minimum nor maximum on $I$.

8. Redo problem 6, but now give an explicit formula for each example.

*9. Redo problem 7, but now give an explicit formula for each example. (Answer for part d.)

10. The text proved that if $f(\bar{x}) \geq f(x)$ for all $x$ in an interval $I$, then $\bar{x}$ is either an endpoint of $I$ or a critical point of $f$. Prove the same conclusion for $\underline{x}$ if $f(\underline{x}) \leq f(x)$ for all $x$ in $I$. (This completes the proof of the Critical Point Theorem.)

11. Find all critical points, and the maximum and minimum, of $f(x) = \dfrac{x}{1 + |x|}$ on the interval $[-2, 2]$.

# 5.2

## THE MEAN VALUE THEOREM

The subject of this section is a pivotal theorem which justifies many of the principles used in applications of calculus. Here are the proof and just a few applications; more are found in the next section and in the following chapters.

Figure 1 illustrates a special case of the Mean Value Theorem: Between two successive zeroes of $f$, there must be a zero of $f'$. This is not true for all functions, however; it may be false if $f'(x)$ fails to exist somewhere between the two zeroes (Fig. 2) or if $f$ is discontinuous at one of the zeroes (Fig. 3). When these possibilities are ruled out, then $f'$ must indeed have a zero. The precise statement is:

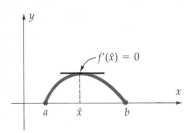

**FIGURE 1**
Rolle's Theorem.

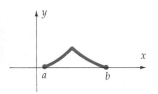

**FIGURE 2**
No point $\hat{x}$ where $f'(\hat{x}) = 0$.

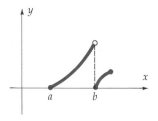

**FIGURE 3**
$f(a) = f(b) = 0$, but $f'(x) \neq 0$, for all $x$ in $(a, b)$.

---

### ROLLE'S THEOREM

Suppose that

   **(i)**   $f'(x)$ exists for $a < x < b$
  **(ii)**   $f$ is continuous for $a \leq x \leq b$
 **(iii)**   $f(a) = f(b) = 0$.

Then $f'(\hat{x}) = 0$ for some point $\hat{x}$ in the open interval $(a, b)$.

---

The idea of the proof is to show that $f$ has an extreme value $f(\hat{x})$, with $f'(\hat{x}) = 0$, as in Fig. 1. Here are the details.

From hypothesis (ii) and the Extreme Value Theorem, $f$ has a maximum and a minimum value on the closed finite interval $[a, b]$. Now consider two cases.

**CASE I.** *The maximum and minimum are both* 0. In this case, $f(x) = 0$ for every $x$ in $a \leq x \leq b$, so $f$ is constant, and $f'(\hat{x}) = 0$ for *every* point $\hat{x}$ between $a$ and $b$.

**CASE II.** *One of the extreme values is not* 0. Let $f(\hat{x})$ be an extreme value (a maximum or minimum) such that $f(\hat{x}) \neq 0$. By the Critical Point Theorem,. at least one of the following is true:

$$\hat{x} \text{ is an endpoint,} \quad \text{or} \quad f'(\hat{x}) \text{ does not exist,} \quad \text{or} \quad f'(\hat{x}) = 0. \qquad (1)$$

But $\hat{x}$ is not an endpoint $a$ or $b$, since $f(\hat{x}) \neq 0$ while $f(a) = 0$ and $f(b) = 0$, by hypothesis (iii). So $a < \hat{x} < b$; it then follows from hypothesis (i) that $f'(\hat{x})$ *does* exist, ruling out the second possibility in (1). The only remaining possibility is that $f'(\hat{x}) = 0$, and this completes the proof of Rolle's Theorem.

---

**EXAMPLE 1**   The function $f(x) = x^3 - x$ satisfies

$$f(0) = 0, \qquad f(1) = 0.$$

The other hypotheses (i) and (ii) are satisfied on the interval $[a, b] = [0, 1]$, so Rolle's Theorem applies: There is a number $\hat{x}$ between 0 and 1 such that

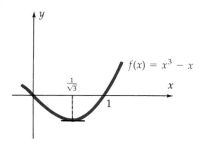

**FIGURE 4**
$f'(1/\sqrt{3}) = 0$.

$f'(\hat{x}) = 0$. In fact

$$f'(x) = 3x^2 - 1$$

and one of the zeroes of $f'$ is at $\hat{x} = 1/\sqrt{3}$, which *is* between 0 and 1 (Fig. 4).

Figure 5 illustrates the Mean Value Theorem. Draw the straight line $L$ through points $(a, f(a))$ and $(b, f(b))$ on the graph of $f$. Then somewhere between these two points, the tangent to the graph is parallel to $L$; that is, the slope of the tangent line equals the slope of $L$:

$$f'(\hat{x}) = \frac{f(b) - f(a)}{b - a}.$$

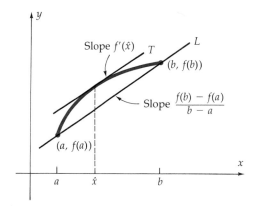

**FIGURE 5**

Mean Value Theorem. Line $T$ is parallel to line $L$, so $f'(\hat{x}) = \dfrac{f(b) - f(a)}{b - a}$.

This is like Rolle's Theorem, but without assuming that $f(a) = f(b) = 0$. The precise statement is:

---

**THE MEAN VALUE THEOREM**

Suppose that

**(i)**   $f'(x)$ exists on the open interval $a < x < b$,
**(ii)**   $f$ is continuous on the closed interval $[a, b]$.

Then there is at least one point $\hat{x}$ in the open interval $a < \hat{x} < b$, where

$$f'(\hat{x}) = \frac{f(b) - f(a)}{b - a}. \qquad (2)$$

---

Why should there be such a point $\hat{x}$? Imagine a line $\hat{L}$ parallel to $L$ and above the graph of $f$ (Fig. 6). Gradually lower $\hat{L}$ until it rests on the graph; at the point of contact $(\hat{x}, f(\hat{x}))$, the line is tangent to the graph. Notice that

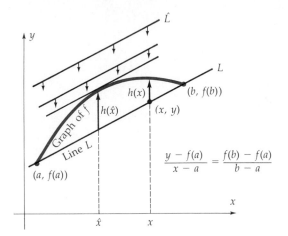

**FIGURE 6**

On $L$, $y = f(a) + (x - a)\dfrac{f(b) - f(a)}{b - a}$.

$\hat{x}$ is precisely the point where the vertical distance between the graph of $f$ and the line $L$ is greatest. So we begin the formal proof by introducing the function $h(x)$ giving the vertical height between $L$ and the graph. The equation of $L$ is (Fig. 6)

$$y = f(a) + (x - a)\frac{f(b) - f(a)}{b - a}. \tag{3}$$

The function $h(x)$ gives the difference between the $y$-coordinate on the graph of $y = f(x)$, and the $y$-coordinate on the line (3):

$$h(x) = f(x) - \left[ f(a) + (x - a)\frac{f(b) - f(a)}{b - a} \right]. \tag{4}$$

Since $f'(x)$ exists for $a < x < b$, the function $h$ has derivative

$$h'(x) = f'(x) - \frac{f(b) - f(a)}{b - a}, \qquad a < x < b. \tag{5}$$

So $h$ satisfies hypothesis (i) of Rolle's Theorem. It also satisfies hypothesis (ii); $h$ is continuous for $a \le x \le b$ because $f$ is, and so are the other terms in (4). Finally, $h(a) = h(b) = 0$; you can see this in Fig. 6, and can check it by evaluating equation (4) at $x = a$ and $x = b$. Since all hypotheses of Rolle's Theorem are satisfied, the conclusion applies: There is a point $\hat{x}$ such that $h'(\hat{x}) = 0$. By (5), this gives

$$0 = f'(\hat{x}) - \frac{f(b) - f(a)}{b - a}.$$

This is equivalent to the conclusion of the Mean Value Theorem, equation (2), so the proof is complete.

**EXAMPLE 2**   The function $f(x) = \sqrt{x}$ satisfies the hypotheses of the Mean Value Theorem on the interval [0, 4]. Find a point $\hat{x}$ such that

$$f'(\hat{x}) = \frac{f(4) - f(0)}{4 - 0}.$$

*SOLUTION*

$$\frac{f(4) - f(0)}{4 - 0} = \frac{\sqrt{4} - 0}{4} = \frac{1}{2}$$

so you need an $\hat{x}$ with $f'(\hat{x}) = 1/2$. The derivative of $f$ is $f'(x) = \dfrac{1}{2\sqrt{x}}$. Clearly $f'(1) = \frac{1}{2}$, so $\hat{x} = 1$ will do.

## Applications of the Mean Value Theorem

Section 2.8 on antiderivatives stated without proof:

> **THEOREM 1**
>
> Any function $f$ whose derivative $f'$ equals 0 throughout an interval $I$ is constant on $I$.

Now we can prove this essential fact. Choose any point $x_0$ in the interval $I$, we will show that $f(x) = f(x_0)$ for every other $x$ in $I$. Since $f' = 0$ throughout $I$, then $f'$ exists at every point in the closed interval from $x$ to $x_0$, and so $f$ is also continuous in that closed interval. Thus the Mean Value Theorem applies:

$$\frac{f(x) - f(x_0)}{x - x_0} = f'(\hat{x}) \tag{6}$$

for some $\hat{x}$ between $x$ and $x_0$. Since $f' = 0$ at *every* point in $I$, then $f'(\hat{x}) = 0$. So (6) implies that $f(x) - f(x_0) = 0$, or $f(x) = f(x_0)$, just what we wanted to prove.

Figure 7 relates the Mean Value Theorem to the linear approximation of $\Delta f$:

$$\Delta f \approx f'(x)\,\Delta x. \tag{7}$$

Apply the Theorem to the interval $[x, x + \Delta x]$:

$$f'(\hat{x}) = \frac{f(x + \Delta x) - f(x)}{(x + \Delta x) - x} = \frac{\Delta f}{\Delta x}$$

or

$$\Delta f = f'(\hat{x})\,\Delta x. \tag{8}$$

Thus, if you could evaluate $f'$ at the appropriate point $\hat{x}$, you would get $\Delta f$ *exactly*, as in (8). But if you replace $\hat{x}$ by the "base point" $x$, you get just the *approximation* (7).

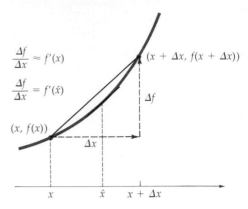

**FIGURE 7**
$\Delta f \approx f'(x)\,\Delta x$; $\Delta f = f'(\hat{x})\,\Delta x$.

Velocity offers another interpretation of the Mean Value Theorem. Let $y$ denote position at time $t$, and

$$\Delta y = y(t + \Delta t) - y(t)$$

be the displacement during the time interval $[t, t + \Delta t]$. Assuming that the derivative $v = dy/dt$ exists throughout this interval, we conclude that there is a time $\hat{t}$ such that

$$\frac{\Delta y}{\Delta t} = y'(\hat{t}) = v(\hat{t}).$$

That is, the *average velocity* over the interval equals the *instantaneous velocity* at some time $\hat{t}$ in that interval.

---

**EXAMPLE 3**   A law-abiding car travels with velocity $|v(t)| \le 55$ mph for $0 \le t \le 2$ hours. Is it necessarily true that the car goes $\le 110$ miles in those two hours?

***SOLUTION***   Here $\Delta t = 2$, and for some $\hat{t}$ in the interval $(0, 2)$,

$$\left|\frac{\Delta y}{\Delta t}\right| = \left|\frac{\Delta y}{2}\right| = |v(\hat{t})| \le 55.$$

So $|\Delta y| \le 2 \cdot 55 = 110$; the car does indeed go $\le 110$ miles. [You must assume, of course, that $v(t)$ is defined throughout this time interval.]

---

## SUMMARY

***The Mean Value Theorem (see Fig. 5):***   If $f'(x)$ exists for $a < x < b$ and $f$ is continuous for $a \le x \le b$, then there is a point $\hat{x}$ between $a$ and $b$ such that

$$f'(\hat{x}) = \frac{f(b) - f(a)}{b - a}.$$

## PROBLEMS

### A

1. Each of the following functions has exactly two zeroes. Find them, and find (if possible) a zero of $f'$ that lies between them. If no such zero exists, show which hypothesis of Rolle's Theorem is not satisfied.
   
   a)  $f(x) = x(x - 1)$     b)  $f(x) = x^3 + x^2$
   
   c)  $f(x) = 1 - |x|$     d)  $f(x) = 1 - \dfrac{1}{x^2}$

2. Given $f(x)$, $a$, and $b$, find a point $\hat{x}$ satisfying the conclusion of the Mean Value Theorem. Sketch the graph on the interval $[a, b]$, the line $L$ through its endpoints, and the parallel tangent line or lines. (There may be more than one such line.)
   
   *a)  $f(x) = \sqrt{x}$, $a = 0$, $b = 1$
   
   b)  $f(x) = x^2$, $a = 0$, $b = 1$ (What is the relation of this to part a?)
   
   *c)  $f(x) = x^2 + 2x + 1$, $a = 0$, $b = 1$
   
   *d)  $f(x) = x^3$, $a = -1$, $b = 1$
   
   *e)  $f(x) = x + \dfrac{1}{x}$, $a = \frac{1}{2}$, $b = 2$
   
   f)  $f(x) = \sin x$, $a = 0$, $b = \pi$
   
   g)  $f(x) = \cos x$, $a = 0$, $b = 2\pi$

3. Find a point $\hat{x}$ between $x$ and $x + \Delta x$ where $\Delta f = f'(\hat{x})\,\Delta x$. [By $\Delta f$, we mean $f(x + \Delta x) - f(x)$].
   
   *a)  $f(x) = x^3$; $x = 0$; $\Delta x = 1$
   
   b)  $f(x) = \dfrac{1}{x}$; $x = 1$; $\Delta x = 2$
   
   *c)  $f(x) = \dfrac{1}{x}$; $x = 3$; $\Delta x = -1$
   
   d)  $f(x) = \sin x$; $x = 0$; $\Delta x = \pi$

4. In each case, show that the conclusion of the Mean Value Theorem fails. Which hypotheses are not satisfied? Draw a sketch.
   
   a)  $f(x) = |x|$ on $[-1, 1]$
   
   b)  $f(x) = \begin{cases} 1, & x > 0 \\ 0, & x = 0 \end{cases}$ on $[0, 1]$
   
   c)  $f(x) = 1/x$ on $[-1, 1]$

5. A car enters a turnpike at $12{:}00$ and at $12{:}30$ reaches an exit 35 miles away. The driver claims that he never exceeded the 55 mph limit. Can you prove him wrong?

6. A *Lorenz curve* (see Sec. 4.5) is the graph of a function on $[0, 1]$, concave up, with $f(0) = 0$ and $f(1) = 1$. The *ESC* (equal share coefficient) of $f$ is a value $\bar{x}$ where $f'(\bar{x}) = 1$. Under what conditions on the function $f$ does such a value $\bar{x}$ exist?

### B

7. Show that in the special case where $f$ is a quadratic function, the "mean value point" $\hat{x}$ for $f$ on any interval $[a, b]$ is precisely the midpoint of the interval.

8. Prove that $|\sin \theta - \sin \varphi| \leq |\theta - \varphi|$. (Apply the Mean Value Theorem to $\sin x$ on an appropriate interval.)

9. Prove that for $|a| \leq 1$ and $|b| \leq 1$, $|b^3 - a^3| \leq 3|b - a|$.

10. Prove that for $|\theta| < \pi/2$ and $|\varphi| < \pi/2$, $|\tan \theta - \tan \varphi| \geq |\theta - \varphi|$.

11. Suppose that $f'$ exists on the closed interval $[x, x + \Delta x]$. Does it follow that
   
   $$|\Delta f| \leq \left[ \max_{[x, x+\Delta x]} |f'| \right] \Delta x?$$
   
   Is $|\Delta f| \geq \left[ \min_{[x, x+\Delta x]} |f'| \right] \Delta x?$

12. Suppose that $f'(x)$ exists everywhere, and has exactly one zero. Prove that $f(x)$ has at most two zeroes. (Hint: If $f(x)$ had three zeroes $a < b < c$, you could apply Rolle's Theorem on the interval $[a, b]$, and again on $[b, c]$. *Note*: This same question was asked in Chapter 2 Section 2.4 problem 11. There you had to answer by intuition; here you can quote a theorem.)

13. Let $f(x) = ax^3 + bx^2 + cx + d$ be a polynomial of degree three: that is, $a \neq 0$. Prove the following:
   
   a)  $f''(x)$ is a linear function with nonzero slope.
   
   b)  $f''(x)$ has exactly one zero.
   
   c)  $f'(x)$ has at most two zeroes on the real axis. (See the previous problem.)
   
   d)  $f(x)$ has at most three zeroes on the real axis.

14. Following the lines of the previous problem, explain why a polynomial of degree $n$ has at most $n$ zeroes on the real axis.

**15.** This problem shows that the "mean value point" $\hat{x}$ can be anywhere in the interval $(a, b)$. For each positive real number $p$, define the function

$$f_p(x) = x^p(x - 1).$$

a) Find the mean value point $\hat{x}$ for $f_p$ on the interval $[0, 1]$.

b) Show that for *any* point $c$ in $(0, 1)$, there is a positive $p$ such that $c$ is the "mean value point" for $f_p$.

c) For what constants $p$ does the function $f_p$ on $[0, 1]$ satisfy the hypotheses of the Mean Value Theorem?

**C**

**16.** *Cauchy's Mean Value Theorem*: Suppose that $f$ and $g$ both satisfy the hypotheses of the Mean Value Theorem on the interval $[a, b]$, and $g'$ is never zero on $(a, b)$. Then there is a point $\hat{t}$ in $(a, b)$ such that

$$\frac{f'(\hat{t})}{g'(\hat{t})} = \frac{f(b) - f(a)}{g(b) - g(a)}.$$

a) Interpret this equation as a statement about the curve defined by the equations $x = g(t)$, $y = f(t)$.

b) Prove the theorem. (Hint: Apply the Mean Value Theorem to the function

$$h(t) = f(t)[g(b) - g(a)] - g(t)[f(b) - f(a)].$$

**17.** Prove l'Hôpital's Rule

$$\lim_{t \to a^+} \frac{f(t)}{g(t)} = \lim_{t \to a^+} \frac{f'(t)}{g'(t)} \qquad (9)$$

given that $f(a) = \lim_{t \to a^+} f(t) = 0$ and $g(a) = \lim_{t \to a^+} g(t) = 0$, and the limit on the right-hand side of (9) exists. (Use the previous problem. Since $\lim_{t \to a^+} \frac{f'(t)}{g'(t)}$ is assumed to exist, it follows that $f'(t)/g'(t)$ is defined in some interval to the right of $a$, hence $g'(t) \neq 0$ in that interval.)

---

## 5.3

## APPLICATIONS TO GRAPHING AND OPTIMIZATION

This chapter ends with some concepts and theorems related to graphing. Some are familiar, some new, but until now they have not been *proved*; for the proofs rely, directly or indirectly, on the Mean Value Theorem.

The main use of the derivative in graphing rests on:

> **THEOREM 1**
>
> If $f' > 0$ throughout an interval $I$, then $f$ is increasing on $I$. That is, for any points $x_1$ and $x_2$ in $I$,
>
> $$x_1 < x_2 \Rightarrow f(x_1) < f(x_2).$$

You have already seen many applications of this. For the proof, apply the Mean Value Theorem to $f$ on the interval $[x_1, x_2]$; this is justified, for $f'$ exists throughout $I$, and it follows that $f$ is continuous on $[x_1, x_2]$, by Theorem 1, Section 3.5. Hence you can write

$$\frac{f(x_2) - f(x_1)}{x_2 - x_1} = f'(\hat{x}),$$

or

$$f(x_2) - f(x_1) = f'(\hat{x}) \cdot (x_2 - x_1).  \tag{1}$$

Since $f' > 0$ at *every* point of $I$, then in particular $f'(\hat{x}) > 0$; and since $x_2 > x_1$, it follows from (1) that $f(x_2) > f(x_1)$, as was to be proved.

In much the same way, you can prove:

---

**THEOREM 2**

If $f$ is continuous on $[a, b]$ and $f' > 0$ on the open interval $(a, b)$, then $f$ is increasing on $[a, b]$.

---

To apply Theorems 1 and 2, you must know the sign of $f'$ throughout the relevant interval. However, certain features of the graph can be determined by the values of $f'$ and $f''$ at just one point. Figure 1 illustrates the most important of these, the *local maximum and minimum points*; points where $f$ has a maximum or minimum with respect to other *nearby* points.

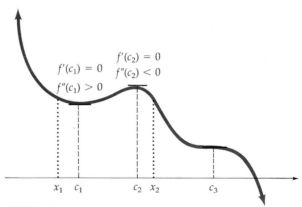

**FIGURE 1**
Local maximum and minimum points.

---

**DEFINITION**

$c$ is a *local maximum point* for $f$ if the domain of $f$ contains an open interval $(x_1, x_2)$ containing $c$, such that

$$f(c) = \max_{(x_1, x_2)} f(x).$$

$c$ is a *strict* local maximum point for $f$ if there is such an interval, with

$$f(c) > f(x)  \qquad \text{for all } x \neq c \text{ in } (x_1, x_2).$$

---

By this definition, an *endpoint* of the domain of $f$ cannot be called a local maximum point; thus a more complete description would be "local *interior* maximum point." Another commonly used term is *relative* maximum, that is, maximum relative to nearby points.

The definitions of local minimum and strict local minimum are similar. In Figure 1

$c_3$ is a local minimum point,
$c_1$ is a strict local minimum point, and
$c_2$ is a strict local maximum point.

The figure suggests, correctly, that strict local maxima and minima can be distinguished by the sign of the second derivative.

---

**THEOREM 3**

If $f'(c) = 0$ and $f''(c) > 0$, then $c$ is a strict local minimum point for $f$.

---

Part of the proof relies on the formal definition of limit in Appendix A, but we can explain most of it here. The assumption $f''(c) > 0$ means that

$$\lim_{x \to c} \frac{f'(x) - f'(c)}{x - c} = f''(c) > 0.$$

It follows that when $x$ is sufficiently close to $c$, then

$$\frac{f'(x) - f'(c)}{x - c} > 0. \tag{2}$$

More precisely, the formal definition of limit guarantees that there is an open interval $(x_1, x_2)$ containing $c$ such that (2) is true for all $x \neq c$ in $(x_1, x_2)$. By assumption, $f'(c) = 0$, so (2) gives

$$\frac{f'(x)}{x - c} > 0 \qquad \text{for } x \neq c \text{ in } (x_1, x_2). \tag{3}$$

when $x > c$, then $x - c > 0$, so (3) implies that

$$f'(x) > 0 \qquad \text{for } c < x < x_2.$$

When $x < c$, then $x - c < 0$, and (3) gives

$$f'(x) < 0 \qquad \text{for } x_1 < x < c.$$

By Theorem 2, $f$ is *decreasing* on $[x_1, c]$ and *increasing* on $[c, x_2]$. It follows that $f(c) < f(x)$ for $x_1 \leq x < c$ and for $c < x \leq x_2$; this proves Theorem 3.

---

**EXAMPLE 1**   For $f(x) = x + \dfrac{1}{x}$, the derivatives are $f'(x) = 1 - 1/x^2$ and $f''(x) = 2/x^3$. Thus

$$f'(1) = 0, \qquad f''(1) = 2 > 0$$
$$f'(-1) = 0, \qquad f''(-1) = -2 < 0.$$

By Theorem 3, 1 is a strict local minimum point, and $-1$ is a strict local maximum point (Fig. 2).

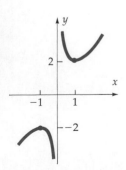

**FIGURE 2**

Local max and min for $f(x) = x + \dfrac{1}{x}$.

**FIGURE 3**
Graph concave up where $f'' > 0$.

With Theorem 3, we can justify the geometric interpretation of the sign of the second derivative (Fig. 3): Where $f'' > 0$ the graph is concave *up*, in the sense that it lies *above* its tangent lines. Precisely:

---

**THEOREM 4**

If $f''(c) > 0$, then there is an open interval $(x_1, x_2)$ containing $c$ such that in $(x_1, x_2)$ the graph of $f$ is above the tangent line at $c$, except at the point of tangency.

---

**PROOF**   Denote by $h$ the difference between the $y$ values on the graph of $f$, and those on the tangent line at $(c, f(c))$:

$$h(x) = f(x) - [f(c) + (x - c)f'(c)].$$

The graph of $f$ is above the tangent line precisely where $h(x) > 0$ (Fig. 4). Now apply Theorem 3 to $h$. You find

$$h'(c) = f'(c) - [0 + f'(c)] = 0$$
$$h''(c) = f''(c) > 0.$$

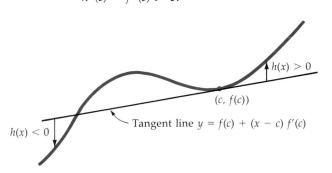

FIGURE 4

So $c$ is a strict local minimum point for $h$, and there is an interval $(x_1, x_2)$ containing $c$ such that $h(x) > h(c) = 0$ for all $x \neq c$ in $(x_1, x_2)$. For these $x$, the graph is above the tangent line, and Theorem 4 is proved.

### Global Extreme Points

Theorem 3 gives only *local* maximum points. In optimization problems, you seek "global" maxima, values $f(c)$ which are greater than or equal to $f(x)$ for all relevant $x$, not just for $x$ near $c$. In general, a local maximum need not be global; Figures 1, 2, and 5a show this. However, Figure 5b suggests a common situation in which a local maximum or minimum is actually global.

(a)
Two critical points; local maximum and minimum are not global.

(b)
One critical point; local maximum is global.

**FIGURE 5**

---

**THEOREM 5**

Suppose that $f$ is continuous on an interval $I$, and $c$ is the *only* critical point of $f$ on $I$. If $c$ is a *local* maximum point, then $f(c)$ is the global maximum of $f$ on the whole interval $I$. In particular, if $c$ is the only critical point and $f''(c) < 0$, then $f(c)$ is a global maximum.

---

**PROOF**    Suppose that $f(c)$ were not a global maximum, that is $f(c) < f(b)$ for some point $b$ in $I$. Suppose, to be definite, that $b > c$, as in Figure 6. (The graph near $c$ is drawn solid, since we know that $c$ is a local maximum; the rest, being hypothetical, is dashed.) Since $c$ is a local maximum, there is an interval $(c, x_2)$ for which

$$f(x) \le f(c) \qquad \text{if } c < x < x_2.$$

Let $a$ be any point in this interval; then $f(a) \le f(c) < f(b)$. By the Intermediate Value Theorem, there is a point $\bar{x}$ between $a$ and $b$ where $f(\bar{x}) = f(c)$ (Fig. 6). Since we chose $c < a$ and supposed that $c < b$, $\bar{x}$ is different from $c$. By the Mean Value Theorem, there is a critical point $\hat{x}$ *between* $c$ and $\bar{x}$ (see problem 28). On the other hand, by hypothesis, $c$ is the *only* critical point of $f$ on $I$. This contradiction arose from supposing that $f(c)$ is not a global maximum; we conclude that $f(c)$ is a global maximum, as was to be proved.

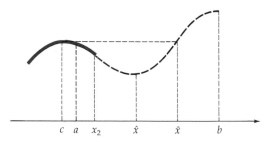

**FIGURE 6**
Illustrating proof of Theorem 5.

---

**EXAMPLE 2**    The function $f(x) = x + 1/x$ in Example 1 has just two critical points: $x = 1$ is a local minimum, and $x = -1$ is a local maximum. On the interval $I = (0, \infty)$, the *only* critical point is the local minimum point $x = 1$; by Theorem 5

$$f(1) = 2 \le f(x) \qquad \text{for } 0 < x < \infty.$$

And on the interval $(-\infty, 0)$ the only critical point is the local maximum point $x = -1$; thus

$$f(-1) = -2 \ge f(x) \qquad \text{for } -\infty < x < 0.$$

---

## PROBLEMS

### A

1. In Theorem 3, what would be the conclusion if $f''(c) > 0$ were replaced by $f''(c) < 0$?

2. In Theorem 5, what would be the conclusion if the assumption "$c$ is a local maximum" were replaced by "$c$ is a local minimum"?

3. Determine all local maximum and minimum points of the given function, using Theorem 3 for the minima and its obvious analog for the maxima.

   *a)  $y = x^2 + 1/x$

   b)  $y = x^3 - 3x$

4. Determine all local maximum and minimum points. In which cases does Theorem 3 apply?
   a) $y = x^2$   b) $y = x^3$   c) $y = x^4$

5. Prove: If $f' < 0$ throughout an interval $I$, then $f$ is decreasing on $I$.

6. Sketch a single continuous function on the interval $(0, 4)$ which has all the following properties: a global maximum; a local maximum which is not a global maximum; a local minimum; no global minimum.

Reduce the following problems to the question of finding a global extreme point of a function $f$ on an interval $I$. Solve the problem, and prove that the extreme point you find is global, either by Theorem 5 or Theorem 1, or by the theorems in Section 5.1.

*7. A rectangle is inscribed in a circle of radius $R$. What dimensions give the rectangle of maximum area?

8. A broken semicircular window of radius $R$ is to be replaced by a rectangular one, inscribed in the original semicircular opening. Give the base and height of the rectangular window of maximum area. (What is the relation of this problem to the previous one?)

9. *a) A triangle is formed by the positive $x$- and $y$-axes and a straight line through $P = (1, 1)$. Find the maximum and minimum area of such a triangle.
   b) Repeat part a if $P = (1, 2)$.

10. A can is made of two rectangles, one $h \times 2\pi r$ for the sides, and another $2r \times 4r$ from which two circles of radius $r$ are cut to form the top and bottom. The can is to contain 1000 cm$^3$. What dimensions $h$ and $r$ minimize the sum of the areas of the two rectangles? Show that for this optional can, the $2r \times 4r$ rectangle has half the area of the $h \times 2\pi r$ rectangle.

*11. A conical cup is to hold 100 cm$^3$. What dimensions require the least surface area?

12. Find the point on the parabola $4y = x^2$ closest to the given point $P$ on the $y$-axis. In each case, show that the line from $P$ to the closest point $Q$ is perpendicular to the parabola at $Q$.
   *a) (0, 1)   b) (0, 2)
   *c) (0, 3)   d) (0, $c$) where $c > 0$
   (Note that (0, 1) is the focus of the parabola.)

13. The stiffness of a rectangular beam of width $w$ and depth $d$ equals $kwd^3$, according to a standard formula from mechanical engineering ($k$ is a constant). Find the dimensions of the stiffest beam that can be cut from a round log of radius $r$, and show that it has the proportions $d/w = \sqrt{3}$.

14. The strength of a rectangular wooden beam of given length is $kwd^2$, where $w$ is width, $d$ is depth, and $k$ is a constant. Find the ratio of depth to width for the strongest beam that can be cut from a circular log. [Strength (resistance to breaking) is different from stiffness (resistance to bending).]

*15. The brightness at a distance $r$ from a point source of light of intensity $I$ is $I/r^2$. Suppose that on the $x$-axis there is a light of intensity $A$ at the origin and a light of intensity $B$ at the point $x = 1$. For what ratio $A/B$ does the darkest spot between the sources occur at $x = 1/3$?

16. An island 20 miles from a relatively straight coast is to arrange permanent car ferry service to a city 50 miles down the coast.
   a) If the ferry travels 15 mph and cars average 45 mph, where should the mainland ferry terminal be located to make the trip as quick as possible?
   b) If the ferry travels $F$ mph and cars average $C$ mph, for what values of $F/C$ should the terminal be located right in the mainland city to make the trip as quick as possible?

17. Suppose it costs $A + Bx$ to produce $x$ units of an item, and they can be sold at a price $C - Dx$ per unit, where $A, B, C, D$ are positive constants.
   a) What value of $x$ maximizes profit, and what is the corresponding price?
   b) If a tax $T$ dollars per unit is added to the production cost, what is the new profit-maximizing price? How much of the tax is "passed on" to the buyer?
   c) What is the situation if tax $T$ is added to the selling price?

*18. Find a point $Q$ on the hyperbola $xy = 1$, which is closest to the given point $P$. In each case, show that the segment from $P$ to $Q$ is perpendicular to the hyperbola at $Q$.
   a) $P = (0, 0)$   b) $P = (0, 1)$

19. A hiker is resting on a hill shaped like the graph of $y = \dfrac{-x^2}{100}$ at the point $(-50, -25)$, dimensions in meters. The hiker's eyes are at $E = (-50, -24)$.

**\*a)** What are the highest and lowest visible points of the hill? (Consider the slope of the line segment from $E$ to points $(x, y)$ on the hill.)

**b)** Show that the line of sight to the highest visible point is tangent to the hill at that point.

**20.** An observer at $(-1, 0)$ sees a ridge shaped like $y = \sqrt{x}$. What part of the ridge is visible to him? Show that the line of sight is tangent to the hill at the farthest visible point.

**21.** Given the two points $F_1 = (1, 0)$ and $F_2 = (-1, 0)$, and the circle $C = \{(x, y): x^2 + y^2 = 4\}$

**a)** Find the points $P$ on $C$ such that $|PF_1| + |PF_2|$ is a minimum.

**b)** Find the points $P$ on $C$ such that $|PF_1| + |PF_2|$ is a maximum.

(If you know about ellipses, this problem has an obvious geometric solution.)

**22.** Given a graph $G$ of a differentiable function $f$, and a point $P_0$ not on $G$, suppose that the distance from $P_0$ to $G$ achieves a minimum at a point $P_1$ on $G$, not an endpoint of $G$. Show that the line through $P_0$ and $P_1$ is perpendicular to $G$ at $P_1$.

**23.** A traveling merchant came to Minsk with his scale out of balance; one side was shorter than the other (Fig. 7). However, he compensated for this by weighing part of every order on each side. If a customer wanted, say, two pounds of sugar, the merchant gave him one "skimpy" pound (Fig. 8a) and one "generous" pound (Fig. 8b). Was this unfair to anyone? If so, to whom? (Hint: By the law of levers, the weight $P_s$ of the "skimpy" pound is given by $l \cdot 1 = xl \cdot P_s$.)

**FIGURE 7**
Unbalanced scale.

(a) Skimpy pound  (b) Generous pound

**FIGURE 8**

**24.** Prove Theorem 2.

**25.** Prove the following version of Rolle's Theorem: If $f$ is continuous on $[a, b]$ and $f(a) = f(b)$, then $f$ has a critical point in the open interval $(a, b)$. [Hint: Consider these cases: i) $f'(x)$ exists throughout $(a, b)$; ii) $f'(x)$ fails to exist at some point of $(a, b)$.]

**26.** Prove: If $f'' > 0$ throughout an interval $I$, and $c$ is in $I$, then the graph of $f$ is above the line tangent to the graph at $(c, f(c))$, at every point of $I$ except the point of tangency.

---

# REVIEW PROBLEMS    CHAPTER 5

(See the optimization problems 7 to 27 in Section 5.3.)

**1. a)** State the Extreme Value Theorem.

**b)** Show by appropriate examples that each hypothesis of the Extreme Value Theorem is necessary.

**2. a)** Define the term "critical point."

**b)** State the Critical Point Theorem.

**c)** Prove the Critical Point Theorem.

**\*3.** Find the critical points.

**a)** $x^3 - x$  **b)** $|x - 1|$  **c)** $x^{1/3}$

**4.** **a)** State the Mean Value Theorem.
   **b)** Find a point $\hat{x}$ as guaranteed by the Mean Value Theorem, for $f(x) = x^3$ on $[0, 1]$.
   **c)** Show by appropriate examples that each hypothesis of the Mean Value Theorem is necessary.

**5.** **a)** Prove that $f(x) = x \sin(x)$ has a maximum and a minimum on the interval $[-100, 100]$.
   **b)** Does $x \sin(x)$ have a maximum or minimum on $(-\infty, +\infty)$?
   **c)** Does $x \sin(x)$ have a maximum or minimum on $(-\pi/2, \pi/2)$?
   **d)** Does $x \sin(x)$ have a maximum or minimum on $(-\pi, \pi)$?

**6.** **a)** Prove: If $f'(x) = 0$ for all $x$ in the interval $(a, b)$, then $f$ is constant on that interval.
   **b)** Suppose that $f'(x) = 0$ for all $x \neq 0$. Is there a constant $C$ such that $f(x) = C$ for all $x \neq 0$? Explain.

**7.** Prove: If $f'(x) = g'(x)$ for all $x$ in the interval $(a, b)$, then there is a constant $C$ such that $f(x) = g(x) + C$ for all $x$ in that interval.

**8.** Suppose that $|f'(x)| \leq 5$ for $0 \leq x \leq 2$. Prove:
   **a)** $|f(2) - f(0)| \leq 10$
   **b)** $|f(1) - f(0)| \leq 5$
   **c)** $|f(x) - f(0)| \leq 5x$ for $0 \leq x \leq 2$

**9.** **a)** State Rolle's Theorem.
   **b)** Prove Rolle's Theorem, assuming the Mean Value Theorem.

**10.** Suppose that $f$ is continuous on a closed finite interval $[a, b]$ and has just two critical points: $x_1$, where $f''(x_1) > 0$, and $x_2$, where $f''(x_2) < 0$. Does it follow that $x_1$ is an absolute minimum for $f$ on $[a, b]$ and $x_2$ an absolute maximum? Explain.

# 6

# INTEGRALS

The two main concepts of calculus are derivatives and *integrals*. For a positive continuous function $f$, the *integral of $f$ from $a$ to $b$* is simply the area lying under the graph of $f$ and over the interval $[a, b]$ on the $x$-axis, as shown in Figure 1. It is denoted $\int_a^b f(x)\, dx$.

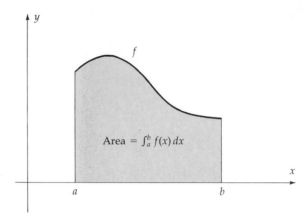

Area $= \int_a^b f(x)\, dx$

**FIGURE 1**

This simple geometric idea has an amazing variety of applications. Integrals represent not just area, but also length, volume, and mass; the distance traveled by a moving particle and the energy required to move it; the total amount of a flow when the rate of flow is known; the revenue that can be achieved by selling at different prices to different consumers; the probability of scoring above a certain percentile on a standardized test; and so on.

Most of these applications are left for the later chapters. Here we explain the main properties of integrals and, particularly, the *Fundamental Theorem of Calculus*, which says, roughly, that an integral is the reverse of a derivative; so integrals can be evaluated by means of antiderivatives!

## 6.1

### THE INTEGRAL AS SIGNED AREA

The integral is easy to define in geometric terms, as in Figure 1. Let $f$ be continuous on the interval $[a, b]$. Suppose at first that $f$ is positive. Then the integral of $f$ from $a$ to $b$, denoted $\int_a^b f(x)\, dx$, is the area between the graph of $f$ and the $x$-axis and between the vertical lines $x = a$ and $x = b$ (Figure 1). The function $f$ is called the **integrand**; the number $a$ is the **lower limit** of *integration* and $b$ is the **upper limit**. Some simple examples:

$\int_1^3 2\, dx$ is the area of a 2 by 2 rectangle (Fig. 2) so

$$\int_1^3 2\, dx = 2 \cdot 2 = 4.$$

$\int_a^b c\, dx$, where $c > 0$, is the area of a rectangle of height $c$ and base $b - a$ (Fig. 3) so

$$\int_a^b c\, dx = c \cdot (b - a).$$

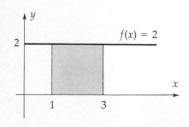

**FIGURE 2**
$\int_1^3 2\, dx = 2 \cdot 2 = 4.$

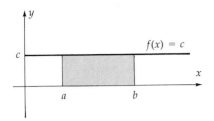

**FIGURE 3**
$\int_a^b c \, dx = c \cdot (b - a)$.

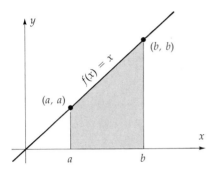

**FIGURE 4**
$\int_a^b x \, dx = \frac{1}{2}b^2 - \frac{1}{2}a^2$.

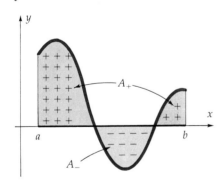

**FIGURE 5**
$\int_a^b f(x) \, dx = A_+ - A_-$.

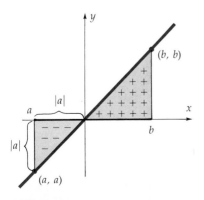

**FIGURE 7**
$\int_a^b x \, dx = \frac{1}{2}b^2 - \frac{1}{2}a^2$.

$\int_a^b x \, dx$, where $0 \leq a < b$, is the area of a right triangle of base $b$ and height $b$, minus a smaller triangle of base $a$ and height $a$ (Fig. 4). The integral is therefore

$$\int_a^b x \, dx = \frac{1}{2}b^2 - \frac{1}{2}a^2.$$

## Integrands with Negative Values

Suppose now that $f$ is continuous, but not necessarily positive, for $a \leq x \leq b$ (Fig. 5). Then part of the region between the graph and the $x$-axis lies *below* the axis. This part is counted as "negative area"; that is, the integral of $f$ equals the area above the axis, minus the area below. This concept of "signed area" seems strange at first. You might think of it in terms of territory. Think of the $x$-axis as the original boundary between two countries (Fig. 6). Suppose the countries arrange a new treaty, called the "$f$-treaty," which moves their boundary from the $x$-axis to the graph of $f$. Take the point of view of the lower country, the one originally below the $x$-axis. Then the areas above the axis count as gains, and the areas below the axis count as losses. The net gain is the area above, minus the area below, that is, the net gain is $\int_a^b f(x) \, dx$.

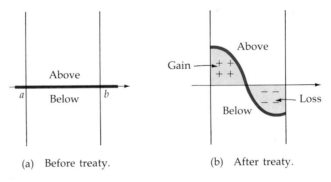

(a) Before treaty.     (b) After treaty.

**FIGURE 6**
Diplomatic view of the integral.

**EXAMPLE 1**   If $a < 0 < b$, then $\int_a^b x \, dx$ is the area of a $b$ by $b$ triangle above the axis, *minus* the area of a triangle below the axis whose dimensions are $|a|$ by $|a|$ (Fig. 7). Hence,

$$\int_a^b x \, dx = \frac{1}{2}b \cdot b - \frac{1}{2}|a| \cdot |a| = \frac{1}{2}b^2 - \frac{1}{2}a^2$$

since $|a| \cdot |a| = a^2$. This is the same formula as for the case where all the area is *above* the axis in Figure 4; counting areas below the axis as negative makes the formulas for these two cases the same.

**EXAMPLE 2**   $\int_{-1}^1 x^3 \, dx = 0$ because the area above the graph for $0 \leq x \leq 1$ cancels the area below the graph for $-1 \leq x \leq 0$ (Fig. 8). The areas cancels exactly, because $x^3$ is an odd function.

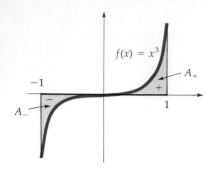

**FIGURE 8**
$\int_{-1}^{1} x^3 \, dx = A_+ - A_- = 0.$

### Riemann Sums

Only very simple integrals can be evaluated by elementary geometric formulas. But any integral can be *approximated*, as in Figure 9. Cut the region between the graph of $f$ and the $x$-axis into vertical strips and approximate each strip by a rectangle. The approximation becomes very accurate if the strips are made very thin. In fact, if you take the *limit* as the strips become infinitely thin and infinitely numerous, you get the *exact value* of the integral.

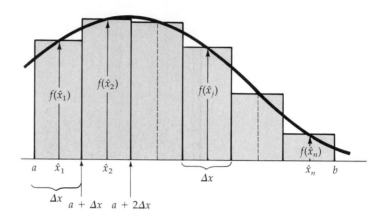

**FIGURE 9**
Forming a Riemann sum.

To form the approximating rectangles, divide the given interval $[a, b]$ into $n$ subintervals of equal length. The original interval has length $b - a$, so each of the $n$ subintervals has length $\dfrac{b - a}{n}$. Denote this by $\Delta x$:

$$\Delta x = \frac{b - a}{n}.$$

The endpoints of the subintervals are (Fig. 9) $a, a + \Delta x, a + 2\,\Delta x, \dots, a + (n - 1)\,\Delta x$, and $b$. The base of each approximating rectangle is $\Delta x$. For the height of the first rectangle, take the *height to the graph* at some point $\hat{x}_1$ in the first subinterval. The height to the graph is $f(\hat{x}_1)$, and the area of the first approximating rectangle is

$$(\text{height}) \times (\text{base}) = f(\hat{x}_1)\,\Delta x.$$

Do the same for each subinterval and add up the terms. The result is called a **Riemann sum** for $f$, denoted $S_n$:

$$S_n = f(\hat{x}_1)\,\Delta x + f(\hat{x}_2)\,\Delta x + \cdots + f(\hat{x}_n)\,\Delta x.$$

The points $\hat{x}_j$ where $f$ is evaluated can be chosen in any convenient way. Three important ways are:

(i)  Choose the *midpoint* of each subinterval (Figs. 9 and 14). The resulting sum is called a **midpoint sum**, which we denote by $MS_n$.

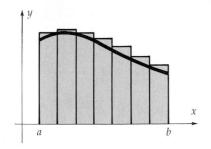

**FIGURE 10**
Upper sum $\bar{S}_n \geq \int_a^b f(x)\,dx$.

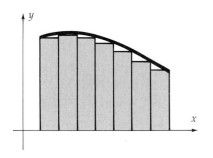

**FIGURE 11**
Lower sum $\underline{S}_n \leq \int_a^b f(x)\,dx$.

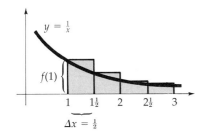

**FIGURE 12**
Upper sum $\bar{S}_4$ for $\int_1^3 \frac{1}{x}\,dx$.

**(ii)** Make $f(\hat{x}_j)$ the *maximum* of $f$ on the $j^{th}$ subinterval (Figs. 10 and 12). The resulting sum is called an **upper sum**, denoted $\bar{S}_n$. When $f$ is positive, the rectangles for an upper sum completely cover the area under the graph, so

$$\bar{S}_n \geq \int_a^b f(x)\,dx.$$

When $f$ is not necessarily positive, a careful consideration of the areas below the axis shows that this inequality remains true in general; each upper sum is greater than or equal to the integral.

**(iii)** Make $f(\hat{x}_j)$ the *minimum* of $f$ on the $j^{th}$ subinterval (Figs. 11 and 13). The resulting sum is called a **lower sum**, denoted $\underline{S}_n$. When $f$ is positive, the rectangles for a lower sum are all contained between the graph, the $x$-axis, and the lines $x = a$ and $x = b$; they don't overlap, so the sum of their areas is less than or equal to the area under the graph:

$$\underline{S}_n \leq \int_a^b f(x)\,dx.$$

This inequality, too, remains valid even when $f$ is not necessarily positive. Thus, for any continuous $f$, the integral lies between each lower sum and upper sum:

$$\underline{S}_n \leq \int_a^b f(x)\,dx \leq \bar{S}_n.$$

**EXAMPLE 3**  For the integral $\int_1^3 \frac{1}{x}\,dx$, form upper and lower sums, with four equal subintervals.

**SOLUTION**  The integral extends from $x = 1$ to $x = 3$ (Fig. 12). Divide this interval into four equal subintervals of length $\Delta x = \dfrac{3-1}{4} = \dfrac{1}{2}$. The endpoints of the intervals are $1$, $1 + \Delta x = 1\frac{1}{2}$, $1 + 2\,\Delta x = 2$, $1 + 3\,\Delta x = 2\frac{1}{2}$, $1 + 4\,\Delta x = 3$.

To form the upper sum, you need the *maximum* of $f(x) = \dfrac{1}{x}$ on each subinterval. Since $f$ is a decreasing function, those maxima occur at the left endpoints (Fig. 12). They are

$$f(1) = \frac{1}{1} = 1; \qquad f\left(1\frac{1}{2}\right) = \frac{1}{1\frac{1}{2}} = \frac{2}{3}; \qquad f(2) = \frac{1}{2}; \qquad f\left(2\frac{1}{2}\right) = \frac{2}{5}$$

and the upper sum is

$$\bar{S}_4 = f(1)\,\Delta x + f(1\tfrac{1}{2})\,\Delta x + f(2)\,\Delta x + f(2\tfrac{1}{2})\,\Delta x$$
$$= 1 \cdot \tfrac{1}{2} + \tfrac{2}{3} \cdot \tfrac{1}{2} + \tfrac{1}{2} \cdot \tfrac{1}{2} + \tfrac{2}{5} \cdot \tfrac{1}{2}$$
$$= \tfrac{1}{2} + \tfrac{1}{3} + \tfrac{1}{4} + \tfrac{1}{5} = 1.2833 \ldots .$$

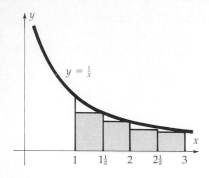

**FIGURE 13**

Lower sum $\underline{S}_4$ for $\int_1^3 \frac{1}{x}\,dx$.

For the lower sum, you need the *minimum* on each interval. In this case, the minima occur at the right endpoint, and the lower sum in Figure 13 is

$$\underline{S}_4 = f(1\tfrac{1}{2})\,\Delta x + f(2)\,\Delta x + f(2\tfrac{1}{2})\,\Delta x + f(3)\,\Delta x$$
$$= \tfrac{2}{3}\cdot\tfrac{1}{2} + \tfrac{1}{2}\cdot\tfrac{1}{2} + \tfrac{2}{5}\cdot\tfrac{1}{2} + \tfrac{1}{3}\cdot\tfrac{1}{2}$$
$$= 0.95.$$

The upper and lower sums together show that

$$0.95 < \int_1^3 \frac{1}{x}\,dx < 1.2833\ldots.$$

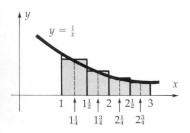

**FIGURE 14**

Midpoint sum $MS_4$ for $\int_1^3 \frac{1}{x}\,dx$.

**EXAMPLE 4**  For $\int_1^3 \frac{1}{x}\,dx$, form the midpoint sum $MS_4$.

**SOLUTION**  As in Example 3, $\Delta x = \tfrac{1}{2}$, but now you evaluate $f$ at the midpoints of the four intervals, halfway between the endpoints in Figure 14:

$$\hat{x}_1 = 1\tfrac{1}{4} = \tfrac{5}{4}, \qquad \hat{x}_2 = 1\tfrac{3}{4} = \tfrac{7}{4}, \qquad \hat{x}_3 = 2\tfrac{1}{4} = \tfrac{9}{4}, \qquad \hat{x}_4 = 2\tfrac{3}{4} = \tfrac{11}{4}.$$

The heights to the graph at these points are

$$f\left(\frac{5}{4}\right) = \frac{1}{5/4} = \frac{4}{5}, \qquad f\left(\frac{7}{4}\right) = \frac{4}{7}, \qquad f\left(\frac{9}{4}\right) = \frac{4}{9}, \qquad f\left(\frac{11}{4}\right) = \frac{4}{11}$$

and the midpoint sum is

$$MS_4 = \tfrac{4}{5}\cdot\tfrac{1}{2} + \tfrac{4}{7}\cdot\tfrac{1}{2} + \tfrac{4}{9}\cdot\tfrac{1}{2} + \tfrac{4}{11}\cdot\tfrac{1}{2} = 1.0897\ldots$$

If you repeat this process with 10 subintervals, you get the midpoint sum

$$MS_{10} = f(1\tfrac{1}{10})\cdot\tfrac{1}{5} + f(1\tfrac{3}{10})\cdot\tfrac{1}{5} + \cdots + f(2\tfrac{9}{10})\cdot\tfrac{1}{5} = 1.097142\ldots.$$

As $n \to \infty$, these sums $MS_n$ come closer and closer to the actual integral, which turns out to be

$$1.0986123\ldots$$

to eight decimal places.

These two examples illustrate the use of Riemann sums. With upper and lower sums, you can pin the integral between two definite numbers. With upper, lower, or any other sums $S_n$, you expect greater accuracy with larger $n$; and if $f$ is continuous, the integral is the *limit* of the Riemann sums as $n \to \infty$:

$$\int_a^b f(x)\,dx = \lim_{n \to \infty} S_n.$$

Of course, we will want to *evaluate* the limit of the Riemann sums; this essential question is left for the next two sections. For now, we use the sums merely as approximations.

*REMARKS*   The notation $\int_a^b f(x)\,dx$ is related to the Riemann sum

$$f(\hat{x}_1)\,\Delta x + \cdots + f(\hat{x}_n)\,\Delta x.$$

The symbol "$\int$" is an elongated *S*, for sum; the "$f(x)\,dx$" is for the terms $f(\hat{x}_j)\,\Delta x$ in the sum; and the numbers *a* and *b* show which interval we are considering.

### Hints on Computing Riemann Sums

It helps to sketch the interval of integration and its subdivision into equal parts by the points

$$a, a + \Delta x, a + 2\,\Delta x, \ldots a + n\,\Delta x = b.$$

For a midpoint sum, the midpoints (Fig. 15) are at

$$a + \tfrac{1}{2}\,\Delta x, a + \Delta x + \tfrac{1}{2}\,\Delta x = a + \tfrac{3}{2}\,\Delta x, a + \tfrac{5}{2}\,\Delta x, \ldots a + (n - \tfrac{1}{2})\,\Delta x.$$

For upper and lower sums, you need to sketch the graph, showing regions of increase and decrease, to find the maximum and minimum of *f* on each subinterval.

A calculator with a "memory +" key is handy for computing these sums. Compute the first term and add it to the memory; repeat for all the other terms; then "recall memory" to show the sum.

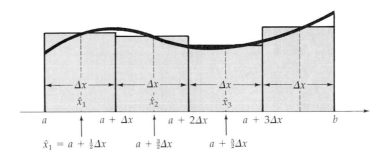

**FIGURE 15**
Midpoints $\hat{x}_1, \hat{x}_2, \ldots$

---

**EXAMPLE 5**   Form upper, lower, and midpoint sums for $\int_0^1 x^2\,dx$ with 10 subdivisions.

*SOLUTION*   Sketch the curve over the interval [0, 1], and divide the interval into ten equal parts (Fig. 16). The upper sum is

$$\begin{aligned}
\bar{S}_{10} &= f(0.1)\cdot\tfrac{1}{10} + f(0.2)\cdot\tfrac{1}{10} + \cdots + f(1)\cdot\tfrac{1}{10} \\
&= (0.1)^2(0.1) + (0.2)^2(0.1) + (0.3)^2(0.1) + \cdots + (1)^2(0.1) \\
&= 0.385,
\end{aligned}$$

and the lower sum is

$$\begin{aligned}
S_{10} &= f(0)\cdot\tfrac{1}{10} + \cdots + f(0.9)\cdot\tfrac{1}{10} \\
&= 0 + (0.1)^2(0.1) + \cdots + (0.9)^2(0.1) \\
&= 0.285.
\end{aligned}$$

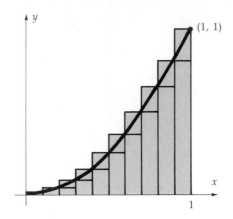

**FIGURE 16**
Upper and lower Riemann sums for $\int_0^1 x^2 \, dx$.

The midpoints of the intervals are at .05, .15, . . . , .95, so the midpoint sum is

$$MS_{10} = (0.05)^2(0.1) + (0.15)^2(0.1) + \cdots + (0.95)^2(0.1)$$
$$= 0.33250.$$

So $\int_0^1 x^2 \, dx$ is between 0.285 and 0.385, probably quite close to 0.33250. Could it be exactly 1/3?

---

**REMARK 1**  In forming Riemann sums $S_n$, we divided the interval $[a, b]$ into $n$ equal parts. This is generally the most convenient way, but it is not necessary—Riemann sums can be formed with partitions into unequal intervals. When it is necessary to emphasize that we are using equal parts, we speak of a "regular" Riemann sum.

**REMARK 2**  We began by defining the integral geometrically, as "net signed area." But what is the real foundation of this definition? How is area defined?

As a matter of fact, the best way to define area is as a limit of sums of rectangular areas, just like the Riemann sums. So in a rigorous course, the integral itself is *defined* as the limit of Riemann sums, and it is *proved* that this limit exists if $f$ is continuous on $[a, b]$. This is the "formal analytic definition" of the integral. The definition we gave, assuming area as a well-defined concept, is the "informal geometric definition."

## SUMMARY

**Geometric Definition of the Integral (see Fig. 5)**

$$\int_a^b f(x) \, dx = A_+ - A_-.$$

The function $f$ is called the *integrand*; $a$ is the *lower limit of integration*, and $b$ is the *upper limit*. $f$ is assumed continuous.

*Riemann Sums (see Figs. 9–11)* have the form

$$S_n = f(\hat{x}_1)\, \Delta x + \cdots + f(\hat{x}_n)\, \Delta x,$$

where

$$\Delta x = \frac{b - a}{n},$$

and the points $\hat{x}_j$ are chosen one from each of $n$ equal subintervals of $[a, b]$. The integral is the limit of these sums:

$$\lim_{n \to \infty} S_n = \int_a^b f(x)\, dx.$$

(This limit is the *analytic* definition of the integral.)

An *upper sum* $\bar{S}_n$ (see Fig. 10) is a Riemann sum where each $f(\hat{x}_j)$ is the *maximum* of $f$ on the $j^{\text{th}}$ interval.

A *lower sum* $\underline{S}_n$ (see Fig. 11) is a Riemann sum where each $f(\hat{x}_j)$ is the *minimum* of $f$ on the $j^{\text{th}}$ interval.

Upper and lower sums "bracket" the integral:

$$\underline{S}_n \leq \int_a^b f(x)\, dx \leq \bar{S}_n.$$

In a *midpoint sum* $MS_n$ (see Fig. 15), $\hat{x}_j$ is the midpoint of the $j^{\text{th}}$ subinterval.

## PROBLEMS

### A

1. Sketch the integrand, and evaluate the integral geometrically. Label the horizontal axis appropriately.

  *a)  $\int_{-1}^{2} 3x\, dx$

  b)  $\int_{-1}^{1} x\, dx$

  *c)  $\int_{-1}^{1} t^5\, dt$

  *d)  $\int_{-1}^{1} \sqrt{1 - z^2}\, dz$

  *e)  $\int_{-1}^{2} |t|\, dt$

  *f)  $\int_{-2}^{2} (s^3 - s)\, ds$

  *g)  $\int_{-1}^{1} |x|\, dx$

  *h)  $\int_{1}^{1} x^2\, dx$

  * i)  $\int_{-1}^{-1} \sqrt{x + 2}\, dx$

2. Evaluate $\int_a^a f(x)\, dx$, for every $f$.

3. If $f$ is an odd function, what can be said about $\int_{-b}^{b} f(x)\, dx$?

4. If $0 \leq f(x) \leq g(x)$ for $a \leq x \leq b$, what is the relation between $\int_a^b f(x)\, dx$ and $\int_a^b g(x)\, dx$? Why?

5. Derive a formula for $\int_a^b cx\, dx$ in each of the following cases. ($c$ is a constant. Make an appropriate sketch for each case.)

  a)  $c > 0, 0 < a < b$

  b)  $c > 0, a < 0 < b$

  c)  $c < 0, 0 < a < b$

  d)  One other case of your choice

6. For each integral $\int_a^b f(x)\, dx$, sketch the graph of $f$ for $a \leq x \leq b$, divide this interval into the given number $n$ of equal parts, sketch the approximating rectangles for the upper sum $\bar{S}_n$, and compute it. Repeat for the lower sum $\underline{S}_n$ and midpoint sum $MS_n$. Verify that

$$\underline{S}_n \leq MS_n \leq \bar{S}_n.$$

  In part a, evaluate the integral and verify that $\underline{S}_n \leq \int_a^b f(x)\, dx \leq \bar{S}_n$.

  *a)  $\int_0^1 x\, dx, \; n = 5$

  *b)  $\int_1^2 \frac{1}{x}\, dx, \; n = 5$

c) $\int_{-1}^{1} \dfrac{1}{1+x^2}\, dx,\ n = 3$

d) $\int_{-1}^{0} x^2\, dx,\ n = 3$

e) $\int_{0}^{1} x^3\, dx,\ n = 10$

*f) $\int_{-1}^{1} x^3\, dx,\ n = 4$

(Notice that when $f(\hat{x}_j) < 0$, the term $f(\hat{x}_j)\,\Delta x$ gives *minus* the area of a rectangle.)

7. Using a calculator, compute midpoint sums $MS_n$ for $\int_0^{\pi/2} \cos x\, dx$, with
   *a) $n = 4$    b) $n = 10$
   What might the integral be, exactly?

## B

8. (Trapezoid sums) Suppose that $f(x)$ is known only at equally spaced points (Fig. 17)
$$x_0 = a,\ x_1 = a + \Delta x,\ x_2 = a + 2\,\Delta x,\ \dots,$$
$$x_n = a + n\,\Delta x = b.$$

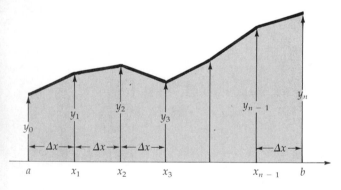

**FIGURE 17**
Trapezoids to approximate an integral.

Then $\int_a^b f(x)\, dx$ can be approximated by the sum of the areas of the $n$ trapezoids in Figure 17.
a) Show that the area of a trapezoid with heights $y_0$ and $y_1$ and base $\Delta x$ is $\frac{1}{2}(y_0 + y_1)\,\Delta x$.
b) Deduce that the sum of the areas of the $n$ trapezoids in Figure 17 is
$$T_n = \Delta x(\tfrac{1}{2}y_0 + y_1 + y_2 + \cdots$$
$$+\ y_{n-1} + \tfrac{1}{2}y_n). \quad (*)$$

[This is called a *trapezoid sum* for the integral $\int_a^b f(x)\, dx$.]

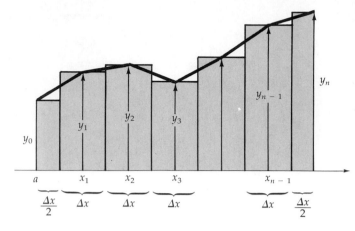

**FIGURE 18**
Trapezoid sum: $T_n = y_0\,\dfrac{\Delta x}{2} + y_1\,\Delta x + \cdots + y_{n-1}\,\Delta x + y_n\,\dfrac{\Delta x}{2}.$

c) Show that the trapezoid sum is the sum of the areas of the $n + 1$ rectangles in Figure 18. (This is a Riemann sum where two of the intervals are smaller than the others.)

*9. A lot between a road and a river is surveyed (Fig. 19) and the width every 10 meters found to be as follows:

| Meters from W end | 0 | 10 | 20 | 30 | 40 | 50 | 60 | 70 | 80 | 90 | 100 |
|---|---|---|---|---|---|---|---|---|---|---|---|
| Width of lot | 41 | 45 | 49 | 52 | 58 | 65 | 71 | 69 | 62 | 55 | 56 |

Estimate the size of the lot in (meters)$^2$ using a trapezoid sum $(*)$.

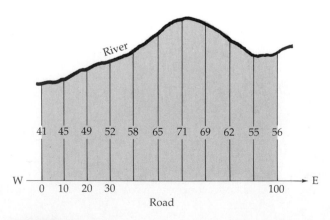

**FIGURE 19**
Measuring an irregular lot.

## 6.2

# SUMMATION NOTATION. LIMITS OF RIEMANN SUMS[1]

Riemann sums are conveniently written with "summation notation," using the letter $\Sigma$ (capital sigma, the Greek version of $S$, for sum):

$$f(x_1)\,\Delta x + f(x_2)\,\Delta x + \cdots + f(x_n)\,\Delta x = \sum_{j=1}^{n} f(x_j)\,\Delta x.$$

The "general term" is $f(x_j)\,\Delta x$; the symbol $\sum_{j=1}^{n}$ instructs you to add all these terms, beginning with $j = 1$ and ending with $j = n$. Thus,

$$\sum_{j=1}^{n} x_j^2\,\Delta x = x_1^2\,\Delta x + x_2^2\,\Delta x + \cdots + x_n^2\,\Delta x;$$

$$\sum_{k=3}^{5} \sin k = \sin 3 + \sin 4 + \sin 5;$$

$$1 + \frac{1}{2} + \frac{1}{3} + \frac{1}{4} + \frac{1}{5} = \sum_{j=1}^{5} \frac{1}{j}.$$

### Sequence Limits

The Riemann sums $S_n$ form a **sequence of numbers**. Each number $S_n$ is called a **term** of the sequence. The sequence has a **limit** $L$ if the terms $S_n$ can be kept arbitrarily close to $L$, just by making $n$ sufficiently large. Thus, if $S_n = \dfrac{1}{n}$, the limit is

$$\lim_{n \to \infty} \frac{1}{n} = 0$$

because $\dfrac{1}{n}$ can be kept arbitrarily close to 0 by making $n$ sufficiently large.

For example, $\dfrac{1}{n}$ is kept $< .001$ by making $n > 1{,}000$; and $\dfrac{1}{n} < 10^{-6}$ if $n > 10^6$.

The rules for function limits apply to sequence limits as well:

$$\lim_{n \to \infty} (S_n + T_n) = \lim_{n \to \infty} S_n + \lim_{n \to \infty} T_n$$

$$\lim_{n \to \infty} cS_n = c\left[\lim_{n \to \infty} S_n\right]$$

$$\lim_{n \to \infty} S_n T_n = \left[\lim_{n \to \infty} S_n\right] \cdot \left[\lim_{n \to \infty} T_n\right]$$

$$\lim_{n \to \infty} [S_n/T_n] = \left[\lim_{n \to \infty} S_n\right] \Big/ \left[\lim_{n \to \infty} T_n\right], \qquad \text{if } \lim T_n \neq 0.$$

---

[1] This section evaluates certain integrals as limits of Riemann sums. It is not strictly necessary for the rest of the text.

Finally, there is a "Trapping Theorem" for sequences:

If $S_n \leq T_n \leq U_n$, while $\lim\limits_{n \to \infty} S_n = L$ and $\lim\limits_{n \to \infty} U_n = L$, then also

$$\lim_{n \to \infty} T_n = L.$$

For example, you can prove that $\lim\limits_{n \to \infty} \dfrac{1}{n} \cos(n) = 0$ as follows:

$-1 \leq \cos(n) \leq 1$, so

$$-\frac{1}{n} \leq \frac{1}{n} \cos(n) \leq \frac{1}{n}.$$

Since $\lim\limits_{n \to \infty} \dfrac{1}{n} = 0$ and $\lim\limits_{n \to \infty} -\dfrac{1}{n} = 0$, then also $\lim\limits_{n \to \infty} \dfrac{1}{n} \cos(n) = 0$.

### Limits of Riemann Sums

We first evaluate the limit of the Riemann sums for $\int_a^b x \, dx$. For simplicity, suppose that $0 \leq a < b$. Figure 1 illustrates the upper sum $\bar{S}_n$, which gives the area of the shaded trapezoid plus $n$ small triangles. The trapezoid is the difference between a large triangle of area $\frac{1}{2}b^2$ and a smaller one of area $\frac{1}{2}a^2$; and each of the $n$ small triangles has area $\frac{1}{2}(\Delta x)(\Delta x)$, where $\Delta x = \dfrac{b - a}{n}$. Thus

$$\bar{S}_n = \frac{1}{2} b^2 - \frac{1}{2} a^2 + n \cdot \frac{1}{2} (\Delta x)^2$$

$$= \frac{1}{2} b^2 - \frac{1}{2} a^2 + n \cdot \frac{1}{2} \left( \frac{b - a}{n} \right)^2$$

$$= \frac{1}{2} b^2 - \frac{1}{2} a^2 + \frac{(b - a)^2}{2n}.$$

Since $\lim\limits_{n \to \infty} \dfrac{(b - a)^2}{2n} = 0$, it follows that

$$\lim_{n \to \infty} \bar{S}_n = \lim_{n \to \infty} \left[ \frac{1}{2} b^2 - \frac{1}{2} a^2 + \frac{(b - a)^2}{2n} \right] = \frac{1}{2} b^2 - \frac{1}{2} a^2.$$

For the lower sums (Fig. 2) you can check that

$$\underline{S}_n = \frac{1}{2} b^2 - \frac{1}{2} a^2 - \frac{(b - a)^2}{2n}$$

and the limit is

$$\lim_{n \to \infty} \underline{S}_n = \tfrac{1}{2} b^2 - \tfrac{1}{2} a^2.$$

Thus the upper sums *overestimate* the integral, and the lower sums *underestimate* it; but their *limits* give the exact value of the integral.

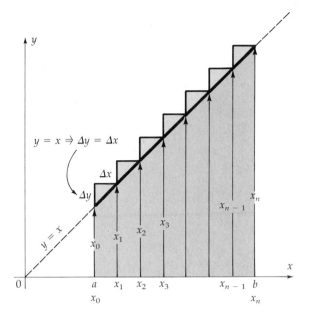

**FIGURE 1**
Upper sum for $\int_a^b x \, dx$,   $x_1 \, \Delta x + \cdots + x_n \, \Delta x$.

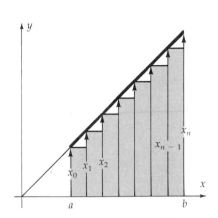

**FIGURE 2**
Lower sum for $\int_a^b x \, dx$,   $x_0 \, \Delta x + \cdots + x_{n-1} \, \Delta x$.

Since the upper and lower sums have a common limit, it follows that *any* sequence of regular Riemann sums $S_n$ for $\int_a^b x \, dx$ will have this same limit. For, any regular Riemann sum $S_n = \sum_{j=1}^{n} \hat{x}_j \, \Delta x$ will lie between the lower sum and the upper sum,

$$\underline{S}_n \le S_n \le \overline{S}_n.$$

Since $\lim_{n \to \infty} \underline{S}_n = \tfrac{1}{2}b^2 - \tfrac{1}{2}a^2 = \lim_{n \to \infty} \overline{S}_n$, it follows by the Trapping Theorem that

$$\lim_{n \to \infty} S_n = \tfrac{1}{2}b^2 - \tfrac{1}{2}a^2.$$

## Evaluating $\int_0^b x^2 \, dx$ (optional)

We conclude this section by evaluating $\int_0^b x^2 \, dx$ as a limit of Riemann sums. In the next section you will find a much easier way to do this; having seen the calculations here, you will appreciate all the more the power of that other method.

Divide the interval $[0, b]$ into $n$ equal parts (Fig. 3), each of length

$$\Delta x = \frac{b - 0}{n} = \frac{b}{n}.$$

The endpoints of these parts are

$$0, \quad x_1 = \frac{b}{n}, \quad x_2 = \frac{2b}{n}, \ldots, x_n = \frac{nb}{n} = b.$$

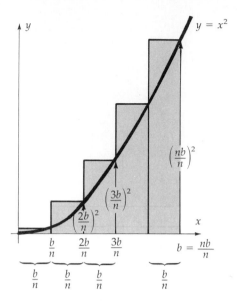

**FIGURE 3**
Upper sum for $\int_0^b x^2 \, dx$.

Since $x^2$ is an increasing function, the upper sum uses the right endpoints (Fig. 3):

$$\bar{S}_n = \left(\frac{b}{n}\right)^2 \cdot \frac{b}{n} + \left(\frac{2b}{n}\right)^2 \cdot \frac{b}{n} + \left(\frac{3b}{n}\right)^2 \cdot \frac{b}{n} + \cdots + \left(\frac{nb}{n}\right)^2 \cdot \frac{b}{n}$$

$$= \frac{b^3}{n^3}(1 + 2^2 + 3^2 + \cdots + n^2) = \frac{b^3}{n^3} \sum_1^n j^2.$$

We will prove below a simple expression for this sum of $n$ squares:

$$\sum_1^n j^2 = \frac{1}{3}n^3 + \frac{1}{2}n^2 + \frac{1}{6}n. \tag{1}$$

Hence,

$$\bar{S}_n = \frac{b^3}{n^3}\left(\frac{1}{3}n^3 + \frac{1}{2}n^2 + \frac{1}{6}n\right) = \frac{b^3}{3} + \frac{b^3}{2n} + \frac{b^3}{6n^2}$$

and

$$\lim_{n \to \infty} \bar{S}_n = \tfrac{1}{3}b^3.$$

Apparently then, $\int_0^b x^2 \, dx = \tfrac{1}{3}b^3$. To make this even more sure, we compute also the limit of the lower sums. $\underline{S}_n$ uses the left endpoints (Fig. 4) and is precisely the same as the upper sum, but without the last term; shift the rectangles for $\bar{S}_n$ one step to the right, then delete the last one, and you have $\underline{S}_n$. So

$$\underline{S}^n = \bar{S}_n - b^2 \cdot \frac{b}{n},$$

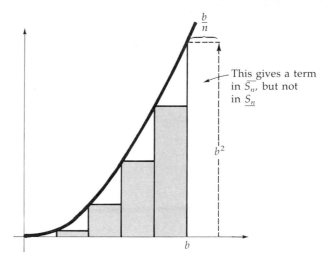

**FIGURE 4**
Shaded rectangles give lower sum $\underline{S}_n$ for $\int_0^b x^2\,dx$.

and

$$\lim_{n\to\infty} \underline{S}_n = \lim_{n\to\infty}\left[\bar{S}_n - \frac{b^3}{n}\right] = \lim_{n\to\infty}\bar{S}_n - \lim_{n\to\infty}\frac{b^3}{n}$$

$$= \frac{1}{3}b^3 - 0 = \frac{1}{3}b^3.$$

Since *both* the upper and lower sums have the same limit, then *all* the Riemann sums have that limit, and

$$\int_0^b x^2\,dx = \tfrac{1}{3}b^3.$$

### Proof of the Sum Formula

It remains to prove (1). Begin with a simpler sum,

$$\sum_1^n j = 1 + 2 + \cdots + (n-1) + n.$$

When you pair the terms as suggested, each pair adds up to $n + 1$. To exploit this, write two of the sums and add:

$$\sum_1^n j = \quad 1 \quad + \quad 2 \quad + \cdots + (n-1) + \quad n$$

$$\sum_1^n j = \quad n \quad + (n-1) + \cdots + \quad 2 \quad + \quad 1$$

$$2\sum_1^n j = (n+1) + (n+1) + \cdots + (n+1) + (n+1)$$

$$= n(n+1),$$

the sum of $n$ terms each equal to $n + 1$. Divide by 2:

$$\sum_1^n j = \frac{1}{2} n(n + 1). \tag{2}$$

The more complicated sum (1) needs a more devious approach, obtaining $j^2$ from the difference of two cubes:

$$j^3 - (j - 1)^3 = j^3 - (j^3 - 3j^2 + 3j - 1) = 3j^2 - 3j + 1.$$

Hence,

$$\sum_1^n [j^3 - (j - 1)^3] = \sum_1^n (3j^2 - 3j + 1) = 3 \sum_1^n j^2 - 3 \sum_1^n j + \sum_1^n 1. \tag{3}$$

The sum on the left "telescopes":

$$\sum_1^n [j^3 - (j - 1)^3] = [1^3 - 0^3] + [2^3 - 1^3] + [3^3 - 2^3] + \cdots$$
$$+ [n^3 - (n - 1)^3] = n^3,$$

for all other terms cancel. Use this on the left in (3), and formula (2) on the right:

$$n^3 = 3 \sum_1^n j^2 - 3 \cdot \frac{1}{2} n(n + 1) + n,$$

for the last sum in (3) is $\sum_1^n 1 = 1 + 1 + \cdots + 1 = n$. Now solve for the sum of squares:

$$\sum_1^n j^2 = \frac{1}{3} \left[ n^3 + \frac{3}{2} n(n + 1) - n \right]$$
$$= \frac{1}{3} n^3 + \frac{1}{2} n^2 + \frac{1}{6} n.$$

This is the formula (1) that we used to evaluate $\int_0^b x^2 \, dx$.

## The Integral $\int_a^b x^k \, dx$

From the three formulas

$$\int_a^b 1 \, dx = b - a, \qquad \int_a^b x \, dx = \tfrac{1}{2} b^2 - \tfrac{1}{2} a^2,$$
$$\int_0^b x^2 \, dx = \tfrac{1}{3} b^3 = \tfrac{1}{3} b^3 - \tfrac{1}{3} 0^3$$

you would naturally suppose that

$$\int_a^b x^k \, dx = \frac{1}{k + 1} b^{k + 1} - \frac{1}{k + 1} a^{k + 1}.$$

The problems show how to verify this for $k = 3$, 4, etc. by evaluating the limits of upper and lower Riemann sums. The case $k = 2$ was worked out by Archimedes about 250 B.C.; the others were done by Cavalieri, Fermat, Pascal, and Roberval about 1635. Each used a somewhat different method. But the next section gives the easiest method of all.

## PROBLEMS

### A

1.   Evaluate the following sums.

   *a) $\displaystyle\sum_{j=1}^{4} \frac{1}{2} j^2$   b) $\displaystyle\sum_{j=0}^{4} 2^{-j}$

   *c) $\displaystyle\sum_{j=0}^{2} \cos(\pi j)$   d) $\displaystyle\sum_{j=0}^{10} \cos(\pi j)$

   e) $\displaystyle\sum_{j=1}^{n} 5$

2.   Write the following sums in "$\Sigma$" notation.
   *a)   $1 + 2 + 3 + 4 + 5$
   b)   $1 + 2 + 3 + 4 + \cdots + n$
   *c)   $\frac{1}{2} + \frac{1}{3} + \frac{1}{4} + \frac{1}{5}$

3.   Explain the following formulas for the lower sums for the integral $\int_a^b x \, dx$. (See Fig. 2.)

   a)   $\displaystyle \underline{S}_n = \sum_{j=1}^{n} x_{j-1} \, \Delta x$

   $\qquad = x_0 \, \Delta x + x_1 \, \Delta x + \cdots + x_{n-1} \, \Delta x$

   b)   $\underline{S}_n = \frac{1}{2} b^2 - \frac{1}{2} a^2 - \frac{1}{2} n(\Delta x)^2$

   c)   $\underline{S}_n = \frac{1}{2} b^2 - \frac{1}{2} a^2 - \frac{(b-a)^2}{2n}$

   d)   $\displaystyle \lim_{n \to \infty} \underline{S}_n = \frac{1}{2} b^2 - \frac{1}{2} a^2$

4.   For $\int_a^b 5x \, dx$,

   a)   Show that

   $$\bar{S}_n = 5 \left[ \frac{1}{2} b^2 - \frac{1}{2} a^2 + \frac{(b-a)^2}{2n} \right].$$

   b)   Compute $\underline{S}_n$.
   c)   Evaluate $\displaystyle\lim_{n \to \infty} \bar{S}_n$ and $\displaystyle\lim_{n \to \infty} \underline{S}_n$.

5.   Evaluate the following integrals by computing the limit of the Riemann sums.

   *a)   $\displaystyle\int_0^2 (5x + 1) \, dx$   b)   $\displaystyle\int_1^3 (2 - x) \, dx$.

### B

6.   For addition and multiplication, we have:

   Commutative laws:   $a + b = b + a$   and
   $\qquad\qquad\qquad\qquad ab = ba$
   Associative laws:   $(a + b) + c = a + (b + c)$
   $\qquad\qquad$ and   $(ab)c = a(bc)$
   Distributive law:   $a(b + c) = ab + ac$.

Verify the following formulas for sums, and in each case explain which of these laws are used.

   a)   $\displaystyle\sum_{j=1}^{n} (ca_j) = c \left[ \sum_{j=1}^{n} a_j \right]$

   *b)   $\displaystyle\sum_{j=1}^{2} a_j + \sum_{j=1}^{2} b_j = \sum_{j=1}^{2} (a_j + b_j)$

   c)   $\displaystyle\sum_{j=1}^{n} a_j + \sum_{j=1}^{n} b_j = \sum_{j=1}^{n} (a_j + b_j)$

7.   Evaluate the following integrals by computing the limit of the Riemann sums.

   *a)   $\displaystyle\int_0^b 7x^2 \, dx$   b)   $\displaystyle\int_0^b (x + x^2) \, dx$

8.   Accept that $\int_0^b x^2 \, dx = \frac{1}{3} b^3$, for every $b > 0$.
   a)   Deduce a formula for $\int_a^b x^2 \, dx$, for $0 < a < b$.
   b)   Deduce a formula for $\int_a^0 x^2 \, dx$, for $a < 0$.
   c)   Deduce a formula for $\int_a^b x^2 \, dx$, for $a < 0 < b$.

9.   a)   Mimic the proof of $\sum_1^n j^2 = \frac{1}{3} n^3 + \frac{1}{2} n^2 + \frac{1}{6} n$ to prove a formula for $\sum_1^n j^3$.
   b)   Compute $\int_0^b x^3 \, dx$ for $b > 0$.
   c)   Compute $\int_a^b x^3 \, dx$ for $0 \le a < b$.

### C

10.   Suppose that $f$ is *increasing* for $a \le x \le b$. Show the following:
   a)   The regular upper and lower sums for $\int_a^b f(x) \, dx$ differ by the amount

   $$\bar{S}_n - \underline{S}_n = [f(b) - f(a)] \frac{b - a}{n}.$$

   (See Fig. 5.)

   b)   $\displaystyle\lim_{n \to \infty} (\bar{S}_n - \underline{S}_n) = 0$.

   c)   $\underline{S}_n \le \displaystyle\int_a^b f(x) \, dx \le \bar{S}_n$, so

   $$0 \le \int_a^b f(x) \, dx - \underline{S}_n \le \bar{S}_n - \underline{S}_n.$$

   d)   $\displaystyle\lim_{n \to \infty} \left[ \int_a^b f(x) \, dx - \underline{S}_n \right] = 0$, so

   $$\lim_{n \to \infty} \underline{S}_n = \int_a^b f(x) \, dx.$$

   e)   $\displaystyle\lim_{n \to \infty} \bar{S}_n = \int_a^b f(x) \, dx$. (Recall parts b and d.)

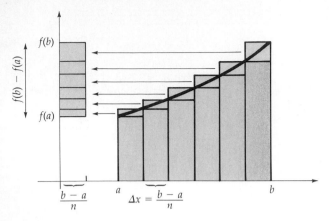

**FIGURE 5**

$$\bar{S}_n - \underline{S}_n = [f(b) - f(a)] \frac{b - a}{n}.$$

**f)** For any regular Riemann, sum $S_n$, $\underline{S}_n \leq S_n \leq \bar{S}_n$.

**g)** $\lim\limits_{n \to \infty} S_n = \int_a^b f(x)\, dx$.

(This argument assumes that $\int_0^b f(x)\, dx$ is defined as an area and then concludes that it is the limit of the Riemann sums. A more subtle argument proves that the Riemann sums have a limit, *without* any appeal to a prior concept of area.)

**11.** Consider the very discontinuous function

$$f(x) = \begin{cases} 1 & \text{if } x \text{ is rational} \\ 0 & \text{if } x \text{ is irrational.} \end{cases}$$

Compute $\bar{S}_n$ and $\underline{S}_n$ for this function on the interval $0 \leq x \leq 1$. Show that $\lim\limits_{n \to \infty} \bar{S}_n = 1$ and $\lim\limits_{n \to \infty} \underline{S}_n = 0$. What can you say about the "area" under this graph?

**12.** This problem evaluates the limit of the Riemann sums for $\int_a^b x^k\, dx$, with $0 \leq a < b$, for $k = 3$,

**4, ....** Explain:

**a)** $x_1^{k+1} - x_0^{k+1}$
$$= \Delta x(x_1^k + x_1^{k-1}x_0 + \cdots + x_0^k);$$
$$\Delta x = x_1 - x_0.$$

**b)** $(k + 1)x_0^k\, \Delta x < x_1^{k+1} - x_0^{k+1}$
$$< (k + 1)x_1^k\, \Delta x.$$

**c)** $\underline{S}_n < \dfrac{1}{k+1} b^{k+1} - \dfrac{1}{k+1} a^{k+1} < \bar{S}_n.$

**d)** $\bar{S}_n - \underline{S}_n = (b^k - a^k)\, \Delta x$ (problem 10a).

**e)** $\left[ \dfrac{1}{k+1} b^{k+1} - \dfrac{1}{k+1} a^{k+1} \right]$
$$- (b^k - a^k)\, \Delta x < \underline{S}_n < \bar{S}_n$$
$$< \left[ \dfrac{1}{k+1} b^{k+1} - \dfrac{1}{k+1} a^{k+1} \right]$$
$$+ (b^k - a^k)\, \Delta x.$$

[Use c) and d).]

**f)** $\int_a^b x^k dx = \lim\limits_{n \to \infty} \bar{S}_n = \lim\limits_{n \to \infty} \underline{S}_n$
$$= \dfrac{1}{k+1} b^{k+1} - \dfrac{1}{k+1} a^{k+1}.$$

**13.** Verify the following formulas by mathematical induction. That is, verify the formula for $n = 1$, and show that if it holds for any particular integer $n$, then it holds also for the next integer $n + 1$.

**a)** $\sum\limits_{k=1}^n k = \dfrac{n(n + 1)}{2}$

**b)** $\sum\limits_{k=1}^n k^2 = \dfrac{n(n + 1)(2n + 1)}{6}$

**c)** $\sum\limits_{j=1}^n j^3 = \left[ \dfrac{n(n + 1)}{2} \right]^2$

**d)** $\sum\limits_{j=1}^n \dfrac{1}{j(j + 1)} = \dfrac{n}{n + 1}$

## 6.3

### THE INTEGRAL OF A DERIVATIVE. THE FIRST FUNDAMENTAL THEOREM OF CALCULUS

We face the question: How to evaluate integrals? The Riemann sums are complicated, and evaluating their limits seems hopeless except in very special cases. Miraculously, the limit is given by a simple formula whenever the in-

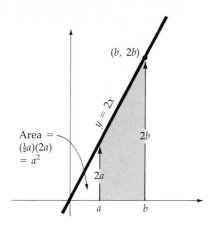

**FIGURE 1**
$\int_a^b 2x\,dx$ = Area of large triangle − area of small triangle = $b^2 - a^2$.

tegrand is a *derivative* $F'$. We will see that

$$\int_a^b F'(x)\,dx = F(b) - F(a). \tag{1}$$

**EXAMPLE 1**   Evaluate $\int_a^b 2x\,dx$.

**SOLUTION**   To apply formula (1), you have to recognize the integrand $2x$ as a derivative $F'(x)$. The obvious choice for $F$ is $F(x) = x^2$; then $F'(x) = 2x = $ the given integrand. So by (1),

$$\int_a^b 2x\,dx = \int_a^b F'(x)\,dx$$
$$= F(b) - F(a)$$
$$= b^2 - a^2,$$

since $F(x) = x^2$. You can check this geometrically (Fig. 1).

The proof of formula (1) uses the Riemann sums for $\int_a^b F'(x)\,dx$:

$$S_n = F'(\hat{x}_1)\,\Delta x + F'(\hat{x}_2)\,\Delta x + \cdots + F'(\hat{x}_n)\,\Delta x, \tag{2}$$

where each point $\hat{x}_1, \hat{x}_2, \ldots, \hat{x}_n$ can be chosen anywhere in the appropriate interval. We will evaluate the integral $\int_a^b F'(x)\,dx = \lim_{n \to \infty} S_n$ by choosing these points judiciously.

Consider the first term, $F'(\hat{x}_1)\,\Delta x$. According to the Mean Value Theorem (Sec. 5.2) there is a point $\hat{x}_1$ between $a$ and $x_1$ such that the tangent line at $(\hat{x}_1, F(\hat{x}_1))$ is parallel to the segment $P_0P_1$ in Figure 2. Then

$$F'(\hat{x}_1) = \frac{F(x_1) - F(a)}{x_1 - a} = \frac{F(x_1) - F(a)}{\Delta x} \tag{3}$$

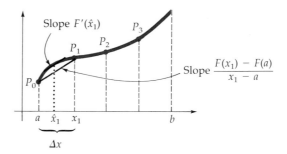

**FIGURE 2**
Graph of $F$.

and hence,

$$F'(\hat{x}_1)\,\Delta x = F(x_1) - F(a).$$

So when $\hat{x}_1$ is chosen in this way, the first term in the Riemann sum (2) gives precisely the *increase in* $F$ over the first interval, from $x = a$ to $x = x_1$.

Choose the second point $\hat{x}_2$ in a similar way, so that

$$F'(\hat{x}_2) = \frac{F(x_2) - F(x_1)}{\Delta x}$$

and the second term in the Riemann sum (2) is

$$F'(\hat{x}_2)\,\Delta x = F(x_2) - F(x_1),$$

the increase in $F$ over the second interval. Do the same for all the remaining intervals. Then *the entire Riemann sum (2) is the sum of the increases in F over all n intervals*, which gives the *increase over the entire interval from a to b*:

$$\begin{aligned}
S_n &= F'(\hat{x}_1)\,\Delta x + F'(\hat{x}_2)\,\Delta x + \cdots + F'(\hat{x}_n)\,\Delta x \\
&= F(x_1) - F(a) + F(x_2) - F(x_1) + \cdots + F(b) - F(x_{n-1}) \\
&= F(b) - F(a)
\end{aligned}$$

since all other terms cancel in pairs. The points $\hat{x}_j$ can be chosen in this way for every $n$, so

$$\begin{aligned}
\int_a^b F'(x)\,dx = \lim_{n \to \infty} S_n &= \lim_{n \to \infty} [F(b) - F(a)] \\
&= F(b) - F(a).
\end{aligned}$$

This is the First Fundamental Theorem of Calculus.

---

**The First Fundamental Theorem of Calculus**

If $F'$ is continuous for $a \leq x \leq b$ then

$$\int_a^b F'(x)\,dx = F(b) - F(a).$$

---

We rewrite this formula to make it easier to apply. The change in $F$ is generally denoted $F(x)\big|_a^b$ or $[F(x)]_a^b$:

$$F(b) - F(a) = F(x)\bigg|_a^b = \left[F(x)\right]_a^b.$$

For example,

$$x^{-1}\bigg|_1^2 = 2^{-1} - 1^{-1} = \tfrac{1}{2} - 1 = -\tfrac{1}{2}$$

$$[x^2 + 2x]_0^1 = [1 + 2] - [0 + 0] = 3.$$

With this notation, the First Fundamental Theorem is written

$$\int_a^b F'(x)\,dx = F(x)\bigg|_a^b. \tag{4}$$

Now, to evaluate an integral $\int_a^b f(x)\,dx$ using (4), you have to set the given integrand $f(x)$ equal to the integrand $F'(x)$ in (4) (as in Example 1). Thus $F'(x) = f(x)$, and $F$ is an antiderivative of $f$,

$$F(x) = D^{-1}f(x).$$

With these substitutions, (4) is rewritten once more as

$$\int_a^b f(x)\,dx = D^{-1}f(x)\Big|_a^b. \tag{5}$$

To apply this, use the antiderivative formulas from Section 2.8:

$$D^{-1}x^n = \frac{1}{n+1}x^{n+1} + C \qquad \text{if } n \neq -1$$

$$D^{-1}(f + g) = D^{-1}f + D^{-1}g$$

$$D^{-1}(cf) = c \cdot D^{-1}f \qquad \text{if } c \text{ is } constant.$$

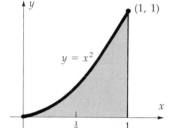

**FIGURE 3**
$\int_0^1 x^2\,dx = \frac{1}{3}.$

**EXAMPLE 2**

$$\int_0^1 x^2\,dx = D^{-1}x^2\Big|_0^1 = \tfrac{1}{3}x^3 + C\Big|_0^1 = [\tfrac{1}{3}\cdot 1 + C] - [0 + C] = \tfrac{1}{3}.$$

This easy calculation gives the area under the parabola in Figure 3. By contrast, the same answer was obtained with great difficulty using Riemann sums in Section 6.2.

*REMARK*   In Example 2, notice that the "$+C$" in the antiderivative formula cancels out when you subtract the value at the lower limit from the value at the upper limit. This cancellation will always occur, so you don't need that "$+C$" in applying formula (5). (You *do* need it, however, in many other uses of antiderivatives.)

**Antiderivative formulas for the trigonometric functions** follow from the corresponding derivative formulas. For example,

$$D \sin \theta = \cos \theta \Rightarrow D^{-1} \cos \theta = \sin \theta + C.$$

We also need $D^{-1} \cos(a\theta)$ for various constants $a$. Since

$$D\left[\frac{1}{a}\sin(a\theta)\right] = \frac{1}{a}D\sin(a\theta) = \cos(a\theta),$$

then

$$D^{-1}\cos(a\theta) = \frac{1}{a}\sin(a\theta) + C.$$

Similarly,

$$D\left[-\frac{1}{a}\cos(a\theta)\right] = \sin(a\theta) \Rightarrow D^{-1}\sin(a\theta) = -\frac{1}{a}\cos(a\theta),$$

$$D\left[\frac{1}{a}\tan(a\theta)\right] = \sec^2 a\theta \Rightarrow D^{-1}\sec^2 a\theta = \frac{1}{a}\tan(a\theta)$$

Notice that the pattern for antiderivatives is not quite the same as for derivatives, because roles are reversed. To be safe, check your antiderivative—differentiate the expression for $D^{-1}f$, to see that you really get $f$.

---

**EXAMPLE 3**

$$\begin{aligned}\int_0^{\pi/2} [\sin\theta + 2\cos 3\theta]\,d\theta &= D^{-1}[\sin\theta + 2\cos 3\theta]\Big|_0^{\pi/2} \\ &= [D^{-1}\sin\theta + 2D^{-1}\cos 3\theta]\Big|_0^{\pi/2} \\ &= \left[-\cos\theta + 2\cdot\left(\frac{1}{3}\sin 3\theta\right)\right]\Big|_0^{\pi/2} \\ &= \left[-\cos\frac{\pi}{2} + \frac{2}{3}\sin 3\frac{\pi}{2}\right] - \left[-\cos 0 + \frac{2}{3}\sin 0\right] \\ &= \left[0 - \frac{2}{3}\right] - [-1 + 0] = \frac{1}{3}.\end{aligned}$$

---

## Reversal of Limits

In $\int_a^b f(x)\,dx$ we have implicity assumed that $a < b$. If $a = b$, then

$$\int_a^a f(x)\,dx = 0$$

since in this case the "area" under the graph is the area of a straight line segment, which is 0 (Fig. 4).

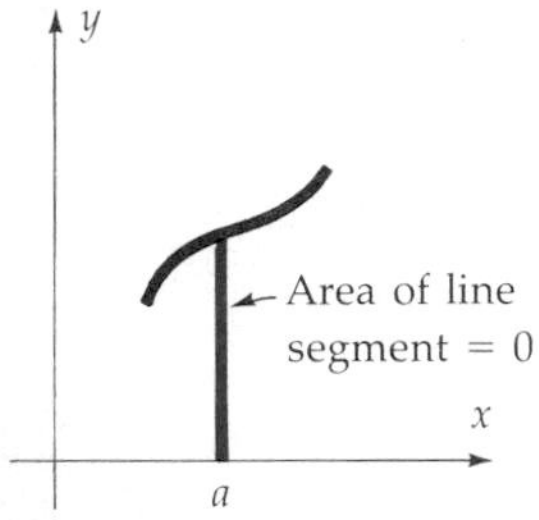

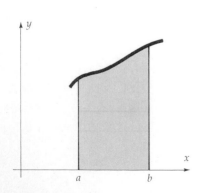

**FIGURE 4**
$\int_a^a f(x)\,dx = 0.$

We also at times reverse the limits of integration. By definition, this *changes the sign of the integral:*

$$\int_b^a f(x)\,dx = -\int_a^b f(x)\,dx.$$

This is consistent with the evaluation formula (1); in $\int_b^a F'(x)\,dx$, the upper limit is $a$ and the lower limit is $b$, so

$$\begin{aligned}\int_b^a F'(x)\,dx &= F(a) - F(b) = -[F(b) - F(a)] \\ &= -\int_a^b F'(x)\,dx.\end{aligned}$$

As a result, for a positive function $f$, $\int_b^a f(x)\,dx$ is *minus* the area under the graph of $f$, if $a < b$ (Fig. 5).

**FIGURE 5**
If $a < b$ and $f \geq 0$,
$\int_b^a f(x)\,dx = -$(area under graph).

## SUMMARY

*The First Fundamental Theorem of Calculus:*

If $F'$ is continuous on $[a, b]$ then

$$\int_a^b F'(x)\, dx = F(b) - F(a) = F(x)\Big|_a^b$$

or

$$\int_a^b f(x)\, dx = D^{-1}f(x)\Big|_a^b.$$

*Reversal of Limits:*    $\int_b^a f(x)\, dx = -\int_a^b f(x)\, dx.$

## PROBLEMS

### A

1.  Evaluate, using the First Fundamental Theorem. Sketch the graph and interval of integration to see that your answer is reasonable.

*a) $\int_0^1 x^3\, dx$

*b) $\int_{-1}^0 x^4\, dx$

*c) $\int_{-1}^1 x\, dx$

*d) $\int_1^2 x^{-2}\, dx$

*e) $\int_2^1 x^{-2}\, dx$

*f) $\int_{-2}^{-1} x^{-2}\, dx$

g) $\int_{-2}^2 (2x^7 + x)\, dx$

h) $\int_0^2 (x + \sqrt{x})\, dx$

i) $\int_0^2 x\sqrt{x}\, dx$

*j) $\int_0^\pi \theta\, d\theta$

*k) $\int_1^{1000} \frac{1}{x^2}\, dx$

l) $\int_0^1 (6t^2 - 2t + 1)\, dt$

m) $\int_{-1}^1 5\sqrt{z}\,(z^2 + 2z)\, dz$

n) $\int_{-1}^1 (x - 2)(x + 2)\, dx$

o) $\int_0^2 (3u - 1)^2\, du$

*p) $\int_0^{\pi/2} \cos\theta\, d\theta$

q) $\int_0^\pi \cos\frac{\theta}{2}\, d\theta$

r) $\int_0^{\pi/2} \sin 2\theta\, d\theta$

*s) $\int_0^\pi \sin 3\theta\, d\theta$

*t) $\int_{-\pi}^\pi \sin\theta\, d\theta$

2.  Evaluate, using the First Fundamental Theorem.

*a) $\int_0^{\pi/2} 2\sec^2\left(\frac{\theta}{2}\right) d\theta$

*b) $\int_{-\pi/3}^{\pi/3} \tan^2\theta\, d\theta \quad [\tan^2\theta = \sec^2\theta - 1]$

c) $\int_0^{\pi/4} 2\sec\theta(\sec\theta + \tan\theta)\, d\theta$

3.  Use the First Fundamental Theorem to show that $\int_a^b c\, dx = c \cdot (b - a)$. Confirm by sketching the graph of $f(x) = c$, $a \le x \le b$, and computing the relevant area.

4.  Evaluate the following integrals by the First Fundamental Theorem of Calculus. Compare the result with the upper and lower Riemann sums $\bar{S}_4$, $\underline{S}_4$, and with the midpoint sum $MS_4$. (Use a calculator to evaluate these sums for part b.)

*a) $\int_1^2 \frac{1}{x^2}\, dx$

b) $\int_0^{\pi/2} \sin\theta\, d\theta$

5.  Compute the area under the graph of $f(x) = x^n$ for $0 \le x \le 1$, where $n \ge 0$. Sketch the region for $n = 1, 2, 3, 4$.

6.  Compute the area under the given curve and over the given interval on the $x$-axis.

a) $y = x^2 + 1$, [0, 2]
*b) $y = 1 - x^2$, [-1, 1]

*c) $y = \cos x - \sin x$, $\left[0, \dfrac{\pi}{4}\right]$

7. Compute the area *above* the given curve and *below* the given interval. Sketch the curve.
   a) $y = x^2 - 1$, [-1, 1]

   *b) $y = \dfrac{1}{x^3}$, [-2, -1]

**B**

8. Compute the area bounded by the following curves:
   a) $y = 4 - x^2$ and the $x$-axis.
   b) $y = 1 - x^{2/3}$ and the $x$-axis.

9. Compute the area which is bounded
   *a) Above by $y = x$ and below by $y = x^2$.
   b) Above by $y = -x^2$ and below by $y = x$.

c) Above by $y = \sin x$ and below by $y = (2x/\pi)^2$.

10. Divide the interval [0, 1] into three equal subintervals and determine points $\hat{x}_1, \hat{x}_2, \hat{x}_3$ such that

$$F'(\hat{x}_j)\,\Delta x = F(x_j) - F(x_{j-1})$$

for each of the following functions $F'$. Then form the Riemann sum using these $\hat{x}_j$ and check that it equals the integral $\int_0^1 f(x)\,dx$.
   a) $F'(x) = x$      b) $F'(x) = x^2$

**C**

11. a) Show that $\frac{1}{2}x|x|$ is an antiderivative of $|x|$. (Check separately $x > 0$, $x = 0$, $x < 0$.)
   b) Evaluate $\int_{-a}^{a} |x|\,dx$ by the First Fundamental Theorem.
   c) Check your answer to part b geometrically.

# 6.4

## THE INTEGRAL OF A RATE OF CHANGE. TRAPEZOID SUMS

The First Fundamental Theorem states that

$$\int_a^b F'(x)\,dx = F(b) - F(a).$$

For functions of time $t$,

$$\int_a^b F'(t)\,dt = F(b) - F(a). \tag{1}$$

The difference $F(b) - F(a)$ is the change in $F$ from time $a$ to time $b$, and $F'(t) = dF/dt$ is the *rate of change* of $F$ with respect to time, so (1) says: *The change in F equals the integral of the rate of change.* Suppose, for instance, that an object is moving on the $y$-axis; its position $y$ at time $t$ is a function of $t$,

$$y = F(t).$$

The rate of change of position is the velocity

$$v = \frac{dy}{dt} = F'(t).$$

In this case, then, the *change in position equals the integral of the velocity*:

$$F(b) - F(a) = \int_a^b v(t)\,dt.$$

In graphical terms, the change in position (or displacement) equals the net signed area between the graph of the velocity $v$ and the $t$-axis.

**EXAMPLE 1**   An object falls with velocity $v(t) = -10t$ m/sec for $0 \leq t \leq 3$. How far does it fall in that time?

*SOLUTION*   The displacement is

$$\int_0^3 v(t)\,dt = \int_0^3 (-10t)\,dt = -5t^2 \Big|_0^3 = -45.$$

The answer *"minus 45"* shows that it *falls* 45 meters from time $t = 0$ to $t = 3$. Figure 1 shows this as the area of a region *below* the $t$-axis. This illustrates once more why it is appropriate to count areas below the axis as negative.

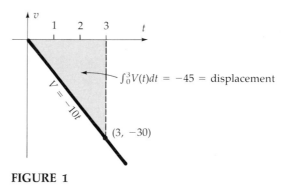

FIGURE 1

Why *should* the displacement be the net signed area between the graph of the velocity and the time axis? This is easy to see when the velocity $v$ is *constant*. For then the displacement during a time interval $a \leq t \leq b$ is simply the (constant) velocity times the elapsed time $b - a$, thus $v \cdot (b - a)$. On the other hand, the graph of a constant $v$ is a horizontal line (Fig. 2) and the product $v \cdot (b - a)$ is the area under the graph of $v$, for $a \leq t \leq b$.

When $v$ is not constant, imagine computing the displacement as follows: Divide the interval $[a, b]$ into $n$ small intervals, each of length $\Delta t = \dfrac{b - a}{n}$ (Fig. 3). In the first small time interval the velocity is nearly constant, so estimate the displacement as

$$(\text{velocity}) \times (\text{elapsed time}) = v(\hat{t}_1)\,\Delta t,$$

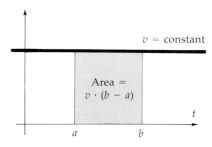

FIGURE 2

FIGURE 3
Partition of $a \leq t \leq b$.

where $\hat{t}_1$ is any convenient point in that first interval. Make the same sort of approximation for each interval, and then approximate the total displacement by the sum

$$S_n = v(\hat{t}_1)\,\Delta t + \cdots + v(\hat{t}_n)\,\Delta t.$$

If the intervals are made shorter and more numerous, the approximations should improve, and you can reasonably expect the actual displacement to be the *limit* of these sums $S_n$ as $n \to \infty$:

$$\text{Displacement} = \lim_{n \to \infty} S_n.$$

Now, the sums $S_n$ are actually Riemann sums approximating the integral $\int_a^b v(t)\,dt$, and their limit is precisely that integral; so

$$\text{Displacement} = \lim_{n \to \infty} S_n = \int_a^b v(t)\,dt. \tag{2}$$

Thus the Riemann sums confirm that the displacement is the integral of the velocity.

Formula (2) brings us back to the Fundamental Theorem of Calculus; for if we denote the position function by $F(t)$, then the displacement is $F(b) - F(a)$, while $v = F'$, so (2) is simply

$$F(b) - F(a) = \int_a^b F'(t)\,dt.$$

Moreover, our derivation of (2) was virtually the same as the proof of the Fundamental Theorem; we gave it just to show this essential idea in the familiar context of motion.

## Trapezoid Sums

When the integrand $f(t)$ is given by a table rather than a formula, you cannot compute $\int_a^b f(t)\,dt$ by an antiderivative formula; but you can *approximate* it by what are called **trapezoid sums**. Suppose that the interval $[a, b]$ is divided into $n$ equal subintervals of length $\Delta t$ by times $t_0 = a,\ t_1,\ t_2,\ \ldots,\ t_n = b$ (Fig. 4). Suppose that at each of those times, $f(t_j) = y_j$ is given. If you plot the corresponding points $(t_j, y_j)$ and connect them by straight line segments, you get an approximation to the graph of $f$; hence the area under these segments approximates the integral $\int_a^b f(t)\,dt$. That area is, in turn, the sum of the areas of the rectangles indicated in the figure, because the "$+$" triangles above the sloping segments are congruent to the "$-$" triangles below, and those triangular areas cancel. The first and last rectangles have base $\dfrac{\Delta t}{2}$, and all the others have base $\Delta t$; thus the approximation is

$$\int_a^b f(t)\,dt \approx y_0\left(\frac{\Delta t}{2}\right) + y_1\,\Delta t + \cdots + y_{n-1}\,\Delta t + y_n\left(\frac{\Delta t}{2}\right)$$

$$= \Delta t\left(\frac{1}{2}y_0 + y_1 + \cdots + y_{n-1} + \frac{1}{2}y_n\right). \tag{3}$$

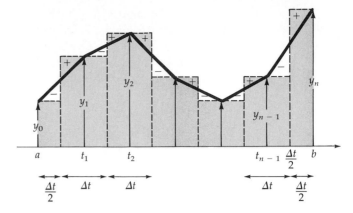

**FIGURE 4**
The area under the sloping segments equals the sum of the areas of the
shaded rectangles, $\left(\dfrac{\Delta t}{2}\right)y_0 + (\Delta t)y_1 + \cdots + (\Delta t)y_{n-1} + \left(\dfrac{\Delta t}{2}\right)y_n.$

This is the *trapezoid sum* approximation to the integral, denoted by $T_n$.
(Another derivation of this formula is outlined in problem 8, Sec. 6.1.)

---

**EXAMPLE 2**   The velocity of a boat on a straight line course is:

| time (minutes) | 0 | 1 | 2 | 3 | 4 | 5 | 6 | 7 | 8 | 9 | 10 |
|---|---|---|---|---|---|---|---|---|---|---|---|
| velocity (miles/hr) | 0 | 5 | 5 | 6 | 8 | 10 | 10 | 10 | 10 | 7 | 0 |

Approximately how many miles did it travel in those ten minutes?

**SOLUTION**   We will measure time in minutes, so the velocities must be
in miles/min; they are 0, 5/60, 5/60, etc.

The displacement is the integral $\int_0^{10} v(t)\,dt$; we approximate it by a
trapezoid sum. In the given table, the time interval $0 \le t \le 10$ has been
partitioned into ten equal subintervals of length $\Delta t = 1$ (Fig. 5). The corresponding trapezoid sum gives

$$\int_0^{10} v(t)\,dt \approx \Delta t(\tfrac{1}{2}v_0 + v_1 + \cdots + v_9 + \tfrac{1}{2}v_{10})$$

$$= 1 \cdot (\tfrac{1}{2} \cdot 0 + \tfrac{5}{60} + \tfrac{5}{60} + \tfrac{6}{60} + \tfrac{8}{60} + \tfrac{10}{60} + \tfrac{10}{60} + \tfrac{10}{60} + \tfrac{7}{60} + \tfrac{1}{2} \cdot 0)$$

$$= \tfrac{71}{60} \approx 1.18 \text{ miles.}$$

This approximates the change in position during those ten minutes.

---

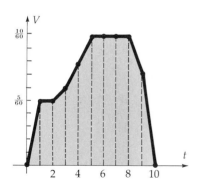

**FIGURE 5**
Trapezoid approximation to integral
of tabulated velocity function.

## Rate of Flow

In diagnosing a patient it can be useful to know how much blood passes
through the heart per minute. To actually collect all that blood would be a
fatal error, but it is feasible to monitor the instantaneous *rate* of flow and to
determine from this the amount flowing through.

Suppose $Q(t)$ is the number of liters that have entered the heart from time 0 to time $t$. Then the rate at which blood enters the heart is the derivative $dQ/dt$. If this rate is known, the actual quantity entering the heart from time $a$ to time $b$ is computed as the integral

$$Q(b) - Q(a) = \int_a^b \frac{dQ}{dt}\, dt.$$

The same formula applies to any fluid flowing through a pipe—the quantity passing a given point for $a \leq t \leq b$ is the integral of the rate of flow, from $t = a$ to $t = b$.

---

**EXAMPLE 3**   Water drains out of a hole in the bottom of a cylindrical tank at a linearly decreasing rate; if $V$ stands for the volume of water in the tank at time $t$, then

$$dV/dt = -3 + \frac{1}{2}\, t\, \frac{\text{liters}}{\text{min}}, \qquad 0 \leq t \leq 6.$$

How much drains out from $t = 0$ to $t = 3$? From $t = 3$ to $t = 6$?

**SOLUTION**   The change in volume $V$ is the integral of the rate of change $dV/dt$. For $0 \leq t \leq 3$ it is

$$V(3) - V(0) = \int_0^3 \frac{dV}{dt}\, dt = \int_0^3 \left(-3 + \frac{1}{2}\, t\right) dt$$

$$= \left[-3t + \frac{1}{4}\, t^2\right]_0^3 = \left[-9 + \frac{9}{4}\right]$$

$$= -\frac{27}{4} = -6\frac{3}{4}\ \text{liters.}$$

So for $0 \leq t \leq 3$, $6\frac{3}{4}$ liters drain out. For $3 \leq t \leq 6$ the change in $V$ is

$$V(6) - V(3) = \int_3^6 \frac{dV}{dt}\, dt = \left[-3t + \frac{1}{4}\, t^2\right]_3^6$$

$$= [-18 + 9] - \left[-9 + \frac{9}{4}\right]$$

$$= -\frac{9}{4} = -2\frac{1}{4}\ \text{liters.}$$

Why does less drain out for $3 \leq t \leq 6$ than for $0 \leq t \leq 3$?

---

*Electrical current* is essentially the flow of a mysterious quantity called electrical charge. Imagine charge moving along a wire. If $Q(t)$ is the amount of charge that has passed a given point $P$ from time 0 to time $t$, then the derivative $dQ/dt$ is called the **current**, denoted $I$:

$$\text{electrical current } I = \frac{dQ}{dt}, \qquad \text{where } Q = \text{charge.}$$

So the amount of charge passing point $P$ for $a \leq t \leq b$ is the integral of the current from time $a$ to time $b$.

*REMARK*   The symbol $dt$ in the integral $\int f(t)\, dt$ is a reminder of the small quantities $\Delta t$ in the Riemann sums

$$S_n = f(\hat{t}_1)\, \Delta t + \cdots + f(\hat{t}_n)\, \Delta t.$$

In applied problems it also keeps the units straight. For example, displacement (in meters) is the integral of velocity (in meters/sec) with respect to time (in seconds):

$$y(b) - y(a) = \int_a^b \frac{dy}{dt}\, dt$$

$$\underset{\text{meters}}{} = \underset{\frac{\text{meters}}{\text{sec}}}{} \cdot \underset{\text{sec}}{}$$

Total flow (liters) is the integral of rate of flow (liters/min) with respect to time $t$ (min):

$$Q(b) - Q(a) = \int_a^b \frac{dQ}{dt}\, dt$$

$$\underset{\text{liters}}{} = \underset{\frac{\text{liters}}{\text{min}}}{} \cdot \underset{\text{min}}{}$$

## SUMMARY

The change in quantity $y$ equals the integral of the rate of change:

$$y(b) - y(a) = \int_a^b \frac{dy}{dt}\, dt.$$

Trapezoid sums for $\int_a^b f(t)\, dt$:

$$T_n = \Delta t(\tfrac{1}{2}y_0 + y_1 + y_2 + \cdots + y_{n-1} + \tfrac{1}{2}y_n),$$

$$\Delta t = \frac{b - a}{n}, \qquad y_j = f(a + j\, \Delta t)$$

## PROBLEMS

**A**

1.  Given the velocity $v$, find the displacement (change in position) during the given time interval.

*a)   $v = -2t; \quad 1 \le t \le 3$

*b)   $v = \dfrac{1}{t^2}; \quad 2 \le t \le 10$

*c)   $v = t^3, \quad -1 \le t \le 1$

d)   $v = \sqrt{t}, \quad 0 \le t \le 10$

e)   $v = t + 3\sec^2 t, \quad 0 \le t \le \dfrac{\pi}{3}$

f)   $v = 2\cos(3t), \quad 0 \le t \le \dfrac{\pi}{6}$

**2.** The velocity of a particle moving on the x-axis is $\dfrac{dx}{dt} = \sin t$. Find the displacement for the given time interval. Sketch a graph of $\dfrac{dx}{dt}$ to explain the answer, to part b.

    **a)**   $0 \le t \le \pi$     **b)**   $0 \le t \le 2\pi$

**\*3.** The velocity of a particle oscillating on the y-axis is $\dfrac{dy}{dt} = \cos(30t)$.

    **a)**   Find the displacement for $0 \le t \le \pi$.

    **b)**   Find the *total distance traveled* back and forth for $0 \le t \le \pi$.

**4.** Given the acceleration $a$, find the change in velocity during the given time interval.

    **a)**   $a = -10$ (constant), $1 \le t \le 5$

    **\*b)**   $a = \sin t - 2$, $-\pi \le t \le \pi$.

**5.** Figure 6 shows a graph of the velocity $v$ of a particle moving in a straight line.

    **\*a)**   Estimate the change in position from $t = 0$ to $t = 3$. (Estimate areas by a trapezoid sum or by counting squares; note the size of the squares.)

    **b)**   Assuming $s(0) = 1$, estimate $s(t)$ for $t = \frac{1}{2}$, $1$, $\frac{3}{2}$, $2$, $\frac{5}{2}$, $3$ and sketch a graph of the position of the particle as a function of time.

**6. \*a)**   The particle in Figure 6 moves to the left for $0 \le t \le 1$, because then $v \le 0$. Estimate how far to the left it moves.

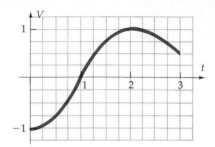

**FIGURE 6**
Graph of velocity $v$.

    **b)**   Estimate how far to the right it moves for $1 \le t \le 3$.

    **c)**   Estimate the *total distance traveled* from $t = 0$ to $t = 3$.

**\*7.** Given below is the velocity of a freight car on a track, every ten seconds. Approximately how far does the car move from its starting place in the given minute? (Use a trapezoid sum.)

| t (seconds) | 0 | 10 | 20 | 30 | 40 | 50 | 60 |
|---|---|---|---|---|---|---|---|
| velocity (mph) | 0 | −5 | −8 | −2 | 5 | 10 | 10 |

**8.** Figure 7 shows the vertical velocity of a rocket in meters/sec. At time 0 seconds it is at ground level ($y = 0$). Estimate the height:

    **\*a)**   At burnout.

    **b)**   At the opening of the parachute.

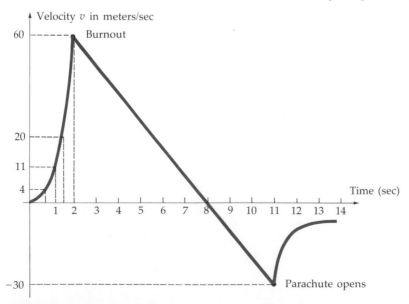

**FIGURE 7**
Velocity function for a rocket. [Adapted from Project CALC]

9. Blood flows through the left coronary artery of a heart with the rate of flow as in Figure 8. From the graph, determine the rate of flow for every tenth of a second, and then determine the total amount flowing through in 0.8 second. (Note the different time units on the two axes.)

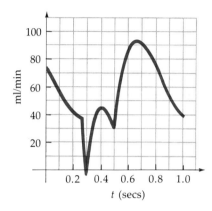

**FIGURE 8**
Rate of flow in left coronary artery in milliliters/min. (one milliliter = $10^{-3}$ liters).

—⁰⁰⁰⁰⁰⁰⁰⟶

**FIGURE 9**
Electrical coil (schematic).

10. Charge flows through an electrical coil (Fig. 9), and the current (from left to right) is measured as

$$I(t) = \frac{dQ}{dt} = 5 \sin t$$

in suitable units. What is the net flow through the coil from left to right, during the given time interval?

a) $0 \le t \le \pi$
b) $\pi \le t \le 2\pi$
c) $0 \le t \le 2\pi$

Sketch the graph of $I$ and reconcile your answers with the graph. When is charge flowing to the right? When to the left?

## 6.5

# PROPERTIES OF INTEGRALS. MEAN VALUE

Integrals, like derivatives, have handy algebraic properties. For example, the integral of a sum equals the sum of the integrals:

$$\int_a^b [f(x) + g(x)] \, dx = \int_a^b f(x) \, dx + \int_a^b g(x) \, dx.$$

To prove this, form Riemann sums for the integrals of $f$ and $g$:

$$S_n = f(\hat{x}_1) \, \Delta x + \cdots + f(\hat{x}_n) \, \Delta x,$$
$$T_n = g(\hat{x}_1) \, \Delta x + \cdots + g(\hat{x}_n) \, \Delta x.$$

Then, if $f$ and $g$ are continuous,

$$\int_a^b f(x) \, dx = \lim_{n \to \infty} S_n \quad \text{and} \quad \lim_{n \to \infty} \int_a^b g(x) \, dx = \lim_{n \to \infty} T_n. \tag{1}$$

The sum of $S_n$ and $T_n$ can be written

$$S_n + T_n = [f(\hat{x}_1) + g(\hat{x}_1)] \, \Delta x + \cdots + [f(\hat{x}_n) + g(\hat{x}_n)] \, \Delta x.$$

This is a Riemann sum for the function $f + g$, and the integral of $f + g$ equals the limit of such sums; so

$$\int_a^b [f(x) + g(x)] \, dx = \lim_{n \to \infty} (S_n + T_n)$$

$$= \lim_{n \to \infty} S_n + \lim_{n \to \infty} T_n$$

$$= \int_a^b f(x) \, dx + \int_a^b g(x) \, dx$$

by (1). This proves the formula for the integral of a sum. A very similar proof shows that the integral of a difference is the difference of the integrals:

$$\int_a^b [f(x) - g(x)] \, dx = \int_a^b f(x) \, dx - \int_a^b g(x) \, dx.$$

Further, any *constant c* can be "factored out" of an integral:

$$\int_a^b cf(x) \, dx = c \int_a^b f(x) \, dx.$$

(Functions, on the other hand, *cannot* be factored out!)

We write these general formulas with the brief notation $\int_a^b f$ in place of $\int_a^b f(x) \, dx$. Thus,

$$\int_a^b (f \pm g) = \int_a^b f \pm \int_a^b g$$

$$\int_a^b cf = c \cdot \int_a^b f \qquad \text{(if } c \text{ is constant).}$$

These properties together are called **linearity**.

---

**EXAMPLE 1**    By the formula for the integral of a difference,

$$\int_0^{\pi/6} [\sec^2 x - \cos x] \, dx = \int_0^{\pi/6} \sec^2 x \, dx - \int_0^{\pi/6} \cos x \, dx$$

$$= \tan x \Big|_0^{\pi/6} - \sin x \Big|_0^{\pi/6} = \frac{1}{\sqrt{3}} - \frac{1}{2}. \tag{2}$$

Figure 1 illustrates formula (2) with areas. On the right-hand side, $\int_0^{\pi/6} \sec^2 x \, dx$ is the area under the graph of $y = \sec^2 x$, and $\int_0^{\pi/6} \cos x \, dx$ is the area under the graph of $y = \cos x$, both for $0 \le x \le \pi/6$. Since $\sec^2 x \ge \cos x$ (Fig. 1), the difference in those areas is just the *area of the region R be-tween the curves* for $0 \le x \le \dfrac{\pi}{6}$ (Fig. 1). This is the right-hand side of (2).

The integral on the left-hand side gives the area between the $x$-axis and the graph of $y = \sec^2 x - \cos x$, $0 \le x \le \pi/6$; apparently *the area under this graph equals the area of R*. The integrand $\sec^2 x - \cos x$ is precisely the vertical distance between the two curves bounding the region $R$ (Fig. 1); so each rectangle in a Riemann sum for $\int_0^{\pi/6} [\sec^2 x - \cos x] \, dx$ can be shifted upward to approximate the corresponding part of region $R$. (This idea is developed systematically in Sec. 7.1)

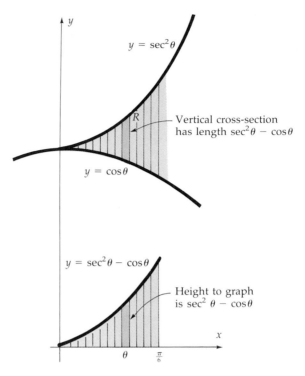

**FIGURE 1**
$\int_0^{\pi/6} (\sec^2 \theta - \cos \theta)\, d\theta = \int_0^{\pi/6} \sec^2 \theta\, d\theta - \int_0^{\pi/6} \cos \theta\, d\theta.$

**EXAMPLE 2**   By the formula for "factoring out" constants,

$$\int_{-1}^{1} 3\sqrt{1 - x^2}\, dx = 3 \int_{-1}^{1} \sqrt{1 - x^2}\, dx$$

$$= 3 \cdot \frac{\pi}{2}; \tag{3}$$

for $\int_{-1}^{1} \sqrt{1 - x^2}\, dx$ is the area under the unit semicircle $y = \sqrt{1 - x^2}$. Figure 2 illustrates this equation geometrically. The integral on the left in (3) gives the area under the graph of $y = 3\sqrt{1 - x^2}$, obtained from the semicircle by stretching it by a factor 3 in the vertical direction (Fig. 2). This triples the height of each approximating rectangle, hence triples its area. Thus, the area under the "stretched" curve is three times the area under the semicircle, $3 \cdot \frac{1}{2}\pi = \frac{3}{2}\pi$. (The "stretched semicircle" is the upper half of an ellipse; the area enclosed by the entire ellipse is $3\pi$.)

Figure 3 illustrates a further property, called **interval additivity**. The integral of $f$ from $a$ to $b$ is the net signed area between the graph of $f$ and the interval $[a, b]$ on the $x$-axis. Add to this the net signed area between the graph and the interval $[b, c]$, and you get the net signed area between the graph and the entire interval $[a, c]$. That is,

$$\int_a^b f + \int_b^c f = \int_a^c f. \tag{4}$$

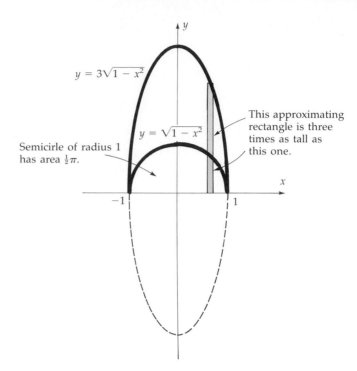

**FIGURE 2**

Area of upper half of ellipse $= \int_{-1}^{1} 3\sqrt{1 - x^2}\, dx = 3 \int_{-1}^{1} \sqrt{1 - x^2}\, dx$

$= 3x \text{ (area of semicircle)} = 3 \cdot \dfrac{\pi}{2}.$

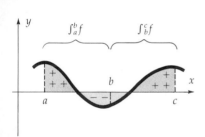

**FIGURE 3**
$\int_{a}^{b} f + \int_{b}^{c} f = \int_{a}^{c} f.$

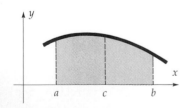

**FIGURE 4**
$\int_{a}^{c} f + \int_{c}^{b} f = \int_{a}^{b} f.$

Figure 3 shows the case where $a < b < c$, but the formula remains true for any order of $a$, $b$, and $c$. For example if $a < c < b$ then Figure 4 shows that

$$\int_{a}^{c} f + \int_{c}^{b} f = \int_{a}^{b} f.$$

Add $-\int_{c}^{b} f$ to both sides,

$$\int_{a}^{c} f = \int_{a}^{b} f - \int_{c}^{b} f.$$

Then recall that $\int_{c}^{b} f = -\int_{b}^{c} f$, so reversing the limits $b$ and $c$ gives

$$\int_{a}^{c} f = \int_{a}^{b} f + \int_{b}^{c} f$$

which is the same as (4).

---

**EXAMPLE 3**    Evaluate $\int_{-1}^{2} |x - 1|\, dx.$

**SOLUTION**    By definition of the absolute value,

$$|x - 1| = \begin{cases} x - 1 & \text{if } x - 1 > 0, \text{ that is, if } x > 1 \\ -(x - 1) & \text{if } x - 1 < 0, \text{ that is, if } x < 1. \end{cases} \tag{5}$$

Since the integrand has different formulas for $x < 1$ and for $x > 1$, we split the integral accordingly in two parts:

$$\int_{-1}^{2} |x - 1| \, dx = \int_{-1}^{1} |x - 1| \, dx + \int_{1}^{2} |x - 1| \, dx$$

$$= \int_{-1}^{1} -(x - 1) \, dx + \int_{1}^{2} (x - 1) \, dx \qquad \text{[using (5)]}$$

$$= \left[ -\frac{x^2}{2} + x \right]_{-1}^{1} + \left[ \frac{x^2}{2} - x \right]_{1}^{2}$$

$$= \left[ \left( -\frac{1}{2} + 1 \right) - \left( -\frac{1}{2} - 1 \right) \right] + \left[ \left( \frac{4}{2} - 2 \right) - \left( \frac{1}{2} - 1 \right) \right]$$

$$= 2 + \frac{1}{2} = 2\frac{1}{2}.$$

You can check this geometrically by sketching the graph.

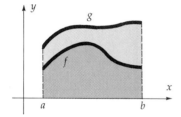

**FIGURE 5**
If $f \leq g$ then $\int_a^b f \leq \int_a^b g$.

Figure 5 illustrates the **Comparison Theorem** for integrals. If $f \leq g$ throughout the interval $[a, b]$, then the area under the graph of $f$ is less than or equal to the area under the graph of $g$, that is,

$$\int_a^b f(x) \, dx \leq \int_a^b g(x) \, dx.$$

The picture illustrates only the case $f \geq 0$, but the same principle remains true in any case (see problem 15).

**The Comparison Theorem**

If $f$ and $g$ are continuous, and

$$f(x) \leq g(x) \qquad \text{for } a \leq x \leq b,$$

then

$$\int_a^b f(x) \, dx \leq \int_a^b g(x) \, dx.$$

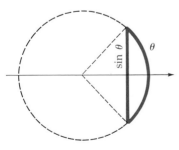

**FIGURE 6**
$\sin \theta \leq \theta$.

**EXAMPLE 4**    As Figure 6 reminds us, $\sin \theta \leq \theta$ for $\theta \geq 0$. So if $b \geq 0$ then, by the Comparison Theorem,

$$\int_0^b \sin \theta \, d\theta \leq \int_0^b \theta \, d\theta.$$

Evaluate the integrals:

$$-\cos b + 1 \leq \tfrac{1}{2}b^2.$$

You can now "solve" for $\cos b$, by adding $\cos b - \tfrac{1}{2}b^2$ to each side of the inequality:

$$1 - \tfrac{1}{2}b^2 \leq \cos b, \qquad b \geq 0.$$

Such inequalities lead to accurate calculations of sine and cosine (see problem 14).

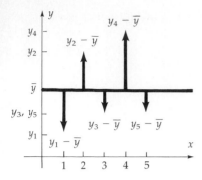

**FIGURE 7**
$(y_1 - \bar{y}) + \cdots + (y_5 - \bar{y}) = 0.$

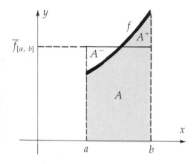

**FIGURE 8**
$A^+ = A^-$, so $\int_a^b f(x)\, dx =$
$A + A^+ = A + A^- = (b - a)\bar{f}_{ab}.$

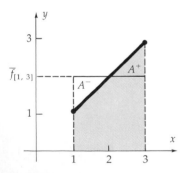

**FIGURE 9**
Mean Value of $f(x) = x$ on $[1, 3]$.

**The mean value** (or average) of $n$ numbers $y_1, y_2, \ldots, y_n$ is

$$\bar{y} = \frac{1}{n}[y_1 + \cdots + y_n]. \tag{6}$$

Suppose we plot the $n$ points $(1, y_1), (2, y_2), \ldots, (n, y_n)$ and the line $y = \bar{y}$ as in Figure 7 and indicate by an appropriate arrow the displacement of each number $y_j$ above or below the mean $\bar{y}$. Then the combined lengths of all the "up arrows" equals the combined lengths of all the "down arrows." For, the length of each "up arrow" is $y_j - \bar{y}$ and of each "down arrow" is $\bar{y} - y_k$; so the sum of all the "ups" *minus* all the "downs" is

$$(y_1 - \bar{y}) + (y_2 - \bar{y}) + \cdots + (y_n - \bar{y}) = [y_1 + \cdots + y_n] - n\bar{y}$$

and it follows from formula (6) defining $\bar{y}$ that this sum is zero.

The mean value of a *function f* on an interval $[a, b]$ is a number denoted $\bar{f}_{[a,b]}$ and is defined analogously (Fig. 8): The area $A^+$ *above* the line $y = \bar{f}_{[a,b]}$ and below the graph of $f$ equals the area $A^-$ *below* that line and above the graph. This implies that the integral of $f$ is exactly the same as the area of the rectangle of base $b - a$ and height $\bar{f}_{[a,b]}$; that is,

$$\int_a^b f(x)\, dx = (b - a)\bar{f}_{[a,b]}.$$

It follows that

$$\bar{f}_{[a,b]} = \frac{1}{b - a}\int_a^b f(x)\, dx. \tag{7}$$

This formula defines the mean value of $f$ on $[a, b]$.

---

**EXAMPLE 5**    The mean value of $f(x) = x$ on $[1, 3]$ is

$$\frac{1}{3 - 1}\int_1^3 x\, dx = \frac{1}{2}\left[\frac{x^2}{2}\right]_1^3 = \frac{1}{2}\left[\frac{9}{2} - \frac{1}{2}\right] = 2,$$

which looks reasonable (Fig. 9).

---

For functions, as for numbers, the mean lies between the extremes:

**THEOREM 1**

If $f$ is continuous on $[a, b]$, then

$$\min_{[a,b]} f \le \frac{1}{b - a}\int_a^b f(x)\, dx \le \max_{[a,b]} f.$$

**PROOF** Min $f$ and max $f$ are respectively the minimum and maximum of
$\underset{[a,b]}{} \quad \underset{[a,b]}{}$
$f$ on the interval $[a, b]$. So for each $x$ in $[a, b]$,

$$\min_{[a,b]} f \leq f(x) \leq \max_{[a,b]} f.$$

By the Comparison Theorem,

$$\int_a^b \min_{[a,b]} f \, dx \leq \int_a^b f(x) \, dx \leq \int_a^b \max_{[a,b]} f \, dx. \tag{8}$$

Now $\min_{[a,b]} f$ is a constant, and the integral of a constant $c$ is

$$\int_a^b c \, dx = (b - a) \cdot c;$$

so $\int_a^b \min_{[a,b]} f \, dx = (b - a) \cdot \min_{[a,b]} f$. It thus follows from (8) that

$$(b - a) \min_{[a,b]} f \leq \int_a^b f(x) \, dx \leq (b - a) \max_{[a,b]} f.$$

Divide by $(b - a)$, and Theorem 1 is proved.

Since the mean value of a continuous function lies between its maximum and minimum values, we can apply the Intermediate Value Theorem and conclude that the mean value is one of the function values $f(\hat{x})$ (Fig. 10). This fact is called:

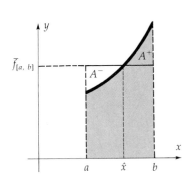

**FIGURE 10**
Illustrating Mean Value Theorem for integrals: $\bar{f}_{[a,b]} = f(\hat{x})$.

---

### THE MEAN VALUE THEOREM FOR INTEGRALS

If $f$ is continuous on $[a, b]$ then there is a number $\hat{x}$ in the interval $[a, b]$ such that

$$\frac{1}{b - a} \int_a^b f(x) \, dx = f(\hat{x}). \tag{9}$$

---

*REMARK 1* Suppose that $f$ has an antiderivative $F$. Then the left-hand side of equation (9) is $\dfrac{1}{b - a} [F(b) - F(a)]$ and the right-hand side is $F'(\hat{x})$, so (9) says

$$\frac{F(b) - F(a)}{b - a} = F'(\hat{x}).$$

This is precisely the Mean Value Theorem for derivatives (see. Sec. 7.2) applied to $F$. The two mean value theorems are virtually the same!

*REMARK 2* The mean value of the velocity is precisely the average velocity, as defined in Section 2.7. To see why, let $y(t)$ be position at time $t$, and

$v = dy/dt$ the velocity. The mean value of the velocity is

$$\frac{1}{b-a} \int_a^b v(t)\, dt = \frac{1}{b-a} [y(b) - y(a)],$$

and the right-hand side is, by definition, the average velocity from time $t = a$ to $t = b$.

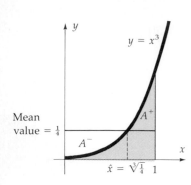

**FIGURE 11**
Mean value of $f(x) = x^3$ on $[0, 1]$
is $\frac{1}{4} = f(\sqrt[3]{\frac{1}{4}})$.

**EXAMPLE 6**   The mean value of $f(x) = x^3$ on $[0, 1]$ is

$$\frac{1}{1-0} \int_0^1 x^3\, dx = 1 \cdot \left[\frac{1}{4} x^4\right]_0^1 = \frac{1}{4}.$$

According to the Mean Value Theorem for Integrals, this is a function value $f(\hat{x})$ with $0 \le \hat{x} \le 1$; in fact (Fig. 11)

$$\tfrac{1}{4} = (\hat{x})^3 \qquad \text{if } \hat{x} = \sqrt[3]{1/4}.$$

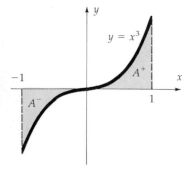

**FIGURE 12**
Mean value of $f(x) = x^3$ on $[-1, 1]$
is 0.

**EXAMPLE 7**   The mean value of $f(x) = x^3$ on $[-1, 1]$ is 0, as you can see in Figure 12. This is borne out by the calculation:

$$\frac{1}{1-(-1)} \int_{-1}^1 x^3\, dx = \frac{1}{2} \left[\frac{1}{4} x^4\right]_{-1}^1 = 0.$$

According to the Mean Value Theorem for integrals, the mean value 0 is a function value $f(\hat{x})$ with $-1 \le \hat{x} \le 1$; in fact,

$$0 = (\hat{x})^3 \qquad \text{if } \hat{x} = 0.$$

## SUMMARY

***Linearity:***   $\displaystyle\int_a^b [f \pm g] = \int_a^b f \pm \int_a^b g$

$$\int_a^b cf = c \int_a^b f$$

***Interval Additivity:***   $\displaystyle\int_a^b f(x)\, dx + \int_b^c f(x)\, dx = \int_a^c f(x)\, dx$

***Comparison Theorem:***   If $f(x) \le g(x)$ for $a \le x \le b$ then

$$\int_a^b f(x)\, dx \le \int_a^b g(x)\, dx.$$

***Mean Value*** of $f$ on the interval $[a, b]$:

$$\bar{f}_{[a,b]} = \frac{1}{b-a} \int_a^b f(x)\, dx.$$

*Mean Value Theorem for Integrals:*   If $f$ is continuous then

$$\frac{1}{b-a} \int_a^b f(x)\, dx = f(\hat{x})$$

for some $\hat{x}$ between $a$ and $b$.

## PROBLEMS

**A**

**1.** Given $\int_a^b f(x)\, dx = 5$ and $\int_a^b g(x) = 3$, compute (if possible!)

**\*a)** $\displaystyle\int_a^b [f(x) + g(x)]\, dx.$

**\*b)** $\displaystyle\int_a^b [f(x) - g(x)]\, dx.$

**\*c)** $\displaystyle\int_a^b 2f(x)\, dx.$

**\*d)** $\displaystyle\int_a^b -3g(x)\, dx.$

**e)** $\displaystyle\int_a^b [3f(x) - 2g(x)]\, dx.$

**\*f)** $\displaystyle\int_a^b f(x)g(x)\, dx.$

**g)** $\displaystyle\int_a^b f(x)\, dx.$

**h)** $\displaystyle\int_a^b g(x)\, dx.$

**i)** $\displaystyle\int_{-a}^{-b} f(x)\, dx.$

**2.** **\*a)** Given $\int_1^2 f(x)\, dx = 3$ and $\int_2^5 f(x)\, dx = -2$, find $\int_1^5 f(x)\, dx$.

**b)** Given $\int_0^{10} g(t)\, dt = 8$ and $\int_5^{10} g(t)\, dt = 6$, find $\int_0^5 g(t)\, dt$.

**\*3.** Evaluate $\int_0^4 |x - 2|\, dx$

**a)** Using interval additivity and the definition of absolute value.

**b)** Geometrically (sketch the graph of $|x - 2|$).

**4.** Compute the mean value and determine a value $\hat{x}$ as in the Mean Value Theorem for Integrals. Sketch the function.

**a)** $x^2$ on $[0, 1]$

**b)** $x^2$ on $[-1, 1]$

**c)** $1 + x^2$ on $[-1, 1]$

**\*d)** $\dfrac{1}{x^3}$ on $[-3, -1]$

**e)** $\sin x$ on $[0, \pi/4]$

**5.** Compute the mean value of the velocity $v(t) = 2 + 3 \sin t$ over the interval. Sketch the curve to see that the answers are reasonable.

**a)** $[0, 2\pi]$     **b)** $[-\pi, 0]$

**\*6.** Water flows out of a tank at the rate $dV/dt = 5 - 10t$, $0 \le t \le 2$. Compute the average rate of flow for that time interval.

**7.** Figure 13 shows the rate of flow in a coronary artery. Compute the average rate of flow for $0 \le t \le 0.8$. (From the figure, this is about the length of one heartbeat. Approximate the integral by a trapezoid sum, Sec. 6.4.)

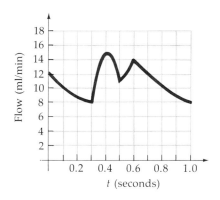

**FIGURE 13**
Rate of flow in right coronary artery.

**8.** The graph of $\dfrac{x^2}{9} + \dfrac{y^2}{4} = 1$ is an ellipse, whose area you will now compute.

**a)** Find the intersections of the ellipse with each coordinate axis and sketch it.

**b)** Solve the equation of the ellipse for $y$.

**c)** Show that the area of the upper half of the ellipse is $\int_{-3}^{3} \frac{2}{3}\sqrt{9 - x^2}\, dx$.

**d)** Compute the area of the ellipse.

**B**

9. Following the lines of the previous problem, compute the area inside the ellipse

$$\frac{x^2}{a^2} + \frac{y^2}{b^2} = 1.$$

10. For the mean value $\bar{f}_{[a,b]} = \frac{1}{b-a} \int_a^b f$, prove that $\int_a^b f(x)\, dx = \int_a^b \bar{f}_{[a,b]}\, dx$.

11. In the Mean Value Theorem for Integrals we assumed that $a < b$. Prove that the theorem is also true if $b < a$. (Hint: Reverse the limits of integration.)

12. Prove the interval additivity formula $\int_a^b f = \int_a^c f + \int_c^b f$ in the following cases:
    a) $c < a < b$    b) $a = c$

13. Use Riemann sums and limits to show that
    a) $\int_a^b cf(x)\, dx = c \int_a^b f(x)\, dx.$
    b) $\int_a^b [f(x) - g(x)]\, dx = \int_a^b f(x)\, dx - \int_a^b g(x)\, dx.$

14. a) Example 4 proved that $\cos x \geq 1 - \frac{1}{2}x^2$ for $x \geq 0$. Integrate this from 0 to $b$ to deduce that $\sin b \geq b - \frac{1}{6}b^3$ for $b \geq 0$.
    b) Deduce that $\cos b \leq 1 - \frac{1}{2}b^2 + \frac{1}{24}b^4$.
    *c) Use part b and Example 4 to approximate $\cos \frac{1}{10}$.

15. This problem outlines a proof of the Comparison Theorem without assuming that both functions $f$ and $g$ are positive. Suppose only that they are continuous and that $f(x) \leq g(x)$ for $a \leq x \leq b$.
    a) Why is $f(\hat{x}_1)\, \Delta x \leq g(\hat{x}_1)\, \Delta x$?
    b) Deduce that corresponding Riemann sums $S_n$ for $\int_a^b f(x)\, dx$ and $T_n$ for $\int_a^b g(x)\, dx$ satisfy $S_n \leq T_n$ and conclude that $\int_a^b f(x)\, dx \leq \int_a^b g(x)\, dx$.

16. We propose nine general properties of integrals. *Some are true for all continuous functions, and some are not.* In each case, decide whether the given property is true for all such functions or not. If it is true, explain why; if you can give no good reason, or are not sure whether it is true or false, try a few examples—if the property fails in any one example, it is not always true!

a) $\int_a^b (f + g + h) \overset{?}{=} \int_a^b f + \int_a^b g + \int_a^b h$

b) $\int_a^b (cf + g) \overset{?}{=} c \int_a^b f + \int_a^b g$

   ($c$ is constant.)

c) $\int_a^b fg \overset{?}{=} \left(\int_a^b f\right)\left(\int_a^b g\right)$

d) $\int_a^b fg \overset{?}{=} f \int_a^b g$

e) $\int_a^b fg \overset{?}{=} f \cdot \int_a^b g + g \cdot \int_a^b f$

f) $\int_a^b f^2 \overset{?}{=} \left(\int_a^b f\right)^2$

g) $\int_a^b \left(\frac{f}{g}\right) \overset{?}{=} \frac{\int_a^b f}{\int_a^b g}$

h) $\int_{-a}^0 f + \int_0^a f \overset{?}{=} 2\int_0^a f$

i) $\int_a^b f' \overset{?}{=} f(b) - f(a)$

   (Assume $f'$ is continuous.)

17. The formula for the mean value of a continuous function $f$ on an interval $[a, b]$ can be explained using Riemann sums. Divide the interval $a \leq x \leq b$ into $n$ equal subintervals; in each subinterval choose a point $\hat{x}_j$. The numbers

$$f(\hat{x}_1), f(\hat{x}_2), \ldots, f(\hat{x}_n)$$

give a sample of the values of $f$. Denote the mean value of this sample by $M_n$:

$$M_n = \frac{1}{n}[f(\hat{x}_1) + f(\hat{x}_n) + \cdots + f(\hat{x}_n)].$$

As $n \to \infty$, the sample gets "better and better," so we define the mean value of $f$ on the interval $[a, b]$ as the *limit*

$$\lim_{n \to \infty} M_n.$$

Compare $M_n$ with a typical Riemann sum $S_n$ for $f$ on $[a, b]$ and show that $M_n = \frac{1}{b-a} S_n$. Deduce that the mean value is

$$\lim_{n \to \infty} M_n = \frac{1}{b-a} \int_a^b f(x)\, dx.$$

## 6.6

## THE SECOND FUNDAMENTAL THEOREM OF CALCULUS

The Second Fundamental Theorem of Calculus answers a basic question: Does every continuous function $f$ have an antiderivative? For many functions the answer is obviously "yes"—for example, $f(x) = x^2$ has the antiderivative

$$D^{-1}(x^2) = \tfrac{1}{3}x^3 + C.$$

But for others, such as $f(x) = \sqrt{1 + x^3}$, the answer is not so easy—no combination of familiar functions yields an antiderivative for this. (Just try to write one that actually works!)

Nevertheless, it *is* true that every continuous function $f$ has an antiderivative, *given by an integral.* Consider the function of a variable $u$ defined by

$$I(u) = \int_a^u f(x)\, dx.$$

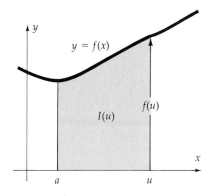

**FIGURE 1**
$I(u) = \int_a^u f(x)\, dx.$

The lower limit $a$ is any fixed number, while the upper limit $u$ varies. When $u > a$, then $I(u)$ is the area under the graph of $f(x)$ for $a \le x \le u$ (Fig. 1). When $u = a$ (Fig. 2) then

$$I(a) = \int_a^a f(x)\, dx = 0.$$

As $u$ moves to the right, the area $I(u)$ grows. We will see that *the rate of increase $I'(u)$ is exactly $f(u)$, the height to the graph at $u$.*

Why? consider a small change $\Delta u$ in $u$, and the resulting change

$$\Delta I = I(u + \Delta u) - I(u)$$

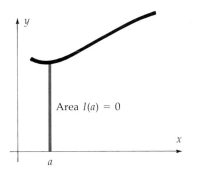

**FIGURE 2**
$I(a) = \int_a^a f(x)\, dx = 0.$

in the area $I$. We will verify what Figures 3 and 4 show:

$$\Delta I = \text{area of thin strip in Figure 3} = \int_u^{u + \Delta u} f(x)\, dx \qquad (1)$$

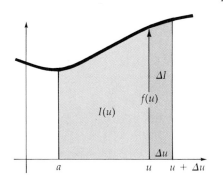

**FIGURE 3**
$\Delta I = $ (area from $a$ to $u + \Delta u$) − (area from $a$ to $u$) = area of shaded strip.

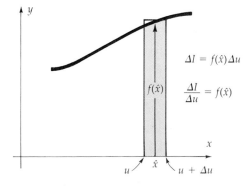

**FIGURE 4**
Area of strip $\Delta I = $ area of rectangle $= f(\hat{x})\, \Delta u.$

and (Fig. 4)

$$\Delta I = f(\hat{x})\,\Delta u \qquad \text{for some } \hat{x} \text{ between } u \text{ and } u + \Delta u. \tag{2}$$

Hence,

$$I'(u) = \lim_{\Delta u \to 0} \frac{\Delta I}{\Delta u} = \lim_{\hat{x} \to u} f(\hat{x}) = f(u). \tag{3}$$

To verify (1), note in Figure 3 that $I(u + \Delta u)$ is the area from $a$ to $u + \Delta u$, while $I(u)$ is the area from $a$ to $u$. The difference $\Delta I = I(u + \Delta u) - I(u)$ is then the area of the thin strip from $u$ to $u + \Delta u$, which is the integral $\int_u^{u+\Delta u} f(x)\,dx$, as claimed in (1).

Now the ratio $\dfrac{\Delta I}{\Delta u} = \dfrac{1}{\Delta u} \displaystyle\int_u^{u+\Delta u} f(x)\,dx$ is the mean value of $f$ on the interval from $u$ to $u + \Delta u$. By the Mean Value Theorem for Integrals (Sec. 6.5), this mean value equals the height to the graph at some point $\hat{x}$ in that interval:

$$\frac{\Delta I}{\Delta u} = \frac{1}{\Delta u} \int_u^{u+\Delta u} f(x)\,dx = f(\hat{x}).$$

This implies (2) and the limit in (3) follows. For as $\Delta u \to 0$, then $u + \Delta u \to u$; since $\hat{x}$ lies between $u$ and $u + \Delta u$, it follows that $\hat{x} \to u$ as well, and

$$I'(u) = \lim_{\Delta u \to 0} \frac{\Delta I}{\Delta u} = \lim_{\hat{x} \to u} f(\hat{x}) = f(u).$$

Thus the integral $I(u) = \int_a^u f(x)\,dx$ has the derivative $I'(u) = f(u)$. We state this fundamental result in Leibniz notation:

---

**THE SECOND FUNDAMENTAL THEOREM OF CALCULUS**

If $f$ is continuous on an interval containing $a$ and $u$, then

$$\frac{d}{du} \int_a^u f(x)\,dx = f(u).$$

---

**EXAMPLE 1**    By the Second Fundamental Theorem, $\dfrac{d}{du} \displaystyle\int_2^u x^2\,dx = u^2$. You can check this by evaluating the integral:

$$\int_2^u x^2\,dx = \tfrac{1}{3}x^3 \Big]_2^u = \tfrac{1}{3}u^3 - \tfrac{8}{3}.$$

So the derivative of the integral is

$$\frac{d}{du} \int_2^u x^2\,dx = \frac{d}{du}\left(\frac{1}{3}u^3 - \frac{8}{3}\right) = \frac{1}{3}\cdot 3u^2$$

$$= u^2.$$

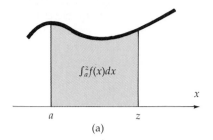

$$\int_a^z f(x)\,dx$$

(a)

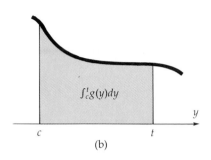

$$\int_c^t g(y)\,dy$$

(b)

**FIGURE 5**

The Second Fundamental Theorem gives the derivative of an integral, considered as a function of the upper limit of integration—the derivative is simply the integrand, evaluated at that upper limit. This can be expressed using various letters:

$$\frac{d}{dz}\int_a^z f(x)\,dx = f(z)$$

$$\frac{d}{dt}\int_c^t g(y)\,dy = g(t)$$

and so on. In each case, try to visualize the integral with appropriate labels on the axes (Fig. 5). Notice that the variable for the upper limit of the integral must be different from the variable of integration, since they play different roles.

---

**EXAMPLE 2**   $\dfrac{d}{dx}\displaystyle\int_0^x \sqrt{1 + t^3}\,dt = \sqrt{1 + x^3}$. There is no independent way to verify this, for there is no antiderivative for $\sqrt{1 + x^3}$, except an integral with a variable limit of integration.

---

**EXAMPLE 3**   Compute first and second derivatives of the function

$$F(x) = \int_0^x \frac{1}{1 + t^2}\,dt.$$

*SOLUTION*   By the Second Fundamental Theorem

$$F'(x) = \frac{1}{1 + x^2}$$

and therefore,

$$F''(x) = \frac{d}{dx}F'(x) = \frac{d}{dx}\left(\frac{1}{1 + x^2}\right) = \frac{-2x}{(1 + x^2)^2}.$$

With this information you can sketch a graph of $F$. First of all,

$$F(0) = \int_0^0 \frac{1}{1 + t^2}\,dt = 0;$$

for $\displaystyle\int_0^0 \frac{1}{1 + t^2}\,dt$ is just the "area" of a line segment (Fig. 6). Second,

$$F'(x) = \frac{1}{1 + x^2} > 0 \qquad \text{for all } x,$$

so the graph is always rising to the right. Third,

$$F''(x) = \frac{-2x}{(1 + x^2)^2}$$

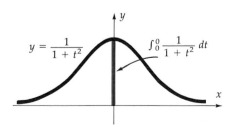

$$y = \frac{1}{1 + t^2}$$

$$\int_0^0 \frac{1}{1 + t^2}\,dt$$

**FIGURE 6**

$$\int_0^0 \frac{1}{1 + t^2}\,dt = \text{area of segment} = 0.$$

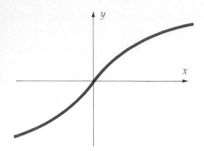

**FIGURE 7**

Graph of $F(x) = \int_0^x \dfrac{dt}{1 + t^2}$.

$\left(\textit{Not the graph of } \dfrac{1}{1 + t^2}!\right)$

is *positive* when $x < 0$ and *negative* when $x > 0$, so the graph is concave *up* for $x < 0$, and concave *down* for $x > 0$. Figure 7 exhibits all these features.

The signs of $F'$ and $F''$ do not determine whether $F(x)$ rises to $+\infty$ or levels off at an asymptote. In fact, it does level off—see problem 6.

**EXAMPLE 4**  $\dfrac{d}{dx} \displaystyle\int_0^{x^2} \sin(t^2)\, dt = ?$

*SOLUTION*   This is a chain rule problem. Let $u = x^2$. Then

$$\frac{d}{dx} \int_0^{x^2} \sin(t^2)\, dt = \frac{d}{dx} \int_0^u \sin(t^2)\, dt$$

$$= \left[\frac{d}{du} \int_0^u \sin(t^2)\, dt\right] \cdot \left(\frac{du}{dx}\right) \qquad \text{(chain rule)}$$

$$= \sin(u^2)\, \frac{du}{dx} \qquad \text{(Second Fundamental Theorem)}$$

$$= \sin((x^2)^2) \cdot 2x \qquad (u = x^2)$$

$$= 2x\, \sin(x^4).$$

**EXAMPLE 5**  $\dfrac{d}{dx} \displaystyle\int_x^b \sin(t^2)\, dt = ?$

*SOLUTION*   The integral $\int_x^b \sin(t^2)\, dt$ is presented as a function of the *lower limit*, whereas the Second Fundamental Theorem gives the derivative with respect to the upper limit. No problem—just reverse the limits and change the sign:

$$\frac{d}{dx} \int_x^b \sin(t^2)\, dt = \frac{d}{dx}\left[-\int_b^x \sin(t^2)\, dt\right]$$

$$= -\frac{d}{dx} \int_b^x \sin(t^2)\, dt = -\sin(x^2).$$

*REMARK*   The First Fundamental Theorem can be proved using the Second. Suppose you want to evaluate $\int_a^b f(x)\, dx$. Let

$$I(u) = \int_a^u f(x)\, dx.$$

By the Second Fundamental Theorem, $I'(u) = f(u)$, so $I$ is an antiderivative of $f$. If $F$ is any other antiderivative of $f$, then

$$I(u) = F(u) + C \qquad\qquad\qquad (4)$$

for a constant $C$. Since $I(a) = \int_a^a f = 0$, setting $u = a$ in (4) gives

$$0 = F(a) + C.$$

So $F(a) = -C$, and from (4), $I(u) = F(u) - F(a)$. Hence,

$$\int_a^b f(x)\, dx = I(b) = F(b) - F(a).$$

This is precisely the First Fundamental Theorem.

**FINAL REMARK**   Both Fundamental Theorems say, roughly, that *integration is the reverse of differentiation*. The First Fundamental Theorem can be written as

$$\int_a^b F'(x)\, dx = F(b) - F(a).$$

So if you start with $F$, and differentiate to get $F'$, then integrate from $a$ to $b$, you recover the original function $F$—more precisely, you get $F(b) - F(a)$. In this sense, integrating "undoes" the derivative.

The Second Fundamental Theorem says that if you start with $f$, integrate to get

$$\int_a^x f(t)\, dt,$$

then differentiate with respect to the upper limit of integration $x$, you recover $f(x)$. So differentiating "undoes" the integral.

## SUMMARY

$$\frac{d}{du} \int_a^u f(x)\, dx = f(u)$$

if $f$ is continuous on an interval containing $a$ and $u$.

## PROBLEMS

### A

1. Compute the derivative in two ways: (i) by the Second Fundamental Theorem, and (ii) by evaluating the integral, then taking the derivative. In each case, sketch and label a rough graph showing the integral.

   *a) $\dfrac{d}{du} \displaystyle\int_0^u 2\, dx$

   b) $\dfrac{d}{dx} \displaystyle\int_1^x \dfrac{1}{\sqrt{t}}\, dt$

   *c) $\dfrac{d}{dt} \displaystyle\int_0^t s^2\, ds$

   d) $\dfrac{d}{du} \displaystyle\int_0^u \cos\theta\, d\theta$

   *e) $\dfrac{d}{dx} \displaystyle\int_1^{1/x} (1/t^2)\, dt$

   f) $\dfrac{d}{dx} \displaystyle\int_x^0 t^2\, dt$

   *g) $\dfrac{d}{dx} \displaystyle\int_{-x}^x t^2\, dt$

   h) $\dfrac{d}{dx} \displaystyle\int_{-x}^x t^3\, dt$

   i) $\dfrac{d}{dx} \displaystyle\int_a^x F'(t)\, dt$

2. Compute the derivative.

   a) $\dfrac{d}{dx} \displaystyle\int_2^x \sqrt{1+z^2}\, dz$

   *b) $\dfrac{d}{dz} \displaystyle\int_0^z \cos\left(-\dfrac{x^2}{2}\right) dx$

   *c) $\dfrac{d}{dt} \displaystyle\int_0^t \varphi(s)\, ds$

3. Compute the derivative.

*a) $\dfrac{d}{dx} \displaystyle\int_2^{x^2} \sin(\theta^2)\, d\theta$

b) $\dfrac{d}{dx} \displaystyle\int_{x^3}^0 \sin(\theta^2)\, d\theta$

*c) $\dfrac{d}{dx} \displaystyle\int_{x^3}^{x^2} \sin(\theta^2)\, d\theta$

d) $\dfrac{d}{d\theta} \displaystyle\int_1^{\cos\theta} \dfrac{dx}{1+x}$

4. a) Show that $\frac{2}{3}[1 + x^3]^{3/2} + C$ is *not* an antiderivative of $[1 + x^3]^{1/2}$.

b) Write an antiderivative for $[1 + x^3]^{1/2}$.

5. Compute the first and second derivative. Evaluate the function at one point, and sketch an appropriate graph, as in Example 3.

*a) $I(x) = \displaystyle\int_0^x \dfrac{1}{t+1}\, dt, \quad x > -1$

b) $G(x) = \displaystyle\int_0^x \dfrac{dt}{\sqrt{1 - t^2}}, \quad |x| < 1$

c) $H(x) = \displaystyle\int_{-1}^x \sqrt{1 + t^3}\, dt, \quad x \geq -1$

## B

6. Let $F(x) = \displaystyle\int_0^x \dfrac{dt}{1 + t^2}$, as in Example 3 and Figure 7. Show:

a) $F(-x) = -F(x)$. (Interpret the integrals as signed areas.)

b) For $x \geq 1$, $F(x) < \displaystyle\int_0^1 1\, dt + \int_1^x \dfrac{1}{t^2}\, dt$.

c) $-2 < F(x) < 2$. (Hence, $F(x)$ does not $\to +\infty$ as $x \to +\infty$.)

7. Suppose that $f'(x) > 0$ for $x > 2$, $f'(x) < 0$ for $x < 2$, and $f(2) = 0$. For the function $F(x) = \int_1^x f(t)\, dt$ is it necessarily true that:

a) $F(2) = 0$?

b) $F(1) = 0$?

c) $F'(x) \geq 0$ for all $x$?

d) The graph of $F$ has an inflection point at $x = 2$?

8. Let $f$ be a continuous function with period 2; that is, $f(x + 2) = f(x)$, for all $x$. Show that $\int_a^{a+2} f(x)\, dx$ has the same value for all numbers $a$, in two ways:

a) By taking a derivative.

b) By an appropriate sketch.

9. Discover and prove a formula for

$$\dfrac{d}{dx} \int_{-x}^x f(t)\, dt,$$

where $f$ is continuous. Illustrate the formula by an appropriate sketch.

## C

10. Suppose that $\varphi_1$ and $\varphi_2$ are differentiable functions (that is, $\varphi_1'$ and $\varphi_2'$ exist) and $f$ is continuous on the range of $\varphi_1$ and $\varphi_2$. Prove that

$$\dfrac{d}{dx} \int_{\varphi_1(x)}^{\varphi_2(x)} f(t)\, dt = f(\varphi_2(x))\varphi_2'(x) - f(\varphi_1(x))\varphi_1'(x).$$

## 6.7

# INDEFINITE INTEGRALS

By the Second Fundamental Theorem of Calculus, every continuous function $f$ has an antiderivative $F$, given by an integral with a variable upper limit:

$$F(u) = \int_a^u f(x)\, dx.$$

This suggests the use of an integral sign to denote the antiderivatives of $f$. The notation that has been adopted is due to Leibniz:

$$\int f(x)\, dx,$$

an integral sign without limits of integration. For example,

$$\int x\, dx = \frac{1}{2} x^2 + C \qquad \left[ \text{since } \frac{d}{dx}\left(\frac{1}{2}x^2 + C\right) = x \right].$$

More generally,

$$\int x^n\, dx = \frac{1}{n+1} x^{n+1} + C, \qquad \text{if } n \neq -1, \tag{1}$$

$$\int \cos(a\theta)\, d\theta = \frac{1}{a}\sin(a\theta) + C, \qquad \int \sin(a\theta)\, d\theta = -\frac{1}{a}\cos(a\theta) + C \tag{2}$$

and so on. (The Summary gives a more complete list.)

These are called **indefinite integrals**. Given $f$, the indefinite integral $\int f(x)\, dx$ is a function $F(x)$, or rather a whole collection of functions $F(x) + C$, where $C$ is any constant.

By contrast, the other notion of integral, $\int_a^b f(x)\, dx$, is called a **definite integral**. Given the function $f$, along with the constants $a$ and $b$, the definite integral $\int_a^b f(x)\, dx$ is a *number*, the net signed area between the graph of $f$ and the fixed interval $[a, b]$.

The antiderivative formulas in Section 2.8, for sums and constant multiples, can be rewritten with the indefinite integral notation:

$$\int cf(x)\, dx = c \int f(x)\, dx \quad (c \text{ is constant.}) \tag{3}$$

$$\int [f(x) + g(x)]\, dx = \int f(x)\, dx + \int g(x)\, dx. \tag{4}$$

We use these formulas, along with algebraic manipulations, to reduce a given integral to known standard indefinite integrals, such as (1) and (2).

---

**EXAMPLE 1** $\displaystyle \int \frac{2x+1}{\sqrt{x}}\, dx = ?$

*SOLUTION*   Rewrite this as a sum of indefinite integrals of powers $x^n$:

$$\begin{aligned}
\int (2x+1)x^{-1/2}\, dx &= \int (2x^{1/2} + x^{-1/2})\, dx & \text{(algebra)} \\
&= \int 2x^{1/2}\, dx + \int x^{-1/2}\, dx & [\text{by (4)}] \\
&= 2\int x^{1/2}\, dx + \int x^{-1/2}\, dx & [\text{by (3)}] \\
&= 2 \cdot \frac{x^{3/2}}{3/2} + \frac{x^{1/2}}{1/2} + C & [\text{by (1)}] \\
&= \frac{4}{3} \cdot x^{3/2} + 2x^{1/2} + C.
\end{aligned}$$

---

**EXAMPLE 2**

$$\int (2 \cos \theta - 3 \sin \theta)\, d\theta = 2 \int \cos \theta\, d\theta - 3 \int \sin \theta\, d\theta$$
$$= 2 \sin \theta + 3 \cos \theta + C.$$

In evaluating *definite* integrals by the First Fundamental Theorem, you can use the indefinite integral notation:

$$\int_a^b f(x)\, dx = D^{-1} f(x) \Big|_a^b$$
$$= \int f(x)\, dx \Big|_a^b \tag{5}$$

**EXAMPLE 3**

$$\int_{-1}^1 x\, dx = \int x\, dx \Big|_{-1}^1$$
$$= \tfrac{1}{2} x^2 \Big|_{-1}^1$$
$$= \tfrac{1}{2}(1)^2 - \tfrac{1}{2}(-1)^2$$
$$= 0.$$

## Rate of Growth

Suppose that a quantity $Q(x)$ is to be determined, given a specific value $Q(a)$ and given the rate of growth $Q'(x)$ for all $x$. Then $Q$ is an indefinite integral of $Q'$:

$$Q(x) = \int Q'(x)\, dx.$$

The indefinite integral contains a constant $C$, which is determined by the given value $Q(a)$.

**EXAMPLE 4**  At time $t = 0$, a container holds 2 liters. Water is added at the rate of $t - t^2$ liters/second for $0 \le t \le 1$. How much water does it hold at time $t = 1$?

**SOLUTION**  Let $Q(t)$ be the amount of water at time $t$. Then $Q(0) = 2$, and $Q'(t) = t - t^2$ for $0 \le t \le 1$. Hence,

$$Q(t) = \int (t - t^2)\, dt$$
$$= \tfrac{1}{2} t^2 - \tfrac{1}{3} t^3 + C, \qquad 0 \le t \le 1.$$

Since $Q(0) = 2$, then $C = 2$ and

$$Q(t) = \tfrac{1}{2} t^2 - \tfrac{1}{3} t^3 + 2, \qquad 0 \le t \le 1.$$

At time $t = 1$, the amount of water is

$$Q(1) = \tfrac{1}{2} - \tfrac{1}{3} + 2 = 2\tfrac{1}{6} \text{ liters.}$$

### Functions with Discontinuities

In an indefinite integral of a function with a discontinuity (like $x^{-2}$), the constant of integration must be handled with care. In writing

$$\int x^{-2}\, dx = -\frac{1}{x} + C,$$

$C$ is constant *in each interval where $x^{-2}$ is continuous*, but may differ from one interval to the other. For example, the function

$$F(x) = \begin{cases} -\dfrac{1}{x} + 1, & x > 0 \\[2mm] -\dfrac{1}{x} - 2, & x < 0 \end{cases}$$

is an indefinite integral of $x^{-2}$; you can check that $F'(x) = x^{-2}$, $x \neq 0$.

As for definite integrals, *we have defined them only when the integrand $f$ is continuous throughout the interval of integration*. If you use the evaluation formula (5) on an interval containing a discontinuity, you may get nonsense. (See problem 22.) The appropriate way to deal with discontinuities is taken up in Section 7.4.

## SUMMARY

The indefinite integral $\int f(x)\, dx$ denotes the antiderivatives of $f(x)$.

$$\int cf(x)\, dx = c \int f(x)\, dx$$

$$\int [f(x) + g(x)]\, dx = \int f(x)\, dx + \int g(x)\, dx$$

$$\int x^n\, dx = \frac{1}{n+1} x^{n+1} + C, \qquad n \neq -1$$

$$\int \sin(ax)\, dx = -\frac{1}{a} \cos(ax) + C$$

$$\int \cos(ax)\, dx = \frac{1}{a} \sin(ax) + C$$

$$\int \sec^2 ax\, dx = \frac{1}{a} \tan(ax) + C$$

$$\int \csc^2 ax\, dx = -\frac{1}{a} \cot(ax) + C$$

$$\int \sec(ax) \tan(ax)\, dx = \frac{1}{a} \sec(ax) + C$$

$$\int \csc(ax) \cot(ax)\, dx = -\frac{1}{a} \csc(ax) + C$$

## PROBLEMS

### A

In problems 1–16, evaluate the definite and indefinite integrals.

*1. $\displaystyle\int -x^3\, dx$

*2. $\displaystyle\int (5t^2 + 1)\, dt$

*3. $\displaystyle\int (2\sqrt{t} + 3\sqrt[3]{t})\, dt$

*4. $\displaystyle\int [2\cos\theta + 3]\, d\theta$

*5. $\displaystyle\int (\sec^2\theta - 1)\, d\theta$

6. $\displaystyle\int \tan^2\theta\, d\theta$

7. $\displaystyle\int 3\cos(\pi\theta)\, d\theta$

*8. $\displaystyle\int_0^1 3\sin(\pi\theta)\, d\theta$

9. $\displaystyle\int_0^1 \sec^2\left(\frac{\pi\theta}{3}\right) d\theta$

10. $\displaystyle\int_1^2 z^{-2}\, dz$

*11. $\displaystyle\int_{-2}^{-1} \frac{1}{x^2}\, dx$

12. $\displaystyle\int_1^2 \frac{x^2 + 2x + 1}{\sqrt{x}}\, dx$

*13. $\displaystyle\int_2^3 \frac{E}{RT^2}\, dT$     (E and R are constants.)

14. $\displaystyle\int \frac{\sin^2 2\theta}{\cos^2 2\theta}\, d\theta$

*15. $\displaystyle\int \sec\theta(\sec\theta + \tan\theta)\, d\theta$

16. $\displaystyle\int \frac{\sin\theta}{1 - \sin^2\theta}\, d\theta$

17. Compute the area of the region *below* the graph of $y = x$ and *above* the graph of $y = x^2$.

*18. Compute the area under one arch of the graph of $y = \cos(\pi\theta)$.

19. Check the following indefinite integrals by differentiating.

a) $\displaystyle\int x\sqrt{1 + x^2}\, dx = \frac{1}{3}(1 + x^2)^{3/2} + C$

b) $\displaystyle\int x^2(1 + x^3)^{99}\, dx = \frac{1}{300}(1 + x^3)^{100} + C$

20. At time $t = 3$, a capacitor has a charge $Q = 10$ volts. Charge flows in at the rate $[8 + 170\cos(120\pi t)]$ volts/sec. What is the charge at $t = 5$ seconds?

21. Is there a constant $k$ such that:

a) $\displaystyle\int \theta\sin(\theta^2)\, d\theta = k\cos(\theta^2) + C$?

b) $\displaystyle\int \theta^2\cos(5\theta^3)\, d\theta = k\sin(5\theta^3) + C$?

c) $\displaystyle\int \sin(\theta^2)\, d\theta = k\cos(\theta^2) + C$?

d) $\displaystyle\int \cos(5\theta^3)\, d\theta = k\sin(5\theta^3) + C$?

### B

22. a) Is this calculation legitimate?

$$\int_{-1}^1 x^{-2}\, dx = \frac{1}{-1}x^{-1}\Big|_{-1}^1$$
$$= (-1) - (1) = -2$$

b) Explain geometrically why $\int_{-1}^1 x^{-2}\, dx$, if it exists at all, should not be negative.

23. a) Explain geometrically why $\int_{-1}^1 x^{-3}\, dx$, if it exists at all, should be 0.

b) Is this calculation legitimate?

$$\int_{-1}^1 x^{-3}\, dx = \frac{1}{-2}x^{-2}\Big|_{-1}^1$$
$$= \left(-\frac{1}{2}\right) - \left(-\frac{1}{2}\right) = 0$$

24. Is the following function an antiderivative (indefinite integral) of $f(x) = 0$, on the whole real line?

$$F(x) = \begin{cases} 1, & x > 0 \\ -1, & x \le 0 \end{cases}$$

Why, or why not? (An antiderivative of a function $f$ on an interval $I$ is a function $F$ such that $F'(x) = f(x)$ for every $x$ in $I$.)

25. Is the equation

$$\int \frac{1}{1 + x^2}\, dx = \int_0^x \frac{1}{1 + t^2}\, dt + C$$

valid?

# 6.8

## DIFFERENTIALS AND SUBSTITUTION

Each derivative formula has a companion formula for indefinite integrals. Three basic "pairs" are:

|  *Derivative* |  *Integral* |
| --- | --- |

$$\frac{d[u(x)]^{n+1}}{dx} = (n+1)[u(x)]^n u'(x) \qquad \int u(x)^n u'(x)\, dx = \frac{1}{n+1} u(x)^{n+1} + C.$$

$$\frac{d\,\sin(u(x))}{dx} = \cos(u(x)) \cdot u'(x) \qquad \int \cos(u(x))u'(x)\, dx = \sin(u(x)) + C.$$

$$\frac{d\,\cos(u(x))}{dx} = -\sin(u(x)) \cdot u'(x) \qquad \int \sin(u(x))u'(x)\, dx = -\cos(u(x)) + C.$$

and in general

$$\frac{dF(u(x))}{dx} = F'(u(x)) \cdot u'(x) \qquad \int F'(u(x))u'(x)\, dx = F(u(x)) + C.$$

On the left, the term $u'(x)$ comes from the Chain Rule. On the right, the term $u'(x)$ must be present, or the formula does not apply! Thus $\int \cos(x^2) \cdot 2x \cdot dx = \sin(x^2) + C$   since   $\dfrac{d\,\sin(x^2)}{dx} = \cos(x^2) \cdot 2x;$   but

$$\int \cos(x^2) \cdot dx \neq \sin(x^2) + C \quad \text{since} \quad \frac{d\,\sin(x^2)}{dx} \neq \cos(x^2).$$

**Differential notation** makes the proper use of the Chain Rule for indefinite integrals automatic. The expression $u'(x)\, dx$ is called the **differential** of $u$, denoted $du$ (see Sec. 4.1):

$$du = u'(x)\, dx.$$

This is consistent with the Leibniz notation for the derivative, $\dfrac{du}{dx} = u'(x)$. For example, if $u(x) = x^2$ then

$$\frac{du}{dx} = 2x, \quad \text{and} \quad du = 2x\, dx.$$

The basic integral formulas above can thus be abbreviated, replacing $u'(x)\, dx$ by $du$:

$$\int u^n\, du = \frac{u^{n+1}}{n+1} + C, \qquad n \neq -1 \tag{1}$$

$$\int \cos(u)\, du = \sin u + C \tag{2}$$

$$\int \sin(u)\, du = -\cos(u) + C, \tag{3}$$

where it must be remembered that $du$ stands for $u'(x)\, dx$!

Many integrals come with the necessary part $u'(x)\,dx$, *except for a constant factor*. This can always be supplied.

---

**EXAMPLE 1** $\int \sin(x^2) \cdot x\,dx$ resembles $\int \sin(u)\,du$ with $u = x^2$, but it is not quite the same. If $u = x^2$, then

$$du = 2x\,dx.$$

The given integral has $x\,dx$, but lacks the factor 2. So we supply that missing factor and compensate by multiplying by $\frac{1}{2}$:

$$\int \sin(x^2) \cdot x\,dx = \int \tfrac{1}{2} \underbrace{\sin(x^2)}_{\sin(u)} \cdot \underbrace{2x\,dx}_{du}$$

$$= \tfrac{1}{2} \int \sin(u) \cdot du$$
$$= \tfrac{1}{2}[-\cos(u) + C]$$
$$= -\tfrac{1}{2}\cos(x^2) + \tfrac{1}{2}C.$$

The constant $\frac{1}{2}C$ can be relabeled as $c$:

$$\int \sin(x^2) \cdot x\,dx = -\tfrac{1}{2}\cos(x^2) + c.$$

---

The method in Example 1 is called **substitution**. You reduce a complicated integral to a basic known integral by identifying some part of the given integrand as a function $u(x)$. There are two essential points to remember:

You must relate $dx$ to $du$ by the formula

$$du = u'(x)\,dx.$$

You can "factor" *constants* out of the integral, but not functions.

---

**EXAMPLE 2** $\int \cos(3x)\,dx$ suggests $\int \cos(u)\,du$, with $u = 3x$. This implies $du = 3\,dx$, or $dx = \frac{1}{3}\,du$, so

$$\int \cos 3x\,dx = \int (\cos u) \cdot (\tfrac{1}{3}\,du)$$

$$= \tfrac{1}{3} \int \cos u\,du$$
$$= \tfrac{1}{3}[\sin u + C]$$
$$= \tfrac{1}{3}\sin u + c,$$

where $c = \frac{1}{3}C$.

---

**EXAMPLE 3** The integral $\int \dfrac{dx}{(2x + 1)^2}$ resembles formula (1) with $n = -2$, if you set

$$u = 2x + 1.$$

Then

$$du = 2\ dx, \qquad \text{so } dx = \tfrac{1}{2}\ du$$

and

$$\int \frac{dx}{(2x + 1)^2} = \int \frac{\frac{1}{2}\ du}{u^2} = \frac{1}{2} \int u^{-2}\ du$$

$$= \frac{1}{2}\left(\frac{1}{-1}\right)u^{-1} + C = \frac{-1}{2u} + C$$

$$= \frac{-1}{2(2x + 1)} + C.$$

Check this by differentiating.

---

**EXAMPLE 4**   $\int \sqrt[3]{x^3 + 1}\ x^2\ dx = ?$

**SOLUTION**   The most prominent part of this integral is the radical, which can be written

$$\sqrt[3]{x^3 + 1} = u^{1/3}$$

by setting $u = x^3 + 1$. Then

$$du = 3x^2\ dx. \qquad\qquad (4)$$

Fortunately, $x^2\ dx$ appears in the given integral, so the substitution will work:

$$\int (x^3 + 1)^{1/3}\ x^2\ dx = \int u^{1/3}\ x^2\ dx$$

$$\qquad\qquad\qquad\qquad\qquad \text{[by (4)]}$$

$$= \int u^{1/3}\left(\frac{1}{3}\ du\right) \qquad \text{(now factor out constant } \tfrac{1}{3}\text{)}$$

$$= \frac{1}{3}\int u^{1/3}\ du = \frac{1}{3}\,\frac{1}{4/3}\,u^{4/3} + C = \frac{1}{4}u^{4/3} + C$$

$$= \frac{1}{4}(x^3 + 1)^{4/3} + C.$$

---

Even if it is not obvious that a substitution will work, it is legitimate to *try*—write everything in terms of the new variable $u$ and see if the result can be integrated.

---

**EXAMPLE 5**   $\int x\sqrt{x - 1}\ dx = ?$

**SOLUTION**   Set $u = x - 1$, so $\dfrac{du}{dx} = 1$, $du = dx$. Also $x = u + 1$ and

$\sqrt{x-1} = u^{1/2}$, so

$$\int x\sqrt{x-1}\,dx = \int (u+1) \cdot u^{1/2}\,du$$

$$= \int (u^{3/2} + u^{1/2})\,du$$

$$= \frac{u^{5/2}}{5/2} + \frac{u^{3/2}}{3/2} + C$$

$$= \frac{2}{5}(x-1)^{5/2} + \frac{2}{3}(x-1)^{3/2} + C.$$

This can be simplified by factoring out $(x-1)^{3/2}$:

$$\int x\sqrt{x-1}\,dx = \tfrac{2}{5}(x-1)^{3/2}(x-1) + \tfrac{2}{3}(x-1)^{3/2} + C$$

$$= \tfrac{2}{15}(x-1)^{3/2}(3x+2) + C.$$

The next example contains a warning.

**EXAMPLE 6**   $\int \sqrt{x^2+1}\,dx = ?$

**SOLUTION**   Try $u = x^2 + 1$, so $\dfrac{du}{dx} = 2x$ and $du = 2x\,dx$. Then

$$\int \sqrt{x^2+1}\,dx = \int u^{1/2}\,\frac{1}{2x}\,du.$$

Now beware! You *cannot* factor out $\dfrac{1}{2x}$ because it is not a constant:

$$\int u^{1/2}\,\frac{1}{2x}\,du \neq \frac{1}{2x}\int u^{1/2}\,du!$$

As we noted before, only constants, not functions, can be "factored out" of an integral. In fact, this particular antiderivative cannot be evaluated by such a simple substitution. We'll return to it in Chapter 9.

## Shorthand Substitution

Once familiar with the method, you can make simple substitutions without actually writing the new variable $u$. For example, the substitution $u = 3x$ in

$$\int \cos(3x)\,dx = \int \cos u\,\tfrac{1}{3}\,du = \tfrac{1}{3}\sin u + C = \tfrac{1}{3}\sin(3x) + C$$

would be abbreviated

$$\int \cos(3x)\,dx = \tfrac{1}{3}\int \cos(3x)\,d(3x) = \tfrac{1}{3}\sin(3x) + C.$$

Another example:

$$\int \sin(\theta^3)\theta^2 \, d\theta = \int \sin(\theta^3)\tfrac{1}{3}(3\theta^2 d\theta)$$

$$= \tfrac{1}{3}\int \sin(\theta^3) \, d(\theta^3) = -\tfrac{1}{3}\cos(\theta^3) + C.$$

### Substitution with Definite Integrals

In the notation $\int_a^b f(x) \, dx$, the limits of integration $a$ and $b$ refer to $x$; if you substitute a new variable $u$, it will usually have different limits. Your notation must make it clear to which variable the limits apply.

---

**EXAMPLE 7**  $\int_0^1 2x\sqrt{x^2 + 1} \, dx$ calls for the substitution $u = x^2 + 1$, $du = 2x \, dx$. There are three ways to handle the limits of integration.

*First Way: Indicate the Variable to Which the Limits Apply*  Substitute the new integrand, but indicate that the limits are for $x$, not $u$:

$$\int_0^1 2x\sqrt{x^2 + 1} \, dx = \int_{x=0}^{x=1} u^{1/2} \, du \qquad \text{(now integrate)}$$

$$= \tfrac{2}{3}u^{3/2}\Big|_{x=0}^{x=1} \qquad \text{(now express integral in terms of } x)$$

$$= \tfrac{2}{3}(x^2 + 1)^{3/2}\Big|_0^1 = \tfrac{2}{3}(2^{3/2} - 1).$$

*Second Way: Convert the Limits on $x$ to the Corresponding Limits on $u$*  Since $u = x^2 + 1$, then $x = 0 \Rightarrow u = 0^2 + 1 = 1$, and $x = 1 \Rightarrow u = 1^2 + 1 = 2$. Substitute the new integrand *and the corresponding limits for $u$*:

$$\int_0^1 2x\sqrt{x^2 + 1} \, dx = \int_1^2 u^{1/2} \, du \qquad \text{(now integrate)}$$

$$= \tfrac{2}{3}u^{3/2}\Big|_1^2 \qquad \text{(now use the } u\text{-limits)}$$

$$= \tfrac{2}{3}(2^{3/2} - 1).$$

With this method, you do not return from "$u$" to "$x$."

*Third Way: Indicate Unknown Limits on the $u$-Integral, and Return to $x$ to Evaluate the Limits*

$$\int_0^1 2x\sqrt{x^2 + 1} \, dx = \int_?^? u^{1/2} \, du = \tfrac{2}{3}u^{3/2}\Big|_?^?$$

$$= \tfrac{2}{3}(x^2 + 1)^{3/2}\Big|_0^1 = \tfrac{2}{3}(2^{3/2} - 1).$$

As you see, the three methods are equivalent.

---

The effect of substitution in a definite integral can be seen geometrically. Figure 1 illustrates a simple case: evaluating $\int_a^b f(2x) \, dx$ by the substitution $u = 2x$. Then $dx = \tfrac{1}{2}du$. For the limits, $x = a \Rightarrow u = 2a$, $x = b \Rightarrow u = 2b$, so

$$\int_a^b f(2x) \, dx = \tfrac{1}{2}\int_{2a}^{2b} f(u) \, du. \tag{5}$$

Figure 1a shows the original integral, and Figure 1b the transformed one. The graph of $y = f(u)$ in the $(u, y)$ plane is like the graph of $y = f(2x)$ in the $(x, y)$

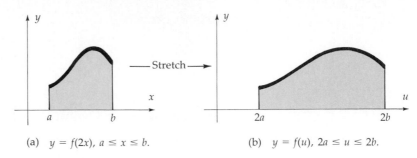

(a) $y = f(2x)$, $a \le x \le b$.     (b) $y = f(u)$, $2a \le u \le 2b$.

**FIGURE 1**

plane, but stretched by a factor 2 in the horizontal direction. This doubles the area, so $\int_{2a}^{2b} f(u) \, du$ is twice $\int_a^b f(2x) \, dx$. This confirms (5); the factor 1/2 corrects for the stretching.

## SUMMARY

The substitution $u = u(x)$ in an integral requires $du = u'(x) \, dx$.

With substitutions in a definite integral, the limits of integration must be treated correctly.

## PROBLEMS

**A**

1. Find $dy$, $dv$, $df$, etc. as appropriate.
   *a) $y = 1/x$          *b) $y = 2x^3$
   c) $y = 5$            d) $v = 9.8t + 3$
   e) $f(x) = \dfrac{12}{x^2 + 1}$     f) $u = \sqrt{x}$
   *g) $y = \sin(3\theta)$      h) $z = \theta \csc(2\theta)$
   *i) $x = t^2 \cot(t/5)$

2. Evaluate the integrals, using the given substitution.
   *a) $\displaystyle\int (2x - 1)^3 \, dx$; $u = 2x - 1$
   *b) $\displaystyle\int x^2(x^3 + 1)^4 \, dx$; $u = x^3 + 1$
   c) $\displaystyle\int_{-1}^1 x\sqrt{x + 1} \, dx$; $u = x + 1$
   *d) $\displaystyle\int \dfrac{dx}{\sqrt{2x + 1}}$; $u = 2x + 1$
   e) $\displaystyle\int \dfrac{dx}{\sqrt{2x + 1}}$; $u = \sqrt{2x + 1}$

*f) $\displaystyle\int \cos(\pi t - 3) \, dt$; $\theta = \pi t - 3$

3. Evaluate $\int \cos \theta \sin \theta \, d\theta$ by the given method. What is the relation between the answers?
   a) Substitute $u = \sin \theta$.
   b) Substitute $u = \cos \theta$.
   c) Use $\cos \theta \sin \theta = \frac{1}{2} \sin 2\theta$.

4. Evaluate $\int \sec^2\theta \tan \theta \, d\theta$ by the given substitution. What is the relation between the two answers?
   a) $u = \tan \theta$     b) $u = \sec \theta$

5. Evaluate the integrals.
   *a) $\displaystyle\int \dfrac{x}{(x^2 + 1)^2} \, dx$
   b) $\displaystyle\int t^2 \sqrt{t^3 - 3} \, dt$
   c) $\displaystyle\int \sqrt{4v + 1} \, dv$
   *d) $\displaystyle\int_0^1 \dfrac{1}{(3x + 1)^3} \, dx$
   *e) $\displaystyle\int_2^{10} (1 - x)^{10} \, dx$

**\*f)** $\displaystyle\int_{-1}^{1} x\sqrt{x+1}\; dx$

**g)** $\displaystyle\int_{2}^{5} \frac{1}{\sqrt{5t+1}}\; dt$

**h)** $\displaystyle\int x^2\sqrt{x+1}\; dx$

**\*i)** $\displaystyle\int \frac{x+1}{(x^2+2x+5)^2}\; dx$

**j)** $\displaystyle\int_{0}^{1} \frac{x}{\sqrt{1+x^2}}\; dx$

**k)** $\displaystyle\int_{0}^{1} x^2\sqrt{1+x^3}\; dx$

**\*l)** $\displaystyle\int x^5\sqrt{1+x^3}\; dx$

**m)** $\displaystyle\int \theta\cos(\theta^2)\; d\theta$

**\*n)** $\displaystyle\int \theta\sin(3\theta^2+1)\; d\theta$

**\*o)** $\displaystyle\int 2\sec^2(5\theta)\; d\theta$

**p)** $\displaystyle\int \sec^2\theta\,\tan^2\theta\; d\theta$

**\*q)** $\displaystyle\int \csc^2\theta\,\cot^2\theta\; d\theta$

**r)** $\displaystyle\int \sin 3\theta\,\cos 3\theta\; d\theta$

**s)** $\displaystyle\int \frac{\sin\theta}{\cos^2\theta}\; d\theta$

**6.** At time $t = 0$, a particle is at $x = 2$ and has velocity $v(0) = 1$. Its acceleration is $a(t) = \sin(3t)$ for $t \geq 0$.
   **a)** Find its velocity $v(t)$ for $t \geq 0$.
   **\*b)** Find its position $x$ as a function of $t$, for $t \geq 0$.

**7. \*a)** Find the area above the $x$-axis and below the first "arch" of the curve $y = x\sin(\pi x^2)$, from $x = 0$ to $x = 1$.
   **b)** Find the corresponding area for the second "arch," from $x = \sqrt{2}$ to $x = \sqrt{3}$.

**8.** Find the mean value of $f(x) = \dfrac{x}{[1+x^2]^2}$ over the interval $[0, 1]$.

**B**

**9. a)** Evaluate $\int_{-1/2}^{1/2}\sqrt{1-4x^2}\; dx$, using $u = 2x$.
   **b)** Sketch the graphs of $y = \sqrt{1-4x^2}$ and $y = \sqrt{1-u^2}$, and compare the areas geometrically.

**10.** Given $\dfrac{dy}{d\theta} = \dfrac{\cos(\theta/2)}{\sin^2(\theta/2)}$ and $y = 0$ when $\theta = \pi$, find $y$ for $0 < \theta < 2\pi$. (Why is it impossible to determine $y$ outside this interval?)

**11. a)** Show that $\int_{-a}^{0} f(x)\, dx = \int_{0}^{a} f(-x)\, dx$.
   **b)** Show (nongeometrically) that if $f$ is even then $\int_{-a}^{a} f(x)\, dx = 2\int_{0}^{a} f(x)\, dx$.
   **c)** Obtain a formula for $\int_{-a}^{a} f(x)\, dx$ if $f$ is odd.

**12. a)** Compare $\int_{0}^{4} f(x)\, dx$ with $\int_{0}^{4} f(4-u)\, du$.
   **b)** Given $y = f(x)$ in Figure 2, sketch $y = f(4-u)$, and explain geometrically the relation between the integrals. (Compute $y$ for $u = 0, 1, 2, 3, 4$.)

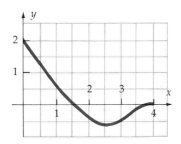

**FIGURE 2**
$y = f(x)$.

**13. a)** Compare $\int_{a}^{b} f(x)\, dx$ with $\int_{a-c}^{b-c} f(u+c)\, du$.
   **b)** Given the graph of $y = f(x)$ in Figure 3, sketch $y = f(u+2)$. (Compute $y$ for $u = a-2, a+1-2, a+2-2, b-2$.)

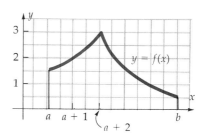

**FIGURE 3**

**\*14.** What is wrong here?
   In evaluating $\int_{-1}^{1} x^2\, dx$, set $u = x^2$, $du = 2x\, dx$, $dx = \dfrac{du}{2x} = \dfrac{du}{2\sqrt{u}}$; thus $\int_{-1}^{1} x^2\, dx =$
   $\int_{1}^{1} u \cdot \dfrac{du}{2\sqrt{u}} = 0$, since $\int_{a}^{a} f(u)\, du = 0$ for any $f$.
   On the other hand, $\int_{-1}^{1} x^2\, dx = \frac{1}{3}x^3\big|_{-1}^{1} = 2/3$.

## REVIEW PROBLEMS CHAPTER 6

### A

*1. For $\int_{-\pi/2}^{\pi/2} \cos^2\theta \, d\theta$, write
   a) The upper sum $\bar{S}_3$.
   b) The lower sum $\underline{S}_3$.
   c) The midpoint sum $MS_3$.

*2. For $\int_1^2 \dfrac{dx}{x}$ write
   a) The upper sum $\bar{S}_{10}$.
   b) The lower sum $\underline{S}_{10}$.
   c) The midpoint sum $MS_{10}$.

3. From the results of the previous problem, which of the following can be deduced?

   a) $\displaystyle\int_1^2 \frac{dx}{x} < 0.72$  b) $\displaystyle\int_1^2 \frac{dx}{x} < 0.7$

   c) $\displaystyle\int_1^2 \frac{dx}{x} > 0.66$  d) $\displaystyle\int_1^2 \frac{dx}{x} > 0.69$

   (Hint: Since $1/x$ is concave up, does the midpoint sum *overestimate* or *underestimate*?)

4. Evaluate.

   *a) $\displaystyle\int \sin \pi x \, dx$  b) $\displaystyle\int \sec^2 x \, dx$

   *c) $\displaystyle\int \sec^2 5\theta \, d\theta$  d) $\displaystyle\int \sec u \tan u \, du$

   *e) $\displaystyle\int_{-1}^1 (x^5 + 1)^2 \, dx$  f) $\displaystyle\int_1^2 \left(t + \frac{1}{t}\right)^2 dt$

   *g) $\displaystyle\int_{-3}^{-1} \frac{dx}{3x^5}$  h) $\displaystyle\int \frac{(x+1)^2}{\sqrt{x}} \, dx$

   *i) $\displaystyle\int (x^3 + 1)^4 x^2 \, dx$  j) $\displaystyle\int x^3 \cos(x^4) \, dx$

5. Find the area of the region under the given curve and above the given interval.
   *a) $y = 3x^2$; [1, 3]
   *b) $y = \sin 3\theta$; [0, $\pi/6$]

   c) $y = \dfrac{1}{(x+2)^3}$; [0, 2]

6. Write out the given sum, without using "$\Sigma$" notation.

   *a) $\displaystyle\sum_{j=1}^3 2^j$  *b) $\displaystyle\sum_{j=0}^4 r^j$  c) $\displaystyle\sum_{k=0}^3 k2^{-k}$

7. Write using "$\Sigma$" notation.
   *a) The sum of the first $n$ positive integers.

*b) The sum of the reciprocals of the first $m$ positive integers.
   c) The reciprocal of the sum of the first $n$ positive integers.

8. Compute the mean value of the given function on the interval $[-1, 1]$.
   *a) 1  *b) $x$  *c) $x^2$
   d) $x^3$  e) $x^n$, $n = 4, 5, 6, \ldots$

*9. The sun's heat is absorbed by a certain south-facing window at a varying rate which is approximated by

$$\frac{dC}{dt} = 100 - 70 \cos \frac{2\pi(t - 10)}{365}$$

calories per day, on day $t$ of the year.
   a) On what day is the rate least? Greatest?
   b) How much heat is absorbed in a year?
   c) What is the average daily rate of absorption for the year? For the month of January?

*10. Figure 1 gives (approximately) the rate of blood flow through an aorta for the length of one heartbeat, about 0.8 seconds. Compute
   a) The total amount of blood flowing through in one heartbeat.
   b) The average rate of flow in ml/sec. (Use a trapezoid sum.)

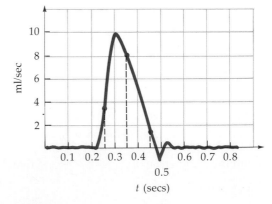

**FIGURE 1**
Rate of flow in an aorta.

11. Given $F(x) = \displaystyle\int_2^x \frac{dt}{t}$.

*a) Compute $F'$ and $F''$.

*b) Estimate $F(3)$ with a midpoint sum $MS_5$.

*c) Estimate $F(1)$ using a midpoint sum $MS_5$.

d) Graph $F$ for $1 \leq x \leq 3$. Plot one point exactly, plot points for $x = 1$ and $x = 3$ using your midpoint approximations, and use the signs of $F'$ and $F''$.

**B**

12. a) Compute $\dfrac{d}{dx} \displaystyle\int_1^x 3t^2 \tan(2t^3) \, dt$.

b) Compute $\dfrac{d}{dx} \displaystyle\int_1^{x^3} \tan(2u) \, du$.

c) Show that the integrals in parts a and b are equal, using the derivatives.

d) Show that the integrals in parts a and b are equal, using a substitution.

13. Write an expression for a function $f$ with $f(5) = 0$ and $f'(x) = \sqrt{1 + x^4}$.

14. Suppose that $f$ is continuous. Let $T$ stand for a fixed constant.

a) Compute the derivative of $\int_a^{a+T} f(t) \, dt$ with respect to $a$.

b) Suppose that $f$ has period $T$; that is, $f(t + T) = f(t)$ for all $t$. What can be deduced about the integral in part a?

15. Prove that $\int_0^x [\int_0^u f(t) \, dt] \, du = \int_0^x f(u) \cdot (x - u) \, du$. (Hint: Express the right-hand side as the sum of two integrals, then compute the derivative of each side.)

# 7

# SOME APPLICATIONS OF INTEGRALS

# 7.1

## AREAS BY INTEGRATION

Figure 1 shows a region $R$ bounded by the graphs of two continuous functions $f$ and $g$, for $a \le x \le b$. Denote by $L(x)$ the length of the vertical cross section of the region at $x$. From the figure,

$$L(x) = [y \text{ value on upper curve}] - [y \text{ value on lower curve}]$$
$$= f(x) - g(x).$$

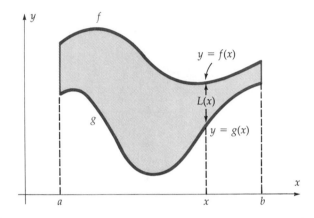

**FIGURE 1**
Area between graphs of $f$ and $g$, $a \le x \le b$. $L(x) = $ length of vertical cross section at $x$.

The *area* of the region is the *integral of the cross-sectional length*, $\int_a^b L(x)\, dx$.

To derive this integral, divide the interval $[a, b]$ into $n$ equal parts, and choose a point $\hat{x}_j$ in each part. Then form $n$ rectangles approximating the region, as suggested in Figure 2. The height of the first approximating rectangle is $L(\hat{x}_1)$, and the base is $\Delta x$, so its area is $L(\hat{x}_1)\, \Delta x$. The sum of the

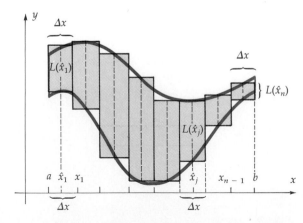

**FIGURE 2**
$S_n = L(\hat{x}_1)\, \Delta x + \cdots + L(\hat{x}_n)\, \Delta x$ approximates the area of the region.

areas of the $n$ approximating rectangles is thus

$$S_n = L(\hat{x}_1)\,\Delta x + \cdots + L(\hat{x}_n)\,\Delta x = \sum_{j=1}^{n} L(\hat{x}_j)\,\Delta x.$$

If $n$ is very large, then the rectangles are very thin and numerous, and we expect them to approximate the region quite closely. We thus *define* the area as the limit of these sums, as $n \to \infty$:

$$\text{area of } R = \lim_{n \to \infty} S_n.$$

But $S_n$ is a Riemann sum for the integral of the function $L$, and the limit of the Riemann sums is the integral, so

$$\text{area of } R = \int_a^b L(x)\,dx.$$

---

**EXAMPLE 1**   Find the area between the graphs of $y = x$ and $y = 1/x^2$, for $1 \le x \le 2$.

**SOLUTION**   Sketch the two graphs and the region (Fig. 3). Choose a "typical" $x$ between 1 and 2 and sketch the vertical cross section of the region at position $x$. This reaches from the lower curve $y = 1/x^2$ to the upper curve $y = x$, so its length is

$$L(x) = x - \frac{1}{x^2}.$$

The required area is therefore

$$\int_1^2 L(x)\,dx = \int_1^2 \left( x - \frac{1}{x^2} \right) dx = \left[ \frac{1}{2}x^2 + \frac{1}{x} \right]_1^2$$
$$= \left[ 2 + \frac{1}{2} \right] - \left[ \frac{1}{2} + 1 \right] = 1.$$

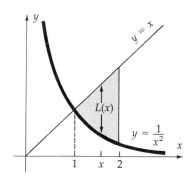

**FIGURE 3**
$L(x) = x - 1/x^2$.

---

**WARNING**   The formula for $L(x)$ must be valid for *all* $x$ between 1 and 2. There is a dangerous tendency to compute a special cross-sectional length (at one of the endpoints, say, or the maximum length) and use this for all $x$. You must compute $L(x)$ for a "typical" $x$, not a special one!

---

**EXAMPLE 2**   The graphs of $y = x^2 - 2$ and $y = x$ bound a finite region. Find its area.

**SOLUTION**   Sketch the two graphs—in this case a parabola and a straight line (Fig. 4). The graphs intersect at points where the two functions $x^2 - 2$ and $x$ agree:

$$x^2 - 2 = x \Leftrightarrow x^2 - x - 2 = 0 \Leftrightarrow x = -1 \text{ or } 2.$$

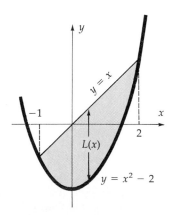

**FIGURE 4**
$L(x) = x - (x^2 - 2)$.

So the region in question lies between $x = -1$ and $x = 2$, and these are the limits of integration:

$$\text{area} = \int_{-1}^{2} L(x)\, dx.$$

You need a formula for $L(x)$, valid for every $x$ from $-1$ to $2$. The vertical cross section at position $x$ reaches from $y = x^2 - 2$ up to $y = x$, so the cross-sectional length is

$$L(x) = \underbrace{(x)}_{\substack{\text{upper} \\ y \text{ value}}} - \underbrace{(x^2 - 2)}_{\substack{\text{lower} \\ y \text{ value}}} = x - x^2 + 2.$$

The area is

$$\int_{-1}^{2} (x - x^2 + 2)\, dx = \left. \frac{x^2}{2} - \frac{x^3}{3} + 2x \right|_{-1}^{2}$$
$$= \left( \frac{4}{2} - \frac{8}{3} + 4 \right) - \left( \frac{1}{2} - \frac{-1}{3} - 2 \right) = \frac{9}{2}.$$

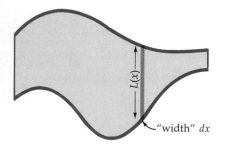

**FIGURE 5**
A useful fiction: Infinitesimal slice has length $L(x)$, width $dx$, area $L(x)\, dx$.

## "Infinitesimal Slices"

The area integral $\int_a^b L(x)\, dx$ is the limit of the sums $\sum_1^n L(\hat{x}_j)\, \Delta x$, where $L(\hat{x}_j)$ is the area of a rectangle of height $L(\hat{x}_j)$ and base $\Delta x$. The corresponding expression in the integral for the area $L(x)\, dx$, has no precise geometric meaning; but it is useful to pretend that it does. If $\Delta x$ is small enough, then the individual rectangles cannot really be seen, and the difference between the sum and its limit, the integral, is negligible. So we adopt a useful fiction: Imagine a so-called "infinitesimal slice" (Fig. 5) of height $L(x)$ and "infinitesimal width" $dx$; the "infinitesimal area" of this slice is $L(x)\, dx$, and the area of the entire region is the "sum" of these infinitesimal areas, the integral $\int_a^b L(x)\, dx$. (This is very close to Leibniz's original view of the integral.)

## Horizontal Sections

It is sometimes better to use horizontal cross sections (Fig. 6). In this case, let $L(y)$ be the length of such a cross section at height $y$. The horizontal "infinitesimal slice" has length $L(y)$ and "infinitesimal width" $dy$; its "infinitesimal area" is $L(y)\, dy$, and the area of the region is the "sum" of these infinitesimal areas, the integral

$$\int_c^d L(y)\, dy,$$

with appropriate limits of integration $c$ and $d$ (Fig. 6).

A skeptic may wonder whether, for a given region $R$, the "horizontal strip" integral necessarily gives the same value as the "vertical strip" integral. For a region bounded by the graphs of continuous functions, the two methods *do* agree; the proof is left to another course.

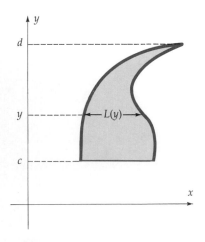

**FIGURE 6**
$L(y) = $ length of cross section at $y$, area $= \int_c^d L(y)\, dy$.

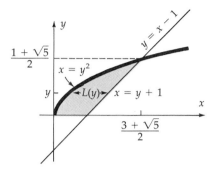

**FIGURE 7**
$L(y) = (y + 1) - y^2$.

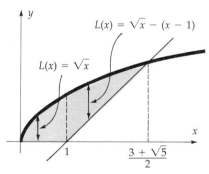

**FIGURE 8**
$L(x)$ has two different formulas.

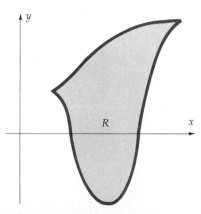

**FIGURE 9**

**EXAMPLE 3**   Compute the area above the $x$-axis, to the right of $y = \sqrt{x}$, and to the left of $y = x - 1$.

*SOLUTION*   Sketch the curves and the region (Fig. 7). The length $L(y)$ of the horizontal cross section at height $y$ is the $x$ value at the right end, minus the $x$ value at the left end. To obtain the formula for $L(y)$, you must express each of those $x$ values in terms of $y$. The right end of the cross section is on the curve $y = x - 1$, so there

$$x = y + 1.$$

The left end is on the curve $y = \sqrt{x}$, so there

$$x = y^2.$$

Hence,

$$L(y) = (y + 1) - y^2$$

and the area is $\int_c^d (y + 1 - y^2) \, dy$, with appropriate limits of integration $c$ and $d$. From Figure 7, $c = 0$; and $d$ is the $y$ value at the point where the curves intersect, that is where

$$y - 1 = y^2$$

or

$$y^2 - y - 1 = 0$$

or, by the quadratic formula,

$$y = \frac{1 \pm \sqrt{1 + 4}}{2} = \frac{1 \pm \sqrt{5}}{2}.$$

You want the positive solution, $y = (1 + \sqrt{5})/2$, so

$$\text{area} = \int_0^{(1+\sqrt{5})/2} (y + 1 - y^2) \, dy = (7 + 5\sqrt{5})/12.$$

To do this with vertical sections, you would need two integrals (Fig. 8):

$$\text{area} = \int_0^1 \sqrt{x} \, dx + \int_1^{(3+\sqrt{5})/2} [\sqrt{x} - (x - 1)] \, dx.$$

*REMARK*   The integral for the area of the region in Figure 1 could be derived "by subtraction":

$$\text{area of } R = [\text{area under graph of } f] - [\text{area under graph of } g]$$
$$= \int_a^b f(x) \, dx - \int_a^b g(x) \, dx$$
$$= \int_a^b [f(x) - g(x)] \, dx$$
$$= \int_a^b L(x) \, dx.$$

The longer derivation using Riemann sums was given for two reasons. First, "subtraction" does not apply directly to a region that crosses the $x$-axis (Fig. 9). Even more important, the use of sums here provides a model for the other integral formulas derived in this chapter. In each case, the quantity in question is divided into small parts, and each part is approximated; the whole quantity is thus approximated by a sum. As the parts become smaller and more numerous, the approximating sums improve. The quantity is the limit of these approximating sums, which are recognized as Riemann sums for an appropriate integral.

## SUMMARY

*Area of region $R = \int_a^b L(x)\, dx$, (see Fig. 1):*

$$L(x) = \text{length of vertical cross section of } R \text{ at position } x,$$

$x = a$ gives left edge of $R$,     $x = b$ gives right edge.

## PROBLEMS

### A

*1. Compute the area of the following regions, using vertical strips.
  a) Between $y = x$ and $y = x^2$.
  b) Above $y = x^3$ and below $y = x$, for $x \geq 0$.
  c) Above $y = x^{-2}$, below $y = 2$, and to the left of $x = 2$.

2. Recompute each of the areas in problem 1, using horizontal strips.

3. Compute the area of the given region by any convenient (legitimate) method.
  *a) Above $y = x^2$ and below $y = 1 - x^2$.
  b) Above $y = x^3$ and below $y = x^2$, for $x \geq 0$.
  *c) Between $x = y^2$ and the vertical line $x = 4$.
  d) Between $y = x^4 - 2x^2$ and $y = 2x^2$.
  e) Between $y = 2x - x^2$ and $y = -3$.
  *f) Between $\sqrt{x} + \sqrt{y} = 1$ and the coordinate axes.
  g) Above $x^{-2}$ and below $x^{-3}$, for $x \geq 1/2$.
  h) Below $x^{-2}$ and above $x^{-4}$, for $|x| \leq 2$.
  *i) Above the $x$-axis, to the right of $y = \sqrt{x}$, and to the left of $y = x - 2$.
  j) To the right of $x = 2y^2$ and to the left of $x = (y - 1)^2$.
  *k) Above the line $y = 1/2$ and below one arch of the graph of $y = \sin\theta$.
  l) To the right of the $y$-axis, above $y = \sin\theta$, and below $y = \cos\theta$.

*m) Below $y = \sin(\pi x/2)$, above $y = x$, for $x \geq 0$.
  n) Any one of the finite regions between the graph of $y = \sec^2 2\theta$ and $y = 2$.

4. Compute by integration the area of the triangle with the given vertices.
  a) $(0, 0)$, $(1, 1)$, $(0, 3)$.
  b) $(0, 0)$, $(1, 2)$, $(3, 0)$.
  c) $(0, 0)$, $(1, 2)$, $(2, -1)$.

*5. Estimate the area of the lot in Figure 10, using the trapezoid rule (Sec. 6.4).

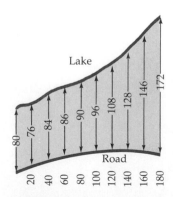

**FIGURE 10**
Compute the area of this irregular lot. (Distances in feet.)

**B**

**6.** A triangle has vertices at $(0, 0)$, $(b, 0)$, and $(c, h)$, with $b > 0$ and $h > 0$. Compute the area by integration.

**7.** A straight line $S$ through points $(a, a^2)$ and $(b, b^2)$ cuts off a region $R$ from the inside of the parabola $y = x^2$ (Fig. 11).

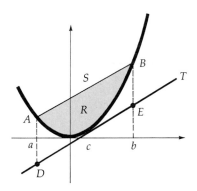

**FIGURE 11**

**a)** At what point $(c, c^2)$ on the parabola is the tangent line parallel to $S$?

**b)** Show that the area of $R$ is exactly two thirds of the area of the parallelogram $ABDE$ in the figure, formed by the straight line $S$, the parallel tangent line, and the vertical lines $x = a$ and $x = b$.

**8.** The Lorenz curve (Sec. 4.5) displays the distribution of income or resources in a given population (Fig. 12); any Lorenz curve $y = f(x)$ is concave upward, with $f(0) = 0$ and $f(1) = 1$. *The Gini index of inequality, $g$, is twice the area between the Lorenz curve and $y = x$, for $0 \leq x \leq 1$.*

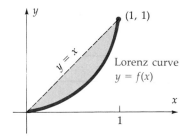

**FIGURE 12**
Gini index $g$ = Twice shaded area.

**a)** Why is $f(x) \leq x$ for $0 \leq x \leq 1$?

**b)** Show that $0 \leq g < 1$. For what curve is $g = 0$, and when is $g$ near 1?

**c)** Compute the Gini index for $y = x^p$, $p \geq 1$.

**d)** Estimate the Gini index for the distribution of land in Bolivia in 1964. (See Fig. 3, Sec. 4.5.)

**e)** Estimate the Gini index for the distribution of land in Denmark (1969). (See Fig. 3, Sec. 4.5.)

---

## 7.2
## VOLUMES BY CROSS SECTIONS

The volume of a solid is given by an integral, derived roughly as follows. Cut the solid into thin slices, as if it were a salami; approximate each slice; add the approximations; and take the limit as the slices become infinitely thin and infinitely numerous.

The approximations will be made by right cylinders with appropriate cross section, or *base* (Fig. 1). If the base is a circle, the cylinder is called circular; but a box (rectangular base), prism (triangular base), and tube (annular base) are also cylinders. In any case, the *volume* of a cylinder is the area of the base, times the height perpendicular to the base.

Now consider a solid $S$ of more general shape, lying between planes perpendicular to the $x$-axis at $x = a$ and $x = b$ (Fig. 2). The cross sections of $S$ perpendicular to the axis vary; denote by $A(x)$ the area of the cross section at $x$ (Fig. 2). Suppose that the cross sections vary continuously; then $A(x)$ is a continuous function. Now cut the solid into $n$ slices of thickness $\Delta x$ by planes perpendicular to the $x$-axis at points $a$, $x_1, \ldots, x_{n-1}$, $b$ (Fig. 3). Approximate the first slice by a cylindrical slab of thickness $\Delta x$, whose base

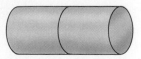

Circular

Rectangular

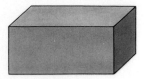

Prism

Tube

Other

**FIGURE 1**
Right cylinders.

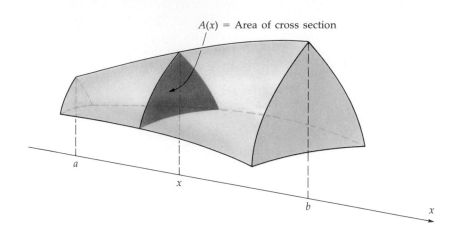

$A(x)$ = Area of cross section

**FIGURE 2**

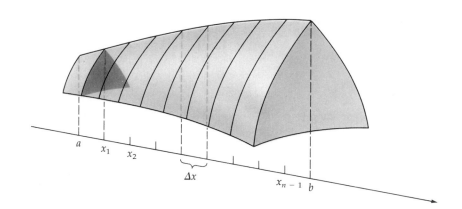

**FIGURE 3**
Solid cut in slices.

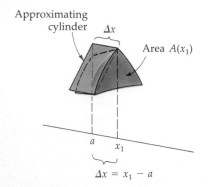

**FIGURE 4**
Volume of first approximating
cylinder = $A(x_1) \, \Delta x$.

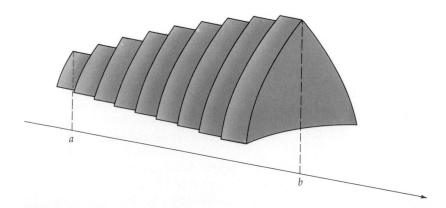

**FIGURE 5**
$V_n = \displaystyle\sum_{j=1}^{n} A(x_j) \, \Delta x.$

coincides with the cross section of $S$ at $x_1$ (Fig. 4). The volume of this slab is

$$(\text{area of base}) \times (\text{thickness}) = A(x_1)\,\Delta x.$$

Approximate the other slices analogously (Fig. 5); the sum of the approximating volumes is

$$V_n = \sum_{j=1}^{n} A(x_j)\,\Delta x.$$

If $n$ is very large then the slices are very thin, and we expect these approximations to be very accurate. We thus define the volume to be their *limit*:

$$V = \lim_{n \to \infty} V_n = \lim_{n \to \infty} \sum_{j=1}^{n} A(x_j)\,\Delta x$$

if the limit exists. But $\sum_{j=1}^{n} A(x_j)\,\Delta x$ is a Riemann sum for the integral $\int_a^b A(x)\,dx$, so the limit *does* exist, and it equals this integral. Thus

$$\text{volume of } S = \int_a^b A(x)\,dx, \tag{1}$$

where $A(x) = $ area of cross section of $S$ at $x$.

You can think of $A(x)\,dx$ as the volume of an "infinitesimal slice" of $S$ (Fig. 6), with cross-sectional area $A(x)$, "infinitesimal thickness" $dx$, and "infinitesimal volume" $A(x)\,dx$. The entire volume is the sum (integral) of these infinitesimal volumes. In this interpretation, the dimensions work out: $A(x)$ has dimensions of area, and $dx$ has dimensions of length, so $A(x)\,dx$ has dimensions of volume.

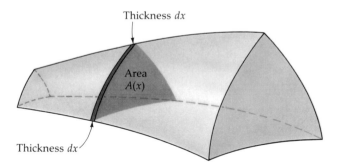

**FIGURE 6**
Interpreting $A(x)\,dx$ as an infinitesimal volume.

Notice how closely the volume integral (1) resembles the integral for the area of a plane region:

$$\text{Area of } R = \int_a^b L(x)\,dx$$

$$L(x) = \text{length of cross section of } R \text{ at } x.$$

**EXAMPLE 1**   *Volume of a pyramid.* A pyramid of height $h$ with square base of edge $b$ lies with the $x$-axis through its center (Fig. 7a). The cross section

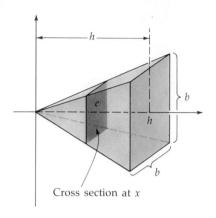

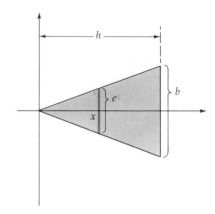

(a) Perspective view.

(b) Schematic side view.

Cross section at $x$

**FIGURE 7**
Pyramid.

at $x$ is a square. The edge $e$ of the square is computed from similar triangles (Fig. 7b):

$$\frac{e}{b} = \frac{x}{h}, \qquad \text{so } e = \frac{bx}{h}.$$

Hence the area of the cross section at $x$ is

$$A(x) = e^2 = \left(\frac{b}{h}\right)^2 x^2,$$

and the volume of the pyramid is

$$\int_0^h A(x)\,dx = \int_0^h \left(\frac{b}{h}\right)^2 x^2\,dx = \left(\frac{b}{h}\right)^2 \frac{h^3}{3} = \frac{1}{3}\,b^2 h.$$

This pyramid would fit neatly in a rectangular box of base $b \times b$ and height $h$. The volume of the box is $b^2 h$, and the volume of the pyramid is exactly $1/3$ of that, which seems reasonable.

Revolving triangle
forms a cone.

Revolving semicircle
forms a sphere.

**FIGURE 8**
Solids of revolution.

A **solid of revolution** is formed by revolving a plane region about an axis. A right triangle revolved about one of its legs forms a cone (Fig. 8); a semicircle revolved about its diameter forms a sphere.

If the region under the graph of a continuous function $f \geq 0$ on an interval $[a, b]$ is revolved about the $x$-axis (Fig. 9) then the cross section at $x$ is a circle of radius $f(x)$. The cross-sectional area is $A(x) = \pi f(x)^2$, so the volume of such a solid is $\int_a^b \pi f(x)^2\,dx$. In this integral, $f(x)$ is the *radius* of the circular cross section at $x$, so we write it $r(x)$; then the volume is

$$\int_a^b \pi r(x)^2\,dx. \tag{2}$$

This is the "disk method" for volumes. An extremely thin slice of the solid is essentially a disk of radius $r(x)$ and infinitesimal thickness $dx$; the infinitesi-

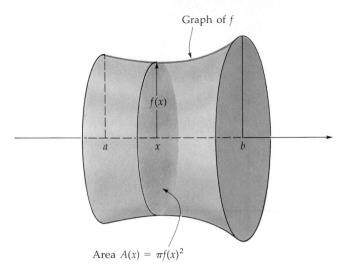

Graph of $f$

$f(x)$

$a$   $x$   $b$

Area $A(x) = \pi f(x)^2$

**FIGURE 9**
Solid formed by revolving the region under the graph of
$f$ on $[a, b]$.

mal volume of this disk is $\pi r(x)^2\, dx$, and the volume of the entire solid is
the sum of these infinitesimal volumes, the integral (2).

**EXAMPLE 2**   The graph of $y = x^2$, $0 \le x \le 1$ (Fig. 10) is revolved about
the $x$-axis, enclosing a solid of revolution. Rather than sketch the entire solid,
we show only the circular base, (obtained by revolving the endpoint $(1, 1)$
about the $x$-axis) and the cross section at $x$—a circle of radius $y = x^2$, with
area

$$A(x) = \pi y^2 = \pi x^4.$$

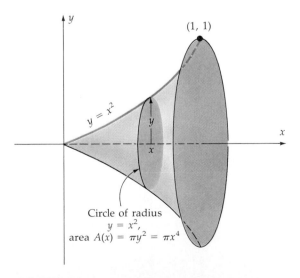

$y$

$(1, 1)$

$y = x^2$

$y$

$x$

$x$

Circle of radius
$y = x^2$,
area $A(x) = \pi y^2 = \pi x^4$

**FIGURE 10**

The volume in question is the integral of this cross-sectional area:

$$\int_0^1 A(x)\,dx = \int_0^1 \pi x^4\,dx = \pi\left.\frac{x^5}{5}\right|_0^1 = \frac{\pi}{5}.$$

**EXAMPLE 3**   The region defined by $x^2 \le y \le 1$ is revolved about the axis $y = 1$ (Fig. 11). Compute the volume generated.

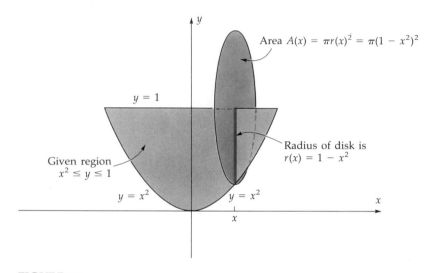

Area $A(x) = \pi r(x)^2 = \pi(1 - x^2)^2$

$y = 1$

Radius of disk is $r(x) = 1 - x^2$

Given region $x^2 \le y \le 1$

$y = x^2$     $y = x^2$

**FIGURE 11**
Cross section at $x$ is a circle of radius $r(x) = 1 - x^2$.

*SOLUTION*   Each slice of this region perpendicular to the $x$-axis is a circle *with center on the line* $y = 1$, the axis of revolution. The radius $r(x)$ of the slice at $x$ is the distance from $y = x^2$ to the axis of revolution:

$$r(x) = 1 - x^2.$$

The area of the slice is $\pi r(x)^2 = \pi(1 - x^2)^2$, and the volume of revolution is

$$\int_{-1}^1 \pi(1 - x^2)^2\,dx = \pi\left[x - \tfrac{2}{3}x^3 + \tfrac{1}{5}x^5\right]_{-1}^1 = \tfrac{16}{15}\pi.$$

## Solids with a Hole

Take the region between the graphs of two continuous functions $0 \le f \le g$ on an interval $[a, b]$, as in Figure 12a, and revolve it about the $x$-axis (Fig. 12b). Then the cross section at $x$ is a circle of radius $g(x)$ with a hole of radius $f(x)$ (Fig. 12c). The area of this cross section is

$$A(x) = \left(\begin{array}{c}\text{area of}\\\text{full circle}\end{array}\right) - \left(\begin{array}{c}\text{area of}\\\text{hole}\end{array}\right)$$
$$= \pi g(x)^2 - \pi f(x)^2$$

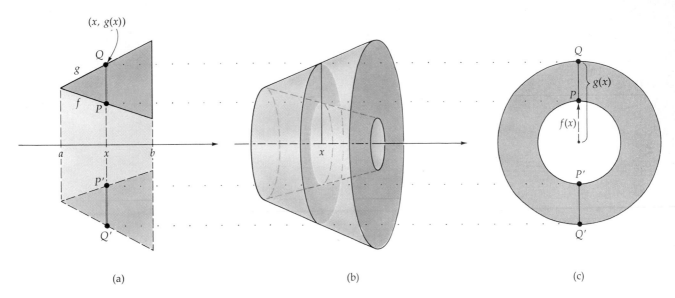

**FIGURE 12**
Region between graphs of $f$ and $g$, revolved about $x$-axis.
(a) The intersection of the solid with the $x$-$y$ plane, including the region between the graphs, and its reflection in the $x$-axis.
(b) Perspective view of the solid, showing section at $x$—an annulus.
(c) The section at $x$ laid flat. Points $P$, $Q$, $P'$, $Q'$ in (a) coincides with those in (c). In (a), segments $PQ$ and $P'Q'$ are the intersection of the cross section at $x$ with the $x$-$y$ plane.

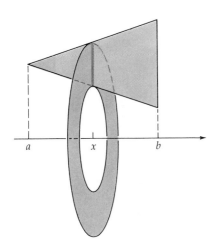

**FIGURE 13**
Region generating a solid of revolution, and annular cross section of solid.

and the volume is $\int_a^b \pi[g(x)^2 - f(x)^2]\, dx$. Here $g(x)$ is the **outer radius** of the cross section, denoted $r_{out}(x)$; and $f(x)$ is the **inner radius**, $r_{in}(x)$. The volume is then

$$\int_a^b \pi[r_{out}(x)^2 - r_{in}(x)^2]\, dx. \qquad (3)$$

This is the "washer method" for volumes. A very thin slice of the solid forms a "washer" with radii $r_{out}(x)$ and $r_{in}(x)$ and infinitesimal thickness $dx$. The volume of the washer is $[\pi r_{out}(x)^2 - \pi r_{in}(x)^2]\, dx$, and the sum of these is the integral (3).

In setting up such an integral, it is useful to draw the solid as in Figure 12; but what you really need is a clear picture of the cross section, a circle with a concentric hole (called an **annulus**). The annulus in Figure 12b is generated by revolving about the $x$-axis the segment $PQ$, the vertical cross section of the given plane region at $x$. To set up the integral, a simple sketch such as Figure 13 suffices; show the generating region, its vertical cross section at $x$ and the annulus formed by revolving that cross section about the $x$-axis. Determine the inner and outer radii of the annulus, and thus its area $A(x)$; then compute the volume $\int_a^b \pi A(x)\, dx$.

**EXAMPLE 4**   Tires, doughnuts, and bagels are examples of the solid called a *torus*. You can form a torus by revolving about the $x$-axis the circle in Figure 14a,

$$x^2 + (y - R)^2 = r^2. \qquad (4)$$

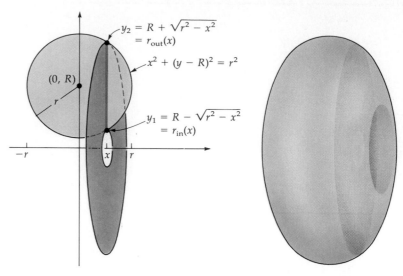

(a) The disk and the cross section of the solid at $x$.

(b) The torus.

**FIGURE 14**

Revolving disk $x^2 + (y - R)^2 \leq r^2$ about $x$-axis generates a torus.

Figure 14b shows the resulting solid. Assume that $R > r$, so the circle lies entirely above the axis. Figure 14a shows the given circle, the vertical cross section at $x$, and the annulus generated by revolving that cross section about the $x$-axis; this annulus is the cross section of the torus at $x$, and you need a formula for its area $A(x)$. Denote the outer radius of the annulus by $y_2$ and the inner radius by $y_1$; the area of the annulus is then

$$A(x) = \pi y_2^2 - \pi y_1^2.$$

To evaluate the volume integral $\int A(x)\, dx$ you must express $y_2$ and $y_1$ in terms of $x$, [obtaining the $r_{out}(x)$ and $r_{in}(x)$ in the general formula (3)]. Solve the equation of the circle (4) for $y$:

$$(y - R)^2 = r^2 - x^2 \Leftrightarrow y - R = \pm\sqrt{r^2 - x^2}$$
$$\Leftrightarrow y = R \pm \sqrt{r^2 - x^2}.$$

So the inner radius of the annulus is $y_1 = R - \sqrt{r^2 - x^2} = r_{in}(x)$, the outer radius is $y_2 = R + \sqrt{r^2 - x^2} = r_{out}(x)$, and the area is

$$A(x) = \pi r_{out}(x)^2 - \pi r_{in}(x)^2$$
$$= \pi[R + \sqrt{r^2 - x^2}]^2 - \pi[R - \sqrt{r^2 - x^2}]^2$$
$$= 4\pi R\sqrt{r^2 - x^2}$$

as you can check. The torus extends from $x = -r$ to $x = r$ (Fig. 14a), so the volume is

$$\int_{-r}^{r} A(x)\, dx = \int_{-r}^{r} 4\pi R\sqrt{r^2 - x^2}\, dx$$
$$= 4\pi R \int_{-r}^{r} \sqrt{r^2 - x^2}\, dx \qquad (5)$$

since the constant $4\pi R$ can be "factored out" of the integral. Now, how to evaluate the integral? The appropriate indefinite integral is hard to discover

(Chapter 9 shows how), but the *definite* integral $\int_{-r}^{r} \sqrt{r^2 - x^2} \, dx$ can be evaluated geometrically; it is the area under the graph of $y = \sqrt{r^2 - x^2}$ from $x = -r$ to $x = r$. The graph is a semicircle of radius $r$, and the area of the region is $\frac{1}{2}\pi r^2$. So the volume integral (5) works out to

$$4\pi R \cdot \tfrac{1}{2}\pi r^2 = (\pi r^2) \cdot (2\pi R).$$

You can interpret this as the *area of the circle of radius r* in Figure 14a, times the *perimeter of a circle of radius R*; the second circle is generated by revolving the center of the first circle about the $x$-axis.

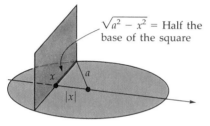

$\sqrt{a^2 - x^2}$ = Half the base of the square

**FIGURE 15**
Cross section of a solid with circular base.

**EXAMPLE 5**   The base of a certain solid is a circle of radius $a$. The sections perpendicular to one diameter of the circle are squares. Compute the volume of the solid.

**SOLUTION**   *Draw* the circular base, one of its diameters, and a square cross section perpendicular to that diameter (Fig. 15). The diameter will be the axis of integration; label it an $x$-axis with 0 at the center.

To compute the area of the square cross section, you need the length of its side. Half of the base of the square forms one leg of a right triangle; the hypotenuse is $a$ and the other leg is $|x|$, so half the base of the square is $\sqrt{a^2 - x^2}$, and the area of the square cross section is

$$A(x) = [2\sqrt{a^2 - x^2}]^2 = 4(a^2 - x^2).$$

The sections run from $x = -a$ to $x = a$, so the volume is

$$\int_{-a}^{a} A(x) \, dx = \int_{-a}^{a} 4(a^2 - x^2) \, dx = \tfrac{16}{3}a^3$$

as you can check. Notice that the dimensions are correct: The radius $a$ has dimensions of length, and the volume $\frac{16}{3}a^3$ has dimensions of (length)$^3$.

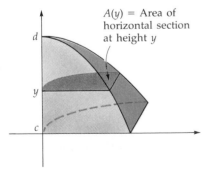

$A(y)$ = Area of horizontal section at height $y$

**FIGURE 16**
$V = \int_c^d A(y) \, dy.$

**REMARK 1**   For a solid lying between planes perpendicular to the $y$-axis at $y = c$ and $y = d$, it is in some cases more natural to take cross sections perpendicular to the $y$-axis (Fig. 16). If $A(y)$ denotes the area of the horizontal cross section of the solid at height $y$, then the volume is

$$\int_c^d A(y) \, dy.$$

The complicated proof that this integral equals the one taken with respect to the $x$-axis is left for another course.

**REMARK 2**   The essential idea of the volume integral was stated, in a different form, by Bonaventura Cavalieri (1598–1647), a professor at Bologna. Calculus as we know it did not yet exist, but the basic intuitions were developing. Among them was *Cavalieri's principle*: Two solids of equal altitude have the same volume if plane cross sections at equal height have the same area. With this principle and some geometric ingenuity, Cavalieri obtained many of the volume formulas that we now deduce from the integral (1). In honor of this, it is appropriate to refer to this integral formula as Cavalieri's principle.

## SUMMARY

*Volume of Solid S:*

$$V = \int_a^b A(x)\, dx,$$

$A(x)$ = area of cross-section of $S$ at position $x$.

When cross section is a circle of radius $r(x)$,

$$V = \int_a^b \pi r(x)^2\, dx.$$

When cross section is an annulus of radii $r_{out}(x)$ and $r_{in}(x)$,

$$V = \int_a^b \pi[r_{out}(x)^2 - r_{in}(x)^2]\, dx.$$

## PROBLEMS

### A

*1. The region between the $x$-axis and the graph of $y = x$, $0 \leq x \leq 1$, is revolved about the $x$-axis. Describe the solid formed and compute its volume.

2. The region between the $x$-axis and the graph of $y = \sqrt{r^2 - x^2}$ for $-r \leq x \leq r$ is revolved about the $x$-axis. Describe the solid generated and compute its volume.

*3. The region between the $x$-axis and the graph of $y = a\sqrt{x}$, $0 \leq x \leq b$, is revolved about the $x$-axis, generating part of a "paraboloid of revolution."
   a) Find the volume.
   b) The solid in part a can be inscribed in a right circular cylinder of radius $a\sqrt{b}$ and height $b$. What fraction of the volume of the cylinder is occupied by the paraboloid?

*4. A diagonal slice is cut off the end of a right circular cylinder of radius $r$, making angle $\theta$ with the end of the cylinder, and cutting through a diameter of the end (Fig. 17).
   a) Show that the area of the cross section in Figure 17 is $\frac{1}{2}(r^2 - x^2)\tan\theta$.
   b) Compute the volume of the slice.

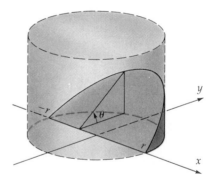

**FIGURE 17**
Slice of a cylinder of radius $r$.

*5. The region between the $x$-axis and the graph of $y = \sqrt{x}$, $0 \leq x \leq 1$, is revolved about the $x$-axis.
   a) Sketch the region and approximate it by five rectangles of equal base, as in forming an upper Riemann sum for the area of the region.
   b) Sketch the cylinder obtained by revolving the third rectangle about the $x$-axis, and compute the volume of this cylinder.
   c) Compute the sum $V_5$ of the volumes of the five approximating cylinders.

**d)** Compute the volume of revolution exactly, and compare it to $V_5$.

**6.** Repeat the previous problem, replacing the graph of $y = \sqrt{x}$ by $y = 1/x$, $1 \leq x \leq 3$.

**7.** A gold bracelet has the shape of the solid formed by revolving the region $\{(x, y): 0 \leq y \leq \sqrt{1 - x^2}\}$ about the line $y = -4$; dimensions are in cm. Compute the volume.

**8.** The region between the $x$-axis, $y = -\sqrt{x}$, and $x = 1$ is revolved about a line $L$. Compute the volume generated if
  **\*a)** $L$ is the line $y = R$, with $R \geq 0$.
  **b)** $L$ is the line $x = R$, with $R > 1$.

**9.** A certain tent has the shape in Figure 18: The floor is an isosceles triangle of base 4 feet and altitude 8 feet; the top is 3 feet high at the front, and at ground level at the back. Assuming no sag in the sides, compute the volume in the tent.

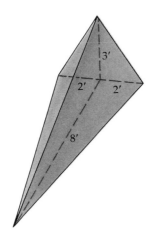

**FIGURE 18**
A simple tent.

**10.** The base of a solid is a circle of radius $R$. Compute the volume of the solid if the sections perpendicular to a given diameter are
  **\*a)** Semicircles.
  **b)** Rectangles, with height half the base.
  **\*c)** Rectangles of fixed height $h$.
  **d)** Equilateral triangles.
  **\*e)** Isosceles triangles of fixed height $h$.
  **f)** Isosceles right triangles, with hypotenuse in the given circular base.
  **\*g)** Isosceles right triangles, with one leg in the given circular base.

**11.** The base of a certain solid is a square of edge $e$. Compute the volume of the solid if the sections perpendicular to a given diagonal of the base are
  **\*a)** Semicircles.
  **b)** Squares.
  **\*c)** Equilateral triangles.
  **d)** Rectangles of fixed height $h$.

**B**

**12.** Compute the volume of a right circular cone of radius $r$ and height $h$.

**\*13.** A spherical tank of radius $r$ is filled to depth $h$, measured from the bottom of the tank. Compute the volume of the fluid in the tank. Check your formula in the special cases $h = 0$, $h = r$, and $h = 2r$, where the volume is known.

**\*14.** A sphere of radius $r$ has a cylindrical hole of radius $b$ cut through the center; $b < r$. What volume remains? (Be careful about the limits of integration.)

**15.** The graph of $\dfrac{x^2}{a^2} + \dfrac{y^2}{b^2} = 1$ is an ellipse (Fig. 19).

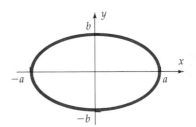

**FIGURE 19**
Ellipse $\dfrac{x^2}{a^2} + \dfrac{y^2}{b^2} = 1$.

  **a)** Verify that the intercepts shown are correct, and that on the graph, $|x| \leq a$ and $|y| \leq b$.
  **\*b)** Compute the volume enclosed by the football-shaped ellipsoid formed by revolving the ellipse about the $x$-axis.
  **c)** Compute the volume enclosed by the ellipsoid formed by rotating the ellipse about the $y$-axis.
  **d)** What familiar shape is the ellipsoid when $a = b$? Do your formulas agree in this case?

**16.** Two cylinders of radius $r$ have axes intersecting at right angles. Compute the volume of the region contained within both cylinders.

**\*17.** An ancient pot with delicate painting cannot safely be immersed in water, so the volume is to be determined by measurements. The pot has circular cross sections, and the diameter at selected heights is given in the following table. Estimate the volume.

| height (cm) | 0 | 1 | 2 | 3 | 4 | 5 | 6 | 7 | 8 | 9 | 10 | 11 |
|---|---|---|---|---|---|---|---|---|---|---|---|---|
| diameter (cm) | 0 | 7 | 9 | 10 | 10.7 | 11 | 10.5 | 9 | 7 | 7.5 | 8.5 | 9 |

# 7.3

## VOLUMES BY CYLINDRICAL SHELLS. FLOW IN PIPES

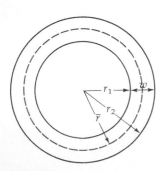

**FIGURE 1**
Cylindrical shell of height $h$, width $w = r_2 - r_1$, and average radius $\bar{r} = \frac{1}{2}(r_1 + r_2)$.

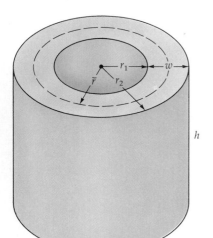

**FIGURE 2**
Annulus. Area $= (2\pi\bar{r})w$.

The integral of the cross-sectional area gives, in principle, the volume of any solid. But an alternative, the method of *cylindrical shells*, leads in some cases to a simpler integral.

What is the volume of the cylindrical shell in Figure 1? The base is an *annulus*, a circle of radius $r_2$ with a concentric hole of radius $r_1$. The area of the circle minus the hole is $\pi r_2^2 - \pi r_1^2$. You can express this in terms of the *average radius*

$$\bar{r} = \tfrac{1}{2}(r_2 + r_1)$$

and the *width* $w = r_2 - r_1$:

$$\begin{aligned} \pi r_2^2 - \pi r_1^2 &= \pi(r_2 + r_1)(r_2 - r_1) \\ &= 2\pi[\tfrac{1}{2}(r_2 + r_1)](r_2 - r_1) \\ &= 2\pi\bar{r}w. \end{aligned}$$

Thus the area of the annulus is the product of the length $2\pi\bar{r}$ of the "average circle," and the width $w$ (Fig. 2). This is reasonable; for if the annulus is cut and straightened (Fig. 3), it forms a rectangle of length $2\pi\bar{r}$ and width $w$. (Of course, the straightening requires some deformation. We suppose that the inner part is stretched and the outer part compressed, while the center maintains its length.)

As for the shell in Figure 1, its volume is

$$(\text{area of base}) \times (\text{height}) = (2\pi\bar{r}w)h.$$

Now we derive the volume integral. Consider a plane region $R$, lying between $x = a$ and $x = b$, revolved about a vertical axis outside $R$ (Fig. 4). Divide the interval $[a, b]$ into $n$ equal subintervals of length $\Delta x$, thus dividing $R$ into vertical strips (parallel to the axis of revolution). Denote by $\hat{x}_j$ the midpoint of the $j^{\text{th}}$ interval, and by $L(\hat{x}_j)$ the length of the cross section at $\hat{x}_j$ (Fig. 5). Approximate the $j^{\text{th}}$ strip by a rectangle of base $\Delta x$ and height $L(\hat{x}_j)$. Revolve this rectangle about the given axis; it forms a shell of average radius $r(\hat{x}_j)$, height $L(\hat{x}_j)$, and width $\Delta x$. The volume of this approximating shell is

$$(\text{area of base}) \times (\text{height}) = (2\pi r(\hat{x}_j)\,\Delta x)L(\hat{x}_j).$$

The sum of these volumes

$$V_n = \sum_{j=1}^{n} 2\pi r(\hat{x}_j)L(\hat{x}_j)\,\Delta x$$

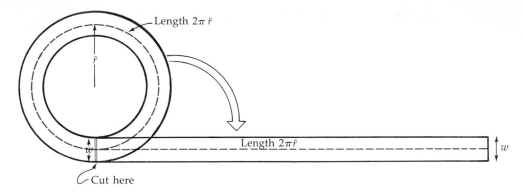

**FIGURE 3**
Area of annulus $= (2\pi \bar{r})w$.

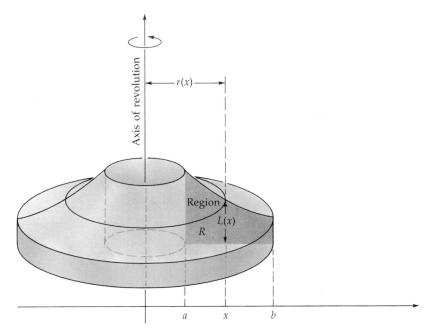

**FIGURE 4**
Region $R$ revolved about a vertical axis (not necessarily the $y$ axis).

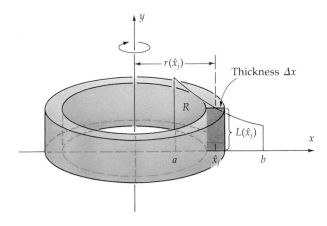

**FIGURE 5**
Cylindrical shell generated by approximating rectangle has volume
$[2\pi r(\bar{x}_j)\,\Delta x]L(\hat{x}_j)$.

approximates the volume of the solid, and it can be proved that the volume is in fact the limit of the approximations $V_n$:

$$\text{volume of solid} = \lim_{n \to \infty} V_n = \lim_{n \to \infty} \sum_{j=1}^{n} 2\pi r(\hat{x}_j)L(\hat{x}_j) \, \Delta x.$$

We recognize $V_n$ as a Riemann sum for the function $2\pi r(x)L(x)$ on the interval $[a, b]$, so

$$\text{volume of solid} = \int_a^b 2\pi r(x)L(x) \, dx, \tag{1}$$

where $r(x)$ is the distance from the section at $x$ to the axis of revolution.

To remember this integral, imagine the solid as formed of infinitesimal shells (Fig. 6). The shell at distance $r(x)$ from the axis has circumference $2\pi r(x)$, height $L(x)$, and "infinitesimal thickness" $dx$. Cut and laid out flat, it would form a slab $2\pi r(x)$ by $L(x)$ by $dx$, with "infinitesimal volume" $2\pi \cdot r(x) \cdot L(x) \cdot dx$. The entire volume is the "sum" of these shells, the integral (1).

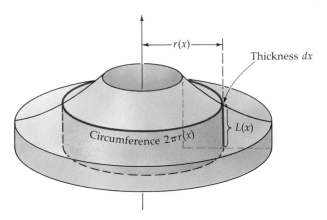

**FIGURE 6**
Volume of infinitesimal shell is $2\pi r(x)L(x) \, dx$.

In setting up the integral, it *helps* to sketch the solid; but you *must* draw the region being revolved, the vertical axis of revolution, and a cross section of the region *parallel to the axis*. From the sketch, determine the distance $r(x)$ from the section to the axis and the length $L(x)$ of the section. Use this in the "shell integral" (1) and evaluate.

---

**EXAMPLE 1** The region between the graphs of $y = x^2$ and $y = \frac{1}{2} - x^2$ is revolved about the $y$-axis to form a lens. Compute the volume of the lens.

**SOLUTION** *Sketch the curves* (Fig. 7). They intersect at the two points $(\pm\frac{1}{2}, \frac{1}{4})$. The lens is formed by revolving just the region to the right of the $y$-axis about that axis. *Draw a cross section* of the region, parallel to the axis of revolution. The distance from this cross section to the axis is $r(x) = x$. The length of the cross section at $x$ is

$$L(x) = [\text{upper } y \text{ value}] - [\text{lower } y \text{ value}]$$
$$= [\tfrac{1}{2} - x^2] - [x^2] = \tfrac{1}{2} - 2x^2.$$

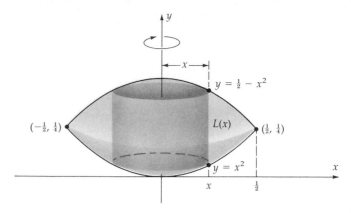

**FIGURE 7**

Region to be revolved about $y$-axis, forming a lens. Red lines indicate cylindrical surface formed by revolving section $L(x)$ about axis.

Imagine this cross section to have width $dx$, hence area $(\frac{1}{2} - 2x^2)\,dx$. Revolved about the $y$-axis, it forms a shell with volume

$$2\pi x L(x)\,dx = \pi(x - 4x^3)\,dx.$$

The lens consists of all these shells for $0 \le x \le 1/2$ (Why not $-\frac{1}{2} \le x \le \frac{1}{2}$?), so its volume is

$$\int_0^{1/2} 2\pi x L(x)\,dx = \int_0^{1/2} \pi(x - 4x^3)\,dx$$
$$= \pi\left[\frac{1}{2}x^2 - x^4\right]_0^{1/2} = \frac{\pi}{16}.$$

This volume can also be computed as an integral of cross-sectional areas. (See problem 12.)

---

If the axis of revolution is horizontal, then you take a horizontal section of the region and integrate with respect to $y$:

$$\text{Volume} = \int_c^d 2\pi r(y) L(y)\,dy$$

where $L(y)$ is the length of the horizontal section at $y$, and $r(y)$ is the distance from this section to the axis of revolution.

---

**EXAMPLE 2**   This region bounded by $x = y^2$ and $x = 4$ is revolved about the line $y = 3$. Compute the volume generated.

*SOLUTION*   Sketch the region and the axis of revolution $y = 3$ (Fig. 8). Draw a cross section of the region parallel to that axis. The distance to the axis from any point in this section is $r(y) = 3 - y$. The length of the section is

$$L(y) = [\text{right-hand } x] - [\text{left-hand } x]$$
$$= 4 - y^2.$$

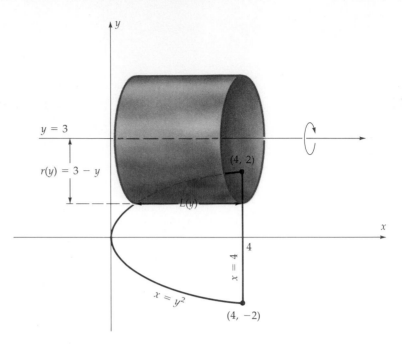

**FIGURE 8**
Region to be revolved about line $y = 3$. Red lines indicate cylindrical surface
formed by revolving section $L(y)$ about axis.

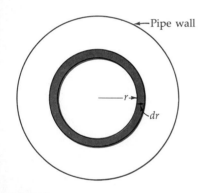

**FIGURE 9**
Cross section of pipe. Infinitesimal
annular ring has area $2\pi r\, dr$.

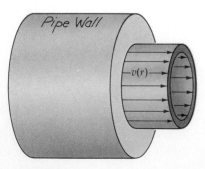

**FIGURE 10**
Fluid passing through annular ring in
one second.

Imagine the cross section to have thickness $dy$, hence area $(4 - y^2)\, dy$.
Revolved about the axis, it forms a shell of perimeter $2\pi r = 2\pi(3 - y)$; the
volume of this shell is

$$2\pi(3 - y)(4 - y^2)\, dy.$$

The region runs from $y = -2$ to $y = 2$ (Fig. 8), so the entire volume is

$$\int_{-2}^{2} 2\pi(3 - y)(4 - y^2)\, dy = 2\pi \int_{-2}^{2} (12 - 4y - 3y^2 + y^3)\, dy$$
$$= 64\pi.$$

## Flow through a Pipe

Suppose that fluid flows through a long straight pipe of radius $R$. At low
speeds, the fluid generally moves parallel to the wall of the pipe, and its
velocity depends only on the distance from the center; denote by $v(r)$ the
velocity at distance $r$ from the center of the pipe, in cm/sec. [Typically
$v(r) = 0$ along the wall of the pipe, and $v(r)$ is maximum at the center.] What
is the rate of flow through the pipe, in cm$^3$/sec?

Consider a plane cross section of the pipe, and compute the volume of
fluid that passes through this cross section in one second. Take a thin annular
ring of the cross section (Fig. 9) with radius $r$ and width $dr$. The velocity of
flow through this ring is $v(r)$, so the fluid that was at the cross section at
time $t = 0$ has advanced a distance $v(r)$ cm by time $t = 1$. Thus the fluid
passing through this ring in one second (Fig. 10) forms a cylindrical shell of
volume $2\pi r \cdot v(r) \cdot dr$. The fluid passing through the entire cross section in

one second is the "sum" of these shells, the integral

$$\int_0^R 2\pi r \cdot v(r) \cdot dr \quad \text{cm}^3$$

where $R$ is the radius of the pipe. The *rate* of flow is then

$$\int_0^R 2\pi r v(r) \, dr \quad \text{cm}^3/\text{sec}.$$

Problem 18 applies this integral.

## SUMMARY

Volume generated by revolving about a vertical axis:

$$V = \int_a^b 2\pi r(x)L(x) \, dx,$$

$L(x) = $ length of cross section of the region to be revolved,

$r(x) = $ distance from the axis of revolution.

For revolving about a horizontal axis, replace $x$ by $y$.

## PROBLEMS

### A

*1.  The region between the graphs of $y = x^2$ and $y = \frac{1}{8} + \frac{1}{2}x^2$ (Fig. 11) is revolved about the $y$-axis to form a lens. Compute the volume of the lens.

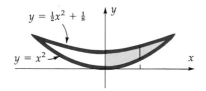

$y = \frac{1}{2}x^2 + \frac{1}{8}$

$y = x^2$

**FIGURE 11**
Cross section of lens.

In problems 2–9, compute the volume of the solid formed by revolving the given plane region about the given axis; use either shells or plane cross sections, as convenient. Describe the solid.

*2.  The region between the graphs of $y = 1 - x$ and $y = x - 1$ for $0 \le x \le 1$, about the $y$-axis.

*3.  The rectangle with corners $(r_1, 0)$, $(r_2, 0)$, $(r_1, h)$, $(r_2, h)$, where $0 < r_1 < r_2$; about
  a)  The $y$-axis.
  b)  The $x$-axis.

4.  The region between the $x$-axis and the graph of $y = \frac{1}{2}\sqrt{4 - x^2}$ for $0 \le x \le 2$, about
  *a)  The $y$-axis.
  b)  The $x$-axis.
  *c)  The line $y = -1$.
  d)  The line $x = -1$.

*5.  The region between the $x$-axis and the graph of $y = \sqrt{r^2 - x^2}$, for $a \le x \le r$, where $0 \le a$; about the $y$-axis.

6.  The region between the graph of $y^2 = 1 + x^2$, the $y$-axis, and the line $x = b$; about
  a)  The $x$-axis.      b)  The $y$-axis.

*7.  The region between the curves $y = x^2$ and $y = x$, about
  a)  The line $x = -2$.
      The line $y = 3$.

8. The region below $y = x^2$ and above $y = x^4$, about
   *a) The $y$-axis.
   b) The line $x = -2$.
   c) The line $y = 1$.

9. The triangle with vertices $(0, 0)$, $(0, 2)$, $(3, 0)$, about
   a) The line $x = 4$.  b) The line $y = 4$.
   (Check your answers by using *both* plane sections *and* cylindrical shells.)

B

10. Compute the volume of a sphere of radius $R$, using cylindrical shells.

11. Compute the volume of a right circular cone of height $h$ and base of radius $R$, using cylindrical shells.

12. Compute the volume of the lens in Example 1 by taking plane cross sections perpendicular to the $y$-axis.

13. The disk $x^2 + y^2 \leq a^2$ is revolved about the line $y = R$, where $R \geq a$. Describe the solid formed, and compute its volume.

14. Find the volume of the solid formed by revolving the region bounded by $x^{1/2} + y^{1/2} = a^{1/2}$ and the coordinate axes, about
   a) The $x$-axis.
   b) The $y$-axis.
   *c) The line $y = 2a$.

15. Find the volume of the solid formed by revolving the graph of $x^{2/3} + y^{2/3} = a^{2/3}$ about the $x$-axis.

16. A hole of radius $a$ is to be drilled through the center of a sphere of radius $R$, so that 10% of the sphere's volume is removed. What is the relation between $R$ and $a$?

*17. A cylindrical pail of water is rotated about its axis at a constant angular velocity of $\omega$ radians/sec. The surface of the water then forms a parabola of revolution generated by revolving the graph of $y = \omega^2 x^2/2g$ about the $y$-axis. (The constant $g$ is the acceleration due to gravity.) If the pail has radius $R$, and the volume of the water is $V$, how deep is the water at the center of the pail? How deep is it at the edge? (Assume that $V \geq \pi\omega^2 R^4/4g$. What complication arises otherwise?)

18. *Poiseuille's law* describes reasonably well the flow of fluid at low speed in water pipes, fuel lines, blood vessels, and so on. The formula determined experimentally by Poiseuille in 1840 gives the velocity $v(r)$ of the flow as a function of the distance $r$ from the center. Let $R$ be the radius of the pipe, $L$ the length of the pipe, $p$ the difference in pressure between the two ends of the pipe, and $k$ a constant measuring the *viscosity* of the fluid. Then the velocity at distance $r$ from the center is

$$v(r) = \frac{p}{4kL}(R^2 - r^2) \text{ cm}^3/\text{sec}, \qquad 0 \leq r \leq R.$$

a) Where is $v$ maximum, and where minimum? Is that reasonable? Why?
b) Why is pressure $p$ in the numerator and length $L$ in the denominator?
*c) Compute the rate of flow through the pipe.
d) The viscosity of a fluid is measured by forcing it with known pressure difference through a pipe with known dimensions. Suppose that volume $V$ passes through the pipe in $T$ seconds. Express the viscosity $k$ in terms of $T$, $V$, $R$, $L$, and $p$. [For a further discussion of Poiseuille's law see Philip Tuchinsky, "Viscous Fluid Flow and the Integral Calculus," UMAP Module 210, available from COMAP, 60 Lowell St., Arlington, MA 02174.

## 7.4

### ENERGY AND WORK

When a force moves an object, work is done. Suppose at first that the force $F$ is a constant, directed along a coordinate line, and that it moves an object from point $a$ to point $b$ on the line (Fig. 1). The work done by that force is defined to be the product of the force $F$ times the displacement $b - a$. Denote

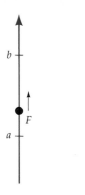

**FIGURE 1**
$W_a^b = F \cdot (b - a)$.

the work in moving from $a$ to $b$ by $W_a^b$; then

$$W_a^b = F \cdot (b - a).\tag{1}$$

This seems sensible; if you double the force $F$, you double the work; and if you double the displacement, you again double the work.

---

**EXAMPLE 1**   An old-fashioned clock is driven by a weight of 200 gm, suspended on a chain. If you raise the weight 30 cm, you exert an upward force of 200 gm (so $F = +200$) through an upward displacement of 30 cm, and the work done is

$$W_0^{30} = 200 \cdot 30 = 6{,}000 \text{ gm-cm}.$$

This work represents **stored energy**, which can drive the clock.

---

Suppose now that a variable force $F$ moves an object along the $x$-axis; for example a spring, as it is stretched, requires an increasing force. Denote by $F(x)$ the force at position $x$. Then the product (1) is replaced by an integral

$$W_a^b = \int_a^b F(x)\, dx.$$

To see why, divide the interval $[a, b]$ into $n$ subintervals of length $\Delta x$ (Fig. 2). In the first subinterval $[a, x_1]$, let $F(\underline{x}_1)$ be the minimum force and $F(\overline{x}_1)$ the maximum. The force in the first subinterval varies between $F(\underline{x}_1)$ and $F(\overline{x}_1)$, while the displacement is $x_1 - a = \Delta x$, so the work done is between $F(\underline{x}_1)\,\Delta x$ and $F(\overline{x}_1)\,\Delta x$:

$$F(\underline{x}_1)\,\Delta x \le W_a^{x_1} \le F(\overline{x}_1)\,\Delta x.$$

**FIGURE 2**

There is a similar inequality for each subinterval:

$$F(\underline{x}_1)\,\Delta x \le W_a^{x_1} \le F(\overline{x}_1)\,\Delta x$$
$$F(\underline{x}_2)\,\Delta x \le W_{x_1}^{x_2} \le F(\overline{x}_2)\,\Delta x$$
$$\vdots$$
$$F(\underline{x}_n)\,\Delta x \le W_{x_{n-1}}^{b} \le F(\overline{x}_n)\,\Delta x.$$

The sum of the terms in the middle is $W_a^b$, the work done in going from $a$ to $b$. The terms on the left add up to a lower sum $\underline{S}_n$ for the integral $\int_a^b F(x)\, dx$, and the terms on the right give the upper sum $\overline{S}_n$, so

$$\underline{S}_n \le W_a^b \le \overline{S}_n.$$

As $n \to \infty$, the upper and lower sums have the limit $\int_a^b F(x)\, dx$, so by the Trapping Theorem

$$W_a^b = \int_a^b F(x)\, dx. \tag{2}$$

In the case of a spring, for moderate displacements, the force on each end increases in proportion to the amount of stretching (Fig. 3)—when it is stretched by an amount $x$, the force required to hold it in that position is

$$F = kx.$$

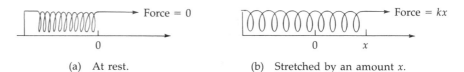

        Force = 0                             Force = $kx$

          0                                  0     $x$

(a)  At rest.                    (b)  Stretched by an amount $x$.

**FIGURE 3**
Hooke's law.

This is Hooke's law; $k$ is the **stiffness constant** of the spring.
    The work done to pull the end from $x = 0$ to $x = L$ is

$$W = \int_0^L F(x)\, dx = \int_0^L kx \cdot dx$$
$$= k\frac{x^2}{2}\Big|_0^L = \frac{1}{2}kL^2.$$

This represents energy stored in the spring, so-called **potential energy**; it could pull something from position $L$ back toward position 0.

## Units

A common unit of force is the **Newton**, the force required to accelerate one kilogram at the rate of 1 meter/sec$^2$; so a Newton has units of (kg)(m)/(sec)$^2$. These units all balance out in Newton's law $F = ma$:

$$\underbrace{(kg)(m)/(sec)^2}_{} = (kg)\ \underbrace{(m/sec^2)}_{}$$

unit of force       unit of   unit of
                           mass    acceleration

By a similar balancing, we determine that when force is measured in Newtons, and distance in meters, then the constant $k$ in Hooke's law

$$F = kx$$

     (kg)(m)/(sec)$^2$     (m)
     or Newton

has units of Newtons/meter, or kg/sec$^2$. If $x$ were in centimeters, then $k$ would be Newtons/cm.

Since work = (force) × (distance), it is often expressed in terms of (Newtons) × (meters). These units are called **Joules**, in honor of an English physicist J. P. Joule; one Joule is the work done by one Newton through a displacement of one meter.

In English units, a pound represents the force exerted by gravity on a certain mass at the surface of the earth; the corresponding unit of work is the foot-pound.

---

**EXAMPLE 2**    A spring is stretched 3 cm from its relaxed state. To hold it in this position requires a force of 5 Newtons. How much work would it take to stretch it 4 cm more?

*SOLUTION*    First, determine the stiffness constant $k$ in Hooke's law. The force required to stretch it 3 cm is 5 Newtons, so $F = kx$ gives

$$5 = k \cdot 3, \qquad \text{hence } k = 5/3 \quad \text{Newton/cm}$$

The work done in stretching 4 cm more, from $x = 3$ to $x = 7$, is

$$\int_3^7 \tfrac{5}{3}x \, dx = \tfrac{100}{3} \text{ Newton cm} = \tfrac{1}{3} \text{ Joule.}$$

---

**Kinetic energy** is energy stored in the *motion* of an object. For example, when you swing a hammer, the motion of the hammerhead stores energy which is used up when you hit the nail, driving it into the wood.

The faster the object moves, the more energy is stored in the motion; and the heavier it is, the more energy there is. The exact relation between the kinetic energy, the mass, and the velocity can be deduced using Newton's second law and the work integral (2), as follows.

Consider an object of mass $m$ moving along a coordinate line; at time $t$ its position is $x(t)$, and its velocity is $v(t)$. The total force $F$ acting on the mass is given by Newton's law

$$F = ma = m\frac{dv}{dt}.$$

From time $t = a$ to $t = b$, the object moves from $x(a)$ to $x(b)$; the work done is

$$W = \int_{x(a)}^{x(b)} F \, dx = \int_{x(a)}^{x(b)} m\frac{dv}{dt} \, dx. \tag{3}$$

Even though the position function is not given, you can "evaluate" this integral by substitution! Set

$$x = x(t), \qquad dx = \frac{dx}{dt} \, dt = v \, dt.$$

While $x$ varies from $x(a)$ to $x(b)$, time $t$ varies from $a$ to $b$, and (3) gives

$$W = \int_a^b \left( m\frac{dv}{dt} \right) \cdot v \, dt. \tag{4}$$

Make another substitution:

$$v = v(t), \qquad dv = \frac{dv}{dt} \, dt.$$

While $t$ varies from $a$ to $b$, the velocity $v$ varies from $v(a)$ to $v(b)$, so (4) gives

$$W = \int_{v(a)}^{v(b)} mv \, dv = \tfrac{1}{2} mv^2 \Big|_{v(a)}^{v(b)}$$
$$= \tfrac{1}{2} mv(b)^2 - \tfrac{1}{2} mv(a)^2.$$

Conclusion: The work done in moving the mass $m$ equals the change in the quantity $\tfrac{1}{2} mv^2$. This quantity is called the energy of motion, or *kinetic energy*, denoted $KE$:

$$KE = \tfrac{1}{2} mv^2.$$

It is the work done by *all forces* acting on the object to move it from rest.

---

**EXAMPLE 3**    At time $t = 0$, a mass $m$ on the end of a spring is at rest at position $x = 0$. Then it is moved, stretching the spring, and comes to rest at time $t = 1$, at position $x = L$. The work done by *all* forces $F$ is then

$$W = \tfrac{1}{2} mv(1)^2 - \tfrac{1}{2} mv(0)^2 = 0 - 0 = 0.$$

Notice the contrast with Example 2. There, we computed the work done only by the force pulling against the spring. But the *total* force on the mass $m$ is that pulling force *combined with* the force exerted by the spring! The work done by the pulling force exactly cancels the work done by the spring.

---

*REMARK*    Heat is actually a form of kinetic energy. A warm object consists of zillions of spinning and vibrating molecules; its heat is the combined kinetic energy of all these particles.

## Other Work Integrals

The basic integral (2) applies when one object is moved from $a$ to $b$. The next example illustrates a different kind of work integral, where you can think of the total work as the sum of the work done on many different parts.

---

**EXAMPLE 4**    Water is pumped from a lake into the cylindrical tank in Figure 4. The base of the tank is 5 meters above lake level. Find the work required to fill it to height $h$.

*SOLUTION*    Think of the water in the tank as a collection of many thin layers, of thickness $dy$. Each layer is a cylinder of volume

$$\pi r^2 \, dy.$$

If the density of water is $w$ gm/m$^3$, then this layer weighs

$$wg \cdot \pi r^2 \, dy.$$

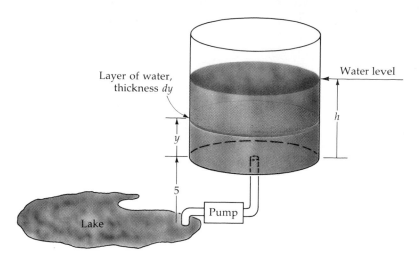

**FIGURE 4**

This is the *force* required to raise the layer of water. The *displacement* of the layer at level $y$ is $5 + y$ meters: 5 meters up to the base of the tank, and $y$ meters above the base. So the work done on the layer at level $y$ is

$$\text{force} \times \text{distance} = wg\pi r^2 dy \, (5 + y)$$
$$= \pi r^2 wg(5 + y)dy.$$

The total work is the sum of the work on all these layers from $y = 0$ to $y = h$:

$$W = \int_0^h \pi r^2 wg(5 + y) \, dy = \pi r^2 wg(5y + \tfrac{1}{2}y^2)\Big]_0^h$$
$$= \pi r^2 wg(5h + \tfrac{1}{2}h^2) = \pi r^2 hwg(5 + \tfrac{1}{2}h).$$

This makes sense: $\pi r^2 h$ is the volume of water, so $\pi r^2 hwg$ is the weight of the water; and $5 + \tfrac{1}{2}h$ is the average height through which it was raised.

If the tank in the preceding example were not a cylinder, the answer would not be so transparent—it would not be obvious what to use as "average height"—so you would have to work out the integral. In general, the layer at height $y$ has area $A(y)$, thickness $dy$, hence volume $A(y) \, dy$, just as in setting up a volume integral. But to compute the work done, this "infinitesimal" volume must be multiplied by the weight per unit volume (to get the force) and then by the distance through which it is raised (to get the work). Finally, integrate the work done on all the layers to get the total work.

## SUMMARY

When a variable force $F$, directed along a coordinate line, moves an object from point $a$ to point $b$ on the line, then the work done by that force is

$$W_a^b = \int_a^b F(x) \, dx.$$

The *kinetic energy* of an object of mass $m$, moving with velocity $v$, is

$$KE = \tfrac{1}{2}mv^2.$$

When different parts of an object are displaced different distances, the total work done is the integral of the work done on the individual parts.

## PROBLEMS

### A

1. A weight of 12 pounds is raised 8 feet. How many pound-feet of work are done?

*2. A spring with stiffness constant $k = 5$ kg/sec$^2$ is stretched from rest position to .06 meters elongation. How much work is done?

3. A spring with stiffness $k = 8$ Newtons/meter is first elongated to .02 meters and then to .05 meters. How much work is done in stretching from .02 to .05 meters?

*4. The end of a relaxed spring is stretched from $x = 0$ to $x = 0.1$ meters and held there by a force of 3 Newtons. Determine the stiffness constant $k$ and the work done.

5. A spring is compressed 5 cm and held by a force of 7 Newtons. How much work is required to compress it 1 cm more?

*6. The force of gravity on an object of mass $m$ at height $x$ kilometers above the surface of the earth is $F(x) = mgR^2(x + R)^{-2}$ Newtons; $g$ is the acceleration due to gravity in m/sec$^2$ at the surface of the earth, and $R$ is the radius of the earth in meters.
   a) Compute the force $F(x)$ at the surface of the earth, where $x = 0$.
   b) How much work is required to raise an object of mass $m$ from the surface of the earth to 1 m above the surface? To 1000 m above the surface? To $\infty$?

7. (Coulomb's law). Two negative (or two positive) charges, of strength $q_1$ and $q_2$, repel each other with a force of $kq_1q_2/r^2$, where $r$ is the distance between them, and $k$ is a constant. Initially they are 2 cm apart. How much work is required to bring them within 1/2 cm? (The answer contains $k$, $q_1$, $q_2$.)

*8. A proton and an electron separated by $r$ meters attract each other with a force of about $2.3 \times$ $10^{-28}/r^2$ Newtons. In a hydrogen atom, they are about $5 \times 10^{-11}$ meters apart. How much work is required to separate them to a distance of 1 m?

*9. The top of a conical tank has radius $R$, the depth of the tank is $H$, and the vertex at the bottom is $V$ meters above a water source (Fig. 5). How much work is required to fill the tank to height $Y$, raising water from the source? (Determine the area of the cross section at height $y$ above the vertex. One m$^3$ of water has a mass of $10^3$ kg, and exerts a force of $9.8 \cdot 10^3$ (kg)m/sec$^2$.)

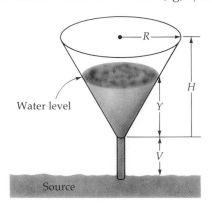

**FIGURE 5**
Conical tank.

10. A pyramid with square base $b$ by $b$, and height $h$, is constructed of stone of density $\delta$ kg/m$^3$. How much work is required to raise all this stone from ground level to its position in the pyramid?

*11. A spherical tank of radius $R$ meters is full of gasoline, which is to be pumped out of the top of the tank. If gasoline has density $w$ kg/m$^3$, how much work is required?

12. A chain 10 ft long coiled on the ground weighs 0.7 lbs/ft. One end is raised, leaving some of the chain on the ground, until the whole chain is in

the air. How much work is done if the top end is raised

a)  5'    b)  10'    c)  15'?

13.  A mechanic lifts a motor weighing 100 lbs by a chain passing over a pulley. One end of the chain is attached to the motor at a point $A$. The chain weighs 4 lbs/ft, and is long enough that it reaches up from the motor, over the pulley and down to the floor, with several feet of chain lying on the floor.

   a)  What force is required to keep point $A$ $x$ feet above the floor?

   *b)  How much work is required to raise point $A$ from 3 ft to 4 ft?

**B**

14.  a)  Suppose that the object in problem 6 has velocity $v_0$ at the surface of the earth and velocity $v_1$ at height $h$. Suppose that gravity is the only force acting on it. Compute $v_1^2 - v_0^2$. (Since gravity acts *downward*, and

the object moves *up*, the work done by gravity is *negative*.)

   b)  Suppose that the object in problem 6 has velocity $v_0$ at the surface, velocity 0 at height $\infty$, and gravity is the only force acting. Find $v_0$. (This is called the *escape velocity* for an object at the surface of the earth.)

15.  A mass $m$ hangs on a spring. By a simple theory that ignores friction, the height $y$ of the mass at time $t$ satisfies the equation

$$m\frac{d^2y}{dt^2} = -ky, \tag{5}$$

where $k$ is the spring constant in Hooke's law.

   a)  Find the work done by the spring in raising the mass from height 0 to height $y$. (This is the negative of the *potential energy* of the mass.)

   b)  Using only equation (5), show that the sum of the potential and the kinetic energy is constant.

=== **7.5** ===

## IMPROPER INTEGRALS

So far, in all our integrals $\int_a^b f(x)\,dx$, the integrand $f$ has been continuous, and the interval $[a, b]$ finite. But some applications require "improper integrals," where the integrand is unbounded, or the interval is infinite. For example, the force of attraction between a proton at the origin and an electron at point $x$ on the $x$-axis is $2.3 \times 10^{-28}x^{-2}$ Newtons ($x$ is in meters). So the work done in moving the electron from $x = a$ to $x = b$ is $\int_a^b kx^{-2}\,dx$, where the constant $k = 2.3 \times 10^{-28}$. If you want to move it along the axis from $x = -1$ to $x = 1$, the work is $\int_{-1}^{1} kx^{-2}\,dx$, an "improper" integral since the function $kx^{-2}$ has a vertical asymptote at $x = 0$ (where the electron meets the proton). If you want to remove the electron from its usual position in the hydrogen atom (at $x = 5 \times 10^{-11}$) and take it completely away from the proton (all the way to $x = \infty$), the work required is $\int_{5 \times 10^{-11}}^{\infty} kx^{-2}\,dx$; this integral is improper, since the interval of integration is infinite.

Improper integrals are handled by taking limits.

In Figure 1, $f$ is discontinuous at the left endpoint $a$, but continuous on the remainder of the interval of integration. So for every $\alpha > a$, the integral $\int_{\alpha}^b f(x)\,dx$ is defined; we then define the *improper integral* $\int_a^b f(x)\,dx$ as the *limit* (Fig. 1)

$$\int_a^b f(x)\,dx = \lim_{\alpha \to a^+} \int_{\alpha}^b f(x)\,dx. \tag{1}$$

If the limit exists, the improper integral is called **convergent**; otherwise, **divergent**.

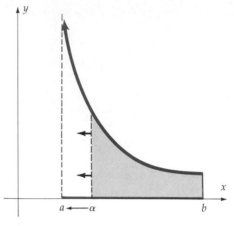

**FIGURE 1**

$$\int_a^b f(x)\, dx = \lim_{\alpha \to a^+} \int_\alpha^b f(x)\, dx.$$

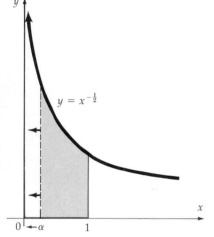

$y = x^{-\frac{1}{2}}$

**FIGURE 2**

$$\int_0^1 x^{-1/2}\, dx = \lim_{\alpha \to 0^+} \int_\alpha^1 x^{-1/2}\, dx.$$

**EXAMPLE 1**  $\int_0^1 x^{-1/2}\, dx$ is an improper integral, because the integrand $x^{-1/2}$ is discontinuous at 0 (Fig. 2). However, if $\alpha > 0$ then $\int_\alpha^1 x^{-1/2}\, dx$ is defined:

$$\int_\alpha^1 x^{-1/2}\, dx = \left.\frac{x^{1/2}}{1/2}\right|_\alpha^1 = 2 - 2\sqrt{\alpha}.$$

The given improper integral is defined to be the limit of this as $\alpha \to 0^+$:

$$\int_0^1 x^{-1/2}\, dx = \lim_{\alpha \to 0^+} \int_\alpha^1 x^{-1/2}\, dx = \lim_{\alpha \to 0^+} (2 - 2\sqrt{\alpha}) = 2.$$

So this improper integral is convergent. Apparently, even though the region is unbounded, the area under the graph is finite!

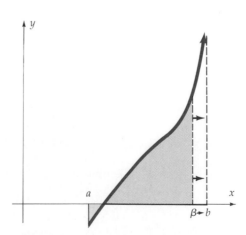

**FIGURE 3**

$$\int_a^b f(x)\, dx = \lim_{\beta \to b^-} \int_a^\beta f(x)\, dx.$$

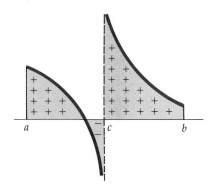

**FIGURE 4**
$\int_a^b f = \int_a^c f + \int_c^b f.$

If $f$ is continuous on $[a, b)$ but not at the right endpoint $b$, then (Fig. 3)

$$\int_a^b f(x)\, dx = \lim_{\beta \to b^-} \int_a^\beta f(x)\, dx.$$

If $f$ has an asymptote at an interior point $c$ (Fig. 4), then the integral is split in two parts at $c$:

$$\int_a^b f(x)\, dx = \int_a^c f(x)\, dx + \int_c^b f(x)\, dx.$$

Each of the integrals on the right may have a discontinuity at the endpoint $c$; if both are convergent, then the original integral $\int_a^b f(x)\, dx$ is called convergent; if either part is divergent, then so is the original integral.

**EXAMPLE 2** $\int_{-1}^1 x^{-2}\, dx = ?$

*SOLUTION* The integrand $x^{-2}$ has a discontinuity at $x = 0$ (Fig. 5), which is inside the interval of integration. Break the interval of integration at this discontinuity:

$$\int_{-1}^1 x^{-2}\, dx = \int_{-1}^0 x^{-2}\, dx + \int_0^1 x^{-2}\, dx$$

$$= \lim_{\beta \to 0^-} \int_{-1}^\beta x^{-2}\, dx + \lim_{\alpha \to 0^+} \int_\alpha^1 x^{-2}\, dx$$

$$= \lim_{\beta \to 0^-} -x^{-1}\Big|_{-1}^\beta + \lim_{\alpha \to 0^+} -x^{-1}\Big|_\alpha^1$$

$$= \lim_{\beta \to 0^-} (-\beta^{-1} - 1) + \lim_{\alpha \to 0^+} (-1 + \alpha^{-1})$$

$$= (+\infty) + (+\infty).$$

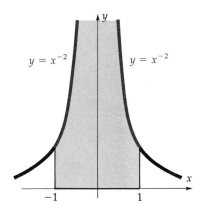

**FIGURE 5**
$\int_{-1}^1 x^{-2}\, dx = +\infty.$

$y = x^{-2}$     $y = x^{-2}$

Apparently, the area under the graph (Fig. 5) is infinite, and this integral diverges.

*WARNING* If you apply the usual antiderivative formula $\int_a^b f(x)\, dx = D^{-1}f(x)\big|_a^b$ directly on the given interval $[-1, 1]$, ignoring the discontinuity at $x = 0$, you get

N O N S E N S E
$$\int_{-1}^1 x^{-2}\, dx = -x^{-1}\big|_{-1}^1 = -1 - (1) = -2.$$
N O N S E N S E

This must be wrong—the area under graph in Figure 5, if defined at all, is surely positive! The trouble is that the integrand $x^{-2}$ has a discontinuity at 0. Moreover, the function $-x^{-1}$ is not an antiderivative of $x^{-2}$ on the entire interval $[-1, 1]$, since neither function is defined at 0; so the Fundamental Theorem of Calculus does not apply.

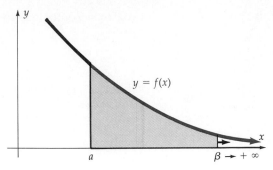

**FIGURE 6**

$$\int_a^\infty f(x)\, dx = \lim_{b \to +\infty} \int_a^b f(x)\, dx.$$

Another kind of improper integral involves an *infinite interval*. This, too, is handled as a limit: If $f$ is continuous on $[a, \infty)$ then (Fig. 6)

$$\int_a^\infty f = \lim_{b \to +\infty} \int_a^b f.$$

**EXAMPLE 3**

$$\int_0^\infty \frac{dx}{(1+x)^2} = \lim_{b \to +\infty} \int_0^b \frac{dx}{(1+x)^2}$$

$$= \lim_{b \to +\infty} \frac{-1}{1+x}\Big|_0^b$$

$$= \lim_{b \to +\infty} \left[ \frac{-1}{1+b} - \frac{-1}{1+0} \right] = 1.$$

Since the integral converges, we conclude that the area under the graph of $y = \dfrac{1}{(1+x)^2}$ on $[0, \infty)$ is finite (Fig. 7).

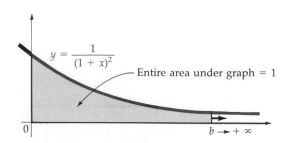

**FIGURE 7**

$$\int_0^\infty \frac{dx}{(1+x)^2} = 1.$$

## Multiple Improprieties

A single integral may display several improprieties. For example, the integrand may have more than one vertical asymptote, or there may be both a vertical asymptote and an infinite interval. In such cases, break the interval into parts, each having just one impropriety, and sum the results.

---

**EXAMPLE 4**  $\int_0^\infty x^{-1/2} \, dx$ has both an infinite interval, and a discontinuity (at 0). So break the integral in two convenient pieces, one from 0 to 1, the other from 1 to $\infty$:

$$\int_0^\infty x^{-1/2} \, dx = \int_0^1 x^{-1/2} \, dx + \int_1^\infty x^{-1/2} \, dx$$

$$= \lim_{a \to 0^+} \int_a^1 x^{-1/2} \, dx + \lim_{b \to +\infty} \int_1^b x^{-1/2} \, dx$$

$$= \lim_{a \to 0^+} 2[1 - \sqrt{a}] + \lim_{b \to +\infty} 2[\sqrt{b} - 1].$$

The first limit equals 2, but the second is infinite, so the given integral diverges.

Making the "break" at $x = 1$ was arbitrary. You would get the same result ("divergent") by breaking at $x = 2$, or at any other point between 0 and $\infty$.

---

## The Comparison Test

For a *positive* function $f$, the question of convergence or divergence can be put geometrically: Is the area under the graph finite or infinite? You can sometimes determine that the area is finite by comparing it to the area under the graph of a larger but simpler function $g$ (Fig. 8). This is the basis of:

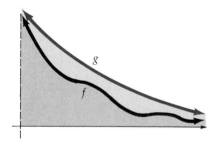

**FIGURE 8**
If $0 \leq f \leq g$ and $\int_a^b g$ converges, then $\int_a^b f$ converges too.

> ### THE COMPARISON TEST FOR IMPROPER INTEGRALS
>
> *Suppose that $f$ is continuous on $(a, b)$, and $0 \leq f \leq g$. If $\int_a^b g(x) \, dx$ converges, then so does $\int_a^b f(x) \, dx$. This holds if $f$ is discontinuous at $a$ or $b$, and if the interval $(a, b)$ is infinite.*

---

**EXAMPLE 5**    Decide the convergence or divergence of

$$\int_0^\infty x^{-1/2}(1 + x)^{-2} \, dx.$$

*SOLUTION*    In this case, an antiderivative of the given integral is hard to find, so we compare it to simpler functions. There are two trouble spots: 0 (where $x^{-1/2}$ becomes infinite) and $\infty$. So break the integral in two parts:

$$\int_0^\infty x^{-1/2}(1 + x)^{-2} \, dx = \int_0^1 x^{-1/2}(1 + x)^{-2} \, dx + \int_1^\infty x^{-1/2}(1 + x)^{-2} \, dx.$$

In the first integral on the right, $x$ lies between 0 and 1, so $(1 + x)^{-2} \leq 1$, hence $x^{-1/2}(1 + x)^{-2} \leq x^{-1/2}$, and

$$\int_0^1 x^{-1/2}(1 + x)^{-2} \, dx \leq \int_0^1 x^{-1/2} \, dx = 2 \qquad \text{(as in Example 4).}$$

It follows that $\int_0^1 x^{-1/2}(1 + x)^{-2} \, dx$ converges, and the value of this integral is some number less than 2.

In the second integral $\int_1^\infty x^{-1/2}(1 + x)^{-2} \, dx$, $x$ is $\geq 1$, so $x^{-1/2} = 1/x^{+1/2} \leq 1$, and $x^{-1/2}(1 + x)^{-2} \leq (1 + x)^{-2}$. Hence,

$$\int_1^\infty x^{-1/2}(1 + x)^{-2} \, dx \leq \int_1^\infty (1 + x)^{-2} \, dx = 1/2.$$

$$\uparrow$$
(check this)

It follows that $\int_1^\infty x^{-1/2}(1 + x)^{-2} \, dx$ converges to some number less than $1/2$.

The two results combine to show that $\int_0^\infty x^{-1/2}(1 + x)^{-2} \, dx$ converges to some number less than $2 + 1/2$.

---

**Integrands with jump discontinuities** are not called "improper," but do require special treatment—split the integral at each jump. Suppose that $f$ is continuous on $[a, c)$ and on $(c, b]$, but has a finite "jump" at $c$ (Fig. 9). Then (just as with vertical asymptotes) we define

$$\int_a^b f(x) \, dx = \int_a^c f(x) \, dx + \int_c^b f(x) \, dx.$$

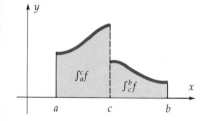

**FIGURE 9**
$\int_a^b f = \int_a^c f + \int_c^b f.$

Usually, the jump corresponds to some change in the formula defining $f$; in the sum on the right, each part would use just one formula, and you could hope to find an antiderivative for it.

---

**EXAMPLE 6** $\int_{-1}^1 \dfrac{x}{|x|} \, dx$ has a discontinuity at $x = 0$. So break the integral in two parts, at $x = 0$:

$$\int_{-1}^1 \frac{x}{|x|} \, dx = \int_{-1}^0 \frac{x}{|x|} \, dx + \int_0^1 \frac{x}{|x|} \, dx. \qquad (2)$$

Since $\dfrac{x}{|x|} = -1$ for $x < 0$ and $= +1$ for $x > 0$, you find

$$\int_{-1}^1 \frac{x}{|x|} \, dx = \int_{-1}^0 (-1) \, dx + \int_0^1 1 \, dx$$

$$= -1 + 1 = 0. \qquad (3)$$

A sketch of the graph will confirm this.

## SUMMARY

$$\int_a^\infty f(x)\,dx = \lim_{b\to\infty} \int_a^b f(x)\,dx.$$

Similarly, if $f$ is discontinuous at some point in the interval $[a, b]$, then $\int_a^b f(x)\,dx$ is defined as an appropriate limit, or sum of limits.

## PROBLEMS

### A

In problems 1–12, evaluate or show to be divergent.

*1. $\displaystyle\int_0^\infty (1+x)^{-3}\,dx$

2. $\displaystyle\int_0^\infty (a+x)^{-3}\,dx$     [$a > 0$ is constant.]

*3. $\displaystyle\int_0^2 x^{-1/2}\,dx$     4. $\displaystyle\int_{-2}^1 x^{-1/3}\,dx$

*5. $\displaystyle\int_{-1}^2 x^{-5/3}\,dx$     6. $\displaystyle\int_0^\infty \sin\theta\,d\theta$

*7. $\displaystyle\int_0^{\pi/2} \sec^2\theta\,d\theta$     *8. $\displaystyle\int_{-\pi/4}^{\pi/4} \csc^2\theta\,d\theta$

*9. $\displaystyle\int_0^1 \frac{x\,dx}{\sqrt{1-x^2}}$     10. $\displaystyle\int_{-1}^0 \frac{dx}{\sqrt{1+x}}$

*11. $\displaystyle\int_0^2 f(x)\,dx$ where $f(x) = \begin{cases} 0, & 0 \le x \le 1 \\ 1, & 1 < x \le 2 \end{cases}$
(Do it both graphically, and as in Example 6.)

12. $\displaystyle\int_{-1}^2 f(x)\,dx$ where $f(x) = \begin{cases} -x^2, & x \le 0 \\ x^2, & x > 0 \end{cases}$

*13. A proton at the origin attracts an electron at point $x$ on the $x$-axis with a force of magnitude $kx^{-2}$ Newtons, $k = 2.3 \times 10^{-28}$. How much work is required to move the electron along the axis from $x = -1$ to $x = 1$?

14. In the hydrogen atom, a proton and an electron are usually separated by about $5 \times 10^{-11}$ meters. How much work is required to remove the electron all the way to $\infty$? (See the previous problem.)

15. The force of gravity on an object of mass $m$ at height $x$ meters above the surface of the earth is $mgR^2(x+R)^{-2}$ Newtons, where $R$ is the radius of the earth. How much work is required to raise an object from the surface of the earth, "all the way to $\infty$"?

16. Let $R$ be the region under the graph of $y = x^{-2/3}$, $1 \le x < \infty$.
   *a) Compute the area of $R$.
   b) Compute the volume of the solid generated by revolving $R$ about the $x$-axis.

17. Let $R$ be the region under the graph of $y = x^{-3/2}$, $1 \le x < \infty$.
   *a) Compute the area of $R$.
   b) Compute the volume of the region generated by revolving $R$ about the $y$-axis.

18. a) Compare the graphs of $y = x^{-2}$ and $y = x^{-1/2}$ on $0 < x \le 1$; which is higher?
   b) Compare $\int_0^1 x^{-2}\,dx$ with $\int_0^1 x^{-1/2}\,dx$.

19. a) Compare the graphs of $y = x^{-1.1}$ and $y = x^{-0.9}$ on $0 < x \le 1$; which is higher?
   b) Compare $\int_0^1 x^{-1.1}\,dx$ with $\int_0^1 x^{-0.9}\,dx$.

20. a) Compare the graphs of $y = x^{-2}$ and $y = x^{-1/2}$ on $1 \le x < \infty$; which is higher?
   b) Compare $\int_1^\infty x^{-2}\,dx$ with $\int_1^\infty x^{-1/2}\,dx$.

### B

21. For what constants $p \ne 1$ is $\int_0^1 x^{-p}\,dx$ convergent? (See problem 24 for the case $p = 1$.)

22. For what constants $p \ne 1$ is $\int_1^\infty x^{-p}\,dx$ convergent? (See problem 24 for the case $p = 1$.)

23. For what constants $p \ne 1$ is $\int_0^\infty x^{-p}\,dx$ convergent? (See previous problems for clues, and problem 24 for $p = 1$.)

### C

24. a) Show by the substitution $u = ax$ that
$$\int_1^2 \frac{dx}{x} = \int_a^{2a} \frac{dx}{x}.$$

**b)** Show that $\int_1^2 \dfrac{dx}{x} = \int_2^4 \dfrac{dx}{x} = \int_4^8 \dfrac{dx}{x} = \dots$.

**c)** Is $\int_1^\infty \dfrac{dx}{x}$ convergent?

**d)** Show that $\int_{1/2}^1 \dfrac{dx}{x} = \int_{1/4}^{1/2} \dfrac{dx}{x} = \dots$, and determine whether $\int_0^1 \dfrac{dx}{x}$ is convergent.

**a)** $\int_1^\infty \dfrac{1}{1 + x^2}\, dx$

**b)** $\int_0^\infty \dfrac{1}{1 + x^2}\, dx$

**c)** $\int_{-1}^1 \dfrac{dx}{\sqrt{1 - x^2}}$ $\left[\text{Compare with } \int_{-1}^0 \dfrac{dx}{\sqrt{1 + x}}\right.$ and $\left. \int_0^1 \dfrac{dx}{\sqrt{1 - x}}.\right]$

**25.** Use the Comparison Test to determine the convergence or divergence of the following integrals.

# REVIEW PROBLEMS  CHAPTER 7

## A

**\*1.** A region $R$ is bounded by the curves $y = x^4$ an $y = x$. Compute
   **a)** The area of $R$.
   **b)** The volume generated by revolving $R$ about the $x$-axis.
   **c)** The volume generated by revolving $R$ about the $y$-axis.
   **d)** The volume generated by revolving $R$ about the line $y = 2$.

**\*2.** A footbridge has a single suspension cable hanging in the shape of the graph of $y = 3 + x^2/40$, $|x| \le 10$, $x$ in meters. From this main cable, many auxiliary cables support a rectangular walkway along the $x$-axis, 2 meters wide, directly under the main cable. Compute the volume enclosed between the system of cables and the walkway.

**\*3.** The circle $x^2 + y^2 = 1$, revolved about the line $x = 2$, forms a so-called *torus*. Compute the volume enclosed by the torus.

**\*4.** A certain spring obeys Hooke's law, approximately, for small displacements, but gradually loses its resilience as it is stretched farther; when stretched by an amount $x$, the force required is $x - x^3$, $0 \le x \le 1$. Compute the work done in stretching it
   **a)** From $x = 0$ to $x = 1/2$.
   **b)** From $x = 1/2$ to $x = 1$.

**\*5.** An artificial pond has the shape of the graph of $y = x^3/400$, $0 \le x \le 10$, revolved about the $y$-axis; $x$ and $y$ are in feet. If it is full, what work is required to empty it by pumping all the water over the edge? [Denote by $w$ the weight (in pounds) of 1 ft$^3$ of water.]

**\*6.** A region $R$ in the first quadrant is bounded by the two axes and the graph of $y = (1 + x)^{-4/3}$. Compute (if finite)
   **a)** The area of $R$.
   **b)** The volume formed by revolving $R$ about the $x$-axis.
   **c)** The volume formed by revolving $R$ about the $y$-axis.

**7.** A region $R$ in the first quadrant is bounded by the $y$-axis, the $x$-axis, the graph of $y = x^{-4/3}$, and the line $x = 1$. Compute (if finite)
   **a)** The area of $R$.
   **b)** The volume formed by revolving $R$ about the $x$-axis.
   **c)** The volume formed by revolving $R$ about the $y$-axis.

**\*8.** An electron at position $x$ on the $x$-axis is attracted to a proton at the origin by a force $kx^{-2}$; $k$ is a constant. Compute the work done
   **a)** By an agent removing the electron from $x = 5$ to $x = +\infty$.
   **b)** By the force between proton and electron in moving the electron from $x = 5$ to the origin.

## B

**9.** For what constants $p \ne 1$ is the given integral convergent?

a) $\displaystyle\int_2^\infty x^{-p}\,dx$  b) $\displaystyle\int_0^2 x^{-p}\,dx$

c) $\displaystyle\int_{-2}^2 |x|^{-p}\,dx$  d) $\displaystyle\int_0^\infty x^{-p}\,dx$

10. *Torricelli's law* describes the velocity of water flowing out of a hole in a container (Fig. 1); the velocity equals that which would be achieved by falling from height $h$, where $h$ is the height from the hole to the surface of the water.

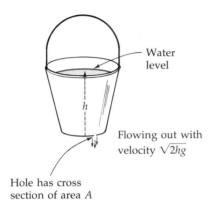

**FIGURE 1**
Water flowing out of a container with a hole.

a) Show that the velocity is $v = \sqrt{2hg}$, where $g$ is the acceleration due to gravity.
b) Suppose that the cross section of the hole

has area $A_H$. Explain why the rate of flow through the hole is $\sqrt{2hg}\,A_H$.

11. According to Torricelli's law (previous problem), when water flows out through a small hole of cross section $A_H$ (Fig. 1), the volume rate of flow is $\sqrt{2gh}\,A_H$; $h$ is the height from the hole to the surface of the water. Compute the rate at which the height $h$ falls, if the hole is in the bottom of
a) A cylinder of radius $r$, with vertical axis.
b) A sphere of radius $r$.
c) A cone with vertex down, its sides making an angle $\theta$ with the vertical.

12. A *water clock* consists of a water tank with a small hole at the bottom, designed so that the water level drops at a constant rate. According to Torricelli's law [see the preceeding problems], when the tank is filled to height $h$, then the water runs out at the rate of $(\sqrt{2g}\,A_H)\sqrt{h}$ cm³/sec; $A_H$ is the area of the cross section of the hole.

Suppose that the tank has the shape formed by revolving the graph of $x = f(y)$ about the (vertical) $y$-axis; $f \geq 0$, and the hole is at $y = 0$.
a) Let $V(h)$ denote the volume of water in the tank when it is filled to height $h$. Compute $dV/dh$.
b) Determine the function $f(y)$ so that the water level $h$ falls at the constant rate of 1 cm/hr.

# 8

# EXPONENTIALS, LOGARITHMS, AND INVERSE FUNCTIONS

Exponentials and logarithms arise in every area of mathematical application: physics, chemistry, biology, demography, statistics, and so on. We shall compute the derivatives of these functions, and of the inverse trigonometric functions, and show some of the applications.

We assume that you are familiar with the laws of exponents, reviewed in Appendix B:

$$b^{x+y} = b^x b^y$$

$$(b^x)^y = b^{xy}, \qquad a^x b^x = (ab)^x$$

$$b^0 = 1, \qquad b^{-x} = \frac{1}{b^x}.$$

## 8.1

## THE NATURAL EXPONENTIAL FUNCTION

Each number $b > 0$ determines a function

$$f(x) = b^x$$

called the *exponential function with base b*. If the base $b = \frac{1}{2}$, then

$$f(-1) = \left(\frac{1}{2}\right)^{-1} = \frac{1}{1/2} = 2, \qquad f(0) = \left(\frac{1}{2}\right)^0 = 1, \qquad f(1) = \left(\frac{1}{2}\right)^1 = \frac{1}{2},$$

and the graph is decreasing as in Figure 1a. If $b = 1$, then $f(x) = 1^x = 1$ for all $x$, and the graph is horizontal (Fig. 1b). If $b = 2$, then

$$f(-1) = 2^{-1} = \tfrac{1}{2}, \qquad f(0) = 2^0 = 1, \qquad f(1) = 2^1 = 2,$$

and the graph is increasing as in Figure 1c. If $b = 4$, then it increases even more rapidly (Fig. 1d).

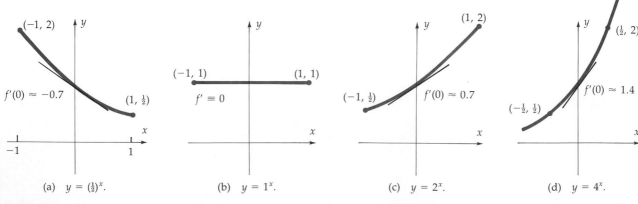

(a)  $y = (\tfrac{1}{2})^x$.  (b)  $y = 1^x$.  (c)  $y = 2^x$.  (d)  $y = 4^x$.

FIGURE 1
$y = b^x$.

The four graphs in Figure 1 illustrate that if $b < 1$ then $b^x$ is a decreasing function [see $(\frac{1}{2})^x$], and if $b > 1$ it is an increasing function. The figure suggests further that each graph has a tangent line at every point, and $f'(x)$ is defined for every $x$. This is true, but hard to prove; in this section we *assume* that $f'(x)$ exists and try to compute it.

Look first at $f'(0)$, the slope at the origin. For $(1/2)^x$, the figure shows $f'(0) < 0$; for $1^x$, $f'(0) = 0$; for $2^x$, $f'(0) > 0$; and for $4^x$, $f'(0)$ is still larger. In fact, by estimating the appropriate limits (problems 1 and 2) you can show that

$$\text{for } (\tfrac{1}{2})^x, \qquad f'(0) \approx -0.7$$
$$\text{for } 1^x, \qquad f'(0) = 0$$
$$\text{for } 2^x, \qquad f'(0) \approx 0.7$$
$$\text{for } 4^x, \qquad f'(0) \approx 1.4.$$

Notice that as the base $b$ is increased, the graph gets steeper at the origin, and $f'(0)$ is larger. Look particularly at $2^x$, giving

$$f'(0) \approx 0.7 < 1;$$

and $4^x$, giving

$$f'(0) \approx 1.4 > 1.$$

Thus at the origin, $2^x$ has slope $<1$ and $4^x$ has slope $>1$. Apparently, for some base between 2 and 4, $f'(0)$ is *exactly equal to 1. This base is denoted e* (for exponential). The function $e^x$ is called the **natural exponential function**. Its special property is (Fig. 2):

$$\text{If} \quad f(x) = e^x, \quad \text{then} \quad f'(0) = 1.$$

We rewrite this as a limit relation. By definition

$$f'(0) = \lim_{h \to 0} \frac{f(h) - f(0)}{h - 0}. \tag{1}$$

In the present case $f(h) = e^h$, $f(0) = e^0 = 1$, and $f'(0) = 1$, so (1) gives

$$\lim_{h \to 0} \frac{e^h - 1}{h} = 1. \tag{2}$$

Now, using the limit (2) and the first law of exponents, we compute the derivative of $e^x$ at all other points $x$:

$$\frac{de^x}{dx} = f'(x) = \lim_{h \to 0} \frac{f(x + h) - f(x)}{h} = \lim_{h \to 0} \frac{e^{x+h} - e^x}{h}$$
$$= \lim_{h \to 0} \frac{e^x e^h - e^x}{h} = e^x \lim_{h \to 0} \frac{e^h - 1}{h}.$$

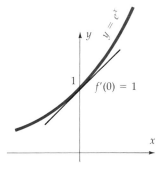

**FIGURE 2**
$f(x) = e^x$, $f'(0) = 1$.

(You can take $e^x$ outside the limit because only $h$, not $x$, varies in the limit process.) Thus from the limit relation (2),

$$\frac{de^x}{dx} = e^x. \tag{3}$$

*REMARK*   You could compute the derivative of the "common" exponential $g(x) = 10^x$ the same way:

$$\lim_{h \to 0} \frac{10^{x+h} - 10^x}{h} = \lim_{h \to 0} 10^x \frac{10^h - 1}{h}$$

$$= 10^x \lim_{h \to 0} \frac{10^h - 1}{h}.$$

It turns out that $\lim_{h \to 0} \dfrac{10^h - 1}{h} = 2.3025851$, to eight figures, so

$$\frac{d10^x}{dx} \approx (2.3025851)10^x.$$

This is certainly inconvenient, compared to formula (3); so in calculus, $e$ is the natural base to use. After all, the base 10 is "common" simply because we have five fingers on each hand!

---

**EXAMPLE 1**

$$\frac{d(x + e^x)}{dx} = \frac{dx}{dx} + \frac{de^x}{dx} = 1 + e^x$$

$$\frac{d(3e^x)}{dx} = 3\frac{de^x}{dx} = 3e^x$$

$$\frac{dx^2 e^x}{dx} = \left(\frac{dx^2}{dx}\right)e^x + x^2\frac{de^x}{dx} = 2xe^x + x^2 e^x = (2x + x^2)e^x$$

$$\frac{d\dfrac{e^x}{x}}{dx} = \frac{x\dfrac{de^x}{dx} - e^x\dfrac{dx}{dx}}{x^2} = \frac{xe^x - e^x}{x^2} = e^x\left(\frac{x - 1}{x^2}\right).$$

---

**Composite Functions $e^u$**

If $u$ is a function of $x$, then by the Chain Rule $\dfrac{de^u}{dx} = \dfrac{de^u}{du}\dfrac{du}{dx} = e^u\dfrac{du}{dx}$; briefly

$$\frac{de^u}{dx} = e^u\frac{du}{dx}.$$

**EXAMPLE 2**

$$\frac{de^{-3t}}{dt} = e^{-3t}\frac{d(-3t)}{dt} = -3e^{-3t}$$

$$\frac{de^{kt}}{dt} = e^{kt}\frac{dkt}{dt} = ke^{kt}, \qquad \text{for any constant } k.$$

**EXAMPLE 3**   $\dfrac{de^{x^2}}{dx} = ?$

*SOLUTION*   Before taking the derivative, make sure what the notation means; $e^{x^2}$ is $e^{(x^2)}$, so

$$\frac{de^{x^2}}{dx} = e^{x^2}\frac{dx^2}{dx} = e^{x^2}(2x)$$

$$= 2xe^{x^2}.$$

Question: What is the derivative of $(e^x)^2$? [Answer: $2e^{2x}$].

## Alternate Notation

The "natural exponential function" $e^x$ is sometimes denoted $\exp(x)$; by definition

$$\exp(x) = e^x.$$

Thus $\dfrac{d\exp(u)}{dx} = \exp(u)\dfrac{du}{dx}$. For example, $\dfrac{d\exp(x^2)}{dx} = \exp(x^2)\cdot 2x.$

**EXAMPLE 4**

$$\frac{d\exp(e^{x^2})}{dx} = \exp(e^{x^2})\frac{de^{x^2}}{dx} = \exp(e^{x^2})(e^{x^2})(2x).$$

## Limits

The graph of $y = e^x$ in Figure 2 suggests that

$$\lim_{x \to +\infty} e^x = +\infty. \tag{4}$$

Figure 3 illustrates a proof. The graph of $e^x$ is concave up, since

$$\frac{d^2 e^x}{dx^2} = \frac{d}{dx}\left(\frac{de^x}{dx}\right) = \frac{d}{dx}(e^x) = e^x > 0.$$

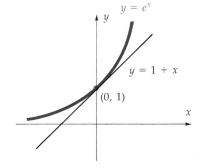

**FIGURE 3**
$e^x > 1 + x$ if $x \neq 0$.

Hence, the entire graph lies above its tangent line at $(0, 1)$. The slope at $x = 0$ is 1, so the tangent line there is $y = 1 + x$. Since the entire graph lies above this line,

$$e^x > 1 + x, \qquad \text{for } x \neq 0. \tag{5}$$

But $\lim\limits_{x \to +\infty} (1 + x) = +\infty$; since $e^x$ is even larger than $1 + x$, it follows that

$$\lim_{x \to +\infty} e^x = +\infty$$

as well. Taking reciprocals, you find

$$\lim_{x \to +\infty} e^{-x} = \lim_{x \to +\infty} \frac{1}{e^x} = 0. \tag{6}$$

Replacing $x$ by $-x$ in (4) and (6) gives

$$\lim_{x \to -\infty} e^{-x} = +\infty \tag{7}$$

$$\lim_{x \to -\infty} e^x = 0. \tag{8}$$

Many other limits follow, by l'Hôpital's Rule (Sec. 4.3).

---

**EXAMPLE 5**   Evaluate $\lim\limits_{x \to +\infty} xe^{-x}$.

**SOLUTION**   To apply l'Hôpital, you need to write the expression as a quotient:

$$\lim_{x \to +\infty} xe^{-x} = \lim_{x \to +\infty} \frac{x}{e^x} \qquad \left[ \frac{\infty}{\infty}, \text{ so you can l'Hôpitalize} \right]$$

$$= \lim_{x \to +\infty} \frac{1}{e^x} \qquad \left[ \frac{\text{finite}}{\infty} \right]$$

$$= 0.$$

---

The limit

$$\lim_{x \to \infty} \frac{x}{e^x} = 0$$

in Example 5 can be viewed as a contest between $x$ and $e^x$. Both go to $+\infty$, but apparently the denominator $e^x$ goes faster than the numerator $x$, since $\dfrac{x}{e^x}$ goes to 0. In fact, $e^x \to \infty$ faster than *any* power $x^n$. That is,

$$\lim_{x \to \infty} \frac{x^n}{e^x} = 0, \qquad \text{for every } n$$

(see problems 5 and 15).

## Calculation of $e$

We defined $e$ indirectly, by requiring that the slope of the graph of $e^x$ at $x = 0$ be 1. How, then, can this fundamental number be computed? One method uses the inequality (5):

$$e^x > 1 + x \qquad \text{for } x \neq 0.$$

Since we know the slope of the graph at $x = 0$, it seems reasonable to use this inequality for $x$ *near* 0. Specifically, set $x = 1/n$ to find

$$e^{1/n} > 1 + \frac{1}{n}.$$

Raise both sides to the $n^{\text{th}}$ power:

$$e > \left(1 + \frac{1}{n}\right)^n. \tag{9}$$

By a similar argument, using (5) with $x = -\dfrac{1}{n+1}$, you can find

$$e < \left(1 + \frac{1}{n}\right)^{n+1}. \tag{10}$$

Now use (9) and (10) with various values of $n$, to get the following estimates for $e$:

$n = 2$:     $(1 + \frac{1}{2})^2 < e < (1 + \frac{1}{2})^3$, or $2.25 < e < 3.375$

$n = 5$:     $(1 + \frac{1}{5})^5 < e < (1 + \frac{1}{5})^6$, or $2.48 \ldots < e < 2.98 \ldots$

$n = 100$:     $(1.01)^{100} < e < (1.01)^{101}$, or $2.70 \ldots < e < 2.73 \ldots$

$n = 1024$:     $\ldots \ldots \ldots 2.716 \ldots < e < 2.719 \ldots$

So, rounded off to two decimals,

$$e \approx 2.72.$$

These calculations suggest that $e$ is the *limit* of the numbers $e_n = \left(1 + \dfrac{1}{n}\right)^n$ as $n \to \infty$. This can be proved, using the Trapping Theorem. Rewrite (9) as

$$e_n < e. \tag{11}$$

Likewise, rewrite (10) as $e < \left(1 + \dfrac{1}{n}\right)^n \left(1 + \dfrac{1}{n}\right) = e_n\left(1 + \dfrac{1}{n}\right)$, which implies

$$\frac{e}{1 + \dfrac{1}{n}} < e_n. \tag{12}$$

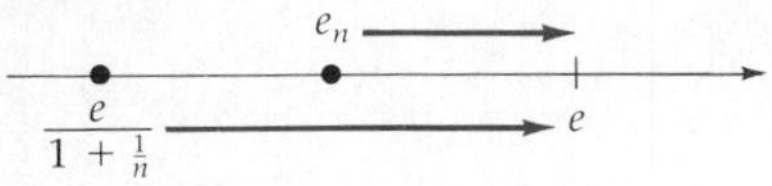

**FIGURE 4**

$$\lim_{n \to \infty} \frac{e}{1 + 1/n} = e, \text{ so } \lim_{n \to \infty} e_n = e.$$

Combining (11) and (12) gives $\dfrac{e}{1 + \dfrac{1}{n}} < e_n < e$ as shown in Figure 4. As

$n \to \infty$ then $\dfrac{1}{n} \to 0$, so $1 + \dfrac{1}{n} \to 1$ and

$$\frac{e}{1 + \dfrac{1}{n}} \to \frac{e}{1} = e.$$

That is, as $n \to \infty$, the left-hand point in Figure 4 approaches the right-hand point $e$. By the Trapping Theorem, the numbers $e_n = \left(1 + \dfrac{1}{n}\right)^n$ must also approach $e$ as $n \to \infty$, and this is what we wanted to prove:

$$e = \lim_{n \to \infty} \left(1 + \frac{1}{n}\right)^n. \tag{13}$$

### Compound Interest

John Napier, the inventor of logarithms, asked himself the following question: If you earn interest at 100% per year, compounded $n$ times a year, and $n$ is very large, can you possibly triple your investment in one year? The answer uses the inequality (9).

First, recall the notions of simple and compound interest. At *simple* interest, with rate $r$ per year, an investment $P_0$ (the "principal") earns interest of $rP_0$ in one year; added to the principal, this gives a total of $P_0 + rP_0 = (1 + r)P_0$ in your account after one year. But if interest is *compounded monthly*, then the rate $r$ per year is considered $r/12$ per month; a first payment of $r/12$ is made after one month, so at that time the balance in your account is

$$P_0 + \frac{r}{12} P_0 = \left(1 + \frac{r}{12}\right) P_0 = \begin{bmatrix} \text{balance at end} \\ \text{of first month} \end{bmatrix}.$$

This new balance earns interest at the rate $r/12$ per month; so after the second month, you earn additional interest of $\left(\dfrac{r}{12}\right)\left(1 + \dfrac{r}{12}\right)P_0$, giving

$$\begin{bmatrix} \text{balance at start} \\ \text{of second month} \end{bmatrix} + \begin{bmatrix} \text{interest earned} \\ \text{in second month} \end{bmatrix}$$

$$= \left(1 + \frac{r}{12}\right)P_0 + \frac{r}{12}\left(1 + \frac{r}{12}\right)P_0$$

$$= \left(1 + \frac{r}{12}\right)\left[\left(1 + \frac{r}{12}\right)P_0\right]$$

$$= \left(1 + \frac{r}{12}\right)^2 P_0 = \begin{bmatrix} \text{balance at end} \\ \text{of second month} \end{bmatrix}.$$

Thus at the end of each month, the balance in your account is multiplied by $(1 + r/12)$. After twelve months, compounding monthly at rate $r$ will yield a balance of $\left(1 + \dfrac{r}{12}\right)^{12} P_0$.

By a similar analysis, compounding $n$ times at rate $r$ per year yields, after one year, the balance

$$\left(1 + \frac{r}{n}\right)^n P_0;$$

the principal $P_0$ is multiplied by the factor $\left(1 + \dfrac{r}{n}\right)^n$.

Return now to Napier's question about 100% interest. In this case, the interest rate is $r = 1$; so compounding $n$ times per year multiplies your principal by $\left(1 + \dfrac{1}{n}\right)^n$ by the end of a year. To triple the principal you need $\left(1 + \dfrac{1}{n}\right)^n \geq 3$. But the inequality (9) says that $\left(1 + \dfrac{1}{n}\right)^n < e$, and we found that $e < 2.72$, so $\left(1 + \dfrac{1}{n}\right)^n$ cannot possibly be $\geq 3$. This is what Napier concluded; at 100% interest, compounded as often as you wish, you still multiply your principal by less than $e \approx 2.72$.

## SUMMARY

$$\frac{de^u}{dx} = e^u \frac{du}{dx}$$

$$\lim_{x \to +\infty} e^x = +\infty \qquad \lim_{x \to -\infty} e^x = 0$$

$$\exp(x) = e^x.$$

$$e = \lim_{n \to \infty} \left(1 + \frac{1}{n}\right)^n$$

## PROBLEMS

### A

1. Estimate $f'(0)$ for each of the following functions by computing the difference quotients (Fig. 5)

$$\frac{f(x) - f(0)}{x - 0}$$

both with $x = 0.1$ and with $x = -0.1$. One of these estimates is larger than the true value of $f'(0)$, and the other is smaller. Which of the two is larger? What feature of the graph explains this?

a)  $f(x) = (1/2)^x$    b)  $f(x) = 1^x$

*c)  $f(x) = 2^x$    *d)  $f(x) = 3^x$

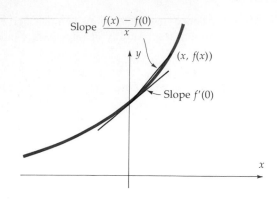

Slope $\dfrac{f(x) - f(0)}{x}$

$(x, f(x))$

Slope $f'(0)$

**FIGURE 5**

$\dfrac{f(x) - f(0)}{x} \approx f'(0)$ when $x$ is small.

e)   $f(x) = 4^x$

    $(2^{0.1} \approx 1.07177, \qquad 2^{-0.1} \approx 0.9330,$
    $\phantom{(}3^{0.1} \approx 1.116, \qquad\quad 3^{-0.1} \approx 0.8960,$
    $\phantom{(}4^{0.1} \approx 1.1487, \qquad 4^{-0.1} \approx 0.8706)$

2.  Estimate $f'(0)$ for $f(x) = 10^x$, using the difference quotients with
   *a)  $x = \pm 0.01$      b)  $x = \pm 0.001$

3.  Compute the first and second derivatives.
   *a)  $2xe^x$               *b)  $e^{x^2}$

   *c)  $(e^x)^2$            *d)  $x^2e^x + \dfrac{1}{x}$

    e)  $e^{5x}$              f)  $\sqrt{1 + 2e^{3x}}$
   *g)  $(2 + e^{3t})^5$      h)  $e^{5t} - 2e^{-5t}$
    i)  $e^{\sqrt{1+x}}$        j)  $e^{-x^2}$

    k)  $\dfrac{e^{2x}}{x}$           *l)  $\exp(1 + x^2)$

    m)  $\exp(2x^3 - x)$   *n)  $\exp(e^x)$
    o)  $\exp(\exp(e^x))$

4.  Compute first and second derivatives.
   *a)  $e^{\sin\theta}$        b)  $e^{\cos(2t)}$
   *c)  $e^{-t}\sin t$     d)  $e^{-2t}\cos 7t$
   *e)  $e^{at}\sin(bt)$

5.  Graph, using the signs of $y$, $dy/dx$, and $d^2y/dx^2$, and the limits as $x \to +\infty$ and $x \to -\infty$.
   *a)  $xe^x$     b)  $xe^{-x}$   *c)  $x^2e^x$
    d)  $x^2e^{-x}$   *e)  $e^{-x^2}$    f)  $xe^{-x^2}$

6.  From $\lim\limits_{x\to+\infty} e^{-x} = 0$ and $\lim\limits_{x\to+\infty} e^x = +\infty$ deduce the following limits.
   *a)  $\lim\limits_{t\to+\infty} e^{-2t}$     b)  $\lim\limits_{t\to+\infty} (2 + 3e^{2t})$
   *c)  $\lim\limits_{t\to-\infty} (3 + 2e^{-5t})$   d)  $\lim\limits_{t\to+\infty} (3 + 2e^{-5t})$

   *e)  $\lim\limits_{t\to+\infty} (e^t + e^{-t})$   *f)  $\lim\limits_{t\to+\infty} \dfrac{e^t - e^{-t}}{e^t + e^{-t}}$
    g)  $\lim\limits_{x\to+\infty} e^{-x^2}$     *h)  $\lim\limits_{x\to-\infty} e^{-x^2}$
    i)  $\lim\limits_{t\to+\infty} \dfrac{1}{e^{2t}}$     *j)  $\lim\limits_{t\to+\infty} \dfrac{1}{e^{-2t}}$

7.  Evaluate, or explain why the limit does not exist.
   *a)  $\lim\limits_{t\to+\infty} e^{-2t}\sin 3t$
   *b)  $\lim\limits_{t\to+\infty} e^{3t}\sin(-2t)$
    c)  $\lim\limits_{t\to+\infty} e^{-at}\cos bt$   $(a > 0, b$ any constant$)$
    d)  $\lim\limits_{t\to+\infty} e^{at}\cos bt$   $(a > 0, b$ constant$)$

8.  Graph the following, showing regions of increase and decrease, and horizontal asymptotes.
   *a)  $x = 1 + e^{-t}$     b)  $x = 2 + 3e^{2t}$
    c)  $x = 3 - 2e^{3t}$     d)  $y = -1 + 2e^{-3t}$

*9.  The rate of a certain reaction depends on temperature. If $T$ is the temperature in degrees Kelvin and $k$ is the rate, then

$$k = A \exp(-E_A/RT),$$

where $A$, $E_A$, and $R$ are constants. At a certain time, $T = 300°$ and is increasing at $3°/\text{sec}$. How fast is $k$ changing?

10.  The hydrogen atom consists of a heavy nucleus and an electron, which we imagine as orbiting around the nucleus. When the atom is in its "unexcited" state, then the probability of finding the electron at distance $r$ from the nucleus is described by the function

$$f(r) = 4r^2 e^{-2r},$$

where $r$ is measured in units of the *Bohr radius*, about $5.3 \times 10^{-11}$ meters. For what value of $r$ is this function maximum? (You can think of this value of $r$ as the most likely distance from nucleus to electron. See Sec. 10.4 for the precise meaning of $f$.)

*11.  The function

$$P(v) = 4\pi \frac{M}{(2\pi kT)^{3/2}} v^2 e^{-mv^2/2kT}$$

describes the probability that a gas molecule will have a speed very near $v$. The letters $M$, $m$, $k$, $T$ denote constants describing the nature and state of the gas. What is the most probable speed of such a gas molecule?

**12.** Suppose that $f(t) = Ce^{kt}$, where $C$ and $k$ are constants. Prove that $f'(t) = kf(t)$.

**B**

**13.** Suppose that $f$ is a function that satisfies the condition $f'(t) = kf(t)$, for all $t$.
   a) Prove that $e^{-kt}f(t)$ is a constant. (Hint: Take the derivative of $e^{-kt}f(t)$, using the Product Rule and the given condition on $f$.)
   b) From part a, deduce that $f(t) = Ce^{kt}$ for some constant $C$.

**14.** A rotating molecule has different possible energy levels $E_1, E_2, E_3, \ldots, E_j, \ldots$. In a gas consisting of such molecules, the amount of rotational energy at level $E_j$ is

$$f(j) = \frac{h^2}{2IkT}(2j + 1)\exp(-j(j + 1)h^2/2IkT).$$

The symbol $h$ denotes Planck's constant, and $I$, $k$, $T$ are constants giving the nature and state of the gas. For what value of $j$ is $f(j)$ largest? [The letter $j$ stands for a positive integer, but treat it as a real number varying in the interval $(0, \infty)$.]

**15.** a) Graph $y = x^{n+1}e^{-x}$ for $x \geq 0$, using the signs of $y$ and $dy/dx$. Assume $n$ is an integer $\geq 1$.
   b) From your graph in part a, deduce that $0 \leq x^{n+1}e^{-x} \leq (n + 1)^{n+1}e^{-n-1}$.
   c) From part b, deduce that

   $$0 \leq x^n e^{-x} \leq \frac{1}{x}(n + 1)^{n+1}e^{-n-1},$$

   hence $\lim_{n \to \infty} x^n e^{-x} = 0$. (This means that $e^{-x} \to 0$ faster than $x^n \to \infty$.)
   d) From part b, deduce that $e^x \geq x^n$ if $x \geq \left(\dfrac{n + 1}{e}\right)^{n+1}$. (This means that as $x \to +\infty$, $e^x$ is eventually larger than any given power of $x$.)

**16.** Prove the inequality (10), $e < \left(1 + \dfrac{1}{n}\right)^{n+1}$.

   $\left[\text{Use inequality (5) with } x = -\dfrac{1}{n + 1}.\right]$

**17.** Prove the following, using inequality (5).
   a) $e > (1 + h)^{1/h}$
   b) $e^{-h/(1+h)} > \dfrac{1}{1 + h}$

   c) $e < (1 + h)^{1+1/h}$

   d) $\dfrac{e}{1 + h} < (1 + h)^{1/h} < e$

   e) $\lim_{h \to 0}(1 + h)^{1/h} = e$

**18.** Use the limit proved in the previous problem to show that

$$\lim_{n \to \infty}\left(1 + \frac{x}{n}\right)^n = e^x.$$

$\left[\text{Hint: }\left(1 + \dfrac{x}{n}\right)^n = \left(\left(1 + \dfrac{x}{n}\right)^{n/x}\right)^x.\right]$

**19.** a) Explain: When $r$ is small, then $(e^r - 1)/r$ is close to 1, so $e^r - 1$ is very close to $r$; that is, $e^r$ is very close to $1 + r$.
   b) Use a calculator to check this when $r = .1$ and $r = .01$. (On some calculators, you get $e^x$ by entering $x$, then pressing "INV" and "ln". The fact is, $e^x$ is the inverse of the natural logarithm $\ln x$; see Sec. 8.2.)

**C**

The remaining problems give a sophisticated, efficient way to compute $e^x$.

**20.** For $x > 0$, show that
   a) $e^x < 1 + xe^x$ (Hint: Both functions equal 1 when $x = 0$. Compare their derivatives for $x > 0$.)

   b) $1 + x < e^x < 1 + x + \dfrac{x^2}{2}e^x$ (Again, compare the derivatives.)

   c) $1 + x + \dfrac{x^2}{2} < e^x < 1 + x + \dfrac{x^2}{2} + \dfrac{x^3}{3 \cdot 2}e^x$

   d) $1 + x + \cdots + \dfrac{x^n}{n!} < e^x$

   $< 1 + x + \cdots + \dfrac{x^n}{n!} + \dfrac{x^{n+1}}{(n + 1)!}e^x$

   where $n! = n(n - 1) \ldots (2)(1)$

   e) $1 + x + \cdots + \dfrac{x^n}{n!} < e^x < \left(1 + x + \dfrac{x^2}{2}\right.$

   $\left. + \cdots + \dfrac{x^n}{n!}\right)\bigg/(1 - x^{n+1}/(n + 1)!)$

**21.** Use problem 20e to show
   a) $2.5 < e < 3$ (Use $n = 2$.)
   b) $2.71666 \ldots < e < 2.721$ (Use $n = 5$.)

c) $2.7182818 < e < 2.7182818 + 8 \times 10^{-8}$
[This method gives quicker results than the method in the text,
$$\left(1 + \frac{1}{n}\right)^n < e < \left(1 + \frac{1}{n}\right)^{n+1}.\right]$$

22. Use problem 20e to estimate the following:
a) $e^{0.5}$ (Use $n = 3$.)
b) $e^{0.1}$ Determine this to 8 decimal places, using whatever value of $n$ does the job.

23. a) Show that if $x < 0$ then
$$1 + x + \cdots + \frac{x^{2k-1}}{(2k-1)!} < e^x$$
$$< 1 + x + \cdots + \frac{x^{2k}}{(2k)!}.$$

b) Compute $e^{-1}$ to four decimal places.

## 8.2

## INVERSE FUNCTIONS

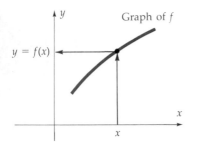

FIGURE 1

To read a graph, you start at a point $x$ on the horizontal axis, go up to the graph, then over to the vertical axis, arriving at the point $y = f(x)$ (Fig. 1). Now, read the graph *backward* as in Figure 2. Start at $y$, go over to the graph, then down to the horizontal axis; you arrive at a point $x$ which is a function of $y$, $x = g(y)$. This is called the **inverse** of the function $f$. The two equations

$$y = f(x) \quad \text{and} \quad x = g(y)$$

are equivalent; if you solve $y = f(x)$ for $x$, you get $x = g(y)$.

You can't do this with every function $f$: If some $y$ value comes from two different values $x_1$ and $x_2$ (Fig. 3), then is $g(y) = x_1$, or $g(y) = x_2$? An *inverse function for $f$ is defined if and only if each $y$ in the range of $f$ comes from just one $x$ in the domain.* A function $f$ with this property is called **one-to-one**.

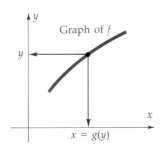

FIGURE 2
$g$ is the inverse of $f$.

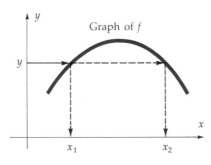

FIGURE 3
$f$ is not one-to-one.

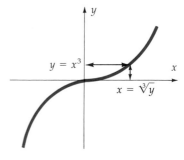

FIGURE 4
$y = x^3$ and its inverse, $x = \sqrt[3]{y}$.

**EXAMPLE 1** The function $f(x) = x^3$ is one-to-one (Fig. 4). Find the inverse function to $f$.

**SOLUTION** Solve $y = x^3$ for $x$, and you get $x = \sqrt[3]{y}$. The inverse of the cube function $f(x) = x^3$ is the cube root, $g(y) = \sqrt[3]{y} = y^{1/3}$.

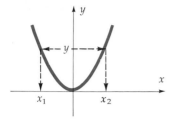

**FIGURE 5**
$f(x) = x^2$, $-\infty < x < \infty$, is not
one-to-one.

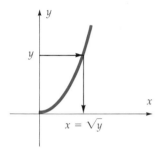

**FIGURE 6**
$f(x) = x^2$, $0 \le x < \infty$, is one-to-one.

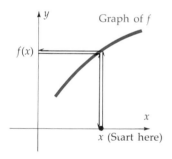

**FIGURE 7**
$x = g(f(x))$.

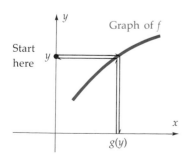

**FIGURE 8**
$y = f(g(y))$.

**EXAMPLE 2**  $f(x) = x^2$, $-\infty < x < \infty$, does not have an inverse, because each $y > 0$ comes from two different $x$'s (Fig. 5). But you can remove the left side of the graph by restricting the domain to $x \ge 0$ (Fig. 6). The restricted function

$$f(x) = x^2, \qquad x \ge 0$$

*is* one-to-one, so it has an inverse. Solving $y = x^2$ for $x \ge 0$ gives $x = \sqrt{y}$, so the inverse is the square root function

$$g(y) = \sqrt{y} = y^{1/2}.$$

**EXAMPLE 3**  The exponential function $f(x) = e^x$ is one-to-one; its inverse is the *natural logarithm*, studied in detail in the next section.

Figure 7 shows how the inverse $g$ "undoes" the original function $f$. Start at point $x$, read the graph forward to get $f(x)$, then read it backward to get $g(f(x))$. You end up where you started, so

$$g(f(x)) = x. \tag{1}$$

And Figure 8 shows that $f$ "undoes" $g$:

$$f(g(y)) = y. \tag{2}$$

For the function $f(x) = x^3$, these equations are

$$\sqrt[3]{x^3} = x \quad \text{and} \quad (\sqrt[3]{y})^3 = y.$$

For $f(x) = x^2$, $x \ge 0$, they are

$$\sqrt{x^2} = x, x \ge 0 \quad \text{and} \quad (\sqrt{y})^2 = y, y \ge 0.$$

The restriction $x \ge 0$ here is essential. For example, $\sqrt{(-1)^2} = \sqrt{1} = +1$. Generally, $\sqrt{x^2} = -x$ if $x$ is negative; in brief, $\sqrt{x^2} = |x|$.

### The Graph of the Inverse Function

When you read the graph of $f$ backward (Fig. 9), the inputs $y$ are on the vertical axis, and the outputs on the horizontal axis. To get a proper graph of the inverse function $x = g(y)$, you have to exchange the positions of the $x$- and $y$-axes; reflect the entire picture across the 45° line $y = x$, as in Figure 10.

In Figure 10, the horizontal axis is now the $y$-axis; this doesn't really matter. But if you want, you can exchange the $x$ and $y$ labels (Fig. 11); the equation of the graph of $g$ is then $y = g(x)$. To obtain Figure 11 directly from Figure 9, reflect only the *graph* across the 45° line, leaving the axes alone.

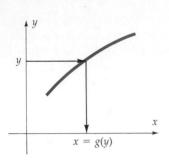

**FIGURE 9**

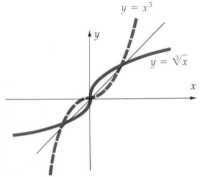

**FIGURE 12**

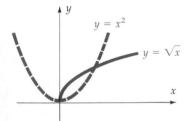

**FIGURE 13**
Graph of the square root.

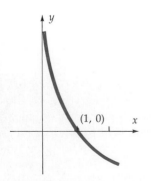

**FIGURE 14**
$y = \dfrac{1}{2x} - \dfrac{x}{2}$, $x > 0$.

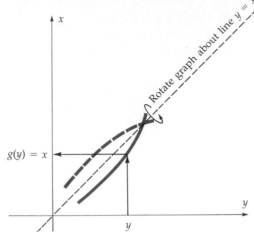

**FIGURE 10**
Graph of $x = g(y)$.

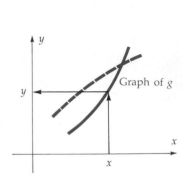

**FIGURE 11**
Graph of $y = g(x)$.

Thus the graph of the cube root $y = x^{1/3}$ (Fig. 12) is the reflection of the graph of $y = x^3$. And the graph of the square root $y = \sqrt{x}$ (Fig. 13) is half of a parabola, the reflection of the right half of the graph of $y = x^2$.

**EXAMPLE 4** The function

$$f(x) = \frac{1}{2x} - \frac{x}{2}, \qquad x > 0$$

is one-to-one (Fig. 14). Find, if possible, a formula for its inverse, and sketch the graph of the inverse.

**SOLUTION** By solving $y = f(x)$ for $x$, you get the inverse $x = g(y)$. In the present example, $y = \dfrac{1}{2x} - \dfrac{x}{2}$ can be solved by rewriting the equation as a quadratic in $x$:

$$y = \frac{1}{2x} - \frac{x}{2} \Leftrightarrow 2xy = 1 - x^2 \Leftrightarrow x^2 + 2xy - 1 = 0$$

$$\Leftrightarrow x = \frac{-2y \pm \sqrt{4y^2 + 4}}{2} = -y \pm \sqrt{y^2 + 1}.$$

Since $x > 0$ in the given definition of $f$, then in the "$\pm$" we must choose the sign making $x$ positive. Thus

$$g(y) = -y + \sqrt{y^2 + 1} = \sqrt{y^2 + 1} - y.$$

This is graphed in Figure 15. Exchange the letters $x$ and $y$, and you get the relation

$$y = \sqrt{x^2 + 1} - x$$

graphed in Figure 16.

### The Derivative of the Inverse Function

Suppose that $f$ is defined on an interval and has a derivative $f'$, which is positive everywhere. Then $f$ is increasing (Fig. 17), so it is one-to-one, and has an inverse $g$. We will compute $g'$.

Figure 18 is obtained from Figure 17 by reflecting both graph and axes across the line $y = x$. In both figures, $\Delta x$ and $\Delta y$ represent corresponding changes in $x$ and $y$. But in Figure 18 the slope is $\dfrac{\Delta x}{\Delta y}$ $\left(\text{not } \dfrac{\Delta y}{\Delta x}\right)$ because the $x$-axis is vertical and the $y$-axis horizontal; so

$$g'(y) = \lim_{\Delta y \to 0} \frac{\Delta x}{\Delta y} = \frac{1}{\displaystyle\lim_{\Delta y \to 0} \frac{\Delta y}{\Delta x}}.$$

As $\Delta y \to 0$ then $\Delta x \to 0$ too (try to visualize this in Fig. 18), so

$$g'(y) = \frac{1}{\displaystyle\lim_{\Delta x \to 0} \frac{\Delta y}{\Delta x}} = \frac{1}{f'(x)}.$$

That is

$$g'(y) = \frac{1}{f'(x)}, \qquad \text{where } y = f(x) \text{ and } x = g(y). \tag{3}$$

In Leibniz notation, $g'(y) = \dfrac{dx}{dy}$ and $f'(x) = \dfrac{dy}{dx}$, so formula (3) is

$$\frac{dx}{dy} = \frac{1}{dy/dx}.$$

We began by supposing that $f'$ is positive everywhere. If, on the other hand, $f'$ is *negative* everywhere, then $f$ is decreasing, so again it is one-to-one, and has an inverse $g$, with derivative given by (3).

The ideas sketched here are stated precisely as follows:

---

**THE INVERSE FUNCTION THEOREM**

Suppose that the domain of $f$ is an interval, and $f' > 0$ everywhere, or $f' < 0$ everywhere. Then $f$ has an *inverse* $g$, defined on an interval, such that

$$g(f(x)) = x \quad \text{and} \quad f(g(y)) = y$$

and

$$g'(y) = \frac{1}{f'(x)} \quad \text{with} \quad y = f(x). \tag{3}$$

---

The rigorous proof is outlined in the problems of Appendix A.5.

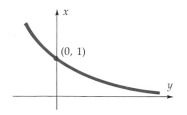

**FIGURE 15**
$x = \sqrt{y^2 + 1} - y$.

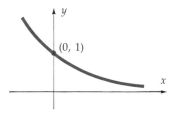

**FIGURE 16**
$y = \sqrt{x^2 + 1} - x$.

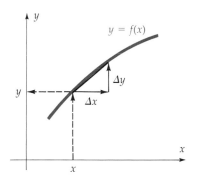

**FIGURE 17**

On graph of $y = f(x)$, slope $= \dfrac{\Delta y}{\Delta x}$.

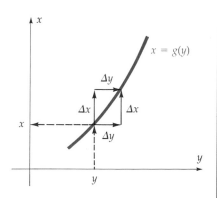

**FIGURE 18**

On graph of $x = g(y)$, slope $= \dfrac{\Delta x}{\Delta y}$.

*REMARK* In (3), $g'$ is evaluated at $y$, while $f'$ is evaluated at $x$, not $y$. Since $x = g(y)$, you can rewrite (3) as

$$g'(y) = \frac{1}{f'(g(y))}.$$

This equation follows directly from the formula

$$f(g(y)) = y. \tag{4}$$

Differentiate each side of (4) with respect to $y$, by the Chain Rule:

$$f'(g(y))g'(y) = 1, \qquad \text{so } g'(y) = \frac{1}{f'(g(y))}.$$

---

**EXAMPLE 5** *The derivative of $g(y) = y^{1/n}$.* For any $n = 1, 2, \ldots$, the function $g(y) = y^{1/n}$ is defined as the inverse of

$$f(x) = x^n, \qquad 0 \le x < \infty.$$

Its derivative can be computed by differentiating $(y^{1/n})^n = y$, with respect to $y$:

$$n(y^{1/n})^{n-1} \frac{d(y^{1/n})}{dy} = 1.$$

Solve this for the desired derivative, and use the laws of exponents:

$$\frac{d(y^{1/n})}{dy} = \frac{1}{n(y^{1/n})^{n-1}} = \frac{1}{ny^{(n-1)/n}} = \frac{1}{n} y^{(1-n)/n}$$

Rewrite the last exponent to find

$$\frac{dy^{1/n}}{dy} = \frac{1}{n} y^{(1/n)-1},$$

confirming the Power Rule.

---

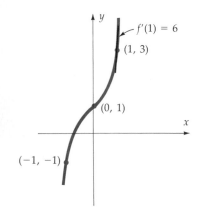

**FIGURE 19**
$f(x) = x^5 + x + 1$.

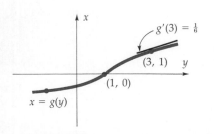

**FIGURE 20**
$g$ is inverse to $f$ in Figure 19.

**EXAMPLE 6** Show that the function $f(x) = x^5 + x + 1$ is one-to-one, so it has an inverse $g$. Compute $g'(3)$.

*SOLUTION* $f'(x) = 5x^4 + 1 \ge 1$, since $x^4 \ge 0$. Since $f'(x)$ exists and is positive for all $x$, the function $f$ is increasing (Fig. 19), hence one-to-one, and has an inverse $g$ (Fig. 20).

To compute $g'(3)$, notice that the point $(3, 1)$ on the graph of $g$ (Fig. 20) corresponds to the point $(1, 3)$ on the graph of $f$ (Fig. 19); for

$$f(1) = 1^5 + 1 + 1 = 3, \qquad \text{so } g(3) = 1.$$

Hence, $g'(3) = 1/f'(1)$. Since $f'(x) = 5x^4 + 1$, then $f'(1) = 6$, so $g'(3) = 1/6$ (Fig. 20).

In this example, we have no explicit formula for the inverse function. The equation $y = f(x) = x^5 + x + 1$ is of fifth degree in $x$, and there is no general method to solve such equations.

### Notation for the Inverse Function

The inverse to a function $f$ is often denoted $f^{-1}$:

$$y = f(x) \Leftrightarrow x = f^{-1}(y).$$

Unfortunately, this conflicts with the use of superscripts as powers; $f^{-1}(y)$ can also denote the *reciprocal* $1/f(y)$, not the inverse! To avoid this ambiguity, we have been using $g$ to stand for the function inverse to $f$. However, when we come to deal with a trigonometric function such as $y = \sin \theta$, the inverse function is commonly denoted $\theta = \sin^{-1} y$.

## SUMMARY

*Inverse Functions:*   If $f$ is defined on an interval $I$, and $f' > 0$ everywhere on $I$, or $f' < 0$ everywhere on $I$, then:

   **a)**   $f$ is one-to-one; each $y$ in the range of $f$ comes from just one $x$ in $I$.

   **b)**   $f$ has an *inverse* $g$, defined by

$$x = g(y) \Leftrightarrow y = f(x).$$

   **c)**   $f(g(y)) = y$ and $g(f(x)) = x$.

   **d)**   $g'(y) = \dfrac{1}{f'(x)}$ if $y = f(x)$.

In Leibniz notation, $\dfrac{dx}{dy} = \dfrac{1}{dy/dx}$.

The graph of $x = g(y)$ is obtained by reflecting the graph of $y = g(x)$, and the axes, across the line $y = x$.

## PROBLEMS

A

1.   Find a formula for the inverse of each of the following, solving $y = f(x)$ for $x$.

   *a)   $f(x) = 2x + 1$

   b)   $f(x) = x^3 + 1$

   *c)   $f(x) = (x + 1)^3 + 3$

   d)   $f(x) = \dfrac{1}{x} - x, \quad x > 0$

   *e)   $f(x) = \sqrt{1 - x^2}, \quad 0 \le x \le 1$

   f)   $f(x) = \sqrt{x^2 - 1} - x$

2.   The function $f(x) = 32 + \frac{9}{5}x$ gives $y =$ degrees Fahrenheit in terms of $x =$ degrees Celsius. What does the inverse function give? What is the formula for the inverse?

3.   Let $f$ be any linear function, not a constant; that is, $f(x) = mx + b$, with $m \ne 0$. Show that the inverse $g$ is linear, and determine its slope. Relate this to the formula $\dfrac{dx}{dy} = \dfrac{1}{dy/dx}$.

4. On a calculator, enter a positive number of your choice, then press "$\sqrt{\phantom{x}}$" followed by "$x^2$". What happens? Why? What if you press first "$x^2$" and then "$\sqrt{\phantom{x}}$"? What if you enter a negative $x$?

5. Why can $x^{1/n}$ be defined for all $x$ if $n$ is an odd integer, but only for $x \geq 0$ if $n$ is even?

6. Sketch the graph of the given function, and evaluate the required limit, or explain why it does not exist.

*a) $\displaystyle\lim_{x \to 0^+} \sqrt[4]{x}$     *b) $\displaystyle\lim_{x \to 0} \sqrt[4]{x}$

*c) $\displaystyle\lim_{x \to +\infty} \sqrt[4]{x}$     *d) $\displaystyle\lim_{x \to 0} \sqrt[5]{x}$

e) $\displaystyle\lim_{x \to +\infty} \sqrt[5]{x}$     f) $\displaystyle\lim_{x \to -\infty} \sqrt[5]{x}$

*g) $\displaystyle\lim_{x \to 0^+} x^{-1/2}$     h) $\displaystyle\lim_{x \to +\infty} x^{-1/2}$

7. *a) Write the equation of the line tangent to the graph of $y = \sqrt{x}$ at $(4, 2)$.

b) What is the relation between the slope of this line and the slope of the line tangent to the graph of $y = x^2$ at $(2, 4)$? Why?

8. a) Sketch the graph of $f(x) = \dfrac{1}{x + 1}$. Is $f$ one-to-one?

b) Write the equation of the line tangent to the graph of $f$ at $(1, 1/2)$.

c) Write the equation of the line tangent to the graph of the inverse function $g$ at $(1/2, 1)$. What is the relation of this line to the one in part b?

**B**

9. Suppose that $q = f(p)$ describes the quantity $q$ of widgets[1] that can be sold at price $p$ dollars per widget.

a) Normally, would you expect that $f$ is an increasing function, or decreasing, or neither?

b) What is the interpretation of the inverse function $p = g(q)$?

10. a) Show that the function $f(x) = x^3 + x$ is increasing on $-\infty < x < \infty$, hence has an inverse $x = g(y)$. (Do not try to write an explicit formula for $g(y)$.)

b) Sketch graphs of $f$ and $g$.

*c) Compute $g'(-2)$, $g'(0)$, and $g'(2)$.

11. Let $f(x) = \dfrac{1}{x^2} - x$, with domain $0 < x < \infty$.

a) Sketch the graph of $f$.

b) Sketch the graph of the inverse function $g$.

c) Compute $g'(0)$.

d) Compute $g'(-7/4)$.

---

# 8.3

## NATURAL LOGARITHMS

Logarithms were invented by John Napier (1550–1617) to help men like Kepler with their astronomical calculations. Henry Briggs simplified Napier's system by relating it to the exponential function and in 1627 published the first widely used tables of logarithms. Such tables remained important practically to the middle of the twentieth century. Now logarithms do their computational magic unseen, inside calculators and computers; but the logarithm function remains an essential theoretical tool in all the sciences.

The logarithm to the base $a$ is the inverse function of the exponential with base $a$. That is,

$$y = \log_a x \quad \text{means that} \quad x = a^y.$$

---

[1] A "widget" is a fictitious product used in illustrations of marketing principles.

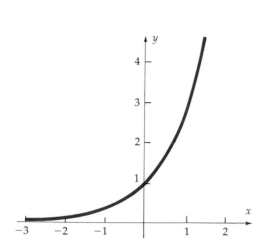

**FIGURE 1**
$y = e^x$.

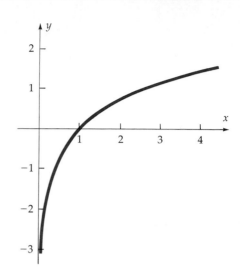

**FIGURE 2**
$y = \ln x$.

The **natural logarithm** is the *logarithm to the base e*. It is denoted ln:

$$\ln x = \log_e x.$$

Thus

$$y = \ln x \quad \text{means} \quad x = e^y. \tag{1}$$

The natural exponential $x = e^y$ is an increasing function with domain $-\infty < y < \infty$ and range $0 < x < \infty$ (Fig. 1). Hence its inverse function $y = \ln x$ (Fig. 2) is also increasing, and has domain $0 < x < \infty$ and range $-\infty < y < \infty$. It follows (Fig. 2) that

$$\lim_{x \to 0^+} \ln x = -\infty$$

$$\lim_{x \to +\infty} \ln x = +\infty.$$

The defining relation (1) has two simple consequences. First, set $x = e^y$ in the equation $y = \ln x$; you get

$$y = \ln(e^y). \tag{2}$$

That is, the natural logarithm "undoes" the power $e^y$. For example, $\ln(e^{-1}) = -1$, and $\ln(1) = \ln(e^0) = 0$ (as in Fig. 2).

Set $y = \ln x$ in the equation $x = e^y$, and you get the relation

$$x = e^{\ln x}; \tag{3}$$

the power with base $e$ "undoes" the natural logarithm. For example,

$$10 = e^{\ln 10}.$$

For another example,

$$\ln x = -1 \Leftrightarrow e^{\ln x} = e^{-1}$$
$$\Leftrightarrow x = 1/e.$$

The laws of logarithms (Appendix B) apply, of course, to natural logarithms:

$$\ln(xy) = \ln x + \ln y \tag{4}$$

$$\ln(x/y) = \ln x - \ln y \tag{5}$$

$$\ln(x^y) = y \ln x. \tag{6}$$

### The Derivative of ln $x$

The function

$$y = \ln x$$

is the inverse of

$$x = e^y.$$

We know the derivative of this exponential:

$$\frac{dx}{dy} = \frac{de^y}{dy} = e^y.$$

The derivative of $y = \ln x$ is therefore

$$\frac{dy}{dx} = \frac{1}{dx/dy} \qquad \text{(Sec. 8.2)}$$

$$= \frac{1}{e^y}$$

$$= \frac{1}{x} \qquad \text{(since } x = e^y\text{).}$$

That is,

$$\frac{d \ln x}{dx} = \frac{1}{x}. \tag{7}$$

---

### EXAMPLE 1

$$\frac{d(x \ln x)}{dx} = \left(\frac{dx}{dx}\right) \ln x + x \frac{d \ln x}{dx} = \ln x + x \cdot \frac{1}{x} = \ln x + 1.$$

$$\frac{d(e^x + 3 \ln x)}{dx} = e^x + 3 \frac{d \ln x}{dx} = e^x + \frac{3}{x}.$$

---

## Derivative of ln|x|

Combined with the Chain Rule, formula (7) gives

$$\frac{d \ln u}{dx} = \frac{d \ln u}{du} \cdot \frac{du}{dx} = \frac{1}{u} \cdot \frac{du}{dx}.$$

If we take $u = -x$, then $\ln(-x)$ is defined for $x < 0$, and the derivative is

$$\frac{d \ln(-x)}{dx} = \frac{1}{-x} \cdot \frac{d(-x)}{dx} = \frac{1}{-x} \cdot (-1) = \frac{1}{x},$$

the same as the derivative of ln $x$. Now recall that

$$|x| = \begin{cases} x, & x > 0 \\ -x, & x < 0 \end{cases}.$$

Thus

$$\frac{d \ln|x|}{dx} = \frac{1}{x}, \qquad \text{both for } x > 0 \text{ and for } x < 0.$$

Combine *this* with the Chain Rule, and you get

$$\frac{d \ln|u|}{dx} = \frac{1}{u}\frac{du}{dx}. \tag{8}$$

---

**EXAMPLE 2**

$$\frac{d \ln(x^2 + 1)}{dx} = \frac{1}{x^2 + 1} \cdot \frac{d(x^2 + 1)}{dx} = \frac{2x}{x^2 + 1}$$

$$\frac{d \ln(e^{2x} + x)}{dx} = \frac{1}{e^{2x} + x} \cdot \frac{d(e^{2x} + x)}{dx}$$

$$= \frac{1}{e^{2x} + x} (2e^{2x} + 1).$$

$$\frac{d \ln\left|\dfrac{2x + 1}{1 - 3x}\right|}{dx} = \frac{d[\ln|2x + 1| - \ln|1 - 3x|]}{dx}$$

$$= \frac{1}{2x + 1} \cdot 2 - \frac{1}{1 - 3x} \cdot (-3)$$

$$= \frac{2}{2x + 1} + \frac{3}{1 - 3x}.$$

[Notice the simplification achieved by applying the formula $\log \dfrac{a}{b} = \log(a) - \log(b)$ *before* differentiating. In this case, it avoided an application of the Quotient Rule.]

In limits involving logarithms, l'Hôpital's Rule is a natural tool.

---

**EXAMPLE 3**   Compute $\lim\limits_{x \to 0^+} x \cdot \ln x$.

**SOLUTION**   As $x \to 0^+$ then $\ln x \to -\infty$, so the product $x \cdot \ln x$ is indeterminate; would the limit be 0, or $-\infty$, or something in between? The question is settled by rewriting the product as a quotient, and applying l'Hôpital's Rule:

$$\lim_{x \to 0^+} x \ln x = \lim_{x \to 0^+} \frac{\ln x}{x^{-1}} \qquad \left( \text{algebra, giving an } \frac{\infty}{\infty} \text{ form} \right)$$

$$= \lim_{x \to 0^+} \frac{1/x}{-x^{-2}} \qquad \text{(l'Hôpital)}$$

$$= \lim_{x \to 0^+} -\frac{1}{x} \cdot x^2$$

$$= \lim_{x \to 0^+} (-x) \qquad \text{(algebra)}$$

$$= 0.$$

---

*REMARK*   In Example 3 we found

$$\lim_{x \to 0^+} x \ln x = 0.$$

This is a contest between $x$ (which $\to 0$) and $\ln x$ (which $\to -\infty$). Apparently $x$ wins. In fact, in the problems you will see that for any positive power $x^a$,

$$\lim_{x \to 0^+} x^a \ln x = 0 \qquad \text{if } a > 0.$$

So *in a contest between a power and a log, the power wins*. Similarly, the examples in Section 8.1 showed that *in a contest between a power and an exponential, the exponential wins*. Briefly:

logs are weaker than powers; powers are weaker than exponentials.

Of course, limits should actually be evaluated by some legitimate method such as l'Hôpital's Rule; but it helps to have this "rule of thumb" to predict the outcome.

The final example is an exercise in careful graphing.

---

**EXAMPLE 4**   Graph $y = x \ln x$.

**SOLUTION**   Assemble the basic facts:

*Domain*   $x > 0$, since $\ln x$ is defined only for $x > 0$.

*Derivatives*

$$\frac{dy}{dx} = \frac{d(x \ln x)}{dx} = \ln x + 1 \qquad \text{(Example 1)}$$

$$\frac{d^2y}{dx^2} = \frac{1}{x} > 0 \qquad \text{(since } x > 0)$$

Thus the graph is *concave up.*

*Zeroes of y*

$$x \ln x = 0 \Leftrightarrow \ln x = 0 \qquad \text{(since } x > 0)$$
$$\Leftrightarrow x = 1. \qquad \text{(see Fig. 2)}$$

*Zeroes of $\dfrac{dy}{dx}$* (critical points):

$$1 + \ln x = 0 \Leftrightarrow \ln x = -1 \Leftrightarrow e^{\ln x} = e^{-1}$$

$$\Leftrightarrow x = e^{-1} = \frac{1}{e} \approx \frac{1}{2.7} \approx 0.37.$$

*Function Value at Critical Point*   The value of $y = x \ln x$ at $e^{-1}$ is

$$e^{-1} \ln(e^{-1}) = \left(\frac{1}{e}\right)(-1) = -1/e.$$

*Limits as $x \to 0^+$* (the endpoint of the domain)

$$\lim_{x \to 0^+} y = \lim_{x \to 0^+} x \ln x = 0 \qquad \text{(Example 3)}$$

$$\lim_{x \to 0^+} \frac{dy}{dx} = \lim_{x \to 0^+} (1 + \ln x) = -\infty \qquad \text{(since } \lim_{x \to 0^+} \ln x = -\infty, \text{ Fig. 2).}$$

Record all this information in a chart:

| $x$ | $0^+$ | $1/e$ | $1$ |
|---|---|---|---|
| $y$ | $0$ | $-1/e$ | $0$ |
| $\dfrac{dy}{dx}$ | $-\infty$ | $0$ | $1$ |

The entries under $0^+$ are limits, of course. Transfer the information to a graph (Fig. 3) and fill in a compatible curve.

*Notice:*

a.   At $(0, 0)$ we have an open dot, since that point is a *limit point,* not actually on the graph.

b.   As the graph approaches the point $(0, 0)$, it becomes tangent to the $y$-axis, since the limiting slope there is $-\infty$.

c.   The short segments at $\left(\dfrac{1}{e}, -\dfrac{1}{e}\right)$ and $(1, 0)$ indicate the slope at each of those points.

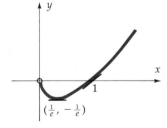

**FIGURE 3**
$y = x \ln x.$

**d.** The graph is concave up everywhere, since $d^2y/dx^2 = 1/x > 0$.

**e.** The curve is smooth everywhere, since $dy/dx$ is continuous.

## SUMMARY

*Natural Logarithms* are logarithms to the base $e$ and denoted by $\ln x$:

$$\ln x = \log_e x$$

$$e^{\ln x} = x \qquad \ln(e^y) = y$$

$$\ln(ab) = \ln a + \ln b$$

$$\ln\left(\frac{a}{b}\right) = \ln a - \ln b$$

$$\ln(a^c) = c \cdot \ln a$$

*Derivative:* $\qquad \dfrac{d\ln x}{dx} = \dfrac{1}{x},$

$$\frac{d\ln|u|}{dx} = \frac{1}{u}\frac{du}{dx}$$

*Limits:* $\qquad \lim\limits_{x \to +\infty} \ln x = +\infty \qquad \lim\limits_{x \to 0^+} \ln x = -\infty$

## PROBLEMS

**A**

**1.** Simplify or rewrite.

**\*a)** $\ln(\sqrt{e})$
**b)** $\ln e^{(x^2)}$
**\*c)** $e^{\ln x^2}$
**\*d)** $e^{x \ln x}$
**e)** $\ln(1/e^2)$
**\*f)** $e^{x + \ln x}$
**g)** $e^{-\ln x}$
**\*h)** $e^{\ln a - \ln b}$
**\*i)** $\ln(x^3 e^{-x^2})$

**2. a)** Given $\ln 2 = 0.69$, $\ln 3 = 1.10$ (to two decimals), compute

$$\ln(\tfrac{1}{2}), \quad \ln 4, \quad \ln(\tfrac{1}{4}), \quad \ln(\tfrac{3}{4}), \quad \ln 1, \quad \ln(1\tfrac{1}{2}).$$

**b)** Sketch the graph of $y = \ln x$ carefully, plotting points for

$$x = 1/4, \ 1/2, \ 3/4, \ 1, \ 1\tfrac{1}{2}, \ 2, \ 3.$$

**3.** Solve for $x$.
**\*a)** $\ln x = 1$
**b)** $\ln x = 0$

**\*c)** $\tfrac{1}{2}\ln x + 1 = 0$
**d)** $e^{x+1} = e^2$
**\*e)** $e^{3x} = 5$

**4.** Solve for $x$.
**\*a)** $6^x = 3^x$
**b)** $x^x = 3^x$
**\*c)** $3^x = 2^{x-1}$
**d)** $e^x = 2^{x^2}$
**\*e)** $\ln(x^2 + 1) = 1$

**5.** Compute the derivative.
**\*a)** $y = x^2 \ln x$
**b)** $y = 2x \ln x - 2x$
**\*c)** $y = (\ln x)^4$
**d)** $y = \ln(x^4)$
**\*e)** $y = e^x \ln x$
**f)** $y = e^{2 \ln x}$
**g)** $x = (\ln t)/e^t$

**6.** Find the first and second derivative. (In some cases, simplify by applying the laws of logarithms before differentiating.)
**\*a)** $y = \ln(x^2 + x)$
**\*b)** $y = \ln(-x), \ x < 0$

**\*c)**  $y = \ln \dfrac{1 + x}{1 - x}$

**d)**  $y = \ln(2x \sqrt[3]{x + 1})$

**e)**  $y = \dfrac{1}{5} \ln\left(\dfrac{x}{1 + 2x}\right)$

**f)**  $y = \ln\left(\sqrt{\dfrac{x}{1 + x}}\right)$

**7.**  Compute the derivative.
  **\*a)**  $y = \ln(\cos x)$    **\*b)**  $y = \ln(\sec 3x)$
  **\*c)**  $z = \sin(\ln(t^2))$    **d)**  $y = \cos(\ln(t^3 - 1))$

**8.**  Compute the second derivative.
  **\*a)**  $x = \ln(\sin 3t)$    **\*b)**  $y = \ln(x \csc x)$
  **\*c)**  $y = e^{\ln(\cos 3t)}$    **d)**  $y = e^{2 \ln|\sec x|}$

**9.**  Graph $y = x^{1/2} \ln x$, $x > 0$, using $y$, $dy/dx$, and $d^2y/dx^2$, and showing the limits of $y$ as $x \to 0^+$ $x \to 0^+$, and show these on the graph.

**10.**  Graph $y = (\ln x)/x$, $x > 0$, using $y$, $dy/dx$, and $d^2y/dx^2$, and showing the limits of $y$ as $x \to 0^+$ and $x \to +\infty$.

**B**

**11.**  Let $a > 0$. Graph $y = x^a \ln x$, $x > 0$, using the signs of $y$ and of $dy/dx$, and showing $\lim\limits_{x \to 0^+} x^a \ln x$ and $\lim\limits_{x \to +\infty} x^a \ln x$.

**12.**  Let $a > 0$. Graph $y = x^{-a} \ln x$, $x > 0$, using the signs of $y$ and of $dy/dx$, and showing $\lim\limits_{x \to 0^+} x^{-a} \ln x$ and $\lim\limits_{x \to +\infty} x^{-a} \ln x$.

**13.**  Graph

$$y = p \ln p + (1 - p) \ln(1 - p), \qquad 0 < p < 1,$$

showing any maxima, minima, and limits at the ends of the domain. (This function defines the "entropy" of a situation with two possible outcomes, one with probability $p$, the other with probability $1 - p$. Entropy is a measure of the uncertainty of the outcome.)

**14.**  A certain calculator has no $e^x$ key. It *does* have keys marked INV, log, ln $x$, and $y^x$. INV gives the *inverse* of the next function pressed; log is $\log_{10}$. How would you calculate $e^x$ on such a calculator? How would you get $e$? Explain.

**\*15.**  In chemistry, the progress of a certain kind of reaction is described by

$$kt = x_e \ln\left(\dfrac{x_e}{x_e - x}\right), \qquad (*)$$

where $x$ is the number of moles reacted at time $t$, $k$ is a constant describing the rate of the reaction, and $x_e$ is a constant describing the amount reacted "at equilibrium". (The "$e$" in $x_e$ is for "equilibrium," not for $2.71828\ldots$.)
  **a)**  Solve equation $(*)$ for $x$.
  **b)**  Find $\lim\limits_{t \to +\infty} x$.
  **c)**  Differentiate $(*)$ with respect to $t$, and express $\dfrac{dx}{dt}$ in terms of $x$, $x_e$, and $k$.

**16.**  **a)**  Write the linear approximation to $\ln(1 + x)$, based at $x = 0$.
  **b)**  Use part a and the second derivative of $\ln(1 + x)$ to deduce that $\ln(1 + x) < x$ for $-1 < x < 0$ and $0 < x$.

**17.**  *The birthday problem.* With $n$ birthdays chosen at random from 365 possibilities (leap years left out of consideration), the probability that *no two* are the same is

$$P_n = \dfrac{(365)(364) \cdots (365 - n + 1)}{(365)(365) \cdots (365)}$$

$$= (1)\left(1 - \dfrac{1}{365}\right)\left(1 - \dfrac{2}{365}\right) \cdots \left(1 - \dfrac{n - 1}{365}\right).$$

The "birthday problem" asks: For which $n$ is $P_n < \frac{1}{2}$? That is, how large must $n$ be to have a *less* than 50% chance of no coincident birthdays? It is difficult to guess the answer, and tedious to compute many "experimental" values of $P_n$. This problem shows a quicker way to find the answer, using logarithms and the linear approximation in the previous problem.
  **a)**  Explain why the formula for $P_n$ is correct for $n = 1$; $n = 2$; $n = 3$.
  **b)**  The birthday problem amounts to this: For what $n$ is $\ln(P_n) < \ln(\frac{1}{2})$? What property of the logarithm justifies this restatement?
  **c)**  $\ln(P_n) = \ln\left(1 - \dfrac{1}{365}\right) + \ln\left(1 - \dfrac{2}{365}\right)$ $+ \cdots + \ln\left(1 - \dfrac{n - 1}{365}\right)$, so you can use the approximation $\ln(1 + x) \approx x$, for small $x$ (see the previous problem). Show that in fact

$$\ln(P_n) < -\dfrac{1}{365} - \cdots - \dfrac{n - 1}{365} = -\dfrac{n(n - 1)}{2 \cdot 365}.$$

[Hint.

$$2[1 + 2 + \cdots (n - 1)]$$
$$= 1 + 2 + \cdots + (n - 2) + (n - 1)$$
$$\quad + (n - 1) + (n - 2) + \cdots + 2 + 1$$
$$= n(n - 1).]$$

*d)   In a group of 30 people with birthdays unknown to each other, is it a good bet at even money that no two birthdays are the same?

e)   Determine the least integer $n$ such that
$$-\frac{n(n - 1)}{2 \cdot 365} < \ln \frac{1}{2} \approx -0.69. \text{ (For this } n,$$
$P_n < \frac{1}{2}.)$

f)   The calculations through part e can be done by hand, but now we require a calculator. Using the original formula for $P_n$, with the number $n$ in part e, show that $P_n < \frac{1}{2}$ and $P_{n-1} > \frac{1}{2}$. [Avoid "overflow" in your calculator; compute $P_n$ as $(\ldots (((364 \div 365) \times 363) \div 365) \ldots).]$

## C

Problems 18 and 19 give a sophisticated, efficient way of computing natural logarithms.

18.   Prove the following for $-1 < x < 1$.

a)   $1 + x^2 + x^4 + \cdots + x^{2n} < \dfrac{1}{1 - x^2}$

$$= 1 + x^2 + x^4 + \cdots + x^{2n} + \frac{x^{2n+2}}{1 - x^2}$$

(Multiply by $1 - x^2$.)

b)   $x + \dfrac{x^3}{3} + \cdots + \dfrac{x^{2n+1}}{2n + 1} < \dfrac{1}{2} \ln\left(\dfrac{1 + x}{1 - x}\right)$

$$< x + \cdots + \frac{x^{2n+1}}{2n + 1} + \frac{1}{1 - x^2} \frac{x^{2n+3}}{2n + 3}$$

if $0 < x < 1$. (All three functions are 0 when $x = 0$; compare their derivatives.)

c)   $x + \cdots + \dfrac{x^{2n+1}}{2n + 1} + \dfrac{1}{1 - x^2} \dfrac{x^{2n+3}}{2n + 3}$

$$< \frac{1}{2} \ln\left(\frac{1 + x}{1 - x}\right) < x + \frac{x^3}{3} + \cdots + \frac{x^{2n+1}}{2n + 1}$$

if $-1 < x < 0$.

19.   a)   Use problem 18b with $x = 1/3$ and $n = 1$ to show that

$$\frac{56}{81} < \ln 2 < \frac{56}{81} + \frac{1}{540}$$

or

$$.6913 < \ln 2 < .69321.$$

Show that this is consistent with the value of $\ln 2$ from a calculator.

b)   Approximate $\ln 2$ using $n = 2$ instead of $n = 1$.

c)   Approximate $\ln(1.5)$, using $n = 2$.

d)   Approximate $\ln(0.5)$, using problem 18c with $n = 1$.

20.   The average $\dfrac{a + b}{2}$ of $a$ and $b$ is called their *arithmetic mean*. For positive $a$ and $b$, there is also a *geometric mean* $\sqrt{ab}$, and a *logarithmic mean* $\dfrac{b - a}{\ln b - \ln a}$. Show

a)   $\text{Min}(a, b) \leq \sqrt{ab} \leq \text{max}(a, b)$, and
$$\text{min}(a, b) \leq \frac{a + b}{2} \leq \text{max}(a, b).$$

b)   The natural log of the geometric mean is the arithmetic mean of $\ln a$ and $\ln b$.

c)   From Figure 4, if $a < b$,
$$e^{1/2(\ln a + \ln b)} (\ln b - \ln a)$$
$$< \int_{\ln a}^{\ln b} e^x \, dx < \frac{b + a}{2} (\ln b - \ln a).$$

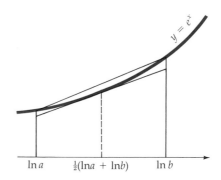

FIGURE 4

(Use the concavity of $e^x$ and compare two trapezoidal areas to the area under the graph.)

d)   $\sqrt{ab} < \dfrac{b - a}{\ln b - \ln a} < \dfrac{1}{2}(a + b)$   if   $a \neq b$
(not just for $a < b$).

e)   $\text{Min}(a, b) < \dfrac{b - a}{\ln b - \ln a} < \text{max}(a, b)$ if
$a \neq b$.

# 8.4

## OTHER BASES. LOGARITHMIC DIFFERENTIATION. INDETERMINATE FORMS

To deal with logarithms or exponentials with an arbitrary base $b$, you can rewrite everything to the base $e$. For logarithms, use the formula

$$\log_b x = \frac{\ln x}{\ln b}, \tag{1}$$

which you are asked to prove in the problems.

---

**EXAMPLE 1**

$$\frac{d \log_{10} x}{dx} = \frac{d}{dx} \frac{\ln x}{\ln 10} \qquad \text{[by (1)]}$$

$$= \frac{d}{dx}\left(\frac{1}{\ln 10} \ln x\right) = \frac{1}{\ln 10} \cdot \frac{1}{x}.$$

$$\frac{d \log_2(t^3 + 1)}{dt} = \frac{d}{dt} \frac{\ln (t^3 + 1)}{\ln 2}$$

$$= \frac{1}{\ln 2} \frac{d}{dt} \ln(t^3 + 1)$$

$$= \frac{1}{\ln 2} \frac{3t^2}{t^3 + 1}.$$

---

For variable powers with a base other than $e$, you can rewrite everything to the base $e$, using the formula

$$b^a = e^{\ln(b^a)} = e^{a \ln b}. \tag{2}$$

---

**EXAMPLE 2**

$$\frac{d10^x}{dx} = \frac{de^{x \ln 10}}{dx} \qquad \text{[by (2)]}$$

$$= e^{x \ln 10} \frac{d(x \ln 10)}{dx} \qquad \left[\frac{de^u}{dx} = e^u \frac{du}{dx}\right]$$

$$= e^{x \ln 10} \cdot \ln 10$$

$$= 10^x \ln 10.$$

Recall from Section 8.1 that

$$\frac{d10^x}{dx} \approx (2.3025851)10^x.$$

Now we find that the complicated constant in this formula is an approximation of $\ln 10$!

---

**Logarithmic differentiation** is another way to attack these problems— take the natural logarithm of the given expression $y$, then differentiate and solve for $dy/dx$. The logarithm simplifies the problem by turning powers into products, products into sums, and quotients into differences.

---

**EXAMPLE 3**  Compute $\dfrac{d(x2^x)}{dx}$.

**SOLUTION**  Let $y = x2^x$. Take ln on both sides:

$$\ln y = \ln x + \ln(2^x) = \ln x + x \ln 2.$$

Differentiate each side *with respect to x*, using the Chain Rule where needed:

$$\frac{1}{y}\frac{dy}{dx} = \frac{1}{x} + \ln 2.$$

Hence,

$$\frac{dy}{dx} = y\left(\frac{1}{x} + \ln 2\right) = x2^x\left(\frac{1}{x} + \ln 2\right).$$

---

**Limits of exponentials** $f(x)^{g(x)}$ can often be evaluated by taking the natural logarithm.

---

**EXAMPLE 4**  $\lim\limits_{x \to 0^+} x^x = ?$

**SOLUTION**  The function $x^x$ is defined for $x > 0$, but not for $x = 0$; that's why the limit question arises. Since the problem involves powers, take logs. Let $y = x^x$. Then

$$\ln y = \ln(x^x) = x \ln x$$

and

$$\lim_{x \to 0^+} \ln y = \lim_{x \to 0^+} x \ln x = 0 \qquad \text{(Sec. 8.3, Example 3).}$$

So,

$$\lim_{x \to 0^+} y = \lim_{x \to 0^+} e^{\ln y} = e^{\lim\limits_{x \to 0^+} \ln y}$$

(because $e^u$ is a continuous function of $u$)

$$= e^0 = 1.$$

---

In Example 4, you might think: "Obviously, $\lim\limits_{x \to 0^+} x^x = 1$; $x^0 = 1$ for any $x \neq 0$; why not for $x = 0$ as well?" On the other hand, $0^x = 0$ for any $x > 0$; why not for $x = 0$ as well? There is a contest here, between the *power* (which suggests a limit of 1) and the *base* (which suggests a limit of 0). We call such a limit, where both base and exponent have limit 0, an "indeterminate form

$0^0$." It is indeterminate, because knowing the limits of the base and the exponent does not determine the limit of the function; only further analysis can settle the issue.

Other exponential indeterminate forms are $\infty^0$ (such as $\lim\limits_{x \to 0} x^{1/x}$) and $1^\infty$ (such as the important limit $\lim\limits_{h \to 0} (1 + h)^{1/h} = e$); the problems ask you to evaluate these.

## SUMMARY

For bases other than $e$, use

$$b^a = e^{a \ln b}$$

$$\log_b a = \frac{\ln a}{\ln b}.$$

Exponential indeterminate forms: $0^0$, $\infty^0$, $1^\infty$.

## PROBLEMS

### A

1. Differentiate.
   *a) $5^x$          *b) $x^{10}$
   *c) $x^x$           d) $x^{1+x}$
   e) $x^{\ln x}$        f) $2^{x^2}$
   g) $a^x \ (a > 0)$    h) $\log_a x \ (a > 0)$
   *i) $x^{\sin x}$      j) $(\ln x)^x$
   k) $x^{1/x}$

2. Evaluate.
   *a) $\lim\limits_{x \to 0^+} x^{1+x}$       *b) $\lim\limits_{x \to 0^+} (2x)^x$

   *c) $\lim\limits_{h \to 0} (1 + h)^{1/h}$    d) $\lim\limits_{x \to +\infty} \left(1 + \frac{1}{x}\right)^x$

   e) $\lim\limits_{x \to 0^+} \left(1 + \frac{a}{x}\right)^{1/x} \ (a > 0)$

   *f) $\lim\limits_{h \to 0} (1 + ah)^{1/h}$   g) $\lim\limits_{x \to +\infty} x^{1/x}$

   *h) $\lim\limits_{x \to +\infty} (e^x + x)^{1/x}$   *i) $\lim\limits_{x \to 0} (e^x + x)^{1/x}$

   j) $\lim\limits_{x \to +\infty} (\ln x)^{1/x}$   k) $\lim\limits_{x \to 0^+} x^{\ln x}$

3. Compute the derivative by logarithmic differentiation.
   a) $y = xe^x$

   *b) $y = x^x$
   c) $y = x\sqrt{x^2 + 1}$
   d) $y = \sqrt{x^2 + 1} \div \sqrt[3]{x^2 - 1}$
   *e) $y = \dfrac{x^2 \sqrt[3]{7x - 14}}{(1 + x^2)^4}$

4. Graph $y = x^x, x > 0$, using $dx/dx$, and showing any appropriate limits.

5. Suppose that $u$ and $v$ are positive functions of $x$. By logarithmic differentiation, deduce
   a) The Product Rule, for the derivative of $uv$.
   b) The Quotient Rule, for the derivative of $u/v$.

6. Prove the formula $\log_b a = \dfrac{\ln a}{\ln b}$. (Start with $a = b^{\log_b a}$, see Appendix B.)

### B

7. Define a function $f$ by the formula
   $$f(x) = \lim\limits_{h \to 0^+} |x|^h.$$
   a) Evaluate $f(x)$ for every real $x$.
   b) Describe any discontinuity of $f$.

## 8.5

## EXPONENTIAL AND LOGARITHMIC INTEGRALS

The derivative formulas

$$\frac{de^u}{du} = e^u, \qquad \frac{d\ln|u|}{du} = \frac{1}{u}$$

imply the integrals

$$\int e^u \, du = e^u + C, \qquad \int \frac{1}{u} \, du = \ln|u| + C.$$

This last integral fills a gap in the general formula for the integral of a power:

$$\int u^n \, du = \frac{1}{n+1} u^{n+1} + C, \qquad \text{if } n \neq -1.$$

When $n = -1$, then $\int u^{-1} \, du = \int \frac{1}{u} \, du = \ln|u| + C.$

---

**EXAMPLE 1**

$$\int \frac{x \, dx}{x^2 + 1} = \int \frac{(1/2) \, du}{u} \qquad (u = x^2 + 1, \, du = 2x \, dx, \, x \, dx = \tfrac{1}{2} \, du)$$

$$= \frac{1}{2} \int \frac{du}{u} = \frac{1}{2} \ln|u| + C$$

$$= \frac{1}{2} \ln(x^2 + 1) + C.$$

(We drop the absolute value sign, since $x^2 + 1 > 0$ for all $x$.)

---

**EXAMPLE 2** $\quad \int xe^{-x^2} \, dx = \int e^u(-\tfrac{1}{2} \, du) = -\tfrac{1}{2}e^{-x^2} + C.$

---

### The Logarithm as an Integral

For any $x > 0$,

$$\int_1^x \frac{1}{t} \, dt = \ln|t| \Big|_1^x = \ln|x| - \ln|1| = \ln x.$$

Thus for $x > 1$, $\ln x$ is the area under the graph of $y = 1/t$, from $t = 1$ to $t = x$ (Fig. 1a). For $0 < x < 1$,

$$\ln x = \int_1^x \frac{1}{t} \, dt = -\int_x^1 \frac{1}{t} \, dt,$$

which is *minus* the area under the graph, from $t = x$ to $t = 1$.

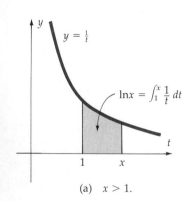

(a) $x > 1$.

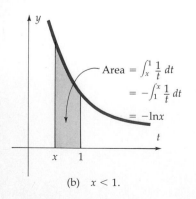

(b) $x < 1$.

**FIGURE 1**
$\ln x$ as an area.

This integral gives an alternate way to introduce the natural logarithm, namely, to *define* it as an integral:

$$\ln x = \int_1^x \frac{1}{t}\, dt, \qquad \text{for } x > 0.$$

Starting with this definition, the derivative is determined by the Second Fundamental Theorem of Calculus:

$$\frac{d}{dx} \int_a^x f(t)\, dt = f(x),$$

so

$$\frac{d \ln x}{dx} = \frac{d}{dx} \int_1^x \frac{1}{t}\, dt = \frac{1}{x}.$$

From this same defining integral, one can prove the laws of logarithms (see problem 8). Then the **exponential function** can be defined as the *inverse of the logarithm*, and its properties derived from those of the logarithm. This approach would avoid many of the assumptions we made without proof, for example, that the function $f(x) = b^x$ is defined for all $b > 0$ and all real $x$; that it has a derivative at $x = 0$; and that $f'(0) = 1$ for exactly one base $b$, namely $b = e$. On the other hand, any standard calculus course is full of plausible but unproved assertions, and there is no special reason to be more rigorous in defining logarithms and exponentials than, say, the trigonometric functions.

## SUMMARY

$$\int e^u\, du = e^u + C, \qquad \int \frac{du}{u} = \ln|u| + C.$$

## PROBLEMS

### A

1. Evaluate.

*a) $\displaystyle \int e^{2x}\, dx$

*b) $\displaystyle \int \sin x\, e^{\cos x}\, dx$

*c) $\displaystyle \int \frac{dx}{x-1}$

*d) $\displaystyle \int \frac{dx}{2-x}$

*e) $\displaystyle \int \frac{x\, dx}{x^2 + 4}$

*f) $\displaystyle \int \frac{2x\, dx}{3x^2 - 2}$

*g) $\displaystyle \int \frac{dx}{x \ln x}$

h) $\displaystyle \int \frac{\sin x\, dx}{\cos x}$

*i) $\displaystyle \int \tan \theta\, d\theta$

j) $\displaystyle \int \frac{x^2 - 1}{x}\, dx$

*k) $\displaystyle \int \frac{x + 4}{x}\, dx$

l) $\displaystyle \int \frac{dx}{\sqrt{x+1}}$

m) $\displaystyle \int \frac{x^2\, dx}{1 - x^3}$

n) $\displaystyle \int \frac{x + 3}{x^2 + 6x + 1}\, dx$

o) $\displaystyle \int e^{3t + 1}\, dt$

p) $\displaystyle \int (x + 1)e^{x^2 + 2x}\, dx$

*q) $\displaystyle \int (e^x + 1)^2\, dx$

r) $\displaystyle \int \frac{(e^x + 1)^2}{e^x}\, dx$

s) $\int \dfrac{e^x}{1 + e^x}\,dx$  *t) $\int 2^x\,dx$

u) $\int \dfrac{dx}{x\,\log_3 x}$

2. Evaluate.

*a) $\int_0^1 \dfrac{dx}{2x + 1}$  *b) $\int_0^1 e^{-3x}\,dx$

*c) $\int_{-2}^{-1} \dfrac{dx}{x}$  d) $\int_2^e \dfrac{dx}{x(\ln x)^2}$

e) $\int_1^2 \dfrac{x + 1}{x}\,dx$

3. Find the area of the region bounded by
   a) The graphs of $y = e^{-x}$ and $y = -e^{-x}$, and the lines $x = 0$ and $x = 3$.
   *b) The graph of $y = 1/x$, $x > 0$, and the lines $y = x$ and $y = 0$.

4. Evaluate the following improper integrals, or explain why they are divergent.

*a) $\int_{-1}^1 \dfrac{dx}{x}$  b) $\int_0^1 \dfrac{dx}{x}$

*c) $\int_0^\infty e^{-2x}\,dx$  *d) $\int_0^\infty xe^{-x^2}\,dx$

e) $\int_{-\infty}^\infty xe^{-x^2}\,dx$  f) $\int_{-\infty}^\infty e^{-x}\,dx$

*g) $\int_{-\infty}^\infty e^{-|x|}\,dx$  h) $\int_0^1 \dfrac{dx}{x^2 - x}$

*i) $\int_0^2 \dfrac{dx}{x - 1}$  j) $\int_2^\infty \dfrac{dx}{x(\ln x)^3}$

*k) $\int_0^\infty e^{-ax}\,dx$, $a > 0$

*5. Electrical charge $Q$ flows into a capacitor at a rate $\dfrac{dQ}{dt} = 5e^{-3t}$, $t > 0$. How much charge flows in for $0 \le t \le 10$? For $10 \le t \le 100$? For $t \ge 0$?

*6. The graph of $y = e^{-x}$ for $0 \le x < \infty$ is revolved about the $x$-axis. Compute the volume enclosed.

**B**

7. Determine whether the following integrals are convergent or divergent.

a) $\int_1^\infty e^{-x^2}\,dx$  b) $\int_1^\infty e^{-x^2}\,dx$

c) $\int_{-\infty}^\infty e^{-x^2}\,dx$

8. Use the formula $\ln(x) = \int_1^x \dfrac{1}{t}\,dt$, $x > 0$, to show

a) $\ln(1) = 0$.
b) $\frac{1}{2} < \ln(2) < \frac{3}{4}$.
c) $-\frac{1}{2} > \ln(\frac{1}{2}) > -\frac{3}{4}$.
d) $\ln(ab) - \ln(b) = \ln(a)$. [In the integral $\ln(a) = \int_1^a \dfrac{1}{t}\,dt$, set $t = s/b$ and determine the corresponding limits for $s$.]

# 8.6

## PROPORTIONAL GROWTH AND DECAY

Imagine a quantity $A(t)$ varying with time $t$. The derivative $dA/dt$ gives its rate of change with time. In some important cases, this rate is proportional to the amount present:

$$\frac{dA}{dt} = kA, \tag{1}$$

where $k$ is a constant.

Suppose, for instance, that $A$ is the mass of a certain bacteria colony. As long as food and space are abundant, the colony grows at a rate proportional to its size (the more parents, the more offspring)—that is, equation (1) holds. The number $k$ is called the **growth constant**. When $A$ is increasing (as in this example) then $\dfrac{dA}{dt} > 0$, so the growth constant $k$ is positive.

In another important example, $A$ is the amount of some radioactive element present in a certain body at time $t$. The element decays at a rate proportional to the amount present (the more unstable atoms there are, the more will decay in a given time interval) so once again $dA/dt = kA$. In this case $A$ is decreasing, so $dA/dt < 0$, hence the constant of proportionality $k$ is negative. Its absolute value $|k|$ is called the **decay constant** for the element.

Suppose that the varying amount $A$ is positive, $A > 0$. Then we can solve the proportional growth equation as follows:

Divide both sides of (1) by $A$:

$$\frac{1}{A}\frac{dA}{dt} = k. \tag{2}$$

Now recall that

$$\frac{d\ln u}{dt} = \frac{1}{u}\frac{du}{dt}, \quad \text{so} \quad \frac{d\ln A}{dt} = \frac{1}{A}\frac{dA}{dt}.$$

Equation (2) can thus be written

$$\frac{d\ln A}{dt} = k. \tag{3}$$

Take antiderivatives on each side of (3):

$$\ln A = kt + c.$$

(Only one constant of integration is needed; if constants $c_1$ and $c_2$ were added, one on each side, they could be combined into a single constant $c_2 - c_1$ on the right-hand side.) This gives $\ln A$; to get $A$ itself, "undo" the logarithm:

$$A = e^{\ln A} = e^{kt+c} = e^c e^{kt}. \tag{4}$$

Thus $A(0) = e^c e^0 = e^c$, $A(1) = e^c e^k$, and so on. The quantity $A(0) = e^c$ is the *initial amount*, denoted $A_0$. Putting $e^c = A_0$ in (4) gives

$$A = A_0 e^{kt}, \qquad \text{where } A_0 = A(0). \tag{5}$$

This is called *exponential growth with growth constant k*. We have proved that *if a positive quantity A statisfies the law of proportional growth* (1), *then A has exponential growth* (5). Conversely, if $A$ is given by (5) then differentiating both sides shows that $dA/dt = kA$.

The growth law (5) contains two constants $A_0$ and $k$, which vary from case to case. Typically, one piece of information determines one of the constants; and two pieces determine both constants, hence determine the function $A$ completely.

**EXAMPLE 1**  A colony of bacteria grows from 1 to 3 gms in one day. Let $t = 0$ at the beginning of the day, $t = 1$ at the end (24 hours later). Assuming exponential growth

a.  Find the constants $A_0$ and $k$ in the growth law.
b.  How many grams are there after two days?
c.  How many hours does it take the colony to double its weight? (This is called the "doubling time.")

*SOLUTION*

a.  Assume exponential growth, $A(t) = A_0 e^{kt}$. We are given $A(0) = 1$ and $A(1) = 3$, so

$$A_0 e^{k \cdot 0} = 1, \qquad A_0 e^{k \cdot 1} = 3.$$

The first equation gives $A_0 = 1$; so put $A_0 = 1$ in the second, and find $e^k = 3$, hence $k = \ln 3$. Therefore

$$A(t) = A_0 e^{kt} = e^{(\ln 3)t},$$

which we simplify to

$$A(t) = (e^{\ln 3})^t = 3^t. \tag{6}$$

Now it is easy to answer any further questions:

b.  How many grams are there after two days? From (6),

$$A(2) = 3^2 = 9.$$

The colony tripled the first day, and tripled again the second day; hence it is multiplied by $3 \cdot 3 = 9$.

c.  How many hours does it take to double its weight? Let $D$ be the "doubling time"; if the weight is $A(t)$ at time $t$, then $D$ hours later it is twice as much,

$$A(t + D) = 2A(t).$$

By (6), this says $3^{(t+D)} = 2 \cdot 3^t$, or $3^t 3^D = 2 \cdot 3^t$, or $3^D = 2$. Solve this for $D$ by taking ln on each side:

$$D \ln 3 = \ln 2,$$

$$D = \frac{\ln 2}{\ln 3} \text{ days} = .63 \ldots \text{ days}$$

$$\approx 15 \text{ hours}.$$

**EXAMPLE 2**  The isotope Radon $C$ has a half-life of about 20 minutes; of a given amount, half of it decays into lead in 20 minutes, then half the remainder decays into lead in the next 20 minutes, and so on.

a.  Find the decay constant $|k|$; measure time $t$ in minutes.
b.  At time $t = 0$, a radiation counter shows that it is decaying at the rate of .003 gm/min. How much Radon $C$ is present at that time?

*SOLUTION*

a.  Assuming exponential decay,

$$A(t) = A_0 e^{kt},$$

where $k < 0$. If half decays in 20 minutes, then $A(20) = \frac{1}{2}A_0$, that is

$$A_0 e^{k \cdot 20} = \frac{1}{2}A_0$$

which gives $e^{20k} = 1/2$, $20\,k = \ln(1/2) = -\ln 2$, $k = -\frac{1}{20}\ln 2$. Hence,

$$A(t) = A_0 e^{-(1/20 \ln 2)t} \qquad (7)$$

which simplifies to

$$A(t) = A_0 (e^{\ln 2})^{-t/20} = A_0 \cdot 2^{-t/20}.$$

b.  The rate of decay at time 0 is given as .003 gm/min, that is,

$$\left.\frac{dA}{dt}\right|_{t=0} = -.003. \qquad (8)$$

You can also compute $dA/dt$ from (7), and compare it to (8):

$$\left.\frac{dA}{dt}\right|_{t=0} = -\left(\frac{1}{20}\ln 2\right) A_0 e^{-(1/20 \ln 2)t}\bigg|_{t=0}$$

$$= -\frac{1}{20}(\ln 2)A_0. \qquad (9)$$

From (8) and (9), $-\dfrac{1}{20}(\ln 2)A_0 = -.003$ so $A_0 = \dfrac{.06}{\ln 2} \approx .086$ gms.

## Carbon Dating

Living organisms constantly absorb carbon from their surroundings. A certain proportion of this carbon is a radioactive type called Carbon-14, or $^{14}C$. When the organism dies, no more $^{14}C$ is absorbed, and the amount already present decays, with a half-life of about 5600 years. This gives a standard technique to determine the age of organic relics.

**EXAMPLE 3**   A relic is found to contain 40% of the $^{14}C$ that it presumably had when alive. How old is it?

*SOLUTION*   Let $A(t)$ be the amount of $^{14}C$ present $t$ years after the death of the organism. Then

$$A(t) = A_0 e^{kt}. \qquad (10)$$

At the time $t$ when 40% is left, $A(t) = .4A_0$, so

$$.4A_0 = A_0 e^{kt}.$$

Solving for $t$,

$$.4 = e^{kt}$$

$$\ln(.4) = kt$$

$$t = \ln(.4)/k. \qquad (11)$$

It remains to determine the constant $k$. The half-life is 5600 years, so from (10)

$$\tfrac{1}{2}A_0 = A(5600) = A_0 e^{5600k}.$$

Divide by $A_0$ and take ln:

$$\ln(1/2) = 5600k$$

$$k = -(\ln 2)/5600.$$

Returning to (11), the age of the relic is

$$t = \ln(.4)/k = -5600 \ln(.4)/\ln 2$$
$$= 7402.797 \ldots \text{ years}$$

(Actually, the accuracy of the given data does not justify an answer more precise than "7400 years." The half-life is given as "about 5600 years," and the 40% presumably means *about* 40%.)

## SUMMARY

The *equation of proportional growth or decay* is

$$\frac{dA}{dt} = kA.$$

Its solution is

$$A(t) = A_0 e^{kt},$$

where $A_0 = A(0)$ is the amount at time $t = 0$.

## PROBLEMS

### A

*1. At time $t = 0$ there is .03 gm of a decaying radioactive substance, and at time $t = 5$ there is .02 gm. Assuming exponential decay, how much is there at time $t$? What is the half-life?

2. At time $t = 0$ there are 5 gms of bacteria, and at time $t = 3$ there are 9 gms. Assuming exponential growth, find the doubling time. How many grams are there at $t = 12$?

*3. At time $t = 0$ days a radioactive substance is decaying at .006 gm/day, and 5 days later it is decaying at .005 gm/day. What is the half-life? How much was there at time $t = 0$?

*4. A deposit earns interest at 5% per year, compounded continuously; the amount after $t$ years is $e^{.05t}P_0$, where $P_0$ is the initial deposit. What is the effective annual interest $\bar{r}$, that is, what simple interest paid once per year would give the same yield at the end of one year?

*5. A savings bond was bought for $12.00 and redeemed for $20.00 ten years later. What was the interest rate, if interest was compounded
a) Continuously? b) Annually?

*6. The cave paintings in Lascaux, France, have apparently lost about 85% of their Carbon-14. How old are the paintings?

7. A piece of charcoal at a prehistoric campsite contains about 45% of the Carbon-14 that it presumably had when it was burned. How long ago was that?

*8. The rate of radiation emitted from a piece of charcoal is measured, and it is found that the Carbon-14 in the piece is decaying at the rate of $R$ gm/yr. How much Carbon-14 is present?

**B**

9. a) A certain substance decays with half-life $H$. Show that the amount $A$ at time $t$ is related to the amount $A_0$ at time 0 by $A = A_0 2^{-t/H}$.

   b) A colony grows exponentially with doubling time $D$. Find a formula for the size of the colony at time $t$, analogous to the formula in part a.

10. Show that
    a) If $k > 0$, every solution of $dA/dt = kA$ has doubling time $D = \dfrac{\ln 2}{k}$.

    b) If $k < 0$, every solution of $dA/dt = kA$ has half-life $H = \dfrac{\ln 2}{|k|}$.

11. For an object traveling at low speed in a fluid (like water, air, or oil) the frictional force is nearly proportional to the velocity, and oppositely directed:

$$F = -kv.$$

Assuming such a friction is the only force at work, use Newton's Law $F = m\dfrac{dv}{dt}$ to express the velocity $v$ in terms of $k$, the mass $m$, and the initial velocity $v_0 = v(0)$.

12. a) Find by "trial and error" a function $A(t)$ that satisfies the equation

$$\frac{d^2 A}{dt^2} = A.$$

   b) Find a function that satisfies $\dfrac{d^2 A}{dt^2} = k^2 A$, where $k$ is a constant.

13. The text proved that if $dA/dt = kA$, then $A = A_0 e^{kt}$, assuming that $A > 0$. This problem outlines a proof that requires no such assumption. Suppose that

$$\frac{dA}{dt} = kA, \qquad (*)$$

where $k$ is constant. Prove that
a) $e^{-kt} A$ is a constant. [Use the Product Rule and (*).]
b) $e^{-kt} A = A_0$ [where $A_0 = A(0)$].
c) $A = A_0 e^{kt}$.

# 8.7

## SEPARABLE DIFFERENTIAL EQUATIONS

An equation involving one or more derivatives of a function $y(t)$, which is to be valid for all $t$ in some interval, is called a **differential equation for $y$**. A function $y$ that satisfies the equation is called a **solution**.

An example is the equation for the position function $y(t)$ of an object falling with constant acceleration $-g$:

$$\frac{d^2 y}{dt^2} = -g.$$

We found the solution (Sec. 2.8)

$$y = y_0 + tv_0 - \tfrac{1}{2}gt^2,$$

where $y_0 = y(0)$ is the initial position and $v_0 = y'(0)$ the initial position. Another example is the proportional growth equation $\dfrac{dA}{dt} = kA$ (Sec. 8.6). We solved this by *separating the variables* $A$ and $t$, then integrating:

$$\frac{1}{A}\frac{dA}{dt} = k \quad \Rightarrow \quad \int \frac{1}{A}\frac{dA}{dt}\,dt = \int k\,dt \quad \Rightarrow \quad \ln|A| = kt + C.$$

This method applies to any differential equation where the variables can be separated, and the equation can be written in the form

$$f(y)\frac{dy}{dt} = g(t). \tag{1}$$

Suppose that $y$ is a function of $t$ that satisfies equation (1). Then the *integral* of the left-hand side with respect to $t$ equals the integral of the right-hand side:

$$\int f(y)\frac{dy}{dt}\,dt = \int g(t)\,dt. \tag{2}$$

By the rule for substitution, $\dfrac{dy}{dt}\,dt = dy$, so

$$\int f(y)\,dy = \int g(t)\,dt. \tag{3}$$

So every solution $y$ of (1) satisfies the relation (3), obtained by integrating the functions $f$ and $g$.

In passing from (1) to (3) you can bypass (2); simply transpose $dt$ from the left side of (1) to the right, leaving both $y$ and $dy$ on the left, and $t$ and $dt$ on the right. Then put integral signs on both sides.

---

**EXAMPLE 1**

    **a.** Solve the equation $\dfrac{dA}{dt} = 2(7 - A)$.

    **b.** Find the solution with $A(0) = 10$, and compute $\displaystyle\lim_{t \to +\infty} A(t)$.

    **c.** Find the solution with $A(0) = 5$, and compute $\displaystyle\lim_{t \to +\infty} A(t)$.

*SOLUTION*

    **a.** Separate variables, putting $A$ and $dA$ on the left, $t$ and $dt$ on the right:

$$\frac{dA}{7 - A} = 2\,dt.$$

Integrate:

$$\int \frac{dA}{7 - A} = \int 2 \, dt,$$

$$-\ln|7 - A| = 2t + c.$$

Multiply by $-1$ and exponentiate:

$$|7 - A| = e^{-2t - c} = e^{-c} e^{-2t}.$$

Hence

$$7 - A = \pm e^{-c} e^{-2t} = Ce^{-2t} \quad [C = \pm e^{-c}],$$

and

$$A(t) = 7 - Ce^{-2t}.$$

**b.** $A(0) = 10 \quad \Rightarrow \quad 10 = 7 - Ce^0 = 7 - C \quad \Rightarrow \quad C = -3 \quad \Rightarrow$
$A(t) = 7 + 3e^{-2t}$. Then

$$\lim_{t \to +\infty} A(t) = 7 + 0 = 7.$$

**c.** $A(0) = 5 \quad \Rightarrow \quad A(t) = 7 - 2e^{-2t}$, and $\lim_{t \to +\infty} A(t) = 7 - 0 = 7.$

**REMARK** The given equation $\dfrac{dA}{dt} = 2(7 - A)$ describes a growth rate which is *positive* when $A < 7$, *negative* when $A > 7$, and *zero* when $A = 7$. Thus it is no surprise that, in both parts b and c, the population $A(t)$ approaches the limiting value 7; when $A < 7$ then $A$ increases, and when $A > 7$ then $A$ decreases.

## Slope Fields

A separable differential equation $f(y) \dfrac{dy}{dt} = g(t)$ specifies the slope of the graph of $y$ versus $t$:

$$\frac{dy}{dt} = \frac{g(t)}{f(y)}.$$

We can present the equation graphically by drawing short line segments with the specified slope at many points in the plane. (This tedious task is best done with a computer program; several are available.) Figure 1 does this for the equation in Example 1, $\dfrac{dA}{dt} = 2(7 - A)$. The slope field shows clearly that every solution has the same limit as $t \to +\infty$:

$$\lim_{t \to +\infty} A(t) = 7.$$

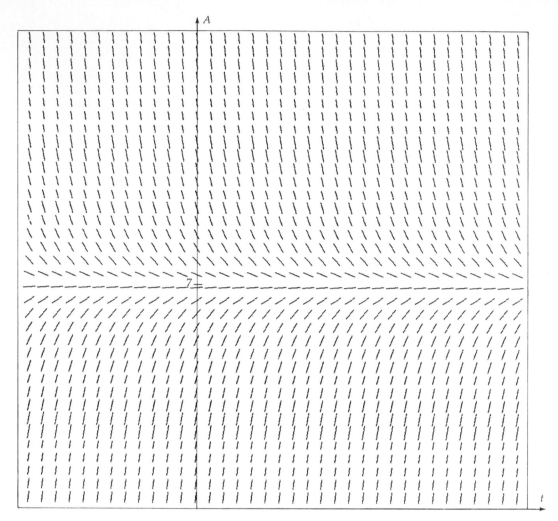

**FIGURE 1**

$\dfrac{dA}{dt} = 2(7 - A)$, $0 \le A \le 15$, $-1 \le t \le 2$.

Figure 2 shows two solutions; at each point $(t, A)$ on the graph of either one, the slope matches that prescribed by the equation $\dfrac{dA}{dt} = 2(7 - A)$.

We conclude with one more example.

---

**EXAMPLE 2**

   **a.**   Solve $\dfrac{dy}{dx} = xy$.

   **b.**   Find a solution such that $y = -1$ when $x = 0$.

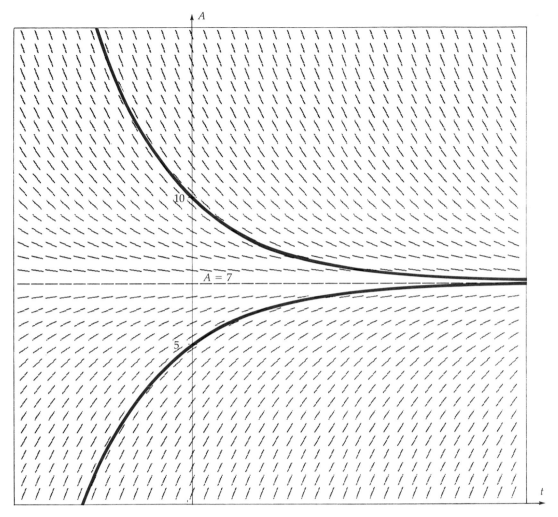

**FIGURE 2**

Slope field and solutions for $\dfrac{dA}{dt} = 2(7 - A)$, $A \geq 0$.

### SOLUTION

**a.** The variables $x$ and $y$ separate:

$$\frac{1}{y} \, dy = x \, dx$$

$$\int \frac{1}{y} \, dy = \int x \, dx$$

$$\ln|y| = \tfrac{1}{2}x^2 + c$$

$$|y| = e^{(1/2)x^2 + c} = e^c e^{(1/2)x^2}.$$

If $y > 0$, then $y = |y| = e^c e^{x^2/2}$; if $y < 0$, then $y = -|y| = -e^c e^{x^2/2}$. Introduce a new constant $k$ with $k = e^c$ if $y > 0$, $k = -e^c$ if $y < 0$, and get in either case the solution

$$y = ke^{x^2/2}. \tag{4}$$

**b.** Now find a solution with $y = -1$ when $x = 0$. (This condition determines the constant $k$). Set $x = 0$ and $y = -1$ in (4):

$$-1 = ke^0 = k.$$

So $k = -1$, and (4) becomes

$$y = -e^{x^2/2}.$$

Figure 3 illustrates the two steps in the solution. *First*, solve the differential equation by separating the variables. This gives many solutions

$$y = ke^{x^2/2},$$

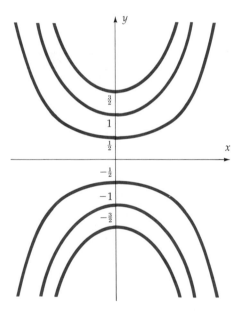

**FIGURE 3**
$y = ke^{x^2/2}$ for $k = 0, \pm\frac{1}{2}, \pm 1, \pm\frac{3}{2}$.

one for each real number $k$. Figure 3 shows the solutions for assorted values of $k$; you can imagine how the rest fit into the picture. *Next*, determine *one particular* solution by the given condition "$y = -1$ when $x = 0$." Geometrically, this means that you pick the solution passing through the point $(0, -1)$, the red one in Figure 3. The condition "$y = -1$ when $x = 0$" is called an **initial condition**. Think of the graph of the solution as starting at $(0, 1)$ and proceeding from there (in both directions) along the appropriate solution curve $y = -e^{x^2/2}$.

Figure 4 shows the slope field for $\dfrac{dy}{dx} = xy$, and the solution with

$y(0) = -1$.

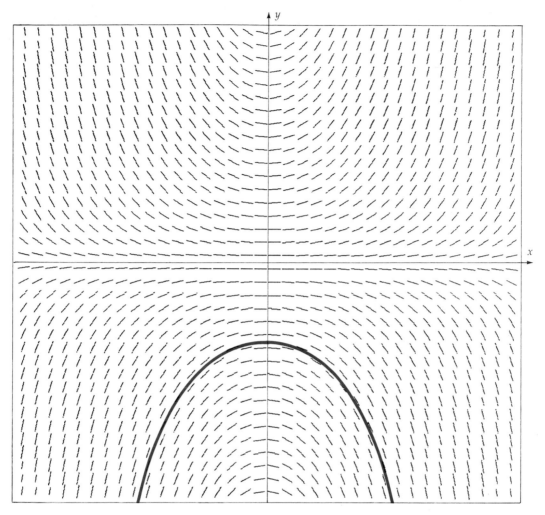

**FIGURE 4**

Slope field for $\dfrac{dy}{dx} = xy$; solution with $y(0) = -1$.

## SUMMARY

An equation which can be written in the form

$$f(y) \frac{dy}{dt} = g(t)$$

is called a *separable differential equation* for the function $y$. It can be solved by "separating the variables"

$$f(y) \, dy = g(t) \, dt$$

and integrating:

$$\int f(y) \, dy = \int g(t) \, dt.$$

## PROBLEMS

### A

1. Solve by separation of variables. Find the solution that satisfies the given initial condition.

   a) $\dfrac{dy}{dx} = 3x^2 + 4$, $y = 1$ when $x = 0$.

   *b) $\dfrac{dy}{dx} = \dfrac{x}{y}$, $x = 2$ when $y = 1$.

   *c) $\dfrac{dy}{dx} = \dfrac{x}{y}$, graph contains the point $(-1, 1)$.

   d) $\sqrt{2xy}\,\dfrac{dy}{dx} = 1$, graph contains the point $(1, 1)$.

   *e) $\dfrac{dy}{dx} = e^{y-x}$, $y = 0$ when $x = 1$.

   *f) $\dfrac{dy}{dx} = \dfrac{x(3 - 2y)}{x^2 + 1}$, $y = 0$ when $x = 0$.

2. Solve the given differential equation, and sketch on a single set of coordinate axes the solutions that pass through the following points: $(0, 0)$, $(0, \pm 1)$, $(0, \pm 2)$, $(0, \pm 3)$.

   *a) $dy/dx = -2xy/(1 + x^2)$

   b) $dy/dx = -xy$

   c) $dy/dx = -x/y$

3. In a "second order" chemical reaction, the concentration $a(t)$ of reactant $A$ varies according to the law $a'(t) = -ka(t)^2$.

   a) Show that for a second order reaction

   $$a(t) = \frac{a_0}{1 + a_0 kt},$$

   where $a_0 = a(0)$.

   b) Compute $\displaystyle\lim_{t \to +\infty} a(t)$.

   *c) Determine $t_{1/2}$, the time at which half the reactant remains.

4. *Newton's law of cooling* applies to an object at temperature $T$ in surroundings of temperature $T_s$. The object loses temperature at a rate proportional to the difference $T - T_s$:

   $$\frac{dT}{dt} = -k(T - T_s), \qquad k > 0. \qquad (5)$$

If $T > T_s$ (for example, $T$ is a hot cup of coffee in a cool room), then $\dfrac{dT}{dt}$ is negative; if $T < T_s$ (a cool drink in a hot room), then $\dfrac{dT}{dt}$ is positive.

   a) Suppose that $T_s = 20°$, $k = 10$. Solve equation (5).

   *b) In part a, find the solution such that $T = 50°$ when $t = 0$. (This is the hot cup of coffee.) Find $\displaystyle\lim_{t \to +\infty} T$. Sketch the graph for $t \geq 0$.

   c) In part a, find the solution such that $T = 0°$ when $t = 0$. (This is a cool drink.) Find $\displaystyle\lim_{t \to +\infty} T$. Sketch the graph for $t \geq 0$.

   d) Solve (5) for arbitrary $k$ and $T_s$, assuming only that $k > 0$. Evaluate $\displaystyle\lim_{t \to +\infty} T(t)$.

5. Temperature $T$ affects the rate constant $k$ for a chemical reaction; the relation between $k$ and $T$ is described by the equation

   $$\frac{dk}{dT} = \frac{kA}{RT^2},$$

   where $A$ and $R$ are constants. Express $k$ as a function of $T$.

6. According to Clapeyron, the vapor pressure $P$ above a liquid is related to the absolute temperature $T$ of the vapor by the equation

   $$\frac{dP}{dT} = \frac{Q}{TV},$$

   where $Q$ is a constant (the molar heat of vaporization) and $V$ is the molar volume of the vapor.

   a) Assume further that $PV = RT$ for a constant $R$. Eliminate $V$ from Clapeyron's equation, solve it, and show that the graph of $\ln P$ versus $T^{-1}$ is a straight line.

   *b) From the graph of $\ln P$ versus $T^{-1}$, what can be determined about the constants $R$ and $Q$?

*7. It follows from Torricelli's law (Chapter 7, Review Problems 10 and 11) that when water flows out of a hole in a cylindrical container, the height

$h$ of the water level above the hole satisfies

$$\frac{dh}{dt} = -\frac{\sqrt{2hg}}{\pi r^2} A, \qquad (6)$$

where $r$ is the radius of the container, $A$ the area of the hole, and $g$ the acceleration due to gravity. Suppose that $h = h_0$ (a given constant) when $t = 0$. Find $h$ as a function of $t$. How long does it take for $h$ to fall to 0?

**8.** Water flows out of a small hole in the bottom of a conical tank, at a rate proportional to the height of water in the tank:

$$\frac{dV}{dt} = -\sqrt{2hg}A, \qquad (7)$$

where $V$ is the volume of water, $h$ the height, $g$ the acceleration due to gravity, and $A$ the area of the hole. The axis of the cone is vertical, with the vertex down.
**a)** Show that $V = \frac{1}{3}\pi h^3 \tan^2\alpha$, where $\alpha$ is the angle between the side of the cone and its axis.
**b)** Solve equation (7) and express $h$ as a function of $t$.

**9.** Water flows out of a small hole in a cylindrical tank (as in problem 7), while more water is poured in at the constant rate $R$.
**a)** What is the new equation corresponding to (6)?
**\*b)** Solve it.
**c)** Show that $\lim\limits_{t \to +\infty} h = R^2/2gA^2$. (This is harder.)

**B**

**10.** Most of the remaining problems involve equations of the form

$$a\frac{dy}{dt} + by = c, \qquad (8)$$

where $a$, $b$, $c$ are constants, with $a \neq 0$ and $b \neq 0$. Derive the solution $y = \frac{c}{b} + ke^{-bt/a}$ ($k$ is an arbitrary constant) in two ways:
**a)** Separate variables.
**b)** Deduce from the equation (8) that $e^{bt/a}(by - c)$ is a constant.
Compare the virtues of the two methods.

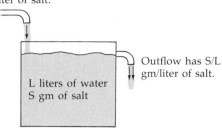

Intake water has Cgm/liter of salt.

Outflow has S/L gm/liter of salt.

L liters of water
S gm of salt

**FIGURE 5**

**\*11.** *Mixing.* At time $t = 0$, a tank contains $L$ liters of water with $S_0$ grams of salt dissolved in it. The tank is being flushed out with slightly salty water, containing $C$ grams of salt per liter. This water flows in at the rate of $v$ liters/sec, is constantly mixed with the water in the tank, and the mixture flows out at $v$ liters/sec. (Fig. 5) Let $S(t)$ be the mass of salt in the tank at time $t$. Then the concentration is $S/L$, so each liter that flows out carries $S/L$ grams of salt with it, and the rate at which salt flows out is $(S/L)v$. Salt flows in at the rate $Cv$, so

$$\frac{dS}{dt} = Cv - \frac{S}{L}v,$$

where $C$, $v$, and $L$ are constants. Solve this equation and find $\lim\limits_{t \to +\infty} S(t)$.

**12.** If glucose is fed intravenously at a constant rate $G$, and transforms into other molecules at a rate proportional to its concentration $c$, then the concentration changes at a rate

$$\frac{dc}{dt} = \frac{G}{V} - kc, \qquad (9)$$

where $V$ is the volume of blood in the body. Solve this equation for the variable concentration $c$. Find the limits $\lim\limits_{t \to +\infty} c(t)$ and $\lim\limits_{t \to +\infty} \frac{dc}{dt}$. (The limit of $c$ is called the *steady state* concentration. Show that it satisfies (9) with $\frac{dc}{dt} = 0$. Why is that reasonable?)

**13.** *Electrical circuits.* In a simple circuit containing a resistance $R$, a capacitance $C$, and a voltage $E$ (Fig. 6), the charge $Q$ in the capacitor satisfies

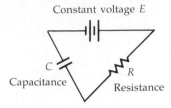

**FIGURE 6**
Circuit.

the differential equation

$$R\frac{dQ}{dt} + \frac{Q}{C} = E.$$

Assume the voltage $E$ is constant.

a) Solve the equation.

b) Sketch the graph of $Q$, for $t \geq 0$, assuming $R = C = E = 1$ and $Q(0) = 2$.

c) Sketch it again, assuming $Q(0) = 0$.

d) Find $\lim\limits_{t \to +\infty} \dfrac{dQ}{dt}$ and $\lim\limits_{t \to +\infty} Q$.

e) Given the limit of $dQ/dt$, could you have determined the limit of $Q$ directly from the differential equation for $Q$? (After a long time, $Q = CE$, and $C = Q/E$. This is why $C$ is called *capacitance*; it is the amount of charge $Q$ held in the capacitor per unit voltage $E$.)

*14. *Electrical circuits.* In a simple circuit containing a resistance $R$, a coil with inductance $L$, and a voltage $E$, the current $I$ satisfies the equation

$$L\frac{dI}{dt} + RI = E.$$

Assuming $E$ is constant, solve the equation; find $\lim\limits_{t \to +\infty} \dfrac{dI}{dt}$ and $\lim\limits_{t \to +\infty} I$. Relate these limits to the differential equation for $I$. (Remark: The current $I$ in a given part of the circuit equals the rate of change of the charge $Q$ in that part, $I = dQ/dt$.)

15. *Electrical circuits.* The equations in problems 13 and 14 can be combined to describe a circuit containing both capacitance and inductance,

$$L\frac{d^2 Q}{dt^2} + R\frac{dQ}{dt} + \frac{Q}{C} = E.$$

Suppose that $E = 0$, and try to find a solution of the equation which has the form $Q = e^{kt}$, for some constant real number $k$. (Substitute $Q = e^{kt}$

into the equation, and solve for $k$.) Show that there are two such solutions if $R^2 > \dfrac{4L}{C}$, one if $R^2 = \dfrac{4L}{C}$, and none if $R^2 < \dfrac{4L}{C}$. (In this last case, the solution requires trigonometric functions; see Sec. 19.5.)

*16. *Falling with friction.* The velocity of an object moving up or down through the atmosphere is affected by air resistance. Assume that this resistance is proportional to the velocity, and opposite in sign, so Newton's law $F = ma = m\,dv/dt$ becomes

$$-mg - kv = m\,dv/dt,$$

where $k$ is a positive constant. (This is apparently realistic for relatively light objects—feathers, or leaves.)

a) Solve this equation for $v$.

b) Find the "limiting velocity" $\lim\limits_{t \to +\infty} v(t)$.

c) Let $v = dy/dt$. Express $y$ in terms of the initial position $y_0 = y(0)$ and initial velocity $v_0 = v(0)$.

17. Let $E$ be the strength of a stimulus applied to a nerve, and $\varepsilon(t)$ be the level of excitement caused in the nerve $t$ seconds after the stimulus began. In a simple model of nerve response,

$$d\varepsilon/dt = k_1 E - k_2\varepsilon(t), \tag{10}$$

where $k_1$ and $k_2$ are positive constants. The term $k_1 E$ represents the growth of the response as the stimulus continues to be applied, while $-k_2\,\varepsilon(t)$ represents a natural "decay" in the response.

a) Solve equation (10), assuming the stimulus $E$ is constant.

b) Find the solution such that $\varepsilon(0) = 0$.

c) Find $\lim\limits_{t \to +\infty} \varepsilon(t)$.

d) Sketch the graph of $\varepsilon(t)$ for $t \geq 0$, using parts b and c.

e) The nerve will cause a response from the brain if and only if $\varepsilon(t)$ reaches a certain "threshold value" $h$. How large must be the stimulus $E$ in order for $\varepsilon(t)$ to reach the threshold value $h$? How long does it take to reach this value?

(For related material see Horelick, Koont, and Gottlieb, "Modeling the Nervous System," UMAP Module 67, available from COMAP, 60 Lowell St., Arlington, MA 02174.)

C

**18.  a)** Solve $x^3 \dfrac{dy}{dx} = y$.

**b)** Explain why the constant of integration for $x < 0$ may be different from the constant for $x > 0$.

**c)** Let

$$f(x) = \begin{cases} c_- e^{-2/2x^2} & \text{if } x < 0 \\ 0 & \text{if } x = 0 \\ c_+ e^{-2/2x^2} & \text{if } x > 0, \end{cases}$$

where $c_-$ and $c_+$ are two constants. Show that $f$ is continuous at 0; that $f'(x) \to 0$ as $x \to 0$; that $f'(0) = 0$; and that $y = f(x)$ is a solution of the equation in part a, for $-\infty < x < \infty$.

**19.** For the function $f(x)$ in the previous problem show that

**a)** $\displaystyle\lim_{x \to 0} x^{-n}f = 0, \quad n = 1, 2, \ldots$

**b)** $f^{(n)}(x) = P_n(1/x)f(x), \quad x \neq 0$, where $P_n$ is a polynomial.

# 8.8

## INVERSE TRIGONOMETRIC FUNCTIONS

Aside from their natural place in trigonometry, the inverse trigonometric functions play a large role in finding antiderivatives, as you will see in the next chapter.

### The Inverse Sine

Given a number $u$ between $-1$ and $+1$, there is exactly one number $\theta$ between $-\pi/2$ and $+\pi/2$ such that $\sin \theta = u$ (Fig. 1). This number $\theta$ is called the **inverse sine** of $u$, denoted $\sin^{-1}u$:

$$\theta = \sin^{-1}u \quad \text{means} \quad \sin \theta = u \quad \text{and} \quad -\frac{\pi}{2} \le \theta \le \frac{\pi}{2}. \tag{1}$$

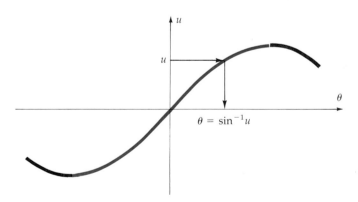

**FIGURE 1**

$u = \sin \theta, \ -\dfrac{\pi}{2} \le \theta \le \dfrac{\pi}{2}; \ \theta = \sin^{-1}u.$

Think of $\sin^{-1}u$ as "the angle (in radians) whose sine is $u$"; the angle in question is to lie between $-\pi/2$ and $\pi/2$. Figure 2 illustrates particular values of

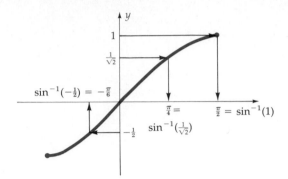

**FIGURE 2**

$\sin^{-1}u$:

$$\sin^{-1}1 = \pi/2, \qquad \sin^{-1}(0) = 0,$$

$$\sin^{-1}(1/\sqrt{2}) = \pi/4, \qquad \sin^{-1}(-1/2) = -\pi/6.$$

The function $\theta = \sin^{-1}u$ is the *inverse* of the function

$$u = \sin\theta, \qquad -\frac{\pi}{2} \leq \theta \leq \frac{\pi}{2} \tag{2}$$

in Figure 1; so the graph of $\theta = \sin^{-1}u$ is obtained from Figure 1 by exchanging the $\theta$ and $u$ axes, thereby rotating the graph around the 45° line $u = \theta$ (Fig. 3); don't rotate it about the origin!

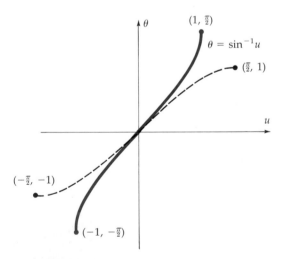

**FIGURE 3**
Graph of inverse sine.

*NOTE 1* $\sin^{-1}u$ is the inverse of the sine *restricted to the interval* $-\pi/2 \leq \theta \leq \pi/2$. *Without* this restriction, the sine is not one-to-one, so it has no well-defined inverse. Restricting to the interval $-\pi/2 \leq \theta \leq \pi/2$ is called "choosing principal values for $\sin^{-1}u$."

*NOTE 2*   The endpoints of the restricted sine graph (Fig. 1) are $\left(-\dfrac{\pi}{2}, -1\right)$ and $\left(\dfrac{\pi}{2}, 1\right)$; so the endpoints of the graph of $\sin^{-1}u$ (Fig. 3) are $\left(-1, -\dfrac{\pi}{2}\right)$ and $(1, \pi/2)$. It is essential to label these when graphing $\theta = \sin^{-1}u$.

## The Derivative of $\sin^{-1}u$

The function $u = \sin\theta$ in Figure 1 has a derivative $\dfrac{du}{d\theta} = \cos\theta$, which is nonzero except at the endpoints $\theta = \pm\pi/2$, where the graph is horizontal. Therefore, by the Inverse Function Theorem (Sec. 8.2), the inverse $\theta = \sin^{-1}u$ has a derivative

$$\frac{d\theta}{du} = \frac{1}{du/d\theta} = \frac{1}{\cos\theta} \tag{3}$$

except at the endpoints $u = -1$ and $u = +1$, where the graph is vertical (Fig. 3). Since $\theta$ is a function of $u$, we express the derivative $d\theta/du$ in terms of $u$, as follows: $u = \sin\theta$, and $\sin^2\theta + \cos^2\theta = 1$, so

$$\cos\theta = \pm\sqrt{1 - \sin^2\theta} = \pm\sqrt{1 - u^2}.$$

Which sign is correct? Since $\theta = \sin^{-1}u$ must lie between $-\pi/2$ and $\pi/2$, then $\cos\theta \geq 0$, so the $+$ sign is correct: $\cos\theta = \sqrt{1 - u^2}$. Hence (3) gives

$$\frac{d\theta}{du} = \frac{1}{\sqrt{1 - u^2}}.$$

(The $+$ sign can also be inferred from Figure 3, which shows that $d\theta/du > 0$.) Recalling that $\theta = \sin^{-1}u$, we have

$$\frac{d\sin^{-1}u}{du} = \frac{1}{\sqrt{1 - u^2}}.$$

In most applications, $u$ is a function of some other variable, say $x$; then by the Chain Rule

$$\frac{d\sin^{-1}u}{dx} = \frac{1}{\sqrt{1 - u^2}}\frac{du}{dx}.$$

**EXAMPLE 1**

$$\frac{d\sin^{-1}(x/4)}{dx} = \frac{1}{\sqrt{1 - (x/4)^2}}\frac{d(x/4)}{dx}$$

$$= \frac{1/4}{\sqrt{1 - x^2/16}} = \frac{1}{\sqrt{16 - x^2}}.$$

*WARNING* $\sin^{-1}u$ does *not* usually mean $(\sin u)^{-1}$ as you might expect! (There is already a special notation for $(\sin u)^{-1}$; it is cosec $u$.) Two alternate notations for the inverse sine avoid this possible confusion. One is *arcsin u*, and the other (used on some calculators) is *invsin u*.

## The Inverse Tangent

Given any real number $u$, there is exactly one number $\theta$ between $-\pi/2$ and $\pi/2$ such that $\tan\theta = u$ (Fig. 4). This number is called the **inverse tangent** of $u$, denoted $\tan^{-1}u$:

$$\theta = \tan^{-1}u \quad \text{means} \quad \tan\theta = u, \quad \text{and} \quad -\frac{\pi}{2} < \theta < \frac{\pi}{2}.$$

This is the inverse of the function $u = \tan\theta$, restricted to the interval $-\pi/2 < \theta < \pi/2$ (Fig. 4). The vertical asymptotes in Figure 4, at $\theta = \pm\pi/2$, become horizontal asymptotes in Figure 5, because axes are exchanged.

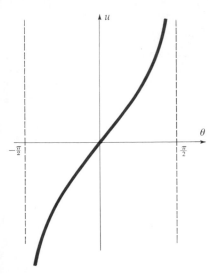

**FIGURE 4**

$u = \tan\theta, \ -\dfrac{\pi}{2} < \theta < \dfrac{\pi}{2}.$

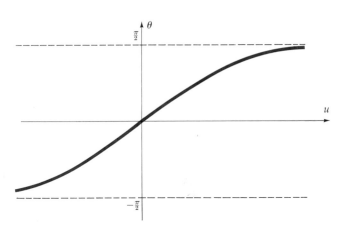

**FIGURE 5**

$\theta = \tan^{-1}u, \ -\infty < u < \infty.$

The derivative of $\theta = \tan^{-1}u$ is

$$\frac{d\theta}{du} = \frac{1}{du/d\theta} = \frac{1}{\sec^2\theta} \qquad (\text{because } u = \tan\theta)$$

$$= \frac{1}{\tan^2\theta + 1} = \frac{1}{u^2 + 1}.$$

Thus $\dfrac{d\tan^{-1}u}{du} = \dfrac{1}{1 + u^2}$ and, by the Chain Rule,

$$\frac{d\tan^{-1}u}{dx} = \frac{1}{1 + u^2}\frac{du}{dx}.$$

**EXAMPLE 2** A picture of height $c$ hangs on a wall, with its base $b$ units above eye level (Fig. 6). Compute the angle $\theta$ subtended by the picture at an eye which is $a$ units away from the wall.

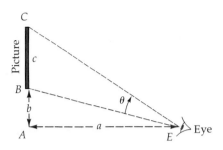

**FIGURE 6**

*SOLUTION* $\theta = \angle CEA - \angle BEA$, and

$$\tan(\angle CEA) = \frac{b+c}{a}, \qquad \tan(\angle BEA) = \frac{b}{a}.$$

By definition of the inverse tangent,

$$\angle CEA = \tan^{-1}\left(\frac{b+c}{a}\right), \qquad \angle BEA = \tan^{-1}\left(\frac{b}{a}\right).$$

Hence

$$\theta = \tan^{-1}\left(\frac{b+c}{a}\right) - \tan^{-1}\left(\frac{b}{a}\right).$$

*The other inverse trigonometric functions* can be defined in a similar way. The summary does this for the cosine and the secant; the cotangent and cosecant are left as problems.

### Evaluating the Inverse Functions

For *positive* inputs $u$, the values of $\sin^{-1}u$, $\tan^{-1}u$, and so on can be interpreted as the radian measure of an angle in an appropriate right triangle. For example, in Figure 7

$$\sin \theta = b/c, \qquad \cos \theta = a/c, \qquad \tan \theta = b/a, \qquad \sec \theta = c/a,$$

so

$$\theta = \sin^{-1}\left(\frac{b}{c}\right) = \cos^{-1}\left(\frac{a}{c}\right) = \tan^{-1}\left(\frac{b}{a}\right) = \sec^{-1}\left(\frac{c}{a}\right).$$

From the triangles in Figure 8,

$$\cos^{-1}(1/\sqrt{2}) = \pi/4, \qquad \sec^{-1}2 = \pi/3, \qquad \tan^{-1}(1/\sqrt{3}) = \pi/6.$$

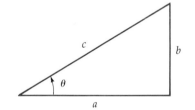

**FIGURE 7**

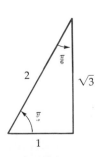

**FIGURE 8**

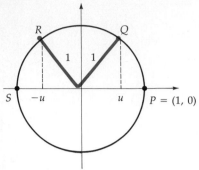

**FIGURE 9**
$\cos^{-1}(-u) = \overset{\frown}{PQR} = \pi - \overset{\frown}{SR}$
$= \pi - \overset{\frown}{PQ} = \pi - \cos^{-1}u.$

To compute the inverse trigonometric functions with *negative* inputs, you can use the following identities:

$$\sin^{-1}(-u) = -\sin^{-1}(u) \qquad (4)$$

$$\tan^{-1}(-u) = -\tan^{-1}(u) \qquad (5)$$

$$\cos^{-1}(-u) = \pi - \cos^{-1}(u) \qquad (6)$$

$$\sec^{-1}(-u) = \pi - \sec^{-1}(u). \qquad (7)$$

The identities (4) and (5) are obvious from the graphs (Figs. 3 and 5); $\sin^{-1}$ and $\tan^{-1}$ are odd functions. Formula (6) is explained in Figure 9; $\cos^{-1}(-u)$ is represented on the upper half of the unit circle [since $0 \le \cos^{-1}(-u) \le \pi$] and the rest of the proof is explained in the figure. The proof of (7) is nearly the same.

---

**EXAMPLE 3**

$$\cos^{-1}(-1/\sqrt{2}) = \pi - \cos^{-1}(1/\sqrt{2}) = \pi - \pi/4 = 3\pi/4;$$
$$\tan^{-1}(-1/\sqrt{3}) = -\tan^{-1}(1/\sqrt{3}) = -\pi/6.$$

---

**Combining trigonometric with inverse trigonometric functions** is best done by drawing an appropriate right triangle.

---

**EXAMPLE 4**  Simplify $\cos(\sin^{-1}x)$.

*SOLUTION*  Draw a right triangle with an angle $\theta = \sin^{-1}x$; Figure 10 does this, since in that figure $\sin\theta = x$. The side adjacent to $\theta$ is computed as $\sqrt{1 - x^2}$, from the Pythagoras Theorem. Thus the triangle shows that

$$\cos(\sin^{-1}x) = \cos\theta = \sqrt{1 - x^2}.$$

The figure is valid for $0 < x < 1$. But in any case, $\theta = \sin^{-1}x$ lies between $-\pi/2$ and $\pi/2$, by definition, so $\cos\theta \ge 0$, and

$$\cos(\sin^{-1}x) = \cos\theta = +\sqrt{1 - \sin^2\theta} = \sqrt{1 - x^2}.$$

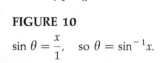

**FIGURE 10**
$\sin\theta = \dfrac{x}{1},$   so $\theta = \sin^{-1}x.$

---

**Relations between the Inverse Trigonometric Functions**

In the right triangle in Figure 11, $u = \cos\left(\dfrac{\pi}{2} - \alpha\right)$, and $\alpha = \sin^{-1}u$, so

$$\cos^{-1}u = \frac{\pi}{2} - \alpha = \frac{\pi}{2} - \sin^{-1}u.$$

This same identity can be proved formally, using the cofunction identity $\sin\left(\dfrac{\pi}{2} - \theta\right) = \cos\theta$; just keep careful track of the range of angles required

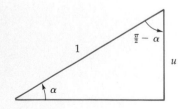

**FIGURE 11**
$\cos^{-1}u = \dfrac{\pi}{2} - \alpha = \dfrac{\pi}{2} - \sin^{-1}u.$

by the inverse functions at each step:

$$\theta = \cos^{-1}u \quad \Leftrightarrow \quad \cos\theta = u \quad \text{and} \quad 0 \le \theta \le \pi$$

$$\Leftrightarrow \quad \sin\left(\frac{\pi}{2} - \theta\right) = u \quad \text{and} \quad \frac{-\pi}{2} \le \frac{\pi}{2} - \theta \le \frac{\pi}{2}$$

$$\Leftrightarrow \quad \frac{\pi}{2} - \theta = \sin^{-1}u \qquad \text{[by (1)]}$$

$$\Leftrightarrow \quad \theta = \frac{\pi}{2} - \sin^{-1}u,$$

which proves that

$$\cos^{-1}u = \frac{\pi}{2} - \sin^{-1}u. \tag{8}$$

Other cofunction identities are

$$\cot^{-1}u = \frac{\pi}{2} - \tan^{-1}u \tag{9}$$

$$\csc^{-1}u = \frac{\pi}{2} - \sec^{-1}u. \tag{10}$$

Finally, the relations

$$\tan\theta = 1/\cot\theta, \qquad \sec\theta = 1/\cos\theta$$

yield formulas

$$\cot^{-1}u = \tan^{-1}(1/u) \tag{11}$$

$$\sec^{-1}u = \cos^{-1}(1/u). \tag{12}$$

**REMARK**   Some books define $\sec^{-1}u$ differently; but not all these identities remain valid with those other definitions.

## SUMMARY

***Definitions and Graphs:***   See Figures 12–15.

***Derivatives:***

$$\frac{d\,\sin^{-1}u}{du} = \frac{1}{\sqrt{1-u^2}}, \qquad \frac{d\,\cos^{-1}u}{du} = \frac{-1}{\sqrt{1-u^2}}$$

$$\frac{d\,\tan^{-1}u}{du} = \frac{1}{1+u^2}, \qquad \frac{d\,\sec^{-1}u}{du} = \frac{1}{|u|\sqrt{u^2-1}}$$

***Symmetries:***

$$\sin^{-1}(-u) = -\sin^{-1}(u), \qquad \cos^{-1}(-u) = \pi - \cos^{-1}(u)$$

$$\tan^{-1}(-u) = -\tan^{-1}(u), \qquad \sec^{-1}(-u) = \pi - \sec^{-1}(u)$$

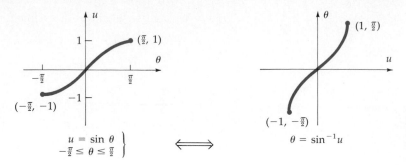

$$u = \sin \theta \left.\right\} \qquad \Longleftrightarrow \qquad \theta = \sin^{-1} u$$
$$-\tfrac{\pi}{2} \le \theta \le \tfrac{\pi}{2}$$

**FIGURE 12**

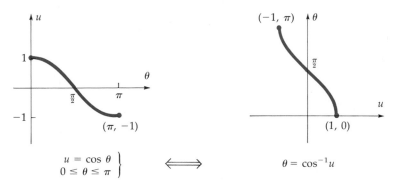

$$u = \cos \theta \left.\right\} \qquad \Longleftrightarrow \qquad \theta = \cos^{-1} u$$
$$0 \le \theta \le \pi$$

**FIGURE 13**

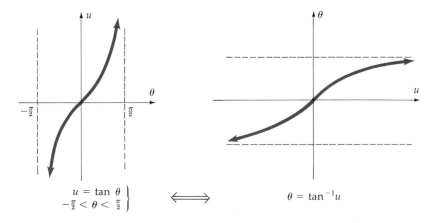

$$u = \tan \theta \left.\right\} \qquad \Longleftrightarrow \qquad \theta = \tan^{-1} u$$
$$-\tfrac{\pi}{2} < \theta < \tfrac{\pi}{2}$$

**FIGURE 14**

*Cofunction Identities:*

$$\cos^{-1} u = \frac{\pi}{2} - \sin^{-1} u, \qquad \cot^{-1} u = \frac{\pi}{2} - \tan^{-1} u$$

$$\sec^{-1} u = \frac{\pi}{2} - \csc^{-1} u$$

*Reciprocals:*

$$\sec^{-1} u = \cos^{-1}(1/u), \qquad \cot^{-1} u = \tan^{-1}(1/u)$$

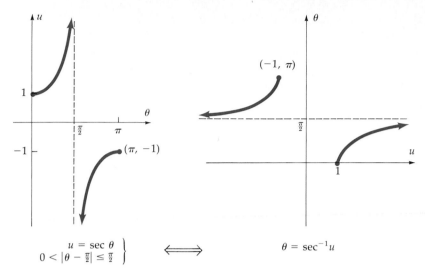

$$u = \sec \theta$$
$$0 < |\theta - \tfrac{\pi}{2}| \le \tfrac{\pi}{2}$$
$$\Longleftrightarrow$$
$$\theta = \sec^{-1} u$$

**FIGURE 15**

## PROBLEMS

### A

*1. Evaluate.
  a) $\sin^{-1}(\tfrac{1}{2})$
  b) $\cos^{-1}(-1/2)$
  c) $\tan^{-1}(1)$
  d) $\tan^{-1}(-\sqrt{3})$
  e) $\tan^{-1}(1/\sqrt{3})$
  f) $\tan^{-1}0$
  g) $\cos^{-1}0$
  h) $\sec^{-1}(\sqrt{2})$
  i) $\sec^{-1}(-\sqrt{2})$

2. Evaluate.
  a) $\sin^{-1}(-1)$
  b) $\tan^{-1}(-1)$
  c) $\sec^{-1}(1)$
  d) $\cos^{-1}(-1)$
  e) $\cot^{-1}(1)$
  f) $\tan^{-1}(\sqrt{3})$
  g) $\sin^{-1}(-1/2)$

3. If $\theta = \sin^{-1}(1/2)$, find the numerical values of $\sin \theta$, $\cos \theta$, $\tan \theta$, $\sec \theta$. (Draw a triangle.)

*4. If $\theta = \tan^{-1}(3/4)$ find the numerical values of $\sin \theta$, $\cos \theta$, $\tan \theta$, $\cot \theta$, $\sec \theta$. (Draw a triangle.)

5. If $\theta = \sin^{-1}(-1/3)$, find the numerical values of $\sin \theta$, $\cos \theta$, and $\tan \theta$.

6. Give the domain and the range.
  a) $\sin^{-1}u$
  b) $\cos^{-1}u$
  c) $\tan^{-1}u$
  d) $\cot^{-1}u$
  e) $\sec^{-1}u$
  f) $\csc^{-1}u$

7. Find.
  a) $\sin^{-1}(\sin \pi/5)$
  b) $\sin^{-1}[\sin(\pi/5 + 2\pi)]$
  *c) $\sin^{-1}(\sin(9\pi/5))$
  *d) $\sin(\cos^{-1}x)$

  e) $\sec(\tan^{-1}x)$
  f) $\tan(\sec^{-1}x)$

8. For which values of $\theta$ or $u$ is
  *a) $\sin^{-1}(\sin \theta) = \theta$?
  *b) $\sin(\sin^{-1}u) = u$?
  c) $\cot^{-1}(\cot \theta) = \theta$?
  d) $\cos(\cos^{-1}u) = u$?

9. Take the derivative.
  *a) $\sin^{-1}(2x)$
  *b) $\tan^{-1}(x^2)$
  *c) $\cos^{-1}(2x)$
  *d) $x \tan^{-1}x$
  e) $x \sin^{-1}x$
  *f) $\sin(\cos^{-1}x)$
  *g) $\cot(\tan^{-1}x)$
  *h) $\sec^{-1}(e^x)$
  i) $\tan^{-1}(\cot x)$
  j) $x \sin^{-1}x + \sqrt{1-x^2}$

10. Obtain the formula for the derivative.
  a) $\cos^{-1}u$
  b) $\sec^{-1}u$ (Consider separately $u > 0$ and $u < 0$; use the principal values in Fig. 15.)
  c) $\cot^{-1}u$
  d) $\csc^{-1}u$

11. Prove the following identities, for $u > 0$, using a right triangle as in Figure 7 or Figure 11.

  a) $\cot^{-1}u = \dfrac{\pi}{2} - \tan^{-1}u$

  b) $\csc^{-1}u = \dfrac{\pi}{2} - \sec^{-1}u$

  c) $\cot^{-1}u = \tan^{-1}(1/u)$
  d) $\sec^{-1}u = \cos^{-1}(1/u)$

**12.** Give a formal proof for the identity $\sec^{-1}u = \cos^{-1}(1/u)$.

**13.**
**a)** Obtain the formula for the derivative of $\cos^{-1}u$, using formula (8).
**b)** Obtain the formula for the derivative of $\sec^{-1}u$, using formula (12) and part a. (Hint: $|u| = \sqrt{u^2}$.)

**14.** Define and graph the function.
**a)** $\cot^{-1}u$ **b)** $\csc^{-1}u$

**15.** In Example 2, take $b$ and $c$ as constants, and $a$ as variable. For what value of $a$ is the subtended angle $\theta$ a maximum?

**16.** Find $dy/dx$ by implicit differentiation.
*a)** $x^3 + x\tan^{-1}y = e^y$
**b)** $\sin^{-1}(x - y) = \cos^{-1}(xy)$

**17.** A compass has a scale giving the radius of the circle to be drawn (Fig. 16).

**a)** If the legs have length 5 cm, at what angle $\theta$ should radius 3 cm be marked?
**b)** If the legs have length $l$, at what angle $\theta$ should radius $r$ be marked?

**B**

**18.**
**a)** Express the area under graph of $y = \sqrt{1-x^2}$ for $0 \le x \le u$ (Fig. 17) in terms of an inverse trigonometric function. (The area is the sum of a circular sector and a triangle. Express the angle of the circular sector in terms of $u$.)
**b)** Differentiate your expression in part a, and relate the derivative to the equation of the graph.

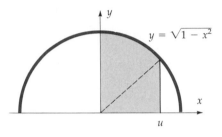

**FIGURE 17**
$\int_0^u \sqrt{1-x^2}\, dx.$

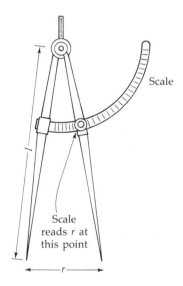

Scale

Scale reads $r$ at this point

$r$

**FIGURE 16**

# 8.9

## OSCILLATIONS (optional)

This section studies the modeling of simple oscillations by sine and cosine functions. The link between the physical oscillations and these functions is a differential equation.

The graph of $y = \cos\theta$ oscillates smoothly up and down (Fig. 1), repeating its motion again and again; it is a natural tool for modeling oscillatory phenomena. For example, the alternating voltage in a typical electrical outlet

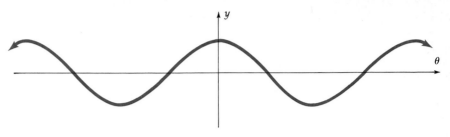

**FIGURE 1**
$y = \cos \theta$.

is given by

$$V = 170 \cos(120\pi t).$$

Since $\cos \theta$ varies between $+1$ and $-1$, the voltage $V$ varies between 170 and $-170$; the **amplitude** is 170. The maximum voltage is achieved whenever $\cos(120\pi t) = 1$, thus when

$$120\pi t = 0, \ \pm 2\pi, \ \pm 4\pi, \ldots$$

or

$$t = 0, \ \frac{\pm 1}{60}, \ \frac{\pm 2}{60}, \ldots$$

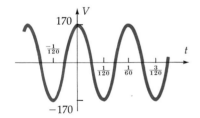

**FIGURE 2**
Alternating voltage
$V = 170 \cos(120\pi t)$.

as shown in Figure 2. The pattern of oscillation repeats every 1/60 second; it has **period** $T = \frac{1}{60}$. The part of the graph corresponding to one of these periods is called a **cycle**. There are 60 cycles per second; this is the **frequency** of the oscillation.

Figure 3 shows the graph of a *general* cosine wave

$$y = A \cos(\omega t - \varphi), \qquad A > 0.$$

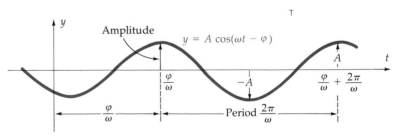

**FIGURE 3**
Cosine wave. $y = A \cos(\omega t - \varphi)$.

The graph is obtained from this formula as follows. Note that $A \cos(\omega t - \varphi)$ achieves its maximum $A$ when $\cos(\omega t - \varphi) = 1$, thus when

$$\omega t - \varphi = 0, \ \pm 2\pi, \ \pm 4\pi, \ldots \quad \text{or} \quad t = \frac{\varphi}{\omega}, \ \frac{\varphi \pm 2\pi}{\omega}, \ \frac{\varphi \pm 4\pi}{\omega}, \ldots.$$

The minimum $-A$ is achieved when $\cos(\omega t - \varphi) = -1$, and this occurs when

$$\omega t - \varphi = \pi, \quad \pi \pm 2\pi, \quad \pi \pm 4\pi, \ldots \quad \text{or} \quad t = \frac{\varphi + \pi}{\omega}, \quad \frac{\varphi + \pi \pm 2\pi}{\omega}, \ldots.$$

These values of $t$ give the points plotted in Figure 3.

The maximum $A$ is the *amplitude* of the wave. The time interval between "peaks" is the *period* $T$, the time required for one complete cycle of the oscillation; as shown in Figure 3,

$$\text{period } T = \frac{2\pi}{\omega}.$$

Since the period is the time for one cycle, its reciprocal is the number of cycles per unit per time, called the *frequency*:

$$\text{frequency} = \frac{1}{T} = \frac{\omega}{2\pi}.$$

Finally, the constant $\varphi$ is the **phase shift**; if $\varphi = 0$, the graph peaks when $t = 0$, and in general it peaks when $t = \varphi/\omega$.

### The Differential Equation of the Cosine Wave

Suppose that the cosine wave

$$y = A \cos(\omega t - \varphi) \tag{1}$$

in Figure 3 describes the motion of an object oscillating up and down on the $y$-axis. The velocity of this motion is

$$v = \frac{dy}{dt} = A[-\sin(\omega t - \varphi)] \frac{d(\omega t - \varphi)}{dt} = -\omega A \sin(\omega t - \varphi).$$

Differentiating once more gives the acceleration

$$a = \frac{d^2 y}{dt^2} = -\omega^2 A \cos(\omega t - \varphi).$$

Compare this acceleration to the given equation (1) for $y$, and you see that the cosine wave satisfies the relation

$$\frac{d^2 y}{dt^2} = -\omega^2 y. \tag{2}$$

This differential equation, relating $y$ to its second derivative, has a simple graphical interpretation: When $y$ is positive then $d^2y/dt^2$ is negative. Thus where the graph in Figure 3 is above the $t$-axis ($y > 0$), it is concave down ($d^2y/dt^2 < 0$); and where the graph is below the axis, it is concave up.

There is a closely related physical interpretation. Suppose that our oscillating object has mass $m$. By (2), and Newton's law $F = ma$, the force acting on the object is

$$F = m \frac{d^2y}{dt^2} = -m\omega^2 y. \tag{3}$$

When $y > 0$, this force $F$ is negative, pushing the object back down toward $y = 0$; and when $y < 0$, then $F$ is positive, pushing the object back up. This is called a **restoring force**, since it always pushes the object back toward $y = 0$. Moreover, according to (3), the restoring force is proportional to the displacement $y$; the constant of proportionality is $-m\omega^2$. Summing up: *The cosine wave $y = A\cos(\omega t - \varphi)$ models the motion of an object of mass $m$, subject to a restoring force $F = -m\omega^2 y$, which is proportional to the displacement $y$.*

### The Oscillating Spring

A classic example of oscillation is the up-and-down motion of a mass on a spring (Fig. 4). To understand the motion, we analyze the forces acting on the mass (ignoring friction, for simplicity). At rest, the spring force exactly cancels the force of gravity, and the net force $F$ acting on the mass is 0. We set up a vertical axis, with $y = 0$ at the rest position of the mass. If the mass is raised to a position $y > 0$, then the spring relaxes and no longer completely cancels the force of gravity, so there is a net *downward* force. How large is it? According to Hooke's law for elastic springs, the decrease in the spring force is proportional to the displacement $\Delta y$. Since the net force $F$ is 0 when $y = 0$, and its change is proportional to the change in $y$, it follows that

$$F = -ky; \tag{4}$$

$k$ is the constant of proportionality, called the spring constant. The minus sign indicates that $F$ is downward when $y > 0$. Thus we have a restoring force proportional to the displacement, and can model the motion with a cosine wave.

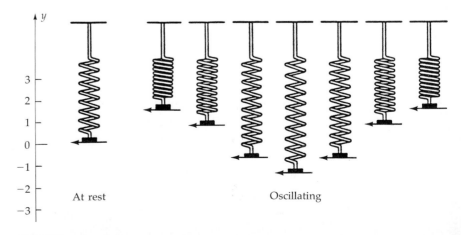

| | | |
|---|---|---|
| At rest | | Oscillating |

**FIGURE 4**
Mass on a spring.

We combine the force equation (4) with Newton's law $F = ma = my''$, obtaining

$$my'' = -ky. \tag{5}$$

This is the differential equation for the position $y$, called the **equation of motion**; it accounts for the forces acting on the mass in the course of the oscillation.

This equation alone does not describe the motion completely; we also need to know how it starts: At time $t = 0$, how far is it above equilibrium? How fast is it moving? That is, along with the equation of motion (5), we need the *initial position*

$$y(0) = y_0 \tag{6}$$

and *initial velocity*

$$y'(0) = v_0. \tag{7}$$

We will prove in Theorem 1 that the equation of motion and the initial position and velocity determine a unique cosine wave; this wave describes the motion of the mass on the spring (ignoring friction).

---

**EXAMPLE 1**  A mass of 50 g hangs on a spring with spring constant $k = 33$. It is raised to 7 cm above equilibrium and at time 0 is released with velocity 0. Find a cosine wave satisfying the corresponding equation of motion (5) and initial conditions (6) and (7).

*SOLUTION*  The equation of motion (5), with $m = 50$ and $k = 33$, gives

$$y'' = -\tfrac{33}{50}y. \tag{8}$$

The initial conditions are

$$y(0) = 7 \qquad \text{(raised to height 7)} \tag{9}$$

$$y'(0) = 0 \qquad \text{(released with velocity 0)}. \tag{10}$$

You are to satisfy these with a cosine wave

$$y = A\cos(\omega t - \varphi). \tag{11}$$

As we saw in (2), this cosine wave satisfies

$$y'' = -\omega^2 y.$$

Compare this to the equation of motion (8), and you see that we need $\omega^2 = \tfrac{33}{50}$; so take

$$\omega = \sqrt{\tfrac{33}{50}}. \tag{12}$$

Turn now to the initial conditions (9) and (10). For the cosine wave (11),

$$y(0) = A\cos(0 - \varphi) = A\cos\varphi,$$

since cosine is an even function. Thus to obtain the desired initial position (9), you need

$$A\cos\varphi = 7. \tag{13}$$

The initial velocity of the cosine wave is

$$y'(0) = -\omega A \sin(0 - \varphi) = \omega A \sin \varphi$$

since sine is an odd function. Thus to obtain the desired initial velocity (10) you need

$$\omega A \sin \varphi = 0. \qquad (14)$$

From (12) and (13), $\omega \neq 0$ and $A \neq 0$, so equation (14) requires $\sin \varphi = 0$; this is satisfied if

$$\varphi = 0.$$

Then $\cos \varphi = \cos 0 = 1$, so (13) gives

$$A = 7.$$

Now all the conditions are satisfied by the cosine wave (11) with these values of $\omega$, $\varphi$, and $A$:

$$y = 7 \cos(\sqrt{\tfrac{33}{50}}t).$$

The period is $T = 2\pi/\omega = 2\pi\sqrt{50/33}$, the amplitude is 7, and the phase shift is 0.

---

It was easy to check that $A \cos(\omega t - \varphi)$ solves the equation $y'' = -\omega^2 y$. But could there possibly be some other type of solution as well? The answer is "no."

---

**THEOREM 1**

Suppose that $y$ is any function satisfying the equation

$$y''(t) = -\omega^2 y(t), \qquad \omega \text{ constant} \qquad (15)$$

for all $t$ in some interval $I$. Then there are constants $A$ and $\varphi$ such that

$$y(t) = A \cos(\omega t - \varphi) \qquad (16)$$

for all $t$ in the interval $I$.

---

The proof consists of some reasonable manipulations which lead, magically, to the trigonometric solution.

First, we rewrite the given equation $y'' = -\omega^2 y$ as an equation in the velocity $v = dy/dt$; for

$$y'' = \frac{dv}{dt} = \frac{dv}{dy} \cdot \frac{dy}{dt} = v \cdot \frac{dv}{dy}$$

so equation (15) gives

$$v \cdot \frac{dv}{dy} = -\omega^2 y.$$

This is solved by separating variables:

$$\int v\,dv = -\omega^2 \int y\,dy$$

$$\Rightarrow$$

$$\tfrac{1}{2}v^2 = -\tfrac{1}{2}\omega^2 y^2 + c.$$

Since $2c = v^2 + \omega^2 y^2$, this constant cannot be negative, and it is legitimate to write it as a square. For future convenience set $2c = C^2$; then

$$v^2 = C^2 - \omega^2 y^2.$$

Next, replace $v$ with $dy/dt$ and solve for $y$:

$$\frac{dy}{dt} = v = \pm\sqrt{C^2 - \omega^2 y^2}$$

$$\Rightarrow$$

$$\int \frac{dy}{\sqrt{C^2 - \omega^2 y^2}} = \pm\int dt = \pm t + k. \tag{17}$$

The integral on the left recalls the derivative of the inverse sine:

$$\frac{d\,\sin^{-1}u}{du} = \frac{1}{\sqrt{1 - u^2}}.$$

You can reduce the integral (17) to this form. Factor out $C$ from the square root and substitute $u = \omega y/C$:

$$\int \frac{dy}{\sqrt{C^2 - \omega^2 y^2}} = \frac{1}{C}\int \frac{dy}{\sqrt{1 - \omega^2 y^2/C^2}} = \frac{1}{\omega}\int \frac{du}{\sqrt{1 - u^2}}$$

$$= \frac{1}{\omega}\sin^{-1}u = \frac{1}{\omega}\sin^{-1}(\omega y/C).$$

Thus (17) gives

$$\frac{1}{\omega}\sin^{-1}(\omega y/C) = \pm t + k.$$

Solve for $y$:

$$\sin^{-1}(\omega y/C) = \pm\omega t + k\omega,$$

$$\frac{\omega y}{C} = \sin(\pm\omega t + k\omega),$$

$$y = \frac{C}{\omega}\sin(\pm\omega t + k\omega).$$

With the "$+$" sign, this reduces to

$$y = \frac{C}{\omega}\cos\left[\omega t - \left(\frac{\pi}{2} - k\omega\right)\right].$$

and with the "$-$" sign to

$$y = \frac{C}{\omega} \cos\left[\omega t - \left(k\omega - \frac{\pi}{2}\right)\right].$$

In either case, we find the desired form (16).

*REMARK*   There is a detail missing in this proof: The equation (17) might at some point switch from "$+t + k$" to "$-t + \tilde{k}$", where $\tilde{k}$ is a constant different from $k$; in fact, this *does* happen when the oscillation passes through its maximum. Nevertheless, the theorem is true; the problems outline a more sophisticated proof with no missing details.

## Other Forms of the Cosine Wave

The addition formula

$$\cos(\theta - \varphi) = \cos\theta\cos\varphi + \sin\theta\sin\varphi$$

gives an alternate way to write the cosine wave:

$$
\begin{aligned}
A\cos(\omega t - \varphi) &= A[\cos(\omega t)\cos\varphi + \sin(\omega t)\sin\varphi] \\
&= (A\cos\varphi)\cos\omega t + (A\sin\varphi)\sin\omega t.
\end{aligned}
$$

We introduce new constants

$$a = A\cos\varphi \quad \text{and} \quad b = A\sin\varphi \tag{18}$$

and obtain the cosine wave in the form

$$y = a\cos\omega t + b\sin\omega t. \tag{19}$$

Thus every cosine wave can be written as a combination of $\sin\omega t$ and $\cos\omega t$, as in (19). Conversely, every function of the form (19) can be written as a cosine wave by solving the equations (18) for $A$ and $\varphi$. $A$ is determined by squaring and adding,

$$a^2 + b^2 = A^2(\cos^2\varphi + \sin^2\varphi) = A^2, \qquad \text{so } A = (a^2 + b^2)^{1/2};$$

then the angle $\varphi$ is determined by

$$\cos\varphi = a/A, \qquad \sin\varphi = b/A. \tag{20}$$

---

**EXAMPLE 2**   Write

$$y = 2\cos 3t + 2\sin 3t$$

as a cosine wave.

*SOLUTION*   The amplitude is

$$A = (a^2 + b^2)^{1/2} = (2^2 + 2^2)^{1/2} = 2\sqrt{2}.$$

The phase shift $\varphi$ satisfies [see (20)]

$$\cos \varphi = \frac{a}{A} = \frac{2}{2\sqrt{2}} = \frac{1}{\sqrt{2}}, \qquad \sin \varphi = \frac{b}{A} = \frac{2}{2\sqrt{2}} = \frac{1}{\sqrt{2}}.$$

You can recognize $(1/\sqrt{2}, 1/\sqrt{2})$ as the point $\pi/4$ units along the unit circle, so

$$1/\sqrt{2} = \cos \pi/4, \qquad 1/\sqrt{2} = \sin \pi/4.$$

Thus $\varphi = \pi/4$, $A = 2\sqrt{2}$, and the desired form is

$$y = 2\sqrt{2} \cos(3t - \pi/4).$$

(As a check, expand this by the addition formula, to be sure that you get the given equation $y = 2 \cos 3t + 2 \sin 3t$.)

---

Since every cosine wave can be written in the alternate form (19), Theorem 1 can be rephrased: *Every solution of the equation $y''(t) = -\omega^2 y(t)$ can be written in the form*

$$y = a \cos \omega t + b \sin \omega t, \qquad \text{where } a \text{ and } b \text{ are constants.}$$

---

**EXAMPLE 3** Find the solution of

$$y'' = -49y \qquad \text{(differential equation)}$$

$$y(0) = 3, \qquad y'(0) = 28 \qquad \text{(initial conditions)}.$$

*SOLUTION* Apply Theorem 1 with $\omega^2 = 49$, $\omega = 7$; any solution of the differential equation $y'' = -49y$ can be written in the form

$$y = a \cos 7t + b \sin 7t.$$

We compare the initial position and velocity for this solution to the desired initial conditions and thus determine the constants $a$ and $b$. The initial position and velocity for $y = a \cos 7t + b \sin 7t$ are

$$y(0) = a \cos 0 + b \sin 0 = a \qquad [\cos 0 = 1, \sin 0 = 0]$$

$$y'(0) = -7a \sin 0 + 7b \cos 0 = 7b.$$

Compare this with the given initial conditions $y(0) = 3$, $y'(0) = 28$ and find

$$a = 3 \quad \text{and} \quad 7b = 28, \qquad \text{or } b = 4.$$

The desired solution is thus

$$y = 3 \cos 7t + 4 \sin 7t.$$

The amplitude of the waves is $A = \sqrt{a^2 + b^2} = \sqrt{3^2 + 4^2} = 5$, and the phase $\varphi$ is determined by

$$\cos \varphi = a/A = 3/5, \qquad \sin \varphi = b/A = 4/5.$$

## SUMMARY

*Differential Equation of Cosine Wave:*

$$y'' = -\omega^2 y \Leftrightarrow y = A \cos(\omega t - \varphi)$$

$\omega$ = angular velocity;   $2\pi/\omega$ = period;   $\omega/2\pi$ = frequency;

$|A|$ = amplitude;   $\varphi$ = phase angle.

*Alternate Form of Solution:*

$$A \cos(\omega t - \varphi) = a \cos \omega t + b \sin \omega t$$

with

$$A \cos \varphi = a, \qquad A \sin \varphi = b, \qquad A^2 = a^2 + b^2.$$

## PROBLEMS

### A

1. Sketch; show two successive maxima, the period, and the amplitude.
   a) $3 \cos(\pi t)$
   *b) $2 \cos(\pi t - 1)$
   c) $(1/2) \cos(2t - \pi/4)$
   d) $3 \sin(t/2 + 1)$

2. Write the following functions in the form of a cosine wave $A \cos(\omega t - \phi)$. Determine the amplitude, the period $T$, and the frequency $1/T$. Sketch the graph, showing at least two successive maxima.
   a) $\sin t$
   *b) $\cos 3t - \sin 3t$
   *c) $5 \sin(2\pi t + 1)$
   d) $2 \cos \pi t + 3 \sin \pi t$

3. A simple electrical generator for alternating current contains coils of wire rotating through a magnetic field. The voltage $E$ produced is very nearly a sine wave with amplitude 170 volts and frequency 60 cycles/second. Write the equation for $E$. (The phase shift is not determined by the given information; leave it as "$\varphi$".)

4. Solve.
   a) $y'' = -y$;   $y(0) = 1, y'(0) = 0$
   b) $y'' = -y$;   $y(0) = 0, y'(0) = 1$
   *c) $y'' = -4y$;   $y(0) = 1, y'(0) = 2$
   d) $y'' = -5y$;   $y(0) = 1, y'(0) = 1$

5. A mass of 30g hangs on a spring with spring constant $k = 17$.
   a) Solve the appropriate equation of motion, and determine the period of the oscillation. (The amplitude $A$ and phase $\varphi$ are not determined by the given information; leave them in your solution as unknown constants.)
   *b) Given the initial conditions $y(0) = 0$, $y'(0) = 1$, determine the amplitude and phase of the oscillation.

6. A mass of 5g hangs on a spring with spring constant $k = 10$.
   a) Solve the appropriate equation of motion, and determine the period of the oscillation. (The amplitude $A$ and phase $\varphi$ are not determined by the given information; leave them in your solution as unknown constants.)
   b) Given the initial conditions $y(0) = 2$, $y'(0) = 1$, determine the amplitude and phase of the oscillation.

7. An electrical oscillator has capacitance $C$ and "inductance" $L$ in series with negligible resistance. The charge $Q$ in the capacitor satisfies the equation

$$\frac{L d^2 Q}{dt^2} + \frac{Q}{C} = 0.$$

**a)** Solve the equation and find the frequency of the oscillation.

**b)** Suppose that $L = 1$. What value of $C$ gives a frequency of 102,500 cycles/second? (This is the frequency of the carrier signal for FM station WCRB in Waltham, Mass.)

**B**

**8.** A mass of $m$ grams hangs on a spring with spring constant $k$.

**a)** Solve the equation of motion, and determine the period $T$.

**b)** If the mass is increased does the period increase or decrease?

**c)** If the mass is unchanged, but the spring is replaced by a stiffer one, is the period increased or decreased?

**9.** Prove that if $dy/dt$ is a cosine wave, then $y$ itself is a cosine wave plus a constant.

**10.** In quantum mechanics, the time-independent behavior of a particle moving in the interval $[0, \pi]$ is described by a function $\psi \neq 0$ solving the three equations

$$\frac{d^2\psi}{dx^2} + \frac{2mE}{\hbar^2}\psi = 0 \qquad \textbf{(21)}$$

$$\psi(0) = 0 \qquad \textbf{(22)}$$

$$\psi(\pi) = 0; \qquad \textbf{(23)}$$

$x$ is the position of the particle, $m$ its mass, $E$ its energy, $2\pi\hbar$ is "Planck's constant," and $\psi(x)$ describes the probability of finding the particle near point $x$.

**a)** Find a function $\psi$ satisfying equations (21) and (22).

**b)** Show that the functions in part a can also satisfy equation (23) only for certain values of the energy $E$, namely

$$E = \frac{\hbar^2}{2m},\ \ 4\frac{\hbar^2}{2m},\ \ 9\frac{\hbar^2}{2m}, \cdots.$$

**11.** Show that every system of equations

$$y'' = -\omega^2 y, \qquad y(0) = y_0, \qquad y'(0) = v_0$$

has a cosine wave solution. ($\omega \neq 0$, $y_0$, $v_0$ are constants.)

**12.** Show that every cosine wave $A\cos(\omega t - \varphi)$ is also a sine wave $A\sin(\omega t - \theta)$, for some constant $\theta$.

**C**

The remaining problems combine to prove that every solution of $y'' = -\omega^2 y$ is a cosine wave.

**13.** Suppose that $y$ is a function satisfying $y'' = -\omega^2 y$. Is $(\omega y)^2 + (y')^2$ necessarily constant? [You cannot assume that $y$ is a cosine wave; assume only the given equation $y'' = -\omega y^2$ and the general rules of calculus. If you think that the answer is "no," give a specific example; if you think "yes," give a proof.]

**14.** Suppose that $y$ is a function satisfying $y'' = -\omega^2 y$, and $y(0) = 0$, $y'(0) = 0$. One possible solution is $y(t) = 0$ for all $t$. Is there any other solution? Why, or why not? (See the preceding problem.)

**15.** Suppose that $y_1(t)$ and $y_2(t)$ both solve

$$y'' = -\omega^2 y$$

$$y(0) = y_0, \qquad y'(0) = v_0$$

with the same constants $\omega$, $y_0$, and $v_0$. What are the corresponding equations for the difference $y = y_1 - y_2$? What must the difference be? (See the preceding problem.)

**16.** Let $y_1$ be *any* solution of the equations in the preceding problem. Let $y_2$ be the cosine wave solution of the same problem (shown to exist in problem 11). Show that $y_1 = y_2$. (See the preceding problem.)

## 8.10

## THE HYPERBOLIC FUNCTIONS

The identities for the trigonometric functions are closely mimicked by certain combinations of exponentials, called the **hyperbolic functions**. This section defines these functions and outlines their essential properties; the straightforward verifications are left as problems.

## Definitions

The hyperbolic sine of any number $t$, denoted sinh($t$) ("sinch $t$") is defined by

$$\sinh(t) = \tfrac{1}{2}(e^t - e^{-t}).$$

The hyperbolic cosine cosh($t$) ("kosh $t$") is defined by

$$\cosh(t) = \tfrac{1}{2}(e^t + e^{-t}).$$

The other hyperbolic functions are defined in terms of these, using the same ratios as for the trigonometric functions:

$$\tanh t = \frac{\sinh t}{\cosh t} = \frac{e^t - e^{-t}}{e^t + e^{-t}},$$

$$\operatorname{sech} t = \frac{1}{\cosh t} = \frac{2}{e^t + e^{-t}}.$$

The hyperbolic cotangent and cosecant are similarly defined, but seldom used.

Figures 1 and 2 show the graphs of the four main functions. The graph of cosh $t$ has the shape assumed by a chain hanging freely from two separated supports (Fig. 3); this shape is called a *catenary*, since "catena" is Latin for chain.

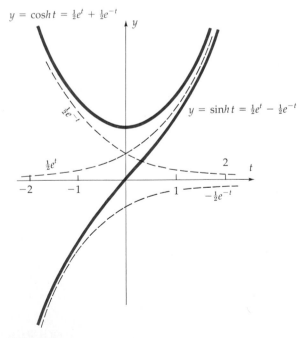

**FIGURE 1**
Sinh $t$ and cosh $t$.

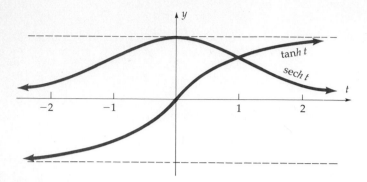

**FIGURE 2**
tanh $t$ and sech $t$; $\tanh^2 t + \operatorname{sech}^2 t = 1$.

**FIGURE 3**
Catenary.

**Hyperbolic identities** are like the familiar trigonometric ones, but with crucial changes in sign:

$$\cosh^2 t - \sinh^2 t = 1 \tag{1}$$

$$1 - \tanh^2 t = \operatorname{sech}^2 t. \tag{2}$$

The first identity shows the relation between these functions and the hyperbola. Imagine a point moving in the plane, so that its position $(x, y)$ at time $t$ is

$$x = \cosh t, \qquad y = \sinh t.$$

Identity (1) implies then that $x^2 - y^2 = 1$, that is, the point moves on the *hyperbola* in Figure 4; it stays on one branch, since $x = \cosh t > 0$.

The hyperbolic addition formulas are

$$\sinh(x \pm y) = \sinh x \cdot \cosh y \pm \cosh x \cdot \sinh y, \tag{3}$$

$$\cosh(x \pm y) = \cosh x \cdot \cosh y \pm \sinh x \cdot \sinh y. \tag{4}$$

From these follow the "double angle" formulas

$$\sinh(2x) = 2 \sinh x \cdot \cosh x, \tag{5}$$

$$\cosh(2x) = \cosh^2 x + \sinh^2 x, \tag{6}$$

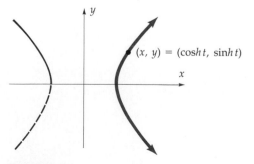

**FIGURE 4**
Moving particle with $x = \cosh t$,
$y = \sinh t$. Path is one branch of
hyperbola $x^2 - y^2 = 1$.

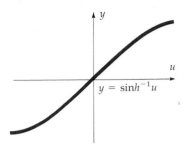

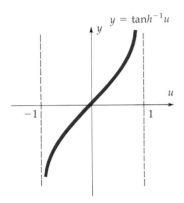

**FIGURE 5**

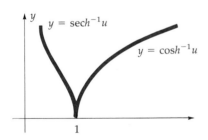

**FIGURE 6**

and hence, in view of (1),

$$\sinh^2 x = \tfrac{1}{2}(\cosh 2x - 1), \tag{7}$$

$$\cosh^2 x = \tfrac{1}{2}(\cosh 2x + 1). \tag{8}$$

**The derivatives** of the hyperbolic functions are computed directly from their definitions:

$$\sinh' x = \cosh x, \tag{9}$$

$$\cosh' x = \sinh x, \tag{10}$$

$$\tanh' x = \operatorname{sech}^2 x, \tag{11}$$

$$\operatorname{sech}' x = -\tanh x \cdot \operatorname{sech} x. \tag{12}$$

Again, these formulas mimick those for the trigonometric functions *except for signs*.

### Inverse Hyperbolic Functions

Since sinh and tanh are strictly increasing functions (Figs. 1 and 2), they are one-to-one, and have well-defined inverses, graphed in Figure 5. The cosh and sech, on the other hand, must be restricted to positive values in order to define an inverse (Fig. 6). Since the hyperbolic functions themselves are defined in terms of exponentials, one might hope that their inverses can be expressed in terms of logarithms. Indeed they can; for example,

$$\sinh^{-1} u = \ln(u + \sqrt{u^2 + 1}) \tag{13}$$

$$\tanh^{-1} u = \tfrac{1}{2} \ln\!\left(\frac{1 + u}{1 - u}\right), \qquad |u| < 1. \tag{14}$$

**EXAMPLE 1**   Prove (13).

*SOLUTION*   We set $y = \sinh^{-1} u$, obtain an equation for $u$ in terms of $y$, and solve it for $y$. Thus

$$u = \sinh y = \tfrac{1}{2}(e^y - e^{-y}).$$

This can be written as a quadratic in $e^y$. Multiply through by $2e^y$:

$$2e^y u = (e^y)^2 - 1$$

or

$$(e^y)^2 - 2u(e^y) - 1 = 0.$$

By the quadratic formula,

$$e^y = \frac{2u \pm \sqrt{4u^2 + 4}}{2} = u \pm \sqrt{u^2 + 1}.$$

Since $e^y$ must be positive, and $\sqrt{u^2 + 1} > u$, we need the "+" sign:

$$e^y = u + \sqrt{u^2 + 1}.$$

Thus

$$\sinh^{-1} u = y = \ln(u + \sqrt{u^2 + 1}).$$

### Derivatives of the inverse hyperbolic functions
any inverse function. They are

$$\frac{d \sinh^{-1} u}{dx} = \frac{1}{\sqrt{1 + u^2}} \frac{du}{dx} \tag{15}$$

$$\frac{d \tanh^{-1} u}{dx} = \frac{1}{1 - u^2} \frac{du}{dx} \tag{16}$$

$$\frac{d \operatorname{sech}^{-1} u}{dx} = \frac{du/dx}{u \sqrt{1 - u^2}}. \tag{17}$$

This completes our outline.

## PROBLEMS

### A

Each of problems 1–17 is to prove the corresponding formula in the text: Problem 1 is to prove formula (1), and so on. Problem 13, to prove formula (13), is worked out in Example 1.

### B

18. **a)** Show that the graphs of $y = \cosh(\omega t)$ and $y = \cos(\omega t)$ are tangent to each other at $t = 0$.

    **b)** Show that the graphs of $y = \sinh(\omega t)$ and $y = \sin(\omega t)$ are tangent to each other at $t = 0$.

19. Find a differential equation (involving $y''$) solved by

$$y = a \cosh(\omega t) + b \sinh(\omega t);$$

$a$, $\omega$, and $b$ are constants.

## REVIEW PROBLEMS   CHAPTER 8

1. Sketch.
    **a)** $y = \ln(x)$        **b)** $y = e^x$
    **c)** $x = 2 + 3e^{-2t}$  **d)** $\theta = \sin^{-1} x$
    **\*e)** $\theta = \cos^{-1}(x/2)$  **f)** $\theta = \tan^{-1}(2x)$

    **e)** $\tan^{-1}(x/5)$    **f)** $e^{\sin 3t}$
    **g)** $\cos^{-1}(e^{t^2 + 1})$  **h)** $10^{x^2 + x}$
    **i)** $\log_2(e^x)$    **j)** $\dfrac{x^3 \sqrt{x^2 + 1}}{(1 + 4x)^2}$

2. Simplify.
    **\*a)** $\ln(2^{3x+1})$    **\*b)** $e^{(\ln 3 - 5 \ln 2)}$
    **\*c)** $\cos(\tan^{-1} 3)$  **\*d)** $\sin(\tan^{-1}(-2))$
    **\*e)** $\tan(\sin^{-1}(1/3))$ **f)** $\sin(2 \tan^{-1} 3)$

4. Evaluate.
    **\*a)** $\lim\limits_{x \to 0^+} x^2 \ln x$

    **\*b)** $\lim\limits_{x \to +\infty} x^{-2} \ln(1 + x^2)$

    **\*c)** $\lim\limits_{x \to +\infty} e^{-x} \ln x$

**\*3.** Differentiate.
    **a)** $\ln(x^2 + 1)$    **b)** $x e^{(x^3)}$
    **c)** $\exp(\exp(x^2 + 1))$  **d)** $\sin^{-1}(x/3)$

**\*d)** $\lim\limits_{h \to +\infty} (1 + a/h)^h$

**e)** $\lim\limits_{t \to +\infty} t^{1/t}$

**5.** Graph, using $y'$, $y''$, and any appropriate limits.

   **\*a)** $y = x^{-1} \ln |x|$    **b)** $y = xe^{-x^2/2}$

**\*6.** Evaluate.

   **a)** $\displaystyle\int_0^1 \frac{x\,dx}{4 + x^2}$    **b)** $\displaystyle\int_0^{10} xe^{-x^2}\,dx$

   **c)** $\displaystyle\int_{-1}^1 \frac{dx}{1 + x^2}$    **d)** $\displaystyle\int \frac{dt}{\sqrt{1 - 4t^2}}$

**\*7.** Solve, and sketch the solution.

   **a)** $\dfrac{dy}{dx} = -xy;$   $y = 1$ when $x = 0$.

   **b)** $\dfrac{dy}{dx} = -xy;$   $y = -1$ when $x = 0$.

   **c)** $\dfrac{dA}{dt} = 3A - 2;$   $A(0) = 0$.

   **d)** $\dfrac{dy}{dt} = y \sin t;$   $y(0) = 1$.

**\*8.** Solve $R\dfrac{dQ}{dt} + \dfrac{Q}{C} = E,$   $Q(0) = Q_0,$ with constants $R > 0$, $C > 0$, and $E$. Compute $\lim\limits_{t \to +\infty} Q$ and $\lim\limits_{t \to -\infty} Q$.

**\*9.** **a)** Solve $\dfrac{d^2y}{dt^2} = -4y,$ $y(0) = 1,$ $y'(0) = -1$.

   **b)** Sketch the solution. Determine its period and amplitude, and indicate them on the sketch.

**\*10.** In a study of the economics of dike building in Holland (Report of the Delta Commission, 1960), it was estimated that the construction cost of raising the dikes to height $x$ was $A + Bx$; and the expected long-run losses due to floods occurring when the sea rose above height $x$ would be $Ce^{-\alpha x}$. The constants $A$, $B$, $C$, and $\alpha$

were empirically determined. According to this model, what height $x$ minimizes the total cost, $A + Bx + Ce^{-\alpha x}$?

**\*11.** In thermodynamics there is a concept called *entropy*. Under certain conditions, the rate of change of entropy $S$ with respect to temperature $T$ is

$$\frac{dS}{dt} = \frac{C(T)}{T},$$

where

$$C(T) = 6 + 2 \times 10^{-3}T - 2 \times 10^{-7}T^2.$$

What is the change in entropy on heating from $T = 100°K$ to $110°K$?

**12.** Derive the formula for $de^x/dx$, assuming only that the graph of $y = e^x$ has slope 1 when $x = 0$.

**\*13.** Let $g$ be the inverse of the function $f(x) = x + x^5$.

   **a)** Determine the domain of $g$.

   **b)** Compute $g'(2)$.

**14.** Derive the formula for the derivative of $\cos^{-1}u$. Explain the sign carefully.

**15.** An oscillating pendulum of length $l$ satisfies the equation $l\dfrac{d^2\theta}{dt^2} = -g \sin \theta;$ $\theta$ is the angle between the pendulum and the vertical and $g$ is the acceleration due to gravity.

   **a)** Explain why, for small oscillations, $\dfrac{d^2\theta}{dt^2}$ is approximately $-\dfrac{g}{l}\theta$.

   **\*b)** Solve the equation $\dfrac{d^2\theta}{dt^2} = -\dfrac{g}{l}\theta$.

   **\*c)** Determine the period of oscillation in the solution of part b.

   **\*d)** In a certain experiment, the period of a pendulum of length 1 meter is found to be 2 seconds. According to this experiment, what is $g$?

# 9

# TECHNIQUES OF
# INTEGRATION

Computing derivatives is a straightforward process, like taking a puzzle apart. Computing integrals is like putting a puzzle back together—much trickier! Worse yet, many integrals are literally impossible to evaluate explicitly; an example is $\int e^{x^2}\, dx$. This cannot be expressed in terms of functions already familiar to us, using just the basic operations of algebra and composition of functions. It is like a "puzzle" where the pieces do not fit together into a familiar picture.

But many useful integrals *can* be evaluated. This chapter shows the most common types and techniques for evaluating them. Along the way, you will exercise and reinforce your mastery of many formulas learned up to now.

Each formula for derivatives implies a corresponding one for integrals. For instance, the Chain Rule implies the formula for substitution

$$\int f(u(x))u'(x)\, dx = \int f(u)\, du.$$

This is the first integration method to keep in mind; you might review it (Sec. 6.8) before proceeding to the techniques in this chapter.

## 9.1

### INTEGRATION BY PARTS

Integration by parts is the integral formula derived from the Product Rule

$$(fg)' = f'g + fg'.$$

Apply the Fundamental Theorem of Calculus, $F(x)\big|_a^b = \int_a^b F'(x)\, dx$, with $F = fg$:

$$f(x)g(x)\Big|_a^b = \int_a^b (fg)'(x)\, dx = \int_a^b [f'(x)g(x) + f(x)g'(x)]\, dx$$
$$= \int_a^b f'(x)g(x)\, dx + \int_a^b f(x)g'(x)\, dx.$$

Rearrange terms, and you have the formula for integration by parts:

$$\int_a^b f(x)g'(x)\, dx = f(x)g(x)\Big|_a^b - \int_a^b g(x)f'(x)\, dx. \tag{1}$$

Starting with the product $fg'$ on the left-hand side, you integrate one part (the factor $g'$) and differentiate the other part (the factor $f$), obtaining the product $gf'$ on the right-hand side. You hope, of course, that the new product will be easier to integrate than the old!

For indefinite integrals, a similar argument gives

$$\int f(x)g'(x)\, dx = f(x)g(x) - \int g(x)f'(x)\, dx. \tag{2}$$

This is usually applied in Leibniz notation. Set $u = f(x)$ and $v = g(x)$. Then

$$du = f'(x)\, dx \quad \text{and} \quad dv = g'(x)\, dx$$

so (2) becomes

$$\int u \, dv = uv - \int v \, du.$$

(3)

**EXAMPLE 1**  Evaluate $\int \ln x \, dx$ and $\int_1^2 \ln(x) \, dx$.

**SOLUTION**

**First**  To use (3), write $\int \ln x \, dx$ in the form $\int u \, dv$. The most obvious way is to set

$$u = \ln x \quad \text{and} \quad dv = dx.$$

**Second**  Determine $du$ and $v$. Get $du$ by differentiating $u$, and $v$ by integrating $dv$:

$$du = d \ln x = \frac{1}{x} \, dx,$$

$$v = \int dv = \int dx \qquad \text{[since } dv = dx\text{]}$$
$$= x + C.$$

*Any $v$ with $dv = dx$ will do, so drop the "$+C$" and set $v = x$.*

**Finally**  Apply the integration by parts formula (3):

$$\underbrace{\int \ln x \, dx}_{u \quad dv} = \underbrace{[\ln x][x]}_{u \quad v} - \int \underbrace{[x]}_{v} \underbrace{\left[ \frac{1}{x} \, dx \right]}_{du}$$

$$= x \ln x - \int 1 \cdot dx$$

$$= x \ln x - x + c.$$

For the definite integral the process is the same, but you include the limits of integration at every step:

$$\int_1^2 \ln x \, dx = x \ln x \Big|_1^2 - \int_1^2 x \frac{1}{x} \, dx$$

$$= [2 \ln 2 - 1 \ln 1] - \int_1^2 dx$$

$$= [2 \ln 2 - 0] - 1 = 2 \ln 2 - 1.$$

**EXAMPLE 2**  $\int xe^{2x} \, dx = ?$

**SOLUTION**  In the product $xe^{2x}$, integrate the factor $e^{2x}$ and differentiate the factor $x$. Thus set

$$x = u, \qquad e^{2x} \, dx = dv.$$

Then

$$du = dx, \quad \text{and} \quad v = \int dv = \int e^{2x} \, dx = \tfrac{1}{2} e^{2x} + C.$$

Drop the "$+C$" and apply (3):

$$\int \underbrace{x}_{u}\ \underbrace{e^{2x}\,dx}_{dv} = \underbrace{x}_{u} \cdot \underbrace{\tfrac{1}{2}e^{2x}}_{v} - \int \underbrace{\tfrac{1}{2}e^{2x}}_{v}\ \underbrace{dx}_{du} = \tfrac{1}{2}xe^{2x} - \tfrac{1}{4}e^{2x} + C.$$

You can check the answer by differentiating it; you should get $xe^{2x}$. In the process you apply the Product Rule, underscoring the fact that integration by parts is the "reverse" of the Product Rule.

---

**REMARK**  Two principles guide the choice of $u$ and $dv$. First, for $dv = g'(x)\,dx$ you must choose something that can be integrated to find $v = g(x)$; and for $u$, choose something that does not grow complicated when differentiated.

---

**EXAMPLE 3**  $\int x \ln x \, dx = ?$

**SOLUTION**  Here $\ln x$ simplifies to $\dfrac{1}{x}$ when differentiated; and $x\,dx$ integrates easily to $\tfrac{1}{2}x^2$. So it is reasonable to try

$$u = \ln x, \qquad du = \frac{1}{x}\,dx,$$

$$dv = x\,dx, \qquad v = \int x\,dx = \tfrac{1}{2}x^2.$$

This gives

$$\int x \ln x \, dx = (\ln x)\left(\frac{1}{2}x^2\right) - \int \frac{1}{2}x^2 \cdot \frac{1}{x}\,dx$$

$$= \frac{1}{2}x^2 \ln x - \int \frac{1}{2}x\,dx$$

$$= \frac{1}{2}x^2 \ln x - \frac{1}{4}x^2 + c.$$

---

Sometimes one integration by parts does not completely solve the problem, and the process must be repeated.

---

**EXAMPLE 4**  $\int x^2 e^x \, dx = ?$

**SOLUTION**  Since $x^2$ simplifies to $2x$ when differentiated, and $e^x$ is easy to integrate, set $u = x^2$, $dv = e^x\,dx$:

$$u = x^2 \Rightarrow du = 2x\,dx, \qquad dv = e^x\,dx \Rightarrow v = \int e^x\,dx = e^x,$$

so

$$\int \underbrace{x^2}_{u}\ \underbrace{e^x\,dx}_{dv} = \underbrace{x^2}_{u}\ \underbrace{e^x}_{v} - \int \underbrace{e^x}_{v}\ \underbrace{2x\,dx}_{du} = x^2 e^x - 2\int xe^x\,dx. \tag{4}$$

To complete the evaluation, integrate $\int xe^x \, dx$ by parts again to reduce the power of $x$ still further. Take $u = x$ and $dv = e^x \, dx$, and find

$$\int \underbrace{x}_{u} \underbrace{e^x \, dx}_{dv} = \underbrace{x \, e^x}_{u \ v} - \int \underbrace{e^x \, dx}_{v \ du} = xe^x - e^x + c.$$

Hence from (4),

$$\int x^2 e^x \, dx = x^2 e^x - 2[xe^x - e^x + c]$$
$$= x^2 e^x - 2xe^x + 2e^x + C,$$

where $C = -2c$ is a new constant of integration.

---

Example 4 suggests that for any integer $n > 0$, the integral $\int x^n e^x \, dx$ can be evaluated by integrating by parts $n$ times. The essence of the calculation is given conveniently in the following *reduction formula*:

$$\int x^n e^x \, dx = x^n e^x - n \int x^{n-1} e^x \, dx. \tag{5}$$

This follows directly from the basic integration by parts formula (3), with $u = x^n$ and $dv = e^x \, dx$. It reduces the problem of integrating $\int x^n e^x \, dx$ to that of integrating $\int x^{n-1} e^x \, dx$, which is one step simpler. This reduction formula is listed for reference in the integral tables inside the covers of this book; it is also easy to derive, when needed.

---

**EXAMPLE 5**

$$\int x^3 e^x \, dx = x^3 e^x - 3 \int x^2 e^x \, dx \qquad \text{[by (5), with } n = 3]$$

$$= x^3 e^x - 3 \left( x^2 e^x - 2 \int xe^x \, dx \right) \qquad \text{[by (5), with } n = 2]$$

$$= x^3 e^x - 3x^2 e^x + 6 \left( xe^x - \int e^x \, dx \right) \qquad \text{[by (5), with } n = 1]$$

$$= x^3 e^x - 3x^2 e^x + 6xe^x - 6e^x + c.$$

---

**The integral $\int e^{ax} \cos bx \, dx$** can be evaluated by integrating by parts twice, then solving algebraically for the integral. Set $u = \cos bx$ and $dv = e^{ax} \, dx$, so

$$du = -b \sin bx \, dx \quad \text{and} \quad v = \frac{1}{a} e^{ax}.$$

Thus

$$\int e^{ax} \cos bx \, dx = \frac{1}{a} e^{ax} \cos bx + \frac{b}{a} \int e^{ax} \sin bx \, dx. \tag{6}$$

This looks like no improvement, but don't give up. Integrate by parts once more! In $\int e^{ax} \sin bx \, dx$ set $u = \sin bx$ and $dv = e^{ax} \, dx$. Then $du = b \cos bx \, dx$, $v = (1/a)e^{ax}$, so

$$\int e^{ax} \sin bx \, dx = \frac{1}{a} e^{ax} \sin bx - \frac{b}{a} \int e^{ax} \cos bx \, dx. \tag{7}$$

Substitute (7) for the integral on the right-hand side of (6):

$$\int e^{ax} \cos bx \, dx = \frac{1}{a} e^{ax} \cos bx + \frac{b}{a^2} e^{ax} \sin bx - \frac{b^2}{a^2} \int e^{ax} \cos bx \, dx. \tag{8}$$

Now view this as *an equation to be solved for* $\int e^{ax} \cos bx \, dx$. In (8), collect all the integrals on the left by adding $\dfrac{b^2}{a^2} \int e^{ax} \cos bx \, dx$ to both sides:

$$\left(\frac{b^2}{a^2} + 1\right) \int e^{ax} \cos bx \, dx = \frac{1}{a} e^{ax} \cos bx + \frac{b}{a^2} e^{ax} \sin bx.$$

Divide by the constant $\dfrac{b^2}{a^2} + 1$, that is, by $\dfrac{b^2 + a^2}{a^2}$, and simplify, obtaining

$$\int e^{ax} \cos bx \, dx = (e^{ax}) \frac{a \cos bx + b \sin bx}{a^2 + b^2} + C. \tag{9}$$

We had to add a constant of integration $C$ in (9); at the beginning [line (6)] it was not needed, since there was an indefinite integral sign on each side of the equation.

Formula (9) is listed for reference in the integral tables; the derivation is too long to be repeated with each application!

### Some General Patterns

It is crucial to make good choices of $u$ and $dv$. The most common integrals yielding to integration by parts fall into two classes:

**I.**  $\int x^n e^{ax} \, dx$, $\int x^n \sin(ax) \, dx$, $\int x^n \cos(ax) \, dx$, $n = 1, 2, 3, \ldots$ . In these, the aim is to *reduce the power of $x$*; choose $u = x^n$, so that $du = nx^{n-1} \, dx$ has a lower power. If $n$ is 2 or more, it is wise to look up (or derive) the appropriate reduction formula.

**II.**  $\int x^n \ln x \, dx$, $\int x^n \sin^{-1} x \, dx$, $\int x^n \tan^{-1} x \, dx$. Here, "simplify" the inverse function by differentiating it; choose $u = \ln x$ [or $\sin^{-1} x$, or $\tan^{-1} x$] so that $du/dx$ is an algebraic function.

**A common source of error** is in computing antiderivatives of $e^{ax}$, $\sin(ax)$, and $\cos(ax)$; derivatives get mixed up with integrals. To avoid this, keep the *derivative* formulas as the basis of your calculations, as in the following examples.

**EXAMPLE 6**

**a.**   $dv = e^{-3x}\,dx \Rightarrow v$ is a constant times $e^{-3x}$. So write

$$v = (\quad)e^{-3x}$$

leaving the parentheses blank. Mentally differentiate $e^{-3x}$ (getting $-3e^{-3x}$) to see that you must cancel a factor $-3$; then fill in the necessary $\dfrac{1}{-3}$ in the parentheses:

$$v = \left(\frac{1}{-3}\right)e^{-3x}.$$

**b.**   $dv = \sin(ax)\,dx \Rightarrow v$ is a constant times $\cos(ax)$. Write

$$v = (\quad)\cos(ax).$$

Mentally differentiate $\cos(ax)$ [getting $-a\sin(ax)$] to see that you must cancel a factor $-a$; then fill in the necessary $-1/a$ in the parentheses, getting $v = (-1/a)\cos(ax)$.

The formula $\int v\,du = uv - \int u\,dv$ has been immortalized in an apocryphal ditty:

"The integral of $v\,du$, as now you clearly see,
is $uv$ less the integral of $u\,d\,v$.
If you learn this little ditty and you take it to your hearts,
then you always will remember how to integrate by parts."

## SUMMARY

*Integration by Parts Formulas:*

$$\int u\,dv = uv - \int v\,du$$

$$\int_a^b f(x)g'(x)\,dx = f(x)g(x)\Big|_a^b - \int_a^b f'(x)g(x)\,dx.$$

## PROBLEMS

**NOTE**   The Appendix to this section applies integration by parts to solve some interesting differential equations.

**A**

*1.   $\displaystyle\int \ln ax\,dx$

2.   $\displaystyle\int x \ln ax\,dx$

*3.   $\displaystyle\int x \ln x^2\,dx$

4.   $\displaystyle\int x^2 \ln x\,dx$

*5.   $\displaystyle\int_1^e (\ln x)^2\,dx$

6.   $\displaystyle\int x\,(\ln x)^2\,dx$

*7.   $\displaystyle\int \sqrt{t}\,\ln t\,dt$

8.   $\displaystyle\int xe^{3x}\,dx$

*9.   $\displaystyle\int_0^1 x^2 e^{-10x}\,dx$

10.   $\displaystyle\int x \sin x\,dx$

*11. $\int x \cos 2x \, dx$

12. $\int_0^{\pi/2} x(1 + \cos x) \, dx$

*13. $\int_0^{\pi} \theta^2 \sin \theta \, d\theta$

14. $\int_0^{\pi} \theta^2 \sin(2\theta) \, d\theta$

*15. $\int_0^{\infty} xe^{-x} \, dx$

16. $\int_0^{\infty} xe^{-2x} \, dx$

*17. $\int \tan^{-1}x \, dx$

18. $\int \arcsin x \, dx$

## B

19. $\int \cos(\ln x) \, dx$ (First substitute $w = \ln x$.)

*20. $\int e^{\sqrt{x}} \, dx$ (First substitute $u = \sqrt{x}$.)

21. $\int \sin \sqrt{x} \, dx$

22. $\int \cos(x^{1/3}) \, dx$

In problems 23–27, *prove* the reduction formulas.

23. $\int x^n e^{ax} \, dx = \dfrac{1}{a} x^n e^{ax} - \dfrac{n}{a} \int x^{n-1} e^{ax} \, dx$

24. $\int x^n \sin ax \, dx =$

$$-\frac{1}{a} x^n \cos ax + \frac{n}{a} \int x^{n-1} \cos ax \, dx$$

25. $\int x^n \cos ax \, dx = ?$ (Discover a formula analogous to problem 24.)

26. $\int x^m (\ln x)^n \, dx =$

$$\frac{1}{m+1} x^{m+1}(\ln x)^n - \frac{n}{m+1} \int x^m (\ln x)^{n-1} \, dx$$

27. $\int x^n e^{-ax^2} \, dx = ?$

28. Derive the formula for $\int e^{ax} \sin bx \, dx$.

29. Is there any fallacy in this evaluation of $\int \ln(x+1) \, dx$? Take $u = \ln(x+1)$, $dv = dx$, and $du = \dfrac{dx}{x+1}$, $v = x+1$, so

$$\int \ln(x+1) \, dx = (x+1)\ln(x+1) - \int \frac{x+1}{x+1} \, dx$$

$$= (x+1)\ln(x+1) - x + C.$$

In problems 30–37 evaluate the improper integral, or show its divergence.

*30. $\int_0^{\infty} xe^{-ax} \, dx$, $a > 0$

31. $\int_0^{\infty} x^2 e^{-ax} \, dx$, $a > 0$

*32. $\int_0^{\infty} xe^{-x^2} \, dx$

*33. $\int_{-\infty}^{\infty} xe^{-x^2} \, dx$

34. $\int_{-\infty}^{\infty} xe^{-x} \, dx$

35. $\int_0^{\infty} x^{-2} e^{-x} \, dx$

*36. $\int_0^{\infty} e^{-ax} \sin bx \, dx$, $a > 0$

37. $\int_0^{\infty} e^{-ax} \cos bx \, dx$, $a > 0$

38. Compute $\lim\limits_{a \to +\infty} \int_0^{\infty} xe^{-ax} \, dx$. Could you predict the result from the behavior of the integrand $xe^{-ax}$ as $a \to +\infty$?

39. Compute $\lim\limits_{a \to \pm\infty} \int_0^1 x \sin ax \, dx$. Could you predict the result from the behavior of the integrand as $a \to \pm\infty$?

## C

40. For integers $n > 0$, the *factorial* $n!$ is defined by

$$n! = n(n-1)(n-2) \cdots (2)(1).$$

Leonhard Euler devised a way of defining $x!$ for all real numbers $x > -1$, by means of an improper integral, as here outlined. Show

a) $\int_0^{\infty} t^n e^{-t} \, dt = n \int_0^{\infty} t^{n-1} e^{-t} \, dt$, and $\int_0^{\infty} t^0 e^{-t} \, dt = 1$.

b) $\int_0^{\infty} t^n e^{-t} \, dt = n!$ (by repeated application of part a).

c) $\int_0^{\infty} t^x e^{-t} \, dt$ is a convergent improper integral for each $x > -1$.

As a result of parts b and c, we define

$$x! = \int_0^{\infty} t^x e^{-t} \, dt \qquad \text{for } x > -1.$$

d) Show that by this definition, $0! = 1$.

e) For $x > -1$, $(x+1)! = (x+1)(x!)$.

f) $\lim\limits_{x \to -1^+} x! = +\infty$.

**REMARK** One often works not with $x!$, but with the *Gamma function*, defined as follows:

$$\Gamma(x) = \int_0^{\infty} t^{x-1} e^{-t} \, dt, \qquad x > 0.$$

[$\Gamma$ is the Greek capital letter gamma.] Thus $x! = \Gamma(x+1)$.

## 9.1

## APPENDIX: THE DIFFERENTIAL EQUATION
$$\frac{dy}{dt} + by = f(t) \quad \text{(optional)}$$

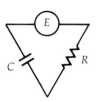

**FIGURE 1**
Electrical circuit.

The circuit in Figure 1 has a voltage $E$, capacitance $C$, and resistance $R$. The charge $Q$ in the capacitor satisfies the differential equation

$$\frac{dQ}{dt} + \frac{Q}{RC} = \frac{E}{R}. \tag{1}$$

Normally, $R$ and $C$ are constants, but the voltage $E$ may vary; for instance, the voltage from a typical power outlet is $E(t) = 170 \cos(120\pi t)$.

Equation (1) is an example of the general equation

$$\frac{dy}{dt} + by = f(t). \tag{2}$$

If $b$ and $f$ are both constants, this can be solved by separating the variables (Sec. 8.6). But if $f$ is *not* constant, separation is no longer possible. In that case, the equation can be solved by an ingenious algebraic manipulation. Multiply both sides of (2) by $e^{bt}$:

$$e^{bt}\frac{dy}{dt} + be^{bt}y = e^{bt}f(t).$$

Now the left-hand side is precisely the derivative of the product $e^{bt}y$ (check this); so the equation can be written

$$\frac{d}{dt}(e^{bt}y) = e^{bt}f(t).$$

To solve for $y$, you must first "undo" the derivative $\dfrac{d}{dt}$. Integrate both sides:

$$e^{bt}y = \int e^{bt}f(t)\, dt + C;$$

the term "$+C$" is the constant of integration for the indefinite integral $\int e^{bt}f(t)\, dt$. Now multiply by $e^{-bt}$, and find the solution of (2):

$$y = e^{-bt}\left[\int e^{bt}f(t)\, dt + C\right]. \tag{3}$$

### KEEP IN MIND

**1.**   This solution is for the case that $b$ is constant.

**2.**   The factor $e^{bt}$ *inside* the integral in (3) does not cancel the factor $e^{-bt}$ outside; $e^{bt}$ is a function, not a constant, and cannot be brought out of the integral to combine with $e^{-bt}$.

**EXAMPLE 1**    Equation (1) with $R = 1$, $C = 1/2$, and alternating voltage $E = 170 \cos(\omega t)$ is

$$\frac{dQ}{dt} + 2Q = 170 \cos \omega t. \tag{4}$$

(For 60 cycle voltage, $\omega = 120\pi$.) This equation is like (2) with $b = 2$ and $f(t) = 170 \cos \omega t$. Multiply both sides by $e^{bt} = e^{2t}$:

$$e^{2t} \frac{dQ}{dt} + 2e^{2t}Q = 170 \, e^{2t} \cos \omega t.$$

Check that the left hand side is the derivative of $e^{2t}Q$; so

$$\frac{d}{dt}(e^{2t}Q) = 170 \, e^{2t} \cos \omega t.$$

Hence

$$e^{2t}Q = \int 170 \, e^{2t} \cos \omega t \, dt$$

$$= 170 \, e^{2t} \frac{2 \cos \omega t + \omega \sin \omega t}{4 + \omega^2} + C$$

from formula (9), Section 9.1. Thus the charge at time $t$ is

$$Q = 170 \, \frac{2 \cos \omega t + \omega \sin \omega t}{4 + \omega^2} + Ce^{-2t}.$$

As $t \to +\infty$, the term $Ce^{-2t} \to 0$, and we obtain the *steady state*

$$Q = \frac{170}{4 + \omega^2} [2 \cos \omega t + \omega \sin \omega t].$$

This is an oscillation with period $2\pi/\omega$, the same as for the "driving voltage" $f(t) = 170 \cos \omega t$ in the original equation (4). But it is "out of phase"; its maximum occurs when $t = \frac{1}{\omega} \tan^{-1} \frac{\omega}{2}$, not at $t = 0$ (Fig. 2).

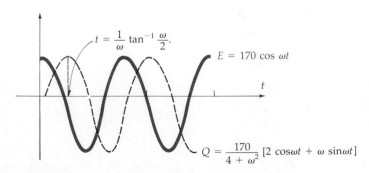

**FIGURE 2**
Peak charge $Q$ is later than peak voltage $E$.

## PROBLEMS

### A

1.   Solve equation (4) with $Q(0) = 0$.

2.   *a)   Solve $\dfrac{dQ}{dt} + \dfrac{Q}{2} = t$.

   b)   Find the solution with $Q(0) = 0$.

3.   Solve equation (1) with a linearly increasing voltage $E = kt$, with constant $R$ and $C$, and with
   *a)   $Q(0) = 0$.      b)   $Q(0) = -1$.

4.   Solve the differential equation for the current $I$ in the circuit in Figure 3, $L\dfrac{dI}{dt} + RI = E$, given that

**FIGURE 3**
Circuit with inductance $L$ and resistance $R$.

a)   $E = E_0$ (constant)      b)   $E = E_0 t$
c)   $E = E_0 \cos \omega t$

### B

5.   Solve equation (1) with $R = C = 1$ (for simplicity), with $Q(0) = 0$, and with

$$E(t) = \begin{cases} 1 - t, & 0 \le t \le 1 \\ 0, & t \ge 1 \end{cases}.$$

[This equation describes a circuit where the voltage is turned down steadily from 1 to 0. Hint: Solve it for $0 \le t \le 1$, determine $Q(1)$, then solve it for $1 \le t < \infty$.]

---

## 9.2

## POWERS AND PRODUCTS OF SINES AND COSINES

The integrals in this section are reduced, by substitution and trigonometric identities, to just three familiar cases:

$$\int \sin ax \, dx = -\frac{1}{a} \cos ax + C,$$

$$\int \cos ax \, dx = \frac{1}{a} \sin ax + C,$$

and the old reliable

$$\int u^n \, du = \frac{1}{n+1} u^{n+1} + C.$$

We discuss three types:

**I.** $\int \sin^m ax \, \cos^n ax \, dx$, with *either* $m$ or $n$ an *odd* positive integer.
   Elementary substitutions reduce such an integral to $\int u^n \, du$. Two examples suggest the general procedure.

**EXAMPLE 1** $\int \sin^2 ax \cos ax \, dx$ is reduced to $\int u^2 \, du$ by setting

$$u = \sin ax, \qquad du = a \cos ax \, dx.$$

So

$$\int \sin^2 ax \cos ax \, dx = \int u^2 \left(\frac{1}{a} \, du\right) = \frac{1}{3a} u^3 + C$$

$$= \frac{1}{3a} \sin^3 ax + C.$$

**EXAMPLE 2** $\int \sin^3 \theta \, d\theta$. Here, split off one factor of $\sin \theta$ and convert the others to $\cos \theta$, by the Pythagorean identity:

$$\sin^3 \theta = (\sin^2 \theta)(\sin \theta) = (1 - \cos^2 \theta) \sin \theta.$$

Then substitute $u = \cos \theta$, $du = -\sin \theta \, d\theta$:

$$\int \sin^3 \theta \, d\theta = \int (1 - \cos^2 \theta) \sin \theta \, d\theta$$

$$= \int (1 - u^2)(-du) = -\int du + \int u^2 \, du$$

$$= -u + \tfrac{1}{3} u^3 + C$$

$$= -\cos \theta + \tfrac{1}{3} \cos^3 \theta + C.$$

To summarize the moral of these two examples: If the sine occurs to an odd power $2k + 1$, split off one factor and convert the rest to cosine:

$$\sin^{2k+1} \theta = (\sin^2 \theta)^k \sin \theta = (1 - \cos^2 \theta)^k \sin \theta;$$

then substitute $u = \cos \theta$. If it is the cosine that occurs to an odd positive power, split off a factor of $\cos \theta$ and convert the rest to $\sin \theta$.

The remaining integrals use some trigonometric identities, which we quickly review. The *addition formulas* are

$$\sin(\theta \pm \varphi) = \sin \theta \cos \varphi \pm \cos \theta \sin \varphi, \tag{1}$$

$$\cos(\theta \pm \varphi) = \cos \theta \cos \varphi \mp \sin \theta \sin \varphi. \tag{2}$$

From these follow the *double angle* formulas

$$\sin(2\theta) = 2 \sin \theta \cos \theta \tag{3}$$

$$\cos(2\theta) = \cos^2 \theta - \sin^2 \theta. \tag{4}$$

This last, combined with the Pythagorean identity, gives

$$1 + \cos 2\theta = (\cos^2 \theta + \sin^2 \theta) + (\cos^2 \theta - \sin^2 \theta) = 2 \cos^2 \theta;$$

hence

$$\cos^2\theta = \tfrac{1}{2}(1 + \cos 2\theta). \tag{5}$$

Similarly, $1 - \cos 2\theta = 2\sin^2\theta$, so

$$\sin^2\theta = \tfrac{1}{2}(1 - \cos 2\theta). \tag{6}$$

Now we are ready for the second type of integral:

**II.** $\int \sin^m ax \cos^n ax \, dx$ with $m$ and $n$ both even, and not negative.

---

**EXAMPLE 3**    $\int \cos^2\theta \, d\theta$ is handled easily by (5):

$$\int \cos^2\theta \, d\theta = \int \tfrac{1}{2}(1 + \cos 2\theta) \, d\theta = \tfrac{1}{2}\int 1 \, d\theta + \tfrac{1}{2}\int \cos 2\theta \, d\theta$$
$$= \tfrac{1}{2}\theta + \tfrac{1}{4}\sin 2\theta + C.$$

This is a convenient form of answer; for future reference, we list another useful form, using (3):

$$\int \cos^2\theta \, d\theta = \tfrac{1}{2}\theta + \tfrac{1}{2}\sin\theta\cos\theta + C.$$

---

**EXAMPLE 4**    $\int \sin^2 3t \, dt$ uses (6):

$$\int \sin^2 3t \, dt = \int \tfrac{1}{2}(1 - \cos 6t) \, dt = \tfrac{1}{2}t - \tfrac{1}{12}\sin 6t + C$$
$$= \tfrac{1}{2}t - \tfrac{1}{6}\sin 3t \cos 3t + C.$$

---

**EXAMPLE 5**

$$\int \cos^2\theta \sin^2\theta \, d\theta = \int \tfrac{1}{2}(1 + \cos 2\theta)\tfrac{1}{2}(1 - \cos 2\theta) \, d\theta$$
$$= \tfrac{1}{4}\int (1 - \cos^2 2\theta) \, d\theta = \tfrac{1}{4}\theta - \tfrac{1}{4}\int \cos^2 2\theta \, d\theta.$$

Now rewrite $\cos^2 2\theta$, using (5) once more:

$$\int \cos^2 2\theta \, d\theta = \int \tfrac{1}{2}(1 + \cos 4\theta) \, d\theta = \tfrac{1}{2}\theta + \tfrac{1}{8}\sin 4\theta + C.$$

Thus,

$$\int \cos^2\theta \sin^2\theta \, d\theta = \tfrac{1}{4}\theta - \tfrac{1}{4}[\tfrac{1}{2}\theta + \tfrac{1}{8}\sin 4\theta] + C'$$
$$= \tfrac{1}{8}\theta - \tfrac{1}{32}\sin 4\theta + C'.$$

This could be rewritten in terms of $\cos\theta$ and $\sin\theta$, using (3) and (4).

*REMARK* An alternate method for these integrals uses a reduction formula proved in the problems.

A third important type of integral is:

**III.** $\int \sin ax \sin bx \, dx, \int \sin ax \cos bx \, dx, \int \cos ax \cos bx \, dx.$ These arise in the study of periodic phenomena, using Fourier series. Their special feature is that one of the factors is evaluated at one point $ax$, while the other is evaluated at a *different* point $bx$. To integrate them, use identities derived from the addition formulas (1) and (2). For the first integral on the list, use

$$\cos(ax - bx) = \cos ax \cos bx + \sin ax \sin bx$$
$$\cos(ax + bx) = \cos ax \cos bx - \sin ax \sin bx.$$

Subtract and divide by 2:

$$\sin ax \sin bx = \tfrac{1}{2}[\cos(ax - bx) - \cos(ax + bx)]. \tag{7}$$

Thus

$$\int \sin ax \sin bx \, dx = \frac{1}{2} \int [\cos((a - b)x) - \cos((a + b)x)] \, dx$$

$$= \frac{\sin(a - b)x}{2(a - b)} - \frac{\sin(a + b)x}{2(a + b)} + C.$$

The other two integrals of this type use similar identities:

$$\cos ax \cos bx = \tfrac{1}{2}[\cos(a + b)x + \cos(a - b)x] \tag{8}$$
$$\sin ax \cos bx = \tfrac{1}{2}[\sin(a + b)x + \sin(a - b)x]. \tag{9}$$

### Special Definite Integrals

Many applications involve the integral of $\sin^2\theta$ or $\cos^2\theta$ over a "period" $0 \le \theta \le 2\pi$, or a half or quarter period. By method II, you can easily show that, *for these special intervals*, the integral is half the length of the interval of integration:

$$\int_0^{\pi/2} \sin^2\theta \, d\theta = \int_0^{\pi/2} \cos^2\theta \, d\theta = \frac{\pi}{4},$$

$$\int_0^{\pi} \sin^2\theta \, d\theta = \int_0^{\pi} \cos^2\theta \, d\theta = \frac{\pi}{2}.$$

More generally, $\cos ax$ and $\sin ax$ have period $2\pi/a$; the integral of their squares over the quarter period $\pi/2a$ is

$$\int_0^{\pi/2a} \sin^2 ax \, dx = \int_0^{\pi/2a} \cos^2 ax \, dx = \pi/4a,$$

with similar results for integrals over any integer multiple of a quarter period: The integral equals half the length of the interval of integration.

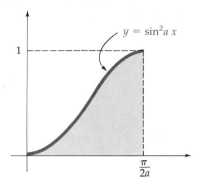

**FIGURE 1**

$$\int_0^{\pi/2a} \sin^2 ax \, dx = \frac{1}{2}\frac{\pi}{2a}.$$

Figure 1 shows a graphic way to remember this; the graph of $y = \sin^2 ax$ divides the rectangle $0 \le x \le \pi/2a$, $0 \le y \le 1$ into two congruent pieces, each of area $\pi/4a$. (The proof that the pieces are congruent uses $\sin^2\theta = \dfrac{1 - \cos 2\theta}{2}$, and the symmetry of $\cos\theta$; see problem 38.)

### An Application

In physics, *power* is the rate at which work is done, or energy is dissipated:

$$\text{power} = \frac{dW}{dt}.$$

The power dissipated by a voltage $E$ across a resistance of $R$ ohms is $E^2/R$ watts. Suppose that $R$ is constant. For *direct* current, the voltage is also constant $E_0$, and the energy dissipated in one second is simply $E_0^2/R$. But for a typical alternating current supply,

$$E = 170 \sin(120\pi t), \tag{10}$$

with $t$ in seconds. As $t$ varies from 0 to 1, then $120\pi t$ varies from 0 to $120\pi = 60 \cdot 2\pi$, sixty periods of the sine function. Thus the voltage in (10) oscillates with a frequency of 60 cycles per second. The power is

$$\frac{dW}{dt} = E^2/R = [170 \sin(120\pi t)]^2/R,$$

and the energy dissipated in one second is

$$\int_0^1 \frac{dW}{dt} \, dt = \frac{(170)^2}{R} \int_0^1 \sin^2(120\pi t) \, dt.$$

The integral extends over 60 whole periods of $\sin(120\pi t)$, so it equals half the length of the interval of integration $[0, 1]$; the energy dissipated in one second is thus $\dfrac{(170)^2}{2R}$. To dissipate energy at the same rate with a *constant* voltage $E_0$ would require

$$\frac{E_0^2}{R} = \frac{(170)^2}{2R}, \quad \text{or} \quad E_0 = \frac{170}{\sqrt{2}} \approx 120.2.$$

The oscillating voltage $170 \sin(120\pi t)$ has a maximum of 170, but its power (averaged over one second) is equivalent to a constant voltage of about 120 volts.

## SUMMARY

$$\int \sin ax \, dx = -\frac{1}{a} \cos ax + C, \qquad \int \cos ax \, dx = \frac{1}{a} \sin ax + C$$

*For $\int \sin^m ax \cos^n ax \, dx$:*

 **I.** If sin $ax$ is raised to an *odd* positive power, split off one power of sin $ax$, convert the rest to cosine, and set $u = \cos ax$.

 If cos $ax$ is raised to an odd positive power, split off one power of cos $ax$, convert the rest to sine, and set $u = \sin ax$.

 **II.** If both powers are even, and $\geq 0$, use

$$\sin^2\theta = \frac{1 - \cos 2\theta}{2}, \qquad \cos^2\theta = \frac{1 + \cos 2\theta}{2}$$

*Other Identities:*

$$\sin ax \sin bx = \tfrac{1}{2}[\cos(a - b)x - \cos(a + b)x]$$

$$\sin ax \cos bx = \tfrac{1}{2}[\sin(a + b)x + \sin(a - b)x]$$

$$\cos ax \cos bx = \tfrac{1}{2}[\cos(a + b)x + \cos(a - b)x]$$

*Reduction Formulas:* See problems 32 and 33.

## PROBLEMS

### A

In problems 1–22, evaluate the integral.

*1. $\displaystyle\int \sin 2\theta \, d\theta$

2. $\displaystyle\int_0^1 \cos 3\pi t \, dt$

*3. $\displaystyle\int \sin^2\theta \, d\theta$

4. $\displaystyle\int \cos^2\pi t \, dt$

*5. $\displaystyle\int \sin^2 x \cos x \, dx$

6. $\displaystyle\int \sqrt{\cos x} \, \sin x \, dx$

*7. $\displaystyle\int \cos^2 ax \, dx$

8. $\displaystyle\int \sin^2 ax \, dx$

*9. $\displaystyle\int \cos^8 3x \sin 3x \, dx$

10. $\displaystyle\int \sin^8 5x \cos^3 5x \, dx$

*11. $\displaystyle\int \sin^2 2\theta \cos^2 2\theta \, d\theta$

12. $\displaystyle\int \cos^4 x \, dx$

*13. $\displaystyle\int \sin^3\theta \, d\theta$

14. $\displaystyle\int_{-\pi}^{\pi} \sin^3\theta \, d\theta$ (Do it geometrically, without calculation.)

*15. $\displaystyle\int_0^{\pi} \cos^2 2\theta \, d\theta$

16. $\displaystyle\int_0^1 \sin^2\pi t \, dt$

*17. $\displaystyle\int_0^1 \cos^2 120\pi t \, dt$

18. $\displaystyle\int \sin^4 5x \, dx$

*19. $\displaystyle\int \sin x \sin 3x \, dx$

20. $\displaystyle\int_0^{\pi/2} \cos x \sin 2x \, dx$

*21. $\displaystyle\int \sin t \sqrt{1 + \cos t} \, dt$

22. $\displaystyle\int e^{\cos x} \sin x \, dx$

23. The variation of the normal mean air temperature through the year at Fairbanks, Alaska, can be approximated by the sine curve

$$T = 37 \sin\left(\frac{2\pi}{365}(x - 101)\right) + 25,$$

which has a minimum of $-37 + 25 = -12$ on day 10 (January 10) and a maximum of $37 + 25 = 62$ on day 192 (July 11).

a) Compute the "degree-days" of heating required to keep a building at $65°$ all year,

$$\int_0^{365} (T_{inside} - T_{outside}) \, dx =$$

$$\int_0^{365} 65 - \left[37 \sin\left(\frac{2\pi}{365}(x - 101)\right) + 25\right] dx.$$

b)   Suppose that the building is heated only to 55°; when the outside temperature is above 55°, no heat is supplied. What then should replace the integral in part a? Evaluate that.

*24.   Compute the volume enclosed by revolving one arch of the curve $y = \sin\theta$ around the $\theta$ axis. (One arch is the part between two successive zeroes.)

*25.   Solve $\dfrac{dy}{dx} = \cos^2 x \cdot \sec y$,   $y(0) = 0$.

26.   Prove
a)   $\sin ax \cos bx = \frac{1}{2}[\sin(a - b)x + \sin(a + b)x]$.
b)   $\displaystyle\int \sin ax \cos bx \, dx =$

$$\frac{\cos((a - b)x)}{2(b - a)} - \frac{\cos((a + b)x)}{2(a + b)} + C,$$

if $a^2 \neq b^2$. (Why is it necessary that $a^2 \neq b^2$?)
c)   The corresponding integral formula if $b = \pm a$.

27.   Prove a general formula for $\int \cos ax \cos bx \, dx$, $a^2 \neq b^2$. (Why is it necessary that $a^2 \neq b^2$?)

28.   Suppose that $m$ and $n$ are integers. Evaluate the following integrals (essential in the theory of Fourier series).
a)   $\displaystyle\int_0^{2\pi} \sin mx \cos nx \, dx$
b)   $\displaystyle\int_0^{2\pi} \sin mx \sin nx \, dx, \, m \neq n$
c)   $\displaystyle\int_0^{2\pi} \sin mx \sin nx \, dx, \, m = n$
d)   $\displaystyle\int_0^{\pi} \sin mx \sin nx \, dx, \, m \neq n$
e)   $\displaystyle\int_0^{\pi} \cos mx \cos nx \, dx, \, m = n$

**B**

29.   Evaluate $\int x \sin^2 x \, dx$.

30.   Evaluate $\int e^{ax} \cos^2 bx \, dx$.

31.   Evaluate $\int \sqrt{a^2 - x^2} \, dx$ by the substitution $x = a \sin\theta$.

32.   Prove

$$\int \sin^n ax \, dx = -\frac{\sin^{n-1} ax \cos ax}{an}$$

$$+ \frac{n-1}{n} \int \sin^{n-2} ax \, dx.$$

(Integrate by parts, with $dv = \sin ax \, dx$. In the new integral, use $\cos^2\theta = 1 - \sin^2\theta$, and then solve for $\int \sin^n ax \, dx$.)

33.   Prove

$$\int \cos^n ax \, dx = \frac{\sin ax \cos^{n-1} ax}{an}$$

$$+ \frac{n-1}{n} \int \cos^{n-2} ax \, dx.$$

34.   Evaluate $\int_0^{\pi/2} \sin^n\theta \, d\theta$ [using Problem 32] for
a)   $n = 1$.                    b)   $n = 2$.
c)   $n = 3$.                    d)   $n = 4$.
e)   Arbitrary odd $n$.     f)   Arbitrary even $n$.
[These formulas were discovered by John Wallis (1616–1703).]

35.   Evaluate, for $a > 0$.
*a)   $\displaystyle\int_0^{\infty} e^{-ax} \sin^2 bx \, dx$
b)   $\displaystyle\int_0^{\infty} e^{-ax} \cos^2 bx \, dx$

36.   Compute $\displaystyle\lim_{a \to \pm\infty} \int_0^1 \sin^2 ax \, dx$. Could you predict the result from the behavior of $\sin^2 ax$ as $a \to \pm\infty$?

37.   The Schrödinger "wave functions" for a particle in a "box" of length $a$ are $\psi_n(x) = A \sin(n\pi x/a)$, where $n$ is an integer, and $A$ an appropriate constant. The wave function is "normalized" if $\int_0^a \psi_n(x)^2 \, dx = 1$. Determine $A$ so that $\psi_n$ is normalized.

38.   Prove that the graph of $y = \sin^2\theta$ divides the rectangle in Figure 2 into two congruent parts.
$$\left[\text{Show that } \sin^2\theta = 1 - \sin^2\left(\frac{\pi}{2} - \theta\right).\right]$$

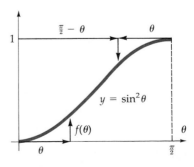

**FIGURE 2**
$y = \sin^2\theta$ divides the rectangle in two congruent parts.

## 9.3

## POWERS OF TANGENT AND SECANT

Any integral involving tangents and secants can be rewritten in terms of sine and cosine. For example

$$\int \tan \theta \, d\theta = \int \frac{\sin \theta}{\cos \theta} \, d\theta = -\int \frac{d \cos \theta}{\cos \theta}$$

since $d \cos \theta = -\sin \theta \, d\theta$. This gives the integral of $\tan \theta$:

$$
\begin{aligned}
\int \tan \theta \, dx &= -\int \frac{d \cos \theta}{\cos \theta} = -\ln|\cos \theta| + C \\
&= \ln \left| \frac{1}{\cos \theta} \right| + C \qquad \left( \ln \frac{1}{a} = -\ln a \right) \\
&= \ln|\sec \theta| + C.
\end{aligned}
$$

Usually, however, it is better to take advantage of special properties of the tangent and secant. Two integrals follow directly from the derivative formulas:

$$\int \sec^2 \theta \, d\theta = \tan \theta + C$$

$$\int \sec \theta \tan \theta \, d\theta = \sec \theta + C.$$

Generally, for integers $m \geq 0$ and $n \geq 0$,

$$\int \sec^m \theta \tan^n \theta \, d\theta \tag{1}$$

is reasonably easy if either

    **(i)**   $\sec \theta$ is raised to a positive even power (then substitute $u = \tan \theta$)

or

    **(ii)**   $\tan \theta$ is raised to a positive odd power (then substitute $u = \sec \theta$).

In each case, the Pythagorean identity $\tan^2 \theta + 1 = \sec^2 \theta$ may come into play.

---

**EXAMPLE 1**   $\int \tan^n \theta \sec^2 \theta \, d\theta$ has $\sec \theta$ to an even power. This is case (i), so set $u = \tan \theta$, $du = \sec^2 \theta \, d\theta$:

$$
\begin{aligned}
\int \tan^n \theta \sec^2 \theta \, d\theta &= \int u^n \, du = \frac{1}{(n+1)} u^{n+1} + C \\
&= \frac{1}{(n+1)} \tan^{n+1} \theta + C.
\end{aligned}
$$

---

**EXAMPLE 2**   $\int \sec^m ax \tan ax \, dx$ has $\tan \theta$ to an odd power as in case (ii). Set $u = \sec ax$, $du = a \sec ax \tan ax \, dx$; and split off one factor from $\sec^m ax$, to combine it with $\tan ax \, dx$:

$$\int \sec^m ax \tan ax \, dx = \int (\sec^{m-1} ax)(\sec ax \tan ax \, dx)$$

$$= \int u^{m-1} \left(\frac{1}{a} du\right) = \frac{1}{ma} u^m + C$$

$$= \frac{1}{ma} \sec^m ax + C.$$

**EXAMPLE 3**

$$\int \tan^2 ax \, dx = \int (\sec^2 ax - 1) dx = \int \sec^2 ax \, dx - \int 1 \, dx$$

$$= \frac{1}{a} \tan ax - x + C.$$

**EXAMPLE 4**   $\int \tan^3 \theta \, d\theta$ has $\tan \theta$ to an odd positive power. To prepare for the substitution $u = \sec \theta$, rewrite it as

$$\int (\sec^2 \theta - 1)\tan \theta \, d\theta = \int \sec \theta(\sec \theta \tan \theta) \, d\theta - \int \tan \theta \, d\theta$$

$$= \int \sec \theta \, d(\sec \theta) - \int \frac{\sin \theta}{\cos \theta} \, d\theta$$

$$= \frac{1}{2} \sec^2 \theta + \ln|\cos \theta| + C.$$

**EXAMPLE 5**   $\int \sec \theta \tan^3 \theta \, d\theta$. This is (1) with an odd power of tangent, so aim for the substitution $u = \sec \theta$, $du = \sec \theta \tan \theta \, d\theta$. Split off one factor of $\tan \theta$ to go with $\sec \theta \, d\theta$:

$$\int \sec \theta \tan^3 \theta \, d\theta = \int (\tan^2 \theta)(\sec \theta \tan \theta \, d\theta).$$

The last parentheses gives $du$; the rest of the integrand, $\tan^2 \theta$, must be expressed in terms of $u = \sec \theta$. By the Pythagorean identity

$$\tan^2 \theta = \sec^2 \theta - 1,$$

and the integral is

$$\int (\sec^2 \theta - 1)(\sec \theta \tan \theta \, d\theta) = \int (u^2 - 1)(du) \quad (u = \sec \theta)$$

$$= (\tfrac{1}{3} u^3 - u) + C$$

$$= \tfrac{1}{3} \sec^3 \theta - \sec \theta + C$$

It remains to consider

$$\int \sec^m\theta \, \tan^n\theta \, d\theta \qquad \text{when } m \text{ is odd and } n \text{ is even.} \tag{2}$$

The simplest of these cases is done by an algebraic trick:

$$\int \sec\theta \, d\theta = \int \sec\theta \, \frac{\tan\theta + \sec\theta}{\sec\theta + \tan\theta} \, d\theta$$

$$= \int \frac{\sec\theta\,\tan\theta + \sec^2\theta}{\sec\theta + \tan\theta} \, d\theta$$

$$= \int \frac{d(\sec\theta + \tan\theta)}{\sec\theta + \tan\theta}$$

$$= \ln|\sec\theta + \tan\theta| + C. \tag{3}$$

The other cases of (2) can be reduced to powers of $\sec\theta$; for when $n$ is even, $\tan^n\theta$ can be rewritten using $\tan^2\theta = \sec^2\theta - 1$. The resulting powers of $\sec\theta$ can then be integrated by the reduction formula

$$\int \sec^n ax \, dx = \frac{\sec^{n-2} ax \, \tan ax}{a(n-1)} + \frac{n-2}{n-1} \int \sec^{n-2} ax \, dx, \tag{4}$$

proved in the problems.

---

**EXAMPLE 6**

$$\int \sec\theta\,\tan^2\theta \, d\theta = \int \sec\theta(\sec^2\theta - 1)\,d\theta$$

$$= \int \sec^3\theta \, d\theta - \int \sec\theta \, d\theta$$

$$= \left[ \frac{\sec\theta\,\tan\theta}{3-1} + \frac{3-2}{3-1}\int \sec\theta\,d\theta \right] - \int\sec\theta\,d\theta$$

(by the reduction formula)

$$= \frac{\sec\theta\,\tan\theta}{2} - \frac{1}{2}\int\sec\theta\,d\theta$$

$$= \frac{1}{2}\sec\theta\,\tan\theta - \frac{1}{2}\ln|\sec\theta + \tan\theta| + C.$$

---

*REMARK* The story for powers of $\csc\theta$ and $\cot\theta$ is essentially the same, with an occasional change in sign.

**The Mercator projection** is one of many ways to represent the round earth on a flat map (Fig. 1). The vertical circles of longitude on the globe are represented as vertical lines on the map. This distorts the horizontal circles

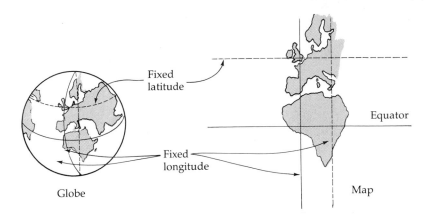

**FIGURE 1**

of latitude; those near the poles are stretched, compared to those near the equator. The Mercator projection stretches the longitude lines, too, by the same factor; as you move further from the equator, the longitude lines are stretched more and more, always by the same factor as the corresponding latitude lines. This vastly exaggerates lengths and areas near the poles compared to those near the equator, but it has two good features:

(i)   Each *small* piece of the map is scaled the same in the N-S direction as in the E-W direction, so that "in the small" there is no distortion.

(ii)   A straight line drawn on a Mercator map corresponds to a course on the globe with a *fixed* compass heading.

Figure 2 shows how to calculate the stretching. On the left is a globe of radius 1 and an equator of length $2\pi$; on the right, a Mercator map with equator also of length $2\pi$. Point $P$ at latitude $\theta$ is mapped to point $\tilde{P}$ at distance $y$ from the map's equator. First, we compute the stretching of the horizontal circle at latitude $\theta$. This circle has radius $\cos\theta$, so its length is

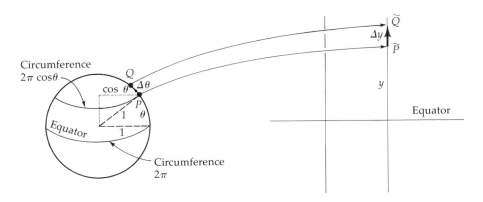

**FIGURE 2**
$\Delta\theta$ is the length of arc between $P$ and $Q$.   $\theta$ is the length of arc from the equator to $P$.

$2\pi \cos \theta$. It is mapped to a line segment of length $2\pi$, parallel to the map's equator; hence the horizontal circle is stretched by a factor $\dfrac{2\pi}{2\pi \cos \theta} = \sec \theta$. So the vertical stretching factor must also be $\sec \theta$, which varies with $\theta$. To interpret this variable stretching factor, consider a point $Q$ at latitude $\theta + \Delta\theta$, mapped to $\tilde{Q}$ at distance $y + \Delta y$ from the map's equator. The arc from $P$ to $Q$ is stretched by a factor $\Delta y / \Delta\theta$, and this should be nearly equal to $\sec \theta$:

$$\frac{\Delta y}{\Delta\theta} \approx \sec \theta.$$

As $\Delta\theta \to 0$, this becomes an equality:

$$\frac{dy}{d\theta} = \sec \theta.$$

This determines $y$:

$$y = \int \sec \theta \, d\theta = \ln|\sec \theta + \tan \theta| + C.$$

Since $y = 0$ when $\theta = 0$, you can check that $C = 0$. Thus on the Mercator map, latitude $\theta$ is plotted at distance $\ln|\sec \theta + \tan \theta|$ from the equator.

## SUMMARY

$$\int \sec^2 ax \, dx = \frac{1}{a} \tan ax + C$$

$$\int \sec ax \tan ax \, dx = \frac{1}{a} \sec ax + C$$

$$\int \tan ax \, dx = -\frac{1}{a} \int \frac{d \cos ax}{\cos ax} = \frac{1}{a} \ln|\sec ax| + C$$

$$\int \sec ax \, dx = \frac{1}{a} \ln|\sec ax + \tan ax| + C$$

$\int \sec^m ax \tan^n ax \, dx$ is easy if either

(i)   the power of secant is positive and *even* (substitute $u = \tan ax$)

or

(ii)   the power of tangent is positive and *odd* (substitute $u = \sec ax$ after appropriate rewriting).

For other cases, use reduction formula (4).

## PROBLEMS

### A

*1. $\int \tan(x/2)\, dx$

2. $\int t^2 \sec^2(t^3)\, dt$

*3. $\int_{-\pi/4}^{\pi/4} \sec\theta\, d\theta$

4. $\int_0^{\pi/3} \sec^2\theta\, d\theta$

*5. $\int_0^{\pi/3} \tan^2\theta\, d\theta$

6. $\int_0^{\pi} \sec(x/2)\, dx$

*7. $\int \sec^2 x \tan^2 x\, dx$

8. $\int \tan^3 2t \sec^3 2t\, dt$

*9. $\int \dfrac{dx}{\cos^2 x}$

10. $\int \dfrac{d\theta}{\sin^2 2\theta}$

*11. $\int \dfrac{\sin^2 t\, dt}{\cos^2 t}$

12. $\int \sec^3 3t\, dt$

*13. $\int \sec \pi x \tan^2 \pi x\, dx$

14. $\int \sec^5 \pi x\, dx$

*15. $\int \sec^3 \pi x \tan^2 \pi x\, dx$

16. $\int \sin^2\theta\, d\theta$

*17. $\int \cos^3 2t \cdot dt$

18. $\int \tan^5\theta\, d\theta$

*19. $\int \sin^2\theta \cos^2\theta\, d\theta$

20. $\int \sin^4\theta \cos^2\theta\, d\theta$

*21. $\int e^{\cos\theta} \sin\theta\, d\theta$

22. $\int \dfrac{\cos x\, dx}{\sin^3 x}$

*23. $\int \cot^3 2x \csc^2 2x\, dx$

24. $\int \tan\theta \sec^2\theta\, d\theta$

*25. $\int \sec^3 x \tan x\, dx$

26. $\int \dfrac{\cos x\, dx}{1 + \sin x}$

*27. $\int x \cos x^2\, dx$

28. $\int_0^{\pi/4} \dfrac{dx}{\sqrt{1 - \sin^2 x}}$

*29. $\int \dfrac{dx}{\cos x}$

30. $\int \dfrac{dx}{\cos^2 x}$

*31. $\int \dfrac{\sec^2\theta}{1 + \tan\theta}\, d\theta$

32. $\int \dfrac{dx}{\sin^4 x}$

*33. $\int \csc\theta\, d\theta$

34. $\int_0^{\pi} \sec x\, dx$   [Improper!]

35. $\int \csc^3 x \cot x\, dx$

36. $\int \csc^2 x \cot^2 x\, dx$

37. $\int \dfrac{\sec\theta\, d\theta}{\tan^2\theta}$

38. $\int x \sec^2 2x\, dx$

39. $\int \theta \tan^2\theta\, d\theta$

*40. The curve $y = \tan\theta$, $0 \le \theta \le \pi/4$, is revolved around the $\theta$-axis. Compute the volume enclosed.

41. Solve $\dfrac{dy}{d\theta} = \sec^2\theta \cos^3 y$.

### B

42. Prove the reduction formulas.

a) Formula (4) in the text. (Split off a factor of $\sec^2\theta$, and integrate by parts.)

b) $\int \sec^n\theta \tan^m\theta\, d\theta = \dfrac{\sec^{n-2}\theta \tan^{m+1}\theta}{(m + n - 1)}$

$\qquad + \dfrac{n - 2}{m + n - 1} \int \sec^{n-2}\theta \tan^m\theta\, d\theta.$

c) $\int \sec^n\theta \tan^m\theta\, d\theta = \dfrac{\sec^n\theta \tan^{m-1}\theta}{m + n - 1}$

$\qquad - \dfrac{m - 1}{m + n - 1} \int \sec^n\theta \tan^{m-2}\theta\, d\theta.$

43. Prove reduction formulas for $\int \csc^n\theta \cot^m\theta\, d\theta$
a) Reducing $n$.   b) Reducing $m$.

## 9.4

## TRIGONOMETRIC SUBSTITUTION

Many integrals containing a *sum or difference of squares* can be evaluated by trigonometric substitutions, exploiting the Pythagorean identities. We give an example, then discuss the general principles.

**EXAMPLE 1** $\displaystyle\int \frac{dx}{\sqrt{4-x^2}}$.

**SOLUTION**  It is tempting to substitute $u = 4 - x^2$, but then $du = -2x\,dx$, and the given integral has no "$x$" to go with the "$dx$." Another method is required—trigonometric substitution.

The expression $4 - x^2 = 2^2 - x^2$ calls to mind a right triangle with hypotenuse 2 and leg $x$ (Fig. 1). The figure suggests that you *substitute* a new variable $\theta$, related to $x$ by

$$\sin \theta = \frac{x}{2}, \quad \text{or} \quad x = 2\sin\theta. \tag{1}$$

The substitution (1) replaces the *difference* of squares $4 - x^2$ by a *perfect square*:

$$4 - x^2 = 4 - 4\sin^2\theta = 4(1 - \sin^2\theta) = 4\cos^2\theta$$

so

$$\sqrt{4 - x^2} = 2\cos\theta. \tag{2}$$

And (1) implies $dx/d\theta = 2\cos\theta$, so

$$dx = 2\cos\theta\,d\theta. \tag{3}$$

With (2) and (3) the original integral becomes

$$\int \frac{dx}{\sqrt{4-x^2}} = \int \frac{2\cos\theta\,d\theta}{2\cos\theta}$$

$$= \int d\theta = \theta + C.$$

Now express the result in terms of $x$, using (1):

$$\theta = \sin^{-1}(x/2),$$

so

$$\int \frac{dx}{\sqrt{4-x^2}} = \sin^{-1}(x/2) + C.$$

Check this by differentiating $\sin^{-1}(x/2)$:

$$\frac{d\,\sin^{-1}(x/2)}{dx} = \frac{1}{\sqrt{1 - (x/2)^2}}\,\frac{2(x/2)}{dx} = \frac{1}{\sqrt{1 - (x/2)^2}} \cdot \frac{1}{2}$$

$$= \frac{1}{\sqrt{(1 - (x/2^2) \cdot 2^2}} = \frac{1}{\sqrt{4 - x^2}}.$$

**NOTE**  In deducing (2), we tacitly assumed that $\cos\theta \geq 0$; this is justified on the next page.

Now, for the general description of the method, we use $u$ to stand for a variable and $a$ for a positive constant. We consider integrals with three

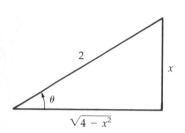

**FIGURE 1**
$\sin\theta = x/2$.

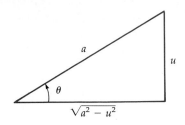

**FIGURE 2**
$u = a \sin \theta$, $\sqrt{a^2 - u^2} = a \cos \theta$.

types of expression:

**Type I:** $\sqrt{a^2 - u^2}$. Inspired by Figure 2, set

$$u = a \sin \theta, \qquad -\frac{\pi}{2} \le \theta \le \frac{\pi}{2}.$$

Then, from the figure,

$$\theta = \sin^{-1}\left(\frac{u}{a}\right) \quad \text{and} \quad \sqrt{a^2 - u^2} = a \cos \theta.$$

$$\left(-\frac{\pi}{2} \le \theta \le \frac{\pi}{2} \Rightarrow \cos \theta \ge 0, \text{ so } \sqrt{a^2 - u^2} = a \cos \theta, \text{ not } -a \cos \theta.\right)$$

**Type II:** $\sqrt{a^2 + u^2}$ **(or even $a^2 + u^2$).** Inspired by Figure 3, set

$$u = a \tan \theta, \qquad -\frac{\pi}{2} < \theta < \frac{\pi}{2}.$$

Then

$$\theta = \tan^{-1}\left(\frac{u}{a}\right) \quad \text{and} \quad \sqrt{a^2 + u^2} = a \sec \theta.$$

$$\left(-\frac{\pi}{2} < \theta < \frac{\pi}{2} \Rightarrow \sec \theta > 0, \text{ so } \sqrt{a^2 + u^2} = a \sec \theta, \text{ not } -a \sec \theta.\right)$$

**Type III:** $\sqrt{u^2 - a^2}$. Inspired by Figure 4, set

$$u = a \sec \theta, \qquad 0 \le \theta < \frac{\pi}{2} \text{ or } \frac{\pi}{2} < \theta \le \pi.$$

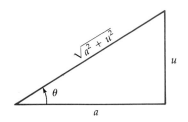

**FIGURE 3**
$u = a \tan \theta$, $\sqrt{a^2 + u^2} = a \sec \theta$.

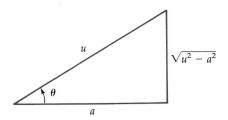

**FIGURE 4**
$u = a \sec \theta$, $\sqrt{u^2 - a^2} = a \tan \theta$.

Then

$$\theta = \sec^{-1}\left(\frac{u}{a}\right) \quad \text{and} \quad \sqrt{u^2 - a^2} = \begin{cases} a \tan \theta, & u \ge a \\ -a \tan \theta, & u \le -a. \end{cases}$$

(The complications with signs are discussed on pages 433 and 434.)

*REMARKS* Each sum or difference of squares suggests the Pythagorean Theorem, applied to an appropriately labeled triangle, as in Figures 1–4. This triangle shows the appropriate substitution and the relations needed to carry it through.

In each case, the range of $\theta$ is chosen to fit the definition of the corresponding inverse trigonometric function (Sec. 8.7). In Types I and II, this gives the appropriate sign for the square root, but in Type III it does not. The examples suggest how to deal with this annoying complication.

These substitutions should be mastered, as part of your understanding of calculus; but they are tedious! Most of the integrals that can be done this way are listed in tables inside the covers of this book (and in other more extensive integral tables, such as those of the Chemical Rubber Company). For all but the simplest of these integrals, it is much easier to use the tables.

---

**EXAMPLE 2** $\int \sqrt{3 + x^2} \, dx = ?$

**SOLUTION** $\sqrt{3 + x^2}$ has a sum of squares $(\sqrt{3})^2 + (x)^2$, calling to mind a right triangle with legs $\sqrt{3}$ and $x$ (Fig. 5). The figure suggests the tangent substitution

$$x = \sqrt{3} \tan \theta, \qquad \sqrt{3 + x^2} = \sqrt{3} \sec \theta.$$

Then $dx = \sqrt{3} \sec^2\theta \, d\theta$, so

$$\int \sqrt{3 + x^2} \, dx = \int \sqrt{3} \sec \theta \sqrt{3} \sec^2\theta \, d\theta$$

$$= 3 \int \sec^3\theta \, d\theta.$$

This calls for the reduction formula of Section 9.3:

$$\int \sec^3\theta \, d\theta = \frac{\sec \theta \tan \theta}{2} + \frac{1}{2} \int \sec \theta \, d\theta$$

$$= \frac{\sec \theta \tan \theta}{2} + \frac{1}{2} \ln|\sec \theta + \tan \theta| + C.$$

So

$$\int \sqrt{3 + x^2} \, dx = 3 \int \sec^3\theta \, d\theta$$

$$= \tfrac{3}{2} \sec \theta \tan \theta + \tfrac{3}{2} \ln|\sec \theta + \tan \theta| + C.$$

Now use Figure 5 to express the answer in terms of $x$:

$$\int \sqrt{3 + x^2} \, dx = \frac{3}{2} \frac{\sqrt{3 + x^2}}{\sqrt{3}} \frac{x}{\sqrt{3}} + \frac{3}{2} \ln \left| \frac{\sqrt{3 + x^2}}{\sqrt{3}} + \frac{x}{\sqrt{3}} \right| + C$$

$$= \frac{x}{2} \sqrt{3 + x^2} + \frac{3}{2} \ln|\sqrt{3 + x^2} + x| + \left[ C - \frac{3}{2} \ln\sqrt{3} \right]$$

$$= \frac{x}{2} \sqrt{3 + x^2} + \frac{3}{2} \ln(\sqrt{3 + x^2} + x) + C'.$$

In the last step, $C'$ is a new constant; and we do not need an absolute value sign since $\sqrt{3 + x^2} + x > 0$ (Why?).

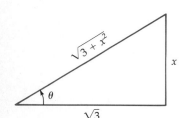

**FIGURE 5**
$x = \sqrt{3} \tan \theta.$

**EXAMPLE 3**   $\displaystyle\int \frac{du}{\sqrt{u^2 - a^2}} = ?$

**SOLUTION**   In $\sqrt{u^2 - a^2}$, the difference of squares $u^2 - a^2$ suggests a right triangle with hypotenuse $u$ and leg $a$ (Fig. 4). This is Type III, so we have to be careful with signs. Suppose at first that $u$ is positive, as in the figure. Set $u = a \sec \theta$, so $du = a \sec \theta \tan \theta \, d\theta$, and $\sqrt{u^2 - a^2} = a \tan \theta$. Hence,

$$\int \frac{du}{\sqrt{u^2 - a^2}} = \int \frac{a \sec \theta \tan \theta \, d\theta}{a \tan \theta}$$

$$= \int \sec \theta \, d\theta$$

$$= \ln|\sec \theta + \tan \theta| + C.$$

Convert back to $u$, using Figure 4:

$$\int \frac{du}{\sqrt{u^2 - a^2}} = \ln\left|\frac{u}{a} + \frac{1}{a}\sqrt{u^2 - a^2}\right| + C$$

$$= \ln\left|u + \sqrt{u^2 - a^2}\right| + C' \tag{4}$$

where $C' = C - \ln a$. This integral, though derived only for $u \geq a$, is valid for $u \leq -a$ as well; for on differentiating the right-hand side, the absolute value sign disappears, and the derivative formula is the same for negative as for positive $u$:

$$\frac{d}{du}\ln\left|u + \sqrt{u^2 - a^2}\right| = \frac{1}{u + \sqrt{u^2 - a^2}} \cdot \left(1 + \frac{u}{\sqrt{u^2 - a^2}}\right)$$

$$= \frac{1}{u + \sqrt{u^2 - a^2}} \frac{\sqrt{u^2 - a^2} + u}{\sqrt{u^2 - a^2}} = \frac{1}{\sqrt{u^2 - a^2}},$$

for all $|u| \geq a$.

---

**EXAMPLE 4**   $\displaystyle\int \frac{\sqrt{x^2 - 1}}{x} \, dx$   calls   for   $x = \sec \theta$,   $dx = \sec \theta \tan \theta \, d\theta$.

Again, we must keep track of the signs. If $x \geq 1$, then $\sqrt{x^2 - 1} = \tan \theta$; you should carry out the integration as an exercise, and find the result

$$\int \frac{\sqrt{x^2 - 1}}{x} \, dx = \tan \theta - \theta + C$$

$$= \sqrt{x^2 - 1} - \sec^{-1}x + C, \qquad x \geq 1. \tag{5}$$

This, however, is not correct for $x \leq -1$, because the derivative $\dfrac{d \sec^{-1}x}{dx} =$

$\dfrac{1}{|x|\sqrt{x^2 - 1}}$ has an absolute value sign, which makes the algebra different for negative $x$. Instead of reworking the problem for $x \leq -1$, observe that the integrand $\dfrac{\sqrt{x^2 - 1}}{x}$ is an *odd* function of $x$; hence, if we adjust the answer

(5) to make it an *even* function of $x$, it will be correct for all $|x| \geq 1$; for the derivative of an even function is odd. Thus we conclude that

$$\int \frac{\sqrt{x^2 - 1}}{x}\, dx = \sqrt{x^2 - 1} - \sec^{-1}|x| + C.$$

You can check this, using

$$\frac{d \sec^{-1}|u|}{du} = \frac{1}{u\sqrt{u^2 - 1}}. \tag{6}$$

## Final Remark on the Substitution $u = a \sec \theta$

These two examples suggest a practical way to deal with this nasty substitution:

(i)   Carry it out for positive $u$, using the relations from the right triangle.

(ii)   In the answer, if $\sec^{-1}\left(\dfrac{u}{a}\right)$ appears, replace it by $\sec^{-1}\left|\dfrac{u}{a}\right|$.

(iii)   Check the result, using (6).

Since we haven't *proved* that steps (i) and (ii) work in all cases, the check (iii) is essential. But formula (6) suggests why this should work; the derivative has no absolute value sign, so the algebra works out the same for negative as for positive $x$. The resulting formulas with $\sec^{-1}|u/a|$ also avoid any confusion in the definition of $\sec^{-1} x$ for $x \leq -1$; we take it in the range $\dfrac{\pi}{2} < \theta \leq \pi$, but some take it in the range $\pi \leq \theta < \dfrac{3\pi}{2}$.

## Use of Tables

You see how laborious these methods are. It is much easier, whenever possible, to reduce your integral to a standard form in a table, such as the forms in Examples 3 and 6 and problem 1.

**EXAMPLE 5**   $\displaystyle\int \frac{dt}{\sqrt{3t^2 - 2}}$ is close to the standard form (4). To match it to that form, set

$$u^2 = 3t^2, \quad \text{or} \quad u = \sqrt{3}\,t.$$

Then

$$\int \frac{dt}{\sqrt{3t^2 - 2}} = \int \frac{du/\sqrt{3}}{\sqrt{u^2 - 2}} = \frac{1}{\sqrt{3}} \int \frac{du}{\sqrt{u^2 - 2}}.$$

This matches (4) with $a = \sqrt{2}$; that formula gives

$$\int \frac{dt}{\sqrt{3t^2 - 2}} = \frac{1}{\sqrt{3}} \ln\left|\sqrt{3}\,t + \sqrt{3t^2 - 2}\right| + C.$$

*The computation of arc length* leads to integrals solvable by trigonometric substitution. You could study that topic (Sec. 10.1) at this point.

## SUMMARY

| *For Integrals With* | *Substitute* | *Use* |
|---|---|---|
| $a^2 - u^2$ (see Fig. 2) | $u = a \sin \theta, \ -\dfrac{\pi}{2} \le \theta \le \dfrac{\pi}{2}$ | $\sqrt{a^2 - u^2} = a \cos \theta,$ $\theta = \sin^{-1}(u/a)$ |
| $a^2 + u^2$ (see Fig. 3) | $u = a \tan \theta, \ -\dfrac{\pi}{2} < \theta < \dfrac{\pi}{2}$ | $\sqrt{a^2 + u^2} = a \sec \theta,$ $\theta = \tan^{-1}(u/a)$ |
| $u^2 - a^2$ (see Fig. 4) | $u = a \sec \theta$ | $\sqrt{u^2 - a^2} = a \tan \theta$ $\theta = \sec^{-1}|u/a|$ |

## PROBLEMS

**A**

**1.** Derive the following by trigonometric substitution.

a) $\displaystyle \int \frac{du}{a^2 + u^2} = \frac{1}{a} \tan^{-1}(u/a) + C$

b) $\displaystyle \int \frac{du}{\sqrt{a^2 + u^2}} = \ln(u + \sqrt{a^2 + u^2}) + C$

c) $\displaystyle \int \sqrt{a^2 + u^2} \, du = \frac{1}{2} u \sqrt{a^2 + u^2}$
$$+ \frac{a^2}{2} \ln(u + \sqrt{a^2 + u^2}) + C$$

d) $\displaystyle \int \sqrt{a^2 - u^2} \, du = \frac{u}{2} \sqrt{a^2 - u^2}$
$$+ \frac{a^2}{2} \sin^{-1}(u/a) + C, \ a > 0$$

e) $\displaystyle \int \frac{du}{\sqrt{a^2 - u^2}} = \sin^{-1}\left(\frac{u}{a}\right) + C$

f) $\displaystyle \int \frac{du}{\sqrt{u^2 - a^2}} = \ln|u + \sqrt{u^2 - a^2}| + C$

g) $\displaystyle \int \frac{du}{u^2 \sqrt{u^2 - a^2}} = \frac{\sqrt{u^2 - a^2}}{a^2 u} + C$

**2.** Evaluate the following, using problem 1 when it applies.

*a) $\displaystyle \int \sqrt{2 - 3x^2} \, dx$

b) $\displaystyle \int_0^1 \frac{dx}{1 + x^2}$

*c) $\displaystyle \int_0^1 \frac{dx}{\sqrt{4 - (x - 1)^2}}$

d) $\displaystyle \int_0^2 \frac{dx}{\sqrt{4 + x^2}}$

*e) $\displaystyle \int_2^4 \frac{dt}{\sqrt{2t^2 - 3}}$

f) $\displaystyle \int_0^1 \sqrt{1 + 4x^2} \, dx$

**3.** Evaluate the following by appropriate substitutions (not necessarily trigonometric).

*a) $\displaystyle \int \frac{x}{\sqrt{9 - x^2}} \, dx$

b) $\displaystyle \int \frac{du}{(a^2 - u^2)^{3/2}}$

*c) $\displaystyle \int \frac{u^2}{a^2 + u^2} \, du$

d) $\displaystyle \int \frac{dx}{x^2 \sqrt{4 - 9x^2}}$

*e) $\displaystyle \int \frac{u \, du}{\sqrt{a^2 + u^2}}$

f) $\displaystyle \int \frac{\sqrt{4 + x^2}}{x} \, dx$

*g) $\displaystyle \int \frac{du}{u^2 \sqrt{u^2 + a^2}}$

h) $\displaystyle \int \frac{t^2 \, dt}{\sqrt{t^2 - 9}}$

*i) $\displaystyle \int \frac{dz}{(4z^2 - 1)^{3/2}}$

j) $\displaystyle \int \frac{x^2 \, dx}{1 + x^2}$

**\*k)** $\displaystyle\int \frac{dx}{(1 + x^2)^2}$ 　　 **l)** $\displaystyle\int \frac{x \, dx}{1 + x^2}$

**\*m)** $\displaystyle\int \frac{x^3 \, dx}{(1 + x^2)^2}$ 　 **n)** $\displaystyle\int \frac{x^4 \, dx}{(1 + x^2)^2}$

**4.** Compute the area under the graph of $y = \sqrt{r^2 - x^2}$, $0 \le x \le a$. (Do it by evaluating an integral, not by geometry.)

**5.** A cylindrical tank of radius $r$ and length $b$, set on its side, contains gasoline to a depth $h$. Find the volume of the gasoline.

**6.** A cylindrical tank of length $l$ and radius $a$, lying on its side, is half full of a fluid of density $\delta$. Compute the work required to pump the fluid out of a hole in the top of the tank.

**\*7.** Evaluate

**a)** $\displaystyle\int_0^\infty \frac{dx}{1 + x^2}$

**b)** $\displaystyle\int_{-0}^\infty \frac{dx}{1 + x^2}$

**8.** Prove formula (6).

# 9.5

## SOME ALGEBRAIC METHODS

We show how to reduce certain integrals to simpler cases already treated. These devices form part of a general method for integrating rational functions, given in the next section.

### Completing the Square

A sum or difference of squares, $u^2 + a^2$ or $a^2 - u^2$, can be reduced to a single square by trigonometric substitution. The *general quadratic* $ax^2 + bx + c$ can, in turn, be reduced to these cases by *completing the square* (see Sec. 12.1 for a review of that).

---

**EXAMPLE 1** $\displaystyle\int \frac{dx}{4x^2 + 4x + 2} = ?$

*SOLUTION*

$$4x^2 + 4x + 2 = 4(x^2 + x) + 2$$
$$= 4(x^2 + x + \tfrac{1}{4}) - 4(\tfrac{1}{4}) + 2$$
$$= 4(x + \tfrac{1}{2})^2 + 1.$$

This is a sum of squares $u^2 + a^2$, with $u = 2(x + \tfrac{1}{2})$, $a = 1$. Substitute $u = 2(x + \tfrac{1}{2}) = 2x + 1$. Then $du = 2 \, dx$, so

$$\int \frac{dx}{4x^2 + 4x + 2} = \int \frac{\tfrac{1}{2} \, du}{u^2 + 1}$$

$$= \frac{1}{2} \tan^{-1} u + C = \frac{1}{2} \tan^{-1}(2x + 1) + C.$$

---

Notice that the first step in completing the square is to factor out the coefficient of $x^2$. To simplify the notation, in the rest of this section we will suppose that this has already been done, and consider quadratics of the form $x^2 + bx + c$, where the coefficient of the squared term is 1.

---

## EXAMPLE 2

$$\int \frac{u\,du}{\sqrt{u^2 - 2bu}} = \int \frac{u\,du}{\sqrt{(u-b)^2 - b^2}} \qquad \text{(completing the square)} \qquad (1)$$

$$= \int \frac{(v+b)\,dv}{\sqrt{v^2 - b^2}} \qquad [v = u - b]$$

$$= \int \frac{v\,dv}{\sqrt{v^2 - b^2}} + b \int \frac{dv}{\sqrt{v^2 - b^2}}. \qquad (2)$$

The first integral on the right in (2) should *not* be done by trigonometric substitution; set $v^2 - b^2 = w$ and get

$$\int \frac{v\,dv}{\sqrt{v^2 - b^2}} = \int \frac{\frac{1}{2}\,dw}{w^{1/2}} = \frac{1}{2}\int w^{-1/2}\,dv = w^{1/2} + C_1$$

$$= \sqrt{v^2 - b^2} + C_1 = \sqrt{u^2 - 2bu} + C_1,$$

since $v = u - b$. The last integral in (2) was done in Example 3, Section 9.4:

$$b \int \frac{dv}{\sqrt{v^2 - b^2}} = b \ln\left|v + \sqrt{v^2 - b^2}\right| + C_2$$

$$= b \ln\left|u - b + \sqrt{u^2 - 2bu}\right| + C_2.$$

Hence

$$\int \frac{u\,du}{\sqrt{u^2 - 2bu}} = \sqrt{u^2 - 2bu} + b \ln\left|u - b + \sqrt{u^2 - 2bu}\right| + C. \qquad (3)$$

---

## Proper Rational Functions

A positive fraction $p/q$ is called **proper** if the numerator $p$ is less than the denominator $q$; thus $2/3$ is proper, and $5/3$ is not. Any improper fraction can be written as an integer plus a proper fraction; for instance

$$\tfrac{5}{3} = 1 + \tfrac{2}{3}.$$

A *rational function* $P/Q$ is called proper if the *degree* of the numerator $P$ is less than the *degree* of the denominator $Q$; briefly, if $\deg(P) < \deg(Q)$. Thus $\dfrac{x}{x^2 + 1}$ is proper, but $\dfrac{x^2}{x^2 + 1}$ and $\dfrac{x^5 + 2x}{x^2 + 1}$ are not.

Every improper rational function can be rewritten as a *polynomial* plus a *proper* rational function:

$$\frac{P}{Q} = D + \frac{R}{Q}, \qquad (4)$$

where $D$ and $R$ are polynomials, with $\deg(R) < \deg(Q)$. For example, you can write

$$\frac{x^2}{x^2 + 1} = \frac{(x^2 + 1) - 1}{x^2 + 1} = \frac{x^2 + 1}{x^2 + 1} - \frac{1}{x^2 + 1} = 1 + \frac{-1}{x^2 + 1},$$

which has the required form

$$\frac{P}{Q} = \ldots\ldots\ldots\ldots\ldots\ldots\ldots\ldots\ldots\ldots\ldots\ldots\ldots = D + \frac{R}{Q}.$$

In this form, the function is easily integrated:

$$\int \frac{x^2 \, dx}{x^2 + 1} = \int \left[ 1 + \frac{-1}{x^2 + 1} \right] dx = x - \tan^{-1}x + C.$$

More complicated examples of improper rational functions can be reduced to proper form by *long division*, as in the following example.

---

**EXAMPLE 3**  Write $\dfrac{2x^3 + x^2 + 1}{x^2 + 1} = \dfrac{P}{Q}$ in the form (4), and integrate the result.

*SOLUTION*  Set up the division, filling in any missing powers in the numerator:

$$Q \to x^2 + 1 \,\big|\, 2x^3 + x^2 + 0x + 1 \leftarrow P.$$

Divide $x^2$, the highest term of $Q$, into $2x^3$, the highest term of $P$, and record the result above $2x^3$; $x^2$ into $2x^3$ goes $2x$ times, so write

$$
\begin{array}{r}
2x \\
x^2 + 1 \,\big|\, \overline{2x^3 + x^2 + 0x + 1}.
\end{array}
$$

Now multiply $x^2 + 1$ by $2x$ and subtract from $P$:

$$
\begin{array}{r}
2x \phantom{xxxxxxxx} \\
x^2 + 1 \,\big|\, \overline{2x^3 + x^2 + 0x + 1} \\
2x^3 \phantom{xxx} + 2x \phantom{xx} \\
\hline
x^2 - 2x + 1
\end{array}.
$$

Repeat the process, dividing $x^2 + 1$ into the remainder $x^2 - x + 1$:

$$
\begin{array}{r}
2x \phantom{xx} + 1 \leftarrow D \phantom{xxxxx} \\
Q \to x^2 + 1 \,\big|\, \overline{2x^3 + x^2 + 0x + 1} \leftarrow P \\
2x^3 \phantom{xxx} + 2x \phantom{xxxxx} \\
\hline
x^2 - 2x + 1 \phantom{xx} \\
x^2 \phantom{xxx} + 1 \phantom{xx} \\
\hline
-2x \phantom{xxxx} \leftarrow R.
\end{array}
$$

The resulting remainder $R = -2x$ has $\mathrm{degree}(R) < \mathrm{degree}(Q)$, so stop at this point.

Think about the operations just performed; they show that

$$(2x^3 + x^2 + 1) - \underbrace{2x(x^2 + 1)}_{\text{first subtraction}} - \underbrace{1(x^2 + 1)}_{\text{second subtraction}} = -2x.$$

Divide by $x^2 + 1$ to find

$$\frac{2x^3 + x^2 + 1}{x^2 + 1} - 2x - 1 = \frac{-2x}{x^2 + 1}$$

or

$$\frac{2x^3 + x^2 + 1}{x^2 + 1} = 2x + 1 + \frac{-2x}{x^2 + 1}.$$

This has the desired form

$$\frac{P}{Q} = D + \frac{R}{Q}$$

where $R/Q$ is proper. Now integrate:

$$\int \frac{2x^3 + x^2 + 1}{x^2 + 1}\, dx = \int (2x + 1)\, dx - \int \frac{2x}{x^2 + 1}\, dx$$

$$= x^2 + x - \ln(x^2 + 1) + C.$$

### Reducible and Irreducible Quadratics

A quadratic $x^2 + bx + c$ is called **reducible** if it is a product of real linear factors:

$$x^2 + bx + c = (x - r)(x - s).$$

For example,

$$x^2 - 1 = (x - 1)(x + 1)$$

is reducible. But $x^2 + 1$ is not; for if it could be factored as

$$x^2 + 1 = (x - r)(x - s)$$

with real linear factors, then $r$ and $s$ would be real zeroes of $x^2 + 1$, whereas $x^2 + 1 = 0$ has *no* real solutions.

   Rational functions with reducible quadratic denominators can be decomposed into simpler parts, called "partial fractions." For example,

$$\frac{1}{x^2 - 1} = \frac{1/2}{x - 1} - \frac{1/2}{x + 1}. \tag{5}$$

Then integration is easy:

$$\int \frac{dx}{x^2 - 1} = \int \frac{(1/2)\, dx}{x - 1} - \int \frac{(1/2)\, dx}{x + 1} = \frac{1}{2}\ln|x - 1| - \frac{1}{2}\ln|x + 1| + C.$$

The next section develops this into a general method of decomposing rational functions into simpler parts for integration. Among the "simpler" parts, we find integrals of the form

$$\int \frac{Bx + C}{(x^2 + bx + c)^n} \, dx \tag{6}$$

with $x^2 + bx + c$ *irreducible*. These require four standard formulas. Three are easy:

$$\int \frac{du}{u^2 + a^2} = \frac{1}{a} \tan^{-1}\left(\frac{u}{a}\right) + C \tag{7}$$

$$\int \frac{u \, du}{u^2 + a^2} = \int \frac{(1/2)d(u^2 + a^2)}{u^2 + a^2} = \frac{1}{2}\ln(u^2 + a^2) + C \tag{8}$$

$$\int \frac{u \, du}{(u^2 + a^2)^n} = \int (u^2 + a^2)^{-n} \frac{1}{2} d(u^2 + a^2)$$

$$= \frac{1}{2(1 - n)}(u^2 + a^2)^{1-n} + C, \qquad n \neq 1. \tag{9}$$

But the remaining form is not so easy: $\int \dfrac{du}{(u^2 + a^2)^n}$ for $n > 1$. This can be done by setting $u = a \tan \theta$, but it is quicker to use the reduction formula (see problem 8)

$$\int \frac{du}{(u^2 + a^2)^{n+1}} = \frac{u}{2na^2(u^2 + a^2)^n} + \frac{2n - 1}{2na^2} \int \frac{du}{(u^2 + a^2)^n}. \tag{10}$$

Now, consider the integral (6) with a general quadratic $x^2 + bx + c$. First, test to see whether the quadratic has real zeroes $r$ and $s$ (not necessarily distinct); this can be determined from the quadratic formula. If it does, then it is reducible; in fact

$$x^2 + bx + c = (x - r)(x - s).$$

This case, as we said, is handled in the next section. But if $x^2 + bx + c$ has *no* real zeroes, then complete the square, and reduce the integral to formulas (7)–(10).

---

**EXAMPLE 4** $\displaystyle\int \frac{x \, dx}{(x^2 + x + 1)^2} = ?$

*SOLUTION*  Test $x^2 + x + 1$ for reducibility. The zeroes

$$x = \frac{-1 \pm \sqrt{1 - 4}}{2}$$

are not real, so the quadratic is irreducible. Complete the square:

$$x^2 + x + 1 = (x + \tfrac{1}{2})^2 + \tfrac{3}{4}.$$

Now the substitution $u = x + \frac{1}{2}$ reduces the integral to formulas (7)–(10):

$$\int \frac{x\, dx}{(x^2 + x + 1)^2} = \int \frac{(u - 1/2)\, du}{(u^2 + 3/4)^2}$$

$$= \int \frac{u\, du}{(u^2 + 3/4)^2} - \frac{1}{2} \int \frac{du}{(u^2 + 3/4)^2}$$

$$= -\frac{1}{2}\left(u^2 + \frac{3}{4}\right)^{-1} - \frac{1}{2}\left[\frac{u}{2(3/4)(u^2 + 3/4)} + \frac{1}{2(3/4)} \int \frac{du}{u^2 + 3/4}\right] \qquad \text{[by (9) and (10)]}$$

$$= \frac{-1}{2(u^2 + 3/4)} - \frac{u}{3(u^2 + 3/4)} - \frac{1}{3} \cdot \frac{2}{\sqrt{3}} \tan^{-1}\left(\frac{2u}{\sqrt{3}}\right) + C$$

$$= \frac{-1}{2(x^2 + x + 1)} - \frac{x + 1/2}{3(x^2 + x + 1)} - \frac{2}{3\sqrt{3}} \tan^{-1}\left(\frac{2x + 1}{\sqrt{3}}\right) + C.$$

## SUMMARY

A rational function $P/Q$ is *proper* if $\deg(P) < \deg(Q)$. Any *improper* rational function is the sum of a polynomial and a *proper* rational function:

$$\frac{P}{Q} = D + \frac{R}{Q}, \qquad \deg(R) < \deg(Q).$$

The polynomials $D$ and $R$ can be found by long division. A quadratic $x^2 + bx + c$ is *reducible* if it is a product of real linear factors,

$$x^2 + bx + c = (x - r)(x - s).$$

This happens if and only if it has real zeroes $r$ and $s$, not necessarily distinct. Otherwise, it is *irreducible*.

## PROBLEMS

### A

1. Evaluate by completing the square and using tables where convenient.

   *a) $\displaystyle\int_0^2 \frac{dx}{x^2 - 2x + 4}$

   b) $\displaystyle\int \frac{x\, dx}{\sqrt{x^2 - 2x + 4}}$

   *c) $\displaystyle\int \frac{x\, dx}{\sqrt{2x - x^2}}$

   d) $\displaystyle\int \frac{dx}{\sqrt{x^2 - 2x - 15}}$

   *e) $\displaystyle\int \frac{(x + 1)\, dx}{\sqrt{x^2 - 4x + 3}}$

   f) $\displaystyle\int \frac{x\, dx}{x^2 + 4x + 8}$

2. Write the following quotients $P/Q$ in the form $D + R/Q$, where $D$ and $R$ are polynomials, and $\deg(R) < \deg(Q)$.

   *a) $\displaystyle\frac{x^2 + 2}{x + 2}$

   b) $\displaystyle\frac{x^2}{(x - 1)(x - 2)}$

   c) $\displaystyle\frac{x^5}{x^3 - x^2 - x + 1}$

   d) $\displaystyle\frac{x^4}{(x + 1)^2}$

*e) $\dfrac{x^5}{(x^2 + 1)^2}$    *f) $\dfrac{x}{(x^2 + 9)^2}$

*g) $\dfrac{2x^3 + x^2 + 2x + 1}{2x^2 + x + 1}$

3. Integrate the functions in problems *2a, *e, *f, g.

4. Compute the volume enclosed by revolving about the $x$-axis the entire graph of $y = \dfrac{1}{a^2 + x^2}$.

**B**

5. a) Factor $x^2 - 3x + 2$.
   b) Determine constants $A$ and $B$ so that
   $$\frac{1}{x^2 - 3x + 2} = \frac{A}{x - 2} + \frac{B}{x - 1}.$$

   *c) Evaluate $\displaystyle\int \frac{dx}{x^2 - 3x + 2}$.

*6. A wire along the $x$-axis from $x = a$ to $x = b$ carries a uniform charge density $\gamma$; an infinitesimal piece of wire of length $dx$, at position $x$, has charge $\gamma\, dx$. This creates at point $(x_0, y_0)$ an electric potential

   $\varphi = \dfrac{\gamma\, dx}{\sqrt{(x - x_0)^2 + y_0^2}}$. Compute the potential

   created by the entire wire, $\displaystyle\int_a^b \frac{\gamma\, dx}{\sqrt{(x - x_0)^2 + y_0^2}}$.

7. A wire along the $x$-axis from $x = a$ to $x = b$ carries a uniform current $I$. According to Ampere's law, an infinitesimal piece of wire of length $dx$, at position $x$, creates at $(x_0, y_0)$ a magnetic field perpendicular to the $xy$ plane, of strength $I \cdot y_0[(x - x_0)^2 + y_0^2]^{-3/2}\, dx$.
   a) Compute the field strength at $(x_0, y_0)$ created by the entire segment.
   b) Compute the field strength at $(x_0, y_0)$ created by an infinitely long wire carrying this current.

8. Prove the reduction formula (10). [Suggestion: Integrate $\displaystyle\int \frac{du}{(a^2 + u^2)^n}$ by parts; then use $u^2 = (a^2 + u^2) - a^2$.]

9. A certain proton repelled from a nucleus moves along the positive $x$-axis with acceleration $d^2x/dt^2 = \frac{1}{2}x^{-2}$, in suitable units. To derive an equation for the motion, show

   a) The velocity $v = dx/dt$ satisfies $\dfrac{dv}{dx} \cdot \dfrac{dx}{dt} =$
   $$v\frac{dv}{dx} = \frac{1}{2}x^{-2}.$$

   b) $\frac{1}{2}v^2 = -\frac{1}{2}x^{-1} + C$, for a *positive* constant $C$.

   c) $\dfrac{dx}{dt} = \pm\sqrt{a^2 - x^{-1}}$ with $a^2 = 2C$.

   d) As the proton recedes from the nucleus, $\dfrac{dx}{dt} \geq 0$. Using the $+$ sign in part c and formula (3) in the text, derive the relation
   $$\frac{\sqrt{a^2x^2 - x}}{a^2} + \frac{1}{2a^3}\ln\left|ax - \frac{1}{2a} + \sqrt{a^2x^2 - x}\right| = t + C_1,$$

   where $C_1$ is a constant.

   e) Suppose that $x_0 > 0$ is the closest approach of the proton to the nucleus. Determine the constant $a$ in part c. (What is $v$, when $x = x_0$?)

   f) From parts d and e, deduce that $x \to +\infty$ as $t \to +\infty$. (Assume that $x \geq x_0 > 0$, and $\dfrac{dx}{dt} > 0$; if $x$ remains bounded, what of the left hand side of the formula in part d?)

   *g) From parts c, e, and f, express the limiting velocity $\displaystyle\lim_{t \to +\infty} \frac{dx}{dt}$, and the limit of the kinetic energy $\frac{1}{2}mv^2$, in terms of the closest approach $x_0$.

═══ **9.6** ═══

# INTEGRATING RATIONAL FUNCTIONS BY PARTIAL FRACTIONS

Every rational function can be written as a sum of so-called *partial fractions*, each of which is easily integrated. For example,

$$\frac{2x + 1}{x(x + 1)} \text{ can be rewritten as } \frac{1}{x} + \frac{1}{x + 1}.$$

(verify this by combining the sum of fractions on a common denominator). So

$$\int \frac{2x+1}{x(x+1)} \, dx = \int \left( \frac{1}{x} + \frac{1}{x+1} \right) dx = \ln|x| + \ln|x+1| + C$$
$$= \ln|x(x+1)| + C.$$

The question is: Given a rational function $P/Q$, *how* do you write it as a sum of easily integrated partial fractions?

**First:  Compare the degrees of $P$ and $Q$.**   If $\deg(P) < \deg(Q)$ then $P/Q$ is proper, and you proceed to the next step. If not, then reduce $P/Q$ (as in Sec. 9.5):

$$\frac{P}{Q} = D + \frac{R}{Q} \quad \text{with} \quad \deg(R) < \deg(Q).$$

The polynomial $D$ is easily integrated, while $R/Q$ is proper, so you can proceed to:

**Second:  Factor the denominator $Q$** into a product of *real linear* and *irreducible quadratic* factors. Collect all like terms so that the denominator is expressed as a constant times *distinct* factors of the form

$$(x + a)^m \quad \text{and} \quad (x^2 + bx + c)^n$$

with $x^2 + bx + c$ irreducible. These factors determine the partial fractions to be introduced, as follows:

**Third:  Express the proper rational function** as a sum of terms, following these two "recipes":

  **I.**   *For each factor $(x + a)^m$, include $m$ terms*

$$\frac{A_1}{x+a} + \frac{A_2}{(x+a)^2} + \cdots + \frac{A_m}{(x+a)^m}$$

  where $A_1, \ldots, A_m$ are constants to be determined.

  **II.**   *For each factor $(x^2 + bx + c)^n$ include $n$ terms*

$$\frac{B_1 x + C_1}{x^2 + bx + c} + \cdots + \frac{B_n x + C_n}{(x^2 + bx + c)^n}$$

  where again, the $B_j$ and $C_j$ are constants to be determined.

**Fourth:  Determine the constants $A_j$, $B_j$, $C_j$** just introduced. (The examples show how.)

**Fifth:  Integrate** each partial fraction.

---

**EXAMPLE 1**   $\displaystyle \int \frac{2x+1}{x^2(x-1)} \, dx = ?$

*SOLUTION* This is a proper rational function, and the denominator is already factored, so we proceed to determine the partial fractions to be introduced. In the denominator $Q(x) = x^2(x - 1)$, the repeated factor $x^2$ has the form $(x + 0)^2$ as in recipe I, so this requires two terms

$$\frac{A_1}{x} + \frac{A_2}{x^2}.$$

The single factor $x - 1$ requires just one term, $\dfrac{A_3}{x - 1}$. So the entire integrand can be expressed as

$$\frac{2x + 1}{x^2(x - 1)} = \frac{A_1}{x} + \frac{A_2}{x^2} + \frac{A_3}{x - 1}. \tag{1}$$

To clear the denominators, multiply by $x^2(x - 1)$:

$$2x + 1 = A_1 x(x - 1) + A_2(x - 1) + A_3 x^2.$$

From this equation, there are various ways to determine the constants $A_1$, $A_2$, $A_3$. The most straightforward way is to *multiply out and combine like powers*:

$$\begin{aligned} 2x + 1 &= A_1 x^2 - A_1 x + A_2 x - A_2 + A_3 x^2 \\ &= (A_1 + A_3)x^2 + (A_2 - A_1)x - A_2. \end{aligned}$$

*Compare the coefficients of like powers* on the two sides of this equation:

$$\begin{aligned} 0 &= A_1 + A_3 && \text{(coefficients of } x^2) \\ 2 &= A_2 - A_1 && \text{(coefficients of } x) \\ 1 &= -A_2 && \text{(coefficients of } x^0). \end{aligned}$$

These three simultaneous equations can be solved, beginning with the last:

$$A_2 = -1, \qquad A_1 = A_2 - 2 = -3, \qquad A_3 = -A_1 = 3.$$

Use these in (1):

$$\frac{2x + 1}{x^2(x - 1)} = \frac{-3}{x} - \frac{1}{x^2} + \frac{3}{x - 1}.$$

Now integrate:

$$\int \frac{2x + 1}{x^2(x - 1)}\, dx = \int \left( \frac{-3}{x} - \frac{1}{x^2} + \frac{3}{x - 1} \right) dx$$

$$= -3 \ln|x| + \frac{1}{x} + 3 \ln|x - 1| + C.$$

*REMARK*  In Example 1, you might hope to obtain $\dfrac{2x + 1}{x^2(x - 1)}$ by combining just two terms: $\dfrac{A_2}{x^2}$ and $\dfrac{A_3}{x - 1}$. But the solution shows that this is impossible; we needed the term $-3/x$ as well. You need, in principle, *all* the terms listed in recipes I and II of the outline.

In those recipes, the number of constants introduced for each factor equals the degree of that factor; thus the total number of constants introduced equals the degree of the denominator $Q(x)$. Call this degree $q$. The number of constants in the typical numerator $P(x)$ is also $q$; for $P(x)$ has degree less than $q$, so

$$P(x) = c_0 + c_1 x + \cdots + c_{q-1} x^{q-1}$$

has $q$ coefficients. The process of equating like powers of $x$ (illustrated in Example 3) gives $q$ simultaneous linear equations for the $q$ constants that were introduced. This suggests why the method works as it does.

---

**EXAMPLE 2**  Evaluate $\displaystyle\int \frac{dx}{x(x^2 + 1)}$.

*SOLUTION*  The denominator has a linear $x$ and an irreducible quadratic $x^2 + 1$. The linear factor requires $A/x$ and, according to recipe II, the quadratic $x^2 + 1$ requires a *first* degree numerator $Bx + C$. So try

$$\frac{1}{x(x^2 + 1)} = \frac{A}{x} + \frac{Bx + C}{x^2 + 1}. \tag{2}$$

Clear denominators and combine like powers:

$$1 = A(x^2 + 1) + (Bx + C)x$$
$$= (A + B)x^2 + Cx + A.$$

Equate coefficients of like powers:

$$0 = A + B \qquad (\text{from } x^2)$$
$$0 = C \qquad (\text{from } x^1)$$
$$1 = A \qquad (\text{from } x^0).$$

This system has a unique solution $A = 1$, $B = -1$, $C = 0$, so from (2)

$$\frac{1}{x(x^2 + 1)} = \frac{1}{x} - \frac{x}{x^2 + 1}.$$

Now integrate:

$$\int \frac{dx}{x(x^2 + 1)} = \int \left( \frac{1}{x} - \frac{x}{x^2 + 1} \right) dx$$
$$= \ln|x| - \frac{1}{2} \ln(x^2 + 1) + C.$$

## A Shortcut Determination of the Coefficients

In these examples, we determined the constants for the partial fractions by the "coefficients of like powers" method. The next example uses the "special values of $x$" method, which in many cases is quicker.

---

**EXAMPLE 3** $\displaystyle\int \frac{(x+1)\,dx}{x^2 - 4x} = ?.$

*SOLUTION*   The denominator factors:

$$x^3 - 4x = x(x^2 - 4) = x(x-2)(x+2).$$

All factors are linear, so according to recipe I we try

$$\frac{x+1}{x^3 - 4x} = \frac{A}{x} + \frac{B}{x-2} + \frac{C}{x+2}. \tag{3}$$

Clear denominators, multiplying by $x(x-2)(x+2)$:

$$x + 1 = A(x-2)(x+2) + Bx(x+2) + Cx(x-2). \tag{4}$$

The partial fractions theory guarantees that there *are* constants $A$, $B$ and $C$ making the two fractions in (3) identical—you just have to *find* them! Since the functions in (3) are to be identical, so are the polynomials in (4), and *they must give the same result for each value of $x$*. In particular, both sides of (4) give the same result when $x = 0$:

$$0 + 1 = A(-2)(2) + B \cdot 0 + C \cdot 0$$

so $1 = 4A$, and

$$A = 1/4.$$

You see why we set $x = 0$; that eliminates all terms on the right except the one containing $A$. Now set $x = 2$ in (4), to eliminate all terms except the one with $B$:

$$2 + 1 = A \cdot 0 + B \cdot 2 \cdot 4 + C \cdot 0$$

so $3 = 8B$ and

$$B = 3/8.$$

Finally, set $x = -2$ in (4):

$$-2 + 1 = A \cdot 0 + B \cdot 0 + C(-2)(-4)$$

so $-1 = 8C$ and $C = -1/8$. Use these values of $A$, $B$, $C$ in (3), and integrate:

$$\int \frac{x+1}{x^3 - 4x}\,dx = \int \left( \frac{1/4}{x} + \frac{3/8}{x-2} + \frac{-1/8}{x+2} \right) dx$$

$$= \frac{1}{4}\ln|x| + \frac{3}{8}\ln|x-2| - \frac{1}{8}\ln|x+2| + c.$$

---

This shortcut method gives all the constants, if there are no multiple or quadratic factors; in any case, it may give some of the constants quickly, making it easier to find the others. But it requires care. For instance, in Example 3, if the form of the decomposition (3) had been wrongly set up, you might still determine constants by using special values of $x$; but the resulting decomposition would not necessarily be valid for other values of $x$. It is *essential* to start with a decomposition of the correct form. (See problem 6.)

When the denominator has no real zeroes, there is no shortcut.

---

**EXAMPLE 4**   $\displaystyle \int \frac{x^3}{(x^2 + 2x + 2)^2}\, dx = ?.$

**SOLUTION**   The quadratic $x^2 + 2x + 2$ is irreducible; according to recipe ll,

$$\frac{x^3}{(x^2 + 2x + 2)^2} = \frac{B_1 x + C_1}{x^2 + 2x + 2} + \frac{B_2 x + C_2}{(x^2 + 2x + 2)^2}. \tag{5}$$

Multiply both sides by $(x^2 + 2x + 2)^2$ and collect like powers:

$$x^3 = B_1 x^3 + (2B_1 + C_1)x^2 + (2B_1 + 2C_1 + B_2)x + 2C_1 + C_2.$$

Equate the coefficients of like powers:

$$\left. \begin{array}{l} 1 = B_1 \\ 0 = 2B_1 + C_1 \\ 0 = 2B_1 + 2C_1 + B_2 \\ 0 = 2C_1 + C_2 \end{array} \right\} \Rightarrow \left\{ \begin{array}{l} B_1 = 1 \\ C_1 = -2 \\ B_2 = 2 \\ C_2 = 4 \end{array} \right.$$

Use these values in (5):

$$\int \frac{x^3 dx}{(x^2 + 2x + 2)^2} = \int \frac{x - 2}{x^2 + 2x + 2}\, dx + \int \frac{2x + 4}{(x^2 + 2x + 2)^2}\, dx. \tag{6}$$

To integrate these terms, complete the square and proceed as in Section 9.5. Thus

$$x^2 + 2x + 2 = (x + 1)^2 + 1,$$

so set $u = x + 1$ and get

$$\int \frac{x - 2}{x^2 + 2x + 2}\, dx = \int \frac{(u - 3)du}{u^2 + 1} = \frac{1}{2} \ln(u^2 + 1) - 3 \tan^{-1} u + C_1$$

$$= \frac{1}{2} \ln(x^2 + 2x + 2) - 3 \tan^{-1}(x + 1) + C_1,$$

$$\int \frac{2x + 4}{(x^2 + 2x + 2)^2}\, dx = \int \frac{2u + 2}{(u^2 + 1)^2}\, du = \frac{-1}{u^2 + 1} + 2 \int \frac{du}{(u^2 + 1)^2}$$

$$= \frac{-1}{x^2 + 2x + 2} + 2 \int \frac{du}{(u^2 + 1)^2}.$$

For the last integral, the recursion formula in Section 9.5 gives

$$2 \int \frac{du}{(u^2 + 1)^2} = 2 \left[ \frac{u}{2(u^2 + 1)} + \frac{1}{2} \tan^{-1} u \right] + C_2$$

$$= \frac{x + 1}{x^2 + 2x + 2} + \tan^{-1}(x + 1) + C_2.$$

Finally, summing up all the parts,

$$\int \frac{x^3 \, dx}{(x^2 + 2x + 2)^2} = \frac{1}{2} \ln(x^2 + 2x + 2) + \frac{x}{x^2 + 2x + 2} - 2 \tan^{-1}(x + 1) + C.$$

---

*WARNING*    Don't waste time trying to "expand" expressions that are already partial fractions!

---

**EXAMPLE 5**    $\displaystyle\int \frac{x \, dx}{(x^2 + x + 1)^2} = ?.$

*SOLUTION*    The integrand $\dfrac{x}{(x^2 + x + 1)^2}$ is already a fraction of the form on the right-hand side of the expansion II,

$$\frac{Bx + C}{(x^2 + x + 1)^2}$$

with $B = 1$, $C = 0$; so proceed with the integration. This was done in Example 4, Section 9.5.

---

We conclude with an applied example.

---

**EXAMPLE 6**    The differential equation

$$\frac{dx}{dt} = kx(b - x), \, b > 0 \tag{7}$$

describes a growth rate which is "jointly proportional" to $x$ and to $b - x$. When $x$ is near 0, this is like the simple proportional growth equation

$$\frac{dx}{dt} = kbx;$$

but as $x$ grows toward $b$, the factor $b - x$ tends to 0, and so does the growth rate. This describes growth where $b$ is the maximum possible size of $x$, as, for example, the amount $x$ of bacteria in a culture might be restricted by the size of the container or the food supply.

Solve (7) by separating variables (Sec. 8.6):

$$\frac{dx}{x(b - x)} = k \, dt. \tag{8}$$

By partial fractions,

$$\frac{1}{x(b-x)} = \frac{1/b}{x} + \frac{1/b}{b-x}$$

so (8) gives

$$\int \frac{1/b}{x}\, dx + \int \frac{1/b}{b-x}\, dx = \int k\, dt$$

$$\frac{1}{b}\ln|x| - \frac{1}{b}\ln|b-x| = kt + c$$

$$\ln\left|\frac{x}{b-x}\right| = bkt + bc,$$

$$\left|\frac{x}{b-x}\right| = e^{bkt+bc} = e^{bc}e^{bkt},$$

$$\frac{x}{b-x} = (\pm e^{bc})e^{bkt},$$

$$\frac{x}{b-x} = Ce^{bkt} \tag{9}$$

where $C = \pm e^{bc}$. To solve for $x$, multiply through by $b-x$ and separate the $x$ terms from the constant terms; you get

$$x = \frac{bCe^{bkt}}{1 + Ce^{bkt}}$$

or, multiplying numerator and denominator by $e^{-bkt}/C$,

$$x = \frac{b}{(e^{-bkt}/C) + 1}. \tag{10}$$

In practical applications the constants $b$ and $k$ are positive, and $x$ lies between 0 and $b$. This has the following consequences:

**a)**   $dx/dt > 0$, from (7).

**b)**   The constant $C > 0$, from (9), since $0 < x < b$.

**c)**   $\lim\limits_{t\to+\infty} e^{-bkt} = 0$ and $\lim\limits_{t\to-\infty} e^{-bkt} = +\infty$, since $bk > 0$.

**d)**   $\lim\limits_{t\to+\infty} x = b$ and $\lim\limits_{t\to-\infty} x = 0$, from (10) and c).

Figure 1 shows the slopes for $\dfrac{dx}{dt} = 4x(1-x)$ and one of the "S-shaped" solution curves, illustrating the limits as $t \to \pm\infty$. The quantity $x$ grows slowly at first (because there is not much of it, so it lacks "internal resources"), then more rapidly, then again more slowly (now it is running out of *external* resources).

The curve in Figure 1 is called a *logistic curve*. (For more background and examples, see Horelick, Koont, and Gottlieb, "Population Growth and the Logistic Curve," UMAP Module 68, COMAP, 60 Lowell St., Arlington, MA 02174.)

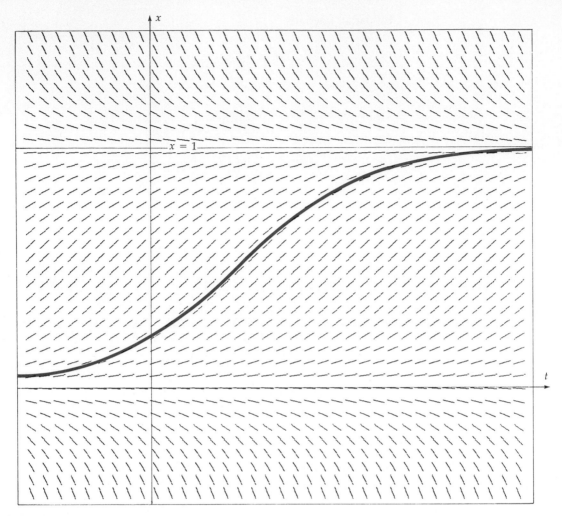

**FIGURE 1**

Slope field for $\dfrac{dx}{dt} = 4x(1 - x)$, with a solution curve.

## SUMMARY

To integrate a rational function $P/Q$:

1. *Divide if improper* $[\deg(P) \geq \deg(Q)]$.

2. *Factor the denominator.*

3. *Write as sum of partial fractions:*
   *I. For each factor $(x + a)^m$, include $m$ terms*

$$\frac{A_1}{x + a} + \frac{A_2}{(x + a)^2} + \cdots + \frac{A_m}{(x + a)^m}$$

II. *For each factor $(x^2 + bx + c)^n$ include $n$ terms*

$$\frac{B_1 x + C_1}{x^2 + bx + c} + \cdots + \frac{B_n x + C_n}{(x^2 + bx + c)^n}$$

4. *Determine the constants $A_j$, $B_j$, $C_j$.*

5. *Integrate.*

## PROBLEMS

### A

**\*1.** For the given denominator $Q$, determine the appropriate form of the partial fractions decomposition of a proper rational function $P/Q$.

a) $Q(x) = x^2 - 1$

b) $Q(x) = x(x^2 - 1)^2$

c) $Q(x) = x^2(x - 1)(x - 2)$

d) $Q(x) = x^2 - 5x + 6$

**\*2.** Integrate the given function.

a) $\dfrac{x + 2}{x^2 - 1}$

b) $\dfrac{x}{x^2 - 1}$ (partial fractions not needed)

c) $\dfrac{1}{x(x^2 - 1)^2}$

d) $\dfrac{x}{x^2 - 5x + 6}$

**3.** For the given denominator $Q$, determine the appropriate form of the partial fractions decomposition for a proper rational function $P/Q$.

**\*a)** $(x^2 + 1)(x^2 - 4)$

**\*b)** $x^2(x^2 + 1)$

c) $x^4 - 1$

d) $(x^2 + 1)^2$

e) $(x^2 + 2x + 2)^3$

f) $x^3 - x^2 - x + 1$ [Hint: $Q(1) = 0$, so $x - 1$ is a factor of $Q$. Divide $Q$ by $x - 1$, and factor the quotient.]

**4.** Integrate the given function.

**\*a)** $\dfrac{x + 2}{x(x^2 + 1)}$

b) $\dfrac{x}{x^2 + 1}$

**\*c)** $\dfrac{2x^2}{x^2 - 1}$

d) $\dfrac{x}{x^4 - 1}$

**\*e)** $\dfrac{x^2 + 2}{(x^2 + 1)^2}$

f) $\dfrac{x}{(x^2 + 1)^2}$

**\*g)** $\dfrac{x^2 + x - 1}{x^3 - x^2 - x + 1}$ (see problem 3f)

h) $\dfrac{x^2}{x^2 - 5x + 5}$

**\*i)** $\dfrac{e^{2t}}{e^{2t} + e^t - 2}$

j) $\dfrac{4x^3 + 10x}{x^4 + 5x^2 + 4}$

k) $x \tan^{-1} x$

**5.** Write the following quotients $P/Q$ in the form $D + R/Q$, where $D$ and $R$ are polynomials, and $\deg(R) < \deg(Q)$. Then integrate.

a) $\dfrac{x^2}{(x - 1)(x - 2)}$

**\*b)** $\dfrac{x^5}{x^3 - x^2 - x + 1}$

c) $\dfrac{x^4}{(x + 1)^2}$

**6.** Suppose that $\dfrac{2x + 1}{x^2(x + 1)}$ could be written as a sum $\dfrac{A}{x^2} + \dfrac{B}{x + 1} = \dfrac{A(x + 1) + Bx^2}{x^2(x + 1)}$.

a) Compare the numerators $2x + 1$ and $A(x + 1) + Bx^2$ for $x = 0$ and $x = -1$, and thus "determine" $A$ and $B$.

b) Check whether the resulting sum $\dfrac{A}{x^2} + \dfrac{B}{x + 1}$ is really equal to $\dfrac{2x + 1}{x^2(x + 1)}$. Why does it fail?

**7.** Solve $\dfrac{dx}{dt} = (1 - x)(3 - x)$ with $x(0) = 2$.

### B

**8.** In certain reactions where two chemicals $A$ and $B$ combine to form compound $x$, the rate of formation of the compound is described by

$$\frac{dx}{dt} = k(a - x)(b - x). \qquad (*)$$

Here $k > 0$ is the *reaction rate constant*, $a$ is the original amount of $A$ (measured in "moles"), $b$ the original amount of $B$, and $x$ the moles of compound $X$ at time $t$. Each mole of $X$ uses one mole of $A$ and one mole of $B$, so $a - x$ is the amount of $A$ remaining at time $t$, and $b - x$ is the amount of $B$ remaining. According to (∗), the reaction rate is "jointly proportional" to these two amounts. Use equation (∗) to answer the following:

**a)** Assuming $x < a < b$, show that $x$ *increases* with time.

**b)** Show that if $x = a$ or $x = b$, then the reaction is stopped. Explain why, using the chemical interpretation of (∗).

**\*c)** Solve (∗), assuming $a \neq b$.

**d)** Show that $\lim_{t \to +\infty} x = a$ if $a < b$. Explain why, using the chemical interpretation of (∗).

**e)** Solve (∗) assuming $a = b$, and evaluate $\lim_{t \to +\infty} x$.

**\*f)** Assuming $a \neq b$, solve (∗) with $x = 0$ when $t = 0$.

**\*g)** Assuming $a = b$, solve (∗) with $x = 0$ when $t = 0$. Graph the solution.

**h)** Show that as $b \to a$, the solution in part f has as its limit the solution in part g.

**9.** Suppose that $r_1 \neq r_2$. Show that any proper rational function of the form

$$\frac{\alpha x + \beta}{(x - r_1)(x - r_2)}$$

can be written as $A/(x - r_1) + B/(x - r_2)$, with uniquely determined constants $A$ and $B$. (Express $A$ and $B$ in terms of $\alpha$ and $\beta$.) Show that this is false if $r_1 = r_2$.

**10.** Show that any proper rational function of the form $\dfrac{\alpha x + \beta}{(x - r)^2}$ can be written as $A/(x - r) + B/(x - r)^2$ for uniquely determined constants $A$ and $B$. Show that "most" functions $\dfrac{\alpha x + \beta}{(x - r)^2}$ can *not* be written as $A/(x - r) + B/(x - r)$, or as $A/(x - r)^2$.

# 9.7

## RATIONALIZING SUBSTITUTIONS

Some integrands, though not themselves rational, can be converted to rational integrands by appropriate substitutions.

### Rational Functions of $x$ and a Radical $(ax + b)^{1/n}$

Such integrands can be rewritten as rational functions of a new variable $u = (ax + b)^{1/n}$. For then $x = \dfrac{1}{a}(u^n - b)$ is a rational function of $u$, and so is

$$dx = \frac{n}{a} u^{n-1} \, du.$$

---

**EXAMPLE 1**  The integrand in $\int x(ax + b)^{2/3} \, dx$ is a rational function of $x$ and $(ax + b)^{1/3}$. Set

$$u = (ax + b)^{1/3}, \ u^3 = ax + b, \ \text{ so } \ x = \frac{1}{a}(u^3 - b) \text{ and } dx = \frac{3}{a} u^2 \, du.$$

Then

$$\int x(ax + b)^{2/3} \, dx = \int \frac{1}{a}(u^3 - b) \cdot u^2 \cdot \frac{3}{a} \, du$$

is a polynomial, easily integrated.

---

A similar approach works with some other algebraic integrands, as you will see in the problems.

**Rational integrands of sin $\theta$ and cos $\theta$** can be rewritten as rational integrands of the ingeniously chosen new variable

$$u = \frac{\sin \theta}{1 + \cos \theta} = \tan\left(\frac{\theta}{2}\right).$$

You can check that

$$\cos \theta = \frac{1 - u^2}{1 + u^2}, \qquad \sin \theta = \frac{2u}{1 + u^2}, \quad \text{and} \quad d\theta = \frac{2\,du}{1 + u^2}. \tag{1}$$

All these are rational functions of $u$; so any integrand which is rational in $\sin \theta$ and $\cos \theta$ is converted to a (possibly complicated) rational integrand in $u$.

Since every trigonometric function is itself a rational function of $\sin \theta$ and $\cos \theta$, this method is very general; keep it in mind for integrands where the simpler methods in the earlier sections do not apply.

---

**EXAMPLE 2**   $\displaystyle\int \frac{d\theta}{\sin \theta + \cos \theta} = ?$

**SOLUTION**   From (1),

$$\sin \theta + \cos \theta = \frac{2u + 1 - u^2}{1 + u^2}.$$

The reciprocal of this, times $d\theta$ $\left[ = \dfrac{2\,du}{1 + u^2} \right]$, gives the new integral $\displaystyle\int \frac{-2\,du}{u^2 - 2u - 1}$. The integrand is a proper rational function, so factor the denominator:

$$u^2 - 2u - 1 = (u - r)(u - s).$$

The constants $r$ and $s$ in this expression stand for the zeroes of $u^2 - 2u - 1$, given by the quadratic formula:

$$r = 1 + \sqrt{2} \quad \text{and} \quad s = 1 - \sqrt{2}. \tag{2}$$

In forming partial fractions, we continue to abbreviate these as $r$ and $s$:

$$\frac{-2}{u^2 - 2u - 1} = \frac{A}{u - r} + \frac{B}{u - s} = \frac{A(u - s) + B(u - r)}{(u - r)(u - s)}. \tag{3}$$

From the particular values $u = r$ and $u = s$, we find $A(r - s) = -2$, $B(s - r) = -2$. From (2), $r - s = 2\sqrt{2}$ and $s - r = -2\sqrt{2}$, so

$$A = \frac{-2}{2\sqrt{2}} = \frac{-1}{\sqrt{2}}, \qquad B = \frac{1}{\sqrt{2}}.$$

Use these values in (3), integrate, and recall that $u = \tan(\theta/2)$; you find

$$\int \frac{d\theta}{\sin\theta + \cos\theta} = \int \frac{-2\,du}{u^2 - 2u - 1} = \frac{1}{\sqrt{2}} \ln\left|\frac{u - 1 + \sqrt{2}}{u - 1 - \sqrt{2}}\right| + C$$

$$= \frac{1}{\sqrt{2}} \ln\left|\frac{\tan(\theta/2) - 1 + \sqrt{2}}{\tan(\theta/2) - 1 - \sqrt{2}}\right| + C.$$

## PROBLEMS

### A

Evaluate.

*1. $\displaystyle\int x^3\sqrt{3x - 2}\,dx$

2. $\displaystyle\int x^2(ax + b)^{4/3}\,dx$

*3. $\displaystyle\int x^3(x^2 + 1)^{1/3}\,dx$

4. $\displaystyle\int x^2(1 - x)^{3/2}\,dx$

*5. $\displaystyle\int \frac{dt}{t^{1/2} + t^{1/4}}$

6. $\displaystyle\int \frac{dx}{x\sqrt{ax + b}}$

*7. $\displaystyle\int \frac{d\theta}{\sin\theta - \cos\theta}$

8. $\displaystyle\int \frac{d\theta}{\sin^2\theta + \cos^2\theta}$

*9. $\displaystyle\int \frac{d\theta}{1 + \sin\theta}$

10. $\displaystyle\int_0^{\pi/2} \frac{d\theta}{1 + \cos\theta}$

*11. $\displaystyle\int \frac{\sin\theta\,d\theta}{2 + \cos\theta}$

## REVIEW PROBLEMS    CHAPTER 9

Before doing these, it is useful to copy onto one or two reference sheets the integration methods in the summaries for the chapter sections. Use tables or not, depending on the type of test you are preparing for.

In problems 1–32 evaluate.

*1. $\displaystyle\int_0^1 x\sin 3x\,dx$

*2. $\displaystyle\int_0^\pi x^2\cos 5x\,dx$

*3. $\displaystyle\int_0^\infty xe^{-ax}\,dx,\ a > 0$

*4. $\displaystyle\int \tan^{-1}x\,dx$

*5. $\displaystyle\int_0^1 x\sin^{-1}x\,dx$

*6. $\displaystyle\int \cos\theta\sin^2\theta\,d\theta$

*7. $\displaystyle\int_0^\pi \cos^{100}\theta\sin^3\theta\,d\theta$

*8. $\displaystyle\int_0^\pi \cos^2 3\theta\,d\theta$

*9. $\displaystyle\int_0^1 \sin^2(t/2)\,dt$

*10. $\displaystyle\int \sin^2\theta\cos^2\theta\,d\theta$

*11. $\displaystyle\int \sec\pi x\,dx$

*12. $\displaystyle\int \sec^4\theta\,d\theta$

*13. $\displaystyle\int \tan^3 2x\,dx$

*14. $\displaystyle\int \frac{\cos\theta\,d\theta}{1 + \sin^2\theta}$

*15. $\displaystyle\int \frac{dx}{\sqrt{3 + x^2}}$

*16. $\displaystyle\int \sqrt{3 - 2x^2}\,dx$

*17. $\displaystyle\int \sqrt{3x^2 - 4}\,dx$

*18. $\displaystyle\int \sqrt{x^2 + 2x}\,dx$

*19. $\displaystyle\int \frac{dx}{\sqrt{2x^2 + x}}$

*20. $\displaystyle\int_0^\pi \cos x\sin 2x\,dx$

*21. $\displaystyle\int_0^\pi \cos(mx)\cos(nx)\,dx,\ n \neq m$, both integers.

*22. $\displaystyle\int_0^\pi \cos mx\cos nx\,dx$ if $m = n$, both integers.

*23. $\displaystyle\int \frac{du}{u^2 - 1}$

*24. $\displaystyle\int \frac{(2x + 1)\,dx}{(x^2 + x + 1)^2}$

25. $\displaystyle\int \frac{x\,dx}{(x^2 + x + 1)^2}$

*26. $\displaystyle\int \frac{x^3\,dx}{x^2 + 1}$

*27. $\displaystyle\int \frac{(x + 1)\,dx}{x^2(x^2 + 1)}$

*28. $\displaystyle\int \frac{dx}{1 + e^x}$

*29. $\displaystyle\int \frac{d\theta}{2 - 3\sin\theta}$

*30. $\displaystyle\int_0^{\pi/2} \frac{d\theta}{1 + \sin\theta + \cos\theta}$

31. $\displaystyle\int x^2\ln(3x)\,dx$

32. $\displaystyle\int (\ln x)^2\,dx$

In problems 33–40 prove an appropriate reduction formula, reducing the power of $n$.

33. $\int x^n e^{-ax}\, dx$

34. $\int x^n \cos ax\, dx$ and $\int x^n \sin ax\, dx$

35. $\int (\ln x)^n x^k\, dx$

36. $\int \sin^n ax\, dx$ $\quad (u = \sin^{n-1} ax,\ dv = \sin ax\, dx)$

37. $\int \cos^n ax\, dx$

38. $\int \sec^n ax\, dx$

39. $\int \dfrac{du}{(a^2 + u^2)^n}$ $\left[\text{Integrate } \int \dfrac{du}{(a^2 + u^2)^{n-1}} \text{ by}\right.$

parts, then use $u^2 = (a^2 + u^2) - a^2.\Big]$

*40. $\int_0^\infty x^n e^{-ax}\, dx$, $a > 0$, $n \geq 1$.

Solve the following differential equations:

*41. $\dfrac{dy}{dx} = (x/y)e^{2t}$, $y(0) = 1$.

*42. $\dfrac{dy}{dx} = x\sqrt{1 - y^2}$, $y(0) = 1/2$.

43. The fall of a reasonably dense object (such as a raindrop) encounters a frictional resistance proportional to the square of the velocity. So in Newton's law $F = ma$, there is a downward force of gravity $-mg$ *together with* an upward force of air resistance $kv^2$, with a constant $k$:

$$m\frac{d^2y}{dt^2} = -mg + kv^2.$$

a) Rewrite this as an equation for $v(= dy/dt)$. Abbreviate $k/mg$ as $\beta^2$, and solve the equation with $v(0) = 0$, to find $v = \dfrac{1}{\beta}\dfrac{e^{-\beta gt} - e^{\beta gt}}{e^{-\beta gt} + e^{\beta gt}}$. [Be careful to use correct signs when removing absolute value signs; be consistent with $v(0) = 0$.]

b) Find the *terminal velocity* $\lim\limits_{t \to +\infty} v$.

c) Find the *position function* $y$, given $y(0) = y_0$.

# 10

# MORE APPLICATIONS
# OF INTEGRALS

# 10.1

## ARC LENGTH

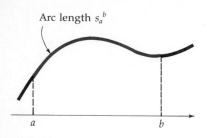

**FIGURE 1**
Arc length.

*Arc length* means length measured along a circular arc or any other curve (Fig. 1). To determine the arc length of the graph of a function $f$ on an interval $[a, b]$, first approximate the graph by straight-line segments, as in Figure 2. Divide $[a, b]$ into $n$ equal subintervals of length $\Delta x$. In each subinterval, approximate the graph by a straight line segment; the sum $L_n$ of the lengths of these $n$ segments approximates the actual length of the graph. If these approximations $L_n$ have a limit as $n \to \infty$, that limit is defined to be the arc length of the graph, denoted $s_a^b$:

$$s_a^b = \lim L_n.$$

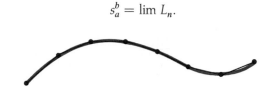

**FIGURE 2**
$L_n$ = sum of lengths of $n$ approximating straight-line segments.

If the derivative $f'$ is continuous, then the approximations can be written as Riemann sums, so their limit, the arc length, is an integral. What is the integrand?

Consider the first approximating segment (Fig. 3); its length is

$$\sqrt{(\Delta x)^2 + (\Delta y)^2}.$$

By the Mean Value Theorem (Sec. 5.2) there is a point $\hat{x}_1$ between $a$ and $x_1$ such that

$$f'(\hat{x}_1) = \frac{\Delta y}{\Delta x}$$

or

$$\Delta y = f'(\hat{x}_1)\,\Delta x.$$

Hence the length of that first segment is

$$\sqrt{(\Delta x)^2 + f'(\hat{x}_1)^2 (\Delta x)^2} = \sqrt{1 + f'(\hat{x}_1)^2}\,\Delta x.$$

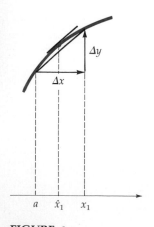

**FIGURE 3**
$f'(\hat{x}_1) = \dfrac{\Delta y}{\Delta x}.$

The approximation $L_n$ is the sum of $n$ such terms:

$$L_n = \sum_{j=1}^{n} \sqrt{1 + f'(\hat{x}_j)^2}\,\Delta x.$$

This is a Riemann sum for the integral $\int_a^b \sqrt{1 + f'(x)^2}\, dx$, so the arc length is

$$s_a^b = \lim_{n \to \infty} L_n = \int_a^b \sqrt{1 + f'(x)^2}\, dx. \tag{1}$$

**EXAMPLE 1**   Compute the arc length of the graph of $f(x) = x^{3/2}$ for $0 \le x \le 1$.

*SOLUTION*

$$s_0^1 = \int_0^1 \sqrt{1 + (\tfrac{3}{2}x^{1/2})^2}\, dx$$
$$= \int_0^1 \sqrt{1 + \tfrac{9}{4}x}\, dx.$$

Set $1 + \tfrac{9}{4}x = u$. Then $du = \tfrac{9}{4}dx$; and $x = 0 \Rightarrow u = 1$, while $x = 1 \Rightarrow u = 13/4$, so

$$s_0^1 = \int_1^{13/4} u^{1/2} \cdot \frac{4}{9}\, du$$
$$= \frac{4}{9} \frac{u^{3/2}}{3/2} \Big|_1^{13/4} = \frac{8}{27} \left[ \left(\frac{13}{4}\right)^{3/2} - 1 \right]$$
$$\approx 1.43971.$$

(Compare this to the length of the single approximating segment in Figure 4, which is $\sqrt{2} \approx 1.414$. The curve is slightly longer than the straight segment, as of course it must be; this is a check on the calculation.)

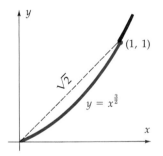

**FIGURE 4**
$s_0^1 > \sqrt{2}$.

*REMARK*   The function in Example 1 was carefully chosen so that the indefinite integral $\int \sqrt{1 + f'(x)^2}\, dx$ required only a simple substitution. Generally, the arc length integral cannot be evaluated so easily. When antidifferentiation fails, you can approximate the length directly by the sums $L_n$ or approximate the arc length integral by an appropriate Riemann sum. More refined approximations of integrals are given at the end of this chapter and in Chapter 11 on infinite series.

**The interpretation of the arc length integral** by infinitesimals is shown in Figure 5. An infinitesimal piece of the curve is practically a straight-line segment, with infinitesimal length denoted by $ds$. This forms the hypotenuse of an infinitesimal right triangle, with legs $dx$ and $dy$, where $dy = f'(x)\, dx$. The Pythagorean Theorem for this infinitesimal triangle gives

$$ds = \sqrt{(dx)^2 + (dy)^2}$$
$$= \sqrt{(dx)^2 + [f'(x)\, dx]^2}$$
$$= \sqrt{1 + f'(x)^2}\, dx. \tag{2}$$

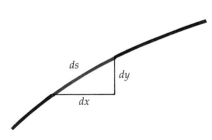

**FIGURE 5**
$(ds)^2 = (dx)^2 + (dy)^2$.

The length of the entire curve is the sum of these infinitesimal lengths, the integral $\int_a^b \sqrt{1 + f'(x)^2}\, dx$.

If the curve in question is the graph of $x = g(y)$, then $dx = g'(y)\,dy$ and the infinitesimal relation (2) gives

$$ds = \sqrt{[g'(y)\,dy]^2 + [dy]^2} = \sqrt{g'(y)^2 + 1}\,dy.$$

Hence the arc length is $\int_c^d \sqrt{g'(y)^2 + 1}\,dy$, where $c$ and $d$ are the appropriate limits on $y$. [The same conclusion would be reached more formally by reversing the roles of $x$ and $y$ in the derivation of the integral (1) above.]

---

**EXAMPLE 2** Compute the length of $y = x^{2/3}$ from the origin to $(8, 4)$ (Fig. 6).

*SOLUTION* From $y = x^{2/3}$ you get

$$\frac{ds}{dx} = \sqrt{1 + (dy/dx)^2} = \sqrt{1 + (4/9)x^{-2/3}}.$$

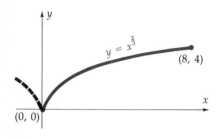

$y = x^{\frac{2}{3}}$

$(8, 4)$

$(0, 0)$

**FIGURE 6**

This becomes infinite as $x \to 0$, and it also looks difficult to integrate, so rewrite it with $y$ as the variable of integration: $y = x^{2/3} \Leftrightarrow x = y^{3/2}$ (since $x \geq 0$ on the given part of the curve), hence

$$\frac{ds}{dy} = \sqrt{1 + \frac{9}{4}y}.$$

Since we integrate with respect to $y$, the limits of integration are $0 \leq y \leq 4$:

$$s = \int_0^4 \sqrt{1 + \tfrac{9}{4}y}\,dy = \tfrac{8}{27}(1 + \tfrac{9}{4}y)^{3/2}\Big|_0^4$$
$$= \tfrac{8}{27}[10^{3/2} - 1] \approx 9.07.$$

As a check, compare this to the straight line length from $(0, 0)$ to $(8, 4)$, which is $\sqrt{64 + 16} = 8.9$.

---

**(Optional) A catenary** is a curve formed by a uniform chain hanging freely between two points of attachment (Fig. 7). The shape of such a curve

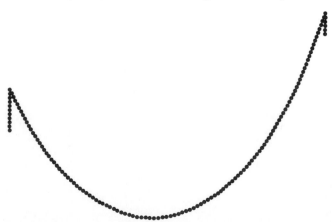

**FIGURE 7**
Hanging chain.

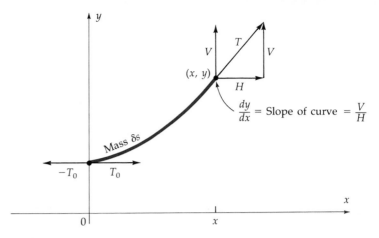

**FIGURE 8**
Portion of hanging chain.

can be determined from the balance of forces in Figure 8, showing the part of the chain between its lowest point (which we take at $x = 0$) and a general point $(x, y)$. The arrow marked $T$ denotes the force of tension in the chain at $(x, y)$; since the chain is completely flexible, this force must be *tangent to the chain* at $(x, y)$. Thus in the right triangle shown,

$$\frac{V}{H} = \frac{dy}{dx}. \tag{3}$$

Moreover, by a basic principle of physics, we can regard the force $T$ as the sum of a horizontal force $H$ and a vertical force $V$. The horizontal force must balance the force $T_0$ at the bottom of the chain, since these are the only two horizontal forces acting on this section of chain:

$$H = T_0. \tag{4}$$

The vertical force must balance the force of gravity acting on this part of the chain. Since the chain is uniform, its mass is the length $s$ times the linear density $\delta$ = mass per unit length; so the force of gravity is $-s\delta g$, with $g$ the acceleration due to gravity. Hence the vertical force required to hold up the part of the chain from 0 to $x$ is

$$V = s\delta g = \delta g \int_0^x \sqrt{1 + f'(t)^2}\, dt, \tag{5}$$

where $y = f(x)$ is the (as yet unknown) equation of the curve. Combined with (3) and (4), this gives

$$T_0 \frac{dy}{dx} = \delta g \int_0^x \sqrt{1 + f'(t)^2}\, dt, \qquad y = f(x).$$

Differentiate each side with respect to $x$, using the Second Fundamental Theorem of Calculus:

$$T_0 \frac{d^2 y}{dx^2} = \delta g \sqrt{1 + f'(x)^2} = \delta g \sqrt{1 + (dy/dx)^2}. \tag{6}$$

You can solve this differential equation (problem 8) to find the slope $dy/dx$, hence the shape of the curve $y$. The result is

$$y = \frac{1}{2k}(e^{kx} + e^{-kx}) + C$$

with $k = \delta g/T_0$, and $C$ a constant depending on the height of the chain above the $x$-axis. This can be expressed in terms of the **hyperbolic cosine** (Sec. 8.9) as

$$y = \frac{1}{k}\cosh(kx) + C.$$

## SUMMARY

**Arc Length of the Graph of $f$ on $[a, b]$:**

$$s_a^b = \int_a^b \sqrt{1 + f'(x)^2}\, dx.$$

**Arc Length Differential (Fig. 5):**

$$s = \int ds, \qquad ds = \sqrt{(dx)^2 + (dy)^2}$$

## PROBLEMS

### A

**1.** Find the arc length of the graph of $y = 2x$ from $(-1, -2)$ to $(1, 2)$
   **a)** By the arc length integrand.
   **b)** By the Theorem of Pythagoras.

**2.** Find the arc length of the given graph over the given interval.
   **\*a)** $y = 3x^{3/2} - 1, \quad 0 \le x \le 1$
   **b)** $y = \frac{2}{3}(x - 1)^{3/2}, \quad 1 \le x \le 5$
   **\*c)** $y = \frac{x^4}{16} + \frac{1}{2x^2}, \quad 2 \le x \le 3$
   **d)** $y = \frac{x^3}{6} + \frac{1}{2x}, \quad a \le x \le b$
   **e)** $y = mx + b, \quad 0 \le x \le 1$

**\*3.** Compute the length of the "astroid of Roemer" in Figure 9, $x^{2/3} + y^{2/3} = 1$. Compare your result to the circumference of the unit circle.

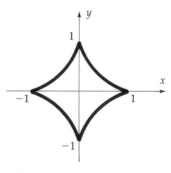

**FIGURE 9**
Astroid, $x^{2/3} + y^{2/3} = 1$.

**4.** Compute the length of the given graph. (As you go down the list, you gradually need more advanced integration techniques from Chapter 9.)
   **\*a)** $e^y = \sec\theta, \quad |\theta| \le \pi/4$
   **b)** $y = ax^2, \quad 0 \le x \le b$
   **\*c)** $x^2 + y^2 = r^2, \quad 0 \le x \le b \quad (b < r)$

d)  $y = e^x, \quad 0 \le x \le b$
(Substitute $u^2 = 1 + e^{2x}$.)

*e)  $y = \ln x, \quad a \le x \le b$

## B

5.  a)  The cable for a suspension bridge with a flat roadway hangs more or less in the shape of a parabola. Compute the length of cable for a bridge of span $2S$ (Fig. 10) with the cable ends at height $H$ above the cable's lowest point. Consider only the part between the towers from which the cable hangs.

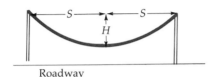

**FIGURE 10**
Suspension bridge.

b)  A proposed bridge across the straits of Messina between Italy and Sicily would be about five miles long. How long would the cable be, if the height $H$ were one-quarter mile?

6.  Compute the length of the graph of $y = x^{2/3}$ for $-1 \le x \le 8$. (Warning: If you express the length as an integral with respect to $y$, you will need to do it in two parts.)

7.  At each point $P$ on the graph of $f$, let $\theta$ denote the angle between the tangent to the graph and the horizontal line through $P$. Assuming that $f'$ is continuous, show that the arc length is $\int \sec\theta \, dx$.

8.  This problem solves the equation (6) for the hanging chain in Figure 8. Since the equation involves $\dfrac{dy}{dx}$ and its derivative, but not $y$ itself, we introduce a new letter for $\dfrac{dy}{dx}$; the traditional letter is $p$. We also abbreviate $\delta g/T_0$ as a single letter $k$. So the equation (6) is $\dfrac{dp}{dx} = k\sqrt{1 + p^2}$.

a)  Solve this for $p$ as a function of $x$. [You can use the integral tables and obtain your answer in terms of exponentials; or else consult the inverse hyperbolic derivative formulas (Sec. 8.9) and obtain the answer in terms of the hyperbolic sine. Determine the constant of integration by evaluating $p(0)$ from Figure 8.]

b)  Solve for the shape of the chain, $y$ as a function of $x$.

9.  (Limit of arc length over chord length) Let $f$ be a function with continuous derivative $f'$. The *chord length* $ch_a^b$ is the length of the straight line segment from $(a, f(a))$ to $(b, f(b))$. Compute the limits.

a)  $\lim\limits_{b \to a^+} \dfrac{s_a^b}{b - a}$  (Use the Second Fundamental Theorem of Calculus.)

b)  $\lim\limits_{b \to a^+} \dfrac{ch_a^b}{b - a}$  (Use the definition of derivative.)

c)  $\lim\limits_{b \to a^+} \dfrac{ch_a^b}{s_a^b}$  (Use the previous two parts.)

# 10.2

# AREA OF A SURFACE OF REVOLUTION

When a curve is revolved about an axis, it generates a **surface of revolution** (Fig. 1). We deduce an integral for the area of such a surface, using the interpretation of $ds$ as an infinitesimal length of arc; a more formal derivation is outlined in the problems.

Consider an "infinitesimal" circular strip of the surface, surrounding the axis of revolution, as in Figure 1. The strip is a sort of belt, with "infinitesimal width" $ds$, and length (circumference) $2\pi r$, where $r$ is the radius of the circular strip. The infinitesimal area of the strip is thus

$$2\pi r \, ds,$$

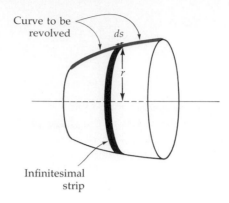

**FIGURE 1**
Surface of revolution.

and the area of the entire surface is the integral

$$\int_a^b 2\pi r \, ds \tag{1}$$

with appropriate limits of integration $a$ and $b$. To apply this formula, choose a variable of integration (usually $x$ or $y$), express both the radius $r$ and the infinitesimal arc length $ds$ in terms of this, and determine the proper limits of integration for the chosen variable.

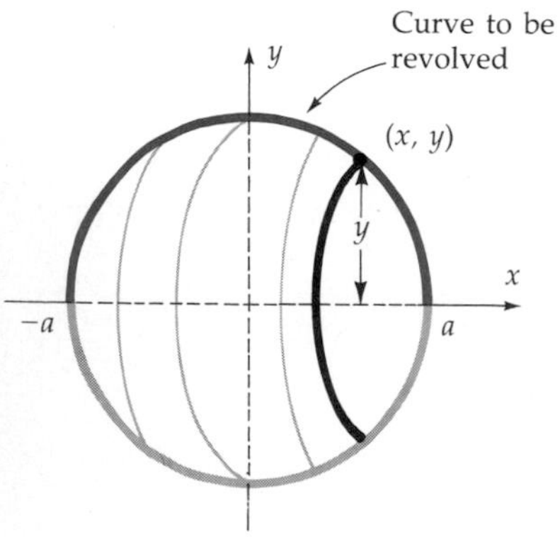

**EXAMPLE 1**   The graph of $y = \sqrt{a^2 - x^2}$, $-a \le x \le a$, is a semicircle of radius $a$ (Fig. 2). Revolve it around the $x$-axis, and you get a *sphere* of radius $a$. Compute the area of this sphere.

***SOLUTION***   In the integral (1), choose $x$ as the variable of integration. Then $r$ and $ds$ must be expressed in terms of $x$ and $dx$. The infinitesimal strip surrounds the axis of rotation, the $x$-axis (Fig. 2); the radius of the strip is precisely the height to the graph,

$$r = y = \sqrt{a^2 - x^2}. \tag{2}$$

The width of the strip, $ds$, is arc length on the graph of $y = \sqrt{a^2 - x^2}$, so

$$dy = \frac{-x \, dx}{\sqrt{a^2 - x^2}}$$

and

$$(ds)^2 = (dx)^2 + (dy)^2 = (dx)^2 + \left[ \frac{-x \, dx}{\sqrt{a^2 - x^2}} \right]^2$$

$$= \left[ 1 + \frac{x^2}{a^2 - x^2} \right] (dx)^2 = \frac{a^2}{a^2 - x^2} (dx)^2.$$

This gives

$$ds = \frac{a}{\sqrt{a^2 - x^2}} \, dx. \tag{3}$$

**FIGURE 2**
$y = \sqrt{a^2 - x^2}$ revolved about $y$-axis.

The limits of integration are seen in Figure 2—add all the strips from $x = -a$ to $x = a$. Thus, using (2) and (3) in the area integral (1), you get

$$\int_{-a}^{a} 2\pi r \, ds = \int_{-a}^{a} 2\pi [\sqrt{a^2 - x^2}] \left[ \frac{a}{\sqrt{a^2 - x^2}} \, dx \right]$$

$$= 2\pi a \int_{-a}^{a} dx = 4\pi a^2.$$

This is the surface area of a sphere of radius $a$.

**EXAMPLE 2**  The graph of $y = x^2$, $0 \le x \le 1$, is revolved about the $y$-axis. Compute the area of the surface generated.

*SOLUTION*  We choose $x$ as the variable of integration (though $y$ would also work). In this example, the strips surround the $y$-axis (Fig. 3), and the radius of the typical strip is simply $x$; so in the area integral (1),

$$r = x.$$

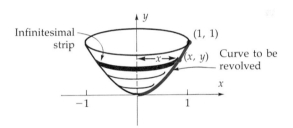

**FIGURE 3**
$y = x^2$, $0 \le x \le 1$, revolved about $y$-axis.

Express the arc length in terms of $x$ and $dx$; $y = x^2$, so $dy = 2x \, dx$, and

$$ds = \sqrt{(dx)^2 + (dy)^2} = \sqrt{(dx)^2 + (2x \, dx)^2}$$
$$= \sqrt{1 + 4x^2} \, dx.$$

The limits of integration must be determined with care. Each infinitesimal strip in Figure 3 corresponds to an $x$ between 0 and 1; so these are the limits of integration, and the surface area is

$$\int_{x=0}^{x=1} 2\pi r \, ds = \int_0^1 2\pi x \sqrt{1 + 4x^2} \, dx.$$

Substitute $u = 1 + 4x^2$, so $du = 8x \, dx$, and the integral becomes

$$\int_1^5 2\pi u^{1/2} \frac{1}{8} \, du = \frac{\pi}{4} \frac{u^{3/2}}{3/2} \Big|_1^5 = \frac{\pi}{6} (5^{3/2} - 1)$$

$$\approx 5.33.$$

As a check, note that the surface in question is not vastly different from a hemisphere of radius 1, whose surface area is $2\pi \cdot 1^2 \approx 6.28$.

## SUMMARY

### Area of a Surface of Revolution (see Fig. 1):

$$\int_a^b 2\pi r \, ds.$$

## PROBLEMS

### A

1. The graph of $y = x^2$, $0 \le x \le 1$, is revolved about the $y$-axis. Compute the surface area by integration with respect to $y$. Compare the result to Example 2.

2. The given graph is revolved about the $x$-axis. Sketch the resulting surface, and compute its area.
   a) $y = \sqrt{4 - x^2}$, $-2 \le x \le 2$
   *b) $y = \sqrt{4 - x^2}$, $-1 \le x \le 2$
   c) $y = 3x$, $0 \le x \le h$
   *d) $y = mx$, $a \le x \le b$ ($m$ is constant.)

3. The given graph is revolved about the $y$-axis. Sketch the resulting surface, and compute its area.
   a) $y = mx$, $a \le x \le b$
   b) $y = \sqrt{9 - x^2}$, $0 \le x \le 3$
   *c) $y = \sqrt{a^2 - x^2}$, $0 \le x \le a/2$
   d) $y = \sqrt{a^2 - x^2}$, $a/2 \le x \le a$

4. The given graph is revolved about the $x$-axis. Find the area of the resulting surface.
   a) $y = x^2$, $0 \le x \le b$
   *b) $y = e^x$, $0 \le x \le b$

5. Compute the surface area generated when the graph of $y = x^3$, $0 \le x \le b$, is revolved about
   a) The $x$ axis.   *b) The $y$-axis.

6. The astroid $x^{2/3} + y^{2/3} = 1$ (in Fig. 9 of Sec. 10.1) is revolved about the $y$-axis. Compute the area of the resulting surface. Compare this area to that of the unit sphere and to that of two unit disks.

*7. The graph of $y = ax^2$ is a parabola with focus at $\left(0, \dfrac{1}{4a}\right)$. Revolve the parabola about its axis of symmetry, and compute the surface area lying below the horizontal plane through the focus.

### B

8. The graph of $y = mx$ for $a \le x \le b$ is revolved about the $x$-axis, generating a *frustum of a cone*. Show that the area is $2\pi rs$, where $r$ is the mid-point radius and $s$ the slant height (Fig. 4), in two ways
   a) Using the integral for surface area.
   b) By elementary geometry. (Cut open the frustum, lay it out flat, and recognize it as the difference of two circular sectors.)

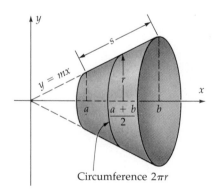

**FIGURE 4**
Area of frustum = $2\pi rs$.

9. Archimedes discovered that when two parallel planes cut a sphere of radius $a$, the area of the spherical surface between the planes is $2\pi ad$, where $d$ is the distance between the planes. Verify this.

### C

10. Suppose that $f'(x)$ is continuous for $a \le x \le b$. The integral for the area of the surface generated by revolving the graph of $f$ about the $x$-axis is $\int_a^b 2\pi f(x)\sqrt{1 + f'(x)^2} \, dx$. This can be

derived by approximating the surface with conical frusta (Fig. 5). The area of a frustum (problem 8) can be deduced by elementary geometry. Show that

a)   The area of the first conical approximation is $\pi(y_1 + y_0)\sqrt{(\Delta x)^2 + (y_1 - y_0)^2}$.

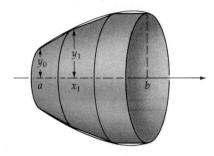

**FIGURE 5**
Surface of revolution from Fig. 1, approximated by conical frusta.

b)   $\sqrt{(\Delta x)^2 + (y_1 - y_0)^2} = \sqrt{1 + f'(\hat{x}_1)^2}\,\Delta x$ for some $\hat{x}_1$ between $a$ and $x_1$.

c)   If $M = \max\limits_{[a,b]} |f'(x)|$, then

$$|y_1 + y_0 - 2f(\hat{x}_1)|$$
$$= \big|[y_1 - f(\hat{x}_1)] + [y_0 - f(\hat{x}_1)]\big|$$
$$\leq M(x_1 - \hat{x}_1) + M(\hat{x}_1 - a) = M\,\Delta x.$$

d)   $\pi(y_1 + y_0)\sqrt{(\Delta x)^2 + (y_1 - y_0)^2} =$ $2\pi f(\hat{x}_1)\sqrt{1 + f'(\hat{x}_1)^2}\,\Delta x + e_1$ where $|e_1| \leq 2\pi M\sqrt{1 + M^2}\,(\Delta x)^2$.

e)   The sum of the $n$ approximations is

$$S_n = \sum_{j=1}^{n} 2\pi f(\hat{x}_j)\sqrt{1 + f'(\hat{x}_j)^2}\,\Delta x + \sum_{j=1}^{n} e_j$$

and $\left|\sum_{j=1}^{n} e_j\right| \leq 2\pi M\sqrt{1 + M^2}(b - a)\,\Delta x$.

f)   $\lim\limits_{n \to \infty} S_n = \int_a^b 2\pi f(x)\sqrt{1 + f'(x)^2}\,dx$.

# 10.3
## CENTER OF MASS

Imagine two masses $m_1$ and $m_2$ on a thin balance arm of negligible mass, resting on a pivot $P$ (Fig. 1). The system is not necessarily in equilibrium when the masses $m_1$ and $m_2$ on both sides are the same; what must be the same on both sides is the *product* of the mass $m$ by its distance $d$ from the pivot point:

$$m_1 d_1 = m_2 d_2. \tag{1}$$

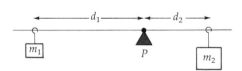

**FIGURE 1**
Balance.

The reason for this is found in the concept of **torque** (turning force):

$$\text{torque} = \pm(\text{force}) \times (\text{distance from pivot}).$$

The $+$ or $-$ sign is determined by taking clockwise torques as positive, and counterclockwise torques as negative. The choice of sign becomes automatic if we assign coordinates to the points on the balance arm, with zero at the pivot, positive numbers to the right, and negative to the left. Then (Fig. 2a)

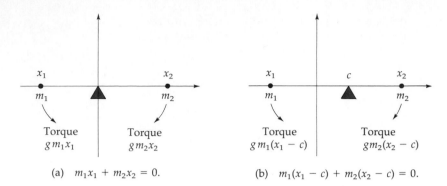

(a)  $m_1x_1 + m_2x_2 = 0.$          (b)  $m_1(x_1 - c) + m_2(x_2 - c) = 0.$

**FIGURE 2**
Equilibrium conditions.

*a mass m at position x exerts a torque gmx; g is the acceleration due to gravity,* and $gm$ the force of gravity on the mass $m$. When $x > 0$ then $m$ is to the right of the pivot and exerts a clockwise (positive) torque $gmx$; when $x < 0$ then $m$ is to the left of the pivot and exerts a counterclockwise (negative) torque $gmx$.

With this convention, consider again the two masses, $m_1$ at position $x_1 = -d_1$ and $m_2$ at position $x_2 = d_2$. The torques are $gm_1x_1$ and $gm_2x_2$; *the system is in equilibrium precisely when the algebraic sum of these torques is zero,*

$$gm_1x_1 + gm_2x_2 = 0.$$

Since $x_1 = -d_1$ and $x_2 = d_2$, you can see that this is the same as condition (1).

More generally, if there are $n$ masses $m_1, \ldots, m_n$ at positions $x_1, \ldots, x_n$, the total torque is

$$g(m_1x_1 + \cdots + m_nx_n),$$

and the system is balanced when this sum is zero.

Suppose now that the pivot is moved from $x = 0$ to $x = c$ (Fig. 2b). Now, a mass $m$ at position $x$ exerts a (clockwise) torque of $gm(x - c)$; and $n$ masses $m_1, \ldots, m_n$ at positions $x_1, \ldots, x_n$ are in balance when

$$g[m_1(x_1 - c) + m_2(x_2 - c) + \cdots + m_n(x_n - c)] = 0.$$

Finally, instead of a simple rod with point masses, consider a horizontal plate of uniform thickness and density, covering a region $R$ in the plane (Fig. 3), and pivoted along the line $x = c$. What is the condition for balance in this case?

Imagine the plate divided into infinitesimal strips of width $dx$, parallel to the pivot line $x = c$. Denote by $L(x)$ the length of the cross section of $R$ at $x$. Then each strip has infinitesimal area $L(x)\,dx$, and its infinitesimal mass is $\delta \cdot L(x)\,dx$; $\delta$ denotes the mass of the plate per unit area. The infinitesimal strip thus exerts a torque

$$g \cdot \delta \cdot L(x)\,dx \cdot (x - c).$$

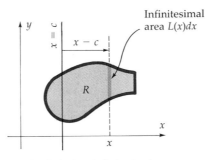

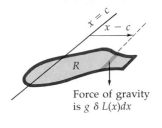

(a)   Strip has infinitesimal
       mass $\delta L(x)dx$.

(b)   Strip contributes torque
       $g\,\delta\,L(x)dx \cdot (x - c)$.

**FIGURE 3**
Computing torque of a plate of density $\delta$ covering region $R$.

The torque of the entire plate is the "sum" of these infinitesimal torques, the integral

$$\int_a^b g\delta L(x)(x - c)\,dx = g\delta \int_a^b L(x)(x - c)\,dx,$$

where $a$ and $b$ are the smallest and largest values of $x$ in the region $R$. The condition for equilibrium is that this torque be zero:

$$\int_a^b L(x)(x - c)\,dx = 0.$$

This is equivalent to

$$\int_a^b L(x) \cdot x\,dx - c \int_a^b L(x)\,dx = 0.$$

The integral $\int_a^b L(x)\,dx$ is precisely the *area* $A$ of the region, and of the plate that covers it; thus the balancing condition is

$$\int_a^b xL(x)\,dx - cA = 0.$$

Solving this equation for $c$, we find that the plate balances when $c$ equals the number $\bar{x}$ given by

$$\bar{x} = \frac{\displaystyle\int_a^b xL(x)\,dx}{A}; \qquad A \text{ is the area of the plate.}$$

We next compute the torque with respect to a line $y = c$ parallel to the $x$-axis. The center of our infinitesimal strip is at its midpoint; denote the $y$-coordinate of this midpoint by $\tilde{y}(x)$ (Fig. 4). The lever arm for computing the torque exerted by this strip about the line $y = c$ is the distance from that line to the midpoint of the strip; taking sign into account, the torque is

$$g\delta L(x)\,dx[\tilde{y}(x) - c].$$

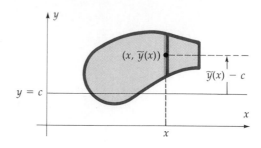

**FIGURE 4**
$\tilde{y}(x)$ gives midpoint of vertical cross section at $x$. Lever arm for strip is $\tilde{y}(x) - c$.

So the torque of the entire plate about the line $y = c$ is

$$\int_a^b g\delta L(x)[\tilde{y}(x) - c]\, dx.$$

The plate is in balance about the line $y = c$ when $c$ has the value $\bar{y}$ given by

$$\bar{y} = \frac{\int_a^b L(x)\tilde{y}(x)\, dx}{A}.$$

The function $\tilde{y}(x)$ gives the midpoint of the vertical cross section at $x$; the number $\bar{y}$ is a sort of average of the function values $\tilde{y}(x)$.

The point $(\bar{x}, \bar{y})$ is called the **center of mass** of the homogeneous plate, or the **centroid** of the region $R$ that it covers. It can be shown (Sec. 17.3) that the plate balances about *any* line through this point. If suspended by a string attached at this point, it is again in balance.

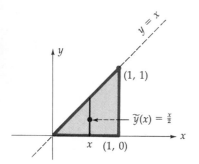

**FIGURE 5**
Computing center of mass of a triangle.

**EXAMPLE 1** Compute the center of mass of the triangle $T$ with vertices $(0, 0)$, $(1, 0)$, $(1, 1)$ (Fig. 5).

**SOLUTION** The top edge of the triangle lies in the line $y = x$; so the vertical cross section of $T$ runs from $y = 0$ to $y = x$, and its length is

$$L(x) = x.$$

The area of the region is $A = \frac{1}{2}$. Hence

$$\bar{x} = \frac{\int_0^1 xL(x)\, dx}{1/2} = \frac{(1/3)x^3\big|_0^1}{1/2} = \frac{2}{3}.$$

The *midpoint* of the cross section at $x$ is (Fig. 5)

$$\tilde{y}(x) = \tfrac{1}{2}x$$

so

$$\bar{y} = \frac{\int_0^1 \tilde{y}(x)L(x)\, dx}{1/2} = \frac{(1/6)x^3\big|_0^1}{1/2} = \frac{1}{3}.$$

The center of gravity is $(\bar{x}, \bar{y}) = (\frac{2}{3}, \frac{1}{3})$.

## SUMMARY

*Center of mass of homogeneous plate is* $(\bar{x}, \bar{y})$, *with*

$$\bar{x} = \frac{\int_a^b xL(x)\,dx}{A},$$

$$\bar{y} = \frac{\int_a^b \tilde{y}(x)L(x)\,dx}{A};$$

$A$ *is the area of the plate;* $\tilde{y}(x)$ *gives the midpoint of the cross section at* $x$; $L(x)$ *is the length of the cross section at* $x$.

## PROBLEMS

### A

Find the center of mass of a homogeneous plate covering the given region.

1. The triangle with vertices $(0, 0)$, $(1, 0)$, $(1, 2)$.

*2. Between the $x$-axis and the parabola $y = 4 - x^2$.

3. The semicircle $0 \le y \le \sqrt{a^2 - x^2}$.

4. In the first quadrant, inside the circle $x^2 + y^2 = a^2$.

*5. In the first quadrant, outside $x^2 + y^2 = a^2$, below $y = a$, and to the left of $x = a$.

6. Between the $y$-axis and $x = y - y^3$, $0 \le y \le 1$.

*7. Between the $x$-axis and $y = \sin x$, for $0 \le x \le \pi$.

8. Between the $x$-axis, the graph of $y = e^x$, and the lines $x = 0$ and $x = 2$.

*9. Between $y = \sqrt{1 + x^2}$, the $x$-axis, and the lines $x = -1$ and $x = 1$.

10. The upper half of the ellipse $\dfrac{x^2}{a^2} + \dfrac{y^2}{b^2} = 1$.

*11. The set $\{(x, y): -b \le y \le \sqrt{a^2 - x^2}, |x| \le a\}$, consisting of a rectangle surmounted by a semicircle.

### B

12. In Example 1, determine whether $\bar{x}$ and $\bar{y}$ are respectively the averages of the $x$- and $y$-coordinates of the vertices of the triangle $T$.

13. a) Prove the *Theorem of Pappus*: The volume of the solid of revolution formed by re-

volving a plane region $R$ about an axis in its plane, but not intersecting the interior of $R$, is $2\pi r A$, where $A$ is the area of the region, and $r$ is the distance from the axis of revolution to the centroid of the region. (Take the $y$-axis as axis of revolution, and recall the computation of volume by the method of cylindrical shells.)

b) Use Pappus' Theorem to write a formula for the volume enclosed by an inner tube.

14. (Center of mass of a homogeneous wire) Imagine a uniform wire bent into the shape of a plane curve. Denote by $\delta$ the linear density of the wire, the mass per unit length; thus a small segment of wire of length $ds$ has mass $dm = \delta\,ds$. The torque exerted by this small segment about the line $x = c$ is

$$(x - c)g\delta\,ds,$$

and the torque exerted by the entire wire is

$$\int_a^b (x - c)g\delta\,ds,$$

where $a$ and $b$ are the appropriate limits of integration corresponding to the endpoints of the wire.

a) Show that the torque about $x = c$ is zero precisely when $c = \dfrac{\int_a^b x\,ds}{L}$, where $L$ is the length of the wire. (This value is called $\bar{x}$.)

b) Determine $\bar{y}$ so that the torque about $y = c$ is zero precisely when $c = \bar{y}$.

*c) Compute the center of mass $(\bar{x}, \bar{y})$ of a wire with constant density $\delta$, in the shape of the semicircle $y = \sqrt{a^2 - x^2}$, $|x| \le a$. Does $(\bar{x}, \bar{y})$ lie *on* the wire?

d) A semicircular wire shaped as in part c is suspended from a string attached to one end

of the wire. What is the angle between the string and the end of the wire?

e) State and prove a Theorem of Pappus relating center of mass to surface area. (See the previous problem.)

---

## 10.4

# PROBABILITY DENSITY FUNCTIONS

The concepts of probability are applied all across the spectrum from marketing to quantum mechanics. To understand the significant applications requires specialized study; but calculus explains some of the basic ideas and formulas and makes the relevant calculations.

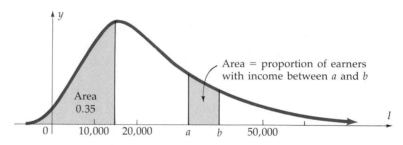

**FIGURE 1**
Distribution of income in country $C$.

The basic idea is illustrated in Figure 1, describing the distribution of income in country $C$. Income I is plotted along the horizontal axis, and the curve is constructed so that *areas* give *proportions*. For example, 35% of potential earners have incomes under $15,000; a small proportion of them actually have *negative* income. Generally:

> The *proportion of people with incomes greater than $a$ and less than $b$ equals the area under the curve for $a < I < b$.*

To put this in brief symbols, denote by $P(a < I < b)$ the percentage of people with incomes between $a$ and $b$; and denote by $f$ the function graphed in Figure 1, called the income density function. Then the long phrase displayed above, relating proportion to area, is simply

$$P(a < I < b) = \int_a^b f(I)\, dI.$$

Similarly,

$$P(I \le 15{,}000) = \int_{-\infty}^{15{,}000} f(I)\, dI = 0.35.$$

These proportions can also be interpreted as probabilities; $P(a < I < b)$ is the probability that a person "chosen at random" from this population will have income between $a$ and $b$. With this interpretation, the function $f$ graphed in Figure 1 is called the **probability density function** for this distribution of incomes.

Figure 2 shows an *exponential* probability density, typically used to describe equipment failure rates, waiting times, and such phenomena. For example, it could describe the probable waiting time before the Amalgamated Airlines reservations agent answers the phone. Time $t$ (seconds) is on the horizontal axis, and the probability of waiting more than $a$ but less than $b$ seconds is the area under the curve from $a$ to $b$:

$$P(a < t < b) = \int_a^b f(t)\, dt.$$

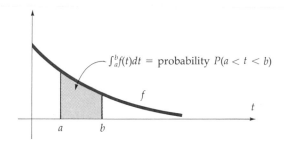

**FIGURE 2**
An exponential density function.

For $t > 0$, $f$ is a decaying exponential, while for $t < 0$ $f$ is $0$; the reservations agent won't answer before you dial.

Figure 3 shows a *normal* probability density used, for example, to describe the results of repeated measurements of the same quantity. (Because of random imperfections in the measuring process, the measurements don't all agree.) This is the famous "bell-shaped" normal curve.

Every probability density $f(x)$ has two essential properties. We don't contemplate negative probabilities, so

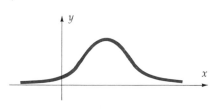

**FIGURE 3**
A normal probability density function.

$$f(x) \geq 0 \qquad \text{for all } x. \tag{1}$$

And the result must surely be *somewhere* between $-\infty$ and $+\infty$, so

$$P(-\infty < x < +\infty) = \int_{-\infty}^{+\infty} f(x)\, dx = 1. \tag{2}$$

In addition, since we want to integrate $f$, we will assume that it is continuous except, perhaps, for isolated jumps or vertical asymptotes; and at any asymptotes, the improper integral must be convergent.

The function in Figure 4 meets these requirements; it is the density for the process of choosing a real number at random from the interval $[0, 1]$.

The exact form of an *exponential* probability density is determined by condition (2). The general decaying exponential function is $Ae^{-kt}$, with constants $A$ and $k$. Since the density is $0$ for negative $t$, we have

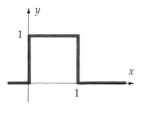

**FIGURE 4**
$$f(x) = \begin{cases} 1, & 0 \leq x \leq 1 \\ 0 & \text{otherwise.} \end{cases}$$

$$f(t) = \begin{cases} Ae^{-kt}, & t \geq 0 \\ 0, & t < 0. \end{cases}$$

Since condition (2) must be satisfied, then

$$1 = \int_{-\infty}^{+\infty} f(t)\, dt = \int_{0}^{\infty} Ae^{-kt}\, dt = \lim_{T \to +\infty} \int_{0}^{T} Ae^{-kt}\, dt$$

$$= \lim_{T \to +\infty} \left. -\frac{A}{k} e^{-kt} \right|_{0}^{T} = \lim_{T \to +\infty} \frac{A}{k}(1 - e^{-kT})$$

$$= A/k.$$

Thus $A/k = 1$, or $A = k$; so every exponential density function has the form

$$f(t) = \begin{cases} ke^{-kt}, & t \geq 0 \\ 0, & t < 0. \end{cases} \tag{3}$$

## The Mean of a Probability Density

Return to the income density function in Figure 1; we are going to compute the average, or *mean* income for country $C$. Suppose the graph represents a total population of $N$ potential wage earners, and that their incomes all lie between a minimum $a$ and a maximum $b$. Divide the interval $[a, b]$ into $n$ equal parts of length $\Delta I$, and let $\hat{I}_j$ denote the midpoint of the $j^{\text{th}}$ interval (Fig. 5)

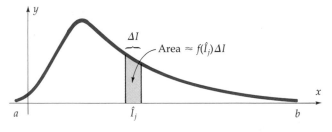

**FIGURE 5**

The proportion of people with incomes in the $j^{\text{th}}$ interval is the area over that interval, approximately $f(\hat{I}_j)\, \Delta I$. With a total population $N$, the actual number of earners with incomes in this interval is $N \cdot f(\hat{I}_j)\, \Delta I$; their incomes are all approximately $\hat{I}_j$, so their combined income is about

$$\hat{I}_j[Nf(\hat{I}_j)\, \Delta I] = N\hat{I}_j f(\hat{I}_j)\, \Delta I.$$

The income of the whole country is approximately

$$\sum_{j=1}^{n} N\hat{I}_j f(\hat{I}_j)\, \Delta I = N \sum_{j=1}^{n} \hat{I}_j f(\hat{I}_j)\, \Delta I.$$

As we let $n \to \infty$ and $\Delta I \to 0$, we obtain the total income as the integral

$$N \int_{a}^{b} If(I)\, dI.$$

The *mean* income is this total divided by the population $N$:

$$\text{Mean income} = \int_{a}^{b} If(I)\, dI.$$

A similar analysis applies to any probability density $f(x)$; its *mean*, denoted $\mu_f$, is given by the integral

$$\mu_f = \int_{-\infty}^{+\infty} xf(x)\, dx. \tag{4}$$

---

**EXAMPLE 1**   The density in Figure 4 is for choosing a number at random from the interval [0, 1]. Compute its mean.

*SOLUTION*   From the figure, $f(x) = 1$ if $0 \le x \le 1$, and $f(x) = 0$ otherwise. Thus

$$\mu_f = \int_{-\infty}^{+\infty} xf(x)\, dx = \int_0^1 x \cdot 1\, dx = \tfrac{1}{2}.$$

This is certainly a reasonable mean for numbers chosen at random from [0, 1].

---

## The Mean as a Center of Mass

Formula (4) for the mean resembles the integral used for the center of mass (Sec. 10.3). Figure 6 shows a typical probability density function $f$. Imagine a thin homogeneous plate bounded by the $x$-axis and the graph of $f$. The $x$-coordinate of the center of mass is

$$\bar{x} = \frac{\displaystyle\int_{-\infty}^{+\infty} xf(x)\, dx}{\displaystyle\int_{-\infty}^{+\infty} f(x)\, dx}.$$

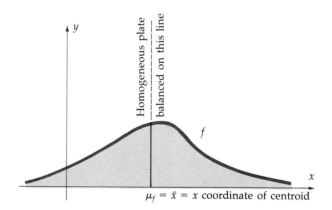

**FIGURE 6**
Region under graph of probability density $f$ balances on line through mean $\mu_f$.

Since $f$ is a probability density, the denominator in this formula is 1, so

$$\bar{x} = \int_{-\infty}^{+\infty} xf(x)\, dx = \mu_f.$$

Thus: *The mean $\mu_f$ of the probability density $f$ is the $x$-coordinate of the centroid of the region bounded by the $x$-axis and the graph of $f$.* The plate would balance on the vertical line $x = \bar{x}$. In particular, if the graph of $f$ is symmetric about a

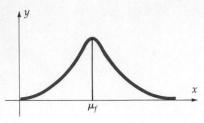

**FIGURE 7**
The mean of a symmetric density of $f$ is right at the line of symmetry.

vertical line $x = c$, then the mean $\mu_f$ is precisely $c$ (Fig. 7). This gives a quick way to do Example 1; the density in Figure 4 is symmetric about the line $x = 1/2$, so $\mu_f = 1/2$.

For the exponential density (3), the mean is

$$\mu_f = \int_{-\infty}^{+\infty} tf(t)\, dt = \int_{0}^{+\infty} tke^{-kt}\, dt = 1/k,$$

as you can check. Thus we can replace the constant $k$ with $1/\mu$, and write the *exponential density function with mean* $\mu$ as

$$f(t) = \begin{cases} \mu^{-1}e^{-t/\mu}, & t \geq 0 \\ 0, & t < 0. \end{cases}$$

Figure 8 shows these densities for three values of $\mu$. For the larger mean $\mu$, there is less "probability" near 0 and more further to the right.

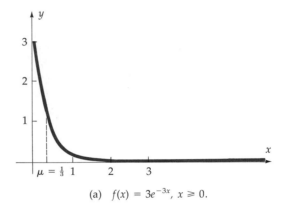

(a)   $f(x) = 3e^{-3x}$, $x \geq 0$.

**FIGURE 8a**
$f(x) = 3e^{-3x}$, $x \geq 0$.

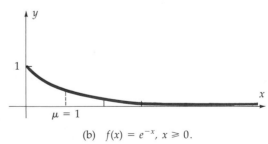

(b)   $f(x) = e^{-x}$, $x \geq 0$.

**FIGURE 8b**
$f(x) = e^{-x}$, $x \geq 0$.

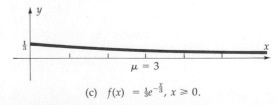

(c)   $f(x) = \frac{1}{3}e^{-\frac{x}{3}}$, $x \geq 0$.

**FIGURE 8c**
$f(x) = \frac{1}{3}e^{-x/3}$, $x \geq 0$.

**EXAMPLE 2** Some standard light bulbs are labeled to have an average lifetime of 1000 hours. It would be reasonable to model the probability of failure of such bulbs by an exponential density with mean $\mu = 1000$; with this model, the probability that the bulb fails by time $T$ is

$$P(0 \le t \le T) = \int_0^T \frac{1}{1000} e^{-t/1000} \, dt.$$

Compute the probability of
    **a.** Failing within the first 100 hours.
    **b.** Burning for at least 1000 hours.

*SOLUTION*
    **a.** The probability of failure within the first 100 hours is

$$P(0 \le t \le 100) = \int_0^{100} \frac{1}{1000} e^{-t/1000} \, dt$$

$$= -e^{-t/1000}\Big|_0^{100} = 1 - e^{-1/10} \approx 0.095.$$

Roughly 10% would fail so quickly.
    **b.** The probability of burning for at least 1000 hours is the probability of failure some time *after* $t = 1000$:

$$P(1000 \le t < \infty) = \int_{1000}^{\infty} \frac{1}{1000} e^{-t/1000} \, dt$$

$$= -e^{-t/1000}\Big|_{1000}^{\infty} = 0 - (-e^{-1}) \approx 0.368.$$

Thus, even though the *average* lifetime is 1000 hours, only about one third of the bulbs burn 1000 hours or longer. The reason is that some bulbs burn *much* longer—as much as 3000 or 4000 hours—and they bring up the average.

**The variance** of a probability density measures how closely it is concentrated near the mean. In Figure 9, the density (b) has a greater variance than (a). Like the mean, the variance is defined by an integral, namely

$$\int_{-\infty}^{+\infty} (x - \mu_f)^2 f(x) \, dx. \tag{5}$$

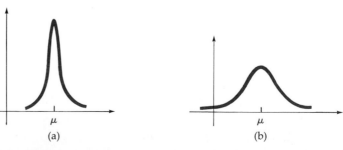

        (a)                    (b)

**FIGURE 9**
Density (b) has greater variance than (a).

In Figure 9a, the area under the graph is concentrated near $x = \mu$, and there $(x - \mu_f)^2$ is very small; so the integral (5) is quite small for this $f$. In Figure 9b, on the other hand, more of the area is farther away from $x = \mu_f$, in the region where $(x - \mu_f)^2$ is larger; so the variance integral (5) is larger in (b) than in (a).

The square root of the variance is called the **standard deviation**, denoted $\sigma_f$; so the variance is $\sigma_f^2$,

$$\sigma_f^2 = \int_{-\infty}^{+\infty} (x - \mu_f)^2 f(x)\, dx. \tag{6}$$

**EXAMPLE 3**  In Figure 4, the mean $\mu_f = 1/2$, so the variance is

$$\sigma_f^2 = \int_{-\infty}^{+\infty} (x - \mu_f)^2 f(x)\, dx = \int_0^1 (x - \tfrac{1}{2})^2 \cdot 1 \cdot dx$$
$$= \tfrac{1}{3}(x - \tfrac{1}{2})^3 \Big|_0^1 = \tfrac{1}{12}.$$

The equation for a normal probability density (as in Fig. 3) is

$$f(x) = \frac{1}{\sqrt{2\pi}\,\sigma}\, e^{-(x-\mu)^2/2\sigma^2}.$$

The parameters $\mu$ and $\sigma$ are precisely the mean and standard deviation of this $f$ (see problems). When such a density $f$ describes the random errors in unbiased measurements of a fixed quantity, then $\mu$ is the true value of the measured quantity; and $\sigma$ describes, in a sense, the average size of the errors made.

## SUMMARY

*Conditions for a Probability Density:*

$$f(x) \geq 0; \qquad \int_{-\infty}^{+\infty} f(x)\, dx = 1.$$

*Mean:*  $\mu_f = \int_{-\infty}^{+\infty} x f(x)\, dx.$

*Variance:*  $\sigma_f^2 = \int_{-\infty}^{+\infty} (x - \mu_f)^2 f(x)\, dx.$

$\sigma_f$ is the standard deviation.

*Exponential Distribution with Mean $\mu$:*

$$f(t) = \begin{cases} \mu^{-1} e^{-t/\mu}, & t \geq 0 \\ 0, & t < 0. \end{cases}$$

*Normal Density with Mean μ and Standard Deviation σ:*

$$f(x) = \frac{1}{\sqrt{2\pi}\,\sigma} \, e^{-(x-\mu)^2/2\sigma^2}.$$

## PROBLEMS

### A

**1.** For the exponential probability model in Example 2, with mean 1000, compute the probability of
  **\*a)** Failure between 100 and 200 hours.
  **b)** Failure within the first hour.
  **\*c)** Burning for at least 2000 hours.
  **d)** Burning for at least 3000 hours.

**2.** The lifetime of a certain type of flashlight battery is modeled by an exponential density with mean 200 hours. Compute the probability of
  **\*a)** Failure within the first hour.
  **b)** Failure within 200 hours.
  **c)** Burning for at least 500 hours.

**3.** For the given density $f$, compute the mean and the probability $P(0 \le t \le 1)$.
  **\*a)** $f(t) = 2e^{-2t}$, $t \ge 0$; $f(t) = 0$; $t < 0$
  **b)** $f(t) = \frac{1}{2}e^{-t/2}$, $t \ge 0$; $f(t) = 0$; $t < 0$
  **\*c)** $f(t) = \begin{cases} 1/4, & 0 \le t \le 4 \\ 0, & \text{otherwise} \end{cases}$
  **d)** $f(t) = \begin{cases} 1/4, & -2 \le t \le 2 \\ 0, & \text{otherwise} \end{cases}$

**\*4.** A probability density can have the form
$$f(x) = \begin{cases} k\sin(x), & 0 \le x \le \pi \\ 0, & \text{otherwise} \end{cases}$$
for an appropriate constant $k$.
  **a)** Determine $k$.
  **b)** Sketch $f$, and determine $\mu_f$ "by inspection."
  **c)** Compute $\mu_f$.
  **d)** Compute the variance.

**5.** A probability density can have the form $f(x) = k/(1 + x^2)$ for an appropriate constant $k$.
  **a)** Determine $k$.
  **b)** Sketch $f$, and determine the mean $\mu_f$ "by inspection."
  **c)** Is the integral defining the mean convergent?

**6.** Write a formula for the probability density for choosing at random a real number between 1 and 10.

**7.** Compute the mean of
  **a)** The "uniform density"
$$f(x) = \begin{cases} (b-a)^{-1}, & a \le x \le b \\ 0, & \text{otherwise} \end{cases}.$$
  **b)** The normal density
$$\frac{1}{\sqrt{2\pi}} e^{-x^2/2}, \qquad -\infty < x < \infty.$$

**8. \*a)** Compute the variance of the uniform density in problem 7a.
  **b)** Compare the variance with $a = 0$, $b = \pi$, to the variance in problem 4d.

**9.** Compute the variance of the exponential density with mean $\mu$.

**10.** In its normal state, the electron in a hydrogen atom is not at a fixed distance from the nucleus; the probability that it lies between distances $a$ and $b$ from the nucleus is
$$P(a < r < b) = \int_a^b kr^2 \exp(-2r/a_0)\,dr,$$
where $k$ is a constant to be determined, and $a_0 \approx 5.29 \times 10^{-11}$ meters. So we have a probability density $f(r) = kr^2 \exp(-2r/a_0)$ for $r \ge 0$, and otherwise $f(r) = 0$. Determine
  **\*a)** The constant $k$ (in terms of $a_0$).
  **b)** The maximum of $f$.
  **\*c)** $\mu_f$.
  **d)** $\sigma_f$.
  (Suggestion: First, prove a reduction formula for $\int_0^\infty r^n e^{-br}\,dr$.)

**11.** Find the distance between the inflection points of the normal density function with mean $\mu$ and standard deviation $\sigma$.

**B**

12. Assuming that $\int_{-\infty}^{+\infty} \frac{1}{\sqrt{2\pi}} e^{-x^2/2} \, dx = 1$, show that

a) $\int_{-\infty}^{+\infty} \frac{1}{\sqrt{2\pi}\,\sigma} e^{-x^2/2\sigma^2} \, dx = 1.$

b) $\int_{-\infty}^{+\infty} \frac{1}{\sqrt{2\pi}\,\sigma} e^{-(x-\mu)^2/2\sigma^2} \, dx = 1.$

13. Verify that the normal density in the Summary has mean $\mu$. [Hint: First compute $\int_{-\infty}^{+\infty} (x - \mu)e^{-(x-\mu)^2/2\sigma}\, dx$ with a change of variable $u = x - \mu$.]

14. Verify that $\sigma^2$ is the variance of the normal probability density in the Summary. [Hint: Change variables to reduce to the case $\mu = 0$, $\sigma = 1$; then integrate by parts, and use condition (2).]

**C**

15. Distribution of income can be displayed graphically by a Lorenz curve [Sec. 4.5]. The population is ranked according to income, and the Lorenz curve plots $y$ versus $x$, with $y =$ proportion of income earned by lowest $x$ proportion of earners. Let $f(I)$ be an "income density function" as in Figure 1.
    a) Write appropriate formulas (in terms of $f$) for the following:
       i) The percent of people with income $I \le$ a given level $I_0$.
       ii) The total income earned by those with income $\le I_0$.
       iii) The proportion of income earned by those with income $\le I_0$.
    b) Describe a process for constructing the Lorenz curve from the income density function.
    c) Would it be possible to reconstruct the income density function from the Lorenz curve?

# 10.5

## NUMERICAL INTEGRATION

The evaluation formula

$$\int_a^b f(x)\, dx = \int f(x)\, dx \Big|_a^b$$

is great, if you can find the indefinite integral $\int f(x)\, dx$. But no matter how many methods or tricks you may learn, you can't evaluate every indefinite integral. Try, for example, the integral $\int \sqrt{1 + 1/x^4}\, dx$ for the arc length along the hyperbola $y = 1/x$ (Sec. 10.1), or $\frac{1}{\sqrt{2\pi}} \int e^{-x^2/2}\, dx$, the integral of the normal probability density (Sec. 10.4). Even worse, the integrand may be given not by a formula, but by experimental measurements; then an antiderivative formula is clearly impossible.

When antiderivatives fail, approximations are used. You can use Riemann sums—midpoint sums are not too bad. But the usual methods of approximation are those given here. *Trapezoid sums* are simple and natural for integrating experimental or tabular data of limited accuracy. *Simpson sums* are slightly more complicated, but much more accurate, and are appropriate when the integrand itself is accurately known. Even greater accuracy is achieved with a *Modified Trapezoid Rule*, which uses the first derivative $f'$, as well as the integrand $f$ itself.

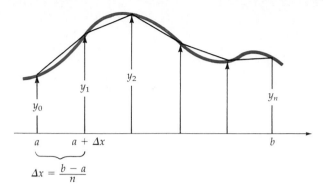

**FIGURE 1**
Approximation by trapezoids.

**Trapezoid sums** are illustrated in Figure 1. Think of $\int_a^b f(x)\,dx$ as an area. Cut the area into $n$ strips of width $\Delta x = (b - a)/n$, and approximate each by a trapezoid. The sum of these trapezoidal areas is denoted $T_n$, and its formula can be derived as follows: The area of the first trapezoid is computed in Figure 2; it is the base times the average height $\dfrac{y_0 + y_1}{2}$. Add to this the areas of the other trapezoids and find

$$T_n = \Delta x\left(\frac{y_0 + y_1}{2}\right) + \Delta x\left(\frac{y_1 + y_2}{2}\right) + \cdots + \Delta x\left(\frac{y_{n-1} + y_n}{2}\right)$$

$$= \Delta x\left[\left(\frac{1}{2}y_0 + \frac{1}{2}y_1\right) + \left(\frac{1}{2}y_1 + \frac{1}{2}y_2\right) + \cdots + \left(\frac{1}{2}y_{n-1} + \frac{1}{2}y_n\right)\right].$$

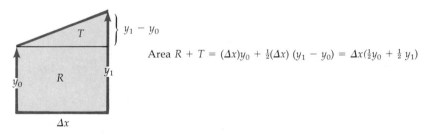

Area $R + T = (\Delta x)y_0 + \frac{1}{2}(\Delta x)(y_1 - y_0) = \Delta x(\frac{1}{2}y_0 + \frac{1}{2}y_1)$

**FIGURE 2**
Area of a trapezoid.

The two terms $\frac{1}{2}y_1$ combine; in fact, all terms combine in pairs except the first and last. Thus

$$T_n = \Delta x[\tfrac{1}{2}y_0 + y_1 + \cdots + y_{n-1} + \tfrac{1}{2}y_n]. \tag{1}$$

(This is derived differently in Sec. 6.4.)

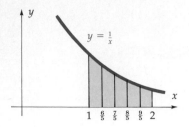

**FIGURE 3**

Approximating $\int_1^2 \dfrac{1}{x}\, dx$.

**EXAMPLE 1** Approximate $\displaystyle\int_1^2 \dfrac{dx}{x}$, using $T_5$.

**SOLUTION** (Fig. 3) In the Trapezoid Rule (1), take $a = 1$, $b = 2$, and $n = 5$. Then $\Delta x = \dfrac{2 - 1}{5} = \dfrac{1}{5}$, and

$$T_5 = \tfrac{1}{5}(\tfrac{1}{2}y_0 + y_1 + y_2 + y_3 + y_4 + \tfrac{1}{2}y_5).$$

The numbers $y_0, y_1, \ldots, y_5$ are the heights to the graph of the integrand at the points

$$1, \tfrac{6}{5}, \tfrac{7}{5}, \tfrac{8}{5}, \tfrac{9}{5}, 2.$$

The integrand is $f(x) = 1/x$, so

$$y_0 = 1, y_2 = \tfrac{5}{6}, \ldots, y_5 = \tfrac{1}{2},$$

and the trapezoid approximation $T_5$ is

$$T_5 = \tfrac{1}{5}(\tfrac{1}{2} + \tfrac{5}{6} + \tfrac{5}{7} + \tfrac{5}{8} + \tfrac{5}{9} + \tfrac{1}{2} \cdot \tfrac{5}{10})$$
$$= 0.69563 \ldots .$$

*Question:* Why bother to approximate $\displaystyle\int_1^2 \dfrac{dx}{x}$, when we know that it is precisely ln 2? It makes a good example, since the numbers come out neatly; and it also has value as a means of *computing* ln 2.

### Accuracy of the Trapezoid Rule

The Trapezoid Rule approximates the integral $\int_a^b f(x)\, dx$; the difference between the actual integral and the approximation is called the remainder, $R(T_n)$:

$$R(T_n) = \int_a^b f(x)\, dx - T_n. \tag{2}$$

This remainder is positive when $f$ is concave down (Fig. 4a), that is, when $f'' < 0$; and when $f''$ is identically zero (Fig. 4b) then $f$ is linear, so the trapezoid rule gives the exact area, and the remainder is zero. This suggests (cor-

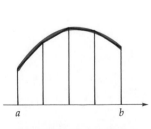

(a) When $f'' < 0$ then $\int_a^b f(x)\, dx > T_n$.

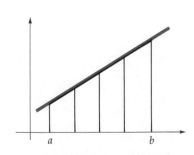

(b) When $f'' \equiv 0$ then $\int_a^b f(x)\, dx = T_n$.

**FIGURE 4**

rectly) that $f''$ tells a lot about the remainder. In fact, the Appendix to this section shows that the remainder can be estimated using the *maximum* of $|f''|$:

$$|R(T_n)| = \left| \int_a^b f(x)\, dx - T_n \right| \leq \max_{[a,b]} |f''(x)| \frac{(b-a)^3}{12n^2}. \tag{3}$$

**EXAMPLE 2**   Estimate the remainder $R(T_5)$ for the integral $\int_1^2 \frac{dx}{x}$ in Example 1.

*SOLUTION*   The integrand is $f(x) = 1/x$, and the second derivative is $f''(x) = 2x^{-3}$. This is a positive decreasing function on the interval $[1, 2]$, so the maximum of its absolute value occurs at the left endpoint $x = 1$:

$$\max_{[1,2]} |f''| = \max_{[1,2]} |2x^{-3}| = 2.$$

Since $b - a = 2 - 1 = 1$, you get from (3)

$$|R(T_5)| \leq 2 \cdot \frac{1^3}{12 \cdot 5^2} = \frac{1}{150} \approx 0.007. \tag{4}$$

This shows that $T_5$ is accurate to about two decimal places. In this case we can verify the accuracy from the known value of the integral $\int_1^2 \frac{dx}{x} = \ln 2 = .693147 \ldots$. The remainder after making the approximation $T_5 = 0.69563 \ldots$ is

$$R(T_5) = \text{true value} - \text{approximation } T_5 = .693147 \ldots - .69563 \ldots$$
$$= -.0025 \ldots,$$

which is in the range given by (4).

### Simpson's Rule

Trapezoid sums replace sections of the graph by straight lines; Simpson sums use parabolas, the graphs of quadratics. To derive the appropriate formula, we first compute the integral of a quadratic $Q(x) = ax^2 + bx + c$ passing through three given points $(-h, y_-)$, $(0, y_0)$, and $(h, y_+)$ as in Figure 5. The integral of $Q(x)$ is

$$\int_{-h}^h (ax^2 + bx + c)\, dx = \frac{a}{3}x^3 + \frac{b}{2}x^2 + cx \Big|_{-h}^h$$
$$= \frac{2ah^3}{3} + 2ch. \tag{5}$$

We need to express this in terms of the three heights $y_-$, $y_0$, and $y_+$ (Fig. 5). The graph of $Q$ passes through $(-h, y_-)$, $(0, y_0)$, and $(h, y_+)$, so

$$\left. \begin{aligned} y_- &= Q(-h) = ah^2 - bh + c \\ y_0 &= Q(0) = c \\ y_+ &= Q(h) = ah^2 + bh + c. \end{aligned} \right\} \tag{6}$$

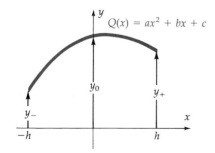

**FIGURE 5**

$\int_{-h}^h Q(x)\, dx = \dfrac{h}{3}(y_- + 4y_0 + y_+).$

Equations (6) give $c = y_0$, and adding the expressions for $y_-$ and $y_+$ leads to

$$2ah^2 = y_- + y_+ - 2c = y_- - 2y_0 + y_+.$$

From (5) then,

$$\int_{-h}^{h} Q(x)\, dx = 2ah^2 \cdot \frac{h}{3} + 2ch$$

$$= (y_- - 2y_0 + y_+)\frac{h}{3} + 2y_0 h$$

$$= \frac{1}{3} h(y_- + 4y_0 + y_+).$$

Now, to approximate $\int_a^b f(x)\, dx$, divide the interval $[a, b]$ into an even *number, n,* of subintervals of length $\Delta x = (b - a)/n$ (Fig. 6). Use one quadratic in the first two subintervals, another in the next two, and so on. Comparing Figures 5 and 6, you see that the integral of the first approximating quadratic is

$$\tfrac{1}{3}\Delta x(y_0 + 4y_1 + y_2).$$

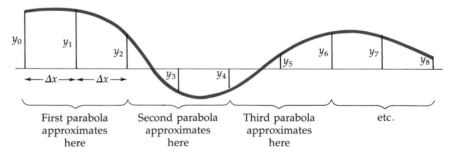

**FIGURE 6**
Simpson's scheme.

Add to this the approximations for the other intervals, and you get the formula for the Simpson sums, called **Simpson's Rule**:

$$SR_n = \tfrac{1}{3}\Delta x[(y_0 + 4y_1 + y_2) + (y_2 + 4y_3 + y_4) + \cdots + (y_{n-2} + 4y_{n-1} + y_n)]$$

or

$$SR_n = \tfrac{1}{3}\Delta x(y_0 + 4y_1 + 2y_2 + 4y_3 + \cdots + 2y_{n-2} + 4y_{n-1} + y_n), \quad (7)$$

where

$$y_0 = f(a),\ y_1 = f(a + \Delta x),\ y_2 = f(a + 2\,\Delta x),\ \ldots,\ y_n = f(b).$$

We denote the remainder in this approximation by

$$R(SR_n) = \int_a^b f(x)\, dx - SR_n.$$

It can be estimated using the *fourth* derivative:

$$\left|R(SR_n)\right| \le \max_{[a,b]} \left|f^{(4)}(x)\right| \cdot \frac{(b-a)^5}{180 \, n^4}. \tag{8}$$

(See the Appendix to this section.)

---

**EXAMPLE 3**   Approximate $\int_1^2 \frac{dx}{x}$ by $SR_4$, and estimate the remainder.

*SOLUTION*   Apply (7): $\Delta x = \dfrac{b-a}{4} = \dfrac{1}{4}$ and

$$y_0 = 1, \qquad y_1 = \tfrac{4}{5}, \qquad y_2 = \tfrac{2}{3}, \qquad y_3 = \tfrac{4}{7}, \qquad y_4 = \tfrac{1}{2}.$$

So the Simpson sum is

$$SR_4 = \tfrac{1}{12}\left(1 + (4)(\tfrac{4}{5}) + (2)(\tfrac{2}{3}) + 4(\tfrac{4}{7}) + (\tfrac{1}{2})\right) = .69325 \ldots .$$

Now estimate the remainder. The fourth derivative $f^{(4)}(x) = 24x^{-5}$ is positive and decreasing for $1 \le x \le 2$, so the maximum of its absolute value on $[1, 2]$ is at $x = 1$:

$$\max_{[1,2]} \left|24x^{-5}\right| = 24.$$

Hence from (8)

$$\left|R(SR_4)\right| \le 24 \cdot \frac{1^5}{(180) \cdot (4^4)} \approx 0.00052.$$

So $SR_4$ is accurate to three decimals. In Example 2, we found for the same integral that $T_5$ was accurate to only two decimals; this illustrates the greater accuracy of Simpson's Rule.

---

Realistically, we often face the problem of approximating an integral to a *preassigned accuracy*; then we have to decide how many terms are needed in the approximating sum. As an illustration:

---

**EXAMPLE 4**   Determine the number $n$ of subdivisions required to approximate $\int_1^2 \frac{dx}{x}$ by Simpson's Rule, with a remainder known to be less than $10^{-6}$.

*SOLUTION*   From Example 3, $\max\limits_{[1,2]} \left|f^{(4)}\right| = 24$, so

$$\left|R(SR_n)\right| \le 24 \cdot \frac{1^5}{180n^4} = \frac{2}{15n^4}.$$

Since $\left|R(SR_n)\right|$ is to be less than $10^{-6}$, we seek an $n$ such that

$$\frac{2}{15n^4} < 10^{-6} \quad \text{or} \quad n^4 > \frac{2 \times 10^6}{15} \approx 1.3 \times 10^5.$$

Clearly, we need $n$ greater than 10 (since $10^4 < 1.3 \times 10^5$), but not greater than 20 (since $20^4 = 2^4 \cdot 10^4 = 16 \times 10^4 > 1.3 \times 10^5$). The calculator shows that $18^4 \approx 1.04 \times 10^5$, so 18 is too small. Since'$n$ must be *even* for a Simpson sum, we settle on $n = 20$.

**Modified trapezoid sums** $MT_n$ are even more accurate than Simpson's. They consist of the trapezoid sums *plus* a correction term using the first derivative of the integrand at the endpoints of the interval of integration:

$$MT_n = T_n - \frac{(\Delta x)^2}{12} [f'(b) - f'(a)].$$

The trapezoid sum $T_n$ approximates $f$ in each subinterval by a first degree polynomial matching $f$ at each end of the subinterval (Fig. 1). The modified trapezoid sum $MT_n$ approximates $f$ with a *cubic* polynomial matching $f$ *and* $f'$ at each end of the subinterval (Fig. 7). This makes sense algebraically. A cubic polynomial

$$p(x) = a + bx + cx^2 + dx^3 \qquad\qquad (9)$$

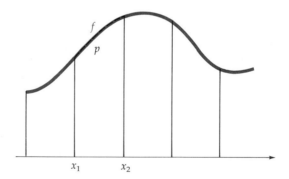

**FIGURE 7**
$p(x)$ is a cubic approximating $f$ in $[x_1, x_2]$; $p(x_1) = f(x_1)$, $p'(x_1) = f'(x_1)$; $p(x_2) = f(x_2)$, $p'(x_2) = f'(x_2)$.

has four coefficients $a$, $b$, $c$, $d$; and matching both $f$ and $f'$ at each of two points gives a total of four conditions to be satisfied.

To derive the formula for $MT_n$, we first approximate $\int_0^h f(x)\, dx$ by the integral of a cubic $p(x)$ matching $f$ and $f'$ at $x = 0$ and $x = h$:

$$f(0) = p(0), \qquad f'(0) = p'(0), \qquad f(h) = p(h), \qquad f'(h) = p'(h).$$

From these equations, you can solve for the coefficients $a$, $b$, $c$, $d$, in (9), finding

$$a = f(0), \qquad b = f'(0),$$
$$c = 3h^{-2}[f(h) - f(0)] - h^{-1}[f'(h) + 2f'(0)]$$
$$d = -2h^{-3}[f(h) - f(0)] + h^{-2}[f'(h) + f'(0)].$$

With these coefficients, the approximating integral works out to

$$\int_0^h p(x)\,dx = ah + \tfrac{1}{2}bh^2 + \tfrac{1}{3}ch^3 + \tfrac{1}{4}dh^4$$
$$= \tfrac{1}{2}h[f(0) + f(h)] + \tfrac{1}{12}h^2[f'(0) - f'(h)]. \tag{10}$$

To simplify the notation, set $f(0) = y_0$, $f'(0) = y_0'$, $f(h) = y_1$, and $f'(h) = y_1'$; and denote the length of the interval, $h$, by $\Delta x$. Then (10) is rewritten as

$$\tfrac{1}{2}\Delta x[y_0 + y_1] + \tfrac{1}{12}(\Delta x)^2[y_0' - y_1']. \tag{11}$$

In this form, the formula gives our approximation to the integral of $f$ over *any* interval of length $\Delta x$.

Now, to approximate $\int_a^b f(x)\,dx$, divide $[a, b]$ into $n$ equal parts of length $\Delta x = \dfrac{b - a}{n}$. Denote the values of $f$ and $f'$ at the ends of these subintervals by

$$y_0, y_1, \ldots, y_n \quad \text{and} \quad y_0', y_1', \ldots, y_n'.$$

Apply (11) on each subinterval, and add the results:

$$\begin{aligned}
\int_a^b f(x)\,dx &\approx \tfrac{1}{2}\Delta x(y_0 + y_1) + \tfrac{1}{12}(\Delta x)^2(y_0' - y_1') \\
&\quad + \tfrac{1}{2}\Delta x(y_1 + y_2) + \tfrac{1}{12}(\Delta x)^2(y_1' - y_2') + \cdots \\
&\quad + \tfrac{1}{2}\Delta x(y_{n-1} + y_n) + \tfrac{1}{12}(\Delta x)^2(y_{n-1}' - y_n') \\
&= \Delta x[\tfrac{1}{2}y_0 + y_1 + \cdots + \tfrac{1}{2}y_n] + \tfrac{1}{12}(\Delta x)^2(y_0' - y_n').
\end{aligned}$$

The sum in brackets times $\Delta x$ is precisely the trapezoid sum $T_n$; and $y_0' = f'(a)$, $y_n' = f'(b)$, so the modified trapezoid sum is

$$MT_n = T_n + \tfrac{1}{12}(\Delta x)^2[f'(a) - f'(b)]. \tag{12}$$

The remainder $R(MT_n)$ for this approximation is

$$|R(MT_n)| \le \max_{[a,b]} |f^{(4)}(x)| \frac{(b - a)^5}{720n^4}, \tag{13}$$

just one-fourth of the remainder estimate for the Simpson sums. This is proved in the Appendix to this section.

---

**EXAMPLE 5**   We approximate $\ln 2 = \displaystyle\int_1^2 \frac{1}{x}\,dx$ using $MT_n$ with 20 subdivisions. In this example, $f'(a) = f'(1) = -1$, and $f'(b) = f'(2) = -1/4$. Formula (12) with $n = 20$ gives

$$\begin{aligned}
MT_n &= \frac{1}{20}\left[\frac{1}{2}\cdot 1 + \frac{1}{1.05} + \frac{1}{1.1} + \cdots + \frac{1}{1.95} + \frac{1}{2}\cdot\frac{1}{2}\right] \\
&\quad + \frac{1}{12}\left(\frac{1}{20}\right)^2\left[-1 + \frac{1}{4}\right] \\
&= 0.693303381 - 0.00015625 = 0.693147131
\end{aligned}$$

to nine places. The remainder estimate uses $f^{(4)}(x) = 24x^{-5}$; its maximum over the interval $[1, 2]$ is 24, so

$$|R(MT_n)| \leq 24 \frac{1}{(720)(20)^4} = 0.000083333.$$

Thus accuracy is guaranteed nearly to four decimal places. (The calculator gives

$$\ln 2 \approx 0.69314718,$$

so the approximation is in fact accurate to seven decimal places, much greater than the guarantee!)

---

## Round-off Error

In these computations, each term of the sum was replaced by a finite decimal expansion, which is itself an approximation. If each of 20 terms is approximated to 8 decimal places, say, then each may have a "round-off" error of $5 \times 10^{-9}$. These errors cannot be expected to cancel each other, so the sum may inherit a round-off error of $20 \times 5 \times 10^{-9} = 10^{-7}$. This must be taken into account, particularly for sums with many terms.

## Choice of Method

For an explicitly given function, it is generally easy to compute the derivatives $f'(a)$ and $f'(b)$, so the modified trapezoid sums are efficient. For tabular data, these derivatives are usually unavailable; then you would use either the trapezoid sums (for rough data) or Simpson sums (for more precise results).

---

## SUMMARY

*Trapezoid Sums:*

$$\int_a^b f(x)\, dx \approx T_n = \Delta x(\tfrac{1}{2}y_0 + y_1 + y_2 + \cdots + y_{n-1} + \tfrac{1}{2}y_n)$$

$$\Delta x = \frac{b-a}{n}, \qquad y_0 = f(a), \qquad y_j = f(a + j\,\Delta x), \qquad y_n = f(b).$$

*Remainder Estimate:*

$$|R(T_n)| = \left| \int_a^b f(x)\, dx - T_n \right| \leq \max_{[a,b]} |f''| \cdot \frac{(b-a)^3}{12n^2}.$$

*Simpson's Rule:*

$$\int_a^b f(x)\, dx \approx SR_n = \frac{\Delta x}{3}(y_0 + 4y_1 + 2y_2 + 4y_3 + 2y_4 + \cdots + 4y_{n-1} + y_n).$$

$\Delta x,\ y_0, \ldots, y_n$ as above; $n$ must be an *even* number.

*Remainder Estimate:*

$$|R(SR_n)| = \left| \int_a^b f(x)\, dx - SR_n \right| \leq \max_{[a,b]} |f^{(4)}| \frac{(b-a)^5}{180n^4}.$$

*Modified Trapezoid Sum:*

$$\int_a^b f(x)\, dx \approx MT_n = T_n + \frac{(\Delta x)^2}{12}[f'(a) - f'(b)]$$

*Remainder Estimate:*

$$\left| R(MT_n) \right| = \left| \int_a^b f(x)\, dx - MT_n \right| \leq \max_{[a,b]} \left| f^{(4)} \right| \frac{(b-a)^5}{720n^4}.$$

## PROBLEMS

### A

1. Approximate the following, using the Trapezoid Rule with the given value of $n$.

   *a) $\displaystyle\int_1^3 \sqrt{4 + x^3}\, dx, \; n = 4$

   b) $\displaystyle\int_0^1 \frac{1}{1 + x^2}\, dx, \; n = 6$

*2. Approximate the integrals in problem 1 using Simpson's Rule, with the same value of $n$ as before. (Answer for part a.)

3. Estimate $\ln 3 = \displaystyle\int_1^3 \frac{dx}{x}$ using

   *a)  $T_4$.        b)  $T_{10}$.
   *c)  $SR_4$.       d)  $SR_{10}$.

*4. Estimate the remainders in problem 3. (Answers for parts a and c.)

5. Approximate the arc length of the following curves, using $SR_4$:
   a)  $y = x^3, 0 \leq x \leq 1$.

   b)  $y = \dfrac{1}{x}, 1 \leq x \leq 3$.

   *c)  $y = \sin x, 0 \leq x \leq \pi/2$.

*6. The hourly temperatures for Alphton for the morning of January 1, 1982, were

| Time (AM) | 6 | 7 | 8 | 9 | 10 | 11 | 12 |
|-----------|----|----|----|----|----|----|----|
| Temp | 10 | 11 | 13 | 15 | 18 | 21 | 23 |

   Estimate the average morning temperature by the Trapezoid Rule.

7. A corrugated sheet is to be made, 100 cm by 100 cm, with cross section $y = 5 \sin(\pi x/5), 0 \leq x \leq 100$. Approximate the surface area of the sheet, using $SR_4$ for an appropriate integral on $0 \leq x \leq 1.25$. (The accuracy is difficult to estimate, because the derivatives are unwieldy.)

8. In approximating $\ln 3 = \displaystyle\int_1^3 \frac{dx}{x}$

   *a)  Determine $n$ so that $\left| R(T_n) \right| < 5 \times 10^{-4}$.

   b)  Determine $n$ so that $\left| R(SR_n) \right| < 5 \times 10^{-4}$.

   *c)  Determine $n$ so that $\left| R(MT_n) \right| < 5 \times 10^{-4}$.

   d)  Compute $\ln 3$ to three decimal places (with an error $< 5 \times 10^{-4}$).

9. With a normal probability density, the probability of a measurement within one standard deviation from the mean is $\dfrac{1}{\sqrt{2\pi}} \displaystyle\int_{-1}^1 e^{-x^2/2}\, dx = $

   $\sqrt{2/\pi} \displaystyle\int_0^1 e^{-x^2/2}\, dx$. For most applications, it is enough to know this to two decimals, that is, with an error $< 5 \times 10^{-3}$. Determine a value of $n$ which guarantees that the last integral is evaluated to this accuracy, and then make the approximation, using
   a)  Trapezoid sums $T_n$.
   b)  Simpson sums $SR_n$.
   c)  Modified trapezoid sums $MT_n$.

10. a)  Evaluate $\displaystyle\int_0^1 \frac{dx}{1 + x^2}$ exactly.

   b)  Approximate the integral with $MT_{10}$, and thus approximate $\pi$.

   c)  Estimate the remainder. [For $f(x) = \dfrac{1}{1 + x^2}$, you should find

   $$f^{(4)}(x) = 24 \cdot \frac{5x^4 - 10x^2 + 1}{(1 + x^2)^5},$$

   and $\left| f^{(4)}(x) \right| \leq 24$ for $0 \leq x \leq 1$.]

**11.** In approximating $\dfrac{\pi}{4} = \displaystyle\int_0^1 \dfrac{dx}{1 + x^2}$, determine
$n$ so that

*a) $\left|R(T_n)\right| < 5 \times 10^{-9}$.

b) $\left|R(SR_n)\right| < 5 \times 10^{-9}$. (See problem 10 for $f^{(4)}$.)

c) $\left|R(MT_n)\right| < 5 \times 10^{-9}$.

**B**

**12.** Let $f(x) = Ax^2 + Bx + C$.

a) Show by direct calculation that

$$\int_a^{a+h} f(x)\, dx - T_1 = -2A \frac{h^3}{12}.$$

b) Show that in this case, the remainder estimate in the summary is exactly equal to the absolute value of the remainder.

**13.** From a sketch, show that $R(T_n) < 0$ when $f$ is concave up $[f'' > 0]$.

**14.** a) Explain geometrically why $\int_a^b f(x)\, dx = T_n$ exactly, if $f$ is linear.

b) Use (3) to show again that $\int_a^b f(x)\, dx = T_n$ if $f$ is linear.

**15.** Show that if $f$ is a polynomial of degree $\leq 3$, then $\int_a^b f(x)\, dx = SR_n = MT_n$ exactly.

**16.** Determine coefficients $a$, $b$, $c$, $d$ such that $p(x) = a + bx + cx^2 + dx^3$ satisfies the four given equations

$$p(0) = y_0, \qquad p'(0) = y'_0,$$
$$p(h) = y_1, \qquad p'(h) = y'_1.$$

**C**

**17.** Problem 12 shows that for every *quadratic* $f$, the difference $\int_a^{a+h} f(x)\, dx - T_1$ depends only on the length of the interval, $h$, not on its endpoint $a$. Prove that this is true *only* for quadratics. That is, if

$$\int_a^{a+h} f(x)\, dx - h\frac{f(a+h) + f(a)}{2} = \varphi(h) \quad (*)$$

is independent of $a$, then $f$ is quadratic. [Hint: Differentiate (∗) twice with respect to $h$, then once with respect to $a$. Assume that $f''$ exists.]

## 10.5

### APPENDIX: PROVING THE REMAINDER ESTIMATES

**Estimating the trapezoid remainder** $R(T_n)$ is difficult, but not impossible. Denote by $R(h)$ the remainder in approximating $\int_0^h f(x)\, dx$ by a *single* trapezoid (Fig. 1):

$$R(h) = \int_0^h f(x)\, dx - \frac{h}{2}[f(h) + f(0)]. \tag{1}$$

We aim to relate this to the maximum of $\left|f''(x)\right|$ for $0 \leq x \leq h$. To do that, consider a "growing remainder" $R(t)$, where $t$ increases from 0 to $h$; speci-

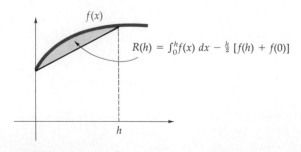

**FIGURE 1**

fically, consider

$$R(t) = \int_0^t f(x)\, dx - \frac{t}{2}[f(t) + f(0)].$$

We compute $R'$, then $R''$, which turns out to be closely related to $f''$.

The first derivative $R'$ is found from the Second Fundamental Theorem of Calculus:

$$R'(t) = f(t) - \frac{1}{2}[f(t) + f(0)] - \frac{t}{2}f'(t)$$

$$= \frac{1}{2}[f(t) - f(0)] - \frac{t}{2}f'(t). \tag{2}$$

Differentiate once more:

$$R''(t) = -\frac{t}{2}f''(t). \tag{3}$$

Thus, knowing the maximum of $|f''|$ tells something about $R''$, and hence about $R$ itself. Specifically, suppose that $|f''(t)| \le M$ for $0 \le t \le h$. Then also $|-f''(t)| \le M$, which is the same as

$$-M \le -f''(t) \le M \qquad \text{for } 0 \le t \le h.$$

Multiply each term by $t/2$. Since $t/2 \ge 0$, this preserves the inequalities:

$$-\frac{t}{2}M \le -\frac{t}{2}f''(t) \le \frac{t}{2}M,$$

and so from (3)

$$-\frac{t}{2}M \le R''(t) \le \frac{t}{2}M, \qquad 0 \le t \le h.$$

Integrate this from 0 to $x$, where $0 \le x \le h$;

$$-\frac{x^2}{4}M \le \int_0^x R''(t)\, dt \le \frac{x^2}{4}M, \qquad 0 \le x \le h. \tag{4}$$

By the First Fundamental Theorem of Calculus, $\int_0^x R''(t)\, dt = R'(x) - R'(0)$; and $R'(0) = 0$ from (2), so $\int_0^x R''(t)\, dt = R'(x)$. Therefore (4) gives

$$-\frac{x^2}{4}M \le R'(x) \le \frac{x^2}{4}M, \qquad 0 \le x \le h. \tag{5}$$

Integrate once more, from 0 to $h$; you get

$$-\frac{h^3}{12}M \le R(h) \le \frac{h^3}{12}M. \tag{6}$$

This is the remainder in approximating by *one* trapezoid of width $h$.

For a general trapezoid sum $T_n$, there are $n$ trapezoids of width $h = (b - a)/n$. Let $M = \max\limits_{[a,b]} |f''(x)|$. Then

$$-M \leq -f''(x) \leq M$$

on the entire interval $[a, b]$, so (6) is true for the remainder in approximating by each of the $n$ trapezoids. The total remainder $R(T_n)$ is the sum of these $n$ remainders, so

$$-n\frac{h^3}{12} M \leq R(T_n) \leq n\frac{h^3}{12} M.$$

Since $h = (b - a)/n$, this gives

$$-M \cdot \frac{(b - a)^3}{12n^2} \leq R(T_n) \leq M \cdot \frac{(b - a)^3}{12n^2}, \tag{7}$$

which is equivalent to $|R(T_n)| \leq M\dfrac{(b - a)^3}{12n^2}$. Since $M = \max\limits_{[a,b]} |f''|$, this is precisely the remainder estimate stated in Section 10.5.

### Estimating the Remainder $R(SR_n)$

This is just a little longer than the proof for the trapezoid remainder. Let $R(h)$ be the difference between $\int_{-h}^{h} f$ and its approximation $SR_2$:

$$R(h) = \int_{-h}^{h} f(x)\, dx - \frac{h}{3}[f(-h) + 4f(0) + f(h)].$$

Then (by what theorem?)

$$R'(t) = \frac{2}{3}[f(t) + f(-t) - 2f(0)] - \frac{t}{3}[f'(t) - f'(-t)] \tag{8}$$

$$R''(t) = \frac{1}{3}[f'(t) - f'(-t)] - \frac{t}{3}[f''(t) + f''(-t)] \tag{9}$$

$$R'''(t) = -\frac{t}{3}[f'''(t) - f'''(-t)].$$

By the Mean Value Theorem

$$f'''(t) - f'''(-t) = (t - (-t))f^{(4)}(\hat{t}) = 2tf^{(4)}(\hat{t})$$

with $-t < \hat{t} < t$. Hence

$$R'''(t) = -\frac{2t^2}{3} f^{(4)}(\hat{t}).$$

If $-M \leq -f^{(4)} \leq M$, then

$$-\frac{2t^2}{3} M \leq R'''(t) \leq \frac{2t^2}{3} M.$$

Integrate from 0 to $x$, noting from (9) that $R''(0) = 0$:

$$-\frac{2x^3}{9}M \le R''(x) \le \frac{2x^3}{9}M.$$

Integrate from 0 to $t$, noting from (8) that $R'(0) = 0$:

$$-\frac{t^4}{18}M \le R'(t) \le \frac{t^4}{18}M.$$

Finally, integrate from 0 to $h$:

$$-\frac{h^5}{90}M \le R(h) \le \frac{h^5}{90}M.$$

This estimates the remainder with one parabola approximating in two intervals of width $h$. With $n/2$ parabolas approximating in $n$ intervals of width $(b-a)/n$, the remainder $R(SR_n)$ satisfies

$$-\frac{n}{2}\left(\frac{b-a}{n}\right)^5\frac{M}{90} \le R(SR_n) \le \frac{n}{2}\left(\frac{b-a}{n}\right)^5\frac{M}{90}$$

or

$$-\frac{(b-a)^5}{180n^4}M \le R(SR_n) \le \frac{(b-a)^5}{180n^4}M, \tag{10}$$

where now $M$ is the maximum of $\left|f^{(4)}\right|$ on $[a, b]$. This is the inequality (8) in Sec. 10.5.

## PROBLEMS

### A

1.  Show that for a quadratic $f(x) = Ax^2 + Bx + C$,

$$\int_a^b f(x)\,dx - T_n = -2A\frac{(b-a)^3}{12n^2}$$

$$= -f''(x)\frac{(b-a)^3}{12n^2}.$$

So in this case, $\left|R(T_n)\right| = \left|\max f''\right|(b-a)^3/12n^2$.

2.  a)  Prove that $SR_n = \int_a^b f(x)\,dx$ if $f$ is a cubic polynomial. [Use (10).]
    b)  Show that if $f$ is a polynomial of degree four, then $f^{(4)}$ is a constant, and $R(SR_n) = -f^{(4)}\frac{(b-a)^5}{180n^4}$ exactly.

3.  The *Midpoint Rule* approximates $\int_a^b f(x)\,dx$ by its midpoint sum

$$MS_n = \Delta x \sum_{j=1}^n f\left(a - \frac{\Delta x}{2} + j\,\Delta x\right),$$

$$\Delta x = \frac{b-a}{n}.$$

a)  Show that if $m \le f''(x) \le M$, then the remainder $R(MS_n) = \int_a^b f(x)\,dx - MS_n$ satisfies

$$m\frac{(b-a)^3}{24n^2} \le R(MS_n) \le M\frac{(b-a)^3}{24n^2}.$$

(half as large as the trapezoid remainder).

b)  Find a function for which this remainder estimate is an equality, not an inequality.

4. This problem sketches a proof of the remainder estimate for the modified trapezoid rule. Define the "error"

$$E(h) = \int_0^h f(x)\, dx - \frac{h}{2}[f(0) + f(h)]$$

$$+ \frac{h^2}{12}[f'(0) - f'(h)];$$

this gives the difference between the integral and $MT_1$. Show that

a) $E''(h) = \frac{h^2}{12} f^{(4)}(h)$.

b) $E(0) = E'(0) = E''(0) = 0$.

c) If $m_4 \le f^{(4)}(x) \le M_4$ for $0 \le x \le h$, then

$$\frac{h^5}{720} m_4 \le E(h) \le \frac{h^5}{720} M_4.$$

d) If $m_4 \le f^{(4)}(x) \le M_4$ for $a \le x \le b$, then

$$\frac{(b - a)^5}{720 n^4} m_4 \le \int_a^b f(x)\, dx - MT_n$$

$$\le \frac{(b - a)^5}{720 n^4} M_4.$$

# REVIEW PROBLEMS    CHAPTER 10

1. Denote by $C$ the graph of $y = e^x$, $0 \le x \le 1$. Set up an integral for
   a) The area below $C$, and above the $x$-axis.
   b) The arc length of $C$.
   c) The area of the surface formed by revolving $C$ about the $x$-axis. (On a sketch, indicate the appropriate radius $r$.)
   d) The area of the surface formed by revolving $C$ about the $y$-axis. (On a sketch, indicate the appropriate $r$.)

2. Evaluate the integral in
   a) Problem 1a.
   *b) Problem 1b (Substitute $u^2 = 1 + e^{2x}$).
   *c) Problem 1c.

*3. Compute the centroid of the region in the first quadrant lying inside the ellipse $\dfrac{x^2}{a^2} + \dfrac{y^2}{b^2} = 1$.

4. a) Show that the function in Figure 1 is a probability density function.

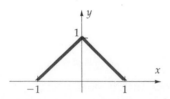

FIGURE 1

*b) Compute its mean and variance.

*5. A probability density is defined by $f(x) = kxe^{-ax}$, $x \ge 0$, with $f(x) = 0$ for $x < 0$; $a$ and $k$ are positive constants. Determine.
   a) The constant $k$    b) $P(0 \le x \le 1)$

   c) The mean $\mu_f$    d) $P(0 \le x \le \mu_f)$
   e) The variance

6. Estimate $\ln(1.5) = \displaystyle\int_1^{1.5} \frac{dx}{x}$ with
   *a) The trapezoid sum $T_4$.
   b) The Simpson sum $SR_4$.
   c) The modified trapezoid sum $MT_4$.

7a*, b*, c. Estimate the error in the corresponding part of the previous problem.

8. How large need $n$ be to approximate $\displaystyle\int_1^{1.5} \frac{dx}{x}$ within $\dfrac{1}{2} \times 10^{-6}$, using
   *a) $T_n$.    b) $SR_n$.    c) $MT_n$.

9. This problem concerns setting up an integral in a new situation. The *demand function* $Q(p)$ tells what quantity of a certain commodity can be sold at price $p$. Thus $Q(p) - Q(p + \Delta p)$ is the number of items that could be sold at price $p$ to people who would not buy at a price $> p + \Delta p$. Suppose that it can be arranged to sell all items in this fashion, selling no items to any buyers at less than the maximum they are willing to pay.
   a) Estimate the revenue from those items sold at a price between $p$ and $p + \Delta p$.
   b) Derive $\displaystyle\int_0^\infty -p \frac{dQ}{dp}\, dp$ as the total revenue earned by selling in this way.
   c) Show that this integral is $\int_0^\infty Q\, dp$, assuming that $\lim_{p \to \infty} pQ(p) = 0$.

# 11

INFINITE SEQUENCES
AND SERIES

The many applications of logarithms, exponentials, and trigonometric functions suggest a fundamental question: How can you actually compute these functions? In practice, you use tables or a calculator. But the tables were not found chiseled in stone on a mountain top, like the Ten Commandments; and calculators are not produced by mysterious elves. The secret of the tables, and of the calculators, lies in efficient, controlled approximations. An early example is the approximation of logarithms by polynomials devised by Gerhard Mercator in 1666. (To construct his famous map, he had to compute $\ln(\sec \theta + \tan \theta)$; see Sec. 9.3). About thirty years later, Johann Bernoulli discovered how to approximate *all* the basic functions of calculus by polynomials. These approximations are now called Taylor polynomials (Oh, fickle fortune!), in honor of Brook Taylor, who seems to have come across them another twenty years later.

Careful work with these approximations leads naturally to a rigorous definition of the *limit of an infinite sequence* of numbers, and also to the powerful technique called *infinite series*.

## 11.1

### TAYLOR POLYNOMIALS

The Taylor polynomials for a function $f$ are a refinement on the idea of the tangent line; they approximate $f(x)$ very well for $x$ near a chosen *base point* $a$. To simplify notation, we begin by taking the base point $a = 0$.

The equation of the tangent line at $(0, f(0))$ is $y = f(0) + f'(0) \cdot x$; the line has $y$-intercept $f(0)$ and slope $f'(0)$, matching the slope of the graph of $f$ at the point of tangency (Fig. 1). This gives the formula for the first degree Taylor polynomial, denoted $P_1(x)$:

$$P_1(x) = f(0) + f'(0) \cdot x. \tag{1}$$

So the graph of $P_1$ is precisely the tangent line; $P_1(x)$ gives the *linear approximation* to $f$, based at 0.

The second degree Taylor polynomial $P_2(x)$ takes into account not only the function value $f(0)$ and the slope $f'(0)$, but also $f''(0)$, the *rate of change*

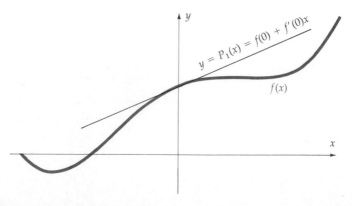

**FIGURE 1**
First degree Taylor polynomial $P_1(x)$ gives linear approximation to $f(x)$.

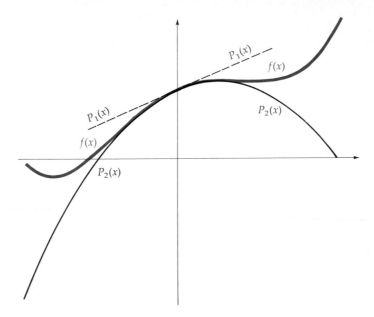

**FIGURE 2**

Approximating $f(x)$ with $P_2(x) = f(0) + f'(0)x + \dfrac{f''(0)}{2} x^2$.

of the slope at 0; it is defined so that $P_2''(0) = f''(0)$. The formula for $P_2$ is

$$P_2(x) = f(0) + f'(0) \cdot x + \frac{f''(0)}{2} x^2. \tag{2}$$

The graph of this quadratic is a parabola, which is tangent to the graph of $f$ (Fig. 2) since it has the same function value and slope at $x = 0$: $P_2(0) = f(0)$, and

$$P_2'(x) = f'(0) + f''(0) \cdot x, \quad \text{so} \quad P_2'(0) = f'(0).$$

Differentiate once more, and you find

$$P_2''(x) = f''(0), \quad \text{so} \quad P_2''(0) = f''(0).$$

Thus at the base point 0, the second degree Taylor polynomial agrees with $f$ in *function value, slope,* and *second derivative.*

The third degree Taylor polynomial is

$$P_3(x) = f(0) + f'(0) \cdot x + \frac{f''(0)}{2} \cdot x^2 + \frac{f'''(0)}{3 \cdot 2} \cdot x^3, \tag{3}$$

and it matches the first three derivatives of $f$ at the base point zero. It is clear from (3) that $P_3(0) = f(0)$; and

$$P_3'(x) = f'(0) + f''(0) \cdot x + \frac{f'''(0)}{2} x^2, \quad \text{so} \quad P_3'(0) = f'(0),$$

$$P_3''(x) = f''(0) + f'''(0)x, \quad \text{so} \quad P_3''(0) = f''(0),$$

$$P_3'''(x) = f'''(0), \quad \text{so} \quad P_3'''(0) = f'''(0),$$

**EXAMPLE 1** Find Taylor polynomials $P_1$, $P_2$, $P_3$ approximating $f(x) = e^x$ with the base point $a = 0$.

**SOLUTION**

$$f(x) = e^x, \qquad f'(x) = e^x, \quad \text{and} \quad f''(x) = f'''(x) = e^x,$$

so

$$f(0) = f'(0) = f''(0) = f'''(0) = e^0 = 1.$$

Hence from (1)–(3),

$$P_1(x) = 1 + x, \qquad P_2(x) = 1 + x + \tfrac{1}{2}x^2,$$

$$P_3(x) = 1 + x + \frac{1}{2}x^2 + \frac{1}{3 \cdot 2}x^3.$$

Figure 3 shows the function $f$ and the approximating polynomials.

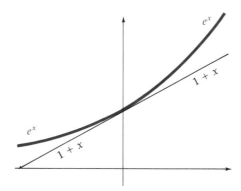

**FIGURE 3a**
$e^x$ and $P_1(x) = 1 + x$.

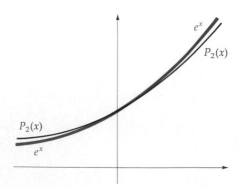

**FIGURE 3b**
$e^x$ and $P_2(x) = 1 + x + x^2/2$.

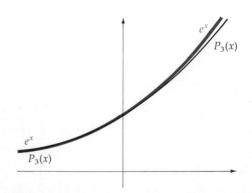

**FIGURE 3c**
$e^x$ and $P_3(x) = 1 + x + x^2/2 + x^3/6$.

The **Taylor polynomial of degree** $n$ is constructed so that at the base point, the *first* $n$ derivatives of $P_n$ agree with the first $n$ derivatives of $f$. With base point 0,

$$P_n(0) = f(0), \; P_n'(0) = f'(0), \; P_n''(0) = f''(0), \ldots, P_n^{(n)}(0) = f^{(n)}(0),$$

where $f^{(n)}$ denotes the $n^{\text{th}}$ derivative of $f$. To write a formula for $P_n$, we need *factorial notation*, $n!$ (read "$n$ factorial"):

$$0! = 1, \qquad 1! = 1, \qquad 2! = 2 \cdot 1, \qquad 3! = 3 \cdot 2 \cdot 1, \ldots$$

$$n! = n \cdot (n-1) \cdot \cdots \cdot (2) \cdot (1).$$

Briefly, $n!$ is the product of all integers from 1 through $n$, for $n \geq 1$; and $0! = 1$, by definition.

With this notation, the $n^{\text{th}}$ degree Taylor polynomial of $f$, based at 0, is

$$P_n(x) = f(0) + f'(0) \cdot x + \frac{f''(0)}{2} \cdot x^2 + \cdots + \frac{f^{(n)}(0)}{n!} \cdot x^n$$

$$= \sum_{k=0}^{n} \frac{f^{(k)}(0)}{k!} x^k. \tag{4}$$

(These Taylor polynomials with base point 0 are also called *Maclaurin polynomials*; everybody gets into the act!)

---

**EXAMPLE 2**   Write the Taylor polynomials $P_n$ for $f(x) = e^x$, based at 0.

*SOLUTION*   For this function, the derivatives are all the same: $f^{(n)}(x) = e^x$, so $f^{(n)}(0) = 1$, for all $n$. Hence from (4)

$$P_n(x) = 1 + x + \frac{1}{2} x^2 + \cdots + \frac{1}{n!} x^n.$$

---

To use these approximations effectively, we need to estimate the *difference* between the polynomials $P_n$ and the approximated function $f$. Consider first the approximation by $P_0(x) = f(0)$. Then the difference $f(x) - P_0(x)$ is simply $f(x) - f(0)$. By the Mean Value Theorem (Fig. 4), there is a number $\hat{x}$ somewhere between 0 and $x$ such that

$$\frac{f(x) - f(0)}{x} = f'(\hat{x}), \quad \text{or} \quad f(x) - f(0) = f'(\hat{x}) \cdot x.$$

Thus

$$f(x) - P_0(x) = f(x) - f(0) = f'(\hat{x}) \cdot x \tag{5}$$

for some $\hat{x}$ between 0 and $x$.

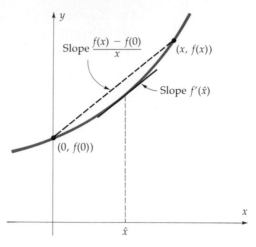

**FIGURE 4**
$$\frac{f(x) - f(0)}{x} = f'(\hat{x}).$$

For the $n^{\text{th}}$ degree Taylor polynomial, there is a similar formula (proved in the Appendix to this section):

$$f(x) - P_n(x) = \frac{f^{(n+1)}(\hat{x})}{(n+1)!} x^{n+1}, \tag{6}$$

where again $\hat{x}$ is somewhere between the base point 0 and the evaluation point $x$. The right-hand side in (6) is easy to remember, for it is almost the same as the $x^{n+1}$ term in the Taylor polynomial; the difference is that in (6), $f^{(n+1)}$ is evaluated at an unspecified point $\hat{x}$, not at the base point 0.

Although $\hat{x}$ is not known exactly, we can nonetheless use (6) to *estimate* the difference $f(x) - P_n(x)$. Denote by $M_{n+1}$ the *maximum* of $\left| f^{(n+1)}(\hat{x}) \right|$ for *all* $\hat{x}$ between 0 and $x$. Then

$$\left| f(x) - P_n(x) \right| = \frac{\left| f^{(n+1)}(\hat{x}) \right|}{(n+1)!} |x|^{n+1} \leq \frac{M_{n+1}}{(n+1)!} |x|^{n+1}. \tag{7}$$

This is the basic estimate for determining the accuracy of the approximation $P_n(x)$.

---

**EXAMPLE 3** Approximate $e$, using the polynomial $P_4$ in Example 2, and estimate the error.

*SOLUTION* In Example 2, $f(x) = e^x$, so $e = e^1 = f(1)$. We approximate this by

$$P_4(1) = 1 + 1 + \frac{1}{2!} + \frac{1}{3!} + \frac{1}{4!} = 2.70833 \ldots . \tag{8}$$

The error is estimated by (7), with $n = 4$:

$$\left|e^1 - P_4(1)\right| \leq \frac{M_5}{5!} \cdot 1^5.$$

The constant $M_5$ is the maximum of $\left|f^{(5)}(\hat{x})\right| = \left|e^{\hat{x}}\right| = e^x$, for $\hat{x}$ between 0 and 1. This maximum is $e^1$ (since $e^{\hat{x}}$ increases as $\hat{x}$ goes from 0 to 1) so $M_5 = e^1 = e < 3$; and then

$$\left|e^1 - P_4(1)\right| \leq \frac{e}{5!} < \frac{3}{5!} = \frac{1}{40} = .025.$$

So the sum (8) approximates $e$ to one decimal place.

---

**REMARK**  Two numbers $a$ and $b$ are said to agree to one decimal place if they differ by less than $0.05 = \frac{1}{2} \times 10^{-1}$. For instance, 3.1 agrees with $\pi$ to one decimal place, since $\left|\pi - 3.1\right| < 0.05$. This is expressed by the suggestive notation $\pi = 3.1 \pm 0.05$. More generally, two numbers $a$ and $b$ agree to $k$ decimal places if $\left|a - b\right| < \frac{1}{2} \times 10^{-k}$. Each digit in the $k^{\text{th}}$ decimal place is worth $10^{-k}$; we require the difference to be less than half of that.

---

**EXAMPLE 4**  Approximate $e$ to five decimal places, using the polynomials $P_n$ in Example 2.

**SOLUTION**  For accuracy to five decimal places, the error must be less than $\frac{1}{2} \times 10^{-5}$; so first determine a value of $n$ such that $\left|e^1 - P_n(1)\right| < \frac{1}{2} \times 10^{-5}$. Use the basic estimate (7):

$$\left|e^1 - P_n(1)\right| \leq \frac{M_{n+1}}{(n+1)!} 1^{n+1} = \frac{e}{(n+1)!} < \frac{3}{(n+1)!} \tag{9}$$

since $M_{n+1} = \max_{[0,1]} \left|e^x\right| = e^1 < 3$, just as in Example 3. By trial and error you find

$$\frac{3}{9!} = 0.000008 \ldots \approx 8 \times 10^{-6} > 5 \times 10^{-6} = \frac{1}{2} \times 10^{-5};$$

but

$$\frac{3}{10!} = 0.0000008 \ldots < 5 \times 10^{-6} = \frac{1}{2} \times 10^{-5},$$

sufficient for five decimal accuracy. So in equation (9), take $n + 1 = 10$, thus $n = 9$, and approximate $e$ by

$$P_9(1) = 1 + 1 + \frac{1}{2!} + \cdots + \frac{1}{9!}$$

$$= 2.7182815 \ldots .$$

So to five decimal places, $e \approx 2.71828$.

### Taylor Polynomials at a General Base Point $a$

The first degree polynomial based at $a$ is precisely the linear approximation to $f$ based at $a$:

$$P_1(x) = f(a) + f'(a)(x - a).$$

The graph of $P_1$ is the tangent line at $(a, f(a))$, the best possible approximation to the graph near that point by a straight line (Fig. 5).

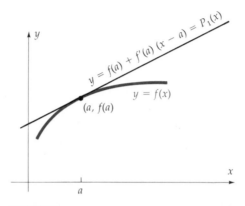

**FIGURE 5**
$P_1(x)$ gives the tangent line at $[a, f(a)]$.

The Taylor polynomial of degree $n$ is

$$P_n(x) = f(a) + f'(a)(x - a) + \frac{f''(a)}{2}(x - a)^2 + \cdots + \frac{f^{(n)}(a)}{n!}(x - a)^n. \tag{10}$$

The difference between the given function $f(x)$ and this approximation is expressed by the **remainder formula**

$$f(x) - P_n(x) = \frac{f^{(n+1)}(\hat{x})}{(n+1)!}(x - a)^{n+1} \tag{11}$$

with $\hat{x}$ between $a$ and $x$. This implies the **remainder estimate**

$$\left| f(x) - P_n(x) \right| \le \frac{M_{n+1}}{(n+1)!} \left| x - a \right|^{n+1} \tag{12}$$

where $M_{n+1}$ denotes the maximum of $\left| f^{(n+1)}(\hat{x}) \right|$ for all $\hat{x}$ between $a$ and $x$.

**REMARK**  The remainder $f(x) - P_n(x)$ can be expressed in various useful ways; the particular expression in (11) is called the **Lagrange form** of the remainder.

Our next (and last) example gives the polynomials used by Mercator.

**EXAMPLE 5**

   **a.** Compute the Taylor polynomial $P_3(x)$ for $f(x) = \ln(x)$, based at $x = 1$.

   **b.** Use it to approximate $\ln(1.1)$.

   **c.** Estimate the error in the approximation.

*SOLUTION*

   **a.** $P_3$, based at 1, uses formula (10) with $n = 3$ and $a = 1$; so you need $f(1)$, $f'(1)$, $f''(1)$, and $f'''(1)$:

$$f(x) = \ln(x) \qquad f(1) = 0$$
$$f'(x) = x^{-1}, \qquad f'(1) = 1$$
$$f''(x) = -x^{-2}, \qquad f''(1) = -1$$
$$f'''(x) = 2x^{-3}, \qquad f'''(1) = 2.$$

So

$$P_3(x) = f(1) + f'(1)(x - 1) + \frac{f''(1)}{2}(x - 1)^2 + \frac{f'''(1)}{3 \cdot 2}(x - 1)^3$$

$$= 0 + 1(x - 1) + \frac{-1}{2}(x - 1)^2 + \frac{2}{3 \cdot 2}(x - 1)^3,$$

$$P_3(x) = (x - 1) - \tfrac{1}{2}(x - 1)^2 + \tfrac{1}{3}(x - 1)^3.$$

[Fig. 6 shows $\ln x$, $P_1(x)$, $P_2(x)$, and $P_3(x)$.]

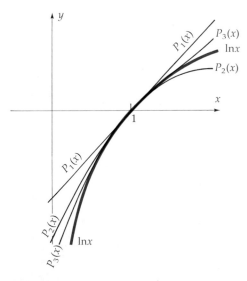

**FIGURE 6**
$\ln x$ together with $P_1$, $P_2$, $P_3$ based at 1.

   **b.** Since $P_3(x)$ approximates $\ln x$, we approximate $\ln(1.1)$ by

$$P_3(1.1) = (1.1 - 1) - \tfrac{1}{2}(1.1 - 1)^2 + \tfrac{1}{3}(1.1 - 1)^3$$
$$= 0.1 - 0.005 + 0.000333 \ldots = 0.0953333 \ldots . \qquad \text{(13)}$$

**c.** Estimate the error, using (12) with $n = 3$. Then $M_{n+1} = M_4$ is the maximum of $\left|f^{(4)}(\hat{x})\right|$, for $\hat{x}$ between the base point $a = 1$ and the evaluation point $x = 1.1$. In the present case

$$\left|f^{(4)}(\hat{x})\right| = \left|-6\hat{x}^{-4}\right| = 6\hat{x}^{-4}$$

is decreasing as $\hat{x}$ varies from 1 to 1.1, so its maximum is at $x = 1$:

$$M_4 = \max_{[1, \, 1.1]} 6\hat{x}^{-4} = 6.$$

So by the remainder estimate (12)

$$\left|\ln(1.1) - P_3(1.1)\right| \leq \frac{6}{4 \cdot 3 \cdot 2}\left|1.1 - 1\right|^4 = 0.000025.$$

This guarantees that the approximation (13) is accurate to four decimal places. [In fact, a calculator gives $\ln(1.1) = 0.09531 \ldots$.]

---

*REMARKS ON EXAMPLE 5* The main factor making the error small is $\left|1.1 - 1\right|^4 = 10^{-4}$. This is small because the evaluation point 1.1 is close to the base point 1. Generally, Taylor polynomials are accurate *close* to the base point.

We chose the base point $a = 1$ because, for the function $f(x) = \ln x$, that is where we already know $f$ and its derivatives with perfect accuracy.

How, then would you compute $\ln x$ with $x$ far from 1? Exploit the basic algebraic property of the logarithm: $\ln(ab) = \ln(a) + \ln(b)$. If $\ln(1.1)$ is known with great accuracy, then so is $2\ln(1.1) = \ln((1.1)^2) = \ln(1.21)$, and likewise $3\ln(1.1) = \ln(1.331)$, and so on. With such algebraic tricks, the computation of $\ln x$ in general is reduced to computations near $x = 1$.

## SUMMARY

*Factorial Notation:* $0! = 1$, $1! = 1$, $\ldots$, $n! = n(n-1)\ldots(2)(1)$.

*Taylor Polynomial of f Based at a:*

$$P_n(x) = \sum_{k=0}^{n} \frac{f^{(k)}(a)}{k!}(x - a)^k$$

*Taylor Formula with Lagrange Remainder:*

$$f(x) = P_n(x) + \frac{f^{(n+1)}(\hat{x})}{(n+1)!}(x - a)^{n+1}, \qquad \hat{x} \text{ between } a \text{ and } x$$

*Remainder Estimate:*

$$\left|f(x) - P_n(x)\right| \leq M_{n+1}\frac{\left|x - a\right|^{n+1}}{(n+1)!}$$

where $M_{n+1} = $ maximum of $\left|f^{(n+1)}(\hat{x})\right|$ for $\hat{x}$ between $a$ and $x$.

# PROBLEMS

## A

*1. Figure 7 shows $\cos(x)$, $-2\pi \leq x \leq 2\pi$, together with the Taylor polynomials $P_2(x)$, $P_4(x)$, and $P_6(x)$. Compute those polynomials.

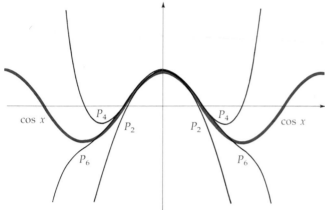

**FIGURE 7**
$\cos x$ and $P_2$, $P_4$, $P_6$.

2. Write the Taylor polynomial $P_4(x)$, based at 0, for
   *a) $f(x) = e^{2x}$
   b) $f(x) = \sin x$
   *c) $f(x) = \ln(1 - x)$

3. In the corresponding part of problem 2, carefully graph the function $f$, and the Taylor polynomials $P_1$, $P_2$, $P_3$. (Use a graphing computer program, if available.)

4. Why is $|e^x| = e^x$? (This was used in Example 3.)

5. Approximate the following, using the Taylor polynomial $P_4$ based at 0, and estimate the error.
   *a) $e^{1/2}$   b) $e^{0.1}$   *c) $e^{-1}$

6. Compute $e^{0.1}$ to eight decimal places, and prove the accuracy.

7. Make the following approximations by Taylor polynomials based at 0, and estimate the error.
   *a) $\sin(1)$ by $P_4(1)$
   b) $\cos(1)$ by $P_5(1)$
   *c) $\sin(0.1)$ by $P_4(0.1)$
   d) $\cos(0.1)$ by $P_5(0.1)$

8. Show that the fourth-degree Taylor polynomial $P_4(x)$, based at $a$, satisfies $P_4(a) = f(a)$, $P_4'(a) = f'(a)$, $P_4''(a) = f''(a)$, $P_4'''(a) = f'''(a)$, $P_4^{(4)}(a) = f^{(4)}(a)$.

9. For $f(x) = \sin x$, compute the following Taylor polynomials based at $\pi/6$.
   *a) $P_2$   b) $P_4$   c) $P_n$

10. Let $P_n(x)$ be the $n^{\text{th}}$ Taylor polynomial of $\sin x$ based at 0. (Fig. 8 shows three of these.) Show that
    a) $P_2 = P_1$, $P_4 = P_3$, $P_6 = P_5$, etc.
    b) For $|x| < \pi/4$, $|\sin x - P_5(x)| < 0.000042$. [Hence, a four-place table of sines for angles $\leq \pi/4$ can be constructed using $P_5(x) = x - x^3/6 + x^5/120$. Hint: In estimating the error, you can safely assume that $\pi < 3.2$, so $\pi/4 < 0.8$.]

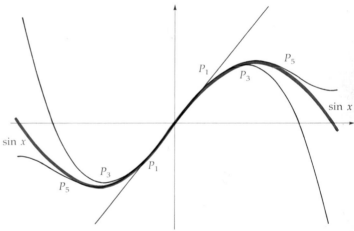

**FIGURE 8**
$\sin x$ and $P_1$, $P_3$, $P_5$.

*11. a) Approximate $\ln(0.9)$, using $P_4(x)$ based at $x = 1$.
    b) Estimate the error in the approximation.

*12. Use an appropriate Taylor polynomial to approximate $\ln(1.1)$ to six decimals.

13. In this problem, you approximate $\sqrt{10}$. Use the function $f(x) = \sqrt{x}$ and the base point 9, since $\sqrt{x}$ and its derivatives are easily evaluated at $x = 9$.
    a) Compute $P_4(x)$.
    b) Use $P_4(x)$ to approximate $\sqrt{10}$.
    c) Estimate the error in the approximation.

14. Continuing the previous problem,
    a) Determine $n$ so that $P_n(10)$ approximates $\sqrt{10}$ with an error $< \frac{1}{2} \times 10^{-8}$.
    b) Make the approximation.

**B**

15. For $f(x) = \int_0^x e^{-t^2/2}\, dt$:
    a) Form the Taylor polynomial $P_3$ based at 0.
    b) Estimate the error $f(x) - P_3(x)$ for $|x| \le \frac{1}{2}$. [To estimate $f^{(4)}(\hat{x})$, show that $|3x - x^3| \le \frac{11}{8}$ for $|x| \le \frac{1}{2}$.]

    *REMARK* This is the integral evaluated in tables of the normal probability distribution. It is difficult to carry the approximation much further with Taylor polynomials; but the method of Taylor *series*, later in this chapter, makes it easy.

**C**

16. This problem shows that $e$ is an irrational number.

    a) Show that $0 < e - \left(1 + 1 + \cdots + \dfrac{1}{n!}\right) <$

    $\dfrac{e}{(n+1)!} < \dfrac{3}{(n+1)!}$.

    b) Supposing that $e = p/q$ with $p$ and $q$ positive integers, show that $q \ge 2$.
    c) Supposing that $e = p/q$, show that

    $$q!\left[\frac{p}{q} - \left(1 + 1 + \frac{1}{2} + \cdots + \frac{1}{q!}\right)\right] \text{ is an}$$

    *integer*, greater than 0, and less than 1.

17. Suppose that $f'' < 0$ throughout an interval $I$. Prove that on $I$, the graph of $f$ lies *below* each of its tangent lines. [Use $P_1(x)$ based at $a$, with remainder.]

18. Suppose that $f$ is a polynomial of degree $n$, and $P_n$ is its $n^{\text{th}}$-degree Taylor polynomial based at any point $a$.
    a) Use the Lagrange remainder to show that $f(x) = P_n(x)$ for all $x$.
    b) Deduce that $f(x) = f(a) + (x - a)Q(x)$ for a polynomial $Q$ of degree $< n$.
    c) For the $Q$ in part b, express $Q(a)$ and $Q'(a)$ in terms of $f$ and its derivatives.

19. Suppose $f'(a) = 0$, $f''(a) = 0$, but $f'''(x) > 0$ in an open interval $I$ containing $a$. Show that $f(a)$ is neither a maximum nor a minimum of $f$ on $I$.

---

## 11.1

### APPENDIX: THE REMAINDER FOR TAYLOR POLYNOMIALS

We derive two expressions for the difference $f(x) - P_n(x)$ between $f$ and its $n^{\text{th}}$ Taylor polynomial. We assume that $f, f', f'', \ldots, f^{(n+1)}$ are all continuous. For $n = 0$, the difference is

$$f(x) - P_0(x) = f(x) - f(a).$$

From the Mean Value Theorem, $f(x) - f(a) = f'(\hat{x})(x - a)$ for some $\hat{x}$ between $a$ and $x$. This gives

$$f(x) - P_0(x) = f'(\hat{x})(x - a), \tag{1}$$

which is the *Lagrange form* of the remainder. On the other hand, from the Fundamental Theorem of Calculus, $f(x) - f(a) = \int_a^x f'(t)\, dt$, so

$$f(x) - P_0(x) = \int_a^x f'(t)\, dt. \tag{2}$$

This is called the **integral form** of the remainder.

For $P_n$, we first derive the integral form of the remainder, then the Lagrange form. We begin with (2) and integrate by parts, differentiating $u = f'$ and integrating $dv = dt$. But to get our formula, we don't take the obvious $v = t$, but rather $v = t - x$!

$$f(x) - f(a) = f'(t)(t - x)\Big|_a^x - \int_a^x f''(t)(t - x) \, dt$$

$$= -f'(a)(a - x) + \int_a^x f''(t)(x - t) \, dt.$$

Transpose to find

$$f(x) - [f(a) + f'(a)(x - a)] = \int_a^x f''(t)(x - t) \, dt. \tag{3}$$

The left-hand side is precisely $f(x) - P_1(x)$; the right-hand side gives the integral form of the remainder for $f - P_1$. Continue integrating by parts, and you find the *integral form of the remainder for* $P_n$:

$$f(x) - P_n(x) = \int_a^x f^{(n+1)}(t) \frac{(x - t)^n}{n!} \, dt. \tag{4}$$

From this we deduce the more easily remembered Lagrange form. To simplify notation, take $x > a$. Denote the *minimum* of $f^{(n+1)}(t)$ on $a \le t \le x$ by $m$, and the *maximum* by $M$; then

$$m \le f^{(n+1)}(t) \le M \qquad \text{for } a \le t \le x.$$

So

$$m \frac{(x - t)^n}{n!} \le f^{(n+1)}(t) \frac{(x - t)^n}{n!} \le M \frac{(x - t)^n}{n!},$$

and hence

$$\int_a^x m \frac{(x - t)^n}{n!} \, dt \le \int_a^x f^{(n+1)}(t) \frac{(x - t)^n}{n!} \, dt \le \int_a^x M \frac{(x - t)^n}{n!} \, dt.$$

The two extreme terms are easily integrated, and the middle term is given in (4), so you find

$$m \frac{(x - a)^{n+1}}{(n + 1)!} \le f(x) - P_n(x) \le M \frac{(x - a)^{n+1}}{(n + 1)!}.$$

Multiply by the positive number $(n + 1)!/(x - a)^{n+1}$:

$$m \le \frac{(n + 1)!}{(x - a)^{n+1}} [f(x) - P_n(x)] \le M. \tag{5}$$

Now, $m$ and $M$ are the *minimum* and *maximum* of $f^{(n+1)}$ on the interval $[a, x]$. Since $f^{(n+1)}$ is assumed *continuous*, and the middle term in (5) lies between the minimum and maximum of $f^{(n+1)}$ on that interval, there must be a point $\hat{x}$ somewhere in $[a, x]$ where $f^{(n+1)}(\hat{x})$ is exactly *equal* to that middle term:

$$f^{(n+1)}(\hat{x}) = \frac{(n + 1)!}{(x - a)^{n+1}} [f(x) - P_n(x)]. \tag{6}$$

This gives the Lagrange form of the remainder:

$$f(x) - P_n(x) = \frac{f^{(n+1)}(\hat{x})}{(n+1)!}(x - a)^{n+1}, \qquad \hat{x} \text{ between } a \text{ and } x.$$

## PROBLEMS

**B**

1. a) Integrate (3) by parts to obtain the integral form of the remainder for $f(x) - P_2(x)$.
   b) Repeat, for $f(x) - P_3(x)$.
   c) Prove (4) by induction.

2. What theorem was used in passing from (5) to (6)?

**C**

3. [Error estimates for Newton's method (Sec. 2.6.)] Let $\bar{x}$ be a solution of $f(\bar{x}) = 0$, and $x_n = x_{n-1} - \dfrac{f(x_{n-1})}{f'(x_{n-1})}$ be its $n^{\text{th}}$ approximation by Newton's method. Show that
   a) $0 = f(\bar{x}) = f(x_n) + (\bar{x} - x_n)f'(\hat{x})$ for some $\hat{x}$ between $\bar{x}$ and $x_n$.
   b) If $|f'| \geq m_1$ between $\bar{x}$ and $x_n$, then $|\bar{x} - x_n| \leq |f(x_n)|/m_1$.
   c) $0 = f(x_n) + (\bar{x} - x_n)f'(x_n) + \frac{1}{2}(\bar{x} - x_n)^2 f''(\hat{x})$ for some $\hat{x}$ between $\bar{x}$ and $x_n$.
   d) $x_{n+1} - \bar{x} = \dfrac{f''(\hat{x})}{2f'(x_n)}(\bar{x} - x_n)^2.$

   e) If $|f''| \leq M_2$ and $|f'| \geq m_1$ in an interval containing $\bar{x}$, $x_{n-1}$, and $x_n$, then $|x_{n+1} - \bar{x}| < (M_2/2m_1)(\bar{x} - x_n)^2 < (M_2/2m_1^3)f(x_n)^2$, and, shifting indices,
   $$|x_n - \bar{x}| < (M_2/2m_1^3)f(x_{n-1})^2.$$

   **REMARK** Frequently, one computes $x_1, x_2, \ldots, x_n, x_{n+1}$, until the difference $x_{n+1} - x_n$ is negligible. The estimate e) gives a *guaranteed* estimate for $x_n - \bar{x}$, using $f(x_{n-1})$, which was already computed to get $x_n$.

4. a) Show that $x^3 + x + 1 = 0$ has precisely one solution $\bar{x}$.
   b) Approximate $\bar{x}$ by Newton's method; use $x_0 = -1$ and compute $x_1$ and $x_2$.
   c) Estimate the error in $x_2$, using the previous problem. (Let $M_2 = \max_{[-1,0]} |f''|$ and $m_1 = \min_{[-1,0]} |f'|$.)

## 11.2

## LIMIT OF A SEQUENCE DEFINED

The first sequence encountered in mathematics is the positive integers

$$1, 2, 3, \ldots . \tag{1}$$

Later on, you meet decimal sequences, such as the decimal approximations of $1/3$:

$$0.3, 0.33, 0.333, \ldots . \tag{2}$$

The numbers in this sequence come closer and closer to $1/3$. We say that they *converge* to $1/3$, and that $1/3$ is the *limit* of the sequence.

The main point of this section is to define precisely the concept of limit of a sequence; using this definition, we prove some basic limit theorems.

An infinite sequence of real numbers

$$a_1, a_2, a_3, \ldots, a_n, \ldots \tag{3}$$

is, from the formal point of view, a rule assigning a real number $a_n$ to each positive integer $n$. In other words, it is a *real-valued function whose domain is the positive integers*; the value of the function at the point $n$ is $a_n$. The sequence as a whole can be indicated using dots as in (1)–(3), or within brackets, as $\{a_n\}_1^\infty$, or simply $\{a_n\}$. The sequence of positive integers is denoted $\{n\}_1^\infty$ or $\{n\}$. The decimal sequence in line (2) can be denoted $\{d_n\}$, where $d_n =$ the $n$-place decimal expansion of $1/3$. Other examples:

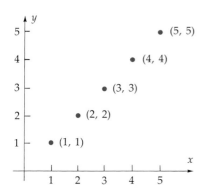

FIGURE 1a
$a_n = n$.

FIGURE 1b
$d_n = n^{\text{th}}$ decimal expansion of $1/3$.

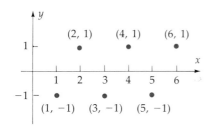

FIGURE 1c
$\{(-1)^n\}_1^\infty$.

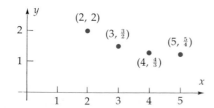

FIGURE 1d
$\left\{\dfrac{n}{n-1}\right\}_2^\infty$.

$$\{(-1)^n\}_1^\infty = -1, 1, -1, 1, \ldots \tag{4}$$

$$\left\{\frac{n}{n-1}\right\}_2^\infty = \frac{2}{1}, \frac{3}{2}, \frac{4}{3}, \ldots \tag{5}$$

The sequence (5) starts with $n = 2$, not $n = 1$, because $\dfrac{n}{n-1}$ is undefined when $n = 1$. We oversimplified in stating that an infinite sequence is defined for all the positive integers; it may start with any integer $n$, not necessarily $n = 1$.

Like any function, a sequence $\{a_n\}$ has a *graph*, consisting of the points $(1, a_1), (2, a_2), (3, a_3), \ldots$. Figure 1 graphs the four sequences given above.

The graphs illustrate different behaviors as $n$ grows larger. The sequence $\{n\}_1^\infty$ increases without bound. The sequence $\{d_n\}$ increases, but converges toward a fixed "limit" $1/3$. The sequence $\{n/(n-1)\}$ decreases toward the limit 1. The sequence $\{(-1)^n\}$ oscillates back and forth between $-1$ and $+1$, and does not converge to any fixed limit; the points on the graph do not approach any one horizontal line as $n$ grows larger.

In calculus, we are interested mainly in sequences that converge to a limit. Often, the limit is some quantity that we want to calculate, and the sequence is a means of calculation, as for example a sequence of Riemann sums is a means of calculating an integral, or a sequence of Taylor polynomials $P_n(x)$ approximates a function $f(x)$. In fact, the idea of limit of a sequence is a cornerstone on which the theory of calculus can be built, so it deserves a precise definition.

The definition of convergence is based on the idea of approximation. In the decimal sequence $\{d_n\}$, the $n^{\text{th}}$ term approximates $1/3$ with an error less than $10^{-n}$:

| | | |
|---|---|---|
| $d_1 = 0.3,$ | $\|d_1 - 1/3\| < 0.1$ | (error $< 10^{-1}$) |
| $d_2 = 0.33,$ | $\|d_2 - 1/3\| < 0.01$ | (error $< 10^{-2}$) |
| $d_3 = 0.333,$ | $\|d_3 - 1/3\| < 0.001$ | (error $< 10^{-3}$) |
| $d_4 = 0.3333,$ | $\|d_4 - 1/3\| < 0.0001$ | (error $< 10^{-4}$) |

and so on.

Different applications will generally require different degrees of accuracy in the approximation; say in one case the error must be no more than $10^{-2}$, or in another it must be no more than $10^{-5}$. Suppose that in a given application, the maximum allowable error is $\varepsilon$ ("epsilon," Greek "e," for error). No matter how small $\varepsilon$ may be, you can approximate 1/3 with an error $< \varepsilon$ simply by taking a term $d_n$ with $n$ sufficiently large. Just how large $n$ must be depends of course on the allowable error $\varepsilon$; when $\varepsilon$ is very small then $n$ may have to be very large.

The definition of limit of a sequence is a precise analytic statement of this idea.

---

### DEFINITION 1

A sequence $\{a_n\}$ converges to $A$ if and only if, for each positive number $\varepsilon$, there is a number $N$ such that

$$n > N \quad \Rightarrow \quad |a_n - A| < \varepsilon. \tag{6}$$

The number $A$ is called the **limit** of the sequence $\{a_n\}$, and we write

$$\lim_{n \to \infty} a_n = A, \quad \text{or} \quad \lim a_n = A, \quad \text{or} \quad a_n \to A.$$

If $\lim a_n = A$ for some number $A$, we say the sequence $\{a_n\}$ *has a limit* or *is* **convergent**. If there is no such limit $A$, the sequence $a_n$ is called **divergent**.

---

Figure 2 illustrates the definition. The inequality $|a_n - A| < \varepsilon$ in (6) means that $a_n$ is within distance $\varepsilon$ of $A$, that is, $a_n$ lies between $A - \varepsilon$ and $A + \varepsilon$. The crucial statement (6) says that all terms with $n > N$ must remain in this interval. On the graph (Fig. 2) the strip between $y = A - \varepsilon$ and $y = A + \varepsilon$ contains the entire "tail" of the sequence, beginning with the term $a_{N+1}$. Briefly, the tail of the sequence remains in an "$\varepsilon$-band" about $A$.

Definition 1 speaks of *the* limit of a sequence, implying that a given sequence $\{a_n\}$ cannot have more than one limit $A$. This is true; if the numbers $a_n$ remain arbitrarily close to a certain number $A$, they cannot at the same

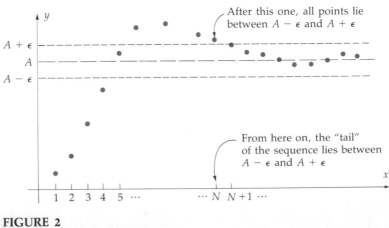

After this one, all points lie between $A - \epsilon$ and $A + \epsilon$

From here on, the "tail" of the sequence lies between $A - \epsilon$ and $A + \epsilon$

**FIGURE 2**

$n > N \quad \Rightarrow \quad |a_n - A| < \varepsilon.$

time remain arbitrarily close to a different number $B$. (A formal proof, using Definition 1, is outlined in the problems.)

Here are two examples applying the definition.

---

**EXAMPLE 1**   Prove that $\lim \dfrac{1}{n} = 0$.

**SOLUTION**   Intuitively, $1/n \to 0$ because $1/n$ is very close to 0 when $n$ is very large. But to *prove* this, you have to show that Definition 1 applies, with $a_n = 1/n$ and $A = 0$. Suppose that an $\varepsilon > 0$ is given, for instance $\varepsilon = 1/10$. Then you have to show that there is an $N$ such that

$$n > N \quad \Rightarrow \quad |a_n - A| < \varepsilon$$

that is

$$n > N \quad \Rightarrow \quad \left| \frac{1}{n} - 0 \right| < 1/10.$$

But $\left| \dfrac{1}{n} - 0 \right| = \left| \dfrac{1}{n} \right| = \dfrac{1}{n}$, so you need an $N$ such that

$$n > N \quad \Rightarrow \quad \frac{1}{n} < \frac{1}{10}.$$

Clearly $N = 10$ will do, since (Fig. 3) $n > 10 \quad \Rightarrow \quad \dfrac{1}{n} < \dfrac{1}{10}$.

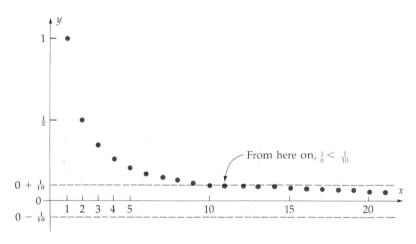

**FIGURE 3**

$$n > 10 \quad \Rightarrow \quad \left| \frac{1}{n} - 0 \right| < \frac{1}{10}.$$

What we did with $\varepsilon = 1/10$ can be done with any $\varepsilon > 0$:

$$n > N \quad \Rightarrow \quad \left| \frac{1}{n} - 0 \right| < \varepsilon$$

is equivalent to

$$n > N \quad \Rightarrow \quad \frac{1}{n} < \varepsilon$$

and this is satisfied with $N = 1/\varepsilon$:

$$n > \frac{1}{\varepsilon} \quad \Rightarrow \quad \frac{1}{n} < \varepsilon.$$

So for each $\varepsilon$, there is a corresponding $N$ (in this case, $N = 1/\varepsilon$) satisfying (6).

---

**EXAMPLE 2**    Guess $\displaystyle\lim_{n \to \infty} \frac{1}{2^n}$, and prove the answer.

*SOLUTION*    The terms of the sequence $\left\{\dfrac{1}{2^n}\right\}$ are

$$\frac{1}{2^1} = \frac{1}{2}, \quad \frac{1}{2^2} = \frac{1}{4}, \quad \frac{1}{2^3} = \frac{1}{8}, \quad \frac{1}{16}, \quad \frac{1}{32}, \ldots$$

They grow smaller, and seem to approach 0 as a limit. To prove this, you must show that for each $\varepsilon > 0$ there is an $N$ such that

$$n > N \quad \Rightarrow \quad \left| \frac{1}{2^n} - 0 \right| < \varepsilon.$$

But $\left| \dfrac{1}{2^n} - 0 \right| = \left| \dfrac{1}{2^n} \right| = \dfrac{1}{2^n}$, so the required inequality is $\dfrac{1}{2^n} < \varepsilon$. This is equivalent to each of the following:

$$1 < 2^n \varepsilon$$

$$\frac{1}{\varepsilon} < 2^n$$

$$\ln\left(\frac{1}{\varepsilon}\right) < n \ln 2 \qquad \text{(because } \ln x \text{ is an increasing function of } x\text{)}$$

$$\frac{\ln(1/\varepsilon)}{\ln 2} < n \qquad \text{(because } \ln 2 > 0\text{)}.$$

So for the desired $N$, take $N = \ln(1/\varepsilon)/\ln 2$, fulfilling condition (6):

$$n > \frac{\ln(1/\varepsilon)}{\ln 2} \quad \Rightarrow \quad \frac{1}{2^n} < \varepsilon \quad \Rightarrow \quad \left| \frac{1}{2^n} - 0 \right| < \varepsilon.$$

---

The definition is the ultimate recourse in settling questions about convergence, but it is clumsy and awkward to use except in simple cases like Examples 1 and 2. Rather than apply it directly in each case, we use it first to prove a few basic limits (for example, $\lim \dfrac{1}{n} = 0$) and then to prove general facts that help to evaluate the more complicated cases. One of these general facts is:

---

**THEOREM 1**

If $\lim\limits_{n \to \infty} a_n = A$, and $c$ is any constant, then

$$\lim_{n \to \infty} ca_n = c \lim_{n \to \infty} a_n = cA.$$

---

**EXAMPLE 3**   Prove that $\lim\limits_{n \to \infty} \dfrac{-1}{n} = 0$.

*SOLUTION*   You could use Definition 1, but Theorem 1 is easier:

$$\lim\left(-\frac{1}{n}\right) = \lim\left[(-1) \cdot \frac{1}{n}\right] \qquad \text{(algebra)}$$

$$= (-1) \lim \frac{1}{n} \qquad \text{(Theorem 1, } c = -1)$$

$$= (-1) \cdot 0 \qquad \text{(Example 1)}$$

$$= 0.$$

---

Figures 4 and 5 illustrate the proof of Theorem 1 for the case where $c = 2$.

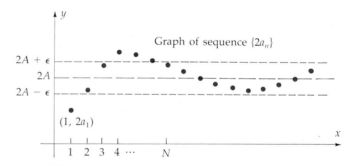

**FIGURE 4**
Wanted: An $N$ such that $\left|2a_n - 2A\right| < \varepsilon$ whenever $n > N$.

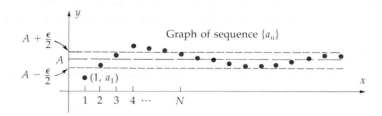

**FIGURE 5**
Since $\lim\limits_{n \to \infty} a_n = A$, there is an $N$ such that $\left|a_n - A\right| < \varepsilon/2$ whenever $n > N$.

For each $\varepsilon > 0$, it must be shown that there is an $N$ such that

$$n > N \quad \Rightarrow \quad \left|2a_n - 2A\right| < \varepsilon$$

as in Figure 4. But $\left|2a_n - 2A\right| = \left|2(a_n - A)\right| = 2\left|a_n - A\right|$, so

$$\left|2a_n - 2A\right| < \varepsilon$$

is equivalent to $2\left|a_n - A\right| < \varepsilon$, or (dividing by 2)

$$\left|a_n - A\right| < \varepsilon/2.$$

Since $\lim_{n \to \infty} a_n = A$ by hypothesis, there *is* a tail of the sequence $\{a_n\}$ which lies entirely in the $\varepsilon/2$-band about $A$ (Fig. 5). That is, there *is* an $N$ such that

$$n > N \quad \Rightarrow \quad \left|a_n - A\right| < \varepsilon/2.$$

For this same $N$,

$$n > N \quad \Rightarrow \quad \left|2a_n - 2A\right| < \varepsilon.$$

This completes the proof, for $c = 2$.

If $c$ is any nonzero constant, the proof is nearly the same. Since $\lim_{n \to \infty} a_n = A$, and $\varepsilon/|c|$ is a positive number, there is an $N$ such that

$$n > N \quad \Rightarrow \quad \left|a_n - A\right| < \varepsilon/|c|. \tag{7}$$

But $\left|a_n - A\right| < \varepsilon/|c|$ is equivalent to $|c| \cdot \left|a_n - A\right| < \varepsilon$, and $|c| \cdot \left|a_n - A\right| = \left|c(a_n - A)\right| = \left|ca_n - cA\right|$, so (7) is equivalent to

$$n > N \quad \Rightarrow \quad \left|ca_n - cA\right| < \varepsilon.$$

This completes the proof of Theorem 1 when $c \neq 0$; and when $c = 0$, the conclusion of the Theorem is obvious.

Similar proofs establish three more algebraic properties: The limit of a sum, product, or quotient is the sum, product or quotient of the limits, provided the quotient has no zero denominator. Stated formally:

---

**THEOREM 2**

If $\{a_n\}$ and $\{b_n\}$ are convergent sequences, then so are $\{a_n + b_n\}$ and $\{a_n b_n\}$, and

$$\lim(a_n + b_n) = \lim a_n + \lim b_n$$

$$\lim(a_n b_n) = (\lim a_n)(\lim b_n).$$

If $\lim b_n \neq 0$, then the quotient $a_n/b_n$ is convergent, and

$$\lim \frac{a_n}{b_n} = \frac{\lim a_n}{\lim b_n}.$$

---

## EXAMPLE 4

$$\lim_{n \to \infty} \frac{1}{n^2} = \lim_{n \to \infty} \frac{1}{n} \cdot \frac{1}{n} = \left( \lim \frac{1}{n} \right)\left( \lim \frac{1}{n} \right) = 0 \cdot 0 = 0$$

$$\lim_{n \to \infty} \left( \frac{1}{n} + \frac{2}{n^2} \right) = \lim \frac{1}{n} + \lim \frac{2}{n^2} = 0 + 2 \lim \frac{1}{n^2} = 2 \cdot 0 = 0.$$

$$\lim_{n \to \infty} \frac{n^2}{n^2 + 1} = \lim \frac{1}{1 + 1/n^2} \qquad \text{(algebra)}$$

$$= \frac{\lim 1}{\lim(1 + 1/n^2)} = \frac{1}{1 + \lim(1/n^2)} = \frac{1}{1 + 0} = 1.$$

Figure 6 illustrates another basic theorem, the "Trapping Theorem." There is an "upper sequence" $\{c_n\}$, a "lower sequence" $\{a_n\}$, and a "middle sequence" $\{b_n\}$ trapped between the two.

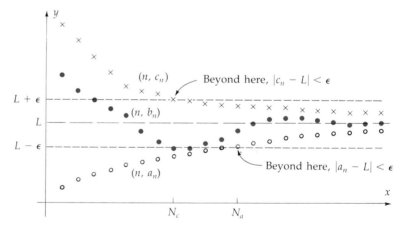

**FIGURE 6**
The Trapping Theorem. $\{a_n\}$ marked $\circ$, $\{b_n\}$ marked $\bullet$, $\{c_n\}$ marked $\times$.

---

### THEOREM 3

If $a_n \leq b_n \leq c_n$, while $\{a_n\}$ and $\{c_n\}$ have a common limit $L$, then $\{b_n\}$ has that same limit $L$.

---

**PROOF**   Suppose that $\varepsilon > 0$. Since $\lim_{n \to \infty} c_n = L$, there is a number $N_c$ such that (Fig. 6)

$$n > N_c \quad \Rightarrow \quad |c_n - L| < \varepsilon. \tag{8}$$

Similarly, there is a number $N_a$ such that

$$n > N_a \quad \Rightarrow \quad |a_n - L| < \varepsilon. \tag{9}$$

Let $N = \max(N_a, N_c)$. If $n > N$ then both $a_n$ and $c_n$ lie in the $\varepsilon$-band about $L$ (Fig. 6), hence $b_n$ does, too. That is,

$$n > N \quad \Rightarrow \quad |b_n - L| < \varepsilon.$$

This completes the proof.

---

**EXAMPLE 5**  $\displaystyle \lim_{n \to \infty} \frac{1}{n^n} = 0$, because

$$0 \le \frac{1}{n^n} \le \frac{1}{n}.$$

The upper sequence $\left\{ \dfrac{1}{n} \right\}$ has limit 0, and so does the lower sequence $\{0\}$.

Hence the middle sequence $\left\{ \dfrac{1}{n^n} \right\}$ has the same limit.

The argument is indicated briefly as follows, using the symbol $\therefore$ to stand for "therefore":

$$0 \le \frac{1}{n^n} \le \frac{1}{n}$$
$$\downarrow \quad \therefore \downarrow \quad \downarrow$$
$$0 \quad\;\; 0 \quad\;\; 0$$

---

**EXAMPLE 6**  $\displaystyle \lim_{n \to \infty} \frac{\cos n}{n} = 0$ because

$$-1 \le \cos n \le 1$$

so

$$-\frac{1}{n} \le \frac{\cos n}{n} \le \frac{1}{n}.$$
$$\downarrow \quad \therefore \downarrow \quad \downarrow$$
$$0 \quad\;\; 0 \quad\;\; 0$$

---

## SUMMARY

*Definition*    $\displaystyle \lim_{n \to \infty} a_n = A$ if and only if, for every $\varepsilon > 0$, there is a number $N$ such that

$$n > N \quad \Rightarrow \quad |a_n - A| < \varepsilon.$$

"$a_n$ *converges to* $A$", or $a_n \to A$, means $\displaystyle \lim_{n \to \infty} a_n = A$.

If $a_n \to A$ and $a_n \to B$, then $A = B$.
$\{a_n\}$ *diverges* means that $a_n$ does *not* converge to any finite number $A$.

**Basic Limits:**

$$\lim_{n \to \infty} 1/n^k = 0, \qquad \text{if } k > 0$$

$$\lim_{n \to \infty} x^n = 0, \qquad \text{if } |x| < 1$$

**Algebra of Limits:**   If $\lim_{n \to \infty} a_n$ and $\lim_{n \to \infty} b_n$ exist, then

$$\lim_{n \to \infty} (ca_n) = c \lim_{n \to \infty} a_n \qquad \text{(constant factor } c)$$

$$\lim_{n \to \infty} (a_n + b_n) = \lim_{n \to \infty} a_n + \lim_{n \to \infty} b_n \qquad \text{(limit of a sum)}$$

$$\lim_{n \to \infty} (a_n b_n) = \left( \lim_{n \to \infty} a_n \right) \left( \lim_{n \to \infty} b_n \right) \qquad \text{(limit of a product)}$$

$$\lim_{n \to \infty} \left( \frac{a_n}{b_n} \right) = \frac{\lim_{n \to \infty} a_n}{\lim_{n \to \infty} b_n} \qquad \text{if } \lim_{n \to \infty} b_n \neq 0. \qquad \text{(limit of a quotient)}$$

**Trapping Theorem:**   If $a_n \leq b_n \leq c_n$, and $\{a_n\}$ and $\{c_n\}$ converge to the same limit $L$, then also $\lim_{n \to \infty} b_n = L$.

## PROBLEMS

### A

**1.** Write out the terms for $n = 1, 2, 10, 100$, and guess the limit of the sequence.

a) $\dfrac{1}{n^2}$    b) $\dfrac{1}{n^3}$    c) $\dfrac{1}{(-n)^3}$

d) $\dfrac{1}{3^n}$    e) $\left( \dfrac{9}{10} \right)^n$    f) $\dfrac{n}{n+1}$

**\*2.** For each part of the previous problem, *prove* that your guess is correct, using the definition of limit directly (and not using any theorems derived from the definition). (Answers given for parts b and e.)

**3.** Write out the terms for $n = 1, 2, 10, 100$, and guess the limit of the sequence.

a) $\dfrac{n^3}{n^3 + 1}$    b) $\dfrac{n}{1 + 2n^2}$

c) $\dfrac{3n^2}{1 + 2n^2}$    d) $\dfrac{1}{n!}$

e) $1 - \dfrac{1}{(-n)^n}$    f) $\dfrac{\sin(2n + 1)}{n^2} - 1$

**\*4.** For each part of the previous problem, *prove* that your guess is correct, using the Theorems and Examples of this section. (Answers given for parts a, c, e.)

**5.** Prove the following limits, using the Trapping Theorem, together with appropriate examples in the text, or in earlier problems.

**\*a)** $\lim_{n \to \infty} \dfrac{\sin n}{n}$    b) $\lim_{n \to \infty} \dfrac{(-1)^n}{n^2}$

**\*c)** $\lim_{n \to \infty} \left( \dfrac{\cos n}{n} \right)^n$    d) $\lim_{n \to \infty} \left( -\dfrac{1}{2} \right)^n$

**\*e)** $\lim_{n \to \infty} \dfrac{1}{3^n}$    f) $\lim_{n \to \infty} \dfrac{-1}{n \cdot 2^n}$

### B

**6.** This problem proves that $\{(-1)^n\}$ is not convergent.

**a)** Show that $\lim(-1)^n = L$ is false, for every $L \geq 0$. (Suppose that $L \geq 0$. Let $\varepsilon = 1/2$. Sketch the sequence, pick any $L \geq 0$, and sketch the $\varepsilon$-band about $L$, for $\varepsilon = 1/2$. Explain why there is *no* number $N$ such that for *every* $n > N$,

$$|(-1)^n - L| < \varepsilon.)$$

**b)** Show that $\lim(-1)^n = L$ is false, for every $L \leq 0$.

**c)** Deduce that $\{(-1)^n\}$ is not convergent.

**7.** Suppose that $\lim\limits_{n \to \infty} a_n = A$.

**a)** Prove that $\lim\limits_{n \to \infty} a_{n+1} = A$, that is, for every $\varepsilon > 0$ there is an $N$ such that

$$n > N \Rightarrow |a_{n+1} - A| < \varepsilon.$$

**b)** Prove that $\lim\limits_{n \to \infty} a_{n-1} = A$.

**8.** Prove: $\lim a_n = 0 \Leftrightarrow \lim |a_n| = 0$.

**C**

**9.** (Limit of a sum.) Suppose that $\lim\limits_{n \to \infty} a_n = A$ and $\lim\limits_{n \to \infty} b_n = B$. Explain why

**a)** For every $\varepsilon > 0$, there is a number $N_a$ such that

$$n > N_a \Rightarrow |a_n - A| < \varepsilon/2$$

and a number $N_b$ such that

$$n > N_b \Rightarrow |b_n - B| < \varepsilon/2.$$

**b)** There is a number $N$ such that for $n > N$, both

$$|a_n - A| < \varepsilon/2 \quad \text{and} \quad |b_n - B| < \varepsilon/2.$$

**c)** For the $N$ in part b,

$$n > N \Rightarrow |(a_n + b_n) - (A + B)| < \varepsilon.$$

Hence, $\lim(a_n + b_n) = A + B$. (Note: $|(a_n + b_n) - (A + B)| = |(a_n - A) + (b_n - B)| \leq |a_n - A| + |b_n - B|$.)

**10.** (Limit of a product.) Suppose that $\lim a_n = A$ and $\lim b_n = B$. Explain why each of the following is true:

**a)** $|a_n b_n - AB| \leq |a_n - A| \cdot |b_n| + |A| \cdot |b_n - B|$.

**b)** Given any $\varepsilon_a > 0$ and $\varepsilon_b > 0$, there are numbers $N_a$ and $N_b$ such that

$$n > N_a \Rightarrow |a_n - A| < \varepsilon_a \quad \text{and}$$
$$n > N_b \Rightarrow |b_n - B| < \varepsilon_b.$$

**c)** If $n > N_b$ then $|b_n| \leq |B| + \varepsilon_b$.

**d)** Given any $\varepsilon > 0$, there is an $\varepsilon_b > 0$ and an $\varepsilon_a > 0$ such that

$$|A|\varepsilon_b < \varepsilon/2 \quad \text{and} \quad (|B| + \varepsilon_b)\varepsilon_a < \varepsilon/2.$$

**e)** Given any $\varepsilon > 0$, there is an $N$ such that

$$n > N \Rightarrow |a_n b_n - AB| < \varepsilon.$$

Hence, $\lim a_n b_n = AB$.

**11.** (Limit of a reciprocal.) Suppose that $\lim b_n = B \neq 0$. Explain why

**a)** There is an $N_b$ such that $n > N_b \Rightarrow |b_n - B| < |B|/2$.

**b)** If $n > N_b$ then $|b_n| > |B|/2$.

**c)** If $n > N_b$ then $\left| \dfrac{1}{b_n} - \dfrac{1}{B} \right| < \dfrac{2|B - b_n|}{|B|^2}$.

$$\left( \text{Put } \frac{1}{b_n} - \frac{1}{B} \text{ on a common denominator.} \right)$$

**d)** Given $\varepsilon > 0$, there is an $N \geq N_b$ such that

$$n > N \Rightarrow |b_n - B| < \frac{\varepsilon}{2} |B|^2 \Rightarrow \left| \frac{1}{b_n} - \frac{1}{B} \right| < \varepsilon.$$

Hence, $\lim \dfrac{1}{b_n} = \dfrac{1}{B}$.

**12.** (Limit of a quotient.) Suppose that $\lim\limits_{n \to \infty} a_n = A$, and $\lim\limits_{n \to \infty} b_n = B \neq 0$. Deduce from the previous two problems that

$$\lim\left( \frac{a_n}{b_n} \right) = \frac{A}{B}.$$

**13.** (Uniqueness of limits.) Suppose that $\lim\limits_{n \to \infty} a_n = L$, for a certain number $L$. Prove that if $M \neq L$, then $\lim\limits_{n \to \infty} a_n = M$ is false. (Hint: In case $M > L$, take an $\varepsilon > 0$ such that $2\varepsilon \leq M - L$, as in Figure 7. Since $a_n \to L$, there is an $N_1$ such that $n > N_1 \Rightarrow |a_n - L| < \varepsilon$. Explain why there is *no* number $N_2$ such that $n > N_2 \Rightarrow |a_n - M| < \varepsilon$. Explain also the case where $M < L$.)

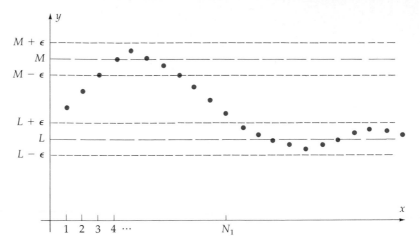

**FIGURE 7**
$|a_n - L| < \varepsilon$ if $n > N_1$.

## 11.3

## MORE ON SEQUENCES

There is a simple relation between *sequence limits* as $n \to \infty$, and *function limits* as $x \to \infty$. Suppose that $f$ is a function for which the limit

$$\lim_{x \to \infty} f(x) = L$$

exists (Fig. 1). This means that as $x \to +\infty$ through all real numbers $x$, the function values $f(x)$ approach the limit $L$. In particular, if we consider $f(n)$ just for the *integer* values $n = 1, 2, 3, \ldots$, then $f(n)$ approaches $L$ as $n \to \infty$. That is:

$$\text{If} \quad \lim_{x \to +\infty} f(x) = L, \quad \text{then} \quad \lim_{n \to \infty} f(n) = L.$$

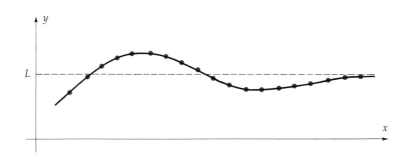

**FIGURE 1**
If $\lim_{x \to \infty} f(x) = L$, then also $\lim_{n \to \infty} f(n) = L$.

In Figure 1, the sequence $\{f(n)\}$ is represented by the dots on the graph; if the whole graph approaches the line $y = L$ as $x \to +\infty$, then so do the dots as $n \to \infty$. The formal proof is left to the problems.

Using this principle, l'Hôpital's Rule can be applied to evaluate certain sequence limits—evaluate $\lim\limits_{x \to +\infty} f(x)$ by l'Hôpital's Rule, and conclude that $\lim\limits_{n \to \infty} f(n)$ is the same.

---

**EXAMPLE 1** Evaluate $\lim\limits_{n \to \infty} \dfrac{1}{n} \ln n$.

**SOLUTION** The function $f(x) = \dfrac{1}{x} \ln x$ has the limit

$$\lim_{x \to +\infty} \frac{\ln x}{x} = \lim_{x \to +\infty} \frac{1/x}{1} = 0 \qquad \text{(l'Hôpital's Rule, since } \frac{\ln x}{x} \text{ is an } \frac{\infty}{\infty}$$
$$\text{form as } x \to +\infty.)$$

Hence $\lim\limits_{n \to \infty} \dfrac{1}{n} \ln n = \lim\limits_{x \to +\infty} \dfrac{1}{x} \ln x = 0$.

A calculator computation suggests the same result:

$$n = 10: \qquad \frac{1}{10} \ln 10 = .230 \ldots$$

$$n = 100: \qquad \frac{1}{100} \ln 100 = .046 \ldots$$

$$n = 10{,}000: \qquad \frac{1}{10{,}000} \ln(10{,}000) = .00092 \ldots.$$

---

Sequence limits can also be evaluated as function limits with $x \to 0+$ (Fig. 2). If $\lim\limits_{x \to 0+} f(x) = L$, then also $\lim\limits_{n \to \infty} f\left(\dfrac{1}{n}\right) = L$; for as $n \to \infty$, then $x = 1/n \to 0^+$.

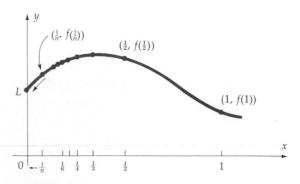

**FIGURE 2**

If $\lim\limits_{x \to 0^+} f(x) = L$, then also $\lim\limits_{n \to \infty} f\left(\dfrac{1}{n}\right) = L$.

**EXAMPLE 2**   Evaluate $\lim\limits_{n \to \infty} n \ln\left(1 - \dfrac{1}{n}\right)$.

*SOLUTION*   Here, it is convenient to replace $1/n$ by $x$, and consider the limit

$$\lim_{x \to 0^+} \left(\frac{1}{x}\right) \ln(1 - x) = \lim_{x \to 0^+} \frac{\ln(1 - x)}{x}$$

$$= \lim_{x \to 0^+} \frac{-1/(1 - x)}{1} = -1 \qquad \text{(l'Hôpital)}.$$

Thus $\lim\limits_{x \to 0^+} \left(\dfrac{1}{x}\right) \ln(1 - x) = -1$. Hence, with $x = 1/n$,

$$\lim_{n \to \infty} n \ln\left(1 - \frac{1}{n}\right) = -1.$$

Another relation concerns a "function of a sequence," $f(a_n)$. Suppose that $\lim\limits_{n \to \infty} a_n = L$, and $f$ is continuous at $L$. Then as $n \to \infty$, the numbers $a_n$ approach $L$ (Fig. 3); since $f$ is continuous at $L$, then $f(a_n)$ approaches $f(L)$ (Fig. 3). Briefly:

$$\lim_{n \to \infty} f(a_n) = f\left(\lim_{n \to \infty} a_n\right) \qquad \text{if } f \text{ is continuous at } \lim_{n \to \infty} a_n. \tag{1}$$

The proof of this requires the formal definition of continuity; it is left to the Appendix.

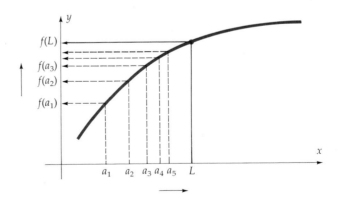

**FIGURE 3**
$f(a_n) \to f(L)$ if $a_n \to L$ and $f$ is continuous at $L$.

**EXAMPLE 3**   Evaluate $\lim\limits_{n \to \infty} \sqrt{\dfrac{n}{n + 1}}$.

*SOLUTION*   The function $f(x) = \sqrt{x}$ is continuous at every point where it is defined, so

$$\lim_{n \to \infty} \sqrt{\frac{n}{n+1}} = \sqrt{\lim_{n \to \infty} \frac{n}{n+1}} \qquad [\lim f(a_n) = f(\lim a_n)]$$

$$= \sqrt{\lim_{n \to \infty} \frac{1}{1 + 1/n}} = \sqrt{1} = 1.$$

The principle (1) justifies the computation of limits by taking logarithms. Suppose that you can evaluate $\lim(\ln a_n)$. Since $a_n = e^{\ln a_n}$, and $e^x$ is a continuous function, then

$$\lim a_n = \lim e^{\ln a_n} \qquad \text{(algebra)}$$
$$= e^{\lim(\ln a_n)} \qquad (e^x \text{ is continuous}).$$

**EXAMPLE 4**   $\displaystyle\lim_{n \to \infty} \left(1 - \frac{1}{n}\right)^n = ?$

*SOLUTION*   This is a problem involving powers, so take logs:

$$\ln\left(1 - \frac{1}{n}\right)^n = n \ln\left(1 - \frac{1}{n}\right).$$

By l'Hôpital's Rule, $\displaystyle\lim_{n \to \infty} n \ln\left(1 - \frac{1}{n}\right) = -1$ (Example 2), so

$$\lim_{n \to \infty} \left(1 - \frac{1}{n}\right)^n = \lim_{n \to \infty} e^{\ln(1 - 1/n)^n} = e^{\lim(\ln(1 - 1/n)^n)} = e^{-1} = 1/e.$$

*WARNING*   You can think of $\displaystyle\lim_{n \to \infty} \left(1 - \frac{1}{n}\right)^n$ as a "$1^\infty$" form. Since $\left(1 - \frac{1}{n}\right)^n < 1$, the infinite exponent suggests that the limit might be 0, while the limiting base 1 suggests the limit 1. *Neither of these is correct!* The actual limit lies somewhere between these naively computed extremes. Indeterminate forms such as $\frac{0}{0}$, $\frac{\infty}{\infty}$, $\infty^0$, $1^\infty$, and $0^0$ must be evaluated carefully, usually using l'Hôpital's Rule.

### Bounded Sequences

The sequences

$$1, 2, 3, \ldots, n, \ldots$$

and

$$-1, 1, -1, 1, \ldots, (-1)^n, \ldots$$

both diverge, but there is an important difference between them; the first one contains arbitrarily large terms, whereas the second one is **bounded**. That is, all the terms of the second sequence have absolute value less than or equal to a fixed constant: $\left|(-1)^n\right| \le 1$, for every $n$.

---

**DEFINITION**

$\{a_n\}$ is *bounded* if there is a number $M$ such that

$$|a_n| \le M \qquad \text{for every } n.$$

Graphically, this means that the sequence stays between $-M$ and $M$ (Fig. 4).

---

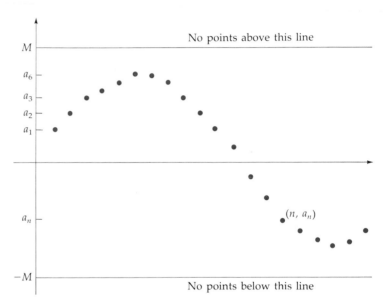

**FIGURE 4**
A bounded sequence: $|a_n| \le M$ means $a_n \le M$ and $a_n \ge -M$.

---

**EXAMPLE 5**   Is the sequence $\{5^n/n!\}$ bounded?

**SOLUTION**   The first few terms of the sequence are

$$5, \quad \frac{5 \cdot 5}{2 \cdot 1} = 12.5, \quad \frac{5 \cdot 5 \cdot 5}{3 \cdot 2 \cdot 1} = 20.8 \ldots, \quad \frac{5 \cdot 5 \cdot 5 \cdot 5}{4 \cdot 3 \cdot 2 \cdot 1} = 26.04 \ldots,$$

$$\frac{5 \cdot 5 \cdot 5 \cdot 5 \cdot 5}{5 \cdot 4 \cdot 3 \cdot 2 \cdot 1} = 26.04 \ldots, \quad \frac{5 \cdot 5 \cdot 5 \cdot 5 \cdot 5 \cdot 5}{6 \cdot 5 \cdot 4 \cdot 3 \cdot 2 \cdot 1} = 21.7 \ldots.$$

The terms $a_n = 5^n/n!$ increase up to $n = 4$, then level off, and apparently decrease thereafter. The reason is that each term $a_n$ equals $5/n$ times the previous term $a_{n-1}$;

$$\frac{5 \cdot 5 \cdot \cdots \cdot 5 \cdot 5}{n \cdot (n-1) \cdots (2)(1)} = \left(\frac{5}{n}\right) \frac{5 \cdot \cdots \cdot 5 \cdot 5}{(n-1) \cdots (2)(1)},$$

that is,

$$a_n = \frac{5}{n} a_{n-1}.$$

When $n < 5$, then $\frac{5}{n} > 1$, so the terms increase up to $n = 5$; and when $n \geq 5$ then $\frac{5}{n} \leq 1$, so the terms decrease:

$$a_n = \frac{5}{n} a_{n-1} \leq a_{n-1}, \qquad \text{if } n \geq 5.$$

It follows that $a_5$ is the largest term:

$$\frac{5^n}{n!} \leq \frac{5^5}{5!} \qquad \text{for every } n. \tag{2}$$

This proves that the sequence $\{5^n/n!\}$ is bounded.

The sequence decreases from the term $a_5$ onward; does it converge to zero? Yes:

$$0 \leq \frac{5^n}{n!} = \frac{5}{n} \cdot \frac{5^{n-1}}{(n-1)!} \leq \frac{5}{n} \frac{5^5}{5!} \qquad \text{[by (2), replacing } n \text{ by } n-1]$$

$$= \left(\frac{5^6}{5!}\right) \frac{1}{n}.$$

But $\left(\dfrac{5^6}{5!}\right) \dfrac{1}{n} \to 0$, so by the Trapping Theorem $\dfrac{5^n}{n!} \to 0$, too.

---

The method used at the end of Example 4 can be stated as a general principle:

> **THEOREM 1**
>
> If $\lim\limits_{n \to \infty} a_n = 0$, and $\{b_n\}$ is a bounded sequence, then $\lim a_n b_n = 0$.

**PROOF**  Since $\{b_n\}$ is bounded, there is a number $M$ such that $|b_n| \leq M$ for every $n$. Suppose that $\varepsilon > 0$ is given. Then also $\varepsilon/M > 0$; since $\lim\limits_{n \to \infty} a_n = 0$, there is an $N$ such that

$$n > N \quad \Rightarrow \quad |a_n - 0| < \varepsilon/M.$$

Further, since $|b_n| \leq M$,

$$|a_n b_n - 0| = |a_n b_n| = |a_n| |b_n| \leq |a_n| \cdot M.$$

Hence

$$n > N \quad \Rightarrow \quad |a_n b_n - 0| \leq |a_n| \cdot M < \frac{\varepsilon}{M} \cdot M = \varepsilon.$$

It follows that $\lim_{n \to \infty} a_n b_n = 0$, as was to be proved.

---

**EXAMPLE 6**   $\lim \dfrac{(-1)^n}{n^2} = 0$, since $\dfrac{(-1)^n}{n^2} = \dfrac{1}{n^2}(-1)^n$, and $\lim_{n \to \infty} \dfrac{1}{n^2} = 0$, while $\{(-1)^n\}$ is bounded.

---

**EXAMPLE 7**   Show that for every number $x$, $\lim_{n \to \infty} \dfrac{x^n}{n!} = 0$.

*SOLUTION*   Follow the pattern of Example 5. The sequence of absolute values

$$\left| \frac{x^n}{n!} \right| = \frac{|x|^n}{n!}$$

increases until $n \geq |x|$, and then decreases, so it is a bounded sequence. Now,

$$\frac{x^n}{n!} = \frac{x}{n} \cdot \frac{x^{n-1}}{(n-1)!},$$

so $x^n/n!$ is the product of a sequence $\left\{ \dfrac{x}{n} \right\}$ that converges to zero, and a bounded sequence $\left\{ \dfrac{x^{n-1}}{(n-1)!} \right\}$. It follows that $x^n/n! \to 0$.

---

We conclude with two theorems for future reference:

---

**THEOREM 2**

Every convergent sequence is bounded.

---

Figure 5 illustrates the proof for the case $A \geq 0$. Suppose that $\lim_{n \to \infty} a_n = A$. Take any $\varepsilon > 0$. Then the tail of the sequence, starting with some term $a_{N+1}$, lies in an $\varepsilon$-band about $A$, so all these terms lie between $A - \varepsilon$ and $A + \varepsilon$; it follows that the absolute values $|a_n|$ are $\leq A + \varepsilon$, for all $n > N$. The remaining absolute values $|a_1|, \ldots, |a_N|$ are finite in number, so there is a largest among them. Hence, for all $n$, $|a_n| \leq M$, where $M$ is the largest of the numbers

$$|a_1|, |a_2|, \ldots, |a_N|, \qquad A + \varepsilon.$$

That is, $\{a_n\}$ is a bounded sequence. (You can modify this proof for the case $A < 0$.)

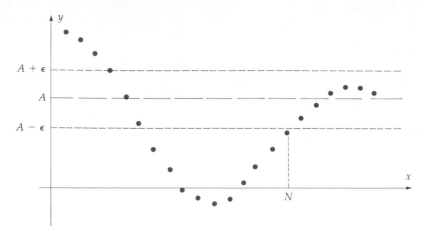

**FIGURE 5**
$A - \varepsilon < a_n < A + \varepsilon$, if $n > N$.

The final theorem concerns *order*:

**THEOREM 3**

If $a_n \le b_n$ for every $n$, and both sequences converge, then

$$\lim_{n \to \infty} a_n \le \lim_{n \to \infty} b_n.$$

This may seem obvious. A proof is suggested in the problems, along with some cautionary examples.

## SUMMARY

*Functions and Sequences:*

$$\lim_{n \to \infty} f(n) = \lim_{x \to +\infty} f(x), \qquad \text{if the latter exists.}$$

$$\lim_{n \to \infty} f\left(\frac{1}{n}\right) = \lim_{x \to 0^+} f(x), \qquad \text{if the latter exists.}$$

$$\lim_{n \to \infty} f(a_n) = f\left(\lim_{n \to \infty} a_n\right), \qquad \text{if } f(x) \text{ is continuous at the point } x = \lim_{n \to \infty} a_n.$$

*Bounded Sequences:*   A sequence $\{a_n\}$ is *bounded* if there is a number $M$ such that $|a_n| \le M$, for every $n$.

**THEOREM 1**

If $\lim_{n \to \infty} a_n = 0$ and $\{b_n\}$ is bounded, then $\lim_{n \to \infty} a_n b_n = 0$.

> **THEOREM 2**
>
> If the sequence $\{a_n\}$ converges, then it is bounded.

*Order:*   If $a_n \le b_n$ then $\lim\limits_{n \to \infty} a_n \le \lim\limits_{n \to \infty} b_n$, provided the limits exist.

*An Important Special Limit:*

$$\text{For every } x, \ \lim_{n \to \infty} \frac{x^n}{n!} = 0.$$

## PROBLEMS

### A

**1.** Compute the limit of the sequence $\{a_n\}$ with the given general term, if it exists.

*a) $\sqrt{\dfrac{n^2}{3n^2 + 2}}$     *b) $e^{1/n}$

*c) $e^{\sqrt{n}}$     d) $\ln\left(\dfrac{n}{n + 1}\right)$

*e) $(n + 1)^{1/n}$     f) $\left(1 + \dfrac{1}{n}\right)^n$

*g) $\left(1 + \dfrac{2}{n}\right)^n$     h) $\left(1 - \dfrac{2}{n}\right)^n$

i) $\left(1 + \dfrac{1}{n^2}\right)^n$     *j) $\left(1 + \dfrac{1}{n}\right)^{n^2}$

*k) $n \sin\left(\dfrac{1}{n}\right)$     l) $(\sin n)/n$

*m) $n^{-n}$     n) $(-1)^{\cos(n\pi/2)}$

o) $\dfrac{2 + (-1)^n n}{e^n}$

**\*2.** Which of the following sequences are bounded?

a) $\left\{\left(1 + \dfrac{1}{n}\right)^n\right\}$     b) $\{(-2)^n\}$

c) $\{2(-1)^n\}$     d) $\{\sin(\pi n/3)\}$

e) $\{e^{\cos n}\}$     f) $\{(-n)^n\}$

g) $\{n^{-n}\}$     h) $\left\{\dfrac{n}{1 + n^2}\right\}$

i) $\left\{\dfrac{n^2}{1 - n}\right\}$

**3.** Define a function $f$ by $f(x) = 1$ for $x \ge 0$, and $f(x) = 0$ for $x < 0$. Is $\lim\limits_{n \to \infty} f(a_n) = f(\lim a_n)$, for every sequence $\{a_n\}$ converging to 0? (Hint: Redraw Figure 3 for this $f$, with $L = 0$.)

**\*4.** Some of the following are universally true, others not. If true, explain why (quoting appropriate theorems or principles). If the statement is not universally true, give an example showing why not.

a) Every bounded sequence converges.

b) Every convergent sequence is bounded.

c) If $a_n < b_n$ for every $n$, and both sequences converge, then $\lim\limits_{n \to \infty} a_n < \lim\limits_{n \to \infty} b_n$.

### B

**5.** a) Suppose that $f(0) = 0$, and $f'(0)$ exists. What is $\lim\limits_{n \to \infty} nf(1/n)$? [Hint: What is the slope of the segment from the origin to the point $\left(\dfrac{1}{n}, f\left(\dfrac{1}{n}\right)\right)$ on the graph of $f$?]

b) Use part a to evaluate $\lim\limits_{n \to \infty} n \tan(1/n)$; $\lim\limits_{n \to \infty} n \ln(1 + 1/n)$; $\lim\limits_{n \to \infty} n(e^{1/n} - 1)$.

### C

**6.** a) Give a definition of $\lim\limits_{x \to +\infty} f(x) = L$, similar to the definition of $\lim\limits_{n \to \infty} a_n = L$.

b) Prove that $\lim\limits_{x \to +\infty} \dfrac{1}{x + 1} = 0$, using your definition.

**c)** Prove that $\lim\limits_{x \to +\infty} f(x) = L$ implies $\lim\limits_{n \to \infty} f(n) = L$, using your definition.

**7.** Suppose that $\lim\limits_{n \to \infty} a_n = A$ and $\lim\limits_{n \to \infty} b_n = B$, while $a_n \leq b_n$. Prove that $A \leq B$. [Hint: Suppose, on the contrary, that $A > B$. Apply the definition of limit with $\varepsilon = \frac{1}{2}(A - B)$, and reach a contradiction.]

**8.** In Theorem 2, for the case $A < 0$, redraw the figure and write out the proof.

**9.** This problem gives an alternate proof that

$$\lim_{n \to \infty} \frac{x^n}{n!} = 0.$$

**a)** Show, by upper and lower Riemann sums, that

$$\int_1^n \ln x \, dx < \ln 2 + \ln 3 + \cdots + \ln n$$
$$< \int_1^{n+1} \ln x \, dx.$$

**b)** Deduce that

$$n^n e^{-(n-1)} < n! < (n+1)^{n+1} e^{-n}.$$

**c)** Deduce that $\lim\limits_{n \to \infty} \dfrac{(n!)^{1/n}}{n} = 1/e$.

**d)** Deduce that

$$\lim_{n \to \infty} \frac{x}{(n!)^{1/n}} = 0, \text{ and } \lim_{n \to \infty} \frac{x^n}{n!} = 0.$$

## 11.4

### INFINITE SERIES

The thought of adding infinitely many numbers is a challenge to the imagination, posed long ago by Zeno of Elea (fifth century B.C.). Suppose, said Zeno, that you want to go from $A$ to $B$, a distance of two paces (Fig. 1). You must first go halfway, to $B_1$; then half the remaining distance, to $B_2$; and so on *ad infinitum*. The total distance of two paces is thus broken up into the sum of infinitely many steps:

$$2 = 1 + 1/2 + 1/4 + 1/8 + 1/16 + \cdots + 1/2^n + \cdots$$

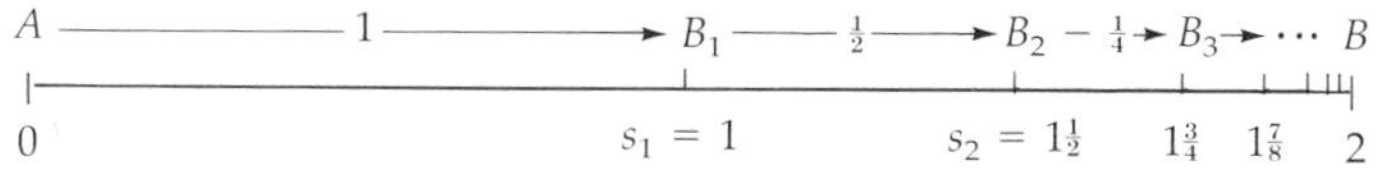

**FIGURE 1**
Illustrating Zeno's paradox.

This could be viewed as a paradox; is it really possible to add up infinitely many positive quantities? Could such a sum be finite?

Now we resolve these difficulties in the following way: Form the so-called *partial sums* (Fig. 1)

$$s_1 = 1$$
$$s_2 = 1 + \tfrac{1}{2}$$
$$s_3 = 1 + \tfrac{1}{2} + \tfrac{1}{4}$$
$$\vdots$$
$$s_n = 1 + \frac{1}{2} + \frac{1}{4} + \cdots + \frac{1}{2^n}.$$

Then *define* the sum of the infinitely many terms as the *limit* of these partial sums. As Figure 1 shows, $\lim_{n \to \infty} s_n = 2$; so by definition,

$$1 + \frac{1}{2} + \frac{1}{4} + \cdots + \frac{1}{2^n} + \cdots = \lim_{n \to \infty} \left( 1 + \frac{1}{2} + \frac{1}{4} + \cdots + \frac{1}{2^n} \right)$$
$$= \lim_{n \to \infty} s_n = 2.$$

The partial sums can be written in "sigma notation" as

$$\sum_{j=0}^{n} \frac{1}{2^j} = \frac{1}{2^0} + \frac{1}{2^1} + \cdots + \frac{1}{2^n}.$$

And we write their limit as $\displaystyle\sum_{j=0}^{\infty} \frac{1}{2^j}$;

$$\sum_{j=0}^{\infty} \frac{1}{2^j} = \lim_{n \to \infty} \sum_{j=0}^{n} \frac{1}{2^j} = 2.$$

This resolution of Zeno's philosophical paradox is just one example of a powerful and efficient means of calculation, called **infinite series**. Suppose you are given an infinite sequence of terms $a_0, a_1, a_2, \ldots$. The notation $\sum_{j=0}^{\infty} a_j$ tells you to add all these terms, as follows: Form the partial sums

$$s_0 = a_0$$
$$s_1 = a_0 + a_1$$
$$\vdots$$
$$s_n = a_0 + a_1 + \cdots + a_n = \sum_{0}^{n} a_j$$

and then take the *limit* of these sums as $n \to \infty$. If this limit exists, then $\sum_{0}^{\infty} a_j$ is called a **convergent infinite series**, and we write

$$\sum_{j=0}^{\infty} a_j = \lim_{n \to \infty} \sum_{0}^{n} a_j. \tag{1}$$

If the limit of the partial sums does not exist, the series is called a **divergent infinite series**.

*WARNING*   There are *two* sequences here:

(i)   The *terms* of the series, $a_0, a_1, a_2, \ldots$, and

(ii)   the *partial sums* $s_0, s_1, s_2, \ldots$, formed by adding more and more of those terms.

With infinite series, it is the limit of the partial sums that concerns us.

**The geometric series** is an important special case. For any number $r$, the sum of powers of $r$

$$\sum_{j=0}^{\infty} r^j$$

is called a geometric series. The partial sums are

$$s_0 = 1$$
$$s_1 = 1 + r$$
$$s_2 = 1 + r + r^2$$

and generally

$$s_n = 1 + r + r^2 + \cdots + r^n = \sum_{j=0}^{n} r^j.$$

We rewrite $s_n$ in a more convenient form using the factorization

$$1 - r^{n+1} = (1 - r)(1 + r + \cdots + r^n).$$

If $r \neq 1$, we can divide by $1 - r$ and find the partial sums

$$s_n = \sum_{j=0}^{n} r^j = 1 + r + \cdots + r^n = \frac{1 - r^{n+1}}{1 - r}, \qquad r \neq 1. \tag{2}$$

From this it follows that the series is convergent for certain values of $r$, and not for others. We treat four cases:

**(i)** If $|r| < 1$ then $\lim_{n \to \infty} r^{n+1} = 0$, so

$$\sum_{j=0}^{\infty} r^j = \lim_{n \to \infty} \sum_{0}^{n} r^j = \lim_{n \to \infty} \frac{1 - r^{n+1}}{1 - r} = \frac{1}{1 - r}.$$

**(ii)** If $|r| > 1$ then $\left| r^{n+1} \right| \to +\infty$, so the sequence $\{s_n\}$ in (2) does not converge.

**(iii)** If $r = -1$, then $r^{n+1} = (-1)^{n+1}$ does not converge, so neither does the sequence (2).

**(iv)** Finally, if $r = 1$ then

$$s_n = \sum_{j=0}^{n} r^n = \sum_{j=0}^{n} 1^j = 1^0 + 1^1 + \cdots + 1^n = n + 1$$

and this too diverges as $n \to \infty$.

In sum:

$$\sum_{j=0}^{\infty} r^j = \frac{1}{1 - r} \qquad \text{if } |r| < 1 \tag{3}$$

and

$$\sum_{j=0}^{\infty} r^j \text{ diverges} \qquad \text{if } |r| \geq 1.$$

The infinite series in Zeno's paradox is the geometric series with $r = 1/2$:

$$\sum_{j=0}^{\infty} \frac{1}{2^j} = \sum_{j=0}^{\infty} \left(\frac{1}{2}\right)^j = \frac{1}{1 - 1/2} = 2.$$

Repeating decimal expansions can also be viewed as geometric series. For example,

$$1.111\ldots = 1 + \frac{1}{10} + \left(\frac{1}{10}\right)^2 + \left(\frac{1}{10}\right)^3 + \cdots$$

$$= \sum_{j=0}^{\infty} \left(\frac{1}{10}\right)^j = \frac{1}{1 - 1/10} = \frac{10}{9} = 1\frac{1}{9}.$$

## Two Operations on Series

A convergent series $\sum_{j=0}^{\infty} a_j$ can be multiplied term-by-term by any constant:

$$c \sum_{j=0}^{\infty} a_j = \sum_{j=0}^{\infty} c a_j. \tag{4}$$

This is proved using the partial sums:

$$c \sum_{j=0}^{n} a_j = c(a_0 + \cdots + a_n) = (ca_0 + ca_1 + \cdots + ca_n)$$

$$= \sum_{j=0}^{n} c a_j.$$

Thus

$$c \sum_{j=0}^{\infty} a_j = c \lim_{n \to \infty} \sum_{j=0}^{n} a_j \qquad \text{[formula (1)]}$$

$$= \lim_{n \to \infty} c \sum_{j=0}^{n} a_j \qquad \text{(limit theorem)}$$

$$= \lim_{n \to \infty} \sum_{j=0}^{n} c a_j \qquad \text{(algebra)}$$

$$= \sum_{j=0}^{\infty} c a_j \qquad \text{[by (1) again]}$$

and this proves (4).

In a very similar way you can prove that if two series $\sum_{j=0}^{\infty} a_j$ and $\sum_{j=0}^{\infty} b_j$ both converge, then

$$\sum_{j=0}^{\infty} a_j + \sum_{j=0}^{\infty} b_j = \sum_{j=0}^{\infty} (a_j + b_j). \tag{5}$$

The geometric series formula $\sum_{j=0}^{\infty} r^j = \frac{1}{1-r}$, $|r| < 1$ can be adapted to series which appear slightly different.

**EXAMPLE 1**

$$\sum_{j=0}^{\infty} 3 \cdot 2^{-j} = 3 \sum_{j=0}^{\infty} \left(\frac{1}{2}\right)^{j} = 3 \frac{1}{1 - \frac{1}{2}} = 6.$$

$$\sum_{n=1}^{\infty} (1/3)^{n} = \frac{1}{3} + \left(\frac{1}{3}\right)^{2} + \left(\frac{1}{3}\right)^{3} + \cdots = \frac{1}{3}\left[1 + \frac{1}{3} + \left(\frac{1}{3}\right)^{2} + \cdots\right]$$

$$= \frac{1}{3} \sum_{j=0}^{\infty} \left(\frac{1}{3}\right)^{j} = \frac{1}{3} \frac{1}{1 - \frac{1}{3}} = \frac{1}{2}$$

or, by another method,

$$\sum_{n=1}^{\infty} \left(\frac{1}{3}\right)^{n} = \left[\sum_{n=0}^{\infty} \left(\frac{1}{3}\right)^{n}\right] - \left(\frac{1}{3}\right)^{0} = \frac{1}{1 - \frac{1}{3}} - 1 = \frac{1}{2}.$$

**Repeating decimals** are those which, from some point on, consist of a certain finite sequence of integers repeated over and over. For example,

$$3.2141414\ldots$$

consists, after the first three digits, of the finite sequence 14 repeated over and over. You can show, as in the following example, that *every repeating decimal is equal to a fraction.*

**EXAMPLE 2**  Express the repeating decimal 3.2141414 . . . as a fraction.

*SOLUTION*  3.21414 . . . is interpreted as the infinite series

$$3.2 + .014 + .00014 + \cdots = 3\frac{1}{5} + 14 \times 10^{-3} + 14 \times 10^{-5} + \cdots$$

$$= 3\frac{1}{5} + \frac{14}{1000}\left[1 + \frac{1}{100} + \left(\frac{1}{100}\right)^{2} + \cdots\right]$$

$$= 3\frac{1}{5} + \frac{14}{1000} \sum_{j=0}^{\infty} \left(\frac{1}{100}\right)^{j} \quad \text{(geometric series!)}$$

$$= 3\frac{1}{5} + \frac{14}{1000} \frac{1}{1 - 1/100} = 3\frac{1}{5} + \frac{14}{10} \cdot \frac{1}{99}$$

$$= 3\frac{106}{495}.$$

In other words, the partial sums

$$3.2, 3.214, 3.21414, 3.2141414, \ldots$$

tend to the limit $3\frac{106}{495}$. This is, by definition, the meaning of the infinite repeating decimal.

The geometric series arises in surprising ways.

**EXAMPLE 3** Suppose that of every dollar received in the United States, 75 cents is again spent in the U.S. If $1,000,000 is "pumped into the economy by government spending," how much total spending is generated?

*SOLUTION* Of every dollar received, 3/4 is spent in the U.S. So of the $10^6$ dollars pumped in, $\frac{3}{4} \cdot 10^6$ is spent in the U.S. But what is spent is also received, so $\frac{3}{4}$ of this $\frac{3}{4} \cdot 10^6$ is again spent in the U.S. Then $\frac{3}{4}$ of the $\frac{3}{4} \cdot \frac{3}{4} \cdot 10^6$ is spent once more in the U.S., and so on ad infinitum. The grand total, including the original $\$10^6$, is therefore

$$10^6 + \frac{3}{4} \cdot 10^6 + \frac{3}{4} \cdot \frac{3}{4} \cdot 10^6 + \frac{3}{4} \cdot \frac{3}{4} \cdot \frac{3}{4} \cdot 10^6 + \cdots$$

$$= 10^6 \left( 1 + \frac{3}{4} + \left(\frac{3}{4}\right)^2 + \cdots \right) \qquad \left( \text{geometric series, } r = \frac{3}{4} \right)$$

$$= 10^6 \cdot \frac{1}{1 - 3/4}$$

$$= 4 \cdot 10^6.$$

Under this "3/4 spending" assumption, the $\$10^6$ pumped in generates $\$4 \cdot 10^6$ of spending, or economic activity. The factor 4 is called the "multiplier." The "3/4 spending" assumption is a highly simplified model of reality, but the multiplier effect is real, and the calculation indicates how significant it is.

The series in the next example has no special significance—its purpose is to illustrate the concepts of partial sums and divergent series.

**EXAMPLE 4** Does $\displaystyle\sum_{j=1}^{\infty} \frac{1}{\sqrt{j}}$ converge?

*SOLUTION* Plot some partial sums (Fig. 2). The terms $1/\sqrt{j}$ tend to zero, but the partial sums $s_n = \sum_{j=1}^{n} 1/\sqrt{j}$ appear to grow rapidly, and it is not clear whether they approach a finite limit, or "march off" to $+\infty$. In fact, they march off:

$$s_n = 1 + \frac{1}{\sqrt{2}} + \cdots + \frac{1}{\sqrt{n}} \geq \underbrace{\frac{1}{\sqrt{n}} + \frac{1}{\sqrt{n}} + \cdots + \frac{1}{\sqrt{n}}}_{n \text{ terms}}$$

$$= n\left(\frac{1}{\sqrt{n}}\right) = \sqrt{n}.$$

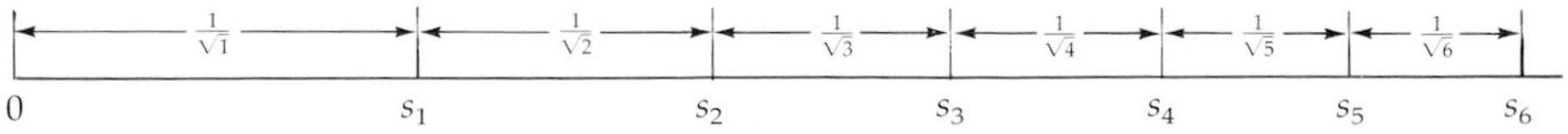

**FIGURE 2**

$\displaystyle\sum_{1}^{\infty} \frac{1}{\sqrt{j}} = \lim_{n \to \infty} S_n = +\infty$. (Note the difference between the *terms of the series*, $1/\sqrt{j}$, and their *partial sums, $s_j$*.)

As $n \to \infty$, then $\sqrt{n} \to \infty$. Since $s_n \geq \sqrt{n}$, then $s_n \to \infty$ too. Since the partial sums diverge, the infinite series $\sum_{j=1}^{\infty} 1/\sqrt{j}$ diverges.

NOTE    The *terms* of the series are the numbers $1/\sqrt{j}$; these converge to the limit 0. But the question concerns the *sums* of these terms, and those sums diverge. Imagine walking along the axis, stepping at $s_1, s_2, s_3, \ldots$. The term $a_j$ gives the length of the $j^{th}$ step, while the partial sum $s_n$ gives the distance gone in $n$ steps. The *size* of the steps tends to 0, but they all add up to a journey of infinite length.

## SUMMARY

*Definition:*

$$\sum_{j=1}^{\infty} a_j = \lim_{n \to \infty} \sum_{j=1}^{n} a_j.$$

If the limit exists, the series $\sum_{1}^{\infty} a_j$ is called *convergent*; otherwise *divergent*.

*Algebra:*

$$\sum_{j=1}^{\infty} ca_j = c \sum_{j=1}^{\infty} a_j,$$

$$\sum_{j=1}^{\infty} (a_j + b_j) = \sum_{j=1}^{\infty} a_j + \sum_{j=1}^{\infty} b_j,$$

assuming in both cases that the series on the right converge.

*The Geometric Series:*

$$\sum_{j=0}^{\infty} r^j = \frac{1}{1-r} \qquad \text{if } |r| < 1.$$

## PROBLEMS

A

1.  Plot the partial sums $s_1, s_2, s_3, s_4$ as in Figure 2.

    a) $\displaystyle\sum_{j=1}^{\infty} 1/j$

    b) $\displaystyle\sum_{j=1}^{\infty} (-1)^j/j$

    c) $\displaystyle\sum_{j=1}^{\infty} (-1)^j/\sqrt{j}$

2.  Plot the partial sums $s_1, s_2, s_3, s_4$, as in Figure 2, for $\sum_{j=0}^{\infty} (4/5)^j$. Evaluate the sum, and plot that on the same axis.

3.  Evaluate.

    *a) $\displaystyle\sum_{j=0}^{\infty} (1/3)^j$    *b) $\displaystyle\sum_{n=1}^{\infty} (1/3)^n$

    *c) $\displaystyle\sum_{j=0}^{\infty} (-1)^j/3^j$   d) $\displaystyle\sum_{n=0}^{\infty} (2)(5^{-n})$

    e) $\displaystyle\sum_{n=1}^{\infty} 2^{1-n}$

    f) $2 + 2/3 + 2/9 + 2/27 + \cdots$

    *g) $5 - 1 + 1/5 - 1/25 + 1/125 - \cdots$

    h) $\displaystyle\sum_{n=0}^{\infty} \frac{2^n + 5^n}{10^n}$

4. Express each repeating decimal as a fraction.
   a) 0.999 . . .
   *b) 2.323232 . . .
   c) 0.78217821 . . .

5. (The "spending multiplier".) A big convention is booked into city $C$, bringing conventioneers who will spend $1,000,000. Suppose that, on the average, each businessman in $C$, great and small, spends 2/3 of his gross income in $C$; the rest is spent elsewhere, or saved. So in addition to the million spent directly by the conventioneers, 2/3 million is spent in $C$ by its own citizens. Then 2/3 of *this* is spent in $C$, and so on. Show that the total amount spent in $C$ as a result of the convention is $3 million.

6. Evaluate the following series, or show their divergence, by expressing their partial sums in simple form. (Write out the partial sums "long-hand" to see how they simplify.)
   *a) $\displaystyle\sum_{j=1}^{\infty}\left(\frac{1}{j}-\frac{1}{j+1}\right)$
   *b) $\displaystyle\sum_{j=1}^{\infty}\ln\left(\frac{j}{j+1}\right)$
   c) $\displaystyle\sum_{j=2}^{\infty}\frac{2}{j^2-1}\quad\left(\frac{2}{j^2-1}=\frac{1}{j-1}-\frac{1}{j+1}\right)$

7. Suppose that $\sum_{j=0}^{\infty}(a_j+b_j)$ converges.
   *a) Do $\sum_{j=0}^{\infty}a_j$ and $\sum_{j=0}^{\infty}b_j$ necessarily converge?
   b) If $\sum_{j=0}^{\infty}a_j$ and $\sum_{j=0}^{\infty}(a_j+b_j)$ converge, can $\sum_{j=0}^{\infty}b_j$ diverge?

**B**

8. A ball is dropped from height $h$ metres. According to an elementary theory of the bounce, each time the ball hits, it bounces up to $r$ times its previous height; $r$ is a constant, called the "coefficient of restitution." Show that, according to this theory
   a) The total distance traversed by the ball (down and up and down and up and . . .) is
   $$h + 2h(r + r^2 + r^3 + \cdots).$$
   b) The total distance is $h\dfrac{1+r}{1-r}$.

9. A ball falling distance $h$, starting at rest, takes $\sqrt{2h/g}$ seconds, where $g$ is a constant. If it bounces to a height $h$ above the floor, that too takes $\sqrt{2h/g}$ seconds. How *long* does it take the ball in the previous problem to make its infinitely many bounces?

10. When a drug is absorbed, its concentration in the blood decays more or less exponentially—an initial concentration $C_0$ decays according to the law
    $$C = C_0 e^{-kt},$$
    where $k > 0$; $t$ is in hours. Suppose that $g_0$ grams are injected every $T$ hours in a patient with $b$ liters of blood; then the initial concentration is $g_0/b$ gms/liter. This decays for $T$ hours, at which time a further $g_0$ grams are injected, and so on.
    a) Explain: Just after $n$ injections [and $(n-1)T$ hours] the concentration is
    $$\frac{g_0}{b}[1 + e^{-kT} + e^{-2kT} + \cdots + e^{-(n-1)kT}]$$
    $$= \frac{g_0}{b}\frac{1 - e^{-nkT}}{1 - e^{-kT}}.$$
    b) Obtain an analogous formula for the concentration just before $n + 1$ injections.
    c) Explain: After very many injections, the concentration just before an injection is nearly
    $$\frac{g_0}{b}\frac{1}{e^{kT} - 1},$$
    and just after an injection is nearly $\dfrac{g_0}{b}\dfrac{e^{kT}}{e^{kT} - 1}$. (This affects the size and frequency of injections—they must reach a certain level to be effective and remain below a certain level to be safe. This problem is adapted from UMAP unit 72.)

*11. In Russian roulette, two players alternately fire a revolver with one of six chambers loaded, until one loses by firing the loaded chamber. (A safer version uses a six-sided die.) The probability that the first player wins is
    $$\frac{5}{6}\cdot\frac{1}{6} + \left[\frac{5}{6}\right]^3\cdot\frac{1}{6} + \left[\frac{5}{6}\right]^5\cdot\frac{1}{6} + \cdots.$$
    Compute this probability. (If you know about probability, try to explain the formula.)

*12. A layer of insulation of a certain kind reflects 2/3 the incident radiant heat and transmits 1/3, regardless of the direction. What fraction of the incident radiant heat is transmitted by two separated layers of this material?

13. Evaluate the following, and state for which values of $x$ the result is valid.

*a) $\displaystyle\sum_{j=0}^{\infty} \left(\frac{1}{x}\right)^j$

b) $\displaystyle\sum_{j=0}^{\infty} (2x)^j$

c) $\displaystyle\sum_{j=0}^{\infty} \left(\frac{1}{1+x}\right)^j$

14. a) Express the partial sums of $\sum_{j=0}^{\infty} jr^{j-1}$ in "closed form," by differentiating formula (2) in the text.

b) Deduce the formula for $\sum_{j=0}^{\infty} jr^{j-1}$.

*c) Evaluate $\displaystyle\sum_{j=1}^{\infty} \frac{j}{2^j}$.

15. a) Obtain a formula for $\sum_{j=0}^{\infty} j(j-1)r^{j-2}$.

b) Obtain a formula for $\sum_{j=1}^{\infty} j^2 r^j$.

c) Evaluate $\sum_{j=1}^{\infty} j^2 2^{-j}$.

## 11.5

## TAYLOR SERIES

Many functions are approximated by Taylor polynomials based at a convenient point $a$:

$$P_n(x) = \sum_{k=0}^{n} \frac{1}{k!} f^{(k)}(a)(x-a)^k.$$

These are precisely the partial sums of the infinite series

$$\sum_{k=0}^{\infty} \frac{1}{k!} f^{(k)}(a)(x-a)^k,$$

called the **Taylor series** of $f$, based at $a$. When the base point $a$ is 0, it is called the **Maclaurin series** of $f$,

$$\sum_{k=0}^{\infty} \frac{1}{k!} f^{(k)}(0)x^k.$$

We generally find that the approximation by Taylor polynomials improves as $n \to \infty$, so: Does the *infinite* Taylor series equal $f$ *exactly*? Often it does, at least for appropriate values of $x$.

---

**EXAMPLE 1**  For $f(x) = \dfrac{1}{1-x}$: a) What is the Maclaurin series? b) For what values of $x$ is the sum of the Maclaurin series equal to $f(x)$?

**SOLUTION**

a.  $f(x) = (1-x)^{-1}$, $f'(x) = (1-x)^{-2}$, $f''(x) = 2(1-x)^{-3}, \ldots,$ $f^{(k)}(x) = k!(1-x)^{-k-1}$,

so

$$f(0) = 1, \ f'(0) = 1, \ f''(0) = 2, \ldots, f^{(k)}(0) = k!$$

and

$$\sum_{k=0}^{\infty} \frac{1}{k!} f^{(k)}(0) x^k = \sum_{k=0}^{\infty} \frac{1}{k!} k! x^k = \sum_{k=0}^{\infty} x^k.$$

**b.**   The Maclaurin series in this case is just the geometric series studied in Section 11.4. For $|x| < 1$, it converges to $\dfrac{1}{1-x}$; for all other $x$, it diverges.

---

**EXAMPLE 2**   For what values of $x$ does $e^x$ equal its Maclaurin series?

**SOLUTION**   First, compute the Maclaurin series. The derivatives of $f(x) = e^x$ are $f^{(k)}(x) = e^x$; so $f^{(k)}(0) = e^0 = 1$, and the Maclaurin series of $e^x$ is

$$\sum_{k=0}^{\infty} \frac{1}{k!} x^k. \tag{1}$$

The partial sums are

$$\sum_{k=0}^{n} \frac{1}{k!} x^k,$$

precisely the Taylor polynomials of $e^x$. Suppose that $x \geq 0$. By the Remainder Estimate of Section 11.1,

$$\left| e^x - P_n(x) \right| \leq \frac{|x|^{n+1}}{(n+1)!} \max_{0 \leq t \leq x} |e^t| = \frac{|x|^{n+1}}{(n+1)!} e^x. \tag{2}$$

Moreover, for each given $x$

$$\lim_{n \to \infty} \frac{|x|^{n+1}}{(n+1)!} = 0 \tag{3}$$

(shown in Sec. 11.3; or see the Remark below). So from (2), $\left| e^x - P_n(x) \right| \to 0$, or $P_n(x) \to e^x$. Since $P_n(x)$ is the $n^{\text{th}}$ partial sum of the Maclaurin series (1), it follows that

$$\sum_{k=0}^{\infty} \frac{1}{k!} x^k = e^x.$$

We have proved this for $x \geq 0$; the case $x < 0$ is left as a problem.

**REMARK**   The limit (3) can be explained briefly as follows. Let $N$ be a fixed integer $\geq |x|$. For $n > N$,

$$\frac{|x|^{n+1}}{(n+1)!} = \frac{|x|}{n+1} \cdot \frac{|x|^n}{n!}$$

$$= \left( \frac{|x|}{n+1} \right) \cdot \underbrace{\left( \frac{|x|}{n} \cdot \frac{|x|}{(n-1)} \cdot \ \cdots \ \cdot \frac{|x|}{(N+1)} \right)}_{n - N \text{ factors, all } < 1, \text{ since } n > N \geq |x|} \cdot \frac{|x|^N}{N!}$$

$$< \frac{|x|}{n+1} \frac{|x|^N}{N!}.$$

Since $|x|$ and $N$ are fixed and $\dfrac{1}{n+1} \to 0$, it follows that $|x|^{n+1}/(n+1)! \to 0$.

**EXAMPLE 3**  Evaluate $\displaystyle\sum_{k=0}^{\infty} \frac{2^k}{k!}$.

**SOLUTION**  This is $\displaystyle\sum_{k=0}^{\infty} \frac{x^k}{k!}$ with $x = 2$. Since $\displaystyle\sum_{k=0}^{\infty} \frac{x^k}{k!} = e^x$ for *all* $x$, then surely for $x = 2$:

$$\sum_{k=0}^{\infty} \frac{2^k}{k!} = e^2.$$

**EXAMPLE 4**  For what values of $x$ does $\sin x$ equal its Maclaurin series?

**SOLUTION**  Compute the derivatives of $f(x) = \sin x$, and evaluate them at $x = 0$:

$$f(0) = 0, \qquad f'(0) = 1, \qquad f''(0) = 0,$$

$$f'''(0) = -1, \qquad f^{(4)}(0) = 0, \qquad f^{(5)}(0) = 1, \ldots$$

The Maclaurin series is therefore

$$0 + \frac{1}{1!} x + 0 + \frac{-1}{3!} x^3 + 0 + \frac{1}{5!} x^5 + 0 + \frac{-1}{7!} x^7 + \cdots.$$

The pattern is clear, but clumsily expressed. Only odd powers of $x$ appear, so we write them as $x^{2k+1}$, where $k = 0, 1, 2, \ldots$. The alternating signs $+1$, $-1$, $+1$, $-1$, $\ldots$ can be given by a factor $(-1)^k$:

$$(-1)^0 = +1, \qquad (-1)^1 = -1, \qquad (-1)^2 = +1, \qquad (-1)^3 = -1, \ldots.$$

The Maclaurin series is therefore

$$\sum_{k=0}^{\infty} \frac{(-1)^k}{(2k+1)!} x^{2k+1} = \frac{(-1)^0}{1!} x^1 + \frac{(-1)^1}{3!} x^3 + \cdots$$

$$= x - \frac{1}{3!} x^3 + \cdots.$$

Does this series equal $\sin x$? The partial sum is

$$\sum_{k=0}^{n} \frac{(-1)^k}{(2k+1)!} x^{2k+1} = x - \frac{x^3}{3!} + \cdots + (-1)^n \frac{x^{2n+1}}{(2n+1)!}. \qquad (4)$$

This is the Taylor polynomial $P_{2n+1}(x)$ for $\sin x$. By the Remainder Estimate,

$$\left|\sin x - P_{2n+1}(x)\right| \le \frac{|x|^{2n+2}}{(2n+2)!} \max\left|\sin^{(2n+2)}(t)\right|.$$

Now, the derivative $\sin^{(2n+2)}(t)$ is $\pm \sin t$, so

$$\max\left|\sin^{(2n+2)}(t)\right| = \max\left|\pm \sin t\right| \le 1.$$

It follows that

$$\left|\sin x - P_{2n+1}(x)\right| \leq \frac{|x|^{2n+2}}{(2n+2)!}. \tag{5}$$

Since $|x|^{2n+2}/(2n+2)! \to 0$ as $n \to \infty$, it follows from (5) that $P_{2n+1}(x) \to \sin x$. But $P_{2n+1}(x)$ is the $n^{\text{th}}$ partial sum of the series (4), so

$$\sum_{k=0}^{\infty} \frac{(-1)^k}{(2k+1)!} x^{2k+1} = \sin x, \qquad \text{for all } x.$$

The series in these examples are among the most important in mathematics:

$$\frac{1}{1-x} = \sum_{k=0}^{\infty} x^k, \qquad |x| < 1 \tag{6}$$

$$e^x = \sum_{k=0}^{\infty} \frac{1}{k!} x^k, \qquad \text{all } x \tag{7}$$

$$\sin x = \sum_{k=0}^{\infty} \frac{(-1)^k}{(2k+1)!} x^{2k+1}, \qquad \text{all } x. \tag{8}$$

The problems ask you to prove another:

$$\cos x = \sum_{0}^{\infty} \frac{(-1)^k}{(2k)!} x^{2k}, \qquad \text{all } x. \tag{9}$$

*Notice* that the Maclaurin series of $f$ does not necessarily equal $f(x)$ for *all* values of $x$; for example, the series for $\dfrac{1}{1-x}$ is valid only for $|x| < 1$. One of the problems shows a more extreme example—a function which is positive when $x \neq 0$, yet its Maclaurin series converges to 0 for every $x$! So in each case, it must be carefully determined for what values of $x$ the Maclaurin series of $f$ actually converges to $f(x)$.

## Operations on Series

Once a series is established, others can be obtained from it by simple algebraic manipulations.

**EXAMPLE 5**  Obtain a series for $\dfrac{1}{1 + 4t^2}$.

*SOLUTION*  This calls to mind the geometric series:

$$\frac{1}{1 + 4t^2} = \frac{1}{1 - x} \qquad \text{if you set } x = -4t^2.$$

So replace $x$ by $-4t^2$ on both sides of (6):

$$\frac{1}{1 + 4t^2} = \frac{1}{1 - (-4t^2)} = \sum_{k=0}^{\infty} (-4t^2)^k, \qquad |-4t^2| < 1.$$

The terms $(-4t^2)^k$ can be rewritten as

$$(-4t^2)^k = (-4)^k(t^2)^k = (-4)^k t^{2k}.$$

And the inequality $\left|-4t^2\right| < 1$ can be simplified, since the absolute value of a product is the product of the absolute values:

$$\left|-4t^2\right| = \left|-4\right| \cdot \left|t\right|^2 = 4\left|t\right|^2$$

so $\left|-4t^2\right| < 1$ gives $4\left|t\right|^2 < 1$, or $\left|t\right|^2 < 1/4$, or $\left|t\right| < 1/2$. Thus

$$\frac{1}{1 + 4t^2} = \sum_{k=0}^{\infty} (-4)^k t^{2k}, \qquad \left|t\right| < 1/2.$$

In these calculations, it is essential to do all the algebra carefully, particularly in handling minus signs!

**EXAMPLE 6**    The "hyperbolic sine" is defined by

$$\sinh x = \frac{e^x - e^{-x}}{2}.$$

Obtain a series for this. (sinh is pronounced "sinch." This function is discussed in Sec. 8.9.)

**SOLUTION**    You could compute $f^{(k)}(0)$ and write the Maclaurin series; but then you would have to determine for what values of $x$ the series is equal to $\sinh x$. It is easier to work with the known series for $e^x$. Replace $x$ by $-x$ in the exponential series (7):

$$e^{-x} = \sum_{k=0}^{\infty} \frac{1}{k!}(-x)^k = \sum_{k=0}^{\infty} \frac{(-1)^k}{k!} x^k$$

$$= 1 - x + \frac{x^2}{2} - \frac{x^3}{3!} + \frac{x^4}{4!} - \cdots.$$

Hence, for all $x$,

$$e^x - e^{-x} = 1 + x + \frac{x^2}{2} + \frac{x^3}{3!} + \frac{x^4}{4!} + \frac{x^5}{5!} + \cdots$$

$$- \left[ 1 - x + \frac{x^2}{2} - \frac{x^3}{3!} + \frac{x^4}{4!} - \frac{x^5}{5!} + \cdots \right]$$

$$= 0 + 2x + 0 + 2\frac{x^3}{3!} + 0 + 2\frac{x^5}{5!} + \cdots$$

$$= 2\left( x + \frac{x^3}{3!} + \frac{x^5}{5!} + \cdots \right)$$

and

$$\sinh x = \frac{1}{2}(e^x - e^{-x}) = x + \frac{x^3}{3!} + \frac{x^5}{5!} + \cdots$$

$$= \sum_{k=0}^{\infty} \frac{x^{2k+1}}{(2k+1)!}.$$

This is like the series (8) for sin $x$, but without the "alternating sign" factor $(-1)^k$.

## SUMMARY

*Taylor Series of f,* based at $a$:

$$\sum_{k=0}^{\infty} \frac{1}{k!} f^{(k)}(a)(x-a)^k$$

*Special Series:*

$$e^x = \sum_{k=0}^{\infty} \frac{x^k}{k!} \qquad \frac{1}{1-x} = \sum_{k=0}^{\infty} x^k, \qquad |x| < 1$$

$$\sin x = \sum_{k=0}^{\infty} (-1)^k \frac{x^{2k+1}}{(2k+1)!} \qquad \cos x = \sum_{k=0}^{\infty} (-1)^k \frac{x^{2k}}{(2k)!}$$

## PROBLEMS

### A

1.   Compute the Maclaurin series of the following functions by computing $f^{(k)}(0)$. Do not try to establish convergence.

a)  $\dfrac{1}{1+x}$     *b)  $\dfrac{1}{2+x}$     *c)  $\ln(1+x)$

d)  $e^{x/2}$     *e)  $e^{-2x}$     f)  $\frac{1}{2}(e^x - e^{-x})$

g)  $\dfrac{1}{(1-x)^2}$

2.   Use known series to derive series for the following. State for which values of $t$ the series converges.

*a)  $e^{2t^2}$     b)  $e^{-t^2/2}$

*c)  $\dfrac{\cos t - 1}{t^2}$     d)  $\dfrac{1}{1-2t}$

*e)  $\dfrac{1}{1+t^2/4}$     f)  $\dfrac{1}{2+t} \left[ = \dfrac{1}{2}\dfrac{1}{1+t/2} \right]$

3.   Evaluate the following.

*a)  $\displaystyle\sum_{k=0}^{\infty} \frac{5^k}{k!}$     b)  $\displaystyle\sum_{k=0}^{\infty} \frac{1}{k!}$

c)  $\displaystyle\sum_{k=0}^{\infty} \frac{(-1)^k}{k!}$     *d)  $\displaystyle\sum_{k=0}^{\infty} (-1)^k/(2k+1)!$

e)  $\displaystyle\sum_{k=0}^{\infty} (-1)^k/(2k)!$     *f)  $\displaystyle\sum_{k=0}^{\infty} (-2)^k/(2k)!$

g)  $\displaystyle\sum_{0}^{\infty} \frac{5}{3^j}$     *h)  $\displaystyle\sum_{1}^{\infty} \frac{5}{3^j}$

4.   The "hyperbolic cosine" is defined by $\cosh x = \frac{1}{2}(e^x + e^{-x})$ (Sec. 8.9).
 a)   Compute the Maclaurin series for $\cosh x$.
 b)   Obtain the series for $\cosh x$ in a different way, using the series for $e^x$.

5.   a)   Obtain the Maclaurin series for $\cos x$.
 b)   Prove that it equals $\cos x$ for all $x$.

### B

6.   a)   From the geometric series, deduce that

$$\frac{-x}{1-x} = \frac{1}{1-1/x} = 1 + \frac{1}{x} + \frac{1}{x^2} + \cdots$$

and

$$\frac{x}{1-x} = x + x^2 + x^3 + \cdots$$

**b)** If you add the two series in part a you get

$$\cdots + x^3 + x^2 + x + 1 + \frac{1}{x} + \frac{1}{x^2} + \cdots$$

$$= \frac{x}{1-x} + \frac{-x}{1-x} = 0.$$

Yet when $x$ is positive, all terms of the series are positive. Where is the error in the calculation?

**C**

**7.** This problem studies a function $f$ whose Maclaurin series $\sum_{k=0}^{\infty} \frac{1}{k!} f^{(k)}(0) x^k$ converges for every $x$, but

*does not converge to $f$.* It is defined by:

$$f(0) = 0, \quad \text{and} \quad f(x) = e^{-1/x^2} \quad \text{if } x \neq 0.$$

(A similar function is in problem 19, Sec. 8.7.) Prove the following:

**a)** For any integer $n$, $x^{-n} e^{-1/x^2} \to 0$ as $x \to 0$.
**b)** $f'(0) = 0$.
**c)** If $x \neq 0$, then $f'(x) = P_1(1/x)e^{-1/x^2}$ for some polynomial $P_1$.
**d)** $f''(0) = 0$.
**e)** If $x \neq 0$, then $f^{(n)}(x) = P_n(1/x)e^{-1/x^2}$ for some polynomial $P_n$.
**f)** $f^{(n)}(0) = 0$ for all $n$.
**g)** $\sum_{k=0}^{\infty} \frac{1}{k!} f^{(k)}(0)x^k = f(x)$ only when $x = 0$.

## 11.6

### OPERATIONS ON POWER SERIES

#### Power Series

The exponential series

$$\sum_{k=0}^{\infty} \frac{1}{k!} x^k = e^x \tag{1}$$

and the geometric series

$$\sum_{k=0}^{\infty} x^k = \frac{1}{1-x}, \qquad |x| < 1 \tag{2}$$

are called **power series**, because each term is a constant times a power of $x$. In general, any series

$$\sum_{k=0}^{\infty} c_k x^k$$

where the $c_k$ are constants, is a power series.

Suppose that a given power series $\sum c_k x^k$ *converges for $|x| < r$.* (For example, the geometric series converges for $|x| < 1$, and the exponential series converges for $|x| < \infty$.) *Then for $|x| < r$, the series can be differentiated and integrated "term by term," as if it were a polynomial:*

$$\frac{d}{dx} \sum_0^{\infty} c_k x^k = \sum_0^{\infty} \frac{d}{dx} [c_k x^k] = \sum_0^{\infty} k c_k x^{k-1}, \qquad |x| < r$$

$$\int_0^x \left( \sum_0^{\infty} c_k t^k \right) dt = \sum_0^{\infty} \left[ \int_0^x c_k t^k \, dt \right] = \sum_0^{\infty} \frac{c_k}{k+1} x^{k+1}, \qquad |x| < r.$$

The proof is left to Appendix A, at the end of the text.

*WARNING*   These operations are *not* valid for *all* kinds of infinite series! So we are especially grateful to the power series, that they are so well behaved.

**EXAMPLE 1**   $\displaystyle\sum_0^\infty \frac{1}{k!} x^k = e^x$ for all $x$, that is, for $|x| < +\infty$. So for all $x$, the series can be differentiated term-by-term:

$$\frac{d}{dx} \sum_0^\infty \frac{1}{k!} x^k = \frac{d}{dx}\left[ 1 + x + \frac{x^2}{2!} + \frac{x^3}{3!} + \cdots + \frac{x^k}{k!} + \cdots \right]$$

$$= 0 + 1 + \frac{2x}{2!} + \frac{3x^2}{3!} + \cdots + \frac{kx^{k-1}}{k!} + \cdots .$$

Note that $\dfrac{2}{2!} = 1$, $\dfrac{3}{3!} = \dfrac{3}{3 \cdot 2 \cdot 1} = \dfrac{1}{2!}$, and similarly $\dfrac{k}{k!} = \dfrac{1}{(k-1)!}$. Thus

$$\frac{d}{dx} \sum_0^\infty \frac{1}{k!} x^k = 1 + x + \frac{x^2}{2!} + \cdots + \frac{x^{k-1}}{(k-1)!} + \cdots$$

$$= \sum_{j=0}^\infty \frac{1}{j!} x^j.$$

This confirms the formula $de^x/dx = e^x$.

**EXAMPLE 2**   (The logarithm.) From the geometric series (2) with $x = -t$, you get

$$\frac{1}{1+t} = \sum_0^\infty (-t)^k = \sum_0^\infty (-1)^k t^k, \qquad |t| < 1.$$

Then

$$\int_0^x \frac{dt}{1+t} = \int_0^x \sum_0^\infty (-1)^k t^k \, dt = \sum_0^\infty (-1)^k \frac{1}{k+1} x^{k+1}, \qquad |x| < 1.$$

Evaluate the integral on the left and rewrite the sum on the right, setting $k + 1 = m$:

$$\ln(1 + x) = \sum_1^\infty (-1)^{m-1} \frac{1}{m} x^m, \qquad |x| < 1.$$

In particular, this is valid with $x = -\frac{1}{2}$. Set $x = -1/2$, note that $\ln(1 - \frac{1}{2}) = \ln(\frac{1}{2}) = -\ln 2$, and multiply through by $-1$, to find a series for $\ln 2$:

$$\ln 2 = -\sum_1^\infty (-1)^{m-1} \frac{1}{m} \left(-\frac{1}{2}\right)^m$$

$$= \frac{1}{2} + \frac{1}{2}\left(\frac{1}{2}\right)^2 + \frac{1}{3}\left(\frac{1}{2}\right)^3 + \frac{1}{4}\left(\frac{1}{2}\right)^4 + \frac{1}{5}\left(\frac{1}{2}\right)^5 + \cdots$$

$$= \frac{1}{2} + \frac{1}{8} + \frac{1}{24} + \frac{1}{64} + \frac{1}{160} + \cdots \tag{3}$$

*REMARK* In Example 2, the theorem on term-by-term operations guarantees the validity of the integrated series only for $|x| < 1$; but in this case, the integrated series is valid for $x = 1$ as well:

$$\ln(1 + x) = \sum_{1}^{\infty} (-1)^{m-1} \frac{1}{m} x^m, \qquad -1 < x \le 1.$$

Problem 10 shows why.

---

**EXAMPLE 3** (The inverse tangent.) Replace $x$ by $-t^2$ in (2):

$$\frac{1}{1 + t^2} = \sum_{0}^{\infty} (-t^2)^k$$

$$= \sum_{0}^{\infty} (-1)^k t^{2k}, \qquad |t| < 1.$$

This is a power series convergent for $|t| < 1$, so it can be integrated term-by-term from 0 to $x$, if $|x| < 1$. The result is a power series for $\tan^{-1} x$:

$$\tan^{-1} x = \int_0^x \frac{dt}{1 + t^2} = \sum_{0}^{\infty} (-1)^k \frac{x^{2k+1}}{2k + 1}, \qquad |x| < 1. \tag{4}$$

(A careful study of this particular example shows that the series is actually valid for $|x| \le 1$; see problem 8.) Since $\tan \dfrac{\pi}{6} = \dfrac{1}{\sqrt{3}}$, the series gives

$$\pi/6 = \tan^{-1}(1/\sqrt{3}) = \sum_{0}^{\infty} (-1)^k \frac{(1/\sqrt{3})^{2k+1}}{2k + 1}$$

$$= \frac{1}{\sqrt{3}} \sum_{0}^{\infty} \frac{(-1/3)^k}{2k + 1}$$

$$= \frac{1}{\sqrt{3}} \left( 1 - \frac{1}{3 \cdot 9} + \frac{1}{5 \cdot 9^2} - \frac{1}{7 \cdot 9^3} + \cdots \right).$$

---

## Power Series are Maclaurin Series

In Example 3 we found the power series (4) for $\tan^{-1} x$, but not by the Maclaurin series method, computing all the derivatives of $\tan^{-1} x$. In fact, it's a messy job to compute those derivatives beyond the first two—the integration method is much easier. This raises a question: Is the series (4) necessarily the Maclaurin series for $\tan^{-1} x$? Yes it is. *If any function $f$ is given by a power series*

$$f(x) = c_0 + c_1 + c_2 x^2 + c_3 x^3 + \cdots, \qquad |x| < r \tag{5}$$

where $r > 0$, *then the coefficients $c_k$ in that series are the same as the coefficients in the Maclaurin series:*

$$c_k = \frac{1}{k!} f^{(k)}(0), \qquad k = 0, 1, 2, \ldots. \tag{6}$$

The proof? First set $x = 0$ in (5) to find

$$f(0) = c_0 + 0 + 0 + \cdots = c_0.$$

proving (6) for $k = 0$. Next differentiate (5) and then set $x = 0$:

$$f'(x) = c_1 + 2c_2x + 3c_3x^2 + 4c_4x^4 + \cdots \tag{7}$$

so $f'(0) = c_1$, proving (6) for $k = 1$. Now differentiate (7), set $x = 0$, and you find

$$f''(0) = 2c_2$$

proving (6) for $k = 2$. Clearly you can continue, to prove (6) for any $k$; so the power series (5) *is* the Maclaurin series for $f$, no matter by what (legitimate) means it was established.

The next example evaluates an otherwise intractible integral, by a power series.

---

**EXAMPLE 4**   The integral $\int_a^b e^{-x^2}\,dx$ cannot be done by finding an antiderivative in "closed form." However, an antiderivative *can* be found by power series. Set $x = -t^2$ in the exponential series, and integrate the result from 0 to $x$:

$$\int_0^x e^{-t^2}\,dt = \int_0^x \sum_{k=0}^{\infty} \frac{1}{k!}(-t^2)^k\,dt$$

$$= \int_0^x \sum_{k=0}^{\infty} \frac{1}{k!}(-1)^k t^{2k}\,dt$$

$$= \sum_{k=0}^{\infty} \frac{1}{k!}(-1)^k \frac{x^{2k+1}}{2k+1}$$

$$= x - \frac{x^3}{3} + \frac{x^5}{5 \cdot 2!} - \frac{x^7}{7 \cdot 3!} + \cdots.$$

In particular, setting $x = 1$,

$$\int_0^1 e^{-t^2}\,dt = 1 - \frac{1}{3} + \frac{1}{5 \cdot 2!} - \frac{1}{7 \cdot 3!} + \cdots. \tag{8}$$

---

*REMARK*   Equation (8) means that you can approximate the integral as accurately as desired by summing enough of the terms on the right-hand side. But how many is enough? If you want accuracy to three decimals, say, how many terms are needed? This is answered in the next section.

### Power Series and Limits

As an alternative to l'Hôpital's Rule, you can use power series to evaluate indeterminate forms.

**EXAMPLE 5** Evaluate $\lim\limits_{x \to 0} \dfrac{\sin x - x}{\sin(x^2)}$, using Maclaurin series.

*SOLUTION* The Maclaurin series of the numerator is

$$\left( x - \frac{x^3}{3!} + \frac{x^5}{5!} + \cdots \right) - x = -\frac{x^3}{3!} + \frac{x^5}{5!} + \cdots$$

$$= x^3 \left( \frac{-1}{3!} + \frac{1}{5!} x^2 - \cdots \right)$$

and for the denominator it is

$$\sin(x^2) = x^2 - \frac{(x^2)^3}{3!} + \cdots = x^2 \left( 1 - \frac{x^4}{3!} + \cdots \right).$$

Hence the quotient in question is written as

$$\frac{\sin x - x}{\sin(x^2)} = \frac{x^3}{x^2} \frac{(-1/3! + x^2/5! - \cdots)}{(1 - x^4/6 + \cdots)}.$$

Since $\lim\limits_{x \to 0} \left[ \dfrac{x^3}{x^2} \right] = \lim\limits_{x \to 0} x = 0$, you find

$$\lim_{x \to 0} \frac{\sin x - x}{\sin(x^2)} = 0 \cdot \frac{-1/3!}{1} = 0.$$

This example illustrates the power series method for these limits: Expand numerator and denominator and in each, factor out the lowest power of $x$; then form the quotient, and take the limit.

**EXAMPLE 6** Evaluate $\lim\limits_{x \to 0} \dfrac{\sin x - \tan^{-1} x}{x \sin x^2}$.

*SOLUTION* l'Hôpital's Rule would require the third derivative of numerator and denominator, a tedious calculation. But from Example 3,

$$\sin x - \tan^{-1} x = \left( x - \frac{x^3}{6} + \frac{x^5}{120} - \cdots \right) - \left( x - \frac{x^3}{3} + \frac{x^5}{5} - \cdots \right)$$

$$= \frac{x^3}{6} - \frac{23}{120} x^5 + \cdots = x^3 (1/6 - 23x^2/120 + \cdots)$$

and

$$x \sin x^2 = x \left( x^2 - \frac{(x^2)^3}{6} + \cdots \right) = x^3 - \frac{x^7}{6} + \cdots$$

$$= x^3 (1 - x^4/6 + \cdots)$$

so

$$\frac{\sin x - \tan^{-1} x}{x \sin x^2} = \frac{x^3}{x^3} \frac{1/6 - 23x^2/120 +}{1 - x^4/6 + \cdots} \to 1/6 \qquad \text{as } x \to 0.$$

## SUMMARY

***Theorem 1:*** If $\sum_{k=0}^{\infty} c_k x^k$ converges for $|x| < r$, then:

$$\frac{d}{dx} \sum_{k=0}^{\infty} c_k x^k = \sum_{k=0}^{\infty} k c_k x^{k-1}, \qquad |x| < r$$

$$\int_0^x \sum_{k=0}^{\infty} c_k t^k \, dt = \sum_{k=0}^{\infty} \frac{c_k}{k+1} x^{k+1}, \qquad |x| < r.$$

***Theorem 2:*** If $f(x) = \sum_{k=0}^{\infty} c_k x^k$, $|x| < r$, where $r > 0$, then this is the Maclaurin series of $f$; that is

$$c_k = f^{(k)}(0)/k!.$$

***Logarithm Series:***

$$\ln(1 + x) = \sum_{k=1}^{\infty} (-1)^{k-1} \frac{1}{k} x^k, \qquad -1 < x \le 1.$$

## PROBLEMS

### A

1. Obtain the Maclaurin series for $(1 - x)^{-2}$ by differentiating the geometric series.

2. **a)** Differentiate the sine series, and show that you get the cosine series.
   **b)** Differentiate the cosine series. What do you get?

3. **a)** Write out three terms of the series $\tan^{-1} x = \sum_0^{\infty} (-1)^k \frac{x^{2k+1}}{2k+1}$.
   **\*b)** Use the series to compute $f(0)$, $f'(0)$, $f''(0)$, $f'''(0)$, $f^{(4)}(0)$, $f^{(5)}(0)$ for the function $f(x) = \tan^{-1} x$.
   **c)** Try computing those same derivatives by repeated differentiation of $\tan^{-1} x$. (You'll see why this is an impractical way to obtain the Maclaurin series for $\tan^{-1} x$.)

4. The series $\sum_0^{\infty} \frac{nx^n}{n+1}$ can be shown to converge for $|x| < 1$. Call the sum of the series $f(x)$. Compute $f(0)$, $f'(0)$, $f''(0)$, $f'''(0)$, $f^{(4)}(0)$.

5. The following series can be shown to converge for all $x$. In each case, determine $f(0)$, $f'(0)$, $f''(0)$, and $f^{(n)}(0)$.

   **a)** $\sum_{n=1}^{\infty} \frac{x^n}{(n!)^2}$  **\*b)** $\sum_{k=0}^{\infty} \frac{x^{2k}}{(k!)^2}$

6. Obtain the Maclaurin series, by any legitimate method, for:

   **\*a)** $\dfrac{1}{1 + x^2}$.  **b)** $\dfrac{1}{a + x} \left( = \dfrac{1}{a} \cdot \dfrac{1}{1 + x/a} \right)$.

   **\*c)** $e^{-x^2/2}$.  **d)** $\dfrac{1}{\sqrt{2\pi}} \displaystyle\int_0^x e^{-t^2/2} \, dt$.

   **\*e)** $xe^x$.  **\*f)** $(\sin x)/x$.

   **\*g)** $\displaystyle\int_0^x \frac{\sin t}{t} \, dt$.

   **\*h)** $\frac{1}{2}(e^x + e^{-x})$. (Notice a similarity to the series for $\cos x$.)

   **i)** $\dfrac{1 - \cos x}{x^2}$.

   (The integral of part d gives the "standard normal" probability distribution, and part g is the Fresnel integral in optics; part h is the "hyperbolic cosine" $\cosh x$.)

7. Evaluate the following limits by power series.

   **\*a)** $\displaystyle\lim_{x \to 0} \frac{1 - \cos x}{x^2}$

**b)** $\lim\limits_{x \to 0} \dfrac{x \ln(1 + x) + 2 \cos x - 2}{x^2 \sin x}$

**\*c)** $\lim\limits_{x \to 0} \dfrac{e^x - e^{-x} - 2 \tan^{-1} x - x^3}{1 - \cos x - x^2/2}$

## B

**8.** This problem i) shows that the Maclaurin series for $\tan^{-1} x$ is valid for $|x| \leq 1$, and ii) estimates the error in using only $N$ terms. Rather than appeal to Theorem 1 on integrating power series in general, we obtain a more complete result by working directly with the explicit formula for the partial sums of the geometric series. Explain the following:

**a)** $\dfrac{1}{1 + t^2} = 1 - t^2 + \cdots + (-1)^N t^{2N}$

$$+ \dfrac{(-1)^{N+1} t^{2N+2}}{1 + t^2}, \text{ all } t.$$

**b)** If $0 \leq x \leq 1$ then

$$\left| \tan^{-1} x - \sum_0^N (-1)^n \frac{x^{2n+1}}{2n+1} \right|$$

$$\leq \int_0^x \frac{t^{2N+2}}{1 + t^2}\, dt \leq \int_0^x t^{2N+2}\, dt$$

$$= \frac{x^{2N+3}}{2N+3}.$$

**c)** $\tan^{-1} x = \sum_0^\infty (-1)^n \dfrac{x^{2n+1}}{2n+1}, \ -1 \leq x \leq 1.$
(Use part b for $0 \leq x \leq 1$, then note that $\tan^{-1} x$ is odd.)

**d)** $\pi/4 = 1 - \frac{1}{3} + \frac{1}{5} - \frac{1}{7} + \frac{1}{9} - \cdots.$

**9.** (Computation of $\pi$)

**a)** Prove $\tan(\alpha + \beta) = \dfrac{\tan \alpha + \tan \beta}{1 - \tan \alpha \tan \beta}.$

**b)** Let $\alpha = \tan^{-1}(1/2)$ and $\beta = \tan^{-1}(1/3)$. Prove $\tan(\alpha + \beta) = 1$, so $\alpha + \beta = \pi/4$.

**c)** Use $\sum_0^3 (-1)^n x^{2n+1}/(2n+1)$ to approximate $\tan^{-1}(1/2)$ and $\tan^{-1}(1/3)$.

**d)** Use problem 8b to estimate the accuracy of the approximations in part c.

**\*e)** Approximate $\pi$, and estimate the accuracy.

**10.** This problem shows that the Maclaurin series for $\ln(1 + x)$ is valid for $-1 < x \leq 1$. Explain:

**a)** $\dfrac{1}{1+t} = 1 - t + \cdots + (-1)^N t^N + \dfrac{(-t)^{N+1}}{1 + t}.$

**b)** If $0 \leq x \leq 1$,

$$\left| \ln(1 + x) - \sum_0^N (-1)^n \frac{x^{n+1}}{n+1} \right|$$

$$= \int_0^x \frac{t^{N+1}}{1 + t}\, dt \leq \int_0^x t^{N+1}\, dt = \frac{x^{N+2}}{N+2}.$$

**c)** $\ln(1 + x) = \sum_0^\infty \dfrac{(-1)^n x^{n+1}}{n+1}$ if $0 \leq x \leq 1.$

**d)** $\ln(1 + x) = \sum_1^\infty (-1)^{n-1} \dfrac{x^n}{n}$ if $-1 < x \leq 1.$
(Use Example 2 for $-1 < x < 1$, and part c for $x = 1$.)

**e)** $\ln 2 = 1 - \frac{1}{2} + \frac{1}{3} - \frac{1}{4} + \cdots.$

**11.** (Probability.) A parts factory tests its product, and finds that in the long run, the proportion of good parts is $p$; thus $p$ is the probability that a tested part is good, and $1 - p$ is the probability that it is bad. If $p$ is nearly equal to 1, then most of the parts are good, and the tester finds long strings of good parts, interrupted by an occasional bad part or two:

$$ggbggggggbbggggggggggggbgggg \ldots.$$

We ask: *What is the average length of a string of good parts?*

Suppose a string of $g$'s (good parts) has just ended, and we have arrived at a $b$ (bad part). Consider the string of $g$'s following this $b$; the string has length zero if and only if the next part is a $b$, and the probability of this is $(1 - p)$. The string of $g$'s has length 1 if the next parts come in the order $gb$, and the probability of this is $p(1 - p)$. (The probability that each of *two* independent events occurs is the product of the probabilities of each individual event.) Generally, the next string of good parts has length $n$ if the parts test like this:

$$\ldots bg \ldots\ldots\ldots gb.$$

$n$ times

end of previous      end of this
string              string

The probability of this event is $p^n(1 - p)$. Hence in all the strings of $g$'s that occur:

$(1 - p)$ is the proportion of strings of length 0,
$p(1 - p)$ is the proportion of strings of length 1,
$\vdots$
$p^n(1 - p)$ is the proportion of strings of length $n$.

Now:

**a)** If $p(n)$ denotes the proportion of strings of length $n$, show that $\sum_0^\infty p(n) = 1$. Why is this reasonable?

**\*b)** To obtain the average length of the "good strings," multiply each string length by the proportion of times it occurs, and add:

$$0(1-p) + 1 \cdot p(1-p) + \cdots + \underbrace{n}_{\text{length of string}} \underbrace{p^n(1-p)}_{\substack{\text{proportion of times} \\ \text{it occurs}}} + \cdots$$

Find the average length of the "good strings."

$$\left[ \sum_{n=0}^\infty nx^n = x\frac{d}{dx}\sum_0^\infty x^n. \right]$$

**c)** What is the average length of a good string if $p = 9/10$? $p = 1/2$? $p = 3/5$?

**d)** Show that the average "good string" length is an *increasing* function of $p$. Why is this reasonable?

*REMARK*  The same analysis applies to many other situations.

*EXAMPLE 1*  $p$ is the probability that a basketball player sinks his foul shots; then $\dfrac{p}{1-p}$ is the average length of a string of good shots.

*EXAMPLE 2*  At a certain stop sign, cars travel on the main street with random gaps between them. Let $p$ be the probability that a gap is too short for a car on the stop street to cross the main street; then $\dfrac{p}{1-p}$ is the average number of cars one must wait for before crossing.

## 11.7

### ESTIMATING REMAINDERS (optional)

Suppose some quantity $S$ is given by a convergent infinite series:

$$S = \sum_0^\infty a_k.$$

You can approximate $S$ by the partial sums of the series, $s_n = \sum_0^n a_k$. The error in this approximation is the sum of all the unused terms, $\sum_{n+1}^\infty a_k$; an important part of the calculation is to estimate this error. We give two methods here, and another in the next section.

### Comparison Estimates

When all the terms $a_k$ are $\geq 0$, the simplest method is to compare the terms $a_k$ to the terms of a series whose sum is known, for example, a geometric series. This is a **comparison estimate**.

---

*EXAMPLE 1*  Since $e^x = \sum_0^\infty \dfrac{x^k}{k!}$, then $e^1 = \sum_0^\infty \dfrac{1^k}{k!}$, that is

$$e = \sum_0^\infty \frac{1}{k!}.$$

Estimate the error in approximating $e$ by five terms of this series.

*SOLUTION*

$$e = \underbrace{\left[\frac{1}{0!} + \frac{1}{1!} + \frac{1}{2!} + \frac{1}{3!} + \frac{1}{4!}\right]}_{\text{The five terms}} + \underbrace{\left[\frac{1}{5!} + \frac{1}{6!} + \frac{1}{7!} + \cdots\right]}_{\text{The remaining terms}}. \tag{1}$$

In approximating by means of the first five terms, you have to estimate the sum of the unused terms

$$\sum_{5}^{\infty} \frac{1}{k!} = \frac{1}{5!} + \frac{1}{6!} + \frac{1}{7!} + \frac{1}{8!} + \cdots. \tag{2}$$

Since $6! = (6)(5!)$, $7! = (7 \cdot 6)(5!)$, $8! = (8 \cdot 7 \cdot 6)(5!)$, etc, you can factor 5! out of each denominator in (2):

$$\sum_{5}^{\infty} \frac{1}{k!} = \frac{1}{5!}\left( 1 + \frac{1}{6} + \frac{1}{7 \cdot 6} + \frac{1}{8 \cdot 7 \cdot 6} + \cdots \right). \tag{3}$$

Don't try to compute this sum exactly; instead *estimate it* by noticing that

$$\frac{1}{7 \cdot 6} < \frac{1}{6 \cdot 6} = \frac{1}{6^2}, \frac{1}{8 \cdot 7 \cdot 6} < \frac{1}{6 \cdot 6 \cdot 6} = \frac{1}{6^3}, \text{ etc.}$$

So from (3),

$$\sum_{k=5}^{\infty} \frac{1}{k!} < \frac{1}{5!}\left( 1 + \frac{1}{6} + \frac{1}{6^2} + \frac{1}{6^3} + \cdots \right) = \frac{1}{5!}\sum_{0}^{\infty}\left(\frac{1}{6}\right)^k = \frac{1}{5!}\frac{1}{1 - (1/6)}$$

$$= \frac{1}{5!} \cdot \frac{6}{5} = \frac{1}{100} = 0.01.$$

Hence the unused terms contribute less than 0.01, and

$$e = 1 + 1 + \tfrac{1}{2} + \tfrac{1}{6} + \tfrac{1}{24} = 2.708333 \dots$$

with an error $< 0.01$.

---

The method in Example 1 applies when all the unused terms $a_{n+1}$, $a_{n+2}, \ldots$ are positive. If you can find modified terms $b_k$ such that $a_k \le b_k$ for $k > n$, and the sum $\sum_{n+1}^{\infty} b_k$ is known, then the sum of the unused terms is less than $\sum_{n+1}^{\infty} b_k$:

$$\text{If} \quad 0 \le a_k \le b_k \text{ for } k > n, \quad \text{then} \quad 0 \le \sum_{n+1}^{\infty} a_k \le \sum_{n+1}^{\infty} b_k.$$

### Estimates for Alternating Series

A series whose terms are alternately positive and negative is called **alternating**. If, in addition, the *absolute values* of the terms decrease monotonically to zero, then the accuracy is easily estimated: The size of the error in approximating by any partial sum is *less than the size of the first unused term*. To show this, write the series as

$$b_1 - b_2 + b_3 - b_4 + \cdots = S,$$

where the $b_j$ are all positive, and $b_1 \ge b_2 \ge b_3 \ge b_4 \ge \ldots$ (since the absolute values decrease monotonically). The partial sums

$$s_1 = b_1, \quad s_2 = b_1 - b_2, \quad s_3 = b_1 - b_2 + b_3, \ldots$$

are illustrated in Figure 1; they oscillate back and forth around their limit $S$. As the figure shows, $0 < S - s_4 < s_5 - s_4 = b_5$, and $0 < s_3 - S < s_3 -$

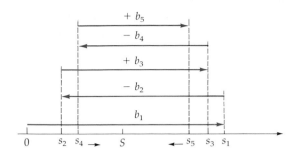

**FIGURE 1**
Partial sums of an alternating series.

$s_4 = b_3$. In general, the distance between $S$ and $s_n$ is less than the distance between $s_{n+1}$ and $s_n$:

$$\left| S - s_n \right| < \left| s_{n+1} - s_n \right| = b_{n+1}.$$

Thus the error in approximating $S$ by $s_n$ is less than the absolute value of the first unused term, $b_{n+1}$.

---

**EXAMPLE 2**   Approximate sin 1, using four terms of the sine series.

**SOLUTION**

$$\sin x = \sum_0^\infty \frac{(-1)^k}{(2k+1)!} x^{2k+1}$$

so

$$\sin 1 = \sum_0^\infty \frac{(-1)^k}{(2k+1)!}$$

$$= 1 - \frac{1}{3!} + \frac{1}{5!} - \frac{1}{7!} + \frac{1}{9!} - \frac{1}{11!} + \cdots.$$

$$\underbrace{\phantom{1 - \frac{1}{3!} + \frac{1}{5!} - \frac{1}{7!}}}_{\text{four terms}} \quad \underbrace{\phantom{\frac{1}{9!} - \frac{1}{11!}}}_{\text{remaining terms}}$$

The series has strictly alternating signs, and the absolute values $1, \frac{1}{3!}, \frac{1}{5!}, \frac{1}{7!}, \ldots$ decrease monotonically to zero, so the error in approximating by four terms is less than the size of the fifth term, $1/9!$:

$$\left| \sin 1 - \left( 1 - \frac{1}{3!} + \frac{1}{5!} - \frac{1}{7!} \right) \right| < \frac{1}{9!} = .00000276 \ldots.$$

So to five decimal places

$$\sin 1 = 1 - \frac{1}{3!} + \frac{1}{5!} - \frac{1}{7!} = .841468 \ldots$$

$$= .84147$$

rounded to five places.

**EXAMPLE 3** Compute the "normal probability integral" $\int_0^1 e^{-x^2/2}\,dx$ to four decimal places.

*SOLUTION* From the exponential series,

$$e^{-x^2/2} = \sum_0^\infty \frac{(-x^2/2)^k}{k!} = \sum_0^\infty \frac{(-1)^k}{k!} \frac{x^{2k}}{2^k}$$

so

$$\int_0^1 e^{-x^2/2}\,dx = \int_0^1 \sum_0^\infty \frac{(-1)^k}{k!} \frac{x^{2k}}{2^k}\,dx$$

$$= \sum_0^\infty \frac{(-1)^k}{k!\,2^k} \int_0^1 x^{2k}\,dx$$

$$= \sum_0^\infty \frac{(-1)^k}{k!\,2^k} \frac{1}{2k+1}$$

$$= 1 - \frac{1}{2\cdot 3} + \frac{1}{2\cdot 4\cdot 5} - \frac{1}{6\cdot 8\cdot 7} + \frac{1}{24\cdot 16\cdot 9}$$

$$\qquad - \frac{1}{120\cdot 32\cdot 11} + \cdots$$

$$= 1 - \frac{1}{6} + \frac{1}{40} - \frac{1}{336} + \frac{1}{3456} - \frac{1}{42240} + \cdots.$$

The signs are strictly alternating, and the absolute values decrease (rapidly!) so the error is less than the size of the first unused term. The sixth term has absolute value

$$\frac{1}{42240} = .000023\ldots < .00003$$

so we can neglect this for four-decimal accuracy:

$$\int_0^1 e^{-x^2/2}\,dx = 1 - \frac{1}{6} + \frac{1}{40} - \frac{1}{336} + \frac{1}{3456} = .85565$$

with an error $< .00003$.

## SUMMARY

*Comparison Estimate:* If $0 \le a_n \le b_n$ for $n > N$ then the difference between the infinite series $\sum_0^\infty a_n$ and the $N^{\text{th}}$ partial sum $\sum_0^N a_n$ is estimated by

$$\sum_{N+1}^\infty a_n \le \sum_{N+1}^\infty b_n.$$

*Alternating Series Estimate:* If $a_n = (-1)^n|a_n|$ and $\{|a_n|\}$ is a *decreasing* sequence for $n > N$, then

$$\left| \sum_0^\infty a_n - \sum_0^N a_n \right| < |a_{N+1}|.$$

## PROBLEMS

### A

1. Estimate the error in approximating $e$ by

   a) $1 + 1 + \dfrac{1}{2!} + \dfrac{1}{3!} + \dfrac{1}{4!} + \dfrac{1}{5!}$.

   *b) $\displaystyle\sum_{0}^{10} \dfrac{1}{k!}$.

*2. Approximate $e^{1/2}$ by three terms of the exponential series. Estimate the error.

*3. Approximate $e^{-1}$ by five terms of the exponential series. Estimate the error.

4. Approximate the given quantity by the given number of (nonzero) terms of an appropriate series, and estimate the error.

   *a) $\displaystyle\int_0^1 \dfrac{\sin x}{x}\,dx$, four terms

   *b) $\displaystyle\int_0^{1/2} e^{x^2}\,dx$, four terms

   c) $\displaystyle\int_0^{1/2} e^{-x^2}\,dx$, three terms

   d) $\displaystyle\int_0^1 \sin x^2\,dx$, three terms

### B

5. Approximate the given quantity to five decimal places, and prove the accuracy.

   *a) $\displaystyle\int_0^{1/2} \dfrac{dx}{1 + x^3}$

   b) $\displaystyle\int_0^{0.1} e^{-x^2/2}\,dx$

   c) $\displaystyle\int_0^{1/2} \dfrac{\sin x}{x}\,dx$

6. Write the power series for $\tan^{-1} x = \displaystyle\int_0^x \dfrac{dt}{1 + t^2}$.

   For each of the following, find a partial sum that has error $< 10^{-8}$ in absolute value. (Do not attempt to calculate the sum; just determine how many terms are needed.)

   a) $\tan^{-1}(0.01)$  b) $\tan^{-1}(1/5)$

   c) $\tan^{-1}(3/5)$  d) $\tan^{-1}(1)$

## 11.8

### CONVERGENCE TESTS. THE INTEGRAL TEST

So far we have dealt with infinite series where the sum is known beforehand, and the series is a means of computing it. But many series come to us without a known sum, and it may not be clear from the outset that the series converges. For example, the differential equation

$$x\,\dfrac{d^2 y}{dx^2} + \dfrac{dy}{dx} + xy = 0$$

arises in the study of vibrating membranes (like drumheads); its solution gives the cross-sectional shape of the drumhead, in its so-called fundamental mode of vibration. This solution is given by a power series

$$\sum_{k=0}^{\infty} \dfrac{(-1)^k}{(k!)^2} \left(\dfrac{x}{2}\right)^{2k}.$$

*Assuming that the series converges,* you can check that it solves the differential equation, and then evaluate it by the methods in the previous section. But first, you must show that it really does converge!

So now we take up an important general problem: *Given an infinite series, decide whether or not it converges.* This is the central problem of the rest of the chapter.

The first prerequisite for the convergence of a series $\sum_1^\infty a_k$ is that the individual terms $a_k$ must tend to zero as $k \to \infty$. This is:

---

**THE ZERO-LIMIT TEST**

    **a.**  If the series $\sum_1^\infty a_k$ converges, then the sequence $\{a_k\}$ converges to zero.

    **b.**  If $\{a_k\}$ does *not* converge to 0, then $\sum_1^\infty a_k$ diverges.

---

Each of these two statements implies the other, as a moment's reflection will show, so we need prove only the first. Suppose that $\sum_1^\infty a_k$ converges. Then the partial sums $s_n = \sum_1^n a_k$ converge to a limit $S$. Now, $a_n$ can be expressed as the difference between $s_n$ and $s_{n-1}$:

$$s_n - s_{n-1} = (a_1 + \cdots + a_{n-1} + a_n) - (a_1 + \cdots + a_{n-1}) = a_n.$$

Since $s_n \to S$, it follows that $s_{n-1} \to S$ as well, so

$$a_n = s_n - s_{n-1} \to S - S = 0.$$

This proves part a, and part b follows.

*REMARK 1*    We stress the distinction between the *terms* $a_k$ and the *partial sums* $s_n$. If the partial sums $s_n$ have a limit $S$ (not necessarily 0), then the terms $a_k$ necessarily have the limit 0. Think of the $a_k$ as lengths laid end-to-end to form the partial sums (Fig. 1), like the successive steps in Zeno's paradox. If the partial sums $s_n$ tend to a limit $S$, then the successive steps $a_k$ must become vanishingly *small*.

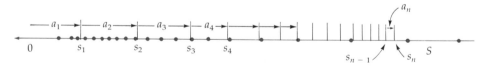

**FIGURE 1**
$s_n = \sum_1^n a_k$. If $s_n \to S$, then $a_n \to 0$.

*REMARK 2*    Read the Test carefully. If $a_k \not\to 0$ then $\sum a_k$ diverges; on the other hand, $a_k \to 0$ *does not guarantee convergence*, as you will see in the examples.

---

**EXAMPLE 1**    Test $\sum_1^\infty \dfrac{k}{k+1}$ for convergence.

*SOLUTION*    Take the limit of the general term:

$$a_k = \frac{k}{k+1} \to 1 \quad \text{as } k \to \infty.$$

The general term does *not* converge to 0, so by the Zero-limit Test the series diverges. Eventually, the successive steps are nearly one unit long; as you

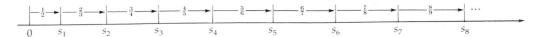

**FIGURE 2**

$$\sum_1^\infty \frac{k}{k+1} = \lim_{n\to\infty} S_n = \infty.$$

add up more and more of them, the partial sums grow arbitrarily large (Fig. 2).

**EXAMPLE 2**   Test $\sum_1^\infty (-1)^k$.

***SOLUTION***   The series is $-1 + 1 - 1 + 1 - \cdots$. The general term $(-1)^k$ oscillates between $+1$ and $-1$, and does not converge to *any* limit. In particular, the general term does not converge to zero so, by the Zero-limit Test, the series $\sum(-1)^k$ diverges. (You could see this without any test; the sequence of partial sums is $-1, -1 + 1 = 0, -1 + 1 - 1 = -1, \ldots$; it oscillates back and forth between $-1$ and $0$, and does not converge.)

**EXAMPLE 3**   Test $\sum_1^\infty \frac{1}{\sqrt{k}}$ for convergence.

***SOLUTION***   The general term $\dfrac{1}{\sqrt{k}}$ converges to 0, so the series "passes" the Zero-limit Test. However, this series diverges (Sec. 11.4, Example 3). The Zero-limit Test is just a preliminary: If the series "fails" this test, it surely diverges. But if it "passes," it might converge $\left[\text{like } \sum \left(\frac{1}{2}\right)^j\right]$ or diverge $\left[\text{like } \sum \dfrac{1}{\sqrt{k}}\right]$; then more testing is required.

The Zero-limit Test can *never* prove that a series *converges*. Tests that do prove convergence all rest on a basic property of the real numbers, illustrated in Figure 3. Consider a sequence of real numbers $s_1, s_2, s_3, \ldots$ that keep increasing, but never pass a certain bound $M$. It seems clear, then, that they must converge to some limit $L \le M$. This is true, but the proof depends on a deep study of the real numbers, well beyond the scope of this book.

**FIGURE 3**

$s_1 \le s_2 \le s_3 \le \cdots \le M$, so $s_n \to L$ for some $L \le M$.

So we state without proof:

---

**The Increasing Sequence Principle**

An increasing sequence with an upper bound $M$ converges to a finite limit $L \leq M$.

---

*Increasing* means, formally, that $s_n \leq s_{n+1}$ for every $n$.

A *decreasing* sequence with a *lower bound* $m$ must also converge to a finite limit $L \geq m$ (Fig. 4).

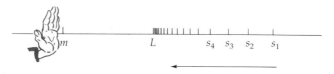

**FIGURE 4**
$s_1 \geq s_2 \geq \cdots \geq m$, so $s_n \to L$ for some $L \geq m$.

## Positive Series

A series $\sum_{k=1}^{\infty} a_k$ is called *positive* if each term $a_k$ is $\geq 0$. For a positive series, the partial sums form an increasing sequence:

$$a_2 \geq 0 \quad \Rightarrow \quad s_2 = a_1 + a_2 \geq a_1 = s_1$$

and generally

$$a_n \geq 0 \quad \Rightarrow \quad s_n = a_1 + \cdots + a_n \geq a_1 + \cdots + a_{n-1} = s_{n-1}.$$

If the sequence of partial sums $\{s_n\}$ is unbounded, then it diverges, and we write $\sum_{k=1}^{\infty} a_k = \infty$. On the other hand, if the partial sums $s_n$ have a fixed bound $M$, independent of $n$, then by the Increasing Sequence Principle $\{s_n\}$ has a limit; and this means, by definition, that $\sum_1^{\infty} a_k$ converges.

How are you to tell whether the partial sums $s_n$ are bounded or not? In some cases you can compare them to a related integral:

---

**THE INTEGRAL TEST**

Suppose that $a_k = f(k)$, where $f$ is a continuous, positive and decreasing function. Then

$$\sum_1^{\infty} f(k) \quad \text{and} \quad \int_1^{\infty} f(x)\, dx \tag{1}$$

either both converge, or both diverge.

---

**PROOF**   Suppose at first that the integral in (1) diverges. Since $f(x) > 0$, this means that

$$\lim_{n \to \infty} \int_1^n f(x)\, dx = +\infty.$$

Now, you can compare the partial sums $s_n$ to the integral $\int_1^{n+1} f(x)\, dx$ as in Figure 5. Each rectangle has base 1, and the heights are $f(1), f(2), \ldots, f(n)$. So the *areas* of the rectangles are $f(1), f(2), \ldots, f(n)$, and the sum of these areas is precisely the partial sum $s_n$. Since the rectangles cover the area under the graph,

$$s_n \geq \int_1^{n+1} f(x)\, dx.$$

If the integral diverges, then, $\int_1^{n+1} f(x)\, dx \to +\infty$, so $s_n \to +\infty$ too. Thus the partial sums $s_n = \sum_1^n f(k)$ do not converge to a finite limit, and the series $\sum_1^\infty f(k)$ diverges, as claimed.

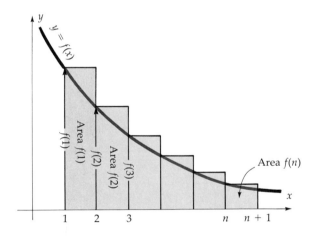

**FIGURE 5**
$$s_n = f(1) + f(2) + \cdots + f(n) \geq \int_1^{n+1} f(x)\, dx.$$

Now for the more interesting part—suppose that $\int_1^\infty f(x)\, dx$ *does* converge to a finite value. This means that the area under the graph of $f(x)$ for $1 \leq x < \infty$ is finite. To show that the partial sums $s_n$ form a bounded sequence, modify Figure 5; draw the partial sums *underneath* the graph of $f$. We can no longer cover the first rectangle (Fig. 6), but we do cover all the others:

$$\sum_2^n f(k) \leq \int_1^\infty f(x)\, dx.$$

Add $f(1)$ to each side of this:

$$s_n = \sum_1^n f(k) \leq f(1) + \int_1^\infty f(x)\, dx.$$

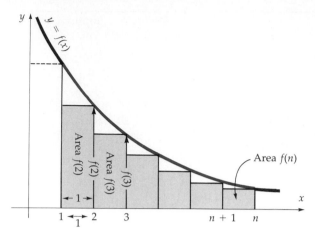

**FIGURE 6**

$$f(2) + \cdots + f(n) \leq \int_1^n f(x)\, dx.$$

Thus $s_n < M$, where $M$ is a fixed number independent of $n$. So the sequence of partial sums $\{s_n\}$ is bounded; and it is increasing (because each $a_k \geq 0$) so it converges to a finite limit $S$. Thus the series $\sum_1^\infty f(k)$ converges to a limit $S$, as was to be proved.

---

**EXAMPLE 4**   Test $\displaystyle\sum_1^\infty \frac{1}{k^2}$ for convergence.

**SOLUTION**   $\dfrac{1}{k^2} = f(k)$ if $f(x) = 1/x^2$. This function $f$ is positive and decreasing, so the Integral Test may be applied. You find

$$\int_1^\infty f(x)\, dx = \int_1^\infty \frac{1}{x^2}\, dx = \lim_{n \to \infty} \int_1^n \frac{1}{x^2}\, dx$$

$$= \lim_{n \to \infty} \left(1 - \frac{1}{n}\right) = 1.$$

The integral converges, so the series also converges: $\displaystyle\sum_1^\infty \frac{1}{k^2} < \infty$.

---

**REMARK 3**   *Note that the sum of the series does not equal the integral*—the relation between the series and the integral is one of *inequality*. In Figure 5, the integral includes only the area under the graph of $f$, while the sum of the series includes as well the "steps" projecting above the graph.

**REMARK 4**   The series $\displaystyle\sum_{k=1}^\infty \frac{1}{k^2}$ is a so-called "$p$ series"; it has the form

$\displaystyle\sum_{k=1}^\infty \frac{1}{k^p}$, with a constant $p$. All these series can easily be tested for conver-

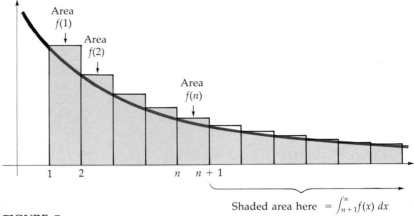

FIGURE 7

$$f(1) + f(2) + \cdots + f(n) + \int_{n+1}^{\infty} f(x)\, dx \le \sum_{1}^{\infty} f(k).$$

gence by the Integral Test. You find that

$$\sum_{k=1}^{\infty} \frac{1}{k^p} \qquad \begin{cases} \text{converges} & \text{if } p > 1, \\ \text{diverges} & \text{if } p \le 1. \end{cases}$$

**Approximating the sum of an infinite series** is easy, when convergence is proved by the Integral Test. In Figure 7, the shaded area is entirely covered by the infinite series of rectangles, and this shows that

$$\sum_{k=1}^{n} f(k) + \int_{n+1}^{\infty} f(x)\, dx \le \sum_{k=1}^{\infty} f(k). \tag{2}$$

Shifting to Figure 8, the infinite series of rectangles is entirely covered by the shaded area, so

$$\sum_{1}^{\infty} f(k) \le \sum_{k=1}^{n} f(k) + \int_{n}^{\infty} f(x)\, dx. \tag{3}$$

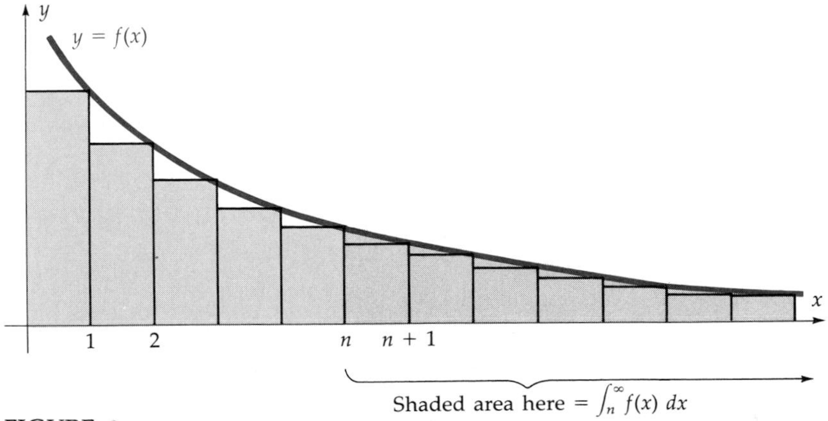

Shaded area here $= \int_{n}^{\infty} f(x)\, dx$

FIGURE 8

$$\sum_{1}^{\infty} f(k) \le \sum_{1}^{n} f(k) + \int_{n}^{\infty} f(x)\, dx.$$

These two inequalities bracket the infinite sum between two computable quantities.

---

**EXAMPLE 5** Approximate the sum of the series $\displaystyle\sum_1^\infty \frac{1}{k^2}$, to two decimal places.

**SOLUTION** Combine (2) and (3), using $f(x) = \dfrac{1}{x^2}$:

$$\sum_1^n \frac{1}{k^2} + \int_{n+1}^\infty \frac{1}{x^2}\,dx < \sum_1^\infty \frac{1}{k^2} < \sum_1^n \frac{1}{k^2} + \int_n^\infty \frac{1}{x^2}\,dx.$$

Evaluate the integrals, to get a *lower* approximation of $\displaystyle\sum_{k=1}^\infty \frac{1}{k^2}$, and an *upper* approximation:

$$\sum_1^n \frac{1}{k^2} + \frac{1}{n+1} < \sum_1^\infty \frac{1}{k^2} < \sum_1^n \frac{1}{k^2} + \frac{1}{n}. \tag{4}$$

For two-decimal accuracy, you need to choose $n$ so that the difference between these two approximations is $< \frac{1}{2} \times 10^{-2}$. The difference is

$$\frac{1}{n} - \frac{1}{n+1} = \frac{1}{n(n+1)}.$$

A little experimentation shows this to be $< \frac{1}{2} \times 10^{-2}$ when $n = 14$. So compute each side of (4) with $n = 14$:

$$1.64266\ldots < \sum_{k=1}^\infty \frac{1}{k^2} < 1.64742\ldots.$$

[By a more sophisticated method (Fourier series) it can be shown that

$$\sum_1^\infty \frac{1}{k^2} = \frac{\pi^2}{6} = 1.644934\ldots.]$$

---

In Example 5, the approximations consist of a partial sum *plus an integral correction*; the corrected sum is much more accurate than the partial sum itself.

---

## SUMMARY

***Zero-limit Test:*** If $\{a_k\}$ does *not* converge to 0, then $\sum_1^\infty a_k$ diverges.

***Increasing Sequence Principal:*** Every bounded increasing sequence converges to a finite limit.

***Integral Test:*** If $f(x)$ is positive and decreasing for $x \ge 1$, then

$$\sum_1^\infty f(k) \quad \text{converges if and only if} \quad \int_1^\infty f(x)\,dx < \infty.$$

*p-series:*

$$\sum_{k=1}^{\infty} \frac{1}{k^p} \quad \begin{cases} \text{converges} & \text{if } p > 1, \\ \text{diverges} & \text{if } p \leq 1. \end{cases}$$

## PROBLEMS

### A

**\*1.** Which of the following series "pass" the Zero-limit Test? Of those that do, decide convergence by the integral test. (Of course, there are many series whose convergence cannot be decided by these two tests; other tests will be given in later sections.)

**a)** $\displaystyle\sum_{1}^{\infty} \frac{k+1}{2k+1}$

**b)** $\displaystyle\sum_{1}^{\infty} \frac{1}{k^3}$

**c)** $\displaystyle\sum_{1}^{\infty} (-1)^{n+1} \frac{n}{n+1}$

**d)** $\displaystyle\sum_{0}^{\infty} \frac{1}{\sqrt{n+1}}$

**e)** $\displaystyle\sum_{1}^{\infty} \frac{\ln k}{k}$

**f)** $1 + \dfrac{1}{2\sqrt{2}} + \dfrac{1}{3\sqrt{3}} + \dfrac{1}{4\sqrt{4}} + \cdots$

**g)** $\displaystyle\sum_{1}^{\infty} \frac{1}{n^{1.01}}$

**h)** $1 + \dfrac{1}{3} + \dfrac{1}{5} + \dfrac{1}{7} + \cdots$

**i)** $\displaystyle\sum_{1}^{\infty} \frac{1}{k^2 + 1}$

**j)** $\displaystyle\sum_{0}^{\infty} \frac{2^j}{j^2 + 1}$

**k)** $\displaystyle\sum_{1}^{\infty} \frac{5}{3k}$

**l)** $\displaystyle\sum_{2}^{\infty} \frac{1}{k \ln k}$

**2.** Repeat problem 1 for the following series.

**a)** $\displaystyle\sum_{2}^{\infty} \frac{1}{m(\ln m)^2}$   **b)** $\displaystyle\sum_{0}^{\infty} \frac{1}{(k+1)^4}$

**c)** $\displaystyle\sum_{0}^{\infty} \frac{k^2}{k^3 + 3}$   **d)** $\displaystyle\sum_{1}^{\infty} n^{1/n}$

**e)** $\displaystyle\sum_{1}^{\infty} 3^n/n^2$   **f)** $\displaystyle\sum_{1}^{\infty} \left(\frac{1}{k} + \frac{1}{k^2}\right)$

**3.** Approximate the following series to two decimal places, using (2) and (3).

**\*a)** $\displaystyle\sum_{k=1}^{\infty} \frac{1}{k^3}$   **b)** $\displaystyle\sum_{0}^{\infty} \frac{k}{(k^2 + 1)^2}$

**4.** Compute $\displaystyle\sum_{k=1}^{\infty} \frac{1}{k^4}$ to four decimal places. (This sum is $\pi^4/90$.)

### B

**5.** Show that

**a)** $\displaystyle\sum_{1}^{\infty} \frac{1}{n^p}$ converges if the constant power $p$ is greater than 1.

**b)** $\displaystyle\sum_{1}^{\infty} \frac{1}{n^p}$ diverges if $p \leq 1$. (Treat separately the cases $p = 1$ and $p < 1$.)

**6.** For what values of $q$ does $\displaystyle\sum_{2}^{\infty} \frac{1}{k(\ln k)^q}$ converge?

**7.** For what values of $p$ and $q$ does $\displaystyle\sum_{2}^{\infty} \frac{1}{k^p(\ln k)^q}$ converge?

### C

**8.** This problem shows how to solve the differential equation

$$xy'' + y' + xy = 0 \quad \text{(Bessel's equation)}$$

by power series. Assume that $y$ is given by such a series:

$$y = a_0 + a_1 x + a_2 x^2 + a_3 x^3 + a_4 x^4 + \cdots$$
$$+ a_n x^n + \cdots. \tag{5}$$

a) Write out the first five terms, and the term in $x^n$, for each of $xy$, $y'$, and $xy''$.

b) Add these terms, and set the coefficient of each power of $x$ equal to 0.

c) Deduce that $a_1 = a_3 = a_5 = \cdots = 0$, while $a_2 = -2^{-2}a_0$, $a_4 = -4^{-2}a_2, \ldots$.

d) Determine $a_2$, $a_4$, $a_6$, $a_{2k}$, in terms of $a_0$.

e) Write the resulting series (5) with $a_0 = 1$; this is the series given at the beginning of this section. We will show in Section 11.11 that it converges for every $x$. Its sum is the *Bessel function* $J_0(x)$, one of the standard solutions of Bessel's equation, just as $\sin x$ is one of the standard solutions of $y'' + y = 0$.

## 11.9

### THE COMPARISON TEST

The Integral Test compares a series to an integral. You can also compare one series to another:

---

**THE COMPARISON TEST**

If

$$0 \le a_k \le b_k \tag{1}$$

and $\sum_1^\infty b_k$ converges, then $\sum_1^\infty a_k$ converges.

---

**PROOF** Because $0 \le a_k$, the partial sums $s_n = \sum_1^n a_k$ form an increasing sequence. Because $a_k \le b_k$,

$$s_n = \sum_1^n a_k \le \sum_1^n b_k. \tag{2}$$

Because $b_k \ge 0$ from (1), and $\sum_1^\infty b_k$ converges, we also have

$$\sum_1^n b_k \le \sum_1^n b_k + \sum_{n+1}^\infty b_k = \sum_1^\infty b_k. \tag{3}$$

By (2) and (3) together, the increasing sequence $s_n$ is bounded by a fixed number $M = \sum_1^\infty b_k$. By the Increasing Sequence Principle, the sequence of partial sums $s_n$ converges to a finite limit $S$. That is, the series $\sum_1^\infty a_k$ converges, and the Comparison Test is proved.

Briefly, for two positive series with $a_k \le b_k$, if the larger one converges, so does the smaller. This implies also that *if the smaller series* $\sum a_k$ *diverges, then so does the larger one*. For if, on the contrary, $\sum_1^\infty b_k$ converged, then so would $\sum_1^\infty a_k$ converge, by the Comparison Test.

---

**EXAMPLE 1** Test $\sum_1^\infty \dfrac{1}{k^2 + k}$ for convergence.

**SOLUTION** Compare $\sum_1^\infty \dfrac{1}{k^2 + k}$ to $\sum_1^\infty \dfrac{1}{k^2}$. The terms $\dfrac{1}{k^2 + k}$ are less than

$\frac{1}{k^2}$, (larger denominators make smaller fractions) and $\sum_1^\infty \frac{1}{k^2}$ converges (by the Integral Test). Thus

$$0 \le \frac{1}{k^2 + k} < \frac{1}{k^2} \qquad \text{for every } k, \text{ and } \sum_1^\infty \frac{1}{k^2} \text{ converges.}$$

So by the Comparison Test, $\sum \frac{1}{k^2 + k}$ converges.

In order to make comparisons, you need a list of positive series whose convergence or divergence is known. A good list contains:

The geometric series: $\quad \sum_0^\infty r^k \qquad \begin{array}{l} \text{converges if } 0 \le r < 1 \\ \text{diverges if } r \ge 1 \end{array}$

The exponential series: $\quad \sum_0^\infty \frac{r^k}{k!} \qquad \text{converges for every } r.$

The "$p$-series": $\quad \sum_1^\infty \frac{1}{k^p} \qquad \begin{array}{l} \text{converges if } p > 1 \\ \text{diverges if } p \le 1. \end{array} \qquad \text{(Sec. 11.8)}$

**EXAMPLE 2**   Test $\sum_1^\infty \frac{1}{2k - 1}$ for convergence.

*SOLUTION*   You could apply the Integral Test, but a comparison is easier:

$$\frac{1}{2k - 1} > \frac{1}{2k} \qquad \text{(since } 2k - 1 < 2k\text{)}$$

and

$$\sum_1^\infty \frac{1}{2k} = \frac{1}{2} \sum_1^\infty \frac{1}{k}$$

is divergent, because $\sum \frac{1}{k}$ is a "$p$-series" with $p = 1$. Since the smaller series $\sum \frac{1}{2k}$ diverges, so does the larger one $\sum \frac{1}{2k - 1}$.

**EXAMPLE 3**   Test $\sum_1^\infty \frac{1}{2k + 1}$ for convergence.

*SOLUTION*   When $k$ is large, the terms $\frac{1}{2k + 1}$ behave much like the simpler terms $\frac{1}{2k}$; so we expect the series to diverge, because $\frac{1}{2} \sum_1^\infty \frac{1}{k}$ diverges. In order to find an appropriate comparison to *prove* divergence, compare the first several terms of the given series with this known divergent series:

$$\text{Given, } \frac{1}{2k+1}: \quad \frac{1}{3}, \ \frac{1}{5}, \ \frac{1}{7}, \ \frac{1}{9}, \ \frac{1}{11}, \dots$$

$$\text{Known, } \frac{1}{k}: \quad 1, \ \frac{1}{2}, \ \frac{1}{3}, \ \frac{1}{4}, \ \frac{1}{5}, \dots$$

This suggests that the term $\dfrac{1}{2k+1}$ is always *at least 1/3 as large* as $1/k$:

$$\frac{1}{3} \cdot \frac{1}{k} \le \frac{1}{2k+1} \qquad \text{for } k = 1, 2, \dots. \tag{4}$$

This is correct, since $\dfrac{2k+1}{k} = 2 + \dfrac{1}{k} \le 3$, for $k = 1, 2$, etc.

The inequality (4) proves that $\displaystyle\sum_{1}^{\infty} \frac{1}{2k+1}$ diverges, by comparison to the known divergent series $\dfrac{1}{3} \displaystyle\sum_{1}^{\infty} \frac{1}{k}$.

---

The essential idea in Example 3 was to see what the terms are like for large $k$; basically, they are like $1/2k$, and it turned out that we could make a comparison to $\displaystyle\sum_{1}^{\infty} \frac{1}{k}$. This idea is systematized in:

---

### THE LIMIT COMPARISON TEST

Suppose that $a_k \ge 0$ and $b_k \ge 0$. If

$$\lim_{k \to \infty} \frac{a_k}{b_k} \text{ is finite and nonzero,}$$

then $\sum a_k$ and $\sum b_k$ either *both converge*, or *both diverge*.

---

To apply this in Example 3, take $a_k = \dfrac{1}{2k+1}$ and $b_k = 1/k$. Then

$$\lim \frac{a_k}{b_k} = \lim \frac{1/(2k+1)}{1/k} = \frac{1}{2}.$$

This limit is finite and nonzero; since the comparison series $\sum b_k = \sum \dfrac{1}{k}$ diverges, so does the given series $\sum a_k = \sum \dfrac{1}{2k+1}$.

**The proof of the limit comparison test** uses Theorem 2, Section 11.3:

Every convergent sequence is bounded.

Since $\lim \dfrac{a_k}{b_k}$ is assumed finite, the sequence $\{a_k/b_k\}$ is bounded by some constant $M$:

$$a_k/b_k \leq M. \tag{5}$$

Thus $a_k \leq Mb_k$, so the convergence of $\sum b_k$ implies that of $\sum a_k$. Moreover, since $\lim \dfrac{a_k}{b_k}$ is not zero, the reciprocal of this sequence also has a finite limit; $\lim \dfrac{b_k}{a_k}$ exists. This implies that $b_k/a_k$ is a bounded sequence, so $b_k \leq Ma_k$ for some constant $M$. Hence the convergence of $\sum a_k$ implies that of $\sum b_k$. This completes the proof.

The Limit Comparison Test is an obvious choice for terms involving only powers and roots of the index $k$; for large $k$, they behave like a certain power of $k$, and can be compared to a $p$-series $\sum 1/k^p$.

---

**EXAMPLE 4**   Test $\sum \dfrac{3k + 2k^{1/3}}{(k^{3/2} + 1)^2}$ for convergence. (The physical world is unlikely to confront you with such a series, but calculus instructors are something else. Be prepared!)

*SOLUTION*   In an expression such as this general term, the behavior for large $k$ is determined by dropping all but the highest power of $k$ in numerator and denominator, getting $\dfrac{3k}{k^3} = 3/k^2$. Hence, compare the given series to the convergent $p$-series $\sum \dfrac{1}{k^2}$, and expect the given series to converge. To *prove* convergence, take $b_k = \dfrac{1}{k^2}$, $a_k =$ the given complicated term, and compute

$$\lim_{k \to \infty} \frac{a_k}{b_k} = \lim_{k \to \infty} \frac{3k + 2k^{1/3}}{(k^{3/2} + 1)^2} \div \frac{1}{k^2} = \lim_{k \to \infty} \frac{3k^3 + 2k^{7/3}}{(k^{3/2} + 1)^2}$$

$$= \lim_{k \to \infty} \frac{3 + 2k^{-2/3}}{(1 + k^{-3/2})^2} \qquad \text{(divide numerator and denominator by } k^3\text{)}$$

$$= 3.$$

The limit is finite and nonzero, and $\sum b_k = \sum \dfrac{1}{k^2}$ converges, so the given series also converges.

---

**REMARK 1** The Limit Comparison Test stated above requires that $\dfrac{a_k}{b_k}$ have a finite, nonzero limit. What if $\dfrac{a_k}{b_k} \to 0$? Then for large $k$, $\dfrac{a_k}{b_k} < 1$, so $a_k < b_k$, and the conclusion of the ordinary Comparison Test holds:

$$\sum b_k \text{ converges} \quad \Rightarrow \quad \sum a_k \text{ converges.}$$

**REMARK 2** Comparisons are useful in estimating the error in approximating a series by its partial sums; see Section 11.7, Example 1, and problem 4b.

## SUMMARY

**Comparison Test:** If $0 \le a_k \le b_k$ then

$$\sum b_k \text{ converges} \quad \Rightarrow \quad \sum a_k \text{ converges}$$
$$\sum a_k \text{ diverges} \quad \Rightarrow \quad \sum b_k \text{ diverges.}$$

**Limit Comparison Test:** If $a_k \ge 0$, $b_k \ge 0$, and $\displaystyle\lim_{k \to \infty} \dfrac{a_k}{b_k}$ is finite and nonzero, then

$$\sum a_k \text{ converges} \quad \Leftrightarrow \quad \sum b_k \text{ converges.}$$

## PROBLEMS

**A**

**1.** Test for convergence by the Comparison Test.

*a) $\displaystyle\sum_{1}^{\infty} \frac{1}{10k - 1}$

b) $\displaystyle\sum_{1}^{\infty} \frac{1}{k^2 + 1}$

*c) $\displaystyle\sum_{1}^{\infty} \frac{5}{n(n + 1)}$

d) $\displaystyle\sum_{1}^{\infty} \frac{1}{k \cdot 2^k}$

*e) $\displaystyle\sum_{1}^{\infty} \frac{(100)^k}{k \cdot (k!)}$

f) $\displaystyle\sum_{1}^{\infty} \frac{2^k}{100}$

**2.** Use the Limit Comparison Test together with an appropriate $p$-series $\sum 1/k^p$ to prove convergence or divergence.

*a) $\displaystyle\sum_{0}^{\infty} \frac{1}{n + 1}$

b) $\displaystyle\sum_{1}^{\infty} \frac{\sqrt{n}}{n + 1}$

*c) $\displaystyle\sum_{0}^{\infty} \frac{1}{\sqrt{n^2 + 1}}$

d) $\displaystyle\sum_{1}^{\infty} \frac{1}{(2n - 3)^2}$

*e) $\displaystyle\sum_{2}^{\infty} \frac{1}{n^2 - n}$

f) $\displaystyle\sum_{0}^{\infty} \frac{1}{\sqrt[3]{n^2 + 1}}$

*g) $\displaystyle\sum_{1}^{\infty} \frac{1}{n\sqrt{2n - 1}}$

**3.** Prove convergence or divergence by any appropriate test.

a) $\displaystyle\sum_{1}^{\infty} \frac{1}{k^{2k}}$

b) $\displaystyle\sum_{0}^{\infty} \frac{1}{(k!)^2}$

c) $\displaystyle\sum_{2}^{\infty} \frac{1}{\ln k}$

d) $\displaystyle\sum \frac{2}{\sqrt{k - 1}}$

e) $\displaystyle\sum \frac{2}{(k + 2)^{3/2}}$

f) $\displaystyle\sum_{2}^{\infty} \frac{k^{1/2}}{(2k^3 - 1)^{2/3}}$

g) $\displaystyle\sum_{0}^{\infty} \frac{3^k}{(k!)^2}$

h) $\displaystyle\sum_{1}^{\infty} \frac{1}{k^k}$

**i)** $\displaystyle\sum_{1}^{\infty} \frac{1}{\sqrt{2n^2 - 1}}$    **j)** $\displaystyle\sum_{2}^{\infty} \frac{1}{k(\ln k)^2}$

**k)** $\displaystyle\sum_{1}^{\infty} \frac{\ln k}{k^2}$

**B**

**4.** The "modified Bessel function" $K_0$ is defined by the series

$$K_0(x) = \sum_{0}^{\infty} \frac{(x/2)^{2k}}{(k!)^2}.$$

**a)** Prove that the series converges for every $x$.

**b)** Approximate $K_0(1)$, using three terms of the series, and estimate the error in the approximation by the comparison method (Sec. 11.7, Example 1).

**5.** Suppose that $a_k > 0$ and $b_k > 0$ for all $k$, and that $\displaystyle\lim \frac{a_k}{b_k} = \infty$. What can be concluded about $\sum a_k$, if

**a)** $\displaystyle\sum b_k$ converges?    **b)** $\displaystyle\sum b_k$ diverges?

# 11.10

## ALTERNATING SERIES

The Integral and Comparison tests apply only to *positive* series. They establish convergence of $\sum a_k$ by showing that the terms $a_k$ go "rapidly" to zero as $k \to +\infty$.

The Alternating Series Test is based, instead, on cancellation between positive and negative terms. Suppose that the terms in a series are *alternately positive and negative*, as in the series

$$1 - \frac{1}{2} + \frac{1}{3} - \frac{1}{4} + \cdots + \frac{(-1)^k}{k} + \cdots$$

and

$$-\frac{1}{3} + \frac{1}{9} - \frac{1}{27} + \cdots + \left(-\frac{1}{3}\right)^k,$$

(but *not* in the series

$$1 - \tfrac{1}{3} - \tfrac{1}{4} + \tfrac{1}{3} - \tfrac{1}{6} - \tfrac{1}{8} + \tfrac{1}{5} - \tfrac{1}{10} - \tfrac{1}{12} + \tfrac{1}{7} - \cdots).$$

The general term of such a series can be written as $a_k = (-1)^k b_k$, where $b_k = |a_k|$ is the absolute values of the given general term. If these absolute values are *decreasing to zero*, then the series converges:

---

**THE ALTERNATING SERIES TEST**

$\sum_{0}^{\infty} (-1)^k b_k$ converges if

$$\text{each } b_k \geq 0 \qquad\qquad (1)$$

$$\text{each } b_k \geq b_{k+1} \qquad\qquad (2)$$

$$b_k \to 0 \quad \text{as} \quad k \to \infty. \qquad (3)$$

---

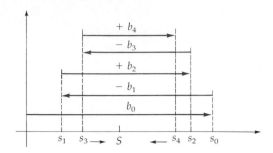

**FIGURE 1**
Proof of the Alternating Series Test.

**PROOF**   As you see in Figure 1, the *odd* partial sums $s_1, s_3, s_5, \ldots$ form an increasing sequence. For, $b_2 - b_3 \geq 0$ by (2), so

$$s_3 = (b_0 - b_1) + (b_2 - b_3) \geq (b_0 - b_1) = s_1.$$

Similarly

$$s_5 = (b_0 - b_1 + b_2 - b_3) + (b_4 - b_5) \geq s_3$$

because $b_4 - b_5 \geq 0$; and so on. Moreover, these odd partial sums are all less than or equal to $b_0$:

$$s_1 = b_0 - b_1 \leq b_0$$
$$s_3 = b_0 - (b_1 - b_2) - b_3 \leq b_0$$

since $b_1 - b_2 \geq 0$ and $b_3 \geq 0$; likewise

$$s_5 = b_0 - (b_1 - b_2) - (b_3 - b_4) - b_5 \leq b_0$$

and so on.

Thus, by the Increasing Sequence Principle, the odd partial sums $s_{2k-1}$ have a limit $S$:

$$s_{2k-1} \to S \quad \text{as} \quad k \to \infty.$$

As for the even sums, they are close to the odd sums:

$$s_{2k} = (b_0 - b_1 + \cdots - b_{2k-1}) + b_{2k}$$
$$= s_{2k-1} + b_{2k}.$$

Since $s_{2k-1} \to S$, and $b_{2k} \to 0$ by hypothesis, then also

$$s_{2k} \to S \quad \text{as} \quad k \to \infty.$$

So the even partial sums $s_{2k}$ and the odd sums $s_{2k-1}$ have the same limit $S$; it follows that the entire sequence $\{s_n\}$ has limit $S$. Thus $\sum_0^\infty (-1)^k b_k$ converges and

$$\sum_0^\infty (-1)^k b_k = S.$$

**EXAMPLE 1**   Test $\sum_{2}^{\infty} (-1)^k \dfrac{\ln k}{k}$ for convergence.

*SOLUTION*   The signs are strictly alternating, and the absolute values

$$\frac{\ln k}{k} \to 0$$

as you can check, by l'Hôpital's Rule. Do the terms $b_k = \dfrac{\ln k}{k}$ form a *decreasing* sequence? Check by differentiating:

$$\frac{d}{dx} \frac{\ln x}{x} = \frac{x \cdot \frac{1}{x} - \ln x \cdot 1}{x^2} = \frac{1 - \ln x}{x^2}.$$

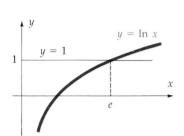

**FIGURE 2**
$\ln x > 1$ if $x > e$.

The derivative is negative if $\ln x > 1$, thus if $x > e$ (Fig. 2), so $\dfrac{\ln x}{x}$ is a decreasing function *if $x > e$*; thus

$$\frac{\ln 3}{3} > \frac{\ln 4}{4} > \frac{\ln 5}{5} > \frac{\ln 6}{6} > \cdots.$$

It follows that $\sum_{3}^{\infty} (-1)^k \ln k/k$ converges; hence the given series $\sum_{2}^{\infty} (-1)^k \ln k/k$ converges, too.

---

As the method in Example 1 shows, it suffices that the sequence of absolute values be decreasing *beyond a certain point*.

### Estimating Remainders

Figure 1 shows that the sum $S$ lies between any *odd* partial sum $s_{2k-1}$ and any *even* partial sum $s_{2k}$. This makes it easy to estimate the error in approximating $S$ by the partial sums $s_n$—*the error is less than the size of the first unused term $b_{n+1}$.* (See "Estimates for Alternating Series" in Sec. 11.7 for more details, examples, and related problems.)

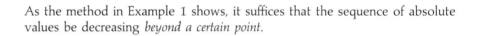

## SUMMARY

*Alternating Series Test:*

$$\sum_{0}^{\infty} (-1)^k b_k \quad \text{converges if} \quad b_k \ge b_{k+1} \ge 0 \quad \text{for every } k, \text{ and } b_k \to 0.$$

Each partial sum is accurate up to the size of the first unused term:

$$\left| \sum_{0}^{\infty} (-1)^k b_k - \sum_{0}^{n} (-1)^k b_k \right| \le b_{n+1}.$$

## PROBLEMS

### A

*1. Decide whether the given series converges.

a) $\displaystyle\sum_{1}^{\infty} (-k)^{-3}$

b) $\displaystyle\sum_{1}^{\infty} (-n)^{-1}$

c) $\displaystyle\sum_{1}^{\infty} (-1)^k/\sqrt{k}$

d) $\displaystyle\sum_{1}^{\infty} (-1)^k \frac{k+1}{k}$

e) $\displaystyle\sum_{0}^{\infty} (-1/2)^n$

f) $\displaystyle\sum_{0}^{\infty} 5^{-n}$

g) $\displaystyle\sum_{1}^{\infty} \frac{3-n}{n^3}$

h) $\displaystyle\sum_{0}^{\infty} (-5)^k$

i) $\displaystyle\sum_{0}^{\infty} \frac{(-1)^{n+1}}{2n+1}$

j) $\displaystyle\sum_{1}^{\infty} (-1)^n (\ln n)^2/n$

k) $\displaystyle\sum_{m=0}^{\infty} (-e)^{-m}$

2. Estimate the error in approximating the infinite sum by the partial sum $s_n$. Compute the partial sum. (To justify this, show that $|a_{k+1}| \le |a_k|$ for $k > n$.

*a) $\displaystyle\sum_{1}^{\infty} (-1)^{k+1}/k; \quad s_6$

b) $\displaystyle\sum_{0}^{\infty} (-1)^k/k!; \quad s_4$

c) $\displaystyle\sum_{0}^{\infty} (-2)^k/k!; \quad s_5$

*d) $\displaystyle\sum_{0}^{\infty} (-1)^k/\sqrt{k+1}; \quad s_{10}$  (Use a calculator.)

### B

3. For what value of $n$ would the partial sum $s_n$ approximate the given series to the given accuracy?

*a) $\displaystyle\sum_{1}^{\infty} (-1)^{k+1}/k, \quad |\text{error}| < 10^{-4}$

b) $\displaystyle\sum_{0}^{\infty} (-1)^k/k!, \quad |\text{error}| < 10^{-8}$

*c) $\displaystyle\sum_{0}^{\infty} (-1)^k/\sqrt{k+1}, \quad |\text{error}| < 10^{-4}$

d) $\displaystyle\sum_{2}^{\infty} (-1)^k/\ln k, \quad |\text{error}| < 10^{-8}$

4. The Bessel function $J_1(x)$ is defined by the series $\displaystyle\sum_{0}^{\infty} (-1)^k \frac{(x/2)^{2k+1}}{k!(k+1)!}$. Prove that the series for $J_1(2)$ converges, and compute it to two decimal places.

5. Is it possible to have two series $\sum a_k$ and $\sum b_k$, both convergent, with $\sum a_k b_k$ divergent?

## 11.11

## ABSOLUTE AND CONDITIONAL CONVERGENCE

Series involving sines and cosines, such as $\displaystyle\sum_{1}^{\infty} \frac{\sin kx}{k^2}$, arise in the analysis of periodic phenomena. The Alternating Series Test does not apply, because the factors $\sin(kx)$ do not have strictly alternating signs, and their absolute values oscillate up and down. The only elementary way to test such a series is to take the *absolute value* of each term. The resulting series has terms $\ge 0$, so may perhaps be amenable to the Integral or Comparison Test; and *if the series of absolute values converges, so does the original series*:

> **ABSOLUTE CONVERGENCE THEOREM**
>
> If $\sum |a_k|$ converges, then so does $\sum a_k$.

The proof is short, using an astute observation: For any real number $a$, $0 \le a + |a| \le 2|a|$. Thus

$$0 \le a_k + |a_k| \le 2|a_k|.$$

By hypothesis, $\sum 2|a_k|$ converges; and by the Comparison Test, so does $\sum(a_k + |a_k|)$. And then so does $\sum a_k$, for it is the difference of two convergent sequences:

$$\sum a_k = \sum \left[(a_k + |a_k|) - |a_k|\right] = \sum (a_k + |a_k|) - \sum |a_k|.$$

---

**EXAMPLE 1**   For what values of $x$ does the series $\sum\limits_{1}^{\infty} \dfrac{\sin kx}{k^2}$ converge?

*SOLUTION*   Consider the absolute values $\sum\limits_{1}^{\infty} \dfrac{|\sin kx|}{k^2}$. Since $0 \le |\sin kx| \le 1$, then

$$0 \le \frac{|\sin kx|}{k^2} \le \frac{1}{k^2}.$$

$\sum \dfrac{1}{k^2}$ is a convergent "$p$-series," so by the Comparison Test $\sum\limits_{1}^{\infty} \dfrac{|\sin kx|}{k^2}$ converges. Hence the given series $\sum \dfrac{\sin kx}{k^2}$ converges, for *every* $x$.

---

## Conditional and Absolute Convergence

The series

$$\sum_{1}^{\infty} \frac{(-1)^k}{k} \tag{1}$$

converges, by the Alternating Series Test; but if you replace each term by its absolute value, you get the divergent series

$$\sum_{1}^{\infty} \frac{1}{k}.$$

A series such as (1) is called **conditionally convergent**. The idea is that it converges only because of the partial cancellation between terms.

On the other hand, if you replace each term of the series

$$\sum_{1}^{\infty} \frac{(-1)^k}{k^2} \tag{2}$$

by its absolute value, you get the convergent series $\sum \dfrac{1}{k^2}$. A series such as (2) is called **absolutely convergent**.

---

### DEFINITION

$\sum a_k$ is called *absolutely convergent* if $\sum |a_k|$ converges.

$\sum a_k$ is called *conditionally convergent* if $\sum a_k$ converges, but $\sum |a_k|$ diverges.

---

For series with mixed signs, you have two options for demonstrating convergence:

(i)  The Alternating Series Test (when it applies), or

(ii)  proving absolute convergence, $\sum |a_k|$ converges.

---

**EXAMPLE 2**  The series

$$\sum_{k=0}^{\infty} \frac{(-1)^k}{(k!)^2} \left( \frac{x}{2} \right)^{2k} \tag{3}$$

is proposed as a solution of the equation $x\dfrac{d^2y}{dx^2} + \dfrac{dy}{dx} + xy = 0$. Decide whether the series converges, and whether *absolutely* or *conditionally*.

**SOLUTION**  Test first for absolute convergence. The absolute values $\left| \dfrac{(-1)^k}{(k!)^2} \left( \dfrac{x}{2} \right)^{2k} \right|$ can be compared to the terms in an exponential series

$$e^r = \sum_{0}^{\infty} \frac{1}{k!} r^k: \tag{4}$$

$$\left| \frac{(-1)^k}{(k!)^2} \left( \frac{x}{2} \right)^{2k} \right| = \frac{1}{(k!)^2} \left| \left( \frac{x}{2} \right)^2 \right|^k$$

$$\leq \frac{1}{k!} \left( \left( \frac{x}{2} \right)^2 \right)^k \tag{5}$$

for $1/(k!)^2 \leq 1/k!$, because $k! \geq 1$. Using (4) with $r = (x/2)^2$ gives the convergent series

$$\sum_{0}^{\infty} \frac{1}{k!} \left( \left( \frac{x}{2} \right)^2 \right)^k = e^{(x/2)^2}.$$

So, by the comparison (5), the series of absolute values of the given series converges for every $x$. Hence the original series (3) converges *absolutely*, for every $x$.

---

**REMARK**  Since the series (3) converges for every $x$, it defines a function of $x$. This particular function is called a Bessel function, and denoted by $J_0$; it is defined by the equation

$$J_0(x) = \sum_{0}^{\infty} \frac{(-1)^k}{(k!)^2} \left( \frac{x}{2} \right)^{2k}.$$

## SUMMARY

***Absolute Convergence:***  If $\sum |a_k|$ converges, then $\sum a_k$ also converges. In this case, we say $\sum a_k$ "*converges absolutely.*"

***Conditional Convergence:***  If $\sum a_k$ converges, but $\sum |a_k|$ diverges, then $\sum a_k$ is *conditionally convergent.*

## PROBLEMS

### A

**1.** Which series are absolutely convergent? Which are conditionally convergent?

*a)  $\displaystyle\sum_{1}^{\infty} (-2)^k/k!$

b)  $\displaystyle\sum_{0}^{\infty} k^{-2} \cos kx$

*c)  $\displaystyle\sum_{1}^{\infty} 2^{-k} \sin kx$

d)  $\displaystyle\sum_{1}^{\infty} \frac{1}{\sqrt{2n^2 - 1}}$

*e)  $\displaystyle\sum_{0}^{\infty} (-1)^k/(2k + 1)$

f)  $\displaystyle\sum (-1)^k \cos(\pi/k)$

*g)  $\displaystyle\sum_{2}^{\infty} (-1)^k (\ln k)/k$

h)  $\displaystyle\sum_{1}^{\infty} \frac{(-1)^n n}{n^2 + 1}$

i)  $\displaystyle\sum_{1}^{\infty} \frac{n}{n^2 + 1}$.

**2.** Prove that the following series converge, for every $x$.

a)  $\displaystyle\sum_{0}^{\infty} \frac{1}{(k!)^2} \left(\frac{x}{2}\right)^{2k}$

b)  $\displaystyle\sum_{1}^{\infty} \frac{2}{k(k!)} x^k$

c)  $\displaystyle\sum_{1}^{\infty} \left(\frac{x}{k}\right)^k$

d)  $\displaystyle\sum_{1}^{\infty} \frac{\cos(k\pi x)}{k^3}$

*3.** Compute $\displaystyle J_0(1) = \sum_{k=0}^{\infty} \frac{(-1)^k}{(k!)^2} \left(\frac{1}{2}\right)^{2k}$ to three decimal places.

### B

**4.** The Bessel function $J_2$ is defined by

$$J_2(x) = \sum_{0}^{\infty} \frac{(-1)^k}{(k!)(k + 2)!} \left(\frac{x}{2}\right)^{2k + 2}$$

Prove that the series converges absolutely, for every $x$.

**5.** Set $\displaystyle f(x) = \sum_{1}^{\infty} \frac{\sin kx}{k^2}$.

a)  Show that $f(x)$ is defined for every $x$ (that is, the series converges for every $x$.)

b)  Is $f(x + 2\pi) = f(x)$, for all $x$?

c)  Define a new series by differentiating the series for $f$, term by term. Does this new series converge for every $x$?

**6.** Prove that the Bessel function in Example 2, $\displaystyle J_0(x) = \sum_{k=0}^{\infty} \frac{(-1)^k}{(k!)^2} \left(\frac{x}{2}\right)^{2k}$, solves the differential equation $xJ_0'' + J_0' + xJ_0 = 0$. (This is its raison d'être. Differentiate the series term by term and collect like powers of $x$.)

**7.** Find a series $\sum a_k$ with the given property, or explain why no such series exists.

a)  $\sum a_k$ converges, but $\sum |a_k|$ diverges.

b)  $\sum a_k$ diverges, but $\sum |a_k|$ converges.

c)  $\sum a_k$ diverges, but $\sum (a_k^2)$ converges.

d)  $\sum a_k$ converges, but $\sum (a_k^2)$ diverges.

e)  $\sum |a_k|$ converges, but $\sum (a_k)^2$ diverges.

# 11.12

## THE RATIO TEST. THE RADIUS OF CONVERGENCE

Power series can be manipulated practically like polynomials, for those values of $x$ where they converge (Sec. 11.6). These values can frequently be determined using the Ratio Test presented here.

To test a series $\sum a_k$ by this method, form the "test ratio"

$$\left| \frac{a_{k+1}}{a_k} \right| = \frac{|a_{k+1}|}{|a_k|}.$$

If this ratio is $< 1$ for all large $k$, then $|a_{k+1}| < |a_k|$, so the terms $a_k$ are decreasing in absolute value. This alone does not guarantee convergence; but if the *limit* of the test ratio is a number $\rho < 1$, then the series *converges absolutely*. This is the main point of:

---

**THE RATIO TEST**

If $\lim_{k \to \infty} \left| \dfrac{a_{k+1}}{a_k} \right| = \rho$, then

   **a.**  $\rho < 1 \Rightarrow \sum a_k$ converges absolutely;

   **b.**  $\rho = 1 \Rightarrow$ no conclusion;

   **c.**  $\rho > 1 \Rightarrow \sum a_k$ diverges.

---

We give some examples, then the proof.

---

**EXAMPLE 1**   Test $\sum_1^\infty k(-2/3)^k$ for convergence.

**SOLUTION**   Here $a_k = k(-2/3)^k$. Form the test ratio

$$\left| \frac{a_{k+1}}{a_k} \right| = \left| \frac{(k+1)(-2/3)^{k+1}}{k(-2/3)^k} \right| = \frac{(k+1)(2/3)^{k+1}}{k(2/3)^k} = \frac{k+1}{k} \cdot \frac{2}{3}.$$

The limit of the test ratio is

$$\lim_{k \to \infty} \left| \frac{a_{k+1}}{a_k} \right| = \lim_{k \to \infty} \frac{k+1}{k} \cdot \frac{2}{3} = \frac{2}{3}.$$

The limit is $< 1$, so by the Ratio Test, the series converges absolutely.

**NOTE**   Of course, the series does not converge to $2/3$; the series converges to the limit

$$\lim_{n \to \infty} \sum_1^n k(-2/3)^k$$

whatever that is, and not to

$$\lim \left| \frac{a_{k+1}}{a_k} \right| = 2/3.$$

The next example shows how the Ratio Test works for power series.

**EXAMPLE 2**   Test $\sum_1^\infty (2x)^k/k$ for convergence.

*SOLUTION*   The test ratio is

$$\left|\frac{a_{k+1}}{a_k}\right| = \left|\frac{(2x)^{k+1}/(k+1)}{(2x)^k/k}\right| = \left|\frac{k}{(k+1)}\, 2x\right| = 2\left(\frac{k}{k+1}\right)|x|.$$

The limit of the test ratio is

$$\lim 2\left(\frac{k}{k+1}\right)|x| = 2|x|.$$

If $2|x| < 1$ the series converges absolutely, by part a of the Ratio Test; and if $2|x| > 1$, the series diverges by part c of the Test. Thus the series *converges* if $|x| < 1/2$ and *diverges* if $|x| > 1/2$.

In this example, the Ratio Test is indecisive where $|x| = \frac{1}{2}$, that is, at $x = \pm\frac{1}{2}$. We investigate these values separately.

If $x = \frac{1}{2}$, the given series is

$$\sum_1^\infty (2 \cdot \tfrac{1}{2})^k/k = \sum_1^\infty 1/k,$$

which diverges by the Integral Test. If $x = -\frac{1}{2}$, the given series is $\sum (-1)^k/k$, which converges by the Alternating Series Test. So the series $\sum_1^\infty (2x)^k/k$ converges if $-\frac{1}{2} \le x < \frac{1}{2}$, and diverges for all other values of $x$. The *interval of convergence* is $[-\frac{1}{2}, \frac{1}{2})$ (Fig. 1).

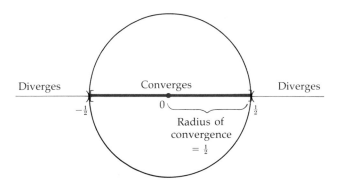

**FIGURE 1**

Interval of convergence for $\displaystyle\sum_1^\infty \frac{(2x)^k}{k}$

is $-\frac{1}{2} \le x < 1/2$.

## Radius of Convergence

The power series in Example 2 converges at all points $x$ inside the open interval $(-\frac{1}{2}, \frac{1}{2})$, and diverges outside the closed interval $[-\frac{1}{2}, \frac{1}{2}]$. These intervals are centered at the origin and have "radius" $\frac{1}{2}$; this is called the **radius of convergence of the power series**.

Every power series $\sum c_k x^k$ has a radius of convergence $r$ (possibly infinite), such that

$$\sum c_k x^k \quad \begin{cases} \text{converges} & \text{if } |x| < r, \\ \text{diverges} & \text{if } |x| > r. \end{cases}$$

The geometric series $\sum_0^\infty x^k$ has radius of convergence $r = 1$; the exponential series $\sum_0^\infty x^k/k!$ has radius of convergence $r = \infty$, since it converges for every $x$.

**The binomial series** is the Maclaurin series of $(1 + x)^p$; the power $p$ can be any constant. The series is

$$1 + px + \frac{p(p - 1)}{2} x^2 + \frac{p(p - 1)(p - 2)}{3} x^3 + \cdots$$

$$= 1 + \sum_{k=1}^\infty \frac{p(p - 1) \cdots (p - k + 1)}{k!} x^k, \tag{1}$$

as you can check. By the Ratio Test, this converges for $|x| < 1$. But the Ratio Test merely proves convergence; it does not prove that the sum of the series is precisely $(1 + x)^p$. That requires a further argument, for example, the following:

Since the series is known to converge to *something* for $|x| < 1$, we can define a function $S$ by

$$S(x) = 1 + \sum_{k=1}^\infty \frac{p(p - 1) \cdots (p - k + 1)}{k!} x^k, \qquad |x| < 1. \tag{2}$$

Our aim is to prove that $S(x) = (1 + x)^p$. Differentiate term by term the series in (2) [or (1)], carefully collect like powers of $x$, and you can show that

$$(1 + x) \frac{dS}{dx} = pS(x).$$

This differential equation for the as yet unknown function $S$ can be solved by separating variables:

$$\frac{dS}{S} = \frac{p}{1 + x} dx.$$

Integrate:

$$\ln|S| = p \ln(1 + x) + c \qquad \text{(Recall that } x > -1, \text{ so } 1 + x > 0.)$$

$$|S| = e^c e^{p \ln(1 + x)} = e^c (1 + x)^p$$

$$S(x) = (\pm e^c)(1 + x)^p = C(1 + x)^p. \tag{3}$$

It remains only to determine $C$. For this, set $x = 0$ in (2) and (3):

$$1 = S(0) = C(1 + 0)^p = C.$$

So $C = 1$; then (3) gives $S(x) = (1 + x)^p$, which proves: *The binomial series (2) converges to $(1 + x)^p$ for $|x| < 1$.*

## Proof of the Ratio Test

Before studying this, recall the definition of limit (Sec. 11.2). In the first part of the Ratio Test, we suppose that

$$\lim_{k \to \infty} \left| \frac{a_{k+1}}{a_k} \right| = \rho < 1,$$

and must prove that $\sum_0^\infty |a_k|$ converges. Choose any positive number $\varepsilon < 1 - \rho$ (Fig. 2). Since $|a_{k+1}/a_k| \to \rho$, there is an integer $N$ such that

$$k \geq N \quad \Rightarrow \quad \left| |a_{k+1}/a_k| - \rho \right| < \varepsilon.$$

Then for $k \geq N$, $|a_{k+1}/a_k|$ lies between $\rho - \varepsilon$ and $\rho + \varepsilon$ (Fig. 2). In particular, for $k \geq N$

$$|a_{k+1}/a_k| < \rho + \varepsilon. \tag{4}$$

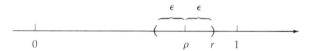

**FIGURE 2**

$$\varepsilon < 1 - \rho. \quad \left| \frac{a_k}{a_{k+1}} - \rho \right| < \varepsilon \Rightarrow \frac{|a_k|}{|a_{k+1}|} < \rho + \varepsilon = r.$$

Let $r = \rho + \varepsilon$ (Fig. 2); then $r < 1$. From (4), $|a_{k+1}|/|a_k| = |a_{k+1}/a_k| < r$ so

$$|a_{k+1}| < r|a_k|.$$

This applies for $k \geq N$, thus for $k = N, N + 1, N + 2, \ldots$, so

$$|a_{N+1}| < r|a_N|$$
$$|a_{N+2}| < r|a_{N+1}| < r(r|a_N|) = r^2|a_N|$$
$$|a_{N+3}| < r|a_{N+2}| < r(r^2|a_N|) = r^3|a_N|$$
$$\vdots$$

It follows that

$$|a_N| + |a_{N+1}| + |a_{N+2}| + |a_{N+3}| + \ldots$$
$$\leq |a_N| + r|a_N| + r^2|a_N| + r^3|a_N| + \ldots$$
$$= |a_N| (1 + r + r^2 + \ldots) = |a_n| \frac{1}{1 - r}.$$

By the Comparison Test, $\sum_N^\infty |a_k|$ converges; hence also $\sum_1^\infty |a_k|$ converges, and $\sum_1^\infty a_k$ converges absolutely. This proves part (a) of the Ratio Test.

*PROOF OF PART C*

Suppose that

$$\lim_{k \to \infty} \left| \frac{a_{k+1}}{a_k} \right| = \rho > 1.$$

Then for all sufficiently large integers $k$,

$$\left| \frac{a_{k+1}}{a_k} \right| \geq 1,$$

which implies that $|a_{k+1}| \geq |a_k| > 0$. So the sequence $\{|a_k|\}$ is *increasing* and cannot converge to zero; hence $\{a_k\}$ does not converge to zero either, and $\sum a_k$ diverges by the Zero-limit Test.

*PROOF OF PART B*

If $\rho = 1$, the series may either converge or diverge. For example, $\rho = 1$ for each of the following series:

$$\sum_{1}^{\infty} 1/k \qquad \text{(diverges)}$$

$$\sum_{1}^{\infty} (-1)^k/k \qquad \text{(converges, but not absolutely)}$$

$$\sum_{1}^{\infty} 1/k^2 \qquad \text{(converges absolutely)}.$$

This completes the proof of the Ratio Test.

## SUMMARY

*The Ratio Test:*   If $\lim \left| \dfrac{a_{k+1}}{a_k} \right| = \rho$, then

   **a.**  $\rho < 1$  $\Rightarrow$  $\sum a_k$ converges absolutely,

   **b.**  $\rho = 1$  $\Rightarrow$  no conclusion,

   **c.**  $\rho > 1$  $\Rightarrow$  $\sum a_k$ diverges.

*The Binomial Series:*

$$(1 + x)^p = 1 + px + \frac{p(p-1)}{2} x^2 + \frac{p(p-1)(p-2)}{3!} x^3 + \dots, \qquad |x| < 1.$$

*Radius of Convergence of a Power Series, r:*

$$\sum c_k x^k \quad \begin{cases} \text{converges} & \text{if } |x| < r \\ \text{diverges} & \text{if } |x| > r. \end{cases}$$

## PROBLEMS

### A

**\*1.** Test for convergence.

a) $\sum 2^n/n^2$    b) $\sum (-3)^n/n!$

c) $\sum 1/(n!)^2$

d) $\sum \dfrac{100 - n}{n!}$

e) $\sum (3n + 1)(-2)^{-n}$

f) $\sum n!/n^n$

g) $\sum \dfrac{1}{n(\ln n)^3}$

**2.** Test for convergence.

a) $\sum n^n/n!$
b) $\sum 3^{n-1} 4^{-n} n^{100}$
c) $\sum (-1)^n/(n^2 + 1)$
d) $\sum (-1)^n/\sqrt{n^2 + 1}$

e) $\sum \dfrac{(-1)^n}{n(\ln n)^2}$

f) $\sum (-1)^n 2^{1/n}$
g) $\sum (1 - 1/n)^n$

h) $\sum \left(\dfrac{n}{3n + 1}\right)^n$

i) $\sum \dfrac{1}{\sqrt{n^3 + 1}}$

**3.** Determine the interval of convergence; test for convergence at each endpoint.

**\*a)** $\sum x^n/n^2$    b) $\sum n^2 x^n$

**\*c)** $\sum \sqrt{k}(2x)^k$    d) $\sum (2x)^n/n!$

**\*e)** $\sum (x/2)^n/(n!)^2$    f) $\sum (x/3)^{2n}/n!$

**\*g)** $\sum \dfrac{\ln n}{n^2} x^n$    h) $\sum \dfrac{(3n)!}{(2n)!} x^n$

**\*i)** $\sum \dfrac{(2n)!}{(3n)!}(-1)^n x^{2n}$

**4.** Determine the radius of convergence.

a) $\sum (nx)^n/n!$
**\*b)** $\sum n!(x/n)^n$
c) $\sum n!(x/10)^n$

**\*5.** Determine the interval of convergence.

a) $\sum (x - 1)^n 3^{-n}$

(Hint: $|x - 1| < 3 \quad \Leftrightarrow \quad -3 < x - 1 < 3$
$\Leftrightarrow \quad -2 < x < 4$.)

b) $\sum n(x + 3)^n$
c) $\sum (2x - 1)^n/(n + 1)$

**6.** Write the Maclaurin series for the following special cases of the binomial series.

**\*a)** $\sqrt{1 + x}$

b) $\dfrac{1}{\sqrt{1 + x}}$

c) $\dfrac{1}{\sqrt{1 - x^2/c^2}}$    (c is constant.)

d) $(1 - x)^p$

**\*e)** $(a + x)^p$

### B

**7.** Prove that the binomial series converges, for $|x| < 1$.

**8.** Use the binomial series to compute $\sqrt[3]{30}$ to three decimals. [Hint: $\sqrt[3]{30} = (27 + 3)^{1/3} = 3(1 + \tfrac{1}{9})^{1/3}$.]

**9.** a) Can you compute $\sqrt{2}$, using the binomial series for $\sqrt{1 + x}$ with $x = 1$?

b) Compute $\sqrt{2} = \tfrac{3}{2}\sqrt{\tfrac{8}{9}}$ to five decimal places, using the binomial series for $\sqrt{1 + x}$.

c) Compute $\sqrt{3}$ to five decimal places.

**10.** a) Use the binomial series to get the series for

$$\sin^{-1} x = \int_0^x \dfrac{dt}{\sqrt{1 - t^2}}.$$

b) Evaluate $\sin^{-1}(0.5)$ to three decimal places.
c) Approximate $\pi$, using part b.

**11.** By term-by-term differentiation, prove that the series $S(x)$ in (2) satisfies

$$(1 + x)S'(x) = pS(x).$$

(Collect like powers of $x$ on the left-hand side. Start with the lowest term $x^0$, and work up to the general term $x^k$.)

## C

**12.** Prove the *Root Test*: Suppose that
$$\lim_{k \to \infty} |a_k|^{1/k} = \rho.$$

   **a)** If $\rho < 1$, then $\sum |a_k|$ converges.

   **b)** If $\rho = 1$, no conclusion follows.

   **c)** If $\rho > 1$ then $\sum a_k$ diverges.

**13.** Test for convergence by the Root Test.

   **a)** $\displaystyle\sum_{1}^{\infty} \frac{n}{2^n}$    **b)** $\displaystyle\sum_{1}^{\infty} \left(\frac{x}{n}\right)^n$

**14.** (Fibonacci numbers.)

   **a)** Show that *if* the function $f(x) = \dfrac{1}{1 - x - x^2}$ has a power series

$$f(x) = \sum a_k x^k \qquad (*)$$

   converging for $|x| < R$ (with $R > 0$), then $a_0 = 1$, $a_1 = 1$, and $a_n = a_{n-1} + a_{n-2}$ for $n \geq 2$. [Hint: $f(x) - xf(x) - x^2 f(x) = 1$. Write this out using the series $(*)$.]

   **b)** Find $a_2, a_3, a_4, a_5, a_6$.

   **c)** Show that $a_{n+1}/a_n \leq 2$.

   **d)** Show that $\sum a_n x^n$ converges at least for $|x| < \frac{1}{2}$.

   **e)** Show that $\sum a_n x^n = f(x)$, wherever the series converges. [Hint: If $g(x)$ is the sum of the series, what is $g - xg - x^2 g$?]

   **f)** Obtain another power series for $f$, by using the partial fractions

$$\frac{1}{1 - x - x^2} = \frac{A}{r_1 - x} + \frac{B}{r_2 - x},$$

   where $r_1$ and $r_2$ are the zeroes of $1 - x - x^2 = 0$, and $A$ and $B$ are appropriate constants. Note further that

$$\frac{1}{r - x} = \frac{1}{r}\left(\frac{1}{1 - x/r}\right) = \frac{1}{r}\sum_{0}^{\infty}(x/r)^n.$$

   **g)** Deduce that

$$a_{n-1} = \frac{(1 + \sqrt{5})^n - (1 - \sqrt{5})^n}{2^n \sqrt{5}}.$$

(The numbers $a_0, a_1, \ldots$ are called Fibonacci numbers, in honor of Fibonacci of Pisa.

Fibonacci hypothesized that a pair of newborn rabbits would bear no offspring the first month, but would bear a new pair of rabbits every month thereafter. Thus, beginning with one pair of bearing rabbits, and denoting by $a_n$ the number of pairs of bearing rabbits in month $n$, we find $a_0 = 1$, $a_1 = 1$, and $a_n = a_{n-1} + a_{n-2}$ for $n > 1$.)

**15.** (Defining the exponential function.) In defining $e^x$ (Sec. 8.1) we took for granted that $a^x$ is defined for all $a > 0$ and all real $x$, and that there is a number $e$ such that the derivative of $e^x$ at the origin equals 1. Power series provide a rigorous way to define an exponential function with these properties. Assume nothing about $e^x$, but accept the general theorems for power series given above. By the Ratio Test, the series $\sum_{0}^{\infty} x^k/k!$ converges for all real $x$, and therefore defines a function which we call $\exp(x)$. Starting with just the infinite series, you can prove the essential properties of the exponential function (if you take them in the right order). Show that

   **a)** $\exp(0) = 1$.

   **b)** $\exp'(x) = \exp(x)$.

   **c)** $\exp(x) \cdot \exp(-x) = 1$, for all $x$. (Use b), not the series.)

   **d)** Given a constant $C$, there is one and only one differentiable function $f$ such that $f' = f$ and $f(0) = C$; it is given by $f(x) = C \exp(x)$. [Suppose $f' = f$, and show that $\exp(-x)f(x)$ is constant.]

   **e)** $\exp(a + x) = \exp(a) \cdot \exp(x)$. (Use part d.)

   **f)** $\exp(x) > 0$ for all $x$.

   **g)** exp is a strictly increasing function.

   **h)** $\lim\limits_{x \to +\infty} \exp(x) = +\infty$. [Hint: When $x \geq 0$, then $\exp(x) > 1 + x$, from the series.]

   **i)** $\lim\limits_{x \to +\infty} \exp(-x) = 0$. (Use parts c and h.)

   **j)** *Define* $e$ as $\exp(1)$. Show that for integer $n$, $\exp(nx) = (\exp(x))^n$. (Use part e.)

   **k)** For rational $m/n$, $e^{m/n} = \exp(m/n)$.

   **l)** *Define* ln as the inverse of exp. Show that $\ln(ab) = \ln(a) + \ln(b)$.

   **m)** We *define* $a^x$ as $\exp(x \cdot \ln(a))$, for any $a > 0$ and all real $x$. Justify this, by showing that $\exp\left(\dfrac{m}{n}\ln(a)\right) = a^{m/n}$, for all rational numbers $m/n$.

   **n)** Show that $a^{x+y} = a^x \cdot a^y$.

## REVIEW PROBLEMS   CHAPTER 11

### A

*1. Write the Taylor polynomial $P_3(x)$ for
   a)  $f(x) = (1 + x)^{2/3}$, based at 0.
   b)  $f(x) = x^{2/3}$, based at 1.

*2. a) Use $P_3(x)$, based at an appropriate point, to approximate $9^{2/3}$.
   b)  Estimate the error in the approximation.

3. a) Define the meaning of "$\lim_{n \to \infty} a_n = L$".

   b)  Prove that $\lim_{n \to \infty} \dfrac{n^2}{2n^2 + 1} = \dfrac{1}{2}$, using your definition.

*4. Compute the limit, or explain why it fails to exist.

   a)  $\lim_{n \to \infty} \dfrac{1 - (9/10)^n}{1/10}$
   b)  $\lim_{n \to \infty} \dfrac{1 - (10/9)^n}{1/10}$

   c)  $\lim_{n \to \infty} (1 + n)^{1/n}$
   d)  $\lim_{n \to \infty} \left(1 - \dfrac{1}{n}\right)^n$

5. a) Define the meaning of "$\sum_1^\infty a_k = A$".

   b)  For what $r$ does $\sum_0^\infty r^k = \dfrac{1}{1 - r}$?

   c)  *Prove* your assertion in part b.

*6. Determine $n$ so that $\sum_0^n \dfrac{x^k}{k!}$ approximates $e^x$ with an error less than $\frac{1}{2} \times 10^{-5}$, for all $x$ in the interval $[-1, 1]$.

7. Write the power series in powers of $x$ for

   *a)  $\cos(x^2)$.
   b)  $\dfrac{\cos(x) - 1}{x}$.

   *c)  $\displaystyle\int_0^x \dfrac{\cos(t) - 1}{t}\, dt$.

*8. Compute $\displaystyle\int_0^1 \dfrac{\cos(t) - 1}{t}\, dt$ to three decimal places, and prove that your answer is that accurate.

*9. Obtain the power series for $(1 + x)^{-3}$ by term-by-term differentiation of the geometric series.

10. a) Write the power series for $\tan^{-1}x$. (Represent $\tan^{-1}x$ as an integral.)

   *b)  Use three terms of the series to estimate (in radian measure) the smallest angle in a right triangle with legs of length 3 and 4.

*c) Estimate the error in the approximation in part b.

*11. Prove convergence or divergence.

   a)  $\displaystyle\sum_2^\infty \dfrac{k}{k + 1}(-1)^k$
   b)  $\displaystyle\sum_{-\infty}^\infty \dfrac{1}{k^2 + 1}$

   c)  $\displaystyle\sum_1^\infty \dfrac{k}{k^2 + 1}$
   d)  $\displaystyle\sum_1^\infty \dfrac{1}{\sqrt{k^2 + 1}}$

   e)  $\displaystyle\sum_1^\infty \dfrac{(-1)^k}{\sqrt{k^2 + 1}}$
   f)  $\displaystyle\sum_0^\infty k^{-k}(k!)^2$

*12. Determine for which $x$ the given series converges.

   a)  $\displaystyle\sum_2^\infty \dfrac{1}{\ln k}\, x^k$
   b)  $\displaystyle\sum_0^\infty \left[\dfrac{x - 2}{3}\right]^k$

   c)  $\displaystyle\sum_1^\infty \dfrac{\sin(kx)}{k^{3/2}}$
   d)  $\displaystyle\sum_0^\infty e^{-n^2 x}$

   e)  $\displaystyle\sum_1^\infty \dfrac{1}{x^k}$
   f)  $\displaystyle\sum_1^\infty k^{-x}$

### B

*13. a) Estimate the error in approximating $\displaystyle\sum_1^\infty \dfrac{1}{k^3}$ by $\displaystyle\sum_1^{10} \dfrac{1}{k^3}$.

   b)  Improve the approximation by adding an appropriate integral, and estimate the error in this.

14. a) Obtain a power series for $\displaystyle\int_0^x \dfrac{dt}{(2 + t^2)^{1/3}}$.
   [Hint: $(2 + t^2)^{1/3} = 2^{1/3}(1 + \frac{1}{2}t^2)^{1/3}$.]

   *b)  What is the radius of convergence of the series?

*15. Is it possible to have two convergent series $\displaystyle\sum_1^\infty a_k$ and $\displaystyle\sum_1^\infty b_k$ such that $\displaystyle\sum_1^\infty a_k b_k$ is divergent?

16. The theorem on term-by-term differentiation and integration of power series justifies these operations *inside* the interval of convergence, but not at the endpoints. Compare the intervals of convergence of the following series:

$$\sum_1^\infty x^n/n, \qquad \sum_1^\infty x^n, \qquad \sum_1^\infty x^{n+1}/n(n + 1).$$

(As these cases suggest, you may *lose* convergence at an endpoint by differentiating and may *gain* it by integrating. The next problem asks you to prove that you can never lose convergence at an endpoint by integrating a power series.)

17. Prove: If $\sum_0^\infty a_k$ converges, then $\sum_0^\infty a_k/(k+1)$ converges. [Big hint: Set $s_N = \sum_0^n a_k$. Check that

$$\sum_0^N a_n/(n+1) = \sum_0^N s_n/(n+1) - \sum_1^N (s_{n-1})/(n+1)$$

$$= \sum_0^N s_n \frac{1}{(n+1)(n+2)} + \frac{s_N}{N+2}.\Big]$$

# 12

## CONIC SECTIONS. POLAR COORDINATES

It is surprising that a purely mathematical study can have great physical significance. A classic example is that of the conic sections, curves studied by the ancient Greeks. Nearly two thousand years later, Johann Kepler discovered that the planetary orbits have the shape of conic sections! This in turn contributed to Newton's theory of mechanics and gravity, which governs the flight of baseballs, satellites, and much more.

The conic sections are obtained as the intersections of a double cone and a plane—hence the name. There are three basic types (Fig. 1): ellipse, parabola, and hyperbola. This chapter gives the main geometric features of these curves, and their equations. Then it presents the polar coordinate system, often used when one point plays a special role, as the sun does for the motion of the planets.

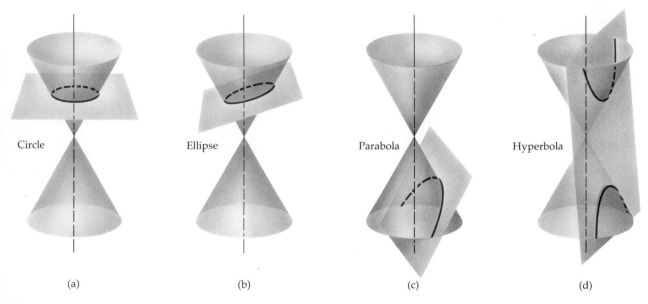

Circle

Ellipse

Parabola

Hyperbola

(a)

(b)

(c)

(d)

**FIGURE 1**

## 12.1

## PARABOLAS. COMPLETING THE SQUARE

A parabola is defined in terms of a given point $F$, called the **focus**, and a given line $L$ not containing $F$, called the **directrix** (Fig. 2).

> **DEFINITION**
>
> The parabola with focus $F$ and directrix $L$ consists of all points $P$ which are equidistant from the focus $F$ and directrix $L$.

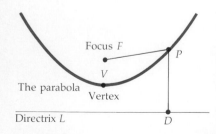

Focus $F$

$P$

$V$

The parabola

Vertex

Directrix $L$

$D$

**FIGURE 2**

Parabola: $|PF| = |PD|$.

Denote by $D$ the foot of the perpendicular from $P$ to the directrix $L$ (Fig. 2); then the defining relation is

$$|PF| = |PD|. \tag{1}$$

We get a simple equation for the parabola by introducing coordinates such that the focus is the point

$$F = (0, p),$$

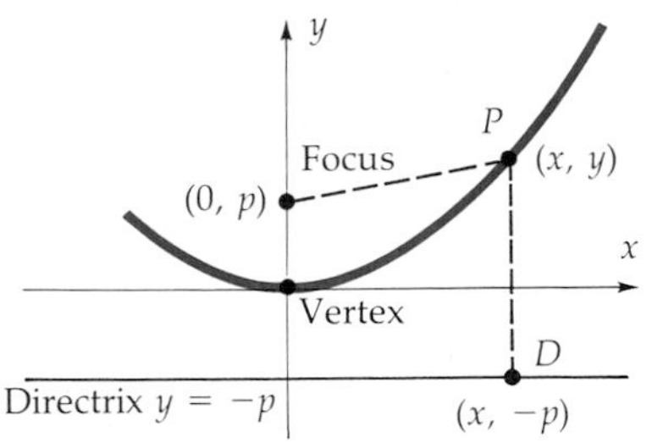

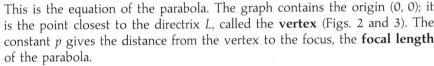

and the directrix is the line $y = -p$ (Fig. 3). By the distance formula, equation (1) can be written

$$\sqrt{(x - 0)^2 + (y - p)^2} = \sqrt{0^2 + (y + p)^2},$$

which is equivalent to each of the following:

$$x^2 + y^2 - 2py + p^2 = y^2 + 2py + p^2,$$

$$x^2 = 4py$$

$$y = \frac{1}{4p} x^2. \tag{2}$$

**FIGURE 3**
Parabola with coordinates.

This is the equation of the parabola. The graph contains the origin $(0, 0)$; it is the point closest to the directrix $L$, called the **vertex** (Figs. 2 and 3). The constant $p$ gives the distance from the vertex to the focus, the **focal length** of the parabola.

The coefficient $\dfrac{1}{4p}$ in equation (2) could be any nonzero number, call it $a$:

$$a = \frac{1}{4p} \quad \text{so} \quad p = \frac{1}{4a}.$$

Then the equation of the parabola (2) is written in the familiar form

$$y = ax^2. \tag{3}$$

Hence, every quadratic $y = ax^2$ is the equation of a parabola: The focal length is $p = 1/4a$, the focus is $F = (0, 1/4a)$, and the directrix is the line $y = -1/4a$.

The point $F$ is called the focus because of the reflecting property of the parabola: Light rays entering parallel to the axis are reflected to the focus (see Sec 2.2).

---

**EXAMPLE 1**   Find the focus and directrix of the parabola $y = 3x^2$.

**SOLUTION**   This is simply equation (3) with $a = 3$; thus the focal length is $\dfrac{1}{4a} = \dfrac{1}{12}$. The focus is at $\left(0, \dfrac{1}{12}\right)$ and the directrix is the line $y = -\dfrac{1}{12}$.

---

### Parabolas with Arbitrary Vertex

The parabola in Figure 4 has a vertical axis, but its vertex $V$ is at a point $(x_0, y_0)$, not the origin. We derive its equation using new coordinates based

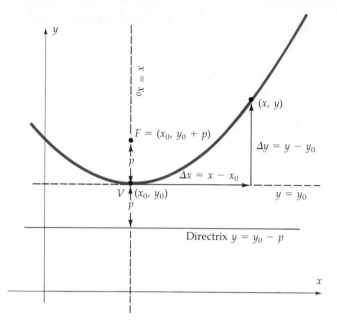

**FIGURE 4**
Graph of $y - y_0 = a(x - x_0)^2$
Focal length $p = 1/4a$.

at $(x_0, y_0)$; we denote them by

$$\Delta x = x - x_0, \qquad \Delta y = y - y_0.$$

Just as $x$ measures signed distance from the vertical line $x = 0$, the new co-ordinate $\Delta x$ measures signed distance from the vertical line $x = x_0$ (Fig. 4). Similarly, $\Delta y$ measures signed distance from the horizontal line $y = y_0$. In these coordinates, the point $(x_0, y_0)$ plays the role of the origin. Thus, the vertical parabola with focal length $p$ and vertex $V = (x_0, y_0)$ has equation

$$\Delta y = \frac{1}{4p} (\Delta x)^2$$

or

$$y - y_0 = \frac{1}{4p} (x - x_0)^2. \tag{4}$$

**EXAMPLE 2**   Write the equation of the parabola with vertex $(1, 3)$ and focal length $p = \frac{1}{2}$ (Fig. 5).

**SOLUTION**   In equation (4), the vertex is $V = (x_0, y_0)$; in this example, $V = (1, 3)$, so $x_0 = 1$, $y_0 = 3$, and the equation of the parabola is

$$y - 3 = \frac{1}{4(1/2)} (x - 1)^2 = \frac{1}{2} (x - 1)^2.$$

The equation can be "simplified" to $y = \frac{1}{2}x^2 - x + \frac{7}{2}$.

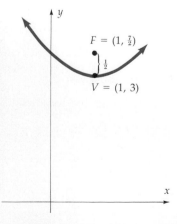

**FIGURE 5**
Vertex $V = (1, 3)$
Focal length $p = 1/2$.

The quadratic in Example 2 turned out to be the graph of the quadratic function $f(x) = \frac{1}{2}x^2 - x + \frac{7}{2}$. Conversely, the graph of *any* quadratic function $f(x) = ax^2 + bx + c$ is a parabola; the proof of this requires a method which we now describe.

**Completing the square** is a useful way to rewrite a quadratic $ax^2 + bx + c$, so that the variable $x$ occurs only once. Begin by factoring the coefficient $a$ out of the first two terms,

$$ax^2 + bx = a\left(x^2 + \frac{b}{a}x\right), \tag{5}$$

and proceed as follows:

$$ax^2 + bx + c = a\left[x^2 + \frac{b}{a}x + \phantom{\left(\frac{1}{2}\cdot\frac{b}{a}\right)^2}\right] + c \tag{6}$$

$$\text{add } \left(\frac{1}{2}\cdot\frac{b}{a}\right)^2 \text{ here} \qquad \text{subtract } a\left(\frac{1}{2}\cdot\frac{b}{a}\right)^2 \text{ here}$$

$$= a\left[x^2 + \frac{b}{a}x + \left(\frac{1}{2}\cdot\frac{b}{a}\right)^2\right] + c - a\cdot\left(\frac{1}{2}\cdot\frac{b}{a}\right)^2.$$

Now recognize that the expression in brackets is a perfect square:

$$x^2 + \frac{b}{a}x + \left(\frac{1}{2}\cdot\frac{b}{a}\right)^2 = \left(x + \frac{1}{2}\cdot\frac{b}{a}\right)^2$$

so

$$ax^2 + bx + c = a\left(x + \frac{b}{2a}\right)^2 + c - a\cdot\left(\frac{b}{2a}\right)^2$$

$$= a\left(x + \frac{b}{2a}\right)^2 + \frac{4ac - b^2}{4a}. \tag{7}$$

This is what we were aiming for; the variable $x$ occurs only once, in a perfect square.

*WARNING* On the right-hand side of (5), notice that the factor $a$ multiplies each term in the parentheses; this is why you get $b/a$ inside, not just $b$. Similarly in (6), the term $\left(\frac{1}{2}\cdot\frac{b}{a}\right)^2$ is added *inside* the brackets, and thus is multiplied by the factor $a$ to the left of the brackets; hence the compensating factor to be subtracted is $a\cdot\left(\frac{1}{2}\cdot\frac{b}{a}\right)^2$, not just $\left(\frac{1}{2}\cdot\frac{b}{a}\right)^2$.

---

**EXAMPLE 3**

$$3x^2 + 2x + 1 = 3[x^2 + \tfrac{2}{3}x + \phantom{(\tfrac{1}{3})^2}] + 1$$

$$= 3[x^2 + \tfrac{2}{3}x + (\tfrac{1}{3})^2] + 1 - 3\cdot(\tfrac{1}{3})^2$$

$$= 3[x + \tfrac{1}{3}]^2 + \tfrac{2}{3}.$$

We now apply this process in graphing.

---

**EXAMPLE 4**   Graph $y = 3x^2 + 2x + 1$.

*SOLUTION*   Complete the square (Example 3) and find
$$y = 3(x + \tfrac{1}{3})^2 + \tfrac{2}{3}.$$
Rewrite this in the form of equation (4):
$$y - \tfrac{2}{3} = 3[x - (-\tfrac{1}{3})]^2. \tag{8}$$
This is the equation of a parabola with vertex $V = (-1/3, 2/3)$. For, with
$$\Delta x = x - (-\tfrac{1}{3}), \qquad \Delta y = y - \tfrac{2}{3},$$
equation (8) takes the form
$$\Delta y = 3(\Delta x)^2.$$
This is like the equation $y = 3x^2$ in Example 1, except that $y$ and $x$ measures distances from the $x$-axis and the $y$-axis, whereas $\Delta y$ measures distance from

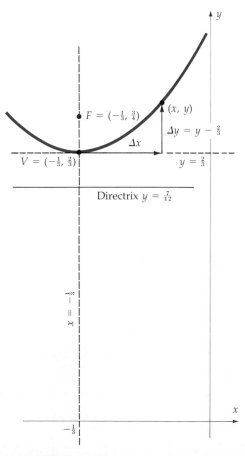

**FIGURE 6**
$y = 3x^2 + 2x + 1 \Leftrightarrow \Delta y = 3(\Delta x)^2.$

the line $y = \frac{2}{3}$, and $\Delta x$ measures distance from the line $x = -\frac{1}{3}$ (Fig. 6). This shows that the graph of (8) is a parabola with vertex $V = (-\frac{1}{3}, \frac{2}{3})$. The focal distance is determined by the coefficient of the square, as in Example 1; it is $p = \dfrac{1}{4 \cdot 3} = \dfrac{1}{12}$. Hence the focus is $1/12$ units above $V$, at the point

$$F = (-\tfrac{1}{3}, \tfrac{2}{3} + \tfrac{1}{12}) = (-\tfrac{1}{3}, \tfrac{3}{4}).$$

The directrix, $1/12$ below $V$, is the line

$$y = \tfrac{2}{3} - \tfrac{1}{12} = \tfrac{7}{12}.$$

---

Example 4 illustrates the general process. By completing the square, the equation

$$y = ax^2 + bx + c \qquad (9)$$

is rewritten in the form

$$y - y_0 = a(x - x_0)^2. \qquad (10)$$

Let $\Delta x = x - x_0$ and $\Delta y = y - y_0$ (Fig. 3). Then equation (10) can be further rewritten as

$$\Delta y = a(\Delta x)^2,$$

which is like $y = ax^2$, except that $y$ and $x$ measure distances from the $x$- and $y$-axes, whereas $\Delta x$ and $\Delta y$ measure distances from the lines $x = x_0$ and $y = y_0$, as in Figure 4. So the graph of (9) is like the graph of $y = ax^2$, with the vertex $V$ shifted to $(x_0, y_0)$; it is a parabola, with focal length $p = \dfrac{1}{4a}$. The focus is the point $F$, $p$ units above $V$; the directrix is the line $L$, $p$ units below $V$. The parabola consists of all points $P$ which are equidistant from $F$ and $L$.

If $a < 0$, then Figure 4 is flipped over; the graph opens downward from $V$.

A quadratic $x = ay^2 + by + c$ is like (9), but with $x$ and $y$ exchanged; its graph is a parabola with *horizontal* axis.

**The quadratic formula** solves the equation $ax^2 + bx + c = 0$, and thus gives the $x$-intercepts of the graph of $y = ax^2 + bx + c$. It is proved by completing the square as in (7):

$$ax^2 + bx + c = 0 \quad \Leftrightarrow \quad a\left(x + \frac{b}{2a}\right)^2 + \frac{4ac - b^2}{4a} = 0$$

$$\Leftrightarrow \quad \left(x + \frac{b}{2a}\right)^2 = \frac{b^2 - 4ac}{4a^2}. \qquad (11)$$

The square $\left(x + \dfrac{b}{2a}\right)^2$ is always $\geq 0$. If $b^2 < 4ac$, the right-hand side of (11) is negative, and the equation has no solution. If $b^2 \geq 4ac$ then the solution is found by taking square roots:

$$x + \frac{b}{2a} = \pm\sqrt{\frac{b^2 - 4ac}{4a^2}} = \pm\frac{\sqrt{b^2 - 4ac}}{2a},$$

$$x = -\frac{b}{2a} \pm \frac{\sqrt{b^2 - 4ac}}{2a} = \frac{-b \pm \sqrt{b^2 - 4ac}}{2a}.$$

This is the famous quadratic formula.

## SUMMARY

A *parabola* consists of all points $P$ equidistant from a given point $F$ (the *focus*) and a given line $L$ (the *directrix*). The point on the parabola nearest to the directrix is the *vertex*.

The graph of any quadratic of the form

$$y = ax^2 + bx + c \quad \text{or} \quad x = ay^2 + by + c$$

is a parabola with axis parallel to a coordinate axis. The vertex, focus, and directrix can be found by completing the square, obtaining either

$$y - y_0 = a(x - x_0)^2 \quad \text{or} \quad x - x_0 = a(y - y_0)^2.$$

The vertex is $V = (x_0, y_0)$ and the focal length is $p = 1/4a$.

## PROBLEMS

### A

1. Complete the square, writing the quadratic in the form $a(x - x_0)^2 + y_0$ (with change of letters where appropriate).
   *a) $x^2 + 2x + 3$    b) $-2x^2 + 5x - 4$
   *c) $-16t^2 + 3t$

2. Sketch the graph, showing the vertex, focus, directrix, and intercepts.
   *a) $y = 2x^2$              b) $y = x(2 - x)$
   c) $2y = x^2 + 1$           d) $x = 2y^2$
   *e) $y^2 + y + x = 0$
   *f) $y = x^2 + 3x + 1$
   g) $2x = -y^2 + 4y + 2$

### B

3. One end of a string of length $s$ is attached to a fixed point $F$, and the other end is attached by a slide to a fixed rod $R$, $r$ units from $F$ (Fig. 7). The

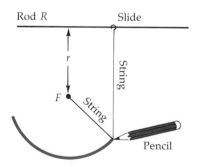

Rod $R$      Slide

**FIGURE 7**

string is held taut by a pencil; as the pencil moves, the string remains perpendicular to $R$. Show that that pencil traces part of a parabola. Find the focal length.

*4. Find an equation of the parabolic arch in Figure 8. Find the focus.

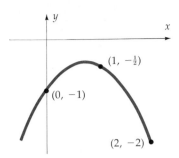

**FIGURE 8**

marches directly toward the grandstand. What curve should it be, if the band is to end up marching in a straight line toward the grandstand? (Note that if the marchers are equally spaced along the semicircle, then they are *unequally* spaced along the straight line. The same effect occurs with light reflected from a parabola; if the source is uniform, then the reflected light is more intense along the axis of the parabola than farther out.)

5.  The members of a marching band are arranged in a semicircle, facing away from a grandstand at the bottom of Figure 9. They march in unison away from the center of the circle, at the same speed. As each reaches a certain curve, she turns and

6.  Prove that the vertex of a parabola is the point closest to the focus.

*7.  A comet moves in a parabolic orbit with the sun as focus (Fig. 10). When it is 20 million miles from the sun, the line from sun to comet makes an angle of $60°$ with the axis of the parabola. How close does it come to the sun?

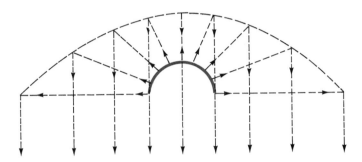

**FIGURE 9**

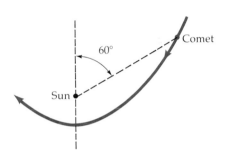

**FIGURE 10**

---

# 12.2
## ELLIPSES

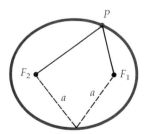

**FIGURE 1**
$|PF_1| + |PF_2| = 2a.$

**DEFINITION**

Given two points $F_1$ and $F_2$ in the plane, and a number $2a > |F_1F_2|$, the set of all points $P$ such that

$$|PF_1| + |PF_2| = 2a \tag{1}$$

is an *ellipse* (Fig. 1). The points $F_1$ and $F_2$ are the *foci* of the ellipse.

*REMARKS*   Foci, the plural of focus, is pronounced "foe-sigh." The given number is called $2a$, not simply $a$, in order to simplify the equation of the ellipse, derived below.

To sketch an ellipse, you could put tacks at the foci $F_1$ and $F_2$ (Fig. 2), attach a string of length $2a$ between them, and move a pencil around the

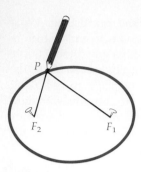

**FIGURE 2**
String length = $2a$.

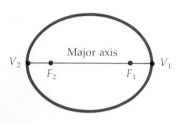

**FIGURE 3**
$V_1V_2$ = major axis, has length $2a$.

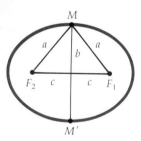

**FIGURE 4**
$MM'$ = minor axis, has length $2b$.

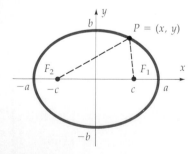

**FIGURE 5**
Ellipse in standard position.
$\dfrac{x^2}{a^2} + \dfrac{y^2}{b^2} = 1$, $a \geq b$, $a^2 = b^2 + c^2$.

foci while holding it taut against the string. As noted in the definition, the string length must be greater than the distance $|F_1F_2|$ between the foci; for if $2a = |F_1F_2|$, then the string goes straight from $F_1$ to $F_2$, and the curve reduces to a line segment; and if $2a < |F_1F_2|$, then the string does not reach from $F_1$ to $F_2$, and there are no points $P$ that satisfy the defining equation (1).

The line segment crossing the ellipse through the foci is called the **major axis** (Fig. 3). Its endpoints are the **vertices**, and its midpoint is the **center** of the ellipse; the curve is symmetric about this point.

The length of the major axis is the same constant $2a$ that appears in the defining equation (1). For in Figure 3, by symmetry, $|V_2F_2| = |V_1F_1|$; hence

$$\text{length of major axis} = |V_2F_2| + |V_1F_2| = |V_1F_1| + |V_1F_2|$$
$$= 2a.$$

[The last step follows from definition (1), since $V_1$ is a point on the ellipse, one of the points $P$ in equation (1).] The number $a$, half the length of the major axis, is called the **semimajor axis**.

The segment crossing the ellipse perpendicular to the major axis at its center (Fig. 4) is the **minor axis**. Half the length of the minor axis is denoted by the letter $b$ and is called the **semiminor axis**.

Consider the point $M$ at one end of the minor axis (Fig. 4). By symmetry, $|MF_1| = |MF_2|$. Since $M$ is on the ellipse, the sum $|MF_1| + |MF_2|$ equals $2a$; hence $|MF_1| = |MF_2| = a$, as shown.

Now let $c$ denote the distance from the center to each focus. From the right triangle in Figure 4

$$a^2 = b^2 + c^2. \tag{2}$$

The numbers $a$, $b$, $c$ are the crucial dimensions of the ellipse:

$$a = \text{semimajor axis.}$$
$$b = \text{semiminor axis.}$$
$$c = \text{distance from center to focus.}$$

Any two determine the shape of the ellipse; the third can be computed from equation (2).

## Equation of the Ellipse

An ellipse is in "standard position" if the major axis lies on the $x$-axis with its center at the origin (Fig. 5); then the foci are at $(\pm c, 0)$. In this case the ellipse has a very simple equation, which we now derive.

Let $P = (x, y)$ denote any point on the ellipse. By the distance formula, the defining equation (1) says

$$\sqrt{(x - c)^2 + y^2} + \sqrt{(x + c)^2 + y^2} = 2a. \tag{3}$$

To simplify this, subtract $\sqrt{(x-c)^2 + y^2}$ from both sides of the equation, then square both sides, obtaining

$$(x+c)^2 + y^2 = 4a^2 - 4a\sqrt{(x-c)^2 + y^2} + (x-c)^2 + y^2.$$

This reduces to

$$cx - a^2 = -a\sqrt{(x-c)^2 + y^2}.$$

Square again, simplify and rearrange terms:

$$a^2(a^2 - c^2) = (a^2 - c^2)x^2 + a^2y^2. \tag{4}$$

From (2) we set $a^2 - c^2 = b^2$. Finally, divide (4) by $a^2b^2$, obtaining

$$\frac{x^2}{a^2} + \frac{y^2}{b^2} = 1. \tag{5}$$

We have shown that every point $(x, y)$ on the ellipse satisfies equation (5). Conversely, any point satisfying (5) is on the ellipse; for the operations leading from (3) to (5) are all reversible. [Squaring is not always reversible: $(-3)^2 = 3^2$ does not imply that $-3 = 3$. But when the two quantities squared are known to be *positive*, then squaring *is* a reversible operation; this is the case in the derivation of (5) from (3).]

If the foci are on the $y$-axis (Fig. 6) then the equation is

$$\frac{x^2}{b^2} + \frac{y^2}{a^2} = 1. \tag{6}$$

In any case, the *larger* of the two denominators is denoted $a^2$ and the smaller one $b^2$; otherwise, equation (2) would have to be relabeled.

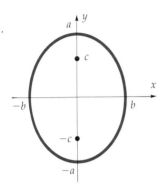

**FIGURE 6**
$\dfrac{x^2}{b^2} + \dfrac{y^2}{a^2} = 1,\ a > b,\ a^2 = b^2 + c^2.$

---

**EXAMPLE 1**   Write the equation of the ellipse in standard position with semimajor axis 5 and semiminor axis 3. Sketch it. Locate the foci.

**SOLUTION**   Use equation (5) with $a = 5$ and $b = 3$:

$$\frac{x^2}{25} + \frac{y^2}{9} = 1.$$

To sketch it, mark the vertices on the $x$-axis at $a = 5$ units from the center $(0, 0)$, and the ends of the minor axis at $b = 3$ units from $(0, 0)$. Then draw in the ellipse as a smooth curve through these four points, symmetric about each axis (Fig. 7).

The focus $F_1$ is determined from the right triangle in Figure 7:

$$c^2 = 5^2 - 3^2 = 16$$

so $c = 4$, and the foci are $(\pm 4, 0)$.

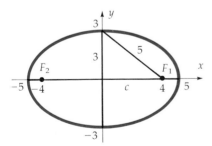

**FIGURE 7**
$\dfrac{x^2}{25} + \dfrac{y^2}{9} = 1.$

---

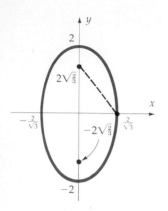

**FIGURE 8**
$3x^2 + y^2 = 4$.

**EXAMPLE 2**  Sketch the graph of $3x^2 + y^2 = 4$, and plot the foci.

**SOLUTION**  The given equation can be written in the standard form for an ellipse—divide through by 4 and rewrite a little:

$$\frac{x^2}{4/3} + \frac{y^2}{4} = 1.$$

The larger denominator is under the $y^2$ term, so the graph is an ellipse of the form (6), not (5). The ends of the two axes are on the coordinate axes, and can be found by the usual method for intercepts. On the $x$-axis, $y = 0$ and the given equation of the ellipse reduces to

$$3x^2 + 0^2 = 4, \quad \text{so} \quad x = \pm \frac{2}{\sqrt{3}}.$$

Similarly the $y$-intercepts are $y = \pm 2$. This gives the graph in Figure 8. The major axis lies on the $y$-axis; so then do the foci. They are at $(0, \pm c)$ where (Fig. 8)

$$c^2 = 2^2 - \left[\frac{2}{\sqrt{3}}\right]^2 = \frac{8}{3}; \quad \text{so} \quad c = 2\sqrt{2/3}.$$

## Reflection in an Ellipse

Imagine a ray of light leaving one focus and reflected from the ellipse. Where does it go? To the other focus! The reason is (Fig. 9) that the tangent line at the point of reflection $P$ makes equal angles with the two lines from $P$ through the foci (see problem 16).

Sound is reflected in much the same way as light, and this has led whimsical architects to design "whispering galleries." The ceiling is a surface of revolution formed by revolving an ellipse about its major axis (Fig. 10). A whisperer $W$ at one focus is heard clearly by an eavesdropper $E$ at the other. This is due partly to the reflective property (all sound from $W$ is reflected to $E$) and partly to the fact that all the reflected paths from $W$ to $E$ have the same length; thus the reflections of a given sound pulse arriving from different parts of the ceiling all arrive *at the same time* and reinforce each other.

A new treatment for kidney stones uses the same idea. A reflector shaped like part of an elliptical surface of revolution has an electrode at one focus. It is placed so that the kidney stone is at the other focus. Shock waves generated by the electrode are focused on the kidney stone, crushing it to harmless bits. Except at the foci, the waves are relatively dispersed and pass harmlessly through the body tissues.

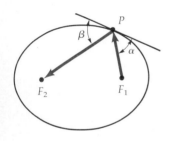

**FIGURE 9**
Reflection in an ellipse: $\alpha = \beta$.

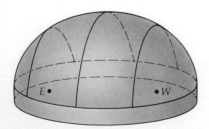

**FIGURE 10**
Whispering gallery. Eavesdropper $E$ hears whisperer $W$.

## Translated Ellipses

The equation

$$\frac{(x - x_0)^2}{a^2} + \frac{(y - y_0)^2}{b^2} = 1, \qquad a > b \tag{7}$$

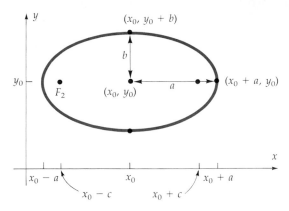

**FIGURE 11**

$$\frac{(x - x_0)^2}{a^2} + \frac{(y - y_0)^2}{b^2} = 1; \ a > b.$$

gives an ellipse with center at the point $(x_0, y_0)$ (Fig. 11). For, the quantities

$$\Delta x = x - x_0 \quad \text{and} \quad \Delta y = y - y_0$$

play exactly the same role in (7) that $x$ and $y$ play in (5). With $a > b$, the major axis is on the horizontal line $y = y_0$. If the larger denominator $a^2$ goes under $(y - y_0)^2$, then the major axis is vertical (Fig. 12).

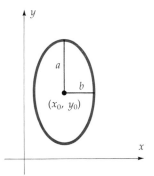

**FIGURE 12**

$$\frac{(x - x_0)^2}{b^2} + \frac{(y - y_0)^2}{a^2} = 1; \ a > b.$$

**EXAMPLE 3**  Show that $9x^2 + 25y^2 - 18x + 100y = 116$ is an ellipse. Find the center, axes, and foci.

*SOLUTION*  Collect $x$ and $y$ terms and complete the squares, to write the equation in the standard form for an ellipse:

$$(9x^2 - 18x) + (25y^2 + 100y) = 116$$

$$9(x^2 - 2x \quad) + 25(y^2 + 4y \quad) = 116.$$

Add terms to complete the squares on the left, and compensating terms on the right to preserve the equality:

$$9(x^2 - 2x + 1) + 25(y^2 + 4y + 4) = 116 + 9 \cdot 1 + 25 \cdot 4$$

$$9(x - 1)^2 + 25(y + 2)^2 = 116 + 9 + 100 = 225$$

$$\frac{(x - 1)^2}{25} + \frac{(y + 2)^2}{9} = 1.$$

This is like (7) with $a = 5$, $b = 3$, $x_0 = 1$, $y_0 = -2$. Since the larger denominator is under the $x$ term, the major axis is horizontal. The center is at $(x_0, y_0) = (1, -2)$, the semiaxes are $a = 5$ and $b = 3$, and the distance from center to focus is

$$c = \sqrt{a^2 - b^2} = \sqrt{25 - 9} = 4.$$

To sketch the graph (Fig. 13) plot the ends of the axes at $(1 \pm 5, -2)$ and $(1, -2 \pm 3)$. Show the foci at $(1 \pm 4, -2)$.

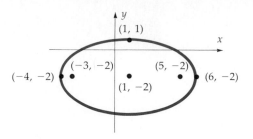

**FIGURE 13**
$9x^2 + 25y^2 - 18x + 100y = 116.$

## SUMMARY

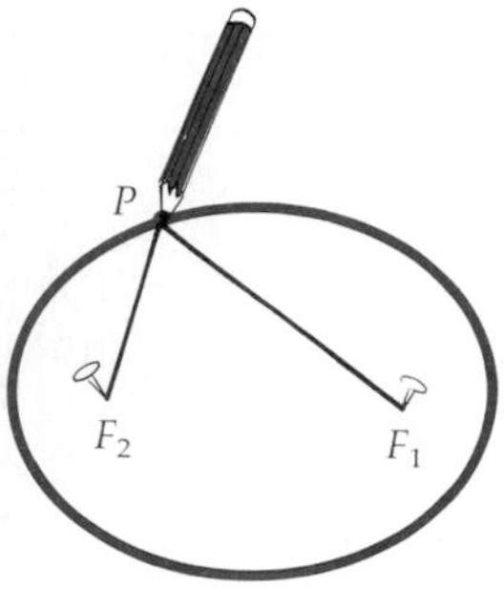

*Geometric Definition of the Ellipse (see Fig. 14):*

$$|PF_1| + |PF_2| = 2a.$$

*Algebraic Equation in Standard Form (see Fig. 15):*

$$\frac{x^2}{a^2} + \frac{y^2}{b^2} = 1$$

$a =$ semimajor axis.

$$a^2 = b^2 + c^2 \qquad b = \text{semiminor axis.}$$

$c =$ distance from focus to center.

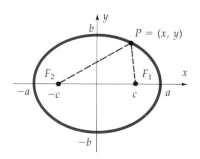

**FIGURE 14**

**FIGURE 15**

## PROBLEMS

**A**

1. Sketch the ellipse, labeling the foci and ends of the axes.

   *a) $\dfrac{x^2}{25} + \dfrac{y^2}{16} = 1$   b) $\dfrac{x^3}{3} + \dfrac{y^2}{4} = 1$

   *c) $x^2 + 2y^2 = 1$   d) $2x^2 + y^2 = 4.$

2. In each part, write the equation of an ellipse in standard position with the given properties.
   a) Semiaxes 2 and 1.
   *b) Major axis of length 10 and foci 6 units apart.
   c) Foci 2 units apart and minor axis of length 2.

**3.** The largest possible ellipse is to be cut from a rectangular piece of wood 4 ft by 8 ft, with axes parallel to the sides of the rectangle.
   **a)** Where are the foci? Make a sketch showing the measurements.
   **b)** Suppose the ellipse is to be drawn using a string attached to two tacks. Where do the tacks go, and how long is the free part of the string (not counting the part in the knots at the ends)?
   **\*c)** Suppose the ellipse is to be drawn using a loop of string, looped over two tacks. How long is the loop?

**\*4.** In each part, write the equation of an ellipse with the given data:
   **a)** Ends of major axis ($\pm 3$, 0) and ends of minor axis (0, $\pm 1$).
   **b)** Foci at ($\pm 1$, 0) and vertices at ($\pm 5$, 0).
   **c)** Foci at (0, 1) and (4, 1) and ends of minor axis at (2, 0) and (2, 2).

**5.** Write the equation of an ellipse with the given properties, and sketch it.
   **a)** Foci (0, 0) and (1, 0); one vertex at (2, 0).
   **b)** Vertices at (0, 0) and ($-3$, 0); passing through ($-3/2$, 1).
   **c)** Vertices at ($-2$, 0) and (4, 0); one focus at the origin.

**6.** Find the center, foci, and ends of the axes. Sketch.
   **\*a)** $\dfrac{(x+1)^2}{16} + \dfrac{(y-3)^2}{9} = 1$
   **b)** $\dfrac{(x-1)^2}{2} + \dfrac{(y+3)^2}{3} = 1$
   **\*c)** $9x^2 + y^2 - 18x + 2y + 1 = 0$
   **d)** $2x^2 + x + y^2 - y = 4$

**7.** Find the equation of the line tangent to
   **\*a)** $\dfrac{x^2}{25} + \dfrac{y^2}{16} = 1$ at the point $\left(3, \dfrac{16}{5}\right)$.
   **b)** $x^2 + 2x + 3y^2 = 0$ at the origin.

**B**

**\*8.** Compute the area of the ellipse $x^2 + 2y^2 = 2$.

**9.** Show that the ellipse with semiaxes $a$ and $b$ has area $\pi ab$. How is this related to the formula for the area of a circle?

**\*10.** The ellipse $2x^2 + y^2 = 4$ is revolved about the $x$-axis. Compute the volume enclosed.

**11.** The ellipse $\dfrac{x^2}{a^2} + \dfrac{y^2}{b^2} = 1$ is revolved about the $x$-

axis. Compute the volume enclosed. Relate your answer to the formula for the volume of a ball.

**\*12.** The ellipse in the previous problem is revolved about the $y$-axis. Compute the volume enclosed, and relate the result to the formula for the volume of a ball.

**13.** Show that the line tangent to the ellipse $b^2x^2 + a^2y^2 = a^2b^2$ at $(x_0, y_0)$ is

$$b^2x_0x + a^2y_0y = a^2b^2.$$

**14.** Figure 16 shows a stiff rod of length $a + b$; a point $P$ is marked on it, $b$ units from the lower end. The upper end rests on the $y$-axis, and the lower end on the $x$-axis. As the lower end is moved from the origin out along the $x$-axis, $P$ traces a quarter of an ellipse. Prove this.

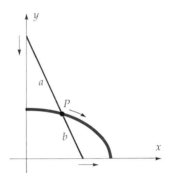

**FIGURE 16**

**\*15.** An ellipse has foci at (1, 1) and ($-1$, $-1$) and vertices at (2, 2) and ($-2$, $-2$). Find the ends of the minor axis. Find the equation of the ellipse. [Imitate the derivation of (5).]

**16.** In Figure 9, prove that $\alpha = \beta$. (A straightforward method is to use the equation of the ellipse and the formula giving the angle $\varphi$ between two lines of slopes $m_1$ and $m_2$: $\tan \varphi = \dfrac{m_2 - m_1}{1 + m_1m_2}$.)

**17.** Prove that the ellipse $\dfrac{x^2}{a^2} + \dfrac{y^2}{b^2} = 1$ is convex; that is, the graph is concave down for $y > 0$ and concave up for $y < 0$.

**18.** Kepler discovered that the orbit of the earth is an ellipse with the sun at one focus. At "perihelion" (closest approach to the sun) it is about 91.4 million miles from the sun; at the other extreme ("aphelion") it is about 94.5 million miles therefrom. Determine the fundamental constants $a$, $b$, and $c$ for the orbit.

## 12.3

## HYPERBOLAS

A hyperbola consists of two disjoint parts, called branches (Fig. 1). It is determined by two foci $F_1$ and $F_2$, together with a positive number $a$. The branch near $F_1$ consists of all points $P$ which are $2a$ units closer to $F_1$ than to $F_2$:

$$|PF_2| - |PF_1| = 2a.$$

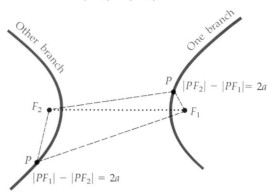

**FIGURE 1**
The two branches of a hyperbola.

The branch near $F_2$ reverses this condition: $|PF_1| - |PF_2| = 2a$, or

$$|PF_2| - |PF_1| = -2a.$$

The two branches together form the hyperbola.

---

**DEFINITION**

Given points $F_1$ and $F_2$ and a number $2a < |F_1 F_2|$, the set of points $P$ such that

$$|PF_2| - |PF_1| = \pm 2a \qquad (1)$$

is a *hyperbola* with foci $F_1$ and $F_2$.

---

The definition requires

$$2a < |F_1 F_2| \qquad (2)$$

since, for the point $P$ on the right-hand branch in Figure 1,

$$2a + |PF_1| = |PF_2| < |F_1 F_2| + |PF_1|.$$

### Equation of the Hyperbola in Standard Position

A hyperbola is in "standard position" if the foci are on the $x$-axis at points $(\pm c, 0)$ with the origin halfway between them (Fig. 2). The distance between the foci is then $|F_1 F_2| = 2c$, so (2) implies that $2a < 2c$, or

$$a < c. \qquad (3)$$

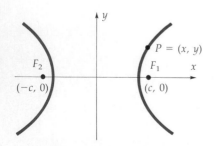

**FIGURE 2**
Hyperbola in standard position.

To derive the equation of a hyperbola thus placed, proceed as with the parabola and ellipse. Let $P = (x, y)$ be any point on the curve; equation (1) becomes

$$\sqrt{(x + c)^2 + y^2} - \sqrt{(x - c)^2 + y^2} = \pm 2a.$$

This is equivalent to the equation obtained by squaring both sides,

$$(x + c)^2 + y^2 - 2\sqrt{(x + c)^2 + y^2}\sqrt{(x - c)^2 + y^2} + (x - c)^2 + y^2 = 4a^2,$$

which can be rewritten

$$x^2 + c^2 + y^2 - 2a^2 = \sqrt{(x + c)^2 + y^2}\sqrt{(x - c)^2 + y^2}. \tag{4}$$

Square again, simplify, and rearrange a few terms to get

$$(c^2 - a^2)x^2 - a^2y^2 = a^2(c^2 - a^2).$$

Finally, divide by $a^2(c^2 - a^2)$:

$$\frac{x^2}{a^2} - \frac{y^2}{c^2 - a^2} = 1. \tag{5}$$

From (3) it follows that $c^2 - a^2 > 0$, so we can define the number

$$b = \sqrt{c^2 - a^2}; \tag{6}$$

then equation (5) becomes

$$\frac{x^2}{a^2} - \frac{y^2}{b^2} = 1. \tag{7}$$

Every point $(x, y)$ on the hyperbola must satisfy this equation; and conversely, every point satisfying (7) is on the hyperbola, because all the steps in deriving (7) are reversible. [The crucial point is that squaring both sides of (4) is reversible, since both sides are positive.]

*SYMMETRY*   The graph of equation (7) is symmetric with respect to both axes, since it is unaffected by replacing $x$ with $-x$, or $y$ with $-y$.

*EXTENT*   Solve equation (7) for $y$:

$$y = \pm\frac{b}{a}\sqrt{x^2 - a^2}. \tag{8}$$

The square root is real only if $x^2 \geq a^2$, so *there are no points on the graph for* $|x| < a$.

On the other hand, solving (7) for $x$ shows: For each $y$ there are two points on the hyperbola, with $x$ coordinate

$$x = \pm\frac{a}{b}\sqrt{y^2 + b^2}.$$

*ASYMPTOTES* For the moment, look just at the first quadrant. There $y > 0$, so we use the $+$ sign in (8):

$$y = \frac{b}{a}\sqrt{x^2 - a^2}, \qquad x \geq a. \tag{9}$$

When $x$ grows large positive, then $a^2$ is insignificant compared to $x^2$; so the graph of (9) should approach the graph of

$$y = \frac{b}{a}\sqrt{x^2} = \frac{b}{a}x. \tag{10}$$

In fact (see problem 11) the difference between (10) and (9) goes to 0 as $x \to +\infty$, so the line (10) is an asymptote of the graph of (9) in the first quadrant (Fig. 3). The hyperbola lies below the asymptote, since

$$\frac{b}{a}\sqrt{x^2 - a^2} < \frac{b}{a}\sqrt{x^2} = \frac{b}{a}x.$$

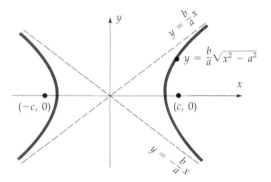

**FIGURE 3**

The asymptotes, $y = \pm\dfrac{b}{a}x$.

Now consider the entire hyperbola. By symmetry, $y = \dfrac{b}{a}x$ is an asymptote as $x \to \pm\infty$, and so is the line $y = -\dfrac{b}{a}x$. These two asymptotes are easily obtained from the equation of the hyperbola; just replace the constant 1 on the right-hand side of (7) with 0:

$$\frac{x^2}{a^2} - \frac{y^2}{b^2} = 0 \quad \Leftrightarrow \quad \left(\frac{x}{a} - \frac{y}{b}\right)\left(\frac{x}{a} + \frac{y}{b}\right) = 0$$

$$\Leftrightarrow \quad y = \frac{b}{a}x \quad \text{or} \quad y = -\frac{b}{a}x.$$

*GRAPHING* Figure 4 shows how to sketch the standard hyperbola

$$\frac{x^2}{a^2} - \frac{y^2}{b^2} = 1. \tag{11}$$

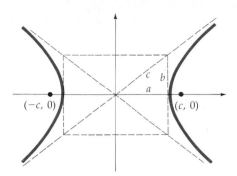

**FIGURE 4**

The box for a hyperbola in standard position, $\dfrac{x^2}{a^2} - \dfrac{y^2}{b^2} = 1$.

Draw the $2a$ by $2b$ rectangle centered at the origin. The diagonals of this rectangle determine the asymptotes of the hyperbola, since both the diagonals and the asymptotes pass through the origin, and both have the same slope $\pm b/a$. The hyperbola lies outside the rectangle, touching it at the two points $(\pm a, 0)$; these are the **vertices**. The line segment connecting them is called the **transverse axis**. The foci are at $(\pm c, 0)$ where, by (6),

$$c = \sqrt{a^2 + b^2}.$$

It follows that the hypotenuse of the right triangle in Figure 4 has length $c$.

Notice the similarities and differences between the standard hyperbola (11) and the standard ellipse

$$\frac{x^2}{a^2} + \frac{y^2}{b^2} = 1. \tag{12}$$

*Similarities:* For both, you can draw a $2a$ by $2b$ rectangle; and both have vertices at $(\pm a, 0)$ and foci at $(\pm c, 0)$.

*Differences:* The equation (12) for the ellipse is a *sum* of squares equal to 1; the larger denominator is denoted $a^2$, and $a \ge b$. The ellipse is *inside* the box (Fig. 5), and so are the foci. The right triangle in the figure has hypotenuse $a$.

The equation (11) for the hyperbola, on the other hand, is a *difference* of squares equal to 1; the denominator of the square with the plus sign is denoted $a^2$, and $a$ need not exceed $b$. The hyperbola is *outside* the box (Fig. 4) and so are the foci. The right triangle in the figure has hypotenuse $c$.

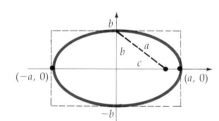

**FIGURE 5**

Ellipse in a box; $\dfrac{x^2}{a^2} + \dfrac{y^2}{b^2} = 1$.

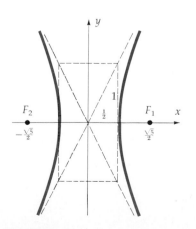

**FIGURE 6**
$4x^2 - y^2 = 1$.

**EXAMPLE 1** Sketch $4x^2 - y^2 = 1$ and plot the foci.

*SOLUTION* The given equation can be written

$$\frac{x^2}{(1/2)^2} - \frac{y^2}{1^2} = 1,$$

which matches the standard hyperbola (11) with $a = 1/2$ and $b = 1$. Draw the $a$ and $b$ rectangle and the asymptotes, and sketch the curve (Fig. 6). The hypotenuse of the right triangle is

$$c = \sqrt{(1/2)^2 + 1^2} = \frac{\sqrt{5}}{2}$$

so the foci are at $\left( \pm \dfrac{\sqrt{5}}{2}, 0 \right)$.

## Hyperbolas out of Standard Position

Any quadratic of the form

$$Ax^2 + Bx + Cy^2 + Dy + E = 0 \tag{13}$$

can be analyzed by completing the squares on $x$ and $y$. If $A$ and $C$ have *opposite signs*, you can then write the equation in one of the three forms

$$\frac{(x - x_0)^2}{a^2} - \frac{(y - y_0)^2}{b^2} = 1 \tag{14}$$

or

$$\frac{(y - y_0)^2}{a^2} - \frac{(x - x_0)^2}{b^2} = 1 \tag{15}$$

or

$$m^2(x - x_0)^2 - (y - y_0)^2 = 0. \tag{16}$$

In the first two, we have a difference of squares equal to 1; the square with the $+$ sign has denominator $a^2$. Equation (14) is just like the standard hyperbola (11), but with the center shifted to $(x_0, y_0)$ (Fig. 7). Equation (15) is similar,

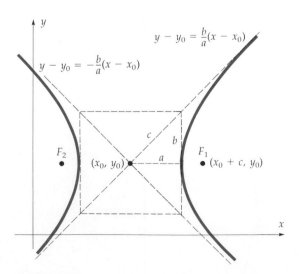

**FIGURE 7**
$$\frac{(x - x_0)^2}{a^2} - \frac{(y - y_0)^2}{b^2} = 1.$$

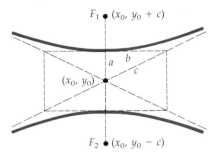

**FIGURE 8**
$$\frac{(y - y_0)^2}{a^2} - \frac{(x - x_0)^2}{b^2} = 1.$$

but with the roles of $x$ and $y$ reversed, so the transverse axis (the segment connecting the vertices) is parallel to the $y$-axis (Fig. 8).

The third equation (16) can be solved for $y$:

$$y = y_0 \pm m(x - x_0).$$

This gives two lines through $(x_0, y_0)$ with slopes $m$ and $-m$. These intersecting lines can be considered a limiting case of a hyperbola, where the vertices have been drawn in to the center.

Return now to the general equation (13), and suppose that $A$ and $C$ have the *same signs*. Then completion of the square yields either an ellipse (as in the previous section), or one of the forms

$$\frac{(x - x_0)^2}{a^2} + \frac{(y - y_0)^2}{b^2} = 0 \qquad \text{[just one point, } (x_0, y_0)\text{]}$$

or

$$\frac{(y - y_0)^2}{a^2} + \frac{(x - x_0)^2}{b^2} = -1 \qquad \text{(empty graph)}.$$

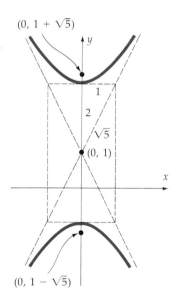

**FIGURE 9**
$$\frac{(y - 1)^2}{4} - \frac{x^2}{1} = 1.$$

**EXAMPLE 2** Graph $4x^2 - y^2 + 2y + 3 + 0$.

**SOLUTION** Complete the square on $y$:

$$4x^2 - (y^2 - 2y) + 3 = 0$$
$$4x^2 - (y - 1)^2 + 1 + 3 = 0$$
$$4x^2 - (y - 1)^2 + 4 = 0.$$

Move the constant term to the right-hand side

$$4x^2 - (y - 1)^2 = -4,$$

divide by the right-hand side and put the positive square first:

$$\frac{(y - 1)^2}{4} - \frac{x^2}{1} = 1.$$

This matches (15) with $a = 2$, $b = 1$, $y_0 = 1$, $x_0 = 0$. The term with $(y - y_0)^2$ has the $+$ sign, so the transverse axis is vertical. The center of the hyperbola is $(x_0, y_0) = (0, 1)$. The distance from center to vertex is $a = 2$; the distance from center to focus is

$$c = \sqrt{a^2 + b^2} = \sqrt{4 + 1} = \sqrt{5}.$$

This gives Figure 9. From the rectangle there, the asymptotes have slope $\pm 2$ and pass through $(0, 1)$; their equations are therefore $y = 1 \pm 2x$.

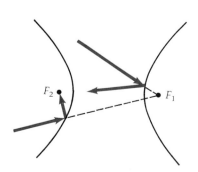

**FIGURE 10**
Reflecting property of the hyperbola.

**The reflecting property** of a hyperbola is that light traveling toward one focus is reflected toward the other (Fig. 10; see problem 16).

### Hyperbolic Orbits

The planets orbit the sun on elliptical paths, held in place by the force of gravity. This same force can generate hyperbolic orbits. In fact, there may well be celestial objects traveling hyperbolic orbits; but any such object can pass the sun only once as it traverses a branch of the hyperbola, and will not return periodically, like a planet or comet.

   The electrical force between an electron and a proton is very similar to the force of gravity, though much stronger. An electron passing a proton travels a hyperbolic orbit with the proton at one focus (Fig. 11). The hyperbola, however, can be so close to its asymptotes that the electron appears to approach the proton on one straight line and leave it on the other, as if bounced from a billiard ball.

**FIGURE 11**
Electron deflected by a proton.

## SUMMARY

*Geometric Definition of the Hyperbola (see Fig. 12):*

$$\left| PF_2 \right| - \left| PF_1 \right| = \pm 2a$$

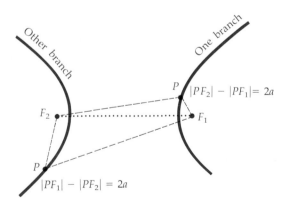

**FIGURE 12**

*Equation of Hyperbola in Standard Position (see Fig. 13):*

$$\frac{x^2}{a^2} - \frac{y^2}{b^2} = 1$$

$$2a = \text{distance between vertices.}$$

$$c^2 = a^2 + b^2 \qquad 2c = \text{distance between foci.}$$

$$\pm b/a = \text{slope of asymptotes.}$$

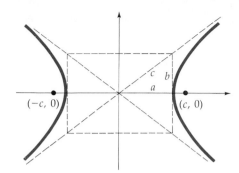

**FIGURE 13**

## PROBLEMS

### A

1. Sketch the graph, showing any vertices, foci, and asymptotes. Write equations for the asymptotes.

   *a) $x^2 - y^2 = 1$     b) $x^2 - \dfrac{y^2}{4} = 1$

   c) $x^2 - y^2 = 4$     *d) $y^2 - 4x^2 = 4$

   e) $x^2 - y^2 = 0$

2. Graph a hyperbola in standard position with the given properties. Show vertices, foci, and asymptotes. Write an equation of the graph.

   *a) Distance between vertices = 2, distance between foci = 3.

   b) Distance between vertices = 4; asymptotes $y = \pm 2x$.

   *c) Distance between foci = 4; asymptotes intersecting at right angles.

   d) Vertices at $(\pm 1, 0)$; graph passes through the point (2, 3).

3. Sketch the graph, showing any vertices, foci, and asymptotes. Write equations for the asymptotes.

   *a) $(x - 1)^2 - (y + 1)^2 = 1$

   b) $(x + 3)^2 - 9(y + 2)^2 = 36$

   *c) $x^2 - 4y^2 + 2x + 8y - 7 = 0$

   d) $4y^2 = x^2 - 4x + 8$

   *e) $4x^2 = y^2 - 4y$

   *f) $x^2 - y^2 - 2x - 2y = 0$

   g) $4x^2 + y^2 + 4y = 0$

   h) $4x^2 + 4y = 0$

   i) $4x^2 + y^2 + 4y + 4 = 0$

   j) $4x^2 + y^2 + 4y + 5 = 0$

### B

4. a) Show that the graphs of $xy = 1$ and $x^2 - y^2 = 1$ intersect at right angles.

   b) Show the same for the graphs of $xy = c$ and $x^2 - y^2 = r$, for any constants $c$ and $r \neq 0$.

5. Show that the line tangent to the graph of $b^2x^2 - a^2y^2 = a^2b^2$ at point $(x_0, y_0)$ on the curve is

$$b^2x_0x - a^2y_0y = a^2b^2.$$

6. A light ray emanates from the focus $(c, 0)$ of a hyperbola in standard position and is reflected from the right-hand branch of the hyperbola. Sketch the reflected ray, and explain why it lies on a line passing through the other focus. (Assume the reflecting property in Fig. 10.)

7. For each constant $a > 0$, the equation $x^2 - y^2 = a^2$ defines a hyperbola with asymptotes $y = \pm x$. Where are the foci? What is the graph like when $a$ is very large? When $a$ is very small? What happens to the graph in the limit as $a \to 0$?

8. a) The foci of a hyperbola are $(a, a)$ and $(-a, -a)$, and the distance between the vertices is $2a$. Show that the hyperbola is the graph of $y = a^2/2x$.

   b) Find the foci of the hyperbola $y = k/x$ if $k > 0$.

*9. a) Two towers $T_1$ and $T_2$, 1000 km apart, transmit synchronized signals at a speed of $3 \times 10^5$ km/sec. A ship at sea receives the signal from $T_2$ $18 \times 10^{-4}$ seconds before the signal from $T_1$. Deduce that the ship lies on a certain hyperbola. [Put $T_1$ at $(-500, 0)$ and $T_2$ at $(500, 0)$.]

**b)** A third tower at the point $(1500, 0)$ also transmits signals, synchronized with those from $T_2$. The signal from $T_2$ is received $10^{-3}$ seconds before the signal from $T_3$. Assuming that the ship is above the $x$-axis, where is it exactly? (This is the basis of the "LORAN" navigation system.)

**10.** A gun at $(c, 0)$ is fired at a target at $(-c, 0)$. Find the set of all points where the sound of the gun is heard simultaneously with the sound of the bullet striking the target. Assume that the speed of sound is $S$ m/sec and that of the bullet is $B$ m/sec, $B > S$. (Where are the two heard simultaneously if $B = S$?)

**11.** Prove the property of the asymptotes of the hyperbola, that

$$\lim_{x \to +\infty} \left[ \frac{b}{a} x - \frac{b}{a} \sqrt{x^2 - a^2} \right] = 0.$$

$$\left[ \text{Use } \sqrt{A} - \sqrt{B} = \frac{A - B}{\sqrt{A} + \sqrt{B}}. \right]$$

**12.** Prove that as $x \to +\infty$, the limiting slope of a hyperbola equals the slope of its asymptote.

**13.** Show that an ellipse and a hyperbola with the same foci intersect at right angles. (This can be proved from the equality of angles implied by the reflective properties of the two curves.)

**14.** Figure 14 shows an arrangement of mirrors in a telescope. Light rays entering parallel to the vertical axis are reflected toward $F_1$, intercepted by

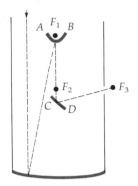

**FIGURE 14**
Reflections in a telescope.

the mirror $AB$ and sent through $F_2$ to mirror $CD$, and reflected there to $F_3$. Describe the conic section containing $AB$, and the section containing $CD$.

**15. a)** For given points $F_1$ and $F_2$, what curve is defined by the equation $|PF_1| = |PF_2|$?

**b)** Relate this to the hyperbola defined by $|PF_1| - |PF_2| = \pm 2a$, for small $a$.

**16.** Let $A$ be a fixed point inside the branch defined by $|PF_2| - |PF_1| = 2a$.

**a)** Among all points $P$ on this branch, for which is the sum $|AP| + |PF_1|$ a minimum? (Use no calculus; there is a simple geometric solution.)

**b)** Let $P_0$ be the intersection of the first branch with the line from $A$ to $F_2$ and $T$ be the line tangent to the first branch at $P_0$. Among all points $Q$ on $T$, for which is the sum $|AQ| + |QF_1|$ a minimum?

**c)** Deduce the reflective property of the hyperbola: Segments $AP_0$ and $F_1P_0$ make equal angles with the tangent line at $P_0$.

---

## 12.4

### GENERAL QUADRATICS. ROTATION OF AXES

The general quadratic equation in two variables has the form

$$Ax^2 + Bxy + Cy^2 + Dx + Ey + F = 0,$$

where $A$, $B$, and $C$ are not all zero. If $B = 0$ and the $xy$ term is missing, then the graph can be analyzed by completing the square.

What if $B \neq 0$ and there *is* an $xy$ term? An example is $xy = 1$. The familiar graph (Fig. 1) appears to be a hyperbola, but with its transverse axis at $45°$ to the coordinate axes. To analyze such graphs, we introduce new

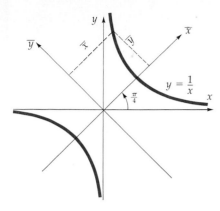

**FIGURE 1**

coordinate axes, appropriately tipped. In Figure 1, these new axes are labeled $\bar{x}$ and $\bar{y}$. Using these new coordinates, the hyperbola will have an equation of the form

$$\frac{\bar{x}^2}{a^2} - \frac{\bar{y}^2}{b^2} = 1,$$

which allows you to identify the vertices, foci, and asymptotes of the graph.

To carry out this idea, we derive the relation between the original $xy$-coordinates and the rotated $\bar{x}\bar{y}$-coordinates. Figure 2 shows a new set of axes labeled $\bar{x}$ and $\bar{y}$, making angle $\varphi$ with the original axes. For a given point $P$, denote by $(x, y)$ its coordinates relative to the original axes, and by $(\bar{x}, \bar{y})$ the coordinates relative to the new set. What is the relation between these coordinates?

Let $r$ denote the distance from the origin to $P$. From the right triangle $OAP$,

$$x = r \cos(\alpha + \varphi) = r \cos \alpha \cos \varphi - r \sin \alpha \sin \varphi.$$
$$y = r \sin(\alpha + \varphi) = r \sin \alpha \cos \varphi + r \cos \alpha \sin \varphi. \tag{1}$$

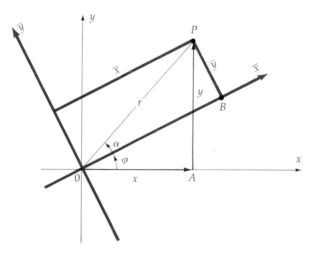

**FIGURE 2**
Axes rotated through angle $\varphi$.

From triangle $OBP$, on the other hand, $\bar{x} = r \cos \alpha$ and $\bar{y} = r \sin \alpha$. Combine this with (1), and the desired relation appears:

$$x = \bar{x} \cos \varphi - \bar{y} \sin \varphi,$$
$$y = \bar{x} \sin \varphi + \bar{y} \cos \varphi. \tag{2}$$

---

**EXAMPLE 1**   Rewrite $xy = 1$, using axes rotated through $45°$.

**SOLUTION**   From the rotation equations (2) with $\varphi = \pi/4$,

$$x = \bar{x}\left(\frac{1}{\sqrt{2}}\right) - \bar{y}\left(\frac{1}{\sqrt{2}}\right), \qquad y = \bar{x}\left(\frac{1}{\sqrt{2}}\right) + \bar{y}\left(\frac{1}{\sqrt{2}}\right)$$

so

$$xy = \left[\frac{\bar{x}}{\sqrt{2}} - \frac{\bar{y}}{\sqrt{2}}\right] \cdot \left[\frac{\bar{x}}{\sqrt{2}} + \frac{\bar{y}}{\sqrt{2}}\right] = \frac{\bar{x}^2}{2} - \frac{\bar{y}^2}{2}.$$

The given equation $xy = 1$ is thus

$$\frac{\bar{x}^2}{2} - \frac{\bar{y}^2}{2} = 1,$$

the equation of a hyperbola with $a^2 = 2$, $b^2 = 2$, $c^2 = a^2 + b^2 = 4$. The vertices are on the $\bar{x}$-axis at

$$\bar{x} = \pm\sqrt{2}, \qquad \bar{y} = 0 \tag{3}$$

and the foci are at

$$\bar{x} = \pm c = \pm 2, \qquad \bar{y} = 0. \tag{4}$$

Convert back to $xy$-coordinates, using (2). From (3) with the $+$ sign, you find a vertex at

$$x = \frac{+\sqrt{2}}{\sqrt{2}} - \frac{0}{\sqrt{2}} = +1, \qquad y = \frac{+\sqrt{2}}{\sqrt{2}} + \frac{0}{\sqrt{2}} = +1$$

in agreement with Figure 3. From (4) with the $+$ sign, you find a focus at

$$x = \frac{+2}{\sqrt{2}} - \frac{0}{\sqrt{2}} = +\sqrt{2}, \qquad y = \frac{+2}{\sqrt{2}} + \frac{0}{\sqrt{2}} = +\sqrt{2}.$$

The other focus is apparently $(-\sqrt{2}, -\sqrt{2})$.

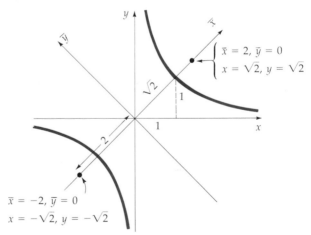

**FIGURE 3**
Foci and vertices for $xy = 1$.

Any quadratic equation

$$Ax^2 + Bxy + Cy^2 + Dx + Ey + F = 0 \tag{5}$$

can, by means of a suitable rotation, be rewritten in the form

$$\bar{A}\bar{x}^2 + \bar{C}\bar{y}^2 + \bar{D}\bar{x} + \bar{E}\bar{y} + \bar{F} = 0 \tag{6}$$

where no "cross term" $\bar{x} \cdot \bar{y}$ appears. In this new form, you can complete the squares on $\bar{x}$ and $\bar{y}$ to analyze the graph as in Sections 12.1 through 12.3. The important question is, what angle of rotation $\varphi$ converts (5) into (6), with no $\bar{x} \cdot \bar{y}$ term?

To determine the right $\varphi$, substitute the rotation equations (2) into the general quadratic (5), multiply out, and combine terms. Denote by $\bar{A}$ the coefficient of $\bar{x}^2$, and so on. You get an equation of the form

$$\bar{A}\bar{x}^2 + \bar{B}\bar{x}\bar{y} + \bar{C}\bar{y}^2 + \bar{D}\bar{x} + \bar{E}\bar{y} + \bar{F} = 0 \tag{7}$$

where

$$\begin{aligned}
\bar{A} &= A \cos^2\varphi + B \cos\varphi \sin\varphi + C \sin^2\varphi \\
&= \tfrac{1}{2}[A(1 + \cos 2\varphi) + B \sin 2\varphi + C(1 - \cos 2\varphi)], \tag{7a} \\
\bar{B} &= B(\cos^2\varphi - \sin^2\varphi) + 2(C - A) \cdot \sin\varphi \cdot \cos\varphi \\
&= B \cos 2\varphi + (C - A) \sin 2\varphi, \tag{7b} \\
\bar{C} &= A \sin^2\varphi - B \sin\varphi \cos\varphi + C \cos^2\varphi \\
&= \tfrac{1}{2}[A(1 - \cos 2\varphi) - B \sin 2\varphi + C(1 + \cos 2\varphi)]. \tag{7c}
\end{aligned}$$

You may check these equations, and compute the other coefficients $\bar{D}$, $\bar{E}$, and $\bar{F}$. We want $\bar{B} = 0$, so (7b) gives

$$0 = B \cos 2\varphi + (C - A) \sin 2\varphi,$$

which reduces to

$$\cot 2\varphi = \frac{A - C}{B}. \tag{8}$$

Conclusion: *The rotated equation has no $\bar{x} \cdot \bar{y}$ term if the angle of rotation $\varphi$ satisfies (8).*

---

**EXAMPLE 2**    Graph $5x^2 + 6xy + 5y^2 - 8 = 0$.

**SOLUTION**    This is a quadratic (5) with $A = 5$, $B = 6$, $C = 5$. Equation (8) gives

$$\cot 2\varphi = \frac{A - C}{B} = 0.$$

Recalling that $\cot 2\varphi = (\cos 2\varphi)/(\sin 2\varphi)$, we see that $\cos 2\varphi = 0$, and take $2\varphi = \pi/2$, $\varphi = \pi/4$. The rotation equations (2) give

$$x = \frac{\bar{x}}{\sqrt{2}} - \frac{\bar{y}}{\sqrt{2}}, \qquad y = \frac{\bar{x}}{\sqrt{2}} + \frac{\bar{y}}{\sqrt{2}}.$$

The given equation becomes

$$5\left(\frac{\bar{x}}{\sqrt{2}} - \frac{\bar{y}}{\sqrt{2}}\right)^2 + 6\left(\frac{\bar{x}}{\sqrt{2}} - \frac{\bar{y}}{\sqrt{2}}\right)\left(\frac{\bar{x}}{\sqrt{2}} + \frac{\bar{y}}{\sqrt{2}}\right) + 5\left(\frac{\bar{x}}{\sqrt{2}} + \frac{\bar{y}}{\sqrt{2}}\right)^2 - 8 = 0.$$

Multiply out and combine terms:

$$8 \cdot \bar{x}^2 + 2 \cdot \bar{y}^2 - 8 = 0 \tag{9}$$

or

$$\bar{x}^2 + \frac{\bar{y}^2}{4} = 1.$$

The graph is an ellipse with major axis on the $\bar{y}$-axis; $a^2 = 4$, $b^2 = 1$, and $c^2 = a^2 - b^2 = 3$. The vertices are at $\bar{y} = \pm 2$, the foci at $\bar{y} = \pm\sqrt{3}$, and the ends of the minor axis at $\bar{y} = 0$, $\bar{x} = \pm 1$ (Fig. 4). Compute the coordinates of these points in the original $xy$-system, using (2) once more. You find

$$\text{vertices} \quad (-\sqrt{2}, \sqrt{2}) \text{ and } (\sqrt{2}, -\sqrt{2}),$$
$$\text{foci} \quad (-\sqrt{\tfrac{3}{2}}, \sqrt{\tfrac{3}{2}}) \text{ and } (\sqrt{\tfrac{3}{2}}, -\sqrt{\tfrac{3}{2}}),$$
$$\text{ends of the minor axis} \quad \left(\frac{1}{\sqrt{2}}, \frac{1}{\sqrt{2}}\right) \text{ and } \left(-\frac{1}{\sqrt{2}}, -\frac{1}{\sqrt{2}}\right).$$

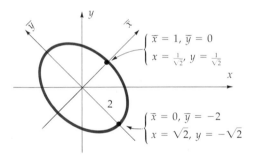

**FIGURE 4**
$5x^2 + 6xy + 5y^2 - 8 = 0.$

When the equation $\cot 2\varphi = \dfrac{A - C}{B}$ gives $2\varphi = 60°, 90°,$ or $120°$, then $\sin \varphi$ and $\cos \varphi$ are easily computed, and you can carry out the rotation beginning with the change of variables (2). But in any case, $\cot 2\varphi$ is given by (8), and this determines $\csc 2\varphi = (1 + \cot^2 2\varphi)^{1/2}$, and with it $\sin 2\varphi = (\csc 2\varphi)^{-1}$ and $\cos 2\varphi = (\cot 2\varphi)(\sin 2\varphi)$. Then the alternate equations in (7) give the new coefficients $\bar{A}, \bar{B}, \bar{C}$. If the original equation also has linear terms $Dx + Ey$, then the new terms $\bar{D}\bar{x} + \bar{E}\bar{y}$ can be computed from (2), using the half angle formulas to determine $\cos \varphi$ and $\sin \varphi$ from $\cos 2\varphi$.

**EXAMPLE 3** Analyze the graph of $32x^2 + 50xy + 7y^2 = 52$.

**SOLUTION**

$$\cot 2\varphi = \frac{A - C}{B} = \frac{1}{2}$$

so

$$\csc 2\varphi = \sqrt{5}/2, \qquad \sin 2\varphi = 2/\sqrt{5}, \qquad \cos 2\varphi = 1/\sqrt{5}.$$

From (7),

$$\bar{A} = \tfrac{1}{2}[32(1 + 1/\sqrt{5}) + 2 \cdot 50/\sqrt{5} + 7(1 - 1/\sqrt{5})]$$
$$= \tfrac{1}{2}(125/\sqrt{5} + 39) \approx 47.45085.$$

$$\bar{B} = 0 \qquad \text{(of course)}.$$

$$\bar{C} = \tfrac{1}{2}[32(1 - 1/\sqrt{5}) - 2 \cdot 50/\sqrt{5} + 7(1 + 1/\sqrt{5})]$$
$$= \tfrac{1}{2}(39 - 125/\sqrt{5}) \approx 8.45085.$$

The new equation is

$$(47.45084967 \ldots)\bar{x}^2 - (8.450849669 \ldots)\bar{y}^2 = 52$$

or

$$\frac{\bar{x}^2}{1.09587 \ldots} - \frac{\bar{y}^2}{6.153227 \ldots} = 1.$$

The graph is a hyperbola, with its transverse axis along the new $\bar{x}$ axis. The angle of rotation $\varphi$ is determined by

$$\cot 2\varphi = \tfrac{1}{2} \quad \Leftrightarrow \quad \tan 2\varphi = 2 \quad \Leftrightarrow \quad 2\varphi = \tan^{-1}2 \quad \Leftrightarrow \quad \varphi \approx 31.7°.$$

## Degenerate Conics

Usually a quadratic equation gives a genuine parabola, ellipse, or hyperbola; but the following cases can also occur:

| Case | Examples |
| --- | --- |
| Empty graph | $x^2 + y^2 = -1; x^2 = -2$ |
| One point | $x^2 + y^2 = 0$ (the origin) |
| One line | $x^2 = 0$ (the $y$-axis) |
| Two parallel lines | $x^2 = 1$ ($x = 1$ and $x = -1$) |
| Two intersecting lines | $xy = 0$ (the two coordinate axes) |

We call these cases "degenerate conics." Thus: *The graph of any quadratic equation (5) is a conic section (possibly degenerate).*

## The Discriminant

Which conic you have can be determined by rotating the axes (if necessary) and completing the square (if needed). However, you can tell a lot without these tedious manipulations, just from the leading coefficients $A$, $B$, $C$, by computing the **discriminant**

$$\delta = B^2 - 4AC.$$

This expression is easy to remember, for it occurs under the radical in the quadratic formula. It turns out that *the discriminant is invariant under rotation of the axes.* That is, if you rotate axes through any angle $\varphi$ and compute the new coefficients $\bar{A}$, $\bar{B}$, $\bar{C}$, then

$$\bar{B}^2 - 4\bar{A} \cdot \bar{C} = B^2 - 4AC.$$

Example 2 illustrates this: $B^2 - 4AC = 6^2 - 4 \cdot 5 \cdot 5 = -64$, and $\bar{B}^2 - 4AC = 0^2 - 4 \cdot 8 \cdot 2 = -64$. The proof in general is a straightforward calculation (see problem 7).

Suppose now that the angle of rotation is chosen so as to make $\bar{B} = 0$. Then the new equation has the form

$$\bar{A}\bar{x}^2 + 0 \cdot \bar{x} \cdot \bar{y} + \bar{C}\bar{y}^2 + \bar{D}\bar{x} + \bar{E}\bar{y} + \bar{F} = 0 \qquad \text{(10)}$$

and the discriminant is

$$\delta = -4\bar{A} \cdot \bar{C} = B^2 - 4AC.$$

Consider three cases:

$\boldsymbol{\delta > 0.}$ Then $\bar{A} \cdot \bar{C} < 0$, so $\bar{A}$ and $\bar{C}$ have opposite sign and the graph of (10) is a hyperbola.

$\boldsymbol{\delta = 0.}$ Then $\bar{A} = 0$ or $\bar{C} = 0$, and the graph is a parabola.

$\boldsymbol{\delta < 0.}$ Then $\bar{A} \cdot \bar{C} > 0$, so $\bar{A}$ and $\bar{C}$ have the same sign, and the graph is an ellipse.

In each case, degenerate conics may occur.

---

**EXAMPLE 4** Classify the graph of

$$x^2 - 4xy + 4y^2 = 4. \qquad \text{(11)}$$

*SOLUTION* $A = 1$, $B = -4$, $C = 4$, so the discriminant is

$$\delta = 16 - 16 = 0.$$

Hence the graph is a parabola, or degenerate. A rotation of axes would reveal that it is in fact two parallel straight lines, a degenerate form of parabola. This can also be seen directly from (11), if you notice that the left-hand side is a perfect square:

$$(x - 2y)^2 = 4, \qquad x - 2y = \pm 2, \qquad y = \tfrac{1}{2}x \pm 1,$$

the equations of two straight lines of slope 1/2.

---

### The Trace

Along with the discriminant $\delta = B^2 - 4AC$, there is another invariant, the **trace** $A + C$. For, after a rotation, $\bar{A} + \bar{C} = A + C$. This invariance provides a simple check on your calculations of $\bar{A}$ and $\bar{C}$; the new trace and discriminant must agree with the old. Thus in Example 2, $A + C = 5 + 5 = 10$, and from (9), $\bar{A} + \bar{C} = 8 + 2 = 10$, which checks.

## SUMMARY

The graph of any quadratic equation

$$Ax^2 + Bxy + Cy^2 + Dx + Ey + F = 0$$

is a (possibly degenerate) conic section.

*The Discriminant* of the equation is $\delta = B^2 - 4AC$:

$$\delta > 0 \Rightarrow \text{the graph is a hyperbola.}$$

$$\delta = 0 \Rightarrow \text{the graph is a parabola.}$$

$$\delta < 0 \Rightarrow \text{the graph is an ellipse.}$$

(In each case, "degenerate" graphs may occur.)

*Rotation of Axes.*   The "cross term" $xy$ is eliminated by rotating axes through an angle $\varphi$ defined by

$$\cot 2\varphi = \frac{A - C}{B}.$$

## PROBLEMS

### A

1. Analyze the given graph by rotating axes. Sketch the graph, and give $xy$-coordinates of any vertices and foci. For the rotated equation, verify that $\bar{B}^2 - 4\bar{A} \cdot \bar{C} = B^2 - 4AC$ and $\bar{A} + \bar{C} = A + C$.
   *a)  $xy = -4$
   b)  $x^2 - xy + y^2 = 2$
   *c)  $2x^2 + 4xy - y^2 = 6$
   *d)  $9x^2 - 24xy + 16y^2 + 50x + 5 = 25y$
   e)  $31x^2 + 10\sqrt{3}xy + 21y^2 = 144$
   *f)  $7x^2 - 6\sqrt{3}xy + 13y^2 = 16$

2. Classify the following conics by the discriminant.
   *a)  $xy = 1$
   b)  $x^2 - y^2 = -1$
   *c)  $2x^2 + y^2 = 1$
   d)  $x^2 + 2xy + 2y^2 = 7$
   e)  $x^2 + 2xy - y^2 = 11$
   *f)  $x^2 = xy$
   g)  $x^2 - xy = 1$

3. Describe the following degenerate conics.

*a)  $y^2 = 1$
*b)  $(x + y)^2 = 0$
*c)  $x^2 - 2xy + y^2 = -1$
d)  $x^2 - y^2 = 0$
e)  $2x^2 + y^2 = 0$
f)  $x^2 + 2y^2 - 2x + 4y + 3 = 0$

4. Under rotation through an angle $\varphi$, to what equation is $x^2 + y^2 = r^2$ transformed?

### B

5. Check formulas (7a)–(7c) for the new coefficients $\bar{A}$, $\bar{B}$, and $\bar{C}$.

6. Determine the new coefficients $\bar{D}$, $\bar{E}$, and $\bar{F}$ in equation (7).

7. After rotating axes through an angle $\varphi$
   a)  Show that $\bar{B}^2 - 4\bar{A} \cdot \bar{C} = B^2 - 4AC$.
   b)  Is $\bar{A} + \bar{C} = A + C$? (Prove or disprove.)
   c)  Is $\bar{A} - \bar{C} = A - C$? (Prove or disprove.)

*8. Show that the graph of $y = ax + b/x$ is a nondegenerate hyperbola if $b \neq 0$. Give its asymptotes.

**9.** Solve equations (2) to express $\bar{x}$ and $\bar{y}$ in terms of $x$ and $y$.

**C**

**10.** Since the discriminant is invariant under rotation, it must be related to some intrinsic geometric property of the graph, not affected by rotations. Relate the discriminant of $Ax^2 + Bxy + Cy^2 = 1$ to the area of the fundamental $2a$ by $2b$ rectangle associated with the graph of an ellipse or a hyperbola. (You can assume that $B = 0$; why?)

**11.** We pulled the discriminant $B^2 - 4AC$ "out of a hat." This problem shows how it might be discovered.

A hyperbola is distinguished from an ellipse by its *extent*; the hyperbola has points $(x, y)$ for all sufficiently large $y$, while the ellipse has points only for a restricted range of $y$. Assuming $A \neq 0$, solve the general quadratic (5) for $x$. Show that if $B^2 - 4AC > 0$ then there are solutions for all sufficiently large $y$; while if $B^2 - 4AC < 0$ then there are solutions for at most a restricted range of $y$.

# 12.5

## POLAR COORDINATES

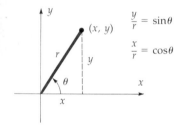

**FIGURE 1**
Polar coordinates $r$ and $\theta$.
$x = r \cos \theta, \; y = r \sin \theta.$

Rectangular coordinates determine points in the plane by ordered pairs of numbers $(x, y)$, where $x$ and $y$ give the distances from the point to two perpendicular axes. *Polar coordinates* determine a point by giving the *distance from the origin* and the *direction from the origin*. The distance is denoted by $r$, and the direction is given by the angle $\theta$ in Figure 1, from the positive $x$-axis to the ray from the origin to the point. The figure shows the relation between these numbers $r$ and $\theta$, and the rectangular coordinates $x$ and $y$:

$$x = r \cos \theta, \qquad y = r \sin \theta. \tag{1}$$

The figure suggests that $r \geq 0$ and $0 \leq \theta \leq 2\pi$, but we do not make these restrictions in general; *any pair of numbers $r$ and $\theta$ satisfying equations (1) are* called **polar coordinates** of the point $(x, y)$.

---

**EXAMPLE 1** Each of the following pairs of numbers $r$ and $\theta$ are polar coordinates of the point $(0, 1)$:

$$r = 1, \quad \theta = \frac{\pi}{2}; \qquad r = -1, \quad \theta = \frac{3\pi}{2}; \qquad r = 1, \quad \theta = -\frac{3\pi}{2}.$$

For each of these pairs, $r \cos \theta = 0$ and $r \sin \theta = 1$ (Fig. 2).

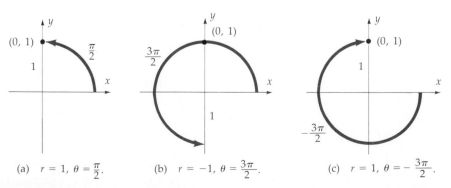

(a) $r = 1, \theta = \frac{\pi}{2}$.      (b) $r = -1, \theta = \frac{3\pi}{2}$.      (c) $r = 1, \theta = -\frac{3\pi}{2}$.

**FIGURE 2**
Three sets of polar coordinates for $(0, 1)$.

The special role of the origin in polar coordinates suggests (correctly) that they serve mainly in applications where one particular point has special significance. For example, in the study of planetary motion we use polar coordinates with the sun at the origin.

Each pair of polar coordinates $r$ and $\theta$ determines a point in the plane, namely $(r \cos \theta, r \sin \theta)$. We denote this point by $(r, \theta)_p$.[1] To plot the point $(r, \theta)_p$, face in the direction of the positive $x$-axis and turn counterclockwise through the angle $\theta$; this determines a certain direction from the origin. Then, if $r > 0$, the point is $r$ units from the origin in that direction; if $r < 0$, the point is $|r|$ units from the origin in the *opposite* direction. Figure 3 shows how to plot $(2, 5\pi/4)_p$ and $(-2, 5\pi/4)_p$. You can check from (1) that the rectangular coordinates of $(-2, 5\pi/4)_p$ are $\left( -2 \cos \dfrac{5\pi}{4}, -2 \sin \dfrac{5\pi}{4} \right) = (\sqrt{2}, \sqrt{2})$ as shown.

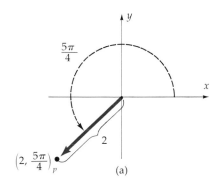
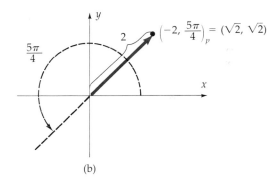

**FIGURE 3**
Plotting in polar coordinates.

As we saw in Example 1, each point has many different pairs of polar coordinates; this is the main disadvantage of the system. From Figure 1

$$r^2 = x^2 + y^2, \tag{2}$$

so $r = \pm\sqrt{x^2 + y^2}$ is determined up to sign. But there is great latitude in choosing $\theta$. In particular, when $r = 0$ then *every* value of $\theta$ gives the same point, the origin; $(0, \theta)_p = (0, 0)$.

---

**EXAMPLE 2**   Plot the points $(1, \pi/2)_p$, $(-1, \pi/2)_p$, $(\sqrt{2}, \pi/4)_p$, $(-\sqrt{2}, 5\pi/4)_p$ directly, without converting to rectangular coordinates. Then find the rectangular coordinates using (1), and check your results. (Try it yourself; then check the solution in Figure 10.)

---

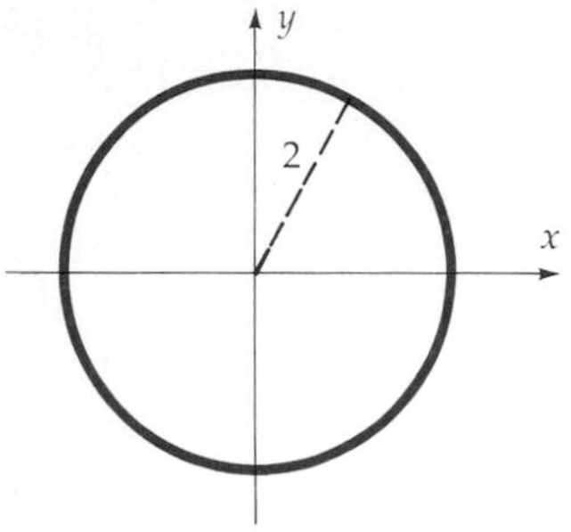

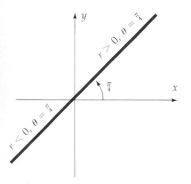

**FIGURE 4**
Polar graph of $r = 2$.

**FIGURE 5**
Polar graph of $\theta = \pi/4$.

**EXAMPLE 3** Find polar coordinates for $(0, 0)$, $(1, -1)$, and $(-1, \sqrt{3})$. Find at least two pairs of polar coordinates for $(1, 1)$. (The solution is in Figure 11.)

## Graphing in Polar Coordinates

An equation in polar coordinates determines a curve in the plane. For example, the equation $r = 2$ says that the distance from the origin to the point $(r, \theta)_p$ is 2; this gives the circle of radius 2 about the origin (Fig. 4). The equation $\theta = \pi/4$ gives the points on the 45° ray, when $r > 0$ (Fig. 5); when $r < 0$, it gives the points in the opposite direction; and with $r = 0$ it gives the origin. Thus the polar coordinates graph of $\theta = \pi/4$ is precisely the line $y = x$.

The next example is less obvious.

**EXAMPLE 4** Graph $r = 2 \sin \theta$.

**SOLUTION** We take a very unsophisticated approach: Calculate many pairs $r$ and $\theta$ satisfying the given equation,

| $\theta$ | 0 | $\pi/6$ | $\pi/3$ | $\pi/2$ | $2\pi/3$ | $5\pi/6$ | $\pi$ | $7\pi/6$ | $4\pi/3$ | $3\pi/2$ |
|---|---|---|---|---|---|---|---|---|---|---|
| $2 \sin \theta$ | 0 | 1 | $\sqrt{3}$ | 2 | $\sqrt{3}$ | 1 | 0 | $-1$ | $-\sqrt{3}$ | $-2$ |

and plot the corresponding points in polar coordinates (Fig. 6a). The points for $0 \le \theta \le \pi$ appear to lie on a circular curve tangent to the $x$-axis; for $\theta > \pi$ you get the *same points again*, but with *different* polar coordinates. Sketch a smooth curve through these points to obtain the circular graph in Figure 6b.

Is it precisely a circle? To check, convert to rectangular coordinates. In this case, the conversion is simplified if you first multiply the given equation $r = 2 \sin \theta$ by $r$:

$$r^2 = 2r \sin \theta.$$

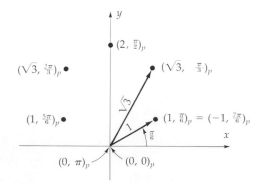

**FIGURE 6a**
Points on polar graph of $r = 2 \sin \theta$.

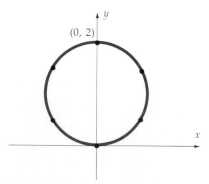

**FIGURE 6b**
Polar graph of $r = 2 \sin \theta$; point $(0, 2)$ is labelled with its rectangular coordinates.

As we noted in Figure 1, $r^2 = x^2 + y^2$, and $r \sin \theta = y$, so

$$x^2 + y^2 = 2y.$$

This is equivalent to $x^2 + (y - 1)^2 = 1$, which is indeed the equation of a circle, with radius 1 and center $(0, 1)$.

To graph a polar equation $r = f(\theta)$ efficiently, use some of the same crucial data as for rectangular coordinates: maxima, minima, and zeros of $f$. The maxima and minima of $r = f(\theta)$ determine how far the curve goes from the origin. The zeros determine its passage through the origin: If $f(\theta_0) = 0$ then the curve passes through the origin *tangent to the line* $\theta = \theta_0$. Figure 7 shows why. The secant line from the origin to the point $P = (r, \theta)_p$ on the curve has slope $m = \tan \theta$. As $\theta \to \theta_0$, then $P$ approaches the origin, since $r = f(\theta) \to f(\theta_0) = 0$. The slope of the curve at the origin is by definition the limiting slope of the secant lines:

$$\text{slope of curve at origin} = \lim_{\theta \to \theta_0} \tan \theta = \tan \theta_0$$

and this is precisely the slope of the line $\theta = \theta_0$. (This argument assumed that $f$ is continuous at $\theta_0$, and $f(\theta) \neq 0$ for $\theta$ near $\theta_0$; these conditions are met in all our examples.)

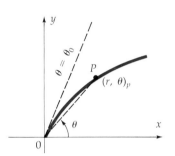

**FIGURE 7**
Segment $OP$ has slope $m = \tan \theta$.

**EXAMPLE 5**   Graph $r = \cos 2\theta$.

**SOLUTION**   The following table includes all the maxima, minima, and zeros of $\cos 2\theta$ for $0 \leq \theta \leq 2\pi$:

| $\theta$ | 0 | $\pi/6$ | $\pi/4$ | $\pi/3$ | $\pi/2$ | $3\pi/4$ | $\pi$ | $5\pi/4$ | $3\pi/2$ | $7\pi/4$ | $2\pi$ |
|---|---|---|---|---|---|---|---|---|---|---|---|
| $r = \cos 2\theta$ | 1 | 1/2 | 0 | $-1/2$ | $-1$ | 0 | 1 | 0 | $-1$ | 0 | 1 |

Plot the corresponding points on the graph. For each of the zeroes $\theta = \dfrac{\pi}{4}$, $\dfrac{3\pi}{4}, \ldots$ sketch a dotted line through the origin at the indicated angle (Fig. 8a). Start sketching the graph at $(0, 1)_p$. As $\theta$ increases from 0 to $\pi/4$, $r$

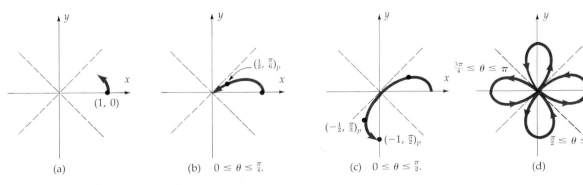

(a)

(b)   $0 \leq \theta \leq \frac{\pi}{4}$.

(c)   $0 \leq \theta \leq \frac{\pi}{2}$.

(d)

**FIGURE 8**
Graphing $r = \cos 2\theta$.

decreases from 1 to 0, and the graph approaches the origin *tangent to the line* $\theta = \pi/4$ (Fig. 8b). Then as $\theta$ increases from $\pi/4$ to $\pi/2$, $r$ decreases further from 0 to $-1$; the angles $\pi/4 \le \theta \le \pi/2$ give directions in the first quadrant, but $r$ is negative, so the graph is actually in the opposite direction (Fig. 8c).

As you continue to follow the curve for $\dfrac{\pi}{2} < \theta < 2\pi$, it loops back and forth through the origin, tracing a sort of four-leaf clover (Fig. 8d). Since the function $\cos 2\theta$ has period $2\pi$, values of $\theta$ outside the interval $[0, 2\pi]$ will give the same points as those already plotted.

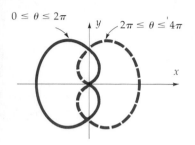

**FIGURE 9**

Polar graph of $r = \sin \dfrac{\theta}{2}$.

**EXAMPLE 6**   Graph $r = \sin(\theta/2)$.

*SOLUTION*   From the maxima, minima, and zeroes you get the graph in Figure 9. The solid part is for $0 \le \theta \le 2\pi$, and the rest for $2\pi \le \theta \le 4\pi$ (or for $-2\pi \le \theta \le 0$). In this case the interval $0 \le \theta \le 2\pi$ is not enough, because $\sin \dfrac{\theta}{2}$ has period $4\pi$, not $2\pi$.

Figure 10 gives the solution for Example 2.

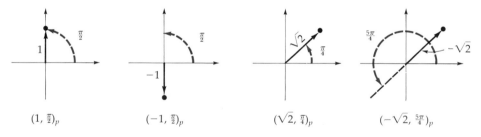

$(1, \frac{\pi}{2})_p$        $(-1, \frac{\pi}{2})_p$        $(\sqrt{2}, \frac{\pi}{4})_p$        $(-\sqrt{2}, \frac{5\pi}{4})_p$

**FIGURE 10**
Solution of Example 2.

Figure 11 gives the solution for Example 3.

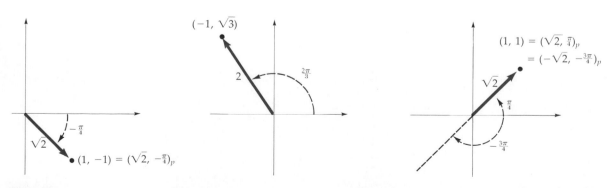

**FIGURE 11**
Solution of Example 3.

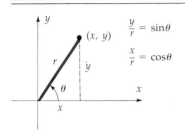

**FIGURE 12**

## SUMMARY

*Polar Coordinates (see Fig. 12):*

$$x = r \cos \theta \qquad r = \pm\sqrt{x^2 + y^2}$$

$$y = r \sin \theta \qquad \tan \theta = \frac{y}{x}$$

## PROBLEMS

### A

1.  Plot the points.
    *a)  $(1, \pi/4)_p$     b)  $(-1, -\pi/4)_p$
    *c)  $(\sqrt{2}, 3\pi/4)_p$    d)  $(\pi, \pi)_p$
    e)  $(1, \pi)_p$      *f)  $(\pi, 1)_p$

2.  Find polar coordinates for each point, with $0 \le \theta < 2\pi$ and $r \ge 0$.
    *a)  $(1, 0)$      b)  $(1, 1)$
    *c)  $(1/2, -1)$   d)  $(-1/2, 1)$
    *e)  $(-1, -1)$   f)  $(-1, 0)$

*3.  Find polar coordinates for each point in problem 2, with $0 \le \theta \le \pi$. (Answers given for parts a, c, e.)

4.  Find all pairs of polar coordinates for the given point.
    *a)  $(0, 0)$   *b)  $(1, 1)$   c)  $(1/2, -1)$

5.  Sketch the graph.
    *a)  $\theta = \pi/2$     b)  $\theta = -\pi/2$
    c)  $\theta = 3\pi/2$    d)  $r = 2$
    *e)  $r = -2$      f)  $r = 0$

6.  Rewrite the following in polar coordinates.
    *a)  $x^2 + y^2 = 1$    *b)  $x = 5$
    c)  $y = 5$      *d)  $xy = 1$
    e)  $x^2 - y^2 = 9$   *f)  $y = x^2$

7.  Identify the graph by converting to rectangular coordinates. (Each graph is a simple geometric figure.)
    *a)  $r \cos \theta = 2$
    b)  $r \sin \theta = 3$
    *c)  $r = \sec \theta$
    *d)  $r \cos\left(\theta - \dfrac{\pi}{2}\right) = 2$

    e)  $r = \dfrac{-1}{\cos \theta + \sin \theta}$
    f)  $r \sin\left(\theta - \dfrac{\pi}{3}\right) = 5$

8.  Identify the graph by converting to rectangular coordinates.
    *a)  $r = 2 \cos \theta$
    *b)  $r = \cos\left(\theta + \dfrac{\pi}{4}\right)$
    c)  $r = 2 \sin(\theta + \pi/3)$
    d)  $r = -2 \cos \theta$
    e)  $r = \sin \theta - \cos \theta$

In problems 9–13, sketch the graph of the given equation in polar coordinates. Show all points where $r$ is maximum or minimum, and all passages through the origin at the proper angle.

9.  *a)  $r = \sin(\theta + \pi/4)$
    b)  $r = 2 \cos(\theta - \pi/3)$
    c)  $r^2 = -\sin \theta$
    *d)  $r^2 = \cos \theta$

10.  (Cardioids)
    a)  $r = 1 + \cos \theta$
    *b)  $r = 1 + \sin \theta$
    c)  $r = 2 - 2 \cos \theta$

11.  (Limacons)
    *a)  $r = 2 + \cos \theta$    *b)  $r = \frac{1}{2} + \cos \theta$
    c)  $r = 2 - \sin \theta$    d)  $r = 1 + 2 \sin \theta$

12.  (Roses)
    *a)  $r = \cos 2\theta$     *b)  $r = \cos 3\theta$
    c)  $r = 5 \sin 2\theta$   d)  $r = \sin 4\theta$
    e)  $r = \sin 3\theta$

13. (Lemniscates)
 *a) $r^2 = \sin 3\theta$  *b) $r^2 = \cos 2\theta$
 c) $r^2 = 4 \sin 2\theta$

14. Graph
 a) A spiral of Archimedes, $r = \theta$.
 b) A "hyperbolic spiral" $r\theta = 1$.
 c) A "lituus" $r^2\theta = 1$.

*15. Identify the graphs by converting to rectangular coordinates.

 a) $r = \dfrac{1}{1 + \cos\theta}$  b) $r = \dfrac{1}{1 + 2\cos\theta}$

 c) $r = \dfrac{2}{2 + \cos\theta}$

**B**

16. a) Show that the graph of $r\cos(\theta - \varphi) = d$ is a straight line. ($\varphi$ and $d$ are constants.)
 *b) On the graph in part a, find the point nearest to the origin.
 c) What is the geometric meaning of $\varphi$ and $d$ in the equation?

*17. Prove that the graph of $r = a\cos\theta + b\sin\theta$ is a circle, if $a^2 + b^2 \neq 0$. Find the center and radius.

18. For the "hyperbolic spiral" $r = 1/\theta$
 a) Compute $\lim_{\theta\to 0^{\pm}} r$, $\lim_{\theta\to\infty} r$, and $\lim_{\theta\to 0} r\sin\theta$.
 b) Show that $-1 < r\sin\theta < 1$.
 c) Sketch the graph.

19. Show that the "lituus" $r^2\theta = 1$ has the horizontal asymptote $y = 0$.

20. Figure 13 gives the graph of $r = f(\theta)$. Sketch the graph of
 *a) $r = -f(\theta)$  b) $r = f(-\theta)$
 c) $r = f(\theta + \pi/4)$  d) $r = f(\theta + \pi)$

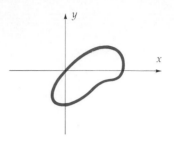

**FIGURE 13**
Graph of $r = f(\theta)$.

21. Compute the square of the distance from $P_1 = (r_1, \theta_1)_p$ to $P_2 = (r_2, \theta_2)_p$. Reduce it to the form $r_1^2 + r_2^2 - 2r_1 r_2 \cos(\theta_2 - \theta_1)$.

22. *a) Sketch the graphs of $r = 2\sin\theta$ and $r = 2\cos\theta$, and find their points of intersection.
 b) The coordinates of the points of intersection do not all satisfy $r = 2\sin\theta = 2\cos\theta$. Why not? [Moral: The points of intersection of the polar coordinate graphs of $r = f(\theta)$ and $r = g(\theta)$ cannot all be found by solving $f(\theta) = g(\theta)$. This is discussed in Sec. 12.7.]

23. a) Sketch together the graphs of $r = 1 + \sin\theta$ and $r^2 = 4\sin\theta$. Observe four points of intersection (three near the origin).
 b) Find the coordinates of the intersections by solving the equations
 $$1 + \sin\theta = \pm 2\sqrt{\sin\theta} \quad \text{and}$$
 $$1 + \sin\theta = \pm 2\sqrt{\sin(\theta + \pi)}.$$
 Why does the second equation give intersections? Why is it needed?

24. a) Indicate *all* the polar coordinates of $(r_0, \theta_0)_p$, if $r_0 \neq 0$.
 b) Indicate all the equations to be solved in seeking intersections of the graphs of $r = f(\theta)$ and $r = g(\theta)$.

## 12.6

### CONIC SECTIONS IN POLAR COORDINATES. ECCENTRICITY

#### Parabolas in Polar Coordinates

Figure 1 shows a parabola with focus $F$ at the origin, and directrix the vertical line $x = -d$. The parabola's equation in polar coordinates is easily derived from the defining relation

$$|PF| = |PD|. \tag{1}$$

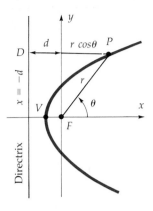

**FIGURE 1**
Parabola: $|PF| = |PD|$
$r = d + r \cos \theta$.

Since the focus $F$ is at the origin, $|PF|$ is simply $r$, the distance from $P$ to the origin. The distance from $P$ to the directrix is, from the figure,

$$|PD| = d + x = d + r \cos \theta.$$

Thus equation (1) gives $r = d + r \cos \theta$; solve this for $r$,

$$r = \frac{d}{1 - \cos \theta}. \tag{2}$$

As a check, note that in (2), the *minimum* of $r$ comes with $\cos \theta = -1$, giving $\theta = \pi$ and $r = d/2$; this gives the vertex $V$ in Figure 1. And $r$ has no maximum; it becomes $+\infty$ as $\theta \to 0$, and this confirms that the parabola encloses the positive $x$-axis.

## More General Conics

Now modify equation (1); instead of requiring that $|PF|$ and $|PD|$ be *equal*, make them *proportional*. The constant of proportionality is denoted by $e$ (*not* the base of the natural logarithms!) Then the modified equation (1) is

$$|PF| = e|PD|. \tag{3}$$

Just as in Figure 1, $|PF| = r$ and $|PD| = d + r \cos \theta$, so

$$r = e(d + r \cos \theta)$$
$$= ed + er \cos \theta. \tag{4}$$

Solve for $r$ to find

$$r = \frac{ed}{1 - e \cos \theta}. \tag{5}$$

When $e = 1$ this is just the parabola in (2). What is it for $e \neq 1$?

Suppose that $0 < e < 1$. Then the denominator $1 - e \cos \theta$ in (5) lies between $1 - e$ (at $\theta = 0$) and $1 + e$ (at $\theta = \pi$); so

$$r_{max} = \frac{ed}{1 - e} \quad [\text{at } \theta = 0] \qquad \text{and} \qquad r_{min} = \frac{ed}{1 + e} \quad [\text{at } \theta = \pi].$$

Figure 2 shows the graph; it looks like an ellipse! Is it, really? To check, convert to rectangular coordinates. Equation (4) gives

$$r = e(d + x), \quad \text{so} \quad r^2 = e^2(d^2 + 2dx + x^2).$$

Use $r^2 = x^2 + y^2$, and complete the square, to find

$$(1 - e^2)\left(x - \frac{e^2 d}{1 - e^2}\right)^2 + y^2 = \frac{e^2 d^2}{1 - e^2}.$$

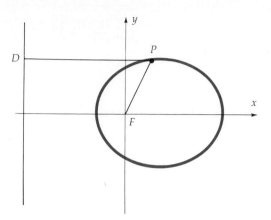

**FIGURE 2**
$|PF| = e|PD|$, with $e = 1/2$.

Denote $\dfrac{e^2d}{1 - e^2}$ by $x_0$, and rewrite this equation in standard form as

$$\frac{(x - x_0)^2}{e^2d^2/(1 - e^2)^2} + \frac{y^2}{e^2d^2/(1 - e^2)} = 1. \tag{6}$$

Since $0 < e < 1$, then also $0 < 1 - e^2 < 1$, so both denominators are positive, and (6) is indeed the equation of an ellipse. We determine its basic dimensions $a$, $b$, $c$. The larger denominator is the one under $(x - x_0)^2$, so that one is $a^2$; thus

$$a = \frac{ed}{1 - e^2}, \qquad b = \frac{ed}{\sqrt{1 - e^2}},$$

and since $c^2 = a^2 - b^2 = e^4d^2/(1 - e^2)^2$,

$$c = \frac{e^2d}{1 - e^2}.$$

The equations for $a$ and $c$ show that $c = ea$, or

$$e = \frac{c}{a}. \tag{7}$$

We can also write

$$e = \frac{2c}{2a} = \frac{\text{distance between foci}}{\text{distance between vertices}}. \tag{8}$$

This ratio is called the **eccentricity** of the ellipse. (Hence the letter $e$ to denote it.) Since it is the ratio of two different dimensions, the eccentricity describes the *proportions* of the ellipse. In fact, any two ellipses with the same eccentricity can differ only in *scale* and *position*.

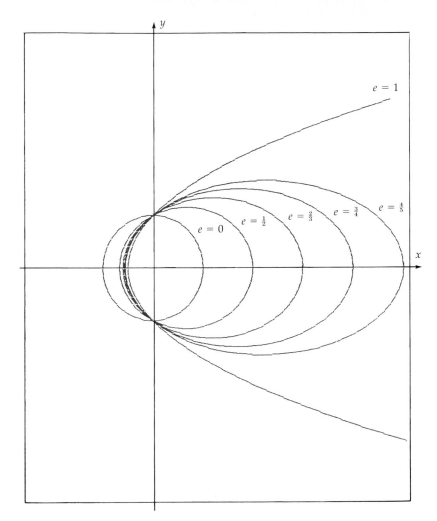

**FIGURE 3**

Five ellipses and a parabola, $r = \dfrac{1}{1 - e \cos \theta}$.

Figure 3 shows the graph of $r = \dfrac{1}{1 - e \cos \theta}$ for various eccentricities $e$. When $e = 0$ it is a circle; this is an "ellipse" with both foci at the center, so the numerator in (8) is zero. As $e$ increases from 0 to 1, the ellipse stretches; in the limit as $e \to 1$, the right-hand vertex goes "off to $\infty$," and the ellipse becomes a parabola!

A similar analysis for the case $e > 1$ shows that then the graph is a *hyperbola*, and equation (8) still holds.

Figure 4 shows some graphs of $r = \dfrac{1}{1 - e \cos \theta}$ for $0 \le e \le 1.5$. As $e$ grows from 0 to 1, the right-hand vertex moves to $x = +\infty$; as $e$ passes 1, this vertex reappears at $x = -\infty$, forming part of the second branch of a hyperbola. As $e$ grows larger, the vertex approaches the origin from the left.

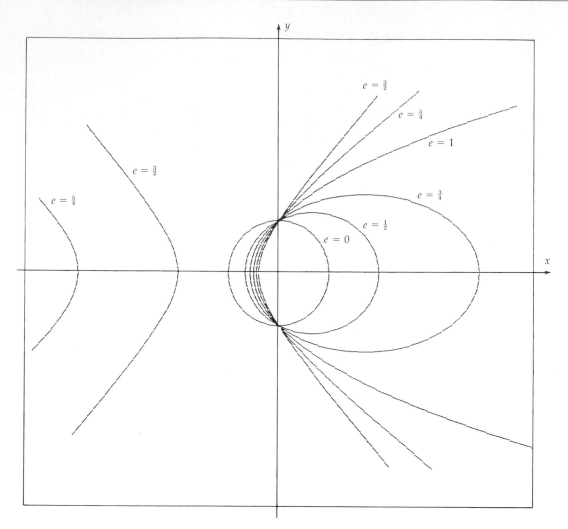

**FIGURE 4**

Conic sections with varying ellipticity,

$$r = \frac{1}{1 - e \cos \theta}.$$

## Directrices

An ellipse or hyperbola can be defined by means of one focus and an associated directrix, by the equation $|PF| = e|PD|$. Because these graphs are symmetric, each has also a second directrix associated with the second focus. The problems ask you to compute the distance from the center of the graph to the directrices; it is $a/e$. Thus, moving out along the major axis from the center of an ellipse, you find (Fig. 5):

| | |
|---|---|
| foci | at distance $ea$, |
| vertices | at distance $a$, |
| directrices | at distance $a/e$. |

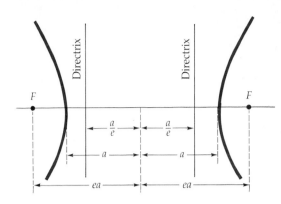

**FIGURE 5**
Foci and directrices of an ellipse of eccentricity $e < 1$.

**FIGURE 6**
Foci and directrices of a hyperbola of eccentricity $e > 1$.

For a *hyperbola* (Fig. 6) with $e > 1$, the order is reversed, but the formulas remain the same:

> directrices   at distance $a/e$,
> vertices     at distance $a$,
> foci       at distance $ea$.

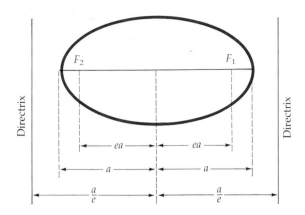

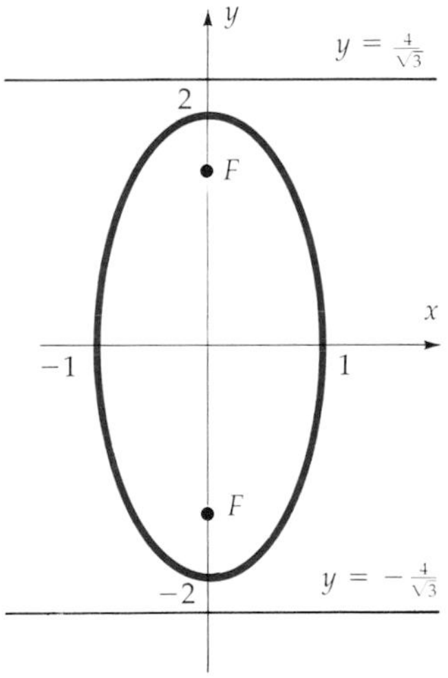

**FIGURE 7**
Foci and directrices of $x^2 + y^2/4 = 1$.

**EXAMPLE 1**   Find the eccentricity and directrices of $x^2 + \dfrac{y^2}{4} = 1$.

**SOLUTION**   Here $a = 2$, $b = 1$, so $c^2 = a^2 - b^2 = 3$. The eccentricity is

$$e = c/a = \sqrt{3}/2.$$

The distance from the center to the directrices is $a/e = \dfrac{4}{\sqrt{3}} \approx 2.3$. Since the major axis is along the $y$-axis, the directrices are the horizontal lines $y = \pm \dfrac{4}{\sqrt{3}}$ (Fig. 7).

**EXAMPLE 2**   Find the eccentricity and directrices of the graph of

$$r = \frac{1}{2 - 4 \cos \theta}.$$

**SOLUTION**   Rewrite the given equation in the standard form (5):

$$r = \frac{1/2}{1 - 2 \cos \theta}. \tag{9}$$

This agrees with (5) when $e = 2$ and $d = 1/4$. So the graph is a hyperbola of eccentricity 2. From Figure 1, the directrix is the vertical line $x = -d$; in this case, $x = -1/4$. The *vertices* of the hyperbola are at $\theta = 0$, $r = -1/2$,

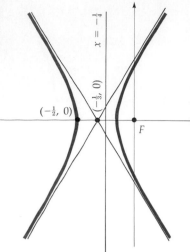

**FIGURE 8**

$r = \dfrac{1/2}{1 - 2\cos\theta}$, with directrix and asymptotes.

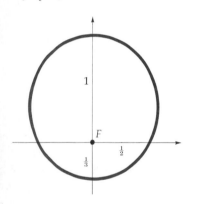

**FIGURE 9**

$r = \dfrac{1}{2 - \sin\theta}$.

and $\theta = \pi$, $r = 1/6$ (Fig. 8); the *center* is midway between them, at $x = -1/3$, $y = 0$. The *asymptotes* of the hyperbola can be found from (9). There, $r \to +\infty$ as $1 - 2\cos\theta \to 0$, thus as $\cos\theta \to 1/2$, or $\theta \to \pm\pi/3$. This gives the angle between the asymptotes and the $x$-axis; their slope is $\pm\tan(\pi/3) = \pm\sqrt{3}$. The asymptotes cross at the center of the hyperbola (Fig. 8); their equations are $y = \pm\sqrt{3}(x + 1/3)$.

---

The standard equation (5) gives a conic section with focus at the origin, and nearest vertex on the negative $x$-axis. If the figure is rotated about the origin, it has an equation very similar to (5).

---

**EXAMPLE 3**    Identify $r = \dfrac{1}{2 - \sin\theta}$ as a conic section.

**SOLUTION**    Try sketching the graph. You find

$$r_{\max} = \frac{1}{2 - 1} = 1, \qquad \text{with } \sin\theta = 1, \quad \text{or} \quad \theta = \pi/2;$$

$$r_{\min} = \frac{1}{2 + 1} = \frac{1}{3}, \qquad \text{with } \sin\theta = -1, \quad \text{or} \quad \theta = -\pi/2.$$

The resulting graph (Fig. 9) appears to be an ellipse with vertices on the $y$-axis. In fact, the given equation is related to the standard one (5) as follows:

$$r = \frac{1/2}{1 - (1/2)\sin\theta} = \frac{1/2}{1 - (1/2)\cos(\pi/2 - \theta)} = \frac{1/2}{1 - (1/2)\cos(\theta - \pi/2)}.$$

This is just (5) with $e = 1/2$ and $d = 2$, *and* with $\theta$ replaced by $\theta - \pi/2$. This has the effect of rotating the "standard" polar ellipse (such as in Fig. 3) through $90°$, obtaining Figure 9.

---

## SUMMARY

*Eccentricity* of a conic section:

$$e = \frac{\text{distance between foci}}{\text{distance between vertices}} = \frac{c}{a}.$$

Given a line (the directrix), a point $F$ not on the directrix, and a number $e > 0$ (the eccentricity); then the set of all points $P$ such that

$$|PF| = e|PD| = e[\text{distance from } P \text{ to directrix}]$$

is a conic section of eccentricity $e$:

$$e = 0 \Rightarrow \text{circle.}$$

$$0 < e < 1 \Rightarrow \text{ellipse.}$$

$$e = 1 \Rightarrow \text{parabola.}$$

$$1 < e \Rightarrow \text{hyperbola.}$$

If the directrix is the vertical line $x = -d$, and the focus $F$ is the origin, then the equation of the conic in polar coordinates is

$$r = \frac{ed}{1 - e \cos \theta}.$$

## PROBLEMS

### A

1. Find eccentricity, vertices, center, the directrix associated with the focus at the origin, and any asymptotes.

   **a)** $r = \dfrac{1}{1 - 2 \cos \theta}$    **\*b)** $r = \dfrac{1}{2 - 3 \cos \theta}$

   **\*c)** $r = \dfrac{5}{3 - \cos \theta}$    **d)** $r = \dfrac{1}{1 + \cos \theta}$

   **\*e)** $r = \dfrac{1}{1 + \sin \theta}$    **f)** $r = \dfrac{2}{3 + 2 \sin \theta}$

2. Find the eccentricity, the foci, and the directrices.

   **\*a)** $y = x^2$    **\*b)** $x^2 + y^2 = 1$

   **c)** $x^2 + 4y^2 = 4$    **d)** $x^2 - y^2 = 1$

   **\*e)** $4x^2 - y^2 = 4$    **f)** $x^2 - 4y^2 = 4$

### B

3. Planets travel in nearly elliptical orbits, with the sun at one focus. The following table gives various data for three of the orbits; under "perihelion" is the distance from planet to sun when it is nearest, and under "aphelion" is the distance when it is farthest. Distances are in millions of miles. For each planet, fill in the missing data. (The headings $2a$ and $2b$ are for the major and minor axes.)

| Planet | Peri-helion | Aphe-lion | $2a$ | $2b$ | Eccen-tricity |
|---|---|---|---|---|---|
| Mercury | . . . | . . . | 71.92 | 70.384 | . . . |
| Earth | 91.45 | 94.56 | . . . | . . . | . . . |
| Pluto | . . . | . . . | 7,328 | . . . | 0.2502 |

4. Recurrent comets, like planets, travel in nearly elliptical orbits with the sun at one focus. However, during most of the orbit they are too far from the sun to be observed. Given the distance at perihelion (in millions of miles) and the eccentricity of the orbit, compute the distance at aphelion (farthest from the sun), and the lengths of the major and minor axes.

| Comet | Perihelion | Eccentricity |
|---|---|---|
| *Halley's | 54 | 0.97 |
| Kohoutek | 12 | 0.999925 |

**\*5.** For an ellipse, which of the following are determined by the eccentricity?

   **a)** The distance between foci.

   **b)** The ratio of axes $b/a$.

   **c)** The angles in the triangle $CFM$, where $C$ is the center, $F$ a focus, and $M$ one end of the minor axis.

6. For a hyperbola, which of the following is deter- by the eccentricity?

   **a)** The angle between the asymptotes.

   **b)** The distance between vertices.

   **c)** The angle between the transverse axis and the asymptotes.

   **d)** The distance between focus and directrix.

7. Demonstrate that if $e > 1$, then the equation

$$|PF| = e \cdot |PD|$$

defines a hyperbola of eccentricity $e$, as claimed in the text. Show that the distance from center to directrix is $a/e$.

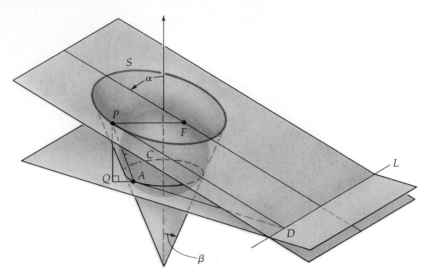

**FIGURE 10**
S is the intersection of a cone with a plane.

**8.** Figure 10 shows a cone and a tipped plane intersecting in a curve $S$. The axis of the cone makes angle $\alpha$ with the plane, and angle $\beta$ with the side of the cone. This problem shows that $S$ is an ellipse, parabola, or hyperbola, according as $\alpha > \beta$, $\alpha = \beta$, or $\alpha < \beta$.

Inscribe a sphere inside the cone and tangent to the tipped plane at a point $F$ (which turns out to be the focus of $S$). The sphere is tangent to the cone in a circle $C$; the plane through $C$ intersects the tipped plane in a line $L$ (which turns out to be the directrix of $S$).

Let $P$ be any point on $S$; $A$, the intersection of circle $C$ with the line from $P$ to the vertex of the cone; $Q$, the foot of the perpendicular from $P$ to the plane through $C$; and $D$, the foot of the perpendicular from $P$ to $L$. Explain why

a) $|PA| = |PF|$
b) $|PQ| = |PA| \cos \beta$
c) $|PQ| = |PD| \cos \alpha$
d) $|PF| = e|PD|$ for a constant $e$
e) $S$ is an ellipse if $\alpha > \beta$, a parabola if $\alpha = \beta$, and a hyperbola if $\alpha < \beta$.

## 12.7

## AREA IN POLAR COORDINATES

### Area of a Circular Sector

The sector in Figure 1 subtends an angle $\Delta\theta$. Since the entire circle subtends an angle $2\pi$, the proportion of the circle lying in the sector is $\dfrac{\Delta\theta}{2\pi}$. The area of the entire circle is $\pi r^2$, so the area of the sector is

$$\frac{\Delta\theta}{2\pi} \cdot \pi r^2 = \frac{1}{2} r^2 \, \Delta\theta.$$

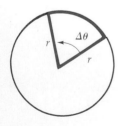

**FIGURE 1**
Area of sector $= \frac{1}{2}r^2 \, \Delta\theta$.

### Area Inside a Polar Graph

Consider now a region as in Figure 2, bounded by the polar graph of $r = f(\theta)$ and the rays $\theta = a$ and $\theta = b$. Assume that $f \geq 0$. Divide the angle from

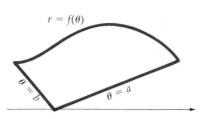

**FIGURE 2**

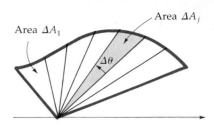

**FIGURE 3**
Region cut into slices.

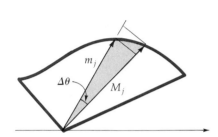

**FIGURE 4**
$\frac{1}{2}m^{j2}\,\Delta\theta \le \Delta A_j \le \frac{1}{2}M_j^2\,\Delta\theta.$

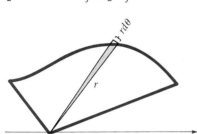

**FIGURE 5**
Area of "infinitesimal triangle"
$= \frac{1}{2}(r)(r\,d\theta) = \frac{1}{2}r^2\,d\theta.$

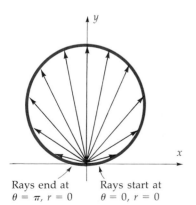

Rays end at          Rays start at
$\theta = \pi, r = 0$    $\theta = 0, r = 0$

**FIGURE 6**
Rays sweeping out area inside
$r = \sin\theta.$

$\theta = a$ to $\theta = b$ into $n$ equal parts, each of angle $\Delta\theta = \dfrac{b-a}{n}$; this slices the region into $n$ parts (Fig. 3). Denote the maximum and minimum values of $r = f(\theta)$ in the $j^{\text{th}}$ slice by

$$M_j = f(\overline{\theta}_j) \quad \text{and} \quad m_j = f(\underline{\theta}_j). \tag{1}$$

Then that slice lies inside a circular sector of radius $M_j$ and contains a sector of radius $m_j$ (Fig. 4), so its area $\Delta A_j$ satisfies

$$\tfrac{1}{2}m_j^2\,\Delta\theta \le \Delta A_j \le \tfrac{1}{2}M_j^2\,\Delta\theta.$$

The total area $A$ of the region is the sum of the sector areas, so

$$\sum_1^n \tfrac{1}{2}m_j^2\,\Delta\theta \le A \le \sum_1^n \tfrac{1}{2}M_j^2\,\Delta\theta$$

or, in view of (1),

$$\sum_1^n \tfrac{1}{2}f(\underline{\theta}_j)^2\,\Delta\theta \le A \le \sum_1^n \tfrac{1}{2}f(\overline{\theta}_j)^2\,\Delta\theta.$$

On the left is a lower Riemann sum for the integral $\int_a^b \frac{1}{2}f(\theta)^2\,d\theta$, and on the right is an upper sum for the same integral. As $n \to \infty$, both sums tend to the integral. Since the area $A$ lies between them, it follows that

$$A = \int_a^b \tfrac{1}{2}f(\theta)^2\,d\theta = \int_a^b \tfrac{1}{2}r^2\,d\theta. \tag{2}$$

Figure 5 illustrates the interpretation of this integral as the "sum" of areas of infinitesimal sectors. The length of an arc of angle $d\theta$ in a circle of radius $r$ is $r\,d\theta$. So the infinitesimal sector of the region is essentially a triangle of base $r\,d\theta$ and altitude $r$. Its "infinitesimal area" is then $\frac{1}{2}(r)\cdot(r\,d\theta) = \frac{1}{2}r^2\,d\theta$; the integral (2) is the sum of these infinitesimal areas.

**EXAMPLE 1**   Compute the area inside the polar graph of $r = \sin\theta$.

**SOLUTION**   *Sketch the graph* (Fig. 6); this is the surest way to determine the limits of integration. Think of rays sweeping out the area inside the curve, illuminating every part of it; they start at $\theta = 0$, continue through $\theta = \pi/4$,

$\theta = \pi/2$, $\theta = 3\pi/4$, and complete the sweep when $\theta = \pi$. Thus the range of integration is $0 \leq \theta \leq \pi$, and the area is

$$\int_0^\pi \frac{1}{2} r^2 \, d\theta = \int_0^\pi \frac{1}{2} \sin^2\theta \, d\theta = \int_0^\pi \frac{1}{4}(1 - \cos 2\theta) \, d\theta = \frac{\pi}{4}.$$

This is reassuring, for the graph of $r = \sin \theta$ is a circle of radius $\frac{1}{2}$; its area is indeed $\pi \left(\frac{1}{2}\right)^2 = \frac{\pi}{4}$.

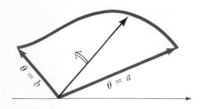

**FIGURE 7**
Rays sweeping out the region inside $r = f(\theta)$, $a \leq \theta \leq b$.

The main difficulty in computing areas in polar coordinates is to determine the limits of integration. For each $\theta$, imagine a ray from the origin to the given graph (Fig. 7); as $\theta$ varies from $a$ to $b$, this ray should sweep out the region exactly once.

**EXAMPLE 2**  Compute the area inside the cardioid $r = 2 + 2 \cos \theta$ and outside the circle $r = 1$.

*SOLUTION*  Graph the two curves (Fig. 8). To compute the required area, you can take the difference between an area inside the cardioid and an area inside the circle, as follows:

(i)  Find the points where the two curves intersect.

(ii)  Choose limits of integration to give the *appropriate part* of the area inside the cardioid and inside the circle.

(iii)  Compute the difference between those areas.

We do this step by step.

(i)  From Figure 8, the curves intersect in the second and fourth quadrants; the corresponding values of $\theta$ are found by equating the for-

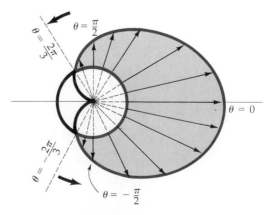

**FIGURE 8**
Shaded region is *outside* $r = 1$ and *inside* $r = 2 + 2 \cos \theta$.

mulas for $r$:

$$r = 2 + 2 \cos \theta = 1,$$

or $\cos \theta = -\dfrac{1}{2}$. There are many solutions for $\theta$:

$$\theta = \frac{2\pi}{3}, \frac{4\pi}{3}, -\frac{2\pi}{3}, \text{ etc.}$$

(ii)  You must choose limits $a \le \theta \le b$ so that the resulting rays sweep out *just the appropriate part* of the region inside each curve. In this case, a natural choice is to start at $\theta = -2\pi/3$, pass through $\theta = -\pi/2$, $\theta = 0$, $\theta = \pi/2$, and stop at $\theta = 2\pi/3$ (Fig. 8).

(iii)  The area in question is the difference

$$A = \int_{-2\pi/3}^{2\pi/3} \frac{1}{2} (2 + 2 \cos \theta)^2 \, d\theta - \int_{-2\pi/3}^{2\pi/3} \frac{1}{2} (1)^2 \, d\theta$$

$$= 2 \int_{-2\pi/3}^{2\pi/3} (1 + 2 \cos \theta + \cos^2\theta) \, d\theta - \frac{2\pi}{3} = \frac{10\pi}{3} + 7\sqrt{3}/2.$$

---

**Finding intersections** of polar curves is tricky, because a given point has many different pairs of coordinates. For example, the circles $r = 2 \sin \theta$ and $r = 2 \cos \theta$ apparently intersect at two points (Fig. 9). Suppose we try to find these intersections algebraically by equating the two formulas for $r$:

$$2 \sin \theta = 2 \cos \theta.$$

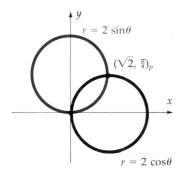

$r = 2 \sin\theta$

$(\sqrt{2}, \frac{\pi}{4})_p$

$r = 2 \cos\theta$

**FIGURE 9**
Intersections of $r = 2 \sin \theta$ with $r = 2 \cos \theta$.

The solutions are $\theta = \pi/4$, $5\pi/4$, $-3\pi/4$, etc. For all these $\theta$, $r = 2 \sin \theta = \pm\sqrt{2}$; they all correspond to the intersection in the first quadrant. So we have not captured the intersection at the origin by solving this equation! Why not? The first curve ($r = 2 \sin \theta$) goes through the origin when $\theta = 0$, $\pm\pi, \ldots$, while the second ($r = 2 \cos \theta$) goes there only when $\theta = \pm\pi/2$, $\pm 3\pi/2, \ldots$.

In general, to find the points of intersection of two curves $r = f(\theta)$ and $r = g(\theta)$, you have to take into account all possible coordinates for each point. First, if $f(\theta)$ has a zero, and $g(\theta)$ has a (possibly different) zero, then the curves intersect at the origin. To find all other intersections, keep in mind that

$$(r, \theta)_p = (r, \theta + 2k\pi)_p$$

and

$$(r, \theta)_p = (-r, \theta + \pi + 2k\pi)_p$$

for all integers $k$. Thus the intersections are found by solving

$$f(\theta) = g(\theta + 2k\pi) \qquad \text{for every integer } k \qquad (3)$$

and

$$f(\theta) = -g(\theta + \pi + 2k\pi) \qquad \text{for every integer } k. \qquad (4)$$

Pragmatically, it is best to sketch a good graph of the two curves together; this will show generally where the intersections are, and you can then determine their polar coordinates by solving equations (3) and (4).

---

**EXAMPLE 3**  Find all intersections of the cardioid $r = 1 + \sin \theta$ with the lemniscate $r^2 = 4 \sin \theta$.

**SOLUTION**  The lemniscate can be written as $r = \pm 2\sqrt{\sin \theta}$, which gives two loops for $0 \leq \theta \leq \pi$; one with $r \geq 0$, one with $r \leq 0$ (Fig. 10). The cardioid appears to intersect this at the origin and three other points. The "obvious" equation for these intersections is

$$1 + \sin \theta = \pm 2\sqrt{\sin \theta},$$

or

$$1 + 2 \sin \theta + \sin^2\theta = 4 \sin \theta \quad \Leftrightarrow \quad (1 - \sin \theta)^2 = 0 \quad \Leftrightarrow \quad \sin \theta = 1,$$

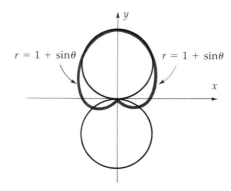

**FIGURE 10**
$r = 1 + \sin \theta$   with   $r^2 = 2 \sin \theta$.

which gives $r = 1 + \sin \theta = 2$. We have thus found the intersection $\left(2, \dfrac{\pi}{2}\right)_p = (0, 2)$ at the top of the figure, but not the one at the origin, or the two below the $y$-axis. The one at the origin comes with $\theta = 0$ on the lemniscate, and $\theta = -\pi/2$ on the cardioid. To find the remaining two, try equation (4) with $k = 0$:

$$r = 1 + \sin \theta = \pm 2\sqrt{\sin(\theta + \pi)} \tag{5}$$

or

$$1 + 2 \sin \theta + \sin^2\theta = 4 \sin(\theta + \pi) = -4 \sin \theta,$$

which gives

$$1 + 6 \sin \theta + \sin^2\theta = 0 \quad \Leftrightarrow \quad \sin \theta = -3 \pm \sqrt{8}.$$

Since $|\sin \theta| \leq 1$, the only solution is

$$\sin \theta = \sqrt{8} - 3 \approx -0.17157.$$

This gives

$$\theta \approx -9.879° \approx -0.1724 \text{ radians}$$

and

$$\theta \approx 189.879° \approx 3.314 \text{ radians.}$$

Since in (5), $r = 1 + \sin \theta$, the corresponding values of $r$ are $r \approx 1 - 0.17157 \approx 0.82843$. Thus the two intersections below the $y$-axis are

$$(0.82843, \ -0.1724 \ldots)_p \quad \text{and} \quad (0.82843, \ 3.314 \ldots)_p.$$

## SUMMARY

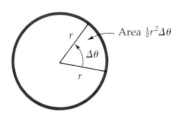

*Area Inside a Polar Curve $r = f(\theta)$:*   $\int_a^b \frac{1}{2} f(\theta)^2 \, d\theta$ (Figs. 11, 12).

*Intersections of Graphs of $r = f(\theta)$ with $r = g(\theta)$:*
   The origin, if both $f$ and $g$ have zeroes.

   Solutions  of  $f(\theta) = g(\theta + 2k\pi)$ and $f(\theta) = -g(\theta + \pi + 2k\pi)$, $k = 0, \pm 1, \ldots$.

**FIGURE 11**

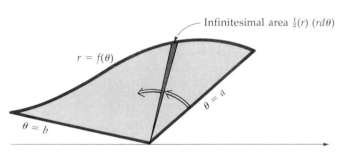

$r = f(\theta)$

Infinitesimal area $\frac{1}{2}(r) \, (r \, d\theta)$

$\theta = b$

$\theta = a$

**FIGURE 12**
Total area $= \displaystyle\int_a^b \frac{1}{2} r^2 \, d\theta = \int_a^b \frac{1}{2} f(\theta)^2 \, d\theta.$

## PROBLEMS

**A**

1.   Compute the area bounded by the given sector of the given curve.
   *a)   The spiral $r = \theta$ for $0 \le \theta \le 2\pi$.
   b)   The part of the cardioid $r = a(1 + \cos \theta)$ lying above the $x$-axis.
   *c)   The entire graph of $r = 2 + \cos \theta$.
   d)   The entire "rose" $r = \cos 3\theta$.
   *e)   One "petal" of the "rose" $r = 3 \cos 4\theta$
   f)   $r = \dfrac{1}{\sin \theta + \cos \theta}$, in the first quadrant.

2.   Find all points of intersection.

   *a)   $r = 1 - 2 \cos \theta$   with   $r = 1$
   b)   $r = 1 + \sin \theta$   with   $r = 1 - \sin \theta$
   *c)   $r = 2 + 2 \cos \theta$   with   $r = 2 - 2 \cos \theta$
   d)   $r = 1 + \sin \theta$   with   $r = 1 - \cos \theta$
   *e)   $r = 1 + \sin \theta$   with   $r = 3 \sin \theta$

3.   Compute the area.
   *a)   Inside both circles $r = \sin \theta$ and $r = \cos \theta$.
   b)   Outside $r = 1 - \cos \theta$ and inside $r = 2$.
   *c)   In the inner loop of the "limacon" $r = 2 + 4 \cos \theta$.
   d)   Inside both the circle $r = 3 \sin \theta$ and the cardioid $r = 1 + \sin \theta$.

**\*e)** Inside all three circles $r = 2$, $r = 4 \sin \theta$, $r = 4 \cos \theta$.

**f)** Inside both $r = 1 - \cos \theta$ and $r = -3 \cos \theta$.

**B**

**\*4.** Compute the area inside the folium $x^3 + y^3 = 3axy$, in the first quadrant. (Use polar coordinates and substitute $z = \tan \theta$. Take $a > 0$.)

**5.** The "spiral of Archimedes" is the graph of $r = a\theta$ for $\theta \geq 0$. This spiral, together with the positive $x$-axis, divides the plane into a sequence of spiral-shaped regions $R_1, R_2, \ldots$. Let $A_j$ denote the area of region $R_j$. Verify the discovery of Archimedes, that $A_2 = 6A_1$, and $A_{n+1} = nA_2$ for $n > 1$.

# REVIEW PROBLEMS   CHAPTER 12

**A**

**\*1.** Find an equation of the parabola with axis parallel to the $y$-axis, and passing through the points $(0, 0)$, $(2, 1)$, $(-6, 3)$. Find its focus and vertex.

**\*2.** Sketch, showing all foci, vertices, and asymptotes.
**a)** $y^2 + 2y + 3x = 5$.
**b)** $x^2 - y^2 + 2x = 0$.
**c)** $x^2 + 2y^2 - y = 0$.

**\*3.** Classify by the discriminant. Find all foci, vertices, and asymptotes.
**a)** $x^2 + 2xy + y^2 = 4\sqrt{2}y$
**b)** $2x^2 + \sqrt{3}xy + y^2 = 2$
**c)** $x^2 + 4xy + y^2 = 1$

**\*4.** Give a specific equation of the form $Ax^2 + Bxy + Cy^2 + Dx + Ey + F = 0$, with $A$, $B$, $C$ not all 0, whose graph is
**a)** Empty.          **b)** A single point.
**c)** One line.       **d)** Two parallel lines.
**e)** Two crossed lines.

**\*5.** Find the angle between the asymptotes of a conic section with eccentricity
**a)** $\sqrt{2}$.        **b)** $e > 1$.

**\*6.** The astroid Icarus moves in an ellipse with the sun at one focus. At perihelion (closest to the sun) it is $17 \times 10^6$ miles from the sun, and at aphelion (furthest from the sun) it is $183 \times 10^6$ miles from it.
**a)** Compute the major and minor axes $2a$ and $2b$, and the distance $2c$ between the foci.
**b)** Compute the eccentricity.

**\*7.** Sketch the following, and identify the graph.
**a)** $r = -2 \sin \theta$     **b)** $r = \dfrac{1}{1 + \sin \theta}$

**c)** $r = \dfrac{1}{2 + \sin \theta}$     **d)** $r = \dfrac{1}{1 - 2 \sin \theta}$

**\*8.** Sketch the polar graph, and compute the area enclosed.
**a)** $r = 1 + \sin \theta$     **b)** $r^2 = \cos 2\theta$.

**9.** **a)** Sketch the limaçon $r = 1 + 2 \sin \theta$.
**\*b)** Compute the area inside its outer loop and outside its inner loop.

**B**

**10.** Derive the equation of the parabola with focus at the origin and directrix the line $x + y = 1$
**\*a)** Using the geometric definition of parabola.
**b)** Using appropriate rotated coordinates $(\bar{x}, \bar{y})$, then converting to $(x, y)$.
**\*c)** In polar coordinates.

**11.** Derive the equation of the ellipse with one focus at the origin and vertices at $(1, 1)$ and $(-2, -2)$
**a)** Using the geometric definition of ellipse.
**b)** Using the appropriate rotated coordinates $(\bar{x}, \bar{y})$, then converting to $(x, y)$.
**c)** In polar coordinates.

**12.** The *latus rectum* of a conic section is its width $L$, measured through the focus, perpendicular to the principal axis. Find the latus rectum of the following graphs.
**a)** $y = x^2$
**b)** $y^2 = Lx + kx^2$, for each $k \geq -1$
**c)** $r = \dfrac{ed}{1 - e \cos \theta}$

For the ellipse and hyperbola, show that $L = 2b^2/a$.

# 13

## PLANE VECTORS

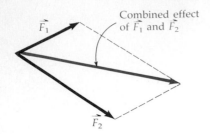

$\vec{F_1}$

Combined effect of $\vec{F_1}$ and $\vec{F_2}$

$\vec{F_2}$

**FIGURE 1**
Adding forces.

The concept of vectors is as old Aristotle; he used arrows to represent forces, as we still do. Suppose, say, that the force is exerted by pulling on a string; then the arrow points in the direction of the string, and its length gives the strength of the pull. These arrows are added by the "parallelogram law" (Fig. 1): The arrows for two forces $\vec{F_1}$ and $\vec{F_2}$ in different directions form two sides of a parallelogram; the combined effect of $\vec{F_1}$ and $\vec{F_2}$ is equivalent to a single force represented by the diagonal of the parallelogram.

From the modern point of view, vectors are algebraic objects; physics and geometry provide the interpretations and applications. This chapter discusses vectors in the plane; the next one develops the subject further, in three-dimensional space.

## 13.1

## THE VECTOR SPACE $R^2$

Ordered pairs $(x, y)$ have a familiar interpretation as points in the plane. *Vectors* provide a new way to work with ordered pairs, giving new interpretations and applications.

The vector given by the ordered pair of numbers $v_1, v_2$ is denoted by

$$\vec{v} = \langle v_1, v_2 \rangle.$$

The numbers $v_1$ and $v_2$ are called the **components** of $\vec{v}$. The **zero vector** $\vec{0}$ has both components equal to zero:

$$\vec{0} = \langle 0, 0 \rangle.$$

### Vector Algebra

We define three operations:

*Addition*

$$\langle v_1, v_2 \rangle + \langle w_1, w_2 \rangle = \langle v_1 + w_1, v_2 + w_2 \rangle$$

*Subtraction*

$$\langle v_1, v_2 \rangle - \langle w_1, w_2 \rangle = \langle v_1 - w_1, v_2 - w_2 \rangle$$

*Scalar multiplication*

$$c\langle v_1, v_2 \rangle = \langle cv_1, cv_2 \rangle.$$

(In contrast to vectors, the real numbers are called "scalars"—hence the name "scalar multiplication" for the product of a real number and a vector.) Thus

$$\langle 2, 3 \rangle + \langle 5, -1 \rangle = \langle 7, 2 \rangle$$
$$\langle 2, 3 \rangle - \langle 5, -1 \rangle = \langle -3, 4 \rangle$$
$$3\langle 1, -2 \rangle = \langle 3, -6 \rangle.$$

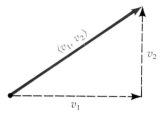

**FIGURE 2**
Vector $\vec{v} = \langle v_1, v_2 \rangle$ with components $v_1$ and $v_2$.

The set of ordered pairs, together with these operations, is called the **vector space** $R^2$ [read "$R$" (for the real numbers) "two" (for the two components)].

Geometrically, $\langle v_1, v_2 \rangle$ is represented as an arrow in the plane (Fig. 2); the first component $v_1$ gives the *difference in the x-coordinates* of tip and tail, while $v_2$ gives the *difference in the y-coordinates*. The arrow can be drawn at any appropriate place. For instance, to represent a force, the arrow is drawn with its tail at the point of application, as if it were a string transmitting the force. A given vector may thus be represented by many different arrows; they are all parallel, point in the same direction, and have the same length (Fig. 3).

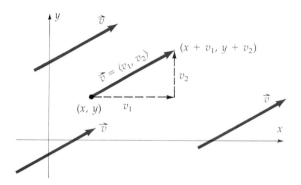

**FIGURE 3**
Arrows, all representing $\vec{v} = \langle v_1, v_2 \rangle$; $|\vec{v}| = \sqrt{v_1^2 + v_2^2}$.

The length of any of the arrows representing $\vec{v}$ is called the **magnitude** of $\vec{v}$, denoted $|\vec{v}|$:

$$|\vec{v}| = \sqrt{v_1^2 + v_2^2}.$$

Figure 4 illustrates the product $c\vec{v} = \langle cv_1, cv_2 \rangle$ of a vector $\vec{v}$ and a number $c$. If $c > 0$ then $c\vec{v}$ points in the same direction as $\vec{v}$, and is $c$ times as long. If $c < 0$, then $c\vec{v}$ points in the direction *opposite* to $\vec{v}$, and is $|c|$ times as long. Thus in any case

$$|c\vec{v}| = |c| \cdot |\vec{v}|. \tag{1}$$

**FIGURE 4**
Scalar multiples of $\vec{v}$.

Figure 5 illustrates the vector sum $\vec{v} + \vec{w}$. Take any arrow representing $\vec{v}$, and draw an arrow for $\vec{w}$ beginning at the tip of the $\vec{v}$-arrow; then the arrow from the tail of $\vec{v}$ to the tip of $\vec{w}$ represents $\vec{v} + \vec{w}$. Think of moving from the tail of $\vec{v}$ to its tip, then on to the tip of $\vec{w}$; the combined motion corresponds to the sum $\vec{v} + \vec{w}$.

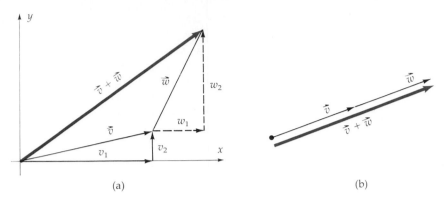

(a)

(b)

**FIGURE 5**
Vector addition.

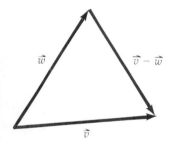

**FIGURE 6**
Vector subtraction.

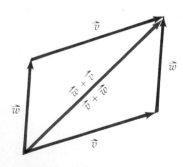

**FIGURE 7**
Parallelogram law of vector addition:
$\vec{v} + \vec{w} = \vec{w} + \vec{v}$.

Since the length of any side of a triangle is less than or equal to the sum of the lengths of the other two sides, we deduce from Figure 5a the **triangle inequality**

$$|\vec{v} + \vec{w}| \le |\vec{v}| + |\vec{w}|. \tag{2}$$

In Figure 5a, the vectors form a genuine triangle, and $|\vec{v} + \vec{w}|$ is strictly less than $|\vec{v}| + |\vec{w}|$; Figure 5b shows a special case where $|\vec{v} + \vec{w}| = |\vec{v}| + |\vec{w}|$.

Figure 6 illustrates the *difference* $\vec{v} - \vec{w}$. The arrows for $\vec{v}$ and $\vec{w}$ are drawn from a common point; then the arrow for $\vec{v} - \vec{w}$ goes from the tip of the $\vec{w}$-arrow to the tip of the $\vec{v}$-arrow. The resulting triangle illustrates that

$$\vec{w} + (\vec{v} - \vec{w}) = \vec{v}.$$

Figure 7 shows $\vec{v} + \vec{w}$ together with the sum taken in the other order, $\vec{w} + \vec{v}$; the various arrows for $\vec{v}$ and $\vec{w}$ form a parallelogram, with $\vec{v} + \vec{w}$ as diagonal; this is the *parallelogram law* for vector addition.

As we see in the figure,

$$\vec{v} + \vec{w} = \vec{w} + \vec{v}; \tag{3}$$

vector addition is "commutative." The formal proof is based on the algebraic definition of addition; compute each side of (3) and compare the results:

$$\vec{v} + \vec{w} = \langle v_1 + w_1, v_2 + w_2 \rangle$$

and

$$\vec{w} + \vec{v} = \langle w_1 + v_1, w_2 + v_2 \rangle.$$

The first components of these two vectors are equal, since $v_1 + w_1 = w_1 + v_1$; this is the commutative law for addition of real numbers. The second components are equal, for the same reason; hence the vectors are equal. Thus the commutative law for vector addition follows from the corresponding law for real numbers.

Vectors inherit further properties from the real numbers—two *associative laws*

$$(\vec{u} + \vec{v}) + \vec{w} = \vec{u} + (\vec{v} + \vec{w}) \tag{4}$$

$$(ab)\vec{v} = a(b\vec{v}), \tag{5}$$

two *distributive laws*

$$a(\vec{v} + \vec{w}) = a\vec{v} + a\vec{w} \tag{6}$$

$$(a + b)\vec{v} = a\vec{v} + b\vec{v}, \tag{7}$$

and three laws for the zero vector

$$\vec{v} - \vec{v} = \vec{0}$$

$$\vec{v} + \vec{0} = \vec{0} + \vec{v} = \vec{v}$$

$$0\vec{v} = \vec{0}.$$

All these, like (3), are proved by computing both sides and comparing components. The formal proofs of laws (1) and (2) consist likewise in computing both sides according to the definition of vector magnitude, and comparing the results.

---

**EXAMPLE 1**  Let $\vec{v} = \langle 2, 1 \rangle$ and $\vec{w} = \langle -1, 1 \rangle$. Compute $\vec{v} + \vec{w}$, $2\vec{w}$, $-\vec{w}$, $-\frac{1}{2}\vec{v}$. Sketch these, along with $\vec{v}$ and $\vec{w}$. Compute the lengths $|\vec{w}|$ and $|2\vec{w}|$.

*SOLUTION*

$$\vec{v} + \vec{w} = \langle 2, 1 \rangle + \langle -1, 1 \rangle = \langle 1, 2 \rangle$$

$$2\vec{w} = 2\langle -1, 1 \rangle = \langle -2, 2 \rangle$$

$$-\vec{w} = -\langle -1, 1 \rangle = \langle 1, -1 \rangle$$

$$-\tfrac{1}{2}\vec{v} = -\tfrac{1}{2}\langle 2, 1 \rangle = \langle -1, -\tfrac{1}{2} \rangle$$

$$|\vec{v}| = \sqrt{2^2 + 1^2} = \sqrt{5}$$

$$|\vec{w}| = \sqrt{(-1)^2 + 1^2} = \sqrt{2}.$$

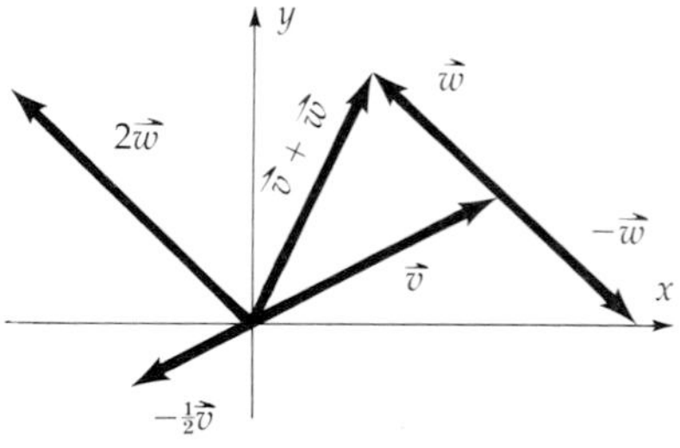

**FIGURE 8**
Vectors in Example 1.

The arrows for these vectors can be drawn with any initial points—Figure 8 shows one way. Note that the arrow drawn for $\vec{w} = \langle -1, 1 \rangle$ does *not end* at $(-1, 1)$; since the arrow begins at $(2, 1)$, it ends at $(2 - 1, 1 + 1) = (1, 2)$.

---

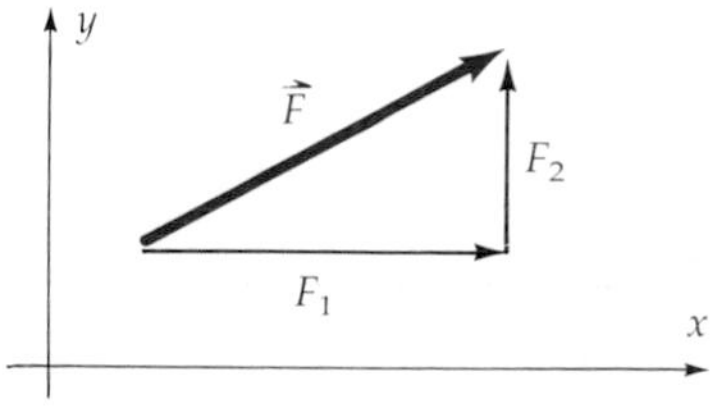

### Force Vectors

In physics, a force $\vec{F}$ can be decomposed into *components* parallel to the co-ordinate axes (Fig. 9). These are precisely the components of the force vector $\vec{F} = \langle F_1, F_2 \rangle$. As Aristotle observed, forces may be combined by the parallelogram law, that is, by vector addition.

**FIGURE 9**
Force $\vec{F}$ has components $F_1$ and $F_2$.

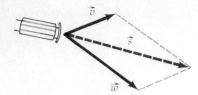

**FIGURE 10**
Parallelogram law for forces.

**EXAMPLE 2**  Victor pulls a sled with force $\vec{v} = \langle 2, 1 \rangle$, and Wendy pulls it with force $\vec{w} = \langle 3, -2 \rangle$ (Fig. 10). Their combined efforts are equivalent to that of a Samson pulling with force

$$\vec{v} + \vec{w} = \langle 2, 1 \rangle + \langle 3, -2 \rangle = \langle 5, -1 \rangle.$$

By the triangle inequality, the magnitude of Samson's force is less than the sum of the exertions of Victor and Wendy:

$$|\vec{v} + \vec{w}| \leq |\vec{v}| + |\vec{w}|.$$

In fact, $|\vec{v} + \vec{w}| = \sqrt{26} \approx 5.1$, whereas $|\vec{v}| + |\vec{w}| = \sqrt{5} + \sqrt{13} \approx 5.83$. Since Victor and Wendy do not pull in exactly the same direction, part of their effort cancels out.

## Unit Vectors

A vector of length 1 is called a **unit vector**. Thus $\left\langle \dfrac{1}{\sqrt{2}}, \dfrac{-1}{\sqrt{2}} \right\rangle$ is a unit vector, since

$$\left| \left\langle \frac{1}{\sqrt{2}}, \frac{-1}{\sqrt{2}} \right\rangle \right| = \sqrt{\left(\frac{1}{\sqrt{2}}\right)^2 + \left(\frac{-1}{\sqrt{2}}\right)^2} = \sqrt{\frac{1}{2} + \frac{1}{2}} = 1.$$

But $\langle 1, 1 \rangle$ is not a unit vector, since

$$|\langle 1, 1 \rangle| = \sqrt{1^2 + 1^2} \neq 1.$$

Figure 11 shows two basic unit vectors denoted $\vec{i}$ and $\vec{j}$, one in the positive $x$ direction, and the other in the positive $y$ direction:

$$\vec{i} = \langle 1, 0 \rangle \quad \text{and} \quad \vec{j} = \langle 0, 1 \rangle.$$

Any vector $\vec{v} = \langle v_1, v_2 \rangle$ can be written in terms of $\vec{i}$ and $\vec{j}$:

$$\vec{v} = v_1 \vec{i} + v_2 \vec{j}$$

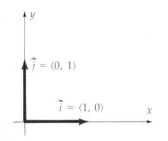

**FIGURE 11**
Basic unit vectors $\vec{i}$ and $\vec{j}$.

since

$$v_1 \vec{i} + v_2 \vec{j} = v_1 \langle 1, 0 \rangle + v_2 \langle 0, 1 \rangle \qquad \text{(definition of } \vec{i} \text{ and } \vec{j}\text{)}$$
$$= \langle v_1, 0 \rangle + \langle 0, v_2 \rangle = \langle v_1, v_2 \rangle = \vec{v}.$$

The use of $\vec{i}$ and $\vec{j}$ is common in physics and engineering.

Given any vector $\vec{v} \neq 0$, you can form the *unit vector in the direction of* $\vec{v}$

$$\frac{1}{|\vec{v}|} \vec{v}.$$

This is simply $\vec{v}$ divided by its length. To check that it really is a unit vector, use formula (1).

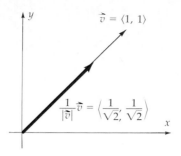

**FIGURE 12**

Unit vector $\left\langle \dfrac{1}{\sqrt{2}}, \dfrac{1}{\sqrt{2}} \right\rangle$ in direction

of $\langle 1, 1 \rangle$.

**EXAMPLE 3**  If $\vec{v} = \langle 1, 1 \rangle$ as in Figure 12, then $|\vec{v}| = \sqrt{2}$, and the unit vector in the direction of $\vec{v}$ is

$$\frac{1}{\sqrt{2}} \vec{v} = \frac{1}{\sqrt{2}} \langle 1, 1 \rangle = \left\langle \frac{1}{\sqrt{2}}, \frac{1}{\sqrt{2}} \right\rangle.$$

Our first applications of vectors are in analytic geometry (right now) and the study of motion (in the next section).

## Position Vectors

With each point $P = (x, y)$ in the plane we associate the **position vector** $\vec{P} = (x, y)$, represented by the arrow from the origin to point $P$ (Fig. 13). The arrow shows how to get to $P$ from the origin, telling how far to go, and in what direction.

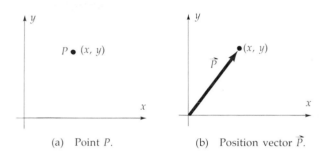

(a)  Point $P$.  (b)  Position vector $\vec{P}$.

**FIGURE 13**

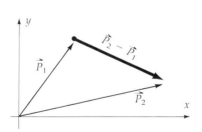

**FIGURE 14**
$\vec{P_2} - \vec{P_1}$ goes from $P_1$ to $P_2$.

## Midpoint of a Segment

Given two points $P_1$ and $P_2$, the difference $\vec{P_2} - \vec{P_1}$ is represented by an arrow from $P_1$ to $P_2$ (Fig. 14). The midpoint $M$ of the segment from $P_1$ to $P_2$ can be reached by going from the origin to $P_1$, then halfway from $P_1$ to $P_2$ (Fig. 15). Expressed in the language of vectors, this is

$$\vec{M} = \vec{P_1} + \tfrac{1}{2}(\vec{P_2} - \vec{P_1}).$$

Now apply the laws of vector algebra, and find

$$\vec{M} = \vec{P_1} + \tfrac{1}{2}\vec{P_2} - \tfrac{1}{2}\vec{P_1} = \tfrac{1}{2}\vec{P_2} + \tfrac{1}{2}\vec{P_1}$$
$$= \tfrac{1}{2}(\vec{P_1} + \vec{P_2}).$$

Thus if $\vec{P_1} = (x_1, y_1)$ and $\vec{P_2} = (x_2, y_2)$, then the midpoint $M$ has position vector

$$\vec{M} = \tfrac{1}{2}(\vec{P_1} + \vec{P_2}) = \tfrac{1}{2}\langle x_1 + x_2, y_1 + y_2 \rangle$$
$$= \langle \tfrac{1}{2}(x_1 + x_2), \tfrac{1}{2}(y_1 + y_2) \rangle.$$

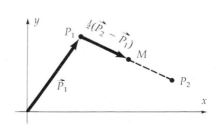

**FIGURE 15**
Getting to the midpoint.

This gives the coordinates of the midpoint of the segment as the *average* of the coordinates of the endpoints.

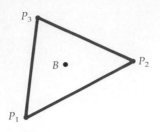

**FIGURE 16**
Barycenter $\vec{B} = \frac{1}{3}(\overrightarrow{P_1} + \overrightarrow{P_2} + \overrightarrow{P_3})$.

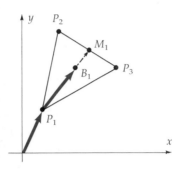

**FIGURE 17**
$\overrightarrow{B_1} = \overrightarrow{P_1} + \frac{2}{3}(\overrightarrow{M_1} - \overrightarrow{P_1})$.

## Barycenter of a Triangle

Three points form the vertices of a triangle. The point obtained by averaging the coordinates of the vertices is, in some sense, the "center" of the triangle; it is called the **barycenter** $B$ (Fig. 16). If the vertices are $P_1$, $P_2$, $P_3$, then the position vector of the barycenter is

$$\vec{B} = \tfrac{1}{3}(\overrightarrow{P_1} + \overrightarrow{P_2} + \overrightarrow{P_3}).$$

Just as a segment will balance if supported at its midpoint, a triangle will balance if supported at its barycenter. (Hence the name, which means essentially center of gravity.) One proof of this depends on a further interesting property: The barycenter is the *intersection of the medians*.

　　A median in a triangle is the segment from a vertex to the midpoint of the opposite side. We will show that *the barycenter is 2/3 of the way from any vertex to the midpoint of the opposite side*. Label the three vertices in any order, as in Figure 17. Let $M_1$ be the midpoint of the side opposite to $P_1$, and $B_1$ be the point 2/3 of the way from $P_1$ to $M_1$. Our claim is that $B_1$ is actually the barycenter $B$; we prove this by showing that the two points have the same position vector. Starting at the origin, you arrive at $B_1$ by going to $P_1$, then 2/3 the way to $M_1$. Hence (Fig. 17) the position vector $\overrightarrow{B_1}$ is

$$\overrightarrow{B_1} = \overrightarrow{P_1} + \tfrac{2}{3}(\overrightarrow{M_1} - \overrightarrow{P_1})$$

Now use the midpoint formula $\overrightarrow{M_1} = \frac{1}{2}(\overrightarrow{P_2} + \overrightarrow{P_3})$:

$$\overrightarrow{B_1} = \overrightarrow{P_1} + \tfrac{2}{3}[\tfrac{1}{2}(\overrightarrow{P_2} + \overrightarrow{P_3}) - \overrightarrow{P_1}]$$
$$= \tfrac{1}{3}\overrightarrow{P_1} + \tfrac{1}{3}\overrightarrow{P_2} + \tfrac{1}{3}\overrightarrow{P_3}.$$

This is precisely the formula giving the barycenter $B$, which is what we wanted to show.

## SUMMARY

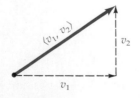

**FIGURE 18**
Vector $\vec{v} = \langle v_1, v_2 \rangle$ with components $v_1$ and $v_2$.

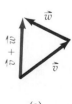

(a)

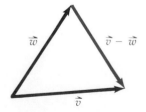

(b)

**FIGURE 19**
$\vec{v} + \vec{w} = \langle v_1 + w_1, v_2 + w_2 \rangle$.

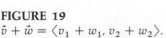

**FIGURE 20**
Vector subtraction.

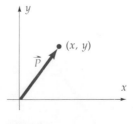

**FIGURE 21**

*Basic Unit Vectors:*　$\vec{i} = \langle 1, 0 \rangle$,　　$\vec{j} = \langle 0, 1 \rangle$　　(see Fig. 22)

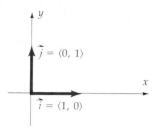

**FIGURE 22**
Basic unit vectors $\vec{i}$ and $\vec{j}$.

**Laws of Magnitude:**

$$|c\vec{v}| = |c| \cdot |\vec{v}| \tag{1}$$

$$|\vec{v} + \vec{w}| \le |\vec{v}| + |\vec{w}| \tag{2}$$

**Commutative Law:**

$$\vec{v} + \vec{w} = \vec{w} + \vec{v} \tag{3}$$

**Associative Laws:**

$$(\vec{u} + \vec{v}) + \vec{w} = \vec{u} + (\vec{v} + \vec{w}) \tag{4}$$

$$(ab)\vec{v} = a(b\vec{v}) \tag{5}$$

**Distributive Laws:**

$$a(\vec{v} + \vec{w}) = a\vec{v} + a\vec{w} \tag{6}$$

$$(a + b)\vec{v} = a\vec{v} + b\vec{v} \tag{7}$$

## PROBLEMS

**A**

1. Sketch each vector twice: once with initial point at the origin and again initiating at the given point.
   *a)   $\langle 1, -2 \rangle$, at $(1, 1)$
   b)   $\langle 7, 6 \rangle$, at $(-1, -3)$
   *c)   $5\vec{i} + 2\vec{j}$, at $(-1, 1)$
   d)   $\langle -1, -2 \rangle$, at $(2, 1)$
   *e)   $-2\vec{i} + \vec{j}$, at $(1, 2)$

*2. Compute the magnitude of each vector in the previous problem. (Answers for parts a, c, and e.)

3. Sketch the vector $\vec{v} = 3\vec{i} - 5\vec{j}$. Then compute and sketch a vector having
   *a)   The same direction as $\vec{v}$ and twice the magnitude.
   b)   The direction opposite to $\vec{v}$ and half the magnitude.

4. Show that
   *a)   $\langle 3, 6 \rangle$ and $\langle 1, 2 \rangle$ have the same direction.
   b)   $\langle 1, -3 \rangle$ and $\langle -2, 6 \rangle$ have opposite directions to each other.
   c)   $\langle 1, 2 \rangle$ and $\langle -2, -1 \rangle$ do not have directions opposite to each other.

5. Find the components of $\vec{P_2} - \vec{P_1}$ and compute the length $|\vec{P_2} - \vec{P_1}|$. Sketch the given points and the vector $\vec{P_2} - \vec{P_1}$.
   *a)   $P_1 = (1, -2)$, $P_2 = (0, 0)$
   b)   $P_1 = (2, -1)$, $P_2 = (-3, 5)$

6. With $\vec{v} = \langle 1, -2 \rangle$ and $\vec{w} = \langle 3, 2 \rangle$, compute and sketch.
   a)   $\vec{v} + \vec{w}$      b)   $3\vec{v}$      c)   $-\vec{v}$
   d)   $\vec{v} - \vec{w}$      e)   $3(\vec{v} - \vec{w})$      f)   $3\vec{v} - 3\vec{w}$

7. For the vector $\vec{v}$ in problem 6, compute.
   a)   $|\vec{v}|$   *b)   $\dfrac{1}{|\vec{v}|} \cdot \vec{v}$   c)   $\left| \dfrac{1}{|\vec{v}|} \vec{v} \right|$

8. For the vectors in problem 6, compute and compare.
   a)   $|\vec{w}|$ and $|-2\vec{w}|$
   *b)   $|\vec{v}| + |\vec{w}|$ and $|\vec{v} + \vec{w}|$

9. a)   Sketch the force vectors $\vec{F_1} = \langle 1, 1 \rangle$ and $\vec{F_2} = \langle 2, -1 \rangle$ from a common initial point.
   b)   Compute and sketch the combined effect of the two forces.
   *c)   What force $F$ will exactly balance the given two?

10. Sketch the vector making the given angle with the positive $x$-axis and having the given length. Compute its components. (Use the signed angle from the positive $x$-axis *to* the vector, as in polar coordinates.)
    *a)   $\pi/4$, length 1      b)   $\pi/4$, length 2
    *c)   $-\pi/4$, length 1      d)   $-\pi/4$, length 1/2
    *e)   $\pi/3$, length 1      f)   $\pi/3$, length 3
    *g)   $5\pi/6$, length 0.3

11. A horse pulls a sleigh, with reins making an angle of 22° with the frozen ground. The tension in the reins is 1000 Newtons (enough to lift about 100 kg). What are the horizontal and vertical components of the force?

12. In the triangle inequality, what is the special relation between $\vec{v}$ and $\vec{w}$ when $|\vec{v} + \vec{w}| = |\vec{v}| + |\vec{w}|$?

13. Is it always true that
    a)   $|\vec{v} - \vec{w}| \le |\vec{v}| + |\vec{w}|$?
    *b)   $|\vec{v} - \vec{w}| \le |\vec{v}| - |\vec{w}|$?

**14.** Given $P_1 = (1, 2)$ and $P_2 = (-1, 3)$, find a point
  a) $\frac{2}{3}$ of the way from $P_1$ to $P_2$.
  b) $3/4$ of the way from $P_2$ to $P_1$.

## B

**15.** Show that every unit vector can be written in the form $\cos\theta\vec{i} + \sin\theta\vec{j}$ for a unique $\theta$ with $-\pi < \theta \leq \pi$.

**16.** Prove the following algebraic properties in the summary:
  a) Associative law (4).
  b) Distributive law (6).
  c) Distributive law (7).

**17.** Solve the following vector equations for $\vec{v}$.
  *a) $\vec{v} + \langle 0, 1 \rangle = \langle 2, -1 \rangle$
  b) $2\vec{v} - \langle 1, 0 \rangle = \langle 3, 1 \rangle$

**18.** Sketch three vectors $\vec{u}, \vec{v}, \vec{w}$ arranged "tip to tail"—the tip of $\vec{u}$ at the tail of $\vec{v}$ and the tip of $\vec{v}$ at the tail of $\vec{w}$. Then draw the arrows representing $\vec{u} + \vec{v}$, $\vec{v} + \vec{w}$, $\vec{u} + (\vec{v} + \vec{w})$, and $(\vec{u} + \vec{v}) + \vec{w}$. Which law of vector algebra is illustrated?

**19.** Prove that if $\vec{v} \neq 0$, then $\dfrac{1}{|\vec{v}|}\vec{v}$ is a unit vector.

**20.** In Figure 15, we found the midpoint $M$ by going to $P_1$ and then halfway from there to $P_2$. Make the corresponding calculation, but going first to $P_2$ and then halfway from there to $P_1$.

**21.** Let $P, P_1,$ and $P_2$ be the vertices of a triangle, with $M_1$ the midpoint of the side opposite $P_2$, and $M_2$ the midpoint of the side opposite $P_1$. Sketch the situation, and show by vector algebra that $\vec{M_2} - \vec{M_1} = \frac{1}{2}(\vec{P_2} - \vec{P_1})$. (In geometric language, the segment joining the midpoints of two sides is parallel to the third side and half as long.)

**22.** Show that the lines joining adjacent midpoints of any quadrilateral form a parallelogram.

**23.** Show that the diagonals of a parallelogram intersect at their midpoints.

**24.** Given a point $P_0$ and a positive number $r$, describe the set of all points $P$ such that $|\vec{P} - \vec{P_0}| = r$.

**25.** Given points $P_1 = (0, 0)$, $P_2 = (1, 1)$, and $P_3 = (2, 1)$, find all points $P$ which, together with the given three, form the vertices of a parallelogram.

**26.** The force exerted by charge $q_1$ at point $P_1$ on charge $q_2$ at point $P_2$ is $\dfrac{1}{4\pi\varepsilon} q_1 q_2 \dfrac{\vec{P_2} - \vec{P_1}}{|\vec{P_2} - \vec{P_1}|^3}$; $\varepsilon$ is a constant depending on the dimensions used.

  a) Show that the magnitude of this force is $\dfrac{q_1 q_2}{4\pi\varepsilon r^2}$, where $r$ is the distance between the charges.

  b) Given three charges in the configuration of Figure 23, compute the total force exerted on $q_3$ by $q_1$ and $q_2$ together.

  c) Compute the force exerted on $q_1$ by $q_2$ and $q_3$ together.

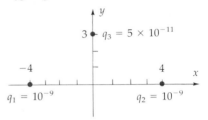

**FIGURE 23**
Three charges.

*27. Figure 24 shows two charges of equal magnitude $q$ and a third on the perpendicular bisector of the line between them, of magnitude $q'$. Where on that perpendicular bisector should $q'$ be placed to experience the least force? The greatest force? (See the previous problem.)

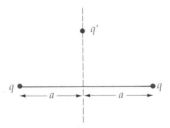

**FIGURE 24**

## C

**28.** Show that a triangle balances on each of its medians (This implies that the center of gravity lies on each median, hence is the intersection of the medians. Hint: Divide the triangle into thin strips parallel to the median, and show by plane geometry that each strip on one side is balanced by a corresponding strip on the opposite side.)

**29.** a) Show that the line through the barycenter of a triangle and any one of its vertices divides the triangle into two parts of equal area.

  b) Does the line through the barycenter, parallel to one of the sides, divide the triangle in two parts of equal area?

## 13.2

# PARAMETRIC AND VECTOR EQUATIONS

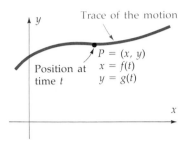

**FIGURE 1**
Moving point.

Imagine a particle moving in the plane (Fig. 1). At time $t$, it is at some point $P = (x, y)$; so the coordinates $x$ and $y$ are functions of time $t$:

$$x = f(t) \quad \text{and} \quad y = g(t). \tag{1}$$

As it moves, the particle traces a curve which we call the **trace** of the motion. The equations (1) are called **parametric equations of the curve**; the variable $t$ is the *parameter*.

---

**EXAMPLE 1**   Galileo discovered the parametric equations for the motion of a projectile. With the $x$-axis horizontal and the $y$-axis vertical, they are

$$x = x_0 + bt, \qquad y = y_0 + ct - \tfrac{1}{2}gt^2,$$

where $g$ is the acceleration due to gravity, and $b$ and $c$ are constants. For example, the equations

$$x = 3t, \qquad y = 2 - 4.9t^2 \tag{2}$$

describe a gently thrown projectile moving to the right at 3 m/sec and simultaneously falling from an initial height of 2 m, pulled down by gravity. Figure 2 plots position $P = (x, y)$ for several times $t$.

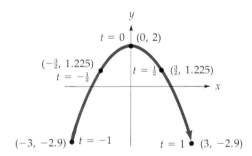

**FIGURE 2**
Trace of $x = 3t$, $y = 2 - 4.9t^2$.

To find a single equation for the curve traced by the projectile, eliminate $t$ from the parametric equations (2). The equation for $x$ gives $t = x/3$, so

$$y = 2 - 4.9t^2 = 2 - (4.9)\left(\frac{x}{3}\right)^2$$

$$= 2 - \frac{4.9}{9}x^2.$$

The projectile traces a parabola.

---

## Vector Functions

The two parametric equations

$$x = f(t), \qquad y = g(t)$$

can be written as a single vector equation

$$\langle x, y \rangle = \langle f(t), g(t) \rangle. \tag{3}$$

This is the **vector equation** of the motion. The expression on the right is a **vector function**; for each input $t$, the resulting output is the vector $\langle f(t), g(t) \rangle$. The function gives the position vector at time $t$, and we denote it by $\vec{P}$:

$$\overrightarrow{P(t)} = \langle f(t), g(t) \rangle.$$

In Example 1,

$$\overrightarrow{P(t)} = \langle 3t, 2 - 4.9t^2 \rangle.$$

### Parametric Equations of a Line

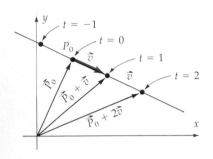

**FIGURE 3**
Particle moving with constant velocity $\vec{v}$.

Imagine a particle moving in space with constant velocity $\vec{v}$ (Fig. 3); the vector $\vec{v}$ points in the direction of motion, and its length is the speed, the distance traveled in unit time. At time $t = 0$, the particle has a certain position $P_0$. Then at $t = 1$ it has moved to the point $P(1)$ with position vector

$$\overrightarrow{P(1)} = \overrightarrow{P_0} + 1 \cdot \vec{v}.$$

At $t = 2$ it has moved twice as far: $\overrightarrow{P(2)} = \overrightarrow{P_0} + 2\vec{v}$. In general, for any time $t$ the position vector is (Fig. 4)

$$\overrightarrow{P(t)} = \overrightarrow{P_0} + t\vec{v}. \tag{4}$$

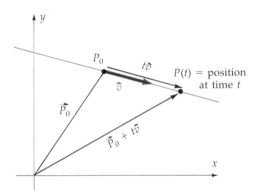

**FIGURE 4**
Line through $P_0$ parallel to $\vec{v}$.

The equation tells how to get to $P(t)$ from the origin: go first to $P_0$, then move with velocity $\vec{v}$ for $t$ units of time. The motion traces the straight line through $P_0$, parallel to the vector $\vec{v}$.

Given the coordinates of $P_0$ and the components of $\vec{v}$,

$$P_0 = (x_0, y_0) \quad \text{and} \quad \vec{v} = \langle a, b \rangle$$

you can write out the vector equation (4) as

$$\langle x, y \rangle = \langle x_0, y_0 \rangle + t\langle a, b \rangle = \langle x_0 + ta, y_0 + tb \rangle.$$

Equating components, you get the parametric equations of the line:

$$x = x_0 + at$$
$$y = y_0 + bt.$$

On the right-hand side, the constant terms $x_0$ and $y_0$ give the coordinates of the point $P_0 = P(0)$, the position at time 0; the coefficients of $t$ give the components $a$ and $b$ of the velocity vector $\vec{v}$.

### EXAMPLE 2

$$x = 1 + 2t$$
$$y = 3 - t$$

are parametric equations of the line through $P_0 = (1, 3)$, parallel to the vector $\vec{v} = \langle 2, -1 \rangle$ (Fig. 5). To sketch the line, plot two points, such as $P_0$ and

$$P(1) = (1 + 2 \cdot 1, \quad 3 - 1) = (3, 2)$$

and sketch the line through them.

To get the "$y = mx + b$" equation, eliminate $t$. The equation for $x$ gives $t = \frac{1}{2}(x - 1)$, so

$$y = 3 - t = 3 - \tfrac{1}{2}(x - 1)$$
$$= -\tfrac{1}{2}x + \tfrac{7}{2}.$$

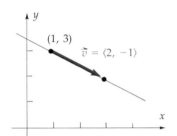

**FIGURE 5**
$x = 1 + 2t, y = 3 - t.$

### Parametric Equations of a Circle

Figure 6 shows a circle with center $P_0$ and radius $r$. To get to a point $P$ on the circle, go first to $P_0$, then on to $P$; in vector terms

$$\vec{P} = \vec{P_0} + (\vec{P} - \vec{P_0}).$$

The vector $\vec{P} - \vec{P_0}$ has components $r \cos \theta$ and $r \sin \theta$ (Fig. 6), as in the equations for polar coordinates. Thus $\vec{P} - \vec{P_0} = \langle r \cos \theta, r \sin \theta \rangle$, and

$$\vec{P} = \vec{P_0} + \langle r \cos \theta, r \sin \theta \rangle. \tag{5}$$

Denoting the coordinates of $P$ and $P_0$ in the usual way, you get

$$\langle x, y \rangle = \langle x_0, y_0 \rangle + \langle r \cos \theta, r \sin \theta \rangle$$
$$= \langle x_0 + r \cos \theta, y_0 + r \sin \theta \rangle.$$

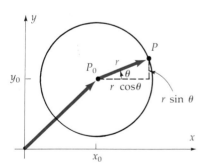

**FIGURE 6**
$\vec{P} = \vec{P_0} + \langle r \cos \theta, r \sin \theta \rangle.$

So, parametric equations of the circle with center $(x_0, y_0)$ and radius $r$ are

$$x = x_0 + r \cos \theta$$
$$y = y_0 + r \sin \theta. \tag{6}$$

The parameter in this case is $\theta$, the angle in Figure 6.

Suppose that a particle moves around this circle, starting with $\theta = 0$ and turning at the rate of $\omega$ radians per second; then $\theta = \omega t$. Using this in the right-hand side of (5) gives the vector equation of a circular motion

$$\overrightarrow{P(t)} = \overrightarrow{P_0} + r\langle \cos \omega t, \sin \omega t \rangle.$$

The parametric equations are

$$x = x_0 + r \cos \omega t$$
$$y = y_0 + r \sin \omega t$$

The constant $\omega$ in these equations is called the **angular velocity** of the motion.

## The Cycloid

Figure 7 shows a wheel of radius $r$ rolling on the $x$-axis. Initially, point $P$ on the wheel is touching the origin, and the center of the wheel is at $(0, r)$. As the wheel rolls, the point traces an interesting curve called a **cycloid**. Parametric equations for the cycloid are derived using Figure 8—we obtain the position vector $\vec{P}$ as the sum $\vec{A} + (\vec{P} - \vec{A})$.

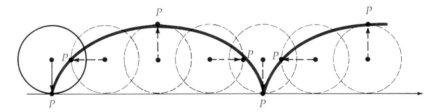

**FIGURE 7**
Cycloid traced by rolling wheel.

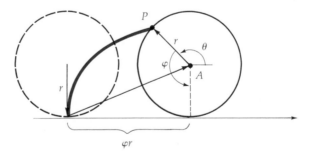

**FIGURE 8**
$\vec{A} = \langle \varphi r, r \rangle$,
$\vec{P} - \vec{A} = \langle r \cos \theta, r \sin \theta \rangle$.

Recall that on a circle of radius $r$, the arc length enclosed in an angle of $\varphi$ radians is $\varphi r$. So when the circle rolls through an angle $\varphi$ radians, an arc of length $\varphi r$ has been in contact with the $x$-axis. The circle rolls without slipping, so the length traversed along the axis equals $\varphi r$; hence the center of the wheel is at $A = (\varphi r, r)$. The vector $\vec{P} - \vec{A}$ in

Figure 8 equals $\langle r \cos \theta, r \sin \theta \rangle$. Since $\varphi + \theta = 3\pi/2$, then $\theta = 3\pi/2 - \varphi$ and

$$\vec{P} - \vec{A} = \left\langle r \cos\left(\frac{3\pi}{2} - \varphi\right), r \sin\left(\frac{3\pi}{2} - \varphi\right) \right\rangle$$
$$= \langle -r \sin \varphi, -r \cos \varphi \rangle.$$

Hence the position is

$$\vec{P} = \vec{A} + (\vec{P} - \vec{A})$$
$$= \langle \varphi r, r \rangle + \langle -r \sin \varphi, -r \cos \varphi \rangle$$
$$= \langle \varphi r - r \sin \varphi, r - r \cos \varphi \rangle$$

and parametric equations of the motion are

$$x = \varphi r - r \sin \varphi, \qquad y = r - r \cos \varphi.$$

The parameter $\varphi$ is the clockwise angle through which the wheel has turned. If the wheel turns with angular velocity $\omega$ radians per second, replace $\varphi$ by $\omega t$ in these parametric equations.

## SUMMARY

***Vector Equation of Motion:*** $\quad \vec{P}(t) = \langle f(t), g(t) \rangle$

***Parametric Equations:*** $\quad x = f(t), \qquad y = g(t)$

## PROBLEMS

**A**

1. For each motion $\overrightarrow{P(t)}$, describe the path, plot the position for the suggested values of $t$, and indicate the direction of motion.

   *a) $\langle 1 - t, 2 + 2t \rangle$, $t = 0, \pm 1, \pm 2$

   b) $(1 + t)\vec{i} + (2 - 2t)\vec{j}$, $t = 0, \pm 1, \pm 2$

   *c) $\langle 3 \cos t, 3 \sin t \rangle$, $t = 0, \pm\dfrac{\pi}{4}, \pm\dfrac{\pi}{2}, \pm\pi$

   d) $\left\langle 3 - \cos\dfrac{\pi t}{2}, 5 - \sin\dfrac{\pi t}{2} \right\rangle$, $t = 0, \pm 1, \pm 2$

   e) $\langle 3 - 2t, 1 + t - 4.9t^2 \rangle$, $t = 0, \pm 1, \pm 2$

   *f) $\langle t \cos t, t \sin t \rangle$, $t = 0, \pm\dfrac{\pi}{4}, \pm\dfrac{\pi}{2}, \pm\pi$

   g) $\left\langle \dfrac{t}{2} - \sin\dfrac{t}{2}, 1 - \cos\dfrac{t}{2} \right\rangle$, for

   $t = 0, \pm\dfrac{\pi}{2}, \pm\pi, \pm\dfrac{3\pi}{2}$

2. For the given position function, write parametric equations, eliminate $t$, and describe the path.

   *a) $\langle 2 + t, 3 - 2t \rangle$

   *b) $\langle t + 1, 2t^2 \rangle$

   *c) $\langle t^2, t^3 \rangle$

   d) $t^3\vec{i} + t^2\vec{j}$

   e) $\langle t^2 + 1, t + 2 \rangle$

   *f) $\langle \sin t, \cos^2 t \rangle$

   g) $(\tan 2t)\vec{i} + (\sec 2t)\vec{j}$

   h) $\langle \sin^2 t, \cos^2 t \rangle$

   i) $\langle \sec t, \tan t \rangle$, $0 \le t < \dfrac{\pi}{2}$

   j) $\langle e^t + e^{-t}, e^t - e^{-t} \rangle$

   k) $\langle \cos 2t, \sin t \rangle$, $0 \le t \le \pi$

3. a) Show that $\vec{P(\theta)} = \langle a\cos\theta,\, b\sin\theta \rangle$ traces an ellipse $x^2/a^2 + y^2/b^2 = 1$.
   b) Sketch the trace of the motion $\vec{P(t)} = \langle 2\sin\pi t/2,\, 3\cos\pi t/2 \rangle$. Plot $P(t)$ for $t = 0$, $\pm 1$, $\pm 2$, $\pm 3$.

*4. A particle moving with constant velocity is at $P_0 = (1, 3)$ when $t = 0$, and at $P_1 = (-1, 1)$ when $t = 1$. Find the velocity $\vec{v}$, the vector equation of motion, and the parametric equations of motion.

5. A particle with constant velocity is at $P_1 = (1, -2)$ when $t = 1$, and at $P_3 = (-1, 2)$ when $t = 3$. Find the velocity $\vec{v}$, the vector equations of motion, and the parametric equations of motion.

6. For the given curve in polar coordinates, write parametric equations for $x$ and $y$, using as parameter the angle $\theta$.
   *a) $r = 2\sin\theta$    b) $r = 1 + \cos\theta$

**B**

7. Let $\vec{P(t)} = \langle 2t,\, t^2 - 1 \rangle$.
   a) Describe the trace of the motion.
   b) Show that if $t$ is an integer, then so are the components of $\vec{P(t)}$, and so is the magnitude $|\vec{P(t)}|$.
   c) Obtain five non-similar right triangles, all of whose sides have integer length.

*8. Describe the following motions. What do they have in common? How do they differ?
   a) $x = t,\ y = t^2$
   b) $x = t^2,\ y = t^4$
   c) $x = \cos t,\ y = \cos^2 t$

9. Describe the following motions. What do they have in common? How do they differ?
   a) $\vec{P(t)} = \langle 0, 1 \rangle + t\langle 1, -1 \rangle$
   b) $\vec{P(t)} = \langle \sin^2 t,\ \cos^2 t \rangle$
   c) $\vec{P(t)} = \langle \sin t,\ 1 - \sin t \rangle$

*10. A Ferris wheel with diameter 10 meters has its center 6 meters above the ground. It rotates once every 20 seconds. Find the angular velocity $\omega$, and write parametric equations for the motion of the point which, at $t = 0$, is at the bottom of the wheel. (Put the $x$-axis on the ground and the $y$-axis through the center of the wheel.)

11. A railroad wheel of radius $r$ has a flange of radius $R > r$. Point $P$ is on the edge of the flange. Write parametric equations for the path of $P$ as the wheel rolls along a rail; use as parameter the angle $\varphi$ through which the wheel has rolled.

12. A wheel of radius 1 is fixed at the origin, and a second wheel of radius 1 rolls counterclockwise around the circumference of the first. Initially, point $P$ on the rolling wheel touches the fixed wheel at $(1, 0)$. Derive parametric equations for the motion of $P$.

**C**

13. A bicycle sprocket with 52 teeth drives a rear sprocket with 26 teeth. The radius of the driving sprocket is 15 cm, and its center is 30 cm above the ground; the rear sprocket is mounted on a wheel of radius 40 cm. Write parametric equations for the motion of one tooth of the driving sprocket as the bicycle is pedaled from left to right along the $x$-axis. Initially, the tooth is on top as it crosses the $y$-axis at $(0, 45)$. Use as parameter the angle through which the driving sprocket has turned from this initial position.

14. The "folium of Descartes" is the graph of $x^3 + y^3 = 3xy$. An effective way to sketch it uses as parameter the slope of the segment from the origin to the point $(x, y)$ on the curve, as outlined here.
   a) Find the point of intersection of the folium with the line $y = mx$.
   b) Write parametric equations of the folium, with parameter the constant $m$ in part a.
   c) Plot points on the curve for $m = 0,\ \frac{1}{4},\ \frac{1}{2},\ \frac{3}{4},\ 1,\ 2,\ 4;\ -4,\ -2,\ -1/2,\ -1/4$.
   d) On your sketch, identify the parts of the folium corresponding to the following ranges: $0 \le m \le 1$; $1 \le m < \infty$; $-\infty < m < -1$; $-1 < m \le 0$.
   e) Show that, on the folium, $\lim\limits_{m \to -1^+} x = -\infty$, $\lim\limits_{x \to -1^-} x = +\infty$, and $\lim\limits_{m \to -1}(x + y) = -1$.
   f) Show that, as $m \to -1$, the folium is asymptotic to the line $y = -x - 1$. (Parametrize the line by $x_{\text{line}} = \dfrac{3m}{1 + m^3}$, $y_{\text{line}} = -\dfrac{3m}{1 + m^3} - 1$, $-\infty < m < -1$ and $-1 < m \le 0$, and consider $\lim\limits_{m \to -1}[y_{\text{folium}} - y_{\text{line}}]$.)

# 13.3

## LIMITS AND DERIVATIVES. VELOCITY AND ACCELERATION

The position function for a motion is a vector function; for each input $t$, the output

$$\vec{P}(t) = \langle x(t),\, y(t)\rangle$$

is a vector. The concepts of calculus carry over to vector functions by considering components. Thus, a vector function $\langle x(t),\, y(t)\rangle$ is *continuous* if the components $x$ and $y$ are continuous. The *limit* of a vector function is computed by taking the limit of each component:

$$\lim_{t \to a} \langle x(t),\, y(t)\rangle = \left\langle \lim_{t \to a} x(t),\, \lim_{t \to a} y(t) \right\rangle. \tag{1}$$

And the derivative is taken by components:

$$\frac{d}{dt} \langle x(t),\, y(t)\rangle = \left\langle \frac{dx}{dt},\, \frac{dy}{dt} \right\rangle. \tag{2}$$

For example, if $\vec{P}(t) = \left\langle \cos t,\, \dfrac{1}{t} \sin t \right\rangle$, then

$$\lim_{t \to 0} \vec{P}(t) = \left\langle \lim_{t \to 0} \cos t,\, \lim_{t \to 0} \frac{1}{t} \sin t \right\rangle = \langle 1,\, 1\rangle;$$

and the derivative of $\vec{P}$ is

$$\vec{P}'(t) = \frac{d\vec{P}}{dt} = \left\langle -\sin t,\, \frac{t \cos t - \sin t}{t^2} \right\rangle.$$

### Velocity

Let $\vec{P}(t)$ be the position function for a motion in the plane. The **average velocity** over a time interval $[t,\, t + \Delta t]$ is the vector

$$\frac{\text{change in position}}{\text{change in time}} = \frac{\vec{P}(t + \Delta t) - \vec{P}(t)}{\Delta t} = \frac{\Delta \vec{P}}{\Delta t}.$$

Figure 1 shows $\Delta \vec{P}$, and Figure 2 the average velocity for $\Delta t = \frac{1}{2}$, and $\Delta t = \frac{1}{4}$. The **instantaneous velocity** is the limit of this average as $\Delta t \to 0$:

$$\vec{v}(t) = \lim_{\Delta t \to 0} \frac{\Delta \vec{P}}{\Delta t} = \lim_{\Delta t \to 0} \left\langle \frac{x(t + \Delta t) - x(t)}{\Delta t},\, \frac{y(t + \Delta t) - y(t)}{\Delta t} \right\rangle$$

$$= \left\langle \lim_{\Delta t \to 0} \frac{x(t + \Delta t) - x(t)}{\Delta t},\, \lim_{\Delta t \to 0} \frac{y(t + \Delta t) - y(t)}{\Delta t} \right\rangle$$

$$= \left\langle \frac{dx}{dt},\, \frac{dy}{dt} \right\rangle$$

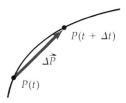

**FIGURE 1**

$\Delta \vec{P} = \vec{P}(t + \Delta t) - \vec{P}(t)$.

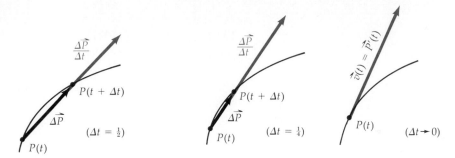

**FIGURE 2**

Velocity $\vec{v}(t) = \vec{P}'(t) = \lim\limits_{\Delta t \to 0} \dfrac{\Delta \vec{P}}{\Delta t}$.

by the definition of derivative for ordinary functions. Looking back at (2), you see that *the velocity is the derivative of the position function*:

$$\vec{v}(t) = \frac{d\vec{P}}{dt} = \vec{P}'(t).$$

In Figure 2, the average velocity vector $\dfrac{\Delta \vec{P}}{\Delta t}$ lies on a secant line cutting the trace of the motion at the two points $P(t)$ and $P(t + \Delta t)$. As $\Delta t \to 0$, the limiting position of this secant line is the line tangent to the trace at $P(t)$. Thus the limit $\vec{v}(t)$, drawn as an arrow with tail at $P(t)$, is *tangent to the trace of the motion*. Moreover, it *points in the direction of motion*, as you can see from Figures 1 and 2. There $\Delta t > 0$, so $P(t + \Delta t)$ is a later position than $P(t)$. Hence $\Delta \vec{P}$ points in the direction of motion, and so does $\dfrac{\Delta \vec{P}}{\Delta t}$ (because $\Delta t > 0$), and so then does the limit $\dfrac{d\vec{P}}{dt}$. This interprets the *direction* of the velocity $\vec{v}$. Its *length* $|\vec{v}|$ is defined to be the *speed* of the motion, as we explain more fully in the next section. Speed tells only "how fast"; velocity tells "how fast, and in what direction."

---

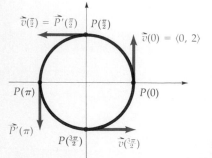

**FIGURE 3**

**EXAMPLE 1** $\vec{P}(t) = \langle 2 \cos t, 2 \sin t \rangle$ gives a circular motion (Fig. 3). The velocity is

$$\vec{v}(t) = \frac{d\vec{P}}{dt} = \langle -2 \sin t, 2 \cos t \rangle.$$

Draw $\vec{v}(t)$ as an arrow with tail at $P(t)$; it is tangent to the circle, and has length $|\vec{v}(t)| = 2$. The point $P$ moves with speed 2, tracing a circle of perimeter $4\pi$ as $t$ varies from 0 to $2\pi$.

---

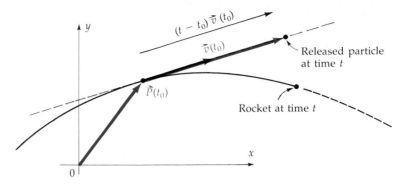

**FIGURE 4**
Interpreting the linear approximation $\vec{L}(t) = \vec{P}(t_0) + (t - t_0)\vec{v}(t_0)$.

**The linear approximation** of a vector function $\vec{P}(t)$, based at a given time $t_0$, is the vector function $\vec{L}$ defined by

$$\vec{L}(t) = \vec{P}(t_0) + (t - t_0)\vec{P}'(t_0) \tag{3}$$
$$= \vec{P}(t_0) + (t - t_0)\vec{v}(t_0). \tag{3a}$$

Think of $\vec{P}(t)$ as the position function of a rocket traveling in space. Suppose that at time $t_0$ a small particle is released from the rocket (Fig. 4). In the absence of any force, it will (by Newton's first law of motion) continue in a straight line with constant velocity $\vec{v}(t_0)$, the velocity it had when released. At time $t_0$ its position is $P(t_0)$, the same as the rocket's; at time $t > t_0$, it has continued from there with velocity $\vec{v}(t_0)$, for time $t - t_0$, so its new position vector is

$$\vec{P}(t_0) + (t - t_0)\vec{v}(t_0).$$

This is precisely the formula (3a) for the linear approximation $\vec{L}(t)$. Thus $\vec{L}(t)$ is the vector equation of the straight line through $P(t_0)$, parallel to the vector $\vec{v}(t_0)$; since this vector is parallel to the trace of $\vec{P}(t)$, *the linear approximation gives the equation of the line tangent to the trace of $\vec{P}$ at $P(t_0)$.*
    Notice how closely $\vec{L}(t)$ resembles the linear approximation of an ordinary function $f$ of one variable $x$, based at $x_0$, which is

$$f(x_0) + (x - x_0)f'(x_0).$$

Replace $f$ by $\vec{P}$ and $x$ by $t$, and you have formula (3).

---

**EXAMPLE 2**   For the motion $\vec{P}(t) = \langle 2\cos t, 2\sin t\rangle$ in Example 1, the linear approximation at time $\pi/2$ is

$$\vec{L}(t) = \vec{P}\left(\frac{\pi}{2}\right) + \left(t - \frac{\pi}{2}\right)\vec{P}'\left(\frac{\pi}{2}\right)$$
$$= \langle 0, 2\rangle + \left(t - \frac{\pi}{2}\right)\langle -2, 0\rangle$$
$$= \langle \pi - 2t, 2\rangle.$$

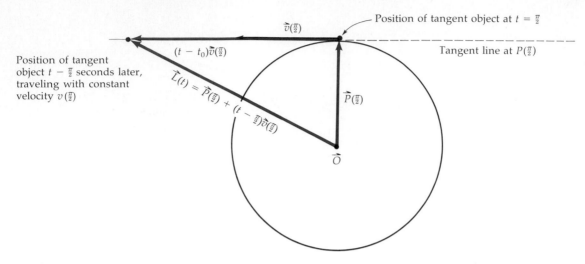

**FIGURE 5**
Equation of linear approximation to $\vec{P}(t) = \langle 2\cos t,\ 2\sin t\rangle$, based at $t = \pi/2$.

This is the vector equation of a line parallel to the vector $\langle -2,\ 0\rangle$ and passing through the point $(0,\ 2)$ at time $t = \pi/2$ (Fig. 5).

---

**The acceleration** $\vec{a}$ of a moving particle is, by definition, the derivative of the velocity:

$$\vec{a} = \vec{v}' = \vec{P}''.$$

---

**EXAMPLE 3**  Motion along the straight line through a point $P_0$ parallel to a fixed vector $\vec{v} = \langle v_1,\ v_2\rangle$ is given by the vector equation

$$\vec{P} = \vec{P}_0 + t\vec{v} = \langle x_0 + tv_1,\ y_0 + tv_2\rangle.$$

The velocity is $\vec{P}' = \langle v_1,\ v_2\rangle = \vec{v}$, a constant vector; so the acceleration (the derivative of the constant $\vec{v}$) is $\vec{0}$.

---

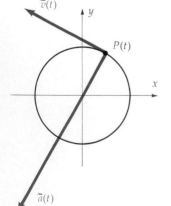

**FIGURE 6**
$\vec{P}(t) = \langle \cos 2t,\ \sin 2t\rangle.$

**EXAMPLE 4**  For the circular motion $\vec{P}(t) = \langle \cos 2t,\ \sin 2t\rangle$, the velocity is

$$\vec{v}(t) = \vec{P}'(t) = \langle -2\sin 2t,\ 2\cos 2t\rangle$$

and the acceleration is

$$\vec{a}(t) = \vec{v}'(t) = \langle -4\cos 2t,\ -4\sin 2t\rangle.$$

Here the *speed* is constant, $|\vec{v}| = 2$. But the *direction* of the velocity varies, so its derivative $\vec{a} = \vec{v}'$ is not $\vec{0}$; $\vec{a}$ is fairly large, and directed toward the center of the circle, perpendicular to the direction of motion (Fig. 6). This is *centripetal* (center-seeking) acceleration.

**Newton's second law of motion** relates the force on a particle to its mass and motion. If the mass $m$ is constant, the law is

$$\vec{F}(t) = m\vec{a}(t),$$

where $\vec{F}(t)$ is the force acting on the particle at time $t$, and $\vec{a}$ the acceleration. Frequently, $\vec{F}$ and $m$ are known; this determines the acceleration by Newton's law. For example, any object of mass $m$ near the surface of the earth experiences a downward gravitational force of magnitude $mg$, where $g$ is a "gravity constant." With the axes in their usual position, the force is directed in the negative $y$ direction; it is given by the vector in Figure 7,

$$\vec{F} = -mg\vec{j}.$$

If this is the only force acting, then the object is in "free fall."

When the acceleration is known, then the velocity is obtained by integration, and a second integration determines the position.

To integrate a vector function such as the velocity, simply integrate each component:

$$\int v_1(t)\vec{i} + v_2(t)\vec{j}\ dt = \left[\int v_1(t)\ dt\right]\vec{i} + \left[\int v_2(t)\ dt\right]\vec{j}.$$

The indefinite integrals will include a separate constant of integration in each component. For example,

$$\int [t\vec{i} + \cos(t)\vec{j}]\ dt = \left[\int t\ dt\right]\vec{i} + \left[\int \cos(t)\ dt\right]\vec{j}$$
$$= [\tfrac{1}{2}t^2 + c_1]\vec{i} + [\sin(t) + c_2]\vec{j}.$$

The two constants can be combined into a single *vector* constant $\vec{c} = c_1\vec{i} + c_2\vec{j}$:

$$\int [t\vec{i} + \cos(t)\vec{j}]\ dt = \tfrac{1}{2}t^2\vec{i} + c_1\vec{i} + \sin(t)\vec{j} + c_2\vec{j}$$
$$= \tfrac{1}{2}t^2\vec{i} + \sin(t)\vec{j} + \vec{c}.$$

The vector constant may be determined by giving the value of the indefinite integral at some particular time.

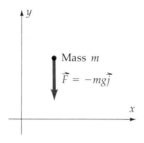

**FIGURE 7**
Force of gravity.

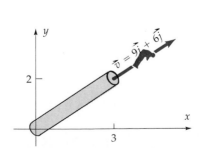

**FIGURE 8**
Human cannonball.

**EXAMPLE 5** A "human cannonball" is shot from a circus cannon (Fig. 8). At time $t = 0$ she emerges from the muzzle of the cannon with velocity

$$\vec{v}(0) = 9\vec{i} + 6\vec{j} \tag{4}$$

in meters/second. The muzzle of the cannon is at position $(3, 2)$.

    **a.** Compute her further path.

    **b.** She is to land in a net two meters above the $x$-axis; where should the center of the net be?

## SOLUTION

a. Between the cannon and the net, she is in free fall and the force is

$$\vec{F} = -mg\vec{j},$$

where $m$ is the (unspecified) mass. By Newton's law,

$$-mg\vec{j} = m\vec{a}$$

so the acceleration is

$$\vec{a} = -g\vec{j}.$$

Then the velocity is

$$\vec{v}(t) = \int \vec{a}(t)\, dt = \int -g\vec{j}\, dt = -gt\vec{j} + \vec{c}, \tag{5}$$

where $\vec{c}$ is the vector constant of integration. The given initial condition (4) determines $\vec{c}$; combining (4) with (5) yields

$$9\vec{i} + 6\vec{j} = \vec{v}(0) = \vec{c}.$$

Substitute this in (5), and find $\vec{v}(t) = 9\vec{i} + (6 - gt)\vec{j}$.

The position $\vec{P}$ is now obtained by integrating this velocity:

$$\vec{P}(t) = \int \vec{v}(t)\, dt = \int [9\vec{i} + (6 - gt)\vec{j}]\, dt$$
$$= 9t\vec{i} + (6t - \tfrac{1}{2}gt^2)\vec{j} + \vec{C},$$

where $\vec{C}$ is a new vector constant of integration. This one is determined by the given position $(3, 2)$ at time 0:

$$3\vec{i} + 2\vec{j} = \vec{P}(0) = \vec{C}$$

hence

$$\vec{P}(t) = 9t\vec{i} + (6t - \tfrac{1}{2}gt^2)\vec{j} + (3\vec{i} + 2\vec{j})$$
$$= (3 + 9t)\vec{i} + (2 + 6t - \tfrac{1}{2}gt^2)\vec{j}. \tag{6}$$

Now we know exactly where she is throughout her "free fall" flight.

b. Where should the net be, to catch her at a height of 2 meters? The height $y$ is the $\vec{j}$ component of the position vector (6),

$$y = 2 + 6t - \tfrac{1}{2}gt^2.$$

From this equation, $y = 2$ when $t = 0$ (as she leaves the muzzle) and again when $t = 12/g$; that is the time when she should strike the net. Her $x$-coordinate then is, from (6),

$$x = 3 + 9t = 3 + \frac{108}{g}.$$

With distance in meters and time in seconds, $g = 9.8$, and the proper location of the center of the net is at

$$x = 3 + \frac{108}{9.8} \approx 14.0 \text{ meters}$$

from the base of the cannon.

## SUMMARY

**Velocity:**   If $\vec{P}(t) = x\vec{i} + y\vec{j}$ is position at time $t$, then the velocity is

$$\vec{v}(t) = \lim_{\Delta t \to 0} \frac{\Delta \vec{P}}{\Delta t} = \vec{P}'(t) = \frac{dx}{dt}\vec{i} + \frac{dy}{dt}\vec{j}.$$

$\vec{v}(t)$ is tangent to the trace of $\vec{P}$ at the point $P(t)$ (Fig. 2).
$|\vec{v}(t)|$ is the speed of motion.

**Linear Approximation:**   $\vec{L}(t) = \vec{P}(t_0) + (t - t_0)\vec{P}'(t_0)$. The trace of $\vec{L}$ is tangent to the trace of $\vec{P}$ at $P(t_0)$.

**Acceleration:**   $\vec{a}(t) = \vec{v}'(t) = \vec{P}''(t)$.

**Newton's Second Law:**   Force $\vec{F} = m\vec{a}$,     if mass $m$ is constant.

**Integration:**   $\int f(t)\vec{i} + g(t)\vec{j}\ dt = \int f(t)\ dt\ \vec{i} + \int g(t)\ dt\ \vec{j}.$

## PROBLEMS

### A

*1.   Compute.

a)   $\lim\limits_{t \to 1} \left\langle \dfrac{t^2 - 1}{t - 1}, e^{1-t} \right\rangle$

b)   $\lim\limits_{t \to +\infty} \left\langle 1 + e^{-t}, \dfrac{\sin t}{t} \right\rangle$

c)   $\lim\limits_{t \to 0} \left( \dfrac{\sin 2t}{t}\vec{i} + t \cos t\vec{j} \right)$

2.   Find the velocity and acceleration of the motion $\vec{P}(t)$. Sketch the trace of the motion, and the velocity and acceleration at the given time $t_0$.
*a)   $\langle 2 \cos 3t, 2 \sin 3t \rangle$; $t_0 = \pi/3$
b)   $\langle 2t, 1 - t^2 \rangle$; $t_0 = 1$
*c)   $e^{2t}\vec{i} + e^{-2t}\vec{j}$; $t_0 = 0$
d)   $2t\vec{i} + (t - 3)\vec{j}$; $t_0 = 1$
*e)   $\langle \sin^2 t, \cos^2 t \rangle$; $t_0 = \pi/2$

*3.   For each motion in problem 2, write the equation of the linear approximation at the given time $t_0$. (Answer to part a.)

4.   Find velocity and position from the given data.
*a)   $\vec{a}(t) = \vec{0}$, $\vec{v}(0) = \vec{i}$, $\vec{P}(0) = \vec{j}$.
*b)   $\vec{a}(t) = e^t\vec{i} + e^{-t}\vec{j}$, $\vec{v}(0) = \vec{0}$, $\vec{P}(0) = \vec{0}$.
*c)   $\vec{a}(t) = \pi \sin \pi t\vec{i} + \pi \cos \pi t\vec{j}$, $\vec{v}(0) = \vec{0}$, $\vec{P}(0) = \vec{0}$.
d)   $\vec{a}(t) = t \cos t\vec{i} + t \sin t\vec{j}$, $\vec{v}(0) = \vec{i}$, $\vec{P}(0) = 2\vec{j}$.

5.   A daredevil cannonballs into the sea from cliffs 10 meters high, with initial velocity $\vec{v}(0) = 3\vec{i} + 2\vec{j}$. What is his velocity as he enters the sea?

*6.   A rocket in space is following the path $\vec{P}(t) = \langle t \cos 2t, t \sin 2t \rangle$. At time $t = 3$, a piece of protective tiling falls off and continues on a straight course with $\vec{0}$ acceleration. Where is the piece at $t = 5$?

### B

7.   A projectile is fired from a gun at time $t = 0$; the muzzle is at position $(x_0, y_0)$, the speed of the projectile as it leaves the muzzle is $s_0$, and the angle of elevation is $\alpha$. Explain why:
a)   $\vec{v}(0) = s_0 \cos \alpha\vec{i} + s_0 \sin \alpha\vec{j}$.
b)   $\vec{P}(t) = \langle x_0 + ts_0 \cos \alpha,$
     $y_0 + ts_0 \sin \alpha - \frac{1}{2}gt^2 \rangle$.
c)   Assuming that the ground coincides with the $x$-axis, the projectile lands at time

$$t = \frac{s_0}{g} \sin \alpha[1 + \sqrt{1 + 2y_0 g s_0^{-2} \csc^2\alpha}].$$

d)   The range (horizontal distance from muzzle to point of landing) is

$$\frac{s_0^2}{2g} \sin 2\alpha[1 + \sqrt{1 + 2y_0 g s_0^{-2} \csc^2\alpha}].$$

**8.** A projectile is fired from the origin with angle of elevation $\alpha = 45°$. It lands 1000 meters away. Find the muzzle speed $s_0$. (See the previous problem.)

**\*9.** A projectile is fired from the origin with muzzle speed $s_0 = 300$ meters/second. What angles of elevation give it a range of exactly 3000 meters? (See problem 7.)

**10.** A particle of mass $m$ moves in a circle of radius $r$ with constant angular velocity $\omega$ radians per second:

$$\vec{P}(t) = r\cos\omega t\,\vec{i} + r\sin\omega t\,\vec{j}.$$

**a)** Find the speed.

**b)** Show that the acceleration (and the force) are "centripetal," that is, directed toward the center of the circle.

**c)** Relate the magnitude of the force to the mass, the speed, and the radius.

**11.** A satellite of mass $m$ moves around a celestial body of mass $M$ in a circular orbit of radius $r$, with constant angular velocity $\omega$. Assume that the mass $M$ is at the origin, so the motion is as in the previous problem. By the law of universal gravitation, the force on the satellite is

$$\vec{F} = -\frac{GMm}{r^3}\,\vec{P}.$$

Show

**a)** $|\vec{F}| = GMm/r^2$.

**b)** The speed $|\vec{v}|$ is $\sqrt{GM/r}$.

**c)** The period $T$ (the time taken to complete one orbit) is $2\pi/\omega$.

**d)** $T = (2\pi/\sqrt{GM})r^{3/2}$. (Thus the period depends on the distance $r$ between body and satellite, but is independent of the mass $m$ of the satellite. Kepler discovered that this applies also to elliptic orbits, such as those of the planets around the sun.)

**\*12. a)** The moon has a nearly circular orbit around the earth, with radius 239,000 miles and period 27.32 days. Using the previous problem, approximate the constant $GM$, where $M$ is the mass of the earth. (Assume that the orbit *is* a circle.)

**b)** An earth satellite is to have a circular orbit with a period of 1 day, so that it remains above a fixed spot on the equator. What should be the radius of its orbit?

**13.** For the motion $\vec{P}(t) = \langle \alpha\cos\omega t,\ \beta\sin\omega t\rangle$, show that

**a)** The path is an ellipse.

**b)** The acceleration is directed toward the center of the ellipse, and proportional to the distance from the center (like the force exerted by a "Hooke's law" spring).

**14.** A point $P$ on the rim of a wheel of radius $r$, rolling along the $x$-axis, follows the motion

$$\vec{P}(t) = \langle r\omega t + r\sin\omega t,\ r + r\cos\omega t\rangle,$$

tracing a cycloid (Sec. 13.2).

**a)** Show that the point $P$ stops once in each revolution of the wheel. Where does it stop?

**b)** What is the speed of the wheel's axle?

**c)** Compare the maximum speed of $P$ to the speed of the axle.

**15.** A point $P$ on the flange of a railroad wheel (Fig. 9) follows the motion

$$\vec{P}(t) = \langle r\omega t + R\sin\omega t,\ r + R\cos\omega t\rangle,$$

with positive constants $r$, $\omega$, and $R$; $r$ is the radius of the part of the wheel riding on the rail (the $x$-axis), $R > r$ is the radius of the flange, and $\omega$ is the angular velocity of the wheel. Show that

**a)** $P$ never stops; the velocity $\vec{v}$ is never $\vec{0}$.

**b)** $P$ moves to the right while above the rail, and to the left while below it.

**c)** The acceleration is the same as for circular motion. (Why is that?)

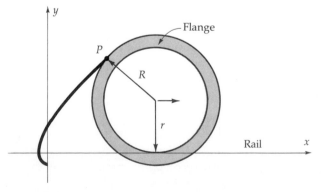

**FIGURE 9**
Point $P$ on flange of railway wheel.

**16.** A rocket is to be fired from $(20, 0)$ with constant velocity $\vec{v}$, to rendezvous with a satellite in the circular orbit $\vec{P}(t) = \langle 10\cos t, 10\sin t\rangle$. At rendezvous, it should have the same position *and velocity* as the satellite.

**\*a)** Where is the rendezvous point?
**\*b)** What is the rocket velocity $\vec{v}$?
**c)** When should the rocket be fired?

**17.** A point moves in the plane with constant non-zero acceleration. Show that it traces either a parabola or a half-line. (By appropriate choice of coordinates, you can assume that $\vec{a} = a\vec{j}$ for a numerical constant $a$.)

**18.** Assuming Newton's second law of motion $\vec{F} = m\vec{a}$, prove his first law: An object acted on by no force moves in a straight line with constant velocity.

**19.** Suppose that $h$ is a scalar function and $\vec{F} = \langle f_1, f_2 \rangle$ is a vector function, both differentiable; that is, $h'$ and $\vec{F}'$ exist. Show that $(h\vec{F})' = h'\vec{F} + h\vec{F}'$.

**20.** Suppose that $\vec{F} = \langle f_1, f_2 \rangle$ is a vector function and $h(s)$ a scalar function, both differentiable.

Prove a formula for the derivative of the composite function $\vec{F}(h(s))$.

**21.** Let $\vec{P}(t) = \left\langle \dfrac{2t}{1+t^2}, \dfrac{1-t^2}{1+t^2} \right\rangle$. Show that $|\vec{P}| = 1$. Compute the limits of $\vec{P}(t)$ and $\vec{P}'(t)$ as $t \to +\infty$ and $t \to -\infty$. Describe the motion for $-\infty < t < \infty$. Show that every point on the unit circle is the limit of a sequence of points $P_n$ on the circle, both of whose coordinates are rational.

**22.** This problem considers whether the tangent line defined for a parametric curve is consistent with the one defined for the graph of a function.
**a)** Compute the slope of the line $\vec{L}(t) = \vec{P}_0 + (t - t_0)\langle a, b \rangle$, where $a \neq 0$.
**b)** Suppose that the curve $\vec{P}(t) = \langle x(t), y(t) \rangle$ lies on the graph of a differentiable function $f$; that is, $y(t) = f(x(t))$. Assume also that $x'(t_0) \neq 0$. Does the tangent line given in vector form by $\vec{L}(t) = \vec{P}(t_0) + (t - t_0)\vec{v}(t_0)$ have slope $f'(x(t_0))$? (Use the Chain Rule.)

# 13.4

## ARC LENGTH

Length measured along a continuous curve is called **arc length**. For a parametric curve

$$\vec{P}(t) = \langle x(t), y(t) \rangle$$

we denote by $L_a^b$ the length traced out as $t$ varies from $a$ to $b$ (Fig. 1). This arc length is defined as follows. Divide the interval $a \leq t \leq b$ into $n$ equal parts of length $\Delta t$ by points $t_1, t_2, \ldots t_{n-1}$. Approximate the curve (as in Fig. 2) by the straight-line segments from $P(a)$ to $P(t_1)$ to $P(t_2) \ldots$ to $P(b)$. The length $L_a^b$ is then approximated by the sum of the lengths of these

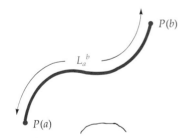

**FIGURE 1**
$L_a^b$ = length along curve as $P(t)$ moves from $P(a)$ to $P(b)$.

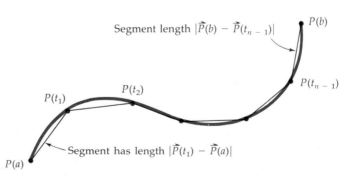

**FIGURE 2**
Approximating length $L_n = |\vec{P}(t_1) - \vec{P}(a)| + |\vec{P}(t_2) - \vec{P}(t_1)| + \cdots + |\vec{P}(b) - \vec{P}(t_{n-1})|$.

segments, which we denote by $L_n$:

$$L_n = \sum_{j=1}^{n} |\vec{P}(t_j) - \vec{P}(t_{j-1})|. \tag{1}$$

(In this sum, $t_0 = a$ and $t_n = b$.) If these approximations have a limit as $n \to \infty$, that limit is defined to be the arc length:

$$L_a^b = \lim_{n \to \infty} L_n.$$

If the derivatives $x'$ and $y'$ are continuous, then the limit does exist, and equals a certain integral, derived as follows. The $j^{\text{th}}$ approximating segment has length

$$|\vec{P}(t_j) - \vec{P}(t_{j-1})| = \sqrt{[x(t_j) - x(t_{j-1})]^2 + [y(t_j) - y(t_{j-1})]^2} \tag{2}$$

Apply the Mean Value Theorem to both $x$ and $y$:

$$x(t_j) - x(t_{j-1}) = x'(\hat{t}_j)\,\Delta t \tag{3}$$

$$y(t_j) - y(t_{j-1}) = y'(t_j^*)\,\Delta t. \tag{4}$$

[We must evaluate $x'$ in (3) and $y'$ in (4) at possibly different points; the Mean Value Theorem doesn't say that the point $\hat{t}_j$ that works for $x$ will be the same as the point $t_j^*$ that works for $y$.] Substitute (3) and (4) into (2):

$$|\vec{P}(t_j) - \vec{P}(t_{j-1})| = \sqrt{[x'(\hat{t}_j)\,\Delta t]^2 + [y'(t_j^*)\,\Delta t]^2}$$
$$= \sqrt{x'(\hat{t}_j)^2 + y'(t_j^*)^2}\,\Delta t.$$

The approximating length $L_n$ is the sum of these lengths:

$$L_n = \sum_{j=1}^{n} \sqrt{x'(\hat{t}_j)^2 + y'(t_j^*)^2}\,\Delta t.$$

Now, if $t_j^*$ were equal to $\hat{t}_j$, this would be a Riemann sum for the integral $\int_a^b \sqrt{x'(t)^2 + y'(t)^2}\,dt$, and we would conclude that

$$L_a^b = \lim_{n \to \infty} L_n = \int_a^b \sqrt{x'(t)^2 + y'(t)^2}\,dt. \tag{5}$$

Even though $L_n$ is not precisely a Riemann sum, the conclusion (5) is correct. For as $n \to \infty$ and $\Delta t \to 0$, the intervals $[t_{j-1}, t_j]$ shrink, and the points $\hat{t}_j$ and $t_j^*$ come arbitrarily close together. From this, it can be proved that the approximating lengths $L_n$ have the same limit as the actual Riemann sums

$$S_n = \sum_{j=1}^{n} \sqrt{x'(\hat{t}_j)^2 + y'(\hat{t}_j)^2}\,\Delta t,$$

and it follows that

$$L_a^b = \lim_{n \to \infty} L_n = \lim_{n \to \infty} S_n = \int_a^b \sqrt{x'(t)^2 + y'(t)^2}\,dt.$$

When the parameter $t$ is time, we recognize the arc length integrand $\sqrt{x'(t)^2 + y'(t)^2}$ as the *speed*, the length of the velocity vector:

$$\sqrt{x'(t)^2 + y'(t)^2} = \left|\langle x'(t), y'(t)\rangle\right|$$
$$= \left|\vec{P}'(t)\right| = \left|\vec{v}(t)\right|.$$

So *arc length equals the integral of speed with respect to time.*

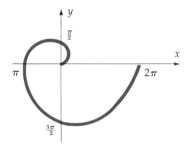

**FIGURE 3**
Spiral $\vec{P}(t) = \langle t\cos t, t\sin t\rangle$,
$0 \le t \le 2\pi$.

**EXAMPLE 1**   The curve $\vec{P}(t) = \langle t\cos t, \sin t\rangle$ is a spiral making one turn as $t$ varies from 0 to $2\pi$ (Fig. 3). Compute the length of this turn of the spiral.

*SOLUTION*

The speed is

$$\left|\vec{P}'(t)\right| = \left|\langle -t\sin t + \cos t, t\cos t + \sin t\rangle\right|$$
$$= \sqrt{1 + t^2}.$$

The length of the curve from $t = 0$ to $t = 2\pi$ is (as you can check by trig substitution, or from the table)

$$\int_0^{2\pi} \sqrt{1 + t^2}\, dt = \frac{t}{2}\sqrt{1 + t^2} + \frac{1}{2}\ln\left|t + \sqrt{1 + t^2}\right| \Big|_0^{2\pi}$$

$$= \pi\sqrt{1 + 4\pi^2} + \frac{1}{2}\ln(2\pi + \sqrt{1 + 4\pi^2}) \approx 21.256.$$

### The Arc Length Function

Any indefinite integral of the arc length integrand is called an **arc length function**, and denoted by $s$:

$$s = \int \sqrt{x'(t)^2 + y'(t)^2}\, dt. \tag{6}$$

The derivative of $s$ is, of course, the integrand

$$\frac{ds}{dt} = \sqrt{x'(t)^2 + y'(t)^2}. \tag{7}$$

Notice that $s$ stands for *length*, not speed; this is an unfortunate but well-established tradition. The speed is the derivative $ds/dt$, the rate at which arc length is traversed.

In differential notation, (7) is

$$ds = \sqrt{x'(t)^2 + y'(t)^2}\, dt = \sqrt{[x'(t)\, dt]^2 + [y'(t)\, dt]^2}$$
$$= \sqrt{(dx)^2 + (dy)^2}$$

so

$$(ds)^2 = (dx)^2 + (dy)^2. \tag{8}$$

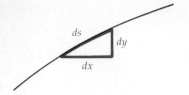

**FIGURE 4**
$(ds)^2 = (dx)^2 + (dy)^2$.

This suggests a convenient way to remember the arc length formula (Fig. 4). A small piece of the curve is practically a straight line segment, the hypotenuse of a right triangle with legs $dx$ and $dy$. So you can think of (8) as the Pythagorean Theorem for an infinitesimally small piece of the curve. The length of arc for $a \leq t \leq b$ is the integral ("sum") of the lengths of these infinitesimal pieces:

$$L_a^b = \int_a^b ds = \int_a^b \sqrt{(dx)^2 + (dy)^2}.$$

*REMARK*    Section 10.1 gave an integral for arc length along the graph of a function $y = f(x)$ and used Figure 4 and formula (8) to remember it:

$$y = f(x), \qquad dy = f'(x)\, dx$$
$$(ds)^2 = (dx)^2 + (dy)^2 = [1 + f'(x)^2](dx)^2$$

so the arc length is

$$\int ds = \int \sqrt{1 + f'(x)^2}\, dx.$$

Problem 8 asks you to show that this integral is consistent with formulas (6) and (7) for arc length on parametric curves.

We have been using time as the parameter just to keep a concrete image in mind; but these differential formulas, and the arc length integral, are equally valid for any parameter.

---

**EXAMPLE 2**    The equations $x = a \cos \theta$, $y = a \sin \theta$ give a circle of radius $a$ (assumed constant); the parameter is not time $t$, but the angle $\theta$ in radians. Compute the length of arc traced for $\theta_1 \leq \theta \leq \theta_2$ (Fig. 5).

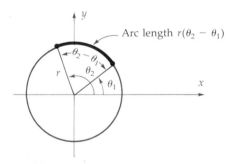

**FIGURE 5**
Arc length on a circle of radius $r$.

*SOLUTION*

$$\frac{dx}{d\theta} = -a \sin \theta \qquad \text{so} \quad dx = -a \sin \theta\, d\theta$$

and likewise $dy = a \cos \theta \, d\theta$. Then

$$(ds)^2 = (dx)^2 + (dy)^2 = a^2 \sin^2\theta (d\theta)^2 + a^2 \cos^2\theta (d\theta)^2$$
$$= a^2 (d\theta)^2.$$

So $ds = a \, d\theta$, and the arc length on the circle for $\theta_1 \le \theta \le \theta_2$ is

$$\int_{\theta_1}^{\theta_2} ds = \int_{\theta_1}^{\theta_2} a \, d\theta = a(\theta_2 - \theta_1).$$

This is the radius a times the radian measure of the angle subtended by the arc (Fig. 5), in accord with the elementary formula for the length of a circular arc.

## Polar Coordinates

An equation

$$r = f(\theta) \tag{9}$$

in polar coordinates defines a parametric curve, with $\theta$ as parameter. The parametric equations are

$$x = r \cos \theta = f(\theta) \cos \theta$$
$$y = r \sin \theta = f(\theta) \sin \theta,$$

so

$$(ds)^2 = (dx)^2 + (dy)^2$$
$$= [f'(\theta) \cos \theta - f(\theta) \sin \theta]^2(d\theta)^2 + [f'(\theta) \sin \theta + f(\theta) \cos \theta]^2(d\theta)^2$$
$$= [f'(\theta)^2 + f(\theta)^2](d\theta)^2$$

(fill in the steps). From (9), $f'(\theta) \, d\theta = dr$, so

$$(ds)^2 = (dr)^2 + r^2(d\theta)^2.$$

Figure 6 shows this as an infinitesimal Pythagorean Theorem. One side of the little right triangle is $dr$, and the other "side" is the arc on a circle of radius $r$ subtending an infinitesimal angle $d\theta$, so its length is $r \, d\theta$.

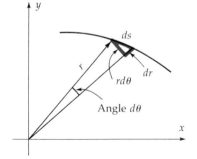

**FIGURE 6**
$(ds)^2 = (dr)^2 + (r \, d\theta)^2.$

## SUMMARY

***Length of Curve***   $\vec{P}(t)$ for $a \le t \le b$ (see Fig. 1)

$$L_a^b = \int_a^b |\vec{P}'(t)| \, dt = \int_a^b \sqrt{x'(t)^2 + y'(t)^2} \, dt$$
$$= \int_a^b \sqrt{(dx)^2 + (dy)^2}.$$

$$\textbf{Arc Length Function:} \quad s = \int \sqrt{\left(\frac{dx}{dt}\right)^2 + \left(\frac{dy}{dt}\right)^2} \, dt$$

$$\frac{ds}{dt} = \sqrt{\left(\frac{dx}{dt}\right)^2 + \left(\frac{dy}{dt}\right)^2}$$

$$(ds)^2 = (dx)^2 + (dy)^2 \qquad \text{[Fig. 4]}.$$

**Polar Coordinates:** (see Fig. 6)

$$(ds)^2 = (dr)^2 + r^2(d\theta)^2$$

$$L_a^b = \int_a^b \sqrt{\left(\frac{dr}{d\theta}\right)^2 + r^2} \, d\theta.$$

## PROBLEMS

### A

1. Find the arc length function, and compute the length of the given part of the curve.
   *a) $\vec{P}(t) = \langle r \cos kt, r \sin kt \rangle$; $L_0^{\pi/k}$
   b) $x = 3t^2$, $y = 2t^3$; $L_0^1$
   *c) $x = 3t^2$, $y = 2t^3$; $L_{-1}^0$ (Caution: $\sqrt{t^2} = t$ only when $t \geq 0$.)
   d) $x = 1$, $y = t^2$; $L_{-1}^1$
   *e) $\vec{P}(t) = e^t \langle \cos t, \sin t \rangle$; $L_0^t$

2. For $\vec{P}(t) = \langle x_0 + at, y_0 + bt \rangle$, show that the arc length for $0 \leq t \leq 1$ equals the speed of motion.

3. Compute the length of the given section of the given curve in polar coordinates.
   a) $r = 2$; $a \leq \theta \leq b$
   *b) $r = a \cos \theta$; $-\dfrac{\pi}{2} \leq \theta \leq \dfrac{\pi}{2}$
   *c) $r = \theta$, $0 \leq \theta \leq \pi/2$
   d) $r = 1/\theta$; $1 \leq \theta \leq 2$.
   *e) $r = \sec \theta$, $0 \leq \theta \leq \pi/4$
   f) $r = 1/\theta$, $\pi \leq \theta \leq 2\pi$
   g) $r = 1/\theta$, $\pi \leq \theta < \infty$
   h) $r = e^\theta$, $0 \leq \theta \leq 1$

4. Compute the length of the cardioid
   $r = a + a \cos \theta$. $\left[ \text{Hint: } 1 + \cos \theta = 2\left(\cos \dfrac{\theta}{2}\right)^2. \right.$
   Be careful; $\sqrt{c^2} = c$ only when $c \geq 0!$ $\Big]$

*5. Compute the length of one arch of the cycloid
   $x = a\theta - a \sin \theta$, $y = a - a \cos \theta$.

$$\left[ \text{Hint: } 1 - \cos \theta = 2\left(\sin \frac{\theta}{2}\right)^2. \right]$$

### B

6. Let $x(t) = t^2$ and $y(t) = t^3$. Find points $\hat{t}$ and $t^*$ such that

$$x(0.5) - x(0.4) = x'(\hat{t}) \cdot (0.1)$$

$$y(0.5) - y(0.4) = y'(t^*) \cdot (0.1).$$

   Show that it is impossible to choose $\hat{t} = t^*$.

7. *a) Compute the length of $x = \cos^2 t$, $y = \sin^2 t$, $0 \leq t \leq 2\pi$.
   b) Show that the curve in part a lies on the line segment $x + y = 1$, $0 \leq x \leq 1$.
   c) Why is the length in part a longer than the segment in part b?

8. Does the arc length integral (5) agree with the integral $\int_a^b \sqrt{1 + f'(x)^2} \, dx$ for the length of a graph? Specifically, suppose that $\vec{P}(t) = \langle x(t), y(t) \rangle$ satisfies

$$y(t) = f(x(t)), \qquad x(\alpha) = a, \qquad x(\beta) = b,$$

   and $x'(t) \geq 0$ for $\alpha \leq t \leq \beta$; thus for $\alpha \leq t \leq \beta$, the parametric curve traces the graph of $f$ from $x = a$ to $x = b$. Can you show that

$$\int_\alpha^\beta \sqrt{x'(t)^2 + y'(t)^2} \, dt = \int_a^b \sqrt{1 + f'(x)^2} \, dx?$$

   What if $x'(t)$ changes sign in the interval $[\alpha, \beta]$?

C

9. Suppose that $\vec{P}(t) = \langle x(t), y(t) \rangle$ where $x'$ and $y'$ are continuous. Evaluate the limiting value of

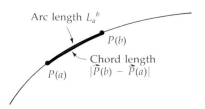

Arc length $L_a^b$

$P(b)$

Chord length
$|\vec{P}(b) - \vec{P}(a)|$

$P(a)$

**FIGURE 7**
Arc length versus chord length.

arc over chord (Fig. 7) as both tend to zero:

$$\lim_{b \to a} \frac{\text{arc length}}{\text{chord length}} = \lim_{b \to a} \frac{L_a^b}{|\vec{P}(b) - \vec{P}(a)|}.$$

(Divide numerator and denominator by $b - a$.)

10. Let $x = t$, $y = t \sin(\pi/2t)$.
   a) Sketch the curve, showing points for $t = 1$, $\frac{1}{2}, \frac{1}{3}, \frac{1}{4}, \frac{1}{5}, \frac{1}{6}$.
   b) Show that the length of the straight-line segments joining these points is greater than $1 + \frac{2}{3} + \frac{2}{5}$. (Use: The hypotenuse of a right triangle is longer than either leg.)
   c) Explain why the length of the curve for $0 < t \le 1$ is infinite.

## 13.5

# PLANETARY MOTION

The burning question of seventeenth-century science was: What holds the moon and planets in their periodic orbits? Over many years, Tycho Brahe had recorded the positions of the planets as seen from this moving and rotating platform, the earth. Tycho's assistant Johann Kepler, after many more years of analysis and some inspired conjectures, distilled from these raw data many "laws" describing the motion of the planets. In 1609 he published these two:

(I)   *Each planet moves in an ellipse, with the sun at one focus.*

(II)  *A given planet sweeps out equal areas in equal times,* as shown in Figure 1. Another way to state the law is this: There is a constant of proportionality $h$ such that (see Fig. 1) if the time of passage from $P$ to $Q$ is $\Delta t$, the area swept out is

$$\Delta A = h \, \Delta t.$$

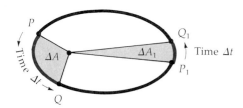

**FIGURE 1**
Kepler's second law: If time of passage $\Delta t$ from $P_1$ to $Q_1$ equals time of passage from $P$ to $Q$, than area $\Delta A_1$ = area $\Delta A$.

A third law (published in 1619) relates the period of motion, $T$, to the major axis of the ellipse, $2a$.

(III) *For all the planets, the ratio $T^2/(2a)^3$ is the same.*

These laws described the planetary motion, but not the forces which compel it. Aristotle had imagined crystal spheres carrying the planets; by 1600, this was a charming but untenable idea. Galileo's discovery that falling bodies appear to travel in parabolic arcs, as if drawn to the earth, made it natural to suppose that some mysterious attractive force likewise draws the planets to the sun. By 1645 it had been conjectured that this was an *inverse square* force, varying as the square of the reciprocal of the distance from planet to sun. Finally (1684), urged on by Hooke (known for his law of elasticity) and Halley (for whom the comet was named), Newton showed how an inverse square force drawing the planets to the sun implies Kepler's laws, and conversely, those laws imply the inverse square force! This fantastic achievement earned for Newton some of the awe once reserved for the heavens and their Creator.

Now, using the methods developed by Newton, Leibniz, and their successors, the relation between Kepler's description of planetary motion and the inverse square force can be explained in a few pages. We will show that Kepler's laws imply the inverse square attraction; the converse is left to the problems.

In the plane of motion, introduce polar coordinates with the sun at the origin. With each point in the plane, we associate two *unit vectors* appropriate to polar coordinates (Fig. 2). At the point $P = (r, \theta)_p$ with $r > 0$, the unit vector

$$\vec{u}_r = \langle \cos \theta, \sin \theta \rangle \tag{1}$$

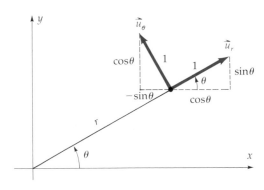

**FIGURE 2**
Unit vectors $\vec{u}_r$ and $\vec{u}_\theta$.

points directly away from the origin, and

$$\vec{u}_\theta = \langle -\sin \theta, \cos \theta \rangle \tag{2}$$

is perpendicular to $\vec{u}_r$, pointing in the direction of increasing $\theta$. The position vector from the origin to $P$ is

$$\vec{P} = \langle x, y \rangle = \langle r \cos \theta, r \sin \theta \rangle = r\vec{u}_r. \tag{3}$$

If $P$ moves, then $r$ and $\theta$ are functions of time $t$. The velocity of the motion is the derivative of the position vector (3):

$$\vec{v} = \frac{d}{dt}\, r\vec{u}_r = \frac{dr}{dt}\, \vec{u}_r + r\, \frac{d}{dt}\, \vec{u}_r.$$

Now

$$\frac{d}{dt}\, \vec{u}_r = \frac{d}{dt}\, \langle \cos \theta, \sin \theta \rangle = \left\langle -\sin \theta\, \frac{d\theta}{dt},\ \cos \theta\, \frac{d\theta}{dt} \right\rangle = \frac{d\theta}{dt}\, \vec{u}_\theta \qquad (4)$$

by the definition of $\vec{u}_\theta$ in (2). So the velocity is

$$\vec{v} = \frac{dr}{dt}\, \vec{u}_r + r\, \frac{d\theta}{dt}\, \vec{u}_\theta. \qquad (5)$$

To get the acceleration, differentiate once more. Just as in (4), you can check that $\dfrac{d}{dt}\, \vec{u}_\theta = -\vec{u}_r\, \dfrac{d\theta}{dt}$, so the product rule applied to (5) gives

$$\vec{a} = \frac{d\vec{v}}{dt} = \left[ \frac{d^2r}{dt^2} - r\left(\frac{d\theta}{dt}\right)^2 \right] \vec{u}_r + \left[ 2\, \frac{dr}{dt} \cdot \frac{d\theta}{dt} + r\, \frac{d^2\theta}{dt^2} \right] \vec{u}_\theta. \qquad (6)$$

This expression for the acceleration is valid for any motion; now we use it to derive the special properties implied by Kepler's laws.

---

**THEOREM 1**

Kepler's law II ("equal areas in equal times") implies that the force acting on each planet is always directed along the line from sun to planet.

---

**PROOF**   Let $r$ and $\theta$ be the polar coordinates of the planet at time $t$, with the sun at the origin. Denote by $A(\theta_1)$ the area swept out by the planet from $\theta = 0$ to $\theta = \theta_1$ [Fig. 2]. The area swept out is, by Section 12.7,

$$A(\theta_1) = \int_0^{\theta_1} \frac{1}{2}\, r^2 \, d\theta.$$

By the Second Fundamental Theorem of Calculus, $\dfrac{dA(\theta_1)}{d\theta_1} = \dfrac{1}{2}\, r^2$. We can drop the subscript on $\theta$ and write $\dfrac{dA(\theta)}{d\theta} = \dfrac{1}{2}\, r^2$; so

$$\frac{dA(\theta)}{dt} = \frac{dA(\theta)}{d\theta} \cdot \frac{d\theta}{dt} = \frac{1}{2}\, r^2\, \frac{d\theta}{dt}. \qquad (7)$$

On the other hand, by Kepler's second law, $\Delta A = h\,\Delta t$, so

$$\frac{dA}{dt} = \lim_{\Delta t \to 0}\frac{\Delta A}{\Delta t} = \lim_{\Delta t \to 0} h = h.$$

Together with (7), this gives

$$\frac{1}{2}r^2\frac{d\theta}{dt} = h. \tag{8}$$

Since $h$ is constant, the derivative of this expression is 0:

$$r\frac{dr}{dt}\frac{d\theta}{dt} + \frac{1}{2}r^2\frac{d^2\theta}{dt^2} = 0. \tag{9}$$

Now look back at formula (6) for the acceleration; the factor multiplying $\vec{u}_\theta$ is just $2/r$ times the expression in (9), so *that factor is zero*. Hence the force is

$$\vec{F} = m\vec{a} = m\left[\frac{d^2r}{dt^2} - r\left(\frac{d\theta}{dt}\right)^2\right]\vec{u}_r. \tag{10}$$

This is a scalar multiple of the vector $\vec{u}_r$ pointing from sun to planet; in other words, it is a central force, as was to be proved. The factor $\dfrac{d^2r}{dt^2} - r\left(\dfrac{d\theta}{dt}\right)^2$ in (10) is the central acceleration.

This proof can be reversed to show that, for any central force, area is swept out at a constant rate.

Now we show that elliptical orbits, together with the equal areas law, imply a central force *varying like* $mk/r^2$.

---

**THEOREM 2**

Kepler's laws I and II together imply that each planet is attracted to the sun by a central force of magnitude $mk/r^2$, where $m$ is the mass of the planet, and $k$ is a constant.

---

**PROOF** Since we assume law II, all the formulas in the proof of Theorem 1 are valid; in particular

$$\vec{F} = m\left[\frac{d^2r}{dt^2} - r\left(\frac{d\theta}{dt}\right)^2\right]\vec{u}_r. \tag{10}$$

We must prove that the factor multiplying $\vec{u}_r$ has the form $-mk/r^2$ for a positive constant $k$.

By Kepler's first law, the orbit is an ellipse with the sun at one focus, so, with suitable axes (Sec. 12.6)

$$r = \frac{ed}{1 - e\cos\theta}. \tag{11}$$

The formula for $1/r$ is simpler than for $r$, so we introduce a new variable

$$w = \frac{1}{r} = \frac{1}{ed} - \frac{1}{d} \cos \theta. \tag{12}$$

We rewrite the central acceleration $\dfrac{d^2r}{dt^2} - r\left(\dfrac{d\theta}{dt}\right)^2$ in terms of $w$, using several times formula (8), repeated here for convenience:

$$r^2 \frac{d\theta}{dt} = 2h. \tag{13}$$

The derivative of $r$ is

$$\frac{dr}{dt} = \frac{dw^{-1}}{dt} = -w^{-2} \frac{dw}{dt} = -r^2 \frac{dw}{d\theta} \frac{d\theta}{dt} = -2h \frac{dw}{d\theta} = \frac{-2h}{d} \sin \theta \tag{14}$$

from (12). The second derivative is

$$\frac{d^2r}{dt^2} = \frac{-2h}{d} \cos \theta \frac{d\theta}{dt}.$$

Use this with (13) to compute the sun's attractive force (10) as

$$m\left[\frac{d^2r}{dt^2} - r\left(\frac{d\theta}{dt}\right)^2\right]\vec{u}_r = m\left[\frac{-2h}{d} \cos \theta \frac{d\theta}{dt} - r\left(\frac{d\theta}{dt}\right)^2\right]\vec{u}_r$$

$$= m\left(\frac{-4h^2}{r^2}\right)\left(\frac{\cos \theta}{d} + \frac{1}{r}\right)\vec{u}_r.$$

But from (12), $\dfrac{\cos \theta}{d} + \dfrac{1}{r} = \dfrac{1}{ed}$; so the magnitude of the central force is

$$\left| m \frac{-4h^2}{r^2} \frac{1}{ed} \vec{u}_r \right| = m \frac{4h^2}{ed} r^{-2}. \tag{15}$$

This is a constant multiple of the inverse square $r^{-2}$, as was to be proved.

The first two laws concern each planet individually; the third shows the relation between the forces exerted by the sun on different planets.

---

**THEOREM 3**

Assuming all three of Kepler's laws, each planet experiences a central force $-mk/r^2$, where $m$ is the mass of the planet, and the constant $k$ is the *same* for each planet.

**PROOF** We must show that the constant $k = 4h^2/ed$ in the central force (15) is the same for each planet; here

$$h = \frac{dA}{dt} = \text{rate of sweeping out area}$$

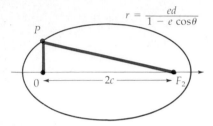

**FIGURE 3**
$|OP| = ed$, $|OP| + |PF_2| = 2a$.

and $ed$ is the numerator in the equation (11) for the elliptical orbit. With the labels in Figure 3, we will show that

$$h = \pi ab/T \qquad (16)$$

and

$$ed = \frac{b^2}{a}. \qquad (17)$$

Thus the constant in (15) is

$$k = \frac{4h^2}{ed} = 4\pi^2 \frac{a^3}{T^2},$$

which can be rewritten as $\dfrac{\pi^2/2}{T^2/(2a)^3}$. By Kepler's third law, $T^2/(2a)^3$ is the same for each planet; hence $k$ is the same for each planet, as was to be proved.

**PROOF OF (16)** Since $h = dA/dt$ and $T = $ period of orbit $=$ time to sweep out the full ellipse, then

$$hT = \text{area of ellipse} = \pi ab \qquad \text{(Sec. 12.2, problems)}$$

and (16) follows.

**PROOF OF (17)** In equation (11) for the ellipse, set $\theta = \pi/2$ to show that Figure 3, $|OP| = ed$. Further, by the fundamental properties of the ellipse (Sec. 12.2)

$$ed + |PF_2| = 2a$$

and

$$b^2 + c^2 = a^2.$$

The first equation gives $ed + \sqrt{(ed)^2 + 4c^2} = 2a$; transpose $ed$, square, and simplify, to find

$$ed = \frac{a^2 - c^2}{a} = \frac{b^2}{a}$$

and (17) follows.

From Kepler's three laws, we have deduced that the force on each planet is toward the sun, and its magnitude is given by the same formula $mkr^{-2}$. The converse, that this force implies Kepler's laws, is outlined in the problems.

## PROBLEMS

### A

1. Suppose that the force acting on an object is central; thus the coefficient of $\vec{u}_\theta$ in (6) is zero. Deduce that the rate $dA/dt$ at which area is swept out is a constant.

2. Suppose that the force attracting an object is central, and inverse square; so $\dfrac{d^2r}{dt^2} - r\left(\dfrac{d\theta}{dt}\right)^2 = -kr^{-2}$ with $k$ constant; and by problem 1, $\dfrac{dA}{dt} = h$ (a constant), or $r^2 \dfrac{d\theta}{dt} = 2h$. Let $w = r^{-1}$. Show that

   a) $\dfrac{d^2w}{d\theta^2} + w = \dfrac{k}{4h^2}$.

   b) $\dfrac{d^2(w - k/4h^2)}{d\theta^2} + \left(w - \dfrac{k}{4h^2}\right) = 0$.

   c) $w - \dfrac{k}{4h^2} = C\cos(\theta - \varphi)$ for constants $C$ and $\varphi$ (see Sec. 8.9).

   d) $r = \dfrac{4h^2}{k + 4h^2 C\cos(\theta - \varphi)}$, the equation of an ellipse if $4h^2 C < k$ (Sec. 12.6).

3. Newton's law of universal gravitation states that the force between particles of masses $m$ and $M$, at distance $r$, is $\dfrac{mM\gamma}{r^2}$, where $\gamma$ is a "universal constant," the same for all particles in the universe.

This implies that the central acceleration of a planet is $-\dfrac{M\gamma}{r^2}\,\vec{u}_r$, where $M$ is the mass of the sun. This is the same for all planets. Deduce Kepler's Third Law. (We are neglecting the attraction of one planet by another, which is very slight compared to that by the sun; however, in a more refined analysis, this too must be taken into account.)

*4. Passive earth satellites are attracted by the earth just as planets are attracted by the sun, so their orbits are periodic and elliptic, with the earth at one focus. Suppose that a satellite is moving in an elliptical orbit about the earth, with distance $r_{\min}$ when it is at "perigee" (Fig. 4) and $r_{\max}$ when at "apogee." The satellite is to be raised to a circular orbit of radius $r_{\max}$, by suddenly increasing its speed when at a certain point in its orbit. Should this be at perigee, or apogee, or somewhere else? (No calculation is required—use just the qualitative facts, that the orbit is periodic and elliptic.)

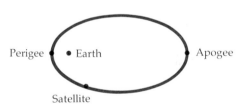

**FIGURE 4**

## REVIEW PROBLEMS   CHAPTER 13

### A

*1. Sketch $\vec{v} = \langle 3, -1 \rangle$ as an arrow beginning at $(1, 2)$; $\vec{w} = -2\vec{i} + 3\vec{j}$ as an arrow beginning at the head of your $\vec{v}$ arrow; and $\vec{v} + \vec{w}$ as an arrow beginning at $(1, 2)$.

*2. Given $P_1 = (3, 5)$, $P_2 = (-1, 2)$, and $P_3 = (0, 1)$
   a) Write parametric equations of the line through $P_1$ and $P_2$.

   b) Find the point $4/5$ of the way from $P_1$ to $P_2$.

   c) Find the point halfway between $P_1$ and the midpoint of the segment from $P_2$ to $P_3$.

3. Prove algebraically.
   *a) $\vec{v} + \vec{w} = \vec{w} + \vec{v}$
   b) $c(\vec{v} + \vec{w}) = c\vec{v} + c\vec{w}$

**4.** An object moving in the plane has position $\vec{P}(t) = \cos 3t\vec{i} + 2 \sin 3t\vec{j}$.

**\*a)** Find the velocity, acceleration, and speed.

**b)** Sketch the position, velocity, and acceleration at time $t = \pi/12$.

**\*c)** At time $t = \pi/12$, a small piece breaks loose from the object, and continues moving for $t \geq \pi/12$ with constant velocity $\vec{v}(\pi/12)$. Where is it at time $t = 1$?

**d)** Show that the object moves on an ellipse.

**\*e)** Write an integral giving the arc length of the ellipse traced out.

**f)** Show that the acceleration is always directed toward the origin and is proportional to the displacement from the origin. (This is a two-dimensional harmonic oscillator.)

**\*5.** An object moving in the plane has acceleration $\vec{a}(t) = \cos 2t\vec{i} + \sin 2t\vec{j}$. At time $t = 0$, it is at rest at the origin. Find its position at time $t$.

**\*6.** The graph of $r = 1/(\theta^2 - 1)$ is a sort of spiral; one turn is formed with $\pi \leq \theta \leq 3\pi$. Compute the arc length of this turn.

**B**

**7. \*a)** Show that the surface area generated by revolving the polar graph of $r = f(\theta)$, $a \leq \theta \leq b$, about the $y$-axis is

$$\int_a^b 2\pi f(\theta)\cos\theta \sqrt{f(\theta)^2 + f'(\theta)^2}\, d\theta.$$

(For surface area, see Sec. 10.2)

**\*b)** Compute the area formed by revolving the polar graph of $r = a \cos\theta$, $|\theta| \leq \pi/2$, about the $y$-axis.

**c)** Derive an integral for the surface area generated by revolving the graph of $r = f(\theta)$, $a \leq \theta \leq b$, about the $x$-axis.

**d)** Check your integral by using it to compute the area formed by revolving the polar graph of $r = a \cos\theta$, $|\theta| \leq \pi/2$, about the $x$-axis. (The surface is familiar, and its area is known.)

**8.** Derive an integral for the area generated by revolving the trace of $x = f(t)$, $y = g(t)$ about

**a)** The $x$-axis.     **b)** The $y$-axis.

**9.** Figure 1 shows a pendulum rod of length $r$, hinged at the origin. (It is an "idealized" rod, completely rigid, and with negligible mass.) The

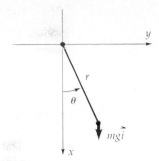

**FIGURE 1**
Mass $m$ on pendulum rod.

$x$-axis points down and the $y$-axis to the right. The force of gravity on a mass $m$ at the end of the pendulum is $mg\vec{i}$. The physical law of conservation of energy declares that the *kinetic energy* $\frac{1}{2}m|\vec{v}|^2$ plus the *potential energy* $-mgx$ is a constant $C$:

$$\frac{1}{2}m|\vec{v}|^2 - mgx = C. \tag{1}$$

The mass moves on the circle of radius $r$, so

$$x = r\cos\theta \quad \text{and} \quad y = r\sin\theta; \tag{2}$$

hence the motion can be described by giving $\theta$ as a function of time $t$.

**a)** Use (1) and (2) to show that $\frac{1}{2}mr^2\left(\dfrac{d\theta}{dt}\right)^2 - mgr\cos\theta = C$.

**b)** Deduce the *pendulum equation* $d^2\theta/dt^2 + (g/r)\sin\theta = 0$.

**c)** For small oscillations, the angle $\theta$ is small, so $\sin\theta \approx \theta$, and the pendulum equation is nearly $d^2\theta/dt^2 + (g/r)\theta = 0$. Determine the period of the solutions of this equation (see Sec. 8.9).

**d)** Suppose that the pendulum swings to a maximum amplitude of $\theta = \alpha$. Show that then in part a, $C = -mgr\cos\alpha$, and

$$\left(\frac{d\theta}{dt}\right)^2 = \frac{2g}{r}(\cos\theta - \cos\alpha).$$

(What is $d\theta/dt$ when the amplitude is maximum?)

**e)** Deduce that the actual period of the pendulum with amplitude $\alpha$ is given by the integral $T(\alpha) = 4\int_0^\alpha \sqrt{r/2g}\,(\cos\theta - \cos\alpha)^{-1/2}\, d\theta$.

**f)** Rewrite the period as $T(\alpha) =$

$$4\sqrt{r/g} \int_0^{\pi/2} \left[ 1 - \left( \sin \frac{\alpha}{2} \sin \varphi \right)^2 \right]^{-1/2} d\varphi,$$

with the substitution $\sin \varphi = \dfrac{\sin(\theta/2)}{\sin(\alpha/2)}$.

**g)** Compare $T(0)$ with the period found in part c.

**h)** Does $T(\alpha)$ increase with $\alpha$, or decrease?

**10.** **a)** For the pendulum of the previous problem, use a linear approximation of $(1 + x)^{-1/2}$ to obtain $T(\alpha) \approx \sqrt{r/g} \left[ 2\pi + \dfrac{\pi}{2} \left( \sin \dfrac{\alpha}{2} \right)^2 \right]$, for small amplitude $\alpha$.

**b)** By a further linear approximation, obtain $T(\alpha) \approx 2\pi \sqrt{r/g} \, (1 + \alpha^2/16)$.

# 14

# SPACE COORDINATES
# AND VECTORS

Real life takes place in three-dimensional space, not a plane, so we extend the concepts of analytic geometry and vectors to space.

## 14.1

### SPACE COORDINATES

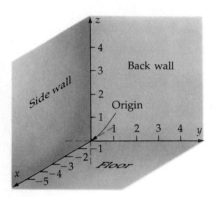

**FIGURE 1**
Coordinate axes.

To describe points in space we use three coordinate axes, like the three lines formed by the intersections of the floor and walls in the corner of a room (Fig. 1). Think of the $y$- and $z$-axes in the plane of the paper, with the $x$-axis projecting out at you. The three axes are mutually perpendicular, intersecting at a point O called the **origin**. On each axis there is a scale of real numbers, with 0 at the origin, and running in the direction of the arrows in Figure 1— on the $z$-axis, numbers increase as you go up; on the $y$-axis, they increase from left to right; and on the $x$-axis, from back to front.

Given the axes, any ordered triple of numbers $(x, y, z)$ determines a point in space, as shown in Figure 2. Starting at the origin, move to position $x$ on the $x$-axis; then move $|y|$ units parallel to the $y$-axis, to the right if $y > 0$, to the left if $y < 0$; and finally move $|z|$ units parallel to the $z$-axis, up if $z \geq 0$, down if $z < 0$. Figure 3 shows a few such points. Comparing this to Figure 1, point $(1, 2, 3)$ is "in the room," and point $(-3, 1, -2)$ is "behind the back wall and below the floor."

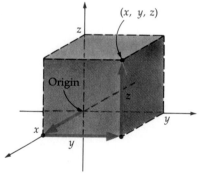

**FIGURE 2**
Going from the origin to point $(x, y, z)$.

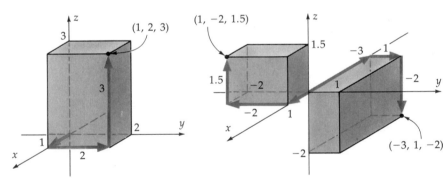

**FIGURE 3**

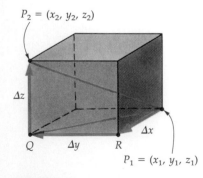

**FIGURE 4**
$|P_1P_2|^2 = (\Delta x)^2 + (\Delta y)^2 + (\Delta z)^2$, Pythagorean Theorem in three dimensions.

The distance $|P_1P_2|$ between two points $P_1 = (x_1, y_1, z_1)$ and $P_2 = (x_2, y_2, z_2)$ is the hypotenuse of the right triangle $P_1QP_2$ in Figure 4, so

$$|P_1P_2|^2 = |P_1Q|^2 + |QP_2|^2. \tag{1}$$

But $|QP_2|^2 = |\Delta z|^2 = (z_2 - z_1)^2$; and $|P_1Q|^2$ is the hypotenuse of the right triangle $P_1RQ$, so

$$|P_1Q|^2 = |\Delta x|^2 + |\Delta y|^2.$$

Use these expressions in (1):

$$|P_1P_2|^2 = |\Delta x|^2 + |\Delta y|^2 + |\Delta z|^2,$$

so the distance is

$$|P_1P_2| = \sqrt{|\Delta x|^2 + |\Delta y|^2 + |\Delta z|^2}$$
$$= \sqrt{(x_2 - x_1)^2 + (y_2 - y_1)^2 + (z_2 - z_1)^2}. \qquad (2)$$

For example, the distance between the points $(1, -2, 1.5)$ and $(-3, 1, -2)$ in Figure 3 is

$$\sqrt{(-3 - 1)^2 + (1 - (-2))^2 + (-2 - 1.5)^2} = \sqrt{37.25}.$$

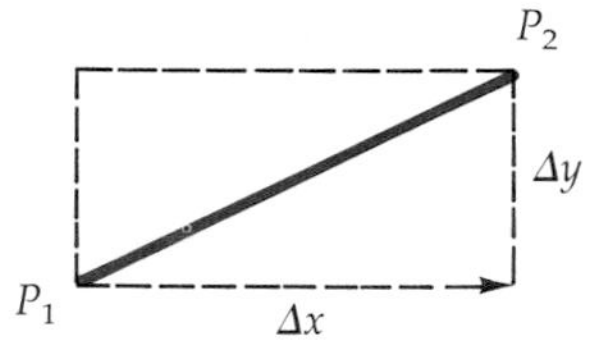

The distance formula (2) is a three-dimensional Pythagorean Theorem. The two-dimensional version gives for a rectangle (Fig. 5)

$$(\text{diagonal})^2 = (\text{length})^2 + (\text{width})^2.$$

The three-dimensional version gives, for the box in Figure 4,

$$(\text{diagonal})^2 = (\text{length})^2 + (\text{width})^2 + (\text{height})^2.$$

**The sphere** with given center $C$ and radius $r$ is the set of all points $P$ at distance $r$ from $C$, the points $P$ satisfying $|PC| = r$, or

$$|PC|^2 = r^2.$$

Denote the coordinates of the center $C$ by $(a, b, c)$ and those of $P$ by $(x, y, z)$. Then the equation for the sphere is written

$$(x - a)^2 + (y - b)^2 + (z - c)^2 = r^2, \qquad (3)$$

and the sphere is the *graph* of this equation.

**EXAMPLE 1**   Describe the graph of $x^2 - 2x + y^2 + 4y + z^2 = 0$.

**SOLUTION**   Complete the square on $x$, and on $y$, to write the given equation in the form (3):

$$[(x - 1)^2 - 1] + [(y + 2)^2 - 4] + z^2 = 0$$

or

$$(x - 1)^2 + (y + 2)^2 + z^2 = 5.$$

This is the equation of the sphere with center $(1, -2, 0)$ and radius $\sqrt{5}$.

## Special Planes

The plane containing the $y$-axis and $z$-axis is called the $yz$-plane; this contains the "back wall" in Figure 1. It consists of all points $(0, y, z)$; the $y$- and $z$-coordinates can be anything, but the $x$-coordinate must be 0. The plane is therefore the graph of the equation $x = 0$.

Move this plane forward by an amount $x_0$, and you get a new plane parallel to the $yz$-plane (Fig. 6). It consists of all points $(x_0, y, z)$ whose

**FIGURE 5**
Pythagorean Theorem in the plane:
$|P_1P_2|^2 = (\Delta x)^2 + (\Delta y)^2.$

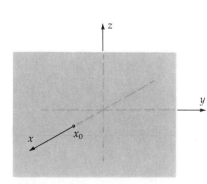

**FIGURE 6**
Plane $\{(x, y, z): x = x_0\}$.

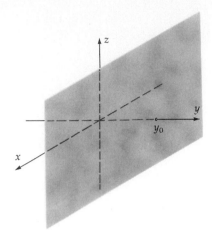

**FIGURE 7**
Plane $\{(x, y, z): y = y_0\}$.

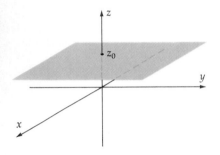

**FIGURE 8**
Plane $\{(x, y, z): z = z_0\}$.

$x$-coordinate equals $x_0$; the $y$-and $z$-coordinates can be anything. This plane is the graph of the equation $x = x_0$; it is perpendicular to the $x$-axis at the point $(x_0, 0, 0)$. Similarly, the graph of the equation $y = y_0$ is a plane (Fig. 7) parallel to the $xz$-plane (the "side wall" in Fig. 1); the graph of $z = z_0$ is a plane (Fig. 8) parallel to the $xy$-plane (the "floor").

Other planes, not parallel to the coordinate planes, are taken up in Section 14.4.

*NOTE* The graph of an equation such as $x = 2$ depends on the context. On the line, $x = 2$ defines just one point. In the plane, $x = 2$ defines a vertical line, the set $\{(x, y): x = 2\}$. In space, $x = 2$ defines a plane perpendicular to the $x$-axis, the set $\{(x, y, z): x = 2\}$.

## SUMMARY

*Distance Formula (see Fig. 4):*

$$|P_1P_2|^2 = |\Delta x|^2 + |\Delta y|^2 + |\Delta z|^2.$$

*Equation of Sphere with Center (a, b, c) and Radius r:*

$$(x - a)^2 + (y - b)^2 + (z - c)^2 = r^2.$$

## PROBLEMS

A

1. Plot the pair of points, and compute the distance between them.
   a) $(0, 0, 0)$ and $(1, 2, 3)$

   *b) $(2, -1, 3)$ and $(2, 1, -3)$
   c) $(-1, 2, -2)$ and $(0, 0, 3)$
   d) $(-1, 2, -2)$ and $(1, -2, 2)$

2. A cube of edge 3 lies in the set where $x \geq 0$, $y \geq 0$, $z \geq 0$ and has one corner at the origin.

Sketch the cube, and label all the corners with the appropriate coordinates.

**\*3.** A cottage is constructed as in Figure 9. Give the coordinates of points $A$ through $F$. (Answers given for points $A$ and $E$.)

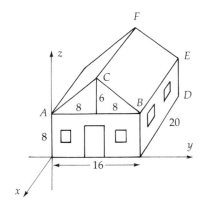

**FIGURE 9**
Coordinate cottage.

**4.** Given are the vertices of a triangle. Determine whether it is isosceles, equilateral, or right. In the latter case, determine which vertex has the right angle, and compute the area.
   **a)** $(7, 1, 3)$, $(4, 2, 5)$, and $(5, 4, 2)$
   **\*b)** $(-2, 2, 1)$, $(-1, 3, 3)$, and $(0, 4, 2)$
   **c)** $(0, 0, 0)$, $(2, 2, 1)$, $(3, 4, 3)$

**5.** Write the equation of the sphere with given center $C$ and given radius $r$.
   **\*a)** $C = (0, 0, 0)$, $r = 1$
   **b)** $C = (0, -1, 2)$, $r = 2$

**6.** Find the center and radius of the given sphere. Which spheres pass through the origin?
   **a)** $x^2 + y^2 + z^2 + 4x - 6y = 0$
   **b)** $x^2 + y^2 + z^2 - 2x + 3y + z - 1 = 0$
   **\*c)** $2x^2 - 4x + 2y^2 + 8y + 2z^2 - 12z + 28 = 0$

**7.** Describe the graph of the given equation in $xyz$-space.
   **a)** $x = 1$   **b)** $y = 3$   **\*c)** $z = -2$
   **\*d)** $xy = 0$   **e)** $xyz = 0$

**8.** Let $P_0 = (2, -1, 3)$. Write the equation of the plane containing $P_0$ which is parallel to
   **a)** The $xy$-plane.   **b)** The $yz$-plane.
   **\*c)** The $xz$-plane.

**9.** Describe the graph of the equation $y = 2$
   **a)** In $xyz$-space.   **b)** In the $xy$-plane.
   **c)** On the $y$-axis.

**10.** Write the equation of a sphere with the given properties.
   **a)** Center $(1, 0, 0)$ and tangent to the $yz$-plane.
   **b)** Opposite ends of a diameter at $(1, 3, 5)$ and $(-1, -3, -5)$.
   **\*c)** Opposite ends of a diameter at $(1, 3, 5)$ and $(2, 4, 1)$.

**11.** Describe the region in $xyz$-space determined by the given inequalities. [If several inequalities are given, the region consists of those points $(x, y, z)$ which satisfy all of them at once.]
   **\*a)** $x > 0$
   **b)** $x > 0$, $y > 0$, and $z > 0$
   **\*c)** $x^2 + y^2 + z^2 \leq 1$
   **d)** $x^2 + y^2 \leq 1$
   **e)** $xyz > 0$
   **f)** $|x| < 1$, $|y| < 1$, and $|z| < 1$
   **\*g)** $1 < x^2 + y^2 + z^2 < 4$
   **h)** $x^2 + y^2 + z^2 \leq 4$ and $x^2 + y^2 \geq 1$

**12.** Describe the given set of points.
   **a)** $\{x : x > 0\}$
   **b)** $\{(x, y) : x > 0\}$
   **c)** $\{(x, y, z) : x > 0\}$
   **d)** $\{(x, y) : xy \geq 0\}$
   **e)** $\{(x, y) : |x| < 1\}$
   **\*f)** $\{(x, y) : |x| < 1 \text{ and } |y| < 1\}$
   **g)** $\{(x, y, z) : |x| < 2, |y| < 2, \text{ and } |z| < 2\}$
   **h)** $\{(x, y, z) : z = \sqrt{1 - x^2 - y^2}\}$

**B**

**13.** Given $A = (1, 0, 0)$ and $B = (2, 3, 5)$, write and simplify the equation for the set of points $P = (x, y, z)$ such that $|PA| = |PB|$. What sort of geometric object is it?

**C**

**14.** Given two distinct points $P_1 = (x_1, y_1, z_1)$ and $P_2 = (x_2, y_2, z_2)$, show that the set of points $P$ such that $|PP_1| = |PP_2|$ has an equation of the form $ax + by + cz = d$, where $a$, $b$, and $c$, are not all 0. Deduce that every plane has such an equation.

## 14.2

### THE VECTOR SPACE $R^3$. THE DOT PRODUCT

The two-dimensional vector space $R^2$ has a three-dimensional analog $R^3$, consisting of all ordered *triples* of real numbers $\vec{v} = \langle v_1, v_2, v_3 \rangle$, $\vec{w} = \langle w_1, w_2, w_3 \rangle$, etc. In $R^3$, the sum, difference, scalar multiple, and magnitude are defined componentwise, just as in $R^2$:

$$\vec{v} + \vec{w} = \langle v_1 + w_1, v_2 + w_2, v_3 + w_3 \rangle$$
$$\vec{v} - \vec{w} = \langle v_1 - w_1, v_2 - w_2, v_3 - w_3 \rangle$$
$$c\vec{v} = \langle cv_1, cv_2, cv_3 \rangle$$
$$|\vec{v}| = \sqrt{v_1^2 + v_2^2 + v_3^2}.$$

The laws of vector algebra established in Section 13.1 remain valid for $R^3$. It is these operations and laws that distinguish the **vector space** $R^3$ from the space considered in analytic geometry, consisting of points $(x, y, z)$. The points just sit there, while the vectors can be added, subtracted, and multiplied by scalars.

Vectors in $R^3$ are represented as arrows in space, beginning at any appropriate point (Fig. 1). The length of the arrow is the magnitude of the vector.

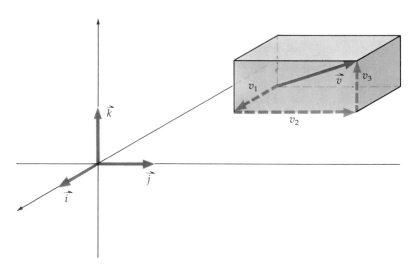

**FIGURE 1**
Vector $\vec{v} = \langle v_1, v_2, v_3 \rangle$.

If $|\vec{v}| = 1$, then $\vec{v}$ is called a **unit vector**. Three basic unit vectors are (Fig. 1)

$$\vec{i} = \langle 1, 0, 0 \rangle, \qquad \vec{j} = \langle 0, 1, 0 \rangle \qquad \vec{k} = \langle 0, 0, 1 \rangle.$$

Any vector in $R^3$ can be written as a combination of these basic three:

$$\vec{v} = \langle v_1, v_2, v_3 \rangle = v_1 \langle 1, 0, 0 \rangle + v_2 \langle 0, 1, 0 \rangle + v_3 \langle 0, 0, 1 \rangle$$
$$= v_1 \vec{i} + v_2 \vec{j} + v_3 \vec{k}.$$

Two vectors $\vec{v}$ and $\vec{w}$ are called **parallel** if one is a scalar multiple of the other, that is, if

$$\vec{v} = c\vec{w} \quad \text{or} \quad \vec{w} = c\vec{v}$$

for some constant $c$. Thus $\langle 1, 2, 3 \rangle$ is parallel to $\langle -2, -4, -6 \rangle$ since

$$\langle -2, -4, -6 \rangle = (-2)\langle 1, 2, 3 \rangle.$$

Any arrow representing $\vec{v}$ is parallel to any arrow representing $c\vec{v}$.

We now define a new operation:

**The dot product** of two vectors $\vec{v} = \langle v_1, v_2, v_3 \rangle$ and $\vec{w} = \langle w_1, w_2, w_3 \rangle$ is formed by multiplying corresponding components and adding the results:

$$\vec{v} \cdot \vec{w} = v_1 w_1 + v_2 w_2 + v_3 w_3. \tag{1}$$

The "factors" $\vec{v}$ and $\vec{w}$ are vectors, but their dot product $\vec{v} \cdot \vec{w}$ is a *number*.

In $R^2$, the dot product is similarly defined:

$$\langle v_1, v_2 \rangle \cdot \langle w_1, w_2 \rangle = v_1 w_1 + v_2 w_2.$$

**EXAMPLE 1**

$$\langle 1, -1 \rangle \cdot \langle 2, 4 \rangle = 1 \cdot 2 + (-1) \cdot 4 = -2$$

$$\langle 2, 3, -5 \rangle \cdot \langle 1, 4, 0 \rangle = 2 \cdot 1 + 3 \cdot 4 + (-5) \cdot 0 = 14.$$

In different notation, this last product is

$$(2\vec{i} + 3\vec{j} - 5\vec{k}) \cdot (\vec{i} + 4\vec{j}) = 2 \cdot 1 + 3 \cdot 4 - 5 \cdot 0 = 14.$$

Notice that the basic vectors $\vec{i}$, $\vec{j}$, and $\vec{k}$ appear in the vectors, but *not* in their product.

The dot product obeys familiar laws:

$$\vec{v} \cdot \vec{w} = \vec{w} \cdot \vec{v} \qquad \text{(commutative law)}$$

$$(c\vec{v}) \cdot \vec{w} = c(\vec{v} \cdot \vec{w}) \qquad \text{(associative law)}$$

$$\vec{u} \cdot (\vec{v} + \vec{w}) = (\vec{u} \cdot \vec{v}) + (\vec{u} \cdot \vec{w}) \qquad \text{(distributive law)}.$$

Moreover, the dot product of any vector with itself equals the magnitude squared:

$$\vec{v} \cdot \vec{v} = |\vec{v}|^2.$$

You can easily check each of these; multiply out each side [using the defining relation (1)] and compare the results.

Geometrically, the dot product is expressed in terms of lengths and angles. To see how, apply the law of cosines (Fig. 2) to the triangle

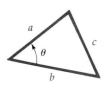

**FIGURE 2**

Law of cosines:
$$c^2 = a^2 + b^2 - 2ab \cos \theta.$$

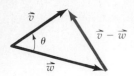

**FIGURE 3**
$|\vec{v} - \vec{w}|^2$
$= |\vec{v}|^2 + |\vec{w}|^2 - 2|\vec{v}|\,|\vec{w}|\cos\theta.$

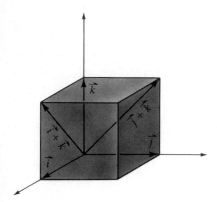

**FIGURE 4**
Diagonals in a cube.

$\vec{v} \cdot \vec{w} > 0$

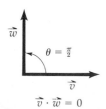

$\vec{v} \cdot \vec{w} = 0$

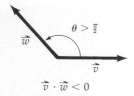

$\vec{v} \cdot \vec{w} < 0$

**FIGURE 5**
Interpreting the sign of $\vec{v} \cdot \vec{w}$.

representing the difference $\vec{v} - \vec{w}$ (Fig. 3):

$$|\vec{v} - \vec{w}|^2 = |\vec{v}|^2 + |\vec{w}|^2 - 2|\vec{v}| \cdot |\vec{w}| \cos\theta$$

or

$$(v_1 - w_1)^2 + (v_2 - w_2)^2 + (v_3 - w_3)^2$$
$$= v_1^2 + v_2^2 + v_3^2 + w_1^2 + w_2^2 + w_3^2 - 2|\vec{v}| \cdot |\vec{w}| \cos\theta.$$

Expand the left-hand side, note that all the squares on the left cancel those on the right, divide by $-2$, and you find

$$v_1 w_1 + v_2 w_2 + v_3 w_3 = |\vec{v}| \cdot |\vec{w}|\cos\theta.$$

The left-hand side is precisely the dot product of $\vec{v}$ and $\vec{w}$; so

$$\vec{v} \cdot \vec{w} = |\vec{v}| \cdot |\vec{w}| \cos\theta \qquad (2)$$

where $\theta$ is the angle between the vectors. This formula gives the dot product its geometric significance. For example, you can compute the *angle* between two nonzero vectors $\vec{v}$ and $\vec{w}$; for (2) implies

$$\cos\theta = \frac{\vec{v} \cdot \vec{w}}{|\vec{v}| \cdot |\vec{w}|}. \qquad (3)$$

---

**EXAMPLE 2** The vectors $\vec{i} + \vec{k}$ and $\vec{j} + \vec{k}$ form the diagonals in two faces of a cube (Fig. 4). Compute the angle between them.

**SOLUTION** The cosine of the angle between the vectors is given by (3):

$$\cos\theta = \frac{(\vec{i} + \vec{k}) \cdot (\vec{j} + \vec{k})}{|\vec{i} + \vec{k}| \cdot |\vec{j} + \vec{k}|} = \frac{1 \cdot 0 + 0 \cdot 1 + 1 \cdot 1}{\sqrt{1^2 + 1^2}\sqrt{1^2 + 1^2}} = \frac{1}{2}.$$

So $\theta = \cos^{-1}(\tfrac{1}{2}) = \pi/3 = 60°$. (This angle can also be found using elementary geometry.)

---

## Orthogonality

Recall that

$$\cos\theta > 0 \qquad \text{if } 0 \le \theta < \pi/2$$

$$\cos\frac{\pi}{2} = 0$$

$$\cos\theta < 0 \qquad \text{if } \pi/2 < \theta \le \pi.$$

So from (2), if neither $\vec{v}$ nor $\vec{w}$ is $\vec{0}$ then (Fig. 5)

$$\vec{v} \cdot \vec{w} > 0 \Leftrightarrow \theta \text{ is acute (less than } \pi/2)$$

$$\vec{v} \cdot \vec{w} = 0 \Leftrightarrow \theta = \frac{\pi}{2}, \text{ and } \vec{v} \text{ is perpendicular to } \vec{w}$$

$$\vec{v} \cdot \vec{w} < 0 \Leftrightarrow \theta \text{ is obtuse (greater than } \pi/2).$$

If one of $\vec{v}$ or $\vec{w}$ is zero, then the angle between the two is not defined. But in this case $\vec{v} \cdot \vec{w} = 0$, and we still say that $\vec{v}$ and $\vec{w}$ are perpendicular, or **orthogonal**. The notation for this is $\vec{v} \perp \vec{w}$. Thus

$$\vec{v} \cdot \vec{w} = 0 \Leftrightarrow \vec{v} \perp \vec{w}.$$

A typical application of the dot product in $R^2$ concerns lines in the plane.

---

**THEOREM**

Let $ax + by + c = 0$ be the equation of any line in the plane. Then:

(i)   The vector $\vec{n} = a\vec{i} + b\vec{j}$ is perpendicular to the line, and
(ii)   the distance from a point $P_0 = (x_0, y_0)$ to the line is

$$\frac{|ax_0 + by_0 + c|}{\sqrt{a^2 + b^2}}. \tag{4}$$

---

**PROOF OF (i)**   Let $P_1 = (x_1, y_1)$ and $P_2 = (x_2, y_2)$ be any two points on the line (Fig. 6). Then they satisfy the equation of the line:

$$ax_1 + by_1 + c = 0$$
$$ax_2 + by_2 + c = 0.$$

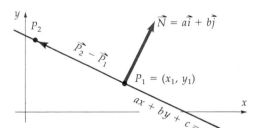

**FIGURE 6**
Interpreting the equation of a line.

Subtract the first equation from the second:

$$a(x_2 - x_1) + b(y_2 - y_1) = 0.$$

The left-hand side is the dot product of the vectors $\vec{n} = \langle a, b \rangle$ and $\vec{P}_2 - \vec{P}_1 = \langle x_2 - x_1, y_2 - y_1 \rangle$ so

$$\vec{n} \cdot (\vec{P}_2 - \vec{P}_1) = 0.$$

That is, $\vec{n}$ is perpendicular to $\vec{P}_2 - \vec{P}_1$. But $\vec{P}_2 - \vec{P}_1$ gives the direction of the line (Fig. 6), so $\vec{n}$ is perpendicular to the line, as claimed. The vector $\vec{n}$ is called a **normal** to the line; for some reason, the word *normal* is used for perpendicular objects.

**PROOF OF (ii), THE DISTANCE FORMULA** Let $P_1$ be any point on the line (Fig. 7), and let $\vec{n}$ be the normal $\langle a, b \rangle$. Suppose that $(x_0, y_0)$ is on the side of the line toward which $\vec{n}$ points. From the right triangle in Figure 7, $\cos \theta = \dfrac{\text{adj}}{\text{hyp}} = \dfrac{D}{|\vec{P_0} - \vec{P_1}|}$, so the distance $D$ to the line is

$$D = |\vec{P_0} - \vec{P}_1| \cos \theta.$$

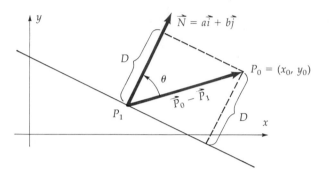

**FIGURE 7**

Distance $D$ from $P_0$ to line: $\cos \theta = \dfrac{\text{adj}}{\text{hyp}} = \dfrac{D}{|\vec{P_0} - \vec{P_1}|}$.

Now use formula (3) for the cosine of the angle between vectors:

$$
\begin{aligned}
D &= |\vec{P_0} - \vec{P}_1| \, \frac{(\vec{P_0} - \vec{P_1}) \cdot \vec{n}}{|\vec{P_0} - \vec{P_1}| \cdot |\vec{n}|} = \frac{(\vec{P_0} - \vec{P_1}) \cdot \vec{n}}{|\vec{n}|} \\
&= \frac{(x_0 - x_1)a + (y_0 - y_1)b}{\sqrt{a^2 + b^2}} \\
&= \frac{ax_0 + by_0 - (ax_1 + by_1)}{\sqrt{a^2 + b^2}}.
\end{aligned}
$$

But $P_1 = (x_1, y_1)$ is on the line, so $ax_1 + by_1 + c = 0$, or $-(ax_1 + by_1) = c$; hence

$$D = \frac{ax_0 + by_0 + c}{\sqrt{a^2 + b^2}}.$$

This proves the distance formula (4) when $P_0$ is on the side of the line toward which $\vec{n}$ points. If $P_0$ is on the other side, then

$$D = -\frac{ax_0 + by_0 + c}{\sqrt{a^2 + b^2}}.$$

In either case, the distance formula (4) is valid.

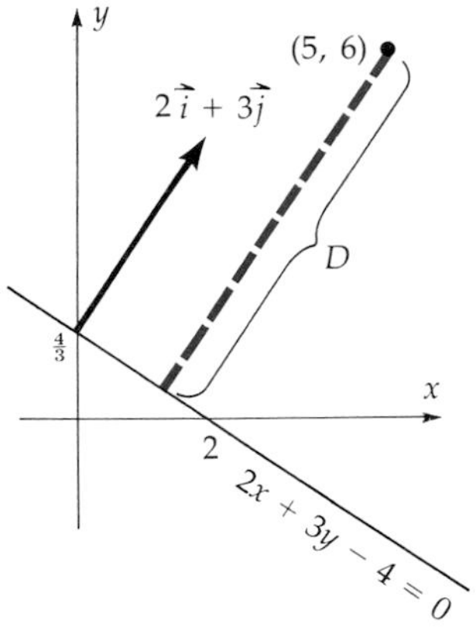


**FIGURE 8**
$D$ = distance from (5, 6) to line.

**EXAMPLE 3** The line $2x + 3y - 4 = 0$ has the normal vector $\vec{n} = \langle 2, 3\rangle$ (Fig. 8). The distance from (5, 6) to the line is

$$D = \frac{2 \cdot 5 + 3 \cdot 6 - 4}{\sqrt{2^2 + 3^2}} = \frac{24}{\sqrt{13}}.$$

## Orthogonal Projection

In geometry, you "drop a perpendicular" from a point $Q$ to a line $L$ (Fig. 9). The corresponding vector concept is the **orthogonal projection** of one vector $\vec{v}$ on another vector $\vec{w}$. Draw $\vec{v}$ and $\vec{w}$ from a common point $P$, and let $Q$ denote the head of the arrow for $\vec{v}$ (Fig. 10). Then the projection of $\vec{v}$ on $\vec{w}$, denoted $\vec{v}_{\vec{w}}$, is represented by the arrow from $P$ to the projection of $Q$ on the line through $\vec{w}$. In Figure 10a, where $0 < \theta < \frac{\pi}{2}$, the *length* of the projection is $|\vec{v}| \cos\theta$, with $\theta$ the angle between $\vec{v}$ and $\vec{w}$. The projection itself is this length times the unit vector in the direction of $\vec{w}$, which is $\frac{1}{|\vec{w}|}\vec{w}$. Thus

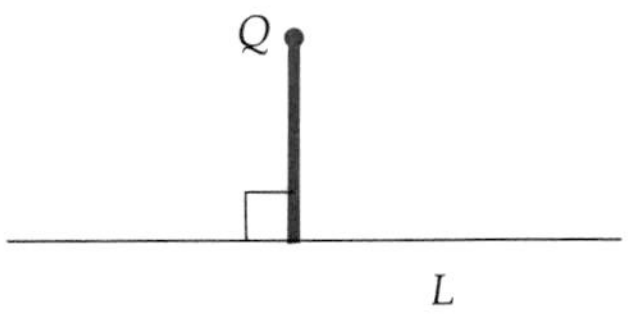


**FIGURE 9**
Dropping a perpendicular from $Q$ to $L$.

$$\vec{v}_{\vec{w}} = (|\vec{v}|\cos\theta)\frac{1}{|\vec{w}|}\vec{w} = |\vec{v}|\frac{\vec{v}\cdot\vec{w}}{|\vec{v}|\cdot|\vec{w}|}\frac{1}{|\vec{w}|}\vec{w}$$

so

$$\vec{v}_{\vec{w}} = \frac{\vec{v}\cdot\vec{w}}{|\vec{w}|^2}\vec{w}.$$

In Figure 10b, where $\theta > \frac{\pi}{2}$, this formula is still correct; there $\vec{v}\cdot\vec{w} < 0$, and the projection $\vec{v}_{\vec{w}}$ has the direction *opposite* to $\vec{w}$.

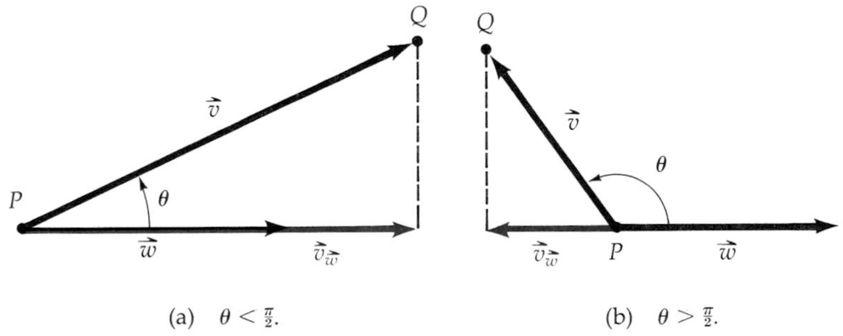


**FIGURE 10**
$\vec{v}_{\vec{w}}$ = Orthogonal projection of $\vec{v}$ on $\vec{w}$.

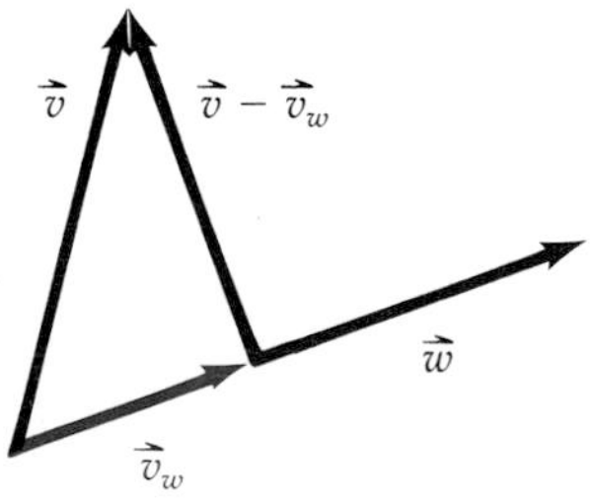


**FIGURE 11**
$\vec{v}_{\vec{w}}$ = projection of $\vec{v}$ on $\vec{w}$
$\vec{v} - \vec{v}_{\vec{w}}$ is perpendicular to $\vec{w}$.

The *magnitude* of the projection is simply $|\vec{v}_{\vec{w}}| = |\vec{v}|\cdot|\cos\theta|$; see Figure 10.

Figure 11 illustrates that the vector $\vec{v}$ is the sum of its projection on $\vec{w}$, and a vector $\vec{v} - \vec{v}_{\vec{w}}$ which is orthogonal to $\vec{w}$. This orthogonality is verified algebraically by computing the dot product:

$$(\vec{v} - \vec{v}_{\vec{w}}) \cdot \vec{w} = \vec{v} \cdot \vec{w} - \vec{v}_{\vec{w}} \cdot \vec{w} = \vec{v} \cdot \vec{w} - \frac{\vec{v} \cdot \vec{w}}{|\vec{w}|^2} \vec{w} \cdot \vec{w}$$
$$= \vec{v} \cdot \vec{w} - \vec{v} \cdot \vec{w} \qquad [\vec{w} \cdot \vec{w} = |\vec{w}|^2]$$
$$= 0.$$

**EXAMPLE 4** A 170 lb cyclist rides a 30 lb bike up a 20° hill (Fig. 12). What force must be exerted to counteract gravity?

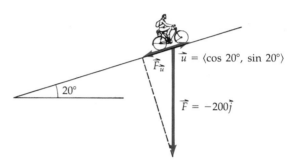

**FIGURE 12**
Uphill exertion $= -\vec{F}_{\vec{u}}$.

*SOLUTION* The force of gravity is 200 lbs, directed downward:

$$\vec{F} = -200\vec{j}.$$

We can express this as the sum of two terms, one perpendicular to the surface of the road, and another parallel to the surface. The force perpendicular to the surface is counteracted by the pressure between the tires and the road; the rest must be counteracted by the exertions of the cyclist. The force of gravity parallel to the road is, from Figure 12,

$$200 \cos 70° = 200 \sin 20° \approx 68.4 \text{ lbs.}$$

This must be counteracted by the cyclist.

We analyze this again, using orthogonal projection. Introduce a unit vector parallel to the road,

$$\vec{u} = \langle \cos 20°, \sin 20° \rangle.$$

Then the component of the force of gravity parallel to the road is the projection of $\vec{F}$ on $\vec{u}$:

$$\vec{F}_{\vec{u}} = \frac{\vec{F} \cdot \vec{u}}{|\vec{u}|^2} \vec{u} = \frac{\langle 0, -200 \rangle \cdot \langle \cos 20°, \sin 20° \rangle}{|\vec{u}|^2} \vec{u}$$
$$= -200 \sin 20° \, \vec{u},$$

since $|\vec{u}| = 1$. The magnitude of this force is

$$|\vec{F}_{\vec{u}}| = 200 \sin 20° \, |\vec{u}| = 200 \sin 20°.$$

Another application of orthogonal projection can be made in the proof of the Theorem above; the distance from the point $P_0$ to the line in Figure 7 is the length of the projection of $\vec{P}_0 - \vec{P}_1$ on the normal vector $\vec{n}$.

## SUMMARY

***Space Vectors (see Fig. 1):***

$$\vec{v} = \langle v_1, v_2, v_3 \rangle = v_1\vec{i} + v_2\vec{j} + v_3\vec{k}$$
$$|\vec{v}| = \sqrt{v_1^2 + v_2^2 + v_3^2}$$

***Dot Product (Fig. 13):***

$$\vec{v} \cdot \vec{w} = v_1w_1 + v_2w_2 + v_3w_3$$
$$= |\vec{v}| \cdot |\vec{w}| \cos \theta$$

$$\vec{v} \cdot \vec{w} = \vec{w} \cdot \vec{v} \qquad \text{(commutative law)}$$

$$(c\vec{v}) \cdot \vec{w} = c(\vec{v} \cdot \vec{w}) \qquad \text{(associative law)}$$

$$\vec{u} \cdot (\vec{v} + \vec{w}) = (\vec{u} \cdot \vec{v}) + (\vec{u} \cdot \vec{w}) \qquad \text{(distributive law).}$$

$$\vec{v} \cdot \vec{v} = |\vec{v}|^2.$$

***Orthogonal Projection of $\vec{v}$ on $\vec{w}$:***

$$\vec{v}_{\vec{w}} = \frac{\vec{v} \cdot \vec{w}}{|\vec{w}|^2} \vec{w}.$$

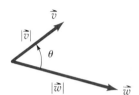

**FIGURE 13**
$\vec{v} \cdot \vec{w} = |\vec{v}| \cdot |\vec{w}| \cos \theta.$

## PROBLEMS

### A

1. Calculate $|\vec{v}|$, $|\vec{w}|$, $\vec{v} \cdot \vec{w}$, and the angle between $\vec{v}$ and $\vec{w}$. Which pairs of vectors form an obtuse angle? Which are orthogonal?
   *a)  $\vec{v} = \langle 1, 1, 1 \rangle$, $\vec{w} = \langle 2, -2, 1 \rangle$
   b)  $\vec{v} = \langle 1/\sqrt{3}, 1/\sqrt{3}, 1/\sqrt{3} \rangle$,
       $\vec{w} = \langle -1/\sqrt{2}, -1/\sqrt{2}, 0 \rangle$
   *c)  $\vec{v} = \vec{i} + \vec{j} + \vec{k}$, $\vec{w} = \vec{k} + \vec{j}$
   d)  $\vec{v} = \langle 0, 1 \rangle$, $\vec{w} = \langle 1, 1 \rangle$
   e)  $\vec{v} = \langle 2, 2, 1 \rangle$, $\vec{w} = \langle 1, -2, 2 \rangle$

2. Given $\vec{a} = \langle 2, 3, 4 \rangle$, $\vec{b} = \langle 1, 2, -2 \rangle$, $\vec{c} = \langle 0, 1, 1 \rangle$, compute
   *i)   $2\vec{a} + \vec{b}$        ii)  $|\vec{a} - 3\vec{b}|$
   *iii)  $(\vec{a} - \vec{b}) \cdot \vec{c}$    iv)  $(\vec{a} - \vec{b}) \cdot (\vec{c} - \vec{b})$

3. a)  Find a vector orthogonal to $\langle 1, 2 \rangle$.

   b)  Find all unit vectors that are orthogonal to $\langle 1, 2 \rangle$.

*4. A triangle has vertices $(0, 2, 2)$, $(1, 7, -5)$, and $(-1, 0, -1)$. Compute the cosine of the interior angle at $(0, 2, 2)$.

5. The "quantity vector" $\vec{q} = \langle 25, 35, 17 \rangle$ gives the numbers of three types of refrigerator stocked by an appliance store. The "price vector" $\vec{p} = \langle 399, 499, 599 \rangle$ gives the prices of those three types. What is the meaning of the dot product $\vec{p} \cdot \vec{q}$?

6. Given $\vec{v} = \langle 1, 1, 1 \rangle$, $\vec{w} = \langle 2, -1, 3 \rangle$, and $\vec{i} = \langle 1, 0, 0 \rangle$, compute the orthogonal projection of:
   a)  $\vec{v}$ on $\vec{i}$      *b)  $\vec{w}$ on $\vec{i}$
   c)  $\vec{v}$ on $\vec{w}$      d)  $\vec{w}$ on $\vec{v}$
   *e)  $\vec{i}$ on $\vec{w}$

7. A boat and trailer weighing 900 lbs are pulled up a 15° ramp. Compute the force $\vec{F}$ required to counteract gravity, and the magnitude $|\vec{F}|$.

The following two problems review concepts from Section 13.1.

8. Given $P = (1, -1, 2)$ and $Q = (2, 3, 1)$, find the point
   a) Midway between $P$ and $Q$.
   *b) Two-thirds of the way from $P$ to $Q$.

9. The barycenter of a triangle with vertices $P, Q,$ and $R$ is the point $B$ defined by

   $$\vec{B} = \tfrac{1}{3}(\vec{P} + \vec{Q} + \vec{R}).$$

   Find the barycenter of the triangle with vertices $(1, 0, 0)$, $(0, 1, 0)$, and $(0, 0, 1)$. Sketch the vertices and the barycenter.

10. For the line $2x - y = 0$:
    a) Find a vector $\vec{n}$ normal to the line. Sketch the line and the normal.
    *b) Compute the distance from the line to the point $(1, 1)$.
    *c) Is $(1, 1)$ on the side of the line toward which $\vec{n}$ points?

11. For the line $x + y = 2$:
    a) Find a normal vector $\vec{n}$. Sketch the line and its normal.
    b) Compute the distance from the origin to the line.

12. The "unit cube" has its faces in the six planes $x = 0$, $x = 1$, $y = 0$, $y = 1$, $z = 0$, and $z = 1$.
    a) Sketch the cube, and find the coordinates of each corner.
    *b) Compute the angle between an edge of the cube and the diagonal from $(0, 0, 0)$ to $(1, 1, 1)$.
    *c) Compute the angle between the diagonal in part b and the diagonal in one face from $(0, 0, 0)$ to $(0, 1, 1)$.

13. a) Determine a constant $c$ such that the four points $(1, 0, 0)$, $(0, 1, 0)$, $(0, 0, 1)$, and $(c, c, c)$ are equidistant. (The four points are then the vertices of a *regular tetrahedron*.)
    b) Compute the angle between adjacent edges of the regular tetrahedron.

14. Verify the following laws for the dot product:
    a) The commutative law $\vec{v} \cdot \vec{w} = \vec{w} \cdot \vec{v}$.
    b) The associative law $(c\vec{v}) \cdot \vec{w} = c(\vec{v} \cdot \vec{w})$.

    c) The distributive law $\vec{u} \cdot (\vec{v} + \vec{w}) = \vec{u} \cdot \vec{v} + \vec{u} \cdot \vec{w}$.
    d) $\vec{v} \cdot \vec{v} = |\vec{v}|^2$.

15. Prove that $|c\vec{v}| = |c| \cdot |\vec{v}|$.

**B**

*16. Find three vectors $\vec{u}, \vec{v},$ and $\vec{w}$ with $\vec{u} \cdot \vec{v} = \vec{u} \cdot \vec{w}$, and $\vec{u} \neq \vec{0}$, but $\vec{v} \neq \vec{w}$. (This shows that you cannot "cancel" a factor from a dot product equation.)

17. *Direction cosines* of a vector $\vec{v}$ are the cosines of the angles between $\vec{v}$ and the three basic vectors $\vec{i}, \vec{j},$ and $\vec{k}$; these angles are generally denoted $\alpha$, $\beta$, and $\gamma$ (Fig. 14). For $v = \langle v_1, v_2, v_3 \rangle$, show that
    a) $\cos \alpha = v_1/|\vec{v}|$, $\cos \beta = v_2/|\vec{v}|$, $\cos \gamma = v_3/|\vec{v}|$.
    b) $\langle \cos \alpha, \cos \beta, \cos \gamma \rangle = \dfrac{\vec{v}}{|\vec{v}|}$, which is the unit vector in the direction of $\vec{v}$.
    c) $\cos^2\alpha + \cos^2\beta + \cos^2\gamma = 1$.

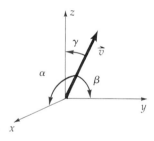

**FIGURE 14**
Angles between $\vec{v}$ and the positive coordinate half-axes.

18. Is it possible to find a unit vector $\vec{u}$ making angle $\pi/3$ with each of the coordinate axes?

19. Under what conditions is
    a) $\vec{v} \cdot \vec{w} = |\vec{v}| \cdot |\vec{w}|$?
    *b) $\vec{v} \cdot \vec{w} = -|\vec{v}| \cdot |\vec{w}|$?
    c) $|\vec{v} + \vec{w}| = |\vec{v}| + |\vec{w}|$?
    d) $|\vec{v} - \vec{w}| = |\vec{v}| - |\vec{w}|$?
    e) $|\vec{v} - \vec{w}|^2 = |\vec{v}|^2 + |\vec{w}|^2$?

20. Prove that
    a) $|\vec{v} + \vec{w}|^2 = |\vec{v}|^2 + |\vec{w}|^2 + 2\vec{v} \cdot \vec{w}$.
    *b) $(\vec{v} + \vec{w}) \cdot (\vec{v} - \vec{w}) = |\vec{v}|^2 - |\vec{w}|^2$.
    c) $|\vec{v} + \vec{w}|^2 + |\vec{v} - \vec{w}|^2 = 2|\vec{v}|^2 + 2|\vec{w}|^2$.
    Draw a sketch illustrating this.

**21.** Show that
  **a)** If $\vec{w}$ is a unit vector then $\vec{v}_{\vec{w}} = (\vec{v} \cdot \vec{w})\vec{w}$.
  **b)** For any vector $\vec{v}$, $\vec{v} = v_{\vec{i}}\vec{i} + v_{\vec{j}}\vec{j} + v_{\vec{k}}\vec{k}$.

**22.** A vertical mast, pivoted at the base, is steadied by three guy wires from the mast top $M$ (Fig. 15). The tension $T_1$ in the wire from $M$ to $P_1$ is 20 Newtons. Denote by $\vec{F}_j$ the vector force exerted on $M$ by the wire from $M$ to $P_j$.
  **a)** Compute $\vec{F}_1$.
  **b)** Why is $\vec{F}_1 + \vec{F}_2 + \vec{F}_3 = \langle 0, 0, z \rangle$ for some number $z$?
  **c)** Compute the tensions $T_2$ and $T_3$ in the wires to $P_2$ and $P_3$.

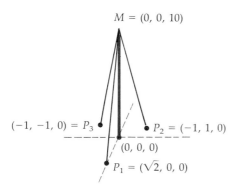

**FIGURE 15**
Mast with three guy wires.

**23.** Suppose that $|\vec{v}| = |\vec{w}|$. Prove algebraically that $(\vec{v} + \vec{w}) \perp (\vec{v} - \vec{w})$. Draw an illustrative sketch.

**24.** The *midpoint* of the segment from $P_1$ to $P_2$ is the point $M$ defined by
$$\vec{M} = \tfrac{1}{2}(\vec{P_1} + \vec{P_2}).$$
Set $P_1 = (x_1, y_1, z_1)$, $P_2 = (x_2, y_2, z_2)$ and verify algebraically that $|P_1 M| = |M P_2| = \tfrac{1}{2}|P_1 P_2|$.

**25.** Prove that the segments joining the midpoints of successive sides of any quadrilateral form a parallelogram.

**26.** Points $A$, $B$, $C$, $D$ in space are the successive vertices of a quadrilateral, not necessarily lying in a plane. Determine whether the segment from $A$ to $C$ necessarily bisects the segment from $B$ to $D$.

**27.** Denote by $\Delta P_1 P_2 P_3$ the triangle with vertices $P_1$, $P_2$, and $P_3$ in space. A *median* of the triangle is the line segment from a vertex to the midpoint of the opposite side. The *barycenter* of the triangle is the point $B$ defined by
$$\vec{B} = \tfrac{1}{3}(\vec{P_1} + \vec{P_2} + \vec{P_3}).$$
Show that the barycenter $B$ is on each median, 2/3 of the way from the vertex to the midpoint of the opposite side.

**28.** Four points $P_1$, $P_2$, $P_3$, $P_4$ not lying in a single plane form the vertices of a *tetrahedron* $T$, a solid bounded by four triangular faces. A *median* of the tetrahedron is the segment from a vertex to the barycenter of the opposite face. (See the previous problem.) The *barycenter* of the tetrahedron is the point $B_T$ defined by
$$\vec{B_T} = \tfrac{1}{4}(\vec{P_1} + \vec{P_2} + \vec{P_3} + \vec{P_4}).$$
Show that $B_T$ lies on each median. (Hint: First guess how far $B_T$ is from the vertex; then imitate the solution of the previous problem.

**29.** Given points $Q = (a, b, c)$ and $R = (d, e, f)$, prove: The set $\{P = (x, y, z): \text{angle } RPQ = \pi/2\}$ is a sphere with diameter $RQ$.

# 14.3

## THE CROSS PRODUCT. DETERMINANTS

In three-dimensional analytic geometry, we often encounter the following problem: Given two vectors $\vec{v}$ and $\vec{w}$, find a vector $\vec{n}$ that is orthogonal to both $\vec{v}$ and $\vec{w}$ (Fig. 1). There is a standard solution to this, called the **cross product** of $\vec{v}$ and $\vec{w}$, which can be derived as follows. Let

$$\vec{v} = \langle v_1, v_2, v_3 \rangle, \qquad \vec{w} = \langle w_1, w_2, w_3 \rangle,$$

and denote by $x$, $y$, and $z$ the as yet unknown coordinates of the desired orthogonal vector $\vec{n}$:

$$\vec{n} = \langle x, y, z \rangle.$$

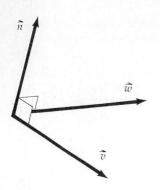

**FIGURE 1**
$\hat{n} \perp \vec{v}$ and $\hat{n} \perp \vec{w}$.

Since $\hat{n}$ is to be orthogonal to both $\vec{v}$ and $\vec{w}$, its dot product with each must be 0:

$$\vec{v} \cdot \hat{n} = v_1 x + v_2 y + v_3 z = 0 \qquad \text{(1)}$$
$$\vec{w} \cdot \hat{n} = w_1 x + w_2 y + w_3 z = 0.$$

Eliminate $z$; multiply the first equation by $w_3$, the second by $v_3$, and subtract:

$$(w_3 v_1 - v_3 w_1)x + (w_3 v_2 - v_3 w_2)y = 0. \qquad \text{(2)}$$

This equation has the form $ax + by = 0$; an obvious solution is $x = b$ and $y = -a$, for then $ax + by = ab + b(-a) = 0$. So we can solve (2) with

$$x = w_3 v_2 - v_3 w_2, \qquad y = -(w_3 v_1 - v_3 w_1).$$

Substitute these in (1) and solve for $z$; you find

$$z = v_1 w_2 - v_2 w_1.$$

Thus, as a vector orthogonal to both $\vec{v}$ and $\vec{w}$, we take

$$\hat{n} = \langle x, y, z \rangle = \langle v_2 w_3 - v_3 w_2, -(v_1 w_3 - v_3 w_1), v_1 w_2 - v_2 w_1 \rangle.$$

This is called the *cross product* of $\vec{v}$ and $\vec{w}$, denoted $\vec{v} \times \vec{w}$:

$$\vec{v} \times \vec{w} = \langle v_2 w_3 - v_3 w_2, -(v_1 w_3 - v_3 w_1), v_1 w_2 - v_2 w_1 \rangle. \qquad \text{(3)}$$

This forbidding formula is easier to remember using the notation of *determinants*. A $2 \times 2$ determinant assigns a number to a square array of four numbers:

$$\det \begin{bmatrix} a & b \\ c & d \end{bmatrix} = ad - bc.$$

It is the product of the entries on the diagonal going down to the right, minus the product of the entries on the diagonal going down to the left. Thus

$$\det \begin{bmatrix} 1 & 2 \\ 3 & 4 \end{bmatrix} = 1 \cdot 4 - 2 \cdot 3 = -2.$$

Using determinants, the cross product formula (3) can be written

$$\vec{v} \times \vec{w} = \left\langle \det \begin{bmatrix} v_2 & v_3 \\ w_2 & w_3 \end{bmatrix}, -\det \begin{bmatrix} v_1 & v_3 \\ w_1 & w_3 \end{bmatrix}, \det \begin{bmatrix} v_1 & v_2 \\ w_1 & w_2 \end{bmatrix} \right\rangle. \qquad \text{(4)}$$

The three determinants are formed from the components of $\vec{v}$ and $\vec{w}$, arrayed in two rows:

$$\begin{bmatrix} v_1 & v_2 & v_3 \\ w_1 & w_2 & w_3 \end{bmatrix}.$$

For the *first* component of $\vec{v} \times \vec{w}$, delete the *first* column of this array and take the determinant of what remains. For each of the other two components, delete the corresponding column and take the determinant of what remains; moreover, *insert a minus sign in front of the middle determinant.*

---

**EXAMPLE 1**   For $\vec{v} = \langle 1, 2, 3 \rangle$ and $\vec{w} = \langle 4, 5, 6 \rangle$, the array is

$$\begin{bmatrix} 1 & 2 & 3 \\ 4 & 5 & 6 \end{bmatrix}.$$

and the cross product is

$$\vec{v} \times \vec{w} = \left\langle \det \begin{bmatrix} & 2 & 3 \\ & 5 & 6 \end{bmatrix}, \ -\det \begin{bmatrix} 1 & & 3 \\ 4 & & 6 \end{bmatrix}, \ \det \begin{bmatrix} 1 & 2 & \\ 4 & 5 & \end{bmatrix} \right\rangle$$
$$= \langle -3, 6, -3 \rangle.$$

(The blank columns indicate which components of the given vectors are deleted in forming the determinant.) A similar calculation gives

$$\vec{w} \times \vec{v} = \langle 3, -6, 3 \rangle = -\vec{v} \times \vec{w}.$$

---

Another formula for the cross product uses *3 × 3 determinants*, which are defined as follows:

$$\det \begin{bmatrix} u_1 & u_2 & u_3 \\ v_1 & v_2 & v_3 \\ w_1 & w_2 & w_3 \end{bmatrix} = u_1 \det \begin{bmatrix} & v_2 & v_3 \\ & w_2 & w_3 \end{bmatrix} - u_2 \det \begin{bmatrix} v_1 & & v_3 \\ w_1 & & w_3 \end{bmatrix} + u_3 \det \begin{bmatrix} v_1 & v_2 & \\ w_1 & w_2 & \end{bmatrix}$$

$$= u_1 \det \begin{bmatrix} v_2 & v_3 \\ w_2 & w_3 \end{bmatrix} - u_2 \det \begin{bmatrix} v_1 & v_3 \\ w_1 & w_3 \end{bmatrix} + u_3 \det \begin{bmatrix} v_1 & v_2 \\ w_1 & w_2 \end{bmatrix}$$

$$= u_1(v_2 w_3 - v_3 w_2) - u_2(v_1 w_3 - v_3 w_1) + u_3(v_1 w_2 - w_1 v_2).$$

The factor multiplying $u_1$ is the determinant of the array formed by deleting the first row and column of the given 3 × 3 array; the factor multiplying $-u_2$ is the determinant of the array formed by deleting the given first row and *second* column; for the third factor, delete the first row and *third* column.

   With this notation, the cross product $\vec{v} \times \vec{w}$ is a "vector determinant," with vectors $\vec{i}, \vec{j}, \vec{k}$ in the first row, the components of $\vec{v}$ in the second row, and the components of $\vec{w}$ in the third:

$$\vec{v} \times \vec{w} = \det \begin{bmatrix} \vec{i} & \vec{j} & \vec{k} \\ v_1 & v_2 & v_3 \\ w_1 & w_2 & w_3 \end{bmatrix}$$

$$= \vec{i} \det \begin{bmatrix} & v_2 & v_3 \\ & w_2 & w_3 \end{bmatrix} - \vec{j} \det \begin{bmatrix} v_1 & & v_3 \\ w_1 & & w_3 \end{bmatrix} + \vec{k} \det \begin{bmatrix} v_1 & v_2 & \\ w_1 & w_2 & \end{bmatrix}$$

$$= (v_2 w_3 - v_3 w_2)\vec{i} - (v_1 w_3 - v_3 w_1)\vec{j} + (v_1 w_2 - v_2 w_1)\vec{k}.$$

**EXAMPLE 2**

$(\vec{i} + 2\vec{j} + 3\vec{k}) \times (4\vec{i} + 5\vec{j} + 6\vec{k})$

$$= \det \begin{bmatrix} \vec{i} & \vec{j} & \vec{k} \\ 1 & 2 & 3 \\ 4 & 5 & 6 \end{bmatrix}$$

$$= \vec{i} \det \begin{bmatrix} 2 & 3 \\ 5 & 6 \end{bmatrix} - \vec{j} \det \begin{bmatrix} 1 & 3 \\ 4 & 6 \end{bmatrix} + \vec{k} \det \begin{bmatrix} 1 & 2 \\ 4 & 5 \end{bmatrix}$$

$$= -3\vec{i} + 6\vec{j} - 3\vec{k}$$

as in Example 1.

**The algebraic properties of the cross product** are partly familiar:

$$(c\vec{v}) \times \vec{w} = c(\vec{v} \times \vec{w}) = \vec{v} \times (c\vec{w})$$

$$\vec{u} \times (\vec{v} + \vec{w}) = \vec{u} \times \vec{v} + \vec{u} \times \vec{w}.$$

But one is quite different:

$$\vec{v} \times \vec{w} = -\vec{w} \times \vec{v},$$

as noted in Example 1. This is an **anticommutative** law. These properties are checked by tedious but straightforward calculation of each side (see problem 11).

There is no associative law for a repeated cross product; except in special cases, $\vec{u} \times (\vec{v} \times \vec{w}) \neq (\vec{u} \times \vec{v}) \times \vec{w}$ [Problem 15].

**The geometric properties of the cross product** explain its importance. First, as we know, it is orthogonal to each factor:

$$\vec{v} \times \vec{w} \perp \vec{v} \quad \text{and} \quad \vec{v} \times \vec{w} \perp \vec{w}. \tag{5}$$

Second, its *length* can be computed (see problem 11); it is

$$|\vec{v} \times \vec{w}|^2 = |\vec{v}|^2 \cdot |\vec{w}|^2 - (\vec{v} \cdot \vec{w})^2.$$

Further, since $\vec{v} \cdot \vec{w} = |\vec{v}| \cdot |\vec{w}| \cos \theta$, with $\theta$ the angle between $\vec{v}$ and $\vec{w}$, we find

$$|\vec{v} \times \vec{w}|^2 = |\vec{v}|^2 \cdot |\vec{w}|^2(1 - \cos^2\theta) = |\vec{v}|^2 \cdot |\vec{w}|^2 \sin^2\theta$$

so

$$|\vec{v} \times \vec{w}| = |\vec{v}| \cdot |\vec{w}| \sin \theta, \qquad \theta = \text{angle between } \vec{v} \text{ and } \vec{w}. \tag{6}$$

The product $|\vec{v}| \cdot |\vec{w}| \sin \theta$ is precisely the *area of the parallelogram* spanned by $\vec{v}$ and $\vec{w}$ (Fig. 2). For you can think of $|\vec{v}|$ as the base of the parallelogram, and

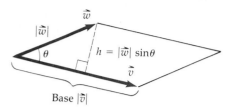

**FIGURE 2**
Area of parallelogram =
(base) × (height) = $|\vec{v}|\,|\vec{w}|\sin\theta$.

$|\vec{w}|\sin\theta$ as the altitude. (A parallelogram is *spanned* by vectors $\vec{v}$ and $\vec{w}$ if arrows representing those two vectors, drawn from a common point, form two sides of the parallelogram.)

---

**EXAMPLE 3**   The area of the parallelogram spanned by the vectors $\vec{v} = \langle 1, 2, 3 \rangle$ and $\vec{w} = \langle 4, 5, 6 \rangle$ in Example 2 is

$$|\vec{v} \times \vec{w}| = |-3\vec{i} + 6\vec{j} - 3\vec{k}| = 3|-\vec{i} + 2\vec{j} - \vec{k}| = 3\sqrt{6}.$$

---

The two properties (5) and (6) *almost* determine the cross product. Imagine $\vec{v}$, $\vec{w}$, and $\vec{v} \times \vec{w}$ all drawn from a common point. If $\vec{v}$ and $\vec{w}$ are not parallel, they determine a plane; the orthogonality property (5) requires that $\vec{v} \times \vec{w}$ lie along the line $L$ perpendicular to that plane (Fig. 3). The length is determined by (6). This leaves two possibilities for $\vec{v} \times \vec{w}$, corresponding to the two directions along $L$: Which way does $\vec{v} \times \vec{w}$ point? The answer is given by the "Right-Hand Rule" (Fig. 4): *If you grasp $\vec{v} \times \vec{w}$ with the right hand, fingers pointing from $\vec{v}$ toward $\vec{w}$, then the thumb points in the direction of $\vec{v} \times \vec{w}$.* (Warning: The orientation of the cross product depends on the orientation of the axes. In this book, we always use "right-handed" axes, where the vector $\vec{k}$ is oriented with respect to $\vec{i}$ and $\vec{j}$ by the Right-Hand Rule. If the axes were drawn with $\vec{k}$ in the opposite direction, then cross products would be "left-handed.")

To prove the Right-Hand Rule requires, first of all, that it be stated more formally. We leave this for a later course.

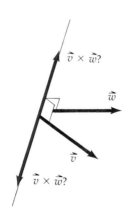

**FIGURE 3**
Vectors orthogonal to both $\vec{v}$ and $\vec{w}$.

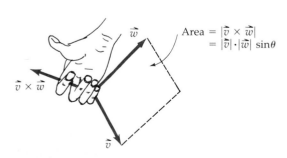

Area = $|\vec{v} \times \vec{w}|$
= $|\vec{v}| \cdot |\vec{w}| \sin\theta$

**FIGURE 4**
Right-Hand Rule for cross product.

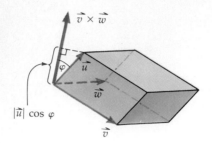

**FIGURE 5**
Volume of parallelepiped = $|\vec{u} \cdot (\vec{v} \times \vec{w})|$.

## 3 × 3 Determinants and Volume

Using the cross product, you can compute the volume of the parallelepiped spanned by three vectors $\vec{u}$, $\vec{v}$, and $\vec{w}$ (Fig. 5). The volume equals the area of the base spanned by $\vec{v}$ and $\vec{w}$, times the height $h$ perpendicular to that base (Fig. 5). The area of the base is the length of the cross product $\vec{v} \times \vec{w}$, while the height is $h = |\vec{u}| \cos \varphi$, with $\varphi$ the angle between $\vec{u}$ and $\vec{v} \times \vec{w}$; thus

$$\text{volume} = (h) \cdot (\text{area of base})$$
$$= (|\vec{u}| \cos \varphi) \cdot |\vec{v} \times \vec{w}|.$$

This is precisely the dot product of $\vec{u}$ with $\vec{v} \times \vec{w}$! So the volume $= \vec{u} \cdot (\vec{v} \times \vec{w})$. Multiply out the dot product, and you find

$$\text{volume} = u_1 \cdot \det\begin{bmatrix} v_2 & v_3 \\ w_2 & w_3 \end{bmatrix} - u_2 \cdot \det\begin{bmatrix} v_1 & v_3 \\ w_1 & w_3 \end{bmatrix} + u_3 \cdot \det\begin{bmatrix} v_1 & v_2 \\ w_1 & w_2 \end{bmatrix}.$$

This, in turn, is the *determinant*

$$\det\begin{bmatrix} u_1 & u_2 & u_3 \\ v_1 & v_2 & v_3 \\ w_1 & w_2 & w_3 \end{bmatrix}.$$

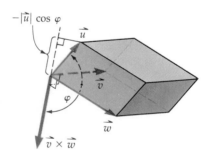

**FIGURE 6**
$\vec{u} \cdot (\vec{v} \times \vec{w}) = -(\text{volume of parellelepiped})$.

This calculation used Figure 5, where the angle $\varphi$ between $\vec{u}$ and $\vec{v} \times \vec{w}$ is less than $\pi/2$. If that angle is $> \pi/2$ (Fig. 6) then $\cos \varphi$ is negative, and $h = -|\vec{u}| \cos \varphi$, so

$$\text{volume} = -(|\vec{u}| \cos \varphi) \cdot |\vec{v} \times \vec{w}| = -\vec{u} \cdot (\vec{v} \times \vec{w})$$
$$= -\det\begin{bmatrix} u_1 & u_2 & u_3 \\ v_1 & v_2 & v_3 \\ w_1 & w_2 & w_3 \end{bmatrix}.$$

In either case, the *absolute value* of the determinant gives the volume spanned, while its *sign* gives the "orientation" of the vectors with respect to each other, determining whether $\vec{u}$ is roughly in the same direction as $\vec{v} \times \vec{w}$, or in the opposite direction.

---

**EXAMPLE 4** The volume of the parallelepiped spanned by $\vec{i} + \vec{j} + \vec{k}$, $\vec{i} + \vec{j} - \vec{k}$, and $\vec{i} - \vec{j} + \vec{k}$ is the absolute value of

$$\det\begin{bmatrix} 1 & 1 & 1 \\ 1 & 1 & -1 \\ 1 & -1 & 1 \end{bmatrix} = 1 \cdot 0 - 1 \cdot 2 + 1 \cdot (-2) = -4.$$

So the volume is 4; the minus sign means that the first vector makes an obtuse angle with the cross product of the second and third.

---

## 2 × 2 Determinants and Signed Area

Just as 3 × 3 determinants give signed volume in space, the 2 × 2 determinant

$$\det \begin{bmatrix} v_1 & v_2 \\ w_1 & w_2 \end{bmatrix}$$

gives the "signed area" of the parallelogram in the plane spanned by vectors

$$\vec{v} = v_1 \vec{i} + v_2 \vec{j} \quad \text{and} \quad \vec{w} = w_1 \vec{i} + w_2 \vec{j}.$$

If the direction of rotation from $\vec{v}$ to $\vec{w}$ is counterclockwise (the same as the direction from $\vec{i}$ to $\vec{j}$, see Fig. 7a), then

$$\det \begin{bmatrix} v_1 & v_2 \\ w_1 & w_2 \end{bmatrix} = \text{area of parallelogram spanned by } \vec{v} \text{ and } \vec{w}.$$

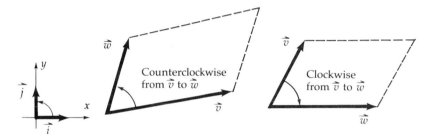

(a)   $\det \begin{bmatrix} v_1 & v_2 \\ w_1 & w_2 \end{bmatrix}$ = Area of parallelogram.   (b)   $\det \begin{bmatrix} v_1 & v_2 \\ w_1 & w_2 \end{bmatrix}$ = −Area of parallelogram.

**FIGURE 7**

But if the rotation is clockwise (Fig. 7b) then

$$\det \begin{bmatrix} v_1 & v_2 \\ w_1 & w_2 \end{bmatrix} = -(\text{area of parallelogram spanned by } \vec{v} \text{ and } \vec{w}).$$

To show this, imagine $\vec{v} = v_1 \vec{i} + v_2 \vec{j}$ and $\vec{w} = w_1 \vec{i} + w_2 \vec{j}$ as vectors in space which happen to lie in the $xy$-plane. Their cross product is easily computed; you find

$$\vec{v} \times \vec{w} = \det \begin{bmatrix} v_1 & v_2 \\ w_1 & w_2 \end{bmatrix} \vec{k}. \tag{7}$$

Hence the area of the parallelogram is the absolute value of this determinant:

$$\text{area} = |\vec{v} \times \vec{w}| = \left| \det \begin{bmatrix} v_1 & v_2 \\ w_1 & w_2 \end{bmatrix} \right|.$$

As for the sign, suppose that the rotation from $\vec{v}$ to $\vec{w}$ is counterclockwise. Then the Right-Hand Rule requires that $\vec{v} \times \vec{w}$ be in the same direction as

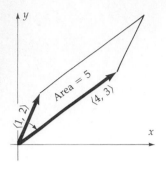

**FIGURE 8**

$$\det \begin{bmatrix} 1 & 2 \\ 4 & 3 \end{bmatrix} = -5.$$

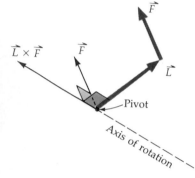

**FIGURE 9**
Torque $\vec{T} = \vec{L} \times \vec{F}$.

$\vec{k}$, so the determinant in (7) is positive. On the other hand, if the rotation is clockwise then the determinant must be negative.

---

**EXAMPLE 5**   The determinant of $\vec{v} = \langle 1, 2 \rangle$ and $\vec{w} = \langle 4, 3 \rangle$ is

$$\det \begin{bmatrix} 1 & 2 \\ 4 & 3 \end{bmatrix} = 1 \cdot 3 - 2 \cdot 4 = -5.$$

The rotation from $\langle 1, 2 \rangle$ to $\langle 4, 3 \rangle$ is clockwise (Fig. 8), and these vectors span a parallelogram of area 5. Exchanging the rows reverses the direction of rotation, and changes the sign:

$$\det \begin{bmatrix} 4 & 3 \\ 1 & 2 \end{bmatrix} = 4 \cdot 2 - 3 \cdot 1 = 5.$$

---

**Torque**  is one of many applications of the cross product in physics. A lever arm $\vec{L}$ is fixed at one end, and a force $\vec{F}$ is applied at the other end (Fig. 9). The force tends to turn the arm around the fastened end. Torque quantifies this tendency to turn the lever arm $\vec{L}$; it is defined by

$$\text{torque } \vec{T} = \vec{L} \times \vec{F}.$$

The magnitude $|\vec{L} \times \vec{F}| = |\vec{L}| \cdot |\vec{F}| \cdot \sin \theta$ gives the strength of the "turning tendency," while the direction of $\vec{L} \times \vec{F}$ gives the "axis of rotation," the axis about which the lever arm would turn if it were free to do so.

## SUMMARY

*Determinants:*

$$\det \begin{bmatrix} v_1 & v_2 \\ w_1 & w_2 \end{bmatrix} = v_1 w_2 - v_2 w_1$$

$$= \pm[\text{area spanned by } \vec{v} \text{ and } \vec{w} \text{ in } R^2].$$

$$\det \begin{bmatrix} u_1 & u_2 & u_3 \\ v_1 & v_2 & v_3 \\ w_1 & w_2 & w_3 \end{bmatrix} = u_1 \det \begin{bmatrix} v_2 & v_3 \\ w_2 & w_3 \end{bmatrix} - u_2 \det \begin{bmatrix} v_1 & v_3 \\ w_1 & w_3 \end{bmatrix} + u_3 \det \begin{bmatrix} v_1 & v_2 \\ w_1 & w_2 \end{bmatrix}$$

$$= \pm[\text{volume of parallelogram spanned by } \vec{u}, \vec{v}, \text{ and } \vec{w} \text{ in } R^3].$$

*Cross Product (see Fig. 4):*

$$\vec{v} \times \vec{w} = \det \begin{bmatrix} \vec{i} & \vec{j} & \vec{k} \\ v_1 & v_2 & v_3 \\ w_1 & w_2 & w_3 \end{bmatrix}$$

$$= \left\langle \det \begin{bmatrix} v_2 & v_3 \\ w_2 & w_3 \end{bmatrix}, -\det \begin{bmatrix} v_1 & v_3 \\ w_1 & w_3 \end{bmatrix}, \det \begin{bmatrix} v_1 & v_2 \\ w_1 & w_2 \end{bmatrix} \right\rangle$$

$\vec{v} \times \vec{w}$ is orthogonal to both $\vec{v}$ and $\vec{w}$.

$$|\vec{v} \times \vec{w}| = \text{area of parallelogram spanned by } \vec{v} \text{ and } \vec{w}$$
$$= |\vec{v}| \cdot |\vec{w}| \sin \theta.$$

**Algebraic Laws for the Cross Product:**

$$\vec{v} \times \vec{w} = -\vec{w} \times \vec{v}$$
$$(c\vec{v}) \times \vec{w} = c(\vec{v} \times \vec{w}) = \vec{v} \times (c\vec{w})$$
$$\vec{u} \times (\vec{v} + \vec{w}) = (\vec{u} \times \vec{v}) + (\vec{u} \times \vec{w})$$
$$(\vec{u} + \vec{v}) \times \vec{w} = (\vec{u} \times \vec{w}) + (\vec{v} \times \vec{w})$$

## PROBLEMS

### A

1. Evaluate the following determinants. Sketch the parallelogram whose area they give, and check that the sign corresponds to the correct direction of rotation.

   *a) $\det \begin{bmatrix} 1 & 0 \\ 0 & 1 \end{bmatrix}$
   
   *b) $\det \begin{bmatrix} 0 & 1 \\ 1 & 0 \end{bmatrix}$

   *c) $\det \begin{bmatrix} 1 & 2 \\ -2 & -1 \end{bmatrix}$
   
   d) $\det \begin{bmatrix} -2 & -1 \\ 1 & 2 \end{bmatrix}$

   e) $\det \begin{bmatrix} -2 & 1 \\ -1 & 2 \end{bmatrix}$
   
   f) $\det \begin{bmatrix} 2 & 3 \\ -4 & -6 \end{bmatrix}$

*2. In Figure 10, which sets of axes are right-handed, and which are left-handed?

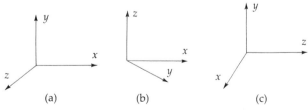

(a)          (b)          (c)

**FIGURE 10**
Which are right-handed? (The slanted axis points toward you, out of the paper.)

3. Evaluate the following determinants. Sketch the parallelepiped whose volume they give, and check that the sign is correct: positive if the first vector makes an acute angle with the cross product of the second and third.

*a) $\det \begin{bmatrix} 1 & 0 & 0 \\ 0 & 1 & 0 \\ 0 & 0 & 1 \end{bmatrix}$
b) $\det \begin{bmatrix} 0 & 1 & 0 \\ 1 & 0 & 0 \\ 0 & 0 & 1 \end{bmatrix}$

*c) $\det \begin{bmatrix} 0 & 1 & 0 \\ 0 & 0 & 1 \\ 0 & 1 & 1 \end{bmatrix}$
d) $\det \begin{bmatrix} 0 & 1 & 1 \\ 0 & -1 & 1 \\ 2 & 0 & 1 \end{bmatrix}$

*e) $\det \begin{bmatrix} 1 & 2 & 3 \\ 4 & 5 & 6 \\ 7 & 8 & 9 \end{bmatrix}$

*4. Compute all nine possible cross products formed from pairs of the basis vectors $\vec{i}$, $\vec{j}$, and $\vec{k}$.

5. For the given vectors, compute $\vec{v} \times \vec{w}$; check that it is orthogonal to $\vec{v}$ and $\vec{w}$.
   *a) $\vec{v} = \langle 4, 5, 6 \rangle$, $\vec{w} = \langle 1, 2, 3 \rangle$
   b) $\vec{v} = \vec{i} + \vec{j}$, $\vec{w} = \vec{j} + \vec{k}$
   *c) $\vec{v} = \langle 1, 2, -3 \rangle$, $\vec{w} = -4\vec{i} + \vec{j} + 2\vec{k}$
   d) $\vec{v} = v_1\vec{i} + v_2\vec{j}$, $\vec{w} = w_1\vec{i} + w_2\vec{k}$

6. Find the area of the parallelogram spanned by $\vec{v}$ and $\vec{w}$.
   a) $\vec{v} = \vec{i} + \vec{k}$, $\vec{w} = \vec{j}$
   *b) $\vec{v} = (1, 2, 0)$, $\vec{w} = (2, 1, 0)$

7. Find the area of the triangle with vertices at the points $P$, $Q$, and $R$. Find a vector $\vec{n}$ which is perpendicular to the plane of the triangle.
   a) $P = (5, -2, 1)$, $Q = (0, 0, 0)$, $R = (5, 1, 3)$
   *b) $P = (-3, 2, 0)$, $Q = (5, 4, 1)$, $R = (9, 7, 2)$

*8. A parallelepiped has three edges representing the vectors $\vec{A} = \langle 1, -1, 2 \rangle$, $\vec{B} = \langle 0, 3, -1 \rangle$, and $\vec{C} = \langle 3, -4, 1 \rangle$. Compute
   a) The area of the face spanned by $\vec{A}$ and $\vec{B}$.
   b) The volume of the parallelepiped.

**c)** The angle between $\vec{C}$ and the face spanned by $\vec{A}$ and $\vec{B}$.

**\*9.** One end of a theoretical weightless lever arm is attached to the origin. To its other end, at $(1, -3, 2)$, is attached a weight exerting a downward force of $-6\vec{k}$.
**a)** Compute the magnitude of the torque.
**b)** Write parametric equations of the axis about which the arm revolves.

**B**

**10.** Problem 4 asked you to compute the cross products $\vec{i} \times \vec{i}, \vec{j} \times \vec{j}, \vec{k} \times \vec{k}, \vec{i} \times \vec{j}, \vec{j} \times \vec{k}, \vec{k} \times \vec{i}$. Now, assuming just these six products and the algebraic laws given in the Summary, deduce the formula for the cross product of $\langle v_1, v_2, v_3 \rangle$ and $\langle w_1, w_2, w_3 \rangle$.

**11.** Show.
**a)** $\vec{v} \times \vec{w} = -\vec{w} \times \vec{v}$
**b)** $(c\vec{v}) \times \vec{w} = c(\vec{v} \times \vec{w}) = \vec{v} \times (c\vec{w})$
**c)** $\vec{u} \times (\vec{v} + \vec{w}) = \vec{u} \times \vec{v} + \vec{u} \times \vec{w}$
**d)** $(\vec{v} + \vec{w}) \times \vec{u} = \vec{v} \times \vec{u} + \vec{w} \times \vec{u}$ (Save time; use parts a and c.)
**e)** $(\vec{v} \times \vec{w}) \cdot (\vec{v} \times \vec{w}) = |\vec{v}|^2 \cdot |\vec{w}|^2 - (\vec{v} \cdot \vec{w})^2$
**f)** $\vec{u} \cdot (\vec{v} \times \vec{w}) = (\vec{u} \times \vec{v}) \cdot \vec{w}$

**12.** Show that a $2 \times 2$ determinant is *reversed in sign* if you exchange the two rows.

**13.** How is a $2 \times 2$ determinant affected if you
**a)** Exchange the two columns?
**b)** Make a new determinant whose *columns* are the rows of the original determinant, in the same order?
**c)** Multiply one row by a constant $c$?
**d)** Make a new determinant whose first row is the original one, and whose second row is the original second row plus the original first row?

**14.** How is a $3 \times 3$ determinant affected if
**a)** Two rows are exchanged?
**b)** One row is multiplied by a constant $c$?

**15.** Find vectors such that $\vec{u} \times (\vec{v} \times \vec{w}) \neq (\vec{u} \times \vec{v}) \times \vec{w}$. (It can be done by choosing $\vec{u}, \vec{v}, \vec{w}$ from among the vectors $\vec{i}, \vec{j}$, and $\vec{k}$.)

**16.** Suppose that $\vec{u}, \vec{v}$, and $\vec{w}$ are *coplanar* (all lie in a single plane). What can be said about their determinant? Is the converse also true?

**17. a)** Explain: For vectors in $R^2$, $\vec{v}$ is parallel to $\vec{w}$ if and only if $\det \begin{bmatrix} v_1 & v_2 \\ w_1 & w_2 \end{bmatrix} = 0$. (Thus the $2 \times 2$ determinant "determines" wheter $\vec{v}$ and $\vec{w}$ are parallel or not.)
**b)** The two equations

$$a_1 x + b_1 y = c_1$$
$$a_2 x + b_2 y = c_2$$

define lines in the plane. Explain why the lines intersect in a unique point if and only if

$$\det \begin{bmatrix} a_1 & b_1 \\ a_2 & b_2 \end{bmatrix} \neq 0.$$

**18.** Four points in the plane form the vertices of a *tetrahedron*, a figure bounded by four triangles. The volume of this tetrahedron is a simple fraction of the volume of a related parallelepiped.
**a)** Guess what the fraction is.
**b)** Use your guess to compute the volume of the tetrahedron with vertices $(0, 0, 0)$, $(1, 0, 0)$, $(0, 1, 0)$, $(0, 0, 1)$.
**c)** Compute the same volume as an integral of cross-sectional areas.

# 14.4

## LINES AND PLANES

**The vector equation of a line** through a given point $P_0$, parallel to a given vector $\vec{v}$, is

$$\vec{P}(t) = \vec{P}_0 + t\vec{v}. \tag{1}$$

To get to point $P(t)$, go first to $P_0$ and then add $t\vec{v}$, a multiple of $\vec{v}$ (Fig. 1). Each different multiple $t$ gives a different point on the line; as $t$ varies over all real numbers, equation (1) gives *all* points on the line.

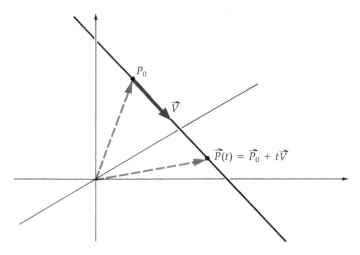

**FIGURE 1**
Equation of the line through $P_0$,
parallel to $\vec{v}$.

To get parametric equations of the line, assign coordinates to $P_0$, $\vec{v}$, and $P$:
$P_0 = (x_0, y_0, z_0)$, $\vec{v} = \langle a, b, c \rangle$, and $\vec{P}(t) = \langle x, y, z \rangle$. Then

$$\langle x, y, z \rangle = \langle x_0 + ta, y_0 + tb, z_0 + tc \rangle,$$

and parametric equations of the line are

$$x = x_0 + ta, \qquad y = y_0 + tb, \qquad z = z_0 + tc.$$

---

**EXAMPLE 1**  The positions of two rockets at time $t$ are
$$\vec{P}_1(t) = \langle 1, 0, 2 \rangle + t \langle 100, 200, 0 \rangle$$

and

$$\vec{P}_2(t) = \langle 11, 10, 6 \rangle + t \langle 100, 100, 40 \rangle.$$

    **a.**  Do the rockets collide?
    **b.**  Do their paths cross?

*SOLUTION*
    **a.**  The rockets collide if they are at the same point at the same time
$t$, that is if

$$\vec{P}_1(t) = \vec{P}_2(t) \tag{2}$$

for some $t$. Write equation (2) in coordinates:

$$1 + 100t = 11 + 100t$$

$$0 + 200t = 10 + 100t$$

$$2 + 0t = 6 + 40t.$$

The first equation has no solution—cancelling $100t$, you get the false equation
$1 = 11$. So at no time do the rockets collide.

**b.** Their paths cross if one rocket is in a certain place at time $t_1$, and the other is at the same place at time $t_2$, that is if

$$\vec{P}_1(t_1) = \vec{P}_2(t_2) \tag{3}$$

for some $t_1$ and $t_2$. In coordinates, equation (3) is

$$1 + 100t_1 = 11 + 100t_2$$
$$200t_1 = 10 + 100t_2$$
$$2 = 6 + 40t_2.$$

From the last equation, $t_2 = \dfrac{-1}{10}$. Then the second equation says

$$200t_1 = 10 - 10 = 0$$

so $t_1 = 0$. These values of $t_1$ and $t_2$ satisfy the last two equations. Do they also satisfy the first? Yes (check this!). So the paths do cross, at the point

$$P_1(0) = (1, 0, 2) = P_2\left(\frac{-1}{10}\right).$$

## Planes

A plane is determined by a point $P_0$ on the plane, and a vector $\vec{n}$ *perpendicular* to the plane (Fig. 2). The vector $\vec{n}$ is called a **normal** to the plane. The plane consists of all points $P$ such that $\vec{P} - \vec{P}_0$ is perpendicular to $\vec{n}$. Its equation is therefore

$$\vec{n} \cdot (\vec{P} - \vec{P}_0) = 0. \tag{4}$$

To write this in coordinates, set $P = (x, y, z)$, $P_0 = (x_0, y_0, z_0)$, and $\vec{n} = \langle a, b, c \rangle$. Then (4) is

$$a(x - x_0) + b(y - y_0) + c(z - z_0) = 0. \tag{5}$$

Multiply this out and introduce a new constant $d = -ax_0 - by_0 - cz_0$, to obtain

$$ax + by + cz + d = 0. \tag{6}$$

This is the *equation of a plane with normal $\vec{n} = a\vec{i} + b\vec{j} + c\vec{k}$*; the *components of the normal* are precisely the *coefficients of $x$, $y$, and $z$ in the equation of the plane*. For example, $x + 2y + 3z + 4 = 0$ is the equation of a plane with normal $\vec{n} = \vec{i} + 2\vec{j} + 3\vec{k}$.

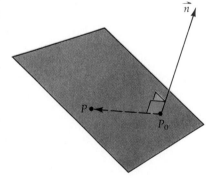

**FIGURE 2**
Plane through $P_0$ with normal $\vec{n}$:
$\vec{n} \cdot (\vec{P} - \vec{P}_0) = 0.$

**EXAMPLE 2**   Write the equation of the plane through $(1, 1, 1)$ perpendicular to the line

$$\vec{P}(t) = \langle 1, 1, 1 \rangle + t\langle 2, 1, 0 \rangle.$$

***SOLUTION***   Visualize the situation (Fig. 3). The vector $\vec{v} = \langle 2, 1, 0 \rangle$ is parallel to the line. Since the plane is perpendicular to the line, this vector $\vec{v}$ can be taken as the normal $\vec{n}$ of the desired plane:

$$\vec{n} = \langle 2, 1, 0 \rangle = 2\vec{i} + \vec{j}.$$

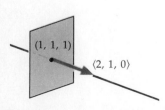

**FIGURE 3**
Line perpendicular to plane. Vector
*parallel* to line is *normal* to plane.

So use equation (5) with $a = 2$, $b = 1$, $c = 0$. The plane passes through the point $P_0 = (1, 1, 1)$, so in (5) take $x_0 = 1$, $y_0 = 1$, $z_0 = 1$:

$$2(x - 1) + 1(y - 1) + 0(z - 1) = 0$$

or

$$2x + y - 3 = 0.$$

**ALTERNATE SOLUTION**   using equation (6). As before, the normal is $\vec{n} = 2\vec{i} + \vec{j}$, so $a = 2$, $b = 1$, $c = 0$. Thus equation (6) is

$$2x + y + 0z + d = 0,$$

and it remains to determine the constant $d$. The plane passes through $(1, 1, 1)$ so

$$2 \cdot 1 + 1 + 0 \cdot 1 + d = 0.$$

Hence $d = -3$, and the equation (6) for our plane is $2x + y - 3 = 0$.

---

**EXAMPLE 3**   Find the intersections of the plane

$$2x + y + z = 2 \tag{7}$$

with the coordinate axes. Sketch the plane.

**SOLUTION**   Any point on the $x$-axis can be written $(x, 0, 0)$. If it is also on the plane (7) then

$$2x + 0 + 0 = 2,$$

so $x = 1$, and the intersection of the plane with the $x$-axis is $(1, 0, 0)$. This same process yields the other two intersections, $(0, 2, 0)$ and $(0, 0, 2)$. These three points determine the plane (Fig. 4).

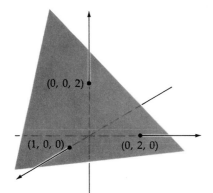

**FIGURE 4**
Portion of plane $2x + y + z = 2$.

---

**EXAMPLE 4**   Find the equation of the plane containing the points $P = (1, 2, 3)$, $Q = (2, 3, 4)$, and $R = (4, 3, 1)$.

**SOLUTION**   Since points $P$ and $Q$ are in the plane, so is the arrow from $P$ to $Q$, representing $\vec{Q} - \vec{P}$ (Fig. 5); so also is the arrow for $\vec{R} - \vec{P}$. Thus the cross product of these is a normal to the plane:

$$\vec{n} = (\vec{R} - \vec{P}) \times (\vec{Q} - \vec{P}) = \langle 3, 1, -2 \rangle \times \langle 1, 1, 1 \rangle$$
$$= \langle 3, -5, 2 \rangle.$$

The plane with this normal, passing through the point $P = (1, 2, 3)$, has the equation

$$3(x - 1) - 5(y - 2) + 2(z - 3) = 0$$

or

$$3x - 5y + 2z = -1.$$

As a check, verify that $P$, $Q$, and $R$ satisfy this equation.

**Remark**   Notice that $\vec{P} \times \vec{Q}$ is *not* an appropriate normal. The arrows for $\vec{P}$

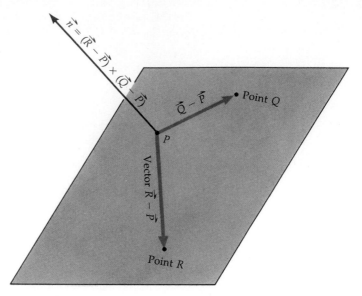

**FIGURE 5**
Normal to plane containing $P$, $Q$,
and $R$.

and $\vec{Q}$ do *not* lie in the plane; they go from the origin to the plane. It is the
arrow for the *difference* $\vec{Q} - \vec{P}$ that lies in the plane (Fig. 5).

**EXAMPLE 5**    Show that the two planes

$$\pi_1: x + y + z = 1 \quad \text{and} \quad \pi_2: x - y + z = 1 \tag{8}$$

intersect in a line. Find parametric equations for the line.

*SOLUTION*    The planes have normals

$$\vec{n}_1 = \vec{i} + \vec{j} + \vec{k} \quad \text{and} \quad \vec{n}_2 = \vec{i} - \vec{j} + \vec{k}.$$

Since these are not parallel, the planes are not parallel either, and so they
intersect in a line $L$. To find equations for $L$, you need a vector $\vec{v}$ parallel to
$L$, and a point $P_0$ on $L$.

Figure 6 shows how to find $\vec{v}$. Since the line $L$ is contained in both
planes, it is perpendicular to both $\vec{n}_1$ and $\vec{n}_2$. But the cross product $\vec{n}_1 \times \vec{n}_2$
is also perpendicular to these two vectors, so it is parallel to the line. Compute

$$\vec{n}_1 \times \vec{n}_2 = 2\vec{i} - 2\vec{k} = 2(\vec{i} - \vec{k}),$$

and conclude that $\vec{v} = \vec{i} - \vec{k}$ is parallel to the line $L$.

It remains to find a point on $L$; any point $(x_0, y_0, z_0)$ satisfying both equations (8) will do. Suppose that we find the point where $L$ crosses the $xy$-plane;
to do so, set $z = 0$ and solve equations (8). You find $x = 1$ and $y = 0$, so
the point where $L$ crosses the $xy$-plane is $(1, 0, 0)$; use this for $P_0$ in the equation
of the line:

$$\vec{P} = \vec{P}_0 + t\vec{v} = \langle 1, 0, 0 \rangle + t \langle 1, 0, -1 \rangle.$$

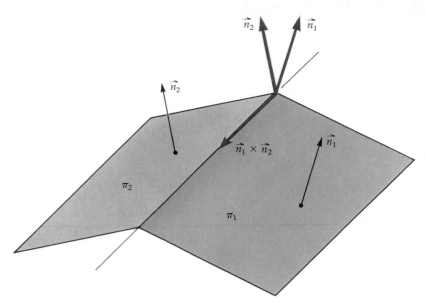

**FIGURE 6**
$\vec{n}_1 \times \vec{n}_2$ is parallel to line of
intersection of $\pi_1$ and $\pi_2$.

Parametric equations are

$$x = 1 + t, \qquad y = 0, \qquad z = -t.$$

[As a final check, substitute these in (8); both equations are satisfied, for all $t$.]

## Symmetric Equations of a Line

The parametric equations for a line

$$x = x_0 + at, \qquad y = y_0 + bt, \qquad z = z_0 + ct \tag{9}$$

can be rewritten by solving each equation for $t$ and equating the results:

$$\frac{x - x_0}{a} = \frac{y - y_0}{b} = \frac{z - z_0}{c}. \tag{10}$$

These are called the **symmetric equations** of the line. The first equation gives

$$b(x - x_0) = a(y - y_0),$$

which determines a plane. The second equation in (10) gives another plane, so the symmetric equations express the line as the intersection of two planes.

   If any of the denominators in (10) is zero, then of course there is a problem. Suppose, for instance, that $a = 0$. Then (9) shows that $x = x_0$ all along the line, and the first equation in (10) can be replaced by $x - x_0 = 0$, again the equation of a plane. Thus in (10), *if any denominator is zero, it is understood that the numerator is also zero.*

## SUMMARY

**Line through $P_0 = (x_0, y_0, z_0)$ Parallel to $\vec{v} = \langle a, b, c \rangle$:**

$$\{P: \vec{P} = \vec{P}_0 + t\vec{v} \text{ for some number } t\}.$$

**Parametric Equations of the Line:**

$$x = x_0 + at, \qquad y = y_0 + bt, \qquad z = z_0 + ct$$

**Plane through $P_0$ with Normal $\vec{n}$:**

$$\{P: \vec{n} \cdot (\vec{P} - \vec{P}_0) = 0\}$$

**Equation of a Plane with Normal $\vec{n} = \langle a, b, c \rangle$:**

$$ax + by + cz + d = 0.$$

## PROBLEMS

### A

**1.** Find a normal to the given plane, and the intersection of the plane with each coordinate axis. Sketch the plane.
   a)   $x + y + z = 1$
   b)   $x + y + z = 0$
   *c)   $x + y + 2z - 1 = 0$
   d)   $x + 2y = 1$
   *e)   $x - z = 2$
   f)   $y + z = 3$

**2.** Write the equation of a plane through the given point $P_0$ with given normal $\vec{n}$.
   *a)   $P_0 = (0, 1, 2), \vec{n} = \vec{i} + \vec{j} - \vec{k}$
   b)   $P_0 = (0, 0, 0), \vec{n} = \langle 1, 0, 0 \rangle$
   c)   $P_0 = (1, 3, 5), \vec{n} = \langle 2, 4, -1 \rangle$
   d)   $P_0 = (0, 0, 1), \vec{n} = \vec{k}$
   *e)   $P_0 = (0, 1, 0), \vec{n} = -\vec{k}$

**\*3.** For the line through $(1, 1, 2)$ parallel to $\langle 1, -3, 2 \rangle$, write
   a)   Parametric equations.
   b)   Symmetric equations.

**\*4.** Find a vector parallel to the line $x = t, y = 1 + t,$ $z = 2 - t$.

**\*5.** Does the line in problem 4 lie in the plane $x + y + 2z = 5$?

**\*6.** Does the line in problem 3 intersect the line in problem 4?

**\*7.** Is the line in problem 4 perpendicular to the plane $x + y - z = 0$?

**8.** Write a vector equation for the line through the two points $(1, 7, 11)$ and $(-2, 3, 0)$.

**\*9.** Write the equation of a plane through the origin and parallel to the plane $x - y + z = 1$.

**10.** Write an equation of the plane through the point $(x_0, y_0, z_0)$ and parallel to
   *a)   The $xy$-plane.      b)   The $yz$-plane.
   c)   The $xz$-plane.

**\*11.** Find an equation for the plane that passes through $(2, -1, 3)$ and is perpendicular to the line $\vec{P}(t) = \langle 1, 0, 0 \rangle + t\langle 3, -2, 4 \rangle$.

**12.** Find parametric equations of the line that contains $(1, -2, -3)$ and is perpendicular to the plane $3x - y - 2z + 5 = 0$.

**13.** Let $L_1$ and $L_2$ be lines with direction vectors $\vec{v}_1$ and $\vec{v}_2$, and let $\pi_1$ and $\pi_2$ be planes with normals $\vec{n}_1$ and $\vec{n}_2$. Then $\pi_1$ is parallel to $\pi_2$ if $\vec{n}_1$ is a scalar multiple of $\vec{n}_2$. Give similar conditions on $\vec{n}$ and $\vec{v}$ for the following; illustrate each situation with an appropriate sketch showing $\vec{n}$ and $\vec{v}$.
   a)   $\pi_1$ is perpendicular to $\pi_2$.
   b)   $\pi_1$ is parallel to $L_1$.
   c)   $\pi_1$ is perpendicular to $L_1$.
   d)   $L_1$ is parallel to $L_2$.
   e)   $L_1$ is perpendicular to $L_2$.

**\*14.** At time $t$, an object is at position $\vec{P}(t) = \langle 1, 2, 3 \rangle + t\langle -1, 0, 1 \rangle$.
   a)   At what time does it cross the plane $x + y + 2z = 1$? Where does it cross?

b) Show that it moves parallel to the plane $x + y + z = 0$.

15. Find the angle of intersection of the planes $x + y + z = 1$ and $2x + y + z = 1$. (The angle between the planes equals the angle between their normals.)

16. Write an equation of the plane containing the three given points.
   *a) $(1, 2, 3)$, $(5, 4, 3)$, $(2, -1, 4)$
   b) $(1, 5, 2)$, $(2, 4, -1)$, $(3, 3, 4)$

17. Write an equation of the plane through the origin and parallel to the two lines

$$x = t, \qquad y = 2 + t, \qquad z = 1 - t$$

$$x = -t, \qquad y = 1 + 2t, \qquad z = 3t.$$

18. Write an equation of the plane through $(-1, 2, 1)$ perpendicular to the line of intersection of the planes $x - y + 2z = 3$ and $2x + y + z = 2$.

19. Given the two planes $x + 2y + 3z = 1$ and $y = z$, find
   a) A point in the intersection of the planes.
   b) A vector parallel to the line of intersection.
   c) Parametric equations of the line of intersection.

20. Given the two planes $x - y - z = 2$ and $3x - 2y - z = 1$, find
   a) Two distinct points in the line of intersection.
   b) A vector equation of the line of intersection.

21. Find a plane through $(2, 3, 1)$ and $(2, 1, 3)$, perpendicular to the plane $x - 2y - 4z = 7$.

22. Show that the lines

$$x = 3 + t, \qquad y = -2 + 2t, \qquad z = -t$$

and

$$x = 2 - t, \qquad y = 3 - 2t, \qquad z = 1 + t$$

are parallel. Find an equation of the plane containing them.

23. Determine whether the lines intersect. If they do, find the point of intersection, and the angle of intersection.
   *a) $x = 4t - 2$, $y = 3$, $z = 2 - t$ and $x = 2t + 2$, $y = 2t + 3$, $z = t + 1$.
   b) $x = 4t$, $y = 3$, $z = 2 - t$ and $x = 2t + 2$, $y = 2t + 3$, $z = t + 1$.

**B**

*24. Find the distance from $(1, 1, 1)$ to the plane $x - y - z = 0$. (See the Theorem on the dis-

tance from a point to a line in the plane, Sec. 14.2.)

25. Show that
   a) The distance from a point $P_0 = (x_0, y_0, z_0)$ to the plane $ax + by + cz + d = 0$ is

$$\left| \frac{ax_0 + by_0 + cz_0 + d}{\sqrt{a^2 + b^2 + c^2}} \right|.$$

   b) The coefficient $d$ in the equation of the plane is $\pm |\vec{n}| \cdot$ (distance from origin to plane), with $\vec{n} = a\vec{i} + b\vec{j} + c\vec{k}$.

26. a) Show that the planes $x + 2y + z = 3$ and $2x + 4y + 2z = 3$ are parallel.
   b) Compute the distance between those planes.

27. Given the lines $x = 1 + t$, $y = 2 - t$, $z = 3$ and $x = -t$, $y = 2 + t$, $z = 1 + 2t$:
   a) Do they intersect?
   b) Are they parallel?
   c) Find the distance between them. (Lines which are neither parallel nor intersecting are called *skew*.)

*28. Show that the entire line $x = t$, $y = 2t$, $z = 1 + t$ lies above the plane $x - y + z = 0$ and below the plane $x - z + 2 = 0$. (Take the $z$-axis pointing up, as usual.)

29. Write the equation in $x$, $y$, and $z$ for the set of points

$$\{P: |\vec{P} - \vec{P_1}|^2 = |\vec{P} - \vec{P_2}|^2\},$$

where $P_1 = (1, 1, 1)$ and $P_2 = (-1, 2, 1)$. Show that the set is a plane.

30. The three equations

$$a_1 x + b_1 y + c_1 z = d_1$$
$$a_2 x + b_2 y + c_2 z = d_2$$
$$a_3 x + b_3 y + c_3 z = d_3$$

define planes with normals $\vec{n}_1$, $\vec{n}_2$, and $\vec{n}_3$. Explain geometrically why
   a) The second two planes intersect in a line precisely when $\vec{n}_2 \times \vec{n}_3 \neq \vec{0}$.
   b) The three planes intersect in a unique point precisely when

$$\det \begin{bmatrix} a_1 & b_1 & c_1 \\ a_2 & b_2 & c_2 \\ a_3 & b_3 & c_3 \end{bmatrix} \neq 0.$$

## 14.5

# VECTOR FUNCTIONS AND SPACE CURVES

This section concerns motion and curves in space, extending the ideas developed in Chapter 13 for motion in the plane. But before discussing motion, we give two rules for differentiating dot products and cross products.

A vector function $\vec{F}(t) = \langle f_1(t), f_2(t), f_3(t) \rangle$ is called **differentiable** if each component $f_j(t)$ has a derivative; then the derivative of $\vec{F}$ is

$$\vec{F}'(t) = \langle f'_1(t), f'_2(t), f'_3(t) \rangle.$$

If $\vec{G}(t)$ is also a differentiable vector function, then the dot product $\vec{G}(t) \cdot \vec{F}(t)$ is a real-valued function of $t$; its derivative can be computed by the familiar-looking Product Rule

$$(\vec{F} \cdot \vec{G})'(t) = \vec{F}'(t) \cdot \vec{G}(t) + \vec{F}(t) \cdot \vec{G}'(t).$$

The proof consists in expressing each side of the equation in terms of the components of $\vec{F}$ and $\vec{G}$. Similarly, for the cross product

$$(\vec{F} \times \vec{G})' = \vec{F}' \times \vec{G} + \vec{F} \times \vec{G}'.$$

In this case *the order of the factors is important,* since the cross product is anticommutative.

Now suppose that the vector function

$$\vec{P}(t) = \langle x(t), y(t), z(t) \rangle$$

gives the position at time $t$ of an object moving in space. The analysis of plane curves in Sections 13.3 and 13.4 carries over word-for-word to space curves. The *velocity* is the derivative of the position with respect to time:

$$\vec{v}(t) = \vec{P}'(t).$$

The acceleration is the derivative of the velocity:

$$\vec{a}(t) = \vec{v}'(t).$$

The velocity vector $\vec{v}(t)$ is tangent to the curve traced out by the motion, and its length $|\vec{v}(t)|$ is the speed along the curve. The arc length function is the integral of the speed,

$$s = \int |\vec{v}(t)| dt,$$

and the derivative of the arc length is $\dfrac{ds}{dt} = |\vec{v}(t)|$. The arc length traversed as $t$ varies from $a$ to $b$ is the integral

$$L_a^b = \int_a^b |\vec{v}(t)| \, dt.$$

**EXAMPLE 1**    Compute the velocity, speed and arc length function for the curve

$$\vec{P}(t) = \langle \cos t, \sin t, t \rangle.$$

Compute the arc length traversed as $t$ varies from 0 to $2\pi$.

*SOLUTION*

Velocity:   $\vec{v} = \vec{P}'(t) = \langle -\sin t, \cos t, 1 \rangle$

Speed:   $|\vec{v}(t)| = \sqrt{(-\sin t)^2 + (\cos t)^2 + 1} = \sqrt{2}$

Arc length function:   $s = \int |\vec{v}(t)| \, dt = \int \sqrt{2} \, dt = \sqrt{2}t + c$

Length for $0 \le t \le 2\pi$:   $L_0^{2\pi} = \int_0^{2\pi} |\vec{v}| \, dt = \int_0^{2\pi} \sqrt{2} \, dt = 2\pi\sqrt{2}.$

Figure 1 shows the curve. The $x$ and $y$ components alone describe a motion around the unit circle in the $xy$-plane:

$$x = \cos t, \, y = \sin t \Rightarrow x^2 + y^2 = 1.$$

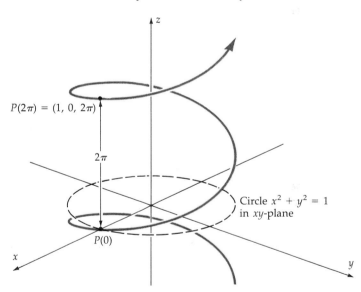

**FIGURE 1**
Helix $\vec{P}(t) = \langle \cos t, \sin t, t \rangle.$

In addition to this circular motion, the curve rises at a steady rate, since $z = t$. These two motions together trace a curve shaped like a coiled spring, called a **helix**, winding around the cylinder of radius 1 surrounding the $z$-axis. As $t$ varies from 0 to $2\pi$, the curve goes once around the cylinder (Fig. 1). Thus the length $L_0^{2\pi}$ that we computed is the length of one turn of the helix, measured along the curve. The straight-line distance between the beginning and end of this turn is $2\pi$ (Fig. 1), while the length along the curve is $L_0^{2\pi} = 2\pi\sqrt{2}$.

Figure 2 shows this turn of the helix "unwound," forming the hypotenuse of a right triangle. The altitude $2\pi$ is the distance between the ends of the turn in Figure 1; the base $2\pi$ is the perimeter of the unit circle. The figure confirms that the length along the curve is $2\pi\sqrt{2}$.

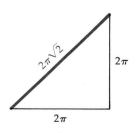

**FIGURE 2**
One turn of the helix unwound.

Helices describe the motion of electrons in a constant magnetic field. In biology, the DNA molecule has a shape like two helices, one surrounding the other.

### The Unit Tangent

The velocity $\vec{v} = d\vec{P}/dt$ gives the direction of the curve *and* its speed. If only the direction concerns us, we form the *unit tangent vector* $\vec{T}$:

$$\vec{T}(t) = \frac{\vec{v}(t)}{|\vec{v}(t)|}.$$

This is defined whenever $\vec{v}(t) \neq \vec{0}$; where $\vec{v}(t) = \vec{0}$, there $\vec{T}(t)$ is not defined, and in fact the curve may not have a well-defined direction.

**EXAMPLE 2**  $\vec{P}(t) = \langle t^3, t^2 \rangle$. This curve is in $R^2$, but the definition of unit tangent still applies. The velocity is

$$\vec{v}(t) = \langle 3t^2, 2t \rangle = t \langle 3t, 2 \rangle,$$

so

$$|\vec{v}(t)| = |t|\sqrt{9t^2 + 4}.$$

The unit tangent is

$$\vec{T}(t) = \frac{\vec{v}(t)}{|\vec{v}(t)|} = \frac{t}{|t|}\frac{1}{\sqrt{9t^2 + 4}}\langle 3t, 2 \rangle, \qquad t \neq 0.$$

Here $\vec{T}(0)$ is not defined, since the denominator $|\vec{v}(0)| = 0$. Moreover, this curve has no well-defined direction at the point $P(0)$ (Fig. 3). As $t$ passes through 0, the moving point comes in to the origin [since $P(0) = (0, 0)$], stops [since $\vec{v}(0) = \vec{0}$], and moves off in a new direction — up, instead of down.

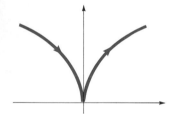

**FIGURE 3**
$\vec{P}(t) = \langle t^3, t^2 \rangle$. At the origin, the unit tangent $\vec{T}$ is not defined.

### Arc Length as Parameter

In the rest of this section, we consider only those parts of the curve where $\vec{v}(t)$ is defined and nonzero. In this case

$$\frac{ds}{dt} = |\vec{v}(t)| > 0,$$

so the arc length function is strictly increasing and has a differentiable inverse (Sec. 8.2). That is, we can express $t$ as a function of $s$,

$$t = g(s)$$

and the derivative is

$$\frac{dt}{ds} = \frac{1}{ds/dt} = \frac{1}{|\vec{v}(t)|}. \tag{1}$$

Since $t = g(s)$, we can consider $\vec{P}$ as a function of arc length $s$;

$$\vec{P}(t) = \vec{P}(g(s)).$$

The derivative is $\dfrac{d\vec{P}}{ds} = \dfrac{d\vec{P}}{dt}\dfrac{dt}{ds} = \dfrac{d\vec{P}/dt}{ds/dt} = \dfrac{\vec{v}(t)}{|\vec{v}(t)|} = \vec{T}(t)$. Thus: *The unit tangent vector is the derivative of position with respect to arc length*:

$$\vec{T} = \frac{d\vec{P}}{ds}.$$

---

**EXAMPLE 3**    For the helix $\vec{P}(t) = \langle \cos t,\ \sin t,\ t \rangle$, the arc length function was found in Example 1:

$$s = \int |\vec{v}|\,dt = \int \sqrt{2}\,dt = \sqrt{2}\,t + C.$$

To get a specific arc length function, take $C = 0$; then $s = \sqrt{2}\,t$. The inverse function is

$$t = \frac{s}{\sqrt{2}}.$$

So position as a function of arc length is found by replacing $t$ with $s/\sqrt{2}$:

$$\vec{P} = \left\langle \cos\frac{s}{\sqrt{2}},\ \sin\frac{s}{\sqrt{2}},\ \frac{s}{\sqrt{2}} \right\rangle.$$

The derivative of position with respect to arc length is

$$\frac{d\vec{P}}{ds} = \left\langle \frac{-1}{\sqrt{2}}\sin\frac{s}{\sqrt{2}},\ \frac{1}{\sqrt{2}}\cos\frac{s}{\sqrt{2}},\ \frac{1}{\sqrt{2}} \right\rangle. \tag{2}$$

This is indeed the unit tangent:

$$\vec{T}(t) = \frac{\vec{v}(t)}{|\vec{v}(t)|} = \frac{\langle -\sin t,\ \cos t,\ 1 \rangle}{\sqrt{2}},$$

which agrees with (2), since $t = s/\sqrt{2}$.

---

## Curvature

Figure 4 shows two curves, one straighter than the other. Along a segment of given length $\Delta s$, the second curve changes direction more rapidly than the first; in Figure 4b), $\Delta\vec{T}$ is longer than in 4a, and hence so is the ratio $\dfrac{\Delta\vec{T}}{\Delta s}$.

The magnitude $\left|\dfrac{\Delta\vec{T}}{\Delta s}\right|$ is called the **average curvature** of the curve along the segment of length $\Delta s$; and the magnitude of the derivative $d\vec{T}/ds$ at a given point $P$ on the curve is defined to be the *curvature* at $P$, denoted $\kappa$:

$$\kappa = |d\vec{T}/ds|. \tag{3}$$

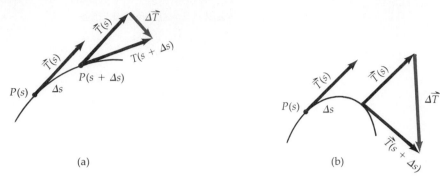

**FIGURE 4**
Two segments of equal length $\Delta s$. The second has greater curvature, and $\Delta \vec{T}$ is greater.

This formula involves the derivative with respect to arc length $s$, but you don't have to represent the curve explicitly as a function of $s$ to compute it; the Chain Rule gives

$$\frac{d\vec{T}}{ds} = \frac{d\vec{T}}{dt}\frac{dt}{ds} = \frac{d\vec{T}/dt}{ds/dt}$$

and thus

$$\kappa = \frac{|d\vec{T}/dt|}{ds/dt}. \tag{4}$$

---

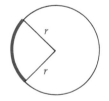

(a) Smaller radius $r$, larger curvature $\kappa$.

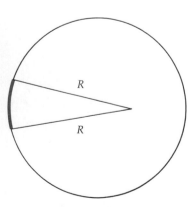

(b) Larger radius $R$, smaller curvature $\kappa$.

**FIGURE 5**

**EXAMPLE 4** Compute the curvature for the circle $\vec{P}(\theta) = \langle r \cos \theta, r \sin \theta \rangle$.

**SOLUTION** The parameter here is $\theta$, not $t$, but the geometry is the same; the derivatives will be with respect to $\theta$ instead of $t$. Thus we set

$$\vec{v} = \vec{P}'(\theta) = \langle -r \sin \theta, r \cos \theta \rangle$$

and find

$$\frac{ds}{d\theta} = |\vec{v}(\theta)| = \sqrt{(-r \sin \theta)^2 + (r \cos \theta)^2} = r.$$

The unit tangent is

$$\vec{T} = \frac{\vec{v}}{|v|} = \frac{\langle -r \sin \theta, r \cos \theta \rangle}{r} = \langle -\sin \theta, \cos \theta \rangle.$$

The derivative of the unit tangent with respect to arc length is

$$\frac{d\vec{T}}{ds} = \frac{d\vec{T}/d\theta}{ds/d\theta} = \frac{\langle -\cos \theta, -\sin \theta \rangle}{r},$$

so the curvature is

$$\kappa = \left| \frac{d\vec{T}}{ds} \right| = \frac{1}{r} |\langle -\cos \theta, -\sin \theta \rangle| = \frac{1}{r}.$$

This is worth noting:

*The curvature of a circle of radius $r$ is $1/r$.*

The smaller the radius, the greater the curvature (Fig. 5).

### The Principal Normal

The derivative $d\vec{T}/ds$, whose magnitude gives the curvature, has a significant direction as well. First of all, it is *orthogonal to* $\vec{T}$. Why? $\vec{T}$ has constant length 1, so

$$1 = |\vec{T}|^2 = \vec{T} \cdot \vec{T}.$$

Take the derivative of each side with respect to $s$, using the Product Rule:

$$0 = \frac{d\vec{T}}{ds} \cdot \vec{T} + \vec{T} \cdot \frac{d\vec{T}}{ds} = 2\vec{T} \cdot \frac{d\vec{T}}{ds}.$$

Thus $\vec{T} \cdot \dfrac{d\vec{T}}{ds} = 0$, and $\dfrac{d\vec{T}}{ds}$ is indeed orthogonal to the unit tangent vector $\vec{T}$.

If $\dfrac{d\vec{T}}{ds}$ is not zero, you can form the unit vector

$$\vec{N} = \frac{d\vec{T}/ds}{|d\vec{T}/ds|} = \frac{1}{\kappa}\frac{d\vec{T}}{ds}. \tag{5}$$

This is called the **principal normal vector**. In the plane, there are just two unit vectors normal to $\vec{T}$; the principal normal $\vec{N}$ is the one pointing in the direction toward which the curve is turning, toward the "inside of the turn." For instance, with the circle in Example 4, the principal normal is

$$\frac{1}{\kappa}\frac{d\vec{T}}{ds} = \langle -\cos\theta, -\sin\theta \rangle;$$

it points toward the center of the circle (Fig. 6).

In space, there is a whole circle of unit vectors orthogonal to $\vec{T}$ (Fig. 7); the *principal* normal gives the direction toward which the curve is turning.

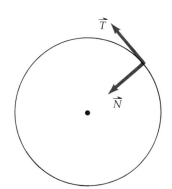

**FIGURE 6**
Principal normal $\vec{N}$ for a circle of radius $r > 1$.

**FIGURE 7**
Circle of unit vectors normal to $\vec{T}$. The *principal* normal $\vec{N}$ shows in which direction the curve is turning.

---

**EXAMPLE 5**   Compute the curvature and principal normal for the helix

$$\vec{P}(t) = \langle \cos t, \sin t, t \rangle.$$

*SOLUTION*   Start with the velocity

$$\vec{v} = \vec{P}'(t) = \langle -\sin t, \cos t, 1 \rangle$$

and find

$$\frac{ds}{dt} = |\vec{v}(t)| = \sqrt{(-\sin t)^2 + (\cos t)^2 + 1} = \sqrt{2}.$$

So

$$\vec{T} = \frac{\vec{v}}{|\vec{v}|} = \frac{\langle -\sin t, \cos t, 1 \rangle}{\sqrt{2}}$$

and

$$\frac{d\vec{T}}{ds} = \frac{d\vec{T}/dt}{ds/dt} = \frac{\langle -\cos t, -\sin t, 0 \rangle}{\sqrt{2} \cdot \sqrt{2}} = \frac{1}{2}\langle -\cos t, -\sin t, 0 \rangle.$$

The curvation is

$$\kappa = |d\vec{T}/ds| = \tfrac{1}{2}$$

and the principal normal is

$$\vec{N} = \frac{d\vec{T}/ds}{|d\vec{T}/ds|} = \langle -\cos t, \ -\sin t, \ 0 \rangle.$$

Drawn at the relevant point $P(t)$ on the curve (Fig. 8), $\vec{N}$ points directly at the $z$-axis around which the curve winds.

This helix wraps around a cylinder of radius 1. The perpendicular cross section of that cylinder is a circle of radius 1, and curvature 1. The helix is not curved as sharply as that circle; its curvature is merely $\tfrac{1}{2}$.

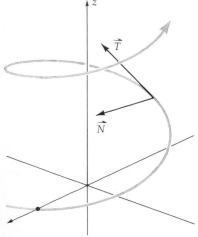

**FIGURE 8**
Unit tangent and principal normal for a helix.

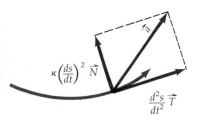

**FIGURE 9**
Normal and tangential components of acceleration.

## Normal and Tangential Components as Acceleration

Acceleration has an interesting expression as the sum of two parts, one in the direction of motion, the other perpendicular to it. To derive this, first write the velocity as the product "speed $|\vec{v}|$ times direction $\vec{T}$":

$$\vec{v} = |\vec{v}| \left( \frac{\vec{v}}{|\vec{v}|} \right) = |\vec{v}|\vec{T} = \frac{ds}{dt}\,\vec{T}. \tag{6}$$

Differentiate this last expression for $\vec{v}$:

$$\vec{a} = \frac{d\vec{v}}{dt} = \frac{d^2 s}{dt^2}\,\vec{T} + \frac{ds}{dt}\frac{d\vec{T}}{dt} \qquad \text{(Product Rule)}$$

$$= \frac{d^2 s}{dt^2}\,\vec{T} + \frac{ds}{dt}\left( \frac{d\vec{T}}{ds}\frac{ds}{dt} \right) \qquad \text{(Chain Rule)}.$$

By the definition (5) of the unit normal, $\dfrac{d\vec{T}}{ds} = \kappa\vec{N}$, so

$$\vec{a} = \frac{d^2 s}{dt^2}\,\vec{T} + \left( \frac{ds}{dt} \right)^2 \kappa\vec{N}. \tag{7}$$

Figure 9 illustrates this equation. The *tangential component of the acceleration* is the factor $\dfrac{d^2 s}{dt^2}$ multiplying the unit tangent $\vec{T}$; it is the rate of change of *speed* $ds/dt$ with respect to time. The *normal component of acceleration* is the factor $(ds/dt)^2\kappa$ multiplying the unit normal $\vec{N}$; this part of the acceleration produces the curvature.

For motion in a circle of radius $r$, the curvature is $\kappa = 1/r$, and the normal component of acceleration is $\left( \dfrac{ds}{dt} \right)^2 \kappa = |\vec{v}|^2 \cdot \dfrac{1}{r}$. You may recognize this from physics as **centripetal acceleration**.

In formula (7), the term $d^2 s/dt^2 = d|\vec{v}|/dt$ may be tedious to compute directly; and calculating $\kappa$ from the defining formula (3) or (4) is even worse, since it requires the derivative of the unit tangent $\vec{T}$. Simplified formulas for these quantities are derived from (7).

First, take the dot product of each side of (7) with the unit tangent $\vec{T}$. Since $\vec{T} \cdot \vec{T} = 1$ (for $\vec{T}$ is a unit vector) and $\vec{N} \cdot \vec{T} = 0$ (for $\vec{N}$ is a normal vector), you find from (7)

$$\vec{a} \cdot \vec{T} = \frac{d^2 s}{dt^2}.$$

And $\vec{T} = \vec{v}/|\vec{v}|$, so

$$\frac{d^2 s}{dt^2} = \frac{\vec{a} \cdot \vec{v}}{|\vec{v}|}. \tag{8}$$

Thus when the acceleration makes an acute angle with the velocity (Fig. 10a), then $\vec{a} \cdot \vec{v} > 0$, so $\frac{d^2 s}{dt^2} > 0$, and the speed increases. But when that angle is obtuse (Fig. 10b), the speed decreases.

We obtain a useful formula for curvature by taking the cross product of each side of (7) with the vector $\vec{T}$:

$$\vec{a} \times \vec{T} = \left[ \frac{d^2 s}{dt^2} \vec{T} \right] \times \vec{T} + \left[ \left( \frac{ds}{dt} \right)^2 \kappa \vec{N} \right] \times \vec{T}.$$

The cross product $\vec{T} \times \vec{T}$ is zero, since the two factors are parallel; and $|\vec{N} \times \vec{T}| = |\vec{N}| \cdot |\vec{T}| \sin \theta = 1$, since the two unit vectors $\vec{N}$ and $\vec{T}$ are perpendicular, making $\sin \theta = 1$. So

$$|\vec{a} \times \vec{T}| = \left( \frac{ds}{dt} \right)^2 \kappa.$$

Thus

$$\left( \frac{ds}{dt} \right)^2 \kappa = |\vec{a} \times \vec{T}| = |\vec{a} \times \vec{v}|/|\vec{v}|.$$

Hence the curvature is $\kappa = \dfrac{|\vec{a} \times \vec{T}|}{(ds/dt)^2} = \dfrac{|\vec{a} \times \vec{v}|/|\vec{v}|}{|\vec{v}|^2}$, or

$$\kappa = \frac{|\vec{a} \times \vec{v}|}{|\vec{v}|^3}. \tag{9}$$

For a curve in the $xy$-plane, $\vec{P} = x\vec{i} + y\vec{j}$, the cross product in (9) reduces to $(x''y' - y''x')\vec{k}$ (where $x' = dx/dt$, etc.). For such curves

$$\kappa = \frac{|x''y' - y''x'|}{[x'^2 + y'^2]^{3/2}}. \tag{10}$$

For the graph of a function $y = f(x)$, we can simplify (10) by taking $x = t$, $y = f(x) = f(t)$; then the curvature formula reduces to

$$\kappa = \frac{|f''(x)|}{[1 + f'(x)^2]^{3/2}}. \tag{11}$$

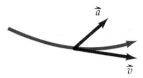

(a)   Speeding up, $\dfrac{d^2 s}{dt^2} > 0$.

(b)   Slowing down, $\dfrac{d^2 s}{dt^2} < 0$.

**FIGURE 10**
$$\frac{d^2 s}{dt^2} = \frac{\vec{a} \cdot \vec{v}}{|\vec{v}|}.$$

## SUMMARY

**Product Rules:**
$$(\vec{F} \times \vec{G})' = \vec{F}' \times \vec{G} + \vec{F} \times \vec{G}'$$
$$(\vec{F} \cdot \vec{G})' = \vec{F}' \cdot \vec{G} + \vec{F} \cdot \vec{G}'.$$

**Motion and Curves:**

$$\text{Velocity } \vec{v} = \frac{d\vec{P}}{dt} \qquad \text{Acceleration } \vec{a} = \frac{d\vec{v}}{dt}$$

$$\text{Speed} = |\vec{v}| \qquad \text{Arc length } s = \int |\vec{v}(t)| \, dt$$

$$\text{Unit tangent } \vec{T} = \frac{\vec{v}}{|\vec{v}|}$$

$$\text{Curvature } \kappa = \left| \frac{d\vec{T}}{ds} \right| = \frac{|\vec{a} \times \vec{v}|}{|\vec{v}|^3}$$

$$\text{Principal normal } \vec{N} = \frac{1}{\kappa} \frac{d\vec{T}}{ds} = \frac{d\vec{T}/dt}{|d\vec{T}/dt|}$$

Normal and tangential components of acceleration:

$$\vec{a} = \frac{d^2s}{dt^2} \vec{T} + \kappa \left( \frac{ds}{dt} \right)^2 \vec{N} = \frac{\vec{v} \cdot \vec{a}}{|\vec{v}|} \vec{T} + \frac{|\vec{a} \times \vec{v}|}{|\vec{v}|} \vec{N}.$$

## PROBLEMS

### A

**1.** Let $\vec{F}(t) = \langle t, t^2, t^3 \rangle$ and $\vec{G}(t) = \langle e^t, 1, t \rangle$.
   **a)** Compute $\vec{F}'$, $\vec{G}'$, $\vec{F} \cdot \vec{G}$, $\vec{F}' \cdot \vec{G}$, $\vec{F} \cdot \vec{G}'$, and $(\vec{F} \cdot \vec{G})'$; check the Product Rule in this case.
   **\*b)** Compute $\vec{F} \times \vec{G}$, $(\vec{F} \times \vec{G})'$, $\vec{F}' \times \vec{G}$, and $\vec{F} \times \vec{G}'$; check the Product Rule in this case.

Problems 2–11 refer to the *helix*

$$\vec{P}(t) = \langle a \cos \omega t, a \sin \omega t, bt \rangle$$

and the *twisted cubic*

$$\vec{P}(t) = \langle 3t, 3t^2, 2t^3 \rangle.$$

**2.** For the helix, compute
   **a)** Velocity and acceleration.
   **\*b)** The arc length $s$ as a function of $t$.
   **\*c)** The arc length of the part of the curve for $0 \le t \le 2\pi/\omega$. Show that this is one turn of the helix.

**\*3.** An object leaves the helix at time $t = 2$, and continues with constant velocity $\vec{v}(2)$ for $t \ge 2$. Where is it at time $t = 5$?

**4.** For the cubic, compute
   **a)** Velocity and acceleration.
   **\*b)** The arc length $s$ as a function of $t$.
   **\*c)** The length of the curve for $0 \le t \le 1$.

**\*5.** Write the vector equation of the line tangent to the twisted cubic at the point $P(1)$.

**\*6.** For the helix, compute.
   **a)** The unit tangent $\vec{T}$.
   **b)** The angle between $\vec{T}$ and the z-axis.

**7.** For the helix,
   **\*a)** Compute the curvature $\kappa$.
   **b)** The helix winds around a cylinder of radius $a$. Is the curvature of the helix greater or less than the curvature of the circular cross section of the cylinder?

8.   For the helix
   *a)   Compute the principal normal $\vec{N}$.
   b)   Show that $\vec{N}(t)$ is parallel to the $xy$-plane and points toward the $z$-axis if drawn as a vector starting at $P(t)$.

9.   For the cubic, compute $\vec{T}$, $\kappa$, and $\vec{N}$. At what point is $\kappa$ maximum?

*10.   For the helix, compute tangential and normal components of acceleration.

11.   For the cubic, compute tangential and normal components of acceleration.

12.   A model electric train traveling at 90 cm/sec passes from a straight section of track onto a circular arc of radius 30 cm. Compute the jump in acceleration (the jolt received by the model passengers).

13.   At time $t = 3$, an object in space has position $P(3) = (1, -2, 5)$ and velocity $\vec{v}(3) = \langle 3, 1, 2 \rangle$. Where is it, approximately, at $t = 3.1$?

14.   Compute the curvature of the cycloid

   $$x = a\theta - a \sin \theta, \qquad y = a - a \cos \theta.$$

   At what part of the curve is $\kappa$ minimum? Maximum?

15.   a)   Compute the curvature of the spiral $\vec{P}(t) = \langle t \cos t, t \sin t \rangle$.
   b)   Show that this spiral lies on the polar curve $r = \theta$.
   c)   Where is the curvature greatest?

16.   a)   Prove that $(\vec{F} \cdot \vec{G})' = \vec{F}' \cdot \vec{G} + \vec{F} \cdot \vec{G}'$
   b)   Prove that $(\vec{F} \times \vec{G})' = \vec{F}' \times \vec{G} + \vec{F} \times \vec{G}'$

B

17.   A car is going 45 mph in a curve posted for 35 mph. By what factor is its centripetal acceleration greater than it would be at 35 mph?

18.   A section of railroad track is to be designed with a straight segment $AB$, a segment $BC$ of gradually increasing curvature, and a circular segment $CD$ of constant curvature. Point $A$ is on the negative $x$-axis, $B$ at the origin, and the segment $BC$ is to have the shape of the graph of $y = ax^3$.
   a)   Show that the curvature is continuous at $B$.
   b)   Show that if $a = 1/\sqrt{45}$ then the maximum curvature of $y = ax^3$ is at $(1, a)$.
   c)   Take $a = 1/\sqrt{45}$, and $C = (1, a)$. What should be the radius of the circular segment $CD$? Where should the center of the circle be?

19.   Use $\kappa = |\vec{a} \times \vec{v}|/|\vec{v}|^3$ to derive formula (10) for the curvature of a plane curve $\vec{P}(t) = x(t)\vec{i} + y(t)\vec{j}$, in terms of $dx/dt$, $dy/dt$, $d^2x/dt^2$, and $d^2y/dt^2$.

20.   The graph of a function $f$ of one variable can be considered as the curve traced by

   $$\vec{P}(t) = t\vec{i} + f(t)\vec{j}.$$

   Use $\kappa = |\vec{a} \times \vec{v}|/|\vec{v}|^3$ to derive formula (11) for the curvature in terms of $f'$, and $f''$.

21.   Show that if $|\vec{F}(t)|$ is constant, then $\vec{F}'(t)$ is orthogonal to $\vec{F}(t)$. (Recall the proof that $d\vec{T}/ds$ is orthogonal to $\vec{T}$.)

22.   Prove or disprove: If $\vec{F}'(t)$ is orthogonal to $\vec{F}(t)$ for all $t$, then $|\vec{F}(t)|$ is constant.

23.   Show that a particle moves with constant speed if and only if the acceleration is always orthogonal to the velocity.

24.   The force exerted by a constant magnetic field $\vec{H}$ on a moving particle with charge $q$ is

   $$\vec{F} = q\vec{v} \times \vec{H}$$

   in appropriate units. Suppose that $\vec{H} = H\vec{k}$, where $H$ is the constant magnetic field strength. Take the position function to be a helix

   $$\vec{P}(t) = a \cos \omega t\vec{i} + a \sin \omega t\vec{j} + bt\vec{k}.$$

   Let the mass of the particle be $m$, and determine the constants $a$, $\omega$, and $b$ so that $\vec{P}$ will satisfy Newton's law $m\vec{a} = \vec{F} = q\vec{v} \times \vec{H}$.

25.   a)   For $0 \le t \le 1$, a car is moving in a circle of radius $r$ meters, with constant speed 20 m/sec [about 55 mph]. Compute the magnitude of the acceleration.
   *b)   At $t = 1$ the brakes are applied, causing a constant deceleration of 5 m/sec$^2$. Compute the magnitude of the acceleration just after the brakes are applied, while the speed is still 20 m/sec.
   c)   Comparing part b with part a, explain why braking could cause the car to skid.

26.   A motorcycle and rider of combined mass $m$ are going around a circular track of radius $r$ at speed $v = r\omega$; the position function is

   $$\vec{P}(t) = r \cos \omega t\vec{i} + r \sin \omega t\vec{j}.$$

   The total force $\vec{F} = m\vec{a}$ is the sum of a gravitational force $\vec{F}_g = -m\gamma\vec{k}$ (where $\gamma$ is a gravity constant) and the force $\vec{F}_{track}$ exerted by the track through its interaction with the tires.
   a)   Compute $\vec{F}_{track}$.

**b)** Compute the angle between $\vec{F}_{track}$ and the vertical direction $\hat{k}$. Express the angle in terms of the speed $v = |\vec{v}|$, the radius $r$, and the gravity constant $\gamma$. (The motorcycle must be tipped at precisely this angle to avoid falling.)

**c)** On a race track, a curve of radius 0.2 miles is intended to carry cars at about 150 mph. How steeply should it be banked to minimize skidding?

**d)** On an icy road, a curve of radius 0.1 mile is banked at an angle $\theta = \tan^{-1}(1/4)$. Cars skid toward the outside of the curve if too fast, and slip toward the inside if too slow. What is the optimal speed to avoid slipping and skidding?

**27.** A curve $\vec{P}(s)$ is given with arc length $s$ as parameter. Show that if the curvature $\kappa$ is zero, then the "curve" is a straight line.

**28.** A particle moves so that the acceleration $\vec{a}$ is always parallel to the position vector $\vec{P}$. (The acceleration is called *centrifugal* if $\vec{a}$ has the same direction as $\vec{P}$, and *centripetal* if it has the opposite direction. A force causing either kind of acceleration is called *central*.) Show that

**a)** $\vec{P} \times \vec{v}$ is constant.

**b)** $\vec{P}$ lies on the plane through the origin with normal $\vec{P} \times \vec{v}$; that is, $\vec{P} \cdot (\vec{P} \times \vec{v}) = 0$. (Thus an object accelerated by a central force moves in a fixed plane.)

**29.** If the unit tangent $\vec{T}$ and principal normal $\vec{N}$ are defined, then the *binormal* $\vec{B}$ is defined by

$$\vec{B} = \vec{T} \times \vec{N},$$

and the *torsion* $\tau$ of the curve is defined by

$$\tau = |d\vec{B}/ds|.$$

**a)** Show that $\vec{B} = \dfrac{\vec{v} \times \vec{a}}{|\vec{v} \times \vec{a}|}$.

**b)** Compute the binormal and the torsion of the helix

$$\vec{P}(t) = \langle a,\ \cos \omega t,\ a \sin \omega t,\ bt \rangle.$$

**c)** Show that, for any curve, if $\tau \equiv 0$ then $\vec{B}$ is constant, and $\vec{P} \cdot \vec{B}$ is constant; so the path of the motion lies in a plane with normal $\vec{B}$.

**30.** The spin of a ball is described by a vector $\vec{S}$; the magnitude $|\vec{S}|$ is the angular velocity of the spin, and the direction of $\vec{S}$ is the spin axis, oriented right-handed. If the velocity is $\vec{v}$, the spin causes an acceleration

$$\vec{a}_{spin} = c\vec{S} \times \vec{v},$$

where $c$ is a constant depending on the nature of the ball and its surrounding medium. Taking gravity into account, but ignoring the slowing effect of friction, the total acceleration is

$$\vec{a} = c\vec{S} \times \vec{v} - g\vec{k}. \qquad (*)$$

Suppose that $\vec{S} = \omega \vec{k}$, with $\omega$ the angular velocity of the spin, assumed constant.

**a)** Show that $\vec{P} = \langle a \cos(c\omega t),\ a \sin(c\omega t),\ bt - gt^2/2 \rangle$ solves the equation of motion $(*)$, if the constant $a$ is appropriately chosen.

**b)** Determine the curvature of the "$xy$" part of the motion, $a \cos(c\omega t)\vec{i} + a \sin(c\omega t)\vec{j}$.

**c)** For a baseball, reasonable values of the constants are $c = .001$ ft/sec, $\omega = 240$ rad/sec; and a reasonable initial velocity is $\vec{v}_0 = \langle 0, 120, 0 \rangle$. For these values, determine $a$ and $b$ in part a, and the curvature in part b.

**\*d)** The value of the constant $a$ in part c starts the ball at $(500, 0, 0)$. If the ball traveled 60' (the distance from the pitching mound to home plate) with its initial velocity $\vec{v}_0 = \langle 0, 120, 0 \rangle$, it would be at $(500, 60, 0)$ in 1/2 second. Where is it at that time, according to the solution in part a?

## 14.6

## QUADRIC SURFACES. CYLINDERS

The graph of an equation of the special form

$$ax + by + cz + d = 0 \qquad (a, b, c \text{ not all } 0)$$

is a plane. Other equations in three variables $(x, y, z)$ usually define curved surfaces. To visualize these surfaces, we take their intersections with convenient planes, as in the following examples.

**Quadric surfaces** are defined by quadratic polynomial equations of the form

$$Ax^2 + By^2 + Cz^2 + Dxy + Exz + Fyz + Gx + Hy + Iz = \text{constant}.$$

The plane sections turn out to be conics, the curves discussed in Chapter 12. The name of each surface is determined by a "majority vote" of the plane sections.

---

**EXAMPLE 1**

$$\frac{x^2}{a^2} + \frac{y^2}{b^2} + \frac{z^2}{c^2} = 1. \tag{1}$$

*The xy-plane* has equation $z = 0$. To intersect the graph of (1) with this plane, set $z = 0$ in (1); the intersection is defined by

$$z = 0, \qquad \frac{x^2}{a^2} + \frac{y^2}{b^2} = 1.$$

By Section 12.2, this curve is an ellipse (Fig. 1).
*The xz-plane* intersected with (1) gives

$$y = 0, \qquad \frac{x^2}{a^2} + \frac{z^2}{c^2} = 1$$

again an ellipse (Fig. 1).
*The yz-plane,* intersected with (1), gives

$$x = 0, \qquad \frac{y^2}{b^2} + \frac{z^2}{c^2} = 1$$

yet another ellipse (Fig. 1).

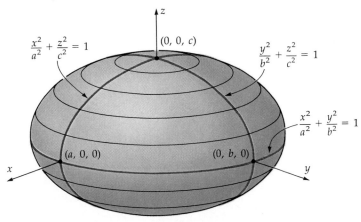

**FIGURE 1**
$\dfrac{x^2}{a^2} + \dfrac{y^2}{b^2} + \dfrac{z^2}{c^2} = 1$, ellipsoid.

All other plane sections are ellipses (or single points). For example, the section by a horizontal plane $z = z_0$ is

$$z = z_0, \qquad \frac{x^2}{a^2} + \frac{y^2}{b^2} = 1 - \left(\frac{z_0}{c}\right)^2,$$

which is an ellipse if $|z_0| < c$, a point if $|z_0| = c$, and empty if $|z_0| > c$. The latter two possibilities are considered "degenerate" ellipses. With this understanding, we can say that all plane sections are ellipses; by a unanimous vote, the surface is called an **ellipsoid**.

EXAMPLE 2   $\dfrac{x^2}{a^2} + \dfrac{y^2}{b^2} - \dfrac{z^2}{c^2} = 1$. The sections in the coordinate planes are:

$$x = 0, \qquad \frac{y^2}{b^2} - \frac{z^2}{c^2} = 1 \qquad \text{(hyperbola in } yz\text{-plane)}$$

$$y = 0, \qquad \frac{x^2}{a^2} - \frac{z^2}{c^2} = 1 \qquad \text{(hyperbola in } xz\text{-plane)}$$

$$z = 0, \qquad \frac{x^2}{a^2} + \frac{y^2}{b^2} = 1 \qquad \text{(ellipse in } xy\text{-plane).}$$

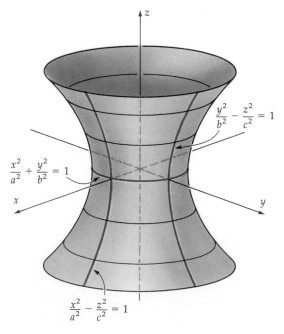

**FIGURE 2**

Hyperboloid of one sheet, $\dfrac{x^2}{a^2} + \dfrac{y^2}{b^2} - \dfrac{z^2}{c^2} = 1$.

Further, all sections parallel to the $xy$-plane are ellipses,

$$z = z_0, \qquad \frac{x^2}{a^2} + \frac{y^2}{b^2} = 1 + \frac{z_0^2}{c^2}.$$

To visualize the graph, sketch these ellipses and "tie them together" with the two hyperbolic sections (Fig. 2).

The sections in two directions out of three are hyperbolas; the majority determines that this surface is called a **hyperboloid** (Fig. 2). In contrast to the next example (Fig. 3) it is a **hyperboloid of one sheet**.

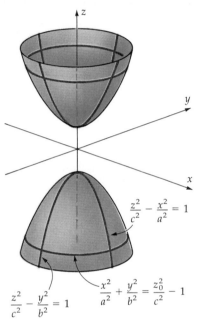

**FIGURE 3**
Hyperboloid of two sheets,
$$-\frac{x^2}{a^2} - \frac{y^2}{b^2} + \frac{z^2}{c^2} = 1.$$

**EXAMPLE 3** $\quad -\dfrac{x^2}{a^2} - \dfrac{y^2}{b^2} + \dfrac{z^2}{c^2} = 1$. The coordinate plane sections are:

$$x = 0, \qquad \frac{z^2}{c^2} - \frac{y^2}{b^2} = 1 \qquad \text{(hyperbola)}$$

$$y = 0, \qquad \frac{z^2}{c^2} - \frac{x^2}{a^2} = 1 \qquad \text{(hyperbola)}$$

$$z = 0, \qquad -\frac{x^2}{a^2} - \frac{y^2}{b^2} = 1 \qquad \text{(empty)}.$$

This last equation has no solution because the left-hand side is $\leq 0$, and cannot equal 1. So try a plane $z = z_0$ *parallel* to the $xy$-plane:

$$z = z_0, \qquad -\frac{x^2}{a^2} - \frac{y^2}{b^2} + \frac{z_0^2}{c^2} = 1 \quad \text{or} \quad \frac{x^2}{a^2} + \frac{y^2}{b^2} = \frac{z_0^2}{c^2} - 1.$$

If $|z_0| < c$ then $\dfrac{z_0^2}{c^2} - 1 < 0$, and there is again no solution. But if $|z_0| > c$ then $\dfrac{z_0^2}{c^2} - 1 > 0$, and the curve is an ellipse.

Since the sections in two out of three standard directions are hyperbolas, this is a **hyperboloid**, of **two sheets**, since the graph has two parts, one for $z \geq c$, another for $z \leq -c$ (Fig. 3).

**EXAMPLE 4** $\quad \dfrac{x^2}{a^2} + \dfrac{y^2}{b^2} - \dfrac{z^2}{c^2} = 0$. The sections in the coordinate planes are:

$$x = 0, \qquad \frac{y^2}{b^2} - \frac{z^2}{c^2} = 0, \quad \text{or} \quad z = \pm\frac{c}{b} y \qquad \text{(two lines)}$$

$$y = 0, \qquad \frac{x^2}{a^2} - \frac{z^2}{c^2} = 0, \quad \text{or} \quad z = \pm\frac{c}{a} x \qquad \text{(two lines)}$$

$$z = 0, \qquad \frac{x^2}{a^2} + \frac{y^2}{b^2} = 0 \qquad \text{[one point, } (0, 0, 0)\text{]}.$$

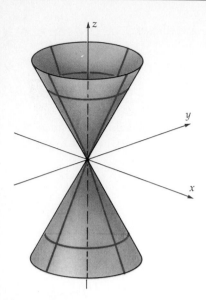

**FIGURE 4**

Double cone $\dfrac{x^2}{a^2} + \dfrac{y^2}{b^2} - \dfrac{z^2}{c^2} = 0$.

This "one-point" intersection is not very revealing. To get a better picture, try other sections parallel to the $xy$-plane:

$$z = z_0, \qquad \frac{x^2}{a^2} + \frac{y^2}{b^2} = \frac{z_0^2}{c^2} \qquad \text{(ellipse).}$$

Sketch these ellipses, tie them together by the straight-line plane sections, and you get a *double cone* with elliptical cross sections (Fig. 4).

In Examples 1–4, all sections parallel to the $xy$-plane are ellipses; but the sections perpendicular to this plane reveal the different natures of the three surfaces. In Examples 1–3, the equations have been "normalized," having sums and differences of squares on the left, and $+1$ on the right. In Example 1, where all signs are $+$, all sections are ellipses, and the surface is an ellipsoid. In Example 2, with *one* minus sign, we have a hyperboloid of *one* sheet. Example 3, with *two* minus signs, gives a hyperboloid of *two* sheets.

Example 4 is different; the right-hand side is 0, not 1, and the surface is a cone. The axis of the cone is revealed by taking sections in the coordinate planes.

## Cylinders

Take the unit circle in the $xy$-plane, and through each point on it pass a vertical line (Fig. 5). The result is a **right circular cylinder**. What is its equation?

To begin, bear in mind that the $xy$-plane in Figure 5 is the plane $z = 0$ in $xyz$ space. So the unit circle in that plane is described by *two* equations

$$z = 0, \qquad x^2 + y^2 = 1.$$

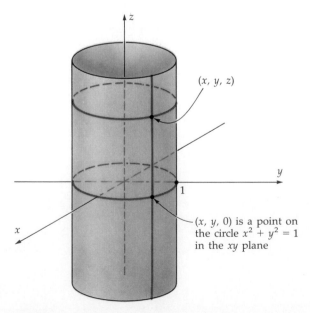

**FIGURE 5**

Cylinder $\{(x, y, z): x^2 + y^2 = 1\}$.

Now take a point $(x, y, 0)$ on this circle. To generate the vertical line through this point, simply let the $z$ coordinate vary, holding $x$ and $y$ fixed; as $z$ varies, you move up and down the vertical line. Algebraically, this means that you drop the restriction $z = 0$, but retain $x^2 + y^2 = 1$, since the original point $(x, y, 0)$ must be on the circle. The equation of the cylinder is thus

$$x^2 + y^2 = 1;$$

the cylinder itself is the set

$$\{(x, y, z): x^2 + y^2 = 1\}.$$

*Any* equation in $(x, y, z)$ where the variable $z$ does not actually appear describes a kind of cylinder, formed by all the vertical lines passing through an appropriate curve in the $xy$-plane. Generally, if any one of the three variables is missing, the graph is a cylinder formed of lines parallel to the axis of the missing variable.

---

**EXAMPLE 5**   $y = x^2$ gives a parabola in the $xy$-plane. Viewed as an equation in $(x, y, z)$, it defines a **parabolic cylinder** (Fig. 6).

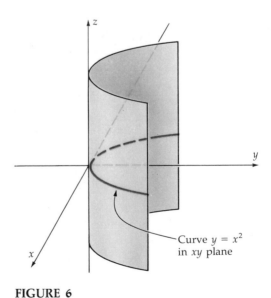

**FIGURE 6**
Cylinder $\{(x, y, z): y = x^2\}$.

---

**EXAMPLE 6**   In $z = e^y$, the variable $x$ is missing; so the graph is the cylinder formed by all lines parallel to the $x$-axis, passing through the exponential curve

$$x = 0, \qquad z = e^y$$

in the $yz$-plane (Fig. 7). (This, of course, is not a quadric surface.)

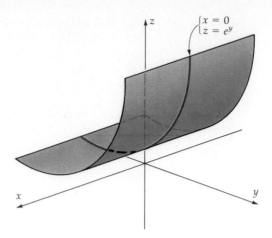

**FIGURE 7**
Cylinder $\{(x, y, z): z = e^y\}$.

## Concluding Remarks

These examples give most of the possible shapes for quadric surfaces (the summary contains two more). Each example could be varied by shifting its center or vertex from the origin to some other point, and then you would need plane sections through that point. It could also be rotated, but then you would need more sophisticated algebraic methods to deduce the nature of the surface from its equation.

## SUMMARY

*Standard Equations of Conic Surfaces*

*Ellipsoid (see Fig. 1):*

$$\frac{x^2}{a^2} + \frac{y^2}{b^2} + \frac{z^2}{c^2} = 1 \qquad \text{(all } + \text{ on left, } +1 \text{ on right)}$$

*Hyperboloid of One Sheet (see Fig. 2):*

$$\frac{x^2}{a^2} + \frac{y^2}{b^2} - \frac{z^2}{c^2} = 1 \qquad \text{(one } - \text{ on left, } +1 \text{ on right)}$$

*Hyperboloid of Two Sheets (see Fig. 3):*

$$-\frac{x^2}{a^2} - \frac{y^2}{b^2} + \frac{z^2}{c^2} = 1 \qquad \text{(two } - \text{ on left, } +1 \text{ on right)}$$

*Double Cone (see Fig. 4):*

$$\frac{x^2}{a^2} + \frac{y^2}{b^2} - \frac{z^2}{c^2} = 0 \qquad \text{(one or two } - \text{ on left, } 0 \text{ on right)}$$

*Elliptic Paraboloid (Fig. 8):*

$$z = \frac{x^2}{a^2} + \frac{y^2}{b^2}$$

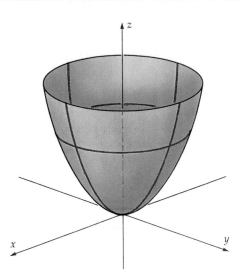

**FIGURE 8**

Elliptic paraboloid $z = \dfrac{x^2}{a^2} + \dfrac{y^2}{b^2}$.

*Hyperbolic Paraboloid (Fig. 9):*

$$z = \frac{x^2}{a^2} - \frac{y^2}{b^2}$$

*Cylinder Parallel to z-axis:*

$$\{(x, y, z)\colon f(x, y) = 0\} \qquad \text{(equation with no } z).$$

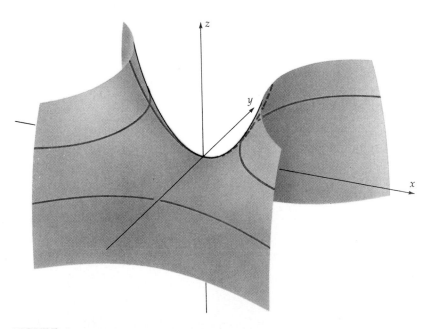

**FIGURE 9**

Hyperbolic paraboloid $z = \dfrac{x^2}{a^2} - \dfrac{y^2}{b^2}$.

## PROBLEMS

### A

**1.** Carefully sketch the section in each coordinate plane, and try to sketch the surface. One is a "prolate spheroid" (football-shaped), the other an "oblate spheroid" (discus-shaped).

a) $x^2 + \dfrac{y^2}{4} + \dfrac{z^2}{4} = 1$

b) $x^2 + 4y^2 + 4z^2 = 1$

**\*2.** Take sections parallel to each coordinate plane, identify them, and name the figure appropriately as an ellipsoid, hyperboloid of one sheet, or hyperboloid of two sheets. Find any intercepts with the coordinate axes, and sketch the surface.

a) $x^2 + y^2 + 4z^2 = 1$
b) $x^2 - y^2 + z^2 = 1$
c) $x^2 + y^2 - z^2 = -9$
d) $x^2 + y^2 - 2z^2 = 5$

**3.** Repeat problem 2 for the following:

a) $x^2 - \dfrac{y^2}{2} + z^2 = 1.$

b) $\dfrac{x^2}{4} + y^2 + z^2 = 3.$

c) $x^2 - y^2 + z^2 = -1.$
d) $x^2 - y^2 + 2y + 2z^2 = 0.$

**4.** The equation $z = x^2 + 2y^2$ defines an *elliptic paraboloid*. Take sections in the planes $x = 0$, $y = 0$, $z = z_0$; sketch the figure; and explain the name.

**5.** The equation $y = x^2 + z^2$ defines a *circular paraboloid*, or *paraboloid of revolution*. Take sections in the planes $x = 0$, $y = y_0$, and $z = 0$; sketch the figure; and explain the names.

**\*6.** Take sections of the graph of $z = xy$ by the vertical planes $y = x$ and $y = -x$, and the horizontal planes $z = \pm 1$. Devise an appropriate name for this surface. Try to sketch it.

In problems 7–20, sketch the surface, and identify it as one of the following types: Plane, cylinder (circular, elliptical, parabolic, hyperbolic, or other), ellipsoid (or sphere), paraboloid (circular, elliptic, or hyperbolic), hyperboloid (of one or two sheets), cone. All are to be considered as equations in $(x, y, z)$ space.

**\*7.** $(x - 1)^2 + y^2 + z^2 = 1$

**8.** $x^2 + (y + 1)^2 = 2$

**\*9.** $y^2 + z^2 = 9$      **10.** $x = y$

**\*11.** $x^2 = y^2$      **\*12.** $y = \cos x$

**13.** $x^2 + y^2 + z^2 = 9$      **14.** $x^2 + y^2 + z = 9$

**\*15.** $2x^2 + y^2 - z = 9$      **\*16.** $xz = -1$

**17.** $x^2 - y^2 = 1$      **18.** $e^{x^2 + y^2 - z^2} = 1$

**19.** $z = x^2 + y^2$      **20.** $z = \sqrt{x^2 + y^2}$

### B

**21.** Given a point $F$ in space and a plane $\pi$ not containing $F$, describe the locus of all points $P$ in space such that $|PF| = |PD|$, where $D$ is the foot of the perpendicular from $P$ to the plane $\pi$.

**22.** Given two points $F_1$ and $F_2$ in space, and a number $2a > |F_1 F_2|$, describe the locus of all points $P$ in space such that $|PF_1| + |PF_2| = 2a$.

**23.** Given two points $F_1$ and $F_2$ in space, and a number $2a < |F_1 F_2|$, describe the locus of all points $P$ in space such that $|PF_1| - |PF_2| = \pm 2a$.

## REVIEW PROBLEMS    CHAPTER 14

**\*1.** For $\vec{v} = \langle 1, 2, 3 \rangle$ and $\vec{w} = 2\vec{i} + 3\vec{j} + 4\vec{k}$, compute.

a) $\vec{v} - 2\vec{w}$    b) $\vec{v} \cdot \vec{w}$
c) $\vec{v} \times \vec{w}$    d) $|\vec{v} - 2\vec{w}|$
e) The angle between $\vec{v}$ and $\vec{w}$

f) The area of the parallelogram spanned by $\vec{v}$ and $\vec{w}$

**\*2.** Let $P = (-2, 2, 1)$, $Q = (0, 4, 2)$, and $R = (-1, 3, 3)$. Determine

a) The distance from $P$ to $Q$.

**b)** The midpoint of side $PQ$.

**c)** The equation of the line through $P$ and $Q$.

**d)** The intersection of the line through $P$ and $Q$ with the $xy$-plane.

**e)** The cosine of the angle between sides $RP$ and $QP$.

**f)** The area of triangle $PRQ$.

**g)** The equation of the plane through $P$, $Q$, and $R$.

**h)** The distance from the origin to the plane through $P$, $Q$, and $R$.

**i)** The equation of the line through $P$, normal to triangle $PQR$.

**j)** The equation of the sphere with $PQ$ as a diameter.

**k)** Whether $R$ is on the sphere in part j.

**3.** Prove that the given formula is true in all cases, *or else* give an example showing that it is false in some cases.

**\*a)** $\vec{v} \times \vec{w} = \vec{w} \times \vec{v}$

**\*b)** $\vec{v} \cdot \vec{w} = \vec{w} \cdot \vec{v}$

**c)** $\vec{u} \times (\vec{v} \times \vec{w}) = (\vec{u} \times \vec{v}) \times \vec{w}$

**d)** $|\vec{v} \times \vec{w}|^2 = |\vec{v}|^2 \cdot |\vec{w}|^2 - (\vec{v} \cdot \vec{w})^2$

**\*4.** The position function for a bead sliding down a helical wire is $\vec{P}(t) = \langle \sin(gt^2/4),\ \cos(gt^2/4),\ -gt^2/4 \rangle$, $t \geq 0$; $g = 9.8$ is the acceleration due to gravity. Determine

**a)** The position, velocity, and acceleration when $t = 1$.

**b)** The time required to complete one turn, starting from $t = 0$.

**c)** Parametric equations of the line tangent to the bead's motion at $t = 1$.

**d)** The length of the path traversed for $0 \leq t \leq 1$.

**e)** The unit tangent $\vec{T}$, for all time $t \geq 0$.

**f)** $d\vec{T}/ds$.

**g)** The curvature $\kappa$.

**h)** The principal normal vector at time $t$.

**i)** The normal and tangential components of acceleration.

**j)** Whether or not $\frac{1}{2}|\vec{v}|^2 = -gz$. (This equation equates the kinetic energy to the loss in potential energy; it should be satisfied, when friction is ignored.)

**\*5.** Suppose that $\vec{a}(t) = 2\cos(t)\vec{i} + 3\sin(t)\vec{j}$, $\vec{v}(0) = \vec{k}$, and $\vec{P}(0) = \vec{0}$. Find $\vec{P}(t)$ for all $t$.

**6.** Suppose that $\vec{P}(t)$ and $\vec{Q}(t)$ are differentiable functions of $t$. State and prove a formula for

**a)** $\dfrac{d(\vec{P} + \vec{Q})}{dt}$.

**b)** $\dfrac{d(\vec{P} \cdot \vec{Q})}{dt}$.

**c)** $\dfrac{d(\vec{P} \times \vec{Q})}{dt}$.

**7.** Let $\vec{N}(t)$ be the unit principal normal to a curve. Is it true that $d\vec{N}/dt$ is perpendicular to $\vec{N}$? If so, prove it.

**\*8.** Describe the intersections of the graph of the given equation with appropriate planes; from these, deduce the nature of the graph in $(x, y, z)$ space.

**a)** $x^2 + (y + 1)^2 = 3$

**b)** $x + y^2 + z^2 = 5$

**c)** $x^2 + 2y^2 + z^2 = 1$

**d)** $x^2 - 4y^2 + z^2 = 0$

**e)** $x^2 - 4y^2 + z^2 = 1$

**f)** $x^2 - 4y^2 + z^2 = -1$

**g)** $y = \frac{1}{2}\sqrt{x^2 + z^2}$

**h)** $x^2 + y^2 + z = 0$

**i)** $x + y + z = 0$

**j)** $x = y$

**k)** $y = x^2$

**l)** $x^2 + 2y^2 + 3z^2 = 0$

**m)** $x^2 + 2(y - 1)^2 + z^2 = -1$

# 15

# FUNCTIONS OF TWO VARIABLES

Most things depend on more than one variable. For example, the area of a rectangle

$$A = lw$$

is a function of the two variables $l$ (length) and $w$ (width). The temperature $T$ at a point $(x, y, z)$ in space, at time $t$, is a function of the four variables $x, y, z, t$, so we write

$$T = f(x, y, z, t).$$

Functions of several variables present interesting new features not found in the one-variable theory. Most of these are easiest to visualize in the case of two variables, so we begin with that. The next chapter develops these ideas further, and extends them to functions of more variables.

# 15.1

## GRAPHS AND LEVEL CURVES

The *domain* of a function of two variables $x$ and $y$ is the set of all points $(x, y)$ for which the function is defined; to each point in its domain, the function $f$ assigns a definite number $f(x, y)$. The set of all numbers obtained in this way is called the *range* of $f$.

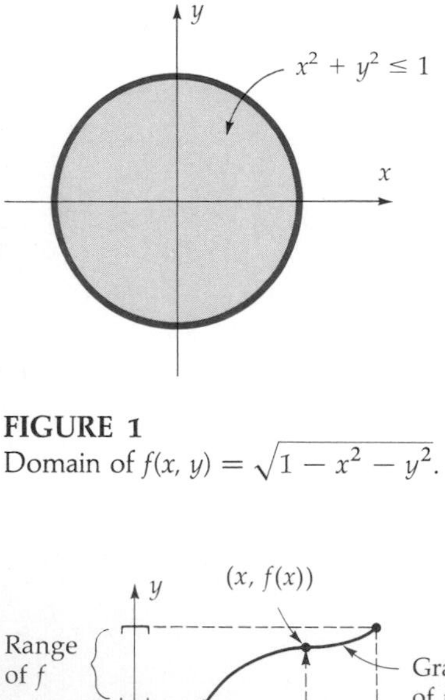

---

**EXAMPLE 1** Let $f(x, y) = \sqrt{1 - x^2 - y^2}$. Compute $f(0, 0)$, $f(1, 0)$, $f(1/2, -1/3)$. Find the domain of $f$.

**SOLUTION**

$$f(0, 0) = \sqrt{1 - 0^2 - 0^2} = 1$$

$$f(1/2, -1/3) = \sqrt{1 - (1/2)^2 - (-1/3)^2} = \sqrt{23/6}.$$

The domain of $f$ consists of all points $(x, y)$ such that $\sqrt{1 - x^2 - y^2}$ is defined. This square root is defined if and only if $1 - x^2 - y^2 \geq 0$, that is, when

$$x^2 + y^2 \leq 1.$$

So the domain is the set of all points $(x, y)$ inside or on the unit circle (Fig. 1). The range of $f$ is revealed in the following discussion of the *graph*.

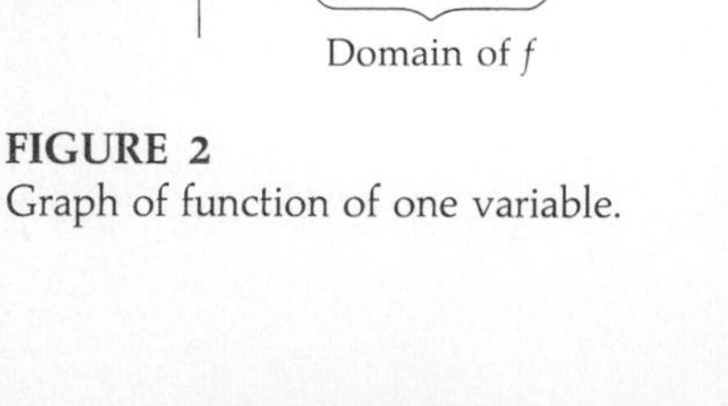

---

**The Graph**

For a function $f(x)$ of one variable, the graph of $f$ consists of all points $(x, f(x))$ in the plane, where $x$ is in the domain of $f$ (Fig. 2). We need one axis for the "inputs" $x$, and another for the "outputs" $y = f(x)$.

To graph a function $f(x, y)$ of *two* variables, we need *two* axes for the inputs $x$ and $y$, and a third (the $z$-axis) for the outputs; the graph consists of all points $(x, y, f(x, y))$, where $(x, y)$ is in the domain of $f$ (Fig. 3). This usually gives some kind of surface in three-dimensional space. In set notation, the graph of $f$ is

$$\{(x, y, z) \colon z = f(x, y)\}.$$

The equation of the graph is $z = f(x, y)$.

**FIGURE 1**
Domain of $f(x, y) = \sqrt{1 - x^2 - y^2}$.

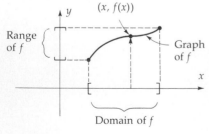

**FIGURE 2**
Graph of function of one variable.

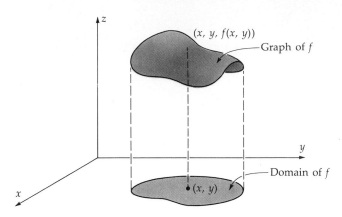

**FIGURE 3**
Domain and graph of $f$.

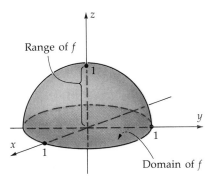

**FIGURE 4**
Graph of $f(x, y) = \sqrt{1 - x^2 - y^2}$.

**EXAMPLE 1 (cont.)**   For $f(x, y) = \sqrt{1 - x^2 - y^2}$, the equation of the graph is

$$z = \sqrt{1 - x^2 - y^2}. \tag{1}$$

This implies that $z^2 = 1 - x^2 - y^2$, or

$$x^2 + y^2 + z^2 = 1.$$

Hence the graph lies on the sphere of radius 1 centered at the origin. It is not, however, the entire sphere; equation (1) implies that $z \geq 0$, so the graph is just the upper half of the sphere, the "northern hemisphere" together with the equator (Fig. 4). The projection of the graph on the $xy$-plane is the domain of $f$, the disk in Figure 1. The projection on the $z$-axis is the range $[0, 1]$.

## Level Curves

It can be a severe artistic challenge to sketch the graph of a function $f(x, y)$; level curves offer a simpler representation. A level curve $f$ is a set in the plane consisting of all those points $(x, y)$ where $f(x, y)$ has a fixed value $c$, the set

$$\{(x, y): f(x, y) = c\}.$$

In abbreviated form, the level curve is denoted $\{f = c\}$. A sketch of these plane curves for several well-chosen equally spaced values $c$ gives a good representation of the function.

**EXAMPLE 2**   Sketch level curves of the function $f(x, y) = 2 - x^2 - y^2$.

**SOLUTION**   The equation of the level curve $\{f = c\}$ is

$$2 - x^2 - y^2 = c, \quad \text{or} \quad x^2 + y^2 = 2 - c.$$

If $2 - c < 0$ there is no solution; if $2 - c > 0$ the curve is a circle of radius $\sqrt{2 - c}$. Figure 5a shows the level curves for $c = 0, \pm 1, \pm 2$; $f(0, 0) = 2$ is a **maximum** value, since the function values decrease as you move away from the point $(0, 0)$.

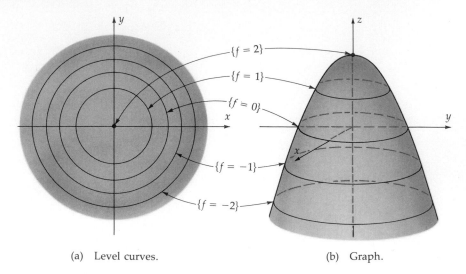

(a)   Level curves.                              (b)   Graph.

**FIGURE 5**
$f(x, y) = 2 - x^2 - y^2$.

The level curves help to visualize the graph of $f$ (Fig. 5b). Raise (or lower) each curve to the given level, and you have a curve which forms a part of the graph. Where the level curves are closer together, the graph is steeper.

## Plane Sections of a Graph

The part of the graph lying over the level curve $\{f = c\}$ is the intersection of the graph with the plane $z = c$ (Fig. 6), a horizontal plane section. Think of the graph as the surface of a hill. Then the horizontal section with $z = c$

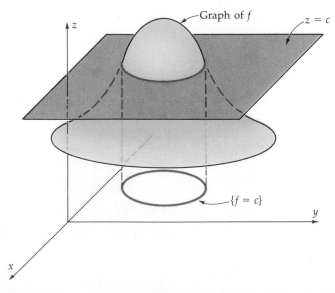

**FIGURE 6**
Level curve and horizontal section of graph.

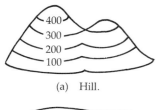

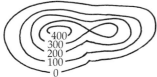

(a)  Hill.

(b)  Topographic map.

**FIGURE 7**

forms a path along the hill at constant elevation $c$; along this path, you walk neither up nor down. The level curve $\{f = c\}$ is the projection of this path on the $xy$-plane. Topographic maps show these curves for equally spaced elevations, say, every 100 meters (Fig. 7).

By combining the horizontal sections with a few well-chosen vertical ones, you get a sort of framework. The graph can then be visualized as a flexible sheet stretched over the frame.

**EXAMPLE 3**  For the function $f(x, y) = 2 - x^2 - y^2$, find the vertical plane sections in the $xz$-plane and the $yz$-plane.

**SOLUTION**  The equation of the $yz$-plane is $x = 0$. To get the section of the graph in this plane, set $x = 0$ in the equation of the graph:

$$z = 2 - 0^2 - y^2 = 2 - y^2.$$

Graph this curve in the $yz$-plane (Fig. 8).

For the section in the $xz$-plane, set $y = 0$ in the equation of the graph:

$$z = 2 - x^2 - 0^2 = 2 - x^2.$$

Graph this curve in the $xz$-plane (Fig. 8); this requires extra thought, since the $xz$-plane is seen at an angle. These two curves, together with a few horizontal sections, give a framework for the graph.

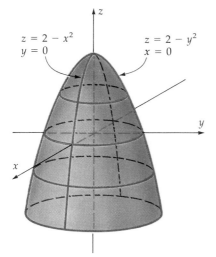

$z = 2 - x^2$
$y = 0$

$z = 2 - y^2$
$x = 0$

**FIGURE 8**
Sections of graph of
$f(x, y) = 2 - x^2 - y^2$.

**EXAMPLE 4**  Sketch level curves for $f(x, y) = xy$. Describe the graph.

**SOLUTION**  The equations of the level curves are

$$xy = c. \qquad (2)$$

When $c = 0$ then either $x = 0$ or $y = 0$, so the level curve $\{f = 0\}$ consists of just the two coordinate axes (Fig. 9). When $c \neq 0$, then $x \neq 0$, so equation (2) can be solved for $y$:

$$y = c/x.$$

These curves are hyperbolas, with the $x$- and $y$-axes as asymptotes (Fig. 9). The value of $f$ is marked on each curve.

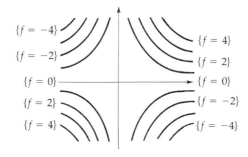

$\{f = -4\}$  $\{f = 4\}$
$\{f = -2\}$  $\{f = 2\}$
$\{f = 0\}$  $\{f = 0\}$
$\{f = 2\}$  $\{f = -2\}$
$\{f = 4\}$  $\{f = -4\}$

**FIGURE 9**
Level curves of $f(x, y) = xy$.

The sketch shows that $f$ has no maximum and no minimum—it is positive in the first and third quadrants, negative in the second and fourth, and zero on the axes. Viewed as a topographic map, this describes a mountain pass. Traversing from lower right to upper left, you climb up to the pass at the origin, then descend on the other side. A graph with this shape is called a **saddle**.

The vertical sections in the $xz$-plane and $yz$-plane are not helpful in this case—they coincide with the $x$-and $y$-axes. Instead, we use the sections in the vertical planes $y = x$ and $y = -x$. In the plane $y = x$,

$$z = xy = x^2,$$

the equation of a parabola opening upward. In the plane $y = -x$,

$$z = xy = -x^2,$$

the equation of a parabola opening downward. Figure 10 shows these two sections, and a number of horizontal ones. The $x$-axis has been drawn at an unusual angle so you can see the top of the graph.

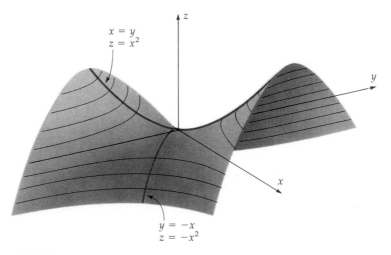

**FIGURE 10**
Graph of $f(x, y) = xy$. The $x$ and $y$ axes lie in the graph.

## Linear Functions of Two Variables

A function of the form

$$f(x) = mx + b \qquad (m \text{ and } b \text{ constant})$$

is called **linear**, since its graph is a straight line. A function of two variables of the form

$$f(x, y) = ax + by + c \qquad (a, b, c \text{ constant}) \tag{3}$$

is also called linear. The equation of the graph is

$$z = ax + by + c \quad \text{or} \quad ax + by - z + c = 0. \tag{4}$$

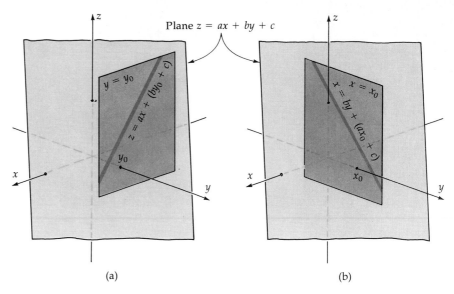

**FIGURE 11a**

$z = ax + by + c$ intersects $y = y_0$ in a line of slope $a$.

**FIGURE 11b**

$z = ax + by + c$ intersects $x = x_0$ in a line of slope $b$.

So the graph is a plane, and all its plane sections are straight lines. (This justifies calling the function linear.) The normal to the graph is given by the coefficients of $x$, $y$, and $z$ in the last equation: $\vec{n} = \langle a, b, -1 \rangle$.

The coefficients $a$ and $b$ have an interpretation as *slopes*. Consider the intersection of the graph of $f$ with the vertical plane $\{y = y_0\}$, parallel to the $x$-axis (Fig. 11a). Setting $y = y_0$ in (4) gives

$$z = ax + (by_0 + c);$$

this is the equation of a line with slope $a$. Thus *every section of the graph of f by a vertical plane parallel to the x-axis is a line of slope a*, the coefficient of $x$ in the formula (3) for $f$. We call this coefficient the "$x$-slope" of the graph.[1] It gives the increase in $z$ for each unit increase in $x$, with $y$ held constant.

The section of the graph in a vertical plane $\{x = x_0\}$ parallel to the $y$-axis is likewise a straight line (Fig. 11b)

$$z = ax_0 + by + c = by + (ax_0 + c).$$

The slope of this line is b, the coefficient of $y$ in the formula (4) defining $f$. We call this the "$y$-slope" of the graph; it gives the increase in $z$ for each unit increase in $y$, with $x$ held constant.

Given a point $P_0 = (x_0, y_0, z_0)$ and an $x$-slope $a$ and $y$-slope $b$, there is a unique plane passing through $P_0$ with these prescribed slopes; it has the equation

$$z = z_0 + a(x - x_0) + b(y - y_0). \tag{5}$$

---

[1] This terminology is nonstandard but, we hope, easily understood.

This is the two-variable version of the familiar equation

$$y = y_0 + m(x - x_0)$$

for the line of slope $m$, through a given point $(x_0, y_0)$.

---

**EXAMPLE 5**  Find a linear function $f(x, y) = ax + by + c$ with $f(2, 1) = 3$, whose graph has $x$-slope 1 and $y$-slope $-1$.

*SOLUTION*  The graph of $f$ is a plane with $x$-slope 1 and $y$-slope $-1$; and the condition $f(2, 1) = 3$ implies that it passes through the point $(2, 1, 3)$. The equation of this plane is, by (5):

$$\overset{x\text{-slope}}{\underset{\underset{f(2,\,1)\,=\,3}{\uparrow}}{z = 3 + \overset{\downarrow}{1}(x - 2) + (\overset{\overset{y\text{-slope}}{\downarrow}}{-1})(y - 1)}}.$$

The linear function with this graph is

$$f(x, y) = 3 + (x - 2) - (y - 1) = x - y + 2.$$

---

## SUMMARY

*Level Curves:*   $\{f = c\}$ is the set in the plane $\{(x, y): f(x, y) = c\}$.

*Plane through $(x_0, y_0, z_0)$ with x-slope a and y-slope b:*

$$z = z_0 + a(x - x_0) + b(y - y_0).$$

---

## PROBLEMS

### A

1.  Evaluate $f(0, 0), f(-1, 2), f(t, t^2), f(u + v, u - v)$.
    *a)   $f(x, y) = 1 + x$
    *b)   $f(x, y) = e^{x+y}$
    c)   $f(x, y) = x^2 + 2y^2$
    d)   $f(x, y) = 1$

2.  Describe the domain of each function. If the domain has a boundary, use a dashed line to indicate any part of the boundary that is excluded from the domain, and a solid line to indicate any included part.

    *a)   $\sqrt{2 - x^2 - y^2}$         b)   $x/y$

    *c)   $\sqrt{y/x}$                    d)   $\dfrac{xy}{x^2 + y^2}$

    *e)   $\dfrac{x^2 + y^2}{xy}$         *f)   $\ln xy$

    g)   $\ln(x + y)$                     h)   $\ln |x + y|$

3.  Find the $x$-slope and $y$-slope of the given linear function. Sketch the section in each vertical coordinate plane, and sketch the graph.

*a)   $f(x, y) = x - y + 1$
b)    $f(x, y) = x + 2$
*c)   $f(x, y) = 3$
d)    $f(x, y) = -2x + y - 1$

4.   Write the equation of the linear function $f(x, y) = ax + by + c$ having the given $x$-slope and $y$-slope, and the given value at the given point.
*a)   $x$-slope 3, $y$-slope $-1$, $f(0, 0) = 2$
b)    $x$-slope $-1$, $y$-slope 2, $f(-1, 1) = -1$

5.   Sketch level curves for levels $-2, -1, 0, 1, 2$. Then describe the graph. Imagine yourself standing at the origin, looking around—in which directions does the ground rise? In which directions is it level? Where are there ridges or valleys?
*a)   $x^2 y$          b)   $x^2 + y^2$
*c)   $x^2 - y^2$      d)   $2 - x^2 - y^2$
*e)   $y^2 - x$        f)   $x + y$
g)    $\sqrt{x^2 + y^2}$   h)   $x^2 + y^2 - 2x$

6.   Let $f(x, y) = \dfrac{2x}{1 + x^2 + y^2}$. Show that the level curve $\{f = c\}$ is
a)   A point if $c = 1$ or $c = -1$.
b)   A line if $c = 0$.
c)   A circle if $0 < |c| < 1$.
d)   Empty if $|c| > 1$.
Sketch level curves for $c = \pm 1, \pm 2/3, \pm 1/3, 0$, and describe the graph, noting any "hills" or "hollows."

7.   For the given function, sketch the sections of the graph in the $xz$-plane and $yz$-plane, and several horizontal sections. Describe the graph.
*a)   $x + y$          *b)   $x^2 + y^2$
*c)   $x^2 - y^2$       d)   $1 + x^2 - y^2$
e)    $2x - y + 1$     *f)   $\sqrt{x^2 + y^2}$
g)    $2 - x^2 - y^2$   h)   $y^2 - x$
*i)   $(x^2 + y^2)^{-1}$  j)   $x^2 + y^2 - 2y$
k)    $2x^2 - y^2$

8.   Sections of a graph in the $xz$-plane and $yz$-plane are given, and the level curves are described. Describe the shape of the resulting graph, and sketch it.

|      | Section in $xz$-plane | Section in $yz$-plane | Level curves |
|------|------|------|------|
| *a)  | $z = x$ | $z = y$ | straight lines |
| b)   | $z = |x|$ | $z = |y|$ | circles centered at $(0, 0)$ |
| *c)  | $z = x^2$ | $z = y^2$ | circles centered at $(0, 0)$ |
| d)   | $z = x^3$ | $z = -y^3$ | straight lines |

**B**

*9.   For each graph in problem 8, obtain a formula for $f(x, y)$. (Answers given for parts a and d.)

10.  Sketch level curves for levels $0, \pm 1, \pm 2$. Describe the graph.
a)   $x/y$      *b)   $e^{-x} \cos y$

11.  Let $f(x, y) = \dfrac{xy}{x^2 + y^2}$. Show that the level curves of $f$ are straight lines through the origin (but with the origin deleted). Find the maximum and minimum values of $f$. (Polar coordinates are useful here, but not necessary.)

12.  The level curves of $f(x, y) = x^3 - 3xy^2$ are hard to sketch, but the vertical plane sections of the graph are easy. Sketch the sections in the $xz$-plane, the $yz$-plane, and the planes $x = \pm 1, \pm 2, \pm 3$. The graph is called a "monkey saddle"—why?

*13.  Analyze the graph in the previous problem using polar coordinates. What angles $\theta$ correspond to $f > 0$, and what angles to $f < 0$? (Hint: show that $\cos 3\theta = \cos^3 \theta - 3 \sin^2 \theta \cos \theta$.)

**C**

14.  (Do problems 12 and 13 first.) Saddles are available for animals with any number of appendages hanging down, for example, the turtle (4, considering the tail insignificant), horse (5), elephant or spider (6). Provide a saddle for each of these as the graph of a polynomial $f(x, y)$ in two variables. [Show that the functions $r^n \sin(n\theta)$ in polar coordinates are polynomials in $x$ and $y$.]

15.  a)   Suppose that $f(x, y) = g(x - y)$, where $g$ is a given function of one variable. Describe the level curves of $f$.
     b)   Suppose that the level curves of $f$ are all straight lines of slope 1. Explain why $f(x, y) = g(x - y)$ for some function $g$ of one variable.

16.  Suppose that the level curves of $f$ are straight-line segments from the origin to $\infty$, with the origin deleted. Explain why $f(r \cos \theta, r \sin \theta) = g(\theta)$ for a function $g$ of period $2\pi$.

17.  Find the maximum value of the function $f(x, y) = x + 2y$ over the unit disk $x^2 + y^2 \le 1$. (Hint: Sketch and label several level curves of $f$.)

## 15.2
## CONTINUITY AND LIMITS

Before proceeding further, we need at least an intuitive idea of limits and continuity for functions of several variables.

Figures 1 and 2 illustrate typical **discontinuities**. In Figure 1 the function values $f(x, y)$ become infinite as $(x, y)$ approaches the origin; so $f$ is discontinuous at $(0, 0)$. In Figure 2 there is a "jump" in the function values all along the positive $x$-axis and $y$-axis; that function is discontinuous at those points.

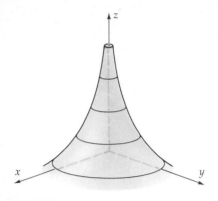

**FIGURE 1**

$f(x, y) = \dfrac{1}{x^2 + y^2}$ is discontinuous at $(0, 0)$.

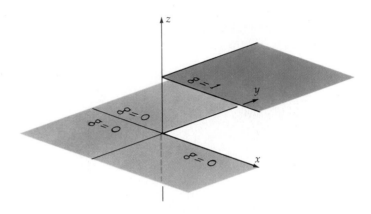

**FIGURE 2**

$g(x, y) = \begin{cases} 1 & \text{in first quadrant} \\ 0 & \text{elsewhere.} \end{cases}$

[Figure rotated 90° for clarity.]

These discontinuities are clear from the graphs; but how could you detect them, given just the definitions of the functions? In the first case,

$$f(x, y) = \frac{1}{x^2 + y^2},$$

the clue is that $f(0, 0) = \dfrac{1}{0^2 + 0^2}$ is undefined; division by zero "explains" the discontinuity at $(0, 0)$ in Figure 1. As for the function $g$ in Figure 2, this is given by *two different* formulas, one for the first quadrant and another for the other three quadrants. The two formulas disagree where the first quadrant meets the others, and this creates a discontinuity.

The rest of this section fills in the technical details underlying the idea of continuity for functions of two variables.

### Limits in the Plane

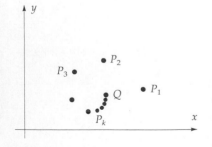

**FIGURE 3**

$\lim_{k \to \infty} P_k = Q$.

A sequence of points in the plane $P_1, P_2, P_3, \ldots, P_k, \ldots$ converges to a limit $Q$ (Fig. 3) if the coordinates of $P_k$ converge to the coordinates of $Q$. That is:

$$(x_k, y_k) \to (a, b)$$

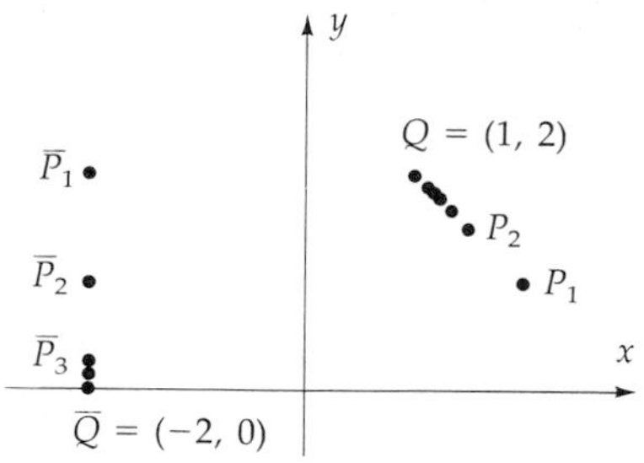

**FIGURE 4**

$$P_k = \left(1 + \frac{1}{k}, \frac{2k-1}{k}\right) \to (1, 2);$$

$$\bar{P}_k = \left(-2, \frac{2}{k^2}\right) \to (-2, 0).$$

means that

$$x_k \to a \quad \text{and} \quad y_k \to b \quad \text{as } k \to \infty.$$

For example (Fig. 4)

$$\left(1 + \frac{1}{k}, \frac{2k-1}{k}\right) \to (1, 2)$$

$$\left(-2, \frac{2}{k^2}\right) \to (-2, 0).$$

## Function Limits

The statement

$$\lim_{(x,y) \to (a,b)} f(x, y) = L$$

means roughly this: As the point $(x, y)$ approaches $(a, b)$, the function values $f(x, y)$ approach $L$. We require that $f$ be defined for all points $(x, y)$ sufficiently close to $(a, b)$, except perhaps at $(a, b)$ itself; and the same limit $L$ must be obtained, no matter *how* $(x, y)$ approaches $(a, b)$. This idea can be expressed in terms of sequence limits.

---

**DEFINITION 1**

$\lim\limits_{(x,y) \to (a,b)} f(x, y) = L$ if and only if

(i) $f(x, y)$ is defined for all points in some disk centered at $(a, b)$, except perhaps at $(a, b)$; and

(ii) for every sequence $(x_k, y_k)$ converging to $(a, b)$, with $(x_k y_k) \neq (a, b)$,

$$\lim_{k \to \infty} f(x_k, y_k) = L.$$

---

(A similar definition can be made for functions of one variable. We didn't do that in Chapter 3, because sequence limits were not rigorously defined until Section 11.2.)

---

**EXAMPLE 1** Show that $\lim\limits_{(x,y) \to (2,1)} xe^{xy} = 2e^2$.

*SOLUTION* Condition (i) is satisfied, since $xe^{xy}$ is defined everywhere. Now check condition (ii). If $(x_k, y_k) \to (2, 1)$, then $\lim\limits_{k \to \infty} x_k = 2$ and $\lim\limits_{k \to \infty} y_k = 1$, so the theorems on sequence limits give

$$\lim_{k \to \infty} x_k e^{x_k y_k} = (\lim x_k)(\lim e^{x_k y_k}) \qquad \text{(limit of a product)}$$

$$= (\lim x_k)e^{\lim x_k y_k} \qquad \text{($e^x$ is a continuous function.)}$$

$$= 2e^2 \qquad \text{($\lim x_k = 2$ and $\lim y_k = 1$).}$$

**EXAMPLE 2** For the function $g$ in Figure 2, $\lim\limits_{(x,y)\to(0,0)} g(x, y)$ does not exist, because different sequences $(x_k, y_k)$ converging to $(0, 0)$ yield different limits for $g(x_k, y_k)$. For instance, the points

$$(x_k, y_k) = \left(\frac{1}{k}, \frac{1}{k}\right)$$

lie in the first quadrant, so $g\left(\dfrac{1}{k}, \dfrac{1}{k}\right) = 1$, and $\lim\limits_{k\to\infty} g\left(\dfrac{1}{k}, \dfrac{1}{k}\right) = 1$. On the other hand, the points $\left(-\dfrac{1}{k}, \dfrac{1}{k}\right)$ lie in the second quadrant, so $g\left(-\dfrac{1}{k}, \dfrac{1}{k}\right) = 0$, and $\lim\limits_{k\to\infty} g\left(-\dfrac{1}{k}, \dfrac{1}{k}\right) = 0$. Since different sequences converging to $(0, 0)$ give different function limits, the limit $\lim\limits_{(x,y)\to(0,0)} g(x, y)$ does not exist.

**EXAMPLE 3** The function $f$ in Figure 1 has no limit as $(x, y) \to (0, 0)$. For example, the sequence $\left(\dfrac{1}{k}, 0\right)$ converges to $(0, 0)$, but

$$f\left(\frac{1}{k}, 0\right) = \frac{1}{(1/k)^2 + 0^2} = k^2$$

has no limit as $k \to \infty$.

**Continuity** is now defined just as for one-variable functions.

---

**DEFINITION 2**

$f$ is continuous at $(a, b)$ if and only if $f(a, b)$ is defined, and

$$\lim\limits_{(x,y)\to(a,b)} f(x, y) = f(a, b).$$

---

Recalling the definition of limit, this means:

(i) $f(x, y)$ is defined throughout some disk centered at $(a, b)$, and

(ii) for every sequence $(x_k, y_k)$ converging to $(a, b)$,

$$\lim\limits_{k\to\infty} f(x_k, y_k) = f(a, b).$$

Thus, for example, the function $f(x, y) = xe^{xy}$ is continuous everywhere; for it is defined everywhere, and if $\lim x_k = a$ and $\lim y_k = b$ then the familiar properties of sequence limits give

$$\lim f(x_k, y_k) = \lim x_k e^{x_k y_k} = ae^{ab} = f(a, b).$$

In a similar way, you can prove that *any function defined by a single formula with a finite number of algebraic operations and compositions, possibly using the trigonometric and exponential functions and their inverses, is continuous wherever defined.* Moreover, the sum or product of two continuous functions is itself continuous; so is the quotient, at points where the denominator is not zero. Further, if $x(t)$ and $y(t)$ are continuous at $t_0$, and $f(x, y)$ is continuous at $(x(t_0), y(t_0))$, then the composition $f(x(t), y(t))$ is also continuous at $t_0$. If $f(x, y)$ is continuous at $(x_0, y_0)$ and $Q(z)$ is continuous at $z_0 = f(x_0, y_0)$, then the composition $Q(f(x, y))$ is continuous at $(x_0, y_0)$.

The function $f$ in Figure 1 is *discontinuous* at $(0, 0)$, for two reasons: First, $f(0, 0)$ is not defined, and second

$$\lim_{(x,y) \to (0,0)} f(x, y) = \lim_{(x,y) \to (0,0)} \frac{1}{x^2 + y^2}$$

does not exist.

The function $g$ in Figure 2 is discontinuous at the origin, and at every point on the positive $x$-axis and positive $y$-axis; for if $(a, b)$ is any of those points, then $\lim_{(x,y) \to (a,b)} g(x, y)$ does not exist.

Here is a more difficult example.

---

**EXAMPLE 4**   Determine whether the following function is continuous everywhere:

$$f(x, y) = \begin{cases} \dfrac{xy}{\sqrt{x^2 + y^2}}, & (x, y) \neq (0, 0) \\ 0, & (x, y) = (0, 0) \end{cases}$$

***SOLUTION***   If $(a, b) \neq (0, 0)$, and $(x_k, y_k) \to (a, b)$, then

$$\lim_{k \to \infty} f(x_k, y_k) = \lim \frac{x_k \cdot y_k}{\sqrt{x_k^2 + y_k^2}} = \frac{ab}{\sqrt{a^2 + b^2}} = f(a, b)$$

by the usual limit theorems; so $f$ is continuous except perhaps at the origin. At the origin, $f(0, 0) = 0$, so we must determine whether

$$\lim_{(x,y) \to (0,0)} f(x, y) = 0.$$

Let $(x_k, y_k) \to (0, 0)$. Introduce polar coordinates, setting

$$x_k = r_k \cos \theta_k, \qquad y_k = r_k \sin \theta_k.$$

Then $r_k = \sqrt{x_k^2 + y_k^2} \to 0$, and

$$\lim_{k \to \infty} \frac{x_k y_k}{\sqrt{x_k^2 + y_k^2}} = \lim_{k \to \infty} \frac{(r_k \cos \theta_k)(r_k \sin \theta_k)}{\sqrt{r_k^2 \cos^2 \theta_k + r_k^2 \sin^2 \theta_k}}$$

$$= \lim_{k \to \infty} r_k(\cos \theta_k)(\sin \theta_k) = 0$$

since $\lim r_k = 0$ and $|\cos \theta_k \sin \theta_k| \leq 1$. Thus $f$ is continuous at the origin, too.

---

## SUMMARY

**Definition 1:** $\lim\limits_{(x,y)\to(a,b)} f(x, y) = L$ if and only if

(i) $f(x, y)$ is defined for all points in some disk centered at $(a, b)$, except perhaps at $(a, b)$; and

(ii) For every sequence $(x_k, y_k)$ converging to $(a, b)$, with $(x_k, y_k) \neq (a, b)$,

$$\lim_{k\to\infty} f(x_k, y_k) = L.$$

**Definition 2:** $f$ is continuous at $(a, b)$ if and only if $f(a, b)$ is defined and

$$\lim_{(x,y)\to(a,b)} f(x, y) = f(a, b).$$

**Theorem 1:** If $f$ and $g$ are continuous at $(a, b)$, then so are $cf$, $f + g$, $fg$, and also $f/g$ if $g(a, b) \neq 0$.

**Theorem 2:** a) If $x(t)$ and $y(t)$ are continuous at $t = t_0$, and $f$ is continuous at $(x(t_0), y(t_0))$, then $f(x(t), y(t))$ is continuous at $t_0$.
 b) If $f$ is continuous at $(a, b)$, and $Q$ is a function of one variable, continuous at $f(a, b)$, then $Q(f(x, y))$ is continuous at $(a, b)$.

**Theorem 3:** Any function defined by a single formula with a finite number of algebraic operations, possibly using the trigonometric and exponential functions and their inverses, is continuous wherever defined.

## PROBLEMS

### A

1. Evaluate the limit of the sequences as $n \to +\infty$.

 *a) $\left(\dfrac{n+1}{n}, \left(\dfrac{1}{2}\right)^n\right)$    b) $\left(\left(1 + \dfrac{1}{n}\right)^n, n^{1/n}\right)$

2. Evaluate:

 *a) $\lim\limits_{(x,y)\to(0,0)} \sqrt{x^2 + y^2}$

 b) $\lim\limits_{(x,y)\to(1,1)} \ln(x/y)$

 *c) $\lim\limits_{(x,y)\to(0,\pi)} e^x \cos y$

3. Find all discontinuities of the function.

 *a) $f(x, y) = \dfrac{1}{\sqrt{x^2 + y^2}}$

 *b) $f(x, y) = \dfrac{\sin(x^2 + y^2)}{x^2 + y^2}$

 *c) $g(x, y) = \dfrac{1}{xy}$

 *d) $G(x, y) = \ln(x + y)$

 e) $h(x, y) = \ln|x + y|$

 *f) $i(x, y) = (4 - x^2 - y^2)^{1/3}$

 g) $i(x, y) = (x - y)^{-1/3}$

 h) $f(x, y) = \begin{cases} 1, & xy > 0 \\ 0 & \text{otherwise} \end{cases}$

4. Let $g(x, y)$ be the function in Figure 2.
 a) Find two sequences $(x_k, y_k)$ and $(\overline{x}_k, \overline{y}_k)$ converging to $(1, 0)$ such that

$$\lim_{k\to\infty} g(x_k, y_k) = 1$$

 while

$$\lim_{k\to\infty} g(\overline{x}_k, y_k) = 0.$$

 [Thus $\lim\limits_{(x,y)\to(1,0)} g(x, y)$ does not exist, and $g$ is not continuous at $(1, 0)$.]
 b) Show that $g$ is not continuous at $(0, 1)$.

**5.** Prove that the given function is continuous wherever defined, *using Definition 2*.
  **a)** $g(x, y) = \ln(x + y)$
  **b)** $h(x, y) = (xy)^{1/3}$
  **c)** $f(x, y) = \sin(x^2 + y^2)$

**B**

**\*6.** At what points is the function $g$ continuous?

$$g(x, y) = \begin{cases} xy, & \text{if } xy > 0 \\ 0, & \text{if } xy \leq 0 \end{cases}$$

**7.** Discover and prove any discontinuities of the function

$$f(x, y) = \begin{cases} \ln(x^2 + y^2) & \text{if } (x, y) \neq (0, 0) \\ 0 & \text{if } (x, y) = (0, 0) \end{cases}$$

**\*8.** Evaluate $\lim\limits_{(x,y)\to(0,0)} \dfrac{x^2 - y^2}{\sqrt{x^2 + y^2}}$. (Use polar coordinates.)

**9.** Does $\lim\limits_{(x,y)\to(0,0)} \dfrac{x^2 - y^2}{x^2 + y^2}$ exist? (Use polar coordinates.)

# 15.3

## PARTIAL DERIVATIVES. THE TANGENT PLANE AND THE GRADIENT

For a function $f(x, y)$ of two variables, both $x$ and $y$ may vary. The **partial derivatives** of $f$ are obtained by holding one variable constant and taking the derivative with respect to the other. Holding $y$ constant, we take the derivative with respect to $x$, obtaining the *partial derivative of f with respect to x*, denoted $f_x$. Likewise, holding $x$ constant, we take the partial derivative with respect to $y$, denoted $f_y$. Alternate notations are $\dfrac{\partial f}{\partial x}$ (read "partial $f$ partial $x$") and $\dfrac{\partial f}{\partial y}$. You can think of the symbol $\partial$ as a "partial $d$," part of the script version of the letter $d$.

---

**EXAMPLE 1**  $f(x, y) = x^2 + xy + 3y^2$. To compute $f_x$, treat $y$ as a constant and take the derivative with respect to $x$:

$$f_x = \frac{\partial f}{\partial x} = \frac{\partial(x^2 + xy + 3y^2)}{\partial x} = 2x + y.$$

Similarly, holding $x$ constant and differentiating with respect to $y$ gives

$$f_y = \frac{\partial f}{\partial y} = \frac{\partial(x^2 + xy + 3y^2)}{\partial y} = x + 6y.$$

The *values* of these derivatives at the point $(-1, 2)$ are

$$f_x(-1, 2) = \frac{\partial f}{\partial x}(-1, 2) = (2x + y)\Big|_{(-1,2)} = -2 + 2 = 0$$

$$f_y(-1, 2) = \frac{\partial f}{\partial y}(-1, 2) = (x + 6y)\Big|_{(-1,2)} = -1 + 12 = 11.$$

---

When $z$ stands for the function values, $z = f(x, y)$, you can also denote the partial derivatives by $\dfrac{\partial z}{\partial x}$ or $z_x$, $\dfrac{\partial z}{\partial y}$ or $z_y$.

---

**EXAMPLE 2**   If $z = \sqrt{1 - x^2 - y^2}$, then

$$z_x = \frac{\partial z}{\partial x} = \frac{\partial \sqrt{1 - x^2 - y^2}}{\partial x} = \frac{-x}{\sqrt{1 - x^2 - y^2}}$$

$$z_y = \frac{\partial z}{\partial y} = \frac{\partial \sqrt{1 - x^2 - y^2}}{\partial y} = \frac{-y}{\sqrt{1 - x^2 - y^2}}.$$

---

Figure 1 illustrates the partial derivatives. Holding $y$ fixed, say $y = y_0$, restricts us to the intersection of the graph with the vertical plane $y = y_0$. In that intersection is the graph of $f(x, y_0)$, a function of just one variable $x$, for $y_0$ is held fixed. The derivative $\dfrac{\partial f}{\partial x}(x_0, y_0)$ gives the slope of that graph at $(x_0, y_0)$; we call it the "$x$-slope" of the graph of $f$ at the point $(x_0, y_0)$. Similarly, $\dfrac{\partial f}{\partial y}(x_0, y_0)$ gives the "$y$-slope" at that point; this is the slope of the graph of $f(x_0, y)$, a function of the one variable $y$.

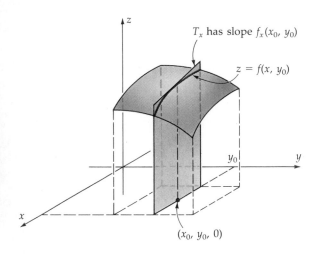

(a)  Section in plane $y = y_0$.

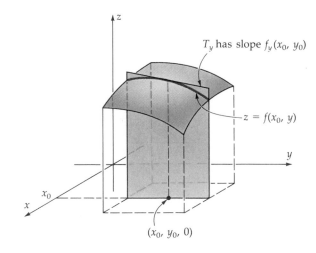

(b)  Section in plane $x = x_0$.

**FIGURE 1**
Lines $T_x$ and $T_y$ are tangent to the graph of $f$.

## The Tangent Plane

For a one-variable function $f$, if the derivative $f'(x_0)$ exists, then the equation

$$y = f(x_0) + f'(x_0)(x - x_0) \tag{1}$$

defines a line tangent to the graph of $f$ at $(x_0, f(x_0))$ (Fig. 2).

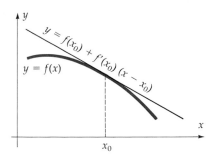

**FIGURE 2**
Tangent line.

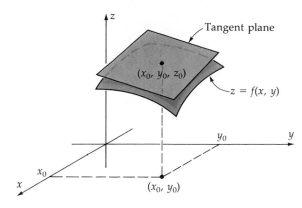

**FIGURE 3**
Plane tangent to graph of $f$ at $(x_0, y_0, z_0)$, with $z_0 = f(x_0, y_0)$.

For a two-variable function $f(x, y)$ we seek a *plane* tangent to the graph at a given point $(x_0, y_0, f(x_0, y_0))$ as in Figure 3. If the partial derivatives $f_x$ and $f_y$ are continuous then there *is* a tangent plane (Sec. 15.4). So assume that $f_x$ and $f_y$ are indeed continuous. Then the tangent plane is determined by the two lines $T_x$ and $T_y$ shown tangent to the graph in Figures 1 and 4. Thus the tangent plane

$$\text{passes through the point } (x_0, y_0, f(x_0, y_0)), \tag{2}$$

$$\text{has } x\text{-slope } f_x(x_0, y_0) \quad \text{(the slope of } T_x), \tag{3}$$

and

$$\text{has } y\text{-slope } f_y(x_0, y_0) \quad \text{(the slope of } T_y). \tag{4}$$

Its equation is therefore (Sec. 15.1)

$$z = f(x_0, y_0) + f_x(x_0, y_0)(x - x_0) + f_y(x_0, y_0)(y - y_0). \tag{5}$$

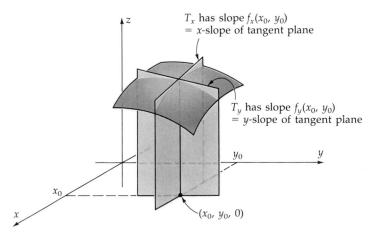

**FIGURE 4**
Tangent plane contains lines $T_x$ and $T_y$.

In this equation, the clumsy symbols $f(x_0, y_0)$, $f_x(x_0, y_0)$, and $f_y(x_0, y_0)$ are all *constants* determined by $f$ and the chosen point $(x_0, y_0)$. Thus (5) has the form of a plane equation

$$z = c + a(x - x_0) + b(y - y_0) = ax + by + (c - ax_0 - by_0).$$

The $x$-slope in (5) is $f_x(x_0, y_0)$ (the coefficient of $x$), and the $y$-slope is $f_y(x_0, y_0)$, as required in (3) and (4); and the remaining condition (2) is easy to check, so (5) satisfies all the conditions for the tangent plane.

The *normal* to the graph of $f$ is, by definition, the normal to the tangent plane. To find it, rewrite (5) as

$$f_x(x_0, y_0)(x - x_0) + f_y(x_0, y_0)(y - y_0) - z + f(x_0, y_0) = 0.$$

The coefficients of $x$, $y$, and $z$ in this equation give the normal to the graph at $(x_0, y_0, f(x_0, y_0))$:

$$\vec{n} = f_x(x_0, y_0)\vec{i} + f_y(x_0, y_0)\vec{j} - \vec{k}. \tag{6}$$

---

**EXAMPLE 3**  Take $f(x, y) = \sqrt{9 - x^2 - y^2}$ and $(x_0, y_0) = (1, 2)$. Then

$$f(1, 2) = \sqrt{9 - 1 - 4} = 2$$

$$f_x(1, 2) = \left. \frac{-x}{\sqrt{9 - x^2 - y^2}} \right|_{(1,2)} = -\frac{1}{2}$$

$$f_y(1, 2) = \left. \frac{-y}{\sqrt{9 - x^2 - y^2}} \right|_{(1,2)} = -1.$$

The point of tangency is $(x_0, y_0, f(x_0, y_0)) = (1, 2, 2)$. From (5), the equation of the plane tangent to the graph at that point is

$$z = f(1, 2) + f_x(1, 2)(x - 1) + f_y(1, 2)(y - 2) = 2 - \tfrac{1}{2}(x - 1) - (y - 2)$$

or

$$\tfrac{1}{2}x + y + z - \tfrac{9}{2} = 0. \tag{7}$$

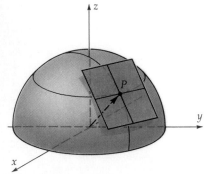

We can check this geometrically; the graph of the given function $f(x, y) = \sqrt{9 - x^2 - y^2}$ is a hemisphere (Fig. 5), so the normal $\vec{n}$ to the tangent plane should be parallel to the vector $\vec{P}$ from the origin to the point of tangency $P = (1, 2, 2)$. In fact, the normal to the tangent plane (4) is

$$\vec{n} = \langle \tfrac{1}{2}, 1, 1 \rangle = \tfrac{1}{2}\langle 1, 2, 2 \rangle = \tfrac{1}{2}\vec{P},$$

parallel to $\vec{P}$.

**FIGURE 5**
On sphere, tangent plane at $P$ is perpendicular to position vector $\vec{P}$.

---

## The Gradient Vector

Figure 6 shows the normal vector $\vec{n}$ to the graph of $f$ as the sum of a vertical part $-\vec{k}$ and a horizontal part $f_x\vec{i} + f_y\vec{j}$. The horizontal part is called the

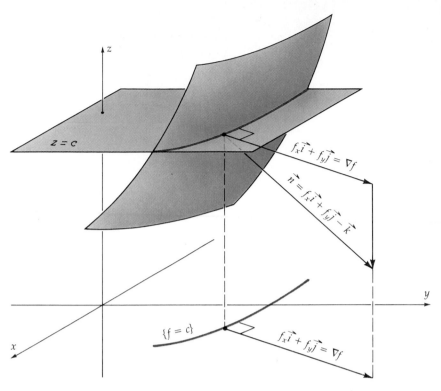

**FIGURE 6**
$f_x\vec{i} + f_y\vec{j} = \nabla f$ is perpendicular to the level curve $\{f = c\}$.

**gradient** of $f$, denoted $\nabla f$:

$$\nabla f = f_x\vec{i} + f_y\vec{j}.$$

(The symbol "$\nabla$" is read "del"; it is an inverted $\Delta$, Greek delta.) The gradient vector plays the main role in the rest of this chapter. First, we relate it to the level curves.

   Again in Figure 6, the graph of $f$ intersects the horizontal plane $\{z = c\}$ in a curve. The plane's normal $\vec{n}$ is orthogonal to this horizontal curve of intersection. It follows that $\nabla f$, the horizontal part of the normal, is also orthogonal to the curve of intersection (see problems 14 and 15). Down in the $xy$-plane, directly below the curve of intersection, lies the level curve $\{f = c\}$, so $\nabla f$ is orthogonal to this level curve:

> The gradient vector $\nabla f(x, y)$ is orthogonal to the level curve
> $\{f = c\}$ passing through the point $(x, y)$.

Moreover, the gradient points in the direction where the graph is rising, toward higher values of $f$. (This conclusion will be confirmed by calculations in Sec. 15.5.) For example, if $T(x, y)$ is the temperature at point $(x, y)$ in the plane, then the gradient vector $\nabla T$ points toward regions of higher temperature.

**EXAMPLE 4** For $f(x, y) = xy$, the partial derivatives are $f_x = y$ and $f_y = x$, so the gradient vector is

$$\nabla f(x, y) = f_x \vec{i} + f_y \vec{j} = y\vec{i} + x\vec{j}.$$

Figure 7 shows the gradient vectors at two points:

$$\nabla f(1, 2) = \langle 2, 1 \rangle \quad \text{and} \quad \nabla f(-3, 2) = \langle 2, -3 \rangle.$$

The arrow for $\nabla f(1, 2)$ begins at $(1, 2)$. At that point, the function value is $f(1, 2) = 2$, so the level curve there is $\{f = 2\}$; $\nabla f(1, 2)$ is orthogonal to this curve. It points toward the level curve $\{f = 4\}$, where $f$ has a higher value.

Similarly, the arrow for $\nabla f(-3, 2)$ begins at $(-3, 2)$ and is orthogonal to the level curve $\{f = -6\}$ which passes through that point. It aims from $\{f = -6\}$ toward higher levels $\{f = -4\}$, $\{f = -2\}$, and so on.

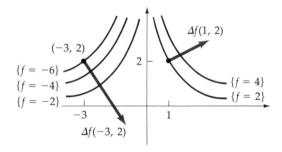

**FIGURE 7**
Gradients and level curves for
$f(x, y) = xy$.

## SUMMARY

**Partial Derivatives of $f(x, y)$:**

$$f_x = \frac{\partial f}{\partial x} = \text{derivative with respect to } x \ (y \text{ held constant}).$$

$$f_y = \frac{\partial f}{\partial y} = \text{derivative with respect to } y \ (x \text{ held constant}).$$

**Gradient Vector:** $\qquad\qquad \nabla f = f_x \vec{i} + f_y \vec{j}$

When $f_x$ and $f_y$ are continuous then $\nabla f(x, y)$ is orthogonal to the level curve through $(x, y)$, and points toward level curves with higher values of $f$ (see Fig. 6).

**Equation of Tangent Plane at $(x_0, y_0, f(x_0, y_0))$:**

$$z = f(x_0, y_0) + f_x(x_0, y_0)(x - x_0) + f_y(x_0, y_0)(y - y_0).$$

**Normal Vector to Tangent Plane, and to Graph:**

$$\vec{n} = f_x \vec{i} + f_y \vec{j} - \vec{k}.$$

## PROBLEMS

### A

1. Compute $f_x$, $f_y$, $\nabla f$.
   a) $x^2 + y^2$      *b) $xe^{xy}$
   c) $xe^{-(x^2+y^2)/2}$      *d) $\ln(x^2 + y^2)$
   *e) $\dfrac{2x}{x^2 + y^2}$      f) $\dfrac{2xy}{x^2 + y^2}$
   g) $(x^2 + 2y^2)e^{-x^2 - y^2}$      h) $e^{x^2 + y^2}\sin(xy)$
   *i) $\cos(x - y)$

2. For $f(x, y) = 25 - x^2 - y^2$, sketch the level curve $\{f = 0\}$. Then, for the given point $P$, compute $\nabla f(x, y)$, and sketch it as an arrow beginning at $P$. Note that it is orthogonal to the level curve.
   *a) $(3, 4)$      b) $(4, -3)$      c) $(-3, 4)$

*3. For $f(x, y) = y/x$, sketch the level curve $\{f = c\}$ for $c = 0, \pm 1, \pm 2$. Sketch $\nabla f(P)$ for $P = (\pm 1, 0)$, $(\pm 1, 1)$, $(\pm 1, 2)$.

*4. Given $f(x, y) = x^2 + y$, compute $\nabla f$ at the given point $P$, and sketch it as an arrow beginning at $P$. Determine $f(P)$, and sketch the level curve $\{f = f(P)\}$ through $P$.
   a) $P = (0, 0)$      b) $P = (1, 1)$
   c) $P = (-2, -1)$

5. Repeat problem 4, but with $f(x, y) = x^2 - y^2$.

6. Write the equation of the plane tangent to the graph of $f$ at the given point. (Check that the point is really on the graph.)
   *a) $f(x, y) = x^2 + y^2$ at $(1, 2, 5)$
   b) $f(x, y) = e^{x-y}\cos(x + y)$ at $(1, -1, e^2)$

7. Write parametric equations of the normal to the graph of $g$ at the given point.
   *a) $g(x, y) = \ln(x/y)$ at $(e, 1, 1)$
   *b) $g(x, y) = x\tan(x^2 + 2y^2)$ at $(0, 0, 0)$
   c) $g(x, y) = xe^{6 - x^2 - 2y^2}$ at $(2, 1, 2)$

*8. A point moves along the intersection of the paraboloid $z = x^2 - 3y^2$ and the vertical plane $\{x = 1\}$. What is the rate of change of $z$ with respect to $y$?

*9. A point moves along the intersection of the ellipsoid $x^2 + 2y^2 + 3z^2 = 10$ and the vertical plane $\{y = 2\}$. What is the rate of change of $z$ with respect to $x$?

10. Find all points on the given graph where the tangent plane is horizontal.
   *a) $z = x^2 y^2$
   *b) $z = ye^{x^2 - y^2}$
   c) $z = (x^2 + y^2)e^{-x^2 - 2y^2}$

11. Show that the equation of the plane tangent to the ellipsoid $\left\{ z = \dfrac{x^2}{a^2} + \dfrac{y^2}{b^2} \right\}$ at $(x_0, y_0, z_0)$ is the graph of
$$z + z_0 = \frac{2x_0 x}{a^2} + \frac{2y_0 y}{b^2}.$$

12. If gas is contained in a flexible vessel, and the total heat in the gas remains constant, then ideally the pressure $P$, volume $V$, and temperature $T$ (in degrees above absolute zero) are related by
$$T = kPV, \qquad (*)$$
where $k$ is a constant depending on the type of gas, its mass, and its heat.
   a) Using $(*)$ to define $T$ as a function of $P$ and $V$, compute $\dfrac{\partial T}{\partial P}$ and $\dfrac{\partial T}{\partial V}$.
   b) Using $(*)$ to define $P$ as a function of $T$ and $V$, compute $\dfrac{\partial P}{\partial T}$ and $\dfrac{\partial P}{\partial V}$.
   c) Using $(*)$ to define $V$ as a function of $T$ and $P$, compute $\dfrac{\partial V}{\partial T}$ and $\dfrac{\partial V}{\partial P}$.
   d) Compute the *coefficient of compressibility*
$$-\frac{1}{V}\left(\frac{\partial V}{\partial P}\right).$$
   *e) Compute and simplify $\dfrac{\partial T}{\partial P} \cdot \dfrac{\partial P}{\partial V} \cdot \dfrac{\partial V}{\partial T}$. (The answer is *not* 1! This shows that a partial derivative such as $\partial T/\partial P$ *cannot* be thought of as an actual fraction with numerator $\partial T$ and denominator $\partial P$.)

*13. Imagine the graph of $f(x, y)$ as an icy hillside, with a small rock of mass $m$ upon it. Gravity pulls down on the rock with force $-mg\vec{k}$. The rock is to be held in equilibrium by an additional horizontal force $\vec{H}$, such that the combined force $\vec{H} - mg\vec{k}$ is perpendicular to the hill; then the icy hill can contribute the final balancing force $mg\vec{k} - \vec{H}$, perpendicular to its surface. What horizontal force $\vec{H}$ is required? (See Fig. 6.)

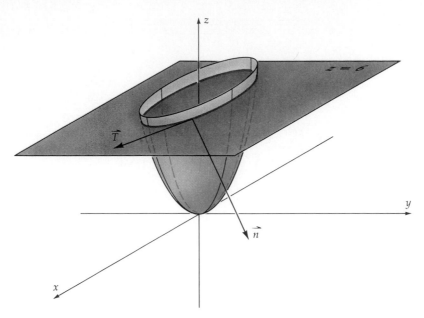

**FIGURE 8**
$z = x^2 + 2y^2$ intersecting plane $z = 6$.

*14. Given $f(x, y) = x^2 + 2y^2$.
a) Find a vector perpendicular to the graph at $(2, 1, 6)$.
b) Find a vector perpendicular to the plane $\{z = 6\}$.
c) Find a vector $\vec{T}$ tanget to the curve of intersection of the given graph and the plane $\{z = 6\}$ at the point $(2, 1, 6)$ (Fig. 8).
d) Check that $\nabla f(2, 1)$ is perpendicular to the vector in c), and thus is perpendicular to the level curve $\{f = 6\}$.

**B**

15. Suppose that $f_x$ and $f_y$ are continuous at $P_0 = (x_0, y_0)$, so the graph of $f$ has a tangent plane at $(x_0, y_0, f(x_0, y_0))$. The graph intersects the horizontal plane $\{z = f(x_0, y_0)\}$ in a curve $C$.
*a) Find a vector $\vec{v}$ tangent to $C$ at $(x_0, y_0, f(x_0, y_0))$. (Use the normals to the plane and to the graph.)
b) Show that $\vec{v}$ is orthogonal to $\nabla f(x_0, y_0)$.

16. Show that the following functions all satisfy the equation $\dfrac{\partial f}{\partial x} = \dfrac{\partial f}{\partial y}$.

a) $(x + y)^3$   b) $e^{x+y}$   c) $\sin(2x + 2y)$
d) $\varphi(x + y)$, where $\varphi$ is any differentiable function of one variable.

17. Let $f(x, y) = \dfrac{xy}{(x^2 + y^2)^2}$ for $(x, y) \neq (0, 0)$, and define $f(0, 0) = 0$.
a) Do $f_x$ and $f_y$ exist at all points $(x, y)$? [Be careful about the point $(0, 0)$; compute the derivative of $f(x, 0)$ with respect to $x$, and of $f(0, y)$ with respect to $y$.]
b) Is $f$ continuous at $(0, 0)$?

$$\overline{\overline{\phantom{====}}}\ \mathbf{15.4}\ \overline{\overline{\phantom{====}}}$$

## THE LINEAR APPROXIMATION. TANGENCY

Figure 1 shows the graph of a function $f$, and its tangent plane at $(x_0, y_0, f(x_0, y_0))$. The equation of the plane is

$$z = f(x_0, y_0) + f_x(x_0, y_0)(x - x_0) + f_y(x_0, y_0)(y - y_0). \tag{1}$$

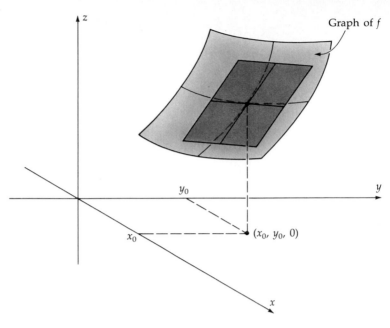

**FIGURE 1**
Tangent plane at $(x_0, y_0, f(x_0, y_0))$.

The function on the right-hand side is called the **linear approximation** *to f, based at* $(x_0, y_0)$. We denote it by $f_{\text{lin}}$:

$$f_{\text{lin}}(x, y) = \underbrace{f(x_0, y_0)}_{\substack{\text{value at} \\ \text{base point}}} + \underbrace{f_x(x_0, y_0)}_{\substack{\text{rate of} \\ \text{change} \\ \text{in } x}} \underbrace{(x - x_0)}_{\substack{\text{change} \\ \text{in } x}} + \underbrace{f_y(x_0, y_0)}_{\substack{\text{rate of} \\ \text{change} \\ \text{in } y}} \underbrace{(y - y_0)}_{\substack{\text{change} \\ \text{in } y}}. \tag{2}$$

We will prove that when $f_x$ and $f_y$ are *continuous* at $(x_0, y_0)$ then $f_{\text{lin}}(x, y)$ is a good approximation to $f(x, y)$, and the plane (1) really is tangent to the graph of $f$.

---

**EXAMPLE 1**   Take $f(x, y) = x^2 + y^2$, with base point $(x_0, y_0) = (3, 2)$. Then

$$f(x_0, y_0) = 13; \qquad f_x(x_0, y_0) = 2x\Big|_{(3,2)} = 6; \qquad f_y(x_0, y_0) = 2y\Big|_{(3,2)} = 4$$

so

$$f_{\text{lin}}(x, y) = 13 + 6(x - 3) + 4(y - 2).$$

For $(x, y)$ near $(3, 2)$

$$f(x, y) \approx f_{\text{lin}}(x, y) = 13 + 6(x - 3) + 4(y - 2).$$

For example,

$$f_{\text{lin}}(3.1, 2.08) = 13 + 6(.1) + 4(.08) = 13.92.$$

The actual function value is

$$f(3.1, 2.08) = (3.1)^2 + (2.08)^2 = 13.9364.$$

You can approximate the *difference* $\Delta f$ in function values at two nearby points

$$\Delta f = f(x, y) - f(x_0, y_0)$$

by the linear difference

$$\begin{aligned}
\Delta f_{\text{lin}} &= f_{\text{lin}}(x, y) - f(x_0, y_0) \\
&= f_x(x_0, y_0)(x - x_0) + f_y(x_0, y_0)(y - y_0).
\end{aligned}$$

To abbreviate the notation, set $\Delta x = x - x_0$, $\Delta y = y - y_0$; then

$$\Delta f_{\text{lin}} = f_x \, \Delta x + f_y \, \Delta y = \frac{\partial f}{\partial x} \Delta x + \frac{\partial f}{\partial y} \Delta y$$

with the understanding that the partial derivatives are evaluated at the base point $(x_0, y_0)$.

## Differential Notation

The linear approximation $\Delta f_{\text{lin}}$ is sometimes denoted by $df$; then the differences $\Delta x$ and $\Delta y$ are denoted by $dx$ and $dy$. Thus

$$\Delta f \approx \Delta f_{\text{lin}} = \frac{\partial f}{\partial x} \Delta x + \frac{\partial f}{\partial y} \Delta y$$

is written

$$\Delta f \approx df = \frac{\partial f}{\partial x} dx + \frac{\partial f}{\partial y} dy.$$

The expression $df = \dfrac{\partial f}{\partial x} dx + \dfrac{\partial f}{\partial y} dy$ is called the **total differential** of $f$.

How does this relate to our earlier use of $dx$ and $dy$, as "infinitesimal increments"? The idea is that when the changes $\Delta x$ and $\Delta y$ are so small that they can be considered "infinitesimals" $dx$ and $dy$, then the resulting increment $\Delta f$ is practically indistinguishable from the "infinitesimal increment" in $f$, $df = \dfrac{\partial f}{\partial x} dx + \dfrac{\partial f}{\partial y} dy$. Other more sophisticated uses of differentials occur in the advanced study of curves and surfaces; Section 18.3 gives some hint of this.

**EXAMPLE 2**  Let $P$, $T$, and $V$ be respectively the pressure, volume, and temperature of a given quantity of mercury. In a reasonable range of temperature (in °C) and pressure (in atmospheres),

$$\frac{\partial V}{\partial T} = (1.8 \times 10^{-4})V \quad \text{and} \quad \frac{\partial V}{\partial P} = (-3.9 \times 10^{-6})V.$$

**a.**    Interpret the signs of these two partial derivatives.

**b.**    Suppose that the temperature is increased by 1.3°C, and the pressure by 0.2 atm. Approximate the resulting change in volume.

*SOLUTION*

**a.**    Since $\partial V/\partial T > 0$, any increase in temperaturre alone will increase the volume. On the other hand, $\partial V/\partial P < 0$, so an increase in pressure will *decrease* the volume.

**b.**    The change in volume is approximately

$$dV = \frac{\partial V}{\partial T}\,dT + \frac{\partial V}{\partial P}\,dP$$

$$= (1.8 \times 10^{-4})V \cdot (1.3) + (-3.9 \times 10^{-6})V \cdot (0.2)$$

$$= (2.34 \times 10^{-4})V - (7.8 \times 10^{-7})V \approx (2.34 \times 10^{-4})V. \tag{3}$$

The effect of increasing the pressure by this amount is negligible, compared to the effect of the temperature change. Even the temperature has a relatively small effect on the volume. That's why thermometers are designed with a very thin column of mercury, with a bulb at the bottom. In the thin column, a small increase in volume causes a relatively large increase in height; and the bulb provides some volume, enlarging the factor $V$ on the right-hand side of (3), hence the increase $dV$ on the left.

---

The main question concerning the linear approximation is: *Under what conditions is it good; and in what sense is it good?* A precise answer is:

---

**The Approximation Lemma**

If $f_x$ and $f_y$ are continuous at $P_0 = (x_0, y_0)$, then

$$f(P) = f(P_0) + f_x(P_0)\,\Delta x + f_y(P_0)\,\Delta y + \varepsilon_1\,\Delta x + \varepsilon_2\,\Delta y \tag{4}$$

where $P = (x, y)$, $\Delta x = x - x_0$, $\Delta y = y - y_0$, and

$$\lim_{P \to P_0} \varepsilon_1(P) = 0 = \lim_{P \to P_0} \varepsilon_2(P). \tag{5}$$

---

Think of the Approximation Lemma as follows: When $\Delta x$ and $\Delta y$ are sufficiently small, then (5) guarantees that $\varepsilon_1$ and $\varepsilon_2$ are small, so $\varepsilon_1\,\Delta x$ and $\varepsilon_2\,\Delta y$ are much smaller than $\Delta x$ and $\Delta y$. Thus

$$f(P) = \quad f(P_0) \quad + [f_x(P_0)\,\Delta x + f_y(P_0)\,\Delta y] + \ [\varepsilon_1\,\Delta x + \varepsilon_2\,\Delta y]$$

$$= \begin{bmatrix} \text{value} \\ \text{at base} \\ \text{point} \end{bmatrix} + \begin{bmatrix} \text{linear approximation} \\ \text{comparable in size} \\ \text{to } \Delta x \text{ and } \Delta y \end{bmatrix} + \begin{bmatrix} \text{error in linear} \\ \text{approximation,} \\ \text{much smaller} \\ \text{than } \Delta x \text{ and } \Delta y \end{bmatrix}.$$

**EXAMPLE 3** Return to the function $f(x, y) = x^2 + y^2$ in Example 1, with base point $P_0 = (3, 2)$. We found

$$f_{\text{lin}}(x, y) = 13 + 6(x - 3) + 4(y - 2)$$
$$= 13 + 6\,\Delta x + 4\,\Delta y.$$

Compare this to the actual function values

$$f(x, y) = f(3 + \Delta x, 2 + \Delta y) = (3 + \Delta x)^2 + (2 + \Delta y)^2$$
$$= 13 + 6\,\Delta x + 4\,\Delta y + (\Delta x)^2 + (\Delta y)^2.$$

The error in the linear approximation is

$$f(x, y) - f_{\text{lin}}(x, y) = (\Delta x)^2 + (\Delta y)^2.$$

The following table compares both the linear approximation $\Delta f_{\text{lin}} = f_x\,\Delta x + f_y\,\Delta y$, and the error $f - f_{\text{lin}}$, to $\Delta x$ and $\Delta y$:

| $\Delta x$ | 0.2 | 0.01 | 0.002 |
|---|---|---|---|
| $\Delta y$ | $-0.1$ | 0.03 | 0.001 |
| $\max(|\Delta x|, |\Delta y|)$ | 0.2 | 0.03 | 0.002 |
| $f_x\,\Delta x + f_y\,\Delta y$ | 0.8 | 0.18 | 0.016 |
| error, $f - f_{\text{lin}}$ | 0.05 | 0.001 | 0.000005 |

When $\Delta x$ and $\Delta y$ are very small, so also is the linear correction $f_x\,\Delta x + f_y\,\Delta y$; but the error $f - f_{\text{lin}} = \varepsilon_1\,\Delta x + \varepsilon_2\,\Delta y$ is very *very* small.

## Tangency

The Approximation Lemma has a geometric interpretation: *If the derivatives $f_x$ and $f_y$ are continuous at $(x_0, y_0)$, then the plane given by the linear approximation is tangent to the graph of $f$ at the point $(x_0, y_0, f(x_0, y_0))$.* Figure 2 illustrates the definition of tangency. There

$$P_0 = (x_0, y_0) \quad \text{and} \quad P = (x, y)$$

are in the $xy$-plane;

$$Q_0 = (x_0, y_0, f(x_0, y_0)) \quad \text{and} \quad Q = (x, y, f(x, y))$$

are on the graph of $f$; while

$$Q_T = (x, y, f_{\text{lin}}(x, y))$$

is on the plane given by the linear approximation. Finally, $\theta$ in Figure 2b is the angle between the horizontal plane $\{z = f(P_0)\}$ and the vector $\overrightarrow{Q_0 Q}$, while $\theta_T$ is the angle between the horizontal plane and the vector $\overrightarrow{Q_0 Q_T}$ in the

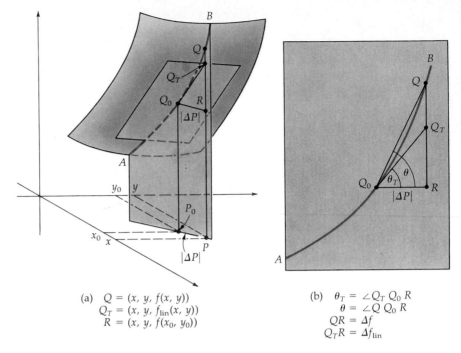

(a)   $Q = (x, y, f(x, y))$
$Q_T = (x, y, f_{\text{lin}}(x, y))$
$R = (x, y, f(x_0, y_0))$

(b)   $\theta_T = \angle Q_T Q_0 R$
$\theta = \angle Q Q_0 R$
$QR = \Delta f$
$Q_T R = \Delta f_{\text{lin}}$

**FIGURE 2**
Tangency: $\theta - \theta_T \to 0$ as $P \to P_0$.

approximating plane. As a definition of tangency, we take the condition

$$\theta - \theta_T \to 0 \quad \text{as} \quad P \to P_0$$

or what is the same,

$$\tan \theta - \tan \theta_T \to 0 \quad \text{as} \quad P \to P_0. \tag{6}$$

We will prove that this is satisfied when $f_x$ and $f_y$ are continuous at $P_0$. From Figure 2b,

$$\tan \theta = \frac{\Delta f}{|\Delta P|} \quad \text{and} \quad \tan \theta_T = \frac{\Delta f_{\text{lin}}}{|\Delta P|}$$

so

$$\tan \theta - \tan \theta_T = \frac{\Delta f}{|\Delta P|} - \frac{\Delta f_{\text{lin}}}{|\Delta P|} = \frac{\Delta f - \Delta f_{\text{lin}}}{|\Delta P|}$$

$$= \frac{[f(P) - f(P_0)] - [f_x(P_0)\,\Delta x + f_y(P_0)\,\Delta y]}{|\Delta P|}.$$

Thus from (4)

$$\tan \theta - \tan \theta_T = \frac{\varepsilon_1\,\Delta x + \varepsilon_2\,\Delta y}{|\Delta P|}$$

$$= \varepsilon_1 \frac{\Delta x}{|\Delta P|} + \varepsilon_2 \frac{\Delta y}{|\Delta P|} \to 0 \quad \text{as} \quad P \to P_0; \tag{7}$$

for $\varepsilon_1 \to 0$ and $\varepsilon_2 \to 0$ as $P \to P_0$, while

$$\frac{|\Delta x|}{|\Delta P|} = \frac{|\Delta x|}{\sqrt{(\Delta x)^2 + (\Delta y)^2}} \leq 1$$

and likewise $\dfrac{|\Delta y|}{|\Delta P|} \leq 1$.

The limit relation (7) proves that the graph of $f_{\text{lin}}$ is tangent to the graph of $f$; we have shown it to hold when $f_x$ and $f_y$ are continuous. When the partial derivatives are *not* continuous, there may be no tangent plane.

---

**EXAMPLE 4**    Consider the function in Figure 3

$$f(x, y) = \begin{cases} -1 & \text{if } x > 0 \text{ and } y > 0 \\ 1 & \text{otherwise.} \end{cases}$$

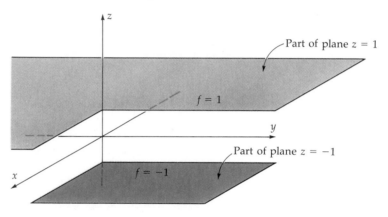

**FIGURE 3**
$f = -1$ in first quadrant, $f = +1$ elsewhere. There is a jump discontinuity along the positive $x$ and $y$ axes.

The graph is flat except along the positive $x$- and $y$-axes, so $f_x = 0$ and $f_y = 0$, except perhaps at those points, which we investigate further. Take a point $(x, 0)$ on the positive $x$-axis. The $x$-section of the graph through this point is a horizontal line, so $f_x(x, 0) = 0$. The $y$-section at $(x, 0)$ has a jump at $y = 0$, so $f_y$ does not exist at this point. Analogously, along the positive $y$-axis $f_y = 0$, and $f_x$ does not exist. At the origin, the $x$-section and $y$-section are both horizontal lines, so $f_x(0, 0) = 0 = f_y(0, 0)$.

Thus the two partial derivatives exist at the origin—both are 0. But they are not *continuous* at the origin; for $f_x$ is not defined on the positive $y$-axis, nor is $f_y$ defined on the positive $x$-axis.

In this example, $f$ is not continuous at the origin, so its graph near that point is not approximated by any one plane. Thus the mere fact that both partial derivatives *exist* at $(0, 0)$ is not very informative—it gives no hint of the discontinuity of $f$ at that point.

## Proof of the Approximation Lemma

Consider the difference $\Delta f = f(x, y) - f(x_0, y_0)$. In moving from $(x_0, y_0)$ to $(x, y)$, *both* variables can change; we rewrite $\Delta f$ as a sum where, in each term, only *one* variable changes:

$$f(x, y) - f(x_0, y_0) = [f(x, y) - f(x, y_0)] + [f(x, y_0) - f(x_0, y_0)].$$

In the last bracket, $f(x, y_0) - f(x_0, y_0)$ can be viewed as a difference in two values of a function $f(x, y_0)$ of just one variable, $x$. By the Mean Value Theorem for functions of one variable,

$$f(x, y_0) - f(x_0, y_0) = f_x(\hat{x}, y_0)\, \Delta x$$

for some $\hat{x}$ between $x_0$ and $x$. Similarly

$$f(x, y) - f(x, y_0) = f_y(x, \hat{y})\, \Delta y$$

for some $\hat{y}$ between $y_0$ and $y$. Thus

$$f(x, y) - f(x_0, y_0) = f_x(\hat{x}, y_0)\, \Delta x + f_y(x, \hat{y})\, \Delta y.$$

We rewrite this in the form (3) given in the Approximation Lemma:

$$
\begin{aligned}
f(x, y) - f(x_0, y_0) &= f_x(x_0, y_0)\, \Delta x + [f_x(\hat{x}, y_0) - f_x(x_0, y_0)]\, \Delta x \\
&\quad + f_y(x_0, y_0)\, \Delta y + [f_y(x, \hat{y}) - f_y(x_0, y_0)]\, \Delta y \\
&= f_x(x_0, y_0)\, \Delta x + f_y(x_0, y_0)\, \Delta y + \varepsilon_1\, \Delta x + \varepsilon_2\, \Delta y,
\end{aligned}
$$

where

$$\varepsilon_1 = f_x(\hat{x}, y_0) - f_x(x_0, y_0) \quad \text{and} \quad \varepsilon_2 = f_y(x, \hat{y}) - f_y(x_0, y_0).$$

As $x \to x_0$ and $y \to y_0$, then $\hat{x} \to x_0$ (since $\hat{x}$ lies between $x_0$ and $x$) and likewise $\hat{y} \to y_0$. By assumption, $f_x$ and $f_y$ are continuous at $(x, y)$, so

$$f_x(\hat{x}, y_0) \to f_x(x_0, y_0) \quad \text{and} \quad f_y(x, \hat{y}) \to f_y(x_0, y_0).$$

Thus $\varepsilon_1 \to 0$ and $\varepsilon_2 \to 0$; this proves (4), the conclusion of the Approximation Lemma.

## SUMMARY

*Linear Approximation Based at $P_0$:*

$$f_{\text{lin}}(P) = f(P_0) + f_x(P_0)\, \Delta x + f_y(P_0)\, \Delta y,$$

$$P = (x, y), \qquad P_0 = (x_0, y_0), \qquad \Delta x = x - x_0, \qquad \Delta y = y - y_0.$$

*Approximation Lemma:* If $f_x$ and $f_y$ are continuous at $P_0$, then

$$f(P) - f_{\text{lin}}(P) = \varepsilon_1\, \Delta x + \varepsilon_2\, \Delta y,$$

where

$$\varepsilon_1 \to 0 \quad \text{and} \quad \varepsilon_2 \to 0 \quad \text{as } P \to P_0.$$

**Tangency:** If $f_x$ and $f_y$ are continuous at $P_0$, then the plane defined by $f_{\text{lin}}$ is tangent to the graph of $f$ at $(x_0, y_0, f(x_0, y_0))$.

**Differential Notation:** $df = f_x\, dx + f_y\, dy$

## PROBLEMS

### A

1. Write the linear approximation $f_{\text{lin}}$ to $f$ at the given base point $(x_0, y_0)$. Use $f_{\text{lin}}$ to approximate $f$ at the given point $(x, y)$.
   *a) $x^2 + y^2$; $(x_0, y_0) = (1, 2)$; $(x, y) = (1.02, 2.01)$
   b) $e^y \sin x$; $(x_0, y_0) = (0, 0)$; $(x, y) = (-.09, 0.12)$
   *c) $x^2 + 2xy - 4x$; $(x_0, y_0) = (1, 2)$; $(x, y) = (1.01, 2.04)$
   *d) $(xy)^{1/3}$; $(x_0, y_0) = (8, 1)$; $(x, y) = (8.2, 0.9)$
   e) $\ln\sqrt{e + xy}$; $(x_0, y_0) = (0, 2)$; $(x, y) = (-0.1, 2.2)$

*2. A rectangle is designed to be 3 by 4, but actually it is 3.1 by 3.99. Estimate the difference between the area of the actual rectangle and the designed one.

3. In problem 2, estimate the difference between the diagonals of the actual rectangle and of the designed one.

*4. In a certain economics model, the cost $C$ of producing a commodity depends on the price of oil ($O$) and the price of labor ($L$). Suppose that currently, $\dfrac{\partial C}{\partial O} = 0.7$ and $\dfrac{\partial C}{\partial L} = 1.3$. Estimate the change in cost if the price of oil declines by 1 and the price of labor increases by 0.5. Does the production cost rise, or fall?

5. The radius of a right circular cylinder is measured as 5 cm, with a possible error of $\pm 0.01$ cm; the height is measured as 12 cm, with a possible error of $\pm 0.005$ cm. Use differentials to estimate the possible error in computing
   a) The volume of the cylinder.
   b) The surface area of the cylinder.

6. The radius of the base of a conical pile of sawdust is measured to be 10 meters, with a possible error of $\pm 0.5$ m; the height is measured as 21 meters, with a possible error of $\pm 0.7$ m. Use differentials to estimate the possible error in computing

a) The volume of the sawdust.
b) The exposed surface area of the pile.

*7. For the gas in a certain flexible container, the pressure $P$, temperature $T$, and volume $V$ are related by

$$P = kT/V,$$

where $k$ is a constant. At a certain time, $V = 2$ and $T = 300$. Use $dP$ to estimate the change in pressure if $V$ rises to 2.01 and $T$ rises to 303.

8. Let $f(x, y) = xy$ and $P_0 = (1, 1)$. Compare $\max(|\Delta x|, |\Delta y|)$ to $\Delta f_{\text{lin}}(P) = f_x(P_0)\,\Delta x + f_y(P_0)\,\Delta y$, and to the error $f(P) - f_{\text{lin}}(P)$, for the following points $P$:
   a) $(1.1, 0.9)$     b) $(1.02, 1.03)$
   c) $(0.9995, 1.0004)$

### B

9. A thermometer is being designed to have a mercury column that rises 1 mm for each temperature increase of $1°C$. The diameter of the column is to be $10^{-2}$ mm. What must then be the volume of mercury in the thermometer? (See Example 2 for the relevant properties of mercury.)

10. Let $f(x, y) = \dfrac{xy}{\sqrt{x^2 + y^2}}$ for $(x, y) \neq (0, 0)$, and $f(0, 0) = 0$.
   a) Show that $f_x(0, 0) = 0 = f_y(0, 0)$, and the graph of the linear approximation is the $xy$-plane.
   b) Determine whether $\dfrac{f(P) - f_{\text{lin}}(P)}{|\Delta P|} \to 0$ as $P \to (0, 0)$.
   c) Does the angle between the $xy$-plane and the segment from $(0, 0, 0)$ to $(x, y, f(x, y))$ go to 0 as $(x, y) \to (0, 0)$? (Check points on the lines $y = x$ and $y = -x$.)

# 15.5

## THE DERIVATIVE ALONG A CURVE.
## THE DIRECTIONAL DERIVATIVE

Suppose that $f(x, y)$ gives the temperature at position $(x, y)$ in the plane. Imagine a point moving in the plane; at time $t$ its position is

$$P(t) = (x(t), y(t)).$$

Then $f(P(t))$ gives the temperature encountered by the moving point at time $t$ (Fig. 1). What is the derivative "along the curve," $\dfrac{df(P(t))}{dt}$? By definition of the derivative,

$$\frac{df(P(t))}{dt}\bigg|_{t=t_0} = \lim_{t \to t_0} \frac{f(P(t)) - f(P(t_0))}{t - t_0}. \tag{1}$$

Set (as in Fig. 2)

$$P_0 = P(t_0), \qquad \Delta x = x(t) - x(t_0), \qquad \Delta y = y(t) - y(t_0), \qquad \Delta t = t - t_0.$$

Assume that the partial derivatives of $f$ are continuous at $P_0$. Then by the Approximation Lemma (Sec. 15.4)

$$f(P) = f(P_0) + f_x(P_0)\, \Delta x + f_y(P_0)\, \Delta y + \varepsilon_1\, \Delta x + \varepsilon_2\, \Delta y$$

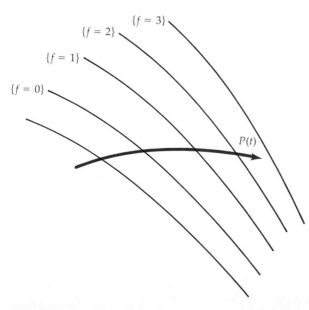

**FIGURE 1**
Moving across the level curves.

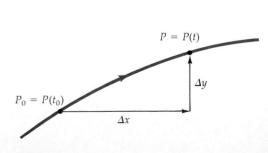

**FIGURE 2**

where $\varepsilon_1 \to 0$ and $\varepsilon_2 \to 0$ as $P \to P_0$. So the difference quotient in (1) can be written

$$\frac{f(P(t)) - f(P(t_0))}{t - t_0} = f_x(P_0)\frac{\Delta x}{\Delta t} + f_y(P_0)\frac{\Delta y}{\Delta t} + \varepsilon_1 \frac{\Delta x}{\Delta t} + \varepsilon_2 \frac{\Delta y}{\Delta t}. \tag{2}$$

In the limit as $t \to t_0$, then $\Delta t \to 0$, so

$$\frac{\Delta x}{\Delta t} \to \frac{dx}{dt} \quad \text{and} \quad \frac{\Delta y}{\Delta t} \to \frac{dy}{dt} \qquad \text{(definition of derivative)};$$

and $P(t) \to P_0$, so $\varepsilon_1 \to 0$ and $\varepsilon_2 \to 0$. Thus the derivative along the curve, the limit of the difference quotient (2), is

$$\frac{df(P(t))}{dt}\bigg|_{t=t_0} = \lim_{t \to t_0} \frac{f(P(t)) - f(P(t_0))}{t - t_0}$$

$$= f_x(P(t_0))\frac{dx}{dt}(t_0) + f_y(P(t_0))\frac{dy}{dt}(t_0).$$

The right-hand side is precisely the dot product of the *gradient* $\nabla f = \langle f_x, f_y \rangle$ and the *velocity vector* $\vec{P}' = \left\langle \dfrac{dx}{dt}, \dfrac{dy}{dt} \right\rangle$:

$$\frac{df(P(t))}{dt}\bigg|_{t=t_0} = \nabla f(P(t_0)) \cdot \vec{P}'(t_0). \tag{3}$$

This is a new version of the Chain Rule. The familiar one-variable version is

$$\frac{df(x(t))}{dt} = \frac{df}{dx}\frac{dx}{dt} = f'(x(t))x'(t),$$

the derivative of the "outside" $f(x)$, times the derivative of the "inside" $x(t)$. The new version (3) has the same form; the derivative of the composite function $f(P(t))$ is the derivative of the "outside" $f$, times the derivative of the "inside" $P(t)$; the derivative of the "outside" is the gradient $\nabla f$, the derivative of the "inside" is the vector derivative $\vec{P}'$, and the product is the dot product.

This Chain Rule has various notations. Dropping the evaluation at $t_0$ and setting $P = (x, y)$, we rewrite (3) as

$$\frac{df(x, y)}{dt} = \langle f_x, f_y \rangle \cdot \left\langle \frac{dx}{dt}, \frac{dy}{dt} \right\rangle = f_x \frac{dx}{dt} + f_y \frac{dy}{dt}. \tag{4}$$

Introducing a variable $z = f(x, y)$, we get the formula in Leibniz notation,

$$\frac{dz}{dt} = \frac{\partial z}{\partial x}\frac{dx}{dt} + \frac{\partial z}{\partial y}\frac{dy}{dt}. \tag{5}$$

In this last form, $\dfrac{\partial z}{\partial x}$ and $\dfrac{\partial z}{\partial y}$ are the partial derivatives of $z$, considered as a function of the two variables $x$ and $y$. Along the given curve, $x$ and $y$ are functions of a single variable $t$, and so then is $z$; $\dfrac{dz}{dt}$ stands for the derivative of $z$ in this sense.

---

**EXAMPLE 1**   $z = x^2 + 2y^2$, $x = \cos t$, $y = \sin t$. By the Chain Rule (5)

$$\frac{dz}{dt} = \frac{\partial z}{\partial x}\frac{dx}{dt} + \frac{\partial z}{\partial y}\frac{dy}{dt}$$
$$= 2x(-\sin t) + 4y(\cos t)$$
$$= 2(\cos t)(-\sin t) + 4(\sin t)(\cos t).$$

You can check this by substituting the given $x$ and $y$ directly in the formula for $z$:

$$z = x^2 + 2y^2 = \cos^2 t + 2\sin^2 t,$$

$$\frac{dz}{dt} = 2(\cos t)(-\sin t) + 4(\sin t)(\cos t).$$

in agreement with the Chain Rule calculation.

---

**EXAMPLE 2**   If $z = xy$ then $\dfrac{\partial z}{\partial x} = y$ and $\dfrac{\partial z}{\partial y} = x$, so the Chain Rule (5) gives

$$\frac{dz}{dt} = y\frac{dx}{dt} + x\frac{dy}{dt},$$

that is,

$$\frac{d(xy)}{dt} = y\frac{dx}{dt} + x\frac{dy}{dt}.$$

This is precisely the Product Rule!

---

*WARNING*   The one-variable Chain Rule

$$\frac{dy}{dt} = \frac{dy}{du}\frac{du}{dt}$$

can be remembered by an informal cancellation of the differential symbols $du$ on the right-hand side. But nothing of the sort is possible in the Chain Rule (5). Instead, think of the two terms on the right in (5) as two contributions to the rate of change of $z$. As $t$ changes, so does $x$, and the term $\dfrac{\partial z}{\partial x}\dfrac{dx}{dt}$ gives the rate of change of $z$ due to this change in $x$; but $y$ changes too, and the rate of change of $z$ due to this is $\dfrac{\partial z}{\partial y}\dfrac{dy}{dt}$. The total rate of change of $z$ is the *sum* of these two terms.

The rest of this section is devoted to just one application of the Chain Rule; many more are given in Section 16.2.

## Directional Derivatives

The partial derivatives $f_x$ and $f_y$ give the rate of change of $f$ in the direction of the two coordinate axes. The *directional derivative* $D_{\vec{u}}f(x, y)$ gives the rate of change in the direction of any *unit vector* $\vec{u}$. Figure 3 illustrates the definition. Consider the straight line $\vec{P}(s) = \langle x, y \rangle + s\vec{u}$ with $s \geq 0$. The point $\vec{P}(s)$ is $s$ units away from $(x, y)$ in the direction of the unit vector $\vec{u}$. Thus the *average rate of change of $f$* with respect to distance in direction $\vec{u}$ is

$$\frac{f(P(s)) - f(x, y)}{s} = \frac{f(P(s)) - f(P(0))}{s}.$$

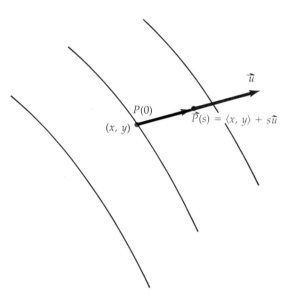

**FIGURE 3**
Straight line through $(x, y)$ in direction of unit vector $\vec{u}$; parameter $s = $ arc length.

The *rate of change* of $f$ with respect to distance in direction $\vec{u}$ is then defined as the limit

$$
\begin{aligned}
D_{\vec{u}}f(x, y) &= \lim_{s \to 0^+} \frac{f(P(s)) - f(P(0))}{s} \\
&= \frac{df(P(s))}{ds}\bigg|_{s=0} && \text{[definition of derivative]} \\
&= \nabla f(P(0)) \cdot \vec{P}'(0) && \text{[Chain Rule (3)].}
\end{aligned}
$$

But $P(0) = (x, y)$ and $\vec{P}'(0) = \dfrac{d}{ds}[\langle x, y \rangle + s\vec{u}] = \vec{u}$, so

$$D_{\vec{u}}f(x, y) = \nabla f(x, y) \cdot \vec{u}. \tag{6}$$

Here $\vec{u}$ is assumed to be a unit vector.

The derivative in the direction of a *nonunit* vector $\vec{v}$ is defined as

$$\nabla f(x, y) \cdot \dfrac{\vec{v}}{|\vec{v}|}; \tag{7}$$

we use the unit vector $\dfrac{\vec{v}}{|\vec{v}|}$ in the direction of $\vec{v}$.

---

**EXAMPLE 3**   Let $f(x, y) = xy$.

**a.**   Compute the directional derivative of $f$ at the point $P_0 = (2, 2)$ in the direction of each of the vectors $\vec{i}, \vec{j}, \vec{i} + \vec{j}, \vec{i} - \vec{j}$.

**b.**   Compute the directional derivative at $P_0$ in the direction toward $(3, 4)$.

*SOLUTION*

**a.**   The gradient is $\nabla f = y\vec{i} + x\vec{j}$, and at $P_0 = (2, 2)$,

$$\nabla f(P_0) = 2\vec{i} + 2\vec{j}.$$

Since $\vec{i}$ and $\vec{j}$ are unit vectors, the derivatives in these directions are

$$\nabla f(P_0) \cdot \vec{i} = 2$$
$$\nabla f(P_0) \cdot \vec{j} = 2.$$

To compute the derivative in the direction $\vec{v} = \vec{i} + \vec{j}$, form the unit vector $\dfrac{\vec{v}}{|\vec{v}|} = \dfrac{\vec{i} + \vec{j}}{\sqrt{2}}$, and find the directional derivative as in (7):

$$\nabla f(P_0) \cdot \dfrac{\vec{v}}{|\vec{v}|} = (2\vec{i} + 2\vec{j}) \cdot \dfrac{\vec{i} + \vec{j}}{\sqrt{2}} = \dfrac{2}{\sqrt{2}} + \dfrac{2}{\sqrt{2}} = 2\sqrt{2}.$$

The derivative in the direction $\vec{i} - \vec{j}$ is

$$\nabla f(P_0) \cdot \dfrac{\vec{i} - \vec{j}}{|\vec{i} - \vec{j}|} = \dfrac{2}{\sqrt{2}} - \dfrac{2}{\sqrt{2}} = 0.$$

To interpret these derivatives, think of the graph of $f$ as a hillside—Figure 4 gives the contour map. The vector $\vec{i} - \vec{j}$ points along a level curve, where $f$ is constant, so the derivative in this direction is 0. The vector $\vec{i} + \vec{j}$ points uphill, so the derivative in this direction is positive.

**b.**   To compute the directional derivative at $P_0 = (2, 2)$ toward $(3, 4)$, you need the vector from $(2, 2)$ to $(3, 4)$,

$$\vec{v} = \langle 3, 4 \rangle - \langle 2, 2 \rangle = \langle 1, 2 \rangle.$$

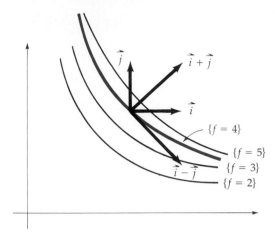

**FIGURE 4**

The unit vector in this direction is

$$\frac{\vec{v}}{|\vec{v}|} = \frac{1}{\sqrt{5}} \langle 1, 2 \rangle.$$

The derivative of $f$ in that direction is

$$\nabla f(P_0) \cdot \frac{1}{\sqrt{5}} \langle 1, 2 \rangle = \langle 2, 2 \rangle \cdot \frac{1}{\sqrt{5}} \langle 1, 2 \rangle = \frac{6}{\sqrt{5}}.$$

---

The directional derivative formula $D_{\vec{u}} f = \nabla f \cdot \vec{u}$ reveals the full meaning of the gradient. Let $\theta$ be the angle between $\nabla f$ and $\vec{u}$ (Fig. 5). Then

$$D_{\vec{u}} f(x, y) = \nabla f(P_0) \cdot \vec{u} = |\nabla f(P_0)| \, |\vec{u}| \cos \theta.$$

Since $\vec{u}$ is a unit vector, we find

$$D_{\vec{u}} f(x, y) = |\nabla f(P_0)| \cos \theta. \tag{8}$$

Now,

$$\cos \theta = \begin{cases} 1 & \text{if } \theta = 0 \text{ (maximum of } \cos \theta) \\ 0 & \text{if } \theta = \pi/2 \\ -1 & \text{if } \theta = \pi \text{ (minimum of } \cos \theta) \end{cases}$$

so from (8)

$$D_{\vec{u}} f = \begin{cases} |\nabla f| & \text{if } \vec{u} \text{ has the direction of } \nabla f \text{ (maximum } D_{\vec{u}} f) \\ 0 & \text{if } \vec{u} \perp \Delta f \\ -|\nabla f| & \text{if } \vec{u} \text{ has the direction of } -\nabla f \text{ (minimum } D_{\vec{u}} f) \end{cases}$$

Figure 5a illustrates the maximum directional derivative: $\vec{u}$ points in the direction of $\nabla f$, directly toward the higher level curves. In Figure 5b $\vec{u}$ still points

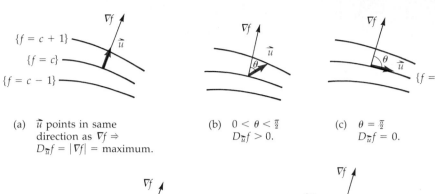

(a)   $\vec{u}$ points in same direction as $\nabla f$ ⇒ $D_{\vec{u}}f = |\nabla f| =$ maximum.

(b)   $0 < \theta < \frac{\pi}{2}$ $D_{\vec{u}}f > 0.$

(c)   $\theta = \frac{\pi}{2}$ $D_{\vec{u}}f = 0.$

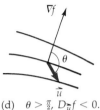

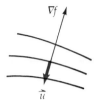

(d)   $\theta > \frac{\pi}{2}$, $D_{\vec{u}}f < 0.$

(e)   $\vec{u}$ opposite to $\nabla f$ ⇒ $D_{\vec{u}}f = -|\nabla f| =$ minimum.

**FIGURE 5**
How $D_{\vec{u}}f$ depends on direction $\vec{u}$.

toward higher levels, but not as directly; the rate of increase in such a direction is positive, but not as large as in the direction of $\nabla f$. In Figure 5c $\vec{u}$ is tangent to the level curve, where $f$ has a constant value; the derivative in this direction is zero. In Figure 5d, $\vec{u}$ points toward *lower* levels, so $D_{\vec{u}}f < 0$. Finally, in the direction directly opposite to $\nabla f$ (Fig. 5e), $\vec{u}$ points directly toward the lower level curves, and $D_{\vec{u}}f$ is as negative as possible.

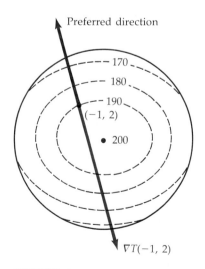

**FIGURE 6**
The flight of the ant from $(-1, 2)$.

**EXAMPLE 4**   An ant in a frying pan seeks rapid relief from the heat. Its position is $(-1, 2)$ and the temperature at $(x, y)$ is

$$T(x, y) = 200 - x^2 - 2y^2.$$

Which way should it run?

**SOLUTION**   The gradient of $T$ is

$$\nabla T(x, y) = -2x\vec{i} - 4y\vec{j}$$

so

$$\nabla T(-1, 2) = 2\vec{i} - 8\vec{j}.$$

This is the direction of maximum *increase* of temperature. The ant should run in the opposite direction, $-2\vec{i} + 8\vec{j}$. Figure 6 shows the level curves and the direction of flight. (Perhaps, with six legs on the pan, the ant can actually determine the direction of decreasing temperature.)

The gradient can be *approximated*, given the function values at appropriate points.

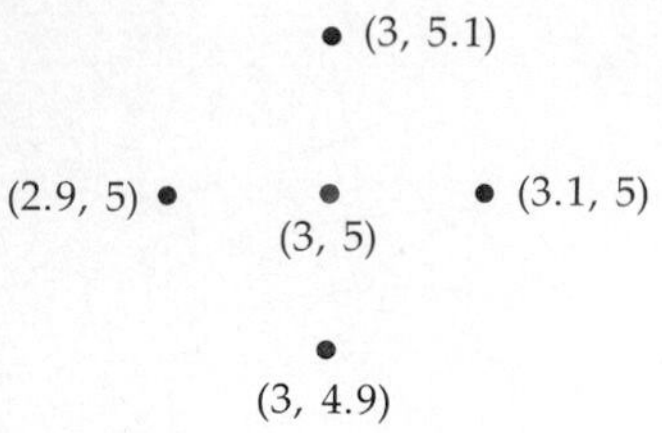

• (3, 5.1)

(2.9, 5) •   •   • (3.1, 5)
        (3, 5)

•
(3, 4.9)

**FIGURE 7**
Four points surrounding (3, 5).

**EXAMPLE 5**  The electric field $\varphi$ in an insulating plate is monitored at four points surrounding the point (3, 5) (Fig. 7):

$$\varphi(3.1, 5) = 5.1 \qquad \varphi(2.9, 5) = 4.8$$
$$\varphi(3, 5.1) = 5.2 \qquad \varphi(3, 4.9) = 5.0.$$

Estimate the quantities $\dfrac{\partial\varphi}{\partial x}$, $\dfrac{\partial\varphi}{\partial y}$, $\nabla\varphi$, and $D_{\vec{u}}\varphi$ for the unit vector $\vec{u} = \dfrac{1}{\sqrt{2}}(\vec{i} + \vec{j})$, all evaluated at the point (3, 5).

**SOLUTION**  The derivative $\dfrac{\partial\varphi}{\partial x}$ gives the rate of change of $\varphi$ in the $x$ direction, with $y$ held constant. To estimate it, compute the *average* rate of change between the two nearby points (2.9, 5) and (3.1, 5), where $x$ varies a little but $y$ is constantly 5:

$$\frac{\partial\varphi}{\partial x}(3, 5) \approx \frac{\varphi(3.1, 5) - \varphi(2.9, 5)}{3.1 - 2.9} = \frac{0.3}{0.2} = 1.5.$$

Likewise

$$\frac{\partial\varphi}{\partial y}(3, 5) \approx \frac{\varphi(3, 5.1) - \varphi(3, 4.9)}{5.1 - 4.9} = \frac{0.2}{0.2} = 1.$$

Hence $\nabla\varphi\langle 3, 5\rangle \approx \langle 1.5, 1\rangle$. The directional derivative in the direction $\vec{u} = \dfrac{1}{\sqrt{2}}(\vec{i} + \vec{j})$ is

$$D_{\vec{u}}\varphi(3, 5) = \nabla\varphi \cdot \vec{u} \approx (1.5)\left(\frac{1}{\sqrt{2}}\right) + (1)\left(\frac{1}{\sqrt{2}}\right) = \frac{2.5}{\sqrt{2}}.$$

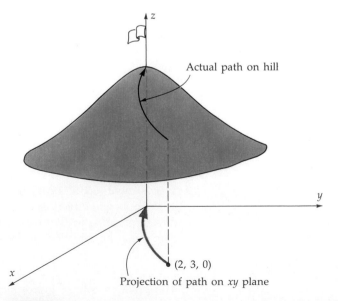

**FIGURE 8**
Steepest route to the top.

**EXAMPLE 6**    A hill has the shape of the graph of $z = e^{-x^2 - 2y^2}$. Starting at the point $(2, 3, e^{-22})$, find the steepest route to the top.

**SOLUTION**    The actual path is on the hill, but we describe it by its projection on the $xy$-plane (Fig. 8); this projection is what a map of the route would show (Fig. 9).

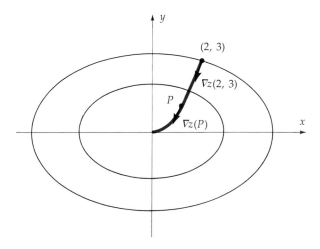

**FIGURE 9**
Map of steepest path from $(2, 3)$ to top. At each point $P$, the path is tangent to $\nabla z(P)$.

Starting at the given point, the steepest route is in the direction of $\nabla z (2, 3)$ (Fig. 9). But if you continue in this direction, you will never reach the top; the direction must change as the path rises. At each point in the $xy$-plane, the gradient $\nabla z$ gives the direction in which $z$ increases most rapidly; so we seek a curve which has *at each point* the direction of the gradient

$$\nabla z = \left\langle \frac{\partial z}{\partial x}, \frac{\partial z}{\partial y} \right\rangle = 2e^{-x^2 - 2y^2}\langle -x, \ -2y \rangle.$$

The *slope* of this vector is the $y$ component divided by the $x$ component (Fig. 10), so

$$\text{slope of desired curve at } (x, y) = \frac{z_y}{z_x} = \frac{2y}{x}.$$

Thus the curve in Figure 9 has slope

$$\frac{dy}{dx} = \frac{2y}{x}.$$

Solve this differential equation:

$$\frac{dy}{y} = \frac{2dx}{x} \Leftrightarrow \ln|y| = 2 \ln|x| + c = \ln(x^2) + c$$

so

$$y = (\pm e^c)x^2 = Cx^2.$$

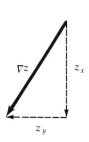

**FIGURE 10**
$\dfrac{z_y}{z_x} = $ slope of $\nabla z$.

Since the curve passes through $(2, 3)$, then $3 = C \cdot 2^2$; so $C = 3/4$ and $y = \frac{3}{4}x^2$. The actual path in space is thus

$$y = \tfrac{3}{4}x^2, \qquad z = e^{-x^2 - 2y^2} = e^{-x^2 - 9x^4/8}.$$

## SUMMARY

**Chain Rule:**

$$\frac{df(P(t))}{dt} = \nabla f(P(t)) \cdot \vec{P}'(t)$$

$$\frac{df(x, y)}{dt} = \frac{\partial f}{\partial x}\frac{dx}{dt} + \frac{\partial f}{\partial y}\frac{dy}{dt}.$$

**Directional Derivative** in the direction of unit vector $\vec{u}$:

$$D_{\vec{u}}f(x, y) = \nabla f(x, y) \cdot \vec{u}.$$

This is *greatest* in the direction of $\nabla f$, *zero* in directions perpendicular to $\nabla f$, and *most negative* in the direction of $-\nabla f$.

## PROBLEMS

### A

**1.** Find $dz/dt$, if $z = f(x, y)$ and $x$ and $y$ are the given functions of $t$. Use the Chain Rule. Then express $z$ directly as a function of $t$, and check the result by computing $dz/dt$ directly.
  **\*a)** $z = xy$, $x = \cos t$, $y = \sin t$
  **b)** $z = e^{x+y}$, $x = t$, $y = t^2$
  **\*c)** $z = \tan^{-1}(y/x)$, $x = t^3$, $y = 2t^3$
  **d)** $z = \cos^{-1}(x/\sqrt{x^2 + y^2})$, $x = \cos 2t$, $y = \sin 2t$

**2.** Compute the directional derivative at the given point $Q$, in the direction of the given vector $\vec{v}$.
  **\*a)** $x^3 + xy^2$, $Q = (-1, -1)$, $\vec{v} = \vec{i} + \vec{j}$
  **b)** $\dfrac{x + y}{x - y}$, $Q = (2, -2)$, $\vec{v} = \vec{i}$
  **\*c)** $\sqrt{x^2 + y^2}$, $Q = (1, -3)$, $\vec{v} = 2\vec{i} + \vec{j}$
  **d)** $x^2 + y^2$, $Q = (0, 0)$, $\vec{v} = 10\vec{i} + 20\vec{j}$

**3.** Compute the directional derivative of $e^{xy}$ at $P = (2, 1)$, in the direction
  **a)** From $P$ toward $(1, 2)$.
  **\*b)** From $P$ toward the origin.

**4.** At the given point $Q$, find the direction in which $f$ increases most rapidly, and compute the rate of change in that direction.
  **\*a)** $3x^2 - 2y^2$, $Q = (2, 1)$
  **b)** $x^2 + xy + y^2$, $Q = (-1, -1)$
  **c)** $\sin^{-1}(xy/4)$, $Q = (1, 2)$

**\*5.** Sheep are grazing in a meadow where the height of the grass is $f(x, y) = x^2 - x + y^2 + 2y$. A sheep at the origin wants to go toward higher grass, nibbling as it goes. Which direction do you recommend?

**6.** A small fish is in a shallow pond where the pollution level at $(x, y)$ is $e^{-xy}$. The fish is at $(1, 2)$. Which direction should it go to reach cleaner water most rapidly?

**\*7.** The electric potential in a copper plate is $e^{3x} \sin 3y$. An electron at the origin seeks higher potential. Which way does it go?

**\*8.** Near a mountain pass at $(0, 0)$, the elevation above sea level is $z = 1500 + 5x^2 - 3y^2$. The $x$-axis points east, and the $y$-axis points north. A climber is at position $(3, -5)$.

**a)** If she goes due south, does she head uphill or downhill?

**b)** Which direction is steepest up? Steepest down?

**c)** In which directions would she follow a level path?

**9.** For any function $f$ with continuous derivatives

**a)** Show that $f_x$ is the directional derivative in direction $\vec{i}$.

**b)** What is the directional derivative in direction $\vec{j}$? In direction $-\vec{j}$?

**\*10.** At a certain time, the radius of the base of a cone is 5 cm and increasing at 2 cm/min, while the altitude is 3 cm and decreasing at 2 cm/min.

**a)** At what rate is the volume of the cone increasing?

**b)** At what rate is the total surface area (including the base) increasing?

**11.** At a certain time, the radius of a cylinder is 6 cm and decreasing at 1 cm/sec, while the height is 7 cm and increasing at 2 cm/sec.

**a)** How fast is the volume increasing?

**b)** How fast is the total surface area increasing?

# B

**12.** The temperature $T(x, y)$ is given at several points. Estimate the partial derivatives, the gradient, and the derivative in the given direction.

**a)**  $T(7.99, 5) = 33.1$   $T(8.01, 5) = 33.2$
 $T(8, 4.99) = 33.2$   $T(8, 5.01) = 33.0$
 Estimate $D_{\vec{u}}T(8, 5)$ in the direction $\vec{u} = \frac{1}{5}(3\vec{i} - 4\vec{j})$.

**\*b)**  $T(85, 13) = 2.1$   $T(86, 13) = 1.9$
 $T(85, 14) = 2.0$
 Plot the three points where $T$ is given. Estimate $D_{\vec{u}}T(85, 13)$ in the direction from $(85, 13)$ toward $(86, 14)$.

**13.** Given $D_{\vec{u}}f(2, 1) = 3$ for $\vec{u} = \frac{1}{5}(3\vec{i} + 4\vec{j})$ and $D_{\vec{v}}f(2, 1) = -1$ for $\vec{v} = \frac{1}{5}(4\vec{i} - 3\vec{j})$, compute the partial derivatives of $f$ at $(2, 1)$.

**\*14.** The temperature at $(x, y)$ is $e^{-x^2 - 3y^2}$. A worm at $(2, 1)$ moves in a path seeking at each instant the most rapid increase in temperature. Find the path of its motion.

**15.** Find a curve in the $xy$-plane passing through $(1, 1)$, and crossing all the level curves $\{(x, y): y - x^2 = c\}$ at right angles.

**\*16.** A ball is released from point $(1, 3, 3)$ on the graph of $z = xy$ ($z$-axis up). It rolls on the graph,

as on a hillside. In what direction does it begin to roll? (Give an appropriate three-dimensional vector $\vec{v}$, tangent to the graph.)

**17.** Prove the following formulas for the gradient. (Compute each side and compare results.)

**a)** $\nabla(f + g) = \nabla f + \nabla g$

**b)** $\nabla(fg) = f\nabla g + g\nabla f$

**c)** $\nabla(cf) = c\nabla f$   ($c$ is constant.)

**d)** $\nabla\left(\dfrac{f}{g}\right) = \dfrac{1}{g^2}[g\nabla f - f\nabla g]$

**18. a)** Prove a formula for $\nabla(f^n)$, where $n$ is a constant and $f > 0$.

**b)** Prove a formula for $\nabla(g(f))$, where $f$ is a function of two variables, and $g$ a function of one variable.

**19.** Suppose that $\nabla f = 0$ all along a differentiable curve $P(t)$. Prove that $f$ is constant along the curve.

**20.** Suppose that $P(t)$ is a curve in space whose points all happen to lie on the graph of a function $f$ with continuous derivatives. Using the Chain Rule, prove that $\vec{P}'(t)$ is perpendicular to the normal to the graph of $f$ at $P(t)$. [If $P = (x, y, z)$ is on the graph, then $z = f(x, y)$. Use this in computing $\vec{P}'(t)$.]

# C

**21.** Suppose you are at a point $P_0$ near, but not on, the level curve $\{f = 0\}$ of a function $f$ with continuous derivatives. Use $f(P_0)$ and $\nabla f(P_0)$ to determine

**a)** A unit vector $\vec{u}$ pointing as nearly as possible toward $\{f = 0\}$.

**b)** The approximate distance from $P_0$ to the level curve.

**\*c)** A vector giving the approximate displacement from $P_0$ to the nearest point on $\{f = 0\}$.

**22.** Suppose that $f$ has continuous derivatives at every point of a circular disk $D$.

**a)** Prove that for any two points $P_1$ and $P_2$ in $D$, there is a point $Q$ on the segment joining $P_1$ and $P_2$ such that $f(P_2) - f(P_1) = (\vec{P}_2 - \vec{P}_1) \cdot \nabla f(Q)$. [Apply the Mean Value Theorem to $f(P(t))$, where $\vec{P}(t) = P_1 + t(\vec{P}_2 - \vec{P}_1)$.]

**b)** Prove that $|f(P_2) - f(P_1)| \leq M|\vec{P}_2 - \vec{P}_1|$, where $M$ is the maximum of $|\nabla f|$ on the disk $D$.

## 15.6

### HIGHER DERIVATIVES. MIXED PARTIALS.
### $C^k$ FUNCTIONS

For any two-variable function $f$, the derivatives $f_x$ and $f_y$ are again functions; *their* derivatives are the **second derivatives** of $f$. There are four of them, denoted as follows:

$$(f_x)_x = f_{xx} \quad \text{or} \quad \frac{\partial}{\partial x}\left(\frac{\partial f}{\partial x}\right) = \frac{\partial^2 f}{\partial x^2}$$

$$(f_x)_y = f_{xy} \quad \text{or} \quad \frac{\partial}{\partial y}\left(\frac{\partial f}{\partial x}\right) = \frac{\partial^2 f}{\partial y \partial x}$$

$$(f_y)_x = f_{yx} \quad \text{or} \quad \frac{\partial}{\partial x}\left(\frac{\partial f}{\partial y}\right) = \frac{\partial^2 f}{\partial x \partial y}$$

$$(f_y)_y = f_{yy} \quad \text{or} \quad \frac{\partial}{\partial y}\left(\frac{\partial f}{\partial y}\right) = \frac{\partial^2 f}{\partial y^2}.$$

For derivatives of order higher than two, the notation is similar.

---

**EXAMPLE 1**   $f(x, y) = xy + x^4 - x^3 y^2$ has *first derivatives*

$$f_x = y + 4x^3 - 3x^2 y^2 \qquad f_y = x - 2x^3 y,$$

second derivatives

$$f_{xx} = 12x^2 - 6xy^2, \qquad f_{xy} = 1 - 6x^2 y,$$
$$f_{yx} = 1 - 6x^2 y, \qquad\qquad f_{yy} = -2x^3,$$

third derivatives

$$f_{xyy} = -6x^2 \qquad f_{yyx} = -6x^2,$$

and so on.

---

In Example 1, the "mixed partials" $f_{xy}$ and $f_{yx}$ came out equal. The same is true for every function $f$, when the derivatives involved are *continuous*. You can think of this as a "commutative law" for partial derivatives:

$$\frac{\partial}{\partial y}\frac{\partial}{\partial x} = \frac{\partial}{\partial x}\frac{\partial}{\partial y}.$$

Partial derivatives can be taken in any order, as long as they are continuous. From this it follows, for instance, that

$$\frac{\partial}{\partial y}\frac{\partial}{\partial y}\frac{\partial}{\partial x} f = \frac{\partial}{\partial x}\frac{\partial}{\partial y}\frac{\partial}{\partial y} f$$

or $f_{xyy} = f_{yyx}$, as in Example 1.

## $C^k$ Functions

The equality $f_{xy} = f_{yx}$ may fail unless those partial derivatives are *continuous*. And recall (Sec. 15.3) that the graph of $f$ may have no tangent plane unless $f_x$ and $f_y$ are continuous. In general, for partial derivatives, mere existence is not enough to guarantee a viable theory; but for continuous derivatives, the theory is fine. There is a shorthand notation for stating the hypothesis of continuity:

> $f$ is a $C^1$ function (a "C-one" function) in a region $R$ if $f_x$ and $f_y$ are continuous at every point of $R$.
>
> $f$ is a $C^2$ function in $R$ if all second derivatives are also continuous in $R$.

Generally, $f$ is a $C^k$ *function* in $R$ if all derivatives of order $\leq k$ are continuous in $R$. Thus if $f$ is a $C^1$ function, the graph of $f$ has a tangent plane. If $f$ is a $C^2$ function, then $f_{xy} = f_{yx}$. If $f$ is a $C^3$ function, then $f_{xyy} = f_{yyx}$; and so on.

## The Proof that $f_{xy} = f_{yx}$

It is difficult to see, geometrically, *why* the mixed partials $f_{xy}$ and $f_{yx}$ must be equal, but it can be proved by representing $f_{xy}$ and $f_{yx}$ as limits of appropriate difference quotients. By the definition of the derivative as a limit,

$$f_{yx}(a, b) = \text{derivative of } f_y(x, b) \text{ at } x = a$$

$$= \lim_{h \to 0} \frac{f_y(a + h, b) - f_y(a, b)}{h}.$$

So $f_{yx}$ is approximated by this difference quotient:

$$f_{yx} \approx \frac{f_y(a + h, b) - f_y(a, b)}{h}. \tag{1}$$

Going one step further, we make the approximations

$$f_y(a, b) \approx \frac{f(a, b + h) - f(a, b)}{h}, \quad f_y(a + h, b) \approx \frac{f(a + h, b + h) - f(a + h, b)}{h}.$$

Use these in (1), obtaining the quotient

$$Q(h) = \frac{f(a + h, b + h) - f(a + h, b) - f(a, b + h) + f(a, b)}{h^2} \tag{2}$$

as an approximation to $f_{yx}(a, b)$.

Now construct the analogous difference quotient for $f_{xy}$. Approximate the derivative of $f_x$ with respect to $y$ by a quotient:

$$f_{xy} \approx \frac{f_x(a, b + h) - f_x(a, b)}{h}. \tag{3}$$

Similarly, use

$$f_x(a, b + h) \approx \frac{f(a + h, b + h) - f(a, b + h)}{h}, \, f_x(a, b) \approx \frac{f(a + h, b) - f(a, b)}{h}$$

to expand (3) into

$$\frac{f(a + h, b + h) - f(a + h, b) - f(a, b + h) + f(a, b)}{h^2}.$$

This approximation for $f_{xy}(a, b)$ is precisely the expression $Q(h)$ in (2), approximating $f_{yx}(a, b)$! We will show that when $f$ is a $C^2$ function, then each mixed partial derivative is in fact the limit of $Q(h)$ as $h \to 0$; so the mixed partials are equal.

---

**THEOREM 1**

If $f_x$, $f_y$, $f_{xy}$, and $f_{yx}$ all exist in a disk centered at $(a, b)$, and are continuous at $(a, b)$, then

$$f_{xy}(a, b) = \lim_{h \to 0} Q(h) = f_{yx}(a, b).$$

---

We'll prove the limit for $f_{xy}$; the other limit is analogous. Define the function of one variable

$$F(x) = f(x, b + h) - f(x, b). \tag{4}$$

Then

$$Q(h) = \frac{F(a + h) - F(a)}{h^2}. \tag{5}$$

By the Mean Value Theorem

$$F(a + h) - F(a) = hF'(\hat{x}) \qquad \text{with } \hat{x} \text{ between } a \text{ and } a + h. \tag{6}$$

Referring to (4), $F'(x) = f_x(x, b + h) - f_x(x, b)$, so (6) gives

$$F(a + h) - F(a) = h[f_x(\hat{x}, b + h) - f_x(\hat{x}, b)].$$

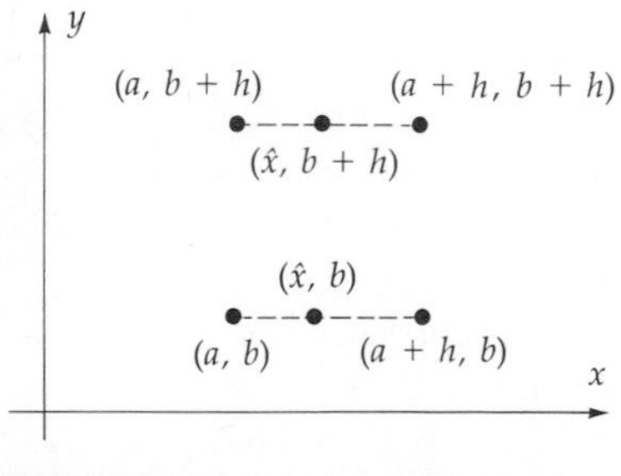

(Figure 1 shows the various points of evaluation introduced so far.) Now apply the Mean Value Theorem once more, to the function $G(y) = f_x(\hat{x}, y)$ on the interval from $b$ to $b + h$. The derivative of $G$ is

$$G'(y) = \frac{\partial}{\partial y} f_x(\hat{x}, y) = f_{xy}(\hat{x}, y)$$

so

$$F(a + h) - F(a) = h[hf_{xy}(\hat{x}, \hat{y})] \qquad \text{with } \hat{y} \text{ between } b \text{ and } b + h.$$

**FIGURE 1**
Points of evaluation for $f_x$.

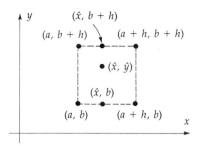

**FIGURE 2**
Point of evaluation for $f_{xy}(\hat{x}, \hat{y})$.
As $h \to 0$, $(\hat{x}, \hat{y}) \to (a, b)$.

[Figure 2 adds $(\hat{x}, \hat{y})$ to the picture.] Finally, from (5),

$$Q(h) = f_{xy}(\hat{x}, \hat{y}). \qquad (7)$$

As $h \to 0$ then $a + h \to a$ and $b + h \to b$, so $\hat{x} \to a$ and $\hat{y} \to b$. Since $f_{xy}$ is assumed continuous at $(a, b)$, it follows from (7) that

$$\lim_{h \to 0} Q(h) = f_{xy}(a, b).$$

Reversing the roles of $x$ and $y$ in the above argument shows further that $\lim_{h \to 0} Q(h) = f_{yx}(a, b)$. So $f_{xy}$ and $f_{yx}$ are given by the same limit, and must be equal.

## SUMMARY

*f is a $C^k$ Function* in a region $R$ if all partial derivatives of $f$, of order $\leq k$, are continuous in $R$.

*Partial derivatives may be taken in any order, if the resulting derivatives are continuous:*

$$f_{xy} = f_{yx}, \qquad f_{xyx} = f_{xxy}, \text{ etc.}$$

## PROBLEMS

**A**

1. Compute all four second order derivatives $f_{xx}$, $f_{xy}$, $f_{yx}$, $f_{yy}$:
   *a)  $xy$
   b)  $x^2 - y^2$
   *c)  $x^2 + 2xy + 3y^2 + x - y$
   d)  $x^3 + 3xy + y^3$
   *e)  $e^x \cos y$

2. Let $f(x, y) = e^x \sin(x + y)$. Compute and compare
   a)  $f_{xy}$ and $f_{yx}$
   b)  $\dfrac{\partial^3 f}{\partial x^2 \, \partial y}$, $\dfrac{\partial^3 f}{\partial x \, \partial y \, \partial x}$, and $\dfrac{\partial^3 f}{\partial y \, \partial x^2}$.

3. For the following functions $f(x, t)$, compare $\dfrac{\partial^2 f}{\partial t^2}$ with $\dfrac{\partial^2 f}{\partial x^2}$.
   *a)  $\sin(x + ct)$    b)  $\sin(x - ct)$

c)  $e^{x + ct}$        d)  $e^{x - ct} + \sin(x + ct)$

**B**

*4. Let $\varphi$ and $\psi$ be $C^2$ functions of one variable. Compare $\dfrac{\partial^2 f}{\partial x^2}$ with $\dfrac{\partial^2 f}{\partial t^2}$ if
   a)  $f(x, t) = \varphi(x + ct)$
   b)  $f(x, t) = \psi(x - ct)$
   c)  $f(x, t) = \varphi(x + ct) - \psi(x + ct)$.
   [Note: $\varphi(x + ct)$ here does not mean "$\varphi$ times $x + ct$"; it means "function $\varphi$ evaluated at $x + ct$," as in the examples in problem 3.]

5. a)  Find a function $f(x, y)$ such that $\nabla f(x, y) = e^x \cos y \vec{i} - e^x \sin y \vec{j}$.
   b)  Show that there is no function $f(x, y)$ such that $\nabla f(x, y) = x^2 y \vec{i} + y^2 x \vec{j}$. [Assuming there were such an $f$, compute $f_{xy}$ and $f_{yx}$.]

**6.** Given $f(2.1, 3.1) = 3.3$, $f(2.0, 3.1) = 3.2$, $f(2.1, 3.0) = 3.1$, $f(2, 3) = 3.0$, approximate $f_x$, $f_y$, and $f_{xy}$ at $(2, 3)$.

**C**

**7.** Obtain the second difference quotient $Q_{hh} = \dfrac{f(a + 2h) - 2f(a + h) + f(a)}{h^2}$ as an approximation to $f''(a)$. Prove that $\lim_{h \to 0} Q_{hh} = f''(a)$ if $f''$ is continuous at $a$.

**8.** One can approximate $f_x$, $f_y$, and $f_{xy}$ at a point $(a, b)$ using values of $f$ at the four points $(a \pm h, b \pm h)$.

**a)** Sketch $(a, b)$ and the four surrounding points.

**b)** Approximate $f_x(a, b)$, using $f$ at all four points.

**c)** Approximate $f_{xy}$ and $f_{yx}$ at $(a, b)$.

**9.** Set $f(x, y) = \dfrac{xy(x^2 - y^2)}{x^2 + y^2}$ if $(x, y) \neq (0, 0)$, and $f(0, 0) = 0$.

**a)** Compute $f_x$ and $f_y$, and show that they are continuous everywhere. (Be careful in computing $f_x(0, 0)$ and $f_y(0, 0)$.)

**b)** Show that $f_{xy}(0, 0) \neq f_{yx}(0, 0)$. Does this violate the theorem on mixed partials?

## 15.7

## CRITICAL POINTS. THE SECOND DERIVATIVE TEST

Partial derivatives are the natural tool for seeking maximum and minimum points for functions of two variables.

A point $P_0$ is called a **local maximum point** for $f$ if

$$f(P_0) \geq f(P)$$

for all points $P$ inside some circle centered at $P_0$ (Fig. 1). Outside the circle $f$ may be undefined, or may assume values greater than $f(P_0)$.

Suppose that $P_0 = (x_0, y_0)$ is a local maximum point, and the partial derivatives $f_x(P_0)$ and $f_y(P_0)$ exist. Setting $y = y_0$, you get a function $f(x, y_0)$ of $x$ alone, which has a local maximum at $x = x_0$; so the derivative with respect to $x$ is zero when $x = x_0$:

$$f_x(x_0, y_0) = 0.$$

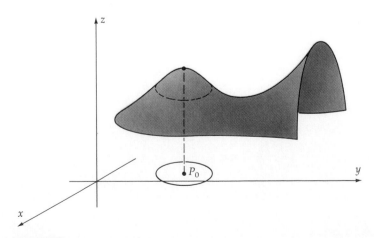

**FIGURE 1**
$P_0$ is a local maximum point.

Likewise, setting $x = x_0$, you get a function $f(x_0, y)$ with a local maximum at $y = y_0$, so

$$f_y(x_0, y_0) = 0.$$

Thus *if the partial derivatives at a local maximum point exist, they equal zero.* The same conclusion would follow for a local minimum point. So you seek local maxima and minima among those points where both partial derivatives equal zero, or where one of them fails to exist. These are called the **critical points** of $f$. Hence:

---

### THE CRITICAL POINT THEOREM

If $P_0$ is a local maximum or minimum point for $f$, then $P_0$ is a critical point; $\nabla f(P_0) = \vec{0}$, or is undefined.

---

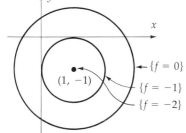

**FIGURE 2**
$(1, -1)$ is a local minimum point.

**EXAMPLE 1**   Find all local maximum or minimum points of the function

$$f(x, y) = x^2 + y^2 - 2x + 2y.$$

**SOLUTION**   Find all critical points, where both $f_x = 0$ and $f_y = 0$:

$$f_x = 2x - 2 = 0 \quad \Leftrightarrow \quad x = 1$$

$$f_y = 2y + 2 = 0 \quad \Leftrightarrow \quad y = -1.$$

Thus both derivatives are zero at just one point, $P_0 = (1, -1)$. The level curves (Fig. 2) show that $(1, -1)$ is a local minimum point.

---

A critical point might be *neither* a maximum nor a minimum. For instance, $f(x, y) = xy$ has a critical point at the origin:

$$f_x(0, 0) = y\Big|_{(0,0)} = 0, \qquad f_y(0, 0) = x\Big|_{(0,0)} = 0.$$

At this point $f(0, 0) = 0$; but this is not a local maximum, since $xy > 0$ everywhere in the first and third quadrants (Fig. 3), nor is it a local minimum, since $xy < 0$ in the other two quadrants. The graph is shaped like a saddle (Fig. 4) and the origin is called a **saddle point**.
   To distinguish between local maxima, minima, and saddle points, there is a simple test involving the second derivatives $f_{xx}, f_{xy}$, and $f_{yy}$ at the critical point. We derive it first for quadratic functions of the form

$$f(x, y) = Ax^2 + Bxy + Cy^2.$$

Suppose that $A \neq 0$, and complete the square on $x$:

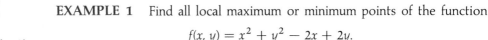

$$f(x, y) = A\left(x + \frac{By}{2A}\right)^2 - \frac{B^2 y^2}{4A} + Cy^2$$

$$= A\left[\left(x + \frac{By}{2A}\right)^2 + \frac{4AC - B^2}{4A^2}y^2\right]. \tag{1}$$

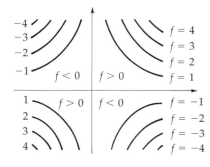

**FIGURE 3**
Level curves, $f(x, y) = xy$; $(0, 0)$ is a saddle point.

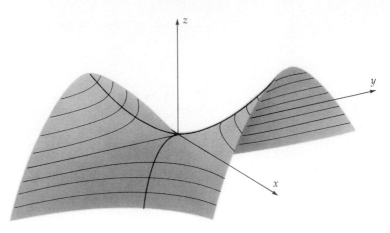

**FIGURE 4**
(0, 0) is a saddle point for $f(x, y) = xy$.

Consider three cases:

*Case I:* $4AC - B^2 > 0$   Then both terms $\left(x + \dfrac{By}{2A}\right)^2$ and $\dfrac{4AC - B^2}{4A^2}\, y^2$ in (1) are positive, except when $y = 0$ and $x = 0$. So the sign of $f$ depends only on the constant $A$:

$$A > 0 \quad \Rightarrow \quad f(x, y) > 0 \text{ for } (x, y) \neq (0, 0), \text{ so } (0, 0) \text{ is a minimum point.} \quad \textbf{(Ia)}$$

$$A < 0 \quad \Rightarrow \quad f(x, y) < 0 \text{ for } (x, y) \neq (0, 0), \text{ so } (0, 0) \text{ is a maximum point.} \quad \textbf{(Ib)}$$

*Case II:* $4AC - B^2 < 0$   Now the two terms $\left(x + \dfrac{By}{2A}\right)^2$ and $\dfrac{4AC - B^2}{4A^2}\, y^2$ have opposite signs, and (0, 0) gives neither a maximum nor a minimum. For when $y = 0$ and $x \neq 0$ then

$$f(x, y) = Ax^2$$

has the same sign as $A$; while if $y \neq 0$ and $x + \dfrac{By}{2A} = 0$ then

$$f(x, y) = A\,\frac{4AC - B^2}{4A^2}\, y^2$$

has the sign opposite to $A$, since $4AC - B^2 < 0$. In this case, the origin is a *saddle point*.

*Case III:* $4AC - B^2 = 0$   Now

$$f(x, y) = A\left(x + \frac{By}{2A}\right)^2$$

is zero all along the line $x + \dfrac{By}{2A} = 0$. This situation is rather special, and in this case the origin is called a **degenerate** critical point.

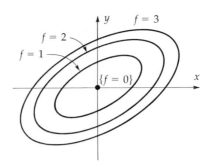

**FIGURE 5**
Level curves of $f(x, y) = Ax^2 + Bxy + Cy^2$, with $4AC - B^2 > 0$ and $A > 0$;
$(0, 0)$ is a minimum point.

Except in the degenerate case, the sign of $4AC - B^2$ distinguishes between a maximum or minimum on the one hand, and a saddle point on the other hand. At the same time, it *determines the nature of the level curves*

$$Ax^2 + Bxy + Cy^2 = \text{constant.} \tag{2}$$

Section 12.4 shows that the level curves (2) are

ellipses if $4AC - B^2 > 0$,      hyperbolas if $4AC - B^2 < 0$.

In the first case, the ellipses surround the critical point, which is either a minimum (Fig. 5), or a maximum. In the second case, the level curve $\{f = 0\}$ consists of two straight lines through the critical point (Fig. 6); the other

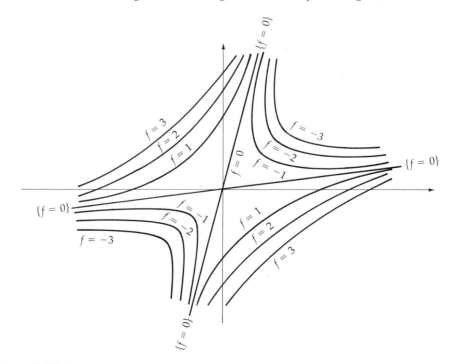

**FIGURE 6**
Level curves of $f(x, y) = Ax^2 + Bxy + Cy^2$, with $4AC - B^2 < 0$, are hyperbolas, and $(0, 0)$ is a saddle point.

level curves are hyperbolas with these straight lines as asymptotes, revealing the "saddle" shape of the graph.

To generalize from the special case of quadratic functions, we rephrase the conditions involving $A$, $B$, $C$, in terms of the second derivatives of the function that we considered:

$$f(x, y) = Ax^2 + Bxy + Cy^2,$$

$$f_{xx} = 2A, \qquad f_{xy} = B, \qquad f_{yy} = 2C.$$

Thus

$$4AC - B^2 = f_{xx}f_{yy} - (f_{xy})^2$$

and the distinction between Cases I, II, and III depends on the sign of this expression; while the distinction between Ia and Ib depends on the sign of $f_{xx}$.

The crucial expression $f_{xx}f_{yy} - f_{xy}^2$ can be remembered as a *determinant*, formed from the four second partial derivatives of $f$:

$$\det \begin{bmatrix} f_{xx} & f_{xy} \\ f_{yx} & f_{yy} \end{bmatrix} = f_{xx}f_{yy} - f_{xy}f_{yx} = f_{xx}f_{yy} - f_{xy}^2$$

since $f_{xy} = f_{yx}$. This is called the **Hessian determinant** of $f$. The position of each derivative in the determinant array depends on the subscripts; the first subscript in each derivative determines the row ($x$ for the first row, $y$ for the second), while the second subscript determines the column.

The conditions on the coefficients $A$, $B$, $C$ distinguishing the various cases can thus be expressed in terms of the Hessian determinant, and the partial derivative $f_{xx}$. In this form, they apply to all $C^2$ functions:

## The Second Derivative Test

Suppose that $f$ is a $C^2$ function, and that $P_0$ is a critical point:

$$f_x(P_0) = 0 = f_y(P_0).$$

Denote by $H$ the Hessian determinant:

$$H = \det \begin{bmatrix} f_{xx} & f_{xy} \\ f_{yx} & f_{yy} \end{bmatrix}.$$

Then:

(I)   If $H(P_0) > 0$ then $P_0$ is
  (Ia)   a local minimum point if $f_{xx}(P_0) > 0$,
  (Ib)   a local maximum point if $f_{xx}(P_0) < 0$.
(II)   If $H(P_0) < 0$ then $P_0$ is a saddle point.
(III)   If $H(P_0) = 0$ then the test gives no conclusion, and $P_0$ is called a *degenerate* critical point.

A proof is given later in this section. Notice that in general, the maxima and minima are *local*, whereas for quadratics they are *absolute*.

In the quadratic case, we linked the nature of the critical point $P_0$ to the shape of the level curves. There is a similar link in the general case: When

$H(P_0) > 0$, then the nearby level curves look like tiny ellipses centered at $P_0$. When $H(P_0) < 0$, then the level curve $\{f = f(P_0)\}$ forms an "×" at $P_0$; and near $P_0$, the other level curves look like hyperbolas asymptotic to this ×.

**EXAMPLE 2**   Classify the critical points of $3xy - x^3 - y^3$.

**SOLUTION**   *Find* the critical points:

$$f_x = 3y - 3x^2 = 0 \iff y = x^2$$
$$f_y = 3x - 3y^2 = 0 \iff x = y^2.$$

Thus each partial derivative is zero along a certain curve (Fig. 7). The critical points are found where *both* $f_x$ and $f_y$ equal zero, at the intersections of these two curves:

$$(0, 0) \quad \text{and} \quad (1, 1).$$

To classify these critical points, take the second derivatives:

$$f_{xx} = -6x, \qquad f_{yy} = -6y, \qquad f_{xy} = 3.$$

At $(0, 0)$, $f_{xx}f_{yy} = 0$ and $(f_{xy})^2 = 9$, so $\det \begin{bmatrix} f_{xx} & f_{xy} \\ f_{yx} & f_{yy} \end{bmatrix} < 0$ as in Case II, and $(0, 0)$ is a saddle point.

At the other critical point $(1, 1)$, $\det \begin{bmatrix} f_{xx} & f_{xy} \\ f_{yx} & f_{yy} \end{bmatrix} = 36 - 9 > 0$, as in Case I, so this is a local maximum or minimum point. Both $f_{xx}$ and $f_{yy}$ are negative at this point, as in Case Ib, so it is in fact a local maximum point.

Figure 8 shows a few level curves. The curve $\{f = 0\}$ forms an × at the origin, which is a saddle point. The level curves near $(1, 1)$ are roughly elliptical. Figure 9 shows the graph; the top of the hill is the point $(1, 1, 1)$.

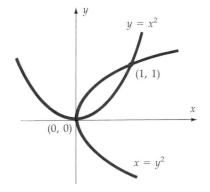

**FIGURE 7**
Simultaneous solution of
$x = y^2$ and $y = x^2$.

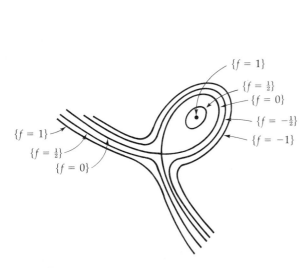

**FIGURE 8**
Level curves of $3xy - x^3 - y^3$.

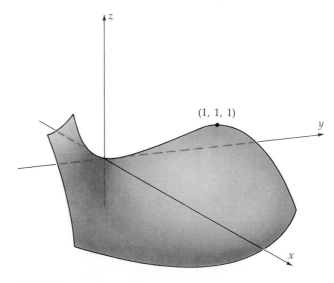

**FIGURE 9**
Graph of $3xy - x^3 - y^3$.

We conclude with an applied example.

---

**EXAMPLE 3**   Determine the shape of the rectangular box containing given volume $V$, and with minimum surface area.

**SOLUTION**   We do not *assume* that any side of the box is a square. Let the dimensions be $x$, $y$, and $z$. Considering all six sides of the box (Fig. 10), you find the surface area

$$S = 2xy + 2yz + 2xz.$$

The volume $V$ is given, so

$$V = xyz$$

is a constant. Solve for $z$ to find $z = V/xy$, and

$$S = 2xy + 2\frac{V}{x} + 2\frac{V}{y}. \tag{3}$$

Find the critical points:

$$\frac{\partial S}{\partial x} = 2y - 2Vx^{-2} = 0, \qquad \text{so } y = Vx^{-2} \tag{4}$$

$$\frac{\partial S}{\partial y} = 2x - 2Vy^{-2} = 0, \qquad \text{so } x = Vy^{-2}.$$

Combine these to find $y = Vx^{-2} = V(Vy^{-2})^{-2} = y^4V^{-1}$, so

$$y - y^4V^{-1} = 0 \quad \text{or} \quad y(1 - y^3V^{-1}) = 0.$$

It follows that $y = 0$, or else $y = V^{1/3}$. The function $S$ in (3) is undefined when $y = 0$, so $y = V^{1/3}$ is the only possibility. Hence from (4)

$$x = V(V^{1/3})^{-2} = V^{1/3}.$$

Finally, $z = V/xy = V^{1/3}$. So

$$x = y = z = V^{1/3}$$

and the critical point of the surface area function $S$ gives dimensions for a cube. It is intuitively clear that this must be a minimum, but we check the criterion for a local minimum just to be sure:

$$S_{xx} = 4Vx^{-3} = 4 \qquad \text{when } x = V^{1/3}$$

$$S_{yy} = 4Vy^{-3} = 4 \qquad \text{when } y = V^{1/3}$$

$$S_{xy} = 2.$$

So $H = S_{xx}S_{yy} - (S_{xy})^2 > 0$, and $S_{xx} > 0$, satisfying the criteria for a local minimum.

---

**REMARK**   Critical points and the second derivative test find *local* extreme points, but not those on the boundary of a region. For example, the maximum of the function $f(x, y) = xy$ in the disk $x^2 + y^2 \le 2$ is $f(-1, -1) = f(1, 1) = 1$;

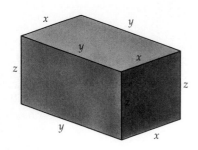

**FIGURE 10**
Box, $x$ by $y$ by $z$.

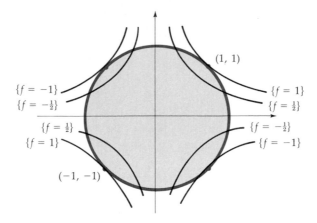

**FIGURE 11**
Maximum of $f(x, y) = xy$ for all $(x, y)$ in the disk $x^2 + y^2 \leq 2$ occurs at $(1, 1)$ and $(-1, -1)$.

you can see this from the level curves (Fig. 11). But $(1, 1)$ is not a critical point. This does not contradict the Critical Point Theorem; $f(1, 1) \geq f(x, y)$ for all points *in the disk* $x^2 + y^2 \leq 2$, but not for points near $(1, 1)$ *outside* that disk.

Extreme points on the boundary of a region are discussed in Section 16.5.

### The Proof of the Second Derivative Test

Suppose that $f$ is a $C^2$ function satisfying the conditions for a minimum at $P_0$:

$$f_x(P_0) = 0, \qquad f_y(P_0) = 0,$$

$$f_{xx}(P_0) > 0, \qquad \det \begin{bmatrix} f_{xx} & f_{xy} \\ f_{yx} & f_{yy} \end{bmatrix}(P_0) > 0. \tag{5}$$

Since the second derivatives are continuous, it can be proved that the inequalities in (5) remain valid for all points $P$ in some disk centered at $P_0$. We will show that $f(P) > f(P_0)$ for all points $P$ in that disk, and hence $P_0$ is a local minimum point.

Denote by $r$ the radius of the disk $D$ (Fig. 12). Then every point $P$ in $D$ can be written as $\vec{P} = \vec{P}_0 + t\vec{u}$ for some unit vector $\vec{u} = \langle a, b \rangle$ and some number $t$ with $|t| < r$. For each fixed $\vec{u}$, we compute the *second* derivative of $f$ along the line $\vec{P} = \vec{P}_0 + t\vec{u}$, given parametrically by $(x, y) = (x_0 + ta, y_0 + tb)$. By the Chain Rule, the first derivative with respect to $t$ is

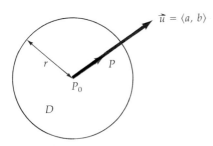

**FIGURE 12**
$\vec{P} = \vec{P}_0 + t\vec{u}, |\vec{u}| = 1, |t| < r,$
gives all points in disk $D$ of radius $r$.

$$\frac{df(x_0 + ta, y_0 + tb)}{dt} = f_x \frac{dx}{dt} + f_y \frac{dy}{dt}$$

$$= f_x(x_0 + ta, y_0 + tb) \cdot a + f_y(x_0 + ta, y_0 + tb) \cdot b. \tag{6}$$

Apply the Chain Rule to each term on the right of (6), and simplify the notation by leaving out the points of evaluation:

$$\frac{d^2f}{dt^2} = (f_x a)_x \frac{dx}{dt} + (f_x a)_y \frac{dy}{dt} + (f_y b)_x \frac{dx}{dt} + (f_y b)_y \frac{dy}{dt}$$
$$= f_{xx}a^2 + f_{xy}ab + f_{yx}ba + f_{yy}b^2$$
$$= f_{xx}a^2 + 2f_{xy}ab + f_{yy}b^2.$$

Complete the square on $a$:

$$\frac{d^2f}{dt^2} = f_{xx}\left[\left(a + \frac{bf_{xy}}{f_{xx}}\right)^2 + \frac{f_{xx}f_{yy} - f_{xy}^2}{f_{xy}^2}b^2\right].$$

The partial derivatives are evaluated at the point $\vec{P} = \vec{P}_0 + t\vec{u}$, with $|t| < r$; hence $P$ is in the disk $D$ in Figure 12. In that disk, (5) holds; hence $f_{xx}f_{yy} - f_{xy}^2 > 0$, and (since $f_{xx} > 0$ by assumption)

$$\frac{d^2f}{dt^2} > 0 \qquad \text{for all } |t| < r.$$

It follows that the critical point at the center of the disk is a minimum along this line segment:

$$f(P) > f(P_0) \text{ for } \vec{P} = \vec{P}_0 + t\vec{u}, \qquad |t| < r.$$

This is true for every unit vector $\vec{u}$, hence for every point $P$ in the disk $D$. This proves the second derivative test for a minimum; the other cases have similar proofs.

## SUMMARY

**$P_0$ is a Critical Point** if $f_x(P_0) = 0 = f_y(P_0)$, or else if either partial derivative fails to exist.

**$P_0$ is a Local Maximum Point** for $f$ if there is a disk $D$ centered at $P_0$ with

$$f(P_0) \geq f(P) \qquad \text{for all points } P \text{ in } D.$$

The definition of local minimum point is similar.

**The Critical Point Theorem:** If $P_0$ is a local maximum or minimum point for $f$, then $P_0$ is a critical point.

**Hessian Determinant:** $H = \det\begin{bmatrix} f_{xx} & f_{xy} \\ f_{yx} & f_{yy} \end{bmatrix}$.

**The Second Derivative Test:** At a critical point $P_0$:

$$H(P_0) > 0 \quad \Rightarrow \quad P_0 \text{ is a } \begin{cases} \text{local minimum if } f_{xx}(P_0) > 0 \\ \text{local maximum if } f_{xx}(P_0) < 0. \end{cases}$$

$$H(P_0) < 0 \quad \Rightarrow \quad P_0 \text{ is a saddle point.}$$

If $H(P_0) = 0$, then $P_0$ is a *degenerate* critical point.

## PROBLEMS

### A

1. Find all the critical points, and classify each as a local maximum, local minimum, saddle, or degenerate. Sketch the level curve passing through each critical point and one or two other level curves.
   *a) $x^2 + 2x + 3y^2 + 4y + 1$
   *b) $x^2 + xy$
   c) $2xy - 2x^2 - 5y^2 - x + 2y + 1$

2. Find and classify all the critical points.
   *a) $x^3 + y^3 + 3xy$   *b) $x^4 + y^4 - 4xy$
   c) $xe^{-x^2 - y^2}$

3. Figure 13 shows the graph of $f(x, y) = (x^2 + 2y^2)e^{-x^2 - y^2}$. Find the coordinates of the five critical points, and classify them.

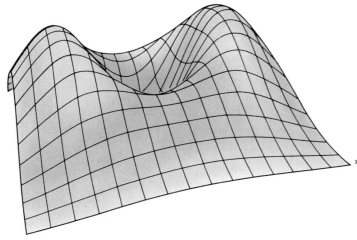

**FIGURE 13**
$z = (x^2 + 2y^2)e^{-x^2 - y^2}$.

4. Show that the origin is a degenerate critical point for each of the following. Determine in each case whether it is a local maximum, minimum, or neither (in the sense that there are both larger and smaller values of $f$ in every disk centered at the origin).
   *a) $x^2$           b) $xy^2$
   *c) $x^2 + y^3$    d) $x^2 + y^4$
   *e) $x^2 - y^4$   *f) $x^2 + 2xy + y^2$

5. Show that the origin is a minimum point for $\sqrt{x^2 + y^2}$. In what way is it a critical point?

6. For $f(x, y) = y(1 - x^2)$
   a) Find and classify all the critical points.

b) Sketch $\{f = c\}$ for $c = 0, \pm 1/4, \pm 1/2$.
c) Describe the graph of $f$.

*7. Find the distance between the lines given by the vector functions
$$\vec{P}(t) = \langle t, 2t, 3t \rangle$$
and
$$\vec{Q}(t) = \langle t + 1, 2t - 1, t \rangle.$$
[Minimize $|\vec{P}(t) - \vec{Q}(s)|^2$.] Deduce that the lines are *skew*, i.e. neither parallel nor intersecting. Redo the problem using the cross product, and compare results.

8. A box is to be constructed of three different materials—the bottom costs $2/m^2$, the top and sides cost $3/m^2$, and the front and back cost $4/m^2$. For a box holding a given volume $V$, find the dimensions giving minimum cost. What fraction of the cost is spent on the front and back together?

### B

*9. Find all points on the plane $x + y + z = 1$ where the function $x^2 + y^2 + z^2$ has a minimum value. Interpret your result geometrically.

*10. On the surface $\{xyz = 1\}$, find the point(s) nearest to the origin.

11. On the paraboloid $\{z = x^2 + 2y^2\}$, find the point(s) nearest to the point $(0, 0, 1)$.

12. Find the dimensions $x$, $y$, and $z$ of a box with with maximum volume, having length plus girth equal to 84 in. (If $z$ is the longest dimension, then the girth is $2x + 2y$, the perimeter measured perpendicular to the length.)

*13. For the function $f(x, y) = (x^2 + y^2)^2 - 4x^2$
   a) Find and classify all critical points.
   b) Sketch $\{f = 0\}$. [Write $f(x, y)$ as $(x^2 + y^2 + 2x)(x^2 + y^2 - 2x)$.]
   c) Show where $f > 0$ and where $f < 0$.
   d) To your sketch of $\{f = 0\}$, add plausible sketches of $\{f = 2\}$ and $\{f = -2\}$.

14. For $f(x, y) = (x^2 + y^2 - 3)^2 - 4x^2$
   a) Find and classify the (five) critical points. Determine any local maximum or minimum values of $f$.

b) Sketch $\{f = 0\}$, noting that $f(x, y) = (x^2 + 2x - 3 + y^2)(x^2 - 2x - 3 + y^2)$.

c) To your sketch of $\{f = 0\}$, add plausible sketches of $\{f = c\}$ for $c = 8, 1, -1$, and $-10$.

15. Let $f(x, y) = (y - x^2)(y - 2x^2)$.
   a) Plot $\{f = 0\}$. Show the regions where $f > 0$ and where $f < 0$.
   b) Show that on every straight line through the origin, $f$ has a minimum at $(0, 0)$. [Compute $f(x, y)$ along the line $x = at$, $y = bt$.]
   c) Is the origin a local minimum point for $f$?

16. The force on an electron in a static electric field can be described by a "potential" $\varphi$; the force at point $P$ is $\vec{F} = \nabla \varphi(P)$, in suitable units. Thus the electron is accelerated toward higher potential. At a *local minimum point* $P_0$ for $\varphi$, the force is zero: $\vec{F}(P_0) = \nabla \varphi(P_0) = \vec{0}$. An electron at $P_0$, with zero velocity, will stay at $P_0$ forever. However, such a minimum point is *unstable*; for if the electron is pushed to any other point nearby, the force drives the electron toward higher potential, away from $P_0$. On the other hand, a local maximum point $P_0$ is stable; if the electron is

moved slightly away from $P_0$, then the force pushes it back toward the higher potential at $P_0$. Near a *saddle point* for $\varphi$, the electron may be driven either toward the saddle or away from it, depending on where it is.

Thus, to understand the nature of the force $\vec{F} = \nabla \varphi$, it is useful to find and classify the critical points of $\varphi$. Realistically, the potential $\varphi$ depends on $x$, $y$, and $z$. For simplicity, in this problem, we consider potentials depending only on $x$ and $y$, and suppose that the motion of the electron is confined to the $xy$-plane.

a) Find and classify the critical points of $\varphi = x^2 + xy - y^2 + x + y$.

b) Find and classify the critical points of $\varphi = \ln|\vec{P} - \vec{P}_1| + \ln|\vec{P} - \vec{P}_2|$, where $P_1 = (1, 0)$ and $P_2 = (0, 1)$.

c) In empty space, a static electric potential depending only on $x$ and $y$ satisfies *Laplace's equation* $\varphi_{xx} + \varphi_{yy} = 0$. Check this equation for the potentials in parts a and b.

d) Show that if $\varphi_{xx} + \varphi_{yy} = 0$, then all critical points must be either *saddle points*, or *degenerate*.

## 15.8

## THE LEAST SQUARES LINE (optional)

Imagine that you have run $n$ experiments, each time recording two related variables $x$ and $y$ (for example, hours studied per week, and grade point average). Your results give $n$ values of $x$ and $y$, say

$$(x_1, y_1), \ldots, (x_n, y_n).$$

These can be plotted as $n$ points in the plane. If there is a linear relation $y = mx + b$ between the two, and if all the measurements are made with perfect precision, then the points will lie precisely on a line. But if the relation is only approximately linear, or if there is some lack of precision, you might still wish to fit a straight line as closely as possible to the plotted points. How to do it? You could "eyeball it"—set down a ruler where it seems to fit best, and draw the line. But it is better to have a systematic way; the most common is the method of least squares.

The vertical deviation of the point $(x_j, y_j)$ from a given line $y = mx + b$ is $mx_j + b - y_j$ (Fig. 1). The *least squares line* is the one that minimizes the sum of the *squares* of these deviations, the sum

$$\sum_{j=1}^{n} (mx_j + b - y_j)^2.$$

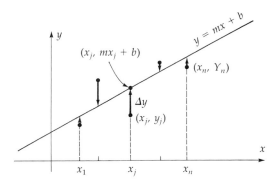

**FIGURE 1**
$\Delta y = mx_j + b - y_j$.

The points $(x_j, y_j)$ are given, so the sum depends on the constants $m$ and $b$ in the as yet undetermined line. That is, the sum of squares is a function of $m$ and $b$:

$$f(m, b) = \sum_{j=1}^{n} (mx_j + b - y_j)^2.$$

At a minimum, the partial derivatives are zero:

$$\frac{\partial f}{\partial m} = \sum_{j=1}^{n} 2(mx_j + b - y_j)x_j = 0 \tag{1}$$

$$\frac{\partial f}{\partial b} = \sum_{j=1}^{n} 2(mx_j + b - y_j) = 0. \tag{2}$$

These simultaneous linear equations determine $m$ and $b$. The second derivative test shows that the critical point does give a minimum. (See problem 7.)

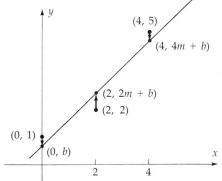

**FIGURE 2**
Fitting a line to three points.

**EXAMPLE 1**   Find the least squares line fitting the points $(0, 1)$, $(2, 2)$ $(4, 5)$.

**SOLUTION** (Fig. 2)   The sum of the squares of the deviations is

$$f(m, b) = (b - 1)^2 + (2m + b - 2)^2 + (4m + b - 5)^2.$$

The partial derivatives are

$$\frac{\partial f}{\partial m} = 4(10m + 3b - 12)$$

$$\frac{\partial f}{\partial b} = 2(6m + 3b - 8).$$

Set both equal to zero, and find $m = 1$, $b = 2/3$. Thus the least squares line is

$$y = x + \tfrac{2}{3}.$$

If you are to do this process frequently, or have data with many points, it simplifies the work to derive general formulas for the coefficients $m$ and $b$.

In equations (1) and (2), divide by 2 and expand, to get

$$m \sum_{j=1}^{n} x_j^2 + b \sum_{j=1}^{n} x_j - \sum_{j=1}^{n} x_j y_j = 0$$

$$m \sum_{j=1}^{n} x_j + nb - \sum_{j=1}^{n} y_j = 0.$$

You can solve these simultaneous equations for $m$ and $b$:

$$m = \frac{n \sum x_j y_j - (\sum x_j)(\sum y_j)}{n \sum x_j^2 - (\sum x_j)^2} \tag{3}$$

$$b = \frac{\sum y_j - m \sum x_j}{n}. \tag{4}$$

A table such as this makes for efficient calculation:

| $j =$ | 1 | 2 | $\cdots$ | $\cdots$ | $n$ | $\sum$ |
|---|---|---|---|---|---|---|
| $x_j$ | $x_1$ | $x_2$ | $\cdots$ | $\cdots$ | $x_n$ | $\sum x_j$ |
| $y_j$ | $y_1$ | $y_2$ | $\cdots$ | $\cdots$ | $y_n$ | $\sum y_j$ |
| $x_j y_j$ | $x_1 y_1$ | $x_2 y_2$ | $\cdots$ | $\cdots$ | $x_n y_n$ | $\sum x_j y_j$ |
| $x_j^2$ | $x_1^2$ | $x_2^2$ | $\cdots$ | $\cdots$ | $x_n^2$ | $\sum x_j^2$ |

Sum the rows as indicated, and plug the results into formulas (3) and (4). (There are hand calculators and "spread sheets" programmed to do all this internally.)

---

**EXAMPLE 1 (redone)**  For the given points $(0, 1)$, $(2, 2)$, and $(4, 5)$, the table is

| $j =$ | 1 | 2 | 3 | $\sum$ |
|---|---|---|---|---|
| $x_j$ | 0 | 2 | 4 | 6 |
| $y_j$ | 1 | 2 | 5 | 8 |
| $x_j y_j$ | 0 | 4 | 20 | 24 |
| $x_j^2$ | 0 | 4 | 16 | 20 |

Formulas (3) and (4) give

$$m = \frac{3 \cdot 24 - 6 \cdot 8}{3 \cdot 20 - (6)^2} = 1$$

$$b = \frac{8 - 1 \cdot 6}{3} = \frac{2}{3}.$$

The least squares line is $y = x + \frac{2}{3}$, as before.

---

## PROBLEMS

### A

1. Find the least squares line for the given data. Plot the points and the line.
   **\*a)** (0, 2), (1, 1), (4, −1)
   **b)** (−2, 1), (0, 3), (2, 4)
   **\*c)** (0, 0), (1, 1), (2, 2), (3, 4)

2. Find the least squares line for the following data; $x$ is the total seasonal depth of water applied through irrigation, and $y$ is the yield in tons of alfalfa per acre.

   | $x$ | 12 | 18 | 24 | 30 | 36 | 42 |
   |---|---|---|---|---|---|---|
   | $y$ | 5.27 | 5.68 | 6.25 | 7.21 | 8.20 | 8.71 |

3. An experiment with freely falling objects obtains the following distances $s$ for the given times of fall $t$:

   | $t$ | 0.2 | 0.4 | 0.6 | 0.8 | 1.0 |
   |---|---|---|---|---|---|
   | $s$ | 0.2 | 0.78 | 1.76 | 3.14 | 4.9 |

   The expected relation is $s = \frac{1}{2}gt^2$.
   **a)** Determine the constant $g$ by doing a least squares fit of the points $(t_j^2, s_j)$.
   **b)** Do a least squares fit for the points $(\ln(t_j), \ln(s_j))$, and derive a relation $\ln(s) = $

$m \ln(t) + b$; then derive the corresponding relation between $s$ and $t$.

### B

4. Derive the formulas (3) and (4) for $m$ and $b$.

5. The averages of the $x_j$ and $y_j$ are

$$\bar{x} = \frac{1}{n}\sum_{j=1}^{n} x_j \quad \text{and} \quad \bar{y} = \frac{1}{n}\sum_{j=1}^{n} y_j.$$

Is $(\bar{x}, \bar{y})$ on the least squares line?

6. Prove or disprove: For a set of just two points $(x_1, y_1)$ and $(x_2, y_2)$ with $x_1 \neq x_2$, the least squares line is the line through those two points.

### C

7. For the function $f(m, b)$ that we minimized:
   **a)** Check that $f_{mm} = 2\sum_{j=1}^{n} x_j^2$, $f_{bb} = 2n$, and $f_{mb} = 2\sum_{j=1}^{n} x_j$.
   **b)** Show that the critical point gives a minimum, unless all $x_j$ are equal. [Hint: For vectors in two or three dimensions, $|\vec{v} \cdot \vec{w}| = |\vec{v}| \cdot |\vec{w}| \cdot |\cos\theta| < |\vec{v}| \cdot |\vec{w}|$ unless $|\cos\theta| = 1$, in which case $\vec{v}$ is parallel to $\vec{w}$. The same is true in $n$ dimensions. Apply this to the vectors $\vec{x} = (x_1, x_2, \ldots, x_n)$ and $\vec{1} = (1, 1, \ldots, 1)$.]

## REVIEW PROBLEMS   CHAPTER 15

**\*1.** Determine the domain of

   **a)** $\log (xy)$   **b)** $\log|xy|$   **c)** $\dfrac{xy}{x^2 + y^2}$

**\*2.** For $f(x, y) = x(y + 1)$:
   **a)** Sketch level curves $\{f = c\}$ for $c = 0, \pm 2, \pm 4$.
   **b)** Describe the graph of $f$, by identifying its horizontal sections, and the sections in the vertical planes $y = x$ and $y = -x$.
   **c)** Sketch $\nabla f(2, 1)$ carefully; note that $(2, 1)$ lies on one of the level curves sketched in part a.
   **d)** Find the directional derivative of $f$ at $(2, 1)$, in the direction from $(2, 1)$ toward $(0, 0)$.
   **e)** Write the equation of the tangent plane at $(2, 1, 4)$.

**\*3.** An ant crawls along the $xy$-plane. At time $t = 1$, its position is $(2, 1)$, and its velocity is $\dfrac{dx}{dt}\vec{i} + \dfrac{dy}{dt}\vec{j} = 2\vec{i} + 3\vec{j}$ mm/sec. The temperature at $(x, y)$ is $T = x(y + 1)$, as in problem 2; $x$ and $y$ are in millimeters.

   **a)** Is the ant's ambient temperature increasing or decreasing? How rapidly, in deg/sec?
   **b)** What is the directional derivative of the temperature in the direction of the ant's motion?
   **c)** Approximately where will the ant be at time $t = 1.1$?

**d)** In which direction *should* the ant move from (2, 1) to find warmer temperatures most rapidly?

**\*4.** At the origin in the $xy$-plane, the pressure $p$ is 5 atmospheres, and the pressure gradient is $\nabla p = \vec{i} - 2\vec{j}$. Approximate the pressure $p$ at (0.05, 0.1).

**\*5.** The van der Waals equation gives the temperature $T$ of a gas as $T = k(V - b)(P + aV^{-2})$; $V$ is volume, $P$ is pressure, and $a$, $b$, $k$ are constants. (This is a refinement of the "ideal gas" equation $T = kPV$.)

**a)** Suppose that $V$ is increased by an amount $\Delta V$, while $P$ is held constant. Estimate $\Delta T$, the increase in $T$.

**b)** Suppose that $V$ is increased by $\Delta V$; what corresponding change $\Delta P$ would keep the temperature $T$ nearly constant?

**\*6.** A function is defined by $f(x, y) = \dfrac{2x^2 y^2}{x^4 + y^4}$ for $(x, y) \neq (0, 0)$, and $f(0, 0) = 0$. Is $f$ continuous at all points of the plane? Explain carefully.

**7.** State the second derivative test for the nature of critical points of a function $f(x, y)$. Give a simple example of each case, and show that it satisfies the corresponding criterion.

**8. \*a)** Find and classify the critical points of $f(x, y) = x(x^2 + y^2 - 1)$.

**b)** Sketch the level set $\{f = 0\}$.

**c)** Fill in level curves for $f$, consistent with the critical points found in part $a$.

**\*9.** Suppose that $u = \varphi(xy)$, where $\varphi$ is a differentiable function of one variable. Express $u_x$ and $u_y$ in terms of $\varphi'$, and show that $x \cdot u_x = y \cdot u_y$.

**\*10.** Show that there is *no* $C^2$ function $f$ such that $f_x = x^2 y$ and $f_y = y^2 x$.

**11.** The force $\vec{F}$ exerted by a static electric field is the gradient of a so-called *potential function* $\varphi$; $\vec{F} = \nabla \varphi$. The level curves of $\varphi$ are called *equipotentials*. Curves crossing every equipotential at right angles are called *lines of force*, since the force $\vec{F} = \nabla \varphi$ is tangent to such curves. Find the lines of force for the given potential.

**\*a)** $\varphi = x^2 - y^2$    **b)** $\varphi = x^2 - 2y^2$

**\*c)** $\varphi = x^2 + y^2$    **d)** $\varphi = x^2 + 2y^2$

**\*e)** $\varphi = e^y \sin(x)$

# 16

# FUNCTIONS OF SEVERAL VARIABLES. FURTHER TOPICS

## 16.1

## FUNCTIONS OF THREE VARIABLES

For a function $f(x, y, z)$ of three variables, the basic formulas are nearly the same as in the two-variable case. The partial derivative $f_x = \partial f/\partial x$ is the derivative of $f$ with respect to $x$, holding both $y$ and $z$ constant; $f_y$ and $f_z$ are similarly defined. For example,

$$\frac{\partial e^{x^2+y^2+z^2}}{\partial y} = 2ye^{x^2+y^2+z^2}.$$

The *gradient* $\nabla f$ has three components,

$$\nabla f = f_x\vec{i} + f_y\vec{j} + f_z\vec{k}.$$

If $\nabla f(P) = \vec{0}$, or if $\nabla f(P)$ is undefined, then $P$ is a *critical point* of $f$. Local maxima and minima of $f$ occur only at critical points.

The *linear approximation* of $f$, based at point $P_0$, is given by

$$f_{\text{lin}}(P) = f(P_0) + f_x(P_0)\,\Delta x + f_y(P_0)\,\Delta y + f_z(P_0)\,\Delta z, \tag{1}$$

where $P_0 = (x_0, y_0, z_0)$, $P = (x, y, z)$, $\Delta x = x - x_0$, $\Delta y = y - y_0$, and $\Delta z = z - z_0$. The resulting approximation to the increment $\Delta f = f(P) - f(P_0)$ is often written in differential notation:

$$\Delta f \approx df = f_x\,dx + f_y\,dy + f_z\,dz,$$

where now $dx = x - x_0$, and so on.

### The Chain Rule

Suppose that $x, y, z$ are functions of $t$, defining a curve in space,

$$P(t) = (x(t), y(t), z(t)).$$

The Chain Rule gives the derivative of $f$ along the curve $P$ as

$$\frac{df(P(t))}{dt} = \nabla f(P(t)) \cdot \frac{d\vec{P}}{dt}$$

or

$$\frac{df(x, y, z)}{dt} = \frac{\partial f}{\partial x} \cdot \frac{dx}{dt} + \frac{\partial f}{\partial y} \cdot \frac{dy}{dt} + \frac{\partial f}{\partial z} \cdot \frac{dz}{dt}.$$

This is valid when $\nabla f$ is continuous and $d\vec{P}/dt$ exists. For a physical interpretation, suppose that $f(x, y, z)$ is the temperature in the ocean at point $(x, y, z)$, and the curve $P(t)$ gives the motion of a skin diver; then $f(P(t))$ is the temperature encountered by the diver at time $t$, and $\dfrac{df(P(t))}{dt}$ tells how fast this temperature is changing.

The **directional derivative** of $f$ at a point $P$ in the direction of a nonzero vector $\vec{v}$ is given by the dot product of $\nabla f$ with the unit vector $\vec{u} = |\vec{v}|^{-1}\vec{v}$:

$$D_{\vec{u}}f(P) = \nabla f(P) \cdot \frac{\vec{v}}{|\vec{v}|}.$$

This is greatest when $\vec{v} = \nabla f(P)$; the derivative in that direction is $|\nabla f(P)|$.

---

**EXAMPLE 1**  If $f(x, y, z) = xy + z^2$, then the gradient of $f$ is

$$\nabla f = y\vec{i} + x\vec{j} + 2z\vec{k}.$$

The derivative of $f$ at $(1, 1, 1)$ in the direction $\vec{v} = 2\vec{i} - \vec{j} + \vec{k}$ is

$$\nabla f(1, 1, 1) \cdot \frac{\vec{v}}{|\vec{v}|} = (\vec{i} + \vec{j} + 2\vec{k}) \cdot \frac{2\vec{i} - \vec{j} + \vec{k}}{\sqrt{6}} = \sqrt{3/2}.$$

Along the curve $x = t$, $y = t^2$, $z = t^3$, the derivative of $f$ is

$$\frac{df(x, y, z)}{dt} = \frac{\partial f}{\partial x}\frac{dx}{dt} + \frac{\partial f}{\partial y}\frac{dy}{dt} + \frac{\partial f}{\partial z}\frac{dz}{dt}$$
$$= y \cdot 1 + x \cdot 2t + 2z \cdot 3t^2$$
$$= t^2 + 2t^2 + 6t^5.$$

---

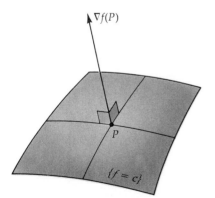

**FIGURE 1**
$\nabla f(P)$ is perpendicular to $\{f = c\}$ at $P$.

## The Geometry of Functions of Three Variables

A function $f(x, y, z)$ has a graph; its equation is

$$w = f(x, y, z).$$

This requires four mutually perpendicular axes: three for the "inputs" $x$, $y$, $z$, and one for the "output" $w$. Since four mutually perpendicular axes cannot be visualized in three-dimensional space, such graphs cannot be sketched by ordinary mortals. But it *is* possible to visualize the *level surfaces*, the sets $\{f = c\}$ where $f$ assumes a fixed value $c$. The gradient $\nabla f(P)$ is perpendicular to the level surface passing through $P$ (Fig. 1; a proof is sketched in problem 20a.)

---

**EXAMPLE 2**  For $f(x, y, z) = x^2 + y^2 - z^2$, describe the level surfaces $\{f = c\}$ with $c = -1$, $0$, and $1$.

**SOLUTION**  These are quadric surfaces (Sec. 14.6); $\{f = -1\}$ is the graph of

$$x^2 + y^2 - z^2 = -1,$$

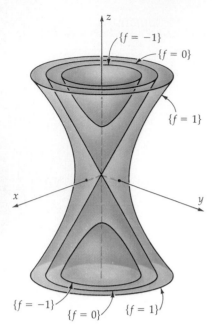

**FIGURE 2**
Level sets of
$f(x, y, z) = x^2 + y^2 - z^2$.

a hyperboloid of two sheets (Fig. 2). And $\{f = 0\}$ is the graph of

$$x^2 + y^2 - z^2 = 0,$$

a cone containing the hyperboloid $\{f = -1\}$; while $\{f = 1\}$ is a hyperboloid of one sheet

$$x^2 + y^2 - z^2 = 1$$

wrapped around the cone (Fig. 2).

The origin is a critical point:

$$\nabla f(0, 0, 0) = \langle 2x, 2y, -2z \rangle \big|_{(0,0,0)} = 0.$$

This is a sort of three-dimensional "saddle", neither a maximum nor a minimum; $f(0, 0, 0) = 0$, but there are higher values of $f$ everywhere outside the cone $\{f = 0\}$, and lower values everywhere inside.

**EXAMPLE 3**   The electrical potential $\varphi$ at point $(x, y, z)$ in a certain box is

$$\varphi(x, y, z) = e^{5x} \cos 3y \sin 4z.$$

The gradient is

$$\nabla \varphi(x, y, z) = e^{5x}(5 \cos 3y \sin 4z \, \vec{i} - 3 \sin 3y \sin 4z \, \vec{j} + 4 \cos 3y \cos 4z \, \vec{k}).$$

At the origin it is

$$\nabla \varphi(0, 0, 0) = 4\vec{k}.$$

The level surface $\{\varphi = 0\}$ passing through the origin includes the $xy$-plane $\{z = 0\}$. The gradient at the origin is perpendicular to this surface, and points directly toward level surfaces with higher values of $\varphi$ (Fig. 3).

The level surfaces of $\varphi$ are called **equipotential surfaces**. An electron in the box experiences a force $\vec{F} = c\nabla\varphi$ ($c$ is constant), proportional to the gradient of the potential. The force on an electron at the origin is $c\nabla\varphi(0, 0, 0) = 4c\vec{k}$, perpendicular to the equipotential surface $\{\varphi = 0\}$, driving the electron toward higher potential.

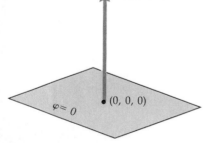

**FIGURE 3**
$\nabla\varphi$ perpendicular to an equipotential surface.

Suppose that $f$ is a $C^1$ function, and the gradient $\nabla f(P)$ is not $\vec{0}$. Being perpendicular to the level surface passing through $P$, it is a *normal to the tangent plane* to the level surface at $P$.

**EXAMPLE 4**   For $f(x, y, z) = x^2 + 2y^2 + z^2$, sketch the level surface passing through the point $P_0 = (1, 1, 2)$. Find the equation of the plane tangent to the level surface at $P_0$.

*SOLUTION*   The value of $f$ at $P_0$ is $1^2 + 2 \cdot 1^2 + 2^2 = 7$, so the level surface in question is the graph of $x^2 + 2y^2 + z^2 = 7$, an ellipsoid (Fig. 4). A

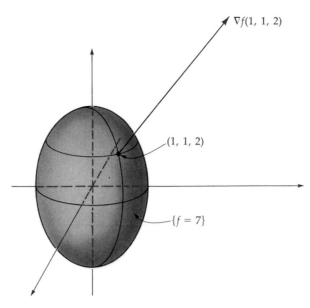

**FIGURE 4**
$\nabla f$ normal to a level surface.

normal to $\{f = 7\}$ at $P_0$ is the gradient $\nabla f(x, y, z) = 2x\vec{i} + 4y\vec{j} + 2z\vec{k}$, evaluated at $(1, 1, 2)$:

$$\nabla f(1, 1, 2) = 2\vec{i} + 4\vec{j} + 4\vec{k}.$$

Since $\nabla f$ is normal to the level surface, any convenient multiple of $\nabla f$ is a normal $\vec{n}$ to the plane tangent to the level surface; we multiply $\nabla f$ by $\frac{1}{2}$, and take the normal

$$\vec{n} = \vec{i} + 2\vec{j} + 2\vec{k}.$$

So an equation of the plane is

$$1(x - 1) + 2(y - 1) + 2(z - 2) = 0.$$

---

**IMPORTANT REMARK**   We now have two types of tangent plane:

I. *For a function $f(x, y)$ of two variables*, the *graph* $z = f(x, y)$ has a tangent plane at $(x_0, y_0, f(x_0, y_0))$ given by the equation of the linear approximation:

$$z = f(x_0, y_0) + f_x(x_0, y_0)(x - x_0) + f_y(x_0, y_0)(y - y_0). \qquad (2)$$

II. *For a function $F(x, y, z)$ of three variables*, the *level surface* $\{F = F(P_0)\}$ passing through $P_0$ has normal $\nabla F(P_0)$; the tangent plane at $P_0$ is

$$\nabla F(P_0) \cdot (\vec{P} - \vec{P}_0) = 0. \qquad (3)$$

The "level surface" method can also be adapted to give the tangent plane to the graph of a function $f(x, y)$ of two variables. For

$$z = f(x, y)$$

is equivalent to

$$f(x, y) - z = 0;$$

so the *graph* of $f(x, y)$, a function of two variables, is a *level surface* for the function of *three* variables, $F(x, y, z) = f(x, y) - z$.

---

**EXAMPLE 5**    Write the equation of the plane tangent to the graph of $f(x, y) = x \sin(2xy)$ at the point $(1, 0, 0)$.

**FIRST SOLUTION**    The equation of the linear approximation of $f$ at $(1, 0)$ is

$$z = f(1, 0) + f_x(1, 0)(x - 1) + f_y(1, 0)(y - 0)$$
$$= 0 + 0 \cdot (x - 1) + 2 \cdot (y - 0)$$

or

$$z = 2y. \tag{4}$$

**SECOND SOLUTION**    The equation of the graph, $z = x \sin(2xy)$, is equivalent to

$$x \sin(2xy) - z = 0.$$

This defines a level surface of the function $F(x, y, z) = x \sin(2xy) - z$. The gradient of $F$ is

$$\nabla F(x, y, z) = (\sin 2xy + 2xy \cos 2xy)\vec{i} + (2x^2 \cos 2xy)\vec{j} + (-1)\vec{k}.$$

At the given point, $\nabla F(1, 0, 0) = 0\vec{i} + 2\vec{j} - \vec{k}$. This is a normal of the tangent plane to the level surface $\{F = 0\}$ at $(1, 0, 0)$, so the equation of the plane is

$$\nabla F(1, 0, 0) \cdot (\vec{P} - \langle 1, 0, 0 \rangle) = 0$$

or

$$(2\vec{j} - \vec{k}) \cdot \langle x - 1, y, z \rangle = 2y - z = 0.$$

This is equivalent to the first solution, (4).

---

## SUMMARY

*The Level Surface* $\{f = c\}$ is the set of all points $(x, y, z)$ where $f(x, y, z) = c$.

*The Gradient* of $f$ is $\nabla f = f_x\vec{i} + f_y\vec{j} + f_z\vec{k}$.
$\nabla f(P)$ is perpendicular to the level surface through $P$, and also to the plane tangent to this surface at $P$.

*$P_0$ is a Critical Point* of $f$ if $\nabla f(P_0) = \vec{0}$, or if $\nabla f(P_0)$ is undefined.

**The Linear Approximation:**   $\Delta f \approx f_x \, \Delta x + f_y \, \Delta y + f_z \, \Delta z.$

**The Chain Rule:**   $\dfrac{df(x, y, z)}{dt} = \dfrac{\partial f}{\partial x}\dfrac{dx}{dt} + \dfrac{\partial f}{\partial y}\dfrac{dy}{dt} + \dfrac{\partial f}{\partial z}\dfrac{dz}{dt}.$

**The Directional Derivative** in the direction of a unit vector $\vec{u}$, at point $P$:

$$D_{\vec{u}}f(P) = \nabla f(P) \cdot \vec{u}.$$

## PROBLEMS

### A

**\*1.** For $f(x, y, z) = e^{3x + 4y} \cos(5z)$ compute
   a) $f(0, 0, 0)$     b) $f(x, x, x)$
   c) $f_x(0, 0, 0)$     d) $f_y(1, 1, 1)$
   e) $f_z(0, 0, 0)$

**\*2.** If $f(x, y, z) = xy + yz$, find the equation of the level surface passing through $(1, -5, 1)$.

**3.** Compute the gradient, and the directional derivative in the given direction $\vec{v}$ at the given point $P$.
   **\*a)** $e^{xyz}$; $\vec{v} = 2\vec{i} - \vec{j}$, $P = (0, 1, 1)$
   **\*b)** $ax + by + cz + d$; $\vec{v} = a\vec{i} + b\vec{j} + c\vec{k}$, $P = (x_0, y_0, z_0)$ ($a, b, c, d$ are constants.)
   **\*c)** $x^2 + 2y^2 - z^2$; $\vec{v} = -\vec{i}$, $P = (1, 1, 1)$.
   **d)** $x^2 + 2y^2 - z^2 + x$; $\vec{v} = \vec{i} + \vec{j} + \vec{k}$, $P = (-1, 2, 3)$.

**\*4.** For each part of problem 3, write the equation of the tangent plane to the level surface passing through $P$. (Answers given to parts a, b, and c.)

**5.** a) Write the equation of the plane tangent to the cone $x^2 + y^2 - z^2 = 0$ at the point $(3, 4, 5)$.
   b) Show that the plane passes through the origin, and explain geometrically why this should be so.

**6.** Use the Chain Rule to find $dw/dt$. Check your result by expressing $w$ explicitly as a function of $t$ and differentiating.
   **\*a)** $w = x^2 + y^2 + z^2$, $x = \cos(t)$, $y = \sin(t)$, $z = t$
   **b)** $w = \sqrt{x^2 + y^2 + z^2}$, $x = t$, $y = 2t$, $z = -3t$
   **\*c)** $w = x^3 y^2 z$, $x = t$, $y = t^2$, $z = t^3$
   **d)** $w = xz^2 \int_0^y \sin(\theta^2) \, d\theta$, $x = 1$, $y = t$, $z = 1$

**\*7.** The concentration of blood in a certain part of the ocean is given by $e^{-x^2 - 2y^2 - z^2}$. A shark at

$(1, 3, -1)$ heads in the direction of maximum increase of concentration. What direction is that?

**8.** An electron in a potential field $\varphi(x, y, z) = e^{5x - 12y} \cos(13z)$ undergoes a force $\vec{F} = q\nabla\varphi$, where $q$ is the charge on the electron, a constant.
   **a)** Compute the force at the origin.
   **\*b)** Is there any place where $\vec{F} = \vec{0}$?
   **c)** Show that $\varphi$ satisfies *Laplace's equation*

$$\frac{\partial^2 \varphi}{\partial x^2} + \frac{\partial^2 \varphi}{\partial y^2} + \frac{\partial^2 \varphi}{\partial z^2} = 0.$$

**9.** For each of the following functions, describe the level surfaces $\{f = c\}$ for $c = -1, 0, 1$.
   **\*a)** $x^2 + y^2 + z^2$     **\*b)** $2x^2 + y^2 - z^2$
   **c)** $x^2 - y^2 - 3z^2$     **d)** $x^2 + y^2 - z$
   **\*e)** $e^{x^2 - y^2 - z^2}$     **f)** $(x^2 + y^2 + z^2)^{-1}$

**10.** Find all critical points.
   **\*a)** $x^2 + x - 2y^2 + 3z^2$
   **\*b)** $\sqrt{x^2 + y^2 + z^2}$
   **\*c)** $f(x, y, z) = x^2 - xy$
   **d)** $xye^{-2x^2 - 2y^2 - 2z^2}$

**\*11.** Use the linear approximation for $xe^{y^2 - z^2}$, based at $(2, 1, 1)$, to approximate $(2.01)e^{(.9)^2 - (1.1)^2}$.

**12.** A box has sides measured as 4, 8, and 10 cm, with a possible error of $\pm 0.02$ cm in each dimension. Use differentials to estimate the possible resulting error in computing
   a) The volume.     b) The surface area.

**13.** In a triangle, sides $a$ and $b$ are measured as $a = 3$ and $b = 5$, with a possible error $\pm 0.03$. The included angle is measured as $31°$, with a possible error of $\pm 1°$. Estimate the resulting possible error in computing
   a) The area of the triangle.
   b) The length of the third side.
   c) The perimeter of the triangle.

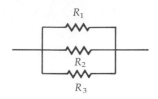

**FIGURE 5**
Three resistances in parallel.

**14.** Three resistances $R_1$, $R_2$, $R_3$ in parallel (Fig. 5) have the effect of a single resistance $R$, where

$$\frac{1}{R} = \frac{1}{R_1} + \frac{1}{R_2} + \frac{1}{R_3}. \qquad (*)$$

**a)** Compute $R$ if $R_1 = 200$, $R_2 = 1000$, $R_3 = 30$.

**b)** Show that $\partial R/\partial R_j = R^2/R_j^2$ for $j = 1, 2, 3$. [Hint: Differentiate each side of (*) with respect to $R_j$.]

**c)** Use the linear approximation to show that

$$\frac{\Delta R}{R} \approx \frac{R}{R_1}\frac{\Delta R_1}{R_1} + \frac{R}{R_2}\frac{\Delta R_2}{R_2} + \frac{R}{R_3}\frac{\Delta R_3}{R_3}.$$

**\*d)** Suppose that each of the three resistances might vary by no more than 5%; $\left|\dfrac{\Delta R_j}{R_j}\right| \le 0.05$. Estimate the maximum percent variation in $R$, that is $\Delta R/R$.

**15.** Show that the function

$$w = \sin(3x)\cos(4y)\cos(5t)$$

solves the *wave equation*

$$w_{xx} + w_{yy} = w_{tt}.$$

**16.** Show that the function $\varphi = e^{5x}\cos 3y \sin 4z$ in Example 3 solves *Laplace's equation*

$$\varphi_{xx} + \varphi_{yy} + \varphi_{zz} = 0.$$

**B**

**17. a)** Show that for any $C^1$ function $f(x, y, z)$, $\nabla(f^2) = 2f\nabla f$.

**b)** Prove a formula for $\nabla(f^p)$, for arbitrary constant $p$.

**c)** Suppose that $g$ is a differentiable function of one variable. Prove a formula for $\nabla g(f(x, y, z))$.

**18.** Find the angle of intersection of the sphere $x^2 + y^2 + z^2 = 1$ with the paraboloid $z = x^2 + y^2$.

**\*19.** Find the angle of intersection of the helix

$$x = \cos(t), \qquad y = \sin(t), \qquad z = t$$

with the sphere $x^2 + y^2 + z^2 = 2$.

**20.** Let $P(t)$ be a differentiable space curve [that is, $\vec{P}'(t)$ exists], and $f$ be a function with $\nabla f$ continuous.

**a)** Suppose that $P(t)$ lies in the surface $\{f = c\}$, that is, $f(P(t)) = c$ for all $t$. Prove that $\nabla f(P(t))$ is perpendicular to $\vec{P}'(t)$. (This shows that $\nabla f$ is perpendicular to $\{f = c\}$.)

**b)** Suppose, conversely, that $\vec{P}'(t)$ is perpendicular to $\nabla f(P(t))$ for every $t$. Prove that the curve $P(t)$ lies in a single level surface of the function $f$.

**21.** Show that the plane $\dfrac{xx_0}{a^2} + \dfrac{yy_0}{b^2} + \dfrac{zz_0}{c^2} = 1$ is tangent to the ellipsoid $\dfrac{x^2}{a^2} + \dfrac{y^2}{b^2} + \dfrac{z^2}{c^2} = 1$ at any point $(x_0, y_0, z_0)$ on the ellipsoid.

**22.** Show that the plane $\dfrac{2xx_0}{a^2} + \dfrac{2yy_0}{b^2} = c(z + z_0)$ is tangent to the paraboloid $\dfrac{x^2}{a^2} + \dfrac{y^2}{b^2} = cz$ at any point $(x_0, y_0, z_0)$ on the paraboloid.

**23.** Write the equation of the plane tangent to the cone $\dfrac{x^2}{a^2} + \dfrac{y^2}{b^2} = \dfrac{z^2}{c^2}$ at the point $(x_0, y_0, z_0)$, in a compact form as in the previous two problems.

**24.** (Least squares planes.) Suppose that $n$ experiments are performed, each yielding three numbers $(x_j, y_j, z_j)$. A linear model $z = ax + by + c$ is to be fitted to these data. Write the equations for the coefficients of the best-fitting plane in the sense of least squares. (See Sec. 15.8 for the analogous case of the least squares line. Write three simultaneous equations for $a$, $b$, $c$. The coefficients in these equations can be written in terms of $|x|^2 = \sum x_j^2$, $|y|^2 = \sum y_j^2$, $x \cdot y = \sum x_j y_j$, $\bar{x} = \dfrac{1}{n}\sum x_j$, etc.)

**25.** In the previous problem, show that if there are just three data points $(x_j, y_j, z_j)$, then the plane contains those three.

**26.** A quadratic $y = ax^2 + bx + c$ is to be fitted to $n$ data points $(x_1, y_1), \ldots, (x_n, y_n)$. Write the equations for $a$, $b$, $c$ which give the best fit, in the sense of least squares.

# 16.2

## THE GENERAL CHAIN RULE

We are going to generalize the Chain Rule proved in Section 15.5,

$$\frac{df(P(t))}{dt} = \nabla f(P) \cdot \frac{d\vec{P}}{dt}$$

$$= \frac{\partial f}{\partial x}\frac{dx}{dt} + \frac{\partial f}{\partial y}\frac{dy}{dt}.$$

First, we introduce a variable $w = f(x, y)$; then $\frac{\partial f}{\partial x}$ is $\frac{\partial w}{\partial x}$, etc., and the Chain Rule is

$$\frac{dw}{dt} = \frac{\partial w}{\partial x}\frac{dx}{dt} + \frac{\partial w}{\partial y}\frac{dy}{dt}. \tag{1}$$

Here $t$ is the *independent variable*; the variables $x$ and $y$ depend on $t$; and $w$ depends in turn on $x$ and $y$. Schematically:

$$t \rightarrow (x, y) \rightarrow w$$

independent $\rightarrow$ intermediate $\rightarrow$ dependent.

There can be more than one independent variable. Suppose, for example, that $w = f(x, y)$, and we introduce polar coordinates

$$x = r \cos \theta, \qquad y = r \sin \theta.$$

Then $w$ depends on $x$ and $y$, which depend in turn on the two independent variables $r$ and $\theta$; the scheme is

$$(r, \theta) \rightarrow (x, y) \rightarrow w. \tag{2}$$

In this case, what does the Chain Rule give for $\partial w/\partial r$? To compute $\partial w/\partial r$ you hold $\theta$ constant, so the Chain Rule is essentially the same as if there were only *one* independent variable, $r$. Thus you can apply (1); replace $t$ by $r$, and the ordinary derivative $\dfrac{d}{dt}$ by the partial derivative $\dfrac{\partial}{\partial r}$:

$$\frac{\partial w}{\partial r} = \frac{\partial w}{\partial x}\frac{\partial x}{\partial r} + \frac{\partial w}{\partial y}\frac{\partial y}{\partial r}. \tag{3}$$

Referring to the scheme (2), we are computing the derivative of the dependent variable $w$ with respect to one of the independent variables, $r$; the sum on the right in (3) includes the derivative of $w$ with respect to *each intermediate* variable $x$ and $y$, times the derivative of that intermediate variable with respect to the independent variable $r$. Both terms are needed, since $r$ affects $w$ through both $x$ and $y$.

In general, a dependent variable $w$ can depend on any number of intermediate variables $(x_1, \ldots, x_n)$ and these in turn can depend on any number of independent variables $(t_1, \ldots, t_m)$. The partial derivatives of $w$ with respect to any of the independent variables $t_j$ must take into account *all the intermediate variables*:

$$\frac{\partial w}{\partial t_j} = \frac{\partial w}{\partial x_1} \cdot \frac{\partial x_1}{\partial t_j} + \cdots + \frac{\partial w}{\partial x_n} \cdot \frac{\partial x_n}{\partial t_j}.$$

**REMARK**    Equation (3) shows that the *partial* derivatives $\dfrac{\partial w}{\partial x}$ and $\dfrac{\partial x}{\partial r}$ *cannot* be treated as fractions; cancelling the symbols $\partial x$ and $\partial y$ on the right-hand side of (3) would give the nonsensical result $\dfrac{\partial w}{\partial r} = \dfrac{\partial w}{\partial r} + \dfrac{\partial w}{\partial r}$! Thus, unlike $dx$, the symbols $\partial w$, $\partial x$, $\partial r$, and so on have no meaning by themselves; they are just *part* of the notation for the partial derivatives.

---

**EXAMPLE 1**    Suppose that $w = f(x, y, z)$. In the system of *spherical coordinates* (to be discussed in Chapter 17) $x$, $y$, and $z$ are expressed in terms of new variables $\rho$, $\varphi$, and $\theta$ by

$$x = \rho \sin \varphi \cos \theta, \qquad y = \rho \sin \varphi \sin \theta, \qquad z = \rho \cos \varphi.$$

Compute $\partial w/\partial \rho$.

**SOLUTION**    Since the problem asks for $\dfrac{\partial w}{\partial \rho}$, $w$ is to be considered a dependent variable, and $\rho$ independent. The equations for $x$, $y$, and $z$ show that the other independent variables are $\varphi$ and $\theta$, while $x$, $y$, and $z$ are intermediate:

$$(\rho, \varphi, \theta) \to (x, y, z) \to w.$$

So

$$\frac{\partial w}{\partial \rho} = \frac{\partial w}{\partial x} \cdot \frac{\partial x}{\partial \rho} + \frac{\partial w}{\partial y} \cdot \frac{\partial y}{\partial \rho} + \frac{\partial w}{\partial z} \cdot \frac{\partial z}{\partial \rho}.$$

From the given formulas for $x$, $y$, and $z$ you can compute $\partial x/\partial \rho$, $\partial y/\partial \rho$, and $\partial z/\partial \rho$, and thus find

$$\frac{\partial w}{\partial \rho} = \left(\frac{\partial w}{\partial x}\right)(\sin \varphi \cos \theta) + \left(\frac{\partial w}{\partial y}\right)(\sin \varphi \sin \theta) + \left(\frac{\partial w}{\partial z}\right)(\cos \varphi).$$

You can go no farther; the function $f$ relating $w$ to $(x, y, z)$ is not given, so $\dfrac{\partial w}{\partial x}, \dfrac{\partial w}{\partial y}$, and $\dfrac{\partial w}{\partial z}$ cannot be evaluated more explicitly.

---

**EXAMPLE 2**    Suppose that $w = f(x - ct)$, where $f$ is a differentiable function of one variable. Relate $w_x$ to $w_t$, and $w_{xx}$ to $w_{tt}$.

*SOLUTION*   Since $f$ is a function of just one variable, you can use the Chain Rule from Chapter 3, the "derivative of the outside" times the "derivative of the inside." To compute $w_x$, treat $t$ as a constant:

$$w_x = \frac{\partial w}{\partial x} = \frac{\partial f(x - ct)}{\partial x} = f'(x - ct) \cdot \frac{\partial(x - ct)}{\partial x} = f'(x - ct) \cdot 1.$$

And for $w_t$, treat $x$ as a constant:

$$w_t = \frac{\partial w}{\partial t} = \frac{\partial f(x - ct)}{\partial t} = f'(x - ct) \frac{\partial(x - ct)}{\partial t} = f'(x - ct) \cdot (-c).$$

Compare the two to see the relation

$$w_t = -cw_x.$$

The second derivatives are

$$w_{xx} = \frac{\partial}{\partial x}(w_x) = \frac{\partial}{\partial x} f'(x - ct) = f''(x - ct) \frac{\partial(x - ct)}{\partial x} = f''(x - ct)$$

and

$$w_{tt} = \frac{\partial}{\partial t}(w_t) = \frac{\partial}{\partial t}(-cf'(x - ct)) = -cf''(x - ct)(-c) = c^2 f''(x - ct).$$

Thus $w_{tt} = c^2 w_{xx}$.

---

*REMARK*   Example 2 concerns the *one-dimensional wave equation*, describing (approximately) the vibrations of a one-dimensional elastic medium, such as a guitar string. Imagine the string stretched along the $x$-axis. As it vibrates, each part of the string moves up and down; $w$ denotes the displacement of the string above or below the axis at position $x$ and time $t$ (Fig. 1). From Newtonian mechanics, one deduces the equation of motion $w_{tt} = c^2 w_{xx}$; the constant $c$ depends on the nature and tightness of the string. Example 2 shows that any function of the form $w = f(x - ct)$ solves this equation. In this solution, when $t = 0$, then $w = f(x - c \cdot 0) = f(x)$, so the graph of $f$ is simply the shape of the string at $t = 0$; suppose this is the shape in Figure 2. What then is the shape of the string at some fixed time $t$? With the particular $f$ in Figure 2, $f(x - ct) = 0$ when $x - ct = 0$, that is $x = ct$ (Fig. 3); you may thus persuade yourself that the graph of $f(x - ct)$ has the same shape as the graph of $f(x)$, shifted $ct$ units to the light. So as $t$ varies, $w = f(x - ct)$ describes a *traveling wave* moving to the right with velocity $c$. The problems give other similar solutions of the wave equation.

String

$w$

$x$

**FIGURE 1**
Vibrating string at time $t$;
$w$ = displacement above or below
axis.

**FIGURE 2**
Traveling wave at $t = 0$;
$w = f(x - c \cdot 0) = f(x)$.

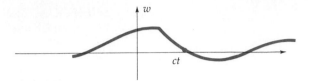

**FIGURE 3**
Traveling wave at time $t$,
$w = f(x - ct)$.

Here is a more difficult (but important) example, computing higher order derivatives by repeated application of the Chain Rule.

---

**EXAMPLE 3**    If $w = f(x, y)$ and $x = r \cos \theta$, $y = r \sin \theta$, relate $w_{\theta\theta}$ to the derivatives of $w$ with respect to $x$ and $y$.

*REMARK*    The conversion from rectangular to polar coordinates in this example is used in the study of heat conduction. It is carried further in the problems.

*SOLUTION*    First, compute

$$w_\theta = \frac{\partial w}{\partial \theta} = \frac{\partial w}{\partial x} \cdot \frac{\partial x}{\partial \theta} + \frac{\partial w}{\partial y} \cdot \frac{\partial y}{\partial \theta}$$
$$= (w_x)(-r \sin \theta) + (w_y)(r \cos \theta)$$
$$= -r \sin \theta\, w_x + r \cos \theta\, w_y. \tag{4}$$

The second derivative $w_{\theta\theta}$ is

$$\frac{\partial^2 w}{\partial \theta^2} = \frac{\partial w_\theta}{\partial \theta} = \frac{\partial}{\partial \theta}[-r \sin\theta\, w_x + r \cos\theta\, w_y]. \tag{5}$$

Now remember the meaning of the symbols; $w$ is a function of $x$ and $y$, so the partial derivatives $w_x$ and $w_y$ are also functions of $x$ and $y$, and thus depend indirectly on $r$ and $\theta$. So on the right-hand side of (5), you need the Product Rule:

$$\frac{\partial^2 w}{\partial \theta^2} = \left[\frac{\partial(-r \sin \theta)}{\partial \theta}\, w_x + (-r \sin \theta)\, \frac{\partial w_x}{\partial \theta}\right]$$
$$+ \left[\frac{\partial(r \cos \theta)}{\partial \theta}\, w_y + r \cos \theta\, \frac{\partial w_y}{\partial \theta}\right]. \tag{6}$$

The derivative $\partial w_x/\partial \theta$ can be handled just as we did $\partial w/\partial \theta$ in (4); in that formula, simply replace $w$ by $w_x$:

$$\frac{\partial w_x}{\partial \theta} = \frac{\partial w_x}{\partial x} \cdot \frac{\partial x}{\partial \theta} + \frac{\partial w_x}{\partial y} \cdot \frac{\partial y}{\partial \theta}$$
$$= (w_{xx})(-r \sin \theta) + (w_{xy})(r \cos \theta).$$

Similarly

$$\frac{\partial w_y}{\partial \theta} = (w_{yx})(-r \sin \theta) + (w_{yy})(r \cos \theta).$$

Substitute these in (6) and rearrange terms (recalling that $w_{xy} = w_{yx}$), to find

$$\frac{\partial^2 w}{\partial \theta^2} = -r \cos\theta\, w_x - r \sin\theta\, w_y + r^2 \sin^2\theta\, w_{xx}$$
$$- 2r^2 \sin\theta \cos\theta\, w_{xy} + r^2 \cos^2\theta\, w_{yy}.$$

## SUMMARY

$$\frac{\partial w}{\partial t_j} = \frac{\partial w}{\partial x_1} \cdot \frac{\partial x_1}{\partial t_j} + \cdots + \frac{\partial w}{\partial x_n} \cdot \frac{\partial x_n}{\partial t_j}.$$

## PROBLEMS

### A

**1.** Find $\partial w/\partial u$ and $\partial w/\partial v$ by the Chain Rule. Check by substituting the given expressions for $x$ and $y$ and differentiating directly.
 **\*a)** $w = x^2 + y^2$, $x = u \cos v$, $y = u \sin v$
 **b)** $w = x^2 - y^2$, $x = u + v$, $y = u - v$

**2.** Find $\partial w/\partial x$, $\partial w/\partial y$, and $\partial w/\partial z$. Evaluate the derivatives when $x = 1$, $y = -1$, $z = 2$.
 **\*a)** $w = u \cdot e^{v^2}$, $u = x + y$, $v = y + z$
 **b)** $w = (u - v)(u + v)$, $u = xz$, $v = xyz$

**3.** Set $w = (x_1)^2 + (x_2)^2 + (x_3)^2 + (x_4)^2$, $x_1 = \rho \sin \varphi \cos \theta$, $x_2 = \rho \sin \varphi \sin \theta$, $x_3 = \rho \cos \varphi$, $x_4 = z$. Express by the Chain Rule.

 **\*a)** $\dfrac{\partial w}{\partial \rho}$   **b)** $\dfrac{\partial w}{\partial \varphi}$   **c)** $\dfrac{\partial w}{\partial z}$

**4.** Suppose that $w = x^2 + y^2 - z^2$, and $x = \rho \sin \varphi \cos \theta$, $y = \rho \sin \varphi \sin \theta$, $z = \rho \cos \varphi$.

 Show that **(a)** $\dfrac{\partial w}{\partial \rho} = \dfrac{2w}{\rho}$ and **(b)** $\dfrac{\partial w}{\partial \theta} = 0$, in two ways:
 **i)** By the Chain Rule.
 **ii)** By expressing $w$ explicitly as a function of $\rho$, $\varphi$, and $\theta$, then differentiating.

**5.** Let $w = f(x, y)$, $x = r \cos \theta$, $y = r \sin \theta$.

 **a)** Express $\dfrac{\partial w}{\partial r}$ and $\dfrac{\partial w}{\partial \theta}$, as explicitly as possible, in terms of $w_x$ and $w_y$.
 **b)** Show that

 $$|\nabla f(x, y)|^2 = \left(\frac{\partial w}{\partial r}\right)^2 + r^{-2}\left(\frac{\partial w}{\partial \theta}\right)^2.$$

**6.** Let $x = r \cos \theta$ and $y = r \sin \theta$, and suppose that $w = f(x, y)$ has continuous second derivatives.

 **a)** Express $\dfrac{\partial^2 w}{\partial r^2}$ in terms of derivatives of $w$ with respect to $x$ and $y$.
 **b)** Show that $\dfrac{\partial^2 w}{\partial r^2} + \dfrac{1}{r} \cdot \dfrac{\partial w}{\partial r} + \dfrac{1}{r^2} \dfrac{\partial^2 w}{\partial \theta^2} = \dfrac{\partial^2 w}{\partial x^2} + \dfrac{\partial^2 w}{\partial y^2}$. (Use Example 3. The sum $w_{xx} + w_{yy}$ is called the *Laplacian* of $w$; if $w$ represents a steady state temperature distribution in a homogeneous plate, then $w_{xx} + w_{yy} = 0$.)

**7.** Suppose that $f$ is a differentiable function of one real variable. Relate $w_x$ to $w_y$ if
 **\*a)** $w = f(xy)$      **b)** $w = f(x + y)$
 **c)** $w = f(x^2 + y^2)$

8. Suppose that $w = f(\rho)$ and $\rho = (x^2 + y^2 + z^2)^{1/2}$.
   a) Prove that $\nabla w(x, y, z)$ is parallel to $x\vec{i} + y\vec{j} + z\vec{k}$, and $|\nabla w|^2 = \left(\dfrac{\partial w}{\partial \rho}\right)^2$.

   b) What shape do the level surfaces $\{f = c\}$ have? Relate this to the direction of $\nabla w$.

9. Suppose that $f$ is a function of three variables, with $\nabla f$ continuous, and $w = f(x - y, y - z, z - x)$. Prove that $w_x + w_y + w_z = 0$.

10. Suppose that $f$ is a differentiable function of one variable, and $w = f(x + ct)$. Relate $\dfrac{\partial w}{\partial x}$ to $\dfrac{\partial w}{\partial t}$.

11. Suppose that $f$ and $g$ are differentiable functions of one variable. Relate $\dfrac{\partial^2 w}{\partial x^2}$ to $\dfrac{\partial^2 w}{\partial t^2}$ if
   a) $w = f(x - ct)$.
   *b) $w = g(x + ct)$.
   c) $w = f(x - ct) + g(x + ct)$.

12. Suppose that $f$ is a $C^2$ function, $w = f(u, v)$, and $u = x - ct$, $v = x + ct$.
   a) Relate $w_x$ and $w_t$ to $w_u$ and $w_v$. Show that $w_x w_t = c(w_v^2 - w_u^2)$.
   b) Relate $w_{xx}$ and $w_{tt}$ to derivatives of $w$ with respect to $u$ and $v$. Show that $w_{tt} - c^2 w_{xx} = -4c^2 w_{uv}$.

**B**

13. Suppose that $f$ is a continuous function of one variable.
   a) Compute $\varphi_u$ and $\varphi_v$ for the function $\varphi(u, v) = \int_u^v f(x)\, dx$.
   [Hint: $\varphi(u, v) = \int_a^v f(x)\, dx - \int_a^u f(x)\, dx$ for a constant $a$.]
   b) Suppose that $a(t)$ and $b(t)$ are differentiable functions of $t$. Show that
   $$\frac{d}{dt} \int_{a(t)}^{b(t)} f(x)\, dx = f(b(t)) \frac{db}{dt} - f(a(t)) \frac{da}{dt}.$$

14. a) Show that $\dfrac{\partial w}{\partial r}$ is the directional derivative in the direction of increasing $r$ ($\theta$ fixed).
   b) What is the directional derivative in the direction of increasing $\theta$ ($r$ fixed)? $\Big($It is not just $\dfrac{\partial w}{\partial \theta}.\Big)$

c) Show that $|\nabla w|^2$ is the sum of the squares of the two directional derivatives in parts a and b.

15. *a) Write the Chain Rule formula for $\dfrac{\partial w}{\partial s}$ involving the links in the chain of dependence $(s, t) \to (u, v) \to (x, y) \to w$.
   b) Write the corresponding formula for $\dfrac{\partial w}{\partial t}$.

**C**

16. The vibrations of a stretched string (as in a guitar or piano) satisfy (approximately) the *wave equation*
   $$w_{tt} = c^2 w_{xx}.$$
   We outline a general method to solve this equation, using new variables $u$ and $v$ as in problem 12.
   a) Suppose that $\partial f(u, v)/\partial v = 0$, for all $u$ and $v$. Prove that $f(u, v) = f(u, 0)$ for all $u$ and $v$, so that $f$ is a function of $u$ alone.
   *b) Suppose that $\partial^2 f(u, v)/\partial u\, \partial v = 0$. Prove that $\partial f/\partial u$ is a function of $u$ alone, and $f$ itself is the sum of a function of $u$ and a function of $v$:
   $$f(u, v) = \varphi(u) + \psi(v).$$
   c) Suppose that $w_{tt} = c^2 w_{xx}$. Prove that there are functions $\varphi$ and $\psi$ such that
   $$w = \varphi(x - ct) + \psi(x + ct).$$
   (Use problem 12.)

17. In this problem, you find some solutions of *Laplace's equation* for heat distribution in the plane,
   $$w_{xx} + w_{yy} = 0.$$
   a) Suppose that $w = \varphi(r)$, where $r = \sqrt{x^2 + y^2}$, and $w$ satisfies Laplace's equation. Prove that $\varphi''(r) + \dfrac{1}{r}\varphi'(r) = 0$. (Use problem 6.)
   b) Show that $\varphi(r) = A + B \cdot \ln r$ for some constants $A$ and $B$. [Hint: $\varphi'' + \dfrac{1}{r}\varphi' = r^{-1}(r\varphi')'$.]

# 16.3

## IMPLICIT FUNCTIONS

Implicit differentiation allows you to compute the slope of a curve given by an equation of the form

$$f(x, y) = c. \tag{1}$$

For example, to compute the slope of

$$x^3 + y^3 - 3xy = 0$$

you differentiate, assuming that $y$ is a differentiable function of $x$:

$$3x^2 + 3y^2 \frac{dy}{dx} - 3y - 3x \frac{dy}{dx} = 0.$$

Then solve for $dy/dx$:

$$\frac{dy}{dx} = \frac{3y - 3x^2}{3y^2 - 3x}.$$

This method raises two questions. First, under what conditions does an equation such as (1) actually define $y$ as a differentiable function of $x$? And when $y$ is thus defined, what is the formula for its derivative, in general?

Suppose at first that the level set defined by (1) contains a $C^1$ parametric curve, that is, $x$ and $y$ are functions of a parameter $t$ such that

$$f(x(t), y(t)) = c$$

for every $t$. Then $\dfrac{df}{dt} = 0$, and by the Chain Rule

$$\frac{\partial f}{\partial x} \cdot \frac{dx}{dt} + \frac{\partial f}{\partial y} \cdot \frac{dy}{dt} = 0. \tag{2}$$

Now suppose that the level set (1) contains the *graph* of a function $y = \varphi(x)$; thus

$$f(x, \varphi(x)) = c. \tag{3}$$

This graph can be represented parametrically by

$$x = t, \qquad y = \varphi(t).$$

Hence $\dfrac{dx}{dt} = 1$ and $\dfrac{dy}{dt} = \dfrac{dy}{dx}$, so by (2)

$$\frac{\partial f}{\partial x} + \frac{\partial f}{\partial y} \cdot \frac{dy}{dx} = 0. \tag{4}$$

If $\dfrac{\partial f}{\partial y} \neq 0$, we can solve for $\dfrac{dy}{dx}$, the slope of the graph lying in the level set:

$$\frac{dy}{dx} = -\frac{\partial f/\partial x}{\partial f/\partial y}. \tag{5}$$

To derive (5), we *assumed* a differentiable function $\varphi$ that satisfies the relation (3); we then found that the derivative $\varphi' = \dfrac{dy}{dx}$ is given by (5), if $\partial f/\partial y \neq 0$. It turns out that this condition $\partial f/\partial y \neq 0$ guarantees that there *really is* a function $\varphi$ defined implicitly by (3).

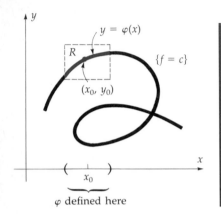

**FIGURE 1**
Near $(x_0, y_0)$, the level set $\{f = c\}$ is the graph of a function $\varphi$.

---

### THE IMPLICIT FUNCTION THEOREM

*Hypotheses:* $f(x_0, y_0) = c$, $f_x$ and $f_y$ are continuous at all points near $(x_0, y_0)$, and $f_y(x_0, y_0) \neq 0$.

*Conclusions:* Inside some rectangle $R$ centered at $(x_0, y_0)$ (Fig. 1), the level curve $\{f = c\}$ coincides with the graph of a function $\varphi(x)$, defined on an open interval containing $x_0$, with derivative

$$\varphi'(x) = \frac{-f_x(x, \varphi(x))}{f_y(x, \varphi(x))}.$$

---

**REMARK 1**  In equation (5), the variable $y$ plays two different roles. In computing $\dfrac{\partial f}{\partial y}$, $x$ and $y$ are independent variables. In $\dfrac{dy}{dx}$, on the other hand, $y$ is a *function of $x$*. We sidestepped this ambiguity by first considering $x$ and $y$ as functions of $t$, allowing a distinction between the role of $y$ in $\dfrac{\partial f}{\partial y}$, and that in $\dfrac{dy}{dt}$. But it is quicker to pass directly from (1) to (4) to (5), keeping in mind the double roles:

$$f(x, y) = c,$$

$$\frac{\partial f}{\partial x} \cdot \frac{dx}{dx} + \frac{\partial f}{\partial y} \cdot \frac{dy}{dx} = 0, \quad \text{or} \quad \frac{\partial f}{\partial x} + \frac{\partial f}{\partial y} \cdot \frac{dy}{dx} = 0,$$

$$\frac{dy}{dx} = -\frac{\partial f/\partial x}{\partial f/\partial y}.$$

**REMARK 2**  If $\dfrac{\partial f}{\partial y} = 0$ but $\dfrac{\partial f}{\partial x} \neq 0$, you can reverse the roles of $x$ and $y$, and express $x$ as a differentiable function of $y$, $x = \psi(y)$. The derivative is found as before; differentiate

$$f(x, y) = c,$$

thinking of $x$ as a function of $y$:

$$\frac{\partial f}{\partial x}\frac{dx}{dy} + \frac{\partial f}{\partial y} = 0,$$

$$\frac{dx}{dy} = -\frac{\partial f/\partial y}{\partial f/\partial x}.$$

**REMARK 3**    The theorem does not guarantee a *formula* for the implicit function—usually there is none. It merely guarantees that there *is* a function, so that any calculations using the derivative $\dfrac{dy}{dx}$ are legitimate.

Proofs of the Implicit Function Theorem can be found in several texts.[1]

---

**EXAMPLE 1**    Determine whether the graph of

$$x^3 + y^3 - 3xy = 0 \tag{6}$$

can be represented as a function graph near each of the points $(3/2, 3/2)$, $(2^{2/3}, 2^{1/3})$, $(0, 0)$. If it can, then compute the slope.

**SOLUTION**    The function $f(x, y) = x^3 + y^3 - 3xy$ in (6) has the derivatives

$$f_x = 3x^2 - 3y, \qquad f_y = 3y^2 - 3x. \tag{7}$$

At the point $(\frac{3}{2}, \frac{3}{2})$

$$f_x = \tfrac{27}{4} - \tfrac{9}{2} = \tfrac{9}{4} \neq 0$$

$$f_y = \tfrac{9}{4} \neq 0.$$

So, near that point, the graph of (6) can be represented in either form (Fig. 2)

$$y = \varphi(x) \quad \text{or} \quad x = \psi(y).$$

The derivatives are

$$\varphi'(x) = \frac{dy}{dx} = -\frac{f_x}{f_y}, \qquad \psi'(y) = \frac{dx}{dy} = -\frac{f_y}{f_x}.$$

In particular, at the point $\left(\dfrac{3}{2}, \dfrac{3}{2}\right)$, the slope is $\dfrac{dy}{dx} = -\dfrac{f_x}{f_y} = -\dfrac{9/4}{9/4} = -1.$

At the point $(2^{2/3}, 2^{1/3})$,

$$f_x = 3 \cdot 2^{4/3} - 3 \cdot 2^{1/3} = 3 \cdot 2^{1/3}(2 - 1) \neq 0$$

$$f_y = 3 \cdot 2^{2/3} - 3 \cdot 2^{2/3} = 0.$$

Since $f_y = 0$, you cannot expect to get $y = \varphi(x)$ near this point. But $f_x \neq 0$, so you *can* get $x = \psi(y)$ (Fig. 3) and

$$\psi'(y) = \frac{dx}{dy} = -\frac{f_y}{f_x}.$$

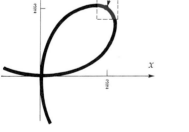

Here $y = \varphi(x)$ and $x = \psi(y)$

**FIGURE 2**
Near $(\frac{3}{2}, \frac{3}{2})$, the graph of $x^3 + y^3 - 3xy = 0$ can be represented either as $y = \varphi(x)$ or $x = \psi(y)$.

---

[1] See, for example, Marsden and Tromba, *Vector Calculus* (Freeman, 1981) or Seeley *Calculus of Several Variables* (Scott Foresman, 1970).

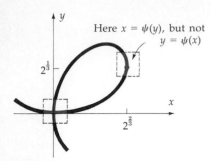

**FIGURE 3**
At $(2^{2/3}, 2^{1/3})$, $f_y = 0$ but $f_x \neq 0$; there, $y$ is not a function of $x$, but $x$ is a function of $y$. Near $(0, 0)$, neither representation is possible.

In the figure: Here $x = \psi(y)$, but not $y = \psi(x)$

In particular, at $(2^{2/3}, 2^{1/3})$ you find $\dfrac{dx}{dy} = -\dfrac{f_y}{f_x} = 0$; at that point, the graph has a vertical tangent.

At $(0, 0)$, both partial derivatives (7) are 0, so you cannot expect either representation. In fact, at this point the graph of (6) consists of two curves that cross each other (Fig. 3) and cannot be represented in either of the two forms $y = \varphi(x)$ or $x = \psi(y)$.

There is an analogous Implicit Function Theorem for equations with three variables

$$f(x, y, z) = c. \tag{8}$$

If $f_x$, $f_y$, and $f_z$ are continuous, then (8) can be solved for $z$,

$$z = \varphi(x, y)$$

near any point $(x_0, y_0, z_0)$ where $f_z(x_0, y_0, z_0) \neq 0$. The partial derivatives of $\varphi$ are formed by differentiating the relation that defines $\varphi$,

$$f(x, y, \varphi(x, y)) = c. \tag{9}$$

Rather than introduce new independent variables, we apply the Chain Rule using double roles of $x$ and $y$, with the scheme $(x, y) \rightarrow (x, y, z) \rightarrow f$. Differentiate both sides of (8):

$$\frac{\partial f}{\partial x}\frac{\partial x}{\partial x} + \frac{\partial f}{\partial y}\frac{\partial y}{\partial x} + \frac{\partial f}{\partial z}\frac{\partial z}{\partial x} = 0.$$

But $\partial x/\partial x = 1$, $\partial y/\partial x = 0$, and $\partial z/\partial x = \partial \varphi/\partial x$ [because $z = \varphi(x, y)$], so

$$\frac{\partial f}{\partial x} + \frac{\partial f}{\partial z}\frac{\partial \varphi}{\partial x} = 0$$

hence

$$\frac{\partial \varphi}{\partial x} = -\frac{\partial f/\partial x}{\partial f/\partial z}. \tag{10a}$$

Similarly

$$\frac{\partial \varphi}{\partial y} = -\frac{\partial f/\partial y}{\partial f/\partial z}. \tag{10b}$$

Such formulas are more useful in deriving general relationships than in computing derivatives with specific formulas, as in Example 1. (In fact, we computed the slopes in Example 1 back in Section 3.3, long before any mention of partial derivatives!) Physical chemistry has important examples of these general relationships.

**EXAMPLE 2** In a given state, a given quantity of fluid has a certain pressure $P$, volume $V$, and temperature $T$. These are not independent, but satisfy a relation

$$f(P, V, T) = 0. \tag{11}$$

For a quantity of "ideal gas" the relation is $PV - kT = 0$ ($k$ is a constant). Generally, $f$ is not explicitly known, but it *is* assumed that the relation (11) defines (in principle) any one of the three variables as a $C^1$ function of the other two. Then an expression like $\partial P/\partial V$ implies that $P$ is considered as a function of $V$ and $T$; and $\partial P/\partial V$ is the rate of change of pressure with respect to volume, holding temperature constant. To make it clear which derivative is held constant, we write

$$\left(\frac{\partial P}{\partial V}\right)_T$$

for the derivative of $P$ with respect to $V$, holding $T$ constant.

One important property of the fluid is its *thermal expansivity*

$$\alpha = \frac{1}{V}\left(\frac{\partial V}{\partial T}\right)_P.$$

This gives the proportional increase in volume per unit increase in temperature, keeping pressure constant. Another is the *isothermal compressivity*

$$\beta = -\frac{1}{V}\left(\frac{\partial V}{\partial P}\right)_T,$$

the proportional compression of the fluid per unit increase in pressure, holding temperature constant. For mercury, in a reasonable range of pressure, volume, and temperature, $\alpha$ and $\beta$ are nearly constant. With pressure measured in atmospheres, volume in liters, and temperature in °C, these quantities for mercury are experimentally determined as

$$\alpha \approx 1.8 \times 10^{-4}, \qquad \beta \approx 3.9 \times 10^{-6}.$$

Suppose now that you have a thermometer where the mercury expands to fill the void completely at 80°C. How much pressure would be generated by raising it to 85°C? Ignoring any expansion or fracture of the thermometer, the volume $V$ is constant and the question concerns the increase in pressure per unit increase in temperature,

$$\left(\frac{\partial P}{\partial T}\right)_V.$$

Is this yet another quantity to be determined, perhaps by new experiments? No—it can be expressed in terms of the quantities $\alpha$ and $\beta$. Compute the various partial derivatives by differentiating the assumed relation (11). For the desired derivative $(\partial P/\partial T)_V$ the two independent variables are $V$ and $T$, so use the Chain Rule scheme

$$(V, T) \rightarrow (P, V, T) \rightarrow f(P, V, T) = 0.$$

Differentiate $f = 0$ with respect to $T$:

$$\left(\frac{\partial f}{\partial P}\right)\left(\frac{\partial P}{\partial T}\right)_V + \left(\frac{\partial f}{\partial V}\right)\left(\frac{\partial V}{\partial T}\right)_V + \left(\frac{\partial f}{\partial T}\right)\left(\frac{\partial T}{\partial T}\right)_V = 0$$

so

$$\left(\frac{\partial f}{\partial P}\right)\left(\frac{\partial P}{\partial T}\right)_V + \left(\frac{\partial f}{\partial T}\right) \cdot 1 = 0$$

and

$$\left(\frac{\partial P}{\partial T}\right)_V = -\frac{\partial f/\partial T}{\partial f/\partial P}. \tag{12}$$

Similarly, the derivatives appearing in the two quantities $\alpha$ and $\beta$ are

$$\left(\frac{\partial V}{\partial T}\right)_V = -\frac{\partial f/\partial T}{\partial f/\partial V} \quad \text{and} \quad \left(\frac{\partial V}{\partial P}\right)_T = -\frac{\partial f/\partial P}{\partial f/\partial V}.$$

In the *ratio* of these two, their denominators cancel, leaving precisely the ratio in (12):

$$\left(\frac{\partial P}{\partial T}\right)_V = -\frac{\partial f/\partial T}{\partial f/\partial P} = -\frac{(\partial V/\partial T)_P}{(\partial V/\partial P)_T}.$$

So from the definitions of $\alpha$ and $\beta$,

$$\left(\frac{\partial P}{\partial T}\right)_V = -\frac{\alpha V}{-\beta V} = \frac{\alpha}{\beta}.$$

This holds for any fluid. In the case of mercury

$$\frac{\alpha}{\beta} = \frac{1.8 \times 10^{-4}}{3.9 \times 10^{-6}} \approx 46.$$

For the unexpanding mercury column, each additional °C above 80°C raises the pressure by 46 atmospheres; at 85°C the pressure would be 230 atmospheres!

## PROBLEMS

### A

In problems 1–6 show that the given point is on the given curve. At each point, decide whether the Implicit Function Theorem guarantees a representation $y = \varphi(x)$, or $x = \psi(y)$, or both, or neither. Find the functions $\varphi$ and $\psi$, if practical.

*1. $x^2 + y^2 = 1$, at

   a) $(1, 0)$   b) $\left(\dfrac{1}{\sqrt{2}}, \dfrac{1}{\sqrt{2}}\right)$   c) $(0, -1)$

2. $x^4 + x^2y^2 - y^4 = 1$, at $(1, 1)$

*3. $x^4 + x^2y^2 - y^4 = 0$, at $(0, 0)$

*4. $x \cos xy = 0$, at $(1, \pi/2)$

5. $x^5 + y^5 + xy = -4$, at $(-2, 2)$

6. $x^5 + y^5 + 2xy = 0$, at
   a) $(0, 0)$
   b) $(-1, -1)$

In problems 7–9 decide whether, at the given point, the given equation can be solved as $z = \varphi(x, y)$, or $y = \varphi(x, z)$, or $x = \varphi(y, z)$. Compute those partial derivatives $\partial z/\partial x$, etc. that exist.

7. $x^2 + y^2 + z^2 = 1$, near
   a) $(0, 0, 1)$   b) $\left(\dfrac{1}{\sqrt{3}}, \dfrac{1}{\sqrt{3}}, \dfrac{1}{\sqrt{3}}\right)$
   c) $(1, 0, 0)$

*8. $x^5 + y^5 + z^5 = 3$, at $(1, 1, 1)$

9. $e^{z+y} \cos(x + y) \sin(x + z) = 0$, at $(1, 1, -1)$

**B**

10. Suppose that $f(x, y, z) = 0$ defines any one of the variables as a function of the other two. Is it true that

$$\left(\frac{\partial x}{\partial y}\right)\left(\frac{\partial y}{\partial z}\right)\left(\frac{\partial z}{\partial x}\right) = -1?$$

11. In *thermodynamics* there are five basic quantities: $T$ (temperature), $V$ (volume), $P$ (pressure), $E$ (energy), and $S$ (an esoteric quantity called *entropy*). It is generally assumed that *any two* of these may be considered as independent variables; the other three are then functions of those two. A symbol such as $\partial S/\partial T$ would imply that $T$ is considered an independent variable, but it would not tell which is the *other* independent variable. This ambiguity is eliminated by writing $(\partial S/\partial T)_V$ for the derivative of $S$ with respect to $T$ when $T$ and $V$ are considered independent, and $(\partial S/\partial T)_P$ for the derivative of $S$ with respect to $T$ when $T$ and $P$ are independent, and so on. With this convention, certain identities are obvious; for example, when $T$ and $V$ are independent we have

$$\left(\frac{\partial T}{\partial T}\right)_V = 1, \qquad \left(\frac{\partial T}{\partial V}\right)_T = 0$$

$$\left(\frac{\partial V}{\partial T}\right)_V = 0, \qquad \left(\frac{\partial V}{\partial V}\right)_T = 1.$$

Beyond these formal identities, the theory of thermodynamics assumes two basic equations:

$$T\left(\frac{\partial S}{\partial T}\right)_V = \left(\frac{\partial E}{\partial T}\right)_V \qquad (*)$$

and

$$T\left(\frac{\partial S}{\partial V}\right)_T = \left(\frac{\partial E}{\partial V}\right)_T + P. \qquad (**)$$

Corresponding relations when other independent variables are chosen can then be deduced by the Chain Rule, as outlined here.

a) Considering $T$ and $V$ as independent variables, differentiate $(*)$ with respect to $V$ and $(**)$ with respect to $T$, and conclude that

$$\left(\frac{\partial P}{\partial T}\right)_V = \left(\frac{\partial S}{\partial V}\right)_T.$$

(Assume that all first and second derivatives are continuous.)

b) Suppose that $T$ and $P$ are chosen as independent. Using the scheme $(T, P) \to (T, V) \to P$, explain why

$$\left(\frac{\partial P}{\partial P}\right)_T = \left(\frac{\partial P}{\partial T}\right)_V \left(\frac{\partial T}{\partial P}\right)_T + \left(\frac{\partial P}{\partial V}\right)_T \left(\frac{\partial V}{\partial P}\right)_T.$$

*c) From part b, deduce that

$$1 = \left(\frac{\partial P}{\partial V}\right)_T \left(\frac{\partial V}{\partial P}\right)_T.$$

d) Compute $(\partial P/\partial T)_P$ by the scheme in part b, and deduce that

$$\left(\frac{\partial P}{\partial T}\right)_V = -\left(\frac{\partial P}{\partial V}\right)_T \left(\frac{\partial V}{\partial T}\right)_P.$$

e) Compute $(\partial S/\partial P)_T$ by the scheme $(T, P) \to (T, V) \to S$. Using parts a, c, and d, show that

$$\left(\frac{\partial S}{\partial P}\right)_T = -\left(\frac{\partial V}{\partial T}\right)_P.$$

f) Prove that

$$\left(\frac{\partial E}{\partial P}\right)_T = -T\left(\frac{\partial V}{\partial T}\right)_P - P\left(\frac{\partial V}{\partial P}\right)_T,$$

using $(*)$ or $(**)$ and any methods or results from parts a–e.

g) Show that $T = \left(\dfrac{\partial E}{\partial S}\right)_V.$

---

# 16.4

## EXACT DIFFERENTIAL EQUATIONS (optional)

Whenever a level curve $\{f(x, y) = c\}$ defines $y$ implicitly as a function of $x$, then the derivative $dy/dx$ satisfies the equation

$$\frac{\partial f}{\partial x} + \frac{\partial f}{\partial y}\frac{dy}{dx} = 0 \qquad (1)$$

(Sec. 16.3). In other words, the level curve $\{f = c\}$ is the graph of a solution of the differential equation (1). For example, $x^2 + xy + y^2 = 1$ defines a curve solving the differential equation

$$(2x + y) + (x + 2y)\frac{dy}{dx} = 0.$$

Can this process be reversed? Suppose that a differential equation is given,

$$M(x, y) + N(x, y)\frac{dy}{dx} = 0. \tag{2}$$

*If* there is a function $f$ such that $M = \dfrac{\partial f}{\partial x}$ and $N = \dfrac{\partial f}{\partial y}$, then equation (2) coincides with (1), and the level curves of $f$ give the solutions of (2). When such an $f$ exists, then equation (2) is called an *exact* differential equation. Precisely:

---

**DEFINITION**

Suppose that $M(x, y)$ and $N(x, y)$ are continuous functions in a region $R$ of the plane. The equation

$$M(x, y) + N(x, y)\frac{dy}{dx} = 0$$

is called **exact in R** if there is a function $f$ defined in $R$ with

$$M = \frac{\partial f}{\partial x} \quad \text{and} \quad N = \frac{\partial f}{\partial y}.$$

Such an $f$ is called a **potential function** for the equation.

---

A potential function exists only under special conditions. In fact, if there *is* an $f$ satisfying (2) then, since "mixed partials are equal," $\dfrac{\partial M}{\partial y} = \dfrac{\partial}{\partial y}\left(\dfrac{\partial f}{\partial x}\right) = \dfrac{\partial}{\partial x}\left(\dfrac{\partial f}{\partial y}\right) = \dfrac{\partial}{\partial x} N$; that is,

$$\frac{\partial M}{\partial y} = \frac{\partial N}{\partial x}. \tag{3}$$

Conversely, when condition (3) is satisfied *in a rectangle R*, then a potential function $f$ *can* be constructed, as in the following example.

---

**EXAMPLE 1**   Find, if possible, a function $f$ with $\partial f/\partial x = x + y$ and $\partial f/\partial y = x - y$.

**SOLUTION**   We seek an $f$ with

$$f_x = x + y, \tag{4a}$$

$$f_y = x - y. \tag{4b}$$

First, check the "mixed partials" condition. If there is such a potential $f$, then

$$f_{xy} = \frac{\partial}{\partial y} f_x = \frac{\partial}{\partial y} (x + y) = 1$$

and

$$f_{yx} = \frac{\partial}{\partial x} f_y = \frac{\partial}{\partial x} (x - y) = 1.$$

These are equal, so such a potential might exist. How to find it?

In equation (4a), the partial derivative $f_x$ is computed by differentiating with respect to $x$, holding $y$ constant. Thus, to solve for $f$, *integrate* (4a) with respect to $x$, again holding $y$ constant:

$$f(x, y) = \int f_x(x, y) \, dx = \int (x + y) \, dx = \tfrac{1}{2}x^2 + xy + c(y). \tag{5}$$

Here, $c(y)$ is the "constant" of integration. For each $y$, we integrate with respect to $x$; the constant of integration for one $y$ need not be the same as for another; it depends on $y$, so we write it as $c(y)$.

The function $f$ in (5) satisfies the first relation (4a):

$$f_x = \frac{\partial}{\partial x} \left[ \frac{1}{2} x^2 + xy + c(y) \right] = x + y.$$

We must now choose $c(y)$ to satisfy (4b). Substitute the $f$ in (5) into the desired relation (4b), to find

$$f_y = \frac{\partial}{\partial y} \left[ \frac{1}{2} x^2 + xy + c(y) \right] = x + c'(y) = x - y.$$

The $x$'s cancel, so this equation is satisfied if and only if $c'(y) = -y$; thus

$$c(y) = -\tfrac{1}{2}y^2 + C.$$

(*This* $C$ is a genuine constant, independent of both $x$ and $y$.) Substitute this expression for $c(y)$ into (5), to obtain the desired function:

$$f(x, y) = \tfrac{1}{2}x^2 + xy - \tfrac{1}{2}y^2 + C.$$

We have now constructed all the functions $f$ that satisfy the two equations (4a and b).

---

Now we illustrate the solution of an exact differential equation by means of a potential function.

---

**EXAMPLE 2**  Solve the equation

$$x + y + (x - y) \frac{dy}{dx} = 0. \tag{6}$$

***SOLUTION***  This has the form $M + N \dfrac{dy}{dx} = 0$, with $M = x + y$ and $N = x - y$. It satisfies the condition for exactness, $\dfrac{\partial M}{\partial y} = \dfrac{\partial N}{\partial x}$, so its solutions are given implicitly as the level curves of a potential function $f$ with

$$f_x = M \quad \text{and} \quad f_y = N.$$

We constructed all such functions in Example 1, finding

$$f(x, y) = \tfrac{1}{2}x^2 + xy - \tfrac{1}{2}y^2 + C.$$

We need only one such function, so take $C = 0$:

$$f(x, y) = \tfrac{1}{2}x^2 + xy - \tfrac{1}{2}y^2.$$

*Every level curve of this $f$ solves the given differential equation (6).* Those level curves are

$$\tfrac{1}{2}x^2 + xy - \tfrac{1}{2}y^2 = c.$$

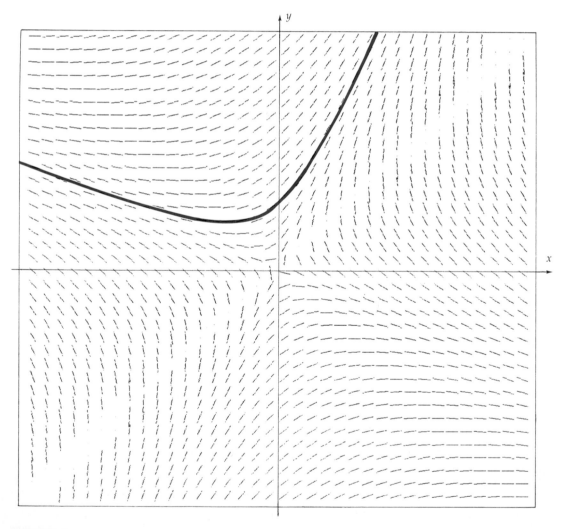

**FIGURE 1**
Slope field for
$$x + y + (x - y)\frac{dy}{dx} = 0,$$
with a solution curve.

Check the solution by implicit differentiation:

$$x + y + x\frac{dy}{dx} - y\frac{dy}{dx} = 0,$$

which is precisely (6). The solutions are conic sections; by the discriminant test (Sec. 12.4) they are hyperbolas.

Figure 1 shows the slope field for equation (6), $\dfrac{dy}{dx} = -\dfrac{x+y}{x-y}$; at many points $(x, y)$ a segment of the prescribed slope $-\dfrac{x+y}{x-y}$ is sketched. These segments are tangent to the various solution curves. (The figure suggests that the solutions are hyperbolas, as indeed they are.)

In conclusion, we prove that a potential function exists whenever the mixed partials condition is satisfied.

---

**THEOREM**

Suppose that $M(x, y)$ and $N(x, y)$ are $C^1$ functions defined in a rectangle $R$, and

$$M_y = N_x. \qquad (7)$$

Then there is a function $f(x, y)$ with $f_x = M$ and $f_y = N$, found as follows:

$$f(x, y) = \int M(x, y)\, dx + c(y) \qquad (8)$$

with $c(y)$ determined by solving

$$f_y = \frac{\partial}{\partial y}\int M(x, y)\, dx + c'(y) = N(x, y). \qquad (9)$$

The level curves of $f$ give solutions of $M + N\dfrac{dy}{dx} = 0$.

---

**PROOF**  Since $M$ is continuous, we may define a function $f$ by equation (8). This function automatically satisfies $f_x = M$:

$$f_x = \frac{\partial}{\partial x}\left[\int M(x, y)\, dx + c(y)\right] = M(x, y) + 0.$$

To satisfy the other condition $f_y = N$, equation (9) must be satisfied; thus

$$c'(y) = N(x, y) - \frac{\partial}{\partial y}\int M(x, y)\, dx. \qquad (10)$$

Since $c'(y)$ depends only on $y$, this equation can be solved only if the right-hand side is independent of $x$. That is, the derivative of the right-hand side of (10) with respect to $x$ must be 0:

$$\frac{\partial}{\partial x}\left[ N(x, y) - \frac{\partial}{\partial y} \int M(x, y)\, dx \right] = 0, \quad \text{or}$$

$$N_x = \frac{\partial}{\partial x} \frac{\partial}{\partial y} \int M(x, y)\, dx. \qquad \text{(11)}$$

By the equality of mixed partials,

$$\frac{\partial}{\partial x} \frac{\partial}{\partial y} \int M(x, y)\, dx = \frac{\partial}{\partial y}\left[ \frac{\partial}{\partial x} \int M(x, y)\, dx \right] = \frac{\partial}{\partial y} M = M_y.$$

By the hypothesis (7), $M_y = N_x$; so (11) *is* satisfied; so the right hand-side of (10) *is* independent of $x$; so (10) *can* be integrated to find $c(y)$. This function $c(y)$ is then substituted in (8) to complete the construction of $f$, with $f_x = M$ and $f_y = N$.

*REMARK*   We carried out the construction in a rectangle $R$, because the theory of antiderivatives applies only to functions defined on a single interval. If $M$ and $N$ were defined, say, in an annular ring (Fig. 2) then some of the integrations would be over two intervals, requiring two "constants" of integration $c_1(y)$ and $c_2(y)$. These complications are discussed in Chapter 18.

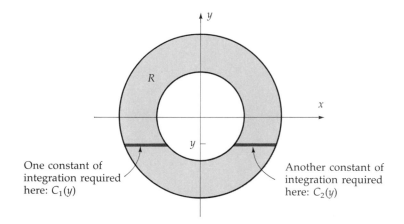

**FIGURE 2**
Complications arising when integrating in a nonrectangular region $R$.

## SUMMARY

A differential equation of the form $M(x, y) + N(x, y)\dfrac{dy}{dx} = 0$ is *exact* in a rectangular region $R$ if

$$M_y = N_x \qquad \text{in } R.$$

Solutions of such an equation are given by the level curves of a *potential func-tion f*, such that $f_x = M$ and $f_y = N$.

## PROBLEMS

### A

**1.** Determine whether the equation is exact. If it is, solve it, and describe the shape of the solution curves.

**\*a)**  $x - y + (y - x)\dfrac{dy}{dx} = 0$

**b)**  $x + y + (y - x)\dfrac{dy}{dx} = 0$

**\*c)**  $\dfrac{x}{x^2 + y^2} - \dfrac{y}{x^2 + y^2}\dfrac{dy}{dx} = 0$

**d)**  $\dfrac{x}{x^2 + y^2}\dfrac{dy}{dx} = \dfrac{y}{x^2 + y^2}$

**\*e)**  $\dfrac{x}{x^2 + y^2} + \dfrac{y}{x^2 + y^2}\dfrac{dy}{dx} = 0$

**f)**  $\sin(y) + \cos(y)\dfrac{dy}{dx} = 0$

**\*g)**  $e^x \sin y + e^x \cos y \dfrac{dy}{dx} = 0$

**2.** Solve.

**\*a)**  $2e^{3x} \cos 2y \dfrac{dy}{dx} = -3e^{3x} \sin 2y,\ y(0) = 1$

**b)**  $x^2 + y^2 + 2xy\dfrac{dy}{dx} = 0;\ y(1) = 1$

**c)**  $e^{x^2 + y^2}\left(1 + 2x^2 + 2xy\dfrac{dy}{dx}\right) = 0;\ y(2) = 1$

**3.** Show that: (i) The two given equations are equiv-alent; and (ii) one is exact, while the other is not. (Thus, whether or not an equation is exact de-pends on how it is written.)

**a)**  $y - x\dfrac{dy}{dx} = 0;\ \dfrac{y}{x^2 + y^2} - \dfrac{x}{x^2 + y^2}\dfrac{dy}{dx} = 0$

**b)**  $\sin(x) - 2\cos(x)\dfrac{dy}{dx} = 0;$

$e^{2y} \sin x - 2e^{2y} \cos x \dfrac{dy}{dx} = 0$

**4.** Differential equations are sometimes written in the form $M\, dx + N\, dy = 0$; this can be interpreted either as $M + N\dfrac{dy}{dx} = 0$, or $M\dfrac{dx}{dy} + N = 0$. Solve.

**\*a)**  $x\, dx + y\, dy = 0$

**b)**  $y\, dx + x\, dy = 0$

**c)**  $y^2 \sec^2 x\, dx + 2y \tan x\, dy = 0$

### B

**5.** A differential equation in the form $\dfrac{dy}{dx} = \dfrac{f(x)}{g(y)}$ is *separable*. Show that every separable equation can be rewritten as an exact equation. (Thus, the method of "exact" equations is more general than separation of variables.)

## 16.5

### LAGRANGE MULTIPLIERS (optional)

Many optimization problems require the maximum or minimum value of a function *f*, subject to some *constraints*. For example, the *x* by *y* rectangle with area 1 and minimum perimeter is found by minimizing the function giving the perimeter,

$$f(x, y) = 2x + 2y,$$  (1)

subject to the constraint that the area be 1:

$$xy = 1. \tag{2}$$

Such problems are common in business. Suppose that a company produces items $A$, $B$, and $C$. Producing $x$ items of $A$, $y$ items of $B$, and $z$ items of $C$ yields a certain profit $f(x, y, z)$ and requires certain financial resources $g(x, y, z)$. With given resources $k$, the company would try to maximize the profit $f(x, y, z)$ subject to the constraint $g(x, y, z) = k$.

The minimum perimeter problem is easily solved using level curves. The constraint $xy = 1$ defines a curve; Figure 1 shows this constraint curve, along with several level curves of the function $f(x, y) = 2x + 2y$ to be minimized, in this case straight lines. The figure shows rather graphically that *along the constraint curve*, the minimum of $f$ is 4; $f(1, 1) = 4$, and $f(x, y) > 4$ for all other points $(x, y)$ on the curve. So the minimum perimeter is with $x = 1$, $y = 1$; the optimal rectangle is a 1 by 1 square.

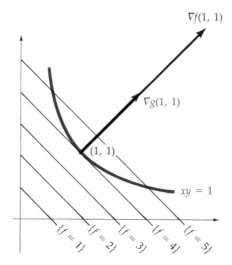

**FIGURE 1**
Along the curve $xy = 1$,
$f(x, y) = 2x + 2y$ has a minimum of
4 at $(x, y) = (1, 1)$.

Notice that at the minimum point $(1, 1)$, the level curve $\{f = 4\}$ is *tangent* to the constraint curve. This tangency can be expressed in terms of gradients. The constraint curve $xy = 1$ is a level curve for the function $g(x, y) = xy$. This level curve is tangent to the level curve of $f$ precisely when the *normals* to the two curves are parallel (Fig. 1); the normals are the gradients of $f$ and $g$, and they are parallel when one is a multiple of the other, that is, when

$$\nabla f = \lambda \nabla g.$$

This is the **Lagrange condition** for a constrained maximum or minimum of $f$ along the curve $\{g = c\}$. The constant $\lambda$ (lambda, Greek *l*) is called a **Lagrange multiplier**, in honor of J. L. Lagrange, the inventor of the method.

Precisely stated:

---

## THE LAGRANGE MULTIPLIER THEOREM

If $P_0$ gives a maximum or minimum of $f(P)$ among all those points $P$ satisfying the constraint

$$g(P) = c,$$

and if $\nabla f$ and $\nabla g$ are continuous at $P_0$, and $\nabla g(P_0) \neq \vec{0}$, then

$$\nabla f(P_0) = \lambda \nabla g(P_0) \tag{3}$$

for some constant $\lambda$.

---

Figure 2 explains why $\nabla f = \lambda \nabla g$ is really necessary at a maximum. Point $P_1$ in the figure cannot give a maximum; for suppose that $P(t)$ is a point moving on the constraint curve $\{g = c\}$ in the direction indicated by the velocity vector $\vec{v} = \dfrac{d\vec{P}}{dt}$ in the figure. Then at $P$,

$$\frac{df(P(t))}{dt} = \nabla f \cdot \frac{d\vec{P}}{dt} = |\nabla f|\left|\frac{d\vec{P}}{dt}\right| \cos \theta_1 > 0$$

since $0 < \theta_1 < \pi/2$; so $f$ increases in the indicated direction along the constraint curve, and $P_1$ is not a maximum. Similarly $P_2$ cannot give a maximum. In fact, whenever the angle $\theta$ between $\nabla f$ and a tangent to the curve is less than $\dfrac{\pi}{2}$, then $\cos \theta > 0$, and we have on the curve a direction in which $f$ is increasing, as at $P_1$ and $P_2$. Thus, any maximum must occur at a point $P_0$ where $\nabla f$ is *perpendicular* to the constraint curve $\{g = c\}$. But $\nabla g$ is perpendicular to this curve; so if $\nabla g \neq \vec{0}$, and $\nabla f$ is also perpendicular to $\{g = c\}$, then $\nabla f$ must be a multiple of $\nabla g$. This is the Lagrange condition (3).

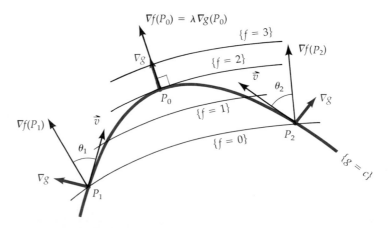

**FIGURE 2**
$f$ is not maximum at $P$ unless $\nabla f(P)$ is perpendicular to $\{g = c\}$.

The vector equation (3) gives two scalar equations, one for each component:

$$f_x(x, y) = \lambda g_x(x, y)$$
$$f_y(x, y) = \lambda g_y(x, y).$$

In addition, the desired point must satisfy the constraint equation

$$g(x, y) = c.$$

These are three equations in the *three* unknowns $x$, $y$, and $\lambda$; if $(x, y, \lambda)$ is a solution, then $(x, y)$ is called a **Lagrange point** for $f$ on the surface $\{g = c\}$. Any maximum or minimum of $f$ on that surface must occur at a Lagrange point, or a point where $\nabla g = 0$, or where $\nabla f$ or $\nabla g$ is discontinuous.

---

**EXAMPLE 1**  Find the points on the curve

$$5x^2 + 5y^2 + 6xy = 4, \tag{4}$$

which are nearest to the origin, and those farthest from it.

*SOLUTION*  The distance from $(x, y)$ to the origin is $\sqrt{x^2 + y^2}$. Maximizing this is equivalent to the simpler problem of maximizing its square, $x^2 + y^2$. So seek maxima and minima of

$$f(x, y) = x^2 + y^2$$

with the constraint that $(x, y)$ lie on the given curve. This constraint is to be written in the form $g = c$; thus we define $g$ to be the function in equation (4), and write the constraint as

$$g(x, y) = 5x^2 + 5y^2 + 6xy = 4.$$

Solve this simultaneously with the vector equation $\nabla f(x, y) = \lambda \nabla g(x, y)$:

$$2x = \lambda(10x + 6y) \tag{5}$$

$$2y = \lambda(10y + 6x). \tag{6}$$

This system of equations is not linear, so there is no standard way to solve it; each system has its own quirks. In this particular case, you can solve (6) for $\lambda$ and use the result in (5), yielding

$$x = \frac{(5x + 3y)y}{5y + 3x}$$

or

$$5xy + 3x^2 = 5xy + 3y^2.$$

Cancel first $5xy$, then 3, to find $x^2 = y^2$, so

$$x = \pm y. \tag{7}$$

This came from (5) and (6); now use it in (4). Of the two cases $x = \pm y$, take first $x = y$:

$$5y^2 + 5y^2 + 6y^2 = 4 \quad \Leftrightarrow \quad y^2 = 1/4 \quad \Leftrightarrow \quad y = \pm 1/2.$$

Since $x = y$, this gives the points

$$(\tfrac{1}{2}, \tfrac{1}{2}) \quad \text{and} \quad (-\tfrac{1}{2}, -\tfrac{1}{2}). \tag{8}$$

Now use (4) with the other case of (7), $x = -y$:

$$5x^2 + 5y^2 - 6y^2 = 4 \quad \Leftrightarrow \quad 4y^2 = 4 \quad \Leftrightarrow \quad y = \pm 1.$$

Since $x = -y$, this gives the points

$$(-1, 1) \quad \text{and} \quad (1, -1). \tag{9}$$

So the equations (4) through (6) yield the four points in (8) and (9).

Figure 3 shows the geometry. The points $\pm(\frac{1}{2}, \frac{1}{2})$ are on a circle inscribed in the curve $\{g = 4\}$; there the normal of the circle is parallel to the normal of the curve:

$$\nabla f = \lambda \nabla g.$$

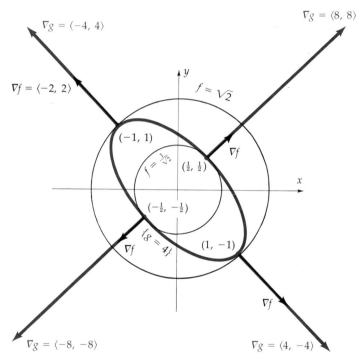

**FIGURE 3**
(Gradients not drawn to scale.)

The other two points $\pm(1, -1)$ lie on a circle circumscribed around the ellipse $\{g = 4\}$, tangent to it at the Lagrange points.

---

For functions of three variables, the Lagrange method is the same; in seeking maxima or minima of $f$ on the level surface $\{g = c\}$, you solve $\nabla f = \lambda \nabla g$, together with the constraint. This gives *four* equations in the four unknowns $(x, y, z, \lambda)$:

$$f_x(x, y, z) = \lambda g_x(x, y, z)$$

$$f_y(x, y, z) = \lambda g_y(x, y, z)$$

$$f_z(x, y, z) = \lambda g_z(x, y, z)$$

$$g(x, y, z) = c.$$

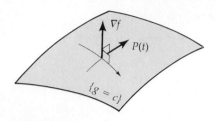

**FIGURE 4**
$\nabla f$ perpendicular to every curve in $\{g = c\}$, so $\nabla f$ is a normal to $\{g = c\}$, and $\nabla f = \lambda \nabla g$.

These conditions can be justified as in the case of two variables, though the picture is harder to see; here is an alternative argument. Suppose that $f(P_0)$ is the maximum of $f(P)$, considering only points $P$ on the surface $\{g = c\}$. To prove that $\nabla f(P_0)$ is perpendicular to $\{g = c\}$ at $P_0$, we show that it is perpendicular to every curve $P(t)$ lying in $\{g = c\}$ and passing through $P_0$ (Fig. 4). Suppose $P(t)$ is such a curve, with $P(t_0) = P_0$. Then $f(P(t))$ has a maximum at $t = t_0$, so its derivative at that point is zero:

$$0 = \left.\frac{df(P(t))}{dt}\right|_{t = t_0} = \nabla f(P(t_0)) \cdot \vec{P}'(t_0).$$

Since $P(t_0) = P_0$,

$$0 = \nabla f(P_0) \cdot \vec{P}'(t_0).$$

Thus $\nabla f(P_0)$ is indeed perpendicular to every curve lying in $\{g = c\}$ and passing through $P_0$. We conclude that it is perpendicular to $\{g = c\}$, hence must be a multiple of $\nabla g$, and the Lagrange condition $\nabla f = \lambda \nabla g$ is established.

---

**EXAMPLE 2** Find the points on the sphere

$$x^2 + y^2 + z^2 = 1 \qquad (10)$$

where the product $xyz$ is maximum.

**SOLUTION** To maximize $f(x, y, z) = xyz$ on the sphere (10), set $\nabla f = \lambda \nabla g$:

$$\left.\begin{array}{l} yz = \lambda \cdot 2x \\ xz = \lambda \cdot 2y \\ xy = \lambda \cdot 2z. \end{array}\right\} \qquad (11)$$

Solve these simultaneously with (10). From (11)

$$2\lambda = \frac{yz}{x} = \frac{xz}{y} = \frac{xy}{z}. \qquad (12)$$

(Why can we divide by $x$, $y$, and $z$? Because, at a maximum of $xyz$ on the sphere, none of $x$, $y$, or $z$ can be zero.) From the second equality in (12), $y^2z = x^2z$ or, since $z \neq 0$,

$$y^2 = x^2.$$

Combined with the last of equations (12), this gives $z^2 = y^2 = x^2$; so from (10)

$$x^2 = \tfrac{1}{3}, \qquad y^2 = \tfrac{1}{3}, \qquad z^2 = \tfrac{1}{3}.$$

This gives eight Lagrange points:

$$\left(\pm\frac{1}{\sqrt{3}}, \pm\frac{1}{\sqrt{3}}, \pm\frac{1}{\sqrt{3}}\right) \qquad (13)$$

with any combination of $+$ or $-$ signs. With one or three $+$ signs, (11) holds with $\lambda = \sqrt{3}/6$; with one or three $-$ signs, (11) holds with $\lambda = -\sqrt{3}/6$.

Now, which of these points give maxima? Taking one or three plus signs in (13) gives $f(x, y, z) = \left(\dfrac{1}{\sqrt{3}}\right)^3$, while one or three minus signs gives $f = -\left(\dfrac{1}{\sqrt{3}}\right)^3$; clearly the former points give the maximum, and the latter points the minimum.

There are other Lagrange points, where some coordinates are zero; they are $(\pm 1, 0, 0)$, $(0, \pm 1, 0)$ and $(0, 0, \pm 1)$. These are saddle points for $f$ on the sphere.

---

*REMARK*  An alternative to the Lagrange method is to use the constraint to reduce the number of variables. For example, in maximizing $f(x, y) = 2x + 2y$ on the curve $xy = 1$, the constraint gives $y = 1/x$, and the function to be maximized is then

$$2x + 2(1/x) \qquad \text{for } x > 0.$$

We solved the problem this way in Chapter 3.

Another alternative is to parametrize the constraint curve. For example, to maximize $xy$ on the circle $x^2 + y^2 = 1$, you can parametrize the circle by $x = \cos t$, $y = \sin t$, $0 \le t \le 2\pi$; then maximize

$$xy = \cos(t)\sin(t), \qquad 0 \le t \le 2\pi.$$

Since $\cos(t)\sin(t) = \tfrac{1}{2}\sin(2t)$, the maximum is $\tfrac{1}{2}$, at $t = \pi/4$ and $t = 5\pi/4$, thus at $x = \cos\dfrac{\pi}{4} = 1/\sqrt{2}$, $y = \sin\dfrac{\pi}{4} = 1/\sqrt{2}$ and at $x = \cos\dfrac{5\pi}{4} = -\dfrac{1}{\sqrt{2}} = y$.

### Existence of Extreme Points

The theorem on the Lagrange method says that *if* there are any extreme points, then they satisfy certain conditions; but it does not guarantee that the desired extremes exist. For example, the problem "Find the maximum of $f(x, y) = 2x + 2y$ on $xy = 1$" has no solution (Fig. 1).

For functions of one variable, the Extreme Value Theorem guarantees the existence of both maximum and minimum for a continuous function $f$ on a *closed, bounded* interval $[a, b]$. With several variables, the analogous theorem guarantees that a continuous function $f$ has a maximum and a minimum over any *closed and bounded set*. Exactly what does this mean? A set $S$ in the plane is *bounded* if it is entirely contained in some (perhaps very large) finite disk. A set $S$ is *closed* if it contains all its boundary points, in the following sense: If any sequence of points $P_k$ in $S$ has a limit point $P$, then also $P$ is in $S$. Thus the disk

$$\{P \colon |P| \le 1\} \tag{14}$$

is a closed set (Fig. 5). For if $|P_k| \le 1$ for all $k$, then also $|\lim P_k| \le 1$, so $\lim P_k$ is in the disk (14). On the other hand, the disk

$$\{P \colon |P| < 1\} \tag{15}$$

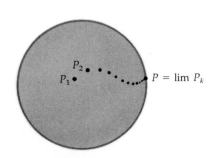

**FIGURE 5**
Disk $\{P \colon |P| \le 1\}$ is closed. If each $P_k$ is in the disk, then $|\lim P_k| \le 1$, so $\lim P_k$ is also in the disk.

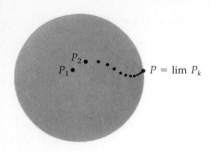

**FIGURE 6**
Disk $\{P: |P| < 1\}$ is *not* closed.
There are sequences $\{P_k\}$ with
$|P_k| < 1$, but $\left|\lim P_k\right| = 1$, so $\lim P_k$
is *not* in this disk.

is not closed (Fig. 6); it can easily happen that $|P_k| < 1$ for all $k$, but $\left|\lim P_k\right| = 1$, so $\lim P_k$ is not in the disk (15).

In general, if $g$ is continuous then any set of the form $\{P: g(P) \le c\}$ is closed; but $\{P: g(P) < c\}$ need not be. Moreover, the *level curves* $\{g = c\}$ are closed; for if $g(P_k) = c$ then also $g(\lim P_k) = \lim g(P_k) = c$. More generally still, any set is closed if it is defined by a finite number of equalities of the type $g = c$, and finitely many inequalities of the type $g \le c$ (not $g < c$!), with continuous functions $g$.

In dimensions higher than 2, the definition of closed set is the same; and a set is *bounded* it it is contained in some finite ball $\{P: |P| \le R\}$.

That said, we can state:

---

## THE EXTREME VALUE THEOREM

If $S$ is a closed and bounded set, and $F$ is continuous at every point of $S$, then $f$ has a maximum and a minimum on $S$.

---

Assuming the set $S$ is closed and bounded, and $f$ is continuous, how are the maximum and minimum to be found? If $S$ is a "level set" $\{g = c\}$, any extreme point must occur at a Lagrange point (where $\nabla f = \lambda \nabla g$) or a point where some hypothesis of the Lagrange Multiplier Theorem is violated: $\nabla g = \vec{0}$, or $\nabla f$ or $\nabla g$ is not continuous.

If $S$ is defined by an inequality $g \le c$, then it has a *boundary* $\{g = c\}$ and an *interior* $\{g < c\}$. (For example, the closed disk $\{P: |P| \le 1\}$ has boundary the circle $\{P: |P| = 1\}$, and interior the open disk $\{P: |P| < 1\}$.) Any extreme points in the interior $\{g < c\}$ are *critical points* of $f$, where $\nabla f = \vec{0}$, or $\nabla f$ fails to exist. Extreme points on the boundary $\{g = c\}$ can be found (in principle) by the methods in this section.

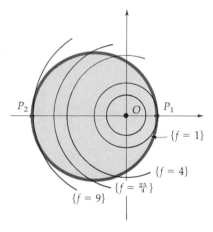

**FIGURE 7**
Level curves of $f(x, y) = x^2 + y^2$ in
the disk $\{(x + 1)^2 + y^2 \le 2\}$. 0 is a
critical point; $P_1$ and $P_2$ are
Lagrange points. $f(0, 0) = 0$ is the
minimum, $f(P_2) = 9$ is the maximum.

**EXAMPLE 3**  Find the maximum and minimum of $f(x, y) = x^2 + y^2$ in the closed disk $\{(x + 1)^2 + y^2 \le 2\}$.

*SOLUTION*  Figure 7 shows the disk and some level curves of $f$. There is just one critical point, at $(0, 0)$, and this gives the minimum since $f(x, y) = x^2 + y^2 \ge 0 = f(0, 0)$. There are two Lagrange points on the boundary, at $(1, 0)$ and $(-3, 0)$. Since $f(1, 0) = 1$ and $f(-3, 0) = 9$, the maximum is 9.

---

**EXAMPLE 4 (Entropy in Probability Theory.)**  Imagine an experiment with two possible outcomes, for example, flipping a coin. Denote by $p$ the probability of the first outcome (heads, say) and by $q$ the probability of the second (tails). Probability theory requires that

$$p + q = 1 \tag{16}$$

and

$$0 \le p, \qquad 0 \le q. \tag{17}$$

The *entropy* is a measure of the uncertainty in the situation; it is defined by

$$E(p, q) = -p \ln(p) - q \ln(q). \tag{18}$$

The function $p \ln p$ is defined only for $p > 0$; but $p \ln p \to 0$ as $p \to 0$, so we take $p \ln p = 0$ when $p = 0$, and $q \ln q = 0$ when $q = 0$. Thus the function $E(p, q)$ is defined for $p \ge 0$ and $q \ge 0$.

If $p = 1$ then there is perfect certainty: heads every time. In this case, we should find the entropy (uncertainty) to be 0, and indeed it is. In view of (16), $p = 1 \Rightarrow q = 0$, and

$$E(1, 0) = 1 \ln 1 + 0 = 0.$$

Similarly, if $p = 0$ then $q = 1$, and there is no uncertainty; $E(0, 1) = 0$.

What situation displays *maximum* uncertainty, in the sense of entropy? In other words, what is the maximum of $E(p, q)$ with the constraints (16) and (17)? These constraints define a line segment in the $pq$-plane (Fig. 8). This is a bounded set, and also closed, since it contains the endpoints $(1, 0)$ and $(0, 1)$; hence $E(p, q)$ must have a maximum on the segment. It is not at either endpoint, since $E = 0$ there, while $E > 0$ at other points of the segment; so the maximum must occur at a Lagrange point. The gradient of the constraint function in (16) is

$$\nabla g = g_p \vec{i} + g_q \vec{j} = \vec{i} + \vec{j}.$$

The gradient of the entropy is

$$\nabla E = E_p \vec{i} + E_q \vec{j} = -(1 + \ln p)\vec{i} - (1 + \ln q)\vec{j}.$$

The Lagrange condition $\nabla E = \lambda \nabla g$ gives

$$-1 - \ln p = \lambda = -1 - \ln q.$$

So $\ln p = \ln q$, hence $p = q$; then from (16)

$$p = q = \tfrac{1}{2}.$$

*CONCLUSION:*   The uncertainty is maximum when both outcomes are equally likely.

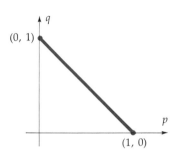

**FIGURE 8**
Segment $p + q = 1$, $p \ge 0$, $q \ge 0$.

## SUMMARY

**The Lagrange Multiplier Method:**   If $P_0$ gives a maximum or minimum of $f(P)$ among all those points $P$ satisfying the constraint

$$g(P) = c, \tag{*}$$

and if $\nabla f$ and $\nabla g$ are continuous at $P_0$, and $\nabla g(P_0) \ne 0$, then

$$\nabla f(P_0) = \lambda \nabla g(P_0). \tag{**}$$

Any such constrained maxima and minima may therefore be found by solving (*) and (**) simultaneously.

## PROBLEMS

### A

**1.** Figure 9 shows a variety of points where the Lagrange condition $\nabla f = \lambda \nabla g$ holds. Which (if any) are maxima of $f$ on $\{g = c\}$? Which are minima, and which are neither?

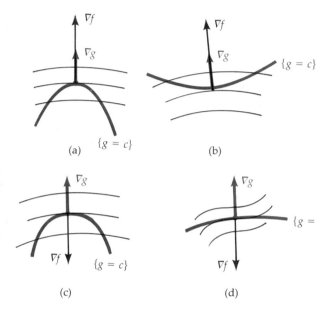

(a)   (b)

(c)   (d)

**FIGURE 9**
Four Lagrange points. Which are max?
Min? Neither?

In problems 2–7, find the maximum or minimum of $f$ on the given set $\{g = c\}$ by the Lagrange method. Illustrate your conclusions by sketching the set and some level curves of $f$.

*2. $f(x, y) = xy$ on $2x + 2y = 4$

*3. $f(x, y) = xy$ on $(x + 1)^2 + y^2 = 1$

4. $f(x, y) = x^2 + y^2$ on $5x^2 + 6xy + 5y^2 = 16$

*5. $f(x, y, z) = x + y + z$ on $x^2 + y^2 + z^2 = 1$

6. $f(x, y, z) = x - y + z$ on $x^2 + y^2 + z^2 = 1$

7. $f(x, y) = y$ on $2x^2 + y^2 = 3$

*8. Find the maximum of $f(x, y) = 3x^2 + 2xy + 4y^2$ on $x^2 + y^2 = 9$ by parametrizing the constraint curve.

In problems 9–13, use Lagrange multipliers to optimize.

**9.** Find the rectangular box with minimum surface area containing given volume $V$.

**10.** Find the rectangular box with maximum volume and given surface area $S$.

**11.** Find the points $P$ on the curve $x^2 - y^2 = 1$ which are closest to the origin.

*12. Find the points $P$ on the cone $x^2 + 2y^2 = z^2$ which are closest to the point $Q = (0, 0, 1)$. Show that at these points, the line through $P$ and $Q$ is perpendicular to the cone. (It can be shown directly from the Lagrange condition.)

**13.** Find the vector $\vec{v}$ of length 3 with the sum of its components maximum.

**14.** Find the maximum and minimum of
*a) $xy$ on the disk $\{(x + 1)^2 + y^2 \le 1\}$.
b) $x^2 + y^2$ on the disk $\{(x + 2)^2 + (y - 3)^2 \le 1\}$.
c) $x^2 + 2y^2$ on the disk $\{x^2 + y^2 \le 1\}$.
d) $x^2 + 2y^2$ on the rectangle $\{(x, y): |x| \le 1, |y| \le 1\}$.

**15.** In a certain economic model, $f(x, y) = x^{1/3}y^{2/3}$ gives the production achieved with $x$ dollars spent on labor, and $y$ dollars on capital. With given resources $x + y = C$, what ratio of $x$ to $y$ maximizes production?

*16. Rework the previous problem with the general "Cobb-Douglas production function"
$$f(x, y) = kx^\alpha y^\beta, \qquad \alpha + \beta = 1, \qquad k \text{ constant.}$$

**17.** Reverse the roles in the previous problem: Minimize the cost $C = x + y$ subject to the constraint that production $f(x, y) = kx^\alpha y^\beta$ has a fixed value.

### B

**18.** On the surface $\{g = c\}$, you seek the point $P$ closest to a given point $P_0$ not on $\{g = c\}$. Interpret the Lagrange condition $\nabla f = \lambda \nabla g$ as an orthogonality relation.

**19.** Fermat's "least time principle" (Sec. 4.4), applied in the study of reflection from a curved surface, leads to the following problem. Given two points $P_1$ and $P_2$, seek a point $P$ on the surface $\{g = c\}$ minimizing $|\vec{P} - \vec{P}_1| + |\vec{P} - \vec{P}_2|$. Show that
a) $\nabla |\vec{P} - \vec{P}_1|$ is the unit vector
$$|\vec{P} - \vec{P}_1|^{-1}(\vec{P} - \vec{P}_1).$$

**b)** If $\vec{u}_1$ and $\vec{u}_2$ are unit vectors, then $\vec{u}_1 + \vec{u}_2$ bisects the angle between $\vec{u}_1$ and $\vec{u}_2$.

**c)** The Lagrange condition for $P$ determines that the normal to $\{g = c\}$ at $P$ lies in the plane of points $P$, $P_1$, and $P_2$, and bisects the angle between $\vec{P} - \vec{P}_1$ and $\vec{P} - \vec{P}_2$.

**20.** An escaped prisoner can walk with velocity $v_1$ along roads $AO$ and $OB$, and with velocity $v_2 \leq v_1$ across the field between them (Fig. 10). Determine $x$ and $y$ to minimize transit time from $A$ to $B$ in the following cases:

**a)** $v_1 = v_2 = 1$.     **b)** $v_1 = 2, v_2 = 1$.
**c)** $v_1 = \sqrt{2}, v_2 = 1$.

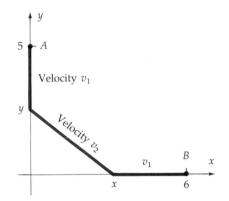

**FIGURE 10**
Escaped convict cutting a corner.

(This problem is adapted from an article by N. L. Silver in the *American Mathematical Monthly*, Vol. 94, pp. 545–47.)

**21.** A certain experiment has three possible outcomes, with probabilities $p_1$, $p_2$, and $p_3$, satisfying

$$p_1 \geq 0, \qquad p_2 \geq 0, \qquad p_3 \geq 0 \qquad (19)$$

and

$$p_1 + p_2 + p_3 = 1. \qquad (20)$$

The *entropy* is

$$E(p_1, p_2, p_3) = -\sum_{j=1}^{3} p_j \ln(p_j).$$

By definition, we take $p \ln p = 0$ when $p = 0$.

**a)** Sketch the set $S$ in $(p_1, p_2, p_3)$ space defined by (19) and (20). Persuade yourself that $S$ is closed and bounded.

**b)** Find the maximum of $E$ on the boundary of $S$.

**c)** Find the maximum of $E$ on the part of $S$ away from the boundary.

**d)** Find the maximum and minimum of $E$ on $S$. (These exist, since $S$ is closed and bounded, and $E$ is continuous.)

**22.** [The geometric mean]. The *arithmetic mean* of three numbers $x$, $y$, $z$ is their average

$$A(x, y, z) = \tfrac{1}{3}(x + y + z).$$

The *geometric mean* of three nonnegative numbers is

$$G(x, y, z) = (xyz)^{1/3}.$$

Show that

**a)** The geometric mean corresponds to averaging the logarithms:

$$\ln G(x, y, z) = A(\ln x, \ln y, \ln z),$$
$$\text{for } x > 0, y > 0, z > 0.$$

**b)** $\min(x, y, z) < G(x, y, z) < \max(x, y, z)$.

**c)** For any given number $a$, the maximum of $G(x, y, z)$ on the set where $A(x, y, z) = a$, $x \geq 0$, $y \geq 0$, $z \geq 0$, equals $a$; and it occurs only where $x = y = z = a$.

**d)** For any nonnegative numbers $x$, $y$, $z$, $G(x, y, z) < A(x, y, z)$ unless $x = y = z$.

**C**

**23.** A triangle with sides $x$, $y$, and $z$ has perimeter $p = x + y + z$.

**\*a)** Prove *Heron's formula* for the area $A$: $A^2 = \tfrac{1}{2}p(\tfrac{1}{2}p - x)(\tfrac{1}{2}p - y)(\tfrac{1}{2}p - z)$.

**b)** Among all triangles with given perimeter $p$, determine the one of maximum area. Is there one of minimum area?

**24.** Show that the solutions of the equations

$$\nabla f(P) = \lambda \nabla g(P), \qquad g(P) = c$$

are the critical points of the function $L(P, \lambda) = f - \lambda(g - c)$.

**25.** Prove that the Lagrange condition is necessary, in the case of $C^1$ functions $f$ and $g$, as follows:

**a)** Suppose that $g_y(x_0, y_0) \neq 0$, and $f(x_0, y_0) \geq f(x, y)$ for all points $(x, y)$ on the level set $\{g = c\}$ near $(x_0, y_0)$. Deduce that at $(x_0, y_0)$

$$f_x - f_y \frac{g_x}{g_y} = 0$$

and hence $\nabla f = \lambda \nabla g$ with $\lambda = \dfrac{f_y}{g_y}$. (Use the Implicit Function Theorem, Sec. 16.3.)

b) Prove an analogous result if $g_x(x_0, y_0) \neq 0$.

26. To optimize $f(x, y, z)$ subject to *two* constraints

$$g_1 = c_1 \quad \text{and} \quad g_2 = c_2 \qquad (21)$$

you solve equations (21) together with

$$\nabla f = \lambda_1 \nabla g_1 + \lambda_2 \nabla g_2;$$

the constants $\lambda_1$ and $\lambda_2$ are among the unknowns. Explain this method geometrically.

27. Find the maximum of $x^2 + y^2 + z^2$ on the curve defined by the two equations $y - x = 1$ and $y^2 - z^2 = 1$. (See the previous problem.)

# REVIEW PROBLEMS    CHAPTER 16

*1. Suppose that the air pressure at point $(x, y, z)$ is $f(x, y, z) = e^{x^2 + y^2 - 2z^2}$.

   a) Describe the level surfaces $\{f = 0\}$, $\{f = \frac{1}{2}\}$, $\{f = 1\}$, $\{f = 2\}$.

   b) Compute the pressure gradient $\nabla f(1, -1, 1)$.

   c) Write the equation of the plane tangent to the "isobarometric surface" $\{f = 1\}$ at $(1, -1, 1)$.

   d) Find the directional derivative of the pressure at $(1, -1, 1)$, in the direction from $(1, -1, 1)$ toward $(2, 3, 2)$.

   e) In what direction is the pressure decreasing most rapidly, at $(1, -1, 1)$?

   f) A bat flies through the point $(1, -1, 1)$ with velocity $2\vec{i} + 5\vec{j} + \vec{k}$. What is the rate of change (with respect to time) of the pressure felt by the bat, at that point? What is the rate of change of pressure (with respect to distance) in the direction of the bat's motion?

*2. A rectangular box is designed with dimensions $l = 4$, $w = 4$, $h = 3$. As produced, each dimension could be larger or smaller by 0.15. Estimate the difference that this makes in

   a) The volume.    b) The surface area.

*3. Show that in polar coordinates $r$ and $\theta$, $|\nabla w|^2 = \left(\dfrac{\partial w}{\partial r}\right)^2 + r^{-2}\left(\dfrac{\partial w}{\partial \theta}\right)^2$.

*4. a) Express $\dfrac{\partial}{\partial x} f[x, y, \varphi(x, y)]$ in terms of derivatives of $f$ and of $\varphi$.

   b) Suppose that $\varphi$ in part a is a function such that $f(x, y, \varphi(x, y)) \equiv 1$ for all $x$ and $y$. Deduce expressions for $\dfrac{\partial \varphi}{\partial x}$ and $\dfrac{\partial \varphi}{\partial y}$ in terms of derivatives of $f$.

*5. Suppose that $f(x, y)$ satisfies the equation $f_x = f_y$. Define a new function $\varphi$ by $\varphi(u, v) = f(u - v, u + v)$.

   a) Express $\varphi_u$ and $\varphi_v$ in terms of derivatives of $f$.

   b) Show that one of the derivatives in part a is a constant. What is the constant?

6. Suppose that $w$ depends on $x$ and $y$, and that $x = u + 2v$, $y = u - 2v$.

   a) Express $w_v$ in terms of $w_x$ and $w_y$, as explicitly as possible.

   *b) Express $w_{vu}$ in terms of $w_{xx}$, $w_{xy}$, and $w_{yy}$.

*7. a) By the Lagrange method, find the point $P$ on $\{x + 2y + z = 1\}$ which is closest to the origin.

   b) Show that $\vec{P}$ is perpendicular to the given plane.

# 17

# MULTIPLE INTEGRALS

The concept of integral extends from one to several variables, with many applications. We begin with a geometric description of the integral of a function of two variables.

# 17.1

## DOUBLE INTEGRALS

Suppose that $R$ is a region in the $xy$-plane, and $f$ is continuous at all points of $R$, including the boundary. Suppose that $f \geq 0$ throughout $R$; then the integral of $f$ over $R$ is the volume lying under the graph of $f$, and over $R$ (Fig. 1). It is denoted $\iint_R f(x, y)\, dA$, and called the **double integral** of $f$ over the region $R$. ["Double" refers to the two variables $(x, y)$; the familiar integral $\int_a^b f(x)\, dx$ is a "single integral."]

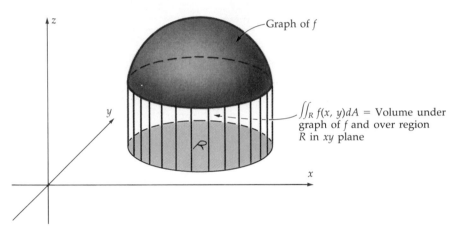

**FIGURE 1**
$\iint_R f(x, y)\, dA$ as volume, for $f \geq 0$.

**EXAMPLE 1**    $R$ is bounded by the unit circle $x^2 + y^2 = 4$, and $f(x, y) \equiv 3$. Then the region above $R$ and below the graph of $f$ is a cylinder of radius two and height 3 (Fig. 2). The volume is $(\pi \cdot 2^2)(3) = 12\pi$, so

$$\iint_R 3\, dA = 12\pi.$$

If $f$ is negative in part of the region $R$ (Fig. 3), then the volume $V_-$ lying *below* the $xy$-plane and above the graph is subtracted from the volume $V_+$ *above* the $xy$-plane and below the graph; the double integral is the *net signed volume*

$$\iint_R f(x, y)\, dA = V_+ - V_-.$$

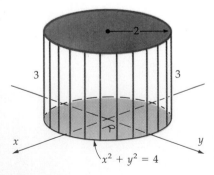

**FIGURE 2**
$\iint_R 3\, dA = 3 \cdot 4\pi = 12\pi.$

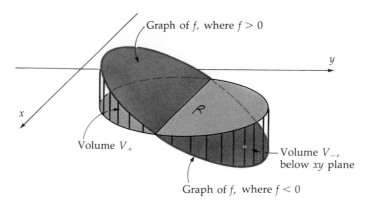

**FIGURE 3**
$\iint_R f(x, y)\, dA = V_+ - V_-$.

**EXAMPLE 2**   Take the region $R$ to be the circle in Example 1, and $f(x, y) = y$ (Fig. 4). Then the volumes $V_+$ and $V_-$ are equal, as you can see from the symmetry of the figure; so

$$\iint_R y\, dA = 0.$$

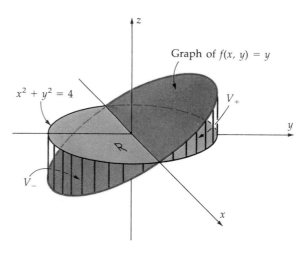

**FIGURE 4**
$\iint_R y\, dA = V_+ - V_- = 0$.

Few integrals can be evaluated by the simple methods in these two examples; but every integral can be *approximated by Riemann sums.* Cover the region $R$ with a grid of horizontal and vertical lines, forming rectangles. Denote those rectangles lying inside the region $R$ by $R_1, \ldots, R_k$. In each rectangle $R_j$, choose a point $P_j$ (Fig. 5). Approximate the volume lying over the rectangle $R_j$ by the volume of a rectangular column of base $R_j$ and height

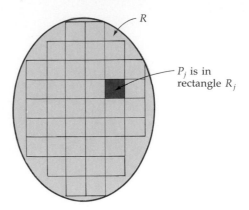

**FIGURE 5**
Shaded region $R$ approximated by
rectangular grid.

$f(P_j)$ (Fig. 6). Let $\Delta A_j$ be the area of $R_j$; then the column has volume

$$f(P_j)\,\Delta A_j.$$

The entire volume under the graph is approximated by the *Riemann sum*

$$\sum_{j=1}^{k} f(P_j)\,\Delta A_j.$$

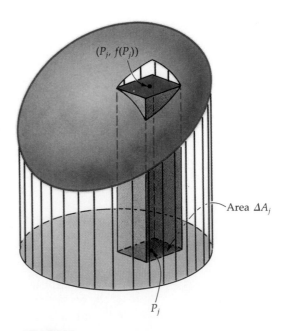

**FIGURE 6**
Volume over $j^{\text{th}}$ rectangle
approximated by rectangular column,
base $\Delta A_j$ and height $f(P_j)$.

If the grid is made finer and finer, so that the length and width of each rectangle $R_j$ tends to zero, then the approximation gets better and better. The integral is the *limit* of such Riemann sums as the lengths and widths go to zero.

If $f(P_j)$ is negative, then $f(P_j)\,\Delta A_j$ is *minus* the volume of the approximating column; so in the Riemann sum, volumes *below* the $xy$-plane are subtracted, just as they are in the integral.

In the notation $\iint_R f(x, y)\,dA$, the "$dA$" is a reminder of the area factors $\Delta A_j$ in the Riemann sum, and the double integral sign $\iint_R$ denotes the integral over a two-dimensional region $R$.

Now, how do you evaluate double integrals? The secret is *Cavalieri's principle*: *The volume of a region is the integral of its cross-sectional area*:

$$\text{Volume} = \int_a^b A(x)\,dx. \tag{1}$$

Figure 7 recalls why. The slice shown there has thickness $\Delta x$, and the face of the slice has area $A(x)$, so its volume is approximately $A(x)\,\Delta x$. The sum $\sum A(x_j)\,\Delta x$ of the volumes of all the slices between $x = a$ and $x = b$ is approximately the total volume $V$. As the slices are made thinner and more numerous, then $\sum A(x_j)\,\Delta x$ approximates the integral $\int_a^b A(x)\,dx$ better. The limit of these sums (as the slice thickness goes to zero) is precisely the volume.

We apply Cavalieri's principle to the double integral

$$\iint_R f(x, y)\,dA.$$

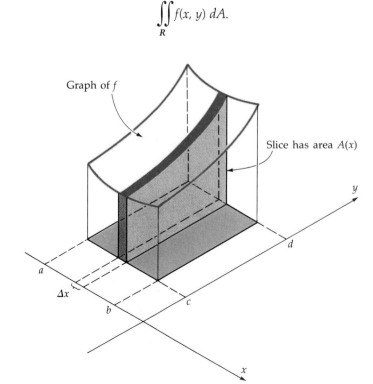

**FIGURE 7**
$\Delta V \approx A(x)\,\Delta x.$

Suppose for simplicity that the region $R$ is a rectangle

$$\{(x, y): a \le x \le b, c \le y \le d\},$$

and that $f \ge 0$, so the integral is simply the volume under the graph (Fig. 7). By Cavalieri's principle (1),

$$\iint\limits_{R} f(x, y) \, dA = \text{volume under graph of } f$$

$$= \int_{a}^{b} A(x) \, dx, \tag{2}$$

where $A(x)$ is the cross-sectional area shown in Figure 8. The top edge of this cross section is the graph of

$$z = f(x, y)$$

as $y$ varies from $c$ to $d$, while $x$ remains constant. The area under that graph is given by the definite integral

$$A(x) = \int_{c}^{d} f(x, y) \, dy. \tag{3}$$

This is a "partial integral" of $f$; in evaluating it, treat $x$ as a constant, and integrate with respect to $y$. Then substitute the result in (2):

$$\iint\limits_{R} f(x, y) \, dA = \int_{a}^{b} \left[ \int_{c}^{d} f(x, y) \, dy \right] dx = \int_{a}^{b} \int_{c}^{d} f(x, y) \, dy \, dx. \tag{4}$$

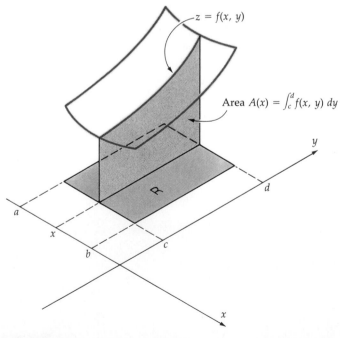

**FIGURE 8**
Cross-sectional area $A(x)$ is an integral.

The expression on the right is called a **repeated integral**, or **iterated integral**. You first evaluate the inner integral (with respect to $y$), then the outer integral (with respect to $x$).

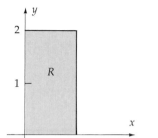

**FIGURE 9**
$R = \{(x, y): 0 \le x \le 1, 0 \le y \le z\}.$

**EXAMPLE 3**   Evaluate $\iint_R (1 - x)\, dA$, where $R$ is the rectangle (Fig. 9)

$$R = \{(x, y): 0 \le x \le 1, 0 \le y \le 2\}.$$

*SOLUTION*   In the given rectangle $R$, for each $x$ between 0 and 1, $y$ varies from 0 to 2. So the repeated integral is

$$\int_0^1 \left[ \int_0^2 (1 - x)\, dy \right] dx. \tag{5}$$

The inner integral is

$$\int_0^2 (1 - x)\, dy = y - xy \Big|_{y=0}^{y=2} = 2 - 2x. \tag{6}$$

In taking this antiderivative, the variable is $y$, while $x$ is constant; the limits of integration 0 and 2 are limits for $y$. The result (6) is a function of $x$, giving the inner integral in (5). Thus the repeated integral is

$$\int_0^1 \overbrace{\int_0^2 (1 - x)\, dy}\ \ \overset{=}{\phantom{xxx}}\ \ dx = \int_0^1 \overbrace{(2 - 2x)}\, dx$$

$$= 2x - x^2 \Big|_0^1 = 2 - 1 = 1.$$

This is the volume under the graph of $f(x, y) = 1 - x$, over the given rectangle $R$ (Fig. 10). As a check, notice that the region is half of a 1 by 1 by 2 box, so its volume is indeed 1.

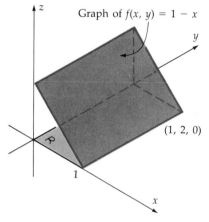

Graph of $f(x, y) = 1 - x$

$(1, 2, 0)$

**FIGURE 10**
$\iint_R (1 - x)\, dA = 1.$

In applying Cavalieri's principle to the double integral $\iint_R f(x, y)\, dA$, you could just as well slice the volume the other way, perpendicular to the $y$-axis (Fig. 11). This gives

$$\iint_R f(x, y)\, dA = \int_c^d A(y)\, dy. \tag{7}$$

The cross section is the region under the graph of $z = f(x, y)$ with $x$ varying from $a$ to $b$, while $y$ is constant; so

$$A(y) = \int_a^b f(x, y)\, dx. \tag{8}$$

Use this in (7):

$$\iint_R f(x, y)\, dA = \int_c^d \int_a^b f(x, y)\, dx\, dy.$$

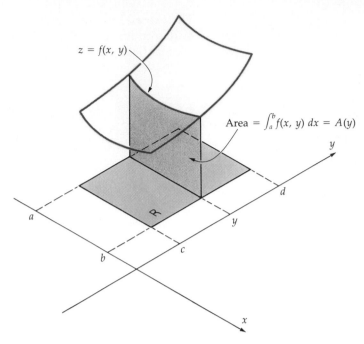

**FIGURE 11**
Area of a slice with fixed $y$, $x$ varying.

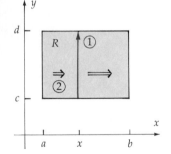

**FIGURE 12**
Scheme for integrating first with respect to $y$, then with respect to $x$:
$\int_a^b \int_c^d f(x, y) \, dy \, dx$.

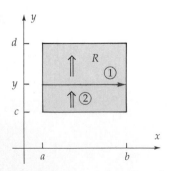

**FIGURE 13**
Scheme for integrating first with respect to $x$, then with respect to $y$:
$\int_c^d \int_a^b f(x, y) \, dx \, dy$.

This is a repeated integral like (4), but in the other order. Both represent the same volume, so they are equal:

$$\iint\limits_R f(x, y) \, dA = \int_a^b \int_c^d f(x, y) \, dy \, dx = \int_c^d \int_a^b f(x, y) \, dx \, dy. \tag{9}$$

Figures 12 and 13 illustrate how the two repeated integrals "sweep out" the rectangle $R$. In Figure 12, first $y$ varies from $c$ to $d$, generating the vertical arrow; then $x$ varies from $a$ to $b$, sweeping the arrow across the entire rectangle $R$. In Figure 13, first $x$ varies, then $y$, and the rectangle is swept out in the other order.

The same idea applies to the Riemann sums $\sum f(P_j) \, \Delta A_j$ approximating the integral (9). Figures 14 and 15 show the rectangle $R$ covered by a grid. Each rectangle $R_j$ in the grid contributes a term $f(P_j) \, \Delta A_j$ to the Riemann sum. You can first form the sum for all the rectangles in each vertical column (Fig. 14), then add those "column sums"; or else form the sum for all rectangles in each horizontal row (Fig. 15) and add the row sums. The first method is analogous to the repeated integral

$$\int_a^b \left[ \int_c^d f(x, y) \, dy \right] dx$$

and the second to the integral

$$\int_c^d \left[ \int_a^b f(x, y) \, dx \right] dy.$$

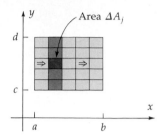

**FIGURE 14**
Summing rectangles over a vertical
column, then summing the columns.

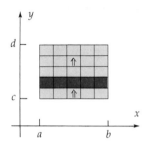

**FIGURE 15**
Summing rectangles over a
horizontal row, then summing the
rows.

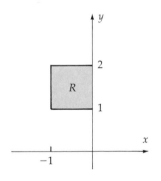

**FIGURE 16**

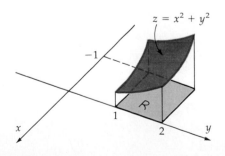

**FIGURE 17**

Formula (9) is the key to evaluating a double integral over a rectangle. In deriving it, we assumed that $f \geq 0$; but it is equally valid when $f$ is negative in any part of the rectangle. In (8), any area where $f$ is negative is subtracted from the area where $f$ is positive; so when you substitute this in (7), the volume where $f$ is negative is subtracted from the volume where $f$ is positive.

**EXAMPLE 4**

**a.** Compute the repeated integral

$$\int_1^2 \int_{-1}^0 (x^2 + y^2) \, dx \, dy. \tag{10}$$

**b.** This integral gives the volume of a solid lying over a certain rectangle $R$; sketch $R$, and sketch the solid.

**c.** Evaluate this same double integral as a repeated integral in the other order.

**SOLUTION**

**a.** The inner integral in (10) is $\int_{-1}^0 (x^2 + y^2) \, dx$, so $x$ varies from $-1$ to $0$ and $y$ is constant:

$$\int_{-1}^0 (x^2 + y^2) \, dx = \tfrac{1}{3}x^3 + y^2 x \Big|_{x=-1}^{x=0} = 0 - [\tfrac{1}{3}(-1)^3 + y^2(-1)]$$
$$= \tfrac{1}{3} + y^2.$$

Use this in (10):

$$\overbrace{\int_1^2 \int_{-1}^0 (x^2 + y^2) \, dx}^{=} \, dy = \int_1^2 (\tfrac{1}{3} + y^2) \, dy$$
$$= \tfrac{1}{3}y + \tfrac{1}{3}y^3 \Big|_1^2 = \tfrac{2}{3} + \tfrac{8}{3} - \tfrac{1}{3} - \tfrac{1}{3}$$
$$= \tfrac{8}{3}.$$

**b.** In (10), for each $y$ from 1 to 2, $x$ varies from $-1$ to $0$. So the rectangle is

$$R = \{(x, y): 1 \leq y \leq 2, \ -1 \leq x \leq 0\}.$$

Figure 16 shows the rectangle, and Figure 17 shows the volume under the graph of $z = x^2 + y^2$ over $R$.

**c.** Reversing the order in (10) gives

$$\int_{-1}^0 \overbrace{\int_1^2 (x^2 + y^2) \, dy}^{=} \, dx = \int_{-1}^0 [x^2 y + \tfrac{1}{3}y^3]_{y=1}^{y=2} \, dx$$
$$= \int_{-1}^0 [2x^2 + \tfrac{8}{3} - (x^2 + \tfrac{1}{3})] \, dx$$
$$= \int_{-1}^0 (x^2 + \tfrac{7}{3}) \, dx$$
$$= \tfrac{1}{3}x^3 + \tfrac{7}{3}x \Big|_{-1}^0 = \tfrac{1}{3} + \tfrac{7}{3} = \tfrac{8}{3}$$

in agreement with the answer to part a.

Integrals over nonrectangular regions are covered in the next section.

**REMARKS**   A more theoretically oriented course would define the double integral $\iint_R f \, dA$ as the *limit of Riemann sums*, and prove (with considerable difficulty) that this limit exists, under appropriate conditions on $f$ and $R$. It would then *define* the volume under the graph as the integral. But we based our description of double integrals on volume, since this gives a direct intuitive understanding of their essential properties.

In applications, however, we often abandon the idea that a double integral is a volume, and think of it variously as the area of a surface, or the mass of a plate, or the torque exerted by a plate about some axis, to name just a few examples.

## SUMMARY

$$\iint_R f(x, y) \, dA = V_+ - V_- \qquad \text{(see Fig. 3)}$$

$$= \text{net signed volume between graph of} \\ f \text{ and region } R \text{ in the } xy \text{ plane}$$

$$= \lim \sum f(P_j) \, \Delta A_j.$$

If $R = \{(x, y): a \le x \le b, c \le y \le d\}$ then

$$\iint_R f(x, y) \, dA = \int_c^d \int_a^b f(x, y) \, dx \, dy$$

$$= \int_a^b \int_c^d f(x, y) \, dy \, dx.$$

## PROBLEMS

### A

**1.** Evaluate the following repeated integrals.

*a)  $\displaystyle\int_0^1 \int_0^1 (x^2 + y^2) \, dx \, dy$

b)  $\displaystyle\int_1^2 \int_2^3 x \, dy \, dx$

*c)  $\displaystyle\int_{-1}^0 \int_{-1}^1 1 \, dx \, dy$

d)  $\displaystyle\int_0^5 \int_0^3 e^{2s+t} \, dt \, ds$

*e)  $\displaystyle\int_{-1}^1 \int_{-2}^2 uve^{-u^2-v^2} \, dv \, du$

f)  $\displaystyle\int_0^{\pi/2} \int_0^{\pi/2} \sin(x + y) \, dy \, dx$

*g)  $\displaystyle\int_0^{\pi} \int_0^{\pi/2} \sin x \cos y \, dx \, dy$

h)  $\displaystyle\int_{-2}^0 \int_1^2 \frac{xy}{2x^2 + y^2} \, dy \, dx$

**2.** For each integral in problem 1, reverse the order of integration and evaluate it again.

**3.** Evaluate the double integral.

*a)  $\iint_R xy \, dA$,
 $R = \{(x, y): -2 \le x \le 2, 0 \le y \le 1\}$

*b)  $\iint_R (x^2 + y^2) \, dA$,
 $R = \{(x, y): |x| \le 1, |y| \le 1\}$

c)  $\iint_R (-1) \, dA$,
 $R = \{(x, y): |x| \le 2, 0 \le y \le 2\}$

**d)** $\iint_R (x^2 - y^2)\, dA,$
$R = \{(x, y) : a \le x \le b,\ a \le y \le b\}$

**4.** Sketch the region whose volume is given by the repeated integral. Compute the volume.

**\*a)** $\displaystyle\int_0^2 \int_{-1}^0 (x^2 + y^2)\, dx\, dy$

**b)** $\displaystyle\int_1^2 \int_{-2}^{-1} (-x)\, dx\, dy$

**5.** Compute the volume enclosed by
**\*a)** The $xy$-plane, the planes $x = 1$, $x = -1$, $y = 1$, $y = -1$, and the surface $z = x^2$.
**b)** The $xy$-plane, the planes $x = 0$, $x = 1$, $y = -1$, $y = 0$, and the surface $z = \sin(\pi x)e^y$.

**B**

**6.** Compute the volume of the finite region bounded
**\*a)** Below by the $xy$-plane and above by the graph of $z = (1 - x^2)(4 - y^2)$.
**b)** Below by the graph of
$z = (x^2 - 4) - (y^2 - 9)$
and above by the $xy$-plane.

**7. \*a)** Approximate the integral $\iint_R (x^2 + y^2)\, dA$ by a Riemann sum, where $R$ is the rectangle $0 \le x \le 1$, $1 \le y \le 3$. Cover $R$ with a grid of $\frac{1}{2}$ by $\frac{1}{2}$ squares; in each square $R_j$, evaluate $f$ at the lower right-hand corner $P_j$.
**b)** Evaluate the double integral, and compare it to your approximation.

**8.** Suppose that $f(-x, y) = -f(x, y)$ for all $x$ and $y$. Explain which of the following integrals is necessarily zero.

**a)** $\displaystyle\int_{-a}^a \int_c^d f(x, y)\, dx\, dy$

**b)** $\displaystyle\int_{-a}^a \int_c^d f(x, y)\, dy\, dx$

**c)** $\displaystyle\int_c^d \int_{-a}^a f(x, y)\, dx\, dy$

**d)** $\displaystyle\int_c^d \int_{-a}^a f(x, y)\, dy\, dx$

**e)** $\iint_R f(x, y)\, dA$
where $R$ is the circle $x^2 + y^2 \le 1$.

**9.** Show that
$$\int_a^b \int_c^d f(x)g(y)\, dy\, dx$$
$$= \left[ \int_a^b f(x)\, dx \right]\left[ \int_c^d g(y)\, dy \right].$$

**C**

**10.** (Differentiating under the integral sign.) If $f(x, y)$ and $f_y(x, y)$ are continuous, then
$$\frac{d}{dy} \int_a^b f(x, y)\, dx = \int_a^b f_y(x, y)\, dx.$$

Prove this as follows:
**a)** Explain why
$$f(x, y) = f(x, y_0) + \int_{y_0}^y f_y(x, v)\, dv.$$
**b)** Use part $a$ in $\int_a^b f(x, y)\, dx$, change order of integration, and differentiate. [Assume the Theorem: If $g(x, y)$ is continuous in $x$ and $y$, then $\int_a^b g(x, y)\, dx$ is continuous in $y$.]

## 17.2

## MORE GENERAL REGIONS

Integrals $\iint_R f(x, y)\, dA$ where $R$ is not a rectangle can be evaluated as repeated integrals, but the limits of integration require special attention. A sketch of the region $R$ is almost essential.

**EXAMPLE 1**  Evaluate $\iint_R x\, dA$, where $R$ is the region between $y = 2x$ and $y = x^2$.

**SOLUTION**  First sketch the region $R$ (Fig. 1). In $R$, $x$ varies from 0 to 2. For each fixed $x$ in this range, imagine a vertical arrow crossing $R$, from bottom to top; it begins where $y = x^2$, and ends where $y = 2x$. Thus, for any given $x$ between 0 and 2, $y$ varies from the lower limit $x^2$ to the upper limit $2x$.

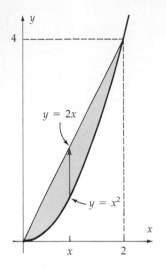

**FIGURE 1**
Region $R$ between $y = x^2$ and $y = 2x$.

Now visualize the volume under the graph of the integrand $f(x, y) = x$ (Fig. 2). The cross-sectional area $A(x)$ is the integral of $f(x, y)$ as $y$ varies from $x^2$ to $2x$:

$$A(x) = \int_{x^2}^{2x} x \, dy = xy \Big|_{y=x^2}^{y=2x} = x \cdot 2x - x \cdot x^2$$
$$= 2x^2 - x^3.$$

The first vertical cross section is at $x = 0$, and the last at $x = 2$, so integrate the cross-sectional areas from 0 to 2:

$$\iint_R x \, dA = \int_0^2 \int_{x^2}^{2x} x \, dy \, dx \tag{1}$$

$$= \int_0^2 (2x^2 - x^3) \, dx$$

$$= \frac{2}{3} x^3 - \frac{1}{4} x^4 \Big|_0^2 = \frac{16}{3} - \frac{16}{4} = \frac{4}{3}. \tag{2}$$

This gives the volume shown in Figure 2.

**NOTE** In the repeated integral (1), the limits of integration on the inner integral are *functions*, not constants! This is because $R$ is not a rectangle.

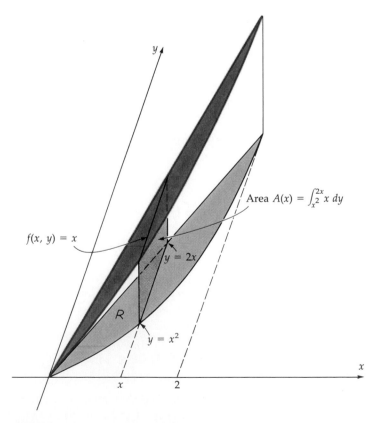

**FIGURE 2**
$\iint_R x \, dA = \int_0^2 A(x) \, dx = \int_0^2 \int_{x^2}^{2x} x \, dy \, dx.$

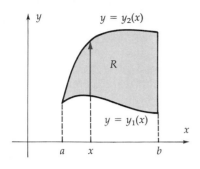

**FIGURE 3**
A "$y$-simple" set: $R = \{(x, y): a \le x \le b, y_1(x) \le y \le y_2(x)\}$.

The region $R$ in Example 1 can be described as the set (Fig. 1)

$$\{(x, y): 0 \le x \le 2, x^2 \le y \le 2x\}.$$

Many regions can be described in a similar way (Fig. 3),

$$R = \{(x, y): a \le x \le b, y_1(x) \le y \le y_2(x)\},$$

where $y_1$ and $y_2$ are continuous functions of $x$. We call such a region "$y$-simple"—crossing it parallel to the $y$-axis, you enter once [at the graph of $y = y_1(x)$] and leave once [at the graph of $y = y_2(x)$]. The graphs may or may not intersect at the ends of the region.

If $f(x, y)$ is a continuous function over such a region $R$, you can form the repeated integral

$$\int_a^b \int_{y_1(x)}^{y_2(x)} f(x, y) \, dy \, dx.$$

It can be proved that this equals the limit of the Riemann sums for $f$ over $R$, so the double integral equals the repeated integral:

$$\iint_R f(x, y) \, dA = \int_a^b \int_{y_1(x)}^{y_2(x)} f(x, y) \, dy \, dx. \tag{3}$$

Given a $y$-simple region $R$, how do you set up the repeated integral (3)? Sketch $R$. Then determine the smallest and largest values of $x$ in $R$; these give the outer limits $a$ and $b$ (Fig. 3). Then choose a "general" $x$ somewhere between the limits $a$ and $b$. The vertical line through $x$ intersects the region $R$ in a segment (Fig. 3); the $y$-coordinate at the lower end of this segment is $y_1(x)$, and at the upper end is $y_2(x)$. Integrate with respect to $y$ between these limits to compute the signed area of the cross section at $x$; then integrate with respect to $x$ between the limits $a$ and $b$ to sweep out the signed volume under the graph; this is the double integral.

Figure 4 shows a region $R$ that is not "$y$-simple"; vertical lines near the right edge intersect $R$ in more than two points. But each *horizontal* line intersects $R$ in at most two points, so it is "$x$-simple"; bounded by the graphs of two continuous functions of $y$, $R$ is the set

$$\{(x, y): c \le y \le d, x_1(y) \le x \le x_2(y)\}.$$

The integral over such a region is

$$\iint_R f(x, y) \, dA = \int_c^d \int_{x_1(y)}^{x_2(y)} f(x, y) \, dx \, dy. \tag{4}$$

If the region is both $x$-simple and $y$-simple, then the two repeated integrals in (3) and (4) are equal, since both give the same double integral. Such regions are called **simple**.

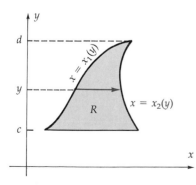

**FIGURE 4**
An "$x$-simple" set: $R = \{(x, y): c \le y \le d, x_1(y) \le x \le x_2(y)\}$.

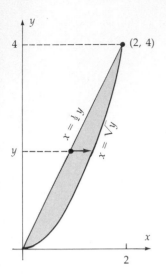

**FIGURE 5**

**EXAMPLE 1 (cont.)**   The region $R$ in Figure 1 is $x$-simple (crossing it parallel to the $x$-axis, you enter just once and leave just once, and the boundary curves are continuous), so the integral over $R$ can be written in the form (4). To do this, write each boundary curve as the graph of an equation $x = x(y)$, as in Figure 5. In $R$, $y$ varies from 0 to 4; for each such $y$, $x$ varies from $y/2$ to $\sqrt{y}$. This means that the cross-sectional area $A(y)$ in Figure 6 is the integral

$$\int_{y/2}^{\sqrt{y}} x \, dx = \frac{1}{2} x^2 \Big|_{y/2}^{\sqrt{y}} = \frac{1}{2}\left(y - \frac{y^2}{4}\right).$$

The region extends from $y = 0$ to $y = 4$, so integrate the cross-sectional areas from 0 to 4:

$$\iint_R x \, dA = \int_0^4 \int_{y/2}^{\sqrt{y}} x \, dx \, dy \qquad (5)$$

$$= \int_0^4 \frac{1}{2}\left(y - \frac{y^2}{4}\right) dy$$

$$= \frac{1}{2}\left[\frac{y^2}{2} - \frac{y^3}{12}\right]_0^4 = \frac{1}{2}\left[8 - \frac{64}{12}\right]$$

$$= \frac{4}{3},$$

the same answer as before.

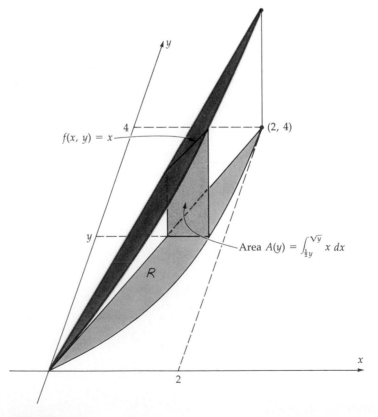

**FIGURE 6**
$\iint_R x \, dA = \int_0^4 \int_{y/2}^{\sqrt{y}} x \, dx \, dy.$

Notice how different the repeated integrals in (2) and (5) appear! There is no simple way to convert one into the other without referring to the sketch of the region of integration.

Sometimes changing the order of integration tames a recalcitrant integral. To make the change, *first* sketch the region of integration, then determine the new limits of integration.

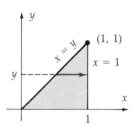

**FIGURE 7**

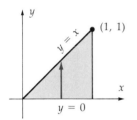

**FIGURE 8**

**EXAMPLE 2**  Evaluate $\int_0^1 \int_y^1 e^{-x^2} \, dx \, dy$.

*SOLUTION*  The given inner integral $\int_y^1 e^{-x^2} \, dx$ cannot be evaluated by the Fundamental Theorem of Calculus, since $e^{-x^2}$ has no elementary anti-derivative. Try changing the order of integration. The region of integration can be determined from the limits on the given integral: $y$ varies from 0 to 1, and for each $y$ in this range, $x$ varies from $x = y$ to $x = 1$; thus the region is bounded by $y = 0$, $x = y$, and $x = 1$ (Fig. 7). Now take it in the other order (Fig. 8); $x$ varies from 0 to 1, and for each $x$ in this range $y$ varies from $y = 0$ to $y = x$. So in the new order, the integral is

$$\int_0^1 \int_0^x e^{-x^2} \, dy \, dx = \int_0^1 y e^{-x^2} \Big|_{y=0}^{y=x} \, dx$$

$$= \int_0^1 x e^{-x^2} \, dx = -\tfrac{1}{2} e^{-x^2} \Big|_0^1$$

$$= \tfrac{1}{2} - \tfrac{1}{2} e^{-1}.$$

Notice again: The limits of integration in the new integral $\int_0^1 \int_0^x e^{-x^2} \, dy \, dx$ bear no simple relation to those in the original integral $\int_0^1 \int_y^1 e^{-x^2} \, dx \, dy$; they must be carefully worked out by sketching the region of integration.

## General Properties of Double Integrals

Several useful properties of double integrals are easily proved using Riemann sums:

$$\iint_R (f \pm g) = \iint_R f \pm \iint_R g. \tag{6}$$

$$\iint_R cf = c \iint_R f \qquad \text{if } c \text{ is constant.} \tag{7}$$

$$\iint_R f \leq \iint_R g \qquad \text{if } f \leq g \text{ throughout } R.$$

These look familiar, since they are also true for single integrals. The proofs are left as problems.

**EXAMPLE 3**  Compute the volume $V$ cut off from the paraboloid $z = x^2 + y^2$ by the plane $z = x$.

*SOLUTION*  Here, the region of integration is not given explicitly, but must be deduced from the given data. You have to visualize the graphs of the two functions: $z = x^2 + y^2$ is a paraboloid with its "nose" at the origin,

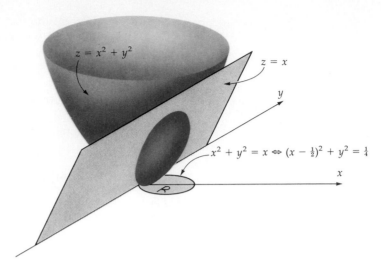

**FIGURE 9**
$z = x^2 + y^2$ cut by $z = x$.

and $z = x$ is a plane cutting the paraboloid right at the nose (Fig. 9). The two surfaces intersect over a curve $C$ in the $xy$-plane. To find the equation of this curve, simply equate the two functions:

$$x^2 + y^2 = x.$$

This gives

$$(x - \tfrac{1}{2})^2 + y^2 = \tfrac{1}{4},$$

a circle of radius $\frac{1}{2}$ with center $(\frac{1}{2}, 0)$. The volume in question lies over the region $R$ inside this circle; it is the volume under the graph of $z = x$, minus the volume under $z = x^2 + y^2$ (Fig. 9), so

$$V = \iint\limits_{R} x \, dA - \iint\limits_{R} (x^2 + y^2) \, dA$$

$$= \iint\limits_{R} (x - x^2 - y^2) \, dA \qquad \text{[using (6)]}$$

$$= \int_0^1 \int_{-\sqrt{x - x^2}}^{+\sqrt{x - x^2}} (x - x^2 - y^2) \, dy \, dx$$

$$= \int_0^1 (x - x^2) 2\sqrt{x - x^2} - \frac{2}{3} (x - x^2)^{3/2} \, dx$$

$$= \frac{4}{3} \int_0^1 (x - x^2)^{3/2} \, dx = \frac{\pi}{128}.$$

(See a Table of Integrals.)

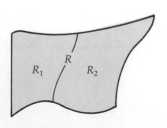

**FIGURE 10**
$\iint_R f = \iint_{R_1} f + \iint_{R_2} f.$ $R$ is cut into two pieces, $R_1$ and $R_2$.

**Additivity over regions** is illustrated in Figure 10. Suppose that region $R$ decomposes into two parts $R_1$ and $R_2$. Then the signed volume over the entire region $R$ equals the part over $R_1$ plus the part over $R_2$:

$$\iint\limits_{R} f = \iint\limits_{R_1} f + \iint\limits_{R_2} f.$$

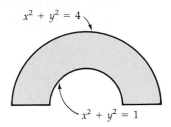

**FIGURE 11**
Annular sector.

**EXAMPLE 4** Evaluate $\iint_R y \, dA$, where $R$ is the annular sector (Fig. 11)

$$1 \leq x^2 + y^2 \leq 4, \qquad y \geq 0.$$

**SOLUTION** The easiest way to do this is by subtraction. Let $R_2$ be the semicircle of radius 2

$$R_2 = \{(x, y): x^2 + y^2 \leq 4, y \geq 0\}$$

and $R_1$ be the semicircle of radius 1

$$R_1 = \{(x, y): x^2 + y^2 \leq 1, y \geq 0\}.$$

Then $R_1$ together with the given sector $R$ forms $R_2$, so

$$\iint_{R_1} f \, dA + \iint_R f \, dA = \iint_{R_2} f \, dA$$

and

$$\iint_R f \, dA = \iint_{R_2} f \, dA - \iint_{R_1} f \, dA.$$

Now

$$\iint_{R_2} y \, dA = \int_{-2}^{2} \int_{0}^{\sqrt{4 - x^2}} y \, dy \, dx$$

$$= \int_{-2}^{2} \frac{4 - x^2}{2} \, dx = \left[ 2x - \frac{x^3}{6} \right]_{-2}^{2} = \frac{16}{3}.$$

Similarly, $\iint_{R_1} y \, dA = \frac{2}{3}$ (check this), so

$$\iint_R y \, dA = \iint_{R_2} y \, dA - \iint_{R_1} y \, dA = \frac{14}{3}.$$

## Still More General Regions

The region $R$ in Figure 12, bounded by two concentric circles, is called an **annulus**. It is neither $x$-simple nor $y$-simple; a vertical or horizontal line crossing the center of the annulus intersects it in four points. Still, a double integral

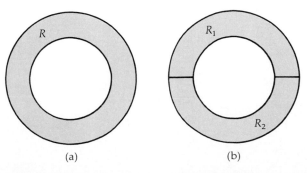

(a)                                  (b)

**FIGURE 12**
Annular ring $R$ cut into two $y$-simple regions $R_1$ and $R_2$.

over $R$ can be expressed as the *sum* of repeated integrals over $y$-simple regions $R_1$ and $R_2$ (Fig. 12b). For example,

$$\iint_R |y| \, dA = \iint_{R_1} |y| \, dA + \iint_{R_2} |y| \, dA$$

$$= \iint_{R_1} y \, dA + \iint_{R_2} (-y) \, dA$$

since $|y| = y$ in $R_1$ (where $y \geq 0$) and $|y| = -y$ in $R_2$. Each of these last integrals can be evaluated as in Example 4; the result is

$$\iint_R |y| \, dA = \frac{14}{3} + \frac{14}{3} = \frac{28}{3}.$$

## SUMMARY

$$\iint_R f(x, y) \, dA = \int_a^b \int_{y_1(x)}^{y_2(x)} f(x, y) \, dy \, dx \qquad \text{if } R \text{ is } y\text{-simple (see Fig. 3).}$$

$$\iint_R f(x, y) \, dA = \int_c^d \int_{x_1(y)}^{x_2(y)} f(x, y) \, dx \, dy \qquad \text{if } R \text{ is } x\text{-simple (see Fig. 4).}$$

$$\iint_R (f + g) \, dA = \int_R f \, dA + \int_R g \, dA$$

$$\int_R cf \, dA = c \int_R f \, dA$$

If $f \leq g$ in $R$ then $\iint_R f \, dA \leq \iint_R g \, dA$.

$$\iint_R f \, dA = \iint_{R_1} f \, dA + \iint_{R_2} f \, dA$$

if $R$ decomposes into $R_1$ and $R_2$ (see Fig. 10).

## PROBLEMS

### A

1. Evaluate the double integral.

   *a) $\displaystyle\int_0^2 \int_0^y y \, dx \, dy$

   *b) $\displaystyle\int_0^1 \int_x^{\sqrt{x}} (x + y) \, dy \, dx$

   c) $\displaystyle\int_2^4 \int_{1+y}^5 (x - y^2) \, dx \, dy$

   *d) $\displaystyle\int_0^3 \int_{-\sqrt{9-x^2}}^{\sqrt{9-x^2}} xy^2 \, dy \, dx$

   *e) $\displaystyle\int_0^2 \int_{y^2}^{2y} (4x - y) \, dx \, dy$

   f) $\displaystyle\int_0^1 \int_{-x-1}^{x-1} (x^2 + y^2) \, dy \, dx$

   g) $\displaystyle\int_0^1 \int_0^x e^{x-y} \, dy \, dx$

   h) $\displaystyle\int_0^1 \int_y^1 \frac{1}{1 + y^2} \, dx \, dy$

   i) $\displaystyle\int_0^1 \int_{x^2}^1 \frac{x}{\sqrt{1 + y}} \, dy \, dx$

*2. For each part of problem 1, sketch the region of integration. Rewrite the given integral with order of integration reversed, and evaluate it once more as a repeated integral, or perhaps a sum of repeated integrals. (Answers given for parts a and b; part f is long.)

3. Evaluate $\iint_R x\, dA$ as a repeated integral (4), where $R$ is the circle $\{P: |P| \leq 1\}$. Sketch the graph of the integrand over $R$, and explain your result geometrically.

4. Evaluate the double integral over the given region.

   *a) $\iint_R xy\, dA$, $R$ bounded by $x = 0$, $y = 2$, $y = \sqrt{x}$

   b) $\iint_R y \cos x\, dA$; $R$ bounded by $y = x$, $x = 0$, $y = \pi$

   *c) $\iint_R \dfrac{1}{1 + x^2}\, dA$;
   $R = \{(x, y): 0 \leq x \leq 2,\ 0 \leq y \leq x\}$

   d) $\iint_R x\, dA$; $R$ bounded by $x^2 + y^2 = 1$ and $x^2 + y^2 = 2$

5. Find the volume enclosed by the given surface(s).

   *a) The cylinder $x^2 + y^2 = 4$ and the planes $z = 0$ and $z = 3 - x$

   b) $z = x^2 + y^2$ and $z = 2 - x^2 - y^2$

   c) The three coordinate planes and the plane
   $$\frac{x}{a} + \frac{y}{b} + \frac{z}{c} = 1$$

   *d) $\dfrac{x^2}{a^2} + \dfrac{y^2}{b^2} + \dfrac{z^2}{c^2} = 1$ (Simplify your integral by substituting $y = bu$ and $x = av$.)

6. Evaluate the integral with order of integration reversed.

   *a) $\displaystyle\int_0^1 \int_{2y}^2 e^{x^2}\, dx\, dy$

   *b) $\displaystyle\int_0^4 \int_{\sqrt{x}}^2 \sin y^3\, dy\, dx$

   c) $\displaystyle\int_0^3 \int_{x^2}^9 x \cos y^2\, dy\, dx$

   d) $\displaystyle\int_0^1 \int_{x^{1/2}}^1 \frac{x}{\sqrt{1 + y}}\, dy\, dx$

7. $\iint_R 1\, dA$ gives the *area* of the region $R$. Why is that?

**B**

8. a) Show that $\int_0^x \int_0^u f(t)\, dt\, du = \int_0^x (x - t)f(t)\, dt$. (Change the order of integration.)

b) Evaluate each side of the equation in part a with $f(t) = e^t$.

c) Evaluate each side of part a with $f(t) = \cos(at)$.

*9. Over which of the following regions $R$ can one show by symmetry that
$$\iint_R \frac{xy}{1 + x^2 + y^2}\, dx\, dy = 0?$$ Give reasons.

   a) $x^2 + y^2 \leq a^2$

   b) $0 \leq x \leq a$, $0 \leq y \leq b$

   c) The rectangle $0 \leq x \leq a$, $0 \leq y \leq b$ together with $-a \leq x \leq 0$ and $-b \leq y \leq 0$

   d) $x^2 + xy + y^2 \leq a^2$

   e) $|x| \leq a$, $0 \leq y \leq b$

   f) $0 \leq x \leq a$, $|y| \leq b$

10. Over which of the regions in the previous problem can one show by symmetry that
$$\iint_R \frac{x + y}{1 + x^2 + y^2}\, dA = 0?$$ Give reasons.

11. Prove formula (6) in the text.

12. Prove formula (7) in the text.

13. $\int_0^\infty \int_0^\infty f(x, y)\, dx\, dy$ is defined as the limit
$$\lim_{N \to \infty} \int_0^N \left[ \lim_{M \to \infty} \int_0^M f(x, y)\, dx \right] dy.$$
Evaluate.

   *a) $\displaystyle\int_0^\infty \int_0^\infty \frac{1}{(1 + x + y)^3}\, dx\, dy$

   *b) $\displaystyle\int_0^\infty \int_0^\infty \frac{1}{(1 + x + y)^2}\, dx\, dy$

   c) $\displaystyle\int_0^\infty \int_0^\infty e^{-x-y}\, dx\, dy$

   d) $\displaystyle\int_0^\infty \int_0^\infty xye^{-x^2-y^2}\, dx\, dy$

**C**

14. a) Evaluate $\int_0^\infty \int_0^\infty e^{-xy} \sin x\, dx\, dy$ as a repeated integral.

   b) Assuming that the order of integration can be reversed in part a, deduce the value of $\int_0^\infty x^{-1} \sin x\, dx$.

15. For the given repeated improper integral, show that

   a) $\displaystyle\int_0^1 \int_0^\pi y^{-1} \cos x\, dx\, dy = 0$.

   b) $\displaystyle\int_0^\pi \int_0^1 y^{-1} \cos x\, dy\, dx$ is undefined.

(If the integrand is unbounded on the region $R$, integration becomes more complicated!)

**16.** For the given repeated improper integral, show that

**a)** $\displaystyle\int_0^\infty \int_{-\infty}^\infty \frac{y}{(1+y^2)^2}\, dy\, dx = 0.$

**b)** $\displaystyle\int_{-\infty}^\infty \int_0^\infty \frac{y}{(1+y^2)^2}\, dx\, dy$ is undefined.

(If $R$ is unbounded, integration becomes more complicated!)

## 17.3

## MASS. CENTER OF MASS. MOMENT OF INERTIA

### Density

Imagine a thin sheet or plate of material extending over a region $R$ in the $xy$-plane. It is called **homogeneous** if a small piece cut out of any part of $R$ is just like a congruent small piece cut out of any other part.

The average density of the plate is its mass per unit area,

$$\text{average density} = \frac{\text{mass}}{\text{area}};$$

so

$$\text{mass} = (\text{average density}) \times (\text{area}).$$

If the plate is homogeneous, all parts have the same density $\delta$; if not, then the density may vary from point to point. The distribution of mass in a non-homogeneous plate is described by a *density function* $\delta(x, y)$. The mass of a very small rectangle containing the point $(x, y)$, with area $dA$, is approximately equal to

$$\delta(x, y)\, dA,$$

the density at $(x, y)$ times the area of the small rectangle (Fig. 1). The mass of the entire plate is then the integral

$$\iint_R \delta(x, y)\, dA.$$

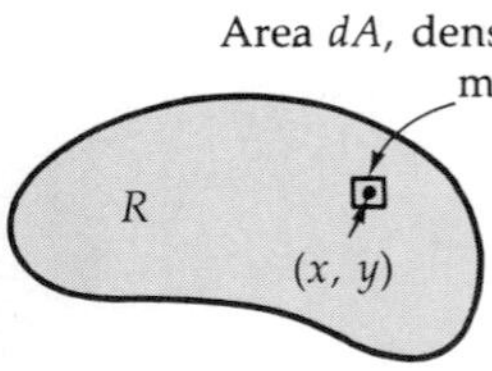

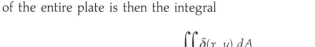

Area $dA$, density $\approx \delta(x, y)$, mass $\approx \delta(x, y)\, dA$

$R$

$(x, y)$

**FIGURE 1**
Mass $= \iint_R \delta(x, y)\, dA.$

**EXAMPLE 1**    A square plate extending over the region

$$R = \{(x, y): 0 \le x \le \pi,\, 0 \le y \le \pi\}$$

has density $\delta(x, y) = \sin x \sin y$. The plate is very light near the edges, for

$$\delta(x, y) = 0 \qquad \text{when } x = 0 \text{ or } \pi, \text{ and when } y = 0 \text{ or } \pi.$$

It is densest in the center, where

$$\delta\left(\frac{\pi}{2}, \frac{\pi}{2}\right) = \sin\frac{\pi}{2} \sin\frac{\pi}{2} = 1.$$

The mass of the plate is

$$\iint_R \sin x \sin y\, dA = \int_0^\pi \int_0^\pi \sin x \sin y\, dx\, dy$$
$$= 4$$

as you can easily check. The average density of the plate is

$$\frac{\text{mass}}{\text{area}} = \frac{4}{\text{area of } \pi \text{ by } \pi \text{ rectangle}} = \frac{4}{\pi^2} \approx 0.4.$$

The average lies between the minimum density, 0, and the maximum density, 1.

## Center of Mass

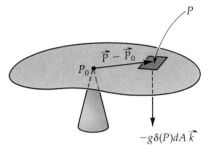

Imagine trying to balance a thin plate on a fulcrum located at $P_0$ (Fig. 2). We will compute the torque $\vec{T}$ exerted by the plate about the fulcrum $P_0$. The plate balances when $\vec{T} = \vec{0}$; this vector equation will determine the "balancing point" $P_0$, called the **center of mass**, or **center of gravity** of the plate.

To compute $\vec{T}$, consider at first a small piece of the plate near point $P$ with area $dA$ and mass $dm = \delta(P)\, dA$. Gravity pulls down on this small piece with force

$$-g\, dm\, \vec{k} = [-g\delta(P)\, dA]\, \vec{k}$$

where $\vec{k}$ is the unit vector along the z-axis. This force exerts a torque (twisting effect) about $P_0$, given by the cross product (Sec. 14.3)

$$(\vec{P} - \vec{P}_0) \times [-g\delta(P)\, dA]\vec{k}. \tag{1}$$

**FIGURE 2**

Gravity exerts torque on small piece at $P$, $d\vec{T} = [\vec{P} - \vec{P}_0] \times [-g\delta(P)\, dA\, \vec{k}]$.

Set $P = (x, y)$ and $P_0 = (\bar{x}, \bar{y})$; then the torque (1) is

$$\begin{vmatrix} \vec{i} & \vec{j} & \vec{k} \\ x - \bar{x} & y - \bar{y} & 0 \\ 0 & 0 & -g\delta(P)\, dA \end{vmatrix} = (y - \bar{y})g\delta(P)\, dA\, \vec{i} + (x - \bar{x})g\delta(P)\, dA\, \vec{j}$$

$$= [-(y - \bar{y})\vec{i} + (x - \bar{x})\vec{j}]g\delta(P)\, dA.$$

The torque over the entire region is the "sum," or integral, of the torques due to all these small pieces:

$$\vec{T} = \iint_R [-(y - \bar{y})\vec{i} + (x - \bar{x})\vec{j}]g\delta(x, y)\, dA.$$

This vector integral is evaluated by integrating each component separately:

$$\vec{T} = \left[\iint_R -(y - \bar{y})g\delta(x, y)\, dA\right]\vec{i} + \left[\iint_R (x - \bar{x})g\delta(x, y)\, dA\right]\vec{j}. \tag{2}$$

Now, suppose that the plate is *balanced* on the point $(\bar{x}, \bar{y})$; then each component of the torque (2) is zero:

$$\iint_R (y - \bar{y})g\delta(x, y)\, dA = 0, \tag{3}$$

$$\iint_R (x - \bar{x})g\delta(x, y)\, dA = 0. \tag{4}$$

The integral (3) can be written as a sum [since $\iint_R (f + g) = \iint_R f + \iint_R g$], and the constants $\bar{y}$ and $g$ can be factored out; then (3) gives

$$g \iint_R y\delta(x, y) \, dA = \bar{y}g \iint_R \delta(x, y) \, dA.$$

Solve this for $\bar{y}$:

$$\bar{y} = \frac{\iint_R y\delta(x, y) \, dA}{\iint_R \delta(x, y) \, dA}. \tag{5}$$

Similarly, from (4),

$$\bar{x} = \frac{\iint_R x\delta(x, y) \, dA}{\iint_R \delta(x, y) \, dA}. \tag{6}$$

These are the coordinates of the balancing point $(\bar{x}, \bar{y})$, the *center of mass* of the plate.

The numerator in (5) is called the **moment about the $x$-axis**, denoted $M_x$:

$$M_x = \iint_R y\delta(x, y) \, dA. \tag{7}$$

The factor $y$ in this integral gives the signed distance from the $x$-axis, the "lever arm" for the mass element $\delta(x, y) \, dA$. Thus, if you hinged the plate along the $x$-axis, it would exert a torque $M_x$ about that axis.

The numerator in (6) is the *moment about the $y$-axis*, denoted $M_y$:

$$M_y = \iint_R x\delta(x, y) \, dA. \tag{8}$$

The factor $x$ gives the lever arm from the $y$-axis.

In terms of mass and moment, the center of gravity $(\bar{x}, \bar{y})$ is

$$\bar{x} = \frac{M_y}{M}, \qquad \bar{y} = \frac{M_x}{M}. \tag{9}$$

Notice an odd feature here: $M_y$ appears in the formula for $\bar{x}$, and $M_x$ in the formula for $\bar{y}$, while a similar reversal takes place in formulas (7) and (8). From this point of view, the formulas (5) and (6) are less confusing; the formula for $\bar{x}$ uses $x$ in the numerator, and the one for $\bar{y}$ uses $y$. Think of $\bar{y}$ in (5) as an "average value" of $y$, while (6) gives an "average" of $x$.

---

**EXAMPLE 2**   A plate of constant density $\delta$ covers the semicircle

$$R = \{(x, y): x^2 + y^2 \leq r^2, y \geq 0\}.$$

Find its center of mass $(\bar{x}, \bar{y})$ (Fig. 3).

**SOLUTION**   Since the density is constant, the mass is

density times area $= \delta \cdot \frac{1}{2}\pi r^2$.

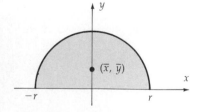

**FIGURE 3**
Center of mass of a semicircle.

The moment about the $x$-axis is

$$M_x = \iint_R y\delta \, dA$$

$$= \int_{-r}^{r} \int_{0}^{\sqrt{r^2 - x^2}} y\delta \, dy \, dx$$

$$= \int_{-r}^{r} \tfrac{1}{2}(r^2 - x^2)\delta \, dx$$

$$= \tfrac{2}{3}r^3\delta.$$

So the $y$-coordinate of the center of gravity $(\bar{x}, \bar{y})$ is

$$\bar{y} = \frac{M_x}{M} = \frac{\tfrac{2}{3}r^3\delta}{\delta \cdot \tfrac{1}{2}\pi r^2} = \frac{4}{3\pi} r \approx 0.4r.$$

As for $\bar{x}$, the symmetry of the figure suggests that $\bar{x} = 0$. Indeed, by symmetry,

$$M_y = \iint_R x\delta \, dA = 0$$

so $\bar{x} = M_y/M = 0$.

---

**The centroid** of a region $R$ is defined to be the center of mass of a plate covering $R$, with constant density $\delta = 1$:

$$\bar{x} = \frac{\iint_R x \, dA}{\iint_R dA}, \qquad \bar{y} = \frac{\iint_R y \, dA}{\iint_R dA}.$$

This would be the center of mass of a plate with *any* constant density $\delta$.

## Units

We do not think of the integrals for $M$, $M_x$, and $M_y$ as volumes. In the mass integral $M = \iint_R \delta \, dA$, the density $\delta$ has units of $\dfrac{\text{mass}}{\text{area}}$, while $dA$ has units of area, so $\delta \, dA$ has units of mass; so then does the integral $\iint_R \delta \, dA$.

In the moment integral $M_x = \iint_R y\delta \, dA$, $y$ is a length, and $\delta \, dA$ a mass, so $M_x$ has units of (length) $\times$ (mass). Then the ratio $\bar{y} = M_x/M$ has units of (length) $\times$ (mass) $\div$ (mass) = length; this is the appropriate unit for $\bar{y}$.

## Moment of Inertia

Imagine a point mass $m$ rotating about an axis (Fig. 4) at distance $r$ from the axis, with angular velocity $\omega$ radians/sec. It has speed $v = r\omega$, and kinetic energy $\tfrac{1}{2}mv^2 = \tfrac{1}{2}mr^2\omega^2$. The quantity $mr^2$ here is called the **moment of inertia** of the mass $m$ with respect to the given axis.

Now imagine a plate of density $\delta$ rotating about the $x$-axis (Fig. 5) with angular velocity $\omega$. Each small piece of mass $dm = \delta \, dA$ is at distance $|y|$ from the axis of rotation (Fig. 5), so the kinetic energy is $\tfrac{1}{2}dm|y|^2\omega^2 = \tfrac{1}{2}\omega^2 y^2\delta \, dA$. The kinetic energy of the entire plate is therefore the integral

$$\iint_R \tfrac{1}{2}y^2\omega^2\delta \, dA = \tfrac{1}{2}\omega^2 \iint_R y^2\delta dA. \tag{10}$$

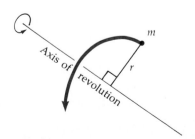

**FIGURE 4**
Point mass $m$ revolving about axis has moment of inertia $I = mr^2$.

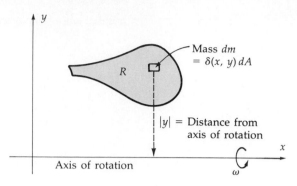

**FIGURE 5**
Kinetic energy of small piece is $\frac{1}{2}(dm)v^2 = \frac{1}{2}\omega^2 y^2 \delta(x, y)\,dA$.

This last integral is called the *moment of inertia* of the plate with respect to the $x$-axis, denoted $I_x$:

$$I_x = \iint_R y^2 \delta\,dA.$$

The kinetic energy due to rotation of the plate about the $x$-axis is thus, from (10),

$$\text{KE} = \tfrac{1}{2}I_x\omega^2.$$

The moment of inertia about the $y$-axis is

$$I_y = \iint_R x^2 \delta\,dA. \tag{11}$$

For rotation about the $z$-axis, the distance to the axis is $r = \sqrt{x^2 + y^2}$, so the relevant moment of inertia uses $r^2 = x^2 + y^2$. It is called the **polar moment** of inertia, denoted $I_0$:

$$I_0 = \iint_R (x^2 + y^2)\delta\,dA. \tag{12}$$

Comparing (12) with (10) and (11), you see that

$$I_0 = I_x + I_y.$$

---

**EXAMPLE 3** The unit square in Figure 6 has constant density $\delta$. Its moment of inertia about the $x$-axis is

$$I_x = \int_0^1 \int_0^1 \delta y^2\,dx\,dy = \frac{\delta}{3}.$$

By symmetry, $I_y$ is the same. Thus the polar moment of inertia is

$$I_0 = I_x + I_y = \frac{2\delta}{3}.$$

If it rotates about the $z$-axis with angular velocity $\omega$, the kinetic energy is $\frac{1}{2}I_0\omega^2 = (\delta/3)\omega^2$.

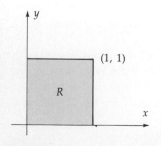

**FIGURE 6**
Unit square $0 \le x \le 1,\ 0 \le y \le 1$.

## SUMMARY

***Mass:*** $\quad M = \iint_R \delta(x, y) \, dA; \quad \delta(x, y)$ is density (mass per unit area).

***Center of Mass*** $(\bar{x}, \bar{y})$: $\quad \bar{x} = \dfrac{\iint_R x\delta \, dA}{\iint_R \delta \, dA}, \quad \bar{y} = \dfrac{\iint_R y\delta \, dA}{\iint_R \delta \, dA}$

***Centroid*** = center of mass with density $\delta \equiv 1$.

***Moments of Inertia:*** $\quad I_x = \iint_R y^2 \delta \, dA, \quad I_y = \iint_R x^2 \delta \, dA,$

$$I_0 = \iint_R (x^2 + y^2)\delta \, dA.$$

With angular velocity $\omega$ and moment of inertia $I$, kinetic energy $= \frac{1}{2}I\omega^2$.

## PROBLEMS

### A

**1.** Compute the mass and center of mass of a plate extending over the given region, with the given mass distribution $\delta(x, y)$.

*a)* $R = \{(x, y): x \geq 0, y \geq 0, x + y \leq 1\}$, $\delta$ is constant.

*b)* $R = \{(x, y): |x| \leq a, |y| \leq b\}$, $\delta(x, y) = e^{x+y}$. (Sec. 17.1, problem 9, simplifies this.)

*c)* $R = \{(x, y): |x| \leq 1, x^2 \leq y \leq 1\}$, $\delta$ constant.

*d)* $R = \{(x, y): |x| \leq 1, x^2 \leq y \leq 1\}$, $\delta(x, y) = y$. (Why is $\bar{y}$ in part d higher than $\bar{y}$ in part c?)

e) $R = \{(x, y): y \geq 0, 1 \leq x^2 + y^2 \leq 4\}$, $\delta \equiv 1$

f) $R = \{(x, y): x \geq 0, a^2 \leq x^2 + y^2 \leq b^2\}$, $\delta \equiv 1$

**2.** Show that the center of mass of the triangle with vertices $(0, 0)$, $(a, 0)$, and $(0, b)$ is the *barycenter* $B = (a/3, b/3)$, the average of the vertices:

$$\vec{B} = \tfrac{1}{3}[\langle 0, 0 \rangle + \langle a, 0 \rangle + \langle 0, b \rangle].$$

**3.** For the rectangle $0 \leq x \leq a, 0 \leq y \leq b$, assume density $\delta \equiv 1$, and compute
a) The center of mass.    b) $I_x$.
c) $I_y$.    d) $I_0$.

**4.** For the triangle with vertices $(0, 0)$, $(a, 0)$, and $(0, b)$, compute
a) $I_x$.    b) $I_y$.    *c)* $I_0$.

**5.** What are the appropriate units for the moment of inertia integrals?

**6.** *Charge density* describes the distribution of electrical charge, just as "mass density" describes the distribution of mass; if the charge density on a plate is $\gamma(x, y)$, then the total charge is

$$C = \iint_R \gamma(x, y) \, dA.$$

The function $\gamma$ is positive where the positively charged protons outnumber the negatively charged electrons. Compute the total charge, given
a) $R = \{(x, y): x^2 + y^2 \leq 1\}$, $\gamma(x, y) = x + y$.

*b)* $R = \{(x, y); |x| < 1 \text{ and } |y| < \pi\}$, $\gamma(x, y) = e^{x/2} \cos(y/2)$.

### B

**7.** Compute the mass and center of mass of a thin plate extending over the entire $xy$-plane with the given density.
*a)* $\delta(x, y) = e^{-|x| - |y|}$
b) $\delta(x, y) = e^{-|x| - |y|} \sin^2 x$

**8.** Show that the center of gravity of the triangle with vertices $(0, 0)$, $(a, 0)$, and $(b, c)$ is the barycenter $B = \left(\dfrac{a + b}{3}, \dfrac{c}{3}\right)$, the average of the three vertices. (See problem 2 for the special case of right triangles.)

**9.** Suppose that a horizontal plate with density function $\delta(x, y)$ is pivoted on a fulcrum at $\hat{P} = (\hat{x}, \hat{y})$, not necessarily the center of gravity. Show that gravity exerts a torque

$$\vec{T} = g(\hat{y}M - M_x)\vec{i} - (\hat{x}M - M_y)\vec{j}$$
$$= Mg[(\hat{y} - \bar{y})\vec{i} - (\hat{x} - \bar{x})\vec{j}]$$
$$= [\hat{P} - \bar{P}] \times [-Mg\vec{k}],$$

where $M$ is the mass, and $\bar{P} = (\bar{x}, \bar{y})$ is the center of mass. (Thus the plate acts like a point mass $M$ concentrated at the center of mass.)

**10. a)** Suppose that $m \leq \delta(x, y) \leq M$ throughout a plate extending over the region $R$. Prove that the average density of the plate lies between $m$ and $M$.

**b)** Let $\delta(x, y)$ be a continuous density over a region $R$. For any rectangle $A$ contained in $R$, let $\delta_A$ be the average density of $A$. Suppose that $A_n$ is a sequence of rectangles shrinking to a point $(x_0, y_0)$ in $R$. What is $\lim_{n \to \infty} \delta_{A_n}$? Explain why.

**11.** Suppose that $f$ and $g$ are functions of one variable with $g \leq f$, and $R$ is the region between the graphs of $f$ and $g$ for $a \leq x \leq b$. Assume a constant density $\delta$.

**a)** Write the double integral formulas for $M_x$ and $M_y$ with appropriate limits of integration.

**b)** Carry out the integration over $y$, and derive formulas involving only integrals with respect to $x$. (Have you seen these formulas before?)

**c)** Repeat parts a and b with a density function $\delta$ depending only on $x$.

# 17.4

## DOUBLE INTEGRALS IN POLAR COORDINATES

Many double integrals are easier to evaluate in polar coordinates, rather than rectangular. We'll see how this is done.

Consider the region $R$ in Figure 1, bounded by two radial lines $\theta = \alpha$ and $\theta = \beta$, and two polar graphs $r = r_1(\theta)$ and $r = r_2(\theta)$, with $0 \leq r_1 \leq r_2$. Suppose you are given a function $f$, continuous on $R$. You can form a "polar Riemann sum" for $f$ over $R$ as follows.

Divide the plane into a "polar grid" (Fig. 2), by radial lines and circles centered at the origin. Number the sections of the grid that lie inside the given region $R$, and in each section choose a point; let $P_k$ be the point chosen from the $k^{th}$ section, and $\Delta A_k$ be the area of that section. The sum

$$\sum f(P_k) \, \Delta A_k$$

is a polar Riemann sum for $f$ over $R$. If you repeat the process with finer and finer grids, and take the limit as the dimensions of the sections tend to zero, you get the *integral* of $f$ over $R$:

$$\iint_R f \, dA = \lim \sum f(P_k) \, \Delta A_k. \tag{1}$$

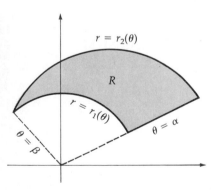

**FIGURE 1**
Region bounded by polar graphs and radial lines.

These Riemann sums lead to a formula for the integral as a *repeated integral in polar coordinates*. The point $P_k$ may be chosen anywhere in the $k^{th}$

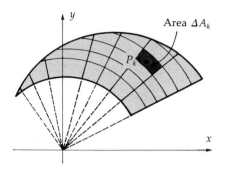

**FIGURE 2**
R covered by a "polar grid."

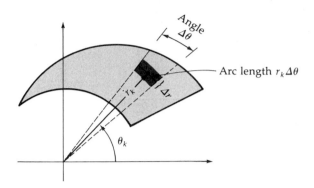

**FIGURE 3**
Area of sector $\Delta A_k = (r_k \, \Delta\theta)(\Delta r)$.

section—take it at the "center," along the ray bisecting the section and halfway between the inner and outer arcs (Fig. 3). Denote the polar coordinates of $P_k$ by $r_k$ and $\theta_k$:

$$P_k = (r_k \cos \theta_k, \, r_k \sin \theta_k).$$

Then the area of the section is exactly (see Fig. 3 and problem 4)

$$\Delta A_k = r_k \, \Delta r \, \Delta\theta. \tag{2}$$

The value of $f(x, y)$ at $P_k$ is $f(r_k \cos \theta_k, \, r_k \sin \theta_k)$, so the polar Riemann sum $\sum f(P_k) \, \Delta A_k$ in (1) is

$$\sum \overbrace{f(r_k \cos \theta_k, \, r_k \sin \theta_k)}^{\substack{\text{height to} \\ \text{graph of } f}} \underbrace{r_k \, \Delta r \, \Delta\theta}_{\substack{\text{area of small piece in} \\ \text{"rectangular grid" (Fig. 4)}}}. \tag{3}$$

Above equation (3) we indicate its derivation as the height to the graph of $f$ times an element of area in polar coordinates for the $xy$-plane. Below the

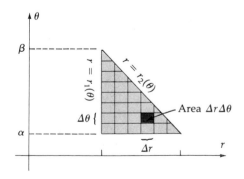

**FIGURE 4**
Region in $r\theta$ plane corresponding to region R in Figures 1–3. Polar grid in Figure 2 gives rectangular grid here.

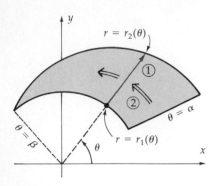

**FIGURE 5**
Setting up limits in polar coordinates: For given $\theta$, $r$ varies from $r_1(\theta)$ to $r_2(\theta)$, then $\theta$ varies from $\theta = \alpha$ to $\theta = \beta$.

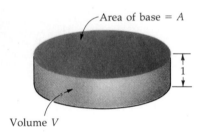

**FIGURE 6**
$V = A \cdot 1 = A.$

equation, we show another way to view it: as an ordinary "rectangular" Riemann sum in the $r\theta$ plane (Fig. 4), with $r$ and $\theta$ plotted on mutually perpendicular axes. There, the area of a small rectangle is simply $\Delta r \, \Delta\theta$. In the sum (3), this small area is multiplied by $f(r_k \cos \theta_k, r_k \sin \theta_k) \cdot r_k$. Hence, when $\Delta r$ and $\Delta\theta$ tend to zero, the limit of the sum (3) is the integral of the function $f(r \cos \theta, r \sin \theta) \cdot r$ over the region in Figure 4:

$$\int_\alpha^\beta \int_{r_1(\theta)}^{r_2(\theta)} f(r \cos \theta, r \sin \theta) r \, dr \, d\theta. \tag{4}$$

The limits of integration in (4) can be deduced from Figure 5: For each fixed $\theta$ between $\alpha$ and $\beta$, $r$ varies from $r_1(\theta)$ to $r_2(\theta)$; then the integrals along these rays are integrated from $\theta = \alpha$ to $\theta = \beta$.

Here is the repeated integral in polar coordinates that we were seeking. The expression $r \, dr \, d\theta$ in (4) corresponds to the area element $dA$ in (1); roughly speaking, $dA = r \, dr \, d\theta$, in analogy with the equation $\Delta A_k = r_k \, \Delta r \, \Delta\theta$ in (2).

When the integrand $f$ is identically equal to 1, then the integral is the volume of a region of uniform thickness 1, so it equals the area of the base (Fig. 6):

$$\iint_R 1 \, dA = \text{area of } R.$$

In polar coordinates, then, the area is

$$\int_\alpha^\beta \int_{r_1(\theta)}^{r_2(\theta)} r \, dr \, d\theta = \int_\alpha^\beta [\tfrac{1}{2} r_2(\theta)^2 - \tfrac{1}{2} r_1(\theta)^2] \, d\theta.$$

This agrees with the formula for area in polar coordinates derived in Section 12.6.

---

**EXAMPLE 1** Compute the moment $M_x = \iint y \, dA$ for the region $R$ outside the circle $r = 1$, inside the circle $r = 2 \cos \theta$, and above the $x$-axis.

**SOLUTION**

*First: Make a sketch to determine the limits of integration* (Fig. 7). The two curves intersect where $2 \cos \theta = 1$, so $\cos \theta = \tfrac{1}{2}$ and $\theta = \pi/3$. So in the given region, $\theta$ varies from 0 (on the $x$-axis) to $\pi/3$ (where the curves intersect). For such a $\theta$, imagine crossing the region on a ray from the origin; you enter where $r = 1$ and leave where $r = 2 \cos \theta$. So the limits of integration are

$$0 \leq \theta \leq \frac{\pi}{3}, \qquad 1 \leq r \leq 2 \cos \theta.$$

*Second: Convert the integrand and area element to polar coordinates.* For the given integral, simply use $y = r \sin \theta$ and $dA = r \, dr \, d\theta$. Hence

$$M_x = \iint_R y \, dA = \int_0^{\pi/3} \int_1^{2 \cos \theta} (r \sin \theta) r \, dr \, d\theta.$$

Notice that the outer limits 0 and $\pi/3$ are constants, while the inner limits may depend on the outer variable $\theta$.

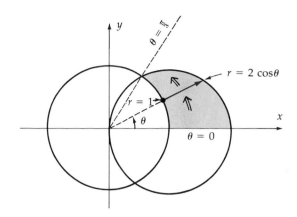

**FIGURE 7**
Region outside $r = 1$, inside $r = 2 \cos \theta$, above $x$-axis. For $0 \leq \theta \leq \pi/3$, $r$ varies from $r = 1$ to $r = 2 \cos \theta$.

*Third: Evaluate the repeated integral.* Combine the two factors of $r$ to find

$$M_x = \int_0^{\pi/3} \int_1^{2 \cos \theta} r^2 \sin \theta \, dr \, d\theta = \int_0^{\pi/3} \sin \theta \, \tfrac{1}{3} r^3 \Big|_1^{2 \cos \theta} \, d\theta.$$

$$= \tfrac{1}{3} \int_0^{\pi/3} \sin \theta \, [8 \cos^3 \theta - 1] \, d\theta = \tfrac{1}{3} [-2 \cos^4 \theta + \cos \theta] \Big|_0^{\pi/3}$$

$$= \tfrac{1}{3} [(-\tfrac{1}{8} + \tfrac{1}{2}) - (-2 + 1)] = \tfrac{11}{24}.$$

**EXAMPLE 2**   Evaluate

$$\iint_R e^{-x^2 - y^2} \, dA \tag{5}$$

where $R$ is the first quadrant.

**SOLUTION**   In rectangular coordinates, you would need an antiderivative for $e^{-x^2}$, which is not easily available. However, polar coordinates are natural for this integral, since both the integrand and the region of integration are easily expressed in terms of $r$ and $\theta$. The integrand is

$$e^{-x^2 - y^2} = e^{-(r \cos \theta)^2 - (r \sin \theta)^2} = e^{-r^2}.$$

In the given region of integration (Fig. 8), $\theta$ varies from $0$ to $\pi/2$; and for each $\theta$ in this range, $r$ varies from $0$ to $\infty$. Set $dA = r \, dr \, d\theta$, and you get the double integral (5) in polar coordinates:

$$\int_0^{\pi/2} \int_0^{\infty} e^{-r^2} r \, dr \, d\theta = \int_0^{\pi/2} \left[ -\frac{1}{2} e^{-r^2} \right]_0^{\infty} d\theta$$

$$= \int_0^{\pi/3} \left[ 0 - \left( -\frac{1}{2} \right) \right] d\theta \qquad (\lim_{r \to \infty} e^{-r^2} = 0)$$

$$= \frac{\pi}{2} \cdot \left( \frac{1}{2} \right) = \frac{\pi}{4}.$$

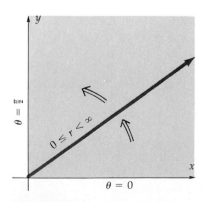

**FIGURE 8**
First quadrant: $0 \leq r < \infty$ and $0 \leq \theta \leq \pi/2$.

In this example, the factor $r$ in the expression $r \, dr \, d\theta$ was a great convenience; $re^{-r^2}$ has a simple antiderivative, but $e^{-r^2}$ does not.

*REMARK* An important quantity in many fields (statistics, for one) is the improper integral

$$I = \int_0^\infty e^{-x^2} \, dx. \tag{6}$$

This can be evaluated by an ingenious method leading to the integral in Example 2. First, replace $x$ by $y$ in (6) to obtain

$$I = \int_0^\infty e^{-y^2} \, dy.$$

Then combine this with (6) to represent $I^2$ as a repeated integral:

$$I^2 = I \cdot \int_0^\infty e^{-x^2} \, dx = \int_0^\infty I e^{-x^2} \, dx$$

$$= \int_0^\infty \left( \int_0^\infty e^{-y^2} \, dy \right) e^{-x^2} \, dx$$

$$= \int_0^\infty \left( \int_0^\infty e^{-y^2} e^{-x^2} \, dy \right) dx$$

$$= \int_0^\infty \int_0^\infty e^{-x^2 - y^2} \, dy \, dx$$

$$= \frac{\pi}{4}$$

as found in Example 2. Since $I^2 = \pi/4$, then

$$I = \int_0^\infty e^{-x^2} \, dx = \frac{\sqrt{\pi}}{2}. \tag{7}$$

This calculation ignored the difficulties inherent in improper multiple integrals; a detailed justification is outlined in problem 7. Essentially, when the integrand is positive and the improper integrals are finite, no problems arise.

## SUMMARY

$$\iint_R f(x, y) \, dA = \int_\alpha^\beta \int_{r_1(\theta)}^{r_2(\theta)} f(r \cos \theta, r \sin \theta) \, r \, dr \, d\theta \qquad \text{(see Fig. 5)}$$

## PROBLEMS

### A

1. Evaluate the repeated integral, and sketch the region of integration $R$ using polar coordinates.

*a) $\displaystyle\int_0^{2\pi} \int_0^1 (r)(r \, dr \, d\theta)$

b) $\displaystyle\int_{-\pi}^\pi \int_0^1 (r)(r \, dr \, d\theta)$

*c) $\displaystyle\int_0^{2\pi} \int_1^2 r \, dr \, d\theta$

d) $\displaystyle\int_{-\pi/2}^{\pi/2} \int_0^{\cos \theta} r \, dr \, d\theta$

**\*e)**  $\int_{-\pi/2}^{\pi/2} \int_0^1 r\cos\theta\, r\, dr\, d\theta$

**\*f)**  $\int_0^\pi \int_0^{\sin\theta} r\sin\theta\, r\, dr\, d\theta$

**g)**  $\int_{-\pi}^{\pi} \int_0^{1+\cos\theta} r\sin\theta\, r\, dr\, d\theta$

**\*h)**  $\int_0^{\pi/2} \int_0^\infty \frac{1}{1+r^2} r\, dr\, d\theta$

**2.**  Sketch the region of integration, express the integral in polar coordinates, and evaluate.

**\*a)**  $\iint_R \frac{1}{x^2+y^2}\, dA,$
$R = \{x, y): 1 \le x^2 + y^2 \le 2\}$

**b)**  $\iint_R x\, dA,$
$R = \{(x, y): 1 \le x^2 + y^2 \le 2, x \ge 0\}$

**\*c)**  $\int_{-a}^{a} \int_0^{\sqrt{a^2-x^2}} y\, dy\, dx$

**\*d)**  $\int_0^2 \int_0^{\sqrt{2x-x^2}} \sqrt{x^2+y^2}\, dy\, dx$

**e)**  $\int_0^2 \int_0^{\sqrt{4-x^2}} \sin(x^2+y^2)\, dy\, dx$

**f)**  $\int_0^\infty \int_0^\infty \frac{dx\, dy}{1+x^2+y^2}$

**3.**  Compute the area of the part of the region inside $r = 2\cos\theta$ which lies outside the circle $r = 1$.

**4.**  Find the center of mass of the given region. (This is fairly long.)
**\*a)**  Inside the curve $r = 3 + \sin\theta$.
**b)**  Inside the cardioid $r = 1 + \cos\theta$.
**c)**  Inside $r = 1 + \cos\theta$ and outside $r = 1$.
**\*d)**  Inside the circle $r = 2a\sin\theta$ and outside the circle $r = a$.
**e)**  Inside the circle $r = a$ and outside the circle $r = a\sin\theta$.
**\*f)**  The part of the annulus $a \le r \le b$ in the first quadrant.

**5.**  Find the volume of the solid bounded by the given surfaces.
**\*a)**  Above $z = 0$, below $z = x^2 + y^2$, and inside $x^2 + y^2 = 2$.
**b)**  Above $z = 0$, below $z = xy$, inside $x^2 + y^2 = 1$.
**\*c)**  Inside the sphere $x^2 + y^2 + z^2 = 4$ and outside the cylinder $x^2 + y^2 = 1$.
**d)**  Inside the sphere $x^2 + y^2 + z^2 = 4$ and the cylinder $x^2 + y^2 - 4x = 0$.

**6.**  Compute the polar moment of inertia $I_0 = \iint_R (x^2 + y^2)\, dA$ for
**\*a)**  The circle of radius $a$ about the origin.
**b)**  The annulus between the circles of radii $a$ and $b$, $b > a$.
**\*c)**  The region inside $r = a(1 + \cos\theta)$ and outside $r = a$.
**d)**  The circle bounded by $r = 2a\cos\theta$.

**7.**  Sketch the region of integration, convert the integral to polar coordinates, and sketch the corresponding region in the $r\theta$-plane.
**a)**  $\iint_R f\, dA$, where $R$ is the unit circle $x^2 + y^2 \le 1$.
**\*b)**  $\int_0^1 \int_0^x f(x, y)\, dy\, dx$.

**B**

**8.  a)**  Show that the centroid of the circular sector $R$ defined by $0 \le r \le a$ and $-b \le \theta \le b$ is $\left(\dfrac{2a\sin b}{3b}, 0\right)$. (Assume that $0 \le b \le \pi$.)
**b)**  Check this formula when $R$ is a circle.
**c)**  Check the formula in the limit as $b \to 0$. (Note that as $b \to 0$, the region $R$ resembles a long thin triangle. What is the centroid of a triangle?)

**9.**  Prove that the area of the annular sector in Figure 9 is $\bar{r}\, \Delta\theta\, \Delta r$, where
$$\Delta\theta = \theta_2 - \theta_1, \qquad \Delta r = r_2 - r_1,$$
$$\text{and} \qquad \bar{r} = \tfrac{1}{2}(r_1 + r_2).$$

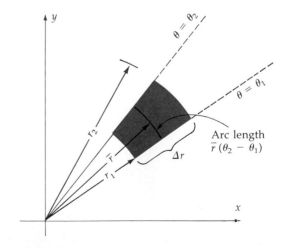

**FIGURE 9**

Area of annular sector is $\bar{r}(\theta_2 - \theta_1)\, \Delta r, \ \bar{r} = \dfrac{r_1 + r_2}{2}.$

**10.** Evaluate $\int_{-\infty}^{\infty} e^{-x^2}\, dx$, using $\int_0^{\infty} e^{-x^2}\, dx = \sqrt{\pi}/2$.

**11.** Evaluate $\int_0^{\infty} e^{-ax^2}\, dx$. (There are at least two ways.)

**12.** This problem justifies the proof of (7). Let $I_b = \int_0^b e^{-x^2}\, dx$.

**a)** Show that $I_b^2 = \int_0^b \int_0^b e^{-x^2-y^2}\, dx\, dy$.

**b)** Show that
$$\int_0^{\pi/2} \int_0^b e^{-r^2} r\, dr\, d\theta \leq \int_0^b \int_0^b e^{-x^2-y^2}\, dx\, dy$$
$$\leq \int_0^{\pi/2} \int_0^{\sqrt{2}b} e^{-r^2} r\, dr\, d\theta.$$

(Sketch the region of integration for each of the three terms.)

**c)** Deduce from the Trapping Theorem that $\lim_{b\to\infty} I_b^2 = \pi/4$, so $\lim_{b\to\infty} I_b = \sqrt{\pi}/2$.

## 17.5

### BIVARIATE PROBABILITY DENSITIES (optional)

A probability density $f(x)$ on the line describes the probabilities of various numerical outcomes of an experiment (Sec. 10.4); the probability of an outcome $x$ in the interval $a \leq x < b$ is given by an integral,

$$P[a \leq x < b] = \int_a^b f(x)\, dx.$$

The density $f$ must satisfy

$$f(x) \geq 0, \quad \text{and} \quad \int_{-\infty}^{\infty} f(x)\, dx = 1. \tag{1}$$

A typical example: $x$ is the number of hours that a certain type of light bulb will burn before failing and, for $a$ and $b \geq 0$,

$$P[a \leq x < b] = \int_a^b \mu^{-1} e^{-x/\mu}\, dx.$$

The density is $f(x) = \mu^{-1} e^{-x/\mu}$ for $x \geq 0$, and 0 for $x < 0$. The constant $\mu$ is the average lifetime of this type of bulb (Sec. 10.4).

A **bivariate** probability density gives probabilities for experiments with *two* numerical outcomes, $x$ and $y$. For example, if two bulbs are lit, then the first fails in $x$ hours, and the second in $y$ hours. Now we can ask for the probabilities of various events involving the two bulbs together, for example:

The first bulb burns for less than $a$ hours, and the second for less than $b$ hours; the probability of this is denoted $P[x < a \text{ and } y < b]$.

A second example:

The first bulb is burned until it fails (at time $x$), then replaced by the second, which is burned until *it* fails (after an additional time $y$). The probability that the two together will fail by time $b$ is denoted $P[x + y < b]$.

Probabilities for events such as these are given as integrals of a **bivariate probability density** $h(x, y)$. The probability in the first example above would be given by an integral

$$P[x < a \text{ and } y < b] = \int_{-\infty}^a \int_{-\infty}^b h(x, y)\, dy\, dx \qquad \text{(Fig. 1)}.$$

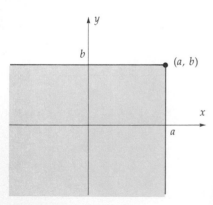

**FIGURE 1**
Region of integration for $P[x < a \text{ and } y < b]$.

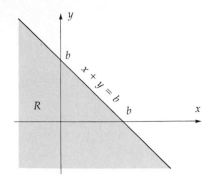

**FIGURE 2**
$R$ is the region of integration for $P[x + y < b]$.

In the second example, the probability would be given by

$$P[x + y < b] = \iint\limits_{R} h(x, y)\, dA, \qquad \text{(Fig. 2)}$$

where $R$ is the set in the plane defined by the inequality $x + y < b$. In general, for any region $R$ in the plane, the probability that $(x, y)$ is in $R$ is given by the integral of the density $h$ over $R$:

$$P[(x, y) \text{ is in } R] = \iint\limits_{R} h(x, y)\, dA.$$

The bivariate probability density $h(x, y)$ must satisfy two conditions analogous to (1):

$$h(x, y) \geq 0 \quad \text{and} \quad \int_{-\infty}^{\infty} \int_{-\infty}^{\infty} h(x, y)\, dx\, dy = 1. \qquad \textbf{(2)}$$

For the case of the two bulbs, we can derive a formula for the density $h(x, y)$. We assume that their lifetimes are *independent* (in the sense of probability). This means that the probability of the two outcomes $a \leq x < b$ and $c \leq y < d$ is the product of the probabilities of the individual outcomes:

$$P[a \leq x < b \text{ and } c \leq y < d] = P[a \leq x < b] \cdot P[c \leq y < d]. \qquad \textbf{(3)}$$

Denote by $f(x)$ the probability density function for the life expectancy of the individual bulbs. Then the product in (3) is

$$\begin{aligned} P[a \leq x < b] \cdot P[c \leq y < d] &= \int_{a}^{b} f(x)\, dx \int_{c}^{d} f(y)\, dy \\ &= \int_{c}^{d} \left[ \int_{a}^{b} f(x)\, dx \right] f(y)\, dy \\ &= \int_{c}^{d} \left[ \int_{a}^{b} f(x)f(y)\, dx \right] dy \\ &= \int_{c}^{d} \int_{a}^{b} f(x)f(y)\, dx\, dy. \end{aligned}$$

Thus (3) gives the integral

$$P[a \leq x < b \text{ and } c \leq y < d] = \int_{c}^{d} \int_{a}^{b} f(x)f(y)\, dx\, dy.$$

In other words, for the rectangle $R = \{(x, y): a \leq x < b \text{ and } c \leq y < d\}$,

$$P[(x, y) \text{ is in } R] = \iint\limits_{R} f(x)f(y)\, dA.$$

This same formula is valid for *any* region $R$ in the plane; for you can approximate $R$ by a collection of rectangles, and add up the results for the rectangles.

More generally, if $f(x)$ and $g(y)$ are probability densities for any two independent experiments, then the bivariate density for the two together is just the product $f(x)g(y)$:

$$P[(x, y) \text{ is in } R] = \iint\limits_{R} f(x)g(y) \, dA.$$

---

**EXAMPLE 1**  Two bulbs are to be burned, the second replacing the first when it fails. For each bulb, the probability of failure by time $b$ is $\int_0^b \mu^{-1}e^{-x/\mu} \, dx$. Compute the probability that *both* fail by time $b$.

**SOLUTION**  The density for the first bulb is

$$f(x) = \begin{cases} \mu^{-1}e^{-x/\mu}, & x > 0 \\ 0, & x < 0, \end{cases}$$

and for the second it is $f(y)$, with the same $f$. The probability that both fail by time $b$ is

$$P[x + y < b] = \iint\limits_{R} f(x)f(y) \, dA$$

with the $R$ in Figure 2. But $f(x) = 0$ when $x < 0$, and $f(y) = 0$ when $y < 0$, so we can replace $R$ with the region $\bar{R}$ in Figure 3. There, $f(x) = \mu^{-1}e^{-x/\mu}$ and $f(y) = \mu^{-1}e^{-y/\mu}$, so

$$P[x + y < b] = \iint\limits_{\bar{R}} \mu^{-1}e^{-x/\mu}\mu^{-1}e^{-y/\mu} \, dA$$

$$= \mu^{-2} \int_0^b \int_0^{b-x} e^{-x/\mu}e^{-y/\mu} \, dy \, dx$$

$$= \int_0^b \frac{e^{-x/\mu} - e^{-b/\mu}}{\mu} \, dx = 1 - \left(\frac{\mu + b}{\mu}\right)e^{-b/\mu}.$$

---

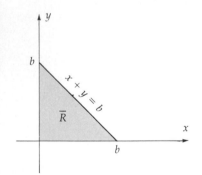

**FIGURE 3**
Alternate region of integration for $P[x + y < b]$, since the integrand is zero for $x < 0$ or $y < 0$.

**EXAMPLE 2**  The "bulb-with-spare" experiment in Example 1 can be viewed as an experiment with a single outcome: the time elapsed from turning on the first bulb to the failure of the second. What is the probability density for that?

**SOLUTION**  Denote by $g(x)$ the desired probability density; so the probability of failure of both bulbs by time $b$ is $\int_{-\infty}^b g(x) \, dx$. On the other hand, this is the probability computed in Example 1, so for $b > 0$,

$$\int_{-\infty}^b g(x) \, dx = 1 - \left(\frac{\mu + b}{\mu}\right)e^{-b/\mu}.$$

Hence for $b > 0$,

$$g(b) = \frac{d}{db} \int_{-\infty}^b g(x) \, dx = \frac{d}{db}\left[1 - \left(\frac{\mu + b}{\mu}\right)e^{-b/\mu}\right] = \mu^{-2}be^{-b/\mu}.$$

For $b < 0$, $g(b) = 0$; so

$$g(x) = \begin{cases} \mu^{-2}xe^{-x/\mu}, & x > 0 \\ 0, & x < 0. \end{cases}$$

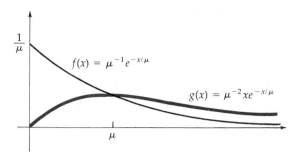

**FIGURE 4**
Density $g$ has a higher mean than $f$.

Figure 4 contrasts this with the density for a single bulb; you see clearly the greater life expectancy of the "bulb-with-spare." The precise calculation of the average life expectancy of this combination is left as a problem.

## PROBLEMS

### A

**1.** The probability density for a certain dart thrower aiming at the origin is $h(x, y) = \dfrac{1}{2\pi} e^{-(x^2 + y^2)/2}$.

Compute the probability that the dart lands
**a)** In the first quadrant.
**\*b)** In a semicircle of radius 1.
**\*c)** In the circle of radius 0.1.
**d)** In the sector $a \le r < b$, $\alpha \le \theta < \beta$, in polar coordinates.

**2.** Suppose that $f(x)$ and $g(y)$ both satisfy the conditions (1). Does the function $h(x, y) = f(x)g(y)$ necessarily satisfy (2)?

**3.** Suppose that $f(x) = \mu^{-1}e^{-x/\mu}$, $x \ge 0$, and $f(x) = 0$ for $x < 0$. Define the density $h(x, y) = f(x)f(y)$, as in Example 1. Compute.
**\*a)** $P[0 \le x < a$ and $0 \le y < b]$
**b)** $P[0 \le x < b$ or $0 \le y < b]$ (Interpretation: The probability that at least one of the bulbs fails by time $b$.)
**\*c)** $P[x \ge b$ and $y \ge b]$ (Interpret this.)
**d)** $P[x \ge b$ or $y \ge b]$ (Interpret this.)

**4.** The probability density for life expectancy of the "bulb-with-spare" combination in Example 2 is

$$g(x) = \begin{cases} \mu^{-2}xe^{-x/\mu}, & x > 0 \\ 0, & x < 0. \end{cases}$$

Compute the expected life of this combination, $\int_{-\infty}^{\infty} xg(x)\, dx$, and compare it to the expected life of the single bulb,

$$\int_{-\infty}^{\infty} xf(x)\, dx = \int_{0}^{\infty} x \cdot \mu^{-1}e^{-x/\mu}\, dx.$$

**5.** Suppose that two independent experiments are done, each described by the standard normal probability density function $f(x) = (1/\sqrt{2\pi})e^{-x^2/2}$.
**a)** Show that the bivariate density for these two experiments is the same as that for the dart "experiment" in problem 1.
**b)** Compute $P[x^2 + y^2 < R^2]$.
**c)** Compute $P[x \le 0]$.

### B

**6.** Continuing the previous problem,
**a)** Compute $P[x + y \le 0]$. (Use the symmetry of $h(x, y)$ to show that this is the same as $P[x \le 0]$.)
**b)** Show that $P[x + y < b] = P[x < b/\sqrt{2}]$.
**c)** Find a function $g(r)$ such that $g(r) = 0$ for $r < 0$, and

$$P[x^2 + y^2 < R^2] = \int_{0}^{R} g(r)\, dr.$$

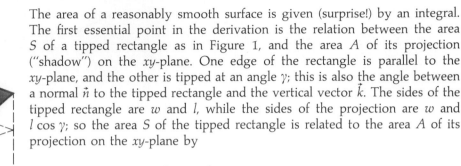

# 17.6

## SURFACE AREA

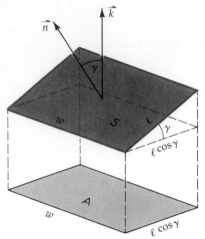

**FIGURE 1**
Tipped rectangle and its projection on $xy$ plane.

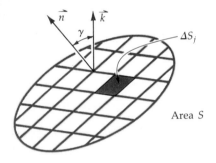

Area $S$

**FIGURE 2**
Plane surface tipped at angle $\gamma$.
$\Delta S_j = (\sec \gamma) \Delta A_j;\ S = (\sec \gamma)A.$

The area of a reasonably smooth surface is given (surprise!) by an integral. The first essential point in the derivation is the relation between the area $S$ of a tipped rectangle as in Figure 1, and the area $A$ of its projection ("shadow") on the $xy$-plane. One edge of the rectangle is parallel to the $xy$-plane, and the other is tipped at an angle $\gamma$; this is also the angle between a normal $\vec{n}$ to the tipped rectangle and the vertical vector $\vec{k}$. The sides of the tipped rectangle are $w$ and $l$, while the sides of the projection are $w$ and $l \cos \gamma$; so the area $S$ of the tipped rectangle is related to the area $A$ of its projection on the $xy$-plane by

$$\frac{S}{A} = \frac{lw}{lw \cos \gamma} = \frac{1}{\cos \gamma} = \sec \gamma$$

or

$$S = (\sec \gamma)A. \tag{1}$$

Now suppose that $S$ is the area of *any* plane region tipped at angle $\gamma$, and $A$ the area of its projection on the $xy$-plane (Fig. 2). Then the relation (1) continues to hold, as we show by a limiting argument. Cover the tipped plane with a grid of lines, those in one direction parallel to the $xy$-plane, the others perpendicular to those, and tipped at the angle $\gamma$. The grid forms many small rectangles; denote by $\Delta S_j$ the area of the $j^{th}$ one. Then the area $S$ is approximated by the sum $\sum \Delta S_j$, including in the sum all those areas completely contained in $S$. Likewise $A$, the projection of $S$, is approximated by the sum $\sum \Delta A_j$ of the projections of the $\Delta S_j$. For each $j$, $\Delta S_j = (\sec \gamma) \Delta A_j$, so

$$\sum \Delta S_j = \sum (\sec \gamma) \Delta A_j = \sec \gamma \sum \Delta A_j.$$

Now refine the grid, and take the limit of these approximating sums, to find

$$S = (\sec \gamma)A. \tag{2}$$

This is valid for the area $S$ of any plane region, tipped at angle $\gamma$.

Finally, consider a curved surface lying over a region $R$ in the plane (Fig. 3). To approximate its surface area $S$, first approximate $R$ by a collection of nonoverlapping rectangles $R_j$, of area $\Delta A_j$. Denote by $P_j$ the point at one corner of $R_j$. Near $P_j$ you can approximate the surface by its tangent plane at $(P_j, f(P_j))$. Denote by $\gamma(P_j)$ the angle between the normal to this plane and the vertical vector $\vec{k}$, and by $\Delta S_j$ the area of the part of this tangent plane lying over $R_j$. By (2),

$$\Delta S_j = \sec \gamma(P_j) \Delta A_j.$$

Now, $\Delta S_j$ is a reasonable approximation to the area of the part of the *curved* surface lying over $R_j$, at least when the rectangles $R_j$ are very small. So we approximate the area of the curved surface by the sum $\sum \Delta S_j$, and find the

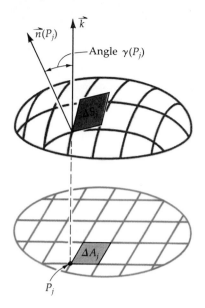

**FIGURE 3**
Small piece of tangent plane
approximating part of surface.
$\Delta S_j = \sec \gamma(P_j) \, \Delta A_j$.

exact area $S$ as the *limit* of these sums:

$$S = \lim \sum \Delta S_j = \lim \sum \sec(\gamma(P_j)) \, \Delta A_j.$$

This last is a Riemann sum for the integral $\iint_R \sec \gamma(x, y) \, dA$, where $\gamma(x, y)$ is the angle between the vertical vector $\vec{k}$ and the upward surface normal $\vec{n}$ at the point on the surface above $(x, y)$ (Fig. 3). This is the desired integral for the surface area:

$$S = \iint_R \sec \gamma \, dA. \tag{3}$$

It is valid when the surface is smooth, in the sense that it can be sufficiently well approximated near each point by its tangent plane, as in Section 15.4.

In the area integral (3), think of $dA$ as the area of a small piece in the $xy$-plane, and $\gamma$ as the "tipping angle" of the corresponding small piece of the curved surface; then $\sec \gamma \, dA$ gives the *area* of the small piece of surface, and the integral (3) gives the entire area.

The evaluation of this integral depends on how the surface is represented: as a graph, or as a level surface.

## Area of a Graph

Suppose that $f$ is a $C^1$ function on a region $R$ in the plane. The graph of $f$ is also the graph of the equation

$$F(x, y, z) = z - f(x, y) = 0.$$

Thus an upward normal to the graph is $\vec{n} = \nabla F = \vec{k} - f_x \vec{i} - f_y \vec{j}$; this points up, since the coefficient of $\vec{k}$ is positive. The cosine of the angle $\gamma$ between $\vec{n}$ and the vertical vector $\vec{k}$ is

$$\cos \gamma = \frac{\vec{n} \cdot \vec{k}}{|\vec{n} \cdot \vec{k}|} = \frac{1}{\sqrt{1 + f_x^2 + f_y^2}}.$$

Hence $\sec \gamma = \dfrac{1}{\cos \gamma} = \sqrt{1 + f_x^2 + f_y^2}$, and the integral (3) gives

$$S = \iint_R \sqrt{1 + f_x^2 + f_y^2} \, dA = \iint_R \sqrt{1 + |\Delta f|^2} \, dA. \tag{4}$$

This resembles the integral for the *arc length* on the graph of a function of *one* variable, $s = \int \sqrt{1 + f'(x)^2} \, dx$.

---

**EXAMPLE 1**  Compute the area of the hemisphere $z = \sqrt{a^2 - x^2 - y^2}$.

**SOLUTION**  The hemisphere is the graph of the function $f(x, y) = \sqrt{a^2 - x^2 - y^2}$ over the disk in Figure 4,

$$R. = \{(x, y): x^2 + y^2 \le a^2\}. \tag{5}$$

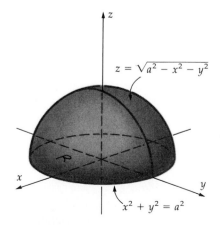

**FIGURE 4**

The surface area integrand in (4) uses

$$1 + |\nabla f|^2 = 1 + f_x^2 + f_y^2$$

$$= 1 + \left[\frac{-x}{\sqrt{a^2 - x^2 - y^2}}\right]^2 + \left[\frac{-y}{\sqrt{a^2 - x^2 - y^2}}\right]^2$$

$$= 1 + \frac{x^2 + y^2}{a^2 - x^2 - y^2} = \frac{a^2}{a^2 - x^2 - y^2}.$$

So the area is

$$\iint_R \sqrt{1 + |\nabla f|^2} \, dA = \iint_R \frac{a}{\sqrt{a^2 - x^2 - y^2}} \, dA,$$

with $R$ the disk (5). This calls for polar coordinates, with $x^2 + y^2 = r^2$ and $dA = r \, dr \, d\theta$; the area is

$$\int_0^{2\pi} \int_0^a \frac{a}{\sqrt{a^2 - r^2}} r \, dr \, d\theta = \int_0^{2\pi} a \left[-\sqrt{a^2 - r^2}\right]_0^a \, d\theta$$

$$= \int_0^{2\pi} a^2 \, d\theta = 2\pi a^2.$$

This is of course the expected result for a hemisphere; the area of a full sphere of radius $a$ is $4\pi a^2$.

---

**The area of a level surface** $\{F = c\}$ of a $C^1$ function $F$ can be computed using (3), as follows. Suppose that the surface projects onto a region $R_{xy}$ in the $xy$-plane, with each point of the surface in question corresponding to just one point of the plane. A normal to the surface is

$$\vec{n} = \nabla F = F_x \vec{i} + F_y \vec{j} + F_z \vec{k}.$$

If $F_z > 0$ this is an upward normal, so $\cos \gamma = \dfrac{\vec{n} \cdot \vec{k}}{|\vec{n}| \cdot |\vec{k}|} = \dfrac{F_z}{|\nabla F|}$. If $F_z < 0$, then $-\vec{n}$ is an upward normal, and $\cos \gamma = \dfrac{-F_z}{|\nabla F|}$. In either case

$$\sec \gamma = \frac{|\nabla F|}{|F_z|}, \quad \text{and} \quad S = \iint_{R_{xy}} \frac{|\nabla F|}{|F_z|} \, dA. \tag{6}$$

The integrand $\dfrac{|\nabla F|}{|F_z|}$ involves $x$, $y$, and $z$; to carry out the integration, you must express $z$ in terms of $x$ and $y$, using the equation of the surface, $F = c$.

---

**EXAMPLE 1 (redone)**  The upper half of the sphere

$$x^2 + y^2 + z^2 = a^2$$

projects onto the disk of radius $a$ in the $xy$-plane; this is the region of integration $R_{xy}$ in (6). For the integrand, we have $F = x^2 + y^2 + z^2$, so

$$|\nabla F| = |2x\vec{i} + 2y\vec{j} + 2z\vec{k}| = 2\sqrt{x^2 + y^2 + z^2} = 2a$$

on the given sphere; and $|F_z| = 2|z|$, so (6) gives

$$S = \iint_{R_{xy}} \frac{2a}{2|z|} \, dA_{xy} = \iint_{R_{xy}} \frac{a}{|z|} \, dA_{xy}.$$

On the hemisphere, $z = \sqrt{a^2 - x^2 - y^2}$ so $S = \iint_{R_{xy}} \dfrac{a \, dx \, dy}{\sqrt{a^2 - x^2 - y^2}}$, the same integral we found before.

---

***REMARK*** For a surface obtained by revolving the graph of $y = f(x)$ about the $x$-axis, Section 10.2 gives the area formula

$$\int_a^b 2\pi f(x)\sqrt{1 + f'(x)^2} \, dx. \tag{7}$$

Is this consistent with the integral given here? To check, represent the upper half of the surface of revolution as the graph of a function of two variables (Fig. 5). Take a point $(x, f(x), 0)$ on the graph of $y = f(x)$, and revolve it about the $x$-axis; it generates a circle of radius $f(x)$, with equation

$$y^2 + z^2 = f(x)^2.$$

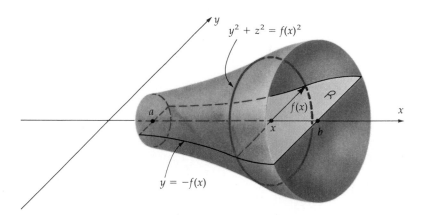

**FIGURE 5**
Graph of $y = f(x)$ revolved about $x$-axis.

So the upper half of the surface of revolution is the graph of $z = \sqrt{f(x)^2 - y^2}$. Then

$$1 + |\nabla z|^2 = 1 + \left[\frac{f(x)f'(x)}{\sqrt{f(x)^2 - y^2}}\right]^2 + \left[\frac{-y}{\sqrt{f(x)^2 - y^2}}\right]^2$$

$$= f(x)^2 \frac{1 + f'(x)^2}{f(x)^2 - y^2}.$$

In the region of integration $R$ [Fig. 5], $x$ varies from $a$ to $b$, and for each such $x$, $y$ varies from $-f(x)$ to $+f(x)$; so the area of this graph is

$$\int_a^b \int_{-f(x)}^{f(x)} f(x)\sqrt{1 + f'(x)^2}\, \frac{1}{\sqrt{f(x)^2 - y^2}}\, dy\, dx$$

$$= \int_a^b f(x)\sqrt{1 + f'(x)^2} \left[ \sin^{-1}\left(\frac{y}{f(x)}\right) \right]_{-f(x)}^{f(x)} dx$$

$$= \int_a^b \pi f(x)\sqrt{1 + f'(x)^2}\, dx.$$

This gives just the upper half of the surface; its double gives the entire surface, in accord with (7).

## SUMMARY

*Area of a Surface Lying over a Plane Region R:*

$$S = \iint_R \sec \gamma\, dA,$$

$\gamma$ = angle between normal to surface and normal to $R$.

For the graph of $z = f(x, y)$

$$S = \iint_R \sqrt{1 + |\nabla f|^2}\, dA.$$

For a level surface $\{F = c\}$,

$$S = \iint_{R_{xy}} \frac{|\nabla F|}{|F_z|}\, dA.$$

## PROBLEMS

### A

1. Compute the area of the given graph.
   *a) $z = 2x + 3y$, $-1 \le x \le 1$, $-1 \le y \le 1$
   b) $z = x^2 + y^2$, $x^2 + y^2 \le 1$
   *c) $z = xy$, $1 \le x^2 + y^2 \le 4$

2. Compute the area of the given graph as an integral. Each is a familiar geometric figure with a known area formula; use this to check your integral.
   a) $z = 3x$, $0 \le x \le 1$, $0 \le y \le 1$
   *b) $z = 3x$, $x^2 + y^2 \le 1$

*c) $z = \sqrt{1 - x^2 - y^2}$, $0 \le x \le 1$, $0 \le y \le \sqrt{1 - x^2}$
   d) $z = \sqrt{1 - x^2}$, $0 \le x \le 1$, $0 \le y \le 1$

3. Compute the area of that part of the plane $z = x - 3y$ lying inside
   *a) The cylinder $x^2 + y^2 = 4$.
   b) The rectangular column $a \le x \le b$, $c \le y \le d$.
   c) The sphere $x^2 + y^2 + z^2 = 1$. (Don't integrate! Does the center of the sphere lie on the plane?)

4. Compute the area of

   *a) The part of the plane $\dfrac{x}{a} + \dfrac{y}{b} + \dfrac{z}{c} = 1$ cut off by the coordinate planes.

   b) The part of the cone $z^2 = 2x^2 + 2y^2$ inside the cylinder $(x - 1)^2 + y^2 = 1$.

   *c) The part of the sphere $x^2 + y^2 + z^2 = 9$ between the planes $z = 1$ and $z = 2$.

   d) The part of the "monkey saddle" $z = \frac{1}{3}x^3 - xy^2$ inside the cylinder $x^2 + y^2 = a^2$.

   e) The part of $z = 1 - x^2 - y^2$ above the $xy$-plane.

*5. A farmer has a sloping lot. On the map, it is a 200' by 400' rectangle, but in the uphill direction it has a slope of 1 in 3; it rises 1 foot for each 3 feet of horizontal displacement. What is the actual surface area of the lot?

**B**

6. A solar collector has an exposed surface of 50 m$^2$. Compare the rate of heat collection when the sun's rays strike it at an angle of 90°, versus an angle of 45°. Explain by a sketch. Ignore reflection.

7. Deduce a formula analogous to (6) for the area of a surface that projects on
   a) The $xz$-plane.     b) The $yz$-plane.

8. Show that over each region in the $xy$-plane, the graphs of $z = 2xy$, $z = x^2 + y^2$, and $z = x^2 - y^2$ have the same surface area.

9. Write a repeated integral for the area of the polar coordinate graph of $z = f(r, \theta)$ over the region $\{(r, \theta)_p : a \le \theta \le b, r_1(\theta) \le r \le r_2(\theta)\}$. (See Sec. 16.2, problem 5, for the expression of $|\nabla z|^2$ in polar coordinates.)

*10. Compute the surface area of a spiral ramp in the shape of $z = c\theta$, $a \le r \le b$, $0 \le \theta \le 2\pi$. (See the previous problem.)

11. The graph of $y = f(x)$, $a \le x \le b$ (where $a \ge 0$) can be revolved about the $y$-axis to generate a surface. For the area of this surface, Section 10.2 gave the integral $2\pi \int_a^b x\sqrt{1 + f'(x)^2}\, dx$. Reconcile this with the surface area formula in this section. (See problem 9.)

## 17.7

## TRIPLE INTEGRALS

You can easily imagine that the integral concept must be extended to functions of three variables. For example, the mass of a solid with variable density will be expressed as the integral of the density function over the region in space occupied by the solid.

In three dimensions there is no simple direct geometric view of the integral, so we begin with the Riemann sums. Suppose that $f$ is a continuous function defined in a rectangular region $W$ as in Figure 1. Subdivide $W$ into many smaller regions $W_1, \ldots, W_n$ with dimensions $\Delta x$, $\Delta y$, $\Delta z$. In each region $W_j$ choose a point $P_j$. Form the sum

$$\sum_{j=1}^{n} f(P_j)\, \Delta V_j \tag{1}$$

where $\Delta V_j = \Delta x\, \Delta y\, \Delta z$ is the volume of $W_j$. Repeat the process with finer and finer subdivisions, such that $\Delta x$, $\Delta y$, and $\Delta z$ all tend to zero; then the Riemann sums (1) have a limit which is independent of the choice of the points $P_j$ (because $f$ is continuous). The limit is the triple integral of $f$ over $W$:

$$\iiint_W f(x, y, z)\, dV = \lim \sum_{j=1}^{n} f(P_j)\, \Delta V_j. \tag{2}$$

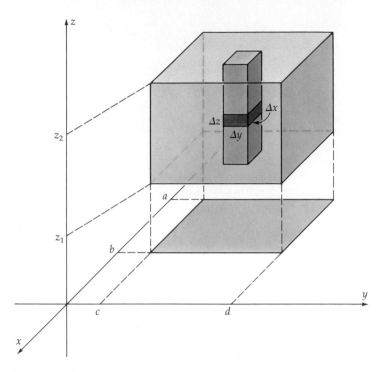

**FIGURE 1**

As we said, a geometric interpretation of this integral is difficult—a single integral gives area in the plane, a double integral gives volume in space, so a triple integral gives some "super-volume" in four-dimensional space—hard to visualize. It is easier to think of the triple integral as *mass*. Suppose that $f(x, y, z)$ is the density function of a substance filling the region $W$. Density is mass per unit volume, so mass is density times volume, and the term $f(P_j) \cdot \Delta V_j$ in the Riemann sum (1) approximates the mass in the small region $W_j$. The Riemann sum approximates the entire mass, and the limit (2) gives the mass exactly. So *the mass of a solid is the integral of its density function.*

You evaluate triple integrals as three successive single integrals. For a rectangular region $W$ such as in Figure 1,

$$\iiint\limits_{W} f(x, y, z) \, dV = \int_a^b \int_c^d \int_{z_1}^{z_2} f(x, y, z) \, dz \, dy \, dx.$$

The inner integral is done first, varying $z$ while holding $x$ and $y$ fixed; this corresponds to the sum over all the small regions in the column shown in Figure 1. Then the integral with respect to $y$ sums these columns from left to right, giving the sum over a slab parallel to the $yz$-plane. Finally, the integral with respect to $x$ sums these slabs from back to front.

If the integrand $f$ is the constant 1, then each term $f(P_j) \Delta V_j$ in the Riemann sum (1) is simply $\Delta V_j$, and the sum gives the entire volume of $W$. Thus

$$\iiint\limits_{W} 1 \, dV = \text{volume of region } W.$$

**EXAMPLE 1**   Evaluate $\int_2^3 \int_1^2 \int_0^1 x \, dz \, dy \, dz$.

*SOLUTION*   First, integrate $\int_0^1 x \, dz$, holding $x$ and $y$ constant:

$$\int_2^3 \int_1^2 \int_0^1 x \, dz \, dy \, dx = \int_2^3 \int_1^2 xz \Big|_{z=0}^{z=1} dy \, dx = \int_2^3 \int_1^2 x \, dy \, dx.$$

Next, integrate with respect to $y$, and finally $x$:

$$\int_2^3 \int_1^2 \int_0^1 x \, dz \, dy \, dx = \int_2^3 \int_1^2 x \, dy \, dx = \int_2^3 xy \Big|_{y=1}^{y=2} dx$$
$$= \int_2^3 x \, dx = \tfrac{1}{2} x^2 \Big|_2^3 = \tfrac{5}{2}.$$

Try it in some other order—you will get the same result.

For more general regions, the limits of integration will not be constants. Figure 2 shows a typical case. $W$ is the region in space between the graphs of two continuous function $z = z_1(x, y)$ and $z = z_2(x, y)$, and the projection of $W$ on the $xy$-plane is a plane region $R_{xy}$. For such a "$z$-simple" region, the triple integral is

$$\iiint\limits_W f(x, y, z) \, dV = \iint\limits_{R_{xy}} \int_{z_1(x,y)}^{z_2(x,y)} f(x, y, z) \, dz \, dA_{xy}. \tag{3}$$

The inner integral with respect to $z$ has variable limits; for each fixed $(x, y)$, $z$ varies from $z_1(x, y)$ to $z_2(x, y)$. You integrate the result of this inner integration over $R_{xy}$.

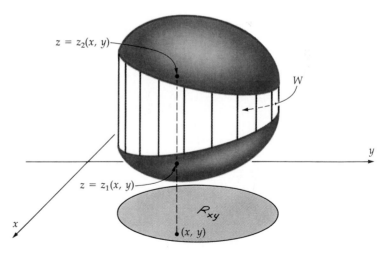

**FIGURE 2**
$z$-simple region;  $W = \{(x, y, z): (x, y) \text{ in } R_{xy}, z_1(x, y) \le z \le z_2(x, y)\}$

To set up the integral (3), you must first visualize the region $W$, and determine the functions $z_1(x, y)$ and $z_2(x, y)$ giving the inner limits of integration. Then sketch the plane projection $R_{xy}$ and determine the appropriate limits of integration for $x$ and $y$, as with any double integral.

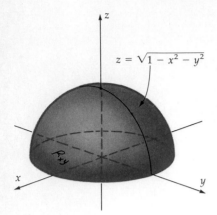

$$z = \sqrt{1 - x^2 - y^2}$$

**FIGURE 3**

**EXAMPLE 2**   Evaluate $\iiint_W z \, dV$, where $W$ is the hemisphere

$$\{(x, y, z): x^2 + y^2 + z^2 \leq 1, z \geq 0\}.$$

**SOLUTION**   Figure 3 shows the region $W$. It seems easiest to project it on the $xy$-plane—the projection is the circle in Figure 4—so you need limits on $z$. In $W$, $z$ varies from $z_1 = 0$ to $z_2 = \sqrt{1 - x^2 - y^2}$ (Fig. 3). So

$$\iiint_W z \, dV = \iint_{R_{xy}} \int_0^{\sqrt{1 - x^2 - y^2}} z \, dz \, dA_{xy}$$

$$= \iint_{R_{xy}} \tfrac{1}{2} z^2 \Big|_0^{\sqrt{1 - x^2 - y^2}} \, dA_{xy}$$

$$= \iint_{R_{xy}} \tfrac{1}{2}(1 - x^2 - y^2) \, dA_{xy}.$$

The integral over $R_{xy}$ can be done in rectangular coordinates, but polar coordinates are easier; $x^2 + y^2 = r^2$ and $dA_{xy} = r \, dr \, d\theta$, so the evaluation continues with

$$\iint_{R_{xy}} \tfrac{1}{2}(1 - x^2 - y^2) \, dA_{xy} = \int_0^{2\pi} \int_0^1 \tfrac{1}{2}(1 - r^2) r \, dr \, d\theta$$

$$= \int_0^{2\pi} \tfrac{1}{4} r^2 - \tfrac{1}{8} r^4 \Big|_0^1 \, d\theta = \tfrac{1}{4}\pi.$$

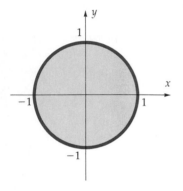

**FIGURE 4**
$R_{xy} = \{(x, y): x^2 + y^2 \leq 1\}.$

Sometimes a different order of integration is more convenient. For example in Figure 5, the region $W$ projects onto the plane region $R_{xz}$, and lies between the graphs of $y = y_1(x, z)$ and $y = y_2(x, z)$. In this case

$$\iiint_W f(x, y, z) \, dV = \iint_{R_{xz}} \int_{y_1(x,z)}^{y_2(x,z)} f(x, y, z) \, dy \, dA_{xz}. \tag{4}$$

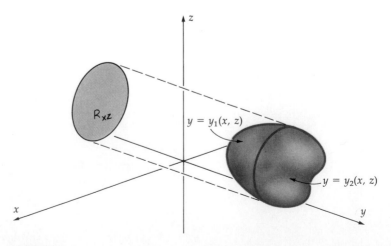

**FIGURE 5**
$y$-simple region; $W = \{(x, y, z): (x, z) \text{ in } R_{xz}, y_1(x, z) \leq y \leq y_2(x, z)\}.$

**EXAMPLE 3**   Evaluate $\iiint_W x\,dV$ where $W$ is the region below the graph of $z = 1 - x^2$ cut off by the $xz$-plane and the plane $z = y$.

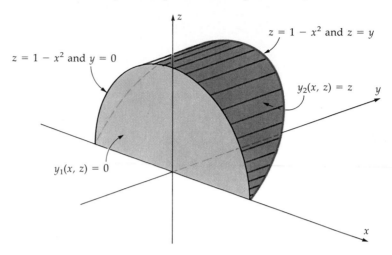

**FIGURE 6**
Left-hand boundary is plane $y = 0$, right-hand boundary is plane $y = z$.

*SOLUTION*   This region (Fig. 6) can be easily projected on any of the three coordinate planes—we'll use the $xz$-plane. Since the function $z = 1 - x^2$ is independent of $y$, its graph consists of lines parallel to the $y$-axis, passing through the curve $z = 1 - x^2$ in the $xz$-plane. So the region $W$ projects onto the plane region $R_{xz}$, which lies between $z = 1 - x^2$ and $z = 0$ (Fig. 7). For each point $(x, z)$ in this region, $y$ varies from $y = 0$ (the equation of the left-hand boundary in Fig. 6) to $y = z$ (the equation of the right-hand boundary). So

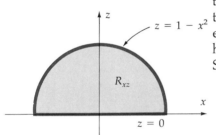

**FIGURE 7**

$$\iiint_W x\,dV = \iint_{R_{xz}} \int_0^z x\,dy\,dA_{xz}$$

$$= \iint_{R_{xz}} xz\,dA_{xz}$$

$$= \int_{-1}^1 \int_0^{1-x^2} xz\,dz\,dx \qquad \text{[Fig. 7]}$$

$$= \int_{-1}^1 x\,\frac{(1-x^2)^2}{2}\,dx = \frac{(1-x^2)^3}{-12}\bigg|_{-1}^1 = 0.$$

All this work, just to get zero? In retrospect, you can see that this integral is 0 just from its symmetry. The region $W$ is symmetric with respect to the $yz$-plane; and the integrand $f(x, y, z) = x$ has the property that

$$f(-x, y, z) = -f(x, y, z).$$

So each contribution $f(x, y, z)\,\Delta V$ to the Riemann sum that comes from a small piece of volume in front of the $yz$-plane is exactly cancelled by the contribution

$$f(-x, y, z)\,\Delta V = -f(x, y, z)\,\Delta V$$

from the symmetrically located small piece behind the $yz$-plane, and the Riemann sum equals zero.

The formulas for the center of mass $(\bar{x}, \bar{y}, \bar{z})$ of an object in space are just like those in two dimensions. If $\delta(x, y, z)$ is the density at $(x, y, z)$, and $M = \iiint_W \delta \, dV$ is the total mass of the object, then

$$\bar{x} = \frac{1}{M} \iiint_W x\delta \, dV, \qquad \bar{y} = \frac{1}{M} \iiint_W y\delta \, dV, \qquad \bar{z} = \frac{1}{M} \iiint_W z\delta \, dV.$$

If you could pivot the object about this point, then gravity would exert no torque, no matter which way you turn the object. (See problem 13 for a derivation of this.)

---

**EXAMPLE 4** Suppose that the hemisphere in Example 2 has constant density $\delta$. Compute the center of mass.

**SOLUTION** Since the density is constant, the mass is $M = \delta \times (\text{volume}) = \frac{2\pi}{3} \delta$. Then

$$\bar{z} = \frac{1}{M} \iiint_W z\delta \, dV = \frac{\delta}{M} \iiint_W z \, dV = \frac{3}{2\pi} \iiint_W z \, dV.$$

Example 2 showed that $\iiint z \, dV = \pi/4$, so

$$\bar{z} = \frac{3}{2\pi} \cdot \frac{\pi}{4} = \frac{3}{8}.$$

This seems reasonable; $\bar{z}$ should be less than $1/2$, since more of the hemisphere lies below $z = \frac{1}{2}$ than above it.

The other coordinates $\bar{x}$ and $\bar{y}$ are zero—you can see this from the symmetry of the figure, and evaluating the integrals bears it out.

---

## SUMMARY

**Triple Integral** in a "z-simple" region $W$:

$$\iiint_W f(x, y, z) \, dV = \iint_{R_{xy}} \int_{z_1(x,y)}^{z_2(x,y)} f(x, y, z) \, dz \, dA$$

Analogous formulas hold for other orders of integration.

**Mass:**

$$M = \iiint_W \delta(x, y, z) \, dV; \ \delta(x, y, z) \text{ is the density at } (x, y, z).$$

**Center of Mass:**

$$\bar{x} = \frac{1}{M} \iiint_W x\delta(x, y, z) \, dV \qquad \bar{y} = \frac{1}{M} \iiint_W y\delta(x, y, z) \, dV$$

$$\bar{z} = \frac{1}{M} \iiint_W z\delta(x, y, z) \, dV$$

## PROBLEMS

### A

1. Evaluate the integral, and describe the region of integration.

*a) $\int_0^1 \int_0^1 \int_0^1 e^{-x-y-z} \, dx \, dy \, dz$

b) $\int_0^1 \int_0^2 \int_{-1}^0 x \, dz \, dy \, dx$

*c) $\int_0^\infty \int_0^\infty \int_0^\infty e^{-x-y-z} \, dx \, dy \, dz$

d) $\int_0^1 \int_0^1 \int_0^{2-x^2-y^2} z \, dz \, dx \, dy$

*e) $\int_0^3 \int_0^{2\sqrt{1-y^2/9}} \int_0^{\sqrt{1-x^2/4-y^2/9}} xz \, dz \, dx \, dy$

(Long, but not hard.)

2. Evaluate $\int_2^3 \int_1^2 \int_0^1 x \, dz \, dy \, dx$ with a different order of integration—first $x$, then $y$, then $z$. Compare the result with Example 1.

3. Evaluate the following integrals over the hemisphere in Example 2, $W = \{(x, y, z): 0 \le z \le \sqrt{1-x^2-y^2}\}$. Explain those that are 0 by appropriate symmetries.

*a) $\iiint_W x \, dV$   *b) $\iiint_W z^2 \, dV$

c) $\iiint_W xy \, dV$

*4. Evaluate $\iiint_W x \, dV$ with $W$ as in Example 3, by projecting the region $W$
   a) On the $xy$-plane.   b) On the $yz$-plane.

5. Evaluate the following integrals over the region $W = \{(x, y, z): x^2 + y^2 + z^2 \le a^2,$
   $$x \ge 0, y \ge 0, z \ge 0\}.$$
   *a) $\iiint_W z \, dV$   b) $\iiint_W x \, dV$

6. Evaluate $\iiint_W xz \, dV$ over the region defined by the given inequalities.
   a) $x \ge 0, y \ge 0, z \ge 0, x + y + z \le 1$ (a tetrahedron) (Long, but easy.)
   b) $x^2 + y^2 + z^2 \le 1, z \le y, z \ge 0$ (a section of a sphere)
   c) $x^2 + y^2 \le z \le 4 - x^2 - 3y^2$

7. Sketch the region whose volume is given by
   a) $\int_0^1 \int_{-1}^0 \int_0^2 dx \, dy \, dz$.
   b) $\int_0^1 \int_{-1}^0 \int_0^2 dz \, dy \, dx$.

*c) $\int_0^1 \int_0^{1-x} \int_0^{1-x-y} dz \, dy \, dx$.

d) $\int_{-2}^2 \int_0^{\sqrt{4-y^2}} \int_0^2 dz \, dx \, dy$.

8. Find the volume of the region bounded by the given surfaces.
   *a) $z = x^2 + y^2$ and $z = 2 - x^2 - y^2$
   b) $z = x^2 + y^2$ and $z = 2 - x^2 - 2y^2$ (Long.)
   *c) $x^2 + z^2 = a^2$ and $y^2 + z^2 = a^2$

9. What is wrong with the following expressions?
   a) $\int_0^x x \, dx$   b) $\int_0^z \int_0^1 \int_x^y z \, dz \, dy \, dx$

10. The following regions have constant density $\delta$. Compute their centers of mass.
    *a) The tetrahedron where $x \ge 0, y \ge 0, z \ge 0,$ $x + y + z \le 1$
    b) The region inside $z = x^2 + y^2$, and below $z = 2$
    *c) The tetrahedron where $x \ge 0, y \ge 0, z \ge 0,$ $x/a + y/b + z/c \le 1$
    *d) The cone where $m\sqrt{x^2 + y^2} \le z \le h$
    e) The region in Example 3
    f) The section of the sphere $x^2 + y^2 + z^2 \le a^2$ where $x \ge 0, z \ge 0$

11. For each region in the previous problem, suppose that the density is $\delta(x, y, z) = z$. Recompute the center of gravity. Which $\bar{x}, \bar{y}, \bar{z}$ are affected? In which direction? Why? (Parts c and e are long.)

12. A storage silo has walls in the form of a vertical cylinder $x^2 + y^2 = r^2$. Grain of constant density $\delta$ fills it from the floor ($z = 0$) to the plane $z = my + b$.
    a) Show that if $b > mr$, then $z = my + b$ does not intersect the floor inside the silo.
    b) If $b > mr$, compute the center of gravity of the pile of grain.

13. For an object with density function $\delta$ extending over the region $W$, the *moment of inertia about the z-axis* is
    $$I_z = \iiint_W (x^2 + y^2)\delta(x, y, z) \, dV$$

    Compute $I_z$ for the following objects, assuming $\delta = 1$.
    *a) The cylinder bounded by $x^2 + y^2 = r^2$, $z = 0, z = h$

**b)** The sphere $x^2 + y^2 + z^2 \le r^2$

*c)** The tetrahedron bounded by the coordinate planes and the plane $x + y + z = 1$ (Long)

**d)** The tetrahedron bounded by the coordinate planes and the plane $\dfrac{x}{a} + \dfrac{y}{b} + \dfrac{z}{c} = 1$ (Long)

*e)** The cylinder $(x - a)^2 + y^2 \le r^2$, $0 \le z \le h$ (Long)

**f)** The cylinder $x^2 + z^2 \le r^2$, $a \le y \le b$ (Long)

**g)** The region inside the sphere $x^2 + y^2 + z^2 = a^2$ and outside the cylinder $x^2 + y^2 = b^2$, where $b < a$

**NOTE** The moment of inertia gives the "resistance to angular acceleration," just as the mass gives the resistance to linear acceleration. (See Sec. 17.3, and problem 15.) If $\omega$ is the angular velocity of rotation about the z-axis, $I_z$ is the moment of inertia, and $T_z$ is the torque about the z-axis, then

$$T_z = I_z \frac{d\omega}{dt} \qquad \left(\text{compare with } \vec{F} = m \frac{d\vec{v}}{dt}\right)$$

and the kinetic energy is $\frac{1}{2} I_z \omega^2$ (compare $\frac{1}{2} m |\vec{v}|^2$).

**C**

**14.** Imagine a rigid body in space pivoted so as to turn freely about a point $P_0$ (which might be inside the body). Imagine a constant gravitational force $\vec{F} = (a\vec{i} + b\vec{j} + c\vec{k})\delta(x, y, z)\, dV$ acting on each small part of the body. [$\delta(x, y, z)\, dV$ is the mass of the small part, and $a\vec{i} + b\vec{j} + c\vec{k}$ is the acceleration due to gravity, pointing in an arbitrary direction; different directions for this gravity vector correspond to different orientations of the body with respect to "up" and "down."] The torque exerted on this small piece is $\vec{F} \times (\vec{P} - \vec{P}_0)$, where $P = (x, y, z)$. Set $P_0 = (x_0, y_0, z_0)$.

**a)** Compute the torque exerted on the entire body.

**b)** Show that this torque is zero for all vectors $a\vec{i} + b\vec{j} + c\vec{k}$ if $P_0$ is the center of mass, $\bar{P}$.

**c)** Show that if $\vec{T} = \vec{0}$, then $P_0$ lies on the line through the center of mass $\bar{P}$, parallel to $a\vec{i} + b\vec{j} + c\vec{k}$.

**d)** Show that if the torque about $P_0$ is nonzero for *any two* nonparallel vectors $a\vec{i} + b\vec{j} + c\vec{k}$, then $P_0$ is the center of mass.

**15.** Suppose that a point of mass $m$ moves in a circle of radius $r$ centered at the origin, with (variable) angular velocity $\omega$. Let $\vec{F}$ be a tangential force and $P$ be the position of the mass; then $\vec{T} = \vec{P} \times \vec{F}$ is the torque exerted by the force $\vec{F}$ about an axis perpendicular to the circle through its center.

**a)** Show that $|\vec{F}| = mr \dfrac{d\omega}{dt}$ and $|\vec{T}| = mr^2 \dfrac{d\omega}{dt}$.

**b)** Deduce that for a rigid body rotating about the z-axis, $|\vec{T}| = I_z\, d\omega/dt$.

**c)** Show that the kinetic energy of the point mass is $\frac{1}{2} mr^2 \omega^2$.

**d)** Deduce that the kinetic energy of a rigid body rotating about the z-axis is $\frac{1}{2} I_z \omega^2$.

# 17.8
## CYLINDRICAL AND SPHERICAL COORDINATES

*Cylindrical coordinates* serve primarily in problems having symmetry with respect to the z-axis. They use the standard z-coordinate, but replace $x$ and $y$ by polar coordinates $r$ and $\theta$ (Fig. 1):

$$x = r \cos \theta$$
$$y = r \sin \theta$$
$$z = z$$

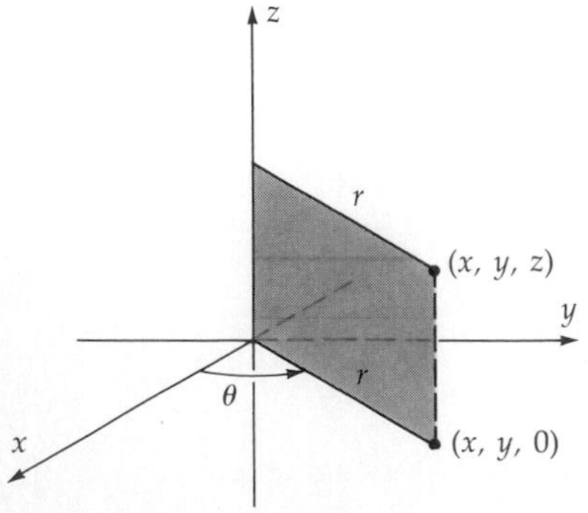

**FIGURE 1**
Cylindrical coordinates $r, \theta, z$.

Triple integrals can be written as repeated integrals in cylindrical coordinates. The inner integral is generally taken with respect to $z$, and then the integral

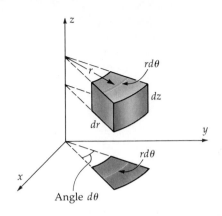

**FIGURE 2**
$dV = (r\,d\theta)(dr)(dz) = r\,dr\,d\theta\,dz.$

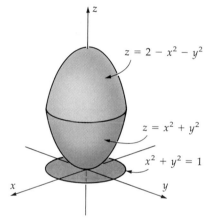

**FIGURE 3**

in $x$ and $y$ is in polar coordinates:

$$\iiint\limits_W f\,dV = \iint\limits_{R_{xy}} \left[ \int_{z_1(r,\theta)}^{z_2(r,\theta)} f\,dz \right] r\,dr\,d\theta$$

$$= \iint\limits_{R_{xy}} \int_{z_1(r,\theta)}^{z_2(r,\theta)} fr\,dz\,dr\,d\theta.$$

In other words, in cylindrical coordinates $dV = r\,dz\,dr\,d\theta$ (Fig. 2).

**EXAMPLE 1**    An object with constant density $\delta = 1$ occupies the region between the graphs of $z = x^2 + y^2$ and $z = 2 - x^2 - y^2$. Compute the *moment of inertia* for rotation about the $z$-axis.

**SOLUTION**    The moment of inertia is defined in Section 17.6, problem 13; it is

$$I_z = \iiint\limits_W (x^2 + y^2)\delta\,dV.$$

Here $\delta = 1$, and in polar coordinates $x^2 + y^2 = r^2$, $dV = r\,dz\,dr\,d\theta$, so

$$I_z = \iiint\limits_W r^2 r\,dz\,dr\,d\theta.$$

The region $W$ determines the limits of integration (Fig. 3). In $W$, $z$ varies from

$$z_1 = x^2 + y^2 = r^2$$

to

$$z_2 = 2 - x^2 - y^2 = 2 - r^2.$$

These two equations define paraboloids, intersecting where $z_1 = z_2$, or

$$z = x^2 + y^2 = 2 - x^2 - y^2.$$

The second equation gives $x^2 + y^2 = 1$, so the intersection is the circle

$$x^2 + y^2 = 1, \qquad z = 1.$$

The region $W$ projects onto a disk in the $xy$-plane,

$$x^2 + y^2 \leq 1.$$

For this disk of radius 1, the polar limits are $0 \leq r \leq 1$, $0 \leq \theta \leq 2\pi$; so the integral is

$$\iiint\limits_W (x^2 + y^2)\,dV = \int_0^{2\pi} \int_0^1 \int_{r^2}^{2-r^2} r^3\,dz\,dr\,d\theta$$

$$= \int_0^{2\pi} \int_0^1 (2 - 2r^2)r^3\,dr\,d\theta$$

$$= \frac{\pi}{3}.$$

**Spherical coordinates** serve well in problems with an apropriate symmetry, for example, symmetry with respect to the origin. The three spherical coordinates of a point $P = (x, y, z)$ are (Fig. 4)

$$\rho = |P| = \text{ the distance from } P \text{ to the origin,}$$

$$\varphi = \text{ the angle between } P \text{ and the positive } z\text{-axis,}$$

$$\theta = \text{ the same as in polar coordinates.}$$

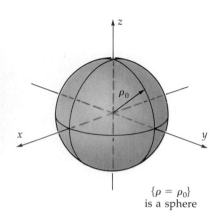

$\{\rho = \rho_0\}$
is a sphere

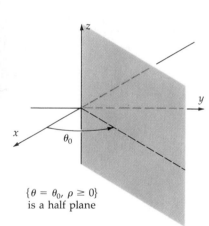

$\{\theta = \theta_0, \rho \geq 0\}$
is a half plane

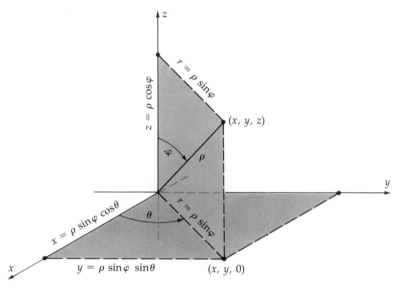

**FIGURE 4**
Spherical coordinates $\rho$, $\varphi$, and $\theta$.

The two "angular" variables $\varphi$ and $\theta$ give the direction from the origin to $P$, and $\rho$ gives the distance. Figure 4 shows that $z = \rho \cos \varphi$, and that the cylindrical coordinate $r$ equals $\rho \sin \varphi$. So

$$x = r \cos \theta = \rho \sin \varphi \cos \theta$$

$$y = r \sin \theta = \rho \sin \varphi \sin \theta$$

$$z = \rho \cos \varphi.$$

It is helpful to visualize the surfaces in which a given variable is constant (Fig. 5):

$\rho = \rho_0$    defines a sphere of radius $\rho_0$,

$\varphi = \varphi_0$    defines a cone making angle $\varphi_0$ with the $z$-axis,

$\theta = \theta_0$    defines a half plane "hinged" on the $z$-axis.

Here we assumed that $\rho \geq 0$; if $\rho < 0$ is allowed, then $\varphi = \varphi_0$ defines a *double cone*, and $\theta = \theta_0$ a complete plane.

You can represent every point in space using spherical coordinates with

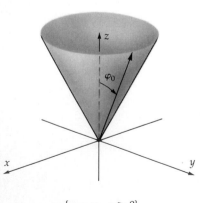

$\{\varphi = \varphi_0, \rho \geq 0\}$
is a single cone

**FIGURE 5**

$$0 \leq \rho < \infty, \qquad 0 \leq \varphi \leq \pi, \qquad 0 \leq \theta < 2\pi.$$

For fixed $\varphi$ and $\theta$, $0 \leq \rho < \infty$ generates a ray through the origin (Fig. 6a). Holding $\theta$ fixed, $0 \leq \varphi \leq \pi$ sweeps this ray from the north pole to the south pole, generating a half plane hinged on the $z$-axis (Fig. 6b). As $\theta$ varies from 0 to $2\pi$, this half plane swings around the $z$-axis, sweeping out all of space. (*Notice* that the range for $\varphi$ is 0 to $\pi$, not 0 to $2\pi$!)

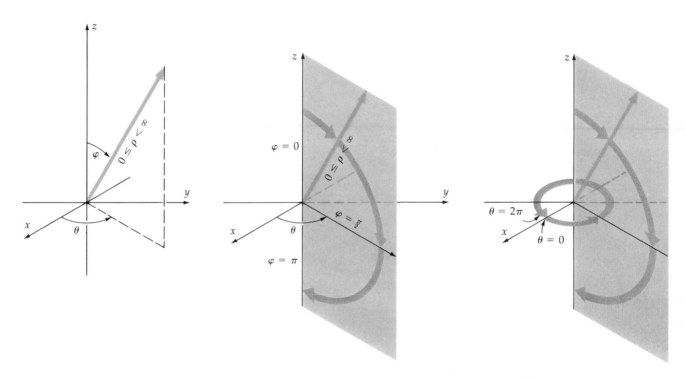

(a)   Line generated with $\theta$ and $\varphi$ fixed, $0 \leq \rho < \infty$.

(b)   Half plane generated with $\theta$ fixed, $0 \leq \rho < \infty$, and $0 \leq \varphi \leq \pi$.

(c)   $R^3$ generated by $0 \leq \rho < \infty$, $0 \leq \varphi \leq \pi$, $0 \leq \theta \leq 2\pi$.

**FIGURE 6**
Sweeping out all of $R^3$. First a ray, then a half plane hinged on the $z$ axis, and finally swing the half plane around 360°.

To use spherical coordinates in triple integrals, you need the appropriate form of the basic volume element $dV$. Think of a small element $dV$ bounded between two spheres, two planes containing the $z$-axis, and two cones (Fig. 7). If the dimensions are very small, then $\rho$ is practically constant throughout the small region, and those edges which are circular arcs are practically straight lines. Four edges are straight lines of length $d\rho$; four are circular arcs of radius $\rho$ and angle $d\varphi$, so their lengths are $\rho\, d\varphi$; the remaining four are circular arcs of radius $\rho \sin \varphi$ and angle $d\theta$, so their lengths are $\rho \sin \varphi\, d\theta$. All the angles are right angles, so the small element is practically a rectangular solid with volume

$$(d\rho)(\rho\, d\varphi)(\rho \sin \varphi\, d\theta) = \rho^2 \sin \varphi\, d\rho\, d\varphi\, d\theta.$$

This is the formula for $dV$ in spherical coordinates. The derivation assumed that $\sin \varphi \geq 0$; thus *the limits of integration for $\varphi$ must be somewhere between* 0 and $\pi$.

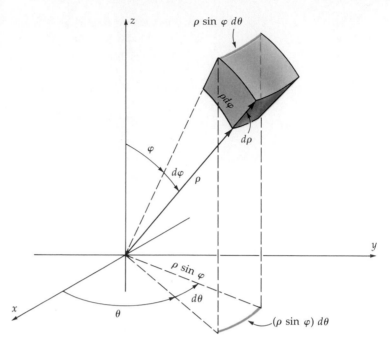

**FIGURE 7**
$dV = (d\rho)(\rho \, d\varphi)(\rho \sin \varphi \, d\theta) = \rho^2 \sin \varphi \, d\rho \, d\varphi \, d\theta.$

---

**EXAMPLE 2**  The density $\delta$ of the atmosphere falls off approximately exponentially with height above the earth:

$$\delta = Ce^{-k\rho} \text{ gms/(meter)}^3$$

where $\rho =$ distance from the center of the earth, in meters, and $C$ and $k$ are constants. Assuming this simplified density function, compute the mass of the entire atmosphere

$$\iiint_W \delta \, dV$$

where $W$ is all of space outside the earth.

**SOLUTION**  The earth is nearly a ball of radius $6.4 \times 10^6$ meters, so $W$ is all of space outside of this ball. In $W$, $\rho$ varies from $\rho_0 = 6.4 \times 10^6$ to $\infty$, while $\varphi$ varies from 0 to $\pi$, and $\theta$ from 0 to $2\pi$:

$$\iiint_W \delta \, dV = \int_0^{2\pi} \int_0^{\pi} \int_{\rho_0}^{\infty} Ce^{-k\rho}\rho^2 \sin \varphi \, d\rho \, d\varphi \, d\theta$$

$$= C \int_0^{2\pi} \int_0^{\pi} \left( \frac{\rho_0^2}{k} + \frac{2\rho_0}{k^2} + \frac{2}{k^3} \right) e^{-k\rho_0} \sin \varphi \, d\varphi \, d\theta$$

$$= C \int_0^{2\pi} 2e^{-k\rho_0} \left( \frac{\rho_0^2}{k} + \frac{2\rho_0}{k^2} + \frac{2}{k^3} \right) d\theta$$

$$= 4\pi Ce^{-k\rho_0} \left( \frac{\rho_0^2}{k} + \frac{2\rho_0}{k^2} + \frac{2}{k^3} \right).$$

The integral here is improper, but the integrand is positive, so no problems arise.

---

*WARNING*   The angles in the spherical coordinate system are similar to those of latitude and longitude on the globe. The angle $\theta$ is just like the longitude east of Greenwich. But the angle $\varphi$ is the *complement* of the latitude; latitude is measured from the equator, while $\varphi$ is measured from the north pole.

## SUMMARY

*Cylindrical Coordinates (see Figs. 1 and 2)*

$$x = r \cos \theta$$
$$y = r \sin \theta \qquad dV = dz\,r\,dr\,d\theta$$
$$z = z$$

*Spherical Coordinates (see Figs. 3 and 4)*

$$x = \rho \sin \varphi \cos \theta$$
$$y = \rho \sin \varphi \sin \theta$$
$$z = \rho \cos \varphi$$
$$dV = \rho^2 \sin \varphi \, d\rho \, d\varphi \, d\theta \qquad (0 \le \varphi \le \pi)$$

## PROBLEMS

### A

1. Describe geometrically the figure given by the following conditions in cylindrical coordinates. Assume $r \ge 0$.
   *a)  $r = 2$           b)  $r = r_0$
   *c)  $r_0 \le r \le r_1$     *d)  $\theta = \pi/2$
   *e)  $\theta = \pi/4$
   f)  $0 \le \theta \le \pi/2$ and $z \ge 0$
   g)  $r_0 \le r \le r_1$ and $z_0 \le z \le z_1$

2. Describe geometrically the figure given by the following conditions in spherical coordinates. Assume $\rho \ge 0$.
   *a)  $\rho = 3$           b)  $\rho = \rho_0$
   *c)  $\rho_0 \le \rho \le \rho_1$     *d)  $\varphi = \pi/4$
   *e)  $\varphi = \pi/2$        *f)  $\varphi = 0$
   g)  $\varphi = \pi/4$ and $\rho = 1$

   *h)  $0 \le \varphi \le \pi/4$
   i)  $\varphi \le \pi/2$
   j)  $\varphi \ge \pi/2$ and $\rho \le 1$

*3. Assume a spherical coordinate system for the earth, with $\theta = 0$ on the north-south line through Greenwich, England, and the positive $z$-axis running from the center of the earth through the North Pole. Find spherical coordinates for New Orleans, which is about $30°$ north of the equator and $90°$ west of Greenwich.

4. Compute $\iiint_W dV$ where $W$ is the unit ball $\{P: |P| \le 1\}$
   a) In cylindrical coordinates.
   b) In spherical coordinates.

*5. Compute $\iiint_W dV$ where $W$ is the region above the cone $z^2 = x^2 + y^2$ and inside the sphere $x^2 + y^2 + z^2 = a^2$

a) In cylindrical coordinates.

b) In spherical coordinates.

6. Compute the moment of inertia $I_z = \iiint_W r^2\, dV$ where $r^2 = x^2 + y^2$ when $W$ is
   *a) The spherical shell $a \le \rho \le b$.
   b) The ball $\rho \le b$.
   *c) The cylindrical shell $r_1 \le r \le r_2$, $0 \le z \le b$.
   d) The cylinder $0 \le r \le a \cos\theta$, $0 \le z \le b$.

7. Compute the center of mass of
   *a) The region cut out of the ball $\rho \le a$ by the cone $\varphi \le \varphi_0$.
   b) The part of the ball $x^2 + y^2 + z^2 \le a^2$ where $x \ge 0$, $y \ge 0$, $z \ge 0$.

8. Show that the area of the part of the cone $z = ar$ (cylindrical coordinates) lying above a region $R$ in the plane is $(\sqrt{1 + a^2}) \times$ (area of $R$).

## B

9. In quantum mechanics, a "wave function" $\psi$ describes the position of an electron with respect to the nucleus. For the "1s orbit" of the hydrogen atom, $\psi = Ne^{-\rho/\alpha}$, where $\rho^2 = x^2 + y^2 + z^2$, $\alpha$ is a constant, and $N$ is a "normalization factor" determined by the condition

$$\iiint_{R^3} \psi^2\, dV = 1.$$

Determine $N$.

10. The probability that an electron in the "1s orbit" of a hydrogen atom lies within a sphere of radius $R$ about the nucleus is

$$P(R) = \iiint_{B_R} \psi^2\, dV,$$

where $B_R$ is the ball of radius $R$ about the origin, and $\psi(x, y, z) = (\pi\alpha^3)^{-1/2}e^{-\rho/\alpha}$ with $\rho^2 = x^2 + y^2 + z^2$ and $\alpha$ constant. Compute $P(R)$.

11. Find the center of mass of the region $W$ defined by the given inequalities, assuming constant density $\delta$.
    a) $\rho \le a$, $0 \le \varphi \le \varphi_0$, $|\theta| \le \theta_0$  (Long)
    b) $r \le a$, $|\theta| \le \theta_0$, $|z| \le z_0$

12. The equation in cylindrical coordinates

$$(r - a)^2 + z^2 = b^2$$

where $0 \le b \le a$, describes a ring-shaped surface called a *torus*. Let $W$ be the region inside this torus. Compute
a) The volume of $W$.
*b) The moment of inertia $I_z = \iiint_W r^2\, dV$, where the density $\delta = 1$). (Long)

13. The gravitational potential at the origin, for a body with density function $\delta(P)$, is

$$\iiint_W \frac{1}{\rho}\, \delta(P)\, dV$$

where $\rho$ is the distance from the origin to $P$. Assume $\delta$ is constant, and compute the gravitational potential of the following bodies.
a) The ball $\rho \le \rho_0$
b) The spherical shell $a \le \rho \le b$
c) The cylinder $r_1 \le r \le r_2$, $z_1 \le z \le z_2$ (Leave your answer as a single integral with respect to $z$.)

14. In studying the motion of the moon and planets, Newton treated them as point masses. He justified this by showing that a homogeneous ball exerts the same gravitational attraction as a point mass at its center, when acting on objects outside the ball. This can be shown by computing the *gravitational potential* of the sphere. The gravitational potential of an object occupying region $W$, with density $\delta$, is a function $\Phi$ defined by the integral

$$\Phi(P_0) = \iiint_W \frac{1}{|P - P_0|}\, \delta(P)\, dV.$$

a) Let $P_0 = (0, 0, c)$, and take $\delta$ as a constant, and $W$ as a ball of radius $a$. Show that $\Phi(P_0) = \frac{4}{3}\pi a^3\delta/c$, the same as the potential of a point mass $\frac{4}{3}\pi a^3\delta$ at the origin. (Let $\rho$, $\varphi$, $\theta$ be spherical coordinates of $P$, and express $|P - P_0|$ in terms of $\rho$, $\varphi$, and $c$, by the law of cosines.)

b) Suppose that $\delta$ is not constant, but depends only on $\rho$, the distance from the origin; then the mass distribution is spherically symmetric. Outside the ball, is the potential the same as for a point mass at the origin?

c) Compute the gravitational potential of a spherically symmetric mass distribution in a "shell" $a < \rho < b$, at points *inside* the shell, that is, where $\rho < a$. (Remember: When $\alpha < 0$ then $\sqrt{\alpha^2} = -\alpha$.)

**d)** For a homogeneous ball ($\delta$ = constant) of radius $a$, show that the gravitational potential at points $(0, 0, c)$ *inside* the sphere (that is, with $c < a$) is a quadratic function of $c$.

**15.** Let $B$ be the unit ball in space, $\{|P| \leq 1\}$. For which powers $q$ is $\iiint_B \rho^q \, dV$ finite?

**16.** Let $W$ be the region outside the unit ball in space. For which powers $p$ is $\iiint_W \rho^p \, dV$ finite?

## 17.9

# GENERAL COORDINATE SYSTEMS. THE JACOBIAN (optional)

Rectangular, polar, cylindrical, and spherical are the most common coordinate systems; but circumstances sometimes require a change to other specially adapted variables. We will see how integrals are transformed under *any* change of variables.

We view a change of variables as a function, or *transformation*, from one space to another. As an example, take the change to polar coordinates for an integral over the annular sector $R$ in Figure 1a:

$$\iint_R f(x, y) \, dA = \int_0^{\pi/2} \int_1^2 f(r \cos \theta, r \sin \theta) r \, dr \, d\theta. \tag{1}$$

On the right in (1) is an integral over a region $R^*$ in the $r\theta$-plane (Fig. 1b), a rectangle. The substitution

$$x = r \cos \theta, \qquad y = r \sin \theta$$

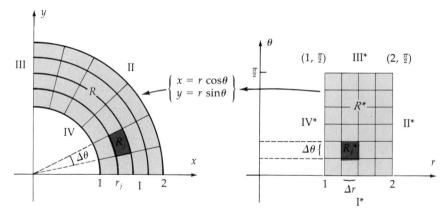

(a)   Annular region $R$ in $xy$ plane. $R_j$ has area $(r_j \Delta\theta)\Delta r$.

(b)   Region $R^*$ in $r\theta$ plane. $R_j^*$ has area $\Delta r \Delta\theta$.

**FIGURE 1**
Polar coordinate transformation.

transforms each point of this rectangle to a corresponding point $(x, y)$ of the annular sector $R$; in particular, the four sides $I^*$, $II^*$, $III^*$, $IV^*$ transform respectively to sides $I$, $II$, $III$, $IV$. What is the role of the factor $r$ in the area element $r \, dr \, d\theta$? Suppose that we cover $R^*$ by a grid of rectangles of area $\Delta r \, \Delta\theta$. Each rectangle is transformed to an annular sector in $R$; the area of that sector is

$$r_j \, \Delta r \, \Delta \theta,$$

where $r_j$ is the value of $r$ at the center of the sector. The factor $r_j$ is the "area multiplication factor" for this transformation; when the area $\Delta r \, \Delta \theta$ in the $r\theta$-plane is transformed to the $xy$-plane, the transformed area is $r_j$ times the original.

To deduce equation (1), we formed a Riemann sum for $\iint_R f(x, y) \, dA$, using the annular sectors instead of rectangles:

$$\iint\limits_{R} f(x, y) \, dA \approx \sum_{j=1}^{n} \underbrace{f(r_j \cos \theta_j, r_j \sin \theta_j)}_{\text{height to graph}} \underbrace{r_j \, \Delta r \, \Delta \theta}_{\text{area of base}}.$$

In the limit as $\Delta r \to 0$ and $\Delta \theta \to 0$, this gives (1).

A general substitution in a double integral is given by two $C^1$ functions

$$x = x(u, v), \qquad y = y(u, v). \tag{2}$$

Suppose that (2) transforms a region $R^*$ in the $uv$-plane to a region $R$ in the $xy$-plane (Fig. 2), such that each point in the interior of $R^*$ is transformed to a different point of $R$; that is, there is no "overlap" in the transformation, except perhaps at the boundary. (A transformation with no "overlap" is called "one-to-one.") Divide $R^*$ into small rectangles $R_j$ of area $\Delta u \, \Delta v$, by a grid of lines. Each line is transformed to a curve in the $xy$-plane, and $R$ is divided into small pieces $R_j^*$, perhaps with curved edges. We approximate these small pieces by parallelograms (Fig. 3). The points $P_j^*$ and $M_j^*$ are transformed to $P_j$ and $M_j$; thus

$$\begin{aligned}
\vec{M}_j - \vec{P}_j &= \langle x(u_j + \Delta u, v_j) - x(u_j, v_j), \, y(u_j + \Delta u, v_j) - y(u_j, v_j) \rangle \\
&\approx \langle x_u(u_j, v_j) \, \Delta u, \, y_u(u_j, v_j) \, \Delta u \rangle \\
&\approx \langle x_u \, \Delta u, \, y_u \, \Delta u \rangle. 
\end{aligned} \tag{3a}$$

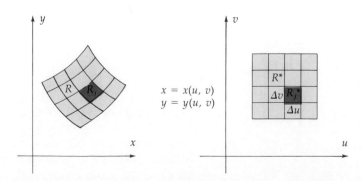

**FIGURE 2**
Transformation from region $R^*$ in $uv$-plane to $R$ in $xy$-plane.

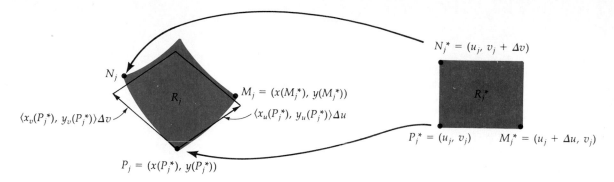

**FIGURE 3**
Approximating $R_j$ by a parallelogram.

[We dropped the point of evaluation $(u_j, v_j)$ to simplify the notation.] Similarly

$$\vec{N}_j - \vec{P}_j \approx \langle x_v \, \Delta v, \, y_v \, \Delta v \rangle. \tag{3b}$$

So $R_j$ is approximated by a parallelogram with sides (3a) and (3b), and its area is approximately that of the parallelogram, given by a determinant (Sec. 14.3):

$$\text{area of } R_j \approx \left| \det \begin{bmatrix} x_u \, \Delta u & y_u \, \Delta u \\ x_v \, \Delta v & y_v \, \Delta v \end{bmatrix} \right| = \left| \det \begin{bmatrix} x_u & y_u \\ x_v & y_v \end{bmatrix} \right| \Delta u \, \Delta v. \tag{4}$$

The determinant in (4) is called the *Jacobian* of the transformation (2) (in honor of C. G. J. Jacobi). It is denoted by $\dfrac{\partial(x, y)}{\partial(u, v)}$:

$$\frac{\partial(x, y)}{\partial(u, v)} = \det \begin{bmatrix} x_u & y_u \\ x_v & y_v \end{bmatrix}.$$

Its absolute value is the "area multiplication factor" for the transformation; from (4)

$$\text{area of } R_j \approx \left| \frac{\partial(x, y)}{\partial(u, v)} \right| \Delta u \, \Delta v. \tag{5}$$

Now form a "warped" Riemann sum for the integral $\iint_R f(x, y) \, dA$, using the small pieces $R_j$. From (5),

$$\sum_{j=1}^{n} f(x_j, y_j) \times [\text{area of } R_j] \approx \sum f(x(u_j, v_j), y(u_j, v_j)) \left| \frac{\partial(x, y)}{\partial(u, v)} \right| \Delta u \, \Delta v. \tag{6}$$

In the limit as $\Delta u \to 0$ and $\Delta v \to 0$, three things happen:

**(i)**   The small pieces $R_j$ shrink to points, and the sum on the left has as its limit the integral $\iint_R f(x, y) \, dA$.

**(ii)**   The parallelograms approximating the small pieces $R_j$ become more accurate, so the sum on the left of (6) has the same limit as the sum on the right.

**(iii)** The sum on the right has the limit

$$\iint_{R*} f(x(u,\, v),\, y(u,\, v)) \left| \frac{\partial(x,\, y)}{\partial(u,\, v)} \right| dA_{uv},$$

where $dA_{uv}$ indicates integration with respect to area in the $uv$-plane.

This gives the general transformation rule:

$$\iint_{R} f(x,\, y)\, dA = \iint_{R*} f(x(u,\, v),\, y(u,\, v)) \left| \frac{\partial(x,\, y)}{\partial(u,\, v)} \right| dA_{uv}. \tag{7}$$

*NOTE* The notation $\dfrac{\partial(x,\, y)}{\partial(u,\, v)}$ tells that the Jacobian is computed from the derivatives of $x$ and $y$ with respect to $u$ and $v$. The essence of formula (7) is the relation between an element of area $dA_{xy} = dx\, dy$ in the $xy$-plane, and a corresponding element $dA_{uv} = du\, dv$ in the $uv$-plane; the relation is

$$dx\, dy = \left| \frac{\partial(x,\, y)}{\partial(u,\, v)} \right| du\, dv.$$

As a rough way to remember this, you can think of "cancelling" $\partial(u,\, v)$ against $du\, dv$ to produce $dx\, dy$.

---

**EXAMPLE 1** In the case of polar coordinates $x = r \cos \theta$, $y = r \sin \theta$, the Jacobian is

$$\frac{\partial(x,\, y)}{\partial(r,\, \theta)} = \det \begin{bmatrix} x_r & y_r \\ x_\theta & y_\theta \end{bmatrix} = \det \begin{bmatrix} \cos \theta & \sin \theta \\ -r \sin \theta & r \cos \theta \end{bmatrix} = r.$$

So (7) reduces to the familiar polar coordinate formula.

---

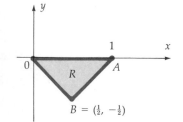

**EXAMPLE 2** Evaluate $\iint_R e^{(x+y)/(x-y)}\, dA$, where $R$ is the triangle in Figure 4, by the substitution

$$u = x + y, \qquad v = x - y. \tag{8}$$

*SOLUTION* You must determine the appropriate region $R*$ in the $uv$-plane, and compute the Jacobian $\dfrac{\partial(x,\, y)}{\partial(u,\, v)}$.

To determine $R*$, look first at the three corners of $R$. Equations (8) transform $A = (1,\, 0)$ in the $xy$-plane to

$$A* = (1 + 0,\, 1 - 0) = (1,\, 1)$$

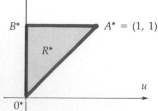

**FIGURE 5**
Region $R*$ in $uv$-plane
corresponding *to* $R$ in Figure 4.

in the $uv$-plane. Similarly, $B$ tranforms to $B* = (0,\, 1)$, and $O$ to $(0,\, 0)$ (Fig. 5). Since the equations (8) are linear, the straight edges of $R$ are transformed to straight line segments in the $uv$-plane; so $R*$ is the triangle $O*A*B*$.

To compute the Jacobian, solve (8) for $x$ and $y$:

$$x = \tfrac{1}{2}(u + v), \qquad y = \tfrac{1}{2}(u - v)$$

so

$$\frac{\partial(x, y)}{\partial(u, v)} = \det \begin{bmatrix} \tfrac{1}{2} & \tfrac{1}{2} \\ \tfrac{1}{2} & -\tfrac{1}{2} \end{bmatrix} = -\tfrac{1}{2}.$$

Hence

$$\iint\limits_{R} e^{(x+y)/(x-y)} \, dA = \iint\limits_{R^*} e^{u/v} \left| -\tfrac{1}{2} \right| \, dA_{uv}$$

$$= \int_0^1 \int_0^v e^{u/v} \tfrac{1}{2} \, du \, dv = \tfrac{1}{2} \int_0^1 v e^{u/v} \Big|_0^v \, dv$$

$$= \tfrac{1}{4}(e - 1).$$

## Orientation

The Jacobian in Example 1 is negative; does that mean anything?

The sign of the Jacobian is related to the *orientations* in the $xy$-plane and the $uv$-plane. In Figure 5, the path $O^* \to A^* \to B^* \to O^*$ around the triangle is counterclockwise, keeping $R^*$ on the left; the transformed path $O \to A \to B \to O$ in Figure 4 is *clockwise*, keeping $R$ on the right. *Transformations that reverse orientation have negative Jacobians.*

### The Jacobian of the Inverse Transformation

Along with any one-to-one transformation

$$x = x(u, v), \qquad y = y(u, v)$$

we have its inverse

$$u = u(x, y), \qquad v = v(x, y).$$

Under the first transformation, any small piece of the $uv$-plane with area $\Delta A_{uv}$ is transformed to a small piece of the $xy$-plane (Fig. 6) with area

$$\Delta A_{xy} \approx \left| \frac{\partial(x, y)}{\partial(u, v)} \right| \Delta A_{uv}. \tag{9}$$

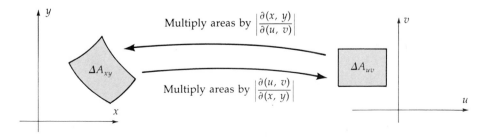

**FIGURE 6**

$$\left| \frac{\partial(x, y)}{\partial(u, v)} \right| = \left| \frac{\partial(u, v)}{\partial(x, y)} \right|^{-1}$$

Under the inverse, the relation between areas is

$$\Delta A_{uv} \approx \left|\frac{\partial(u, v)}{\partial(x, y)}\right| \Delta A_{xy}. \tag{10}$$

Comparing (9) and (10), you see that

$$\left|\frac{\partial(u, v)}{\partial(x, y)}\right| = \left|\frac{\partial(x, y)}{\partial(u, v)}\right|^{-1}.$$

Moreover, if the transformation reverses orientation, so does its inverse; hence both Jacobians have the same sign, and we can drop the absolute value:

$$\frac{\partial(x, y)}{\partial(u, v)} = \left[\frac{\partial(x, y)}{\partial(u, v)}\right]^{-1}.$$

This relation, deduced from the geometric meaning of the Jacobian, can be verified by a laborious calculation with the Chain Rule.

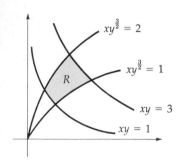

$xy^{\frac{3}{2}} = 2$

$xy^{\frac{3}{2}} = 1$

$R$

$xy = 3$

$xy = 1$

**FIGURE 7**

**EXAMPLE 3** In the theory of heat engines, the work done by the engine in one cycle of operation is the area enclosed by a "Carnot cycle," bounded by curves like those in Figure 7. Compute the area in the case of Figure 7.

*SOLUTION* The figure suggests new variables

$$u = xy, \qquad v = xy^{3/2}.$$

The new region of integration is the rectangle

$$R^* = \{(u, v): 1 \le u \le 3, 1 \le v \le 2\}.$$

It is easy to calculate

$$\frac{\partial(u, v)}{\partial(x, y)} = \det\begin{bmatrix} u_x & v_x \\ u_y & v_y \end{bmatrix} = \det\begin{bmatrix} y & y^{3/2} \\ x & \frac{3}{2}xy^{1/2} \end{bmatrix} = \frac{1}{2}xy^{3/2}.$$

But the Jacobian you need is $\dfrac{\partial(x, y)}{\partial(u, v)} = \left[\dfrac{\partial(u, v)}{\partial(x, y)}\right]^{-1} = \dfrac{2}{xy^{3/2}} = \dfrac{2}{v}.$

So the integral is

$$\iint_{R} dA_{xy} = \iint_{R^*} \left|\frac{\partial(x, y)}{\partial(u, v)}\right| dA_{uv} = \int_1^2 \int_1^3 \frac{2}{v}\, du\, dv = 4 \ln 2.$$

*Triple integrals* can be similarly transformed. The "volume multiplication factor" is

$$\left|\frac{\partial(x, y, z)}{\partial(u, v, w)}\right| = \left|\det\begin{bmatrix} x_u & y_u & z_u \\ x_v & y_v & z_v \\ x_w & y_w & z_w \end{bmatrix}\right|.$$

## SUMMARY

*Jacobian:*

$$\frac{\partial(x, y)}{\partial(u, v)} = \det \begin{bmatrix} x_u & y_u \\ x_v & y_v \end{bmatrix}$$

This is *negative* when the transformation reverses orientation.

*Inverses:*

$$\frac{\partial(u, v)}{\partial(x, y)} = \left[ \frac{\partial(x, y)}{\partial(u, v)} \right]^{-1}$$

*Change of Variable:*

$$\iint_R f(x, y) \, dA_{xy} = \iint_{R^*} f(x(u, v), y(u, v)) \left| \frac{\partial(x, y)}{\partial(u, v)} \right| dA_{uv},$$

where $(u, v) \rightarrow (x(u, v), y(u, v))$ transforms $R^*$ to $R$.

## PROBLEMS

### A

1. Compute the Jacobian. In each case, sketch the image of the unit square $\{(u, v); 0 \le u \le 1, 0 \le v \le 1\}$, and relate its area and orientation to the Jacobian.
   *a) $x = u, y = 2v$
   *b) $x = 2v, y = u$
   c) $x = u, y = u + v$

2. Compute the Jacobian. Sketch the image of the unit square $\{(u, v): 0 \le u \le 1, 0 \le v \le 1\}$, and relate its orientation to the Jacobian.
   *a) $x = u^2 - v^2, y = 2uv$
   b) $x = u^2 - v^2, y = -2uv$

3. The Jacobian for $x = r \cos \theta, y = r \sin \theta$, is $\frac{\partial(x, y)}{\partial(r, \theta)} = r$. Sketch the images of the following rectangles, and reconcile the sign of the Jacobian with any reversal of orientation.
   a) $0 \le r \le 1, \quad 0 \le \theta \le \pi/2$
   b) $-1 \le r \le 0, \quad 0 \le \theta \le \pi/2$

4. Evaluate the integral, using the given change of variable.
   *a) $\iint_R dx \, dy, R = \{(x, y): \frac{x^2}{a^2} + \frac{y^2}{b^2} \le 1\};$
   $x = au, y = bv$
   *b) $\int_0^2 \int_0^{2-y} e^{(y-x)/(y+x)} \, dx \, dy;$
   $u = y - x, v = y + x$

c) $\iint_R dx \, dy; R$ is in the first quadrant and bounded by $xy = 1, xy = 4, y = x, y = 4x;$ $xy = u^2, y/x = v^2$ ($u \ge 0$ and $v \ge 0$)

5. Let $R$ be the region bounded by $xy = a, xy = b, xy^\gamma = c, xy^\gamma = d$. Assume $2 > \gamma > 1, 0 < a < b, 0 < c < d$. Compute.
   a) $\iint_R dA$ *b) $\iint_R y \, dA$

*6. Compute the area of the first-quadrant region bounded by $y = x/2, y = 2x, xy = 1$, and $xy = 2$; use an appropriate change of variable.

7. For the transformation $x = u + v, y = u - v$
   a) Compute $\frac{\partial(x, y)}{\partial(u, v)}$.
   b) Solve for $u$ and $v$ in terms of $x$ and $y$.
   c) Compute $\frac{\partial(u, v)}{\partial(x, y)}$ using part b, and compare with part a.

8. For polar coordinates in the first quadrant
   a) Express $r$ and $\theta$ in terms of $x$ and $y$.
   b) Compute $\frac{\partial(r, \theta)}{\partial(x, y)}$, and compare it to
   $$\frac{\partial(x, y)}{\partial(r, \theta)} = r.$$

9. For spherical coordinates, compute the Jacobian
   $$\frac{\partial(x, y, z)}{\partial(\rho, \varphi, \theta)}.$$

**10.** For cylindrical coordinates, compute $\dfrac{\partial(x, y, z)}{\partial(r, \theta, z)}$.

**\*11.** Compute the area of the surface $z = x^2 - 2y^2$ lying inside the elliptic cylinder $4x^2 + 16y^2 = 16$. [Use new coordinates $x = 2u\cos(v)$, $y = u\sin(v)$.]

**B**

**12.** Suppose that $(x, y)$ are functions of $(u, v)$, and $(u, v)$ in turn are functions of $(s, t)$. Deduce, on geometric grounds, the relation between $\dfrac{\partial(x, y)}{\partial(u, v)}$, $\dfrac{\partial(u, v)}{\partial(s, t)}$, and $\dfrac{\partial(x, y)}{\partial(s, t)}$. Take orientation into account.

**C**

**13.** Use the Chain Rule to prove

**a)** $1 = \dfrac{\partial x}{\partial u}\dfrac{\partial u}{\partial x} + \dfrac{\partial x}{\partial v}\dfrac{\partial v}{\partial x}$, $\quad 0 = \dfrac{\partial x}{\partial u}\dfrac{\partial u}{\partial y} + \dfrac{\partial x}{\partial v}\dfrac{\partial v}{\partial y}$,

$0 = \dfrac{\partial y}{\partial u}\dfrac{\partial u}{\partial x} + \dfrac{\partial y}{\partial v}\dfrac{\partial v}{\partial x}$, $\quad 1 = \dfrac{\partial y}{\partial u}\dfrac{\partial u}{\partial y} + \dfrac{\partial y}{\partial v}\dfrac{\partial v}{\partial y}$.

**b)** $\dfrac{\partial(x, y)}{\partial(u, v)} \cdot \dfrac{\partial(u, v)}{\partial(x, y)} = 1$.

[Try to find an organizing principle for the calculation.]

# REVIEW PROBLEMS   CHAPTER 17

**\*1. a)** Evaluate $\int_0^1 \int_0^x (x + 5y)\, dy\, dx$.
**b)** Change the order of integration, and then re-evaluate the integral.
**c)** Change to polar coordinates, and then re-evaluate the integral.

**\*2.** Rewrite $\int_{\pi/2}^{3\pi/2} \int_0^1 r^2\, dr\, d\theta$ in rectangular coordinates.

**\*3.** Compute the area of the surface $z = x + y^2$ lying over the triangle bounded by $x = 0$, $y = 2x$, and $y = 1$. (Choose the convenient order of integration.)

**\*4.** Compute the area of the part of the cylinder $x^2 + z^2 = 1$ lying inside the cylinder $x^2 + y^2 = 1$. (Use rectangular coordinates, in the convenient order.)

**\*5.** Find the center of mass of a plate covering the quarter circle of radius 1 in the first quadrant, given
**a)** Constant density.
**b)** Density $\delta(x, y) = (1 + x^2 + y^2)^{-2}$.
Which center of mass is closer to the origin? Why?

**\*6.** Find the centroid of the region inside $r = 1$ and $r = 2\cos\theta$, and above the $x$-axis.

**\*7.** Find the center of mass of a solid with constant density, occupying the given region.
**a)** Bounded by the planes $x = 1$, $z = 0$, $y = 0$, $z = x$, $y = x$.
**b)** $x^2 + y^2 \le a^2$, $y \ge 0$, $0 \le z \le y$.

**\*8.** Convert

$$\int_0^1 \int_0^{\sqrt{1-z^2}} \int_0^{\sqrt{1-x^2-z^2}} xe^{-x^2-y^2-z^2}\, dy\, dx\, dz$$

to
**a)** Cylindrical coordinates.
**b)** Spherical coordinates.

**9.** Convert $\int_0^\pi \int_0^{\pi/4} \int_0^1 d\rho\, d\varphi\, d\theta$ to
**a)** Cylindrical coordinates.
**b)** Rectangular coordinates.

**10.** The moment of inertia of a homogeneous plate in the $xy$-plane, with respect to the origin, is

$$I_z = \iint_R (x^2 + y^2)\, dA;$$

$R$ is the region covered by the plate. Evaluate this when $R$ is bounded by the ellipse $\dfrac{x^2}{a^2} + \dfrac{y^2}{b^2} = 1$, using new variables $u$ and $v$ defined by $x = au\cos(v)$, $y = bu\sin(v)$.

**11.** New variables $(u, v)$ are related to $(x, y)$ by $x = uv$, $y = u/v$.
**a)** Sketch a region $R^*$ in the first quadrant of the $(u, v)$ plane, corresponding to the region $R$ in the first quadrant of the $(x, y)$ plane bounded by $xy = 1$, $xy = 4$, $x = y$, $x = 4y$.
**b)** Does this transformation preserve orientation, or reverse it?
**c)** Compute the area of the region $R$ in part a.

# 18

## LINE AND SURFACE INTEGRALS

# 18.1

## WORK AND LINE INTEGRALS

If a constant force $\vec{F}$ moves an object from $P_1$ to $P_2$, then the work done is the dot product of the force $\vec{F}$ by the displacement $\vec{P}_2 - \vec{P}_1$ (Fig. 1):

$$\text{work } W = |\vec{F}| \cos\theta \, |\vec{P}_2 - \vec{P}_1| = \vec{F} \cdot (\vec{P}_2 - \vec{P}_1).$$

The product $|\vec{F}| \cos\theta$ is the *component* of $\vec{F}$ in the direction of the motion, the work done is this force component times the distance traveled.

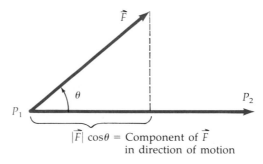

**FIGURE 1**
Work $= |\vec{F}| \cos\theta \, |\vec{P}_2 - \vec{P}_1| = \vec{F} \cdot (\vec{P}_2 - \vec{P}_1).$

If the force varies from point to point, and the path of the object is curved, then this simple product is replaced by an integral.

For a concrete example of a force varying with position, imagine a proton fixed at the origin, attracting an electron with an "inverse square" force directed toward the origin. Denote by $\vec{F}(P)$ the force acting on the electron when it is at position $P = (x, y, z)$ in space. In suitable units, the force is

$$\vec{F}(P) = -\frac{\vec{P}}{|\vec{P}|^3} = \frac{-x\vec{i} - y\vec{j} - z\vec{k}}{(x^2 + y^2 + z^2)^{3/2}}.$$

Figure 2 illustrates this at points $(x, y, 0)$ in the $xy$-plane; the force at $P$ is represented by an arrow beginning at $P$, with the appropriate length and direction. Since vector $\vec{P}$ points from the origin to $P$, the force $\vec{F}(P) = (-|\vec{P}|^{-3})\vec{P}$ points in the opposite direction, from $P$ towards the origin. Its magnitude is

$$|\vec{F}(P)| = |-\vec{P}/|\vec{P}|^3| = |\vec{P}|/|\vec{P}|^3 = 1/|\vec{P}|^2,$$

one over the square of the distance to the origin. So the formula for $\vec{F}$ does indeed describe an inverse square central force, directed toward the origin.

Now, given a variable force $\vec{F}$ such as this, we compute the work done on an object moving along a curve

$$\vec{P}(t) = x(t)\vec{i} + y(t)\vec{j} + z(t)\vec{k}, \qquad a \le t \le b.$$

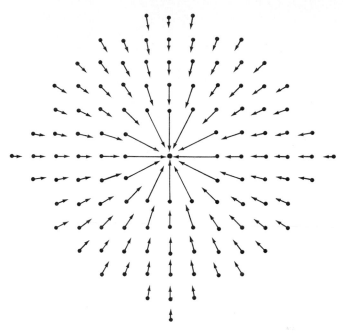

**FIGURE 2**
Inverse square central force.

Divide the time interval $[a, b]$ into $n$ equal parts of length $\Delta t$, thus dividing the path of motion into short segments (Fig. 3). During a short time interval $[t_{j-1}, t_j]$:

**(i)** The force on the particle at time $t$ is $\vec{F}(P(t))$, with $t$ varying from $t_{j-1}$ to $t_j$. Since the interval is short, $t$ is near $t_{j-1}$, and $\vec{F}(P(t))$ is nearly $\vec{F}(P(t_{j-1}))$.

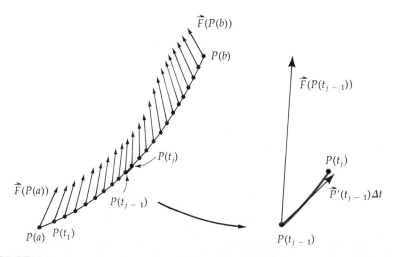

**FIGURE 3**
Curve subdivided in many small pieces, with enlarged view of section from $P(t_{j-1})$ to $P(t_j)$.

**(ii)** The path is nearly straight, and the displacement $\vec{P}(t_j) - \vec{P}(t_{j-1})$ is nearly $\vec{P}'(t_{j-1})\,\Delta t$, so

**(iii)** the work done in this time interval is nearly the dot product

$$(\text{force}) \cdot (\text{displacement}) \approx \vec{F}(P(t_{j-1})) \cdot [\vec{P}(t_j) - \vec{P}(t_{j-1})]$$
$$\approx \vec{F}(P(t_{j-1})) \cdot \vec{P}'(t_{j-1})\,\Delta t.$$

The work done from time $a$ to time $b$ is the sum of the contributions from each interval, so

$$W \approx \sum_1^n \vec{F}(P(t_{j-1})) \cdot \vec{P}'(t_{j-1})\,\Delta t. \tag{1}$$

As $n \to \infty$ and $\Delta t \to 0$, the limit of the sum in (1) gives the work exactly. But this is a Riemann sum; its limit is an integral, and we conclude that

$$W = \int_a^b \vec{F}(P(t)) \cdot \vec{P}'(t)\,dt. \tag{2}$$

This formula applies in two dimensions, as well as three.

---

**EXAMPLE 1** An object moves along a parabola in the $xy$-plane, with position vector

$$\vec{P}(t) = t\vec{i} + (4 - t^2)\vec{j}, \qquad 0 \le t \le 2. \tag{3}$$

Compute the work done on this object by a variable force

$$\vec{F}(x, y) = -x\vec{i} - y\vec{j}.$$

**SOLUTION** Along the given curve (3), $x = t$ and $y = 4 - t^2$, so

$$\vec{F}(P(t)) = -t\vec{i} - (4 - t^2)\vec{j}. \tag{4}$$

The derivative of the position vector (3) is $\vec{P}'(t) = \vec{i} - 2t\vec{j}$, so the work integral (2) is

$$W = \int_0^2 [-t\vec{i} - (4 - t^2)\vec{j}] \cdot [\vec{i} - 2t\vec{j}]\,dt$$
$$= \int_0^2 (7t - 2t^3)\,dt = 6.$$

Figure 4 shows the path, and the force at several points along the way.

---

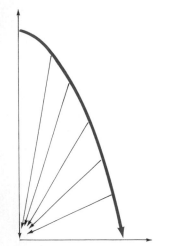

**FIGURE 4**
Force $\vec{F} = -x\vec{i} - y\vec{j}$ along curve $\vec{P}(t) = t\vec{i} + (4 - t^2)\vec{j}, \; 0 \le t \le 2.$

The integral in formula (2) is called the **line integral** of the force $\vec{F}$ over the given curve. We normally denote the curve of motion by a single letter $C$, writing

$$C: \vec{P}(t) = \langle x(t), y(t), z(t) \rangle, \qquad a \le t \le b.$$

To describe the motion, we give the functions $x(t)$, $y(t)$, $z(t)$ (the coordinates at time $t$) and the relevant time interval. In the integral (2) we abbreviate

$\vec{P}'(t)\, dt$ as $d\vec{P}$, and write the line integral as $\int_C \vec{F} \cdot d\vec{P}$:

$$\int_C \vec{F} \cdot d\vec{P} = \int_a^b \vec{F}(P(t)) \cdot \vec{P}'(t)\, dt. \qquad (5)$$

Another notation, convenient for calculations, displays the components of the force. Denote these components by $F_1$, $F_2$, and $F_3$:

$$\vec{F} = F_1\vec{i} + F_2\vec{j} + F_3\vec{k}.$$

On the curve $\vec{P} = x(t)\vec{i} + y(t)\vec{j} + z(t)\vec{k}$, we have $dx = x'\, dt$, etc., so

$$d\vec{P} = \vec{P}(t)\, dt = x'\, dt\,\vec{i} + y'\, dt\,\vec{j} + z'\, dt\,\vec{k}$$
$$= dx\,\vec{i} + dy\,\vec{j} + dz\,\vec{k}.$$

Hence the line integral (5) is written

$$\int_C \vec{F} \cdot d\vec{P} = \int_C F_1\, dx + F_2\, dy + F_3\, dz. \qquad (6)$$

The curve $C$ gives $x$, $y$, and $z$ as functions of time $t$ (or some other parameter). Substitute these functions in the formula for $\vec{F}$, and also use them to express $dx$, $dy$, $dz$ in terms of the parameter on the curve; then evaluate the resulting integral.

---

**EXAMPLE 1 (redone)**    For the force $\vec{F} = -x\vec{i} - y\vec{j}$, the components are $F_1 = -x$, $F_2 = -y$, $F_3 = 0$ so, using (6),

$$\int_C \vec{F} \cdot d\vec{P} = \int_C F_1\, dx + F_2\, dy + F_3\, dz = \int_C -x\, dx - y\, dy.$$

The curve in Example 1 is given by $x = t$, $y = 4 - t^2$, $0 \le t \le 1$, so $dx = dt$, $dy = -2t\, dt$, and

$$\int_C -x\, dx - y\, dy = \int_0^1 (-t)\, dt - (4 - t^2)(-2t\, dt)$$
$$= \int_0^1 (7t - 2t^3)\, dt = 6$$

as before.

---

Yet another notation, useful for interpreting the line integral, involves the unit tangent of the curve (Sec. 14.5),

$$\vec{T} = \frac{1}{|\vec{P}'(t)|}\, \vec{P}'(t),$$

and the arc length function $s$. Recall that $\dfrac{ds}{dt} = |\vec{v}(t)| = |\vec{P}'(t)|$; so $ds = |\vec{P}'(t)|\, dt$.

Hence we can write $\vec{P}'(t)\, dt$ as follows:

$$\vec{P}'(t)\, dt = \left[ \frac{1}{|\vec{P}'(t)|}\, \vec{P}'(t) \right] [|\vec{P}'(t)|\, dt] = \vec{T}\, ds.$$

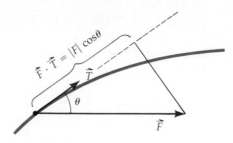

**FIGURE 5**
$\vec{F} \cdot \vec{T}$ = component of $\vec{F}$ in direction of $\vec{T}$.

So the line integral (5) is

$$\int_C \vec{F} \cdot d\vec{P} = \int_a^b \vec{F}(P(t)) \cdot \vec{T} \, ds$$

or, briefly, $\int_C \vec{F} \cdot \vec{T} \, ds$. The dot product $\vec{F} \cdot \vec{T}$ is the component of the force in the direction of motion along the curve (Fig. 5):

$$\vec{F} \cdot \vec{T} = |\vec{F}| \cos \theta |\vec{T}| = |\vec{F}| \cos \theta.$$

And we think of $ds$ as the length of a short piece of the curve, the distance traveled as the object moves just a little. The integrand $\vec{F} \cdot \vec{T} \, ds$ is this short distance times the component of the force in the direction of motion; this gives the little bit of work done along that short piece. The work done along the entire curve is the sum of all these little bits, the integral $\int_C \vec{F} \cdot \vec{T} \, ds$. (Other notations in use include $\int_C \vec{F} \cdot d\vec{s}$ and $\int_C F \cdot d\vec{r}$.)

---

**EXAMPLE 2** Compute the work done by the force $\vec{F} = -\frac{1}{2}y\vec{i} + \frac{1}{2}x\vec{j}$ acting on a particle moving along:

    **a.** The semicircle $C_1$: $\vec{P}(t) = \cos(t)\vec{i} + \sin(t)\vec{j}$, $0 \le t \le \pi$.
    **b.** The same semicircle run backwards,

$$C_2: \vec{P}(t) = \cos(\pi - t)\vec{i} + \sin(\pi - t)\vec{j}, \qquad 0 \le t \le \pi.$$

    **c.** The straight line from $(-1, 0)$ to $(1, 0)$,

$$C_3: \vec{P}(t) = (2t - 1)\vec{i}, \qquad 0 \le t \le 1.$$

**SOLUTION** For the force $\vec{F} = -\frac{1}{2}y\vec{i} + \frac{1}{2}x\vec{j}$, the line integral is $\int_C -\frac{1}{2}y \, dx + \frac{1}{2}x \, dy$, and you can easily compute these integrals using this formula. However, it is more illuminating to deduce the integrals from the relation between the force $\vec{F}$ and the unit tangent $\vec{T}$, as follows.

    **a.** Figure 6 shows the force $\vec{F}$ at various points along the semicircle $C_1$. Apparently the force is always tangent to this curve, and pointing in the direction of motion [from $P(0) = (1, 0)$ to $P(1) = (-1, 0)$]. So the angle $\theta$ between $\vec{F}$ and $\vec{T}$ is 0, and the component of $\vec{F}$ in the direction of motion is

$$\vec{F} \cdot \vec{T} = |\vec{F}| \cos \theta = |\vec{F}|$$
$$= \sqrt{(\tfrac{1}{2}y)^2 + (\tfrac{1}{2}x)^2} \qquad [\vec{F} = -\tfrac{1}{2}y\vec{i} + \tfrac{1}{2}x\vec{j}]$$
$$= \tfrac{1}{2}\sqrt{x^2 + y^2} = \tfrac{1}{2}$$

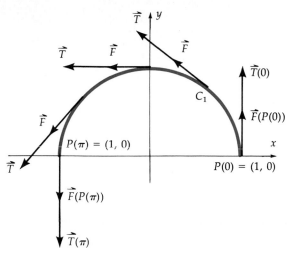

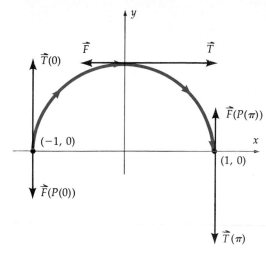

**FIGURE 6**
$\int_{C_1} (-\frac{1}{2}y \, dx + \frac{1}{2}x \, dy) = \int_{C_1} \vec{F} \cdot \vec{T} \, ds = \int_{C_1} \frac{1}{2} \, ds = \frac{1}{2}\pi.$

**FIGURE 7**
$\int_{C_2} \vec{F} \cdot \vec{T} \, ds = \int_{C_2} (-\frac{1}{2}) \, ds = -\frac{1}{2}\pi.$

since, on the given curve $C_1$, $x^2 + y^2 = \cos^2 t + \sin^2 t = 1$. The line integral is therefore

$$\int_{C_1} \vec{F} \cdot \vec{T} \, ds = \int_{C_1} \frac{1}{2} \, ds = \frac{1}{2}(\text{length of } C) = \frac{1}{2}\pi.$$

**b.**   Figure 7 shows $C_2$, which is just $C_1$ run in reverse. Now the angle between $\vec{F}$ and $\vec{T}$ is $\theta = \pi$, so $\cos \theta = -1$ and

$$\vec{F} \cdot \vec{T} = -|\vec{F}| = -\frac{1}{2}.$$

Along this curve the line integral is therefore

$$\int_{C_2} \vec{F} \cdot \vec{T} \, ds = \int_{C_2} (-\frac{1}{2}) \, ds = -\frac{1}{2}\pi.$$

The work done by $\vec{F}$ is negative, since $\vec{F}$ and $\vec{T}$ point in opposite directions and $\vec{F}$ acts *against* the motion.

**c.**   Figure 8 shows $\vec{F}$ along the straight line curve $C_3$. Now $\vec{F}$ is perpendicular to the direction of motion, so $\vec{F} \cdot \vec{T} = 0$, and $\int_{C_3} \vec{F} \cdot \vec{T} \, ds = 0$.

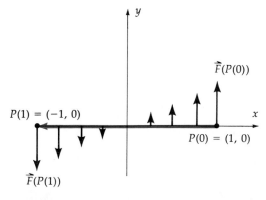

**FIGURE 8**
$\int_{C_3} \vec{F} \cdot \vec{T} \, ds = 0.$

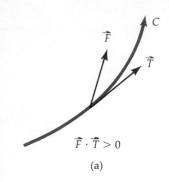

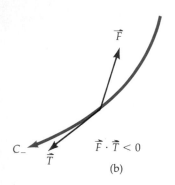

$$\vec{F} \cdot \vec{T} > 0$$

(a)

$$\vec{F} \cdot \vec{T} < 0$$

(b)

**FIGURE 9**
Reversing direction of curve
reverses sign of line integral:
$\int_{C_-} \vec{F} \cdot \vec{T} \, ds = -\int_{C} \vec{F} \cdot \vec{T} \, ds.$

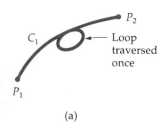

(a)

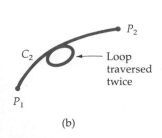

(b)

**FIGURE 10**
$\int_{C_1} \vec{F} \cdot d\vec{P}$ need not equal $\int_{C_2} \vec{F} \cdot d\vec{P}$.

The notation $\int_C \vec{F} \cdot \vec{T} \, ds$ suggests two important properties of line integrals:

**I.** *Reversing the direction of the curve C reverses the sign of the integral.* Example 2 illustrates this; the curve $C_2$ runs in the direction opposite to $C_1$, so the direction of the unit tangent $\vec{T}$ is reversed. This reverses the sign of $\vec{F} \cdot \vec{T}$, and with it the integral $\int_C \vec{F} \cdot \vec{T} \, ds$. Figure 9 illustrates the same fact in general; there, $C_-$ stands for the curve C run backward.

**II.** *The line integral does not depend on the particular parametrization used to represent C, as long as it runs in the proper direction.* For, the unit tangent $\vec{T}$ and the arc length differential $ds$ can be computed equally well with any parametrization. What matters about the curve is *where it starts*, where it *ends*, and the succession of intermediate points. The velocity $\vec{P}'$ does not matter, because the forces we are considering depend only on position P, not on velocity $\vec{P}'$ or time t. Bear in mind, however, that if some part of the curve is traversed more than once, this can affect the line integral. For instance, in Figure 10, the line integral along $C_2$ need not be the same as along $C_1$; the work done going around the loop is counted twice on $C_2$, but only once on $C_1$.

(These two properties are stated and proved more formally in problem 10.)

---

**EXAMPLE 3** Compute $\int_C x \, dy - y \, dx$ along the straight-line segment from $P_1 = (0, 2)$ to $P_2 = (2, 0)$, and again from $P_2$ to $P_1$.

**FIRST SOLUTION** Parametrize the path C from $P_1$ to $P_2$ by the standard formula for the line through given points $P_1$ and $P_2$: $\vec{P} = \vec{P}_1 + t(\vec{P}_2 - \vec{P}_1)$ (Fig. 11a). This gives the curve from $P_1$ to $P_2$ as

$$C: \vec{P}(t) = \langle 0, 2 \rangle + t\langle 2, -2 \rangle = \langle 2t, 2 - 2t \rangle, \qquad 0 \leq t \leq 1.$$

Then

$$\int_C x \, dy - y \, dx = \int_0^1 (2t)(-2 \, dt) - (2 - 2t)(2 \, dt) = \int_0^1 (-4) \, dt = -4.$$

To go in reverse, from $P_2$ to $P_1$, use $\vec{P} = \vec{P}_2 + t(\vec{P}_1 - \vec{P}_2)$ (Fig. 11b). This gives a new curve, which we call $C_-$:

$$C_-: \vec{P}_-(t) = \langle 2, 0 \rangle + t\langle -2, 2 \rangle = \langle 2 - 2t, 2t \rangle, \qquad 0 \leq t \leq 1.$$

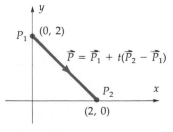

(a) Path from $P_1$ to $P_2$.

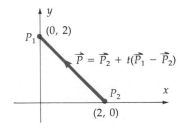

(b) Path from $P_2$ to $P_1$.

**FIGURE 11**

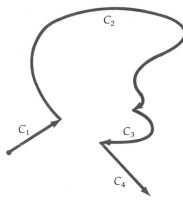

$C_2$

$C_1$   $C_3$

$C_4$

(a)   Chain of curves.

$C_2$

$C_1$

$C_3$

(b)   Closed chain.

**FIGURE 12**
Chain of curves.

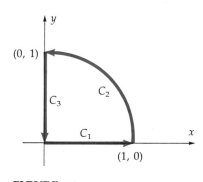

**FIGURE 13**
Chain of curves bounding a quarter circle.

The line integral over $C_-$ is

$$\int_{C_-} x\,dy - y\,dx = \int_0^1 (2 - 2t)(2\,dt) - (2t)(-2\,dt) = \int_0^1 4\,dt = 4,$$

the negative of the integral over the original path $C$ from $P_1$ to $P_2$.

**SECOND SOLUTION**   Represent the curve $C$ as $y = 2 - x$, $0 \le x \le 2$. As $x$ varies from 0 to 2, the curve moves from $P_1 = (0, 2)$ (where $x = 0$) to $P_2 = (2, 0)$ (where $x = 2$); thus $x = 0$ will be the lower limit of integration, and $x = 2$ the upper limit. Since $y = 2 - x$, then $dy = -dx$ and

$$\int_C x\,dy - y\,dx = \int_0^2 x(-dx) - (2 - x)\,dx = \int_0^2 -2\,dx = -4.$$

In the reverse direction, take $y = 2 - x$ again, but now $x$ runs from 2 to 0:

parameter value at terminal point (0, 2)
↓
$$\int_{C_-} x\,dy - y\,dx = \int_2^0 x(-dx) - (2 - x)\,dx = \int_2^0 -2\,dx = +4.$$
↑
parameter value at initial point (2, 0)

*Moral: The lower limit of integration gives the initial point, and the upper limit gives the terminal point.*

---

**A chain of curves** is a finite collection of curves $C_1, C_2, \ldots, C_k$, where the terminal point of each curve is the starting point of the next (Fig. 12). The chain is *closed* if the last link ends where the first one began. A line integral over a chain is, by definition, the sum of the integrals over the links.

---

**EXAMPLE 4**   Integrate $\int_C \vec{F} \cdot d\vec{P}$ where

$$\vec{F}(x, y) = -y\vec{i} + x\vec{j}$$

and $C$ is the boundary of the quarter circle in Figure 13, traversed in the indicated direction.

**SOLUTION**   The boundary is a chain with three links

$$C_1: \vec{P}_1(t) = \langle t, 0 \rangle, \qquad 0 \le t \le 1$$
$$C_2: \vec{P}_2(t) = \langle \cos t, \sin t \rangle, \qquad 0 \le t \le \pi/2$$
$$C_3: \vec{P}_3(t) = \langle 0, 1 - t \rangle, \qquad 0 \le t \le 1.$$

Check that $C_1$ goes in a straight line

from $P_1(0) = (0, 0)$ to $P_1(1) = (1, 0)$;

$C_2$ along the circle $x^2 + y^2 = 1$

from $P_2(0) = (1, 0)$ to $P_2(\pi/2) = (0, 1)$,

and $C_3$ along a straight line

from $P_3(0) = (0, 1)$ to $P_3(1) = (0, 0)$.

The integral of $\vec{F}$ over the chain $C$ consisting of these three curves is

$$\int_C \vec{F} \cdot d\vec{P} = \int_{C_1} \vec{F} \cdot d\vec{P} + \int_{C_2} \vec{F} \cdot d\vec{P} + \int_{C_3} \vec{F} \cdot d\vec{P}$$

$$= \int_0^1 (-0\vec{i} + t\vec{j}) \cdot \vec{i}\, dt$$

$$+ \int_0^{\pi/2} (-\sin t\, \vec{i} + \cos t\, \vec{j}) \cdot (-\sin t\, \vec{i} + \cos t\, \vec{j})\, dt$$

$$+ \int_0^1 (-(1-t)\vec{i} + 0\vec{j}) \cdot (0\vec{i} - \vec{j})\, dt$$

$$= \int_0^1 0\, dt + \int_0^{\pi/2} (\sin^2 t + \cos^2 t)\, dt + \int_0^1 0\, dt$$

$$= \pi/2.$$

## SUMMARY

***The Line Integral*** of $\vec{F} = F_1\vec{i} + F_2\vec{j} + F_3\vec{k}$ over a curve

$$C: \vec{P}(t) = \langle x(t), y(t), z(t)\rangle, \qquad a \leq t \leq b$$

is

$$\int_C \vec{F} \cdot d\vec{P} = \int_a^b \vec{F}(P(t)) \cdot \vec{P}'(t)dt = \int_C \vec{F} \cdot \vec{T}\, ds$$

where $\vec{T} = \dfrac{\vec{P}'}{|\vec{P}'|}$ and $ds = |\vec{P}'|\, dt$. In differential notation

$$\int_C \vec{F} \cdot d\vec{P} = \int_C F_1\, dx + F_2\, dy + F_3\, dz.$$

Reversing the direction of $C$ reverses the sign of the line integral:

$$\int_{C_-} \vec{F} \cdot d\vec{P} = -\int_C \vec{F} \cdot d\vec{P}.$$

If force $\vec{F}$ acts on a particle as it moves along $C$, then the work done by $\vec{F}$ is $\int_C \vec{F} \cdot d\vec{P}$.

## PROBLEMS

### A

1. Compute the following line integrals $\int_C \vec{F} \cdot d\vec{P}$.

*a) $\vec{F} = x^2 y\vec{i} + xy^2\vec{j}$, $\vec{P}(t) = t\vec{i} + t^2\vec{j}$, $0 \leq t \leq 1$

b) $\int_C (x \sin y\, dx + xy^2\, dy)$, $C$: $x = t$, $y = 2t$, $0 \leq t \leq 1$

*c) $\int_C (y + z)\, dx + (z + x)\, dy + (x + y)\, dz$, $C: x = t^2, y = t^3, z = t^4, 0 \leq t \leq 1$

d) $\int_C y^2\, dx + x\, dy$ where $C$ is the graph of $y = x^2$ from $(-1, 1)$ to $(1, 1)$ run from left to right.

*e) $\int_C y^2\, dx + x\, dy$, where $C$ is the graph of $y = x^2$ from $(1, 1)$ to $(-1, 1)$, run from right to left

f) $\int_C \vec{F} \cdot \vec{T}\, ds$ where $\vec{F} = x\vec{i} + y\vec{j} + z\vec{k}$ and $C$ is the helix

$$x = 2\cos t, \quad y = 2\sin t, \quad z = 3t,$$
$$0 \leq t \leq \pi.$$

**\*2.** An electron is pulled away from a proton, along a straight line from one point $(ra, rb, rc)$ to another $(Ra, Rb, Rc)$, where $a^2 + b^2 + c^2 = 1$. The force required is $\vec{F}(P) = \vec{P}/|P|^3$. Compute the work done.

**3.** Compute $\int_C (-y \, dx + x \, dy)$ where $C$ is the given chain. (Each chain is the boundary of a plane region. Do you find any relation between the value of this particular line integral and the area of the region?)

   **\*a)** The boundary of the square $\{(x, y): 0 \leq x \leq 1, 0 \leq y \leq 1\}$ taken counterclockwise.

   **b)** The boundary of the square in (a), but taken clockwise.

   **c)** The straight line from $(-2, 0)$ to $(2, 0)$, followed by the upper half of the circle $x^2 + y^2 = 4$.

**4.** Compute $\int_C -y \, dx + x \, dy$ along the following curves. Note that each curve goes along the unit circle, starting and ending at $(1, 0)$; but some go with different velocities, or in different directions, or traverse their paths more than once.

   **\*a)** $\vec{P}(t) = \cos t \, \vec{i} + \sin t \, \vec{j}, \ 0 \leq t \leq 2\pi$

   **b)** $\vec{P}(t) = \cos \pi t \, \vec{i} + \sin \pi t \, \vec{j}, \ 0 \leq t \leq 2$

   **\*c)** $\vec{P}(t) = \cos t \, \vec{i} + \sin t \, \vec{j}, \ 0 \leq t \leq 4\pi$

   **d)** $\vec{P}(t) = \cos t \, \vec{i} - \sin t \, \vec{j}, \ 0 \leq t \leq 2\pi$

**5.** The force on an electron due to an infinitely long wire with a constant charge density is

$$\vec{F} = \frac{x\vec{i} + y\vec{j}}{x^2 + y^2}.$$

Compute the work done by this force in moving a particle along the given curve. Sketch the force at several points along the curve to interpret the sign of the result.

   **\*a)** The circle $x^2 + y^2 = a^2, z = 0$

   **b)** The straight line from $(1, 0, 0)$ to $(1, 0, 2)$

   **\*c)** The helix $x = a \cos \pi t, y = a \sin \pi t, z = t, 0 \leq t \leq 2$

   **d)** The straight line from $(1, 1, 0)$ to $(2, 2, 0)$

**6.** Let $C$ be the graph of $y = f(x), a \leq x \leq b$. Explain why

   **a)** $\int_C xy \, dx = \int_a^b xf(x) \, dx.$

   **b)** $\int_C (x + y) \, dy = \int_a^b [x + f(x)]f'(x) \, dx.$

   **c)** For any continuous functions $M(x, y)$ and $N(x, y), \int_C M(x, y) \, dx + N(x, y) \, dy = \int_a^b M(x, f(x)) \, dx + \int_a^b N(x, f(x))f'(x) \, dx.$

**7.** The magnetic field around a very long wire carrying a steady electric current is

$$\vec{H} = \frac{-y}{x^2 + y^2} \vec{i} + \frac{x}{x^2 + y^2} \vec{j},$$

in suitable units. Compute the line integral of this vector field along the following paths.

   **\*a)** The circle $x^2 + y^2 = a^2, z = 0$, traversed counterclockwise

   **b)** The circle $x^2 + y^2 = a^2, z = 0$, traversed clockwise

   **\*c)** The straight line from $(a, 0, 0)$ to $(a, 0, 2)$.

   **d)** The helix $x = a \cos \pi t, y = a \sin \pi t, z = t, 0 \leq t \leq 2$

**8.** Show that

   **a)** The curve in part c of the previous problem starts and ends at the same places as the curve in part d.

   **b)** The line integral of $\vec{H}$ along the two curves is not the same.

**B**

**9.** This problem shows that if an object of mass $m$ moves along a path $C$, propelled solely by a force $\vec{F}$ according to Newton's law $\vec{F} = m\vec{a}$, then the work done by $\vec{F}$ equals the increase in kinetic energy $\frac{1}{2}m|\vec{v}|^2$. Show that

   **a)** $\dfrac{d}{dt}(m\vec{v} \cdot \vec{v}) = 2m\vec{a} \cdot \vec{v}.$

   **b)** $m\vec{a} \cdot \vec{v} = \vec{F} \cdot \vec{P}'.$

   **c)** If $C$ is given by $P(t), a \leq t \leq b$, then $\int_C \vec{F} \cdot d\vec{P} = \frac{1}{2}m|\vec{v}(b)|^2 - \frac{1}{2}m|\vec{v}(a)|^2$, the increase in kinetic energy.

**10.** Let $\vec{P}(t), a \leq t \leq b$, be a parametrized curve, and $\vec{F}$ a vector field defined along it. Suppose that a new parameter $\theta$ is related to the original one $t$ by $t = \varphi(\theta), \alpha \leq \theta \leq \beta$. This defines a new parametrization $\tilde{P}(\theta) = \vec{P}(\varphi(\theta)), \alpha \leq \theta \leq \beta$.

   **a)** If $\varphi(\alpha) = a$ and $\varphi(\beta) = b$, then $\varphi$ is *orientation-preserving*. Show that in this case,

$$\int_\alpha^\beta \vec{F}(\tilde{P}(\theta)) \cdot \tilde{P}'(\theta) \, d\theta = \int_a^b \vec{F}(P(t)) \cdot \vec{P}'(t) \, dt.$$

Thus, the line integral is unaffected by an orientation-preserving change of parameter.

   **b)** If $\varphi(\alpha) = b$ and $\varphi(\beta) = a$, then $\varphi$ is *orientation-reversing*. Show that in this case,

$$\int_\alpha^\beta \vec{F}(\tilde{P}(\theta)) \cdot \tilde{P}'(\theta) \, d\theta = -\int_a^b \vec{F}(P(t)) \cdot \vec{P}'(t) \, dt.$$

## 18.2

## GRADIENT FIELDS. CURL

A *vector field* $\vec{F}$ assigns to each point $P$ in space a vector $\vec{F}(P)$. A familiar example is the inverse square central force occurring in gravity and electricity, discussed in Section 18.1:

$$\vec{F}(P) = -\frac{\vec{P}}{|\vec{P}|^3}. \tag{1}$$

Figure 1 shows $\vec{F}$ for points in the $xy$-plane; the arrow beginning at each point $P$ gives the magnitude and direction of the force at $P$.

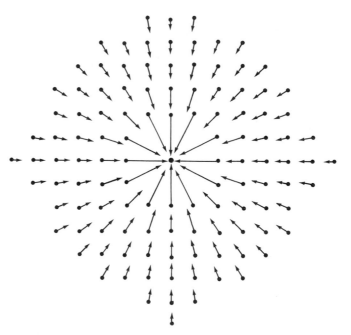

**FIGURE 1**
Inverse square vector field $\vec{F}(P) = -|\vec{P}|^{-3}\vec{P}$.

For another example, imagine a plate covering the plane, rotating counterclockwise around the origin at $\omega$ radians per second. For any $(x, y)$ in the plane, the point on the plate lying over $(x, y)$ moves with velocity

$$\vec{F}(x, y) = -\omega y\vec{i} + \omega x\vec{j}$$

as you may check (problem 16). A sketch of the vector field shows the rotating motion of the plate (Fig. 2); the arrow at each point $P$ represents the velocity at $P$.

$\vec{F}$ is a **gradient field** if it is the gradient of some function $f$; that is, if

$$\vec{F} = \nabla f = f_x\vec{i} + f_y\vec{j} + f_z\vec{k}.$$

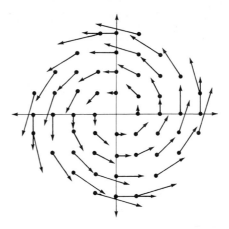

**FIGURE 2**
Rotation velocity field $\vec{F}(x, y) = -\omega y i + \omega x j$.

In this case, the scalar function $f$ is called a **potential** for the vector field $\vec{F}$. The inverse square central force (1) is a gradient, with potential $f(P) = 1/|P|$ (problem 10):

$$\frac{-\vec{P}}{|\vec{P}|^3} = \nabla \frac{1}{|\vec{P}|} \qquad \text{if } \vec{P} \neq \vec{0}.$$

The line integral of a gradient field is easily evaluated. If $\vec{F} = \nabla f$, and $C$ is given by $\vec{P}(t), a \leq t \leq b$, then

$$\int_C \vec{F} \cdot d\vec{P} = \int_a^b \vec{F}(P(t)) \cdot \vec{P}'(t)\, dt = \int_a^b \nabla f(P(t)) \cdot \vec{P}'(t)\, dt$$

$$= \int_a^b \left[\frac{d}{dt} f(P(t))\right] dt \qquad \text{(Chain Rule!)}$$

$$= f(P(b)) - f(P(a)) \qquad \qquad (2)$$

by the Fundamental Theorem of Calculus. Thus (Fig. 3):

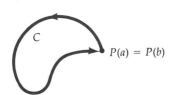

$P(b) = \text{End of } C$

$C$

$P(a) = \text{Beginning of } C$

**FIGURE 3**
$\int_C \nabla f \cdot d\vec{P} = f(P(b)) - f(P(a))$.

---

**THEOREM 1**

The line integral of a gradient field along a curve $C$ depends only on the beginning and end points of $C$. Precisely,

$$\int_C \nabla f \cdot d\vec{P} = f(\text{end of } C) - f(\text{beginning of } C). \qquad (3)$$

---

$C$

$P(a) = P(b)$

**FIGURE 4**
If $C$ is a closed curve, then
$\int_C \nabla f \cdot d\vec{P} = f(P(b)) - f(P(a)) = 0$.

If the curve $C$ is closed (Fig. 4) then the beginning and end are the same point, so the line integral is zero:

---

**THEOREM 2**

The line integral of a gradient field around a closed curve is zero.

**EXAMPLE 1** The inverse square central force $\vec{F}(P) = -|\vec{P}|^{-3}\vec{P}$ is the gradient of the function $f(P) = |\vec{P}|^{-1}$. So if $C$ is given by $\vec{P}(t)$, $a \leq t \leq b$, then

$$\int_C |\vec{P}|^{-3}\vec{P} \cdot d\vec{P} = \int_C \nabla|\vec{P}|^{-1} \cdot d\vec{P}$$
$$= |\vec{P}(b)|^{-1} - |\vec{P}(a)|^{-1}.$$

If an object is moved by this force from $P_1 = (3, 5, 0)$ to $P_2 = (1, 0, 0)$, then the work done is simply $|\langle 1, 0, 0 \rangle|^{-1} - |\langle 3, 5, 0 \rangle|^{-1} = 1 - 1/\sqrt{34}$.

The line integral of a gradient takes a particularly simple form in differential notation:

$$\int_C \nabla f \cdot d\vec{P} = \int_C (f_x\vec{i} + f_y\vec{j} + f_z\vec{k}) \cdot (dx\,\vec{i} + dy\,\vec{j} + dz\,\vec{k})$$
$$= \int_C f_x\,dx + f_y\,dy + f_z\,dz.$$

The integrand here is precisely $df$, the differential of $f$. Thus equation (3) can be written

$$\int_C df = f(\text{end of } C) - f(\text{beginning of } C).$$

**EXAMPLE 2** $d(xy) = x\,dy + y\,dx$, so $\int_C x\,dy + y\,dx = \int_C d(xy)$. If $C$ goes from $(-1, 1)$ to $(1, 1)$ then

$$\int_C d(xy) = xy\Big|_{(1,1)} - xy\Big|_{(-1,1)} = 1 - (-1) = 2.$$

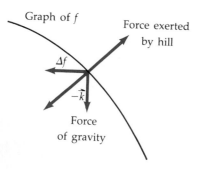

Graph of $f$

Force exerted by hill

$\Delta f$

$-\vec{k}$

Force of gravity

**FIGURE 5**
$\nabla f$ balancing a ball on a hillside.

For any force $\vec{F}$, the line integral gives the work done by $\vec{F}$; for a *gradient* force field $\vec{F} = \nabla f$, equation (2) says: *The work done by $\nabla f$ moving an object along any curve $C$ from $P(a)$ to $P(b)$ is precisely the increase in the potential $f$ from $P(a)$ to $P(b)$.* (In physics, the signs are reversed: A function $\varphi$ is a potential for $\vec{F}$ if $\vec{F} = -\nabla\varphi$; $\varphi$ is called the *potential energy* of the force, and the work done by $\vec{F}$ is the *decrease* in potential energy.)

For a gradient field $\nabla f$ in the plane, the graph of $f(x, y)$ is like a hillside. The gradient points toward the uphill direction; you can think of it as the horizontal force required to balance the pull of gravity on a ball resting on the hillside (Fig. 5; see Sec. 15.3, problem 14]. If the ball is moved by this force from

$$Q_1 = (x_1, y_1, f(x_1, y_1)) \text{ to } Q_2 = (x_2, y_2, f(x_2, y_2))$$

then the work done is $f(x_2, y_2) - f(x_1, y_1)$, precisely the *difference in elevation* between the two positions (Fig. 6). This is the line integral of $\nabla f$ along the curve $C$ in the $xy$-plane lying under the ball's path on the hillside.

We now face two questions: How can you tell whether or not a vector field is a gradient? And if it is, how do you find a potential?

The rotation velocity field

$$\vec{F}(x, y) = -\omega y\vec{i} + \omega x\vec{j}$$

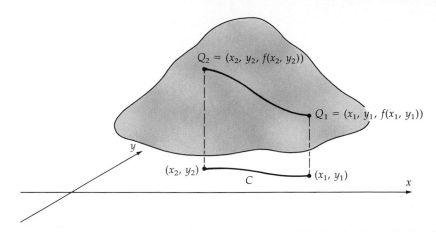

**FIGURE 6**
$\int_C \nabla f \cdot d\vec{P} = f(x_2, y_2) - f(x_1, y_1) = $ Change in elevation from $Q_1$ to $Q_2$.

is *not* a gradient. Indeed, suppose that $\vec{F} = \nabla f$ for some function $f$; then

$$-\omega y \vec{i} + \omega x \vec{j} = f_x \vec{i} + f_y \vec{j}$$

so $f_x = -\omega y$ and $f_y = \omega x$. Hence $f_{xy} = -\omega$ and $f_{yx} = \omega$. But this cannot be, for $f_{xy} = f_{yx}$ when all the derivatives are continuous. So it is impossible that $\vec{F} = \nabla f$ in this case.

This example illustrates that a vector field can be a gradient only if it satisfies certain special conditions, as follows:

---

**THEOREM 3**

If $\vec{F} = F_1 \vec{i} + F_2 \vec{j} + F_3 \vec{k}$ is a gradient, and each component of $\vec{F}$ is a $C^1$ function, then

$$\frac{\partial F_1}{\partial y} = \frac{\partial F_2}{\partial x}, \quad \frac{\partial F_1}{\partial z} = \frac{\partial F_3}{\partial x}, \quad \text{and} \quad \frac{\partial F_2}{\partial z} = \frac{\partial F_3}{\partial y}. \tag{4}$$

---

The reason is simple. If $\vec{F} = \nabla f$ then

$$F_1 \vec{i} + F_2 \vec{j} + F_3 \vec{k} = f_x \vec{i} + f_y \vec{j} + f_z \vec{k}$$

so $F_1 = f_x$, $F_2 = f_y$, $F_3 = f_z$; hence

$$(F_1)_y = f_{xy} = f_{yx} = (F_2)_x,$$

proving the first equation in (4). The other two are similarly proved.

Condition (4) can be expressed in terms of a certain *derivative* of $\vec{F}$, called the **curl**, denoted $\nabla \times \vec{F}$:

$$\nabla \times \vec{F} = \det \begin{bmatrix} \vec{i} & \vec{j} & \vec{k} \\ \partial/\partial x & \partial/\partial y & \partial/\partial z \\ F_1 & F_2 & F_3 \end{bmatrix}$$

$$= \vec{i} \left[ \frac{\partial F_3}{\partial y} - \frac{\partial F_2}{\partial z} \right] - \vec{j} \left[ \frac{\partial F_3}{\partial x} - \frac{\partial F_1}{\partial z} \right] + \vec{k} \left[ \frac{\partial F_2}{\partial x} - \frac{\partial F_1}{\partial y} \right].$$

Compare the components of the curl $\nabla \times \vec{F}$ to the equations (4); you see that (4) means precisely $\nabla \times \vec{F} = \vec{0}$. When this holds, $\vec{F}$ is called "curl-free." So Theorem 1 says:

$$\text{If } \vec{F} \text{ is a gradient, then } \vec{F} \text{ is curl-free:} \quad \nabla \times \vec{F} = \vec{0}.$$

What about the converse? Does every curl-free field $\vec{F}$ have a potential? The answer is "yes, with reservations." The following examples show how to compute the potential of a curl-free $\vec{F}$, and illustrate the reservations about the existence of such a potential.

The first example concerns a vector field in the plane. For such a field, $F_1$ and $F_2$ are independent of $z$, while $F_3 = 0$; thus the last two equations in (4) are automatically satisfied, and only the first comes into play:

$$\text{If } \vec{F}(x, y) = F_1(x, y)\vec{i} + F_2(x, y)\vec{j} \text{ is a gradient field, then } \frac{\partial F_1}{\partial y} = \frac{\partial F_2}{\partial x}.$$

---

**EXAMPLE 3**  Is $\vec{F} = (y + 2)\vec{i} + (x + 1)\vec{j}$ curl-free? If so, try to find a potential $f$. (Note: A very similar problem is worked out in Sec. 16.4; we repeat these details for those who skipped that section.)

*SOLUTION*  This is a plane vector field, so we need check only one equation:

$$\frac{\partial F_2}{\partial x} - \frac{\partial F_1}{\partial y} = \frac{\partial(x + 1)}{\partial x} - \frac{\partial(y + 2)}{\partial y} = 0.$$

So $\vec{F}$ is curl-free, and might have a potential $f$. If $\vec{F} = \nabla f$ then

$$(y + 2)\vec{i} + (x + 1)\vec{j} = f_x\vec{i} + f_y\vec{j}.$$

Equate components:

$$y + 2 = f_x \tag{5}$$

$$x + 1 = f_y. \tag{6}$$

Equation (5) says that $y + 2$ is the partial derivative of $f$ with respect to $x$, holding $y$ constant; so $f$ must be the "partial integral" of $y + 2$ with respect to $x$, treating $y$ as a constant:

$$f(x, y) = \int (y + 2)\, dx = xy + 2x + c(y). \tag{7}$$

The constant of integration can be different for each $y$, so we write it $c(y)$. Equation (7) is a partial solution of the problem; the unknown function $c(y)$ is still to be determined, by the remaining equation (6). To determine $c(y)$, differentiate (7) with respect to $y$; you find $f_y = x + c'(y)$, so (6) gives

$$x + 1 = x + c'(y).$$

The $x$ cancels, leaving $1 = c'(y)$; so

$$c(y) = y + C, \tag{8}$$

where now $C$ is actually a constant, independent of both $x$ and $y$. Insert (8) in (7), and the potential $f$ appears:

$$f(x, y) = xy + 2x + y + C.$$

You can check that $\nabla f = (y + 2)\vec{i} + (x + 1)\vec{j}$, as required.

---

In Example 3, $\vec{F}$ depends on only two variables $(x, y)$, and the potential requires just two partial integrations. For a curl-free vector field in space, you may need three.

---

**EXAMPLE 4**   Is $\vec{F} = (y + z)\vec{i} + (x + z)\vec{j} + (x + y)\vec{k}$ curl-free? If so, is it a gradient?

*SOLUTION*

$$\nabla \times \vec{F} = \det \begin{bmatrix} \vec{i} & \vec{j} & \vec{k} \\ \partial/\partial x & \partial/\partial y & \partial/\partial z \\ y + z & x + z & x + y \end{bmatrix}$$
$$= \vec{i}(1 - 1) - \vec{j}(1 - 1) + \vec{k}(1 - 1) = \vec{0}.$$

So $\vec{F}$ is curl-free, and might possibly be a gradient. If $\vec{F} = \nabla f$, then

$$(y + z)\vec{i} + (x + z)\vec{j} + (x + y)\vec{k} = f_x\vec{i} + f_y\vec{j} + f_z\vec{k}.$$

Equate components:

$$y + z = f_x \tag{9}$$

$$x + z = f_y \tag{10}$$

$$x + y = f_z \tag{11}$$

Integrate (9) with respect to $x$, holding $y$ and $z$ constant:

$$f(x, y, z) = \int (y + z)\, dx$$
$$= yx + zx + c(y, z); \tag{12}$$

the constant of integration $c(y, z)$ can be different for each $y$ and $z$. Next, the function $f$ must also satisfy (10):

$$x + z = f_y.$$

But $f$ is given by (12), so $f_y = x + c_y(y, z)$, and

$$x + z = x + c_y(y, z).$$

The $x$ cancels, leaving $z = c_y(y, z)$; so determine $c(y, z)$ by a partial integration with respect to $y$, holding $z$ constant:

$$c(y, z) = \int c_y(y, z)\, dy$$
$$= \int z\, dy$$
$$= zy + c(z).$$

[The new constant $c(z)$ may depend on $z$, but not on $x$, since $c(y, z)$ is independent of $x$.] Substitute this expression for $c(y, z)$ in (12) to get

$$f(x, y, z) = yx + zx + zy + c(z), \tag{13}$$

satisfying $f_x = y + z$ and $f_y = x + z$. Finally, to satisfy the last equation (11), differentiate both sides of (13) and use (11):

$$x + y = f_z = x + y + c'(z).$$

Here $x$ and $y$ cancel, leaving $c'(z) = 0$; so $c(z) = C$, a genuine constant. At last, we have the desired potential:

$$f(x, y, z) = xy + xz + yz + C,$$

where $C$ is any constant. You can check that $\nabla f = \vec{F}$, as required.

---

*REMARK* The method of successive "partial integrations" in these examples applies to any curl-free vector field defined in a *rectangular* region of $R^2$ or $R^3$; problems 17 and 18 show why. But in a more complicated region, the method may fail. For example, a steady electrical current flowing in a very long wire along the $z$-axis creates a magnetic field

$$\vec{H}(x, y) = \frac{-y\vec{i}}{x^2 + y^2} + \frac{x\vec{j}}{x^2 + y^2},$$

defined in the region $R$ consisting of all points *except* the $z$-axis, where $x^2 + y^2 = 0$; Figure 7 shows the field at points in the $xy$-plane. This magnetic field $\vec{H}$ is curl-free; but it is *not* the gradient of any function $f$ defined in all of the region $R$. For if it were, then the line integral of $\vec{H}$ around any

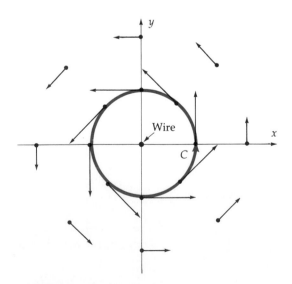

FIGURE 7

Magnetic field $\vec{H}(x, y) = \dfrac{-y\vec{i} + x\vec{j}}{x^2 + y^2}$.

closed curve $C$ in $R$ would be zero, by Theorem 2. On the other hand, if $C$ circles the origin counterclockwise (Fig. 7) then $\vec{H}$ points in the direction of the unit tangent $\vec{T}$ all along $C$, so $\vec{H} \cdot \vec{T} > 0$, and

$$\int_C \vec{H} \cdot d\vec{P} = \int_C \vec{H} \cdot \vec{T} \, ds > 0$$

as you can check (problem 11). So in this case, $\vec{H}$ is curl-free in $R$, but has no potential defined in all of $R$.

### Other Terminology

A gradient field $\vec{F} = \nabla f$ is called **exact**, since it is exactly the gradient of a function $f$.

## SUMMARY

*A Vector Field* $\vec{F}(x, y, z) = F_1 \vec{i} + F_2 \vec{j} + F_3 \vec{k}$ assigns a vector to each point $(x, y, z)$ in its domain. The vector $\vec{F}(x, y, z)$ is plotted as an arrow beginning at the point $(x, y, z)$.

If $\vec{F} = \nabla f$ for some function $f$, then $\vec{F}$ is called a *gradient field*, or an *exact* vector field; and $f$ is a *potential* for $\vec{F}$.

For a gradient field:

$$\int_C \vec{F} \cdot d\vec{P} = \int_C \nabla f \cdot d\vec{P} = \text{increase in potential } f \text{ from beginning of } C \text{ to end of } C.$$

$$\int_C \vec{F} \cdot d\vec{P} = \int_C \nabla f \cdot d\vec{P} = 0 \text{ for every closed curve } C \text{ in the domain of } \vec{F}.$$

*Curl of* $\vec{F}$:

$$\nabla \times \vec{F} = \det \begin{bmatrix} \vec{i} & \vec{j} & \vec{k} \\ \partial/\partial x & \partial/\partial y & \partial/\partial z \\ F_1 & F_2 & F_3 \end{bmatrix}.$$

Every gradient field is *curl-free*: $\nabla \times (\nabla f) = \vec{0}$.

If $\vec{F}$ is curl-free in a rectangular region $R$ of $R^2$ or $R^3$, then it is a gradient in $R$.

## PROBLEMS

### A

1.  **a)**  Sketch the vector field $\vec{F} = y\vec{i} + x\vec{j}$, drawing an appropriate arrow at each of the eight points $(\pm 1, 0)$, $(0, \pm 1)$, $(\pm 1, \pm 1)$.
    **b)**  Show that $\vec{F} = \nabla(xy)$.
    **c)**  Sketch the level sets of $xy$ passing through the eight given points.
    **\*d)**  Compute $\int_C \vec{F} \cdot d\vec{P}$, where $C$ is any curve running from the origin to $(1, 2)$.
    **e)**  Compute $\int_C \vec{F} \cdot d\vec{P}$, with $C$ given by

$$\vec{P}(t) = \langle e^{t^2}, t \rangle, \ 0 \le t \le 1.$$

2. Sketch the following vector fields, drawing an arrow for $\vec{F}(x, y)$ at each of the eight points $(\pm 1, 0)$, $(0, \pm 1)$, $(\pm 1, \pm 1)$.

*a) $\frac{1}{2}x\vec{i} + \frac{1}{2}y\vec{j}$

b) $\vec{i} + \vec{j}$

*c) $x\vec{i} + x\vec{j}$ [Plot this at $(\pm 2, 0)$, $(\pm 2, \pm 2)$, $(0, \pm 2)$ as well.]

d) $y\vec{i}$ [Plot this at $(\pm 2, 0)$, $(\pm 2, \pm 2)$, $(0, \pm 2)$ as well.]

e) $\dfrac{x\vec{i}}{x^2 + y^2} + \dfrac{y\vec{j}}{x^2 + y^2}$

f) $\dfrac{-y\vec{i}}{x^2 + y^2} + \dfrac{x\vec{j}}{x^2 + y^2}$

*3. Compute the curl $\nabla \times \vec{F}$ for each $\vec{F}$ in problem 2. (Answers given for parts a and c.)

4. For each curl-free $\vec{F}$ in problem 2, find a potential function $f$ with $\vec{F} = \nabla f$. (Check your answers.)

5. Evaluate the following line integrals using a potential constructed in the previous problem.

*a) $\int_C \frac{1}{2}x \, dx + \frac{1}{2}y \, dy$; $C$ goes from $(-1, 3)$ to $(2, 4)$.

b) $\int_C dx + dy$ with $C: \vec{P}(t) = \ln|\sec(t)|\vec{i} + \tan^{-1}t\,\vec{j}$, $0 \le t \le 1$.

*c) $\displaystyle\int_C \dfrac{x \, dx + y \, dy}{x^2 + y^2}$ with $C$ the semicircle $x^2 + y^2 = a^2$, $y \ge 0$, run clockwise.

d) $\displaystyle\int_C \dfrac{x \, dx + y \, dy}{x^2 + y^2}$ with $C$ running from $(1, 0)$ to $(0, 5)$.

e) $\displaystyle\int \dfrac{-y \, dx + x \, dy}{x^2 + y^2}$ with $C$ running from $(1, 0)$ to $(5, 5)$.

6. Prove that if $\vec{F} = \nabla f$, then $\dfrac{\partial F_2}{\partial z} = \dfrac{\partial F_3}{\partial y}$ and $\dfrac{\partial F_1}{\partial z} = \dfrac{\partial F_3}{\partial x}$.

7. Compute $\nabla \times \vec{F}$.

a) $x\vec{i} + y\vec{j} + z\vec{k}$

*b) $y\vec{i} + z\vec{j} + x\vec{k}$

c) $yz\vec{i} + zx\vec{j} + xy\vec{k}$

8. For each curl-free $\vec{F}$ in problem 7, find a potential function $f$. (Check your answers.)

9. Evaluate the line integral by finding a potential.

*a) $\int_C x \, dx + y \, dy + z \, dz$ with $C: \vec{P}(t) = \langle t \cos t, t \sin t, t \rangle$, $0 \le t \le \pi$.

b) $\int_C yz \, dx + xz \, dy + xy \, dz$ with $C: \vec{P}(t) = \langle e^{t^3}, e^{t^2}, e^t \rangle$, $0 \le t \le 1$.

10. For the inverse square potential force $\vec{F} = -|\vec{P}|^{-3}\vec{P}$

a) Show that $\vec{F} = \nabla |\vec{P}|^{-1}$. [Write $P = (x, y, z)$, so $|\vec{P}|^{-1} = (x^2 + y^2 + z^2)^{-1/2}$.]

b) Compute $\int_C \vec{F} \cdot \vec{T} \, ds$ with $C$ running from $(1, 0, 0)$ to $(1, 10, 100)$.

c) Compute $\int_C \vec{F} \cdot \vec{T} \, ds$ with $C$ running from $(1, 0, 0)$ "to $\infty$"; specifically, with $\vec{P}(t) = \vec{i} + t\vec{j} + t^2\vec{k}$, $0 \le t \le \infty$.

d) Compute $\int_C \vec{F} \cdot \vec{T} \, ds$ with $C$ running from $(x_0, y_0, z_0)$ to $\infty$, in any fashion whatsoever.

11. a) Show that $\vec{H} = \dfrac{-y\vec{i} + x\vec{j}}{x^2 + y^2}$ is curl-free.

*b) Evaluate $\displaystyle\int_C \dfrac{-y \, dx + x \, dy}{x^2 + y^2}$ for $C: \vec{P}(t) = a \cos t\,\vec{i} + a \sin t\,\vec{j}$, $0 \le t \le 2\pi$.

c) Evaluate $\displaystyle\int_C \dfrac{-y \, dx + x \, dy}{x^2 + y^2}$ for $C: \vec{P}(t) = a \cos t\,\vec{i} + a \sin t\,\vec{j}$, $0 \le t \le \theta$ for any $\theta > 0$.

d) Find a potential for $\vec{H}$ by partial integration, and identify its points of discontinuity. (There are many potentials possible, but all have discontinuities along some line or curve from the origin to $\infty$.)

**B**

12. The magnetic vector field

$$\vec{H} = \dfrac{-y\vec{i}}{x^2 + y^2} + \dfrac{x\vec{j}}{x^2 + y^2}$$

is best understood in polar coordinates $r, \theta$.

a) Express $dx$ and $dy$ in terms of $dr$ and $d\theta$.

b) Express $\vec{H} \cdot d\vec{P}$ in terms of $dr$ and $d\theta$.

c) Conclude that $\int_C \vec{H} \cdot d\vec{P}$ equals the increase in $\theta$ from the beginning of $C$ to the end of $C$. [Note: $\theta$ is not a well-defined function of points $(x, y)$ in the plane; $(x, y)$ determines $\theta$ only up to a constant $2k\pi$. When we speak of "the increase in $\theta$," we require that $\theta$ vary continuously as $C$ is traversed; for example, if $C$ wraps counterclockwise around the origin three times, then $\theta$ will increase by $6\pi$. The next problem gives a more formal view of the situation.]

**d)** Recompute the two integrals in problem 11, using your answer to part c.

**13.** Let $r$ and $\theta$ be polar coordinates of the point $(x, y)$. Show that

**a)** $\theta = \tan^{-1}(y/x)$, $\dfrac{-\pi}{2} < \theta < \dfrac{\pi}{2}$.

**b)** $\theta = \cot^{-1}(x/y)$, $0 < \theta < \pi$.

**c)** $\theta = \tan^{-1}(y/x) + \pi$, $\pi/2 < \theta < 3\pi/2$.

**d)** $\theta = \cot^{-1}(x/y) + \pi$, $\pi < \theta < 2\pi$.

**e)** $\theta = \tan^{-1}(y/x) + 2\pi$, $3\pi/2 < \theta < 5\pi/2$.

**f)** In each case, $\nabla\theta = \dfrac{-y\vec{i} + x\vec{j}}{x^2 + y^2} = \vec{H}$, the magnetic vector field in problems 11 and 12. (Remark: In each part a–e, $\theta$ is a well-defined function in a rectangular region, with $\nabla\theta = \vec{H}$. Moreover, $\theta$ varies continuously from the region in part a to that in part b, and again from b to c, and from c to d, and from d to e. However, in part e, $\theta$ is defined at the same points as in part a, but the two definitions of $\theta$ differ by $2\pi$. This illustrates that a potential can be found in each rectangular region, and even in more general regions; but no *single* potential function can be found in the *entire* region where $\vec{H}$ is defined.)

**14.** Suppose that $f(x, y, z)$ is a $C^2$ function (has continuous second derivatives). Is it always true that $\nabla \times (\nabla f) = \vec{0}$?

**15.** For a vector field $\vec{F} = F_1\vec{i} + F_2\vec{j} + F_3\vec{k}$, the *divergence* is defined as $\nabla \cdot \vec{F} = \dfrac{\partial F_1}{\partial x} + \dfrac{\partial F_2}{\partial y} + \dfrac{\partial F_3}{\partial z}$.
Suppose that $\vec{F}$ is $C^2$. Is it always true that
**a)** $\nabla \times (\nabla \times \vec{F}) = \vec{0}$?
**b)** $\nabla \cdot (\nabla \times \vec{F}) = 0$?

**16.** A point moves on a circle of radius $r$ with angular velocity $\omega$. Show that
**a)** If the initial position is $(r\cos\phi, r\sin\phi)$, then the position at time $t$ is $(x, y) = (r\cos(\omega t + \phi), r\sin(\omega t + \phi))$.
**b)** The velocity is $-\omega y\vec{i} + \omega x\vec{j}$. (This is the velocity field for rotation with angular velocity $\omega$.)

**17.** Suppose that $M(x, y)$ and $N(x, y)$ have continuous partial derivatives, and $M_y = N_x$ in a rectangle $a < x < b$, $c < y < d$. Show that $M\vec{i} + N\vec{j}$ is the gradient of a function of the form $f(x, y) = \int_a^x M(u, y)\, du + c(y)$, as follows. (Assume that all functions arising here are continuous.)

**a)** Show that $\dfrac{\partial f}{\partial x} = M(x, y)$.

**b)** Show that $\dfrac{\partial f}{\partial y} = N(x, y)$ if and only if

$$c'(y) = -\frac{\partial}{\partial y}\int_a^x M(u, y)\, du + N(x, y). \quad (*)$$

**c)** Equation $(*)$ can be solved if and only if the right-hand side is independent of $x$. Show that

$$\frac{\partial}{\partial x}\left[\frac{\partial}{\partial y}\int_0^x M(u, y)\, du - N(x, y)\right] = 0$$

if $N_x = M_y$. $\left(\text{Use } \dfrac{\partial}{\partial x}\dfrac{\partial}{\partial y} = \dfrac{\partial}{\partial y}\dfrac{\partial}{\partial x}.\right)$

**18.** Following the lines of the previous problem, show that if $\nabla \times \vec{F} = \vec{0}$ in a rectangular region $R$ defined by $a_1 < x < b_1$, $a_2 < y < b_2$, $a_3 < z < b_3$, then $\vec{F} = \nabla f$ for an $f$ defined in all of the region $R$.

**19.** (Magnetic field from a long straight wire.) In Figure 8, a current of strength $I$ flowing upward along a small piece of the $z$-axis, of length $dz$, at $Q = (0, 0, z)$, creates at $P = (a, b, c)$ a magnetic field

$$d\vec{H} = \frac{(I\, dz\, \vec{k}) \times (\vec{P} - \vec{Q})}{|\vec{P} - \vec{Q}|^3} \quad \text{(Ampère's law)}.$$

The magnetic field from the entire wire is the integral of this from $z = -\infty$ to $z = +\infty$. Write this integral in terms of $a, b, c, z$, and show that the entire field is

$$\vec{H} = 2I(a^2 + b^2)^{-1}(-b\vec{i} + a\vec{j}).$$

(This is the law of Biot and Savart.)

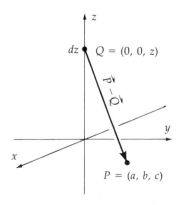

**FIGURE 8**
Current $I$ flowing in short segment $dz$ on $z$-axis creates a magnetic field at each point $P$ outside the wire.

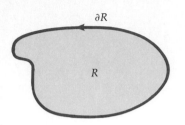

**FIGURE 1**
Region $R$ with positively oriented
boundary curve $\partial R$.

**FIGURE 2**
Positive orientation of $\partial R$: $R$ is on
the left as you follow $\partial R$.

## 18.3

### GREEN'S THEOREM

Recall the First Fundamental Theorem of Calculus:

$$\int_a^b g'(x)\, dx = g(b) - g(a).$$

On the left is the integral of the derivative of $g$ over an interval $[a, b]$. On the right, the terms $g(a)$ and $g(b)$ are the values of $g$ on the boundary of the interval, which consists of the two points $a$ and $b$.

Green's Theorem expresses a certain derivative of a vector field $\vec{F} = M\vec{i} + N\vec{j}$ over a region $R$ in the plane, as the line integral of $\vec{F}$ over the boundary of $R$ (Fig. 1).[1] Denote by $\partial R$ the boundary of $R$, traversed in the positive direction, that is, keeping $R$ on the left (Fig. 2). Then Green's Theorem states:

$$\iint_R \left( \frac{\partial N}{\partial x} - \frac{\partial M}{\partial y} \right) dA = \int_{\partial R} M\, dx + N\, dy, \tag{1}$$

if $\partial N/\partial x$ and $\partial M/\partial y$ are continuous throughout $R$.

The expression $\dfrac{\partial N}{\partial x} - \dfrac{\partial M}{\partial y}$ is a particular kind of derivative of the vector field $\vec{F} = M\vec{i} + N\vec{j}$; precisely, it is the $\vec{k}$ component of the curl

$$\nabla \times \vec{F} = \begin{vmatrix} \vec{i} & \vec{j} & \vec{k} \\ \partial/\partial x & \partial/\partial y & \partial/\partial z \\ M & N & 0 \end{vmatrix} = \left( \frac{\partial N}{\partial x} - \frac{\partial M}{\partial y} \right) \vec{k}.$$

Thus

$$\frac{\partial N}{\partial x} - \frac{\partial M}{\partial y} = (\nabla \times \vec{F}) \cdot \vec{k}$$

is the dot product of $\nabla \times \vec{F}$ with the vector $\vec{k}$, the upward unit normal to the $xy$-plane. So the left-hand side of (1) is $\iint_R (\nabla \times \vec{F}) \cdot \vec{k}\, dA$. The right-hand side is just the line integral of $\vec{F}$ over the boundary $\partial R$, so (1) can be rewritten as

$$\iint_R (\nabla \times \vec{F}) \cdot \vec{k}\, dA = \int_{\partial R} \vec{F} \cdot d\vec{P}.$$

---

**EXAMPLE 1** Check Green's Theorem for $\int_{\partial R} -y\, dx + x\, dy$, where $R$ is the unit circle $x^2 + y^2 \leq 1$.

**SOLUTION (Fig. 3)** Compare the right-hand side of (1) with the given line integral to see that here $M = -y$ and $N = x$. So the left-hand side of

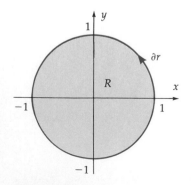

**FIGURE 3**

---

[1] Vector fields in the plane are often denoted $M\vec{i} + N\vec{j}$, rather than $F_1\vec{i} + F_2\vec{j}$; this avoids the use of subscripts.

(1) is

$$\iint\limits_{R}\left(\frac{\partial N}{\partial x}-\frac{\partial M}{\partial y}\right)dA = \iint\limits_{R}\left(\frac{\partial x}{\partial x}-\frac{\partial(-y)}{\partial y}\right)dA$$

$$= \iint\limits_{R} 2\,dA = 2(\text{area of } R)$$

$$= 2\pi.$$

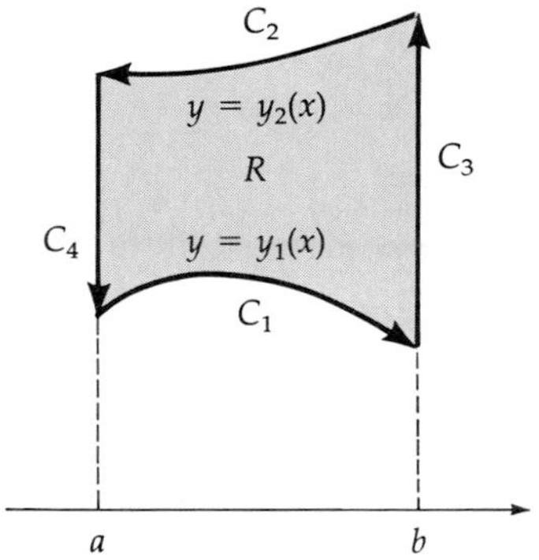

To compute the right-hand side of (1), parametrize the boundary

$$\partial R:\ \vec{P}(t) = \cos t\,\vec{i} + \sin t\,\vec{j}, \qquad 0 \le t \le 2\pi.$$

This curve goes counterclockwise, keeping $R$ on the left. So the right-hand side of (1) is

$$\int_{\partial R} -y\,dx + x\,dy = \int_{0}^{2\pi}(-\sin t)^2\,dt + (\cos t)^2\,dt = \int_{0}^{2\pi} 1\,dt$$

$$= 2\pi,$$

agreeing with the left-hand side.

**FIGURE 4**
A $y$-simple region $R$. $\partial R$ consists of the four curves $C_1$, $C_2$, $C_3$, $C_4$.

Why is Green's Theorem true? Consider at first just the term involving $M$, and suppose that the region $R$ is $y$-simple (Fig. 4). Then

$$\iint\limits_{R} -\frac{\partial M}{\partial y}\,dA = -\int_{a}^{b}\int_{y_1(x)}^{y_2(x)}\frac{\partial M}{\partial y}\,dy\,dx. \tag{2}$$

By the Fundamental Theorem of Calculus,

$$\int_{y_1}^{y_2}\frac{\partial M}{\partial y}\,(x, y)\,dy = M(x, y_2) - M(x, y_1),$$

so

$$\iint\limits_{R} -\frac{\partial M}{\partial y}\,dA = -\int_{a}^{b}[M(x, y_2(x)) - M(x, y_1(x))]\,dx$$

$$= -\int_{a}^{b} M(x, y_2(x))\,dx + \int_{a}^{b} M(x, y_1(x))\,dx. \tag{3}$$

We will show that this is precisely $\int_{\partial R} M\,dx$, where $\partial R$ is the chain with four "links" $C_1$, $C_2$, $C_3$, $C_4$ in Figure 4. The last term in (3) is precisely the line integral $\int_{C_1} M(x, y)\,dx$ over the curve $C_1$; for $C_1$ is given by $y = y_1(x)$, $a \le x \le b$, so

$$\int_{C_1} M(x, y)\,dx = \int_{a}^{b} M(x, y_1(x))\,dx. \tag{4}$$

Moving on to $C_2$, this is the graph of $y = y_2(x)$, $a \le x \le b$, but taken from right to left, from $x = b$ to $x = a$; thus

$$\int_{C_2} M(x, y)\,dx = \int_{b}^{a} M(x, y_2(x))\,dx = -\int_{a}^{b} M(x, y_2(x))\,dx. \tag{5}$$

On each of the curves $C_3$ and $C_4$, $x$ is constant; in evaluating the line integrals over these curves, $dx = 0$, so

$$\int_{C_3} M(x, y) \, dx = 0 \quad \text{and} \quad \int_{C_4} M(x, y) \, dx = 0. \tag{6}$$

Combining (4), (5), (6), and (3), you find

$$\int_{\partial R} M \, dx = \left[ \int_{C_1} + \int_{C_2} + \int_{C_3} + \int_{C_4} \right] M \, dx = \iint_R -\frac{\partial M}{\partial y} \, dA, \tag{7}$$

precisely the statement of Green's Theorem for the case $\vec{F} = M\vec{i}$.

The proof of equation (7) can be summed up briefly as follows: Integrating the double integral $\iint_R -\frac{\partial M}{\partial y} \, dy \, dx$ first with respect to $y$ "cancels" the derivative $\frac{\partial}{\partial y}$, leaving the integral of $M$ with respect to $x$, at the upper and lower limits of the $y$ integration. A careful check shows that this is precisely the line integral $\int_{\partial R} M \, dx$.

A similar argument shows that if $R$ is $x$-simple, then

$$\iint_R \frac{\partial N}{\partial x} \, dA = \int_{\partial R} N \, dy. \tag{8}$$

To prove this, integrate first with respect to $x$, "canceling" $\partial / \partial x$; the remaining integral with respect to $y$ turns out to be $\int_{\partial R} N \, dy$.

If $R$ is both $x$-simple and $y$-simple, then (7) and (8) are both valid; their sum gives Green's Theorem,

$$\iint_R \left( \frac{\partial N}{\partial x} - \frac{\partial M}{\partial y} \right) dA = \int_{\partial R} M \, dx + N \, dy.$$

The theorem remains true for any region $R$ that can be cut up into a finite number of pieces, each of which is both $x$-simple and $y$-simple. For example, the circular ring $R$ in Figure 5 is not $x$-simple and $y$-simple, but it can be cut into four regions $R_1, R_2, R_3, R_4$ that *are* (Fig. 6). The integral of $N_x - M_y$ over $R$ is the sum of the integrals over the four pieces:

$$\iint_R (N_x - M_y) \, dA = \sum_{j=1}^{4} \iint_{R_j} (N_x - M_y) \, dA = \sum_{j=1}^{4} \int_{\partial R_j} M \, dx + N \, dy.$$

In the sum on the right, the radial curve $C_2$ occurs twice, once in $\partial R_1$ and again in $\partial R_2$; the two occurrences are in opposite directions, so they cancel. The same is true of the other radial curves $C_4, C_6, C_9$; each is cancelled by the integral over the same curve run in the opposite direction. This leaves only the four circular curves running counterclockwise around the outer circle, and the four running *clockwise* around the inner circle; taken together, these form the boundary $\partial R$ of the original region $R$, *keeping $R$ on the left*. So for

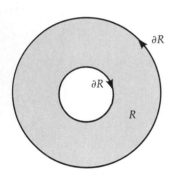

**FIGURE 5**
Annular region $R$ is neither $x$-simple nor $y$-simple. Positively oriented boundary $\partial R$ consists of an outer counterclockwise circle, and an inner clockwise circle, keeping $R$ on the left.

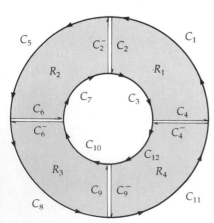

**FIGURE 6**
Annular ring cut into four simple regions.

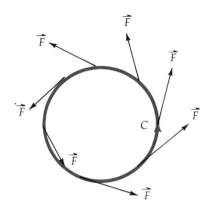

**FIGURE 7**
$\vec{F}$ has positive rotation around $C$.

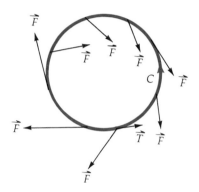

**FIGURE 8**
$\vec{F}$ has negative rotation around $C$.

this region,

$$\int_R (N_x - M_y)\, dx\, dy = \int_{\partial R} M\, dx + M\, dy.$$

A similar "cut-and-patch" method allows you to apply Green's Theorem whenever the region $R$ can be cut into a finite number of simple regions (that is, both $x$-simple and $y$-simple), using a finite number of $C^1$ curves. The vector field $\vec{F} = M\vec{i} + N\vec{j}$ must also be $C^1$.

### The Rotation of a Vector Field

Imagine a particle moving counterclockwise around a circle $C$, in a force field $\vec{F}$. If on the average $\vec{F}$ is in the direction of this motion (Fig. 7) we say "$\vec{F}$ has positive rotation around $C$." Otherwise, the rotation of $\vec{F}$ around $C$ is negative (Fig. 8) or zero.

The precise measure of the rotation of $\vec{F}$ around $C$ is defined as the *line integral* $\int_C \vec{F} \cdot d\vec{P} = \int_C \vec{F} \cdot \vec{T}\, ds$. When $\vec{F}$ points in the direction of the curve, then $\vec{F} \cdot \vec{T} > 0$, making a *positive* contribution to $\int_C \vec{F} \cdot \vec{T}\, ds$; otherwise $\vec{F} \cdot \vec{T} \le 0$, and the contribution is negative or zero.

By Green's Theorem, the rotation of a vector field $\vec{F} = M\vec{i} + N\vec{j}$ around a curve $C$ is

$$\int_C \vec{F} \cdot d\vec{P} = \iint_R (N_x - M_y)\, dA$$

where $R$ is the disk bounded by $C$. So, in any region where $N_x - M_y$ is *positive*, the rotation around every circle is positive; briefly, $\vec{F}$ has *positive rotation*. Where $N_x - M_y < 0$, the rotation is negative. Hence $N_x - M_y$ is called the **rotation** of $M\vec{i} + N\vec{j}$:

$$\operatorname{rot} \vec{F} = \operatorname{rot}(M\vec{i} + N\vec{j}) = N_x - M_y.$$

You can think of rot $\vec{F}$ as the tendency for $\vec{F}$ to form "whirlpools."

### Differential Forms (optional)

The "derivative" $N_x - M_y$ in Green's Theorem can be viewed as the $\vec{k}$ component of the cross product; but it seems unnatural to interpret a two-dimensional problem in three dimensions. Another approach uses "differential forms." By interpreting the symbols appropriately, we can write Green's Theorem as

$$\iint_R d(M\, dx + N\, dy) = \int_{\partial R} (M\, dx + N\, dy) \tag{9}$$

in perfect analogy with the Fundamental Theorem of Calculus

$$\int_a^b dg = \int_a^b g'(x)\, dx = g(b) - g(a),$$

and the formula for the line integral of a gradient

$$\int_C df = f(\text{end of } C) - f(\text{beginning of } C).$$

The expression $M\,dx + N\,dy$ in the line integral $\int_C (M\,dx + N\,dy)$ is called a **differential form**. It is, in a sense, an alternate notation for the vector field $\vec{F} = M\vec{i} + N\vec{j}$; the symbols $dx$ and $dy$ serve the same purpose as $\vec{i}$ and $\vec{j}$, in distinguishing the two components $M$ and $N$. The following table gives further analogies, and defines a *product* of $dx$ and $dy$, similar to the cross product:

| Vectors | Differential forms | |
|---------|--------------------|--|
| $\vec{F} = M\vec{i} + N\vec{j}$ | $\varphi = M\,dx + N\,dy$ | |
| $\nabla f = f_x \vec{i} + f_y \vec{j}$ | $df = f_x\,dx + f_y\,dy$ | (10) |
| $\vec{i} \times \vec{j} = -\vec{j} \times \vec{i}$ | $dx \wedge dy = -dy \wedge dx$ | (11) |
| $\vec{i} \times \vec{i} = 0 = \vec{j} \times \vec{j}$ | $dx \wedge dx = 0 = dy \wedge dy.$ | (12) |

In the "differential forms" column, the product in the last two lines is called the **wedge product**. Notice that, although $\vec{i} \times \vec{j} = \vec{k}$, we do *not* define $dx \wedge dy = dz$; everything is expressed with just the given variables $x$ and $y$.

Using this wedge product, and the *differential* $df = f_x\,dx + f_y\,dy$ of the function $f$, we define the differential of the "form" $M\,dx + N\,dy$:

$$d(M\,dx + N\,dy) = (dM) \wedge dx + (dN) \wedge dy \qquad \text{(we define it this way)}$$
$$= (M_x\,dx + M_y\,dy) \wedge dx + (N_x\,dx + N_y\,dy) \wedge dy$$
$$\text{[by (10)]}$$
$$= M_x\,dx \wedge dx + M_y\,dy \wedge dx + N_x\,dx \wedge dy + N_y\,dy \wedge dy$$
$$\text{(assuming a distributive law for the wedge product)}$$
$$= 0 - M_y\,dx \wedge dy + N_x\,dx \wedge dy + 0$$
$$\text{[by (11) and (12)]}.$$

So (finally)

$$d(M\,dx + N\,dy) = (N_x - M_y)\,dx \wedge dy. \tag{13}$$

The expression in the last line is called a "differential 2-form," because $dx \wedge dy$ is the product of two differentials. This product $dx \wedge dy$ is like the symbol $dx\,dy$ in a double integral, with one important difference: The integral symbol $dx\,dy$ can be reversed to $dy\,dx$ (with the appropriate change in any limits of integration) whereas $dx \wedge dy = -dy \wedge dx$, [see (11)].

Now, we *integrate* the 2-form (13) over a region $R$ in the plane by replacing $dx \wedge dy$ by $dA$:

$$\iint_R (N_x - M_y)\,dx \wedge dy = \iint_R (N_x - M_y)\,dx\,dy.$$

Accepting all this, we have

$$\iint_R d(M\,dx + N\,dy) = \iint_R (N_x - M_y)\,dx \wedge dy$$
$$= \iint_R (N_x - M_y)\,dA$$

so that (9) is precisely Green's Theorem (1).

These intriguing calculations give just a glimpse of the role of differential forms in unifying the basic facts of calculus for functions of one and several variables.

## SUMMARY

**Green's Theorem:**   Suppose that $R$ can be cut into a finite number of simple regions by $C^1$ curves, and that $\vec{F} = M\vec{i} + N\vec{j}$ is a $C^1$ vector field on $R$ and its boundary. Then

$$\int_{\partial R} M\, dx + N\, dy = \iint_R (N_x - M_y)\, dA$$

or

$$\int_{\partial R} \vec{F} \cdot d\vec{P} = \iint_R (\nabla \times \vec{F}) \cdot \vec{k}\, dA.$$

$\int_{\partial R} \vec{F} \cdot d\vec{P}$ denotes the counterclockwise line integral around the boundary of $R$.

**The Rotation** of $\vec{F} = M\vec{i} + N\vec{j}$   is   rot $\vec{F} = N_x - M_y$.

## PROBLEMS

**A**

1.  Evaluate the line integral directly, and again by Green's Theorem.
    *a)  $\int_{\partial R} x\, dy$, where $R$ is the square $|x| \leq 1$, $|y| \leq 1$.
    b)  $\int_C y\, dx$, where $C$ is the circle $x^2 + y^2 = 1$ taken counterclockwise.
    *c)  $\int_C y^2\, dx + x^2\, dy$ where $C$ is the circle $x^2 + y^2 = 1$, taken counterclockwise.
    d)  $\int_{\partial R} y\, dx + x\, dy$ where $R$ is the square $0 \leq x \leq 1, 0 \leq y \leq 1$.

2.  Evaluate the following by Green's Theorem.
    a)  $\int_{\partial R} xy^2\, dy - yx^2\, dx$ where $R$ is the set

    $$\{(x, y): x^2 + y^2 \leq 4, x \geq 0, y \geq 0\}.$$

    *b)  $\int_C \cos y \sin x\, dx + \sin y \cos x\, dy$, where $C$ is the triangle with vertices $(0, 0)$, $(0, \pi)$, and $(\pi, \pi)$, run counterclockwise.
    c)  The same integral as in part b, with $C$ the same triangle, but run clockwise.

3.  A particle goes counterclockwise around the circle $x^2 + y^2 = 4$, pushed by the force $\vec{F}(x, y) = xy\vec{i} + (\frac{1}{2}x^2 + yx)\vec{j}$. Find the work done, using Green's Theorem.

4.  a)  Evaluate $\int_{\partial R} M\, dx + N\, dy$ where

    $$M = \frac{-y}{x^2 + y^2}, \qquad N = \frac{x}{x^2 + y^2},$$

    and $R$ is the unit disk $x^2 + y^2 \leq 1$.
    b)  Show that $N_x - M_y = 0$, except at $(0, 0)$.
    c)  Apparently Green's Theorem does not apply in this case. Why not?

5.  Show that each of the following line integrals gives the *area* of $R$.
    a)  $\int_{\partial R} x\, dy$
    b)  $\int_{\partial R} (-y)\, dx$
    c)  $\int_{\partial R} (\frac{1}{2}x\, dy - \frac{1}{2}y\, dx)$
    d)  $\int_{\partial R} (\frac{1}{2}x - y)\, dy + (x - \frac{1}{2}y)\, dx$

**6.** Use any of the line integrals in the previous problem to compute the area of the region bounded by the following curves.

 **a)** $x = a \cos t$, $y = b \sin t$, $0 \le t \le 2\pi$
 (an ellipse).

 **\*b)** $x = \cos^3 t$, $y = \sin^3 t$, $0 \le t \le 2\pi$
 (an "astroid").

**7.** Suppose that $R$ is "$x$-simple". Prove (using the Fundamental Theorem of Calculus) that

$$\iint_R \frac{\partial N}{\partial x}\, dA = \int_{\partial R} N\, dy.$$

**8.** Suppose that $C$ is the boundary of a region $R$ where Green's Theorem applies, and that $\vec{F}$ is a $C^1$ vector field, and $f$ a $C^2$ function. Show that

 **a)** $\displaystyle\int_C \vec{F} \cdot d\vec{P} = 0$ if $\vec{F}$ is curl-free throughout $R$.

 **b)** $\displaystyle\int_C \nabla f \cdot d\vec{P} = 0$.

 **c)** $\displaystyle\int_C f_y\, dx - f_x\, dy = \iint_R (f_{xx} + f_{yy})\, dx\, dy$.

**B**

**\*9.** Figure 9 shows a curve $C$ surrounding the origin once, counterclockwise. Evaluate

$$\int_C \frac{-y\, dx + x\, dy}{x^2 + y^2}$$

by applying Green's Theorem to the region $R$ between $C$ and a small circle surrounding the origin. (The line integral around the small circle can be computed directly.)

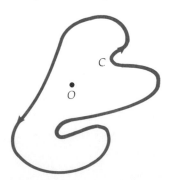

**FIGURE 9**
$C$ surrounds the origin once, counterclockwise.

**10.** The folium of Descartes $x^3 + y^3 = 3xy$ (Fig. 10) can be parametrized by the parameter $t = y/x$;

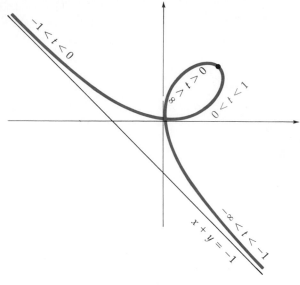

**FIGURE 10**
Folium $x^3 + y^3 = 3xy$; $x = \dfrac{3t}{1 + t^3}$, $y = \dfrac{3t^2}{1 + t^3}$.

thus $y = tx$, and on the folium, $x = \dfrac{3t}{1 + t^3}$,

$y = tx = \dfrac{3t^2}{1 + t^3}$ (Sec. 13.2, problem 14). Figure 10 shows how the parameter runs, and the asymptote of the folium, $x + y = -1$.

 **a)** Compute the area of the first quadrant loop, using $-\int_{\partial R} y\, dx$. For $R$, take the part of the loop enclosed by the folium for $0 \le t \le 1$, together with the line $y = x$. (Evaluate the integral with the substitution $u = 1 + t^3$.)

 **b)** (Much harder) Compute the area between the folium and its asymptote, $x + y = -1$ (Fig. 10). [You can rotate the folium $45°$ clockwise, using $\bar{x} = \dfrac{x - y}{\sqrt{2}}$ and $\bar{y} = \dfrac{x + y}{\sqrt{2}}$.

 Express $\bar{x}$ and $\bar{y}$ in terms of $t$. Check that $\displaystyle\lim_{t \to -1^+} (\bar{x}, \bar{y}) = (-\infty, -1/\sqrt{2})$ and $\displaystyle\lim_{t \to -1^-} (\bar{x}, \bar{y}) = (+\infty, -1/\sqrt{2})$, so $\bar{y} = -1/\sqrt{2}$ is the asymptote. Evaluate the area as $\displaystyle\int_{-\infty}^{+\infty} (\bar{y} + 1/\sqrt{2})\, d\bar{x}$, using

 $\bar{x} = \dfrac{3}{\sqrt{2}} \dfrac{t - t^2}{1 + t^3}$ and $\bar{y} = \dfrac{3}{\sqrt{2}} \dfrac{t + t^2}{1 + t^3}$,

 $-\infty < t < 0$. Notice that $(1 + t)^3$ factors out of both numerator and denominator.

The integral is straightforward, but very tedious.]

**11.** With substitutions in a single integral,

$$dx = \frac{dx}{du}\, du.$$

Show the analog of this for differential 2-forms: If $x = x(u, v)$ and $y = y(u, v)$, then

$$dx \wedge dy = \frac{\partial(x, y)}{\partial(u, v)}\, du \wedge dv.$$

Here $dx = x_u\, du + x_v\, dv$, and $\partial(x, y)/\partial(u, v)$ is the Jacobian determinant (Sec. 17.8).

**12.** Figure 11 shows a *planimeter*, an ingenious device for measuring plane·areas. An arm of radius $\rho$ is pivoted at a fixed point $O$, and on the end $P$ of this arm is pivoted a second arm, of radius $R$. On the second arm, at distance $r$ from the pivot $P$, there is a little knurled wheel $W$. The planimeter measures the area $|A|$ of a region $A$ by recording the net turning of the wheel $W$ as the free end $E$ of the second arm is moved around the boundary of $A$. This problem shows why the turning of $W$ is proportional to the area of $A$.

As $E$ is moved around the boundary of $A$, it traces a closed curve $\bar{C}$, and by problem 5, $|A| = \frac{1}{2}\int_{\bar{C}} (x\, dy - y\, dx)$. At the same time, the pivot $P$ traces another closed curve $C$ lying entirely on the circle of radius $\rho$ about $O$, and the wheel $W$ is dragged over yet another closed curve $\hat{C}$, as in Figure 11. Let $\alpha$ be the angle between the arm $PE$ and the $x$-axis. Then $\vec{u}_1 = \langle \cos \alpha, \sin \alpha \rangle$ is parallel to the arm $PE$, and $\vec{u}_2 = \langle -\sin \alpha, \cos \alpha \rangle$ is perpendicular to it. Since the little knurled wheel is unaffected by motion

parallel to the arm $PE$, but responds to motion perpendicular to $PE$, it turns at a rate proportional to $\vec{u}_2 \cdot \vec{W}'$, where $\vec{W}'$ is the velocity along $\hat{C}$. Thus if $a \le t \le b$ is the interval on which the curves $C$, $\bar{C}$, $\hat{C}$ are defined, the net turning of the wheel is

$$c \int_a^b \vec{u}_2 \cdot \vec{W}'\, dt, \tag{14}$$

where $c$ is a constant of proportionality. Parts a–g show that if the end $E$ is drawn around the boundary of $A$ so that the planimeter returns to its original position, and neither arm swings all the way around in a full circle, then the area $|A| = R \int_a^b \vec{u}_2 \cdot \vec{W}'\, dt$. Thus $|A|$ is proportional to the turning of the wheel, and the planimeter is justified.

**a)** Suppose that $\vec{P}(t)$, $\vec{E}(t)$, $\vec{W}(t)$, are respectively the position vectors of $P$, $E$, and $W$, while $\alpha(t)$ is the angle $\alpha$ at time $t$; and that $t$ varies from $a$ to $b$ as $E$ is drawn around the area $A$. Show that

$$\vec{E} = \vec{P} + \langle R \cos \alpha, R \sin \alpha \rangle,$$
$$\vec{W} = \vec{P} + \langle r \cos \alpha, r \sin \alpha \rangle.$$

**b)** If the second arm returns to its original position, and $-\pi < \alpha < \pi$ throughout the motion, and $\alpha$ is a differentiable function of $t$, then $\int_a^b \alpha'\, dt = 0$. Why? What could happen without the restriction $-\pi < \alpha < \pi$?

**c)** Under the conditions of part b, show that the integral (11) giving the turning of the wheel reduces to

$$c \int_a^b (-\sin \alpha x' + \cos \alpha y')\, dt,$$

where $x(t)$ and $y(t)$ denote the coordinates of point $P$ on the curve $C$.

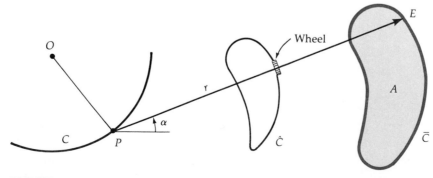

**FIGURE 11**
Planimeter. $|\vec{E} - \vec{P}| = R$.

**d)** Show that $|A| = \frac{1}{2}\int_{\bar{c}} (x\,dy - y\,dx)$ can be "reduced" to

$$\frac{1}{2}\int_{C} (x\,dy - y\,dx) + \frac{R^2}{2}\int_{a}^{b} \alpha'\,dt$$

$$+ \frac{R}{2}\int_{a}^{b} [x(\sin\alpha)' - \frac{dx}{dt}\sin\alpha$$

$$- y(\cos\alpha)' + \frac{dy}{dt}\cos\alpha]\,dt.$$

**e)** If the first arm returns to its original position without swinging all the way around O, then $\int_{C} (x\,dy - y\,dx) = 0$. Why?

**f)** In the expression in part d, show that

$$\int_{a}^{b} x(\sin\alpha)'\,dt = -\int_{a}^{b} \frac{dx}{dt}\sin\alpha\,dt$$

and

$$\int_{a}^{b} -y(\cos\alpha)'\,dt = \int_{a}^{b} \frac{dy}{dt}\cos\alpha\,dt.$$

**g)** Combine parts b–f to show that (14) equals

$$\frac{c}{R}|A|.$$

---

## 18.4

## SURFACE INTEGRALS. GAUSS' THEOREM

Imagine a stream of fluid or gas flowing through a region of space. At a given time, the particle at a given point has a certain velocity; if this velocity is independent of time, the flow is called **steady**. For a steady flow, denote by $\vec{F}(x, y, z)$ the velocity of the particle at point $(x, y, z)$; this gives a vector field $\vec{F}$ in space (Fig. 1).

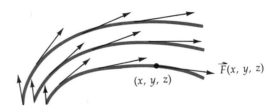

**FIGURE 1**
A flow: $\vec{F}(x, y, z) =$ velocity of particle at $(x, y, z)$.

Imagine the flow passing through a fixed surface $S$, as through a screen; we compute the rate of flow through $S$, given the velocity field.

Suppose at first that the velocity $\vec{F}$ is constant throughout space, and the surface $S$ is a parallelogram (Fig. 2). Denote by $\vec{N}$ a unit normal to $S$. In time $\Delta t$, a particle at point $P$ on the surface is displaced by $\vec{F}\,\Delta t$. Since the flow is the same at all points, the particles that pass through $S$ in time $\Delta t$ form a parallelepiped with volume

$$\text{(altitude)} \times \text{(area of base)} = (|\vec{F}\,\Delta t|\cos\theta) \times \text{(area of } S)$$
$$= (\vec{F} \cdot \vec{N}\,\Delta t) \times \text{(area of } S).$$

So

$$\text{rate of flow} = \frac{\text{flow in time } \Delta t}{\Delta t} = (\vec{F} \cdot \vec{N}) \times \text{(area of } S). \tag{1}$$

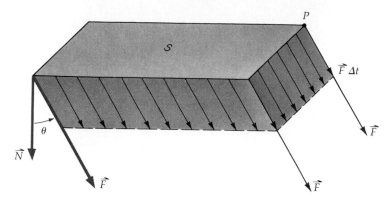

**FIGURE 2**
Flow through $S$ in time $\Delta t$. $\vec{F}$ flows toward side with normal $\vec{N}$;
$\theta < \pi/2$, $\cos \theta > 0$, $\vec{F} \cdot \vec{N} > 0$.

This gives the rate of flow through $S$ *in the direction toward the unit normal*
$\vec{N}$. If $\theta < \pi/2$ as in Figure 2, then fluid flows indeed toward the side in direction $\vec{N}$; correspondingly, the dot product $\vec{F} \cdot \vec{N}$ is positive, and so then is the
rate of flow as given by (1). But if $\theta > \pi/2$ (Fig. 3) then the flow rate is
negative, since the flow is away from direction $\vec{N}$.

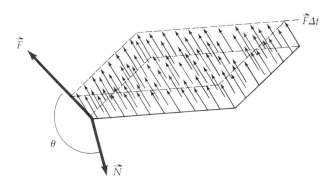

**FIGURE 3**
$\vec{F}$ flows away from side with normal $\vec{N}$; $\theta > \pi/2$, $\vec{F} \cdot \vec{N} < 0$.

Now consider a nonconstant flow $\vec{F}(x, y, z)$ passing through a smooth
curved surface $S$ in space, such as the graph of a $C^1$ function. We suppose
that at each point $P$ of $S$ there is a unit normal $\vec{N}(P)$, varying continuously
with $P$; and that the flow $\vec{F}(x, y, z)$ varies continuously with $(x, y, z)$. Divide the
surface $S$ into small pieces $S_j$, each of which can be approximated by a parallelogram tangent to $S_j$ at a point $P_j$ (Fig. 4). Denote by $\Delta S_j$ the area of the
approximating parallelogram. Since $S_j$ is small and $\vec{F}$ is continuous, $\vec{F}$ can be
approximated by the fixed vector $\vec{F}(P_j)$. Thus the rate of flow through $S_j$ is
approximately

$$\vec{F}(P_j) \cdot \vec{N}(P_j) \, \Delta S_j.$$

The sum $\sum \vec{F}(P_j) \cdot \vec{N}(P_j) \, \Delta S_j$ approximates the rate of flow through the entire
surface. Repeat the process with smaller and smaller pieces; if the approxi-
mating parallelograms are intelligently chosen, then these sums have a limit,

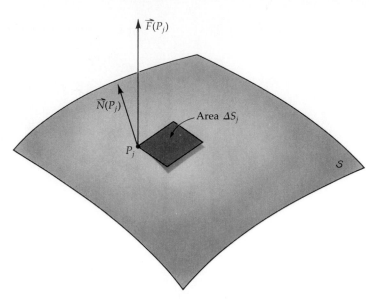

**FIGURE 4**
Approximating a small piece of $S$ by a parallelogram.

called the *integral of $\vec{F}$ over the surface $S$:*

$$\lim \sum \vec{F}(P_j) \cdot \vec{N}(P_j) \, \Delta S_j = \iint\limits_S \vec{F} \cdot \vec{N} \, dS. \tag{2}$$

This surface integral gives the *net rate of flow through $S$* in the direction of the normal vector $\vec{N}$.

How is such an integral computed? Consider the case where $S$ is the graph of a $C^1$ function

$$z = z(x, y)$$

for $(x, y)$ in a region $R$ (Fig. 5). We will identify the sum on the left in (2) as a Riemann sum, and thus represent the surface integral as a double integral

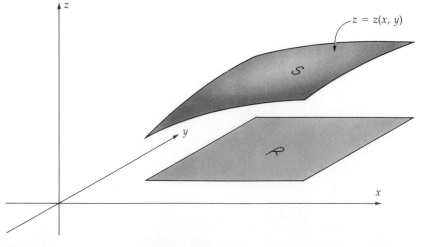

**FIGURE 5**
$S$ is the graph of $z = z(x, y)$, for $(x, y)$ in $R$.

over $R$. Approximate $R$ by small rectangles $R_j$. Let $Q_j = (x_j, y_j)$ be one corner of $R_j$, and let

$$P_j = (x_j, y_j, z_j) \qquad [z_j = z(x_j, y_j)]$$

be the point above $Q_j$ on the surface $S$ (Fig. 6). The small piece $S_j$ above $R_j$ can be approximated by a parallelogram with area $\Delta S_j$. If $\Delta A_j$ denotes the area of $R_j$ then these two areas are related by

$$\Delta S_j = (\sec \gamma_j) \, \Delta A_j, \tag{3}$$

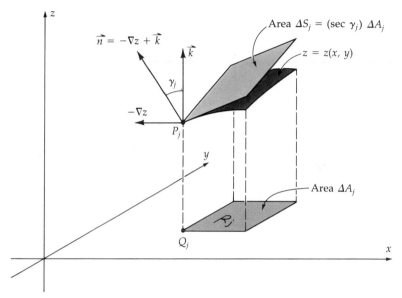

**FIGURE 6**

$$\vec{N} \, \Delta S_j = \frac{\vec{n}}{|\vec{n}|} \, \Delta S_j = \vec{n} \, \Delta A_j.$$

where $\gamma_j$ is the angle between $\vec{k}$ and an upward normal to $S_j$ (see Sec. 17.5). For this upward normal, we take

$$\vec{n} = -z_x \vec{i} - z_y \vec{j} + \vec{k} = -\nabla z + \vec{k}.$$

Then $\cos \gamma_j = \dfrac{\vec{n} \cdot \vec{k}}{|\vec{n}|} = \dfrac{1}{\sqrt{1 + |\nabla z|^2}}$, and the unit upward normal is

$$\vec{N} = \frac{\vec{n}}{|\vec{n}|} = \frac{-z_x \vec{i} - z_y \vec{j} + \vec{k}}{\sqrt{1 + |\nabla z|^2}} = \vec{n} \cos \gamma_j.$$

Together with (3), this gives an expression for the terms of the sum in (2):

$$\vec{F}(P_j) \cdot \vec{N}(P_j) \, \Delta S_j = \vec{F}(P_j) \cdot \vec{n} \, \Delta A_j.$$

Denote the components of $\vec{F}$ by $F_1, F_2, F_3$. Then

$$\vec{F}(P_j) \cdot \vec{N}(P_j) \, \Delta S_j = [-F_1(x_j, y_j, z_j) \, z_x(x_j, y_j) - F_2(x_j, y_j, z_j) \, z_y(x_j, y_j)$$
$$+ F_3(x_j, y_j, z_j)] \, \Delta A_j.$$

This identifies the sum $\sum \vec{F}(P_j) \cdot \vec{N}(P_j) \, \Delta S_j$ as a Riemann sum for an integral over the plane region $R$. Hence the integral of $\vec{F}$ over the surface $S$ is a double integral over $R$:

$$\iint_S \vec{F} \cdot \vec{N} \, dS = \iint_R [-F_1 z_x - F_2 z_y + F_3] \, dA_{xy}. \tag{4}$$

Each component $F_j$ is evaluated at $(x, y, z(x, y))$, and $z_x$, $z_y$ are the partial derivatives $\partial z/\partial x$, $\partial z/\partial y$.

The integrand on the right in (4) can be remembered as

$$[-F_1 z_x - F_2 z_y + F_3] \, dA_{xy} = \vec{F} \cdot (-z_x \vec{i} - z_y \vec{j} + \vec{k}) \, dA_{xy} = \vec{F} \cdot \vec{n} \, dA_{xy}$$

with $\vec{n} = -z_x \vec{i} - z_y \vec{j} + \vec{k}$, the upward normal to the graph that we used in Section 17.5. Thus to remember (4), set

$$\vec{N} \, dS = (-z_x \vec{i} - z_y \vec{j} + \vec{k}) \, dA_{xy}, \tag{5}$$

and take the dot product with $\vec{F}$.

---

**EXAMPLE 1**  For $\vec{F} = x\vec{i} + y\vec{j} + z\vec{k}$, compute the rate of flow up through the hemisphere

$$x^2 + y^2 + z^2 = a^2, \qquad z \geq 0.$$

**SOLUTION**  The surface is the graph of

$$z = \sqrt{a^2 - x^2 - y^2}, \, x^2 + y^2 \leq a^2.$$

The partial derivatives are

$$z_x = \frac{-x}{\sqrt{a^2 - x^2 - y^2}} = \frac{-x}{z} \quad \text{and} \quad z_y = \frac{-y}{\sqrt{a^2 - x^2 - y^2}} = \frac{-y}{z}$$

so by (5), $\vec{N} \, dS = \left\langle \dfrac{x}{z}, \dfrac{y}{z}, 1 \right\rangle \, dA_{xy}$. The components of $\vec{F}$ are $F_1 = x$, $F_2 = y$, $F_3 = z$, so $\vec{F} \cdot \vec{N} \, dS = \left( \dfrac{x^2}{z} + \dfrac{y^2}{z} + z \right) dA_{xy}$ and

$$\iint_S \vec{F} \cdot \vec{N} \, dS = \iint_{x^2+y^2 \leq a^2} \left[ \frac{x^2}{z} + \frac{y^2}{z} + z \right] dA_{xy}$$

$$= \iint_{x^2+y^2 \leq a^2} \frac{x^2 + y^2 + z^2}{z} \, dA_{xy}$$

$$= \iint_{x^2+y^2 \leq a^2} \frac{x^2 + y^2 + (a^2 - x^2 - y^2)}{\sqrt{a^2 - x^2 - y^2}} \, dA_{xy}$$

$$= \int_0^{2\pi} \int_0^a \frac{a^2}{\sqrt{a^2 - r^2}} \, r \, dr \, d\theta$$

$$= \int_0^{2\pi} -a^2 \sqrt{a^2 - r^2} \Big|_0^a \, d\theta = 2\pi a^3.$$

This is positive, since along the hemisphere $S$ the flow is in the direction of the upward normal (Fig. 7).

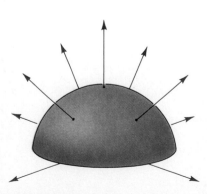

**FIGURE 7**
$\vec{F}(x, y, z) = x\vec{i} + y\vec{j} + z\vec{k}.$

**Gauss' Theorem** expresses the flow out of a region $W$ as a triple integral over $W$. Let $\partial W$ denote the surface bounding $W$, and $\vec{N}$ be the unit normal pointing outward. Then the flow out of $W$ is $\iint_{\partial W} \vec{F} \cdot \vec{N}\, dS$. According to Gauss' Theorem, this equals the integral over $W$ of a certain derivative of $\vec{F}$, the so-called *divergence*, defined by

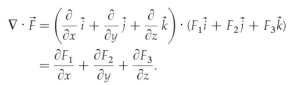

$$\nabla \cdot \vec{F} = \left(\frac{\partial}{\partial x}\vec{i} + \frac{\partial}{\partial y}\vec{j} + \frac{\partial}{\partial z}\vec{k}\right) \cdot (F_1\vec{i} + F_2\vec{j} + F_3\vec{k})$$

$$= \frac{\partial F_1}{\partial x} + \frac{\partial F_2}{\partial y} + \frac{\partial F_3}{\partial z}.$$

Gauss' Theorem is thus:

$$\iint_{\partial W} \vec{F} \cdot \vec{N}\, dS = \iiint_W (\nabla \cdot \vec{F})\, dV. \tag{6}$$

It is assumed that the partial derivatives $\dfrac{\partial F_1}{\partial x}$, $\dfrac{\partial F_2}{\partial y}$, and $\dfrac{\partial F_3}{\partial z}$ are continuous throughout $W$ and its boundary, and that $\partial W$ can be cut into pieces, each represented as the graph of a $C^1$ function, in one of the three forms $z = z(x, y)$, or $y = y(x, z)$, or $x = x(y, z)$.

Gauss' Theorem is like Green's, and the proof is similar. Suppose at first that $F_1 = F_2 = 0$, so $\vec{F} = F_3\vec{k}$. Suppose further that the region $W$ is "z-simple," bounded below and above by two graphs (Fig. 8)

$$z = z_1(x, y), \qquad z = z_2(x, y), \qquad (x, y) \text{ in } R, \tag{7}$$

where $z_1 \leq z_2$. Then

$$\iiint_W \nabla \cdot \vec{F}\, dV = \iint_R \int_{z_1(x,y)}^{z_2(x,y)} \frac{\partial F_3}{\partial z}(x, y, z)\, dA_{xy}$$

$$= \iint_R [F_3(x, y, z_2(x, y)) - F_3(x, y, z_1(x, y))]\, dA. \tag{8}$$

Now consider the left hand side of Gauss' Theorem (6). The boundary $\partial W$ consists of three parts (Fig. 8): the graph of $z_1$, the graph of $z_2$, and a vertical band. Along the vertical band, the normal vector $\vec{N}$ is horizontal. The vector field $\vec{F} = F_3\vec{k}$, on the other hand, is vertical, so $\vec{F} \cdot \vec{N} = 0$, and the integral $\iint \vec{F} \cdot \vec{N}\, dS$ over the vertical band is 0.

On the graph of $z_2$, the outward normal points up, so formula (4) gives the flow out through this part as

$$\iint_R F_3(x, y, z_2(x, y))\, dx\, dy.$$

On the graph of $z_1$, the outward normal points down, the opposite of the normal in (4), so the flow out through this part is

$$-\iint_R F_3(x, y, z_1(x, y))\, dx\, dy.$$

**FIGURE 8**
"z-simple" region.

$z = z_2(x, y)$

$z = z_1(x, y)$

Combining the three parts gives precisely (8), which proves Gauss' Theorem for this special case. Briefly, integrating $\iiint \frac{\partial F_3}{\partial z} \, dV$ first with respect to $z$ "cancels" the derivative $\frac{\partial}{\partial z}$, leaving an integral of $F_3$ with respect to $x$ and $y$; this turns out to be precisely the surface integral $\iint_{\partial W} F_3 \vec{k} \cdot \vec{N} \, dS$.

If $\vec{F} = F_1 \vec{i}$, or $\vec{F} = F_2 \vec{j}$, similar arguments prove (6), supposing that the region $W$ is $x$-simple and $y$-simple. If it is also $z$-simple, then the three results can be added, proving Gauss' Theorem for arbitrary vector fields $\vec{F}$ and simple regions $W$. More generally, if $W$ can be cut up into simple pieces, then the theorem is proved by adding the results for each piece, just as with Green's theorem.

---

**EXAMPLE 2** Compute the rate of flow out through the sphere $x^2 + y^2 + z^2 = a^2$, for the flow $\vec{F} = x\vec{i} + y\vec{j} + z\vec{k}$.

**SOLUTION** The sphere is the boundary $\partial W$ of the solid ball

$$W = \{(x, y, z): x^2 + y^2 + z^2 \leq a^2\}$$

so Gauss' Theorem applies. The divergence of $\vec{F}$ is

$$\nabla \cdot \vec{F} = \frac{\partial F_1}{\partial x} + \frac{\partial F_2}{\partial y} + \frac{\partial F_3}{\partial z}$$

$$= \frac{\partial x}{\partial x} + \frac{\partial y}{\partial y} + \frac{\partial z}{\partial z} = 3.$$

By Gauss' Theorem, the flow out through $\partial W$ is

$$\iint_{\partial W} \vec{F} \cdot \vec{N} \, dS = \iiint_W \nabla \cdot \vec{F} \, dV$$

$$= \iiint_W 3 \, dV$$

$$= 3 \cdot \tfrac{4}{3} \pi a^3 = 4\pi a^3$$

since $W$ is a sphere of radius $a$, with volume $\frac{4}{3}\pi a^3$. In Example 1, we found the flow up through the upper half of $\partial W$ to be $2\pi a^3$; the flow down through the lower half is also $2\pi a^3$, by symmetry, so the two examples concur: the flow through the whole sphere is $4\pi a^3$. But the calculation here is simpler than that in Example 1.

---

Gauss' Theorem explains why $\nabla \cdot \vec{F}$ is called the **divergence**—the integral of $\nabla \cdot \vec{F}$ over a region $W$ is

$$\iiint_W \nabla \cdot \vec{F} \, dV = \iint_{\partial W} \vec{F} \cdot \vec{N} \, ds$$

$$= \text{rate at which fluid flows out through } \partial W.$$

In a region $W$ where $\nabla \cdot \vec{F} > 0$, the integral $\iiint_W \nabla \cdot \vec{F}\, dV$ is positive, so the fluid is expanding and leaving the region, that is, diverging. But if $\nabla \cdot \vec{F} < 0$, the rate at which fluid leaves the region is negative; thus fluid is actually *entering* the region, "converging." In any case, taking sign into account, the divergence $\nabla \cdot \vec{F}(x, y, z)$ measures how rapidly fluid is diverging from the point $(x, y, z)$.

---

**EXAMPLE 3**   The flow $\vec{F} = y\vec{j}$ has divergence

$$\nabla \cdot \vec{F} = \frac{\partial y}{\partial y} = 1.$$

By Gauss' Theorem, fluid leaves any region $W$ at the rate

$$\iiint_W \nabla \cdot \vec{F}\, dV = \iiint_W 1\, dV = \text{volume of } W.$$

Figure 9 illustrates this, with $W$ the cube

$$0 \leq x \leq 1, \qquad \tfrac{1}{2} \leq y \leq \tfrac{3}{2}, \qquad 0 \leq z \leq 1.$$

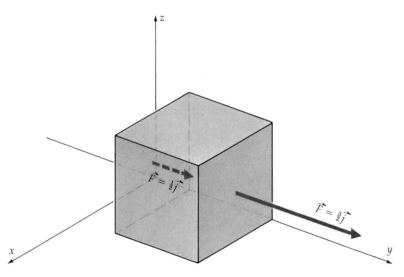

**FIGURE 9**
$\vec{F} = y\vec{j}$ flowing through a cube.

This cube has volume 1, so $\iiint_W \nabla \cdot \vec{F}\, dV = 1$. We compare this with the surface integral $\iint_{\partial W} \vec{F} \cdot \vec{N}\, dS$. The flow $y\vec{j}$ is parallel to the $y$-axis, so there is no flow through $\partial W$ except in the two planes $y = \tfrac{1}{2}$ and $y = \tfrac{3}{2}$. In $y = \tfrac{1}{2}$ the flow is $\vec{F} = \tfrac{1}{2}\vec{j}$, perpendicular to the face. Since the face has area 1, the volume rate of flow inward through this face is $\tfrac{1}{2}$, or $-\tfrac{1}{2}$ outward. On the other face $y = \tfrac{3}{2}$, the flow is $\vec{F} = \tfrac{3}{2}\vec{j}$, and the volume rate of flow is $\tfrac{3}{2}$ outward. The net outward flow is therefore $\iiint_W \vec{F} \cdot \vec{N}\, dS = \tfrac{3}{2} - \tfrac{1}{2} = 1$, in agreement with Gauss' Theorem.

**REMARK**  Throughout this section, we have viewed the vector field $\vec{F}$ as the velocity of a flow. But Gauss' Theorem applies to any $C^1$ vector field, and has many other applications, notably to electric and magnetic fields. Some of these are given in the problems.

## SUMMARY

*Surface Integral:*

$$\iint_S \vec{F} \cdot \vec{N} \, dS = \lim \sum \vec{F}(P_j) \cdot \vec{N}(P_j) \, \Delta S_j$$

$$= \text{rate of flow through } S \text{ in direction toward } \vec{N}. \tag{9}$$

*Surface Integral over Graph* of $z = z(x, y)$ for $(x, y)$ in $R$:

$$\vec{N} \, dS = (-z_x \vec{i} - z_y \vec{j} + \vec{k}) \, dA_{xy}$$

$$\iint_S \vec{F} \cdot \vec{N} \, dS = \iint_R [-F_1 z_x - F_2 z_y + F_3] \, dA_{xy}. \tag{10}$$

*Divergence:*

$$\nabla \cdot \vec{F} = \frac{\partial F_1}{\partial x} + \frac{\partial F_2}{\partial y} + \frac{\partial F_3}{\partial z}$$

*Gauss' Theorem:*  If $\vec{F}$ and $\nabla \cdot \vec{F}$ are continuous in $W$, and $W$ can be cut into a finite number of simple regions by graphs of $C^1$ functions, then

$$\iint_{\partial W} \vec{F} \cdot \vec{N} \, dS = \iiint_W \nabla \cdot \vec{F} \, dV$$

$\vec{N} = $ outer unit normal to $W$.

## PROBLEMS

### A

**1.** Analyze the flow $\vec{F} = y\vec{j}$ through each face of the given cube $W$, along the lines of Example 3, checking Gauss' Theorem in each case.
   **a)** $0 \le x \le 1, 0 \le y \le 1, 0 \le z \le 1$
   **b)** $0 \le x \le 1, -\frac{1}{2} \le y \le \frac{1}{2}, 0 \le z \le 1$

**\*2. a)** Compute the rate of flow upward through the hemisphere

$$z = \sqrt{a^2 - x^2 - y^2}, \qquad x^2 + y^2 \le a^2$$

for the flow $\vec{F} = z\vec{k}$.

   **b)** Sketch the flow at points $(0, 0, a)$,

$$\left(0, \pm\frac{a}{\sqrt{2}}, \frac{a}{\sqrt{2}}\right), \left(\pm\frac{a}{\sqrt{2}}, 0, \frac{a}{\sqrt{2}}\right) \text{ and}$$

$$\left(\pm\frac{a}{\sqrt{2}}, \pm\frac{a}{\sqrt{2}}, 0\right) \text{ on the hemisphere.}$$

**3.** Compute the rate of flow out of the sphere $x^2 + y^2 + z^2 = a^2$, for $\vec{F} = z\vec{k}$, using Gauss' Theorem. Sketch the flow at the points $(0, 0, \pm a)$,

$$\left(0, \pm\frac{a}{\sqrt{2}}, \pm\frac{a}{\sqrt{2}}\right) \text{ and } \left(\pm\frac{a}{\sqrt{2}}, 0, \pm\frac{a}{\sqrt{2}}\right).$$

Relate the answer to this problem with the answer to problem 2.

**4.** The flow $\vec{F} = -y\vec{i} + x\vec{j}$ represents a counterclockwise rotation around the $z$-axis.
   **a)** Compute the divergence $\nabla \cdot \vec{F}$.
   **b)** Explain intuitively why, for this $\vec{F}$, the net outward flow through the boundary of any region should be zero.
   **\*c)** Compute $\iint_S \vec{F} \cdot \vec{N} \, dS$, where $S$ is the square

$$\{(x, y, z): x = 0, 0 \le y \le 1, 0 \le z \le 1\}$$

with unit normal $\vec{N} = \vec{i}$.

**d)** Compute $\iint_S \vec{F} \cdot \vec{N} \, dS$, where $S$ is the rectangle

$$\{(x, y, z): x = 0, |y| \leq 1, 0 \leq z \leq 1\}$$

with unit normal $\vec{N} = \vec{i}$. Relate this answer to the one in part c.

**e)** Compute the *curl* $\nabla \times \vec{F}$.

5. The flow $\vec{F} = z\vec{j}$ represents a "shear" parallel to the $xy$-plane.

**a)** Sketch $\vec{F}$ on each face of the unit cube $0 \leq x, y, z \leq 1$. What is the net rate of flow out of the cube?

**b)** Check your answer to part a by Gauss' Theorem.

**c)** Compute the rate of flow in the direction $\vec{j}$ through the square defined by $y = 0$, $0 \leq x \leq 1, 0 \leq z \leq 1$, and through the rectangle defined by $y = 0, |x| \leq 1, 0 \leq z \leq 1$.

***d)** Compute the rate of flow "up" through the graph of

$$z = x^2 + y^2, 0 \leq x^2 + y^2 \leq 1, y \geq 0.$$

6. In Example 1, show that at every point of the surface $S$, the flow $\vec{F}$ is perpendicular to $S$, and has speed $a$. Using this, explain why the rate of flow through $S$ is $2\pi a^3$.

***7.** Compute the rate of flow out through the surface of the cube $|x| \leq 1, |y| \leq 1, |z| \leq 1$, for the flow $\vec{F} = x^3\vec{i} + y^3\vec{j} + z^3\vec{k}$:

**a)** As a surface integral.

**b)** As a volume integral.

**B**

8. For an electric field $\vec{E}$, the surface integral $\iint_S \vec{E} \cdot \vec{N} \, dS$ is called the *flux* of $\vec{E}$ through the surface $S$. The electric field due to a certain point charge at the origin is $\vec{E} = \dfrac{x\vec{i} + y\vec{j} + z\vec{k}}{x^2 + y^2 + z^2}$.

**a)** Show that $\nabla \cdot \vec{E} = 0$, except at the origin.

**b)** Compute the flux of $\vec{E}$ through the sphere $x^2 + y^2 + z^2 = r^2$, by evaluating the surface integral directly. (Show that $\vec{E} \cdot \vec{N}$ is constant on the sphere.)

**c)** Reconcile parts a and b with Gauss' Theorem.

**d)** Compute the flux of the same field through the "box" bounded by the planes $x = \pm a$, $y = \pm b, z = \pm c$. (Apply Gauss' Theorem to the region between the box and a small sphere centered at the origin.)

9. Except in regions containing electric charges, a stationary electric field $\vec{E}$ has zero divergence. Suppose that two surfaces $S_1$ and $S_2$ together bound a region $W$ in which there is no electric charge. Show that $\iint_{S_1} \vec{E} \cdot \vec{N} \, dS = \iint_{S_2} \vec{E} \cdot \vec{N} \, dS$. (Choose the normal on $S_1$ pointing *out* of $W$, and on $S_2$ pointing into $W$.)

10. An electric charge $q$ at point $P_0$ creates an electric field $\vec{E}(P) = q \dfrac{\vec{P} - \vec{P_0}}{|\vec{P} - \vec{P_0}|^3}$, in suitable units.

**a)** Show that $\nabla \cdot \vec{E} = 0$.

**b)** Suppose that $S$ is a sphere of radius $r$ centered at $P_0$, and $\vec{E}$ is the field due to a point charge at $P_0$. Show that $\iint_S \vec{E} \cdot \vec{N} \, dS = 4\pi q$.

**c)** Suppose that $S$ is a closed surface, and $P_0$ is *outside* $S$. Show that $\iint_S \vec{E} \cdot \vec{N} \, dS = 0$.

**d)** Suppose that there are point charges $q_1$, $q_2, \ldots, q_k$ at points $P_1, P_2, \ldots, P_k$ inside a surface $S$. The electric field $\vec{E}$ due to those $k$ charges is the sum of the fields due to each individual charge. Show that

$$\iint_S \vec{E} \cdot \vec{N} \, dS = 4\pi \sum q_j.$$

(Apply Gauss' Theorem to the region bounded by $S$ together with $k$ small spheres, one surrounding each of the charges. If $\vec{N}$ is the *inner* normal to one of the small spheres $S_j$, then $\iint_{S_j} \vec{E} \cdot \vec{N} \, dS = -4\pi q_j$.)

11. The *Laplacian* of a $C^2$ function $f$ is

$$\Delta f = f_{xx} + f_{yy} + f_{zz}.$$

Show that

**a)** $\iint_{\partial W} \nabla f \cdot \vec{N} \, dS = \iiint_W \Delta f \, dV$. (The expression $\nabla f \cdot \vec{N}$ is the directional derivative of $f$ in the direction of the outward normal $\vec{N}$, called the *outer normal derivative* of $f$, denoted $\partial f / \partial \vec{N}$.)

**b)** $\iint_{\partial W} g \, \nabla f \cdot \vec{N} \, dS = \iiint_W (\nabla g \cdot \nabla f + g \, \Delta f) \, dV.$

**c)** Suppose that $\Delta f = 0$ throughout $W$ and $\nabla f \cdot \vec{N} = 0$ on $\partial W$. Take $g = f$ in part b and deduce that $\iiint_W |\nabla f|^2 \, dV = 0$. (This implies that $\nabla f = \vec{0}$ in $W$; for $|\nabla f|^2 \geq 0$, so any positive contribution to $\iiint_W |\nabla f|^2 \, dV$ from some part of $W$ could not be cancelled by a negative contribution from somewhere else. Moreover, $\Delta f = \vec{0}$ in $W$ implies that $f$ is constant in $W$.)

**d)** Suppose that $\Delta f = 0$ throughout $W$, and $f = 0$ on $\partial W$. Use part b with $f = g$ to show that $\iiint_W |\nabla f|^2 \, dV = 0$. Deduce that $\nabla f = \vec{0}$ in $W$, so $f$ is constant. Why is the constant 0?

**12.** Green's Theorem can be interpreted for flows in the plane. Consider a two-dimensional substance flowing in the plane with a steady velocity $\vec{F}(x, y) = M\vec{i} + N\vec{j}$, varying with position $(x, y)$, but not with time.

**a)** Suppose $\vec{A} = a_1\vec{i} + a_2\vec{j}$ is any vector, and $\vec{B}$ is obtained from $\vec{A}$ by rotating clockwise $90°$. Show with an appropriate sketch that $\vec{B} = a_2\vec{i} - a_1\vec{j}$.

**b)** Suppose that the curve $\vec{P}(t) = x(t)\vec{i} + y(t)\vec{j}$, $a \leq t \leq b$, runs counterclockwise around a region $R$. Show that $\dfrac{dy}{dt}\vec{i} - \dfrac{dx}{dt}\vec{j}$ is an outward normal to $R$.

**c)** Continuing part b, explain why

$$\int_a^b \left( M\frac{dy}{dt} - N\frac{dx}{dt} \right) dt = \int_{\partial R} \vec{F} \cdot \vec{N} \, ds,$$

where

$$\vec{N} = \frac{1}{\sqrt{\left(\dfrac{dx}{dt}\right)^2 + \left(\dfrac{dy}{dt}\right)^2}} \left( \frac{dy}{dt}\vec{i} - \frac{dx}{dt}\vec{j} \right)$$

is a *unit* outer normal to $R$.

**d)** Explain why $\int_{\partial R} M \, dy - N \, dx$ gives the area rate of flow out through $\partial R$.

**e)** By Green's Theorem, show that

$$\int_{\partial R} \vec{F} \cdot \vec{N} \, ds = \iint_R (M_x + N_y) \, dx \, dy$$
$$= \iint_R (\nabla \cdot F) \, dx \, dy.$$

**13.** The surface integral (9) is expressed in the language of differential forms (Sec. 18.3) as

$$\iint_S F_1 \, dy \wedge dz + F_2 \, dz \wedge dx + F_3 \, dx \wedge dy. \quad \text{(11)}$$

The expression $F_1 \, dy \wedge dz + F_2 \, dz \wedge dx + F_3 \, dx \wedge dy$ is called a "differential 2-form"; the "2" refers to the two differentials in each product $dy \wedge dz$, and so on.

**a)** Show that if $z = z(x, y)$ and $dz = z_x \, dx + z_y \, dy$, then (11) reduces to (4), with $dA_{xy}$ replaced by $dx \wedge dy$.

**b)** Show that

$$d(F_1 \, dy \wedge dz + F_2 \, dz \wedge dx + F_3 \, dx \wedge dy)$$
$$= \left( \frac{\partial F_1}{\partial x} + \frac{\partial F_2}{\partial y} + \frac{\partial F_3}{\partial z} \right) dx \wedge dy \wedge dz.$$

**c)** Denote the differential 2-form $F_1 \, dy \wedge dz + F_2 \, dz \wedge dx + F_3 \, dx \wedge dy$ by the single letter $\varphi$. Verify the formula

$$\iint_{\partial W} \varphi = \iiint_W d\varphi.$$

**14.** Suppose that $S$ is a surface not passing through the origin $O$. The *solid angle subtended by $S$ at $O$* is defined to be $\iint_S \vec{\Omega} \cdot \vec{N} \, dS$, where $\vec{\Omega}$ is the vector field $\dfrac{x\vec{i} + y\vec{j} + z\vec{k}}{\rho^3}$, with $\rho = \sqrt{x^2 + y^2 + z^2}$. Compute the solid angle subtended by the following surfaces:

**a)** A sphere of radius $a$, centered at the origin.

**b)** A hemisphere of radius $a$, centered at the origin.

**\*c)** The region where $\rho = a$, $0 \leq \theta \leq \theta_0$ in spherical coordinates.

**d)** The disk $z = 1$, $x^2 + y^2 \leq 1$.

**e)** The disk $z = a$, $x^2 + y^2 \leq R$. What happens as $a \to 0$? $a \to \infty$? $R \to \infty$? Explain geometrically.

# 18.5

## STOKES' THEOREM

Recall Green's Theorem (Sec. 18.3), which can be written

$$\int_{\partial R} \vec{F} \cdot \vec{T} \, ds = \iint_R (\nabla \times \vec{F}) \cdot \vec{k} \, dA,$$

where $R$ is a region in the $xy$-plane with positively oriented boundary $\partial R$ (Fig. 1). The vector $\vec{k}$ in this formula is a unit normal to $R$.

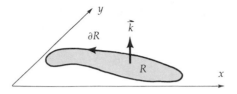

**FIGURE 1**
Region $R$ in $xy$-plane with oriented
boundary $\partial R$ and normal $\vec{k}$.

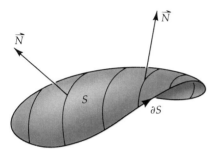

**FIGURE 2**
Surface $S$ in space, with normal
vectors $\vec{N}$ and oriented boundary $\partial S$.

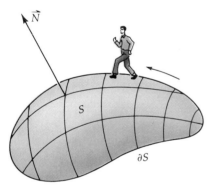

**FIGURE 3**
Positive orientation of $\partial S$.

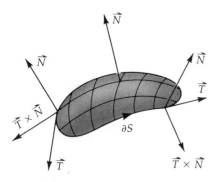

**FIGURE 4**
The boundary of $S$ is positively
oriented when $\vec{T} \times \vec{N}$ points
*away* from the edge.

Stokes' Theorem is a more general version of Green's—the region $R$ is replaced by a (possibly curved) surface $S$ in space (Fig. 2), with boundary curve $\partial S$, and the vector $\vec{k}$ is replaced by a unit normal $\vec{N}$, varying continuously over $S$. The theorem is that

$$\int_{\partial S} \vec{F} \cdot \vec{T} \, ds = \int_{S} (\nabla \times \vec{F}) \cdot \vec{N} \, dS, \tag{1}$$

under appropriate conditions on $\vec{F}$ and $S$.

The normals $\vec{N}$ determine a *positive* side for $S$—the one where the arrows for $\vec{N}$ are drawn. When a positive side is given, $S$ is called **oriented**. This in turn determines an orientation for the boundary curve $\partial S$—if you walk around the boundary of $S$, on the positive side of $S$ (Fig. 3), then the positive direction of $\partial S$ is the one keeping $S$ on your left. This is the orientation used on the left-hand side of (1).

Figure 4 shows another description of the orientation: If $\vec{T}$ is the unit tangent to the oriented boundary $\partial S$, then $\vec{T} \times \vec{N}$ points *away* from the edge of the surface.

Stokes' formula (1) is valid when $\vec{F}$ is a $C^1$ vector field, and the surface $S$ can be cut into a finite number of pieces, each the graph of a $C^1$ function defined in a simple plane region $R$. (For brevity, we call such pieces "simple graphs.") For example, the "balloon" $S$ in Figure 5 can be cut into seven simple graphs: $S_1$ is the graph of a function $z_1(x, y)$, $S_2$ the graph of a function $y_2(x, z)$, $S_4$ the graph of $x_4(y, z)$, and so on.

The proof of Stokes' Theorem has two steps.

I. *Prove it for a graph.* This is a straightforward calculation, best left as a problem.

II. *Prove it in general by adding up the pieces.* The surface integrals over the pieces $S_1, S_2, \ldots$ add up to the integral over the entire surface $S$; this gives the right-hand side of (1). The line integrals along the "cuts" cancel, for each one occurs twice, with opposite directions. [For example, in Figure 5, the cut from $A$ to $B$ occurs once upward (as part of $\partial S_4$) and again downward (as part of $\partial S_6$).] The remaining uncancelled line integrals give the integral along $\partial S$, the left-hand side of (1). (In Figure 5, $\partial S$ consists of the uncancelled parts of $\partial S_4$, $\partial S_5$, $\partial S_6$, and $\partial S_7$.)

**EXAMPLE 1** Compute both sides of Stokes' formula (1), with $\vec{F} = y\vec{i} + z\vec{j} + x\vec{k}$ and $S$ the upper hemisphere $z = \sqrt{1 - x^2 - y^2}$, with outward facing normal.

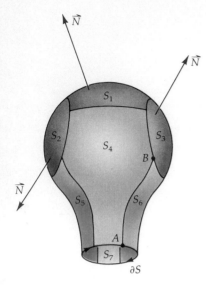

**FIGURE 5**
Surface $S$ cut into seven simple graphs.

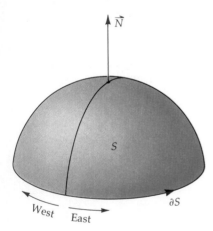

**FIGURE 6**
Northern hemisphere.

**SOLUTION (Fig. 6)** $S$ is the "Northern Hemisphere," the boundary is the equator, and the positive side is the "outside" of the sphere. The positive direction along $\partial S$ is "eastward"—walking east along the equator, the Northern Hemisphere is on your left. So $\partial S$ is the circle

$$\vec{P}(t) = (\cos t, \sin t, 0), \qquad 0 \le t \le 2\pi$$

and the left-hand side of (1) is, in this case,

$$\int_{\partial S} \vec{F} \cdot \vec{T} \, dS = \int_{\partial S} y \, dx + z \, dy + x \, dz = \int_0^{2\pi} -\sin^2 t \, dt = -\pi. \qquad (2)$$

Now compute the right-hand side of (1):

$$\nabla \times \vec{F} = -\vec{i} - \vec{j} - \vec{k}$$

and the surface $S$ is the graph of

$$z = \sqrt{1 - x^2 - y^2}, \qquad (x, y) \text{ in } R$$

where $R$ is the unit disk $x^2 + y^2 \le 1$. On this surface

$$\vec{N} \, dS = (-z_x \vec{i} - z_y \vec{j} + \vec{k}) \, dA_{xy}$$

so

$$\iint_S \nabla \times \vec{F} \cdot \vec{N} \, dS = \iint_R (z_x + z_y - 1) \, dx \, dy$$

$$= \iint_R \left( \frac{-x - y}{\sqrt{1 - x^2 - y^2}} - 1 \right) dx \, dy$$

$$= -\iint_R \frac{x + y}{\sqrt{1 - x^2 - y^2}} \, dx \, dy - \iint_R dx \, dy. \qquad (3)$$

The integrand $\dfrac{x + y}{\sqrt{1 - x^2 - y^2}}$ is odd, and the region of integration is symmetric about the origin, so $\iint_R \dfrac{x + y}{\sqrt{1 - x^2 - y^2}} \, dx \, dy = 0$. In the other integral, $\iint_R dx \, dy = $ area of $R = \pi$, since $R$ is a disk of radius 1. So (3) gives $\iint \nabla \times F \cdot \vec{N} \, dS = -\pi$, the same as (2).

## Gradient Fields (optional)

Stokes' Theorem sheds light on the relation between *curl-free* fields and *gradient* fields. Section 18.2 showed that

$$\vec{F} \text{ is a gradient} \Rightarrow \vec{F} \text{ is curl-free.}$$

But the converse was proved only in a limited sense: If $\vec{F}$ is curl-free in a *rectangular* region $R$, then $\vec{F}$ is a gradient in $R$ (problem 18, Sec. 18.2). This relation is not true for all regions $R$; the magnetic field created by a steady current along the entire $z$-axis is curl-free throughout the region $R$ outside the axis, yet it is not the gradient of any function $f$ defined throughout $R$.

This raises the question: In what type of region $R$ does "$\vec{F}$ is curl-free" imply "$\vec{F}$ is a gradient"? The answer: This is true in regions of the type called

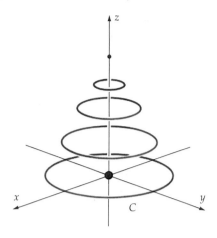

**FIGURE 7**
Region $R_1$ is space $R^3$ with origin deleted. Circle $C$ can be shrunk to a point in $R_1$.

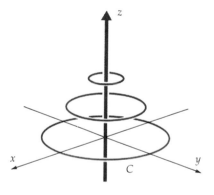

**FIGURE 8**
Region $R_2 = R^3$ with z-axis deleted. Circle $C$ cannot be shrunk to a point in $R_2$.

**simply connected**. Figures 7 and 8 illustrate the meaning of this term. We contrast two regions $R_1$ and $R_2$:

$$R_1 = R^3, \text{ with the origin removed} \qquad \text{(Fig. 7)}$$

and

$$R_2 = R^3, \text{ with the z-axis removed} \qquad \text{(Fig. 8)}.$$

Both regions are *connected*, in the sense that any two points $P$ and $Q$ in $R$ can be joined by a continuous curve $C$ lying entirely in $R$. But there is an important difference between them—in $R_1$, any *closed* curve $C$ can be shrunk to a point, without being opened up, and without touching any point outside of $R_1$. For example, in Figure 7, $C$ is the unit circle in the $xy$-plane. This can be shrunk to a point by raising the curve while the radius is reduced to zero. In $R_2$, on the other hand, a closed curve encircling the z-axis *cannot* be shrunk to a point, *without* ever touching the z-axis (Fig. 8). To distinguish these, the region $R_1$ is called *simply connected* (connected in a simple way) while $R_2$ is not. In sum:

$R$ is *connected* if each pair of points in $R$ can be joined by a continuous curve lying entirely in $R$.

$R$ is *simply connected* if each closed curve in $R$ can be shrunk to a point without being opened up, or touching any point outside $R$.

Now we can state:

---

**THEOREM**

Suppose that $R$ is a region in space which is connected and simply connected; and $\vec{F}$ is a $C^1$ vector field which is curl-free in $R$. Then $\vec{F}$ has a potential in $R$, a function $f$ such that

$$\vec{F} = \nabla f \text{ throughout } R.$$

---

The proof has three steps.

**STEP 1**[1] $\int_C \vec{F} \cdot d\vec{P} = 0$ *for every closed curve $C$ in $R$.* For if $C$ is closed, and $R$ is simply connected, then $C$ can be shrunk to a point in $R$. As it shrinks, $C$ generates a surface $S$, with $C$ as boundary. (For example, in Figure 7 the shrinking curve generates the upper hemisphere, with the equator as boundary.) By Stokes' Theorem,

$$\int_C \vec{F} \cdot \vec{T} \, ds = \iint_C (\nabla \times \vec{F} \cdot \vec{N}) \, dS = 0$$

since $\nabla \times \vec{F} = 0$ in $R$. So $\int_C \vec{F} \cdot \vec{T} \, ds = 0$ for every closed curve in $R$.

**STEP 2** $\int_C \vec{F} \cdot d\vec{P}$ *depends only on the beginning and endpoints of $C$.* For suppose that $C_1$ and $C_2$ are two curves, both beginning at $P_0$ and ending at $P_1$ (Fig. 9). Form the closed curve $C$ consisting of $C_1$ from $P_0$ to $P_1$, then

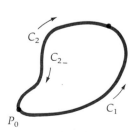

**FIGURE 9**
$C_1$ and $C_{2^-}$ form a closed curve, beginning and ending at $P_0$.

---

[1] This step is presented in a sketchy fashion. A more thorough treatment is properly part of a course in Advanced Calculus.

$C_2$-, which is $C_2$ run backward from $P_1$ to $P_0$. The combined curve $C$ is closed, so Step 1 gives

$$0 = \int_C \vec{F} \cdot d\vec{P} = \int_{C_1} \vec{F} \cdot d\vec{P} + \int_{C_2-} \vec{F} \cdot d\vec{P}. \tag{4}$$

But $\int_{C_2-} \vec{F} \cdot d\vec{P} = -\int_{C_2} \vec{F} \cdot d\vec{P}$, so (4) gives $\int_{C_1} \vec{F} \cdot d\vec{P} = \int_{C_2} \vec{F} \cdot d\vec{P}$. Thus along any two curves from $P_0$ to $P_1$, $\vec{F}$ has the same line integral.

**STEP 3**   *Now define the potential f as a line integral.* Choose a fixed point $P_0$ in R. Then for any point $P$ in R, choose a curve $C_P$ from $P_0$ to $P$, and define

$$f(P) = \int_{C_P} \vec{F} \cdot d\vec{P}. \tag{5}$$

It doesn't matter what curve $C_P$ is chosen from $P_0$ to $P$; for by Step 2, on all these curves, the line integral of $\vec{F}$ is the same.

It is reasonable to define $f$ as in (5), for if a potential $f$ *did* exist, it would satisfy

$$\int_{C_P} \vec{F} \cdot d\vec{P} = f(P) - f(P_0);$$

thus equation (5) should give a potential which has $f(P_0) = 0$.

It remains to show that $\nabla f = \vec{F}$. For the first component, we must show that $f_x = F_1$, more precisely

$$\lim_{h \to 0} \frac{f(x + h, y, z) - f(x, y, z)}{h} = F_1(x, y, z). \tag{6}$$

To prove (6), let $P = (x, y, z)$ and $Q = (x + h, y, z)$, and take the path $C_Q$ in Figure 10, consisting of $C_P$ followed by the straight line segment $C_{PQ}$ from $P$ to $Q$. Then by (5),

$$f(x + h, y, z) - f(x, y, z) = f(Q) - f(P)$$
$$= \int_{C_Q} \vec{F} \cdot d\vec{P} - \int_{C_P} \vec{F} \cdot d\vec{P} \qquad \text{(definition of } f)$$
$$= \left[ \int_{C_P} \vec{F} \cdot d\vec{P} + \int_{C_{PQ}} \vec{F} \cdot d\vec{P} \right] - \int_{C_P} \vec{F} \cdot d\vec{P}$$
$$= \int_{C_{PQ}} \vec{F} \cdot d\vec{P}.$$

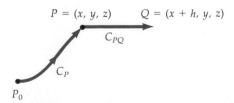

**FIGURE 10**
$C_Q = C_P$ together with $C_{PQ}$.

Hence

$$\lim_{h \to 0} \frac{f(x + h, y, z) - f(x, y, z)}{h} = \lim_{h \to 0} \frac{1}{h} \int_{C_{PQ}} F_1 \, dx + F_2 \, dy + F_3 \, dz$$

$$= \lim_{h \to 0} \frac{1}{h} \int_0^h F_1(x + t, y, z) \, dt$$

$$(dy = dz = 0 \text{ on this segment})$$

$$= F_1(x, y, z),$$

by the Second Fundamental Theorem of Calculus. This proves (6); the proofs that $f_y = F_2$ and $f_z = F_3$ are nearly the same.

## SUMMARY

**Stokes' Theorem:**  If $\vec{F}$ is a $C^1$ vector field, and the surface $S$ can be cut into finitely many simple graphs, then

$$\int_{\partial S} \vec{F} \cdot \vec{T} \, ds = \iint_S \nabla \times \vec{F} \cdot \vec{N} \, dS.$$

## PROBLEMS

### A

Given $\vec{F}$ and $S$, verify Stokes' Formula (1) by computing each side separately.

1. $\vec{F}(x, y, z) = z\vec{i} + x\vec{j} + y\vec{k}$; $S$ is the hemisphere $x^2 + y^2 + z^2 = 4$, $z \geq 0$.

\*2. $\vec{F}(x, y, z) = y^2\vec{i} + z^3\vec{j} + x^2\vec{k}$; $S$ is the triangle $x + y + z = 1$, $x \geq 0$, $y \geq 0$, $z \geq 0$.

3. $\vec{F}(x, y, z) = y\vec{i} + z\vec{j} + x\vec{k}$; $S$ is the part of the paraboloid $z = 4 - x^2 - y^2$ inside the cylinder $x^2 + y^2 = 1$.

\*4. $\vec{F}(x, y, z) = (x + y)\vec{i} + (y + z)\vec{j} + (z + x)\vec{k}$, $S$ is the cone $z = r$, $0 \leq r \leq 1$ in cylindrical coordinates.

### B

5. Let $S$ be the sphere $x^2 + y^2 + z^2 = a^2$, with $\vec{N}$ the outward normal. Let $\vec{F}$ be any $C^1$ vector field defined at every point of $S$.
   a) Show that $\iint_S \nabla \times \vec{F} \cdot \vec{N} \, ds = 0$. (Apply Stokes' Theorem to the upper hemisphere and lower hemisphere.)

   b) Prove the same formula using Gauss' Theorem, by assuming that $\vec{F}$ has continuous *second* derivatives throughout the interior of $S$.

   c) Which proof is more satisfactory?

6. Suppose that $S_1$ and $S_2$ are two oriented surfaces with the same oriented boundary $C$. Prove that $\iint_{S_1} \nabla \times \vec{F} \cdot \vec{N} \, dS = \iint_{S_2} \nabla \times \vec{F} \cdot \vec{N} \, dS$.

7. Suppose that $S$ is a *closed* surface, that is, $\partial S$ is empty. Specifically, suppose that $S$ can be cut into finitely many simple graphs, and each piece of each cut occurs twice, in opposite directions. What then can be said about $\iint_S \nabla \times \vec{F} \cdot \vec{N} \, dS$, for an arbitrary $C^1$ vector field $\vec{F}$?

8. a) If $\vec{F}$ has continuous second partial derivatives, what can be said about $\nabla \cdot (\nabla \times \vec{F})$?
   b) Can $\vec{F}(x, y, z) = x\vec{i} + y\vec{j} + z\vec{k}$ be the curl of some vector field?

9. In each part of this problem, a region is obtained by *removing* a given part of $R^3$. Which of the

regions are connected? Which are simply connected? Remove

**a)** The $xy$-plane.

**\*b)** The $x$-axis.

**c)** The positive $x$-axis.

**\*d)** The unit sphere $x^2 + y^2 + z^2 = 1$.

**e)** The ball $x^2 + y^2 + z^2 \leq 1$.

**f)** The graph of $z = x^2 + y^2$.

**g)** The graph of $z = \sqrt{1 - x^2 - y^2}$.

**h)** The helix traced by $\vec{P}(t) - \langle \cos t, \sin t, t \rangle$.

**10.** (Proof of Stokes' Theorem.) Suppose that $S$ is the graph of a $C^1$ function $z(x, y)$, for $(x, y)$ in $R$, and $R$ is a plane region where Green's Theorem applies.

**a)** Suppose that $\vec{P}(t) = x(t)\vec{i} + y(t)\vec{j}$ gives the oriented boundary of $R$. Explain why $x(t)\vec{i} + y(t)\vec{j} + z(x(t),y(t))\vec{k}$ gives the oriented boundary of $S$, with $\vec{N}$ chosen pointing upward.

**b)** Show that $\int_{\partial S} \vec{F} \cdot \vec{T}\, ds = \int_{\partial R} (F_1 + F_3 z_x)\, dx + (F_2 + F_3 z_y)\, dy$, where $F_1$ denotes $F_1(x, y, z(x, y))$, and similarly for $F_2$ and $F_3$.

**c)** Show Stokes' Theorem for $S$:

$$\iint_S \nabla \times \vec{F} \cdot \vec{N}\, dS = \int_{\partial S} \vec{F} \cdot \vec{T}\, ds.$$

(Use Green's Theorem in part b.)

**11.** For the function $f(P) = \int_{C_P} F_1\, dx + F_2\, dy + F_3\, dz$ in (5) above, prove that $\partial f / \partial y = F_2$.

**12.** In the language of differential forms (Sec. 18.3 and Sec. 18.4, problem 13) Stokes' Theorem follows the same pattern as the theorems of Green and Gauss. Let $\omega = F_1\, dx + F_2\, dy + F_3\, dz$. Show that Stokes' formula (1) reduces to

$$\int_{\partial S} \omega = \int_S d\omega.$$

---

# REVIEW PROBLEMS    CHAPTER 18

**\*1.** Evaluate $\int_C x\, dx + 2xy\, dy$, where $C$ is

**a)** The graph of $y = x^2$, $-1 \leq x \leq 1$, taken from left to right.

**b)** The same graph as in part a, taken from right to left.

**c)** The line segment from $(1, 1)$ to $(-1, 2)$.

**d)** The curve $x = \cos 2t$, $y = \sin 2t$, $0 \leq t \leq 2\pi$.

**\*2.** Sketch the vector field $\vec{F}(x, y) = y\vec{i} - x\vec{j}$.

**\*3.** Evaluate the line integral of the vector field $y\vec{i} - x\vec{j}$ along the given curve.

**a)** The circle $x = \cos \theta$, $y = \sin \theta$, $0 \leq \theta \leq 2\pi$.

**b)** The circle of radius 2 about the origin, taken *clockwise*.

**c)** The graph of $y = e^x$, $0 \leq x \leq 1$.

**d)** The rectangle with corners $(0, 0)$, $(1, 0)$, $(1, 1)$, $(0, 1)$ taken in that order.

**\*4.** Use Green's Theorem to evaluate three of the integrals in the previous problem.

**\*5.** Determine whether the given $\vec{F}$ is a gradient field in the given region. If it is, find a potential function for it in that region.

**a)** $\vec{F} = y\vec{i} - x\vec{j}$, all $(x, y)$

**b)** $\vec{F} = y\vec{i} + x\vec{j}$, all $(x, y)$

**c)** $\vec{F} = [(1 - x)\vec{i} + x\vec{j} + 2x\vec{k}]e^{-x+y+2z}$, all $(x, y, z)$

**d)** $\vec{F} = [x\vec{i} + y\vec{j}](x^2 + y^2)^{-1}$, $(x, y) \neq (0, 0)$

**e)** $\vec{F} = (y\vec{i} - x\vec{j})(x^2 + y^2)^{-1}$, $(x, y) \neq (0, 0)$

**f)** $\vec{F} = (y\vec{i} - x\vec{j})(x^2 + y^2)^{-1}$, $x > 0$

**\*6.** Construct a potential function to evaluate $\int_C yz\, dx + xz\, dy + xy\, dz$ along the given curve $C$.

**a)** The circle $x^2 + y^2 = 1$, $z = 0$, clockwise.

**b)** $\vec{P}(t) = \cos t\, \vec{i} + \sin t\, \vec{j} + t\, \vec{k}$, $0 \leq t \leq 2\pi$.

**c)** The straight line segment from $(0, -1, 2)$ to $(1, 3, -1)$.

**\*7.** A flow velocity is given by $\vec{F} = -y\vec{i} + x\vec{j} - z\vec{k}$. Compute

**a)** The divergence.

**b)** The curl.

**c)** The surface integral over the unit disk $x^2 + y^2 \leq 1$, $z = 0$, with upward unit normal.

**d)** The surface integral over the upper unit hemisphere $x^2 + y^2 + z^2 = 1$, $z \geq 0$, with outward normal. (Include only the spherical part of the surface, not the disk in the $xy$-plane.)

**e)** $\iiint_V \nabla \cdot \vec{F}\, dV$, where $V$ is the solid region inside the upper unit hemisphere in part d.

f)   $\int_C \vec{F} \cdot d\vec{P}$, where $C$ is the unit circle $x^2 + y^2 = 1$, $z = 0$.

g)   The surface integral of $\nabla \times \vec{F}$ over the disk in part c.

h)   The surface integral of $\nabla \times \vec{F}$ over the hemisphere in part d.

**\*8.**   State Gauss' Theorem, and use it to explain the relation between some of the evaluations in problem 7.

**\*9.**   State Stokes' Theorem, and use it to explain the relation between some of the evaluations in problem 7.

**\*10.**   Give an example of a region $R$ in space where the following statement is true, and another example of a region where it is false: "If $\vec{F}$ is a $C^1$ curl-free vector field in $R$, then $\vec{F}$ is a gradient field."

**B**

**11.**   Let $\vec{F} = \dfrac{x\vec{i} + y\vec{j} + z\vec{k}}{(x^2 + y^2 + z^2)^{3/2}}$.

a)   Show that $\iint_S \vec{F} \cdot \vec{N}\, dS = 4\pi$, if $S$ is a sphere centered at the origin, with outer normal $\vec{N}$.

b)   Suppose that $S$ is a surface which, together with a small sphere centered at the origin, bounds a region in which Gauss' Theorem is valid. Show that $\iint_S \vec{F} \cdot \vec{N}\, dS = 4\pi$, if $\vec{N}$ is the outer normal to $S$.

**12.**   (Integration by parts in two variables.) Show, under appropriate conditions

a)   $\displaystyle\iint_R (f_x g + f g_x)\, dA = \int_{\partial R} fg\, dy$.

b)   $\displaystyle\iint_R f_x g\, dA = \int_{\partial R} fg\, dy - \iint_R f g_x\, dA$.

c)   $\displaystyle\iint_R f_y g\, dA = -\int_{\partial R} fg\, dx - \iint_R f g_y\, dA$.

d)   $\displaystyle\iint_R f_x g\, dA = \int_{\partial R} fg\vec{i} \cdot \vec{N}\, ds - \iint_R f g_x\, dA$,
where $\vec{N}$ is the outer normal to $R$.

e)   $\displaystyle\iint_R f_y g\, dA = \int_{\partial R} fg\vec{j} \cdot \vec{N}\, dS - \iint_R f g_y\, dA$
where $\vec{N}$ is the outer normal to $R$.

**13.**   State and prove the analog of the previous problem, for integration by parts in three variables.

# 19

# COMPLEX NUMBERS AND DIFFERENTIAL EQUATIONS

From time immemorial, numbers have been associated with lengths, and it must have been a terrible shock to discover that some lengths are not given by *rational* numbers. But this fact is inescapable, for the diagonal $d$ of a square of side 1 satisfies $d^2 = 2$, whereas *no* rational number $x$ satisfies $x^2 = 2$. The strong intuitive feeling that every length *should* be given by a number led in time to an extension of the rational numbers to form the *real numbers*, which do allow us to solve the equation $x^2 = 2$, and which, in fact, provide all the lengths that geometry requires.

Another extension problem is presented by the quadratic equation $ax^2 + bx + c = 0$, which is solved by the expression

$$x = \frac{-b \pm \sqrt{b^2 - 4ac}}{2a}.$$

(1)

Since a negative number has no real square root, formula (1) does not give a real number $x$ when $b^2 - 4ac < 0$. Shall we nevertheless assign a meaning to $\sqrt{b^2 - 4ac}$ if, say, $b^2 - 4ac = -1$? Or shall we simply say that there are no roots in this case? The answer is, once more, to extend the number system and to include objects like $\sqrt{-1}$. This object is called an *imaginary number*, and the extended system containing combinations of this and the real numbers is called the *complex number system*.

The very names "complex number" and "imaginary number" suggest a certain air of mystery, in sharp contrast to the reassuring "real numbers." The mystery is due, most likely, to the fact that the real numbers satisfy a clear geometric need, while imaginary and complex numbers are introduced for purely algebraic reasons.

It would be natural to expect that complex numbers are useless in applications of mathematics to real physical situations. But in fact, they constitute an extraordinarily powerful tool both in "pure" mathematics and in its applications. This chapter gives a very small sample of these applications and of the underlying theory.

## 19.1

### COMPLEX NUMBERS

The complex number system is obtained from the real numbers by declaring that there *is*, after all, a square root of $-1$. Since none of the "real numbers" can have a negative square, this new root must be an "imaginary number." We denote it by $i$, and boldly write

$$i^2 = -1.$$

(1)

We combine this new number with the real numbers through multiplication and addition, obtaining "complex numbers" such as

$$2i, \qquad 3 + 2i,$$

and so on. It is true (but not obvious) that this can be done, still preserving these basic laws of algebra:

$$ab = ba, \qquad a + b = b + a$$

$$a(bc) = (ab)c, \qquad a + (b + c) = (a + b) + c$$

$$a(b + c) = ab + ac.$$

$$0 + a = 0, \qquad 1 \cdot a = a.$$

Whenever $i^2$ occurs, we replace it with $-1$. Thus,

$$(3 + 2i)(4 - i) = 3(4 - i) + 2i(4 - i) = 12 - 3i + 8i - 2i^2$$
$$= 12 + 5i - 2(-1)$$
$$= 14 + 5i.$$

The results of these operations can always be written in the form $x + yi$, with *real* numbers $x$ and $y$; such numbers form the **complex number system**.

In this system there are *two* square roots of $-1$. One is $i$, and the other is $(-1)i$; for

$$[(-1)i]^2 = (-1)^2 \cdot i^2 = 1 \cdot (-1) = -1.$$

This second root, denoted $-i$, is called the **conjugate** of the first root, $i$.

Generally, for any complex number such as $3 + 5i$, we define a *conjugate* $3 - 5i$, obtained by replacing $i$ with $-i$. The conjugate is denoted with a bar:

$$\overline{3 + 5i} = 3 - 5i.$$

Thus for $z = x + iy$ (with $x$ and $y$ real),

$$\bar{z} = \overline{x + iy} = x - iy. \tag{2}$$

You can check that

$$\overline{z + w} = \bar{z} + \bar{w}$$

and

$$\overline{(zw)} = \bar{z} \cdot \bar{w}.$$

## Geometry of the Complex Numbers

Each complex number $x + iy$ can be plotted as a point $(x, y)$ in the coordinate plane (Fig. 1). The coordinate $x$ is called the **real part** of $x + iy$, denoted $Re(x + iy)$; and $y$ is the **imaginary part**, $Im(x + iy)$. The ordinary real numbers have the form $x + i0$, corresponding to the points $(x, 0)$ on the $x$-axis; this is called the **real axis**. The numbers $0 + iy$ are called "pure imaginary numbers"; they correspond to the points $(0, y)$ on the $y$-axis, which is therefore called the "imaginary axis."

We can also think of $x + iy$ as the vector $\langle x, y \rangle$. Then the addition of complex numbers corresponds to vector addition:

$$(x + iy) + (u + iv) = (x + u) + i(y + v)$$

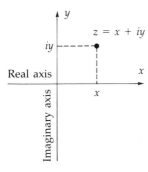

**FIGURE 1**
$x + iy$ plotted as $(x, y)$.

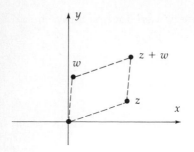

**FIGURE 2**
Parallelogram law for $z + w$.

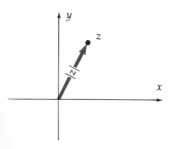

**FIGURE 3**
$|z|$ = Distance from origin to $z$.

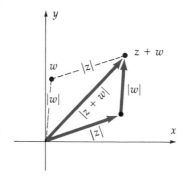

**FIGURE 4**
Triangle inequality:
$|z + w| \le |z| + |w|$.

corresponds to

$$\langle x, y \rangle + \langle u, v \rangle = \langle x + u, y + v \rangle.$$

So when complex numbers are plotted in the plane, addition follows the parallelogram law of vector addition (Fig. 2). The segments from 0 to $z$ and 0 to $w$ form two sides of a parallelogram; in this parallelogram, $z + w$ is the corner opposite 0.

The *distance* from $z = x + iy$ to the origin is called the **absolute value** of $z$, or **modulus** of $z$, denoted $|z|$ (Fig. 3):

$$|z| = |x + iy| = \sqrt{x^2 + y^2}. \tag{3}$$

When $z$ is real, $z = x + i0$, then $|z| = \sqrt{x^2 + 0^2} = |x|$, in accord with the concept of absolute value for real numbers. The absolute value in (3) is also the same as the length of the vector $\langle x, y \rangle$, so the triangle inequality remains true (Fig. 4):

$$|z + w| \le |z| + |w|.$$

Moreover, the distance between two complex numbers $z$ and $w$ is the absolute value of their difference (Fig. 5):

$$|z - w| = \text{distance from } z \text{ to } w.$$

For example, $|z - i|$ is the distance from $z$ to $i$ (Fig. 6); so the set

$$\{z : |z - i| = 1\}$$

is a circle of radius 1, with center $i$.

The product of a complex number and its conjugate has a simple form:

$$(x + iy)(x - iy) = x^2 + iyx - xiy - i^2 y^2 = x^2 - (-1)y^2 = x^2 + y^2.$$

Thus from (2) and (3),

$$z \cdot \bar{z} = |z|^2. \tag{4}$$

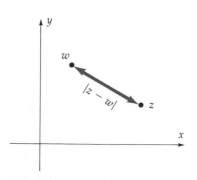

**FIGURE 5**
$|z - w|$ = distance from $z$ to $w$.

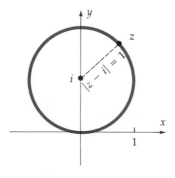

**FIGURE 6**
$|z - i| = 1$ is the equation of a circle of radius 1, centered at $i$.

This formula shows how to construct the reciprocal $z^{-1}$ of any complex number $z \neq 0$. For if $z \neq 0$ then $|z|^2 \neq 0$, and we can divide in (4), obtaining

$$z \cdot \frac{\bar{z}}{|z|^2} = 1.$$

So the reciprocal of $z$ is $z^{-1} = \dfrac{\bar{z}}{|z|^2}$. For example,

$$(3 + 5i)^{-1} = \frac{3 - 5i}{|3 + 5i|^2} = \frac{3 - 5i}{9 + 25} = \frac{3}{34} - \frac{5}{34}\, i.$$

Generally, $(x + iy)^{-1} = \dfrac{x - iy}{x^2 + y^2}.$

**REMARK**  Do the complex numbers share *all* the properties of the real numbers? No; there is no way to extend the *ordering* from the real to the complex numbers, preserving the two basic properties:

   **I.**   For every real number $a$, exactly one of three possibilities holds:

$$a > 0, \quad \text{or} \quad a = 0, \quad \text{or} \quad -a > 0.$$

   **II.**   If $a > 0$ and $b > 0$, then $a + b > 0$ and $ab > 0$.

For if there *were* such an ordering, then by property I, either $i > 0$ or $-i > 0$. Each of these possibilities leads to a contradiction:

   If $i > 0$ then $i^2 = -1 > 0$, by II, and this is false.
   If $-i > 0$, then $(-i)^2 = -1 > 0$, the same contradiction as before.

We can, however, use inequalities, such as the triangle inequality, involving the *absolute values* of complex numbers.

## SUMMARY

For complex numbers $z = x + iy$, $w = u + iv$, and $c = a + ib$,

$$zw = wz, \qquad z + w = w + z$$
$$(zw)c = z(wc), \qquad (z + w) + c = z + (w + c)$$
$$c(z + w) = cz + cw.$$
$$0 \cdot z = 0, \qquad 1 \cdot z = z, \qquad 0 + z = z.$$

$x + iy$ is plotted in the coordinate plane as $(x, y)$.

***Real and Imaginary Parts:***   $\operatorname{Re}(x + iy) = x, \qquad \operatorname{Im}(x + iy) = y.$

*Conjugate:* $\quad \bar{z} = \overline{x + iy} = x - iy.$

*Absolute Value:* $\quad |z| = |x + iy| = \sqrt{x^2 + y^2}.$

$$|z|^2 = z \cdot \bar{z}.$$

*Reciprocal:* $\quad z^{-1} = \dfrac{\bar{z}}{|z|^2}, \qquad (x + iy)^{-1} = \dfrac{x - iy}{x^2 + y^2}.$

## PROBLEMS

### A

1. Compute.
   a) $2 + 3i + (3 - i)$
   *b) $(2 + 3i)(3 - i)$
   c) $(2 - 3i)(2 + 3i)$
   d) $(2 - 3i)(x + iy)$
   *e) $(a + ib)(c + id)$

2. Compute $z^{-1}$, and verify that $z \cdot z^{-1} = 1$.
   *a) $z = (1 + i)$
   *b) $z = i$
   c) $z = 3 - 5i$

3. For each of the following numbers $z$, draw a figure showing $z$, $\bar{z}$, $-z$, and $1/z$.
   a) $1$
   *b) $i$
   c) $-i$
   *d) $-1$
   *e) $1 + i$
   f) $-1 - 2i$
   g) $-3 + \pi i$

4. Sketch the following sets of complex numbers.
   *a) $\{z: |z - 1| = 2\}$
   b) $\{z: |z + i| = 1\}$
   *c) $\{z: \text{Re}(z) = 2\}$
   d) $\{z: \text{Im}(z) = 1\}$

5. a) Show that $\sqrt{-4} = \pm 2i$, in the sense that $(2i)^2 = -4$ and $(-2i)^2 = -4$.
   b) Show that for any real number $a > 0$,
   $$\sqrt{-a} = \pm\sqrt{a}i.$$

6. a) Solve $z^2 - 6z + 34 = 0$, using the quadratic formula.

b) Check that each of the complex numbers obtained really solves the equation.

7. Verify.
   a) $\overline{z + w} = \bar{z} + \bar{w}$
   b) $\overline{zw} = \bar{z} \cdot \bar{w}$
   c) $|z \cdot w| = |z| \cdot |w|$ (Use part b.)

### B

8. Describe the following sets geometrically.
   *a) $\{z: |z - i| = |z - 2|\}$
   *b) $\{z: |z - 1| = \text{Re}(z)\}$
   c) $\{z: |z + 1| = \frac{1}{2}|\text{Re}(z)|\}$
   d) $\{z: |z + i| = 2\text{Im}(z)\}$

9. Describe geometrically the relation between
   a) $z$ and $\bar{z}$.
   b) $z$ and $-z$.

10. *a) Determine all square roots of $i$, by solving $(x + iy)^2 = i$ for $x$ and $y$.
    b) Plot the square roots of $i$.

11. a) Determine all square roots of $1 + i$, as in the previous problem.
    b) Plot the square roots of $1 + i$.

## 19.2

### MULTIPLICATION OF COMPLEX NUMBERS

Our definition of multiplication for complex numbers gives

$$(x + iy)(u + iv) = (xu - yv) + i(yu + xv).$$

But there is a more revealing way to handle these products, based on *complex exponentials*.

The basic algebraic property of the exponential function is

$$e^{a+b} = e^a e^b. \tag{1}$$

We now define *complex* exponentials by the formula

$$e^{x+iy} = e^x(\cos y + i \sin y), \tag{2}$$

and justify this definition by showing that:

**(A)**   The new definition agrees with the old when $x + iy$ happens to be a real number, that is, when $y = 0$.

**(B)**   The basic property (1) remains valid when $a$ and $b$ are complex numbers.

(A) is easily checked. For when $y = 0$, then $\cos y = 1$ and $\sin y = 0$, so (2) gives $e^{x+i0} = e^x$. To check (B), let $a = x + iy$ and $b = u + iv$. Then from (2),

$$e^{a+b} = e^{(x+u)+i(y+v)} = e^{x+u}[\cos(y+v) + i \sin(y+v)]$$
$$= e^{x+u}(\cos y \cos v - \sin y \sin v + i \sin y \cos y + i \cos y \sin v).$$

On the other hand,

$$e^a e^b = e^x(\cos y + i \sin y)e^u(\cos v + i \sin v)$$
$$= e^x e^u(\cos y \cos v - \sin y \sin v + i \sin y \cos v + i \cos y \sin v).$$

The two expressions are indeed the same.

---

**EXAMPLE 1**   Verify the mysterious relation $e^{i\pi} = -1$ between four special numbers. (Do it yourself; the solution is on p. 944.)

---

To explain complex multiplication from a geometric point of view, we use polar coordinates (Fig. 1). If $z = x + iy$, we obtain polar coordinates of the corresponding point $(x, y)$ in the plane by setting

$$r = \sqrt{x^2 + y^2} = |z|$$

and choosing $\theta$ so that

$$r \cos \theta = x, \qquad r \sin \theta = y. \tag{3}$$

Figure 1 illustrates one choice of $\theta$, with $0 \le \theta < 2\pi$; the other possible choices differ from this by integer multiples of $2\pi$.

From (3), it follows that

$$x + iy = r \cos \theta + ir \sin \theta = r(\cos \theta + i \sin \theta) = re^{i\theta}.$$

Any number $\theta$ chosen as in (3) is traditionally called an *argument* of $z$, denoted arg $z$; we think of arg $z$ as giving the angle from the $x$-axis to the line between the origin and $z$, as in Figure 1. Using this notation, and recalling that $r = |z|$, we can write the formula $x + iy = re^{i\theta}$ as

$$z = |z|e^{i \text{ arg } z}. \tag{4}$$

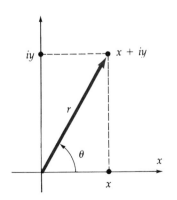

**FIGURE 1**
Polar coordinates; $\theta = \arg(x + iy)$, $r = |z|$.

We note that arg is *not* a function, for it assigns to a given $z$ not one, but many numbers; if $\theta$ is one of these, then the others are $\theta \pm 2\pi, \theta \pm 4\pi, \ldots$.

**EXAMPLE 2**    Find the values of arg 1, arg $i$, arg $(1 - i)$. Find three different values of arg $i$. (Do it yourself; the solution is on p. 944.)

The representation in (4) together with formula (1) for products of exponentials leads immediately to the geometric interpretation of complex products. If we write $z = |z|e^{i \arg z}$ and $w = |w|e^{i \arg w}$, then

$$
\begin{aligned}
zw &= (|z|e^{i \arg z})(|w|e^{i \arg w}) \\
&= |z| \cdot |w|e^{i \arg z}e^{i \arg w} \quad \text{(commutative law)} \\
&= |z| \cdot |w|e^{i(\arg z + \arg w)}. \quad \text{[by (1)]}
\end{aligned}
$$

Thus *to multiply two complex numbers, multiply their absolute values, and add their arguments* (Fig. 2).

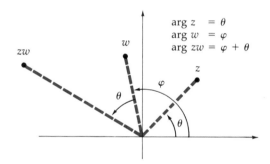

$$
\begin{aligned}
\arg z &= \theta \\
\arg w &= \varphi \\
\arg zw &= \varphi + \theta
\end{aligned}
$$

**FIGURE 2**
Multiplication of complex numbers:
$|zw| = |z| |w|$ and
$\arg zw = \arg z + \arg w$.

**EXAMPLE 3**    Let $w = i$. Then (Fig. 3) $|i| = 1$ and arg $i = \pi/2$, so

$$
iz = |z| \cdot |i|e^{i(\arg z + \arg i)} = |z|e^{i(\arg z + \pi/2)}.
$$

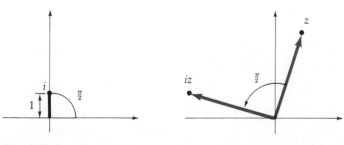

**FIGURE 3**
$\text{Arg}(iz) = \arg z + \arg i$
$\qquad = \arg z + \pi/2.$

Thus to multiply $z$ by $i$, leave the length unchanged, and increase arg $z$ by $\pi/2$; in other words, rotate $z$ counterclockwise about the origin, through an angle $\pi/2$.

**EXAMPLE 4**   Explain the effect of multiplication by $-1$, by observing that $|-1| = 1$ and $\arg(-1) = \pi$. (The solution is on p. 944.)

It is easy to take powers of a complex number $z$ written in the form $re^{i\theta}$:

$$(re^{i\theta})^2 = r^2 e^{2i\theta},$$

and generally

$$(re^{i\theta})^n = r^n e^{ni\theta}.$$

This leads, in turn, to a simple way of extracting the roots of a complex number $z$. Simply write $z$ in the form $re^{i\theta}$, and set

$$z^{1/n} = r^{1/n} e^{i\theta/n}. \tag{5}$$

For then $(r^{1/n} e^{i\theta/n})^n = re^{i\theta} = z$.

**EXAMPLE 5**   To find $\sqrt{-1}$, write $-1$ in "polar form," $-1 = e^{i\pi}$; this leads to $\sqrt{-1} = e^{i\pi/2} = i$. We could also write $-1 = e^{-i\pi}$, leading to $\sqrt{-1} = e^{-\pi i/2} = -i$. Giving both solutions together, $\sqrt{-1} = \pm i$.

**EXAMPLE 6**   We can find $i^{1/3}$ in any of the following ways (Fig. 4):

$$i = e^{i\pi/2}, \qquad i^{1/3} = e^{i\pi/6} = \frac{\sqrt{3}}{2} + \frac{i}{2};$$

$$i = e^{i5\pi/2}, \qquad i^{1/3} = e^{i5\pi/6} = -\frac{\sqrt{3}}{2} + \frac{i}{2};$$

$$i = e^{i9\pi/2}, \qquad i^{1/3} = e^{i9\pi/6} = e^{i3\pi/2} = -i;$$

$$i = e^{i13\pi/2}, \qquad i^{1/3} = i^{i13\pi/6} = \frac{\sqrt{3}}{2} + \frac{i}{2};$$

$$i = e^{-i3\pi/2}, \qquad i^{1/3} = e^{-i\pi/2} = -i;$$

and so on.

Thus $i$ has *three* cube roots; $-i$, and $\pm\dfrac{\sqrt{3}}{2} + \dfrac{i}{2}$.

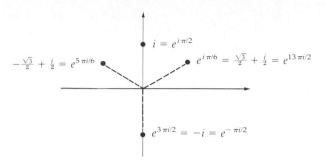

**FIGURE 4**
Three cube roots of $i$.

As Examples 5 and 6 show, the expression (5) produces more than one $n^{th}$ root of a given complex number $z$, because there is more than one way to choose the angle $\theta$ in $z = re^{i\theta}$. The situation is as follows:

---

**THEOREM 1**

Given any complex number $z \neq 0$, there are exactly $n$ different numbers $z_1, \ldots, z_n$ such that $(z_j)^n = z$. The $z_j$ are uniformly spaced around the circle of radius $|z|^{1/n}$ about the origin.

---

We leave the proof as problem 9. Example 6 provides a hint as to why the theorem is true. The various possible choices of arg $z$ all differ from each other by multiples of $2\pi/n$.

Theorem 1 is a good example of the simplification brought about by complex numbers. Compare the neat result stated there to the mess that comes when we deal only with real numbers:

(a) If $x > 0$ and $n$ is even, there are two real $n^{th}$ roots of $x$, denoted $\sqrt[n]{x}$ and $-\sqrt[n]{x}$.

(b) If $x < 0$ and $n$ is even, there are no real $n^{th}$ roots of $x$.

(c) If $n$ is odd, then any real number has exactly one real $n^{th}$ root.

**SOLUTION TO EXAMPLE 1** $\quad e^{i\pi} = \cos \pi + i \sin \pi = -1.$

**SOLUTION TO EXAMPLE 2** From Figure 5 we read arg $1 = 0$, arg $i = \pi/2$, arg$(1 - i) = -\pi/4$. The dashed arc and the dotted arc suggest $-3\pi/2$ and $5\pi/2$ as alternate choices of arg $i$. In general,

$$\arg 1 = 0, \pm 2\pi, \pm 4\pi, \ldots$$

$$\arg i = \pi/2, \pi/2 \pm 2\pi, \ldots$$

$$\arg(1 - i) = -\pi/4, -\pi/4 \pm 2\pi, \ldots$$

**SOLUTION TO EXAMPLE 4** Since $-1 = e^{i\pi}$, multiplication by $-1$ amounts to rotation about the origin through an angle of radian measure $\pi$, that is, through a "straight angle."

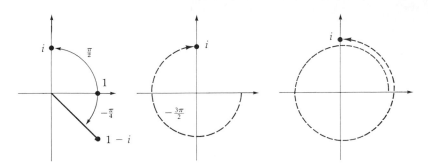

**FIGURE 5**
Solution to Example 2.

# PROBLEMS

## A

1. Find values of arg $z$ for the following numbers.
   *a)  $1 + i$         b)  $1/2 + i/2$
   *c)  $1 + 3i$        d)  $-i$

   e)  $-1 - i$     *f)  $-\dfrac{3}{2} + i\dfrac{\sqrt{3}}{2}$

2. Plot the following numbers as points in the plane.
   a)  $e^{i\pi/3}$           b)  $e^{-i\pi/4}$
   c)  $2e^{i3\pi/4}$         d)  $\frac{1}{2}e^{i5\pi/6}$

*3. Show that $e^{z + 2\pi i} = e^z$.

4. Plot $z$, $w$, and $zw$ in each of the following cases.
   a)  $z = 2i, w = 1 + i$
   b)  $z = -i/2, w = 3 - i$
   c)  $z = 2i, w = e^{\pi i/6}$
   d)  $z = \frac{1}{2}e^{-\pi i/2}, w = 3 + i$

*5. Write each of the four fourth roots of 1 in the form $x + iy$. Plot them.

*6. Write each of the sixth roots of $-2$ in the form $x + iy$. Plot them.

## B

7. a)  Find four solutions of $z^4 - 3z^2 + 2 = 0$. (Hint: First solve for $z^2$.)
   *b)  Find four solutions of $z^4 + 3z^2 + 3 = 0$.

8. Show that $\cos \theta = (e^{i\theta} + e^{-i\theta})/2$. What is the corresponding formula for $\sin \theta$?

9. a)  Find $n$ distinct $n^{\text{th}}$ roots of 1.
   b)  Show that 1 has at most $n$ distinct $n^{\text{th}}$ roots.
   c)  Prove Theorem 1.

10. Prove that $|zw| = |z| \cdot |w|$.

11. Show that $\overline{re^{i\theta}} = re^{-i\theta}$. Illustrate with an appropriate sketch.

12. Prove that $\overline{zw} = \bar{z} \cdot \bar{w}$. (See problem 11.)

13. a)  Let $r$ be a complex number. Prove that $(z - r)(z - \bar{r}) = z^2 + bz + c$, where $b$ and $c$ are *real* numbers.
    b)  Suppose that $P(z) = \sum_0^n a_j z^j$, where the $a_j$ are all *real* numbers. Prove that $P(r) = 0 \Rightarrow P(\bar{r}) = 0$. [Is $\overline{P(r)} = P(\bar{r})$?]

14. a)  Prove *De Moivre's Theorem*:
       $(\cos \theta + i \sin \theta)^n = \cos n\theta + i \sin n\theta$.
    b)  Take real and imaginary parts, with $n = 2$, to derive a formula for $\cos 2\theta$ and $\sin 2\theta$ as polynomials in $\cos \theta$ and $\sin \theta$.
    c)  Derive formulas for $\cos 3\theta$ and $\sin 3\theta$.
    d)  Using De Moivre's Theorem and the Binomial Theorem $(a + b)^n = a^n + na^{n-1}b + \dfrac{n(n-1)}{2} a^{n-2}b^2 + \cdots + nab^{n-1} + b^n$,
       derive formulas for $\cos n\theta$ and $\sin n\theta$. [Note that two complex numbers $z$ and $w$ are equal if $\text{Re}(z) = \text{Re}(w)$ and $\text{Im}(z) = \text{Im}(w)$.]

15. Every nonzero complex number $z$ can be written *uniquely* in the form $z = re^{i\theta}$, where $r = |z|$ and $-\pi < \theta \le \pi$. The number $\theta$ with these restrictions is called the *principal value* of the argument of $z$, denoted Arg $z$. Show from appropriate sketches that

    a)  Arg $(x + iy) = \arctan \dfrac{y}{x}$ if $x > 0$.

**b)** $\text{Arg}\,(x + iy) = \text{arccot}\,\dfrac{x}{y}$ if $y > 0$.

**c)** $\text{Arg}\,(x + iy) = \text{arccot}\,\dfrac{x}{y} - \pi$ if $y < 0$.

**d)** $\text{Arg}\,(x + iy) = \arctan\,\dfrac{y}{x} + \pi$ if $x < 0$ and $y \geq 0$.

**e)** $\text{Arg}\,(x + iy) = \arctan\,\dfrac{y}{x} - \pi$ if $x < 0$ and $y < 0$.

(Recall the principal value restrictions $-\pi/2 < \arctan t < \pi/2$ and $0 < \text{arccot}\,t < \pi$.)

16. For any complex $z$, define the (complex) logarithm by

$$\text{Log}\,z = \log|z| + i\,\text{Arg}\,z$$

(see problem 15 for Arg).
**a)** Compute $\text{Log}\,i$.
**b)** Prove that for all complex $z$, $e^{\text{Log}\,z} = z$.

**c)** For what values of $z$ is $\text{Log}(e^z) = z$?
**d)** Give a general formula for $\text{Log}(e^z)$.

17. Make sense of the equation $i^i = e^{-\pi/2}$.

18. The number $z_n = e^{2\pi i/n}$ is called a *primitive $n^{\text{th}}$ root of unity*.
**a)** Show that $z_n, z_n^2, \ldots, z_n^n$ are all the $n^{\text{th}}$ roots of 1.
**b)** Prove that $z_n + z_n^2 + \cdots + z_n^n = 0$. [Hint: $(1 - r)(1 + r + \cdots + r^{n-1}) = 1 - r^n$.]
**c)** Let $\vec{A}_1, \ldots, \vec{A}_n$ be plane vectors pointing from the origin to the vertices of a regular polygon centered at the origin. Prove that $\vec{A}_1 + \cdots + \vec{A}_n = 0$.

19. The hyperbolic functions are $\cosh x = \frac{1}{2}(e^x + e^{-x})$ and $\sinh x = \frac{1}{2}(e^x - e^{-x})$.
**a)** Derive a formula for $(\cosh\theta + \sinh\theta)^n$, comparable to De Moivre's Theorem (problem 14).
**b)** Relate $\cosh(i\theta)$ and $\sinh(i\theta)$ to $\cos\theta$ and $\sin\theta$.
**c)** Make sense of the formula $\cos(i\theta) = \cosh(\theta)$.

## 19.3

## COMPLEX FUNCTIONS OF A REAL VARIABLE

We studied the differential equation for oscillating motion

$$f''(t) + \omega^2 f(t) = 0, \tag{1}$$

and found the general solution (Sec. 8.8) as a linear combination of sine and cosine,

$$f(t) = A\cos\omega t + B\sin\omega t.$$

The term $\omega^2 f$ in (1) can be thought of as the acceleration due to stretching or compressing a spring. To be more realistic, we should add a frictional force of some kind. If the frictional force is proportional to the velocity $f'$, we get the equation

$$f''(t) + bf'(t) + \omega^2 f(t) = 0, \tag{2}$$

where $b$ is constant.

Equations (1) and (2) are *linear differential equations with constant coefficients*. The general equation of this type has the form

$$a_n f^{(n)}(t) + a_{n-1} f^{(n-1)}(t) + \cdots + a_1 f'(t) + a_0 f(t) = g, \tag{3}$$

where $a_0, \ldots, a_n, g$ are constants, possibly complex. The solution of such equations depends on the use of exponential functions, with either real or complex exponents, as the following examples suggest.

---

**EXAMPLE 1**   Find solutions of $f'' - 4f = 0$.

**SOLUTION**   Try $f(t) = e^{rt}$, where $r$ is a constant to be determined. We get $f' = re^{rt}$, $f'' = r^2 e^{rt}$, so

$$f'' - 4f = (r^2 - 4)e^{rt}.$$

Thus, for functions $f$ of this form $e^{rt}$,

$$f'' - 4f = 0 \quad \Leftrightarrow \quad r^2 = 4, \text{ or } r = \pm 2.$$

We obtain in this way two solutions, $e^{2t}$ and $e^{-2t}$. You can check further that for any constants $A$ and $B$, the linear combination

$$f(t) = Ae^{2t} + Be^{-2t}$$

is a solution.

---

**EXAMPLE 2**   Find solutions of $f'' - 3f' + 2f = 0$.

**SOLUTION**   Try $f(t) = e^{rt}$ again; you get

$$f'' - 3f' + 2f = (r^2 - 3r + 2)e^{rt}.$$

So $e^{rt}$ is a solution $\Leftrightarrow$ $r^2 - 3r + 2 = 0$, which gives $r = 2$ and $r = 1$. You can check that, more generally

$$f(t) = Ae^{2t} + Be^t$$

is a solution for any choice of the constants $A$ and $B$.

---

**EXAMPLE 3**   Find solutions of $f'' + 9f = 0$.

**SOLUTION**   Trying $f(t) = e^{rt}$ as before, you get

$$f'' + 9f = (r^2 + 9)e^{rt}.$$

So $e^{rt}$ is a solution precisely when $r^2 = -9$, or $r = \pm 3i$. We thus get the two functions

$$e^{3it} = \cos 3t + i \sin 3t,$$

$$e^{-3it} = \cos 3t - i \sin 3t.$$

We can easily recognize these as solutions of $f'' + 9f = 0$, *if* complex-valued functions behave as real-valued functions do. This is precisely what we are about to investigate.

A complex function of one real variable has domain on the real line, and assigns to each point of this domain a complex number. The most important examples are the complex polynomials

$$P(t) = a_n t^n + \cdots + a_1 t + a_0$$

(where the numbers $a_0, \ldots, a_n$ may be complex); and the complex exponentials

$$f(t) = e^{rt}$$

(where again the number $r$ may be complex).

Every complex function $f$ can be written as a combination of two real functions $u$ and $v$:

$$f(t) = u(t) + iv(t)$$

with $u(t) = \operatorname{Re} f(t)$ and $v(t) = \operatorname{Im} f(t)$. For example,

$$f(t) = (1 + i)t^2 + t - 3i = (t^2 + t) + i(t^2 - 3)$$

has the desired form $u(t) + iv(t)$, with $u(t) = t^2 + t$ and $v(t) = t^2 - 3$ *real*. Similarly, complex exponentials have this form:

$$f(t) = e^{2it} = \cos 2t + i \sin 2t = u(t) + iv(t),$$

with $u(t) = \cos t$ and $v(t) = \sin 2t$.

The *derivative* of a complex function is taken by differentiating the real and imaginary parts separately:

$$f' = (u + iv)' = u' + iv'.$$

Thus, if $f(t) = e^{2it} = \cos 2t + i \sin 2t$, then

$$f'(t) = -2 \sin 2t + 2i \cos 2t.$$

You can easily check that the usual rules for differentiation apply to complex functions:

$$(f + g)' = f' + g', \tag{4}$$

$$(fg)' = f'g + fg', \tag{5}$$

$$\left(\frac{f}{g}\right)' = \frac{f'g - fg'}{g^2}, \tag{6}$$

and if $h$ is a real-valued function, then the Chain Rule holds:

$$(f \circ h)' = (f' \circ h)h'. \tag{7}$$

In addition to these general results, we need a special formula:

$$\text{If} \quad f(t) = e^{rt} \quad \quad (r \text{ complex})$$
$$\text{then} \quad f'(t) = re^{rt}. \tag{8}$$

The proof is straightforward. Set $r = a + ib$; then

$$f(t) = e^{at+ibt} = e^{at}(\cos bt + i \sin bt),$$

$$f'(t) = e^{at}(-b \sin bt + ib \cos bt) + ae^{at}(\cos bt + i \sin bt)$$
$$= (a + ib)e^{at}(\cos bt + i \sin bt)$$
$$= re^{at+ibt}$$
$$= re^{rt}.$$

When $f$ is a complex function defined on an interval of the real axis, we call any function $F$ such that $F' = f$ an *indefinite integral* of $f$. Just as for real-valued functions, any two indefinite integrals of a given function $f$ differ by a constant. The indefinite integral of $f(t)$ is denoted $\int f(t)\, dt$. Thus, from (8), if $r \neq 0$, we have

$$\int e^{rt}\, dt = \frac{1}{r} e^{rt} + c,$$

where $c$ is any complex number.

The net result of the last few paragraphs is that we calculate with complex-valued functions exactly as with real-valued functions. With this assurance, we return to the differential equation with constant coefficients (3), and take the special case where $g = 0$, (the so-called homogeneous case):

$$a_n f^{(n)} + \cdots + a_0 f = 0. \tag{9}$$

Try an *exponential solution,*

$$f(t) = e^{rt}.$$

Since $f'(t) = re^{rt}, f''(t) = r^2 e^{rt}, \ldots$ and $f^{(n)}(t) = r^n e^{rt}$, we are led from (9) to the equation

$$a_n r^n e^{rt} + \cdots + a_0 e^{rt} = 0.$$

Since $e^{rt}$ is never zero, we may divide and obtain a polynomial equation for $r$,

$$a_n r^n + \cdots + a_0 = 0. \tag{10}$$

Thus *when $r$ satisfies equation (10), then $e^{rt}$ is a solution of (9).* The polynomial $\sum_0^n a_j r^j$ occurring in (10) is called the **characteristic polynomial** of equation (9). The solutions of (10), the zeroes of the characteristic polynomial, are called the **characteristic roots** of the equation.

Finding these exponential solutions of (9) is the first step in solving linear differential equations with constant coefficients; the solution (for second order equations) is completed in the next two sections.

We close with an example illustrating a rather different use of complex functions: the evaluation of integrals.

**EXAMPLE 4** $\int e^t \cos t \, dt = ?$

**SOLUTION** $\cos t = \text{Re}(e^{it})$, so consider the integral

$$\int e^t(\cos t + i \sin t) \, dt = \int e^t e^{it} \, dt = \int e^{(1+i)t} \, dt = \frac{1}{1+i} e^{(1+i)t} + c$$

$$= \frac{1-i}{2} e^t e^{it} + c = e^t \left( \frac{1-i}{2} \right) (\cos t + i \sin t) + c$$

$$= \frac{e^t}{2} [(\cos t + \sin t) + i(\sin t - \cos t)] + c.$$

Taking real parts, you get

$$\int e^t \cos t \, dt = \frac{e^t}{2} (\cos t + \sin t) + c.$$

Taking imaginary parts, you get a bonus:

$$\int e^t \sin t \, dt = \frac{e^t}{2} (\sin t - \cos t) + c.$$

## PROBLEMS

### A

1. For each of the following differential equations, find the roots of the characteristic polynomial and the corresponding exponential solutions.
   *a) $f'' + 4f = 0$
   b) $f'' - 4f = 0$
   *c) $f'' + 2f' + f = 0$
   d) $f'' - 2f' + f = 0$
   *e) $f'' + 2f' + 2f = 0$
   f) $f''' + f = 0$

2. Prove formula (4).

3. Prove formula (5).

4. Prove formula (6).

5. Let $a$ and $b$ be real numbers.
   a) Show that $\int e^{at} \cos bt \, dt = \text{Re} \int e^{(a+ib)t}$, and evaluate the integral.
   b) Evaluate $\int e^{at} \sin bt \, dt$ in a similar way.

6. This problem concerns integrals arising in Fourier series, one of the important topics in advanced calculus. Let $n$ and $m$ be integers.
   a) Show that $\int_0^{2\pi} e^{int} e^{-imt} dt = 0$ if $n \neq m$, and $= 2\pi$ if $n = m$.

   b) Show that if $n \geq 0$ and $m \geq 0$, then $\int_0^{2\pi} e^{int} e^{imt} \, dt = 0$ unless $n = m = 0$.
   c) Take real and imaginary parts of the integrals in part (a) and part (b) to show that for integers $n \geq 0$, $m \geq 0$

   $$\int_0^{2\pi} \cos nt \sin mt \, dt = 0 \text{ for all } n \text{ and } m$$

   $$\int_0^{2\pi} \cos nt \cos mt \, dt = \begin{cases} 0, & n \neq m \\ \pi, & n = m \neq 0 \\ 2\pi, & n = m = 0 \end{cases}$$

   $$\int_0^{2\pi} \sin nt \sin mt \, dt = \begin{cases} 0, & n \neq m \\ \pi, & n = m \neq 0 \\ 0, & n = m = 0. \end{cases}$$

7. Evaluate the following improper integrals, which are important in the study of heat conduction.
   *a) $\int_0^\infty e^{ixt} e^{-t} \, dt$

   b) $\int_{-\infty}^0 e^{ixt} e^t \, dt$

   *c) $\int_{-\infty}^\infty e^{ixt} e^{-|t|} \, dt$

**d)** $\displaystyle\int_0^\infty \cos(xt)e^{-t}\,dt$

**e)** $\displaystyle\int_0^\infty \sin(xt)e^{-t}\,dt$

**f)** $\displaystyle\int_{-\infty}^\infty e^{ixt}e^{-a|t|}\,dt,\ a > 0$

**B**

8. Prove that if $f'$ is identically zero throughout an interval $I$, then $f$ is constant on that interval. (Here, $f$ may be complex, of course.)

9. Prove that $\int (f + g) = \int f + \int g$ and $\int cf = c\int f$, where $f$ and $g$ are any continuous complex functions and $c$ is any complex number.

10. **a)** Prove that $\int fg' = fg - \int f'g$, where $f$ and $g$ are differentiable complex functions.
    ***b)** Find $\int te^{rt}\,dt$, where $r$ is any complex number.
    **c)** Find $\int t^2 e^{rt}\,dt$, where $r$ is any complex number.

11. The complex functions Arg and Log developed in the problems of Section 19.2 can be used to integrate rational functions.
    **a)** Define Log $z$ as in problems 15 and 16, Section 19.2. Let $r = a + ib$ be a complex number, $b \neq 0$, and define $f(t) = \text{Log}(t - r)$.

Show that $f'(t) = \dfrac{1}{t - r}$. (Use problem 15c for $b > 0$, and 15b for $b < 0$.)

**b)** Note that
$$\int \frac{dt}{t^2 + 1} = \int \frac{dt}{(t - i)(t + i)}$$
$$= \frac{1}{2i}\int \frac{dt}{t - i} - \frac{1}{2i}\int \frac{dt}{t + i}.$$

Use part a to evaluate the last two integrals as Logs, and combine them to produce a formula for arctan $t$.

**c)** Use part a to evaluate $\displaystyle\int \frac{dt}{t^2 - 2t + 5}$. [Hint: $t^2 - 2t + 5 = 0$ has solutions $t = 1 \pm 2i$, so it factors into $(t - 1 - 2i)(t - 1 + 2i)$. Use a partial fractions decomposition (Sec. 9.6), and evaluate the results as Logs. Recombine the Logs.]

**d)** Evaluate $\displaystyle\int \frac{dt}{(t^2 + 1)^2}$ using partial fractions and Logs. (The calculations here may be a little longer than those in Section 9.6, but they are more straightforward and do not require any hard-to-remember reduction formulas.)

## 19.4

# DAMPED OSCILLATIONS

Many simple mechanical systems and electrical circuits are described by differential equations of the form

$$ay'' + by' + cy = 0 \tag{1}$$

with constant coefficients $a$, $b$, $c$. These equations are easily solved, using exponential functions; if the solutions are oscillatory, the exponents are complex numbers.

The solution depends on the zeroes of the *characteristic polynomial*

$$P(r) = ar^2 + br + c.$$

These zeroes are called the **characteristic roots** of the equation; denote them by $r_1$ and $r_2$. As seen in Section 19.3, $e^{r_1 t}$ and $e^{r_2 t}$ are solutions of (1), and so is any *linear combination*

$$y = c_1 e^{r_1 t} + c_2 e^{r_2 t}. \tag{2}$$

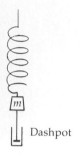

**FIGURE 1**

Oscillating mass $m$, damped by a "dashpot," a sort of shock absorber.

If $r_1 \neq r_2$, then this linear combination gives *all* solutions of (1); it is called the **general solution** of the differential equation. (When $r_1 = r_2$, the general solution has a different form.) We illustrate this with two examples, then prove that (2) really *does* give all solutions, and finally derive the general solution for the case $r_1 = r_2$.

Figure 1 illustrates both examples. A mass $m$ oscillates at the end of a spring of stiffness $k$, and is "damped" by a so-called "dashpot," which contributes a force proportional to the velocity, and opposite in direction. This damping force is $-b\dfrac{dy}{dt}$; the constant $b$ is the damping coefficient. The equation of motion is $m\dfrac{d^2y}{dt^2} = -ky - b\dfrac{dy}{dt}$, or

$$my'' + by' + ky = 0. \tag{3}$$

The nature of the solution depends on the relation between the damping coefficient $b$, the stiffness $k$, and the mass $m$.

---

**EXAMPLE 1** With $m = 1$, $b = 3$, $k = 2$, equation (3) becomes

$$y'' + 3y' + 2y = 0. \tag{4}$$

The characteristic polynomial is $r^2 + 3r + 2$; its zeroes are the characteristic roots:

$$r_1 = -1, \qquad r_2 = -2.$$

The general solution of (4) is

$$y = c_1 e^{-t} + c_2 e^{-2t}.$$

No matter what the constants $c_1$ and $c_2$ may be, the solution tends to zero as $t \to +\infty$, without oscillating. Figure 2 shows the case $c_1 = 1$, $c_2 = -2$.

---

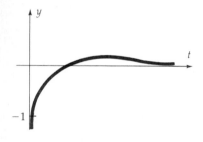

**FIGURE 2**

$y = e^{-t} - 2e^{-2t}$.

**EXAMPLE 2** Take $m = 1$ and $k = 2$ as before, but reduce the damping coefficient to $b = 2$. Now the equation (3) is

$$y'' + 2y' + 2y = 0. \tag{5}$$

The characteristic polynomial is $r^2 + 2r + 2$, with zeroes

$$r = \frac{-2 \pm \sqrt{4-8}}{2} = \frac{-2 \pm \sqrt{4(-1)}}{2} = \frac{-2 \pm 2i}{2}$$

$$= -1 \pm i.$$

The general solution of (5) is

$$y(t) = c_1 e^{(-1+i)t} + c_2 e^{(-1-i)t}. \tag{6}$$

In this form, the solution is hard to interpret; the position function is real-valued, but expressed in terms of complex-valued functions. And since the exponentials are complex, the constants $c_1$ and $c_2$ are presumably complex, as well. To bring the solution back to reality, rewrite (6) in terms of the real-valued functions sine and cosine:

$$e^{(-1+i)t} = e^{-t+it} = e^{-t}(\cos t + i \sin t)$$

and likewise

$$e^{(-1-i)t} = e^{-t}(\cos t + i \sin(-t)) = e^{-t}(\cos t - i \sin t).$$

Use these in (6) and collect like terms:

$$y(t) = e^{-t}[(c_1 + c_2) \cos t + i(c_1 - c_2) \sin t].$$

Relabel the constants, $c_1 + c_2$ as $C_1$, and $i(c_1 - c_2)$ as $C_2$, to get the solution expressed in terms of real functions,

$$y(t) = e^{-t}(C_1 \cos t + C_2 \sin t). \tag{7}$$

This is the product of an oscillating factor

$$C_1 \cos t + C_2 \sin t$$

and an exponential "damping factor" $e^{-t}$. As $t \to +\infty$, the oscillating factor remains bounded, so the product tends to 0 exponentially, as in Figure 3. This exponential decay is the result of the damping force $-b\dfrac{dy}{dt}$, with $b = 2$ in this case.

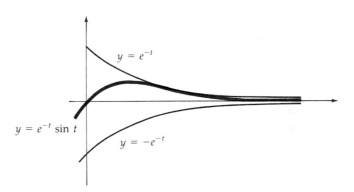

**FIGURE 3**
$y = e^{-t} \sin t \to 0$ as $t \to +\infty$.

In a typical application, the constants $C_1$ and $C_2$ in the general solution are determined by *initial conditions*, specifying the position $y(t_0)$ and velocity $y'(t_0)$ at some particular time $t_0$.

**EXAMPLE 3**   Solve $y'' + 2y' + 2y = 0$, with the initial conditions

$$y(0) = 1, \qquad y'(0) = -2.$$

**SOLUTION**   The previous example gave the general solution:

$$y(t) = e^{-t}(C_1 \cos t + C_2 \sin t).$$

The first initial condition $y(0) = 1$ gives

$$1 = y(0) = e^0(C_1 + C_2 0) = C_1,$$

so $C_1 = 1$. Since

$$y'(t) = -e^{-t}(C_1 \cos t + C_2 \sin t) + e^{-t}(-C_1 \sin t + C_2 \cos t),$$

the second initial condition $y'(0) = -2$ gives

$$-2 = y'(0) = -C_1 + C_2 = -1 + C_2,$$

so $C_2 = -1$. The desired solution is

$$y(t) = e^{-t}(\cos t - \sin t).$$

---

These examples show how to analyze the damped oscillator equation, using exponential solutions. It remains to prove that these give *all* solutions of $my'' + by' + ky = 0$ when the characteristic roots are distinct, and to discover the general solution when those roots are *not* distinct. As a preliminary, we divide by $m$ to get an equivalent equation

$$y'' + \beta y' + \gamma y = 0 \tag{8}$$

with new constants $\beta = b/m$ and $\gamma = k/m$. Then the characteristic polynomial is

$$P(r) = r^2 + \beta r + \gamma. \tag{9}$$

Its zeroes $r_1$ and $r_2$ give the factorization

$$P(r) = (r - r_1)(r - r_2) = r^2 - (r_1 + r_2)r + r_1 r_2. \tag{10}$$

We solve (8) by exploiting the operator notation for the derivative:

$$Dy = y', \qquad D^2 y = D(Dy) = Dy' = y''.$$

Then (8) is written $D^2 y + \beta Dy + \gamma y = 0$, or $(D^2 + \beta D + \gamma)y = 0$, or even

$$P(D)y = 0;$$

$P(D)$ is what you get by replacing $r$ with $D$ in (9). Corresponding to (10), we can *factor* $P(D)$ as

$$P(D) = (D - r_1)(D - r_2). \tag{11}$$

This means that for every function $y$ with second derivatives,

$$P(D)y = (D - r_1)(D - r_2)y.$$

To check this, compute the "operator product" on the right, by applying first $(D - r_2)$, then $(D - r_1)$:

$$(D - r_2)y = Dy - r_2 y = y' - r_2 y,$$

SO

$$(D - r_1)(D - r_2)y = (D - r_1)(y' - r_2 y)$$
$$= D(y' - r_2 y) - r_1(y' - r_2 y)$$
$$= y'' - r_2 y' - r_1 y' + r_1 r_2 y$$
$$= y'' - (r_1 + r_2)y' + r_1 r_2 y$$
$$= D^2 y - (r_1 + r_2)Dy + r_1 r_2 y$$
$$= P(D)y,$$

because of (10). This proves the factorization (11).

Suppose now that $y$ is *any* solution of $P(D)y = 0$ on some interval $I$. Then, factoring $P(D)$ as in (11),

$$(D - r_1)[(D - r_2)y] = 0. \tag{12}$$

We solve (12) by "peeling off" first $D - r_1$, then $D - r_2$. Abbreviate $(D - r_2)y$ by $u$:

$$u = (D - r_2)y. \tag{13}$$

Then by (12), $(D - r_1)u = 0$, or $u' - r_1 u = 0$, or $u' = r_1 u$. The solution of this familiar equation is

$$u = ce^{r_1 t}, \qquad c \text{ a constant.} \tag{14}$$

According to (13), $(D - r_2)y = u$, that is

$$y' - r_2 y = ce^{r_1 t}.$$

It remains to solve this first order equation for $y$ by the method in Section 9.1, Appendix: Multiply each side by $e^{-r_2 t}$, getting

$$e^{-r_2 t}y' - r_2 e^{-r_2 t}y = ce^{(r_1 - r_2)t}.$$

The left-hand side is the derivative of $e^{-r_2 t}y$, so

$$(e^{-r_2 t}y)' = ce^{(r_1 - r_2)t}.$$

Hence

$$e^{-r_2 t}y = \int ce^{(r_1 - r_2)t} \, dt. \tag{15}$$

At this point, we distinguish two cases.

*CASE 1: $r_2 \neq r_1$.*   Then evaluating the integral in (15) gives

$$e^{-r_2 t}y = \frac{c}{r_1 - r_2} e^{(r_1 - r_2)t} + c_2.$$

Multiply through by $e^{r_2 t}$ to find $y = \dfrac{c}{r_1 - r_2} e^{r_1 t} + c_2 e^{r_2 t}$. Relabel the first constant, and you have

$$y = c_1 e^{r_1 t} + c_2 e^{r_2 t}. \tag{16}$$

CASE 2: $r_2 = r_1$.   Then $e^{(r_1 - r_2)t} = 1$, and (15) gives

$$e^{-r_1 t} y = \int c \, dt = ct + c_1.$$

Hence $y = cte^{r_1 t} + c_1 e^{r_1 t}$. Relabel $c$ as $c_2$, and you have

$$y = (c_1 + c_2 t)e^{r_1 t}. \tag{17}$$

To derive the solutions (16) and (17), we assumed only that $y'' + \beta y' + \gamma y = 0$ on an interval. Thus we have proved:

---

**THEOREM**

Suppose that $y$ is any solution of $y'' + \beta y' + \gamma y = 0$ on an interval. Denote by $r_1$ and $r_2$ the zeroes of the characteristic polynomial $r^2 + \beta r + \gamma$.
   If $r_1 \neq r_2$, then there are constants $c_1$ and $c_2$ with

$$y = c_1 e^{r_1 t} + c_2 e^{r_2 t}.$$

If $r_1 = r_2$, then there are constants $c_1$ and $c_2$ with

$$y = (c_1 + c_2 t)e^{r_1 t}.$$

---

This justifies the general solution (2) that we used above for the case of distinct roots, and in addition gives the general solution (17) for the case where $r_1 = r_2$.

---

**EXAMPLE 4**   Find a solution of the equation $y'' - 4y' + 4y = 0$, with initial conditions $y(0) = 1$ and $y'(0) = 2$.

**SOLUTION**   The characteristic polynomial is

$$P(r) = r^2 - 4r + 4 = (r - 2)^2.$$

The characteristic roots are equal, $r_1 = r_2 = 2$. So the general solution is

$$y = (c_1 + c_2 t)e^{2t}.$$

(To check the general theory, verify directly that $te^{2t}$ really is a solution!) The initial conditions determine $c_1$ and $c_2$:

$$1 = y(0) = (c_1 + c_2 \cdot 0)e^0 = c_1,$$

so $c_1 = 1$. And

$$-2 = y'(0) = [c_2 e^{2t} + (c_1 + c_2 t)2e^{2t}]_{t=0} = c_2 + 2c_1 = c_2 + 2$$

so $c_2 = -4$. The required solution is $y = (1 - 4t)e^{2t}$.

---

## SUMMARY

To solve

$$ay'' + by' + cy = 0 \tag{18}$$

find the *characteristic roots* $r_1$ and $r_2$, the zeroes of the characteristic polynomial $ar^2 + br + c$.

  If $r_1 \neq r_2$, the general solution of (18) is

$$y = c_1 e^{r_1 t} + c_2 e^{r_2 t}.$$

  If $r_1 = r_2$, the general solution of (18) is

$$y = (c_1 + c_2 t)e^{r_1 t}.$$

Any complex-valued exponentials can be rewritten in terms of real exponentials, sines, and cosines, using

$$e^{a + i\theta} = e^a e^{i\theta}, \qquad e^{i\theta} = \cos\theta + i\sin\theta.$$

## PROBLEMS

### A

**1.** For the given equation, find the characteristic roots, and the general solution. Write it without complex exponentials.
  **\*a)** $y'' + 4y = 0$
  **b)** $y'' + 2y' + y = 0$
  **\*c)** $y'' - 4y' + 5y = 0$
  **d)** $6y'' + 5y' - 6y = 0$
  **\*e)** $y'' - 4y = 0$
  **f)** $y'' + y' = 0$

**2.** For the equation in the corresponding part of problem 1, find the solution satisfying the given initial conditions.
  **\*a)** $y(0) = 0,\ y'(0) = 1$
  **b)** $y(0) = 1,\ y'(0) = 0$
  **\*c)** $y(0) = 1,\ y'(0) = 1$
  **d)** $y(1) = 0,\ y'(1) = 0$
  **\*e)** $y(1) = 1,\ y'(1) = 0$
  **f)** $y(-1) = 0,\ y'(-1) = 1$

**3.** The charge $Q$ in the capacitor of a simple electrical circuit as in Figure 4 satisfies the equation

$$LQ'' + RQ' + \frac{1}{C}Q = 0.$$

The constants $L$, $R$, $C$ are respectively the *inductance*, *resistance* and *capacitance* of the circuit.

**\*a)** Find the general solution in exponential form, assuming distinct characteristic roots.
**b)** Show that there is an oscillatory solution if and only if $R^2 C < 4L$. Show that the period of the oscillation is $2\pi/\omega_0$, with

$$\omega_0 = \sqrt{1/LC - R^2/4L^2}.$$

**c)** A circuit is to be "tuned" by varying $C$. Is it possible to obtain an arbitrary preassigned period of oscillation $2\pi/\omega_0$ by varying $C$, while keeping $R$ and $L$ fixed?

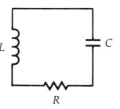

**FIGURE 4**
Circuit with inductance $L$, resistance $R$, capacitance $C$.

### B

**4.** A mass $m = 50$ gm is supported by a spring of stiffness 2 gm/sec². The motion is damped by a

force $-b\dfrac{dy}{dt}$, so

$$50y'' + by' + 2y = 0.$$

a) Show that there is an oscillatory solution if and only if the damping factor $b$ is less than 20. (This is called "undercritical damping.")

b) Determine the period $T$ of the oscillatory solution. As $b \to 20^-$, what is the limit of $T$?

c) For $b = 20$, find the solution with $y(0) = 1$ and $y'(0) = 0$.

d) For $b < 20$, find the solution with $y(0) = 1$ and $y'(0) = 0$.

e) Is the solution in part c the limit of the one in part d, as $b \to 20^-$?

f) For $b > 20$, find the solution with $y(0) = 1$ and $y'(0) = 0$.

g) Is the solution in part c the limit of the one in part f, as $b \to 20^+$?

h) Show that as $t \to +\infty$, the solution in part c, with $b = 20$, tends to zero more rapidly

than the exponential damping factor for the case $b < 20$. $\left[ \text{Compute } \lim_{t \to +\infty} \dfrac{e^{-t/5}(1 + t/5)}{e^{-bt/100}} \right.$

with $b < 20.\Big]$

i) Show that as $t \to +\infty$, the solution in part c tends to zero more rapidly than the solution for $b > 20$. (Compute the limit of the ratio of the two solutions as $t \to +\infty$.)

(In view of parts h and i, the case $b = 20$ is called "critical damping" for the equation $50y'' + by' + 2y = 0$.)

5. a) Suppose that the characteristic roots of $P(D)y = 0$ both have negative real part: $\text{Re}(r_1) < 0$ and $\text{Re}(r_2) < 0$. Prove that all solutions $y$ of the equation satisfy $\lim_{t \to +\infty} y = 0$.

b) What limit is zero if both characteristic roots have positive real part?

---

## 19.5

## FORCED OSCILLATIONS

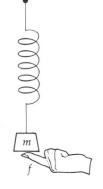

Suppose that a mass $m$, pulled by a spring and linearly damped, experiences a further force $f$, varying with time but not position (Fig. 1). Then Newton's law gives

$$my'' + by' + ky = f; \tag{1}$$

$k$ is the spring constant, and $b$ the damping factor. The resulting oscillations are "forced" by $f$; if $f = 0$, they are called "free."

A similar equation describes the oscillations of an electrical circuit (Fig. 2) with constant resistance $R$, capacitance $C$, and inductance $L$, driven by a possibly varying voltage $E$. This is called an "L-R-C" circuit; the charge $Q$ in the capacitor at time $t$ satisfies

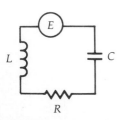

**FIGURE 1**
Mass on a spring, with force $f$.

$$L\frac{d^2Q}{dt^2} + R\frac{dQ}{dt} + \frac{1}{C}Q = E(t). \tag{2}$$

In both equations (1) and (2), we can divide by the leading coefficient and obtain the form

$$y'' + \beta y' + \gamma y = g. \tag{3}$$

This is called a *nonhomogeneous* linear equation, if $g \neq 0$. Its solutions are closely related to those of the *associated homogeneous equation*, obtained by setting the right-hand side equal to 0:

**FIGURE 2**
L-R-C circuit with driving voltage $E$.

$$y'' + \beta y' + \gamma y = 0. \tag{4}$$

**EXAMPLE 1**   The nonhomogeneous equation

$$y'' = 1 \tag{5}$$

is solved by integrating twice:

$$y' = \int 1 \, dt = t + c_1,$$

$$y = \int y' \, dt = \int (t + c_1) \, dt = \tfrac{1}{2}t^2 + c_1 t + c_2. \tag{6}$$

One particular solution is obtained by setting $c_1 = c_2 = 0$; call this solution $y_p$,

$$y_p = \tfrac{1}{2}t^2.$$

The rest of the solution (6), $c_1 t + c_2$, is the general solution of the homogeneous equation associated to (5), $y'' = 0$. We denote this part by $y_h$; then the general solution (6) is the sum of the *particular solution $y_p$* and the *general homogeneous solution $y_h$*:

$$y = y_p + y_h = \tfrac{1}{2}t^2 + c_1 t + c_2.$$

The relation $y = y_p + y_h$ found in Example 1 is true in general.

---

**THEOREM**

Suppose that $y_p$ is one particular solution of the nonhomogeneous equation

$$y'' + \beta y' + \gamma y = g. \tag{7}$$

Then *every* solution of this equation has the form

$$y = y_p + y_h$$

where $y_h$ solves the associated homogeneous equation

$$y'' + \beta y' + \gamma y = 0.$$

---

**PROOF**   By assumption,

$$y_p'' + \beta y_p' + \gamma y_p = g.$$

Suppose that $y$ is any solution of (7). Then $y - y_p$ solves the associated homogeneous equation:

$$(y - y_p)'' + \beta(y - y_p)' + \gamma(y - y_p) = (y'' + \beta y' + \gamma y) - (y_p'' + \beta y_p' + \gamma y_p)$$
$$= g - g = 0.$$

Since $y - y_p$ solves the homogeneous equation, we denote it by $y_h$: $y - y_p = y_h$. Thus $y = y_p + y_h$, and the theorem is proved.

According to this theorem, the solution of the nonhomogeneous linear equation (7) can be separated into two steps:

1.  Construct, in any convenient manner, one particular solution $y_p$.
2.  Add to $y_p$ the general solution $y_h$ of the associated homogeneous equation, to get the general solution $y = y_p + y_h$.

---

**EXAMPLE 2**  Solve the circuit equation (2) with $R = L = C = 1$, and a constant voltage $E = 12$:

$$Q'' + Q' + Q = 12. \tag{8}$$

**SOLUTION**  It is easy to check that a particular solution is $Q_p = 12$, for then $Q_p'' + Q_p' + Q_p = 0 + 0 + 12 = 12$. Look next at the associated homogeneous equation

$$Q'' + Q' + Q = 0. \tag{9}$$

The characteristic polynomial $r^2 + r + 1$ has the characteristic roots $r_{\pm} = -1/2 \pm i\sqrt{3}/2$, so the general homogeneous solution is

$$
\begin{aligned}
Q_h &= c_1 e^{-t/2 + i\sqrt{3}t/2} + c_2 e^{-t/2 - i\sqrt{3}t/2} \\
&= e^{-t/2}\left( C_1 \cos\frac{\sqrt{3}t}{2} + C_2 \sin\frac{\sqrt{3}t}{2} \right).
\end{aligned} \tag{10}
$$

The general solution of the nonhomogeneous equation (8) is thus

$$Q = Q_p + Q_h = 12 + e^{-t/2}\left( C_1 \cos\frac{\sqrt{3}t}{2} + C_2 \sin\frac{\sqrt{3}t}{2} \right).$$

As $t \to +\infty$, then $Q \to 12$; this is the "steady state" charge in a capacitor of capacitance 1 with an applied voltage 12.

---

The next examples illustrate special techniques for constructing particular solutions.

---

**EXAMPLE 3**  Solve the circuit equation (2) with $L = R = C = 1$ and an exponentially decaying voltage $E = e^{-2t}$:

$$Q'' + Q' + Q = e^{-2t}. \tag{11}$$

To find a particular solution, consider that the derivatives of $e^{-2t}$ are again constants times $e^{-2t}$; so we can reasonably hope that (11) has a particular solution of the form

$$Q_p = ce^{-2t}.$$

Try this in (11):

$$(ce^{-2t})'' + (ce^{-2t})' + (ce^{-2t}) = e^{-2t}.$$

The left-hand side reduces to $3ce^{-2t}$, so

$$3ce^{-2t} = e^{-2t}$$

and $c = 1/3$. This gives a *particular solution* $Q_p = \frac{1}{3}e^{-2t}$. The homogeneous equation associated to (11) is the same as in the previous example, with solution (10); so the *general solution* in this case is

$$Q = Q_p + Q_h = \frac{1}{3}e^{-2t} + e^{-t/2}\left(C_1 \cos\frac{\sqrt{3}t}{2} + C_2 \sin\frac{\sqrt{3}t}{2}\right).$$

In this case, as $t \to +\infty$, the charge $Q \to 0$.

---

**EXAMPLE 4**   Take again $L = R = C = 1$, but now with an oscillating voltage $E = \cos \omega t$, as from an AC generator. Then

$$Q'' + Q' + Q = \cos \omega t. \tag{12}$$

To find a particular solution, consider that each derivative of $\cos \omega t$ is a constant times $\cos \omega t$, *or* a constant times $\sin \omega t$. Thus $Q_p = c \cos \omega t$ will not work; but .

$$Q_p = a \cos \omega t + b \sin \omega t \tag{13}$$

might work, with appropriate constants $a$ and $b$. Try it:

$$Q_p'' + Q_p' + Q_p = -a\omega^2 \cos \omega t - b\omega^2 \sin \omega t - a\omega \sin \omega t + b\omega \cos \omega t + a \cos \omega t + b \sin \omega t$$
$$= [a(1 - \omega^2) + b\omega] \cos \omega t + [b(1 - \omega^2) - a\omega] \sin \omega t. \tag{14}$$

To solve (12), this combination must equal $\cos \omega t$; so in (14), the coefficient of $\cos \omega t$ should be 1, and that of $\sin \omega t$ should be 0:

$$a(1 - \omega^2) + b\omega = 1$$
$$-a\omega + b(1 - \omega^2) = 0.$$

The solution is $a = \dfrac{1 - \omega^2}{(1 - \omega^2)^2 + \omega^2}$, $b = \dfrac{\omega}{(1 - \omega^2)^2 + \omega^2}$; so (13) gives the particular solution

$$Q_p = \frac{1 - \omega^2}{(1 - \omega^2)^2 + \omega^2} \cos \omega t + \frac{\omega}{(1 - \omega^2)^2 + \omega^2} \sin \omega t. \tag{15}$$

The general solution of (12) is $Q = Q_p + Q_h$, with $Q_h$ the same homogeneous solution (10) as in the previous two examples.

Consider again the long-term behavior of the circuit. As $t \to +\infty$, then $Q_h \to 0$, so $Q = Q_p + Q_h \to Q_p$; this is the *steady state* solution. It has frequency $\omega/2\pi$, the same as for the driving voltage $\cos \omega t$. As for the amplitude of this steady state, Section 8.9 shows that the amplitude of $a \cos \omega t + b \sin \omega t$ is $A = \sqrt{a^2 + b^2}$. So the amplitude of $Q_p$ in (15) turns out to be

$$A = [(1 - \omega^2)^2 + \omega^2]^{-1/2}.$$

Apparently, the amplitude of the response depends on the frequency $\omega/2\pi$ of the driving voltage. You can check that the maximum amplitude is with $\omega = 1/\sqrt{2}$. The circuit is "tuned" to the frequency $1/2\sqrt{2}\pi$.

The particular solutions $Q_p$ in Examples 3 and 4 were constructed by the method of "undetermined coefficients." In general, for

$$y'' + \beta y' + \gamma y = g(t),$$

if the "forcing term" $g(t)$ is a combination of exponentials, trigonometric functions, and positive integer powers $t^n$, then a particular solution $y_p$ is sought in the form indicated here:

| Forcing term $g(t)$ | Particular solution $y_p$ |
| --- | --- |
| $e^{\alpha t}$ | $ce^{\alpha t}$ |
| $e^{\alpha t}[C_1 \cos \omega t + C_2 \sin \omega t]$ | $e^{\alpha t}[a \cos \omega t + b \sin \omega t]$ |
| $t^n e^{\alpha t}$ | $[a_n t^n + a_{n-1} t^{n-1} + \cdots + a_0]e^{\alpha t}$ |

However, if the forcing term happens to be a solution of the homogeneous equation, the proposed particular solution will not work; it will give 0 instead of something resembling $g(t)$. The situation is saved by multiplying the proposed $y_p$ by $t$, if $g$ corresponds to a simple zero of the characteristic polynomial, or by $t^2$ if it corresponds to a double zero.

## PROBLEMS

### A

1. Find the general solution.
   *a) $y'' - y' - 6y = \cos 3t$
   b) $y'' + 2y' + 2y = t^2 + 2$
   *c) $y'' - y' = \sin t$
   d) $y'' - 7y' + 12y = \sin 2t + \cos 2t$
   *e) $\dfrac{d^2y}{dx^2} - y = e^x + x^2$
   f) $2y'' + 3y' + y = 1 + \cos 2t$

*2. An L-R-C circuit as in Figure 2 has $L = 1$, $R = 20$, $C = 0.002$. At time $t = 0$, the charge $Q$ in the capacitor is 0, and $Q'(0) = 0$ (there is no current). Find the charge at time $t > 0$, and the steady state of the system (as $t \to +\infty$) if the circuit is connected to
   a) A battery with fixed voltage $E = 6$.
   b) A generator with voltage $E = 6 \sin 10t$.

3. For an L-R-C circuit as in Figure 2, find the charge at time $t$ if the circuit is attached to
   a) A battery with fixed voltage $E = E_0$.
   b) A generator with voltage $E = A \cos \omega t$.

*4. A spring with mass $m$, spring constant $k$, and negligible damping is driven by a force $f(t) = c \cos \omega t$. The free vibrations of the spring have frequency $\omega_0/2\pi$, with $\omega_0 = \sqrt{k/m}$. Solve the equation for the forced vibrations, if $\omega \neq \omega_0$.

5. Solve the previous problem for a driving force $c \cos \omega_0 t$ with a frequency $\omega_0/2\pi$ matching that of the free vibrations. Show that the oscillations grow arbitrarily large as $t \to +\infty$. (This is the principle involved in pushing a swing; if the frequency of the pushes matches the natural frequency of the swing, then the oscillations grow ever larger.)

# APPENDIX A

## PROOFS OF THE
## HARD THEOREMS

# A.1

## THE DEFINITION OF LIMIT

From the time of the Greeks, limits were recognized as a powerful but tricky concept. Since it took about 2,000 years to come to an acceptable definition, a few words of introduction are appropriate.

One limit known to the Greeks was the ratio of chord length $x$ to arc length $s$ in a circle (Fig. 1). As $x$ goes to zero, the ratio $x/s$ approaches the "limit" 1. Table 1 illustrates this, showing $x/s$ for a few values of the chord length $x$:

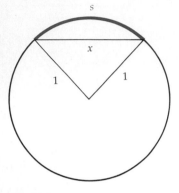

**FIGURE 1**

$\lim\limits_{x \to 0} \dfrac{x}{s} = 1.$

**Table 1**

| $x$ | 1 | 0.5 | 0.1 | 0.01 |
|-----|-----|--------|--------|---------|
| $x/s$ | .959 | 0.9896 | .99958 | .999996 |

This table serves to illustrate a very early attempt to define limits, by d'Alembert, in the first encyclopedia ever written (1765): "One magnitude is said to be the *limit* of another magnitude when the second may approach the first within any given magnitude, however small. . . ." From Table 1, for example, it appears that

$\dfrac{x}{s}$ approaches 1 within 0.5 if $0 < x < 1$;

within 0.011 if $0 < x < 0.05$;

within 0.0005 if $0 < x < 0.1$;

within 0.000005 if $0 < x < 0.01$;

and so on.

d'Alembert's definition contained the germ of the limit idea (which he inherited from Newton and others), but did not provide a clear and unequivocal way to prove things about limits. This came a hundred years later, when Karl Weierstrass finally gave a workable definition which remains the basis of all such proofs.[1] Here is the definition for the limit as $x$ approaches $a$ from the right (Fig. 2):

---

**DEFINITION 1**

$\lim\limits_{x \to a^+} f(x) = L$ if, for every $\varepsilon > 0$, there is a $\delta > 0$ such that

$$|f(x) - L| < \varepsilon \qquad \text{where } a < x < a + \delta. \tag{1}$$

---

The last line (1) says that $f(x)$ approximates the proposed limit $L$ within $\varepsilon$, whenever $x$ approaches $a$ within $\delta$, and $x > a$; we require that such a statement be true for each $\varepsilon > 0$, no matter how small. This expresses precisely what we mean by "$f(x)$ approaches $L$ as $x$ approaches $a$."

On the graph of $f$ (Fig. 2) you can think of the open interval $(L - \varepsilon, L + \varepsilon)$ as a "target." Imagine that the function $f$ takes points $x$ and fires them at the

---

[1] See Chapter 40 in Morris Kline, *Mathematical Thought from Ancient to Modern Times*, (New York: Oxford University Press, 1972).

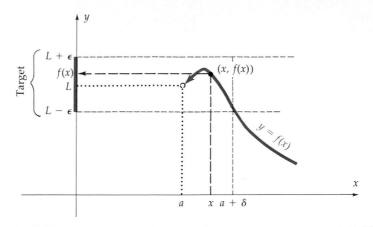

**FIGURE 2**
$a < x < a + \delta \;\Rightarrow\; L - \varepsilon < f(x) < L + \varepsilon.$

target—point $x$ lands at $f(x)$. Then for each target interval $(L - \varepsilon, L + \varepsilon)$, however small, there must be some corresponding "source interval" $(a, a + \delta)$ such that every $x$ in that source interval lands in the target. Then we say that the source interval $(a, a + \delta)$ is *mapped* by $f$ into the image interval $(L - \varepsilon, L + \varepsilon)$.

With Definition 1, you can really prove things about limits. As a first example, we prove something very simple.

---

**EXAMPLE 1**   Prove that $\lim\limits_{x \to a^+} 2x = 2a$.

*SOLUTION*   Figure 3 shows the graph of $f(x) = 2x$, and sets up an arbitrary "target interval" $(2a - \varepsilon, 2a + \varepsilon)$ centered on the proposed limit $2a$.

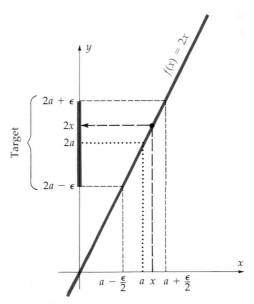

**FIGURE 3**
$a < x < a + \varepsilon/2 \;\Rightarrow\; 2a - \varepsilon < f(x) < 2a + \varepsilon.$

Backtracking from the $y$-axis to the $x$-axis, you see that every $x$ between $a$ and $a + \varepsilon/2$ lands in the target, that is,

$$\left|2a - 2x\right| < \varepsilon \qquad \text{whenever } a < x < a + \varepsilon/2.$$

So condition (1) in the definition is fulfilled, with $\delta = \varepsilon/2$. Hence $\lim\limits_{x \to a^+} 2x = 2a$.

**EXAMPLE 2**    Prove that for any $a > 0$, $\lim\limits_{x \to a^+} \dfrac{1}{x} = \dfrac{1}{a}$.

*SOLUTION*    Figure 4 sets up a target about the proposed limit $\dfrac{1}{a}$; we assume that $\varepsilon < \dfrac{1}{a}$, for otherwise *every* $x > a$ lands in the target. From the figure, every $x$ between $a$ and $\dfrac{a}{1 - a\varepsilon}$ lands in the target interval; so the length of the source interval is

$$\delta = \frac{a}{1 - a\varepsilon} - a = \frac{a^2\varepsilon}{1 - a\varepsilon}.$$

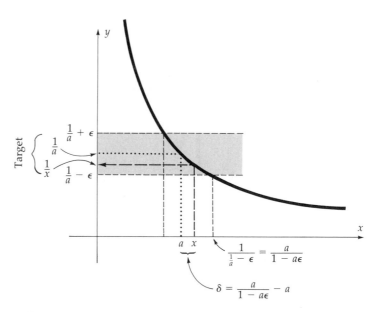

**FIGURE 4**

$$a < x < \frac{a}{1 - a\varepsilon} \quad \Rightarrow \quad \frac{1}{a} - \varepsilon < \frac{1}{x} < \frac{1}{a} + \varepsilon.$$

So Definition 1 is satisfied for every $\varepsilon > 0$. If $0 < \varepsilon < 1/a$ then

$$\left|\frac{1}{x} - \frac{1}{a}\right| < \varepsilon \qquad \text{if } a < x < a + \frac{a^2\varepsilon}{1 - a\varepsilon};$$

and if $\varepsilon > 1/a$, then $\left| \dfrac{1}{x} - \dfrac{1}{a} \right| < \varepsilon$ for every $x > a$. Thus $\lim\limits_{x \to a^+} \dfrac{1}{x} = \dfrac{1}{a}$.

---

You can also use the definition to show that certain limits do *not* exist.

---

**EXAMPLE 3**   Let $f(x) = \begin{cases} 1 & \text{if } x \text{ is rational} \\ 0 & \text{if } x \text{ is irrational.} \end{cases}$

Prove that $\lim\limits_{x \to a^+} f(x)$ does not exist; that is, there is no number $L$ such that $\lim\limits_{x \to a^+} f(x) = L$.

*SOLUTION*   Figure 5 suggests the graph of $f$—for every rational $x$ it has a point on the line $y = 1$, and for every irrational $x$ it has a point on the line $y = 0$.

It seems that no number $L \geq 1/2$ could be the limit of $f$, because $f(x) = 0$ at every irrational $x$; and no number $L < 1/2$ could be the limit of $f$ because $f(x) = 1$ at every rational $x$. This is the general idea; here is the formal proof.

Suppose $L \geq 1/2$. If $\lim\limits_{x \to a^+} f(x) = L$, then for *every* $\varepsilon > 0$ there would be a $\delta > 0$ such that

$$\left| f(x) - L \right| < \varepsilon \qquad \text{if } a < x < a + \delta.$$

In particular, for $\varepsilon = 1/2$ there would be such a $\delta$:

$$\left| f(x) - L \right| < 1/2 \qquad \text{if } a < x < a + \delta. \tag{2}$$

We are supposing $L \geq 1/2$, so $\left| f(x) - L \right| < 1/2$ implies that $f(x) > 0$ (Fig. 6). Hence (2) would imply that

$$f(x) > 0 \qquad \text{if } a < x < a + \delta. \tag{3}$$

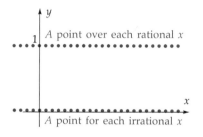

**FIGURE 5**
$f(x) = \begin{cases} 1 & \text{if } x \text{ is rational} \\ 0 & \text{if } x \text{ is irrational.} \end{cases}$

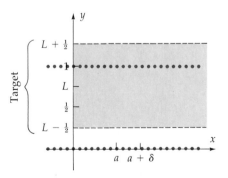

**FIGURE 6**
$\lim\limits_{x \to a} f(x) = L$ is impossible when $L \geq 1/2$.

But any open interval $(a, a + \delta)$ includes some irrational points $x$; at these points, $f(x) = 0$, violating (3). So (3) is false, for every $\delta > 0$. Hence (2) is false for every $\delta > 0$, and condition (1) of the definition is *not* satisfied for any $\delta > 0$. Hence, no number $L \geq 1/2$ is the limit of $f(x)$.

Suppose, on the other hand, that $L < 1/2$; a very similar argument shows again that $L$ is not the limit of $f(x)$. So $f(x)$ has no limit $L \geq 1/2$, and no limit $L < 1/2$ either; it has no limit at all.

---

With the same idea as in Example 2, we can prove an important fact: Limits are unique.

---

**THEOREM 1**

At a given point $a$, a given function $f$ cannot have two different limits from the right.

---

**PROOF**    Suppose, on the contrary, that $f$ *has* two different right-hand limits at some point $a$. Call the smaller one $L$, and the larger one $M$:

$$\lim_{x \to a^+} f(x) = L, \quad \text{and} \quad \lim_{x \to a^+} f(x) = M, \qquad \text{where } L < M.$$

Since $\lim_{x \to a^+} f(x) = L$, there is for every $\varepsilon > 0$ a corresponding $\delta > 0$ such that

$$\left| f(x) - L \right| < \varepsilon \qquad \text{whenever } a < x < a + \delta.$$

Choose $\varepsilon = \frac{1}{2}(M - L)$, as in Figure 7. Then every $x$ in the interval $(a, a + \delta)$ hits the "$L$-target" $(L - \varepsilon, L + \varepsilon)$. Hence *none* of these values of $x$ hits the corresponding "$M$-target" $(M - \varepsilon, M + \varepsilon)$; so there is *no* $\delta > 0$ such that

$$\left| f(x) - M \right| < \varepsilon \qquad \text{whenever } a < x < a + \delta.$$

Hence $\lim_{x \to a^+} f(x) = M$ is false, contrary to our supposition at the beginning of the proof. This contradiction proves the theorem.

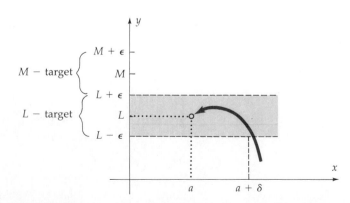

**FIGURE 7**

If $\lim_{x \to a^+} f(x) = L$, then $\lim_{x \to a^+} f(x) \neq M$ for $M > L$.

***REMARK***   A subtle point in this proof is that $f(x)$ is actually defined throughout the interval $(a, a + \delta)$; for otherwise there would be no contradiction. It is important to realize that the statement

$$\left| f(x) - L \right| < \varepsilon \qquad \text{whenever } a < x < a + \delta$$

implies that $f(x)$ is *defined* whenever $a < x < a + \delta$.

## Other Types of Limits as $x$ Approaches $a$

The limit from the left, $\lim\limits_{x \to a^-}$ has a definition very similar to the one for $\lim\limits_{x \to a^+}$ (Fig. 8). The exact statement is left as a problem.

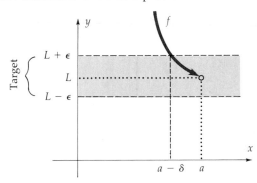

**FIGURE 8**
$\left| f(x) - L \right| < \varepsilon$ if $a - \delta < x < a$; $\lim\limits_{x \to a^-} f(x) = L$.

*The two-sided* limit $\lim\limits_{x \to a}$ is defined using source intervals that extend to each side of $a$ (Fig. 9).

**FIGURE 9**
$\lim\limits_{x \to a} f(x) = L$.

---

### DEFINITION 2

$\lim\limits_{x \to a} f(x) = L$ if, for every $\varepsilon > 0$, there is a $\delta > 0$ such that

$$\left| f(x) - L \right| < \varepsilon \qquad \text{whenever } 0 < \left| x - a \right| < \delta.$$

---

Notice that the inequality $0 < \left| x - a \right|$ rules out $x = a$; nothing is required of $f(a)$—it need not even be defined.

It is not hard to show that $\lim\limits_{x \to a} f(x) = L$ if and only if

$$\lim\limits_{x \to a^+} f(x) = L \quad \text{and} \quad \lim\limits_{x \to a^-} f(x) = L.$$

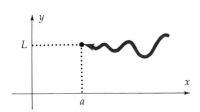

**FIGURE 10**
$f(x)$ can go above and below its "limit."

***REMARK***   The word "limit" suggests a number which may not be exceeded. In fact, this was part of d'Alembert's original definition—"the [second] magnitude may never exceed the magnitude it approaches." But this requirement was found to be inappropriate. We require only that $f(x)$ approach its limit $L$ "within any given magnitude, however small"; $f$ may oscillate above and below $L$ as often as it wants (Fig. 10).

## SUMMARY

*Definition 1*   $\lim\limits_{x \to a^+} f(x) = L$ if, for every $\varepsilon > 0$, there is a $\delta > 0$ such that

$$a < x < a + \delta \quad \Rightarrow \quad |f(x) - L| < \varepsilon.$$

*Definition 2*   $\lim\limits_{x \to a} f(x) = L$ if, for every $\varepsilon > 0$, there is a $\delta > 0$ such that

$$|f(x) - L| < \varepsilon \qquad \text{whenever } 0 < |x - a| < \delta.$$

*Theorem 1*   (Uniqueness of limits) A given function $f$ can have at most one limit as $x \to a^+$. The same is true for limits from the left, and two-sided limits.

## PROBLEMS

### B

1.  Prove the following limits.

    a)  $\lim\limits_{x \to 1^+} 3x = 3$

    b)  $\lim\limits_{x \to a^+} 3x = 3a$

    c)  $\lim\limits_{x \to a^+} \dfrac{1}{x} = \dfrac{1}{a}$, where $a < 0$

    d)  $\lim\limits_{x \to a^+} cx = ca$, where $c > 0$

    e)  $\lim\limits_{x \to a^+} (-x) = -a$

    f)  $\lim\limits_{x \to a^+} cx = ca$, where $c < 0$

    g)  $\lim\limits_{x \to 0^+} \dfrac{1}{x + 1} = 1$

2.  In Example 3, show that $L$ is not a limit if $L < 1/2$.

Problems 3–6 concern limits from the left.

3.  Define the concept $\lim\limits_{x \to a^-} f(x) = L$.

4.  Prove.

    a)  $\lim\limits_{x \to 2^-} \dfrac{1}{x} = \dfrac{1}{2}$

    b)  $\lim\limits_{x \to a^-} \dfrac{1}{x} = \dfrac{1}{a}$ if $a > 0$.

    c)  $\lim\limits_{x \to a^-} \dfrac{1}{x} = \dfrac{1}{a}$ if $a < 0$.

    d)  $\lim\limits_{x \to a^-} cx = ca$, where $c > 0$.

5.  Prove that a given function $f$ cannot have two different limits from the left at a given point $a$.

6.  Define a function $f$ with $\lim\limits_{x \to 0^-} f(x) \neq \lim\limits_{x \to 0^+} f(x)$.

Problems 7–10 concern two-sided limits.

7.  Prove the following two-sided limits (referring to earlier problems where convenient).

    a)  $\lim\limits_{x \to a} \dfrac{1}{x} = \dfrac{1}{a}$ if $a \neq 0$

    b)  $\lim\limits_{x \to a^-} cx = ca$

8.  Prove that a given function $f$ can have at most one two-sided limit as $x \to a$.

9.  Suppose that $\lim\limits_{x \to a} f(x) = L$ in the sense of Definition 2. Prove that $\lim\limits_{x \to a^+} f(x) = L$ and $\lim\limits_{x \to a^-} f(x) = L$.

10. Suppose that $\lim\limits_{x \to a^+} f(x) = L$ and $\lim\limits_{x \to a^-} f(x) = L$. Show that $\lim\limits_{x \to a} f(x) = L$.

# A.2

## SOME LIMIT THEOREMS

With the definition understood, we proceed to prove some basic limit theorems. First (Fig. 1):

---

### THEOREM 1

The Trapping Theorem. If $f(x) \leq g(x) \leq h(x)$ for $x > a$, and $\lim\limits_{x \to a^+} f(x) = L$, and $\lim\limits_{x \to a^+} h(x) = L$, then also $\lim\limits_{x \to a^+} g(x) = L$.

---

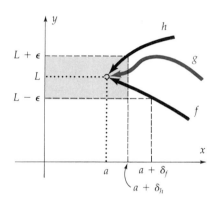

**FIGURE 1**

**PROOF**  Given any "target interval" $(L - \varepsilon, L + \varepsilon)$, we must show that there is a source interval $(a, a + \delta)$ which is mapped into the target by $g$ (Fig. 1). Since $\lim\limits_{x \to a^+} f(x) = L$, there is an interval $(a, a + \delta_f)$ which is mapped into the target by $f$; similarly for $h$, there is an interval $(a, a + \delta_h)$ which is mapped into the target. Let $\delta$ be the smaller of $\delta_f$ and $\delta_h$ (in Fig. 1, $\delta = \delta_h$); then the interval $(a, a + \delta)$ is mapped into the target by *both* $f$ and $h$. Since $g$ lies between $f$ and $h$, $g$ also maps this same interval into the target. That is,

$$|g(x) - L| < \varepsilon \qquad \text{whenever } a < x < a + \delta.$$

This argument works for every $\varepsilon > 0$, so $\lim\limits_{x \to a^+} g(x) = L$, and Theorem 1 is proved.

Next we prove that the limit of a sum equals the sum of the limits.

---

### THEOREM 2

$\lim\limits_{x \to a^+} [f(x) + g(x)] = \lim\limits_{x \to a^+} f(x) + \lim\limits_{x \to a^+} g(x)$, provided that the latter two limits exist.

---

**PROOF**  Let $\lim\limits_{x \to a^+} f(x) = L$ and $\lim\limits_{x \to a^+} g(x) = M$. Given $\varepsilon > 0$, we must show that there is a $\delta > 0$ such that

$$|f(x) + g(x) - (L + M)| < \varepsilon \qquad \text{whenever } a < x < a + \delta. \tag{1}$$

Now, $f(x) + g(x)$ will be within $\varepsilon$ of $L + M$ if $f(x)$ is within $\varepsilon/2$ of $L$ and $g(x)$ is within $\varepsilon/2$ of $M$, so we proceed as follows.

Since $\lim\limits_{x \to a^+} f(x) = L$, and $\varepsilon/2$ is a positive number, there is a $\delta_f > 0$ such that

$$|f(x) - L| < \varepsilon/2 \qquad \text{if } a < x < \delta_f.$$

That is, if $a < x < \delta_f$ then

$$L - \varepsilon/2 < f(x) < L + \varepsilon/2. \tag{2}$$

Similarly, there is a $\delta_g > 0$ such that if $a < x < a + \delta_g$, then

$$M - \varepsilon/2 < g(x) < M + \varepsilon/2. \tag{3}$$

Let $\delta$ be the smaller of $\delta_f$ and $\delta_g$. Then if $a < x < \delta$, *both* inequalities (2) and (3) are true. Adding corresponding terms in the two inequalities, you find that

$$L + M - \varepsilon < f(x) + g(x) < L + M + \varepsilon \qquad \text{if } a < x < \delta.$$

This is the same as the desired statement (1). The argument works for every $\varepsilon > 0$, so $\lim_{x \to a^+} [f(x) + g(x)] = L + M$, and Theorem 2 is proved.

The next theorem asserts that the limit of a product is the product of the limits. The proof uses two facts about absolute values:

$$|ab| = |a| \cdot |b|$$

and the "triangle inequality"

$$|a + b| \le |a| + |b|.$$

You can persuade yourself that these are true.

---

**THEOREM 3**

$$\lim_{x \to a^+} [f(x)g(x)] = \left[ \lim_{x \to a^+} f(x) \right]\left[ \lim_{x \to a^+} g(x) \right], \text{ provided both limits on}$$
the right exist.

---

**PROOF**   Let $\lim_{x \to a^+} f(x) = L$ and $\lim_{x \to a^+} g(x) = M$. Given $\varepsilon > 0$, we must show that there is a $\delta > 0$ such that

$$|f(x)g(x) - LM| < \varepsilon \qquad \text{whenever } a < x < a + \delta. \tag{4}$$

The difference $f(x)g(x) - LM$ is related to $f(x) - L$ and $g(x) - M$ by an equation suggested by Figure 2:

$$f(x)g(x) - LM = [f(x) - L]M + L[g(x) - M] + [f(x) - L][g(x) - M].$$

So

$$|f(x)g(x) - LM| \le |f(x) - L| \cdot |M| + |L| \cdot |g(x) - M|$$
$$+ |f(x) - L| \cdot |g(x) - M|. \tag{5}$$

We will make each of the three terms on the right less than $\varepsilon/3$.

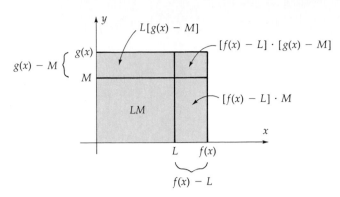

**FIGURE 2**

Because $\lim\limits_{x \to a^+} f(x) = L$ and $\lim\limits_{x \to a^+} g(x) = M$, there are numbers $\delta_1, \delta_2, \delta_3,$ and $\delta_4$ such that:

$$\left.\begin{array}{ll}
|f(x) - L| < \varepsilon & \text{if } a < x < a + \delta_1 \\[2ex]
|g(x) - M| < \dfrac{1}{3} & \text{if } a < x < a + \delta_2 \\[2ex]
|f(x) - L| < \dfrac{\varepsilon}{3|M| + 1} & \text{if } a < x < a + \delta_3 \\[2ex]
|g(x) - M| < \dfrac{\varepsilon}{3|L| + 1} & \text{if } a < x < a + \delta_4.
\end{array}\right\} \tag{6}$$

Let $\delta$ be the minimum of $\delta_1, \delta_2, \delta_3, \delta_4$. Then if $a < x < a + \delta$, all four inequalities on the right of (6) are valid, hence so are the four on the left. Use these in (5):

$$\begin{aligned}
|f(x)g(x) - LM| &\le \frac{\varepsilon}{3|M| + 1}\,|M| + |L|\,\frac{\varepsilon}{3|L| + 1} + \varepsilon \cdot \frac{1}{3} \\[2ex]
&= \varepsilon\,\frac{|M|}{3|M| + 1} + \varepsilon\,\frac{|L|}{3|L| + 1} + \varepsilon\,\frac{1}{3} \\[2ex]
&< \varepsilon \cdot \frac{1}{3} + \varepsilon \cdot \frac{1}{3} + \varepsilon \cdot \frac{1}{3} = \varepsilon.
\end{aligned}$$

This argument is valid for any $\varepsilon > 0$, so $\lim f(x)g(x) = LM$, and Theorem 3 is proved.

*REMARK 1*   It might seem simpler in (6) to use $\dfrac{\varepsilon}{3|M|}$ instead of $\dfrac{\varepsilon}{3|M| + 1}$. But if $M = 0$, $\dfrac{\varepsilon}{3|M|}$ is undefined; so we use $\dfrac{\varepsilon}{3|M| + 1}$ just in case.

*REMARK 2*   A similar proof applies to the limit of a quotient; this is outlined in the problems. But a simpler proof is given in the next section.

Finally, we prove a relation between inequalities and limits.

> **THEOREM 4**
>
> If $f(x) \leq g(x)$ for $x > a$, then $\lim_{x \to a^+} f(x) \leq \lim_{x \to a^+} g(x)$, provided that the limits exist.

**PROOF**   Let $\lim_{x \to a^+} f(x) = L$ and $\lim_{x \to a^+} g(x) = M$. We have to prove that $L \leq M$. Suppose on the contrary that $L > M$ (Fig. 3). Let $\varepsilon = \frac{1}{3}(L - M)$. Then there would be a $\delta_f$ such that $(a, a + \delta_f)$ is mapped by $f$ into the target $(L - \varepsilon, L + \varepsilon)$; and there would be a $\delta_g$ such that $(a, a + \delta_g)$ is mapped by $g$ into the target $(M - \varepsilon, M + \varepsilon)$. Let $\delta = \min(\delta_f, \delta_g)$. Then the interval $(a, a + \delta)$ would be mapped by $f$ into $(L - \varepsilon, L + \varepsilon)$, and by $g$ into $(M - \varepsilon, M + \varepsilon)$. Hence (Fig. 3) $f(x) > g(x)$ for $a < x < a + \delta$. But this contradicts the hypothesis that $f(x) \leq g(x)$ for $x > a$. So supposing (as we did) that $L > M$ leads to a contradiction; it follows that $L \leq M$, and Theorem 4 is proved.

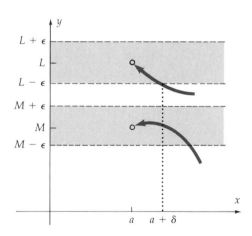

**FIGURE 3**

# PROBLEMS

## B

Assume in each problem that $\lim_{x \to a^+} f(x)$ and $\lim_{x \to a^+} g(x)$ exist.

1.  Prove that $\lim_{x \to a^+} cf(x) = c \lim_{x \to a^+} f(x)$, for any constant $c$. (Treat separately the case where $c = 0$.)

2.  Prove that
$$\lim_{x \to a^+} [f(x) - g(x)] = \lim_{x \to a^+} f(x) - \lim_{x \to a^+} g(x).$$

3.  Prove that $\lim_{x \to a^+} \dfrac{1}{g(x)} = \dfrac{1}{\lim_{x \to a^+} g(x)}$, if $\lim_{x \to a^+} g(x) \neq 0$. (Hints: Let $\lim g(x) = M$, and put $\dfrac{1}{g(x)} - \dfrac{1}{M}$ on a common denominator. Choose $\delta$ so that $|g(x) - M| < |M|/2$, and also $|g(x) - M| < \varepsilon M^2/2$, if $a < x < a + \delta$. Deduce from the first inequality that $|g(x)| > |M|/2$ if $a < x < a + \delta$.)

4. Prove that $\lim\limits_{x \to a^+} \dfrac{f(x)}{g(x)} = \dfrac{\lim\limits_{x \to a^+} f(x)}{\lim\limits_{x \to a^+} g(x)}$ if $\lim\limits_{x \to a^+} g(x) \neq$ 0. (You can use the previous problem.)

5. Prove, using either the definition, or preceding theorems and problems, that

   a) $\lim\limits_{x \to a^+} x = a$.   b) $\lim\limits_{x \to a^+} x^2 = a^2$.

   c) $\lim\limits_{x \to a^+} x^3 = a^3$.

   d) $\lim\limits_{x \to a^+} (c_3 x^3 + c_2 x^2 + c_1 x + c_0)$
      $= c_3 a^3 + c_2 a^2 + c_1 a + c_0$.

6. If you know the method of proof by induction, prove that $\lim\limits_{x \to a^+} x^n = a^n$ for $n = 1, 2, 3, \ldots$

7. Prove.
   a) $\lim\limits_{x \to a^+} P(x) = P(a)$ if $P$ is a polynomial.

   b) $\lim\limits_{x \to a^+} \dfrac{P(x)}{Q(x)} = \dfrac{P(a)}{Q(a)}$ if $P$ and $Q$ are polynomials, and $Q(a) \neq 0$.

8. In Theorem 4, if $f(x) < g(x)$ for $x > a$, does it follow that $\lim\limits_{x \to a^+} f(x) < \lim\limits_{x \to a^+} g(x)$?

## A.3

## CONTINUITY

The definition of continuity is closely related to that of limit.

> **DEFINITION**
>
> $f$ is *continuous at* $a$ if $f(a)$ is defined, and for every $\varepsilon > 0$ there is a $\delta > 0$ such that
>
> $$\left| f(x) - f(a) \right| < \varepsilon \qquad \text{whenever } f(x) \text{ is defined and } |x - a| < \delta.$$
>
> $f$ is called *continuous* if it is continuous at every point where it is defined.

The relation between limits and continuity is spelled out in:

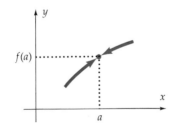

**FIGURE 1**
$f$ is continuous at $a$ if $\lim\limits_{x \to a} f(x) = f(a)$.

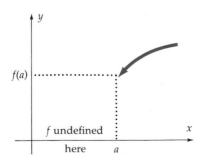

**FIGURE 2**
$f$ is continuous at $a$ if $\lim\limits_{x \to a^+} f(x) = f(a)$.

> **THEOREM 1**
>
> If $f$ is defined in an open interval containing $a$ (Fig. 1), then $f$ is continuous at $a$ if and only if $\lim\limits_{x \to a} f(x) = f(a)$. If $f$ is defined only in an interval $[a, b)$, $[a, b]$, or $[a, \infty)$, as in Figure 2, then $f$ is continuous at $a$ if and only if $\lim\limits_{x \to a^+} f(x) = f(a)$. Similarly, if $a$ is the right endpoint of the domain of $f$, then $f$ is continuous at $a$ if and only if $\lim\limits_{x \to a^-} f(x) = f(a)$.

We leave the proof as a problem.

Most of the functions considered in calculus are continuous at every point where they are defined. To prove this, you begin by showing that a few basic functions are continuous, for example the constant functions, the function $f(x) = x$, and the $n^{\text{th}}$ roots $f(x) = x^{1/n}$. Then you show that the sum and product of two continuous functions are continuous—the proofs are just like those for the limit of a sum or product. Armed with these facts, you can show that every rational function is continuous wherever defined.

To handle a composite function such as $f(x) = \sqrt{1 - x^2}$, you need the following basic result:

---

### THEOREM 2

If $g$ is continuous at $a$, and $f$ is continuous at $g(a)$, then $f \circ g$ is continuous at $a$.

---

Figure 3 illustrates the proof. Given $\varepsilon > 0$, there is a number $\delta_f > 0$ such that

$$\left| f(u) - f(g(a)) \right| < \varepsilon \qquad \text{whenever } \left| g(a) - u \right| < \delta_f.$$

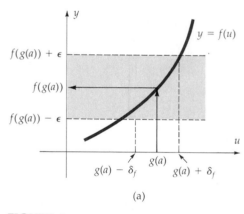

(a)

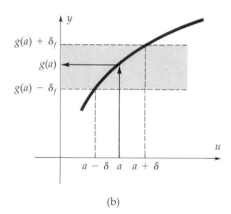

(b)

FIGURE 3

This is true because $f$ is continuous at $g(a)$; it says that the source interval $(g(a) - \delta_f, g(a) + \delta_f)$ is mapped by $f$ into the prescribed target interval $(f(g(a)) - \varepsilon, f(g(a)) + \varepsilon)$, as in Figure 3a.

Now think of $(g(a) - \delta_f, g(a) + \delta_f)$ as a target interval for $g$ (Fig. 3b). Since $g$ is continuous at $a$, there is a source interval $(a - \delta, a + \delta)$ which is mapped by $g$ into the target $(g(a) - \delta_f, g(a) + \delta_f)$. The latter interval, in turn, is mapped by $f$ into the original target $(f(g(a)) - \varepsilon, f(g(a)) + \varepsilon)$. Hence the composition $f \circ g$ maps $(a - \delta, a + \delta)$ into the prescribed target interval $(f(g(a)) - \varepsilon, f(g(a)) + \varepsilon)$ (Fig. 4). This argument is valid for every $\varepsilon > 0$, so $f \circ g$ is continuous at $a$.

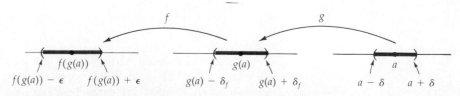

FIGURE 4

> ## COROLLARY 1
>
> If $g$ is continuous at $a$, and $g(a) \neq 0$, then $1/g$ is continuous at $a$.

**PROOF**   The function $f(u) = \dfrac{1}{u}$ is continuous at $u = g(a)$; for $g(a) \neq 0$, and so from Section A.1,

$$\lim_{u \to g(a)} \frac{1}{u} = \frac{1}{g(a)} = f(g(a)).$$

Hence by Theorem 2, the composition

$$f(g(x)) = \frac{1}{g(x)}$$

is continuous at $a$, as claimed in Corollary 1.

> ## COROLLARY 2
>
> If $\lim\limits_{x \to a} g(x) \neq 0$, then $\lim\limits_{x \to a} \dfrac{1}{g(x)} = \dfrac{1}{\lim\limits_{x \to a} g(a)}$.

This can be proved directly from the definition of limit; but here is a simpler proof.

Since the numerical value of $g(a)$ does not affect the existence or the numerical value of $\lim\limits_{x \to a} g(x)$, we can assume that $\lim\limits_{x \to a} g(x) = g(a)$. Then $g$ is continuous at $a$, and by Corollary 1, $1/g$ is continuous at $a$. Hence

$$\lim_{x \to a} \frac{1}{g(x)} = \frac{1}{g(a)}$$

and Corollary 2 is proved.

With a very similar method, one proves:

> ## COROLLARY 3
>
> $\lim\limits_{x \to a} f(g(x)) = f\left( \lim\limits_{x \to a} g(x) \right)$ *provided* that $M = \lim\limits_{x \to a} g(x)$ exists, and that $f$ is defined in an open interval about $M$, and is continuous at $M$.

Theorem 2 has a close analog relating *function limits* and *sequence limits*:

---

**THEOREM 3**

If

$$\lim_{n \to \infty} a_n = a, \qquad (1)$$

$$f(a_n) \text{ is defined for all } n, \qquad (2)$$

and $f$ is continuous at $a$, then

$$\lim_{n \to \infty} f(a_n) = f\left( \lim_{n \to \infty} a_n \right) = f(a). \qquad (3)$$

Conversely, if for every sequence $\{a_n\}$ satisfying (1) and (2), we have (3), then $f$ is continuous at $a$.

---

The proof is left as a problem.

## PROBLEMS

**B**

1. Prove that every rational function is continuous.

2. Prove that $f(x) = \sqrt{x}$ is continuous at every point $a$.

3. Prove that the function $f(x) = \begin{cases} 1, & x \geq 0 \\ 0, & x < 0 \end{cases}$ is discontinuous at 0, but continuous at every other point.

4. Prove Theorem 1.

5. Prove the first half of Theorem 3.

6. Prove the second half of Theorem 3. (If $f$ is *not* continuous at $a$, then for *some* $\varepsilon > 0$ there is *no* $\delta$ as required in the Definition of continuity. Use this to construct a sequence $a_n \nrightarrow a$ with $f(a_n) \nrightarrow f(a)$.

## A.4

### THE INTERMEDIATE VALUE THEOREM

We now need a deep fact about real numbers, discussed in Section 11.8:

**The Increasing Sequence Principle:** Every bounded increasing sequence of real numbers has a limit.

Armed with this, we prove:

---

### THE INTERMEDIATE VALUE THEOREM

If $f$ is continuous on the interval $[a, b]$, and $\hat{y}$ is any number between $f(a)$ and $f(b)$, then there is a point $\hat{x}$ in $[a, b]$ where $f(\hat{x}) = \hat{y}$ (Fig. 1).

---

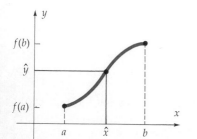

**FIGURE 1**
Intermediate Value Theorem.

The idea of the proof goes back to Simon Stevin (1549–1620). Since $\hat{y}$ is between $f(a)$ and $f(b)$, then either $f(a) \leq \hat{y} \leq f(b)$, or $f(b) \leq \hat{y} \leq f(a)$.

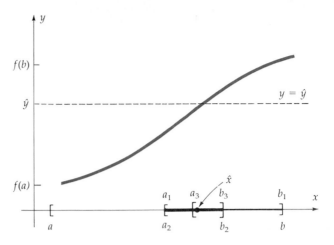

**FIGURE 2**

$a \leq a_1 \leq a_2 \leq a_3$ and
$b \geq b_1 \geq b_2 \geq b_3$.

Suppose, to be definite, that $f(a) \leq \hat{y} \leq f(b)$. Then bisect the interval $[a, b]$ repeatedly (Fig. 2), each time finding a smaller interval $[a_n, b_n]$ where

$$f(a_n) \leq \hat{y} \leq f(b_n). \tag{1}$$

Each interval is half the previous one; the successive intervals shrink down to a point $\hat{x} = \lim_{n \to \infty} a_n$, and we show that $f(\hat{x}) = \hat{y}$. Here are the details.

Let $c$ be the midpoint of $[a, b]$. If $f(c) < \hat{y}$, let $a_1 = c$ and $b_1 = b$ (Fig. 3); then $f(a_1) \leq \hat{y} \leq f(b_1)$. If on the other hand $f(c) \geq \hat{y}$, then let $a_1 = a$ and $b_1 = c$ (Fig. 4); again $f(a_1) \leq \hat{y} \leq f(b_1)$. Either way, (1) is true for $n = 1$; and $[a_1, b_1]$ is one half of $[a, b]$, so

$$a \leq a_1 \leq b_1 \leq b \quad \text{and} \quad b_1 - a_1 = \tfrac{1}{2}(b - a).$$

Carry out the same process with the interval $[a_1, b_1]$ in place of $[a, b]$; you get an interval $[a_2, b_2]$ satisfying (1). Moreover, $[a_2, b_2]$ is one half of $[a_1, b_1]$, so

$$a_1 \leq a_2 \leq b_2 \leq b_1 \quad \text{and} \quad b_2 - a_2 = \tfrac{1}{2}(b_1 - a_1) = \tfrac{1}{4}(b - a).$$

Proceeding ad infinitum, you get sequences $a_1, a_2, a_3, \ldots$ and $b_1, b_2, b_3, \ldots$ such that

$$a \leq a_1 \leq a_2 \leq a_3 \leq \cdots \leq b_3 \leq b_2 \leq b_1 \leq b \tag{2}$$

and

$$b_n - a_n = 2^{-n}(b - a). \tag{3}$$

By (2), $\{a_n\}$ is a bounded increasing sequence, so it has a limit, which we call $\hat{x}$; and $a \leq \hat{x} \leq b$. By (3), $b_n = a_n + 2^{-n}(b - a)$ has the same limit $\hat{x}$, since $2^{-n}(b - a) \to 0$ as $n \to \infty$. By Theorem 3, Section A.3,

$$f(\hat{x}) = \lim_{n \to \infty} f(a_n) \leq \hat{y} \tag{4}$$

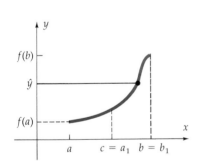

**FIGURE 3**

$f(c) < \hat{y} < f(b)$.

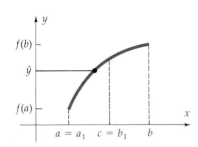

**FIGURE 4**

$f(a) < \hat{y} \leq f(c)$.

because of (1). On the other hand

$$f(\hat{x}) = \lim_{n \to \infty} f(b_n) \geq \hat{y}, \tag{5}$$

again because of (1). The two inequalities (4) and (5) imply that $f(\hat{x}) = \hat{y}$. This proves the Intermediate Value Theorem in the case where $f(a) \leq \hat{y} \leq f(b)$; the proof for the other case is similar.

## PROBLEMS

### B

1. In the proof of the Intermediate Value Theorem, determine $[a_1, b_1]$ and $[a_2, b_2]$ if $f(x) = x^2$, $[a, b] = [0, 1]$, and $\hat{y} = \frac{1}{3}$.

2. Prove the Intermediate Value Theorem in the case where $f(b) \leq \bar{y} \leq f(a)$.

### C

3. Suppose that $f$ is continuous on $[a, b]$, and $a \leq f(x) \leq b$; thus $f$ maps the interval $[a, b]$ into itself. Is it necessarily true that $f(\hat{x}) = \hat{x}$ for some $\hat{x}$ in $[a, b]$? [Consider the function $g(x) = f(x) - x$.]

4. Take for granted that every bounded *increasing* sequence has a limit. Deduce that every bounded *decreasing* sequence has a limit. (Hint: If $\{a_n\}$ is decreasing, what can be said about $\{-a_n\}$?)

## A.5

### THE EXTREME VALUE THEOREM

We prove:

> **THE EXTREME VALUE THEOREM (Sec. 7.1)**
>
> If $f$ is continuous on $[a, b]$, then $f$ has a maximum and a minimum value on $[a, b]$.

We prove the existence of a maximum, and leave the minimum as a problem. Our proof uses Stevin's method of repeated bisections (Fig. 1). Bisect $[a, b]$, and choose the half where $f$ is larger. Then bisect *that* half, and choose again the half where $f$ is larger. Continue the process to obtain an infinite sequence of "nested intervals" that shrink down to a point $\bar{x}$ where $f(\bar{x})$ is maximum. Here are the details.

First, divide the given interval $[a, b]$ in two parts by its midpoint $c$ (Fig. 2):

$$L = [a, c] \quad \text{and} \quad R = [c, b].$$

There are two possibilities:

    **(i)**   For every $x$ in $L$, there is an $x_1$ in $R$ with $f(x_1) \geq f(x)$.
    **(ii)**  For some $\hat{x}$ in $L$, $f(\hat{x}) > f(x)$ for all $x$ in $R$.

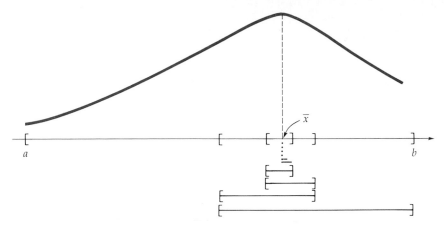

**FIGURE 1**
Nested intervals shrinking down to a point $\bar{x}$ where $f(\bar{x})$ is maximum.

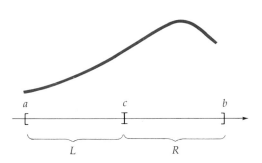

**FIGURE 2**

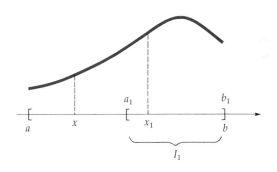

**FIGURE 3**
For every $x$ in $[a, b]$, there is an $x_1$ in $I_1 = [a_1, b_1]$ with $f(x_1) \geq f(x)$.

In case (i) set

$$a_1 = c, \quad b_1 = b, \quad \text{and} \quad I_1 = [a_1, b_1] = R.$$

Then (Fig. 3):

$$\text{For every } x \text{ in } [a, b], \text{ there is an } x_1 \text{ in } I_1 \\ \text{with } f(x_1) \geq f(x). \tag{1}$$

Indeed, if $x$ is in $L$, there is such an $x_1$, by case (i); and if $x$ is in $R$, we can choose $x_1 = x$.

In case (ii) set

$$a_1 = a, \quad b_1 = c, \quad \text{and} \quad I_1 = [a_1, b_1] = L.$$

Then again, (1) is true. Indeed, if $x$ is in $R$, choose $x_1 = \hat{x}$ as in case (ii); and if $x$ is in $L$, choose $x_1 = x$.

Thus, in any case, we have an interval $I_1 = [a_1, b_1]$ with $b_1 - a_1 = \frac{1}{2}(b - a)$, and (1) valid.

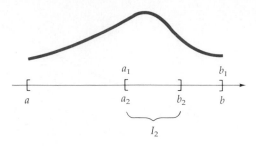

**FIGURE 4**
$a \leq a_1 \leq a_2 \leq b_2 \leq b_1 \leq b$.

Apply the same argument to $I_1$ in place of $[a, b]$, and find an interval $I_2 = [a_2, b_2]$ as in Figure 4, one half of $I_1$, such that

$$a \leq a_1 \leq a_2 \leq b_2 \leq b_1 \leq b, \qquad b_2 - a_2 = 2^{-2}(b - a)$$

and

$$\text{for every } x \text{ in } I_1, \text{ there is an } x_2 \text{ in } I_2 \text{ with } f(x_2) \geq f(x). \tag{2}$$

Continue the process indefinitely, finding $I_n = [a_n, b_n]$ such that

$$a \leq a_1 \leq \cdots a_n \leq b_n \leq \cdots \leq b_1 \leq b, \qquad b_n - a_n = 2^{-n}(b - a) \tag{3}$$

and

$$\text{for every } x \text{ in } I_{n-1} \text{ there is an } x_n \text{ in } I_n \text{ with } f(x_n) \geq f(x). \tag{4}$$

Now let

$$\bar{x} = \lim_{n \to \infty} a_n = \lim_{n \to \infty} b_n. \tag{5}$$

(Why do the two limits exist? Why are they equal?) We claim that $f(\bar{x}) \geq f(x)$, for every $x$ in $[a, b]$. Indeed, given $x$, there is by (1) an $x_1$ in $[a_1, b_1]$ where $f(x_1) \geq f(x)$. Then by (2), there is an $x_2$ in $[a_2, b_2]$ where

$$f(x_2) \geq f(x_1) \geq f(x),$$

and so on. Thus for every $n$ there is an $x_n$ in $[a_n, b_n]$ such that

$$f(x_n) \geq f(x). \tag{6}$$

Moreover, $a_n \leq x_n \leq b_n$, so by (5), $\lim_{n \to \infty} x_n = \bar{x}$. Hence by (6)

$$f(\bar{x}) = \lim f(x_n) \geq f(x).$$

So for every $x$ in $[a, b]$, $f(\bar{x}) \geq f(x)$; thus $f(\bar{x})$ is the maximum value of $f$ on $[a, b]$.

In just the same way you can show that there is a minimum value $f(\underline{x})$, thus completing the proof of the Extreme Value Theorem.

## PROBLEMS

### B

1. In the proof of the Extreme Value Theorem just given, determine $I_1 = [a_1, b_1]$ and $I_2 = [a_2, b_2]$ in each of the following cases:
   a) $f(x) = x^2$ on $[0, 1]$.
   b) $f(x) = -(x - 1)^2$ on $[0, 2]$.

2. In the proof of the Extreme Value Theorem just given, why does $\lim_{n \to \infty} a_n$ exist?

3. In the proof of the Extreme Value Theorem just given, why is $\lim b_n = \lim a_n$?

4. Prove that if $f$ is continuous on $[a, b]$, then there is a point $\underline{x}$ in $[a, b]$ such that $f(\underline{x}) \leq f(x)$ for every $x$ in $[a, b]$:
   a) By imitating the proof just given.
   b) By using the existence of a maximum value for $-f$.

### C

Problems 5–7 outline a proof of the Inverse Function Theorem.

5. Suppose that $f$ is continuous and strictly increasing on $[a, b]$. Prove that $f$ assumes every value in the interval $[f(a), f(b)]$ at exactly one point in $[a, b]$; hence $f$ has an inverse $f^{-1}$ defined on $[f(a), f(b)]$.

6. Continuing problem 5, prove that the inverse $g = f^{-1}$ is continuous. [Hint: Sketch a typical $f$, and read the graph in reverse to get $g$. Given a "target" $(g(\alpha) - \varepsilon, g(\alpha) + \varepsilon)$, consider the interval $(f(g(\alpha) - \varepsilon), f(g(\alpha) + \varepsilon))$.]

7. Continuing problems 5 and 6, suppose that $f'(x) > 0$ everywhere in $(a, b)$. Prove that $g'(y) = 1/f'(g(y))$. [Hint: Given $\bar{y}$ in $(f(a), f(b))$, let $\bar{x} = g(\bar{y})$, and set

$$Q(x) = \begin{cases} \dfrac{f(x) - f(\bar{x})}{x - \bar{x}}, & x \neq \bar{x} \\ f'(\bar{x}), & \text{if } x = \bar{x}. \end{cases}$$

Explain why $Q$ is continuous at $\bar{x}$, and why

$$\frac{g(y) - g(\bar{y})}{y - \bar{y}} = \frac{1}{Q(g(y))} \to \frac{1}{Q(g(\bar{y}))}$$

as $y \to \bar{y}$.]

# A.6

## UNIFORM CONVERGENCE. TERM-BY-TERM OPERATIONS ON SERIES

The evaluation of integrals by series (Sec. 11.6) requires term-by-term integration; and the solution of differential equations by power series (Sec. 11.8) uses term-by-term differentiation. This section justifies these operations, and not just for power series. Roughly speaking, *any* series of functions can be integrated and differentiated term-by-term, as long as all the functions and derivatives are continuous, and all the series involved have a certain property called **uniform convergence**. We begin by discussing this property, and then prove the theorems on term-by-term operations.

Consider a series of functions $\sum_{k=0}^{\infty} u_k(x)$, for example a power series

$$\sum_{k=0}^{\infty} a_k x^k, \quad |x| \leq r \qquad [\text{here } u_k(x) = a_k x^k]$$

or a "trigonometric series"

$$\sum_{k=0}^{\infty} (a_k \cos kx + b_k \sin kx) \qquad [\text{here } u_k(x) = a_k \cos kx + b_k \sin kx].$$

Suppose that all the functions $u_k(x)$ are defined for $x$ in a given interval $I$; and for each $x$ in $I$, the series $\sum_{k=0}^{\infty} u_k(x)$ converges to a sum, which we denote by $s(x)$. For example, the functions $u_k(x) = x^k$ are all defined for $x$ in the interval $I = (-1, 1)$, and the series

$$\sum_{k=0}^{\infty} u_k(x) = \sum_{k=0}^{\infty} x^k = \frac{1}{1-x}, \qquad |x| < 1$$

converges to the sum $s(x) = \dfrac{1}{1-x}$ for all $x$ in $I$.

The series $\sum_{k=0}^{\infty} u_k(x)$ is called **uniformly convergent** on $I$ if, for every $\varepsilon > 0$, there is an integer $N$ such that

$$n > N \quad \Rightarrow \quad \left| s(x) - \sum_{k=0}^{n} u_k(x) \right| < \varepsilon \qquad \text{for all } x \text{ in } I. \tag{1}$$

The convergence is called uniform since the *same* $N$ works for *all* $x$ in $I$.

The defining property (1) can be rewritten as

$$n > N \quad \Rightarrow \quad \left| \sum_{n+1}^{\infty} u_k(x) \right| < \varepsilon \qquad \text{for all } x \text{ in } I. \tag{1a}$$

---

**EXAMPLE 1**    Show that $\sum_{k=0}^{\infty} x^k$ converges uniformly for $|x| \le 1/2$.

*SOLUTION*

$$\sum_{k=0}^{n} x^k = \frac{1 - x^{n+1}}{1 - x},$$

so

$$\left| \frac{1}{1-x} - \sum_{k=0}^{n} x^k \right| = \left| \frac{x^{n+1}}{1-x} \right| \le \frac{(1/2)^{n+1}}{1/2} \qquad \text{if } |x| \le \frac{1}{2}; \tag{2}$$

for $-\frac{1}{2} \le x \le \frac{1}{2}$ implies that $1 - x \ge \frac{1}{2}$. Now, given $\varepsilon > 0$, you can choose $N$ such that

$$n > N \quad \Rightarrow \quad \frac{(1/2)^{n+1}}{1/2} < \varepsilon;$$

in fact, $N = \dfrac{\log \varepsilon}{\log(1/2)}$ will do. Then for $n > N$, line (2) implies that

$$\left| \frac{1}{1-x} - \sum_{0}^{n} x^k \right| < \varepsilon \qquad \text{for all } |x| \le \frac{1}{2}.$$

---

---

**EXAMPLE 2**    $\sum_{k=0}^{\infty} x^k = \dfrac{1}{1-x}, \; |x| < 1$, but the convergence is not *uni-*

*form* in that interval. For if it were, then for each $\varepsilon > 0$ there would be an $N$ such that

$$\left| \frac{1}{1-x} - \sum_{k=0}^{N+1} x^k \right| < \varepsilon \qquad \text{for all } |x| < 1.$$

This is impossible; for $\left| \sum_{k=0}^{N+1} x^k \right| \leq N + 2$ for $|x| < 1$, whereas $\dfrac{1}{1-x}$ is unbounded on that interval.

---

Example 1 was easy because of the simple formula for the partial sums of the geometric series. For more general series, the most common tool to prove uniform convergence is:

---

**THE $M$-TEST**

Suppose that

$$|u_k(x)| \leq M_k \qquad \text{for all } x \text{ in } I \tag{3}$$

and

$$\sum_{k=0}^{\infty} M_k < \infty. \tag{4}$$

Then $\sum_{k=0}^{\infty} u_k(x)$ converges uniformly for $x$ in $I$.

---

**PROOF**  The hypotheses (3) and (4) imply that for each $x$ in the interval $I$, $\sum_{k=0}^{\infty} u_k(x)$ is absolutely convergent; so the series has a well-defined sum, which we call $s(x)$. Then

$$\left| s(x) - \sum_{k=0}^{n} u_k(x) \right| = \left| \sum_{k=0}^{\infty} u_k(x) - \sum_{k=0}^{n} u_k(x) \right|$$

$$= \left| \sum_{n+1}^{\infty} u_k(x) \right| \leq \sum_{n+1}^{\infty} M_k \tag{5}$$

by (3). Hypothesis (4) implies that $\lim\limits_{n \to \infty} \sum_{k=0}^{n} M_k$ exists, and equals the finite quantity $\sum_0^{\infty} M_k$. So by the definition of limit, there is for each $\varepsilon > 0$ an $N$ such that

$$\sum_0^{\infty} M_k - \sum_0^{n} M_k < \varepsilon \qquad \text{if } n > N.$$

Thus by (5), if $n > N$,

$$\left| s(x) - \sum_{k=0}^{n} u_k(x) \right| \leq \sum_{n+1}^{\infty} M_k = \sum_0^{\infty} M_k - \sum_0^{n} M_k < \varepsilon,$$

uniformly for all $x$ in $I$. The $M$-Test is proved.

**EXAMPLE 3**  $\displaystyle\sum_{k=1}^{\infty} \frac{x^k}{k^2}$ converges uniformly for $|x| \leq 1$. For,

$$\left| \frac{x^k}{k^2} \right| \leq \frac{1}{k^2} \qquad \text{if } |x| \leq 1$$

and

$$\sum_{k=1}^{\infty} \frac{1}{k^2} < \infty.$$

So hypotheses (3) and (4) of the $M$-Test are met, with $M_k = \dfrac{1}{k^2}$.

Example 2 showed that the geometric series does not converge uniformly in its entire interval of convergence $|x| < 1$; but (Example 1) it *does* converge uniformly in the smaller interval $|x| \leq \frac{1}{2}$. This illustrates a general fact about power series.

---

**THEOREM 1**

Suppose that $\sum_{k=0}^{\infty} a_k x^k$ converges for $|x| < r$. Then for any $\rho < r$, $\sum_{k=0}^{\infty} a_k x^k$ converges *uniformly* for $|x| \leq \rho$. So does the differentiated series $\sum_{k=1}^{\infty} k \cdot a_k x^{k-1}$.

---

**PROOF**  Given $\rho < r$, choose $x_0$ between $\rho$ and $r$. Then $|x_0| < r$, so $\sum_{k=0}^{\infty} a_k(x_0)^k$ converges. It follows that $a_k(x_0)^k \to 0$, and hence that the sequence $\{a_k(x_0)^k\}$ is bounded; there is an $M$ such that

$$\left| a_k(x_0)^k \right| \leq M \qquad \text{for all } k.$$

Then for $|x| \leq \rho$,

$$\left| a_k x^k \right| \leq |a_k| \rho^k = \left| a_k(x_0)^k \right| \left( \frac{\rho}{x_0} \right)^k$$

$$\leq M \left( \frac{\rho}{x_0} \right)^k. \tag{6}$$

Since $x_0$ is between $\rho$ and $r$, then $\rho/x_0 < 1$, and

$$\sum_{k=0}^{\infty} M \left( \frac{\rho}{x_0} \right)^k < \infty. \tag{7}$$

The inequalities (6) and (7) match the hypotheses (3) and (4) of the $M$-Test, so the power series converges uniformly for $|x| \leq \rho$.

For the differentiated series,

$$\left| k \cdot a_k x^{k-1} \right| \leq M |x_0|^{-1} k \left( \frac{\rho}{x_0} \right)^k. \tag{8}$$

The terms on the right in (8) form a convergent series of constants, by the ratio test, so $\sum_{k=1}^{\infty} k \cdot a_k x^{k-1}$ converges uniformly. This completes the proof of Theorem 1.

Now we prove some of the useful properties of uniformly convergent series.

---

**THEOREM 2**

If $\sum_{k=0}^{\infty} u_k(x)$ converges uniformly on an interval $I$, and each term $u_k(x)$ is continuous on $I$, then the sum $s(x) = \sum_{k=0}^{\infty} u_k(x)$ is also continuous on $I$.

---

**PROOF**   Given $\varepsilon > 0$, and a point $x_0$ in $I$, we must show that there is a $\delta > 0$ such that

$$|x - x_0| < \delta \quad \Rightarrow \quad |s(x) - s(x_0)| < \varepsilon.$$

By uniform convergence, there is an $N$ such that

$$\left| s(x) - \sum_{k=0}^{N} u_k(x) \right| < \varepsilon/3 \qquad \text{for all } x \text{ in } I. \tag{9}$$

Since each $u_k(x)$ is continuous, so is the finite sum $\sum_{k=0}^{N} u_k(x)$; hence there is a $\delta > 0$ such that

$$|x - x_0| < \delta \quad \Rightarrow \quad \left| \sum_{k=0}^{N} u_k(x) - \sum_{k=0}^{N} u_k(x_0) \right| < \varepsilon/3. \tag{10}$$

Then for $|x - x_0| < \delta$,

$$|s(x) - s(x_0)| \leq \left| s(x) - \sum_{k=0}^{N} u_k(x) \right| + \left| \sum_{k=0}^{N} u_k(x) - \sum_{k=0}^{N} u_k(x_0) \right|$$
$$+ \left| \sum_{k=0}^{N} u_k(x_0) - s(x_0) \right|.$$

On the right, the middle term is $< \varepsilon/3$, by (10), and the other two are $< \varepsilon/3$, by (9). Thus for $|x - x_0| < \delta$,

$$|s(x) - s(x_0)| < \varepsilon,$$

as was to be proved.

Next, we prove that uniformly convergent series can be integrated term-by-term.

---

**THEOREM 3**

If each $u_k(x)$ is continuous for $a \leq x \leq b$, and $\sum_{k=0}^{\infty} u_k(x)$ converges uniformly for $a \leq x \leq b$, then

$$\int_a^b \sum_{k=0}^{\infty} u_k(x) \, dx = \sum_{k=0}^{\infty} \int_a^b u_k(x) \, dx.$$

**PROOF** By Theorem 2, $\sum_{k=0}^{\infty} u_k(x)$ is a continuous function on $[a, b]$, so its integral exists. We must show that, given $\varepsilon > 0$, there is an $N$ such that

$$n > N \quad \Rightarrow \quad \left| \int_a^b \sum_{k=0}^{\infty} u_k(x)\, dx - \sum_{k=0}^{n} \int_a^b u_k(x)\, dx \right| < \varepsilon. \tag{11}$$

But the integral of a *finite* sum is the sum of the integrals, so

$$\left| \int_a^b \sum_{k=0}^{\infty} u_k(x)\, dx - \sum_{k=0}^{n} \int_a^b u_k(x)\, dx \right| = \left| \int_a^b \sum_{k=0}^{\infty} u_k(x)\, dx - \int_a^b \sum_{k=0}^{n} u_k(x)\, dx \right|$$

$$= \left| \int_a^b \sum_{n+1}^{\infty} u_k(x)\, dx \right|$$

$$\leq \int_a^b \left| \sum_{n+1}^{\infty} u_k(x) \right| dx. \tag{12}$$

Since the series converges uniformly, there is by (1a) an $N$ such that

$$\left| \sum_{n+1}^{\infty} u_k(x) \right| < \frac{\varepsilon}{b-a} \qquad \text{for all } n \geq N \text{ and all } x \text{ in } [a, b]. \tag{13}$$

Thus $\int_a^b \left| \sum_{n+1}^{\infty} u_k(x) \right| dx < \int_a^b \frac{\varepsilon}{b-a}\, dx = \varepsilon$, and the required inequality (11) follows from (12) and (13). Theorem 3 is proved.

Together with Theorem 1, this justifies the term-by-term integration of power series.

In order to *differentiate* a series $s(x) = \sum_{k=0}^{\infty} u_k(x)$ term-by-term, it is of course necessary that the differentiated series $\sum_{k=0}^{\infty} u_k'(x)$ converge; but to prove that this series converges to $s'(x)$ we assume more, namely: The differentiated series converges *uniformly*.

---

**THEOREM 4**

Suppose that, in a given interval $I$,

$$u_k'(x) \quad \text{is continuous for each } k, \text{ and each } x \text{ in } I,$$

$$\sum_{k=0}^{\infty} u_k'(x) \quad \text{converges uniformly in } I, \text{ and}$$

$$\sum_{k=0}^{\infty} u_k(x_0) \quad \text{converges for at least one } x_0 \text{ in } I.$$

Then for every $x$ in $I$,

$$\frac{d}{dx} \sum_{k=0}^{\infty} u_k(x) = \sum_{k=0}^{\infty} u_k'(x).$$

**PROOF**    By Theorem 2, $\sum_{k=0}^{\infty} u_k'(x)$ is continuous, and by Theorem 3

$$\int_{x_0}^{x} \sum_{k=0}^{\infty} u_k'(t)\, dt = \sum_{k=0}^{\infty} \int_{x_0}^{x} u_k'(t)\, dt$$

$$= \sum_{k=0}^{\infty} u_k(x) - \sum_{k=0}^{\infty} u_k(x_0).$$

Thus

$$\frac{d}{dx} \sum_{k=0}^{\infty} u_k(x) = \frac{d}{dx}\left[\int_{x_0}^{x} \sum_{k=0}^{\infty} u_k'(t)\, dt + \sum_{k=0}^{\infty} u_k(x_0)\right]$$

$$= \sum_{k=0}^{\infty} u_k'(x) + 0,$$

by the Second Fundamental Theorem of Calculus. Theorem 4 is proved.

Now at last we can justify the term-by-term differentiation of power series.

---

**COROLLARY 1**

If $f(x) = \sum_{k=0}^{\infty} a_k x^k$ for $|x| < r$, then

$$f'(x) = \sum_{k=1}^{\infty} k \cdot a_k x^{k-1} \qquad \text{for } |x| < r.$$

---

**PROOF**    We must show that for any number $x_0$ such that $|x_0| < r$,

$$f'(x_0) = \sum_{k=0}^{\infty} k \cdot a_k (x_0)^{k-1}.$$

Given $|x_0| < r$, choose $\rho$ such that $|x_0| < \rho < r$. By Theorem 1, $\sum_{k=1}^{\infty} k \cdot a_k x^{k-1}$ converges uniformly for $|x| \leq \rho$; then by Theorem 4,

$$f'(x) = \frac{d}{dx} \sum_{k=0}^{\infty} a_k x^k = \sum_{k=1}^{\infty} k \cdot a_k x^{k-1}$$

for $|x| \leq \rho$. Since $|x_0| < \rho$, the desired conclusion follows.

---

## PROBLEMS

**B**

**1.**    Prove uniform convergence by the $M$-Test.

a)    $\displaystyle\sum_{k=1}^{\infty} \frac{\sin kx}{k^2}$, all $x$

b)    $\displaystyle\sum_{k=0}^{\infty} ke^{-k} \cos kx$, all $x$

c)    $\displaystyle\sum_{k=0}^{\infty} \frac{x^k}{(k!)^2}$, $|x| \leq 10$

**2.**    Define $u(x, y)$ for $y \geq 0$ to be the sum of the series $\sum_{1}^{\infty} k^{-4} \sin(kx) e^{-ky}$.

a)    Prove that the series converges uniformly for $y \geq 0$.

**b)** Justify term-by-term computation of $u_x$ and $u_y$.

**c)** Prove that $u$ solves *Laplace's equation* for steady-state temperature distribution in the plane, $u_{xx} + u_{yy} = 0$.

**3.** Define $u(x, t) = \sum_{k=1}^{\infty} e^{-k^2t} \sin kx$. Prove that

**a)** For each $t > 0$, the series converges uniformly in $x$.

**b)** For $t > 0$, $u_{xx} = \sum_{k=0}^{\infty} (-k^2) e^{-k^2t} \sin kx$.

**c)** If $\delta > 0$ and $x$ is fixed, the series converges uniformly in $t \geq \delta$.

**d)** If $t_0 > 0$ then

$$u_t(x, t_0) = \sum_{k=0}^{\infty} -k^2 e^{-k^2t} \sin kx = u_{xx}(x, t_0).$$

[The equation $u_t = u_{xx}$ describes the flow of heat in a rod; $t$ is time, $x$ is position along the rod, and $u$ is temperature.]

## C

The following problems show that Theorems 2−4 may fail, if the convergence is *not* uniform.

**4.** Let $s_n(x) = x^{1/n}$, $0 \leq x \leq 1$.

**a)** Determine $\lim_{n \to \infty} s_n(x)$. Is this limit function continuous on $[0, 1]$?

**b)** Define $u_1(x) = x$, and $u_n(x) = s_n(x) - s_{n-1}(x)$

for $n > 1$. Is each $u_n$ continuous for $0 \leq x \leq 1$? Is $\sum_{1}^{\infty} u_n(x)$ continuous?

**5.** Let

$$s_n(x) = \begin{cases} n^2x, & 0 \leq x \leq 1/n \\ -n^2(x - 2/n), & 1/n \leq x \leq 2/n \\ 0, & 2/n \leq x \leq 1. \end{cases}$$

**a)** Sketch $s_2, s_3, s_4$. (Each graph consists of line segments; plot their endpoints.)

**b)** Determine $\lim_{n \to \infty} s_n(x)$. (Treat $x = 0$ and $x > 0$ separately.)

**c)** Compare $\int_0^1 \lim s_n(x) \, dx$ and $\lim \int_0^1 s_n(x) \, dx$. (Evaluate the integrals graphically.)

**d)** Let $u_2(x) = s_2(x)$, and $u_n(x) = s_n(x) - s_{n-1}(x)$ for $n > 2$. Compute $\sum_{2}^{\infty} u_n(x)$, and compare $\int_0^1 \sum_{2}^{\infty} u_n(x) \, dx$ with $\sum_{2}^{\infty} \int_0^1 u_n(x) \, dx$.

**6.** Let $S_n(x) = \int_0^x s_n(t) \, dt$, with $s_n$ as in the previous problem.

**a)** Compute $\lim_{n \to \infty} S_n(x)$ and $\lim_{n \to \infty} S_n'(x)$.

**b)** Show that $\lim_{n \to \infty} S_n'(x) \neq \dfrac{d}{dx} \lim_{n \to \infty} S_n(x)$ for $x = 0$.

**c)** Construct a sequence of functions $U_n(x)$ such that, for some $x$, $\sum_{2}^{\infty} U_n'(x) \neq \dfrac{d}{dx} \sum_{2}^{\infty} U_n(x)$, although both series converge.

# APPENDIX B

## REVIEW OF EXPONENTS AND LOGARITHMS

## Positive Integer Exponents

For any number $a$, and any positive integer $n$, the notation $a^n$ stands for the product with $n$ factors, all equal to $a$:

$$a^n = \underbrace{a \cdot a \cdot \cdots \cdot a}_{n \text{ factors}}$$

This is **exponential notation**. The positive integer $n$ is called the **exponent**, or **power**, and the repeated factor $a$ is the **base**.

These exponentials satisfy three basic "laws":

### First Law of Exponents:

$$a^n \cdot a^m = a^{n+m} \tag{1}$$

since

$$\underbrace{(a \cdots a)}_{n \text{ factors}} \cdot \underbrace{(a \cdots a)}_{m \text{ factors}} = \underbrace{(a \cdots \cdots a)}_{n + m \text{ factors}}.$$

### Second Law of Exponents:

$$(a^n)^m = a^{n \cdot m} \tag{2}$$

since

$$\overbrace{\underbrace{(a \cdots a)}_{n \text{ factors}} \cdot \underbrace{(a \cdots a)}_{n \text{ factors}} \cdots \cdots \underbrace{(a \cdots a)}_{n \text{ factors}}}^{m \text{ groups}} = \underbrace{(a \cdots \cdots a)}_{n \cdot m \text{ factors}}.$$

### Third Law of Exponents:

$$a^n \cdot b^n = (a \cdot b)^n \tag{3}$$

as can be seen from an argument like the previous two. (These informal arguments help to keep the laws straight, in moments of doubt.)

## More General Exponents

If $a \neq 0$, then $a^n$ is defined when $n$ is 0 or a negative integer, in such a way that the three laws remain valid. The first law determines the definition of $a^0$ as follows:

$$a^0 \cdot a^1 = a^{0+1} = a^1;$$

divide through by $a^1$, and you find

$$a^0 = 1. \tag{4}$$

Now from the first law and (4), $a^{-n} \cdot a^n = a^{-n+n} = a^0 = 1$; divide by $a^n$ and you find

$$a^{-n} = \frac{1}{a^n}. \tag{5}$$

If $a > 0$ then $a^x$ is defined for every real number $x$, in such a way that $a^x$ is a continuous function, and preserving laws (1)–(5). The second law implies that the fractional power $a^{1/n}$ is the $n^{\text{th}}$ root $\sqrt[n]{a}$; for this root is defined by the equation $(\sqrt[n]{a})^n = a$, and $a^{1/n}$ satisfies that equation:

$$(a^{\frac{1}{n}})^n = a^{\frac{1}{n} \cdot n} = a^1 = a.$$

Thus

$$a^{1/n} = \sqrt[n]{a}, \qquad a \geq 0. \tag{6}$$

Using the second law again,

$$a^{m/n} = (a^{1/n})^m. \tag{7}$$

For example, $8^{2/3} = (8^{1/3})^2 = (\sqrt[3]{8})^2 = 2^2 = 4$.

When $n$ is odd, then formulas (6) and (7) can be applied even when $a < 0$; for odd roots of negative numbers do exist.

## Logarithms

The logarithm to the base $a$ is defined as the inverse function of the exponential with base $a$. That is,

$$y = \log_a x \quad \text{means that} \quad x = a^y. \tag{8}$$

For example,

$$3 = \log_2 8, \quad \text{since} \quad 8 = 2^3.$$

The defining relation (8) has two simple consequences. First, set $x = a^y$ in the equation $y = \log_a x$; you get

$$y = \log_a(a^y). \tag{9}$$

That is, the logarithm to the base $a$ "undoes" the power $a^y$. For example, $\log_{10}(10^{-1}) = -1$ and $\log_a(a^0) = 0$.

Set $y = \log_a x$ in the equation $x = a^y$, and you get the relation

$$x = a^{\log_a x}. \tag{10}$$

The power with base $a$ "undoes" the logarithm to the base $a$. For example, $7 = 10^{\log_{10} 7}$.

## Laws of Logarithms

Each law of exponents implies a corresponding law of logarithms. As a first example, we derive the formula for the logarithm of a product $xy$. Write

$$\begin{aligned} xy &= a^{\log_a x} \cdot a^{\log_a y} && \text{[by (10)]} \\ &= a^{\log_a x + \log_a y} && \text{[first law of exponents]}. \end{aligned}$$

Take $\log_a$ on each side; this "undoes" the power on the right-hand side, and yields

$$\log_a xy = \log_a x + \log_a y. \tag{11}$$

That is: *The logarithm of a product equals the sum of the logarithms of the factors.* Logarithms simplify multiplication by turning it into addition.

As for division,

$$\frac{x}{y} = \frac{a^{(\log_a x)}}{a^{(\log_a y)}} = a^{(\log_a x)} \cdot a^{-(\log_a y)} = a^{\log_a x - \log_a y}.$$

Take $\log_a$ on each side:

$$\log_a\left(\frac{x}{y}\right) = \log_a x - \log_a y. \tag{12}$$

For powers,

$$x^y = [a^{\log_a x}]^y = a^{(\log_a x)y} = a^{y\,\log_a x}$$

so

$$\log_a(x^y) = y\,\log_a x. \tag{13}$$

We use these laws to solve equations involving powers and logarithms. In the following examples, "log" denotes "$\log_{10}$."

---

**EXAMPLE 1**   Solve $3^{2t} = 5$   for $t$.

**SOLUTION**   You can "get $t$ out of the exponent" of $3^{2t}$ by taking logs. Since $3^{2t} = 5$ then

$$\log(3^{2t}) = \log 5$$

$$(2t)\log 3 = \log 5 \qquad \text{[by (13)]}$$

$$t = \frac{\log 5}{2 \log 3}$$

---

**EXAMPLE 2**   Solve $\log(x^2 + 1) = 1$ for $x$.

**SOLUTION**   Here you have to "get $x$ out of the log." Use formula (10). Since $\log(x^2 + 1) = 1$ then

$$10^{\log(x^2 + 1)} = 10^1$$

Since log denotes $\log_{10}$, formula (10) gives

$$x^2 + 1 = 10$$

so $x = \pm 3$.

**EXAMPLE 3**    A certain country increases its Gross National Product (GNP) by 5% a year. At this rate, how long will it take to double the GNP?

*SOLUTION*    Let $P_0$ be the GNP initially, and $P_1$ be the GNP after 1 year. Since the GNP grows at 5% a year,

$$P_1 = P_0 + .05P_0 = (1.05)P_0.$$

After two years it is

$$P_2 = P_1 + .05P_1 = (1.05)P_1 = (1.05)^2 P_0$$

and after $n$ years it is

$$P_n = (1.05)^n P_0.$$

If it doubles in $n$ years, then

$$2P_0 = P_n = (1.05)^n P_0$$
$$2 = (1.05)^n.$$

To get $n$ out of the exponent, take logs, using (13):

$$\log 2 = n \log(1.05)$$
$$n = \frac{\log 2}{\log(1.05)}.$$

A calculator gives

$$n \approx 14.2 \text{ years.}$$

## Change of Base

Logarithms to a given base $a$ can be converted to a different base $b$ by the formula

$$\log_a b \cdot \log_b c = \log_a c. \tag{14}$$

This is easily remembered; on both sides, $a$ appears as a subscript and $c$ on the line, while the intermediate $b$ "cancels." To prove (14), set $x = \log_b c$; thus

$$c = b^x.$$

Take $\log_a$:

$$\log_a c = \log_a(b^x) = x \log_a b = (\log_b c)(\log_a b).$$

Now (14) follows by exchanging the last two factors.
    As a consequence,

$$\log_b c = \frac{\log_a c}{\log_a b}. \tag{15}$$

**EXAMPLE 4**   Compute $\log_2 7$, using common logs (to the base 10).

*SOLUTION*   Using (15) and a calculator,

$$\log_2 7 = \frac{\log_{10}(7)}{\log_{10}(2)} = 2.807354922\ldots.$$

## SUMMARY

**Laws of Exponents:**   $a^x a^y = a^{x+y}$ $\qquad [a > 0]$

$(a^x)^y = a^{xy}$ $\qquad [a > 0]$

$a^x b^x = (ab)^x$ $\qquad [a > 0, b > 0]$

$a^0 = 1$ $\qquad [a \neq 0]$

$0^0$ is undefined.

$$a^{-x} = \frac{1}{a^x}, \quad a^{1/n} = \sqrt[n]{a}$$

**Definition of Logarithms:**   Let $a > 0$, $a \neq 1$. The *logarithm to the base $a$* is defined by

$$y = \log_a x \quad \text{means} \quad x = a^y$$

Hence

$$a^{(\log_a x)} = x, \qquad \log_a(a^y) = y$$

**Laws of Logarithms:**   $\left.\begin{array}{l} \log_a(xy) = \log_a x + \log_a y \\ \log_a(x/y) = \log_a x - \log_a y \end{array}\right\} x > 0 \text{ and } y > 0$

$$\log_a(x^y) = y \log_a x, \qquad x > 0$$

**Change of Base:**   $\log_a b \cdot \log_b c = \log_a c.$ $\qquad \log_b c = \dfrac{\log_a c}{\log_a b}.$

## PROBLEMS

### A

*1.   Express the following as simply as possible, using only positive integer exponents, fractions, and radicals $\sqrt[n]{\phantom{x}}$

a)   $3^{-2}$
b)   $a^{-2}$
c)   $x^{5/8}$
d)   $t^{-5/8}$
e)   $(-100)^0$
f)   $1^{30}$
g)   $1^{-30}$
h)   $1^{1/8}$
i)   $32^{-3/5}$
j)   $(-1)^{2/3}$

*2.   Explain why $a^n \cdot b^n = (ab)^n$.

3.   Graph the following for $-1 \leq x \leq 1$. Plot points for $x = 0$, $\pm 1$, $\pm 2/3$, $\pm 1/3$. Use a calculator, or tables, or the following 2-decimal approximations:

$$\sqrt[3]{2} \approx 1.26 \qquad \sqrt[3]{3} \approx 1.44 \qquad \sqrt[3]{10} \approx 2.15$$

a) $y = 2^x$     b) $y = 3^x$     c) $y = 10^x$
d) $y = (\frac{1}{2})^x$     e) $y = 1^x$

*4.  Compute
a) $\log_2 64$          b) $\log_3 9$
c) $\log_2(\frac{1}{2})$          d) $\log_2 1$
e) $\log_5(1/25)$       f) $\log_{64}(2)$
g) $\log_{1/2}(2)$       h) $\log_5(\log_2 32)$

5.  Solve for $x$. (log denotes $\log_{10}$.)
*a)  $\log x = 1$          *b)  $\log x = 0$
*c)  $\frac{1}{2}\log x + 1 = 0$     d)  $7^{3x} = 5$
e)  $(x + 1)(\log x - 2) = 0$
f)  $\log(2x + 1) = -1$

6.  Express with common logarithms (and compute with a calculator).
*a)  $\log_2 10$     b)  $\log_5 80$     *c)  $\log_{1/10}(2)$

*7.  The legendary region "Middle Earth" has a GNP growing at 8% a year. In how many years will the GNP double?

*8.  $10,000 is invested at 9% annual interest, compounded yearly; thus the investment grows at 9% a year. In how many years will the investment double?

9.  Sketch the following graphs, plotting points for $x = 1/4, 1/2, 1, 2, 4$.
a)  $y = \log_2 x$     b)  $y = \log_{1/2} x$.

10.  Why is there no function $\log_1 x$?

# APPENDIX C

# REVIEW OF TRIGONOMETRY

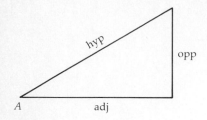

**FIGURE 1**
Labeling sides with respect to angle $A$.

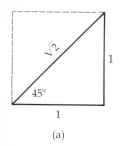

(a)

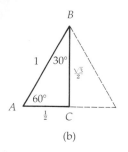

(b)

**FIGURE 2**

**The trigonometric functions of angles** are defined as certain ratios in right triangles. In Figure 1

"opp"   denotes the side opposite angle $A$,
"adj"   denotes the side adjacent to angle $A$,
"hyp"   denotes the hypotenuse, opposite the right angle.

Then the sine and cosine of the angle $A$ are defined by

$$\sin A = \frac{\text{opp}}{\text{hyp}}, \qquad \cos A = \frac{\text{adj}}{\text{hyp}},$$

while the tangent, cotangent, secant, and cosecant are defined by

$$\tan A = \frac{\text{opp}}{\text{adj}} = \frac{\sin A}{\cos A} \qquad \cot A = \frac{\text{adj}}{\text{opp}} = \frac{1}{\tan A}$$

$$\sec A = \frac{\text{hyp}}{\text{adj}} = \frac{1}{\cos A} \qquad \csc A = \frac{\text{hyp}}{\text{opp}} = \frac{1}{\sin A}.$$

The popular cry "SOH CAH TOA" helps to remember that

**S**ine is **O**pposite over **H**ypotenuse      (SOH)
**C**osine is **A**djacent over **H**ypotenuse   (CAH),   and
**T**angent is **O**pposite over **A**djacent      (TOA).

The trigonometric functions of $30°$, $45°$, and $60°$ are determined (and remembered!) from the right triangles in Figure 2. Figure 2a shows a unit square bisected by a diagonal of length $\sqrt{1^2 + 1^2} = \sqrt{2}$, from Pythagoras' Theorem. The diagonal bisects the $90°$ angle at the corner, so the angle $A$ is $45°$; hence

$$\sin 45° = 1/\sqrt{2}, \qquad \cos 45° = 1/\sqrt{2}, \qquad \tan 45° = 1.$$

In Figure 2b, a unit equilateral triangle is bisected by a perpendicular. Since the sum of the angles in any triangle is $180°$, each of the three equal angles in the equilateral triangle is $60°$. Since $BC$ is a bisector, side $AC = \frac{1}{2}$, and then $|BC|$ is computed from Pythagoras' Theorem:

$$(\tfrac{1}{2})^2 + |BC|^2 = 1 \quad \Rightarrow \quad |BC|^2 = \tfrac{3}{4} \quad \Rightarrow \quad |BC| = \sqrt{3}/2.$$

The triangle $ABC$ shows that

$$\sin 60° = \cos 30° = \sqrt{3}/2, \qquad \cos 60° = \sin 30° = \tfrac{1}{2},$$
$$\tan 60° = \sqrt{3}, \qquad \tan 30° = 1/\sqrt{3}.$$

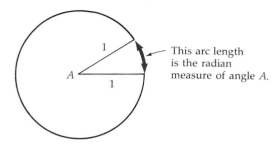

**FIGURE 3**
Radian measure.

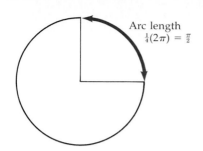

**FIGURE 4**
Radian measure of a right angle.

**The radian measure of an angle** is defined as the length of arc cut off by that angle in a circle of unit radius (Fig. 3). Since a circle of radius $r = 1$ has circumference $2\pi r = 2\pi \cdot 1 = 2\pi$, and a right angle cuts off a quarter of the circle, the radian measure of a right angle is $\dfrac{1}{4}(2\pi) = \dfrac{\pi}{2}$ (Fig. 4). Half a circle gives $180°$ or $(1/2)(2\pi) = \pi$ radians. In general, $(180\alpha)°$ gives $\pi\alpha$ radians, or

$$\alpha \text{ degrees} = \frac{\pi}{180}\,\alpha \quad \text{radians.}$$

In particular (Fig. 5) $60°$, $45°$, and $30°$ are respectively $\pi/3$, $\pi/4$, and $\pi/6$ radians; so Figure 2 shows that

$$\sin\frac{\pi}{6} = \frac{1}{2}, \qquad \sin\frac{\pi}{4} = 1/\sqrt{2}, \qquad \sin\frac{\pi}{3} = \sqrt{3}/2,$$

and so on.

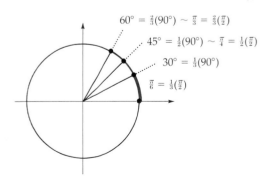

**FIGURE 5**
Radian measure of $30°$, $45°$, $60°$.

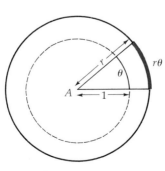

**FIGURE 6**
Arc length on a circle of radius $r$.

In a circle of radius $r$, the length of arc cut off by an angle of $\theta$ radians is $r\theta$ (Fig. 6). For if you enlarge Figure 3 by a factor $r$, the angle $A$ remains the same while all lengths are multiplied by $r$.

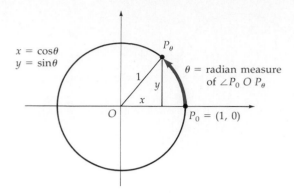

**FIGURE 7**
Sine and cosine defined using unit circle $x^2 + y^2 = 1$.

**The trigonometric functions of numbers** are defined in terms of arc length on the unit circle, the graph of the equation $x^2 + y^2 = 1$ (Fig. 7). Given a real number $\theta$, start at the point $P_0 = (1, 0)$ and go $\theta$ units of arc length *counterclockwise* around the unit circle, arriving at a point $P_\theta = (x, y)$; then

$$x = \cos\theta \quad \text{and} \quad y = \sin\theta. \tag{1}$$

Are these definitions consistent with the previously defined trigonometric functions of angles, using radian measure? Yes; for when $0 < \theta < \dfrac{\pi}{2}$, then $P_\theta$ is in the first quadrant, and $\theta$ is the radian measure of angle $P_\theta O P_0$. Moreover,

$$\cos(\angle P_\theta O P_0) = \frac{\text{adj}}{\text{hyp}} = \frac{x}{1} = x = \cos\theta$$

so the cosine of the number $\theta$ equals the cosine of an angle with radian measure $\theta$. The same holds for the sine,

$$\sin(\angle P_\theta O P_0) = \frac{\text{opp}}{\text{hyp}} = \frac{y}{1} = y = \sin\theta,$$

and for the other trigonometric functions as well.

When $\theta$ is negative, $\sin\theta$ and $\cos\theta$ are defined by means of a sign convention: Go $|\theta|$ units of arc length *backward* (that is, clockwise) around the unit circle, arriving at a point $P_\theta = (x, y)$; then again, by definition, $\cos\theta = x$ and $\sin\theta = y$.

**Table 1**

| $\theta$ | 0 | $\pi/6$ | $\pi/4$ | $\pi/3$ | $\pi/2$ | $3\pi/4$ | $\pi$ | $-\pi/6$ | $-\pi/2$ | $-\pi$ |
|---|---|---|---|---|---|---|---|---|---|---|
| $\cos\theta$ | 1 | $\sqrt{3}/2$ | $1/\sqrt{2}$ | $1/2$ | 0 | $-1/\sqrt{2}$ | $-1$ | $\sqrt{3}/2$ | 0 | $-1$ |
| $\sin\theta$ | 0 | $1/2$ | $1/\sqrt{2}$ | $\sqrt{3}/2$ | 1 | $1/\sqrt{2}$ | 0 | $-1/2$ | $-1$ | 0 |

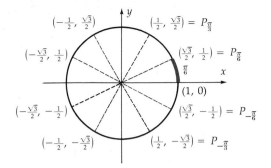

**FIGURE 8**
Coordinates corresponding to integer multiples of $\pi/6$.

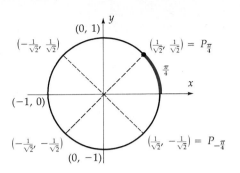

**FIGURE 9**
Coordinates corresponding to integer multiples of $\pi/4$

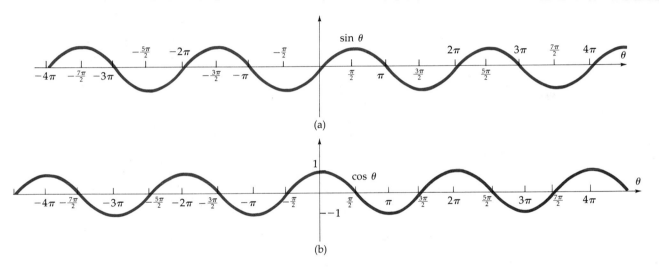

(a)

(b)

**FIGURE 10**

The values in Table 1 can all be read off from Figures 8 and 9. The graphs in Figure 10 are obtained by plotting points like those in the table. Note particularly that

$$\sin 0 = 0 \qquad \sin \frac{\pi}{2} = 1$$

$$\cos 0 = 1 \qquad \cos \frac{\pi}{2} = 0.$$

The other trigonometric functions are defined in terms of sine and cosine:

$$\tan \theta = \frac{\sin \theta}{\cos \theta} \qquad \cot \theta = \frac{\cos \theta}{\sin \theta}$$

$$\sec \theta = \frac{1}{\cos \theta} \qquad \csc \theta = \frac{1}{\sin \theta}.$$

Figure 11 shows the graph of $\tan \theta$; notice the vertical asymptotes at the points where $\cos \theta = 0$.

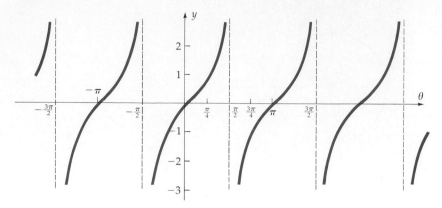

**FIGURE 11**
$y = \tan \theta$

## Trigonometric Identities

The following equations are called identities, since they hold for all values of the numbers $\theta$ and $\varphi$.

**Odd functions:**

$$\sin(-\theta) = -\sin \theta \qquad \csc(-\theta) = -\csc(\theta)$$

$$\tan(-\theta) = -\tan \theta \qquad \cot(-\theta) = -\cot \theta$$

**Even functions:**

$$\cos(-\theta) = \cos \theta \qquad \sec(-\theta) = \sec \theta$$

Figure 12 shows why the sine is odd and the cosine is even. To reach point $P_{-\theta} = (\cos(-\theta), \sin(-\theta))$, go the same distance as to $P_\theta$, but in the *opposite* direction. The two points $P_\theta$ and $P_{-\theta}$ have the same $x$-coordinate, so

$$\cos(-\theta) = \cos \theta;$$

and the $y$-coordinates differ only in sign, so

$$\sin(-\theta) = -\sin \theta.$$

These relations for sine and cosine imply those for the other functions.

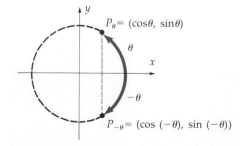

**FIGURE 12**
$\cos(-\theta) = \cos \theta$ and
$\sin(-\theta) = -\sin \theta.$

The graphs in Figure 10 confirm that sine is odd, and cosine even.

**Periodicity:**

$$\sin(\theta + 2\pi) = \sin\theta \qquad \cos(\theta + 2\pi) = \cos\theta$$

$$\tan(\theta + \pi) = \tan\theta \qquad \cot(\theta + \pi) = \cot\theta$$

$$\sec(\theta + 2\pi) = \sec\theta \qquad \csc(\theta + 2\pi) = \csc\theta$$

Figure 13 shows why sine and cosine have period $2\pi$. If you go first to point $P_\theta$, then continue around the circle another $2\pi$ radians, you return to the same point; so $P_{\theta+2\pi} = P_\theta$. Hence

$$(\cos(\theta + 2\pi),\ \sin(\theta + 2\pi)) = (\cos\theta,\ \sin\theta),$$

which means that $\cos(\theta + 2\pi) = \cos\theta$ and $\sin(\theta + 2\pi) = \sin\theta$. This periodicity is seen in the graphs (Fig. 10); the basic shape for $0 \le \theta \le 2\pi$ is repeated ad infinitum in both directions.

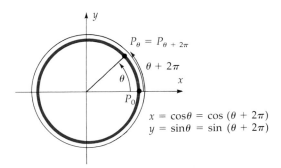

**FIGURE 13**
Arc $\theta$ goes directly from $P_0$ to $P_\theta$; arc $\theta + 2\pi$ goes from $(1, 0)$ once around the circle, then on to $P_\theta$.

The graph of $y = \tan\theta$ (Fig. 11) shows a period $\pi$; the shape for $-\dfrac{\pi}{2} < \theta < \dfrac{\pi}{2}$ is repeated ad infinitum in both directions. The proof that $\tan(\theta + \pi) = \tan\theta$ is left to problem 3.

**Cofunction identities:**

$$\sin\left(\frac{\pi}{2} - \theta\right) = \cos\theta \qquad \cos\left(\frac{\pi}{2} - \theta\right) = \sin\theta$$

$$\tan\left(\frac{\pi}{2} - \theta\right) = \cot\theta \qquad \cot\left(\frac{\pi}{2} - \theta\right) = \tan\theta$$

$$\sec\left(\frac{\pi}{2} - \theta\right) = \csc\theta \qquad \csc\left(\frac{\pi}{2} - \theta\right) = \sec\theta$$

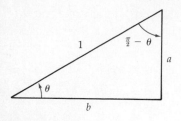

**FIGURE 14**
Cofunction identities:

$$\sin\left(\frac{\pi}{2} - \theta\right) = \frac{b}{1} = \cos\theta;$$

$$\cos\left(\frac{\pi}{2} - \theta\right) = \frac{a}{1} = \sin\theta.$$

Figure 14 illustrates these. Since $\pi/2$ is the radian measure of a right angle, the complement of $\theta$ is $\pi/2 - \theta$; and the sine of the complement of $\theta$ is the cosine of $\theta$, with similar relations holding for the other pairs of functions.

**Pythagorean identities:**

$$\sin^2\theta + \cos^2\theta = 1$$

$$\tan^2\theta + 1 = \sec^2\theta$$

$$1 + \cot^2\theta = \csc^2\theta$$

[Note: $\sin^2\theta$ means $(\sin\theta)^2$.] These identities follow by applying the Pythagorean Theorem to the right triangle in Figure 7, with sides $x = \cos\theta$, $y = \sin\theta$, and hypotenuse 1.

**Addition formulas:**

$$\cos(\theta + \phi) = \cos\theta\cos\phi - \sin\theta\sin\phi$$

$$\sin(\theta + \phi) = \sin\theta\cos\phi + \cos\theta\sin\phi$$

$$\tan(\theta + \varphi) = \frac{\tan\theta + \tan\varphi}{1 - \tan\theta \cdot \tan\varphi}$$

Proofs of these very important formulas are outlined in the problems.

**Double angles:** Set $\varphi = \theta$ in the addition formulas, and you find

$$\sin 2\theta = 2\sin\theta\cos\theta, \qquad \tan 2\theta = \frac{2\tan\theta}{1 - \tan^2\theta}$$

$$\cos 2\theta = \cos^2\theta - \sin^2\theta = 1 - 2\sin^2\theta = 2\cos^2\theta - 1.$$

**Half angles:** Set $\theta = \varphi/2$ in the last two formulas for $\cos 2\theta$, finding

$$\sin\frac{\varphi}{2} = \pm\sqrt{\frac{1 - \cos\varphi}{2}}, \qquad \cos\frac{\varphi}{2} = \pm\sqrt{\frac{1 + \cos\varphi}{2}}.$$

For $\tan\dfrac{\varphi}{2}$ there is a surprisingly simple formula:

$$\tan\frac{\varphi}{2} = \frac{\sin\varphi}{1 + \cos\varphi}$$

(see problem 9).

**The Law of Cosines** (Fig. 15) is like the Pythagorean Theorem, but it applies not just to right triangles:

$$c^2 = a^2 + b^2 - 2ab\cos\theta;$$

$$\theta = \text{angle opposite side } c.$$

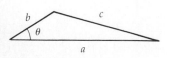

**FIGURE 15**
Law of cosines,
$c^2 = a^2 + b^2 - 2ab\cos\theta.$

Problem 11 outlines a proof.

## PROBLEMS

### A

*1. From Figure 8, evaluate the sine and cosine of:
$2\pi$, $3\pi$, $-\pi$, $-3\pi/2$, $5\pi/6$, $7\pi/6$, $-2\pi/3$.

2. From Figure 9, evaluate the sine and cosine of:
$5\pi/4$, $7\pi/4$, $-3\pi/4$, $-5\pi/4$, $-11\pi/4$.

3. From the addition formulas and Table 1, derive formulas for
   *a) $\sin(\theta + \pi)$       *b) $\cos(\theta + \pi)$
   *c) $\tan(\theta + \pi)$        d) $\sin 2\theta$
   e) $\cos 2\theta$              f) $\tan 2\theta$

   g) $\sin\left(\dfrac{\pi}{2} - \theta\right)$     h) $\cos\left(\dfrac{\pi}{2} - \theta\right)$.

4. Use the addition formulas, and the facts that sine is odd and cosine is even, to prove that
   a) $\sin(\theta - \varphi) = \sin\theta\cos\varphi - \cos\theta\sin\varphi$.
   b) $\cos(\theta - \varphi) = \cos\theta\cos\varphi + \sin\theta\sin\varphi$.

*5. Express $\tan(\alpha - \beta)$ in terms of $\tan\alpha$ and $\tan\beta$.
   (Write $\tan\theta = \dfrac{\sin\theta}{\cos\theta}$, and use the previous problem.)

6. It is easy to fall into the trap of thinking that $\sin(\theta + \varphi)$ equals $\sin\theta + \sin\varphi$. Compute $\sin(\theta + \varphi)$ and $\sin\theta + \sin\varphi$, and compare the results, for the following choices of $\theta$ and $\varphi$:
   a) $\theta = \varphi = \pi/2$.
   b) $\theta = \pi/4$, $\varphi = \pi/2$.

7. Sketch the graphs of the following functions, indicating all zeroes and asymptotes.
   a) $\sec\theta$               b) $\cot\theta$
   c) $\csc\theta$              d) $2\cos 3x$,
   e) $\cos(x/3)$              f) $\tan 2x$
   g) $\tan\pi x$              h) $\cot\pi x$
   i) $-\sec 2\pi x$           j) $\frac{1}{2}\cos(x - \pi/4)$
   *k) $\cos(3x - \pi/4)$      l) $1 + \cos\theta$

8. Prove the following:
   a) $2\cos\alpha\cos\beta = \cos(\alpha + \beta) + \cos(\alpha - \beta)$.
   b) $2\sin\alpha\sin\beta = \cos(\alpha - \beta) - \cos(\alpha + \beta)$.
   c) $\sin(\alpha + \beta) - \sin(\alpha - \beta) = \ldots$ (figure it out).
   d) $2\cos^2\theta = 1 + \cos(2\theta)$.
   e) $2\sin^2\theta = \ldots$ (figure it out).

   f) $\sec\left(\dfrac{\pi}{2} - \theta\right) = \csc\theta$.

g) $\csc\left(\dfrac{\pi}{2} - \theta\right) = \sec\theta$.

9. a) Prove that $\dfrac{\sin 2\theta}{1 + \cos 2\theta} = \tan\theta$.

   b) Deduce that $\tan\dfrac{\varphi}{2} = \dfrac{\sin\varphi}{1 + \cos\varphi}$.

   c) Compute $\tan\dfrac{\pi}{8}$ and $\tan\dfrac{\pi}{12}$.

### B

Problem 10 outlines a proof of the "subtraction" formula

$$\cos(\alpha - \beta) = \cos\alpha\cos\beta + \sin\alpha\sin\beta, \quad (*)$$

based on Figure 16. Problem 11 then deduces the addition formulas.

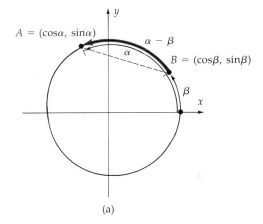

(a)

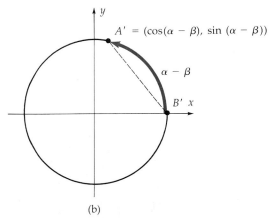

(b)

**FIGURE 16**
Illustrating formula $(*)$.

**10.** In Figure 16a, points $A$ and $B$ are respectively $(\cos \alpha, \sin \alpha)$ and $(\cos \beta, \sin \beta)$. In 16(b), each point has been moved a distance $\beta$ clockwise along the circle; $B$ is moved to $B' = (1, 0)$, and $A$ to $A' = (\cos(\alpha - \beta), \sin(\alpha - \beta))$. Hence

$$|A'B'|^2 = |AB|^2.$$

Use the distance formula to compute each side of this equality, and simplify, to obtain the identity (∗).

**11.** From (∗), deduce:

   **a)** The addition formula for $\cos(\alpha + \beta)$. (Use "odd-even.")

   **b)** The addition formula for $\sin(\alpha + \beta)$.

**12.** (Law of cosines, Fig. 15.)

   **a)** Show that when $\theta = \dfrac{\pi}{2}$, the law of cosines agrees with the Pythagorean Theorem.

   **b)** From Figure 17a, prove the law of cosines when $0 < \theta < \dfrac{\pi}{2}$.

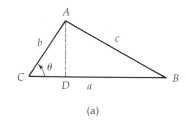

(a)

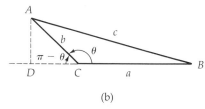

(b)

**c)** From Figure 17(b), prove the law of cosines when $\dfrac{\pi}{2} < \theta < \pi$. [Explain why

$$\cos(\pi - \theta) = -\cos \theta.]$$

## C

The remaining problems show the essential role of the "subtraction formula" (∗) for the trigonometric functions; in fact, one can derive all the other properties from (∗), together with the simple facts that

$$\sin 0 = 0 \qquad \sin(\pm \pi/2) = \pm 1$$
$$\cos 0 = 1 \qquad \cos(\pi/2) = 0. \qquad (**)$$

In each of the remaining problems, assume only (∗), (∗∗), and previous problems from this section C.

**13.** Prove that $1 = \cos^2\theta + \sin^2\theta$.

**14.** Prove that

   **a)** $\cos\left(\dfrac{\pi}{2} - \theta\right) = \sin \theta$.

   **b)** $\sin\left(\dfrac{\pi}{2} - \theta\right) = \cos \theta$.

   **c)** The addition formula for $\sin(\theta + \varphi)$.

**15.** Prove:

   **a)** $\cos(-\theta) = \cos \theta$.

   **b)** $\sin(-\theta) = -\sin \theta$.

   **c)** The addition formula for $\cos(\theta + \varphi)$.

   **d)** The subtraction formula for $\sin(\theta - \varphi)$.

**16.** Prove that

   **a)** $\cos \pi = -1$.

   **b)** $\sin \pi = 0$.

   **c)** $\cos(\theta + \pi) = -\cos \theta$.

   **d)** $\sin(\theta + \pi) = -\sin \theta$.

   **e)** $\cos(\theta + 2\pi) = \cos \theta$.

   **f)** $\sin(\theta + 2\pi) = \sin \theta$.

**FIGURE 17**
Proving the law of cosines. Express $|AD|$ and $|CD|$ in terms of $b$ and $\theta$.

# SELECTED ANSWERS

## Section 1.1

**3a.** $\xrightarrow[\quad 1 \quad\quad 3 \quad]{}$   **3c.** $\xrightarrow[\; -1 \quad\quad 1 \;]{}$   **4a.** $(1, 3]$   **4c.** $[0, 5]$   **4e.** $[-1, 1]$   **5a.** $\{x: -1 \le x < 2\}$

**5c.** $\{x: x^3 + 3x + 1 = 0\}$

**7.** $|P_1P_2| = \sqrt{5}$ in all parts; $m = 2$ in parts a, e, and f; $m = -2$ in parts c and d; $m = 1/2$ in part b

**9a.** center $(-2, 3)$, radius 2   **10.** $\sqrt{10} + \sqrt{13} \approx 6.77 < 7.08$, the length in Example 2

**11.** $\dfrac{12}{50} + \dfrac{\sqrt{34}}{16} \approx 0.604 < 0.6125$, the shortest time in Example 3   **12a.** $4272   **12b.** $3100

## Section 1.2

**1b.** $\pm\sqrt{3}$   **2g.** $m = -1$

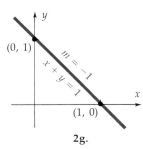

(0, 1)

$x + y = 1$

$m = -1$

(1, 0)

2g.

**3a.** $y = x$   **3b.** $y = -2x + 3$   **3c.** $y = -x$
**4a.** $y = 2x$   **4b.** $y = 5 - x$   **4c.** $y = x/2$

**5a.** $y = \frac{1}{2}x + 2\frac{1}{2}$   **5b.** $y = -\frac{2}{3}x + \frac{4}{3}$   **5c.** $y = 2$   **6a.** $y = -\frac{1}{2}x + \frac{11}{2}$   **6b.** $y = \frac{2}{3}x + \frac{13}{3}$   **7b.** $y = 2$
**8a.** $x = -1$   **8b.** $t = 1/2$   **9a.** $y = x/2$   **11a.** yes   **11b.** yes; yes   **14a.** $10 + (.15)x$   **14b.** $.15 + 10/x$
**15.** $V = 5000 - 500t$; $m = -500$; $V$ decreases when $t$ increases   **16.** $y \approx 0.0541$   **17.** $\ln(7.07) \approx 1.9557$

## Section 1.3

**1a.** $\frac{1}{2}$   **1c.** UND   **1d.** $f(x)$   **1e.** $1/\sqrt{4 - a^2}$   **2b.** Domain $x \ge -1$, range $y \ge 0$
**2c.** Domain $x > -1$, range $y > 0$   **2d.** Domain $x \ne 1$, range $y \ne 0$   **3.** $f(0) = 0$; $f$ is even.

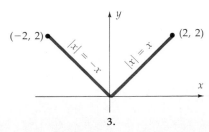

$(-2, 2)$

$|x| = -x$

$|x| = x$

$(2, 2)$

3.

**4a.** Domain [0, 2], range [−1, 1]   **4b.** Domain [−1, 1], range [−2, 2]   **4c.** Domain [0, 4), range {0, 1, 2, 3}
**4d.** Domain (−∞, ∞), range (−π, π)   **5a.** odd   **5c.** neither   **5e.** odd   **5g.** neither   **5i.** even   **7a.** $1 + x^3$
**7b.** $(1 + x)^3$   **7d.** $x$   **7e.** $x^3$   **7f.** (ice)$^3$   **10a.** 4   **11.** $D(x) = \sqrt{x^2 + x^4} = |x|\sqrt{1 + x^2}$   **13.** $s(x) = 7x/11$
**14.** $f = g \neq h$

## Section 1.4

**1a.** $y = \frac{1}{2}(20 - x)$   **1b.** $A = \dfrac{x}{2}(20 - x)$   **1d.** $x = 10, y = 5$   **3a.** $S = 2x^2 + 32/x, 0 < x$

**3b.** In this case, the optimal box appears to be a cube: $x = h = 2$.   **5a.** $S = 2\pi r^2 + 200/r$   **5b.** $r \approx 2.5, h \approx 5$

**7a.** $f(x) = \dfrac{15 - x}{50} + \dfrac{\sqrt{25 + x^2}}{20}, 0 \leq x \leq 15$   **7b.** Optimal $x$ slightly greater than 2.

## Chapter 1 Review

**1.**

   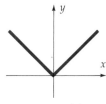   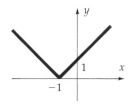

**1a.** $y = x + 1$.   **1b.** $y = |x|$.   **1c.** $y = |x + 1|$.

**2a.** (0, 2]   **2b.** [−1, 1]   **2c.** (−3, 1)   **3a.** $\{x: -1 \leq x < 1\}$   **3b.** $\{x: x^3 + 3x + 1 = 0\}$
**4.** $\sqrt{x^2 + 2x + 5}$   **5a.** $y = 2x - 1$   **5b.** $y = -x + 7$   **5c.** $y = -x + 4$   **6a.** $y = -\frac{1}{2}x$   **6b.** $y = 2x$
**7.** $y = -\frac{1}{2}x + 3/4$   **8a.** $x \neq 1, x \neq -2$   **8b.** $|x| \leq 2$   **8c.** $|x| > 2$   **9a.** odd   **9b,d,f.** even   **9c,e,g.** neither
**10a.** $f(x + 1) = x^3 + 3x^2 + 5x + 3$   **10b.** $f(x) + 1 = x^3 + 2x + 1$   **10c.** $f(g(x)) = (\sqrt{1 + x})^3 + 2\sqrt{1 + x}$
**10d.** $g(f(x)) = \sqrt{1 + x^3 + 2x}$   **10e.** $f(x + h) - f(x) = 3x^2h + 3xh^2 + h^3 + 2h$   **11.** $x/9 + (1/4)\sqrt{x^2 - 10x + 29}$
**12.** $C = 4/h + 16\sqrt{h}$

## Section 2.1

**1a.** $\dfrac{(x + \Delta x)^2 + 2(x + \Delta x) - [x^2 + 2x]}{\Delta x}; 2x + \Delta x + 2; 2x + 2; 6; 0$

**1b.** $\dfrac{(1/2)(x + \Delta x)^2 + 1 - [(1/2)x^2 + 1]}{\Delta x}; x + (1/2)\Delta x; x; 2; -1$   **1d.** $\dfrac{3(x + \Delta x) + 1 - [3x + 1]}{\Delta x}; 3; 3; 3; 3$

**1f.** $f'(x) = 2 - 2x; f'(2) = -2; f'(-1) = 4$   **2a.** $y - 8 = 6(x - 2)$   **2b.** $y - 3 = 2(x - 2)$   **2d.** $y = 3x + 1$
**2e.** $y = 3$   **3b.** $y = -6 - 5(x - 3)$   **4a.** $m$   **4b.** $2ax + b$   **5a.** $\approx -1$   **5b.** $\approx .25$   **5c.** $\approx 1$

## Section 2.2

**1a.** $V = (1/2, 3/4)$   **1b.** $V = (3/2, -13/4)$   **1d.** $V = (5, 25)$   **2a.** $y = 19x - 20$
**3.** 60 meters parallel to the river   **5b.** $x = 2, y = 3/2$   **7b.** $3x^2$   **14a.** $f(x) = 0$ for all $x$ (not actually quadratic)
**14b.** $\frac{3}{2}x^2 - \frac{5}{2}x + 3$

## Section 2.3

**1.** $3x^2 + 2x - 1$   **3.** 0   **4.** 200   **5.** 5   **6.** −6.7   **7.** 459   **9.** $8x - 4$

**13.**

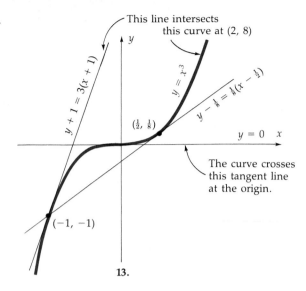

This line intersects this curve at (2, 8)

$y + 1 = 3(x + 1)$

$y = x^3$

$y - \frac{1}{8} = \frac{3}{4}(x - \frac{1}{2})$

$(\frac{1}{2}, \frac{1}{8})$

$y = 0$  $x$

The curve crosses this tangent line at the origin.

$(-1, -1)$

**13.**

## Section 2.4

**2.**

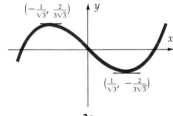

$\left(-\frac{1}{\sqrt{3}}, \frac{2}{3\sqrt{3}}\right)$

$\left(\frac{1}{\sqrt{3}}, -\frac{2}{3\sqrt{3}}\right)$

**2a.**

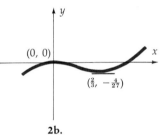

$(0, 0)$

$\left(\frac{2}{3}, -\frac{4}{27}\right)$

**2b.**

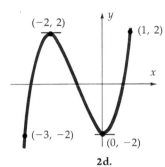

$(-2, 2)$

$(1, 2)$

$(-3, -2)$

$(0, -2)$

**2d.**

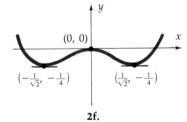

$(0, 0)$

$\left(-\frac{1}{\sqrt{2}}, -\frac{1}{4}\right)$

$\left(\frac{1}{\sqrt{2}}, -\frac{1}{4}\right)$

**2f.**

**4.**  $x = 10, h = 5$; one third  **5.**  $x = h = 10$, one third
**7.**  $r = h = \sqrt{5/\pi}$; 2/3 on sides  **8.**  $x = 28, y = z = 14$

**12a.**  two zeroes, near $-.3$ and 2.

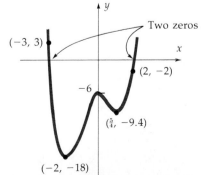

Two zeros

$(-3, 3)$

$(2, -2)$

$-6$

$\left(\frac{5}{4}, -9.4\right)$

$(-2, -18)$

**12a.**

## Section 2.5

**1a.** 0   **1c.** $-6t$   **1e.** $20x^3 - 16$   **1g.** 2
**2a.** $f'' > 0$ for $x < 0$ and for $x > 1$; inflection points at $(0, 1)$ and $(1, 0)$; stationary point at $x = 2$
**2b.** stationary points (approximately): $x = -1.4$, 2, 4; inflection points (approximately): $(-2, -.2)$, $(0, 1)$, $(3, 1.5)$, $(4, 1)$, $(5, .5)$

**4.**

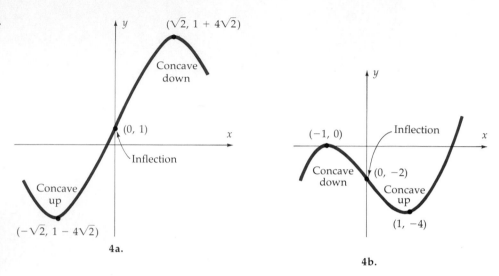

4a.

4b.

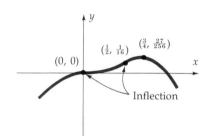

4c.

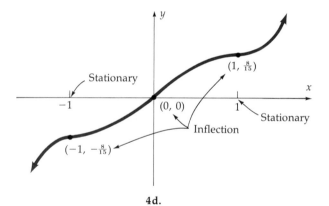

4d.

## Section 2.6

**2a.** $x_1 = -2.5$, $x_2 = -2.3725$, $x_3 = -2.36155$   **2c.** $x_1 = 4/3$, $x_2 = 1.2639$, $x_3 = 1.25993$, $[2^{1/3} = 1.25992]$
**3b.** $1.53 < \bar{x} < 1.54$   **5a.** $E = v^3 - 24.65887v^2 + 3.592v - 0.15327 = 0$
**5b.** $E$ is cubic, and negative at both critical points $\approx 0.073$, 16.4, so $E(v) = 0$ only for $v > 16.4$. Trials give $E(24) < 0$, $E(25) > 0$.
**5c.** $\bar{v} \approx 24.513$

## Section 2.7

**1c.** $v(2) = -64$; $v_{av} = -63.84$   **2b.** $v(1) = -9.8$; $v_{av} = -9.8$
**3a.** up: $t < \frac{1}{2}$; down: $t > \frac{1}{2}$; $y_{max} = 10$; $v(0) = 16$; lands at $t = \frac{1}{2} + \sqrt{10}/4$, with $v = -8\sqrt{10}$
**3b.** up: $0 \le t < 1/\sqrt{3}$; down: $t > 1/\sqrt{3}$; $y_{max} = 2/(3\sqrt{3})$; $v(0) = 1$; lands at $t = 1$, $v(1) = -2$

**4a.** stopped: $t = 3$; to right: $t > 3$; to left: $t < 3$; farthest left: $x(3) = 0$
**4b.** stopped: $t = 0, \pm 3$; to right: $t > 3$ and $-3 < t < 0$; to left: $t < -3$ and $0 < t < 3$; farthest left for $t \geq 0$:
$x(3) = -56$
**6.** $v = -980t$, $a = -980$ **8.** (a) ↔ (iv); (b) ↔ (iii); (c) ↔ (ii); (d) ↔ (i) **11a.** speeding up when $t > 5/16$

## Section 2.8

**1a.** $x^3/3 + x + C$ **1e.** $-16t^2 + 10t + C$ **1g.** $u^{101}/101 + C$ **2a.** $F(x) = (1/3)x^3$
**2b.** $F(x) = (1/3)x^3 + (1/2)x^2 + 7/6$ **3.** $v(\sqrt{6}) = 2\sqrt{6}$ **5.** $y = 1 + 4t - 4.9t^2$; $y_{\max} = 1 + 8/(9.8) \approx 1.816$ meters
**6.** $y = 1 + 4t - 0.8t^2$; $y_{\max} = y(5/2) = 6$ meters! In the air for 5 seconds!

**8.**

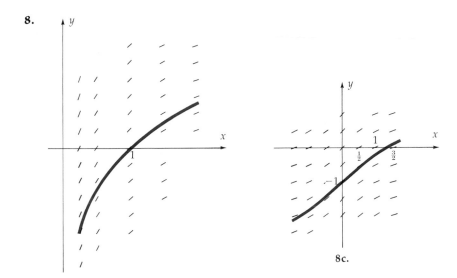

8a.

8c.

**11a.** $y = \frac{3}{4}t^3(t^2 - 5t + \frac{20}{3})$ **11d.** $a_{\max} = 10/\sqrt{3}$

## Chapter 2 Review

**1a.** $\dfrac{\Delta y}{\Delta x} = 4x + \Delta x - 1 \to 4x - 1$ as $\Delta x \to 0$ **1b.** $-2/(2x + 1)^2$ **2a.** $y = 8x + 5$ **2b.** $y = 1 - 2x$
**3a.** stationary at $x = \frac{1}{2}$; concave up **3b.** stationary at $x = 0, -2/3$; inflection at $x = -1/3$
**3c.** stationary at $x = 0, \pm 1$; inflection at $x = \pm 1/\sqrt{3}$ **4.** $r = 1/\sqrt{\pi}$, $h = 5/\sqrt{\pi}$ **5a.** 3 **5b.** 1 **5c.** 2 **5d.** 3
**6a.** $\bar{x} \approx -.6823$ **6b.** $f(-.6823) > 0$, $f(-.6824) < 0$
**7.** $a = 12t - 6$; to left when $0 < t < 1$; stationary when $t = 0, 1$; $v \geq -3/2$; $v(t) \leq 12$, $0 \leq t \leq 2$; speeding up when
$0 < t < \frac{1}{2}$ and when $t > 1$
**8.** $F(x) = x^5/5 - x^2/2 + 3.3$ **9.** $y = 1 + 2t - 3t^2/2$ **10.** $y = t^3/3 + 2t - 7/3$

## Section 3.1

**1a.** 3 **1c.** DNE **2a.** 0 **2c.** DNE **3a.** 2 **3c.** 2 **3e.** $-1$ **4a.** 1 **4b.** DNE **5c.** 0 **7a.** 5
**7b.** 1 **7c.** DNE **7d.** 5 **9a.** DNE **9b.** 0 **10a.** 3 **10b.** DNE **10c.** 1/2 **11c.** DNE **11d.** 1
**13a.** 6 **13c.** 1/2 **13e.** 3 **14a.** 1 **14b.** 0 **15a.** 1

## Section 3.2

**1.**

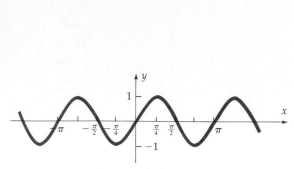

**1a.** $y = \sin 2x$.

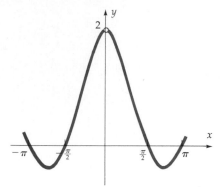

**1c.** $y = \dfrac{\sin 2x}{x},\ |x| < \pi.$

**1b.** 2 **3a.** $\frac{1}{2}$ **3b.** 1 **4a.** 5 **4b.** $\frac{1}{3}$ **5b.** 0 **6.**

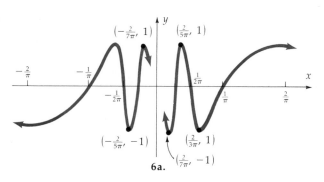

**6a.**

**6b.** odd **6c.** no **8a.** 0 **9a.** 1 **9c.** 0 **10a.** 0.01745 **10b.** 0.84147 **12a.** 0 **12b.** $\frac{1}{2}$

## Section 3.3

**1a.** $-1;\ 1$ **1b.** $0;\ -\infty$ **1c.** DNE; $+\infty$ **3.** $1/2$ **5.** 0 **7.** $+\infty;\ -\infty$ **9.** $2/3$ **11.** 1 **12.** 0
**14.** $+1;\ -1$ **16.** $+1;\ -1$ **18.** 0 **20.** 1 **24a.** 0 **24b.** 0 **24c.** 1 **24d.** 0 **29a.** 0 **29b.** 0

## Section 3.4

**1a.** 1 **1b.** $+\infty$ **1c.** $-\infty$ **1d.** $-\infty$ **2a.** jump at $x = 2$ **2b.** jumps at all the integers
**2c.** oscillatory at $x = a$ **3.** (We list the zeroes, then the discontinuities.) **3a.** none; $-1$ **3b.** none; 0

**3c.** $0;\ -2$ **3d.** 0; none **3g.** 1; 0, 2 **3h.** none; 0, $-1$ **3i.** $-1/2;\ 0,\ -1$ **4a.** $\overset{-\quad +}{\underset{-1}{\rule{1cm}{0.4pt}{+}}}$ **4b.** $\overset{+\quad +}{\underset{0}{\rule{1cm}{0.4pt}}}$

**4c.** $\overset{+\ -\ +}{\underset{-2\ \ 0}{\rule{1.4cm}{0.4pt}}}$ **4d.** $\overset{-\quad +}{\underset{0}{\rule{1cm}{0.4pt}}}$ **4g.** $\overset{-\ +\ -\ +}{\underset{0\ \ 1\ \ 2}{\rule{1.6cm}{0.4pt}}}$ **4h.** $\overset{+\ -\ +}{\underset{-1\ \ 0}{\rule{1.4cm}{0.4pt}}}$ **4i.** $\overset{-\ +\ -\ +}{\underset{-1\ -\frac{1}{2}\ 0}{\rule{1.6cm}{0.4pt}}}$
**5.** (At each discontinuity, we give the limit from the left and from the right.)
**5a.** at $-1$: $-\infty,\ +\infty$ **5b.** at 0: $+\infty,\ +\infty$ **5c.** at $-2$: $+\infty,\ -\infty$ **5d.** no discontinuities
**5g.** at 0: 0, 0; at 2: $-\infty,\ +\infty$ **5h.** at 0: $-\infty,\ +\infty$; at $-1$: $+\infty,\ -\infty$ **5i.** at 0: $-\infty,\ +\infty$; at $-1$: $-\infty,\ +\infty$
**6.** (We give the limit as $x \to -\infty$; as $x \to +\infty$.) **6a.** 0; 0 **6b.** 0; 0 **6c.** 1; 1 **6d.** 0; 0 **6g.** $-\infty;\ +\infty$
**6h.** 0; 0 **6i.** 0; 0 **12.** jumps at $t = 0$ and $t = 3$; continuous at $t = 2$

## Section 3.5

**1a.** $-3$ **1c.** 1 **1e.** 5 **1f.** $-2$ **3a.** $(2x + 3)/5$ **4a.** $(x + 1)^{-2}$ **4b.** $1/3$ **4c.** $-3/x^2$

**4e.** $7(1-x)^{-2}$ **4h.** $2x - 2x^{-3}$ **4i.** $1 - x^{-2}$ **4j.** $300x^{-101}$ **4l.** $-2x(x^2+1)^{-2}$ **5l.** $(6x^2 - 2)(x^2+1)^{-3}$

**6a.** $5/(3x-2)^2$ **6b.** $(x^4 - 6x^2 - 4x - 3)/(x^2 - 1)^2$ **6d.** $3x^2 \dfrac{x^2 - 2}{(x+1)^2}$ **7a.** $-10x^{-3}$ **7c.** $-x^{-2} - 2x^{-3}$

**8a.** $y - \frac{1}{2} = \frac{1}{4}(x-1)$ **9.**

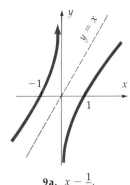

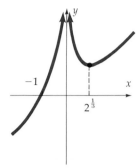

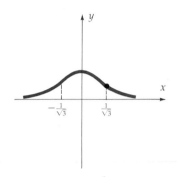

**9a.** $x - \dfrac{1}{x}$. **9c.** $x + \dfrac{1}{x^2}$. **9e.** $\dfrac{1}{x^2 + 1}$.

**12a.** $(1000/3)^{1/2}$ by $(3000)^{1/2}$ **12b.** $1/2$ **13b.** $1/2$ **14.** $4\sqrt{10}$ by $10\sqrt{10}$ **16.** diameter = height

## Section 3.6

**2a.** $\cos\theta - \sin\theta$ **2c.** $-(\csc^2\theta)/5$ **2e.** $-10\csc\theta\,\dfrac{\theta\cot\theta + 1}{\theta^2}$ **2h.** $-\dfrac{\theta(\csc\theta\cot\theta + \csc^2\theta) + 2(\csc\theta + \cot\theta)}{\theta^3}$

**3.** The second derivatives are: **3a.** $-2\sin x - x\cos x$ **3c.** $-4\sin x\cos x$

**3e.** $5\dfrac{\sin^2\theta + 2\cos\theta(\theta\cos\theta - \sin\theta)}{\sin^3\theta}$ **4a.** $f = -f'' = f^{(4)}$ **6a.** $A = \sin\theta(1 + \cos\theta)$ **6b.** $\theta = \pi/3$

**9a.** $2\cos 2\theta$ **10a.** $a\sec^2 a\theta$ **12b.** $.47916 < \sin(\frac{1}{2}) < .47943$

## Section 3.7

**2a.** $f(g(x))$ **2b.** $g(f(x))$ **2c.** $h(f(x))$ **2d.** $f(h(x))$ **3a.** $f(u) = 1/u,\ g(x) = 2x + 1$

**3c.** $f(u) = u^{1/4},\ g(x) = x^2 + 2x + 2$ **3e.** $f(u) = \sqrt{u},\ g(\theta) = \sin\theta$ **4a.** $4(2x+1)$ **5a.** $-2x\sin(1 + x^2)$

**5c.** $1/(2+x)^2$ **6a.** $6(2x+1)^2$ **6b.** $-2(1-x)$ **6c.** $10x(x^2-1)^4$ **6e.** $4(3x^2-1)(x^3 - x + 1)^3$ **7.** $y'' =:$

**7a.** $2\cos x - x\sin x$ **7c.** $2(\sin^2 t - \cos^2 t)$ **7e.** $-\cos(\sin x)\cdot\cos^2 x + \sin(\sin x)\cdot\sin x$ **7g.** $18\sec^2 3x\tan 3x$

**7k.** $x^{-3}(-x^2\sin x - 2x\cos x + 2\sin x)$ **8a.** $\dfrac{-4x}{(x^2+1)^2}$ **8b.** $\dfrac{-30x}{(x^2+1)^4}$ **12a.** $y'' = -9y$

**15.** $f^{(81)}(t) = -8\cdot 3^{81}\cos(3t + 1);\ f^{(402)}(t) = -8\cdot 3^{402}\sin(3t + 1)$

## Section 3.8

**1a.** odd **1b.** even **1c.** neither

**3.**

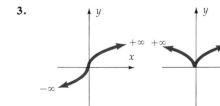

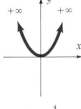

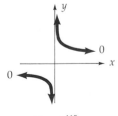

**3a.** $x^{\frac{1}{3}}$. **3b.** $x^{\frac{2}{3}}$. **3c.** $x^{\frac{4}{3}}$. **3f.** $x^{-115}$.

**4a.** $y'' = x(x^2+1)^{-3/2}(2x^2+3)$ **4b.** $-(2/9)(x^2+3)(x^2-1)^{-5/3}$

**5.**

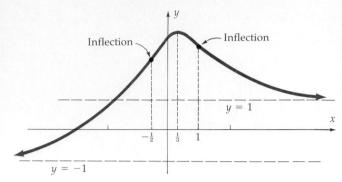

**5a.** $y = (x + 3)(x^2 + 1)^{-\frac{1}{2}}$.

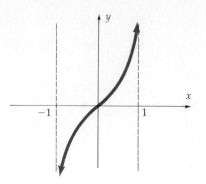

**5c.** $y = x(1 - x^2)^{-\frac{1}{2}}$.

**6a.** $-\dfrac{2x + 3y^4}{1 + 12xy^3}$  **6d.** $\dfrac{26x(x^2 + 2y^2)^{12}}{1 - 52y(x^2 + 2y^2)^{12}}$  **6f.** $\dfrac{y \cos xy}{1 - x \cos xy}$

**7a.** $y - 1 = \dfrac{-7}{25}(x - 2)$  **8a.** $y - 6/7 = (-17/20)(x + 12/7)$  **9b.** $\dfrac{dy}{dx} = y^3/x^3$

**10b.** $y - 1/\sqrt{2} = (3/2)^{3/2}(x - 1/\sqrt{3})$  **11b.** $27/4\sqrt{2}$

**13.** $\dfrac{dy}{dx} = 0$ at $(-2^{1/3}, -2^{2/3})$; $\dfrac{dy}{dx} = \infty$ at $(-2^{2/3}, -2^{1/3})$; $\dfrac{dx}{dy}$ and $\dfrac{dy}{dx}$ undefined at $(0, 0)$

## Chapter 3 Review

**1a.** $\lim\limits_{x \to 0^+} f(x) = +\infty$; $\lim\limits_{x \to 0^-} f(x) = -\infty$; $\lim\limits_{x \to 2} f(x) = 1/2$ ("hole")

**1b.** $\lim\limits_{x \to 0^\pm} f(x) = -\infty$; $\lim\limits_{x \to 2^+} f(x) = 1/2$; $\lim\limits_{x \to 2^-} f(x) = -1/2$ ("jump")

**1c.** $\lim\limits_{x \to 1} g(x) = 1/2$ ("hole"); $\lim\limits_{x \to -1^-} g = -\infty$; $\lim\limits_{x \to -1^+} g = +\infty$

**3a.** $(t + 1)^{-1/2} + t(-1/2)(t + 1)^{-3/2} = (t + 1)^{-3/2}(t/2 + 1)$  **3b.** $(1 + (2x + 1)^{-1/2})\cos(x + \sqrt{2x + 1})$

**3c.** $\dfrac{1 + 1/2\sqrt{z}}{2\sqrt{z + \sqrt{z}}}$  **3d.** $-10 \cos(5\theta + 1) \sin(5\theta + 1)$  **3e.** $3 \sec^2 3t - 3 \csc^2 3t$  **3g.** $\theta^{-2}(2\theta \cos 2\theta - \sin 2\theta)$

**4a.** $(1 - x^2)/(1 + x^2)^2$  **4b.** $\dfrac{1 - x^2}{2x^{1/2}(1 + x^2)^{3/2}}$  **4d.** $-\dfrac{x}{y} \cdot \dfrac{5x^3 + 2y^2}{5y^4 + 2x^2}$  **5a.** $y - 3 = 8(x - 1)$

**5b.** $y - 3 = (-3/14)(x - 1)$  **6a.** $4/3$  **7.** $(40/3)$ by $(40/3)$ by $(10/3)$  **8.** $r = 10(4\pi)^{-1/3}$, $h = 40(4\pi)^{-1/3}$

**9.**

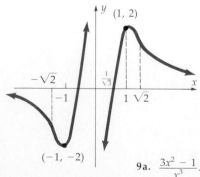

**9a.** $\dfrac{3x^2 - 1}{x^3}$.

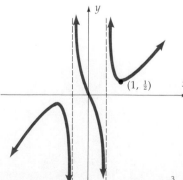

**9b.** $\dfrac{x^3}{3x^2 - 1}$.

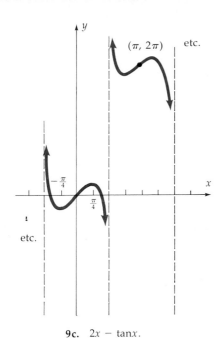

**9c.** $2x - \tan x$.

**10a.** Example: graph of $y = (x - 1)^2$ **10b.** impossible **10c.** impossible **10d.** possible; $f$ discontinuous at 0

## Section 4.1

**1a.** $4 - \frac{1}{8}$ **1b.** $7 + \frac{1}{14}$ **2a.** $-.03$ **2b.** $-.029701$ **6a.** $-1.6\pi$ **6b.** twice as large; $-1/40$ versus $-1/80$
**8a,b** $2\pi(10)^2(0.0003)$ **10a.** $20/\sqrt{3}$ m **10b.** $8\pi/27$ **13a.** $-0.1$ **13b.** $0.1n$
**14.** $P' < 0$ (declining), $P'' > 0$ ($P'$ is increasing toward 0) **17a.** $\dfrac{\Delta V}{V} \approx 3\,\dfrac{\Delta r}{r}$ **18b.** $\dfrac{\Delta S}{S} \approx 2\,\dfrac{\Delta x}{x}$ **19a.** $\dfrac{\Delta P}{P} \approx \dfrac{\Delta x}{x}$

## Section 4.2

**1a.** $13/36\pi$ cm/sec **2.** $26/3$ cm$^2$/sec **3a.** $6\sqrt{5}$ m/sec **4a.** $-50\sqrt{2}$ km/hr **5b.** $-4\sqrt{5}$ mph **6.** $40\sqrt{5}$ mph
**9.** $12; 0; -6; -12$ **11.** $8/5$ m/sec **12a.** $8/3$ ft/sec **13a.** $2500$ m$^2$/sec; $\dfrac{125}{\pi}$ m/sec **15a.** $5/(4\pi)$
**16a.** $-0.0006$ m$^3$/hr **17a.** $90/25(29)^2$ **19.** $3\pi r_0$ in$^3$/hr **21a.** $205\pi/12$ m/hr **21b.** $1025\pi^2/288$ m/hr$^2$

## Section 4.3

**1c.** 2 **1d.** 1 **1e.** 3 **1f.** $+\infty$ **1g.** $\frac{1}{2}$ **1h.** 0 **1i.** $1/3$ **1j.** $-8/3$ **1k.** 0 **1l.** $2/3$
**2a.** $dy/dx = 2$ **2b.** $\tan\theta = -\sqrt{4.39}$ **3a.** $(2\frac{1}{2}, 1\frac{1}{2}); dy/dx = 5/3$ **3b.** $t = 1/2, dy/dx = -5/3$

## Section 4.4

**2.** $20 - \sqrt{5/3}$ miles from $C$
**3a.** Along one bank from $A$ to $P$, $500/\sqrt{3}$ meters short of $B'$ opposite $B$, then straight under the river to $B$.
**5.** $\cos\alpha = 2\cos\beta$

## Section 4.5

**3a.** the dashed line
**3b.** It's not clear. With the solid line, fewer people are disadvantaged, but their disadvantage is greater than in the case of the dashed line.

## Chapter 4 Review

**1.** $2 - 1/40$  **2a.** $dS/dt < 0$, $d^2S/dt^2 < 0$  **3a.** $\Delta g \approx (-g^{3/2}/\pi\sqrt{L})\,\Delta T = -2(g/T)\,\Delta T$  **3b.** A short one.
**4a.** Clockwise  **4b.** $16/3$  **5b.** $-y$  **6.** $2/3$ ft/sec  **7.** 30 m; descending; $50/3$ m/sec  **8.** $0.28/\pi$
**9a.** $100\pi/3 - 50 \approx 55$ m/sec $\approx 200$ km/hr  **9b.** $\pi^2/162$  **10.** $x = \max\left\{ r - \dfrac{Sw}{\sqrt{U^2 - S^2}}, 0 \right\}$  **11a.** $10/3$

**11b.** $5/3$  **11c.** 1  **11d.** 1  **11e.** 2  **11f.** 0

## Section 5.1

**1a.** $\bar{x} = \frac{1}{2}$; min $= -\frac{1}{4}$, max $= 0$  **1c.** $\bar{x} = 1$ [$f'(1)$ does not exist]; min $= 0$, max $= 1$
**1e.** $\bar{x} = 0$ [$f'(0)$ does not exist]; min $= -2$, max $= 2$  **2a.** Applies; min $= 0$, max $= 4$

**2b.** does not apply; no min, max $= 4$  **2c.** applies; min $= \frac{1}{2}$, max $= 1$  **2d.** does not apply; no min, max $= 1$
**3a.** 1 m by 1 m; not solved by Critical Point Theorem (CPT), since Extreme Value Theorem (EVT) does not apply.
**3b.** 1 by 1; EVT applies, so CPT gives guaranteed max.  **4.** EVT applies, so max and min found by CPT.
**4a.** $32/3\sqrt{3}$; occurs at critical points $\pm 2/\sqrt{3}$  **4b.** 0; occurs at endpoints  **6.** all possible

**7.** all possible except (b)  **9d.** $\dfrac{x^2}{1 + x^2} \cos x$ is an example.

## Section 5.2

**2a.** $1/4$  **2c.** $1/2$  **2d.** $\pm 1/\sqrt{3}$  **2e.** 1  **3a.** $1/\sqrt{3}$  **3c.** $\sqrt{6}$  **5a.** $1/\sqrt{3}$  **5b.** $\sqrt{3}$  **5c.** $\sqrt{6}$  **5d.** $\pi/2$

## Section 5.3

**3a.** local min at $x = 2^{-1/3}$  **7.** $\sqrt{2}R$ by $\sqrt{2}R$  **9a.** no max; global min is 2, for an isosceles triangle, by Theorem 5
**11.** $h = (600/\pi)^{1/3}$, $r = h/\sqrt{2}$  **12a.** $(0, 0)$  **15.** $A/B = 1/8$
**18a.** $(1, 1)$ and $(-1, -1)$. Slope of segment $= 1$, slope of tangent at $(1, 1) = -1$.
**18b.** $(x_0, 1/x_0)$, where $x_0$ is the unique solution of $x^4 + x - 1 = 0$ in $x > 0$. Slope of segment is $(1 - x_0)/x_0^2$;
slope of tangent is $-1/x_0^2$; the product is $-1$ precisely when $x_0^4 + x_0 - 1 = 0$.
**19a.** highest, $(-40, -16)$; lowest, $(-60, -36)$

## Chapter 5 Review

**3a.** $\pm 1/\sqrt{3}$  **3b.** 1  **3c.** 0

## Section 6.1

**1a.** $9/2$  **1c.** 0  **1d.** $\pi/2$  **1e.** $5/2$  **1f.** 0  **1g.** 1  **1h.** 0  **1i.** 0  **6a.** $\underline{S}_5 = 0.4$, $MS_5 = 0.5$, $\bar{S}_5 = 0.6$
**6b.** $\underline{S}_5 \approx 0.6456$, $MS_5 \approx 0.6919$, $\bar{S}_5 \approx 0.7456$  **6f.** $\underline{S}_4 = -1/2$, $MS_4 = 0$, $\bar{S}_4 = 1/2$  **7a.** $1.006$  **9.** 5,745 m$^2$

## Section 6.2

**1a.** 15  **1c.** 1  **2a.** $\displaystyle\sum_{1}^{5} j$  **2c.** $\displaystyle\sum_{2}^{5} 1/j$  **5a.** 12  **6b.** commutative law for addition  **7a.** $7b^3/3$

## Section 6.3

**1a.** $1/4$  **1b.** $1/5$  **1c.** 0  **1d.** $1/2$  **1e.** $-1/2$  **1f.** $1/2$  **1j.** $\pi^2/2$  **1k.** $0.999$  **1p.** 1  **1s.** $2/3$
**1t.** 0  **2a.** 4  **2b.** $2\sqrt{3} - 2\pi/3$

**4a.** $\underline{S}_4 = 4(5^{-2} + 6^{-2} + 7^{-2} + 8^{-2}) \approx 0.415$, $\displaystyle\int_1^2 \frac{1}{x^2}\, dx = \frac{1}{2}$, $\bar{S}_4 = 4(4^{-2} + 5^{-2} + 6^{-2} + 7^{-2}) \approx 0.6$  **6b.** $4/3$

**6c.** $\sqrt{2} - 1$  **7b.** $3/8$  **9a.** $1/6$

## Section 6.4

**1a.** $-8$  **1b.** $0.4$  **1c.** $0$  **3a.** $0$  **3b.** $60$  **4b.** $-4\pi$  **5a.** $\approx 7/8$  **6a.** $\approx 5/8$ to the left
**7.** $\approx 0.014$ miles  **8a.** $\approx 32$ meters

## Section 6.5

**1a.** $8$  **1b.** $2$  **1c.** $10$  **1d.** $-9$  **1f.** Cannot be determined from given information.  **2a.** $1$  **3.** $4$
**4a.** $1/3;\ 1/\sqrt{3}$  **4c.** $4/3;\ \pm 1/\sqrt{3}$  **4d.** $-2/9;\ -(9/2)^{1/3}$  **6.** $-5$  **14c.** $.995 < \cos(1/10) < .9950042$

## Section 6.6

**1a.** $2$  **1c.** $t^2$  **1e.** $-1$  **1g.** $2x^2$  **2b.** $\cos(-z^2/2)$  **2c.** $\varphi(t)$  **3a.** $2x \sin x^4$  **3c.** $2x \sin x^4 - 3x^2 \sin x^6$

**5a.** $I'(x) = \dfrac{1}{x+1} > 0$ for $x > -1$; $I''(x) = \dfrac{-1}{(x+1)^2} < 0$, so concave down; $I(0) = 0$.

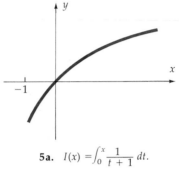

**5a.** $I(x) = \displaystyle\int_0^x \frac{1}{t+1}\,dt.$

## Section 6.7

**1.** $-\dfrac{x^4}{4} + C$  **2.** $\frac{5}{3}t^3 + t + C$  **3.** $\frac{4}{3}t^{3/2} + \frac{9}{4}t^{4/3} + C$  **4.** $2 \sin \theta + 3\theta + C$  **5.** $\tan \theta - \theta + C$  **8.** $6/\pi$
**11.** $1/2$  **13.** $E/6R$  **15.** $\tan \theta + \sec \theta + C$  **18.** $2/\pi$

## Section 6.8

**1a.** $dy = -x^{-2}\,dx$  **1b.** $dy = 6x^2\,dx$  **1g.** $dy = 3 \cos 3\theta\,d\theta$  **1i.** $2t \cot(t/5)\,dt - (t^2/5) \csc^2(t/5)\,dt$
**2a.** $\dfrac{(2x-1)^4}{8} + C$  **2b.** $\dfrac{(x^3+1)^5}{15} + C$  **2d.** $\sqrt{2x+1} + C$  **2f.** $\dfrac{1}{\pi}\sin(\pi t - 3) + C$  **5a.** $\dfrac{-1}{2(x^2+1)} + C$
**5d.** $5/32$  **5e.** $(9^{11}-1)/11$  **5f.** $(1/15)2^{5/2}$  **5i.** $-\frac{1}{2}(x^2 + 2x + 5)^{-1} + C$
**5l.** $\frac{2}{15}(1+x^3)^{5/2} - \frac{2}{9}(1+x^3)^{3/2} + C$  **5n.** $-(1/6)\cos(3\theta^2 + 1) + C$  **5o.** $\frac{2}{5}\tan 5\theta + C$  **5q.** $-(1/3)\cot^3\theta + C$
**6b.** $x = -(1/9)\sin 3t + 4t/3 + 2$  **7a.** $1/\pi$
**14.** With $u = x^2$, $x = \sqrt{u}$ is false in part of the interval of integration $[-1, 1]$.

## Chapter 6 Review

**1a.** $\dfrac{\pi}{3}\left(\dfrac{3}{4} + 1 + \dfrac{3}{4}\right)$  **1b.** $\dfrac{\pi}{3}\left(0 + \dfrac{3}{4} + 0\right)$  **1c.** $\dfrac{\pi}{3}\left(\dfrac{1}{4} + 1 + \dfrac{1}{4}\right)$  **2a.** $(1/10)(1 + \frac{10}{11} + \frac{10}{12} + \cdots + \frac{10}{19}) \approx 0.719$
**2b.** $(\frac{1}{11} + \frac{1}{12} + \cdots + \frac{1}{20}) \approx 0.669$  **2c.** $\frac{1}{10}(\frac{20}{21} + \frac{20}{23} + \cdots + \frac{20}{39}) \approx 0.6928$  **4a.** $-(1/\pi)\cos \pi x + C$
**4c.** $\frac{1}{5}\tan 5\theta + C$  **4e.** $\frac{2}{11} + 2$  **4g.** $\frac{1}{12}(3^{-4} - 1)$  **4i.** $\frac{1}{15}(x^3 + 1)^5 + C$  **5a.** $26$  **5b.** $1/3$
**6a.** $2 + 2^2 + 2^3$  **6b.** $1 + r + r^2 + r^3 + r^4$  **7a.** $\displaystyle\sum_{j=1}^{n} j$  **7b.** $\displaystyle\sum_{j=1}^{m}(1/j)$  **8a.** $1$  **8b.** $0$  **8c.** $1/3$
**9a.** January 10  **9b.** $36{,}500$ calories  **9c.** $100$ cal/day for the year; $100 - \dfrac{35}{31} \cdot \dfrac{365}{\pi}\left(\sin\dfrac{42\pi}{365} + \sin\dfrac{4\pi}{73}\right)$ for January
**10a.** $\approx 1.45$  **10b.** $\approx 1.45/0.8$  **11a.** $F''(x) = -1/x^2$  **11b.** $2(\frac{1}{21} + \frac{1}{23} + \frac{1}{25} + \frac{1}{27} + \frac{1}{29}) \approx 0.4$
**11c.** $-2(\frac{1}{11} + \frac{1}{13} + \frac{1}{15} + \frac{1}{17} + \frac{1}{19}) \approx -0.7$

## Section 7.1

**1a.** $1/6$  **1b.** $1/4$  **1c.** $\frac{9}{2} - 2\sqrt{2}$  **3a.** $2\sqrt{2}/3$  **3c.** $32/3$  **3f.** $1/6$  **3i.** $10/3$  **3k.** $\sqrt{3} - \pi/3$  **3m.** $1/2$
**5.** $\approx 18{,}800 \text{ ft}^2$

## Section 7.2

**1.** cone; $\pi/3$  **3a.** $\dfrac{\pi}{2} a^2 b^2$  **3b.** $1/2$  **4b.** $\frac{2}{3} r^3 \tan \theta$

**5.**

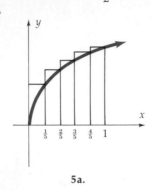

**5a.**

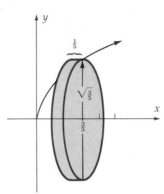

**5b.** $V = \frac{1}{5}\pi(\sqrt{\frac{3}{5}})^2 = 3\pi/25.$

**5c.** $V_5 = \dfrac{\pi}{5}[(1/5) + (2/5) + (3/5) + (4/5) + 1] = 3\pi/5$  **5d.** $V = \int_0^1 \pi x\, dx = \pi/2 < V_5$  **8a.** $\pi(\frac{4}{3}R + \frac{1}{2})$
**10a.** $2\pi R^3/3$  **10c.** $\pi R^2 h$  **10e.** $\int_{-R}^{R} h\sqrt{R^2 - x^2}\, dx = \pi R^2 h/2$  **10g.** $8R^3/3$  **11a.** $\pi e^3/3 \cdot 2^{3/2}$  **11c.** $e^3/\sqrt{6}$
**13.** $\pi h^2\left(r - \dfrac{h}{3}\right)$  **14.** $\dfrac{4\pi}{3}(r^2 - b^2)^{3/2}$  **15b.** $\dfrac{4\pi}{3} ab^2$  **17.** $\approx 720 \text{ cm}^3$

## Section 7.3

**1.** $\pi/64$  **2.** $2\pi/3$  **3a.** $\pi(r_2^2 - r_1^2)h$  **3b.** $\pi(r_2 - r_1)h^2$  **4a.** $8\pi/3$  **4c.** $\pi^2 + 4\pi/3$  **5.** $(2\pi/3)(r^2 - a^2)^{3/2}$
**7a.** $5\pi/6$  **7b.** $13\pi/15$  **8a.** $\pi/6$  **14c.** $3\pi a^3/5$  **17.** $a = V/\pi R^2 - \omega R^2/4g$  **18c.** $\pi p R^4/8kL$

## Section 7.4

**2.** $.009$  **4.** $0.15$ Newton-meter  **6a.** $mg$  **6b.** $\dfrac{mgR}{R + 1}$; $\dfrac{1000mgR}{R + 1000}$; $mgR$  **8.** $\approx (4.6)10^{-18}$ Newton-meter
**9.** $(9.8) \cdot 10^3 \pi (R^2/H^2)(VY^3/3 + Y^4/4)$  **11.** $(9.8)4w\pi R^4/3$  **13b.** $86$ ft-lbs

## Section 7.5

**1.** $1/2$  **3.** $2\sqrt{2}$  **5.** divergent  **7.** $+\infty$  **8.** $+\infty$  **9.** $1$  **11.** $1$  **13.** $\infty$  **16a.** $\infty$  **17a.** $2$

## Chapter 7 Review

**1a.** $3/10$  **1b.** $2\pi/9$  **1c.** $\pi/3$  **1d.** $44\pi/45$  **2.** $76\frac{2}{3}\text{ m}^3$  **3.** $4\pi^2$  **4a.** $7/64$  **4b.** $9/64$  **5.** $1125\pi w/8$
**6a.** $3$  **6b.** $3\pi/5$  **6c.** $\infty$  **8a.** $k/5$  **8b.** $\infty$

## Section 8.1

**1c.** $0.7 > f'(0) > 0.67$  **1d.** $1.16 > f'(0) > 1.104$  **2a.** $2.33 > f'(10) > 2.276$  **3a.** $(2 + 2x)e^x$; $(4 + 2x)e^x$
**3b.** $2xe^{x^2}$; $(4x^2 + 2)e^{x^2}$  **3c.** $2e^{2x}$; $4e^{2x}$  **3d.** $(x^2 + 2x)e^x - 1/x^2$; $(x^2 + 4x + 2)e^x + 2x^{-3}$
**3g.** $15e^{3t}(2 + e^{3t})^4$; $45e^{3t}(2 + e^{3t})^3(2 + 5e^{3t})$  **3l.** $2x \exp(1 + x^2)$; $(4x^2 + 2) \exp(1 + x^2)$
**3n.** $e^x \exp(e^x)$; $(e^x + e^{2x}) \exp(e^x)$  **4a.** $\cos \theta e^{\sin \theta}$; $(\cos^2 \theta - \sin \theta)e^{\sin \theta}$

**4c.** $(a \sin bt + b \cos bt)e^{at}$; $[2ab \cos bt + (a^2 - b^2) \sin bt]e^{at}$ **4e.** $e^{at}(a \sin bt + b \cos bt)$; $e^{at}[(a^2 - b^2) \sin bt + 2ab \cos bt]$

**5.**

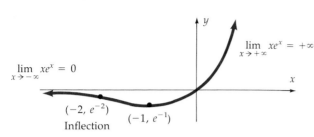

$$\lim_{x \to +\infty} xe^x = +\infty$$
$$\lim_{x \to -\infty} xe^x = 0$$

$(-2, e^{-2})$
Inflection $(-1, e^{-1})$

**5a.** $y = xe^x$.

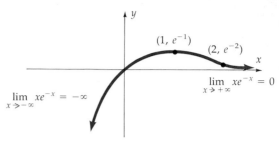

$(1, e^{-1})$
$(2, e^{-2})$
$$\lim_{x \to +\infty} xe^{-x} = 0$$
$$\lim_{x \to -\infty} xe^{-x} = -\infty$$

**5c.** $y = xe^{-x}$

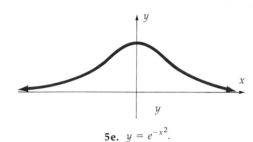

**5e.** $y = e^{-x^2}$.

**6a.** 0 **6c.** $+\infty$ **6e.** $+\infty$ **6f.** 1 **6h.** 0 **6j.** $\infty$
**7a.** 0 **7b.** doesn't exist

**8.**

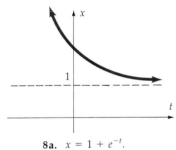

**8a.** $x = 1 + e^{-t}$.

**9.** $3 \cdot (300)^{-2} A \cdot (E_A/R) \exp(-E_A/300R)$ **11.** $v = (2kT/m)^{1/2}$

## Section 8.2

**1a.** $g(y) = (y - 1)/2$ **1c.** $g(y) = (y - 3)^{1/3} - 1$ **1e.** $g(y) = \sqrt{1 - y^2}$, $0 \le y \le 1$ **6a.** 0
**6b.** DNE (undefined for $x < 0$) **6c.** $+\infty$ **6d.** 0 **6g.** $+\infty$ **7a.** $y - 2 = \frac{1}{4}(x - 4)$
**10c.** $g'(-2) = g'(2) = 1/4$, $g'(0) = 1$

## Section 8.3

**1a.** 1/2 **1c.** $x^2$ **1d.** $x^x$ **1f.** $xe^x$ **1h.** $a/b$ **1i.** $3 \ln x - x^2$ **3a.** $e$ **3c.** $e^{-2}$ **3e.** $(\ln 5)/3$ **4a.** 0
**4c.** $-\dfrac{\ln 2}{\ln(3/2)}$ **4e.** $x = \pm\sqrt{e - 1}$ **5a.** $2x \ln x + x$ **5c.** $4(\ln x)^3/x$ **5e.** $e^x(\ln x + 1/x)$
**6a.** $\dfrac{1}{x} + \dfrac{1}{x + 1}$; $\dfrac{-1}{x^2} - \dfrac{1}{(x + 1)^2}$ **6b.** $x^{-1}$; $-x^{-2}$ **6c.** $\dfrac{1}{1 + x} + \dfrac{1}{1 - x}$; $\dfrac{-1}{(1 + x)^2} + \dfrac{1}{(1 - x)^2}$ **7a.** $-\tan x$
**7b.** $3 \tan 3x$ **7c.** $\dfrac{2}{t} \cos(\ln(t^2))$ **8a.** $-9 \csc^2 3t$ **8b.** $\dfrac{-1}{x^2} + \csc^2 x$ **8c.** $-9 \cos 3t$ **15a.** $x = x_e[1 - e^{-kt/x_e}]$
**15b.** $x_e$ **15c.** $\dfrac{dx}{dt} = k(1 - x/x_e)$
**17d.** No. $\ln(P_{30}) < -\dfrac{(30)(29)}{2 \cdot 365} < -1 < \ln\left(\dfrac{1}{2}\right)$, so $P_{30} < \dfrac{1}{2}$; there is a *less than even* chance that no two are the same.

## Section 8.4

**1a.** $5^x \ln 5$  **1b.** $10x^9$  **1c.** $x^x(1 + \ln x)$  **1i.** $x^{\sin x}\left(\cos x \ln x + \dfrac{1}{x}\sin x\right)$  **2a.** $0$  **2b.** $1$  **2c.** $e$  **2f.** $e^a$

**2h.** $e$  **2i.** $e^2$  **3b.** $x^x(\ln x + 1)$  **3e.** $\dfrac{x^2 \sqrt[3]{7x - 14}}{(1 + x^2)^4}\left[\dfrac{2}{x} + \dfrac{1}{3x - 6} - \dfrac{8x}{1 + x^2}\right]$

## Section 8.5

**1a.** $\frac{1}{2}e^{2x} + C$  **1b.** $-e^{\cos x} + C$  **1c.** $\ln|x - 1| + C$  **1d.** $-\ln|2 - x| + C$  **1e.** $\frac{1}{2}\ln(x^2 + 4) + C$
**1f.** $\frac{1}{3}\ln|3x^2 - 2| + C$  **1g.** $\ln|\ln x| + C$  **1i.** $-\ln|\sin \theta| + C$  **1k.** $x + 4\ln|x| + C$  **1q.** $\frac{1}{2}e^{2x} + 2e^x + x + C$

**1t.** $\dfrac{1}{\ln 2}2^x + C$  **2a.** $\frac{1}{2}\ln 3$  **2b.** $\frac{1}{3}(1 - e^{-3})$  **2c.** $\ln \frac{1}{2}$  **3b.** $\infty$  **4a.** diverges

**4c.** $\frac{1}{2}$  **4d.** $\frac{1}{2}$  **4g.** $2$  **4i.** divergent  **4k.** $\frac{1}{a}$
**5.** $Q(10) - Q(0) = \int_0^{10} 5e^{-3t}\, dt = \frac{5}{3}(1 - e^{-30})$; $Q(100) - Q(10) = \frac{5}{3}(e^{-30} - e^{-300})$; $Q(\infty) - Q(0) = 5/3$  **6.** $\pi/2$

## Section 8.6

**1.** $(.03)(3/2)^{-t/5}$; $H = 5 \ln 2/\ln(3/2)$  **3.** $H = 5 \ln 2/\ln(6/5)$; $A_0 = (.03)/\ln (6/5)$  **4.** $e^{.05} - 1 \approx 5.127\%$
**5.** $\frac{1}{10}\ln\frac{5}{3} \approx 5.11\%$; $(\frac{5}{3})^{1/10} - 1 \approx 5.24\%$  **6.** $\approx 15{,}000$ years  **8.** $(5600R)/\ln 2$

## Section 8.7

**1b.** $x = \sqrt{y^2 + 3}$  **1c.** $y = -x$  **1e.** $y = -\ln(e^{-x} + 1 - e^{-1})$  **1f.** $y = \dfrac{3x^2}{2(x^2 + 1)}$

**2.**

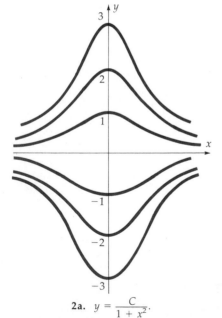

**2a.** $y = \dfrac{C}{1 + x^2}$.

**3c.** $\cdot 1/a_0 k$  **4b.** $T = 20 + 30e^{-10t}$  **6b.** The ratio $Q/R$ is the slope of the graph.
**7.** $h = \left[h_0^{1/2} - \sqrt{\dfrac{g}{2}}\dfrac{At}{\pi r^2}\right]^2$; $t = \sqrt{2h_0/g}\pi r^2/A$  **9b.** $R - \sqrt{2g}A\sqrt{h} - R \ln|R - \sqrt{2g}A\sqrt{h}| = \dfrac{gA^2}{\pi r^2}t + c$

**11.** $S = CL + ke^{-vt/L}$, $\displaystyle\lim_{t \to +\infty} S = CL$  **14.** $I = \dfrac{E}{R} + ke^{-Rt/L}$; $\displaystyle\lim_{t \to +\infty} \dfrac{dI}{dt} = 0$, $\displaystyle\lim_{t \to +\infty} I = E/R$  **16a.** $v = -\dfrac{mg}{k} + ce^{-kt/m}$

**16b.** $-mg/k$   **16c.** $y_0 - mgt/k - \dfrac{km^2 v_0 + m^3 g}{k^3}(1 - e^{-kt/m})$

## Section 8.8

**1a.** $\pi/6$   **1b.** $2\pi/3$   **1c.** $\pi/4$   **1d.** $-\pi/3$   **1e.** $\pi/6$   **1f.** 0   **1g.** $\pi/2$   **1h.** $\pi/4$   **1i.** $3\pi/4$
**4.** 3/5, 4/5, 3/4, 4/3, 5/4   **7c.** $-\pi/5$   **7d.** $\sqrt{1 - x^2}$   **8a.** $-\pi/2 \le \theta \le \pi/2$   **8b.** $-1 \le u \le 1$
**9a.** $2/\sqrt{1 - 4x^2}$   **9b.** $2x/(1 + x^4)$   **9c.** $-2/\sqrt{1 - 4x^2}$   **9d.** $\tan^{-1}x + x/(1 + x^2)$   **9f.** $-x/\sqrt{1 - x^2}$
**9g.** $-1/x^2$   **9h.** $1/\sqrt{e^{2x} - 1}$   **16a.** $y' = \dfrac{3x^2 + \tan^{-1}y}{e^y - x/(1 + y^2)}$

## Section 8.9

**1b.** maxima: $\dfrac{1}{\pi}, \dfrac{1}{\pi} \pm 2, \ldots$; period 2; amplitude 2

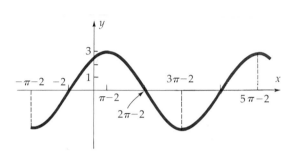

**1d.** $y = 3\sin(\frac{t}{2} + 1)$; period $4\pi$, amplitude 3.

**2b.** $\sqrt{2}\cos(3t + \pi/4)$; amplitude $\sqrt{2}$; period $\dfrac{2\pi}{3}$; frequency $3/2\pi$; maxima at $-\dfrac{\pi}{12}, -\dfrac{\pi}{12} \pm \dfrac{2\pi}{3}, \ldots$

**2c.** $5\cos\left(2\pi t - \left(\dfrac{\pi}{2} - 1\right)\right)$; amplitude 5; period 1; frequency 1; maxima at $\dfrac{1}{4} - \dfrac{1}{2\pi}, \dfrac{1}{4} - \dfrac{1}{2\pi} \pm 1, \ldots$

**4c.** $y = \cos 2t + \sin 2t$   **5b.** $y = \sqrt{30/17}\cos(\sqrt{17/30}t - \pi/2) = \sqrt{30/17}\sin(\sqrt{17/30}t)$

## Chapter 8 Review

**1.**

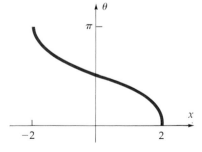

**1e.** $\theta = \cos^{-1}\left(\frac{x}{2}\right)$.

**2a.** $(3x + 1)\ln 2$   **2b.** 3/32   **2c.** $1/\sqrt{10}$   **2d.** $-2/\sqrt{5}$   **2e.** $\sqrt{2}/4$

**3a.** $\dfrac{2x}{x^2 + 1}$   **3b.** $(3x^3 + 1)e^{x^3}$   **3c.** $2x\exp(\exp(x^2 + 1))\exp(x^2 + 1)$   **3d.** $\dfrac{1}{\sqrt{9 - x^2}}$   **3e.** $\dfrac{5}{25 + x^2}$

**3f.** $3\cos 3t\, e^{\sin 3t}$   **3g.** function has empty domain   **3h.** $(2x + 1)\ln(10)10^{x^2 + x}$   **3i.** $\dfrac{1}{\ln 2}$

**3j.** $\dfrac{x^3\sqrt{x^2 + 1}}{(1 + 4x)^2}\left(\dfrac{3}{x} + \dfrac{x}{x^2 + 1} - \dfrac{8}{1 + 4x}\right)$   **4a.** 0   **4b.** 0   **4c.** 0   **4d.** $e^a$

**5a.**

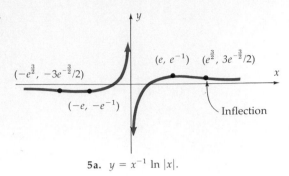

**5a.** $y = x^{-1} \ln|x|$.

**6a.** $\frac{1}{2}\ln(5/4)$ **6b.** $\frac{1}{2}(1 - e^{-100})$ **6c.** $\dfrac{\pi}{2}$ **6d.** $\frac{1}{2}\sin^{-1}(2t) + C$

**7a.** $y = e^{-x^2/2}$ **7b.** $y = -e^{-x^2/2}$ **7c.** $A = \frac{2}{3}(1 - e^{3t})$ **7d.** $y = e^{1 - \cos(t)}$

**8.** $Q = CE + (Q_0 - CE)e^{-t/RC}$; $\lim\limits_{t \to +\infty} Q = CE$, $\lim\limits_{t \to -\infty} Q = -\infty$ if $Q_0 < CE$, $= CE$ if $Q_0 = CE$, $= +\infty$ if $Q > CE$

**9a.** $y = \cos(2t) - \frac{1}{2}\sin(2t)$ **9b.** period $\pi$, amplitude $\sqrt{5}/2$

**10.** $x = -\alpha^{-1}\ln(B/\alpha C)$ **11.** $6\ln(1.1) + .02 - .00021$ **13a.** $(-\infty, \infty)$ **13b.** $1/6$

**15b.** $\theta = A\cos(\sqrt{g/l}\,t - \varphi)$ **15c.** period $2\pi\sqrt{l/g}$ **15d.** $9.87$ m/sec$^2$

## Section 9.1

**1.** $x \ln ax - x + C$ **3.** $\dfrac{x^2}{2}[\ln x^2 - 1] + C$ **5.** $e - 2$ **7.** $\frac{2}{3}t^{3/2}(\ln t - \frac{2}{3}) + C$ **9.** $\frac{1}{500}[1 - 61e^{-10}]$

**11.** $\frac{1}{2}x \sin 2x + \frac{1}{4}\cos 2x + C$ **13.** $\pi^2 - 4$ **15.** $1$ **17.** $x \tan^{-1}x - \frac{1}{2}\ln(1 + x^2) + C$ **20.** $2e^{\sqrt{x}}(\sqrt{x} - 1) + C$

**30.** $a^{-2}$ **32.** $1/2$ **33.** $0$ **36.** $\dfrac{b}{a^2 + b^2}$

## Section 9.1 Appendix

**2a.** $Q = 2t - 4 + ce^{-t/2}$ **3a.** $Q = kCt + kRC^2(e^{-t/RC} - 1)$

## Section 9.2

**1.** $-\frac{1}{2}\cos 2\theta + C$ **3.** $\dfrac{\theta}{2} - \dfrac{\sin 2\theta}{4} + C$ **5.** $\frac{1}{3}\sin^3 x + C$ **7.** $\dfrac{x}{2} + \dfrac{\sin 2ax}{4a} + C$ **9.** $-\frac{1}{27}\cos^9 3x + C$

**11.** $\dfrac{\theta}{8} - \dfrac{\sin 8\theta}{64} + C$ **13.** $-\cos\theta + \frac{1}{3}\cos^3\theta + C$ **15.** $\pi/2$ **17.** $1/2$ **19.** $\dfrac{\sin 2x}{4} - \dfrac{\sin 4x}{8} + C$

**21.** $-\frac{2}{3}(1 + \cos t)^{3/2} + C$ **24.** $\pi^2/2$ **25.** $y = \sin^{-1}\left(\dfrac{x}{2} + \dfrac{\sin 2x}{4}\right)$ **35a.** $\dfrac{1}{2a} - \dfrac{a/2}{a^2 + 4b^2}$

## Section 9.3

**1.** $2\ln\left|\sec\left(\dfrac{x}{2}\right)\right| + C$ **3.** $2\ln(1 + \sqrt{2})$ **5.** $\sqrt{3} - \pi/3$ **7.** $\frac{1}{3}\tan^3 x + C$ **9.** $\tan x + C$ **11.** $\tan t - t + C$

**13.** $\dfrac{1}{2\pi}[\sec \pi x \tan \pi x - \ln|\sec \pi x + \tan \pi x|] + C$ **15.** $\dfrac{\sec^3 \pi x \tan \pi x}{4\pi} - \dfrac{\sec \pi x \tan \pi x}{8\pi} - \dfrac{1}{8\pi}\ln|\sec \pi x + \tan \pi x| + C$

**17.** $\frac{1}{2}\sin 2t - \frac{1}{6}\sin^3 2t + C$ **19.** $\theta/8 - (1/32)\sin 4\theta + C$ **21.** $-e^{\cos\theta} + C$ **23.** $(-1/8)\cot^4 2x + C$

**25.** $\frac{1}{3}\sec^3 x + C$ **27.** $\frac{1}{2}\sin(x^2) + C$ **29.** $\ln|\sec x + \tan x| + C$ **31.** $\ln|1 + \tan\theta| + C$

**33.** $-\ln|\csc\theta + \cot\theta| + C$ **40.** $\pi(1 - \pi/4)$

## Section 9.4

**2a.** $\dfrac{x}{2}\sqrt{2-3x^2}+(1/\sqrt{3})\sin^{-1}(x\sqrt{3/2})+C$    **2c.** $\pi/6$    **2e.** $[(1/\sqrt{2})\ln|\sqrt{2}t+\sqrt{2t^2-3}|]_2^4=\dfrac{1}{\sqrt{2}}\ln\left(\dfrac{4\sqrt{2}+\sqrt{29}}{2\sqrt{2}+\sqrt{5}}\right)$

**3a.** $-\sqrt{9-x^2}+C$    **3c.** $u-a\tan^{-1}(u/a)+C$    **3e.** $\sqrt{a^2+u^2}+C$    **3g.** $-\dfrac{1}{a^2}\dfrac{\sqrt{a^2+u^2}}{u}+C$

**3i.** $-z(4z^2+1)^{-1/2}+C$    **3k.** $\dfrac{1}{2}\left[\tan^{-1}x+\dfrac{x}{1+x^2}\right]$    **3m.** set $u=1+x^2$    **7a.** $\dfrac{\pi}{2}$    **7b.** $\pi$

## Section 9.5

**1a.** $(2/\sqrt{3})\tan^{-1}(1/\sqrt{3})$    **1c.** $\sin^{-1}(x-1)-\sqrt{2x-x^2}$    **1e.** $\sqrt{x^2-4x+3}+3\ln|x-2+\sqrt{x^2-4x+3}|$

**2a.** $x-2+\dfrac{6}{x+2}$    **2e.** $x-\dfrac{2x^3+x}{(x^2+1)^2}$    **2f.** nothing to do    **2g.** $x+\dfrac{1+x}{2x^2+x+1}$

**3a.** $\frac{1}{2}x^2-2x+6\ln|x+2|+C$    **3e.** $\frac{1}{2}x^2-\ln(1+x^2)-\dfrac{1/2}{1+x^2}+C$    **3f.** $\dfrac{-1/2}{x^2+9}+C$

**5c.** $\ln\left|\dfrac{x-2}{x-1}\right|+C$    **6.** $\ln\left|\dfrac{b-x_0+\sqrt{(b-x_0)^2+y_0^2}}{a-x_0+\sqrt{(a-x_0)^2+y_0^2}}\right|$    **9g.** $\lim\limits_{t\to+\infty}v=x_0^{-1/2},\ \lim\limits_{t\to+\infty}\frac{1}{2}mv^2=\frac{1}{2}m/x_0$

## Section 9.6

**1a.** $\dfrac{A}{x-1}+\dfrac{B}{x+1}$    **1b.** $\dfrac{A}{x}+\dfrac{B}{x-1}+\dfrac{C}{(x-1)^2}+\dfrac{D}{x+1}+\dfrac{E}{(x+1)^2}$    **1c.** $\dfrac{A}{x}+\dfrac{B}{x^2}+\dfrac{C}{x-1}+\dfrac{D}{x-2}$

**1d.** $\dfrac{A}{x-2}+\dfrac{B}{x-3}$    **2a.** $\frac{3}{2}\ln|x-1|-\frac{1}{2}\ln|x+1|+C$    **2b.** $\frac{1}{2}\ln|x^2-1|+C$

**2c.** $\ln|x|-\dfrac{1}{2}\ln|x-1|-\dfrac{1/4}{x-1}-\dfrac{1}{2}\ln|x+1|+\dfrac{1/4}{x+1}+C$    **2d.** $-2\ln|x-2|+3\ln|x-3|+C$

**3a.** $\dfrac{A}{x-2}+\dfrac{B}{x+2}+\dfrac{Cx+D}{x^2+1}$    **3b.** $\dfrac{A}{x}+\dfrac{B}{x^2}+\dfrac{Cx+D}{x^2+1}$    **4a.** $\ln\dfrac{x^2}{x^2+1}+\tan^{-1}x+C$    **4c.** $2x+\ln\left|\dfrac{x-1}{x+1}\right|+C$

**4e.** $\dfrac{3}{2}\tan^{-1}x+\dfrac{1}{2}\dfrac{x}{x^2+1}+C$    **4g.** $\dfrac{5}{4}\ln|x-1|-\dfrac{1}{4}\ln|x+1|-\dfrac{1}{2(x-1)}+C$    **4i.** $\frac{2}{3}\ln(e^t+2)+\frac{1}{3}\ln|e^t-1|+C$

**5b.** $x^2+x+2+\dfrac{2x^2+x-2}{x^3-x^2-x+1};\ x^3/3+x^2/2+2x+\frac{9}{4}\ln|x-1|-\frac{1}{2}(x-1)^{-1}-\frac{1}{4}\ln|x+1|+C$

**8c.** $\dfrac{Cae^{(b-a)kt}-b}{Ce^{(b-a)kt}-1}$    **8f.** $ab\dfrac{e^{(b-a)kt}-1}{be^{(b-a)kt}-a}$    **8g.** $\dfrac{a^2kt}{1+akt}$

## Section 9.7

**1.** $\frac{2}{81}\sqrt{3x-2}[\frac{1}{9}(3x-2)^4+\frac{6}{7}(3x-2)^3+\frac{12}{5}(3x-2)^2+\frac{8}{3}(3x-2)]+C$    **3.** $\frac{3}{2}(x^2+1)^{4/3}(x^2/7-\frac{3}{28})+C$

**5.** $2t^{1/2}-4t^{1/4}+4\ln(1+t^{1/4})+C$    **7.** $\dfrac{1}{\sqrt{2}}\ln\left|\dfrac{\tan(\theta/2)+1-\sqrt{2}}{\tan(\theta/2)+1+\sqrt{2}}\right|+C$    **9.** $-2\dfrac{1+\cos\theta}{1+\cos\theta+\sin\theta}+C$

**11.** $-\ln|2+\cos\theta|+C$

## Chapter 9 Review

**1.** $\dfrac{\sin 3-3\cos 3}{9}$    **2.** $\dfrac{-2\pi}{25}$    **3.** $1/a^2$    **4.** $x\tan^{-1}x-\frac{1}{2}\ln(1+x^2)+C$    **5.** $\pi/8$    **6.** $\frac{1}{3}\sin^3\theta+C$

**7.** $\dfrac{4}{(103)(101)}$    **8.** $\pi/2$    **9.** $\frac{1}{2}(1-\sin 1)$    **10.** $\theta/8-(1/32)\sin 4\theta+C$    **11.** $\dfrac{1}{\pi}\ln|\sec\pi x+\tan\pi x|+C$

**12.** $\tan\theta + \frac{1}{3}\tan^3\theta + C$  **13.** $\frac{1}{4}\tan^2 2x + \frac{1}{2}\ln|\cos 2x| + C$  **14.** $\tan^{-1}(\sin\theta) + C$  **15.** $\ln|x + \sqrt{3 + x^2}| + C$

**16.** $\frac{3}{2\sqrt{2}}\sin^{-1}(\sqrt{2}x/\sqrt{3}) + \frac{1}{2}x\sqrt{3 - 2x^2} + C$  **17.** $\frac{x}{2}\sqrt{3x^2 - 4} - (2/\sqrt{3})\ln|\sqrt{3}x + \sqrt{3x^2 - 4}| + C$

**18.** $\frac{x + 1}{2}\sqrt{x^2 + 2x} - \frac{1}{2}\ln|x + 1 + \sqrt{x^2 + 2x}| + C$  **19.** $(1/\sqrt{2})\ln|x + \frac{1}{4} + \sqrt{x^2 + \frac{1}{2}x}| + C$  **20.** $4/3$  **21.** $0$

**22.** $\pi/2$  **23.** $\frac{1}{2}\ln\left|\frac{u - 1}{u + 1}\right| + C$  **24.** $\frac{-1}{x^2 + x + 1} + C$  **26.** $\frac{1}{2}x^2 - \frac{1}{2}\ln(x^2 + 1) + C$

**27.** $\ln|x| - 1/x - \frac{1}{2}\ln(x^2 + 1) - \tan^{-1}x + C$  **28.** $x - \ln(1 + e^x) + C$  **29.** $(1/\sqrt{5})\ln\left|\frac{2\tan(\theta/2) - 3 - \sqrt{5}}{2\tan(\theta/2) - 3 + \sqrt{5}}\right| + C$

**30.** $\ln 2$  **40.** $\frac{n}{a}\int_0^\infty x^{n-1}e^{-ax}\,dx = \frac{n!}{a^{n+1}}$  **41.** $x^2 = \frac{3}{2}e^{-2} - ye^{-2y} - \frac{1}{2}e^{-2y}$  **42.** $y = \sin\left(\frac{1}{2}x^2 + \frac{\pi}{6}\right)$

## Section 10.1

**2a.** $\frac{(85)^{3/2} - 8}{243}$  **2c.** The integrand is $\frac{x^3}{4} + x^{-3}$  **3.** $6$, vs $2\pi$ for the circle.  **4a.** $2\ln(1 + \sqrt{2})$  **4c.** $2r\sin^{-1}(b/r)$

**4e.** $\sqrt{1 + b^2} - \sqrt{1 + a^2} + \frac{1}{2}\ln\left|\frac{(\sqrt{1 + b^2} - 1)(\sqrt{1 + a^2} + 1)}{(\sqrt{1 + b^2} + 1)(\sqrt{1 + a^2} - 1)}\right| = \sqrt{1 + b^2} - \sqrt{1 + a^2} + \ln\frac{b(\sqrt{1 + a^2} + 1)}{a(\sqrt{1 + b^2} + 1)}$

## Section 10.2

**2b.** $12\pi$  **2d.** $\pi m\sqrt{1 + m^2}(b^2 - a^2)$  **3c.** $\pi a^2$  **4b.** $\pi\left[e^b\sqrt{1 + e^{2b}} - \sqrt{2} + \ln\left|\frac{e^b + \sqrt{1 + e^{2b}}}{1 + \sqrt{2}}\right|\right]$

**5b.** $\frac{\pi}{6}[3b^2\sqrt{1 + 9b^4} + \ln|3b^2 + \sqrt{1 + 9b^4}|]$  **7.** $\pi\frac{2\sqrt{2} - 1}{6a^2}$

## Section 10.3

**2.** $(0, 8/5)$  **5.** $\bar{x} = \bar{y} = \frac{2a}{12 - 3\pi} \approx 0.8a$  **7.** $(\pi/2, \pi/8) \approx (\pi/2, 0.4)$  **9.** $\bar{x} = 0, \bar{y} = \frac{4/3}{\sqrt{2} + \ln(1 + \sqrt{2})}$

**11.** $\bar{x} = 0, \bar{y} = 2\frac{2a^2/3 - b^2}{4b + \pi a}$  **14c.** $(0, 2a/\pi)$; no

## Section 10.4

**1a.** $e^{-1/10} - e^{-1/5} \approx 0.086$  **1c.** $e^{-2} \approx 0.135$  **2a.** $\frac{1}{200}\int_0^1 e^{-t/200}\,dt = 1 - e^{-2/100} \approx 0.005$

**3a.** $\mu = 1/2, P(0 \le t \le 1) = 1 - e^{-2} \approx 0.865$  **3c.** $\mu = 2, P(0 \le t \le 1) = 1/4$  **4a.** $1/2$  **4b,c.** $\mu = \pi/2$

**4d.** $\sigma^2 = \pi^2/4 - 2 \approx 0.47$  **8a.** $(b - a)^2/12$  **10a.** $4/a_0^3$  **10c.** $3a_0/2$

## Section 10.5

**1a.** $7.26$  **2a.** $7.22$  **3a.** $1.117$  **3c.** $1.1$  **4a.** $|R| < 0.08$  **4c.** $|R| < 0.017$  **5c.** $1.910$

**6.** $16$ (rounded from $15.75$)  **8a.** $n > 51$  **8c.** $n > 6$  **11a.** $n > 5773$ [How do you prove that $|f''(x)| \le 2$?]

## Chapter 10 Review

**2b.** $\sqrt{1 + e^2} - \sqrt{2} + \ln\left[\frac{e(1 + \sqrt{2})}{1 + \sqrt{1 + e^2}}\right]$  **2c.** $\pi[e\sqrt{1 + e^2} + \ln(e + \sqrt{1 + e^2}) - \sqrt{2} - \ln(1 + \sqrt{2})]$

**3.** $(4a/3\pi, 4b/3\pi)$  **4b.** $\mu = 0, \sigma^2 = 1/6$  **5a.** $k = a^2$  **5b.** $1 - e^{-a} - ae^{-a}$  **5c.** $\mu = 2/a$

**5d.** $1 - 3e^{-2} \approx 0.6$  **5e.** $\sigma^2 = 2a^{-2}$  **6a.** $.406187$  **6b.** $.4054714$  **7a.** $|R(T_4)| \le 1/768$

**7b.** $|R(SR_4)| \le 1/61{,}440$  **8a.** $n > 10^3/2\sqrt{6}$

## Section 11.1

**1.** $P_6(x) = 1 - x^2/2 + x^4/24 - x^6/720$ **2a.** $1 + 2x + 2x^2 + 4x^3/3 + 2x^4/3$ **2c.** $-x - x^2/2 - x^3/3 - x^4/4$
**5a.** $1.6484\ldots$; $|\text{error}| < 0.000521$, since $e^{1/2} < 2$ **5c.** $|e^{-1} - (\frac{1}{2} - \frac{1}{6} + \frac{1}{24})| < 1/120$ **7a.** $|\sin 1 - \frac{5}{6}| < 1/120$

**7c.** $|\sin 0.1 - (0.1 - (0.1)^3/6)| < 8.3 \times 10^{-8}$ **9a.** $\dfrac{1}{2} + \dfrac{\sqrt{3}}{2}(x - \pi/6) - \dfrac{1}{4}(x - \pi/6)^2$

**11a.** $\ln(0.9) \approx -0.105358333\ldots$ **11b.** $|\text{error}| < 1/5 \cdot 9^5 \approx .000003$
**12.** $|\ln(1.1) - (.1 - \frac{1}{2}(.1)^2 + \frac{1}{3}(.1)^3 - \frac{1}{4}(.1)^4 + \frac{1}{5}(.1)^5| < \frac{1}{6}(.1)^6 < \frac{1}{2} \times 10^{-6}$

## Section 11.2

**2b.** $n > \varepsilon^{-1/3} \Rightarrow \left|\dfrac{1}{n^3} - 0\right| < \varepsilon$ **2e.** $\left(\dfrac{9}{10}\right)^n < \varepsilon$ if $n > \dfrac{\ln(1/\varepsilon)}{\ln(10/9)}$ **4a.** $\lim \dfrac{1}{1 + 1/n^3} = \dfrac{\lim 1}{1 + (\lim 1/n)^3} = 1$

**4c.** $\lim \dfrac{3n^2}{1 + 2n^2} = \lim \dfrac{3}{1/n^2 + 2} = \dfrac{\lim 3}{(\lim 1/n)^2 + 2} = 3/2$

**4e.** $1 - \dfrac{1}{n} \le 1 - \dfrac{1}{(-n)^n} \le 1 + \dfrac{1}{n}$; apply the Trapping Theorem **5a.** $-\dfrac{1}{n} \le \dfrac{\sin n}{n} \le \dfrac{1}{n}$

**5c.** $-1/n < -1/n^n < \left(\dfrac{\cos n}{n}\right)^n < 1/n^n < 1/n$ **5e.** $\lim x^n = 0$ if $|x| < 1$

## Section 11.3

**1a.** $\sqrt{1/3}$ **1b.** $1$ **1c.** $+\infty$ (nonexistent) **1e.** $1$ **1g.** $e^2$ **1j.** $\infty$ (nonexistent) **1k.** $1$ **1m.** $0$
**2.** a, c, d, e, g, h are bounded. **4a.** false: $\{(-1)^n\}$ **4b.** true (Theorem 2)
**4c.** false: $-1/n < 1/n$, but $\lim\left(\dfrac{-1}{n}\right) = \lim\left(\dfrac{1}{n}\right)$

## Section 11.4

**3a.** $3/2$ **3b.** $1/2$ **3c.** $3/4$ **3g.** $25/6$ **4b.** $\frac{230}{99}$ **6a.** $1$ **6b.** $-\infty$ (divergent)
**7a.** No; try $a_j = 1$, $b_j = -1$. **11.** $5/11$ **12.** $1/5$ **13a.** $\dfrac{x}{x - 1}$, $|x| > 1$ **14c.** $2$

## Section 11.5

**1b.** $\displaystyle\sum_{n=0}^{\infty} (-1)^n 2^{-n-1} x^n$ **1c.** $\displaystyle\sum_{n=1}^{\infty} \dfrac{(-1)^{n+1}}{n} x^n$ **1e.** $\displaystyle\sum_{k=0}^{\infty} \dfrac{(-2)^k}{k!} x^k$ **2a.** $\displaystyle\sum_{n=0}^{\infty} \dfrac{2^n t^{2n}}{n!}$

**2c.** $\displaystyle\sum_{k=1}^{\infty} \dfrac{(-1)^k}{(2k)!} t^{2k-2} = \sum_{k=0}^{\infty} \dfrac{(-1)^{k+1}}{(2k+2)!} t^{2k}$ **2e.** $\displaystyle\sum_{0}^{\infty} (-4)^{-n} t^{2n}$, $|t| < 2$ **3a.** $e^5$ **3d.** $\sin 1$ **3f.** $\cos 2$ **3h.** $5/2$

## Section 11.6

**3b.** $f^{(5)}(0) = 24$ **5b.** $f^{(2k)}(0) = \dfrac{(2k)!}{(k!)^2}$, $f^{(2k+1)}(0) = 0$ **6a.** $\displaystyle\sum_{n=0}^{\infty} (-1)^n x^{2n}$ **6c.** $\displaystyle\sum_{n=0}^{\infty} \left(\dfrac{-1}{2}\right)^n \dfrac{1}{n!} x^{2n}$

**6e.** $\displaystyle\sum_{n=0}^{\infty} \dfrac{x^{n+1}}{n!}$ **6f.** $\displaystyle\sum_{k=0}^{\infty} \dfrac{(-1)^k}{(2k+1)!} x^{2k}$ **6g.** $\displaystyle\sum_{k=0}^{\infty} \dfrac{(-1)^k x^{2k+1}}{(2k+1)(2k+1)!}$ **6h.** $\displaystyle\sum_{k=0}^{\infty} \dfrac{x^{2k+1}}{(2k+1)!}$ **7a.** $\frac{1}{2}$ **7c.** $0$

**9e.** $\pi \approx 4\left[\dfrac{1}{2} + \dfrac{1}{3} - \dfrac{1}{3}\left(\left(\dfrac{1}{2}\right)^3 + \left(\dfrac{1}{3}\right)^3\right) + \dfrac{1}{5}\left(\left(\dfrac{1}{2}\right)^5 + \left(\dfrac{1}{3}\right)^5\right) - \dfrac{1}{7}\left(\left(\dfrac{1}{2}\right)^7 + \left(\dfrac{1}{3}\right)^7\right)\right] \approx 3.1409$;

$|\text{error}| < 4\left[\dfrac{(1/2)^9}{9} + \dfrac{(1/3)^9}{9}\right] \approx 0.0008$ **11b.** $\dfrac{p}{1 - p}$

## Section 11.7

**1b.** Error $< \dfrac{12}{11(11!)} \approx 5 \times 10^{-8}$ **2.** $e^{1/2} \approx 1.625$, $|\text{error}| < \frac{1}{42} < .03$ **3.** $e^{-1} \approx 9/24 = 0.375$, $|\text{error}| < \dfrac{1}{5!} < .01$

**4a.** $0.946083$, $|\text{error}| < \dfrac{1}{9 \cdot 9!} \approx 3 \times 10^{-7}$ **4b.** $0.54498$, $|\text{error}| < \dfrac{5}{525312} < 10^{-5}$ **5a.** $0.48540$

## Section 11.8

**1.** a,c,j "fail"; b,f,g,i converge **3a.** 1.20

## Section 11.9

**1a.** $\sum a_k > \dfrac{1}{10} \sum \dfrac{1}{k} = \infty$ (divergent) **1c.** $\sum a_k < 5 \sum 1/n^2 < \infty$ (convergent)

**1e.** $\sum a_k < \sum \dfrac{(100)^k}{k!} = e^{100} < \infty <$ (convergent) **2a.** $a_n \sim \dfrac{1}{n}$, divergent **2c.** $a_n \sim \dfrac{1}{n}$, divergent

**2e.** $a_n \sim \dfrac{1}{n^2}$, convergent **2g.** $a_n \sim 1/\sqrt{2} n^{3/2}$, convergent **3.** c,d,i diverge

## Section 11.10

**1.** b,d,h divergent **2a.** $0.617$, $|\text{error}| < 1/7 \approx 0.143$ **2d.** $0.752$, $|\text{error}| < 0.3$ **3a.** $n \geq 10^4 - 1$ **3c.** $n \geq 10^8 - 2$

## Section 11.11

**1a.** converges absolutely **1c.** converges absolutely **1e.** converges conditionally **1g.** converges conditionally
**3.** $\approx 0.765$

## Section 11.12

**1.** a diverges **3a.** $|x| \leq 1$ **3c.** $|x| < \frac{1}{2}$ **3e.** $|x| < \infty$ **3g.** $|x| \leq 1$ **3i.** $|x| < \infty$ **4b.** $e$
**5a.** $-2 < x < 4$ **5b.** $-4 < x < -2$ **5c.** $0 \leq x < 1$ **6a.** $1 + \dfrac{1}{2} x - \dfrac{1}{8} x^2 + \dfrac{\frac{1}{2}(-\frac{1}{2})(-\frac{3}{2})}{3!} x^3 + \cdots$

**6e.** $a^p(1 + x/a)^p = a^p \left( 1 + p\dfrac{x}{a} + \dfrac{p(p-1)}{2} \left(\dfrac{x}{a}\right)^2 + \cdots \right)$

## Chapter 11 Review

**1a.** $1 + \frac{2}{3}x - \frac{1}{9}x^2 + \frac{4}{81}x^3$ **1b.** $1 + \frac{2}{3}(x-1) - \frac{1}{9}(x-1)^2 + \frac{4}{81}(x-1)^3$

**2a.** $9^{2/3} = (8 + 1)^{2/3} \approx 4 + \dfrac{1}{3} - \dfrac{1}{9 \cdot 2^4} + \dfrac{1}{81 \cdot 2^5} \approx 4.3267747$ **2b.** $|\text{error}| < \dfrac{7}{3^5 \cdot 2^{10}} < 3 \times 10^{-5}$ **4a.** 10

**4b.** $-\infty$ (nonexistent) **4c.** $e$ **4d.** $1/e$ **6.** any $n \geq 8$ **7a.** $1 - \dfrac{x^4}{2} + \dfrac{x^8}{4!} - \cdots = \sum_{0}^{\infty} (-1)^k x^{4k}/(2k)!$

**7c.** $\sum_{1}^{\infty} (-1)^k \dfrac{x^{2k}}{(2k)(2k)!}$ **8.** $-\dfrac{1}{4} + \dfrac{1}{4 \cdot 4!}$; $|\text{error}| < \dfrac{1}{6 \cdot 6!} = \dfrac{1}{4320} < \dfrac{1}{2} \cdot 10^{-3}$ (alternating series)

**9.** $\sum_{0}^{\infty} (-1)^j \dfrac{(j+2)(j+1)}{2} x^j$ **10b.** $\approx \dfrac{3}{4} - \dfrac{(3/4)^4}{3} + \dfrac{(3/4)^5}{5}$ **10c.** $|\text{error}| < \dfrac{(3/4)^7}{7} < .04$ **11.** b and e converge

**12a.** $-1 \leq x < 1$ **12b.** $-1 < x < 5$ **12c.** all $x$ **12d.** $x > 0$ **12e.** $|x| > 1$ **12f.** $x > 1$
**13a.** Less than $1/200$ and more than $1/242$ **13b.** $\sum_{1}^{10} k^{-3} + 1/200$; error less than $0.001$
**14b.** radius of convergence $\sqrt{2}$ **15.** Yes; find an example.

## Section 12.1

**1a.** $(x + 1)^2 + 2$   **1c.** $-16(t - \frac{3}{32})^2 + \frac{9}{64}$   **2a.** $V = (0, 0)$, $F = (0, \frac{1}{8})$; $y = -\frac{1}{8}$; intercepts $(0, 0)$
**2e.** $V = (\frac{1}{4}, -\frac{1}{2})$, $F = (0, -\frac{1}{2})$; $x = \frac{1}{2}$; intercepts $(0, 0)$ and $(0, -1)$
**2f.** $V = (-\frac{3}{2}, -\frac{5}{4})$, $F = (-\frac{3}{2}, -1)$; $y = -3/2$; $(-\frac{3}{2} \pm \sqrt{5}/2, 0)$, $(0, 1)$   **4.** $y = -x^2 + \frac{3}{2}x - 1$   **7.** 5 million miles

## Section 12.2

**1a.** foci $(\pm 3, 0)$; ends of axes $(\pm 5, 0)$ and $(0, \pm 4)$   **1c.** $(\pm 1/\sqrt{2}, 0)$; $(\pm 1, 0)$, $(0, \pm 1/\sqrt{2})$   **2b.** $\dfrac{x^2}{25} + \dfrac{y^2}{16} = 1$

**3c.** $2a + 2c = 8 + 4\sqrt{3} \approx 14'11\frac{1}{8}''$   **4a.** $x^2 + 9y^2 = 9$   **4b.** $\dfrac{x^2}{25} + \dfrac{y^2}{24} = 1$   **4c.** $(x - 2)^2 + 5(y - 1)^2 = 5$

**6a.** $(-1, 3)$; $(-1 \pm \sqrt{7}, 3)$; $(-5, 3)$, $(3, 3)$ and $(-1, 0)$, $(-1, 6)$
**6c.** $(1, -1)$; $(1, -1 \pm \sqrt{8})$; $(1, 2)$, $(1, -4)$ and $(0, -1)$, $(2, -1)$   **7a.** $5y + 3x = 25$   **8.** $\pi\sqrt{2}$   **10.** $16\pi\sqrt{2}/3$
**12.** $4\pi a^2 b/3 = 4\pi r^3/3$ when $a = r$, $b = r$   **15.** $7x^2 - 2xy + 7y^2 = 48$

## Section 12.3

**1a.** vertices $(\pm 1, 0)$; foci $(\pm\sqrt{2}, 0)$; asymptotes $y = \pm x$   **1d.** vertices $(0, \pm 2)$; foci $(0, \pm\sqrt{5})$; asymptotes $y = \pm 2x$
**2a.** $5x^2 - 4y^2 = 5$   **2c.** $x^2 - y^2 = 2$
**3a.** vertices $(0, -1)$, $(2, -1)$; foci $(1 \pm \sqrt{2}, -1)$; asymptotes $y = -x$, $y = x - 2$
**3c.** $(-3, 1)$, $(1, 1)$; $(-1 \pm \sqrt{5}, 1)$; $y = 1 \pm \frac{1}{2}(x + 1)$   **3e.** vertices $(0, 0)$, $(0, 4)$; foci $(0, 2 \pm \sqrt{5})$; asymptotes $y = 2 \pm 2x$
**3f.** two lines $y = -x$, $y = x - 2$   **9a.** $\dfrac{x^2}{(270)^2} - \dfrac{y^2}{177{,}100} = 1$   **9b.** $x \approx 666$, $y \approx 949$

## Section 12.4

**1a.** vertices $\pm(2, -2)$; foci $\pm(2\sqrt{2}, -2\sqrt{2})$
**1c.** vertices $(\bar{x}, \bar{y}) = (\pm\sqrt{2}, 0)$, $(x, y) = (\pm\sqrt{8/5}, \pm\sqrt{2/5})$; foci $(\bar{x}, \bar{y}) = (\pm\sqrt{5}, 0)$, $(x, y) = \pm(2, 1)$
**1d.** $(\bar{x} + 1)^2 = \frac{4}{5} - \bar{y}$   **1f.** $\dfrac{\bar{x}^2}{4} + \bar{y}^2 = 1$, vertices $\pm(\sqrt{3}, 1)$, foci $\pm(3/2, \sqrt{3}/2)$   **2a.** $\delta = 1$, hyperbola
**2c.** $\delta = -8$, ellipse   **2f.** $\delta = 1$, hyperbola (but this one is degenerate)   **3a.** two horizontal lines
**3b.** one line   **3c.** empty   **8.** $ax^2 - xy + b = 0$, asymptotes $x = 0$ and $y = ax$

## Section 12.5

**1a.** $(1/\sqrt{2}, 1/\sqrt{2})$   **1c.** $(-1, 1)$   **1f.** $(\pi \cos 1, \pi \sin 1)$   **2a.** $(1, 0)_p$   **2c.** $(\sqrt{5}/2, 2\pi - \tan^{-1}2)_p$
**2e.** $(\sqrt{2}, 5\pi/4)_p$   **3a.** $(1, 0)_p$   **3c.** $(-\sqrt{5}/2, \pi - \tan^{-1}2)$   **3e.** $(-\sqrt{2}, \pi/4)_p$   **4a.** $(0, \theta)_p$ for all $\theta$
**4b.** $\left(\sqrt{2}, \dfrac{\pi}{4} + 2k\pi\right)$ and $\left(-\sqrt{2}, 5\dfrac{\pi}{4} + 2k\pi\right)$, $k = 0, \pm 1, \pm 2, \ldots$

**5.**

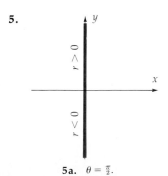

**5a.** $\theta = \frac{\pi}{2}$.

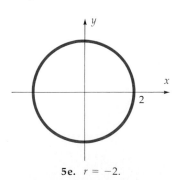

**5e.** $r = -2$.

**6a.** $r = 1$   **6b.** $r = 5 \sec \theta$
**6d.** $r^2 = 2 \csc 2\theta$   **6f.** $r = \sin \theta \sec^2\theta$

**7a.** $x = 2$   **7c.** $x = 1$   **7d.** $y = 2$   **8a.** circle, radius 1   **8b.** circle, radius 1/2

**9.**

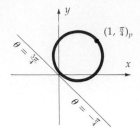

**9a.** $r = \sin(\theta + \frac{\pi}{4})$.

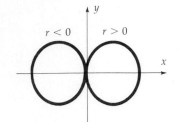

**9d.** $r^2 = \cos\theta$.

**10.**

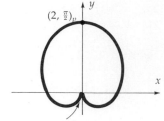

Tangent to $\theta = \frac{3\pi}{2}$

**10b.** $r = 1 + \sin\theta$.

**11.**

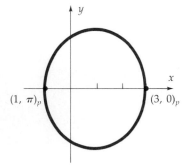

**11a.** $r = 2 + \cos\theta$.

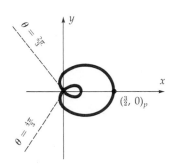

**11b.** $r = \frac{1}{2} + \cos\theta$.

**12.**

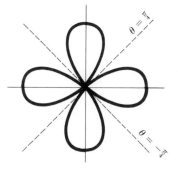

**12a.** $r = \cos 2\theta$.

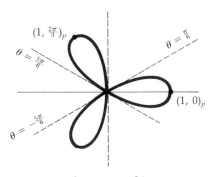

**12b.** $r = \cos 3\theta$.

**13.**

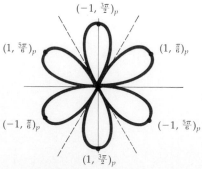

**13a.** $r^2 = \sin 3\theta$.

**13b.** $r^2 = \cos 2\theta$.

**15a.** parabola  **15b.** hyperbola  **15c.** ellipse  **16b.** $(d, \varphi)_p$  **17.** center $(a/2, b/2)$, radius $\frac{1}{2}\sqrt{a^2 + b^2}$

**20.**

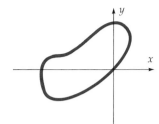

**22a.** $(0, \theta)_p$ and $(\sqrt{2}, \pi/4)_p$

**20a.** This is Figure 13, reflected through the origin.

## Section 12.6

**1b.** $e = 3/2$, vertices $(-1, 0)_p$ and $(1/5, \pi)_p$; center $(-3/5, 0)$; directrix $x = -1/3$; asymptotes $y = \pm(\sqrt{5}/2)(x + 3/5)$
**1c.** $e = 1/3$, vertices $(5/2, 0)_p$ and $(5/4, \pi)_p$, center $(5/8, 0)$, directrix $x = -5$
**1e.** $e = 1$, $V = \left(\dfrac{1}{2}, \dfrac{\pi}{2}\right)_p$, directrix $y = \pm 1$  **2a.** $e = 1$, $F = (0, 1/4)$, directrix $y = -1/4$
**2b.** $e = 0$; $F_1 = F_2 = (0, 0)$; directrices "at $\infty$," since $a/e$ "$= 1/0 = \infty$".  **2e.** $e = \sqrt{5}$, $x = \pm 1/\sqrt{5}$
**4.** (Halley) major axis 3600, minor axis $\approx$ 875, aphelion 3546  **5a.** no  **5b.** $b/a = \sqrt{1 - e^2}$
**5c.** angle $CFM = \sin^{-1}\sqrt{1 - e^2}$

## Section 12.7

**1a.** $\frac{4}{3}\pi^3$  **1c.** $9\pi/2$  **1e.** $9\pi/16$  **2a.** $(1, \pm\pi/2)_p$ and $(1, \pi)_p$  **2c.** $(0, 0)$ and $(2, \pm\pi/2)_p$
**2e.** $(0, 0)$, $(3/2, \pi/6)_p$, $(3/2, 5\pi/6)_p$  **3a.** $\pi/8 - 1/4$  **3c.** $4\pi - 6\sqrt{3}$  **3e.** $5\pi/3 - 2\sqrt{3}$  **4.** $3a^2/2$

## Chapter 12 Review

**1.** $8y = x^2 + 2x$; $F = (-1, 15/8)$, $V = (-1, -1/8)$  **2a.** $(y + 1)^2 + 3x = 6$; $V = (2, -1)$, $F = (5/4, -1)$
**2b.** $(x + 1)^2 - y^2 = 1$; vertices $(0, 0)$, $(-2, 0)$, foci $(-1 \pm \sqrt{2}, 0)$
**2c.** $x^2 + 2(y - \frac{1}{4})^2 = \frac{1}{8}$; vertices $(\pm 1/2\sqrt{2}, 1/4)$, foci $(\pm 1/4, 1/4)$
**3a.** parabola; $V = (3/2\sqrt{2}, 1/2\sqrt{2})$, $F = (1/\sqrt{2}, 1/\sqrt{2})$  **3b.** ellipse $V = \pm(-1, \sqrt{3})$, $F = \pm(2\sqrt{5}, 2\sqrt{3}/5)$
**3c.** hyperbola $V = \pm(1/\sqrt{6}, 1/\sqrt{6})$, $F = \pm(\sqrt{2/3}, \sqrt{2/3})$, $y = (-2 \pm \sqrt{3})x$  **4a.** $x^2 + 1 = 0$  **4b.** $x^2 + y^2 = 0$
**4c.** $x^2 = 0$  **4d.** $x^2 = 1$  **4e.** $xy = 0$  **5a.** $90°$  **5b.** $2\tan^{-1}\sqrt{e^2 - 1}$  **6a.** $2a = 200 \times 10^6$, $2b \approx 112 \times 10^6$,
$2c = 166 \times 10^6$  **6b.** $e = 0.83$  **7a.** circle  **7b.** parabola  **7c.** ellipse, $e = \frac{1}{2}$  **7d.** hyperbola, $e = 2$  **8a.** $3\pi/2$
**8b.** 1 (There are two loops.)  **9b.** $\pi + 3\sqrt{3}$  **10a.** $(x - y)^2 + 2(x + y) = 1$  **10c.** $r = \dfrac{1/\sqrt{2}}{1 - \cos(\theta + 3\pi/4)}$

## Section 13.1

**1.**

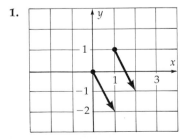

**1a.** $\langle 1, -2 \rangle$.

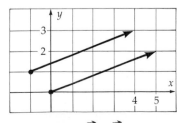

**1c.** $5\vec{i} + 2\vec{j}$.

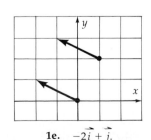

**1e.** $-2\vec{i} + \vec{j}$.

**2.** magnitudes are: a) $\sqrt{5}$ c) $\sqrt{29}$ e) $\sqrt{5}$ **3a.** $\langle 6 - 10 \rangle$ **4a.** $\langle 3, 6 \rangle = 3\langle 1, 2 \rangle$
**5a.** $\vec{P}_2 - \vec{P}_1 = \langle -1, 2 \rangle$ **6a.** $\langle 4, 0 \rangle$ **6c.** $\langle -1, 2 \rangle$ **6e.** $-\langle 6, 12 \rangle$ **7b.** $\langle 1/\sqrt{5}, -2/\sqrt{5} \rangle$
**8b.** $|\vec{v}| + |\vec{w}| = \sqrt{5} + \sqrt{13} \approx 5.84$ **9c.** $\langle -3, 0 \rangle$ **10a.** $(1/\sqrt{2}, 1/\sqrt{2})$ **10c.** $(1/\sqrt{2}, -1/\sqrt{2})$
**10e.** $(1/\sqrt{2}, 3/\sqrt{2})$ **10g.** $(-\sqrt{3}/2, \frac{1}{2})$ **13b.** True only if $\vec{v} = c\vec{w}$ with $c \geq 1$. **14a.** $(-1/3, 8/3)$
**17a.** $\vec{v} = \langle 2, -2 \rangle$ **27.** Least force: At midpoint of segment between the two charges of magnitude $q$. Greatest force: $a/\sqrt{2}$ away from that segment.

## Section 13.2

**1a.** straight line **1c.** circle **1f.** counterclockwise spiral **2a.** $y = -2x + 7$ (straight line)
**2b.** $y = 2(x - 1)^2$ (parabola) **2c.** $x = y^{2/3}$ **2f.** $y = 1 - x^2, |x| \leq 1$ **4.** $\vec{v} = \langle -2, -2 \rangle; \vec{P}(t) = \langle 1, 3 \rangle - 2t\langle 1, 1 \rangle$
**6a.** $x = \sin 2\theta, y = 2\sin^2\theta$ **8a.** All of $y = x^2$, traced once **8b.** $y = x^2$ with $x \geq 0$, traced twice
**8c.** $y = x^2$ with $|x| \leq 1$, traced $\infty$ often **10.** $\omega = \pi/10; x = 5\sin(\pi t/10)$ [or $-5\sin(\pi t/10)$], $y = 6 - 5\cos(\pi t/10)$

## Section 13.3

**1a.** $\langle 2, 1 \rangle$ **1b.** $\langle 1, 0 \rangle$ **1c.** $\langle 2, 0 \rangle$

**2a.** circle; $\vec{v}(\pi/3) = \langle 0, -6 \rangle, \vec{a}(\pi/3) = \langle 18, 0 \rangle$

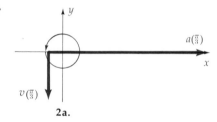

2a.

**2c.** one branch of hyperbola $xy = 1; \vec{a}(0) = \langle 4, 4 \rangle, \vec{v}(0) = \langle 2, -2 \rangle$

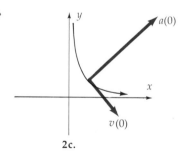

2c.

**2e.** segment $x + y = 1, 0 \leq x \leq 1; \vec{v}(\pi/2) = \langle 0, 0 \rangle, \vec{a}(\pi/2) = \langle -2, 2 \rangle$

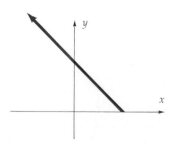

**2e.** $v(\frac{\pi}{2}) = 0, a(\frac{\pi}{2}) = \langle -2, 2 \rangle.$

**3a.** $L(t) = \langle -2, 0 \rangle + (t - \pi/3)\langle 0, -6 \rangle$ **4a.** $\vec{P}(t) = t\vec{i} + \vec{j}$ **4b.** $\vec{P}(t) = (e^t - t - 1)\vec{i} + (e^{-t} + t - 1)\vec{j}$
**4c.** a cycloid (see Sec. 13.2) **6.** $(5\cos 6 - 12\sin 6, 5\sin 6 + 12\cos 6)$ **9.** $9.533°, 80.467°$
**12a.** $7.2209 \times 10^{14}$ (mi)$^3$/(day)$^3$ **12b.** $26,347.8$ miles **16a.** $(5, 5\sqrt{3})$ **16b.** $\langle -5\sqrt{3}, 5 \rangle$

## Section 13.4

**1a.** $\pi r$   **1c.** $2(\sqrt{8}-1)$   **1e.** $\sqrt{2}(e^t-1)$   **3b.** $\pi a$   **3c.** $\dfrac{1}{2}\left[\dfrac{\pi}{2}\sqrt{1+\pi^2/4}+\ln(\pi/2+\sqrt{1+\pi^2/4})\right]$   **3e.** $1$

**5.** $8a$   **7a.** $4\sqrt{2}$

## Section 13.5

**4.** Do it at apogee. If it were done at perigee, the satellite would return to the same perigee, since its orbit is periodic. By Section 13.3, problem 11, the speed should be $\sqrt{GM/r_{\max}}$, with constants $G$ and $M$ described there.

## Chapter 13 Review

**1.**

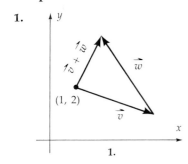

**2a.** $x=3-4t,\ y=5-3t$   **2b.** $(-1/5,\ 13/5)$   **2c.** $(5/4,\ 13/4)$

**3a.** $\vec{v}+\vec{w}=\langle v_1,v_2\rangle+\langle w_1,w_2\rangle=\langle v_1+w_1,v_2+w_2\rangle=\langle w_1+v_1,w_2+v_2\rangle=\vec{w}+\vec{v}$

**4a.** $\vec{v}(t)=\langle-3\sin 3t,6\cos 3t\rangle,\ \vec{a}(t)=\langle-9\cos 3t,-18\sin 3t\rangle,\ \text{speed}=3\sqrt{\sin^2 3t+4\cos^2 3t}$

**4c.** $\langle 1/\sqrt{2},\sqrt{2}\rangle+\left(1-\dfrac{\pi}{12}\right)\left(-\dfrac{3}{2}\sqrt{2},3\sqrt{2}\right)$   **4e.** $3\int_0^{2\pi}\sqrt{\sin^2 3t+4\cos^2 3t}\,dt$

**5.** $\vec{P}(t)=\tfrac14(1-\cos 2t)\vec{i}+\tfrac14(2t-\sin 2t)\vec{j}$   **6.** $\dfrac{1}{2}\left[\dfrac{1}{\pi-1}+\dfrac{1}{\pi+1}-\dfrac{1}{3\pi-1}-\dfrac{1}{3\pi+1}\right]$   **7b.** $(\pi a)^2$   **7d.** $\pi a^2$

## Section 14.1

**1b.** distance $2\sqrt{10}$   **3.** $A=(0,0,8),\ E=(-20,16,8)$
**4b.** $|P_1P_2|^2+|P_2P_3|^2=|P_1P_3|^2$; right angle at $P_2=(-1,3,3)$; area $3\sqrt{2}$   **5a.** $x^2+y^2+z^2=1$
**6c.** center $(1,-2,3)$, radius 0; "degenerate" sphere; does not pass through origin   **7c.** plane parallel to $xy$-plane, 2 units below it   **7d.** union of $yz$- and $xz$-planes   **8c.** $y=-1$   **10c.** $(x-3/2)^2+(y-7/2)^2+(z-3)^2=9/2$
**11a.** region in front of $yz$-plane   **11c.** region inside and on sphere of radius 1
**11g.** region between sphere of radius 1 and sphere of radius 2   **12f.** interior of a square of side 2

## Section 14.2

**1a.** $|\vec{v}|=\sqrt{3},\ |\vec{w}|=3,\ \cos\theta=1/(3\sqrt{3})$; acute   **1c.** $|\vec{v}|=\sqrt{3},\ |\vec{w}|=\sqrt{2},\ \cos\theta=\sqrt{2/3}$; acute   **2i.** $\langle 5,8,6\rangle$
**2iii.** $7$   **4.** $\cos\theta=10/\sqrt{1050}$   **6b.** $2\vec{i}$   **6e.** $\tfrac17\langle 2,-1,3\rangle$   **8b.** $(\tfrac53,\tfrac53,\tfrac43)$   **10b.** $1/\sqrt{5}$
**10c.** Yes, for $\vec{n}=2\vec{i}-\vec{j}$   **12b.** $\cos\theta=1/\sqrt{3},\ \theta\approx 54.7°$   **12c.** $\theta\approx 35.26°$   **16.** Try $\vec{i},\vec{j}$, and $\vec{k}$.
**19b.** $\vec{v}=c\vec{w}$ with $c\le 0$; or one of $\vec{v},\vec{w}$ is zero   **20b.** $(\vec{v}+\vec{w})\cdot(\vec{v}-\vec{w})=\vec{v}\cdot\vec{v}-\vec{v}\cdot\vec{w}+\vec{w}\cdot\vec{v}-\vec{w}\cdot\vec{w}=|\vec{v}|^2-|\vec{w}|^2$

## Section 14.3

**1a.** $1$   **1b.** $-1$   **1c.** $3$   **2.** b and c are left-handed   **3a.** $1$   **3c.** $0$   **3e.** $0$
**4.** $\vec{i}\times\vec{i}=\vec{0},\ \vec{i}\times\vec{j}=\vec{k},\ \vec{i}\times\vec{k}=-\vec{j}$, etc.   **5a.** $\langle 3,-6,3\rangle$   **5c.** $\langle 7,10,9\rangle$   **6b.** $3$
**7b.** $\vec{n}=\pm(\vec{i}+4\vec{j}-16\vec{k})$, area $=\sqrt{273}/2$   **8a.** $\sqrt{35}$   **8b.** $16$   **8c.** $\sin\varphi=16/\sqrt{910}$
**9a.** $\vec{T}=18\vec{i}+6\vec{j},\ |\vec{T}|=6\sqrt{10}$   **9b.** $x=3t,\ y=t,\ z=0$

## Section 14.4

**1c.** $\hat{n} = \langle 1, 1, 2 \rangle$; $(1, 0, 0)$, $(0, 1, 0)$, $(0, 0, 1/2)$   **1e.** $\hat{n} = \langle 1, 0, -1 \rangle$; $(2, 0, 0)$, $(0, 0, -2)$ (plane does not intersect $y$-axis)

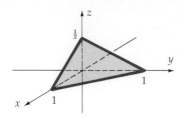

**1c.** The triangle, of course, is just *part* of the entire plane.

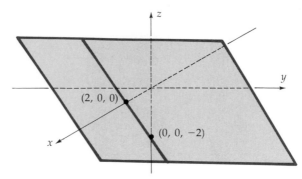

**1e.** The parallelogram is just *part* of the plane.

**2a.** $x + y - z = -1$   **2e.** $z = 0$   **3a.** $x = 1 + t, y = 1 - 3t, z = 2 + 2t$   **3b.** $\dfrac{x-1}{1} = \dfrac{y-1}{-3} = \dfrac{z-2}{2}$

**4.** $\langle 1, 1, -1 \rangle$   **5.** yes   **6.** no   **7.** yes   **9.** $x - y + z = 0$   **10a.** $z = z_0$   **11.** $3x - 2y + 4z = 20$

**14a.** $t = -8$, at $(9, 2, -5)$   **14b.** $\langle -1, 0, 1 \rangle \cdot \langle 1, 1, 1 \rangle = 0$   **16a.** $x - 2y - 7z = -24$

**23a.** intersection $(2, 3, 1)$ with $\cos \theta = 7/3\sqrt{17}$   **24.** $1/\sqrt{3}$

**28.** The line is parallel to both planes; when $t = 0$ it gives $(0, 0, 1)$, which is directly above the point $(0, 0, 0)$ on one plane, and below $(0, 0, 2)$ on the other.

## Section 14.5

**1b.** $(F \times G)' = \langle 0, t^2 e^t + 2t e^t - 1, -e^t(1 + t) \rangle$   **2b.** $s = \sqrt{(a\omega)^2 + b^2}\, t$   **2c.** $2\pi\sqrt{a^2 + (b/\omega)^2}$

**3.** $(a \cos 2\omega - 3a\omega \sin 2\omega, a \sin 2\omega + 3a\omega \cos 2\omega, 5b)$   **4b.** $s = 3t + 2t^3$   **4c.** $5$

**5.** $\vec{P} = \langle 3, 3, 2 \rangle + 3(t - 1)\langle 1, 2, 2 \rangle$   **6a.** $(a^2\omega^2 + b^2)^{-1/2}\langle -a\omega \sin \omega t, a\omega \cos \omega t, b \rangle$   **6b.** $\cos \theta = b/\sqrt{a^2\omega^2 + b^2}$

**7a.** $a\omega^2/(a^2\omega^2 + b^2)$   **8a.** $\langle -\cos \omega t, -\sin \omega t, 0 \rangle$   **10.** $\vec{a} = a\omega^2\vec{N}$   **25b.** $|\vec{a}| = \sqrt{25 + 160000r^{-2}}$

**30d.** $(496.4, 59.9, -4)$

## Section 14.6

**2a.** ellipsoid   **2b and d.** hyperboloid (one sheet)   **2c.** hyperboloid (two sheets)   **6.** hyperbolic paraboloid

**7.** sphere   **9.** circular cylinder   **11.** two planes intersecting in $z$-axis   **12.** "serpentine" cylinder

**15.** elliptic paraboloid   **16.** hyperbolic cylinder   **18.** double cone   **20.** single cone

## Chapter 14 Review

**1a.** $\langle -3, -4, -5 \rangle$   **1b.** $20$   **1c.** $\langle -1, 2, -1 \rangle$   **1d.** $5\sqrt{2}$   **1e.** $\cos^{-1}(20/\sqrt{14 \cdot 29})$   **1f.** $\sqrt{6}$

**2a.** $3$   **2b.** $(-1, 3, 3/2)$   **2c.** $\langle x, y, z \rangle = \langle -2, 2, 1 \rangle + t\langle 2, 2, 1 \rangle$   **2d.** $(-4, 0, 0)$   **2e.** $\cos \theta = \sqrt{2/3}$

**2f.** $3/\sqrt{2}$   **2g.** $x - y = -4$   **2h.** $2\sqrt{2}$   **2i.** $\langle x, y, z \rangle = \langle -2, 2, 1 \rangle + t\langle 1, -1, 0 \rangle$

**2j.** $(x + 1)^2 + (y - 3)^2 + (z - 3/2)^2 = 9/4$   **2k.** Yes   **3a.** false; $\vec{i} \times \vec{j} = \vec{k}, \vec{j} \times \vec{i} = -\vec{k}$   **3b.** true

**4.** Abbreviate $g/4$ as $\gamma$.

**4a.** $\vec{P}(1) = \langle \sin \gamma, \cos \gamma, -\gamma \rangle, \vec{v}(1) = \frac{1}{2}g\langle \cos \gamma, -\sin \gamma, -1 \rangle, \vec{a}(1) = \dfrac{1}{2}g\left\langle \cos \gamma - \dfrac{g}{2} \sin \gamma, -\sin \gamma - \dfrac{g}{2} \cos \gamma, -1 \right\rangle$

**4b.** $t = \sqrt{8\pi/g}$   **4c.** $x = \sin \gamma - \frac{1}{2}g(t - 1) \cos \gamma, y = \cos \gamma - \frac{1}{2}g(t - 1) \sin \gamma, z = -\gamma - \frac{1}{2}g(t - 1)$

**4d.** $g/2\sqrt{2}$   **4e.** $(1/\sqrt{2}) \langle \cos \gamma t^2, -\sin \gamma t^2, -1 \rangle$   **4f.** $\frac{1}{2}\langle -\sin \gamma t^2, -\cos \gamma t^2, 0 \rangle$   **4g.** $1/2$

**4h.** $\vec{N} = \langle -\sin \gamma t^2, -\cos \gamma t^2, 0 \rangle$   **4i.** $\vec{a} = (g/\sqrt{2})\vec{T} + (g^2 t^2/4)\vec{N}$   **4j.** Yes   **5.** $\vec{P} = \langle 2 - 2 \cos t, 3t - 3 \sin t, t \rangle$

**8a.** circular cylinder   **8b.** circular paraboloid   **8c.** ellipsoid   **8d.** double cone   **8e.** hyperboloid, 1 sheet

**8f.** hyperboloid, 2 sheets **8g.** single cone **8h.** circular paraboloid **8i.** plane **8j.** plane
**8k.** parabolic cylinder **8l.** the point $(0, 0, 0)$ **8m.** empty graph

## Section 15.1

**1a.** $f(u + v, u - v) = 1 + u + v$ **1b.** $f(u + v, u - v) = e^{2u}$ **2a.** $\{(x, y): x^2 + y^2 \le 2\}$
**2c.** first and third quadrants, including $x$-axis but excluding $y$-axis **2e.** everything but the two axes
**2f.** first and third quadrants
**3a.** $x$-slope 1, $y$-slope $-1$ **3c.** $x$-slope $= y$-slope $= 0$

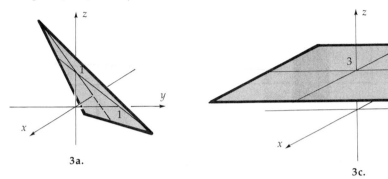

**3a.**

**3c.**

**4a.** $f(x, y) = 3x - y + 2$ **5a.** level ridge along the negative $y$-axis, level valley along the positive $y$-axis
**5c.** From $(0, 0)$: rising in both directions along $x$-axis, falling in both directions along $y$-axis
**5e.** valley along $x$-axis, descending as $x \to +\infty$ **7a.** plane **7b.** circular paraboloid
**7c.** hyperbolic paraboloid, saddle at $(0, 0)$ **7f.** cone **7i.** infinitely high at $(0, 0)$ **8a.** plane
**8c.** circular paraboloid **9a.** $z = x + y$ **9d.** $z = (x - y)^3$
**10b.** Nearly level as $x \to +\infty$. As $x \to -\infty$, ridges rising along lines $y = \pm 2k\pi$, valleys descending along lines
$y = \pm(2k + 1)\pi$.

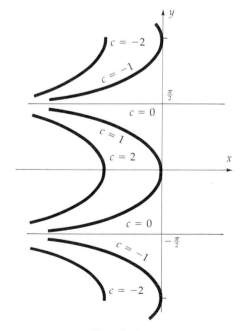

**10b.** $\{e^{-x}\cos y = c\}$

**13.** $z = r^3 \cos 3\theta < 0$ when $\dfrac{\pi}{6} < \theta < \dfrac{\pi}{2}, \dfrac{5\pi}{6} < \theta < \dfrac{7\pi}{6}, \dfrac{9\pi}{6} < \dfrac{11\pi}{6}$. These regions provide for the monkey's legs and tail.

## Section 15.2

**1a.** $(1, 0)$ **2a.** $0$ **2c.** $-1$ **3a.** $(0, 0)$ **3b.** $(0, 0)$ **3c.** the two coordinate axes
**3d.** not defined for $x + y \leq 0$; continuous elsewhere **3f.** continuous everywhere
**6.** All points. This is clear except perhaps on the axes. If $(x_k, y_k) \to (x, 0)$ then $|g(x_k, y_k)| \leq |x_k y_k| \to |x \cdot 0| = 0$, so
$g(x_k, y_k) \to g(x, 0)$; the same is true for $(x_k, y_k) \to (0, y)$. **8.** $0$

## Section 15.3

**1b.** $f_x = e^{xy}(1 + xy), f_y = x^2 e^{xy}, \nabla f = e^{xy}(1 + xy, x^2)$ **1d.** $\nabla f = \dfrac{2x}{x^2 + y^2}\vec{i} + \dfrac{2y}{x^2 + y^2}\vec{j}$

**1e.** $\nabla f = \dfrac{-2x^2 + 2y^2}{(x^2 + y^2)^2}\vec{i} + \dfrac{-4xy}{(x^2 + y^2)^2}\vec{j}$ **1i.** $\nabla f = -\sin(x - y)\vec{i} + \sin(x - y)\vec{j}$

**2a.**

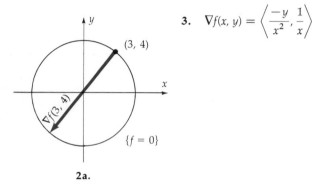

**2a.**

**3.** $\nabla f(x, y) = \left\langle \dfrac{-y}{x^2}, \dfrac{1}{x} \right\rangle$

**4.**

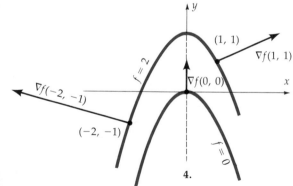

**4.**

**6a.** $z = 5 + 2(x - 1) + 4(y - 2)$

**7a.** $x = e + t/e, y = 1 - t, z = 1 - t$ **7b.** $x = 0, y = 0, z = -t$ **8.** $-6y$ **9.** $-x/3z$
**10a.** all points on the axes **10b.** $(0, \pm 1/\sqrt{2})$ **12e.** $-1$ **13.** $mg\nabla f$ **14a.** $\langle 4, 4, -1 \rangle$ **14b.** $\langle 0, 0, 1 \rangle$
**14c.** $\langle 4, -4, 0 \rangle$ **15a.** $\vec{v} = f_y\vec{i} - f_x\vec{j}$

## Section 15.4

**1a.** $f_{\text{lin}}(x, y) = 5 + 2(x - 1) + 4(y - 2); f_{\text{lin}}(1.02, 2.01) = 5.08$ **1c.** $f_{\text{lin}}(x, y) = 1 + 2(x - 1) + 2(y - 2); 1.1$
**1d.** $f_{\text{lin}}(x, y) = 2 + \frac{1}{12}(x - 8) + \frac{2}{3}(y - 1); 1.95$ **2.** $\Delta A \approx y\,\Delta x + x\,\Delta y = .37$ **4.** $\Delta C \approx -.05$, a fall in cost
**7.** $\Delta P \approx (0.75)k$

## Section 15.5

**1a.** $-y \sin t + x \cos t = \cos^2 t - \sin^2 t$ **1c.** $\left[\dfrac{-y}{x^2 + y^2}\right]3t^2 + \left[\dfrac{x}{x^2 + y^2}\right]6t^2 = 0$ **2a.** $3\sqrt{2}$ **2c.** $-1/5\sqrt{2}$

**3b.** $-4e^2/\sqrt{5}$ **4a.** $\vec{u} = (1/\sqrt{10})\langle 3, -1\rangle$, $D_{\vec{u}}f = 4\sqrt{10}$ **5.** $(-1, 2)$ (one step west and two steps north) **7.** $\vec{j}$

**8a.** downhill ($D_{\vec{u}}z < 0$) **8b.** NE; SW **8c.** NW and SE **10a.** $10\pi/3$ **10b.** $2\pi(10 + 2\sqrt{34} + 10/\sqrt{34})$

**12b.** $T_x \approx -.2$, $T_y \approx -.1$, $D_{\vec{u}}T \approx .3/\sqrt{2}$ **14.** $y = \frac{1}{8}x^3$ **16.** $\dot{v} = -3\vec{i} - \vec{j} - 10\vec{k}$ **21c.** $-f(P_0)|\nabla f(P_0)|^{-2}\nabla f(P_0)$

## Section 15.6

**1a.** $0$; $1$; $1$; $0$ **1c.** $2$; $2$; $2$; $6$ **1e.** $e^x \cos y$; $-e^x \sin y$; $-e^x \sin y$; $-e^x \cos y$

**3a.** $\dfrac{\partial^2 f}{\partial t^2} = -c^2 \sin(x + ct) = c^2 \dfrac{\partial^2 f}{\partial x^2}$ **4.** $\partial^2 f/\partial t^2 = c^2 \cdot \partial^2 f/\partial x^2$

## Section 15.7

**1a.** $(-1, -2/3)$, min **1b.** $(0, 0)$, saddle **2a.** $(-1, -1)$, local max; $(0, 0)$, saddle

**2b.** $(1, 1)$ and $(-1, -1)$, local min; $(0, 0)$, saddle **4a.** local min **4c.** neither **4e.** neither **4f.** local min

**7.** $1/\sqrt{5}$ **9.** $(\frac{1}{3}, \frac{1}{3}, \frac{1}{3})$ **10.** $(1, 1, 1)$, $(-1, -1, -1)$, $(-1, 1, -1)$, and $(-1, -1, 1)$

**13a.** $(0, 0)$ degenerate, $(\pm\sqrt{2}, 0)$ minima **13b.** $\{f = 0\}$ is two circles tangent to the $y$ axis at the origin.

**13c.** $f > 0$ outside $\{f = 0\}$, and $f < 0$ inside $\{f = 0\}$.

**13d.**

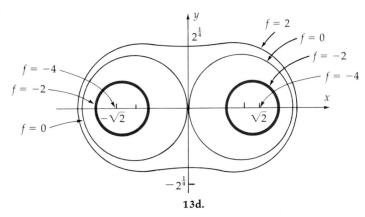

13d.

## Section 15.8

**1a.** $y = -\frac{19}{26}x + \frac{49}{26}$ **1c.** $y = 1.3x - 0.2$

## Chapter 15 Review

**1a.** $xy > 0$ (first and third quadrants) **1b.** $xy \neq 0$ (all but the coordinate axes) **1c.** all $(x, y) \neq (0, 0)$.

**2a.** $\{f = 0\}$ is two perpendicular lines; $\{f = \pm 2\}$ and $\{f = \pm 4\}$ are hyperbolas, asymptotic to $\{f = 0\}$.

**2b.** a hyperbolic paraboloid **2c.** $2\vec{i} + 2\vec{j}$; the arrow goes from $(2, 1)$ to $(4, 3)$, is perpendicular to $\{f = 4\}$.

**2d.** $-6/\sqrt{5}$ **2e.** $z = 4 + 2(x - 2) + 2(y - 1)$ **3a.** increasing at $10°/\text{sec}$ **3b.** $10/\sqrt{13}$ **3c.** $(2.2, 1.3)$

**3d.** $\vec{i} + \vec{j}$ **4.** 4.85 atm **5a.** $\Delta T \approx k(P - aV^{-2} + 2abV^{-3})\Delta V$ **5b.** $0 = \Delta T \Rightarrow \Delta P \approx \dfrac{aV - pV^3 - 2ab}{V^4 - bV^3}\Delta V$

**6.** No. If $(x_k, y_k) \to (0, 0)$ along the line $y = x$ then $\lim f(x_k, y_k) = 1 \neq f(0, 0)$.

**8a.** $(0, \pm 1)$ saddles, $(1/\sqrt{3}, 0)$ a local min, $(-1/\sqrt{3}, 0)$ a local max   **9.**   $x \cdot u_x = xy\varphi'(xy) = y \cdot u_y$   **10.**   $f_{xy} \neq f_{yx}$
**11a.**   $xy = c$   **11c.**   $y = mx$   **11e.**   $y = \ln|\sec x| + c$

## Section 16.1

**1a.**   1   **1b.**   $e^{7x} \cos 5x$   **1c.**   3   **1d.**   $4e^7 \cos 5$   **1e.**   0   **2.**   $xy + yz = -10$
**3a.**   $\nabla f = e^{xyz}(yz\vec{i} + xz\vec{j} + xy\vec{k}); 2/\sqrt{5}$   **3b.**   $\nabla f = a\vec{i} + b\vec{j} + c\vec{k}; \sqrt{a^2 + b^2 + c^2}$   **3c.**   $\nabla f = \langle 2x, 4y, -2z \rangle; -2$
**4a.**   $x = 0$   **4b.**   $a(x - x_0) + b(y - y_0) + c(z - z_0) = 0$   **4c.**   $(x - 1) + 2(y - 1) - (z - 1) = 0$
**6a.**   $-2x \sin t + 2y \cos t + 2z$   **6c.**   $3x^2y^2z + 4x^3yzt + 3x^3y^2t^2$   **7.**   $-\vec{i} - 6\vec{j} + \vec{k}$
**8b.**   No. $\vec{F} = qe^{5x-12y}[(5\vec{i} - 12\vec{j}) \cos 13z - 13\vec{k} \sin 13z]$, so $|\vec{F}| = 13qe^{5x-12y} > 0$.
**9a.**   $c = -1$, empty; $c = 0$, origin; $c = 1$, unit sphere
**9b.**   $c = -1$, hyperboloid of two sheets; $c = 0$, cone; $c = 1$, hyperboloid of one sheet
**9e.**   $c = -1$ and $c = 0$, empty; $c = 1$, cone   **10a.**   $(-\frac{1}{2}, 0, 0)$   **10b.**   $(0, 0, 0)$   **10c.**   The z-axis   **11.**   1.21
**14d.**   5%
**19.**   $\pi/6$ [the complement of the angle between the tangent to the curve and the normal to the sphere at $(\cos 1, \pm \sin 1, \pm 1)$]

## Section 16.2

**1a.**   $\partial w/\partial u = 2x \cos v + 2y \sin v = 2u$, $\partial w/\partial v = -2xu \sin v + 2yu \cos v = 0$
**2a.**   $w_x = e^{v^2}$, $w_y = (1 + 2uv)e^{v^2}$, $w_z = 2uve^{v^2}$   **3a.**   $\partial w/\partial \rho = 2x_1 \sin \varphi \cos \theta + 2x_2 \sin \varphi \sin \theta + 2x_3 \cos \varphi = 2\rho$
**7a.**   $xw_x = yw_y$   **11b.**   $\dfrac{\partial^2 w}{\partial t^2} = c^2 \dfrac{\partial^2 w}{\partial x^2}$   **15a.**   $w_s = w_x x_u u_s + w_x x_v v_s + w_y y_u u_s + w_y y_v v_s$

**16b.**   By 16a), $\partial f/\partial u$ is a function of $u$ alone: $\dfrac{\partial f(u, v)}{\partial u} = g(u)$. Then for each fixed $v$, the function

$f(u, v) = \int_0^u g(z)\, dz$ solves $\dfrac{\partial f}{\partial u} = g(u)$; the general solution is this integral plus a constant of integration, *which may depend*
*on* $v$: $f(u, v) = \int_0^u g(z)\, dz + \psi(v) = \varphi(u) + \psi(v)$.

## Section 16.3

**1a.**   $x = \sqrt{1 - y^2}$   **1b.**   $x = \sqrt{1 - y^2}$ or $y = \sqrt{1 - x^2}$   **1c.**   $y = -\sqrt{1 - x^2}$   **3.**   neither   **4.**   both
**8.**   All the partial derivatives $\dfrac{\partial z}{\partial x}, \dfrac{\partial y}{\partial z}$, etc. are $-1$.
**11c.**   When $T$ and $P$ are the independent variables, then $\partial P/\partial P = 1$ and $\partial T/\partial P = 0$.

## Section 16.4

**1a.**   $\frac{1}{2}(x - y)^2 = c$. Solutions are straight lines of slope 1. $\left(\text{This equation is equivalent to } \dfrac{dy}{dx} = 1.\right)$
**1c.**   not exact
**1e.**   $\frac{1}{2}\ln(x^2 + y^2) = c$. Solutions are circles. (This equation can also be solved by separation of variables.)
**1g.**   $e^x \sin y = c$, or $x = C - \ln|\sin y|$. All solutions are asymptotic to two of the lines $y = 0, \pm\pi, \pm 2\pi, \ldots$ as
$x \to +\infty$. (This, too, can be solved by separation.)
**2a.**   $e^{3x} \sin 2y = \sin 2$   **4a.**   $x^2 + y^2 = c$

## Section 16.5

**2.**   max 1 at $(1, 1)$

**3.** max $3\sqrt{3}/4$ at $(-3/2, -\sqrt{3}/2)$; min $-3\sqrt{3}/4$ at $\left(-\dfrac{3}{2}, -\dfrac{\sqrt{3}}{2}\right)$

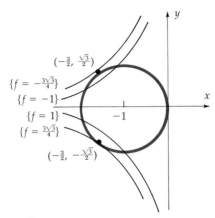

$\left(-\frac{3}{2}, \frac{\sqrt{3}}{2}\right)$

$\{f = -\frac{3\sqrt{3}}{4}\}$
$\{f = -1\}$
$\{f = 1\}$
$\{f = \frac{3\sqrt{3}}{4}\}$

$-1$

$\left(-\frac{3}{2}, -\frac{\sqrt{3}}{2}\right)$

**3.** Maximum and minimum of $f(x, y) = xy$ on circle $(x + 1)^2 + y^2 = 1$.

**5.** max $\sqrt{3}$ at $(1/\sqrt{3}, 1/\sqrt{3}, 1/\sqrt{3})$; min $-\sqrt{3}$ at $\left(\dfrac{-1}{\sqrt{3}}, \dfrac{-1}{\sqrt{3}}, \dfrac{-1}{\sqrt{3}}\right)$

**8.** max at $(3\cos\theta, 3\sin\theta)$ with $\theta = -\dfrac{1}{2}\tan^{-1}2 \pm \dfrac{\pi}{2}$; min with $\theta = -\dfrac{1}{2}\tan^{-1}2$ and $\pi - \dfrac{1}{2}\tan^{-1}2$.

**12.** $(0, \pm\sqrt{2/3}, 2/3)$. $\nabla|\vec{P} - \vec{Q}|^2 = 2(\vec{P} - \vec{Q}) = \lambda\nabla g$ shows that the vector from $Q$ to $P$ is normal to $\{g = c\}$.

**14a.** max $3\sqrt{3}/4$ at $(-3/2, -\sqrt{3}/2)$; min $-3\sqrt{3}/4$ at $(-3/2, \sqrt{3}/2)$ **16.** $x/y = \alpha/\beta$

**23a.** Let $\theta =$ angle opposite $z$. Solve for $\cos\theta$ from the law of cosines, then compute $A = \frac{1}{2}xy \sin\theta$.

## Chapter 16 Review

**1a.** $\{f = 0\}$ is empty, $\{f = 1\}$ is a double cone surrounding the z-axis, $\{f = \frac{1}{2}\}$ is a hyperboloid of two sheets inside the cone, $\{f = 2\}$ is a hyperboloid of one sheet surrounding the cone.

**1b.** $\nabla f(1, -1, 1) = 2\vec{i} - 2\vec{j} - 4\vec{k}$ **1c.** $x - y - 2z = 0$ **1d.** $-10/3\sqrt{2}$ **1e.** $-\vec{i} + \vec{j} + 2\vec{k}$

**1f.** $-10; -10/\sqrt{30}$ **2a.** $\pm 6$ **2b.** $\pm 6.6$ **3.** See Section 16.2, problem 5b

**4a.** $f_u(x, y, \varphi(x, y)) + f_w(x, y, \varphi(x, y))\varphi_x(x, y)$ **4b.** $\partial\varphi/\partial x = -f_u(x, y, \varphi(x, y))/f_w(x, y, \varphi(x, y))$

**5a.** $\varphi_u = f_x(u - v, u + v) + f_y(u - v, u + v)$ **5b.** $\varphi_v = 0$ **6b.** $w_{vu} = 2(w_{xx} - w_{yy})$ **7a.** $\langle \frac{1}{6}, \frac{1}{3}, \frac{1}{6} \rangle$

**7b.** $\langle 1/6, 1/3, 1/6 \rangle = (1/6)\langle 1, 2, 1 \rangle$, and $\langle 1, 2, 1 \rangle$ is a normal to the plane.

## Section 17.1

**1a.** $2/3$ **1c.** $2$ **1e.** $0$ **1g.** $0$ **3a.** $0$ **3b.** $8/3$

**4a.** $V = 10/3$

$(-1, 2, 5)$

$(-1, 2, 0)$

$2$

**5a.** $4/3$ **6a.** $128/9$ **7a.** $8$

**4a.** $V = \frac{10}{3}$.

## Section 17.2

**1a.** $8/3$  **1b.** $3/20$  **1d.** $162/5$  **1e.** $36/5$  **2.**

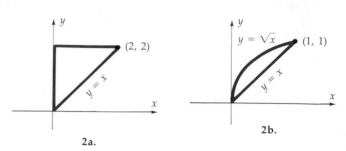

2a.

2b.

**4a.** $16/3$  **4c.** $\frac{1}{2}\ln 5$  **5a.** $12\pi$  **5d.** $\frac{4}{3}\pi abc$  **6a.** $\frac{1}{4}(e^4 - 1)$

**6b.** $\dfrac{1 - \cos\theta}{3}$  **9.** yes for a, e, f  **13a.** $1/2$  **13b.** $+\infty$

## Section 17.3

**1a.** $m = \delta/2$, $(\bar{x}, \bar{y}) = (1/3, 1/3)$  **1b.** $\bar{x} = \dfrac{(a-1)e^a + (a+1)e^{-a}}{(e^a - e^{-a})}$  **1c.** $(\bar{x}, \bar{y}) = (0, 3/5)$  **1d.** $(\bar{x}, \bar{y}) = (0, 5/7)$

**4c.** $\dfrac{ab}{12}(a^2 + b^2)$  **6b.** $8(e^{1/2} - e^{-1/2})$  **7a.** $M = 4$, $(\bar{x}, \bar{y}) = (0, 0)$

## Section 17.4

**1a.** $2\pi/3$  **1c.** $3\pi$  **1e.** $2/3$  **1f.** $\pi/8$  **1h.** $+\infty$  **2a.** $\pi\ln 2$  **2c.** $2a^3/3$  **2d.** $\int_0^{\pi/2}\int_0^{2\cos\theta} r^2\, dr\, d\theta = 16/9$

**4a.** $\bar{y} = 37/40$  **4d.** $\bar{y} = \dfrac{8\pi + 3\sqrt{3}}{4\pi + 6\sqrt{3}} a$  **4f.** $\bar{x} = \bar{y} = \dfrac{4}{3\pi} \cdot \dfrac{b^2 + ab + a^2}{a + b}$  **5a.** $2\pi$  **5c.** $4\sqrt{3}\pi$  **6a.** $Pa^4/2$

**6c.** $(a^4/4)\left(\dfrac{40}{3} + \dfrac{27\pi}{8}\right)$  **7.** $\int_0^{\pi/4}\int_0^{\sec\theta} f(r\cos\theta, r\sin\theta) r\, dr\, d\theta$

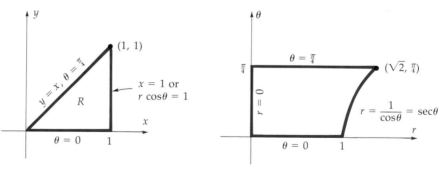

7b.

## Section 17.5

**1b.** $1 - e^{-1/2}$  **1c.** $\frac{1}{2}(1 - e^{-.005})$  **3a.** $(1 - e^{-a/\mu})(1 - e^{-b/\mu})$  **3c.** $e^{-2b/\mu}$

## Section 17.6

**1a.** $4\sqrt{14}$  **1c.** $\dfrac{2\pi}{3}(5^{3/2} - 2^{3/2})$  **2b.** $\pi\sqrt{10}$ (ellipse, $a = \sqrt{10}$, $b = 1$)  **2c.** $\pi/2$ (1/8 of unit sphere)

**3a.** $4\pi\sqrt{11}$ **4a.** $\frac{1}{2}\sqrt{a^2b^2 + b^2c^2 + c^2a^2}$ **4c.** $6\pi$ **5.** $(\sqrt{10}/3)(200)(400)$ ft$^2$

**10.** $2\pi\int_a^b \sqrt{c^2 + r^2}\, dr = \pi[b\sqrt{b^2 + c^2} - a\sqrt{a^2 + c^2} + b^2\ln(b + \sqrt{b^2 + c^2}) - a^2\ln(a + \sqrt{a^2 + c^2})]$

## Section 17.7

**1a.** $(1 - e^{-1})^3$ (cube) **1c.** 1 (octant) **1e.** $\frac{4}{5}$ ($\frac{1}{8}$ of ellipsoid) **3a.** 0 (the part for $x > 0$ cancels the part for $x < 0$.)

**3b.** $2\pi/15$ **4a,b.** 0 **5a.** $\pi a^4/16$ **7c.**

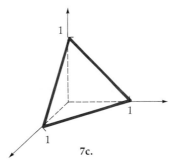

7c.

**8a.** $\pi$ **8c.** $16a^3/3$ **10a.** $(1/4, 1/4, 1/4)$ **10c.** $(a/4, b/4, c/4)$ **10d.** $3h/4$ **13a.** $\dfrac{\pi h R^4}{2}$ **13c.** $1/30$

**13e.** $\dfrac{\pi r^4 h}{2} + \pi r^2 a^2 h = $ (moment about axis of cylinder) + (mass of cylinder) $\times$ (distance from axis of cylinder to $z$ axis)$^2$

## Section 17.8

**1a.** cylinder of radius 2 about $z$-axis **1c.** region between two cylinders **1d.** the part of the $yz$-plane where $y \geq 0$
**1e.** the part of the plane $x = y$ where $x \geq 0$ **2a.** sphere **2c.** spherical shell **2d.** cone **2e.** $xy$-plane

**2f.** positive $z$-axis **2h.** solid cone **3.** $\rho \approx 6400$ km, $\varphi = \pi/3$, $\theta = -\pi/2$ **5.** $\dfrac{2\pi a^3}{3}\left(1 - \dfrac{1}{\sqrt{2}}\right)$

**6a.** $\dfrac{8\pi}{15}(b^5 - a^5)$ **6c.** $\frac{1}{2}\pi b(r_2^4 - r_1^4)$ **7a.** $\dfrac{3a}{8}(1 + \cos\varphi_0)$ **12b.** $(2\pi a)(\pi b^2)\left[\dfrac{3b^2}{4} + a^2\right] = V\left[\dfrac{3b^2}{4} + a^2\right]$

## Section 17.9

**1a.** 2 **1b.** $-2$ **2a.** $4(u^2 + v^2)$ **4a.** $\pi ab$ **4b.** $e - e^{-1}$

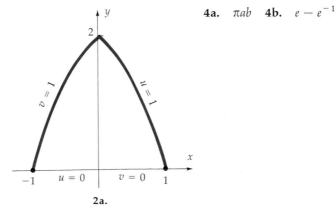

2a.

**5b.** $\dfrac{\gamma - 1}{\gamma - 2}[d^\alpha - c^\alpha][b^{(\gamma - 1)\alpha} - a^{(\gamma - 2)\alpha}]$, wherein $\alpha = 1/(\gamma - 1)$ **6.** $\ln 2$ **11.** $\dfrac{\pi}{12}(17^{3/2} - 1)$

## Chapter 17 Review

**1.** 7/6  **2.** $\int_{-1}^{1} \int_{-\sqrt{1-y^2}}^{0} \sqrt{x^2 + y^2}\, dx\, dy$ or $\int_{-1}^{0} \int_{-\sqrt{1-x^2}}^{\sqrt{1-x^2}} \sqrt{x^2 + y^2}\, dy\, dx$  **3.** $[6^{3/2} - 2^{3/2}]/24$  **4.** 8

**5a.** $\bar{x} = 4/3\pi = \bar{y}$  **5b.** $\bar{x} = 1 - \dfrac{2}{\pi} = \bar{y}$  **6.** $\bar{x} = \frac{1}{2},\ \bar{y} = 5/(8\pi - 3\sqrt{3})$  **7a.** $\bar{x} = 3/4,\ \bar{y} = 3/8$

**7b.** $(0, 3\pi a/16, 3\pi a/32)$  **8a.** $\int_{0}^{\pi/2} \int_{0}^{1} \int_{0}^{\sqrt{1-r^2}} (r\cos\theta)e^{-r^2 - z^2} r\, dz\, dr\, d\theta$

**8b.** $\int_{0}^{\pi/2} \int_{0}^{\pi/2} \int_{0}^{1} \rho \sin\varphi\, \cos\theta\, e^{-\rho^2} \rho^2 \sin\varphi\, d\rho\, d\varphi\, d\theta$

## Section 18.1

**1a.** 17/35  **1c.** 3  **1e.** 26/15  **2.** $1/r - 1/R$  **3a.** 2  **4a.** $2\pi$  **4c.** $4\pi$  **5a.** 0  **5c.** 0  **7a.** $2\pi$  **7c.** 0

## Section 18.2

**1d.** 2  **2.**

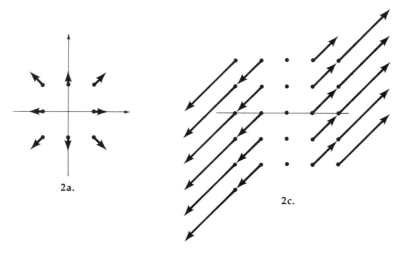

2a.

2c.

**3a.** 0  **3c.** $\vec{k}$  **5a.** 5/2  **5c.** 0  **7b.** $-\vec{i} - \vec{j} - \vec{k}$  **9a.** $\pi^2$  **11b.** $2\pi$

## Section 18.3

**1a.** 4  **1c.** 0  **2b.** 0  **6b.** $3\pi/8$  **9.** $2\pi$

## Section 18.4

**2a.** $\frac{2}{3}\pi a^3$  **2b.**

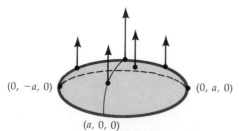

$(0, -a, 0)$  $(0, a, 0)$

$(a, 0, 0)$

2b. $\vec{F} = 0$ at boundary of hemisphere.

**4c.** $\int_{0}^{1} \int_{0}^{1} (-1)\, dy\, dz = -\frac{1}{2}$  **5d.** $-4/5$  **7.** 6  **14c.** $2\theta_0$

## Section 18.5

**2.** $-1$  **4.** $-\pi$  **9b.** connected, not simply connected  **9d.** not connected, but simply connected

# Chapter 18 Review

**1a.** $8/5$ **1b.** $-8/5$ **1c.** $-1/3$ **1d.** $0$ **2.**

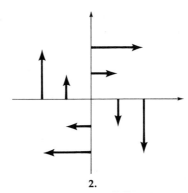

2.

**3a.** $-2\pi$ **3b.** $8\pi$ **3c.** $e - 2$ **3d.** $-2$
**4.** 3a, b,d) can be done by Green's Theorem **5a.** not a gradient **5b.** $\nabla(xy)$ **5c.** $\nabla(xe^{-x+y+2z})$
**5d.** $\nabla(\frac{1}{2}\ln(x^2 + y^2))$
**5e.** There is no potential $f$ valid for all $(x, y) \neq (0, 0)$, since $\int_C \vec{F} \cdot d\vec{P} \neq 0$ for a circle $C$ surrounding $(0, 0)$.
**5f.** $\nabla(\cot^{-1}(y/x))$ **6.** $f = xyz$ **6a.** $0$ **6b.** $0$ **6c.** $-3$ **7a.** $-1$ **7b.** $2\vec{k}$ **7c.** $0$ **7d.** $-2\pi/3$
**7e.** $-2\pi/3$ **7f.** $2\pi$ **7g.** $2\pi$ **7h.** $2\pi$
**8.** Answers to d and c differ by answer to e. Answers g and h are equal, since their difference is the integral of $\nabla \cdot (\nabla \times \vec{F})$ over the inside of the upper hemisphere, and $\nabla \cdot (\nabla \times \vec{F}) = 0$ for any $C^2$ vector field $\vec{F}$.
**9.** Answers g and h are equal to answer f, by Stokes' Theorem.
**10.** true if $R = R^3$; false if $R = R^3$ with one axis removed

# Section 19.1

**1b.** $9 + 7i$ **1e.** $ac - bd + i(bc + ad)$ **2a.** $\frac{1}{2}(1 - i)$ **2b.** $-i$

**3.**

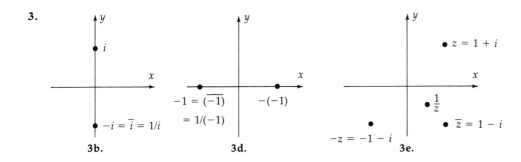

3b. 3d. 3e.

**4a.** circle, radius 2, center 1 **4c.** vertical line $x = 2$ **8a.** perpendicular bisector of segment from $i$ to 2

**8b.** parabola with focus $z = 1$, directrix the imaginary axis **10a.** $\pm\dfrac{1}{\sqrt{2}}(1 + i)$

# Section 19.2

**1a.** $\arg(1 + i) = \dfrac{\pi}{4} \pm 2k\pi$ **1c.** $\arg(1 + 3i) = \arctan 3 \pm 2k\pi$ **1f.** $\arg\left(-\dfrac{3}{2} + i\dfrac{\sqrt{3}}{2}\right) = \dfrac{5\pi}{6} \pm 2k\pi$

**3.** $e^{z+2\pi i} = e^z e^{2\pi i} = e^z(\cos 2\pi + i \sin 2\pi) = e^z$ **5.** $\pm 1, \pm i$ **6.** $\pm 2^{1/6}i, \pm 2^{-5/6}(\sqrt{3} \pm i)$
**7b.** $z^2 = \sqrt{3}e^{i5\pi/6}, \sqrt{3}e^{i7\pi/6}; z = \pm 3^{1/4}e^{i5\pi/12}, \pm 3^{1/4}e^{i7\pi/12}$

## Section 19.3

**1a.** $e^{-2it}$ and $e^{2it}$ **1c.** $e^{-t}$ **1e.** $e^{-t}e^{it}$ and $e^{-t}e^{-it}$ **7a.** $\displaystyle\int_0^\infty e^{ixt}e^{-t}\,dt = \frac{1}{1-ix} = \frac{1+ix}{1+x^2}$

**7c.** $\displaystyle\int_{-\infty}^\infty e^{ixt}e^{-|t|}\,dt = \frac{2}{1+x^2}$ **10b.** $r^{-2}e^{rt}(rt-1)$

## Section 19.4

**1a.** roots $\pm 2i$; $y = C_1\cos 2t + C_2\sin 2t$ **1c.** $r = 2 \pm i$; $y = e^{2t}(C_1\cos t + C_2\sin t)$
**1e.** $r = \pm 2$; $y = C_1 e^{2t} + C_2 e^{-2t}$ **2a.** $y = \frac{1}{2}\sin 2t$ **2c.** $y = e^{2t}(\cos t - \sin t)$ **2e.** $y = \frac{1}{2}(e^{2(t-1)} + e^{-2(t-1)})$
**3a.** $Q = e^{-Rt/2L}[c_1 e^{t\sqrt{(R/2L)^2 - 1/LC}} + c_2 e^{-t\sqrt{(R/2L)^2 - 1/LC}}]$

## Section 19.5

**1a.** $(-1/78)(5\cos 3t + \sin 3t) + c_1 e^{3t} + c_2 e^{-2t}$ **1c.** $\frac{1}{2}\cos t - \frac{1}{2}\sin t + c_1 + c_2 e^t$
**1e.** $\frac{1}{2}xe^x - x^2 - 2 + c_1 e^x + c_2 e^{-x}$ **2a.** $250Q = 3 - \frac{1}{2}e^{-10t}(6\cos 20t + 3\sin 20t)$, steady state $Q = 3/250$
**2b.** $Q = -0.006\cos 10t + 0.012\sin 10t + e^{-10t}(0.006\cos 20t + 0.003\sin 20t)$, steady

state $Q = -0.006\cos 10t + 0.012\sin 10t$ **4.** $\dfrac{c}{m(\omega_0^2 - \omega^2)}\cos\omega t + c_1\cos\omega_0 t + c_2\sin\omega_0 t$

## Appendix B

**1a.** $1/9$ **1b.** $1/a^2$ **1c.** $\sqrt[8]{x^5}$ or $(\sqrt[8]{x})^5$ **1d.** $\sqrt[8]{1/t^5}$, for example **1e.** 1 **1f.** 1 **1g.** 1 **1h.** 1
**1i.** $1/8$ **1j.** 1
**2.** $(a\cdots a)(b\cdots b) = (ab)\cdots(ab)$, by the commutative law $ab = ba$, applied $(n-1) + (n-2) + \cdots + 1$ times
**4a.** 6 **4b.** 2 **4c.** $-1$ **4d.** 0 **4e.** $-2$ **4f.** $1/6$ **4g.** $-1$ **4h.** 1 **5a.** 10 **5b.** 1 **5c.** .01
**6a.** $1/\log 2$ **6c.** $-\log 2$ **7.** $\dfrac{\log 2}{\log(1.08)} \approx 9$ years **8.** $\dfrac{\log 2}{\log(1.09)} \approx 8.04$ years

## Appendix C

**1.** $\cos 2\pi = 1$, $\sin 2\pi = 0$; $\cos 3\pi = \cos(-\pi) = -1$, $\sin 3\pi = \sin(-\pi) = 0$; $\cos(-3\pi/2) = 0$, $\sin(-3\pi/2) = 1$;
$\cos(5\pi/6) = \cos(7\pi/6) = \sqrt{3}/2$, $\sin(5\pi/6) = -\sin(7\pi/6) = \frac{1}{2}$; $\cos(-2\pi/3) = -\frac{1}{2}$, $\sin(-2\pi/3) = -\sqrt{3}/2$

**3a.** $7\sin\theta$ **3b.** $-\cos\theta$ **3c.** $\tan\theta$ **5.** $\dfrac{\tan\alpha - \tan\beta}{1 + \tan\alpha\tan\beta}$ **7.**

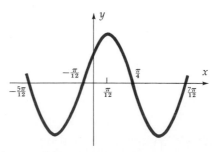

**7k.** Zeros at $\frac{\pi}{4} \pm k\frac{\pi}{3}$.

# INDEX

absolute convergence, 571–72
absolute convergence theorem, 570
absolute value, 2
absolute value, of complex number, 938
absolute value function, 38
acceleration, 73, 654, 706; centripetal, 712
addition formulas, trigonometric, 1006
alternating series, 550, 567ff; test 567
Ampère's law, 907
amplitude (of oscillation), 391
annulus, 307, 841
antiderivative, 77ff, 223; formulas for, 79, 408ff; notation for, 78ff
aphelion, 627
arc length, 458ff; as parameter, 708; in polar coordinates, 663; of parametric curve, 659ff; of space curve, 706
arc length function, 661
Archimedes, 250, 466
area, by integration, 296ff; in polar coordinates, 628ff; of graph, 861; of level surface, 862; of parallelogram, 692–93; of surface of revolution, 463ff
argument, of complex number, 941
Aristotle, 636, 666
arithmetic mean, 360
asymptote, horizontal, 112; of hyperbola, 600; vertical, 121

barycenter, of tetrahedron, 689; of triangle, 642, 644, 688, 849
base of exponential, 336
Bernoulli, Johann, 496
Bessel function, 567, 572–73
binomial series, 576
Biot and Savart, law of, 907
birthday problem, 359
bivariate probability density, 856ff
Bohr radius, 344
Briggs, Henry, 352

$C^k$ functions, 769
capacitance, 380
carbon dating, 369
cardioid, 619, 630, 632

Carnot cycle, 884
catenary, 460ff
Cavalieri, 250, 309
Cavalieri's principle, 309, 829, 831
center of gravity, 845
center of mass, 467ff, 475, 845ff, 870
centripetal acceleration, 712
centroid of a region, 470. 847; of a sector, 855
chain of curves, 895
Chain Rule, 151ff, 758, 788, 795ff
change of base of logarithms, 361, 995
change of variable, in double integral, 879ff
characteristic polynomial, of differential equation, 949, 951
characteristic roots, of differential equation, 949
charge density, 849
circles, 5
Clapeyron's equation, 378
cofunction identities, 388, 1005
comparison estimate, 549, 552
Comparison Test, for improper integrals, 329; for series, 562ff
Comparison Theorem for integrals, 269
completing the square, 436, 587
complex exponential, 940ff
complex functions, 948ff
complex numbers, 936ff; geometry of, 937ff
component of a vector, 636
composite functions, 148ff, 798ff
compound interest, 342ff
concavity, 62, 64, 229
conditional convergence, 571–72
conic sections, 584ff; and planetary orbits, 665ff; degenerate, 611; in polar coordinates, 628ff
connected region, 929
continuity, 120ff, 975ff
continuity, and limits, 975, 977; and sequences, 978; defined, 975; of composite function, 976; of function of two variables, 736ff
convergence, of sequence, 510
convergent improper integral, 325
coordinates, in space, 676ff; in plane, 4; on line, 2
cosine function, 99ff, 1000, 1002
cosine wave, 391ff

critical point, 216; of function of two variables, 773, 788
Critical Point Theorem, 216ff, 773
cross product, 690ff, 696
curl of vector field, 901ff
curl-free vector field, 902, 928ff
curvature, 709, 713–14; of a circle, 710
cycloid, 648
cylinder, 720ff
cylindrical coordinates, 872ff

d'Alembert, 964, 969
damped oscillations, 951ff
damping, critical, 958; undercritical, 958
damping factor, 953
de Moivre's Theorem, 945
decay constant, 367
definite integral, 281
degenerate conics, 611
density, 844, 870
derivative, 32ff, 45ff; and continuity, 132; as limit, 132; numerical approximation of, 37; of complex function, 948; of exponential function, 338; of power, 133ff, 158; of product, 132ff; of quotient, 136; of reciprocal, 135; of vector function, 651; of trigonometric functions, 142ff; second, 61ff; third, 62
Descartes, Rene, 160, 203; folium of, 160ff, 650, 914
determinant, three-by-three, 694, 696; two-by-two, 690, 695, 696
difference quotient, 34
differentiable, 132
differential, of function of one variable, 180; of function of two variables, 750
differential equation, 371ff, 415ff; exact, 808ff; linear with constant coefficients, 946; separable, 377
differential forms, 911ff, 926, 932
differential notation, 180
differentials, and substitution, 285ff
differentiation, under integral sign, 833; of power series, 989; term-by-term, 988
direction cosines, 688
directional derivative, 760–61, 789

directrix, of conic section, 624, 626; of parabola, 584
discontinuity, "blip," 122ff; "hole," 122; oscillatory, 123
discriminant, 611ff
distance, from point to line, 683; in plane, 4; on line, 3
distance formula, 676, 678
divergence, of a vector field, 921, 922ff
divergent improper integral, 325
divergent series, 529
domain, of function of one variable, 18; of function of two variables, 728
Doppler effect, 188, 194
dot product, 681–82, 687
double integral, 826ff
double integral, in polar coordinates, 850ff; properties of, 839ff

e (base of natural exponential), 337
e, computation of, 341, 345ff, 549, 553
eccentricity, 622–23, 627
electrical circuit, 379, 957–58
electron, 897
ellipse, defined, 591; major axis of, 592; minor axis of, 592; semimajor axis of, 592; semiminor axis of, 592
ellipsoid, 718
energy, 319–20; kinetic, 321, 325, 847, 872
entropy, 359, 807, 820
equation, of line, 9ff; of plane, 700ff; point-slope, 11; slope-intercept, 10
escape velocity, 325
Euler, Leonhard, 414
exact differential equation, 808ff
exact vector field, 905
exp(x), 339
exponential, complex, 940–41
exponential function, 336; derivative of, 338
exponential growth, 367
exponential solution, 949
exponents, 992ff
exponents, laws of, 996
extreme value, 214ff
Extreme Value Theorem, 215, 820; proof of, 980ff

Factor Theorem, 52
factorial, 414
Fermat, Pierre de, 26, 32, 38, 206, 250
Fermat's principle, 204
Fibonacci numbers, 580
flow, through a pipe, 316ff; through a surface, 916ff
flux, of an electric field, 925
focus, of hyperbola, 598; of parabola, 584; of ellipse, 591
folium of Descartes, 160ff, 650, 914
frequency (of oscillation), 391
function, 18ff; composite, 148ff; continuous, 120ff; domain of, 18; even, 21; exponential, 336; graph of, 19; increasing or decreasing, 54; inverse, 346ff; odd, 21; quadratic, 40ff; range of, 18–19; rational, 94; trigonometric, 128

function notation, 22
Fundamental Theorem of Calculus (First), 254, 257, 899, 908; (Second), 276

Galileo, 70, 666
gamma function, 414
Gauss' Theorem, 921ff
geometric mean, 360
geometric series, 530, 534
Gini index, 301
global maximum and minimum, 229
gradient field, 898ff 918ff
gradient vector, 744ff, 788
graph, of function of one variable, 19; of function of three variables, 789; of function of two variables, 728
graphing, 53ff; summary of, 168ff; trigonometric functions, 109
Green's Theorem, 908ff; for plane flows, 926
growth constant, 366

Halley, Edmund, 666
helix, 707, 714
Heron of Alexandria, 203
Hessian determinant, 776
homogeneous linear equation, 958
Hooke, Robert, 666
Hooke's Law, 320
hydrogen atom, 878
hyperbola, defined, 598
hyperbolic cosine, 401; functions, 400ff, 946; orbits, 604; sine, 401; sine, series for, 540; tangent, 401
hyperboloid, 718–19

imaginary number, 936
imaginary part of complex number, 937
implicit differentiation, 159ff, 801
implicit function, 801ff
Implicit Function Theorem, 802, 804
improper integral, 325ff
Increasing Sequence Principle, 556, 978
indefinite integral, 280ff; of complex function, 949
indeterminate form, 117, 196ff
index of refraction, 204
inductance, 380
inequality, and limits, 974
infinite series, 528ff; positive, 556
infinitesimal length, 458; slice, 298
inflection point, 62, 64
initial conditions, 953
integral, analytic definition of, 242; definite, 281; double, 826ff; geometric definition of, 236; indefinite, 280ff; iterated, 831; of rate of change, 258ff; repeated, 831; triple, 865ff
integral test, 556ff
integrand, 236, 242; with jump discontinuities, 330
integration by parts, 408ff; numerical, 480ff; of trigonometric functions, 417–29, 453; term-by-term, 987
Intermediate Value Theorem, 123; proof of, 978ff

intersections of polar curves, 631
interval additivity, 267
intervals, 3
inverse function, 346ff; derivative of, 349; hyperbolic, 403; notation, 351; trigonometric, 381ff
Inverse Function Theorem, 349, 983
irreducible quadratic, 439, 441
isothermal compressivity, 805
iterated integral, 831

Jacobian, 881ff, 915; of inverse, 883
Joule (unit of work), 321

Kepler, Johann, 584, 658, 665
Kepler's laws, 665, 670
kinetic energy, 321, 325, 847, 872

l'Hopital's Rule, 195ff, 197, 199, 545
L-R-C circuit, 958
Lagrange, J. L., 814
Lagrange multipliers, 813ff; point, 816; remainder, 502, 504
Laplace's equation, 794, 800, 990
latitude, 877
Law of cosines, 1006, 1008
least squares line, 782ff; plane, 794
Leibniz, Gottfried, 26, 206, 666
Leibniz notation, 46, 153, 179
lemniscate, 620
length of curve, 458ff
level curve, 729; surface, 789
limaçon, 619
limit, 90ff; as x goes to infinity, 111ff; defined, 964, 969; from left, 90ff; from right, 90ff; of exponential functions, 340ff, 362; of function of two variables, 736ff; of logarithmic functions, 356; of product, 972; of rational function, 115; of sequence, 245, 508ff; of sum, 971; of trigonometric function, 99; of vector function, 651
limit comparison test, 564; theorems, 971ff
limits, reversal of, 256–57
line, equation of, 9ff
line integral, 890ff; of gradient field, 899
linear approximation, 178ff; of function of three variables, 788; of function of two variables, 749ff; of vector function, 653
linear functions, 732; interpolation, 14ff; relations, 13
linearity, of integrals, 266
lines, parallel, 12; perpendicular, 12
lituus, 620
ln(x), 353
local maximum point, 227ff, 772, 780
local minimum point, 227ff, 772, 780
logarithm, 352ff, 993ff; as integral, 364; natural, 347, 353; to base a, 352;
logarithmic differentiation, 362; mean, 360
logarithms, change of base, 361, 995; laws of, 996
logistic curve, 449
long division, 438
longitude, 877
LORAN navigation, 606

Lorenz curve, 208ff, 301, 480
lower limit of integration, 236, 242
lower sum, 239

M-test, 985
Maclaurin series, 535, 544
magnetic field, 907
magnitude of vector, 637, 680
mass, 844, 870
maximum value, 214; *See also*, extreme value
mean, of probability density function, 474ff
mean value, of function, 270
Mean Value Theorem, 214, 221ff, 253; applications of, 223ff, 226ff, 253, 770; for integrals, 271
Mercator, Gerhard, 496, 502
Mercator projection, 426ff
midpoint of segment, 641
Midpoint Rule, 493
midpoint sum, 238, 493
minimum value, 214; *See also*, extreme value
mixed partial derivatives, 768ff
modified trapezoid sum, 486ff
moment, about x-axis, 846; about y-axis, 846; of inertia 847, 871–72

Napier, John, 342, 352
natural exponential function, 337ff; derivative of, 338
natural logarithm, 347, 352ff; computation of, 360; derivative of, 354
nested intervals, 980
Newton, Isaac, 26, 206, 666, 878, 964
Newton (unit of force), 320
Newton's Law of Cooling, 378
Newton's method of approximation, 66ff; error estimate, 508
Newton's (Second) Law of Motion, 321, 371, 455, 655, 958
nonhomogeneous linear equation, 958ff
normal component of acceleration, 712
numerical approximation of derivative, 37
numerical integration, 480ff

operator notation, 45, 78–79, 134, 954
optimization, 25ff, 56ff
orientation, and Jacobian, 883; and line integrals, 894, 897
orthogonal projection, 685, 687; vectors, 682
oscillating spring, 393, 952, 958, 962
oscillations, 390ff

p-series, 563
parabola, 41; defined, 584; reflecting property of, 42ff
parallel lines, 12, 704
parallelogram law, for complex numbers, 938; for vectors, 638
parametric equations, 195, 645ff; of circle, 647; of line, 646, 699
partial derivatives, 741ff; mixed, 768ff
partial fractions, 442ff
partial integral, 830, 902, 904
particular solution, of linear equation, 959
Pascal, Blaise, 250

pendulum equation, 672, 673
perihelion, 627
period (of oscillation), 391
periodicity, of trigonometric functions, 1005
Planck's constant, 345
plane, equation of, 700ff, 733
plane sections of graph, 730
planetary motion, 665ff
planimeter, 915
Poiseuille's Law, 318
polar coordinates, 614ff; and arc length, 663; and area, 628ff
position function, 72, 651
position vector, 641
potential, for vector field, 899, 929ff
potential energy, 320, 325, 900
potential function, 808
pound (unit of work), 321
power (and work), 421
power rule, 46, 51ff, 158
power series, 542; differentiation of, 542, 989; integration of, 542, 988
primitive root of unity, 946
principal normal vector, 711
probability density function, 472ff; bivariate, 856ff; exponential, 473; mean of, 474ff; normal, 473; variance of, 477
product, of complex numbers, 942
product rule, 133ff, 408, 706
proper rational function, 437
proportional growth, 366
proton, 897
Ptolemy of Alexandria, 203
Pythagorean identities, 1005
Pythagorean Theorem, 4, 17

quadratic equations in two variables, 606ff
quadratic formula, 41, 589
quadric surfaces, 717ff

radian measure, 99, 1001
radius of convergence, 575, 576
range of function, 18
rate of change, 178
rate of flow, 261
rate of flow, through surface, 916–17
ratio test, 574
rational function, 94, 437; limit of, 115
rational number, 936
real number, 936
real part of complex number, 937
reciprocal, of complex number, 939
reducible quadratic, 439, 441
reduction formula, 411, 414, 423, 429
reflecting property, of ellipse, 594; of hyperbola, 603, 606; of parabola, 42ff
refraction, law of, 26, 203ff
related rates, 186ff
relative error, 182
Remainder Theorem, 52
repeated integral, 831
repeating decimals, 532
Riemann sum, 238ff, 827, 834, 850, 861, 865, 881, 918; regular, 242
right-hand rule, 693
Roberval, 250
Rolle's Theorem, 220

root, of complex number, 943ff
root test, 580
rotation, of vector field, 911; of axes, 606ff
Russian roulette, 535

saddle, 732
saddle point, 773
scalar, 636
Schwarz inequality, 87
secant line, 32
second derivative, 61ff
second derivative test, for functions of two variables, 776ff
separable differential equation, 377
separation of variables, 372, 377
sequence, 508ff; bounded, 522ff; convergent, 510; limit of, 245, 508ff
series, term-by-term operations, 983ff
set-builder notation, 3
sign change principle, 123ff
signed area, 695
simply connected region, 929
Simpson's Rule, 483ff, 492ff
sine function, 99ff, 1000, 1002
skew lines, 705
slope, between two points, 4; of line, 9ff
slope field, 81ff, 373
Snell, Willibrord, 203
Snell's Law, 203ff
solid angle, 926
solid of revolution, 304
speed, 70, 652, 706
spending multiplier, 533, 535
sphere, equation of, 677–78
spherical coordinates, 874ff
spiral of Archimedes, 620
standard deviation, 478
stationary point, 54, 58, 216
step function, 96
Stevin, Simon, 978, 980
stiffness constant of spring, 320
Stokes' Theorem, 927ff
substitution, in definite integral, 289; in integral, 286, 452ff
summation notation, 245
surface area, 860ff; integral, 918ff
symmetric equations of line, 703

tangent line, and magnification, 38; equation of, 36; slope of, 32, 36; to trace of vector function, 653; vertical, 162
tangent plane, 742ff, 752, 791, 792
tangential component of acceleration, 712
Taylor, Brook, 496
Taylor polynomials, 496ff; integral remainder, 507; remainder estimate, 502, 504, 506ff
Taylor series, 536ff; for exponential, 541; for inverse tangent, 544, 548; for logarithm, 548; for sine and cosine, 541
term-by-term operations, 983ff
terminal velocity, 455
tetrahedron, 689, 698
thermal expansivity, 805
thermodynamics, 807
torque, 467, 696, 845, 872
Torricelli's Law, 333, 378
torus, 307

total differential, 750

trace, of motion, 645; of quadratic, 612

transverse axis of hyperbola, 601

trapezoid sum, 260, 481ff, 490ff; modified, 486ff

Trapping Theorem, 105ff, 342, 971; for sequences, 515, 517

triangle inequality, 638

trigonometric addition formulas, 142, 1006

trigonometric functions, derivatives of, 142ff; inverse, 381ff; of angles, 1000; of numbers, 1002

trigonometric identities, 1004ff; limits, 99ff; substitution, 429ff

trigonometry, 1000ff

triple integral, 865ff

twisted cubic, 714

two-form, 912

Tycho Brahe, 665

uniform convergence, 983ff; and continuity, 987; of power series, 986

uniformly convergent, 984

unit tangent vector, 708

unit vector, 640

upper limit of integration, 236, 242

upper sum, 239

van der Waals equation, 141

variance, 477

vector equation, 646, 649; of line, 698

vector field, 898ff; function, 645ff, 646, 706ff; space $R^2$, 637; $R^3$, 680

vectors, 636ff; operations with, 636, 681ff, 689ff

velocity, 70ff, 651, 706; average, 71; instantaneous, 71ff

velocity field, 916

vertex, of ellipse, 592; of hyperbola, 601; of parabola, 585

vertical tangent line, 162

Voltaire, 206

volume, by cross sections, 301ff; by cylindrical shells, 312ff; by disk method, 304; of cylinder, 56, 301

Wallis, John, 423

water clock, 333

wave equation, 794, 797, 800; function, 878

wedge product, 912

Weierstrass, Karl, 964

work, 318ff, 331; and line integrals, 888ff; units of, 320

Wronskian, 141

$x$-simple region, 837

$x$-slope, 734

$y$-simple region, 837, 909

$y$-slope, 734

Zeno's paradox, 528

zero-limit test for divergence, 554

# TABLE OF INTEGRALS

**36.** $\displaystyle\int \frac{du}{u\sqrt{u^2 - a^2}} = \frac{1}{a}\sec^{-1}\left|\frac{u}{a}\right| + C = \frac{1}{a}\cos^{-1}\left|\frac{a}{u}\right| + C$

**37.** $\displaystyle\int \frac{du}{u^2\sqrt{u^2 - a^2}} = \frac{\sqrt{u^2 - a^2}}{a^2 u} + C$

**Forms with $\sqrt{2bu - u^2}$ or $\sqrt{u^2 - 2bu}$:**

**38.** $\displaystyle\int \sqrt{2bu - u^2}\; du = \frac{u - b}{2}\sqrt{2bu - u^2} + \frac{b^2}{2}\sin^{-1}\left(\frac{u - b}{b}\right) + C$

**39.** $\displaystyle\int \frac{du}{\sqrt{2bu - u^2}} = \sin^{-1}\left(\frac{u - b}{b}\right) + C$

**40.** $\displaystyle\int \frac{u\,du}{\sqrt{u^2 - 2bu}} = \sqrt{u^2 - 2bu} + b\ln\left|u - b + \sqrt{u^2 - 2bu}\right| + C.$

**Trigonometric functions:**

**41.** $\displaystyle\int \sin(au)du = -\frac{1}{a}\cos(au) + C$

**42.** $\displaystyle\int \cos(au)\,du = \frac{1}{a}\sin(au) + C$

**43.** $\displaystyle\int \sin^m(ax)\cos^{2n+1}(ax)\,dx = \frac{1}{a}\int u^m(1 - u^2)^n\,du \qquad [u = \sin(ax)]$

**44.** $\displaystyle\int \cos^m(ax)\sin^{2n+1}(ax)\,dx = -\frac{1}{a}\int u^m(1 - u^2)^n\,du \qquad [u = \cos(ax)]$

**45.** $\displaystyle\int \sin^2(au)\,du = \frac{u}{2} - \frac{\sin(2au)}{4a} + C$

**46.** $\displaystyle\int \cos^2(au)\,du = \frac{u}{2} + \frac{\sin(2au)}{4a} + C$

**47.** $\displaystyle\int \sin^n(au)\,du = -\frac{\sin^{n-1}(au)\cos(au)}{na} + \frac{n - 1}{n}\int \sin^{n-2}(au)\,du$

**48.** $\displaystyle\int \cos^n(au)\,du = \frac{\cos^{n-1}(au)\sin(au)}{na} + \frac{n - 1}{n}\int \cos^{n-2}(au)\,du$

*Continued on overleaf*

# TABLE OF INTEGRALS

**49.** If $a^2 \neq b^2$:

   **a.** $\displaystyle \int \sin(au) \cos(bu) \, du = -\frac{\cos(a + b)u}{2(a + b)} - \frac{\cos(a - b)u}{2(a - b)} + C$

   **b.** $\displaystyle \int \sin(au) \sin(bu) \, du = \frac{\sin(a - b)u}{2(a - b)} - \frac{\sin(a + b)u}{2(a + b)} + C$

   **c.** $\displaystyle \int \cos(au) \cos(bu) \, du = \frac{\sin(a - b)u}{2(a - b)} + \frac{\sin(a + b)u}{2(a + b)} + C$

**50.** $\displaystyle \int \tan(au) \, du = -\frac{1}{a} \ln|\cos(au)| + C$

**51.** $\displaystyle \int \cot(au) \, du = \frac{1}{a} \ln|\sin(au)| + C$

**52.** $\displaystyle \int \tan^2(au) \, du = \frac{1}{a} \tan(au) - u + C$

**53.** $\displaystyle \int \cot^2(au) \, du = -\frac{1}{a} \cot(au) - u + C$

**54.** $\displaystyle \int \tan^n(au) \, du = \frac{\tan^{n-1}(au)}{a(n - 1)} - \int \tan^{n-2}(au) \, du, \qquad n \neq 1$

**55.** $\displaystyle \int \cot^n(au) \, du = -\frac{\cot^{n-1}(au)}{a(n - 1)} - \int \cot^{n-2}(au) \, du, \qquad n \neq 1$

**56.** $\displaystyle \int \sec(au) \, du = \frac{1}{a} \ln|\sec(au) + \tan(au)| + C$

**57.** $\displaystyle \int \csc(au) \, du = -\frac{1}{a} \ln|\csc(au) + \cot(au)| + C$

**58.** $\displaystyle \int \tan^m(au) \sec^{2n}(au) \, du = \frac{1}{a} \int u^m(u^2 + 1)^{n-1} \, du \qquad [u = \tan(au)]$

**59.** $\displaystyle \int \sec^m(au) \tan^{2n+1}(au) \, du = \frac{1}{a} \int u^{m-1}(u^2 - 1)^n \, du \qquad [u = \sec(au)]$

**60.** $\displaystyle \int \sec^2(au) \, du = \frac{1}{a} \tan(au) + C$

**61.** $\displaystyle \int \csc^2(au) \, du = -\frac{1}{a} \cot(au) + C$

**62.** $\displaystyle \int \sec^n(au) \, du = \frac{\sec^{n-2}(au) \tan(au)}{a(n - 1)} + \frac{n - 2}{n - 1} \int \sec^{n-2}(au) \, du, \qquad n \neq 1$